CRC

MATERIALS SCIENCE AND ENGINEERING HANDBOOK

THIRD EDITION

CRC

MATERIALS SCIENCE AND ENGINEERING HANDBOOK

THIRD EDITION

James F. Shackelford
Professor of Materials Science and Engineering
Division of Materials Science and Engineering
and
Associate Dean of the College of Engineering
University of California, Davis

William Alexander
Research Engineer
Division of Materials Science and Engineering
University of California, Davis

CRC Press
Boca Raton London New York Washington D C

Library of Congress Cataloging-in-Publication Data

CRC materials science and engineering handbook / [edited by] James F. Shackelford, William Alexander.—3rd ed.
p. cm.
Includes bibliographical references and index.
ISBN 0-8493-2696-6 (alk. paper)
1. Materials—Handooks, manuals, etc. I. Shackelford, James F. II. Alexander, William, 1950 Feb. 13-

TA403.4 .C74 2000
620.1′1—dc21 00-048567

Direct all inquiries to CRC Press LLC, 2000 N.W. Corporate Blvd., Boca Raton, Florida 33431.

International Standard Book Number 0-8493-2696-6
Library of Congress Card Number 00-048567
Printed in the United States of America 1 2 3 4 5 6 7 8 9 0
Printed on acid-free paper

TABLE OF₍ENTS

To

Penelope and Scott

Li-Li and Cassie

CHAPTER 1 Structure of Materials

List of Tables

List of Tables (Continued)

Density

Table 1. ELECTRONIC STRUCTURE OF SELECTED ELEMENTS

At. No.	Element	Sym	Electronic Configuration																	
			1s	2s	2p	3s	3p	3d	4s	4p	4d	4f	5s	5p	5d	5f	6s	6p	6d	7s
1	Hydrogen	H	1																	
2	Helium	He	2																	
3	Lithium	Li	.	1																
4	Beryllium	Be	.	2																
5	Boron	B	.	2	1															
6	Carbon	C	.	2	2															
7	Nitrogen	N	.	2	3															
8	Oxygen	O	.	2	4															
9	Fluorine	F	.	2	5															
10	Neon	N	.	2	6															
11	Sodium	Na	.	.	.	1														
12	Magnesium	Mg	.	.	.	2														
13	Aluminum	Al	.	.	.	2	1													
14	Silicon	Si	.	.	.	2	2													
15	Phosphorus	P	.	.	.	2	3													
16	Sulfur	S	.	.	.	2	4													
17	Chlorine	Cl	.	.	.	2	5													
18	Argon	Ar	.	.	.	2	6													
19	Potassium	K	.	.	.	.	.		1											
20	Calcium	Ca	.	.	.	.	.		2											
21	Scandium	Sc	.	.	.	.	.	1	2											
22	Titanium	Ti	.	.	.	.	.	2	2											
23	Vanadium	V	.	.	.	.	.	3	2											
24	Chromium	Cr	.	.	.	.	.	5	1											
25	Manganese	Mn	.	.	.	.	.	5	2											
26	Iron	Fe	.	.	.	.	.	6	2											
27	Cobalt	Co	.	.	.	.	.	7	2											
28	Nickel	Ni	.	.	.	.	.	8	2											
29	Copper	Cu	.	.	.	.	.	10	1											
30	Zinc	Zn	.	.	.	.	.	10	2											
31	Gallium	Ga	.	.	.	.	.	10	2	1										
32	Germanium	Ge	.	.	.	.	.	10	2	2										
33	Arsenic	As	.	.	.	.	.	10	2	3										
34	Selenium	Se	.	.	.	.	.	10	2	4										
35	Bromine	Br	.	.	.	.	.	10	2	5										
36	Krypton	Kr	.	.	.	.	.	10	2	6										
37	Rubidium	Rb	.	.	.	.	.	.	.	.			1							
38	Strontium	Sr	.	.	.	.	.	.	.	.			2							
39	Yttrium	Y	.	.	.	.	.	.	.	.	1		2							
40	Zirconium	Zr	.	.	.	.	.	.	.	.	2		2							
41	Niobium	Nb	.	.	.	.	.	.	.	.	4		1							
42	Molybdenum	Mo	.	.	.	.	.	.	.	.	5		1							
43	Technetium	Tc	.	.	.	.	.	.	.	.	6		1							
44	Ruthenium	Ru	.	.	.	.	.	.	.	.	7		1							
45	Rhodium	Rh	.	.	.	.	.	.	.	.	8		1							
46	Palladium	Pd	.	.	.	.	.	.	.	.	10									
47	Silver	Ag	.	.	.	.	.	.	.	.	10		1							
48	Cadmium	Cd	.	.	.	.	.	.	.	.	10		2							
49	Indium	In	.	.	.	.	.	.	.	.	10		2	1						
50	Tin	Sn	.	.	.	.	.	.	.	.	10		2	2						
51	Antimony	Sb	.	.	.	.	.	.	.	.	10		2	3						
52	Tellurium	Te	.	.	.	.	.	.	.	.	10		2	5						
53	Iodine	I	.	.	.	.	.	.	.	.	10		2	5						
54	Xenon	Xe	.	.	.	.	.	.	.	.	10		2	6						

At. No.	Element	Sym	1s	2s	2p	3s	3p	3d	4s	4p	4d	4f	5s	5p	5d	5f	6s	6p	6d	7s
55	Cesium	Ce															1			
56	Barium	Ba															2			
57	Lantium	La													1		2			
58	Cerium	Ce										2					2			
59	Praseodymium	Pr										3					2			
60	Neodymium	Nd										4					2			
61	Promethium	Pm										5					2			
62	Samarium	Sm										6					2			
63	Europium	Eu										7					2			
64	Gadolinium	Gd										7			1		2			
65	Terbium	Tb										9					2			
66	Dysprosium	Dy										10					2			
67	Holmium	Ho										11					2			
68	Erbium	Er										12					2			
69	Thulium	Tm										13					2			
70	Ytterbium	Yb										14					2			
71	Lutetium	Lu										14			1		2			
72	Hafnium	Hf										14			2		2			
73	Tantalum	Ta										14			3		2			
74	Tungsten	W										14			4		2			
75	Rhenium	Re										14			5		2			
76	Osmium	Os										14			6		2			
77	Iridium	Ir										14			9					
78	Platinum	Pt										14			9		1			
79	Gold	Au										14			10		1			
80	Mercury	Hg										14			10		2			
81	Thallium	Tl										14			10		2	1		
82	Lead	Pb										14			10		2	2		
83	Bismuth	Bi										14			10		2	3		
84	Polonium	Po										14			10		2	4		
85	Asatine	At										14			10		2	5		
86	Radon	Rn										14			10		2	6		
87	Francium	Fr																		1
88	Radium	Ra																		2
89	Actinium	Ac																	1	2
90	Thorium	Th																	2	2
91	Protoactinium	Pa														2			1	2
92	Uranium	U														3			1	2
93	Neptunium	Np														4			1	2
94	Plutonium	Pu														6				2
95	Americium	Am														7				2
96	Curium	Cm														7			1	2
97	Berkelium	Bk														9				2
98	Californium	Cf														10				2
99	Einsteinium	Es														11				2
100	Fermium	Fm														12				2
101	Mendelevium	Md														13				2
102	Nobelium	No														14				2
103	Lawrencium	Lw														14			1	2

Table 2. AVAILABLE STABLE ISOTOPES OF THE ELEMENTS

(SHEET 1 OF 11)

Element	Mass No.	Natural Abundance (%)
Hydrogen	1	99.985
	2	0.015
Helium	3	0.00013
	4	≈100.0
Lithium	6	7.42
	7	92.58
Beryllium	9	100.0
Boron	10	19.78
	11	80.22
Carbon	12	98.89
	13	1.11
Nitrogen	14	99.63
	15	0.37
Oxygen	16	99.76
	17	0.04
	18	0.20
Fluorine	19	100.0
Neon	20	90.92
	21	0.26
	22	8.82
Sodium	23	100.0
Magnesium	24	78.70
	25	10.13
	26	11.17

Source: Wang, Y., Ed., Handbook of Radioactive Nuclides, The Chemical Rubber Co., Cleveland, 1969, 25.

Table 2. AVAILABLE STABLE ISOTOPES OF THE ELEMENTS

(SHEET 2 OF 11)

Element	Mass No.	Natural Abundance (%)
Aluminum	27	100.0
Silicon	28	92.21
	29	4.70
	30	3.09
Phosphorus	31	100.0
Sulfur	32	95.0
	33	0.76
	34	4.22
	36	0.014
Chlorine	35	75.53
	37	24.47
Argon	36	0.34
	38	0.06
	40	99.60
Potassium	39	93.1
	40[a]	0.01
	41	6.9
Calcium	40	96.97
	42	0.64
	43	0.14
	44	2.06
	46	0.003
	48	0.18
Scandium	45	100.0

Source: Wang, Y., Ed., Handbook of Radioactive Nuclides, The Chemical Rubber Co., Cleveland, 1969, 25.

Table 2. AVAILABLE STABLE ISOTOPES OF THE ELEMENTS

(SHEET 3 OF 11)

Element	Mass No.	Natural Abundance (%)
Titanium	46	7.93
	47	7.28
	48	73.94
	49	5.51
	50	5.34
Vanadium	50	0.24
	51	99.76
Chromium	50	4.31
	52	83.76
	53	9.55
	54	2.38
Manganese	55	100.0
Iron	54	5.82
	56	91.66
	57	2.19
	58	0.33
Cobalt	59	100.0
Nickel	58	67.84
	60	26.23
	61	1.19
	62	3.66
	64	1.08
Copper	63	69.09
	65	30.91

Source: Wang, Y., Ed., Handbook of Radioactive Nuclides, The Chemical Rubber Co., Cleveland, 1969, 25.

Table 2. AVAILABLE STABLE ISOTOPES OF THE ELEMENTS
(SHEET 4 OF 11)

Element	Mass No.	Natural Abundance (%)
Zinc	64	48.89
	66	27.81
	67	4.11
	68	18.57
	70	0.62
Gallium	69	60.4
	71	39.6
Germanium	70	20.52
	72	27.43
	73	7.76
	74	36.54
	76	7.76
Arsenic	75	100.0
Selenium	74	0.87
	76	9.02
	77	7.58
	78	23.52
	80	49.82
	82	9.19
Bromine	79	50.54
	81	49.46
Krypton	78	0.35
	80	2.27
	82	11.56
	83	11.55
	84	56.90
	86	17.37
Rubidium	85	72.15
	87	27.85

Source: Wang, Y., Ed., Handbook of Radioactive Nuclides, The Chemical Rubber Co., Cleveland, 1969, 25.

Table 2. AVAILABLE STABLE ISOTOPES OF THE ELEMENTS

(SHEET 5 OF 11)

Element	Mass No.	Natural Abundance (%)
Strontium	84	0.56
	86	9.86
	87	7.02
	88	82.56
Yttrium	89	100.0
Zirconium	90	51.46
	91	11.23
	92	17.11
	94	17.40
	96	2.80
Niobium	93	100.0
Molybdenum	92	15.84
	94	9.04
	95	15.72
	96	16.53
	97	9.46
	98	23.78
	100	9.63
Ruthenium	96	5.51
	98	1.87
	99	12.72
	100	12.62
	101	17.07
	102	31.61
	104	18.60
Rhodium	103	100.0

Source: Wang, Y., Ed., Handbook of Radioactive Nuclides, The Chemical Rubber Co., Cleveland, 1969, 25.

Table 2. AVAILABLE STABLE ISOTOPES OF THE ELEMENTS

(SHEET 6 OF 11)

Element	Mass No.	Natural Abundance (%)
Palladium	102	0.96
	104	10.97
	105	22.23
	106	27.33
	108	26.71
	110	11.81
Silver	107	51.82
	109	48.18
Cadmium	106	1.22
	108	0.88
	110	12.39
	111	12.75
	112	24.07
	113	12.26
	114	28.86
	116	7.58
Indium	113	4.28
	115	95.72
Tin	112	0.96
	114	0.66
	115	0.35
	116	14.30
	117	7.61
	118	24.03
	119	8.58
	120	32.85
	122	4.72
	124	5.94
Antimony	121	57.25
	123	42.75

Source: Wang, Y., Ed., Handbook of Radioactive Nuclides, The Chemical Rubber Co., Cleveland, 1969, 25.

Table 2. AVAILABLE STABLE ISOTOPES OF THE ELEMENTS

(SHEET 7 OF 11)

Element	Mass No.	Natural Abundance (%)
Tellurium	120	0.09
	122	2.46
	123	0.87
	124	4.61
	125	6.99
	126	18.71
	128	31.79
	130	34.48
Iodine	127	100.0
Xenon	124	0.096
	126	0.090
	128	1.92
	129	26.44
	130	4.08
	131	21.18
	132	26.89
	134	10.44
	136	8.87
Cesium	133	100.0
Barium	130	0.101
	132	0.097
	134	2.42
	135	6.59
	136	7.81
	137	11.30
	138	71.66
Lanthanum	138	0.09
	139	99.91

Source: Wang, Y., Ed., Handbook of Radioactive Nuclides, The Chemical Rubber Co., Cleveland, 1969, 25.

Table 2. AVAILABLE STABLE ISOTOPES OF THE ELEMENTS

(SHEET 8 OF 11)

Element	Mass No.	Natural Abundance (%)
Cerium	136	0.193
	138	0.250
	140	88.48
	142[d]	11.07
Praseodymium		
	141	100.0
Neodymium	142	27.11
	143	12.17
	144	23.85
	146	17.22
	148	5.73
	150	5.62
Samarium	144	3.09
	147[e]	14.97
	148[f]	11.24
	149[g]	13.83
	150	7.44
	152	26.72
	154	22.71
Europium	151	47.82
	153	52.18
Gadolinium	152[h]	0.20
	154	2.15
	155	14.73
	156	20.47
	157	15.68
	158	24.87
	160	21.90
Terbium	159	100.0

Source: Wang, Y., Ed., Handbook of Radioactive Nuclides, The Chemical Rubber Co., Cleveland, 1969, 25.

Table 2. AVAILABLE STABLE ISOTOPES OF THE ELEMENTS

(SHEET 9 OF 11)

Element	Mass No.	Natural Abundance (%)
Dysprosium	156[i]	0.052
	158	0.090
	160	2.29
	161	18.88
	162	25.53
	163	24.97
	164	28.18
Holmium	165	100.0
	186	28.41
Erbium	162	0.136
	164	1.56
	166	33.41
	167	22.94
	168	27.07
	170	14.88
	186	1.59
Thulium	169	100.0
	189	16.1
Ytterbium	168	0.135
	170	3.03
	171	14.31
	172	21.82
	173	16.13
	174	31.84
	176	12.73
Lutetium	175	97.40
	176[j]	2.60

Source: Wang, Y., Ed., Handbook of Radioactive Nuclides, The Chemical Rubber Co., Cleveland, 1969, 25.

Table 2. AVAILABLE STABLE ISOTOPES OF THE ELEMENTS

(SHEET 10 OF 11)

Element	Mass No.	Natural Abundance (%)
Haffiium	174[k]	0.18
	176	5.20
	177	18.50
	178	27.14
	179	13.75
	180	35.24
Tantalum	180	0.012
	181	99.988
Tungsten	180	0.14
	182	26.41
	183	14.40
	184	30.64
Rhenium	185	37.07
	187	62.93
Osmium	184	0.018
	187	1.64
	188	13.3
	190	26.4
	192	41.0
Iridium	191	37.3
	193	62.7
Platinum	190[m]	0.013
	192	0.78
	194	32.9
	195	33.8
	196	25.3
	198	7.2
Gold	197	100.0

Source: Wang, Y., Ed., Handbook of Radioactive Nuclides, The Chemical Rubber Co., Cleveland, 1969, 25.

Table 2. AVAILABLE STABLE ISOTOPES OF THE ELEMENTS
(SHEET 11 OF 11)

Element	Mass No.	Natural Abundance (%)
Mercury	196	0.146
	198	10.02
	199	16.84
	200	23.13
	201	13.22
	202	29.80
	204	6.85
Thallium	203	29.50
	205	70.50
Lead	204	1.48
	206	23.6
	207	22.6
	208	52.3
Bismuth	209	100.0
Thorium	232[n†]	100.0
Uranium	234[o†]	0.0006
	235[p†]	0.72
	238[q†]	99.27

Source: Wang, Y., Ed., Handbook of Radioactive Nuclides, The Chemical Rubber Co., Cleveland, 1969, 25.

[a] half-life = 1.3×10^9 y.
[b] half-life > 10^{15} y
[c] half-life = 5×10^{14} y
[d] half-life = 5×10^{14} y
[e] half-life = 1.06×10^{11} y
[f] half-life = 1.2×10^{13} y
[g] half-life = 1.2×10^{14} y
[h] half-life = 1.1×10^{14} y
[i] half-life = 2×10^{14} y
[j] half-life = 2.2×10^{10} y
[k] half-life = 4.3×10^{15} y
[l] half-life = 4×10^{10} y
[m] half-life = 6×10^{11} y
[n] half-life = 1.4×10^{10} y
[o] half-life = 2.5×10^5 y
[p] half-life = 7.1×10^8 y
[q] half-life = 4.5×10^9 y
[†] naturally occurring.

Table 3. Periodic Table of the Elements

1 IA	2	3	4	5	6	7	8	9	10	11	12	13	14	15	16	17	18 VIIA
1 H	IIA											IIIA	IVA	VA	VIA	VIIA	2 He
3 Li	4 Be											5 B	6 C	7 N	8 O	9 F	10 Ne
11 Na	12 Mg	IIIB	IVB	VB	VIB	VIIB	-----	VIII	-----	IB	IIB	13 Al	14 Si	15 P	16 S	17 Cl	18 Ar
19 K	20 Ca	21 Sc	22 Ti	23 V	24 Cr	25 Mn	26 Fe	27 Co	28 Ni	29 Cu	30 Zn	31 Ga	32 Ge	33 As	34 Se	35 Br	36 Kr
37 Rb	38 Sr	39 Y	40 Zr	41 Nb	42 Mo	43 Tc	44 Ru	45 Rh	46 Pd	47 Ag	48 Cd	49 In	50 Sn	51 Sb	52 Te	53 I	54 Xe
55 Cs	56 Ba		72 Hf	73 Ta	74 W	75 Re	76 Os	77 Ir	78 Pt	79 Au	80 Hg	81 Tl	82 Pb	83 Bi	84 Po	85 At	86 Rn
87 Fr	88 Ra																

57 La	58 Ce	59 Pr	60 Nd	61 Pm	62 Sm	63 Eu	64 Gd	65 Tb	66 Dy	67 Ho	68 Er	69 Tm	70 Yb	71 Lu
89 Ac	90 Th	91 Pa	92 U	93 Np	94 Pu	95 Am	96 Cm	97 Bk	98 Cf	99 Es	100 Fm	101 Md	102 No	103 Lw

Table 4. Periodic Table of Elements in Metallic Materials

1 IA	2 IIA	3 IIIB	4 IVB	5 VB	6 VIB	7 VIIB	8 ----	9 VIII	10 ----	11 IB	12 IIB	13 IIIA	14 IVA	15 VA	16 VIA	17 VIIA	18 VIIA
3 Li	4 Be											5 B					
11 Na	12 Mg											13 Al					
19 K	20 Ca	21 Sc	22 Ti	23 V	24 Cr	25 Mn	26 Fe	27 Co	28 Ni	29 Cu	30 Zn	31 Ga					
37 Rb	38 Sr	39 Y	40 Zr	41 Nb	42 Mo	43 Tc	44 Ru	45 Rh	46 Pd	47 Ag	48 Cd	49 In	50 Sn	51 Sb			
55 Cs	56 Ba		72 Hf	73 Ta	74 W	75 Re	76 Os	77 Ir	78 Pt	79 Au	80 Hg	81 Tl	82 Pb	83 Bi			
87 Fr	88 Ra																

57 La	58 Ce	59 Pr	60 Nd	61 Pm	62 Sm	63 Eu	64 Gd	65 Tb	66 Dy	67 Ho	68 Er	69 Tm	70 Yb	71 Lu
89 Ac	90 Th	91 Pa	92 U	93 Np	94 Pu	95 Am	96 Cm	97 Bk	98 Cf	99 Es	100 Fm	101 Md	102 No	103 Lw

Table 5. PERIODIC TABLE OF ELEMENTS IN CERAMIC MATERIALS

1 IA	2	3	4	5	6	7	8	9	10	11	12	13	14	15	16	17	18 VIIA
	IIA											IIIA	IVA	VA	VIA	VIIA	
3 Li	4 Be											5 B	6 C	7 N	8 O		
11 Na	12 Mg	IIIB	IVB	VB	VIB	VIIB	-----	VIII	-----	IB	IIB	13 Al	14 Si	15 P	16 S		
19 K	20 Ca	21 Sc	22 Ti	23 V	24 Cr	25 Mn	26 Fe	27 Co	28 Ni	29 Cu	30 Zn	31 Ga	32 Ge				
37 Rb	38 Sr	39 Y	40 Zr	41 Nb	42 Mo	43 Tc	44 Ru	45 Rh	46 Pd	47 Ag	48 Cd	49 In	50 Sn	51 Sb			
55 Cs	56 Ba		72 Hf	73 Ta	74 W	75 Re	76 Os	77 Ir	78 Pt	79 Au	80 Hg	81 Tl	82 Pb	83 Bi			
87 Fr	88 Ra																

57 La	58 Ce	59 Pr	60 Nd	61 Pm	62 Sm	63 Eu	64 Gd	65 Tb	66 Dy	67 Ho	68 Er	69 Tm	70 Yb	71 Lu
89 Ac	90 Th	91 Pa	92 U	93 Np	94 Pu	95 Am	96 Cm	97 Bk	98 Cf	99 Es	100 Fm	101 Md	102 No	103 Lw

Table 6. Periodic Table of Elements in Polymeric Materials

1 IA	2 IIA	3 IIIB	4 IVB	5 VB	6 VIB	7 VIIB	8 -----	9 VIII	10 -----	11 IB	12 IIB	13 IIIA	14 IVA	15 VA	16 VIA	17 VIIA	18 VIIA
1 H																	
													6 C	7 N	8 O	9 F	
													14 Si				

Table 7. Periodic Table of Elements in Semiconducting Materials

1 IA	2 IIA	3 IIIB	4 IVB	5 VB	6 VIB	7 VIIB	8 -----	9 VIII	10 -----	11 IB	12 IIB	13 IIIA	14 IVA	15 VA	16 VIA	17 VIIA	18 VIIA
															8 O		
												13 Al	14 Si	15 P	16 S		
											30 Zn	31 Ga	32 Ge	33 As	34 Se		
											48 Cd	49 In	50 Sn	51 Sb	52 Te		
											80 Hg						

Table 8. PERIODIC TABLE OF ELEMENTS IN SUPERCONDUCTING METALS

1	2	3	4	5	6	7	8	9	10	11	12	13	14	15	16	17	18
IA	IIA	IIIB	IVB	VB	VIB	VIIB	-----	VIII	-----	IB	IIB	IIIA	IVA	VA	VIA	VIIA	VIIA
	4 Be																
												13 Al					
			22 Ti	23 V							30 Zn	31 Ga					
			40 Zr	41 Nb	42 Mo	43 Tc	44 Ru				48 Cd	49 In	50 Sn	51 Sb			
				73 Ta	74 W	75 Re	76 Os	77 Ir			80 Hg		82 Pb				

57 La														
	90 Th	91 Pa												

Table 9. ATOMIC AND IONIC RADII OF THE ELEMENTS
(SHEET 1 OF 5)

Atomic Number	Symbol	Atomic Radius (nm)	Ion	Ionic Radius (nm)
1	H	0.046	H^-	0.154
2	He	–	–	–
3	Li	0.152	Li^+	0.078
4	Be	0.114	Be^{2+}	0.054
5	B	0.097	B^{3+}	0.02
6	C	0.077	C^{4+}	<0.02
7	N	0.071	N^{5+}	0.01–0.2
8	O	0.060	0^{2-}	0.132
9	F	–	F^-	0.133
10	Ne	0.160	–	–
11	Na	0.186	Na^+	0.098
12	Mg	0.160	Mg^{2+}	0.078
13	Al	0.143	Al^{3+}	0.057
14	Si	0.117	Si^{4-}	0.198
.			Si^{4+}	0.039
15	P	0.109	P^{5+}	0.03–0.04
16	S	0.106	S^{2-}	0.174
			S^{6+}	0.034
17	Cl	0.107	Cl^-	0.181
18	Ar	0.192	–	–
19	K	0.231	K^+	0.133
20	Ca	0.197	Ca^{2+}	0.106
21	Sc	0.160	Sc^{2+}	0.083

Source: Data from R. A. Flinn and P. K. Trojan, Engineering Materials and Their Applications, Houghton Mifflin Company, Boston, 1975. The ionic radii are based on the calculations of V. M. Goldschmidt, who assigned radii based on known interatomic distances in various ionic crystals.

Table 9. ATOMIC AND IONIC RADII OF THE ELEMENTS

(SHEET 2 OF 5)

Atomic Number	Symbol	Atomic Radius (nm)	Ion	Ionic Radius (nm)
22	Ti	0.147	Ti^{2+}	0.076
			Ti^{3+}	0.069
			Ti^{4+}	0.064
23	V	0.132	V^{3+}	0.065
			V^{4+}	0.061
			V^{5+}	0.04
24	Cr	0.125	Cr^{3+}	0.064
			Cr^{6+}	0.03–0.04
25	Mn	0.112	Mn^{2+}	0.091
			Mn^{3+}	0.070
			Mn^{4+}	0.052
26	Fe	0.124	Fe^{2+}	0.087
			Fe^{2+}	0.067
27	Co	0.125	Co^{2+}	0.082
			Co^{3+}	0.065
28	Ni	0.125	Ni^{2+}	0.078
29	Cu	0.128	Cu^{+}	0.096
30	Zn	0.133	Zn^{2+}	0.083
31	Ga	0.135	Ga^{3+}	0.062
32	Ge	0.122	Ge^{4+}	0.044
33	As	0.125	As^{3+}	0.069
			As^{5+}	~0.04
34	Se	0.116	Se^{2-}	0.191
			Se^{6+}	0.03–0.04
35	Br	0.119	Br^{-}	0.196
36	Kr	0.197	–	–

Source: Data from R. A. Flinn and P. K. Trojan, Engineering Materials and Their Applications, Houghton Mifflin Company, Boston, 1975. The ionic radii are based on the calculations of V. M. Goldschmidt, who assigned radii based on known interatomic distances in various ionic crystals.

Table 9. ATOMIC AND IONIC RADII OF THE ELEMENTS
(SHEET 3 OF 5)

Atomic Number	Symbol	Atomic Radius (nm)	Ion	Ionic Radius (nm)
37	Rb	0.251	Rb^{+}	0.149
38	Sr	0.215	Sr^{2+}	0.127
39	Y	0.181	Y^{3+}	0.106
40	Zr	0.158	Zr^{4+}	0.087
41	Nb	0.143	Nb^{4+}	0.074
			Nb^{5+}	0.069
42	Mo	0.136	Mo^{4+}	0.068
			Mo^{6+}	0.065
43	Tc	–	–	–
44	Ru	0.134	Ru^{4+}	0.065
45	Rh	0.134	Rh^{3+}	0.068
			Rh^{4+}	0.065
46	Pd	0.137	Pd^{2+}	0.050
47	Ag	0.144	Ag^{+}	0.113
48	Cd	0.150	Cd^{2+}	0.103
49	In	0.157	In^{3+}	0.091
50	Sn	0.158	Sn^{4-}	0.215
			Sn^{4+}	0.074
51	Sb	0.161	Sb^{3+}	0.090
52	Te	0.143	Te^{2-}	0.211
			Te^{4+}	0.089
53	I	0.136	I^{-}	0.220
			I^{5+}	0.094
54	Xe	0.218	–	–
55	Cs	0.265	Cs^{+}	0.165

Source: Data from R. A. Flinn and P. K. Trojan, Engineering Materials and Their Applications, Houghton Mifflin Company, Boston, 1975. The ionic radii are based on the calculations of V. M. Goldschmidt, who assigned radii based on known interatomic distances in various ionic crystals.

Table 9. ATOMIC AND IONIC RADII OF THE ELEMENTS

(SHEET 4 OF 5)

Atomic Number	Symbol	Atomic Radius (nm)	Ion	Ionic Radius (nm)
56	Ba	0.217	Ba^{2+}	0.13
57	La	0.187	La^{3+}	0.122
58	Ce	0.182	Ce^{3+}	0.118
			Ce^{4+}	0.102
59	Pr	0.183	Pr^{3+}	0.116
			Pr^{4+}	0.100
60	Nd	0.182	Nd^{3+}	0.115
61	Pm	–	Pm^{3+}	0.106
62	Sm	0.181	Sm^{3+}	0.113
63	Eu	0.204	Eu^{3+}	0.113
64	Gd	0.180	Gd^{3+}	0.111
65	Tb	0.177	Tb^{3+}	0.109
			Tb^{4+}	0.089
66	Dy	0.177	Dy^{3+}	0.107
67	Ho	0.176	Ho^{3+}	0.105
68	Er	0.175	Er^{3+}	0.104
69	Tm	0.174	Tm^{3+}	0.104
70	Yb	0.193	Yb^{3+}	0.100
71	Lu	0.173	Lu^{3+}	0.099
72	Hf	0.159	Hf^{4+}	0.084
73	Ta	0.147	Ta^{5+}	0.068
74	W	0.137	W^{4+}	0.068
			W^{6+}	0.065
75	Re	0.138	Re^{4+}	0.072
76	Os	0.135	Os^{4+}	0.067
77	Ir	0.135	Ir^{4+}	0.066

Source: Data from R. A. Flinn and P. K. Trojan, Engineering Materials and Their Applications, Houghton Mifflin Company, Boston, 1975. The ionic radii are based on the calculations of V. M. Goldschmidt, who assigned radii based on known interatomic distances in various ionic crystals.

Table 9. ATOMIC AND IONIC RADII OF THE ELEMENTS

(SHEET 5 OF 5)

Atomic Number	Symbol	Atomic Radius (nm)	Ion	Ionic Radius (nm)
78	Pt	0.138	Pt^{2+}	0.052
			Pt^{4+}	0.055
79	Au	0.144	Au^{+}	0.137
80	Hg	0.150	Hg^{2+}	0.112
81	Tl	0.171	Tl^{+}	0.149
			Tl^{3+}	0.106
82	Pb	0.175	Pb^{4-}	0.215
			Pb^{2+}	0.132
			Pb^{4+}	0.084
83	Bi	0.182	Bi^{3+}	0.120
84	Po	0.140	Po^{6+}	0.067
85	At	–	At^{7+}	0.062
86	Rn	–	–	–
87	Fr	–	Fr^{+}	0.180
88	Ra	–	Ra^{+}	0.152
89	Ac	–	Ac^{3+}	0.118
90	Th	0.180	Th^{4+}	0.110
91	Pa	–	–	–
92	U	0.138	U^{4+}	0.105

Source: Data from R. A. Flinn and P. K. Trojan, Engineering Materials and Their Applications, Houghton Mifflin Company, Boston, 1975. The ionic radii are based on the calculations of V. M. Goldschmidt, who assigned radii based on known interatomic distances in various ionic crystals.

Table 10. BOND LENGTH VALUES BETWEEN ELEMENTS

(SHEET 1 OF 4)

Elements	Compound	Bond length (Å)		
B-B	B_2H_6	1.770	±	0.013
B-Br	BBF	1.88		
	BBr_3	1.87	±	0.02
B-Cl	BCl	1.715		
	BCl_3	1.72	±	0.01
B-F	BF	1.262		
	BF_3	1.29	±	0.01
B-H	Hydrides	1.21	±	.02
B-H bridge	Hydrides	1.39	±	.02
B-N	$(BClNH)_3$	1.42	±	.01
B-0	BO	1.2049		
	$B(OH)_3$	1.362	±	0.005 (av)
N-Cl	NO_2Cl	1.79	±	0.02
N-F	NF_3	1.36	±	0.02
N-H	$[NH_4]^+$	1.034	±	0.003
	NH	1.038		
	ND	1.041		
	HNCS	1.013	±	0.005
N-N	N_3H	1.02	±	0.01
	N_2O	1.126	±	0.002
	$[N_2]^+$	1.116		
N-O	NO_2Cl	1.24	±	0.01
	NO_2	1.188	±	0.005

To convert Å to nm, multiply by 10^{-1}

Source: from Kennard, O., in Handbook of Chemistry and Physics, 69th ed., Weast, R. C., Ed., CRC Press, Boca Raton, Fla., 1988, F-167.

Table 10. BOND LENGTH VALUES BETWEEN ELEMENTS

(SHEET 2 OF 4)

Elements	Compound	Bond length (Å)		
N=O	N_2O	1.186	±	0.002
	$[NO]^+$	1.0619		
N-Si	SiN	1.572		
O-H	$[OH]^+$	1.0289		
	OD	0.9699		
	H_2O_2	0.960	±	0.005
B-B	B_2H_6	1.770	±	0.013
B-Br	BBF	1.88		
	BBr_3	1.87	±	0.02
B-Cl	BCl	1.715		
	BCl_3	1.72	±	0.01
B-F	BF	1.262		
	BF_3	1.29	±	0.01
B-H	Hydrides	1.21	±	.02
B-H bridge	Hydrides	1.39	±	.02
B-N	$(BClNH)_3$	1.42	±	.01
B-0	BO	1.2049		
	$B(OH)_3$	1.362	±	0.005 (av)
N-Cl	NO_2Cl	1.79	±	0.02
N-F	NF_3	1.36	±	0.02
N-H	$[NH_4]^+$	1.034	±	0.003
	NH	1.038		
	ND	1.041		
	HNCS	1.013	±	0.005

To convert Å to nm, multiply by 10^{-1}

Source: from Kennard, O., in Handbook of Chemistry and Physics, 69th ed., Weast, R. C., Ed., CRC Press, Boca Raton, Fla., 1988, F-167.

Table 10. Bond Length Values Between Elements

(Sheet 3 of 4)

Elements	Compound	Bond length (Å)		
N-N	N_3H	1.02	±	0.01
	N_2O	1.126	±	0.002
	$[N_2]^+$	1.116		
N-O	NO_2Cl	1.24	±	0.01
	NO_2	1.188	±	0.005
N=O	N_2O	1.186	±	0.002
	$[NO]^+$	1.0619		
N-Si	SiN	1.572		
O-H	$[OH]^+$	1.0289		
	OD	0.9699		
	H_2O_2	0.960	±	0.005
O-O	H_2O_2	1.48	±	0.01
	$[O_2]^+$	1.227		
	$[O_2]^-$	1.26	±	0.2
	$[O_2]^{--}$	1.49	±	0.02
P-D	PD	1.429		
P-H	$[PH_4]^+$	1.42	±	0.02
P-N	PN	1.4910		
P-S	$PSBr_3$ (Cl_3,F_3)	1.86	±	0.02
S-Br	$SOBr_2$	2.27	±	0.02
S-F	SOF_2	1.585	±	0.005
S-D	SD	1.3473		
	SD_2	1.345		

To convert Å to nm, multiply by 10^{-1}

Source: from Kennard, O., in Handbook of Chemistry and Physics, 69th ed., Weast, R. C., Ed., CRC Press, Boca Raton, Fla., 1988, F-167.

Table 10. Bond Length Values Between Elements

(Sheet 4 of 4)

Elements	Compound	Bond length (Å)		
S-O	SO_2	1.4321		
	$SOCl_2$	1.45	±	0.02
S-S	S_2Cl_2	2.04	±	0.01
Si-Br	$SiBr_4$	2.17	±	1.01
Si-Cl	$SiCl_4$	2.03	±	1.01 (av)
Si-F	SiF_4	1.561	±	0.003 (av)
Si-H	SiH_4	1.480	±	0.005
Si-O	$[SiO]^+$	1.504		
Si-Si	Si_2Cl_2	2.30	±	0.02

To convert Å to nm, multiply by 10^{-1}

Source: from Kennard, O., in Handbook of Chemistry and Physics, 69th ed., Weast, R. C., Ed., CRC Press, Boca Raton, Fla., 1988, F-167.

Table 11. Periodic Table of Carbon Bond Lengths (Å)

1 IA	2 IIA	3 IIIB	4 IVB	5 VB	6 VIB	7 VIIB	8 -----	9 VIII	10 -----	11 IB	12 IIB	13 IIIA	14 IVA	15 VA	16 VIA	17 VIIA	18 VIIA
H 1.06																	
	Be 1.93											B 1.56	C 1.2	N 1.47	O 1.43	F 1.55	
												Al 2.24	Si 1.8	P 1.87	S 1.81	Cl 1.7	
					Cr 1.92		Fe 1.94	Co 1.93	Ni 18.2				Ge 1.98	As 1.98	Se 1.98	Br 1.9	
					Mo 2.08				Pd 2.27			In 2.16	Sn 2.15	Sb 2.16	Te 2.05	I 2.1	
					W 2.06						Hg 2.07		Pb 2.29	Bi 2.30			

Table 12. Carbon Bond Lengths

(Sheet 1 of 3)

Group No.	Element.	At. No.	Sym.	Bond Length (Å)		Bond Type
1	Hydrogen	1	H	1.056	± 1.115	
2	Beryllium	4	Be	1.93		
6	Chromium	24	Cr	1.92	± 0.04	
	Molybdenum	42	Mo	2.08	± 0.04	
	Tungsten	74	W	2.06	± 0.01	
8	Iron	26	Fe	1.94	± 0.02	
9	Cobalt	27	Co	1.93	± 0.02	
10	Nickel	28	Ni	1.82	± 0.03	
	Palladium	46	Pd	2.27	± 0.04	
12	Mercury	80	Hg	2 .07	± 0.01	
13	Aluminum	13	Al	2.24	± 0.04	
	Boron	5	B	1. 56	± 0.01	
	Indium	49	In	2.16	± 0.04	
14	Carbon	6	C	1.20	± 1.54	Alkyls (CH_3XH_3)
	Germanium	32	Ge	1.98	± 0.03	Alkyls (CH3XH3)
	Lead	82	Pb	2.29	± 0.05	Alkyls (CH3XH3)
	Silicon	14	Si	1.865	± 0.008	Alkyls (CH3XH3)
				1.84	± 0.01	Aryls (C_6H5XH3)
				1.88	± 0.01	Neg. Subst. (CH3XCI3)
	Tin	50	Sn	2.143	± 0.008	Alkyls (CH3XH3)
				2.18	± 0.02	Neg. Subst. (CH3XCI3)
15	Arsenic	33	As	1.98	± 0.02	Paraffinic (CH3)3X
	Bismuth	83	Bi	2.30		Paraffinic (CH3)3X
	Nitrogen	7	N	1 .47	± 1.1	
	Phosphorus	15	P	1.87	± 0.02	Paraffinic (CH3)3X
	Antimony	51	Sb	2.202	± 0.016	Paraffinic (CH3)3X

Source: data from Lide, David R., Ed., CRC Handbook of Chemistry and Physics, CRC Press, Boca Raton, (1990); and "Tables of interatomic distances" Chem. Soc. of London, 1958.

Table 12. CARBON BOND LENGTHS

(SHEET 2 OF 3)

Group No.	Element.	At. No.	Sym.	Bond Length (Å)		Bond Type
16	Oxygen	8	O	1.43	± 1.15	
	Sulfur	16	S	1.81	± 1.55	
	Selenium	34	Se	1 .98	± 1.71	
	Tellurium	52	Te	2.05	± 0.14	
17	Bromine	35	Br	1.937	± 0.003	Paraffinic (mono. substituted) (CH_3X)
			Br	1.937	± 0.003	Paraffinic (disubstituted) (CH_2X_2)
			Br	1.89	± 0.01	Olfinic(CH_2:CHX)
				1.85	± 0.01	Aromatic (C_6H_3X)
				1.79	± 0.01	Acetylenic (HC:CX)
	Chlorine	17	Cl	1.767	± 0.002	Paraffinic (mono. substituted) (CH_3X)
			Cl	1.767	± 0.002	Paraffinic (disubstituted) (CH_2X_2)
			Cl	1.72	± 0.01	Olfinic(CH_2:CHX)
				1.70	± 0.01	Aromatic (C_6H_3X)
				1.79	± 0.01	Acetylenic (HC:CX)
	Fluorine	9	F	1.831	± 0.005	Paraffinic (mono. substituted) (CH_3X)
			F	1.334	± 0.004	Paraffinic (disubstituted) (CH_2X_2)

Source: data from Lide, David R., Ed., CRC Handbook of Chemistry and Physics, CRC Press, Boca Raton, (1990); and "Tables of interatomic distances" Chem. Soc. of London, 1958.

Table 12. CARBON BOND LENGTHS

(SHEET 3 OF 3)

Group No.	Element.	At. No.	Sym.	Bond Length (Å)		Bond Type
	Fluorine con't		F	1.325	± 0.1	Olfinic(CH_2:CHX)
				1.30	± 0.01	Aromatic (C_6H_3X)
				1.635	± 0.004	Acetylenic (HC:CX)
	Iodine	53	I	2.13	± 0.1	Paraffinic (mono. substituted) (CH_3X)
			I	2.13	± 0.1	Paraffinic (disubstituted) (CH_2X_2)
			I	2.092	± 0.005	Olfinic(CH_2:CHX)
				2.05	± 0.01	Aromatic (C_6H_3X)
				1.99	± 0.02	Acetylenic (HC:CX)

Source: data from Lide, David R., Ed., CRC Handbook of Chemistry and Physics, CRC Press, Boca Raton, (1990); and "Tables of interatomic distances" Chem. Soc. of London, 1958.

Table 13. CARBON BOND LENGTHS IN POLYMERS

(SHEET 1 OF 3)

Bond Type	Polymer Type	Bond Length (Å)	
CARBON-CARBON			
Single Bond	Paraffinic	1.541	± 0.003
	In diamond (18°C)	1.54452	± 0.00014

Source: data from CRC Handbook of Chemistry and Physics, David R. Lide, Ed., CRC Press, Boca Raton, (1990) and "Tables of interatomic distances" Chem. Soc. of London, (1958).

Table 13. CARBON BOND LENGTHS IN POLYMERS
(SHEET 2 OF 3)

Bond Type	Polymer Type	Bond Length (Å)	
CARBON-CARBON cont't			
Partial Double Bond	(1) Shortening of single bond in presence of carbon carbon double bond, e.g. $(CH_2),C_3CH_2$; or of aromatic ring e.g. $C_6H_5\ CH_3$	1.53	± 0.01
	(2) Shortening in presence of a carbon oxygen double bond e.g. CH_3CHO	1.516	± 0.005
	(3) Shortening in presence of two carbon oxygen double bonds, e.g. $(CO_2H)_2$	1.49	± 0.01
	(4) Shortening in presence of a carbon oxygen triple bond, e.g. $CH_3C:CH$	1.460	± 0.003
	(5) In compounds with tendency to dipole formation, e.g. C:C.C:N	1.44	± 0.01
	(6) In graphite(at 15 °C)	1.4210	± 0.0001
	(7) In aromatic compounds	1.395	± 0.003
	(8) in presence of a carbon carbon triple bonds, e.g. HC≡C-C≡CH	1.373	± 0.004
Double Bonds	(1) simple	1.337	± 0.006
	(2) Part triple bond, e.g. $CH_2:C:CH_2$	1.309	± 0.005
Triple Bond	(1) Simple, e.g. C_2H_2	1.204	± 0.002
	(2) Conjugated, e.g. $CH_3.(C:C)_2.H$ e.g. C_5H_5N	1.206	± 0.004

Source: data from CRC Handbook of Chemistry and Physics, David R. Lide, Ed., CRC Press, Boca Raton, (1990) and "Tables of interatomic distances" Chem. Soc. of London, (1958).

Table 13. Carbon Bond Lengths in Polymers

(Sheet 3 of 3)

Bond Type	Polymer Type	Bond Length (Å)	
CARBON-HYDROGEN	(1) Paraffinic		
	(a) in methane	1.091	
	(b) in monosubstituted carbon	1.101	
	(c) in disubstituted carbon	1.073	
	(d) in trisubstituted carbon	1.070	
	(2) Olefinic, c.g. CH_2:CH_2	1.07	± 0.01
	(3) Aromatic in C_6H_6	1.094	± 0.006
	(4) Acetylenic, e.g. CH_2:C.X	1.056	± 0.003
	(5) Shortening in presence of a carbon oxygen triple bond, e.g.CH_3CN	1.115	± 0.004
	(6) In small rings, e.g. $(CH_2)_2S$	1.081	± 0.007
CARBON-NITROGEN Single Bond	(1) Paraffinic		
	(a) 4 co-valent nitrogen	1.479	
	(b) 3 co-valent nitrogen	1.472	
	(2) in C-N= e.g. CH_3NO_2	1.475	± 0.010
	(3) Aromatic in $C_6H_5NHCOCH_3$	1.426	± 0.012
	(4) Shortened (partial double bond) in h.heterocyclic systems,	1.352	± 0.005
	(5) Shortened (partial double bond) in N-C=O e.g. HCONH2	1.322	± 0.003
Triple Bond	(1) in R.C:N	1.158	± 0.002

Source: data from CRC Handbook of Chemistry and Physics, David R. Lide, Ed., CRC Press, Boca Raton, (1990) and "Tables of interatomic distances" Chem. Soc. of London, (1958).

Table 14. Bond Angle Values Between Elements

Element	Bond	Compound	Bond angle (°)		
B	H–B–H	B_2H_6	121.5	±	7.5
B	Br–B–Br	BBr_3	120	±	6
B	Cl– B–Cl	BCl_3	120	±	3
B	F–B–F	BF_3	120		
B	O–B–O	$B(OH)_3$	119.7		
N	B–N–B	$(BClNH)_3$	121		
N	F–N–F	NF_3	102.5	±	1.5
N	H–N–C	HNCS	130.25	±	0.25
N	H–N–N'	N_3H	112.65	±	0.5
N	O–N–O	NO_2Cl	126	±	2
N	O–N–O	NO_2	134.1	±	0.25
O	O–O–H	H_2O_2	100	±	2
S	Br–S–Br	$SOBr_2$	96	±	2
S	F–S–F	SOF_2	92.8	±	1
S	O–S–O	SO_2	119.54		

Source: Kennard, O., in Handbook of Chemistry and Physics, 69th ed., Weast, R. C., Ed., CRC Press, Boca Raton, Fla., 1988, F–167.

Table 15. KEY TO TABLES OF CRYSTAL STRUCTURE OF THE ELEMENTS

Table 16. THE SEVEN CRYSTAL SYSTEMS

System	Axial Lengths and Angles	Unit Cell Geometry
Cubic	$a = b = c, \quad \alpha = \beta = \gamma = 90°$	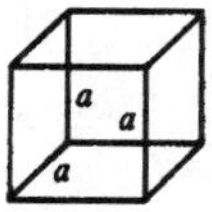
Tetragonal	$a = b \neq c, \quad \alpha = \beta = \gamma = 90°$	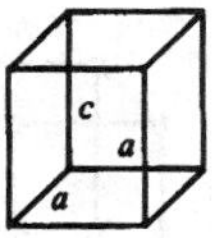
Orthorhombic	$a \neq b \neq c, \quad \alpha = \beta = \gamma = 90°$	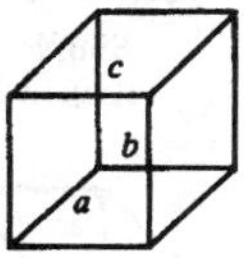
Rhombohedral	$a = b = c, \quad \alpha = \beta = \gamma \neq 90°$	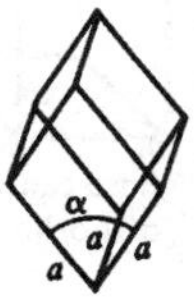
Hexagonal	$a = b \neq c, \quad \alpha = \beta = 90°, \gamma = 120°$	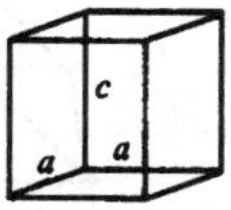
Monoclinic	$a \neq b \neq c, \quad \alpha = \gamma = 90° \neq \beta$	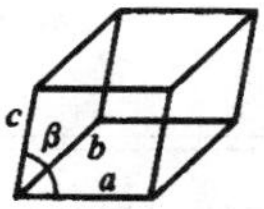
Triclinic	$a \neq b \neq c, \quad \alpha \neq \beta \neq \gamma \neq 90°$	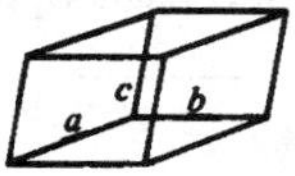

Source: James F. Shackelford, *Introduction to Materials Science for Engineers, 4th ed.*, Prentice-Hall, Upper Saddle River, NJ, 1996.

Table 17. THE FOURTEEN BRAVAIS LATTICES

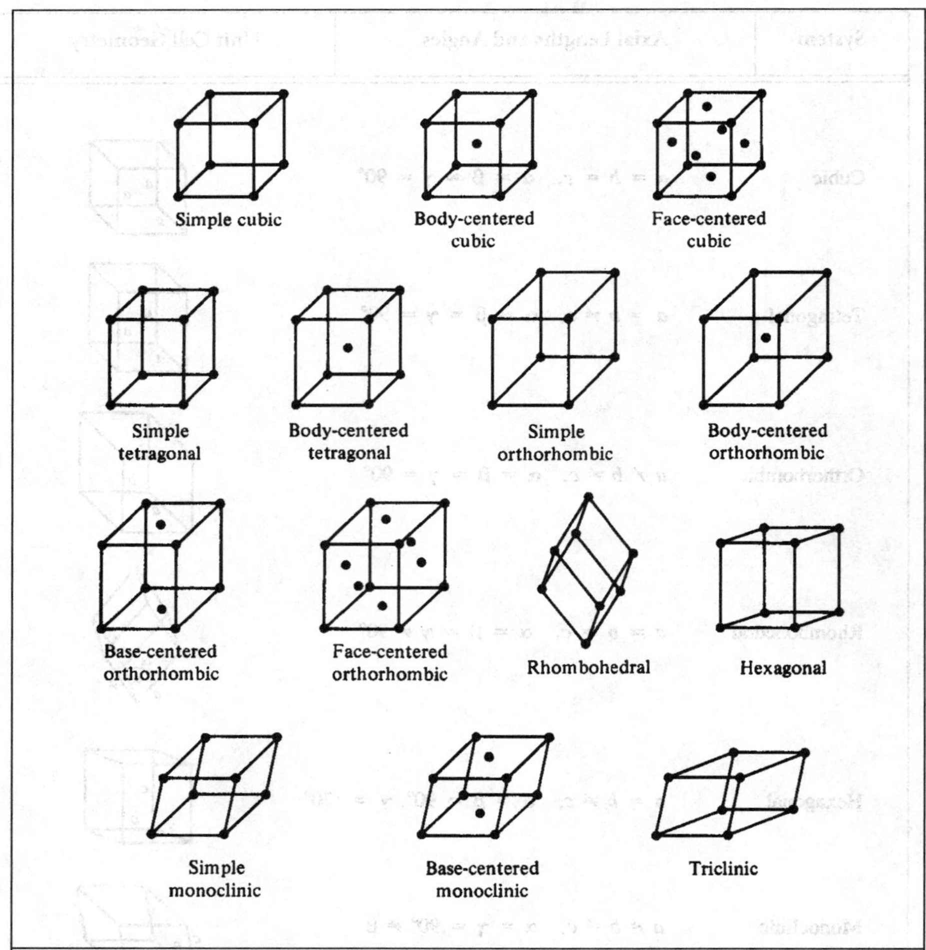

Source: James F. Shackelford, *Introduction to Materials Science for Engineers, 4th ed.*, Prentice-Hall, Upper Saddle River, NJ, 1996.

Table 18. PERIODIC TABLE OF THE BODY CENTERED CUBIC ELEMENTS

1	2	3	4	5	6	7	8	9	10	11	12	13	14	15	16	17	18
IA	IIA	IIIB	IVB	VB	VIB	VIIB	-----	VIII	-----	IB	IIB	IIIA	IVA	VA	VIA	VIIA	VIIA
3 Li																	
11 Na																	
19 K				23 V	24 Cr	25 Mn	26 Fe										
37 Rb				41 Nb	42 Mo												
55 Cs	56 Ba			73 Ta	74 W												
87 Fr	88 Ra																
							63 Eu										

Table 19. Periodic Table of the Face Centered Cubic Elements

1	2	3	4	5	6	7	8	9	10	11	12	13	14	15	16	17	18
IA	IIA	IIIB	IVB	VB	VIB	VIIB	-----	VIII	-----	IB	IIB	IIIA	IVA	VA	VIA	VIIA	VIIA
																	10 Ne
												13 Al	14 Si				18 Ar
	20 Ca								28 Ni	29 Cu			32 Ge				36 Kr
	38 Sr							45 Rh	46 Pd	47 Ag							54 Xe
								77 Ir	78 Pt	79 Au			82 Pb				86 Rn

57 La														
89 Ac														

Table 20. Periodic Table of the Hexagonal Close Packed Elements

1 IA	2 IIA	3 IIIB	4 IVB	5 VB	6 VIB	7 VIIB	8 -----	9 VIII	10 -----	11 IB	12 IIB	13 IIIA	14 IVA	15 VA	16 VIA	17 VIIA	18 VIIA
	4 Be																
	12 Mg																
			22 Ti					27 Co			30 Zn						
		39 Y	40 Zr			43 Tc	44 Ru				48 Cd						
			72 Hf			75 Re	76 Os					81 Tl					
								64 Gd	65 Tb	66 Dy	67 Ho	68 Er	69 Tm		71 Lu		

Table 21. PERIODIC TABLE OF THE HEXAGONAL ELEMENTS

1 IA	2	3	4	5	6	7	8	9	10	11	12	13	14	15	16	17	18 VIIA
	IIA											IIIA	IVA	VA	VIA	VIIA	
													6 C				
		IIIB	IVB	VB	VIB	VIIB	-----	VIII	-----	IB	IIB						
															34 Se		
															52 Te		
	57 La		59 Pr	60 Nd	61 Pm												
							95 Am	96 Cm	97 Bk								

Table 22. STRUCTURE OF CERAMICS

(SHEET 1 OF 6)

Ceramic	Structure
Borides	
Chromium Diboride (CrB_2)	hexagonal, AlB_2 structure (C-32 type) isomorphous with other transition metal diborides a=2.969Å; c=3.066Å; c/a=1.03
Hafnium Diboride (HfB_2)	hexagonal, AlB_2 structure (C-32 type) isomorphous with TiB_2 and ZrB_2 a=3.141 ± 0.002 Å; c=3.470 ± 0.002 Å; c/a=1.105
Tantalum Diboride (TaB_2)	hexagonal, AlB_2 structure (C-32 type) isomorphous with other transition metal diborides a=3.078-3.088Å; c=3.241-3.265Å; c/a=1.06-1.074 low boron composition (64 atom % boron) a=3.097-3.099Å; c=3.244-3.277Å high boron composition :(72 atom % boron) a=3.057-3.060Å; c=3.291-3.290Å
Titanium Diboride (TiB_2)	hexagonal, AlB_2 structure (C-32 type) isomorphous with ZrB_2 a=3.028-3.030Å; c=3.227-3.228Å; c/a=1.064
Zirconium Diboride (ZrB_2)	hexagonal, AlB_2 structure (C-32 type) isomorphous with TiB_2 a=3.1694-3.170Å; c=3.528-3.5365Å; c/a=1.114
Carbides	
Boron Carbide (B_4C)	rhombic, C_3 chains and B_{12} icosahedral in a NaCl structure, extended along a body diagonal

To convert Å to nm, multiply by 10^{-1}.

Source: Data compiled by J.S. Park from No. 1 Materials Index, Peter T.B. Shaffer, Plenum Press, New York, (1964); Smithells Metals Reference Book, Eric A. Brandes, ed., in association with Fulmer Research Institute Ltd. 6th ed. London, Butterworths, Boston, (1983); and Ceramic Source, American Ceramic Society (1986-1991).

Table 22. STRUCTURE OF CERAMICS

(SHEET 2 OF 6)

Ceramic	Structure
Hafnium Monocarbide (HfC)	FCC(B_1), NaCl type isomorphous with HfB and HfN a=4.46-4.643Å
Silicon Carbide (SiC)	low temperature form (β) cubic high temperature form (α) hexagonal β-SiC F43m space group a=4.349-4.358Å α-SiC C6MC space group a=3.073Å; c=15.07Å; c/a=4.899
Tantalum Monocarbide (TaC)	FCC, NaCl type (B_1) a=4.42-4.456Å
Titanium Monocarbide (TiC)	FCC, NaCl type (B_1) isomorphous with TiO and TiN a=4.315-4.3316Å
Trichromium Dicarbide (Cr_3C_2)	orthorhombic D^5_{10} type a=2.82Å, b=5.53Å, c=11.47Å
Tungsten Monocarbide (WC)	Hexagonal a=2.2897-2.90Å

To convert Å to nm, multiply by 10^{-1}.

Source: Data compiled by J.S. Park from No. 1 Materials Index, Peter T.B. Shaffer, Plenum Press, New York, (1964); Smithells Metals Reference Book, Eric A. Brandes, ed., in association with Fulmer Research Institute Ltd. 6th ed. London, Butterworths, Boston, (1983); and Ceramic Source, American Ceramic Society (1986-1991).

Table 22. STRUCTURE OF CERAMICS
(SHEET 3 OF 6)

Ceramic	Structure
Zirconium Monocarbide (ZrC)	FCC(B_1), NaCl type isomorphous with ZrB and ZrN a=4.669-4.694Å
Nitrides	
Aluminum Nitride (AlN)	hexagonal, Wurtzite structure a=3.10-3.114Å; c=4.96-4.981Å
Boron Nitride (BN)	hexagonal (common type) graphite type structure a=2.5038±0.0001Å; c=6.60±0.01Å B-N distance 1.45Å cubic zinc blende structure a=3.615Å B-N distance 1.57Å
Titanium Mononitride (TiN)	cubic a=4.23Å homogeneity range: $TiN_{0.42}$-$TiN_{1.16}$ yields a=4.213 to 4.24Å
Trisilicon tetranitride (Si_3N_4)	α hexagonal a=7.748-7.758Å; c=5.617-5.623Å β hexagonal a=7.608Å; c=2.911Å

To convert Å to nm, multiply by 10^{-1}.

Source: Data compiled by J.S. Park from No. 1 Materials Index, Peter T.B. Shaffer, Plenum Press, New York, (1964); Smithells Metals Reference Book, Eric A. Brandes, ed., in association with Fulmer Research Institute Ltd. 6th ed. London, Butterworths, Boston, (1983); and Ceramic Source, American Ceramic Society (1986-1991).

Table 22. STRUCTURE OF CERAMICS
(SHEET 4 OF 6)

Ceramic	Structure
Zirconium Mononitride (ZrN)	cubic, NaCl type, B1 a=4.567-4.63Å
Oxides	
Aluminum Oxide (Al_2O_3)	hexagonal a=4.785Å; c=12.991Å; c/a=2.72
Beryllium Oxide (BeO)	hexagonal a=2.690-2.698Å; c=4.370-4.380Å
Calcium Oxide (CaO)	cubic, NaCl type a=4.8105Å
Cerium Dioxide (CeO_2)	cubic
Dichromium Trioxide (Cr_2O_3)	trigonal rhombic
Hafnium Dioxide (HfO_2)	monoclinic to 1700 °C tetragonal above 1700 °C a=5.1170Å; b=5.1754Å; c=5.2915Å $\beta = 99.216^o$
Magnesium Oxide (MgO)	cubic, Fm3m space group a=4.313Å
Nickel monoxide (NiO)	face centered cubic, NaCl type
Silicon Dioxide (SiO_2)	hexagonal

To convert Å to nm, multiply by 10^{-1}.

Source: Data compiled by J.S. Park from No. 1 Materials Index, Peter T.B. Shaffer, Plenum Press, New York, (1964); Smithells Metals Reference Book, Eric A. Brandes, ed., in association with Fulmer Research Institute Ltd. 6th ed. London, Butterworths, Boston, (1983); and Ceramic Source, American Ceramic Society (1986-1991).

Table 22. STRUCTURE OF CERAMICS
(SHEET 5 OF 6)

Ceramic	Structure
Thorium Dioxide (ThO_2)	cubic, fluorite type a=5.59525-5.5997Å
Titanium Oxide (TiO_2)	tetragonal (rutile) a=4.594Å; c=2.958Å at 26 °C tetragonal (anatase) rhombic (brookite)
Uranium Dioxide (UO_2)	cubic, fluorite type a=5.471Å
Zircoium Oxide (ZrO_2)	to 1050 °C monoclinic a=5.1505Å; b=5.2031Å; c=5.3154 β=99.194° at room temp. 1050—2100°C tetragonal above 2100°C cubic (stabilized) a=5.132±0.006Å (8.13 mol% Y_2O_3) a=5.145±0.006Å (11.09 mol% Y_2O_3) a=5.146±0.006Å (12.08 mol% Y_2O_3) a=5.153±0.006Å (15.52 mol% Y_2O_3) a=5.162±0.006Å (17.88 mol% Y_2O_3)
Cordierite (2MgO $2Al_2O_3$ $5SiO_2$)	0rthorhombic
Mullite ($3Al_2O_3$ $2SiO_2$)	0rthorhombic a=7.54±0.03Å; b=7.693±0.03Å;c=2.890±0.01
Sillimanite (Al_2O_3 SiO_2)	0rthorhombic

To convert Å to nm, multiply by 10^{-1}.

Source: Data compiled by J.S. Park from No. 1 Materials Index, Peter T.B. Shaffer, Plenum Press, New York, (1964); Smithells Metals Reference Book, Eric A. Brandes, ed., in association with Fulmer Research Institute Ltd. 6th ed. London, Butterworths, Boston, (1983); and Ceramic Source, American Ceramic Society (1986-1991).

Table 22. STRUCTURE OF CERAMICS
(SHEET 6 OF 6)

Ceramic	Structure
Spinel (Al_2O_3 MgO)	cubic a=8.0844Å
Silicides	
Molybdenum Disilicide ($MoSi_2$)	tetragonal, $D_{4h}17$ space group isomorphous with WSi_2 a=3.197-3.20Å; c=7.85-7.871
Tungsten Disilicide (WSi_2)	tetragonal, $D_{4h}17$ space group isomorphous with $MoSi_2$ a=3.212±0.005Å; c=7.880±0.005

To convert Å to nm, multiply by 10^{-1}.

Source: Data compiled by J.S. Park from No. 1 Materials Index, Peter T.B. Shaffer, Plenum Press, New York, (1964); Smithells Metals Reference Book, Eric A. Brandes, ed., in association with Fulmer Research Institute Ltd. 6th ed. London, Butterworths, Boston, (1983); and Ceramic Source, American Ceramic Society (1986-1991).

Table 23. ATOMIC MASS OF SELECTED ELEMENTS

(SHEET 1 OF 4)

At omic Number	Element	Symbol	Atomic Mass
1	Hydrogen	H	1.008
2	Helium	He	4.003
3	Lithium	Li	6.941
4	Beryllium	Be	9.012
5	Boron	B	10.81
6	Carbon	C	12.01
7	Nitrogen	N	14.01
8	Oxygen	O	16.00
9	Fluorine	F	19.00
10	Neon	N	20.18
11	Sodium	Na	22.99
12	Magnesium	Mg	24.31
13	Aluminum	Al	26.98
14	Silicon	Si	28.09
15	Phosphorus (White)	P	30.97
16	Sulfur	S	32.06
17	Chlorine	Cl	35.45
18	Argon	Ar	39.95
19	Potassium	K	39.1
20	Calcium	Ca	40.08
21	Scandium	Sc	44.96
22	Titanium	Ti	47.9
23	Vanadium	V	50.94
24	Chromium	Cr	52.00
25	Manganese	Mn	54.94
26	Iron	Fe	55.85
27	Cobalt	Co	58.93

Source: data from James F. Shackelford, Introduction to Materials Science for Engineers, Second Edition, Macmillian Publishing Company, New York, pp.686-688, (1988).

Table 23. ATOMIC MASS OF SELECTED ELEMENTS
(SHEET 2 OF 4)

At omic Number	Element	Symbol	Atomic Mass
28	Nickel	Ni	58.71
29	Copper	Cu	63.55
30	Zinc	Zn	65.38
31	Gallium	Ga	69.72
32	Germanium	Ge	72.59
33	Arsenic	As	74.92
34	Selenium	Se	78.96
35	Bromine	Br	79.9
36	Krypton	Kr	83.8
37	Rubidium	Rb	85.47
38	Strontium	Sr	87.62
39	Yttrium	Y	88.91
40	Zirconium	Zr	91.22
41	Niobium	Nb	92.91
42	Molybdenum	Mo	95.94
43	Technetium	Tc	98.91
44	Ruthenium	Ru	101.07
45	Rhodium	Rh	102.91
46	Palladium	Pd	106.4
47	Silver	Ag	107.87
48	Cadmium	Cd	112.4
49	Indium	In	114.82
50	Tin	Sn	118.69
51	Antimony	Sb	121.75
52	Tellurium	Te	127.6
53	Iodine	I	126.9
54	Xenon	Xe	131.3
55	Cesium (-10°)	Ce	132.91

Source: data from James F. Shackelford, Introduction to Materials Science for Engineers, Second Edition, Macmillian Publishing Company, New York, pp.686-688, (1988).

Table 23. ATOMIC MASS OF SELECTED ELEMENTS
(SHEET 3 OF 4)

At omic Number	Element	Symbol	Atomic Mass
56	Barium	Ba	137.33
57	Lantium	La	138.91
58	Cerium	Ce	140.12
59	Praseodymium	Pr	140.91
60	Neodymium	Nd	144.24
61	Promethium	Pm	(145)
62	Samarium	Sm	150.4
63	Europium	Eu	151.96
64	Gadolinium	Gd	157.25
65	Terbium	Tb	158.93
66	Dysprosium	Dy	162.5
67	Holmium	Ho	164.93
68	Erbium	Er	167.26
69	Thulium	Tm	168.93
70	Ytterbium	Yb	173.04
71	Lutetium	Lu	174.97
72	Hafnium	Hf	178.49
73	Tantalum	Ta	180.95
74	Tungsten	W	183.85
75	Rhenium	Re	186.2
76	Osmium	Os	190.2
77	Iridium	Ir	192.22
78	Platinum	Pt	195.09
79	Gold	Au	196.97
80	Mercury	Hg	200.59
81	Thallium	Tl	204.37
82	Lead	Pb	207.2
83	Bismuth	Bi	208.98

Source: data from James F. Shackelford, Introduction to Materials Science for Engineers, Second Edition, Macmillian Publishing Company, New York, pp.686-688, (1988).

Table 23. ATOMIC MASS OF SELECTED ELEMENTS
(SHEET 4 OF 4)

At omic Number	Element	Symbol	Atomic Mass
84	Polonium	Po	(~210)
85	Asatine	At	(210)
86	Radon	Rn	(222)
87	Francium	Fr	(223)
88	Radium	Ra	226.03
89	Actinium	Ac	(227)
90	Thorium	Th	232.04
91	Protoactinium	Pa	231.04
92	Uranium	U	238.03
93	Neptunium	Np	237.05
94	Plutonium	Pu	(244)
95	Americium	Am	(243)
96	Curium	Cm	(247)
97	Berkelium	Bk	(247)
98	Californium	Cf	(251)
99	Einsteinium	Es	(254)
100	Fermium	Fm	(257)
101	Mendelevium	Md	(258)
102	Nobelium	No	(259)
103	Lawrencium	Lw	(260)

Source: data from James F. Shackelford, Introduction to Materials Science for Engineers, Second Edition, Macmillian Publishing Company, New York, pp.686-688, (1988).

Table 24. Solid Density of Selected Elements

(Sheet 1 of 3)

Atomic Number	Element	Symbol	Solid Density (Mg/m3)
3	Lithium	Li	0.533
4	Beryllium	Be	1.85
5	Boron	B	2.47
6	Carbon	C	2.27
11	Sodium	Na	0.966
12	Magnesium	Mg	1.74
13	Aluminum	Al	2.7
14	Silicon	Si	2.33
zz			
15	Phosphorus (White)	P	1.82
16	Sulfur	S	2.09
19	Potassium	K	0.862
20	Calcium	Ca	1.53
21	Scandium	Sc	2.99
22	Titanium	Ti	4.51
23	Vanadium	V	6.09
24	Chromium	Cr	7.19
25	Manganese	Mn	7.47
26	Iron	Fe	7.87
27	Cobalt	Co	8.8
28	Nickel	Ni	8.91
29	Copper	Cu	8.93
30	Zinc	Zn	7.13
31	Gallium	Ga	5.91
32	Germanium	Ge	5.32
33	Arsenic	As	5.78
34	Selenium	Se	4.81
37	Rubidium	Rb	1.53
38	Strontium	Sr	2.58

Source: data from James F. Shackelford, Introduction to Materials Science for Engineers, Second Edition, Macmillian Publishing Company, New York, pp.686-688, (1988).

Table 24. SOLID DENSITY OF SELECTED ELEMENTS

(SHEET 2 OF 3)

Atomic Number	Element	Symbol	Solid Density (Mg/m3)
39	Yttrium	Y	4.48
40	Zirconium	Zr	6.51
41	Niobium	Nb	8.58
42	Molybdenum	Mo	10.22
43	Technetium	Tc	11.5
44	Ruthenium	Ru	12.36
45	Rhodium	Rh	12.42
46	Palladium	Pd	12.00
47	Silver	Ag	10.50
48	Cadmium	Cd	8.65
49	Indium	In	7.29
50	Tin	Sn	7.29
51	Antimony	Sb	6.69
52	Tellurium	Te	6.25
53	Iodine	I	4.95
55	Cesium (-10°)	Ce	1.91
56	Barium	Ba	3.59
57	Lantium	La	6.17
58	Cerium	Ce	6.77
59	Praseodymium	Pr	6.78
60	Neodymium	Nd	7.00
62	Samarium	Sm	7.54
63	Europium	Eu	5.25
64	Gadolinium	Gd	7.87
65	Terbium	Tb	8.27
66	Dysprosium	Dy	8.53
67	Holmium	Ho	8.80
68	Erbium	Er	9.04

Source: data from James F. Shackelford, Introduction to Materials Science for Engineers, Second Edition, Macmillian Publishing Company, New York, pp.686-688, (1988).

Table 24. SOLID DENSITY OF SELECTED ELEMENTS
(SHEET 3 OF 3)

Atomic Number	Element	Symbol	Solid Density (Mg/m3)
69	Thulium	Tm	9.33
70	Ytterbium	Yb	6.97
71	Lutetium	Lu	9.84
72	Hafnium	Hf	13.28
73	Tantalum	Ta	16.67
74	Tungsten	W	19.25
75	Rhenium	Re	21.02
76	Osmium	Os	22.58
77	Iridium	Ir	22.55
78	Platinum	Pt	21.44
79	Gold	Au	19.28
81	Thallium	Tl	11.87
82	Lead	Pb	11.34
83	Bismuth	Bi	9.80
84	Polonium	Po	9.2
90	Thorium	Th	11.72
92	Uranium	U	19.05
94	Plutonium	Pu	19.81

Source: data from James F. Shackelford, Introduction to Materials Science for Engineers, Second Edition, Macmillian Publishing Company, New York, pp.686-688, (1988).

Table 25. DENSITY OF IRON AND IRON ALLOYS
(SHEET 1 OF 2)

Class	Metal or Alloy	Density g/cm^3
Iron and Iron Alloys	Pure iron	7.874
	Ingot iron	7.866
	Wrought iron	7.7
	Gray cast iron	7.15
	Malleable iron	7.27
	0.06% C steel	7.871
	0.23% C steel	7.859
	0.435% C steel	7.844
	1.22% C steel	7.830
Low-carbon chromium-molybdenum steels		
	0.5% Mo steel	7.86
	1Cr-0.5Mo steel	7.86
	1.25Cr-0.5Mo steel	7.86
	2.25Cr-1.0Mo steel	7.86
	5Cr-0.5Mo steel	7.78
	7Cr-0.5Mo steel	7.78
	9Cr-1Mo steel	7.67
Medium-carbon alloy steels	1Cr-0.35Mo-0.25V steel	7.86
	H11 die steel (5Cr-1.5Mo-0.4V)	7.79
Other Iron-base alloys	A-286	7.94
	16-25-6 alloy	8.08
	RA-330	8.03
	Incoloy	8.02

Data from *ASM Metals Reference Book, Third Edition*, Michael Bauccio, Ed., ASM International, Materials Park, OH, p152 (1993).

Table 25. DENSITY OF IRON AND IRON ALLOYS

(SHEET 2 OF 2)

Class	Metal or Alloy	Density g/cm^3
Other Iron-base alloys (Con't)	Incoloy T	7.98
	Incoloy 901	8.23
	T1 tool steel	8.67
	M2 tool steel	8.16
	H41 tool steel	7.88
	20W-4Cr-2V-12Co steel	8.89
	Invar (36% Ni)	8.00
	Hipernik (50% Ni)	8.25
	4% Si	7.6
	10.27%Si Si	6.97

Data from *ASM Metals Reference Book, Third Edition*, Michael Bauccio, Ed., ASM International, Materials Park, OH, p152 (1993).

Table 26. DENSITY OF WROUGHT STAINLESS STEELS* (SHEET 1 OF 2)

Type	UNS Designation	Density (Mg/m^3)
201	S20100	7.8
202	S20200	7.8
205	S20500	7.8
301	S30100	8.0
302	S30200	8.0
302B	S30215	8.0
303	S30300	8.0
304	S30400	8.0
304L	S30403	8.0
S30430	S30430	8.0
304N	S30451	8.0
305	S30500	8.0
308	S30800	8.0
309	S30900	8.0
310	S31000	8.0
314	S31400	7.8
316	S31600	8.0
316L	S31603	8.0
316N	S31651	8.0
317	S31700	8.0
317L	S31703	8.0
321	S32100	8.0
329	S32900	7.8
330	N08330	8.0
347	S34700	8.0
384	S38400	8.0
405	S40500	7.8
409	S40900	7.8

Data from *ASM Metals Reference Book, Third Edition*, Michael Bauccio, Ed., ASM International, Materials Park, OH, p360, (1993).

Table 26. DENSITY OF WROUGHT STAINLESS STEELS* (SHEET 2 OF 2)

Type	UNS Designation	Density (Mg/m^3)
410	S41000	7.8
414	S41400	7.8
416	S41600	7.8
420	S42000	7.8
422	S42200	7.8
429	S42900	7.8
430	S43000	7.8
430F	S43020	7.8
431	S43100	7.8
434	S43400	7.8
436	S43600	7.8
440A	S44002	7.8
440C	S44004	7.8
444	S44400	7.8
446	S44600	7.5
PH 13–8 Mo	S13800	7.8
15–5 PH	S15500	7.8
17–4 PH	S17400	7.8
17–7 PH	S17700	7.8

Data from *ASM Metals Reference Book, Third Edition*, Michael Bauccio, Ed., ASM International, Materials Park, OH, p360, (1993).

* Annealed Condition.

Table 27. DENSITY OF STAINLESS STEELS AND HEAT-RESISTANT ALLOYS (SHEET 1 OF 3)

Class	Metal or Alloy	Density g/cm^3
Corrrosion-resistant steel castings	CA-15	7.612
	CA-40	7.612
	CB-30	7.53
	CC-50	7.53
	CE-30	7.67
	CF-8	7.75
	CF-20	7.75
	CF-8M, CF-12M	7.75
	CF-8C	7.75
	CF-16F	7.75
	CH-20	7.72
	CK-20	7.75
	CN-7M	8.00
Heat resistant alloy castings	HA	7.72
	HC	7.53
	HD	7.58
	HE	7.67
	HF	7.75
	HH	7.72
	HI	7.72
	HK	7.75
	HL	7.72
	HN	7.83
	HT	7.92
	HU	8.04
	HW	8.14
	HX	8.14

Data from *ASM Metals Reference Book, Third Edition*, Michael Bauccio, Ed., ASM International, Materials Park, OH, p152-153 (1993).

Table 27. Density of Stainless Steels and Heat-Resistant Alloys (Sheet 2 of 3)

Class	Metal or Alloy	Density g/cm^3
Wrought stainless and heat-resisting steels	Type 301	7.9
	Type 302	7.9
	Type 302B	8.0
	Type 303	7.9
	Type 304	7.9
	Type 305	8.0
	Type 308	8.0
	Type 309	7.9
	Type 310	7.9
	Type 314	7.72
	Type 316	8.0
	Type 317	8.0
	Type 321	7.9
	Type 347	8.0
	Type 403	7.7
	Type 405	7.7
	Type 410	7.7
	Type 416	7.7
	Type 420	7.7
	Type 430	7.7
	Type 430F	7.7
	Type 431	7.7
	Types 440A, 440B, 440C	7.7
	Type 446	7.6
	Type 501	7.7
	Type 502	7.8
	19-9DL	7.97
precipitation-hardening stainless steels	PH15-7 Mo	7.804
	17-4 PH	7.8
	17-7 PH	7.81

Data from *ASM Metals Reference Book, Third Edition*, Michael Bauccio, Ed., ASM International, Materials Park, OH, p152-153 (1993).

Table 27. DENSITY OF STAINLESS STEELS AND HEAT-RESISTANT ALLOYS (SHEET 3 OF 3)

Class	Metal or Alloy	Density g/cm^3
Nickel-base alloys	D-979	8.27
	Nimonic 80A	8.25
	Nimonic 90	8.27
	M-252	8.27
	Inconel	8.51
	Inconel "x" 550	8.30
	Inconel 700	8.17
	Inconel "713C"	7.913
	Waspaloy	8.23
	René 41	8.27
	Hastelloy alloy B	9.24
	Hastelloy alloy C	8.94
	Hastelloy alloy X	8.23
	Udimet 500	8.07
	GMR-235	8.03
Cobalt-chromium-nickel-base alloys	N-155 (HS-95)	8.23
	S-590	8.36
Cobalt-base alloys	S-816	8.68
	V-36	8.60
	HS-25	9.13
	HS-36	9.04
	HS-31	8.61
	HS-21	8.30
Molybdenmun-base alloy	Mo-0.5Ti	10.2

Data from *ASM Metals Reference Book, Third Edition*, Michael Bauccio, Ed., ASM International, Materials Park, OH, p152-153 (1993).

Table 28. DENSITY OF ALUMINUM ALLOYS

(SHEET 1 OF 2)

Class	Metal or Alloy	Density g/cm^3
Pure Aluminum	Aluminum (99.996%)	2.6989
Wrought alloys	EC, 1060 alloys	2.70
	1100	2.71
	2011	2.82
	2014	2.80
	2024	2.77
	2218	2.81
	3003	2.73
	4032	2.69
	5005	2.70
	5050	2.69
	5052	2.68
	5056	2.64
	5083	2.66
	5086	2.65
	5154	2.66
	5357	2.70
	5456	2.66
	6061, 6063	2.70
	6101, 6151	2.70
	7075	2.80
	7079	2.74
	7178	2.82

Data from *ASM Metals Reference Book, Third Edition*, Michael Bauccio, Ed., ASM International, Materials Park, OH, p152 (1993).

Table 28. DENSITY OF ALUMINUM ALLOYS
(SHEET 2 OF 2)

Class	Metal or Alloy	Density g/cm^3
Casting Alloys	A13	2.66
	43	2.69
	108, A108	2.79
	A132	2.72
	D132	2.76
	F132	2.74
	138	2.95
	142	2.81
	195, B195	2.81
	214	2.65
	220	2.57
	319	2.79
	355	2.71
	356	2.68
	360	2.64
	380	2.71
	750	2.88
	40E	2.81

Data from *ASM Metals Reference Book, Third Edition*, Michael Bauccio, Ed., ASM International, Materials Park, OH, p152 (1993).

Table 29. DENSITY OF COPPER AND COPPER ALLOYS

(SHEET 1 OF 3)

Class	Metal or Alloy	Density g/cm^3
Wrought coppers	Pure copper	8.96
	Electrolytic tough pitch copper (ETP)	8.89
	Deoxidized copper, high residual phosphorus (DHP)	8.94
	Free-machining copper, 0.5% Te	8.94
	Free-machining copper, 1.0% Pb	8.94
Wrought alloys	Gilding, 95%	8.86
	Commercial bronze 90%	8.80
	Jewelry bronze, 87.5%	8.78
	Red brass, 85%	8.75
	Low brass, 80%	8.67
	Cartridge brass, 70%	8.53
	Yellow brass	8.47
	Muntz metal	8.39
	Leaded commercial bronze	8.83
	Low-leaded brass (tube)	8.50
	Medium-leaded brass	8.47
	High-leaded brass (tube)	8.53
	High-leaded brass	8.50
	Extra-high-leaded brass	8.50
	Free-cutting brass	8.50
	Leaded Muntz metal	8.41
	Forging brass	8.44
	Architectural bronze	8.47
	Inhibited admiralty	8.53
	Naval brass	8.41
	Leaded naval brass	8.44
	Manganese bronze	8.36
	Phosphor bronze, 5%	8.86
	Phosphor bronze, 8%	8.80

Data from *ASM Metals Reference Book, Third Edition*, Michael Bauccio, Ed., ASM International, Materials Park, OH, p152 (1993).

Table 29. DENSITY OF COPPER AND COPPER ALLOYS
(SHEET 2 OF 3)

Class	Metal or Alloy	Density g/cm^3
Wrought alloys (Con't)	Phosphor bronze, 10%	8.78
	Phosphor bronze, 1.25%	8.89
	Free-cutting phosphor bronze	8.89
	Cupro-nickel, 30%	8.94
	Cupro-nickel, 10%	8.94
	Nickel silver,65-18	8.73
	Nickel silver, 55-18	8.70
	High-silicon bronze	8.53
	Low-silicon bronze	8.75
	Aluminum bronze, 5% Al	8.17
	Aluminum-silicon bronze	7.69
	Aluminum bronze	7.78
	Aluminum bronze	7.58
	Beryllium copper	8.23
Casting alloys	Chromium copper (1% Cr)	8.7
	88Cu-10Sn-2Z	8.7
	88Cu-8Sn-4Zn	8.8
	89Cu-11Sn	8.78
	88Cu-6Sn-1.5Pb-4.5Zn	8.7
	87Cu-8Sn-1Pb-4Zn	8.8
	87Cu-10Sn-1Pb-2Zn	8.8
	80Cu-10Sn-10Pb	8.95
	83Cu-7Sn-7Pb-3Zn	8.93
	85Cu-5Sn-9Pb-1Zn	8.87
	78Cu-7Sn-15Pb	9.25
	70Cu-SSn-2SPb	9.30
	85Cu-5Sn-SPb-SZn	8.80
	83Cu-4Sn-6Pb-7Zn	8.6
	81Cu-3Sn-7Pb-9Zn	8.7
	76Cu-2.5Sn-6.5Pb-15Zn	8.77

Data from *ASM Metals Reference Book, Third Edition,* Michael Bauccio, Ed., ASM International, Materials Park, OH, p152 (1993).

Table 29. DENSITY OF COPPER AND COPPER ALLOYS (SHEET 3 OF 3)

Class	Metal or Alloy	Density g/cm^3
Casting alloys (Con't)	72Cu-1Sn-3Pb-24Zn	8.50
	67Cu-1Sn-3Pb-29Zn	8.45
	61Cu-1Sn-1Pb-37Zn	8.40
	Manganese bronze, 60 ksi	8.2
	Manganese bronze, 65 ksi	8.3
	Manganese bronze, 90 ksi	7.9
	Manganese bronze, 110 ksi	7.7
	Aluminum bronze, Alloy 9A	7.8
	Aluminum bronze, Alloy 9B	7.55
	Aluminum bronze, Alloy 9C	7.5
	Aluminum bronze, Alloy 9D	7.7
	Nickel silver, 12% Ni	8.95
	Nickel silver, 16% Ni	8.95
	Nickel silver, 20% Ni	8.85
	Nickel silver, 25% Ni	8.8
	Silicon bronze	8.30
	Silicon brass	8.30

Data from *ASM Metals Reference Book, Third Edition*, Michael Bauccio, Ed., ASM International, Materials Park, OH, p152 (1993).

Table 30. DENSITY OF MAGNESIUM AND MAGNESIUM ALLOYS

Class	Metal or Alloy	Density g/cm^3
Pure Magnesium	Magnesium (99.8%)	1.738
Casting alloys	AM100A	1.81
	AZ63A	1.84
	AZ81A	1.80
	AZ9lA, B, C	1.81
	AZ92A	1.82
	HK31A	1.79
	HZ32A	1.83
	ZH42, ZH62A	1.86
	ZK51A	1.81
	ZE41A	1.82
	EZ33A	1.83
	EK30A	1.79
	EK41A	1.81
Wrought alloys	M1A	1.76
	A3A	1.77
	AZ31B	1.77
	PE	1.76
	AZ61A	1.80
	AZ80A	1.80
	ZK60A, B	1.83
	ZE10A	1.76
	HM21A	1.78
	HM31A	1.81

Data from *ASM Metals Reference Book, Third Edition*, Michael Bauccio, Ed., ASM International, Materials Park, OH, p153 (1993).

Table 31. DENSITY OF NICKEL AND NICKEL ALLOYS

Class	Metal or Alloy	Density g/cm^3
Pure	Nickel (99.95% Ni+Co)	8.902
	"A" Nickel	8.885
	"D" Nickel	8.78
	Duranickel	8.26
	Cast nickel	8.34
	Monel	8.84
	"K" Mond	8.47
	Monel(cast)	8.63
	"H" Monel(cast)	8.5
	"S" Monel(cast)	8.36
	Inconel	8.51
	Inconel (cast)	8.3
	Ni-o-nel	7.86
Nickel-molybdenum-chromium-iron alloys	Hastelloy B	9.24
	Hastelloy C	8.94
	Hastelloy D	7.8
	Hastelloy F	8.17
	Hastelloy N	8.79
	Hastelloy W	9.03
	Hastelloy X	8.23
Nickel-chromium-molybdenum-copper alloys	Illium G	8.58
	Illium R	8.58
Electrical resistance alloys	80Ni-20Cr	8.4
	60Ni-24Fe-16Cr	8.147
	35Ni-4SFe-20Cr	7.95
	Constantan	8.9

Data from *ASM Metals Reference Book, Third Edition*, Michael Bauccio, Ed., ASM International, Materials Park, OH, p153 (1993).

Table 32. DENSITY OF LEAD AND LEAD ALLOYS

Class	Metal or Alloy	Density g/cm^3
Lead alloys	Chemical lead (99.90+% Pb)	11.34
	Corroding lead (99.73+% Pb)	11.36
	Arsenical lead	11.34
	Calcium lead	11.34
	5-95 solder	11.0
	20-80 solder	10.2
	50-50 solder	8.89
Antimonial lead alloys	1% antimonial lead	11.27
	Hard lead (96Pb-4Sb)	11.04
	Hard lead (94Pb-6Sb)	10.88
	8% antimonial lead	10.74
	9% antimonial lead	10.66
Lead-base babbitt alloys	Lead-base babbitt, SAE 13	10.24
	Lead-base babbitt, SAE 14	9.73
	Lead-base babbitt, Alloy 8	10.04
	Arsenical lead, Babbitt (SAE 15)	10.1
	Arsenical lead, "G" Babbitt	10.1

Data from *ASM Metals Reference Book, Third Edition*, Michael Bauccio, Ed., ASM International, Materials Park, OH, p153 (1993).

Table 33. DENSITY OF TIN AND TIN ALLOYS

Metal or Alloy	Density g/cm^3
Pure tin	7.3
Soft solder (30% Pb)	8.32
Soft solder (37% Pb)	8.42
Tin babbitt, Alloy 1	7.34
Tin babbitt, Alloy 2	7.39
Tin babbitt, Alloy 3	7.46
Tin babbitt, Alloy 4	7.53
Tin babbitt, Alloy S	7.75
White metal	7.28
Pewter	7.28

Data from *ASM Metals Reference Book, Third Edition*, Michael Bauccio, Ed., ASM International, Materials Park, OH, p153 (1993).

Table 34. DENSITY OF WROUGHT TITANIUM ALLOYS

Class	Metal or Alloy	Density (Mg/m^3)
Commercially Pure	99.5Ti	4.51
	99.2Ti	4.51
	99.1Ti	4.51
	99.0Ti	4.51
	99.2 Ti–0.2Pd	4.51
	Ti-0.8Ni-0.3Mo	4.54
Alpha Alloys	Ti-5Al-2.5Sn	4.48
	Ti-5Al-2.5Sn (low O_2)	4.48
Near Alpha Alloys	Ti-8Al-1Mo-1V	4.37
	Ti-11Sn-1Mo-2.25Al-5.0Zr-1Mo-0.2Si	4.82
	Ti-6Al-2Sn-4Zr-2Mo	4.54
	Ti-5Al-5Sn-2Zr-2Mo-0.25Si	4.51
	Ti-6Al-2Nb-1Ta-1Mo	4.48
Alpha-Beta Alloys	Ti-8Mn	4.73
	Ti-3Al-2.5V	4.48
	Ti-6Al-4V	4.43
	Ti-6Al-4V (low O_2)	4.43
	Ti-6Al-6V-2Sn	4.54
	Ti-7Al-4Mo	4.48
	Ti-6Al-2Sn-4Zr-6Mo	4.65
	Ti-6Al-2Sn-2Zr-2Mo-2Cr-0.25Si	4.57
	Ti-10V-2Fe-3Al	4.65
Beta Alloys	Ti-13V-11Cr-3Al	4.84
	Ti-8Mo-8V-2Fe-3Al	4.84
	Ti-3Al-8V-6Cr-4Mo-4Zr	4.82

Data from *ASM Metals Reference Book, Third Edition*, Michael Bauccio, Ed., ASM International, Materials Park, OH, p511, (1993).

Table 35. DENSITY OF TITANIUM AND TITANIUM ALLOYS

Metal or Alloy	Density g/cm^3
99.9% Ti	4.507
99.2% Ti	4.507
99.0% Ti	4.52
Ti-6Al-4V	4.43
Ti-5Al-2.5Sn	4.46
Ti-2Fe-2Cr-2Mo	4.65
Ti-8Mn	4.71
Ti-7Al-4Mo	4.48
Ti-4Al-4Mn	4.52
Ti-4AI-3Mo-1V	4.507
Ti-2.5Al-16V	4.65

Data from *ASM Metals Reference Book, Third Edition*, Michael Bauccio, Ed., ASM International, Materials Park, OH, p153 (1993).

Table 36. DENSITY OF ZINC AND ZINC ALLOYS

Metal or Alloy	Density g/cm^3
Pure zinc	7.133
AG40A alloy	6.6
AC41A alloy	6.7
Commercial rolled zinc 0.08% Pb	7.14
Commercial rolled zinc 0.06 Pb, 0.06 Cd	7.14
Commercial rolled zinc 0.3 Pb, 0.3 Cd	7.14
Copper-hardened, rolled zinc (1% Cu)	7.18
Rolled zinc alloy (1Cu-0.010Mg)	7.18
Zn-Cu-Ti alloy (0.8Cu, 0.15Ti)	7.18

Data from *ASM Metals Reference Book, Third Edition*, Michael Bauccio, Ed., ASM International, Materials Park, OH, p153-154 (1993).

Table 37. DENSITY OF PERMANENT MAGNET MATERIALS

Metal or Alloy	Density g/cm^3
Cunico	8.30
Cunife	8.61
Comol	8.16
Alnico I	6.89
Alnico I	7.09
Alnico III	6.89
Alnico IV	7.00
Alnico V	7.31
Alnico VI	7.42

Data from *ASM Metals Reference Book, Third Edition*, Michael Bauccio, Ed., ASM International, Materials Park, OH, p154, (1993).

Table 38. DENSITY OF PRECIOUS METALS

Metal or Alloy	Density g/cm^3
Silver	10.49
Gold	19.32
70Au-30Pt	19.92
Platinum	21.45
Pt-3.5Rh	20.9
Pt-5Rb	20.65
Pt-lORh	19.97
Pt-20Rb	18.74
Pt-30Rh	17.62
Pt-40Rb	16.63
Pt-5Ir	21.49
Pt-10Ir	2153
Pt-15Ir	2157
Pt-20Ir	21.61
Pt-25Ir	21.66
Pt-30Ir	21.70
Pt-35Ir	21.79
Pt-5Ru	20.67
Pt-10Ru	19.94
Palladium	12.02
60Pd40Cu	10.6
95.5Pd-4.5Ru	12.07
95.5Pd-45Ru	11.62

Data from *ASM Metals Reference Book, Third Edition*, Michael Bauccio, Ed., ASM International, Materials Park, OH, p154, (1993).

Table 39. DENSITY OF SUPERALLOYS

Class	Alloy	Density (Mg/m^3)
Iron-base alloys	Carpenter 20-Cb3	8.055
	Haynes 556	8.23
	Incoloy 800	7.94
	Incoloy 801	7.94
Cobalt-base alloys	Haynes 25(L-605)	9.13
	Haynes 188	9.13
	Stellite 6B	8.38
	UMCo 50	8.05
Nickel-base alloys	Hastelloy B–2	9.21
	Hastelloy C4	8.64
	Hastelloy C–276	8.90
	Hastelloy N	8.93
	Hastelloy S	8.76
	Hastelloy W	9.03
	Hastelloy X	8.23
	Inconel 600	8.42
	Inconel 625	8.44
	Inconel X750	8.25
	René 41	8.25
	Udimet 500	8.14
	Udimet 700	7.92
	Waspaloy	8.20

Data from *ASM Metals Reference Book, Third Edition*, Michael Bauccio, Ed., ASM International, Materials Park, OH, p386, (1993).

Table 40. DENSITY OF SELECTED CERAMICS

(SHEET 1 OF 3)

Class	Ceramic	Density (g/cm^3)
Borides	Chromium Diboride (CrB_2)	5.6
	Hafnium Diboride (HfB_2)	11.2
	Tantalum Diboride (TaB_2)	12.60
	Titanium Diboride (TiB_2)	4.5-4.62
	Zirconium Diboride (ZrB_2)	6.09-6.102
Carbides	Boron Carbide (B_4C)	2.51
	Hafnium Monocarbide (HfC)	12.52-12.70
	Silicon Carbide (SiC)	
	(hexagonal)	3.217
	(cubic)	3.210
	Tantalum Monocarbide (TaC)	14.48-14.65
	Titanium Monocarbide (TiC)	4.92-4.938
	Trichromium Dicarbide (Cr_3C_2)	6.70
	Tungsten Monocarbide (WC)	15.8
	Zirconium Monocarbide (ZrC)	6.44-6.73
Nitrides	Aluminum Nitride (AlN)	3.26-3.30
	Boron Nitride (BN)	
	(cubic)	3.49
	(hexagonal)	2.27
	Titanium Mononitride (TiN)	5.43
	Trisilicon tetranitride (Si_3N_4)	
	(α)	3.184
	(β)	3.187
	Zirconium Mononitride (TiN)	7.349

Source: data compiled by J.S. Park from *No. 1 Materials Index*, Peter T.B. Shaffer, Plenum Press, New York, (1964); *Smithells Metals Reference Book*, Eric A. Brandes, ed., in association with Fulmer Research Institute Ltd. 6th ed. London, Butterworths, Boston, (1983); and *Ceramic Source*, American Ceramic Society (1986-1991).

Table 40. DENSITY OF SELECTED CERAMICS

(SHEET 2 OF 3)

Class	Ceramic	Density (g/cm^3)
Oxides	Aluminum Oxide (Al_2O_3)	3.97-3.986
	Beryllium Oxide (BeO)	3.01-3.03
	Calcium Oxide (CaO)	3.32
	Cerium Dioxide (CeO_2)	7.28
	Dichromium Trioxide (Cr_2O_3)	5.21
	Hafnium Dioxide (HfO_2)	9.68
	Magnesium Oxide (MgO)	3.581
	Nickel monoxide (NiO)	6.8-7.45
	Thorium Dioxide (ThO_2)	9.821
	Titanium Oxide (TiO_2)	
	(anatase)	3.84
	(brookite)	4.17
	(rutile)	4.25
	Uranium Dioxide (UO_2)	10.949-10.97
	Zirconium Oxide (ZrO_2)	
	(monoclinic)	5.56
	(CaO stabilized)	5.5
	(MgO stabilized)	5.43
	(plasma sprayed)	5.6-5.7
	Cordierite ($2MgO\ 2Al_2O_3\ 5SiO_2$)	1.61-2.51
	Mullite ($3Al_2O_3\ 2SiO_2$)	2.6-3.26
	(theoretical)	3.16-3.22
	Sillimanite ($Al_2O_3\ SiO_2$)	3.23-3.24
	Spinel ($Al_2O_3\ MgO$)	3.580
	Zircon ($SiO_2\ ZrO_2$)	4.6

Source: data compiled by J.S. Park from *No. 1 Materials Index*, Peter T.B. Shaffer, Plenum Press, New York, (1964); *Smithells Metals Reference Book*, Eric A. Brandes, ed., in association with Fulmer Research Institute Ltd. 6th ed. London, Butterworths, Boston, (1983); and *Ceramic Source*, American Ceramic Society (1986-1991).

Table 40. DENSITY OF SELECTED CERAMICS

(SHEET 3 OF 3)

Class	Ceramic	Density (g/cm^3)
Silicides	Molybdenum Disilicide ($MoSi_2$)	6.24-6.29
	Tungsten Disilicide (WSi_2)	9.25-9.3

Source: data compiled by J.S. Park from *No. 1 Materials Index*, Peter T.B. Shaffer, Plenum Press, New York, (1964); *Smithells Metals Reference Book*, Eric A. Brandes, ed., in association with Fulmer Research Institute Ltd. 6th ed. London, Butterworths, Boston, (1983); and *Ceramic Source*, American Ceramic Society (1986-1991).

Table 41. DENSITY OF GLASSES

(SHEET 1 OF 10)

Class	Glass	Density (g/cm^3)	Temperature Range of Validity
SiO_2 glass		2.201-2.211	room temp.
	(stabilized)	2.1977	room temp.
	(~1% wt impurity)	2.094	1935°C
	(~1% wt impurity)	2.072	2048°C
	(~1% wt impurity)	2.057	2114°C
	(~1% wt impurity)	2.045	2165°C
	(~1% wt impurity)	1.929	2322°C
	(1300°C for 1 hr then 1000°C for 70 hr)	2.201	
	(1300°C for 1 hr then 1100°C for 22 hr)	2.198	
	(1300°C for 1 hr then 1200°C for 7 hr)	2.201	
	(1300°C for 1 hr then 1400°C for 5 min)	2.201	
SiO_2-Na_2O glass	(5% wt Na_2O)	2.240	20°C
	(10% wt Na_2O)	2.291	20°C
	(14.86% wt Na_2O)	2.334	20°C
	(19.55% wt Na_2O)	2.383	20°C
	(25% wt Na_2O)	2.431	20°C
	(29.20% wt Na_2O)	2.459	20°C
	(35.25% wt Na_2O)	2.498	20°C
	(39.66% wt Na_2O)	2.521	20°C
	(49.20% wt Na_2O)	2.563	20°C

Source: *data compiled by* J.S. Park *from* O. V. Mazurin, M. V. Streltsina and T. P. Shvaiko-Shvaikovskaya, *Handbook of Glass Data, Part A and Part B*, Elsevier, New York, 1983

Table 41. DENSITY OF GLASSES

(SHEET 2 OF 10)

Class	Glass	Density (g/cm^3)	Temperature Range of Validity
SiO_2-Na_2O glass (con't)	(20.1% wt Na_2O)	2.270	987°C
	(20.1% wt Na_2O)	2.240	1249°C
	(20.1% wt Na_2O)	2.220	1388°C
	(30.1% wt Na_2O)	2.270	1004°C
	(30.1% wt Na_2O)	2.230	1252°C
	(30.1% wt Na_2O)	2.205	1400°C
	(45.6% wt Na_2O)	2.260	1044°C
	(45.6% wt Na_2O)	2.225	1243°C
	(45.6% wt Na_2O)	2.190	1413°C
	(50.2% wt Na_2O)	2.250	1075°C
	(50.2% wt Na_2O)	2.215	1259°C
	(50.2% wt Na_2O)	2.180	1421°C
	(55.4% wt Na_2O)	2.245	1105°C
	(55.4% wt Na_2O)	2.205	1258°C
	(55.4% wt Na_2O)	2.165	1412°C
	(60.9% wt Na_2O)	2.250	1052°C
	(60.9% wt Na_2O)	2.190	1243°C
	(60.9% wt Na_2O)	2.145	1413°C
SiO_2-CaO glass	(30% mol CaO)	2.466	1700°C
	(35% mol CaO)	2.475	1700°C
	(39.0% mol CaO)	2.746	20°C
	(40% mol CaO)	2.542	1700°C

Source: *data compiled by* J.S. Park *from* O. V. Mazurin, M. V. Streltsina and T. P. Shvaiko-Shvaikovskaya, *Handbook of Glass Data, Part A and Part B*, Elsevier, New York, 1983

Table 41. DENSITY OF GLASSES

(SHEET 3 OF 10)

Class	Glass	Density (g/cm^3)	Temperature Range of Validity
SiO_2-CaO glass (Con't)	(42.5% mol CaO)	2.555-2.568	1700°C
	(44.6% mol CaO)	2.835	20°C
	(45% mol CaO)	2.590-2.618	1700°C
	(47.5% mol CaO)	2.602-2.604	1700°C
	(50.0% mol CaO)	2.898	20°C
	(50% mol CaO)	2.615-2.617	1700°C
	(52.5% mol CaO)	2.612-2.640	1700°C
	(52.9% mol CaO)	2.918	20°C
	(57.5% mol CaO)	2.953	20°C
	(57.5% mol CaO)	2.641-2.644	1700°C
	(60% mol CaO)	2.661	1700°C
SiO_2-PbO glass	(20.78% mol PbO)	3.6711	room temp.
	(24.90% mol PbO)	3.9606	room temp.
	(29.71% mol PbO)	4.3558	room temp.
	(34.66% mol PbO)	4.7437	room temp.
	(35.0% mol PbO)	5.10	1270K
	(40.2% mol PbO)	5.15	1270K
	(40.80% mol PbO)	5.2543	room temp.
	(44.7% mol PbO)	5.45	1270K
	(45.56% mol PbO)	5.6416	room temp.
	(49.5% mol PbO)	5.85	1270K
	(50.50% mol PbO)	6.0473	room temp.
	(52.7% mol PbO)	5.90	1270K
	(54.45% mol PbO)	6.3322	room temp.
	(58.0% mol PbO)	6.05	1270K
	(59.39% mol PbO)	6.6894	room temp.

Source: *data compiled by* J.S. Park *from* O. V. Mazurin, M. V. Streltsina and T. P. Shvaiko-Shvaikovskaya, *Handbook of Glass Data, Part A and Part B*, Elsevier, New York, 1983

Table 41. DENSITY OF GLASSES

(SHEET 4 OF 10)

Class	Glass	Density (g/cm^3)	Temperature Range of Validity
SiO_2-PbO glass (Con't)	(65.97% mol PbO)	7.0810	room temp.
	(66.7% mol PbO)	6.20	1270K
	(73.0% mol PbO)	6.42	1270K
	(80.0% mol PbO)	6.70	1270K
	(84.9% mol PbO)	7.03	1270K
	(89.0% mol PbO)	7.05	1270K
	(94.2% mol PbO)	7.45	1270K
SiO_2-Al_2O_3 glass	(0.04% wt Al_2O_3 for quintus quartz glass)	2.2000	room temp.
	(0.10% wt Al_2O_3 for Cab-O-Sil glass)	2.2025	room temp.
	(0.37% wt Al_2O_3 for I.R. vitreosil glass)	2.2043	room temp.
	(0.38% wt Al_2O_3 for Cab-O-Sil glass)	2.1977	room temp.
	(0.38% wt Al_2O_3 for quintus quartz glass)	2.1982	room temp.
	(0.41% wt Al_2O_3 for Cab-O-Sil glass)	2.2047	room temp.
	(0.47% wt Al_2O_3 for quintus quartz glass)	2.2048	room temp.
	(0.64% wt Al_2O_3 for I.R. vitreosil glass)	2.2006	room temp.
	(0.77% wt Al_2O_3 for quintus quartz glass)	2.2027	room temp.

Source: *data compiled by* J.S. Park *from* O. V. Mazurin, M. V. Streltsina and T. P. Shvaiko-Shvaikovskaya, *Handbook of Glass Data, Part A and Part B*, Elsevier, New York, 1983

Table 41. DENSITY OF GLASSES

(SHEET 5 OF 10)

Class	Glass	Density (g/cm^3)	Temperature Range of Validity
SiO_2-Al_2O_3 glass (Con't)	(1.22% wt Al_2O_3 for Cab-O-Sil glass)	2.2095	room temp.
	(1.29% wt Al_2O_3 for I.R. vitreosil glass)	2.2072	room temp.
	(2.30% wt Al_2O_3 for I.R. vitreosil glass)	2.2081	room temp.
	(2.34% wt Al_2O_3 for quintus quartz glass)	2.1994	room temp.
	(2.70% wt Al_2O_3 for Cab-O-Sil glass)	2.2031	room temp.
	(5.22% wt Al_2O_3 for quintus quartz glass)	2.2118	room temp.
	(14.82% mol Al_2O_3)	2.319	1707°C
	(14.82% mol Al_2O_3)	2.320	1813°C
	(14.82% mol Al_2O_3)	2.313	1907°C
	(14.82% mol Al_2O_3)	2.302	2008°C
	(30.08% mol Al_2O_3)	2.475	1758°C
	(30.08% mol Al_2O_3)	2.460	1858°C
	(30.08% mol Al_2O_3)	2.448	1909°C
	(30.08% mol Al_2O_3)	2.446	1975°C
	(46.92% mol Al_2O_3)	2.736	1755°C
	(46.92% mol Al_2O_3)	2.724	1803°C
	(46.92% mol Al_2O_3)	2.627	1859°C

Source: *data compiled by* J.S. Park *from* O. V. Mazurin, M. V. Streltsina and T. P. Shvaiko-Shvaikovskaya, *Handbook of Glass Data, Part A and Part B*, Elsevier, New York, 1983

Table 41. DENSITY OF GLASSES

(SHEET 6 OF 10)

Class	Glass	Density (g/cm^3)	Temperature Range of Validity
SiO_2-Al_2O_3 glass (Con't)	(46.92% mol Al_2O_3)	2.625	1910°C
	(46.92% mol Al_2O_3)	2.612	1959°C
	(70.21% mol Al_2O_3)	2.811	1966°C
	(70.21% mol Al_2O_3)	2.791	1995°C
SiO_2-B_2O_3 glasss	(35.1% mol B_2O_3)	2.0436	25°C
	(39.2% mol B_2O_3)	2.0224	25°C
	(44.2% mol B_2O_3)	2.0031	25°C
	(50.8% mol B_2O_3)	1.9865	25°C
	(53.10% mol B_2O_3)	1.892-0.0634 $x10^{-3}T$	1653K< T <1803K
	(58.4% mol B_2O_3)	1.9608	25°C
	(62.40% mol B_2O_3)	1.812-0.0475 $x10^{-3}T$	1553K< T <1733K
	(71.90% mol B_2O_3)	1.785-0.0705 $x10^{-3}T$	1303K<T<1683 K
	(72.7% mol B_2O_3)	1.9135	25°C
	(82.50% mol B_2O_3)	1.737-0.0798 $x10^{-3}T$	1203K< T <1633K
	(83.2% mol B_2O_3)	1.8838	25°C
	(88.6% mol B_2O_3)	1.8682	25°C

Source: *data compiled by* J.S. Park *from* O. V. Mazurin, M. V. Streltsina and T. P. Shvaiko-Shvaikovskaya, *Handbook of Glass Data, Part A and Part B*, Elsevier, New York, 1983

Table 41. DENSITY OF GLASSES
(SHEET 7 OF 10)

Class	Glass	Density (g/cm^3)	Temperature Range of Validity
SiO_2-B_2O_3 glasss (Con't)	(90.00% mol B_2O_3)	1.680-0.0806 x10^{-3}T	1203K< T <1633K
	(92.6% mol B_2O_3)	1.8599	25°C
	(93.91% mol B_2O_3)	1.661-0.0825 x10^{-3}T	1243K< T <1623K
	(100% mol B_2O_3)	1.8453	25°C
B_2O_3 glass		1.844-1.859	25°C
		1.693	411°C
		1.671	450°C
		1.648	500°C
		1.626	550°C
		1.609	600°C
		1.580	700°C
		1.559	800°C
		1.541	900°C
		1.528	1000°C
		1.518	1100°C
		1.509	1200°C
		1.503	1300°C
		1.498	1400°C
B_2O_3-CaO glass	(28.8% mol CaO)	2.475-2.483	25°C
	(31.2% mol CaO)	2.519-2.526	25°C
	(31.2% mol CaO)	2.334-2.341	900°C
	(31.2% mol CaO)	2.279-2.288	1000°C
	(31.2% mol CaO)	2.229-2.231	1100°C
	(31.2% mol CaO)	2.174	1200°C

Source: *data compiled by* J.S. Park *from* O. V. Mazurin, M. V. Streltsina and T. P. Shvaiko-Shvaikovskaya, *Handbook of Glass Data, Part A and Part B*, Elsevier, New York, 1983

Table 41. DENSITY OF GLASSES

(SHEET 8 OF 10)

Class	Glass	Density (g/cm^3)	Temperature Range of Validity
B_2O_3-CaO glass (Con't)	(34.7% mol CaO)	2.583-2.590	25°C
	(34.7% mol CaO)	2.309	1056°C
	(34.7% mol CaO)	2.280	1105°C
	(37.1% mol CaO)	2.622-2.629	25°C
	(37.1% mol CaO)	2.306	1105°C
	(37.1% mol CaO)	2.282	1153°C
	(37.1% mol CaO)	2.259	1200°C
	(42.5% mol CaO)	2.349	1145°C
	(42.5% mol CaO)	2.328	1193°C
	(45.7% mol CaO)	2.403	1106°C
	(45.7% mol CaO)	2.379	1156°C
	(45.7% mol CaO)	2.359	1207°C
	(50.3% mol CaO)	2.417	1156°C
	(50.3% mol CaO)	2.398	1199°C
B_2O_3-Na_2O glass	(3% mol Na_2O)	1.608	920°C
	(3% mol Na_2O)	1.601	1000°C
	(3% mol Na_2O)	1.587	1091°C
	(6% mol Na_2O)	1.705	907°C
	(6% mol Na_2O)	1.691	1000°C
	(6% mol Na_2O)	1.660	1141°C
	(8.21% mol Na_2O)	2.0112	room temp.
	(9% mol Na_2O)	1.794	890°C
	(9% mol Na_2O)	1.773	1000°C
	(9% mol Na_2O)	1.740	1109°C
	(10.33% mol Na_2O)	2.0466	room temp.

Source: *data compiled by* J.S. Park *from* O. V. Mazurin, M. V. Streltsina and T. P. Shvaiko-Shvaikovskaya, *Handbook of Glass Data, Part A and Part B*, Elsevier, New York, 1983

Table 41. DENSITY OF GLASSES

(SHEET 9 OF 10)

Class	Glass	Density (g/cm^3)	Temperature Range of Validity
B_2O_3-Na_2O glass (Con't)	(12% mol Na_2O)	1.872	901°C
	(12% mol Na_2O)	1.842	1000°C
	(12% mol Na_2O)	1.808	1106°C
	(12.03% mol Na_2O)	2.0752	room temp.
	(14.12% mol Na_2O)	2.1053	room temp.
	(15% mol Na_2O)	1.907	934°C
	(15% mol Na_2O)	1.886	1000°C
	(15% mol Na_2O)	1.848	1131°C
	(16.34% mol Na_2O)	2.0466	room temp.
	(18% mol Na_2O)	1.976	882°C
	(18% mol Na_2O)	1.935	1000°C
	(18% mol Na_2O)	1.904	1097°C
	(18.16% mol Na_2O)	2.0752	room temp.
	(20.23% mol Na_2O)	2.1053	room temp.
	(21% mol Na_2O)	2.009	910°C
	(21% mol Na_2O)	1.971	1000°C
	(21% mol Na_2O)	1.921	1136°C
	(22.07% mol Na_2O)	2.2146	room temp.
	(24% mol Na_2O)	2.054	891°C
	(24% mol Na_2O)	2.000	1000°C
	(24% mol Na_2O)	1.958	1106°C
	(24.33% mol Na_2O)	2.2493	room temp.
	(26.18% mol Na_2O)	2.2835	room temp.
	(27% mol Na_2O)	2.043	945°C

Source: *data compiled by* J.S. Park *from* O. V. Mazurin, M. V. Streltsina and T. P. Shvaiko-Shvaikovskaya, *Handbook of Glass Data, Part A and Part B*, Elsevier, New York, 1983

Table 41. DENSITY OF GLASSES

(SHEET 10 OF 10)

Class	Glass	Density (g/cm^3)	Temperature Range of Validity
B_2O_3-Na_2O glass (Con't)	(27% mol Na_2O)	1.992	1077°C
	(27% mol Na_2O)	1.954	1170°C
	(28.17% mol Na_2O)	2.3141	room temp.
	(30% mol Na_2O)	2.059	916°C
	(30% mol Na_2O)	2.018	1000°C
	(30% mol Na_2O)	1.960	1094°C
	(30.68% mol Na_2O)	2.3488	room temp.
	(32.05% mol Na_2O)	2.3591	room temp.
	(33% mol Na_2O)	2.055	909°C
	(33% mol Na_2O)	2.008	1000°C
	(33% mol Na_2O)	1.963	1076°C
	(34.20% mol Na_2O)	2.3755	room temp.
	(36% mol Na_2O)	2.075	885°C
	(36% mol Na_2O)	1.998	1000°C
	(36% mol Na_2O)	1.944	1081°C
	(36% mol Na_2O)	1.944	1081°C

Source: *data compiled by* J.S. Park *from* O. V. Mazurin, M. V. Streltsina and T. P. Shvaiko-Shvaikovskaya, *Handbook of Glass Data, Part A and Part B*, Elsevier, New York, 1983

Table 42. SPECIFIC GRAVITY OF POLYMERS

(SHEET 1 OF 7)

Class	Polymer	Specific Gravity (ASTM D792)
ABS Resins; Molded, Extruded	Medium impact	1.05—1.07
	High impact	1.02—1.04
	Very high impact	1.01—1.06
	Low temperature impact	1.02—1.04
	Heat resistant	1.06—1.08
Acrylics; Cast, Molded, Extruded	Cast Resin Sheets, Rods:	
	General purpose, type I	1.17—1.19
	General purpose, type II	1.18—1.20
	Moldings:	
	Grades 5, 6, 8	1.18—1.19
	High impact grade	1.12—1.16
	Thermoset Carbonate	
	Allyl diglycol carbonate	1.32
	Alkyds; Molded	
	Putty (encapsulating)	2.05—2.15
	Rope (general purpose)	2.20—2.22
	Granular (high speed molding)	2.21—2.24
	Glass reinforced (heavy duty parts)	2.02—2.10
Cellulose Acetate; Molded, Extruded	ASTM Grade:	
	H6—1	
	H4—1	1.29—1.31
	H2—1	1.25—1.31
	MH—1, MH—2	1.24—1.31
	MS—1, MS—2	1.23—1.30
	S2—1	1.22—1.30

Source: *data compiled by* J.S. Park *from* Charles T. Lynch, CRC Handbook of Materials Science, Vol. 3, CRC Press, Boca Raton, Florida, 1975 and Engineered Materials Handbook, Vol.2, Engineering Plastics, ASM International, Metals Park, Ohio, 1988.

Table 42. SPECIFIC GRAVITY OF POLYMERS

(SHEET 2 OF 7)

Class	Polymer	Specific Gravity (ASTM D792)
Cellulose Acetate Butyrate; Molded, Extruded	ASTM Grade:	
	H4	1.22
	MH	1.18—1.20
	S2	1.15—1.18
Cellusose Acetate Propionate; Molded, Extruded	ASTM Grade:	
	1	1.22
	3	1.20—1.21
	6	1.19
	Chlorinated Polymers:	
	Chlorinated polyether	1.4
	Chlorinated polyvinyl chloride	1.54
	Polycarbonates:	
	Polycarbonate	1.2
	Polycarbonate (40% glass fiber reinforced)	1.51
	Chlorinated Polymers	
	Chlorinated polyether	1.4
	Chlorinated polyvinyl chloride	1.54
	Polycarbonates	
	Polycarbonate	1.2
	Polycarbonate (40% glass fiber reinforced)	1.51
Diallyl Phthalates; Molded	Orlon filled	1.31—1.35
	Dacron filled	1.40—1.65
	Asbestos filled	1.50—1.96
	Glass fiber filled	1.55—1.85

Source: *data compiled by* J.S. Park *from* Charles T. Lynch, CRC Handbook of Materials Science, Vol. 3, CRC Press, Boca Raton, Florida, 1975 and Engineered Materials Handbook, Vol.2, Engineering Plastics, ASM International, Metals Park, Ohio, 1988.

Table 42. Specific Gravity of Polymers

(Sheet 3 of 7)

Class	Polymer	Specific Gravity (ASTM D792)
Fluorocarbons; Molded,Extruded	Polytrifluoro chloroethylene (PTFCE)	2.10—2.15
	Polytetrafluoroethylene (PTFE)	2.1—2.3
	Ceramic reinforced (PTFE)	2.2—2.4
	Fluorinated ethylene propylene(FEP)	2.12—2.17
	Polyvinylidene— fluoride (PVDF)	1.77
Epoxies; Cast, Molded, Reinforced	Standard epoxies (diglycidyl ethers of bisphenol A)	
	Cast rigid	1.15
	Cast flexible	1.14-1.18
	Molded	1.80-2.0
	General purpose glass cloth laminate	1.8
	High strength laminate	1.84
	Filament wound composite	2.17-2.18
Epoxies—Molded, Extruded	High performance resins (cycloaliphatic diepoxides)	
	Cast, rigid	1.24
	Molded	1.7
	Glass cloth laminate	1.97
	Epoxy novolacs	
	Cast, rigid	1.22
	Glass cloth laminate	1.97
Melamines; Molded	Filler & type	
	Unfilled	1.48
	Cellulose electrical	1.43—1.50
	Glass fiber	1.8—2.0
	Alpha cellulose and mineral	1.5(a), 1.72(mineral)

Source: *data compiled by* J.S. Park *from* Charles T. Lynch, CRC Handbook of Materials Science, Vol. 3, CRC Press, Boca Raton, Florida, 1975 and Engineered Materials Handbook, Vol.2, Engineering Plastics, ASM International, Metals Park, Ohio, 1988.

Table 42. SPECIFIC GRAVITY OF POLYMERS

(SHEET 4 OF 7)

Class	Polymer	Specific Gravity (ASTM D792)
Nylons; Molded, Extruded	**Type 6**	
	General purpose	1.12—1.14
	Glass fiber (30%) reinforced	1.35—1.42
	Cast	1.15
	Flexible copolymers	1.12—1.14
	Type 8	1.09
	Type 11	1.04
	Type 12	1.01
	6/6 Nylon	
	General purpose molding	1.13—1.15
	Glass fiber reinforced	1.37–1.47
	Glass fiber Molybdenum disulfide filled	1.37—1.41
	General purpose extrusion	1.13–1.15
	6/10 Nylon	
	General purpose	1.07—1.09
	Glass fiber (30%) reinforced	1.3
Phenolics; Molded	Type and filler	
	General: woodflour and flock	1.32—1.46
	Shock: paper, flock, or pulp	1.34—1.46
	High shock: chopped fabric or cord	1.36—1.43
	Very high shock: glass fiber	1.75—1.90
Polyacetals	Homopolymer:	
	Standard	1.425
	20% glass reinforced	1.56
	22% TFE reinforced	1.54
	Copolymer:	
	Standard	1.41
	25% glass reinforced	1.61
	High flow	1.41

Source: *data compiled by* J.S. Park *from* Charles T. Lynch, CRC Handbook of Materials Science, Vol. 3, CRC Press, Boca Raton, Florida, 1975 and Engineered Materials Handbook, Vol.2, Engineering Plastics, ASM International, Metals Park, Ohio, 1988.

Table 42. SPECIFIC GRAVITY OF POLYMERS

(SHEET 5 OF 7)

Class	Polymer	Specific Gravity (ASTM D792)
Phenolics: Molded	Arc resistant—mineral	1.5—3.0
	Rubber phenolic—woodflour or flock	1.24—1.35
	Rubber phenolic—chopped fabric	1.30—1.35
	Rubber phenolic—asbestos	1.60—1.65
	ABS–Polycarbonate Alloy	1.14
	PVC–Acrylic Alloy	
	PVC–acrylic sheet	1.35
	PVC–acrylic injection molded	1.3
Polymides	Unreinforced	1.19—1.47
	Glass reinforced	1.60—1.95
Polyester; Thermoplastic	Injection Moldings:	
	General purpose grade	1.31
	Glass reinforced grades	1.52
	Glass reinforced self extinguishing	1.58
	General purpose grade	1.31
	Glass reinforced grade	1.45
	Asbestos—filled grade	1.46
Polyesters: Thermosets	Cast polyyester	
	Rigid	1.12—1.46
	Flexible	1.06—1.25
Reinforced polyester moldings	High strength (glass fibers)	1.8—2.0
	Heat and chemical resistant (asbestos)	1.5—1.75
	Sheet molding compounds, general purpose	1.65—1.80

Source: *data compiled by* J.S. Park *from* Charles T. Lynch, CRC Handbook of Materials Science, Vol. 3, CRC Press, Boca Raton, Florida, 1975 and Engineered Materials Handbook, Vol.2, Engineering Plastics, ASM International, Metals Park, Ohio, 1988.

Table 42. SPECIFIC GRAVITY OF POLYMERS

(SHEET 6 OF 7)

Class	Polymer	Specific Gravity (ASTM D792)
Phenylene Oxides	SE—100	1.1
	SE—1	1.06
	Glass fiber reinforced	1.21–1.27
Phenylene oxides (Noryl)	Standard	1.24
	Glass fiber reinforced	1.41–1.55
	Polyarylsulfone	1.36
Polypropylene	General purpose	0.900—0.910
	High impact	0.900—0.910
	Asbestos filled	1.11—1.36
	Glass reinforced	1.04—1.22
	Flame retardant	1.2
Polyphenylene sulfide	Standard	1.34—1.35
	40% glass reinforced	1.6—1.64
Polyethylenes; Molded, Extruded	Type I—lower density (0.910—0.925)	
	Melt index 0.3—3.6	0.910—0.925
	Melt index 6—26	0.918—0.925
	Melt index 200	0.91
	Type II—medium density (0.926—0.940)	
	Melt index 20	0.93
	Melt index l.0—1.9	0.930—0.940
	Type III—higher density (0.941—0.965)	
	Melt index 0.2—0.9	0.96
	Melt index 0.l—12.0	0.950—0.955
	Melt index 1.5—15	0.96
	High molecular weight	0.94

Source: *data compiled by* J.S. Park *from* Charles T. Lynch, CRC Handbook of Materials Science, Vol. 3, CRC Press, Boca Raton, Florida, 1975 and Engineered Materials Handbook, Vol.2, Engineering Plastics, ASM International, Metals Park, Ohio, 1988.

Table 42. SPECIFIC GRAVITY OF POLYMERS

(SHEET 7 OF 7)

Class	Polymer	Specific Gravity (ASTM D792)
Olefin Copolymers; Molded	EEA (ethylene ethyl acrylate)	0.93
	EVA (ethylene vinyl acetate)	0.94
	Ethylene butene	0.95
	Propylene—ethylene	0.91
	Ionomer	0.94
	Polyallomer	0.898—0.904
Polystyrenes; Molded	Polystyrenes	
	General purpose	1.04
	Medium impact	1.04—1.07
	High impact	1.04—1.07
	Glass fiber -30% reinforced	1.29
	Styrene acrylonitrile (SAN)	1.04—1.07
	Glass fiber (30%) reinforced SAN	1.35
Polyvinyl Chloride And Copolymers; Molded, Extruded	Nonrigid—general	1.20—1.55
	Nonrigid—electrical	1.16—1.40
	Rigid—normal impact	1.32—1.44
	Vinylidene chloride	1.68—1.75
Silicones; Molded, Laminated	Fibrous (glass) reinforced silicones	1.88
	Granular (silica) reinforced silicones	1.86—2.00
	Woven glass fabric/ silicone laminate	1.75—1.8
Ureas; Molded	Alpha—cellulose filled (ASTM Type l)	1.45—1.55
	Cellulose filled (ASTM Type 2)	1.52
	Woodflour filled	1.45—1.49

Source: *data compiled by* J.S. Park *from* Charles T. Lynch, CRC Handbook of Materials Science, Vol. 3, CRC Press, Boca Raton, Florida, 1975 and Engineered Materials Handbook, Vol.2, Engineering Plastics, ASM International, Metals Park, Ohio, 1988.

Table 43. DENSITY OF 55MSI GRAPHITE/6061 ALUMINUM COMPOSITES

	Reinforcement Content (vol %)	Fiber Orientation	Density (g/cm^3)
55MSI graphite/6061 aluminum composites	34	0°	2.35
55MSI graphite/6061 aluminum composites	34	90°	2.35

Data from *ASM Engineering Materials Reference Book, Second Edition*, Michael Bauccio, Ed., ASM International, Materials Park, OH, p148,(1994).

Table 44. DENSITY OF GRAPHITE FIBER REINFORCED METALS

Composite	Fiber content (vol%)	Density (lb/in^3)
Graphite(a)/lead	41	0.270
Graphite(b)/lead	35	0.280
Graphite(a)/zinc	35	0.191
Graphite(a)/magnesium	42	0.064

(a) Thornel 75 fiber (b) Courtaulds HM fiber

Data from *ASM Engineering Materials Reference Book, Second Edition*, Michael Bauccio, Ed., ASM International, Materials Park, OH, p148,(1994).

Table 45. DENSITY OF Si_3N_4 COMPOSITES

(SHEET 1 OF 2)

Matrix	Dispersed Phase	Density (g/cm^3)
Si_3N_4+ 6 wt % Y_2O_3	None	3.26
Si_3N_4+ 6 wt % Y_2O_3	TiC	3.81
	(Ti, W) C	4.55
	WC	7.70
	TaC	6.87
	HfC	5.74
	SiC	3.24

Containing 30 Vol % of Metal Carbide Dispersoid (2 μm average particle diameter)

Data from *ASM Engineering Materials Reference Book, Second Edition*, Michael Bauccio, Ed., ASM International, Materials Park, OH, p169,(1994).

Table 45. DENSITY OF Si_3N_4 COMPOSITES

(SHEET 2 OF 2)

Matrix	Dispersed Phase	Density (g/cm^3)
Al_2O_3	TiC	4.28

Containing 30 Vol % of Metal Carbide Dispersoid (2 μm average particle diameter)

Data from *ASM Engineering Materials Reference Book, Second Edition*, Michael Bauccio, Ed., ASM International, Materials Park, OH, p169,(1994).

CHAPTER 2

Composition of Materials

List of Tables

Steels and Irons

Copper Alloys

Aluminum Alloys

Tin Alloys

Heat Resistant Alloys

Zinc Alloys

List of Tables (Continued)

Super Alloys

Glass Ceramics

Table 46. COMPOSITION LIMITS OF TOOL STEELS
(SHEET 1 OF 3)

Designations				Composition (%)								
	AISI	SAE	UNS	C	Mn	Si	Cr	Ni	Mo	W	V	Co
Molybdenum high speed steels												
	M2	M2	T11302.	0.78–0.88; 0.95–1.05	0.15–0.40	0.20–0.45	3.75–4.50	0.30 max	4.50–5.50	5.50–6.75	1.75–2.20	
Tungsten high speed steels												
	T1	T1	T12001	0.65–0.80	0.10–0.40	0.20–0.40	3.75–4.00	0.30 max		17.25–18.75	0.90–1.30	
	T15		T12015	1.50–1.60	0.15–0.40	0.15–0.40	3.75–5.00	0.30 max	1.00 max	11.75–13.00	4.50–5.25	4.75–5.25
Chromium hot work steels												
	H11	H11	T20811	0.33–0.43	0.20–0.50	0.80–1.20	4.75–5.50	0.30 max	1.10–1.60		0.30–0.60	
	H13	H13	T20813	0.32–0.45	0.20–0.50	0.80–1.20	4.75–5.50	0.30 max	1.10–1.75		0.80–1.20	

Source: data from *ASM Metals Reference Book*, Second Edition, American Society for Metals, Metals Park, Ohio 44073, p.239, (1984).

Table 46. COMPOSITION LIMITS OF TOOL STEELS

(SHEET 2 OF 3)

Designations			Composition (%)								
AISI	SAE	UNS	C	Mn	Si	Cr	Ni	Mo	W	V	Co
Tungsten hot work steels											
H21	H21	T20821	0.26–0.36	0.15–0.40	0.15–0.50	3.00–3.75	0.30 max		8.50–10.00	0.30–0.60	
H26		T20826	0.45–0.55	0.15–0.40	0.15–0.40	3.75–4.50	0.30 max		17.25–19.00	0.75–1.25	
Air–hardening medium–alloy cold work steels											
A2	A2	T30102	0.95–1.05	1.00 max	0.50 max	4.75–5.50	0.30 max	0.90–1.40		0.15–0.50	
A3		T30103	1.20–1.30	0.40–0.60	0.50 max	4.75–5.50	0.30 max	0.90–1.40		0.80–1.40	
Shock–resisting steels											
S1	S1	T41901	0.40–0.55	0.10–0.40	0.15–1.20	1.00–1.80	0.30 max	0.50 max	1.50–3.00	0.15–0.30	
S5	S5	T41905	0.50–0.65	0.60–1.00	1.75–2.25	0.35 max		0.20–1.35		0.35 max	
S7		T41907	0.45–0.55	0.20–0.80	0.20–1.00	3.00–3.50		1.30–1.80		0.20–0.30	

Source: data from *ASM Metals Reference Book*, Second Edition, American Society for Metals, Metals Park, Ohio 44073, p.239, (1984).

Table 46. COMPOSITION LIMITS OF TOOL STEELS

(SHEET 3 OF 3)

Designations			Composition (%)								
AISI	SAE	UNS	C	Mn	Si	Cr	Ni	Mo	W	V	Co
Low–alloy special–purpose tool steels											
L2		T61202	0.45–1.00	0.10–0.90	0.50 max	0.70–1.20		0.25 max		0.10–0.30	
L6	L6	T61206	0.65–0.75	0.25–0.80	0.50 max	0.60–1.20	1.25–2.00	0.50 max		0.20–0.30	
Water–hardening tool steels											
W1	W108, W109, W110, W112	T72301	0.70–1.50	0.10–0.40	0.10–0.40	0.15 max	0.20 max	0.10 max	0.15 max	0.10 max	

Source: data from *ASM Metals Reference Book*, Second Edition, American Society for Metals, Metals Park, Ohio 44073, p.239, (1984).

Table 47. COMPOSITION LIMITS OF GRAY CAST IRONS

UNS	SAE grade	Composition Limits (%)				
		TC	Mn	Si	P	S
F10004	G1800	3.40 to 3.70	0.50 to 0.80	2.80 to 2.30	0.15	0.15
F10005	G2500	3.20 to 3.50	0.60 to 0.90	2.40 to 2.00	0.12	0.15
F10009	G2500	3.40 min	0.60 to 0.90	1.60 to 2.10	0.12	0.12
F10006	G3000	3.10 to 3.40	0.60 to 0.90	2.30 to 1.90	0.10	0.16
F10007	G3500	3.00 to 3.30	0.60 to 0.90	2.20 to 1.80	0.08	0.16
F10010	G3500	3.40 min	0.60 to 0.90	1.30 to 1.80	0.08	0.12
F10011	G3500	3.50 min	0.60 to 0.90	1.30 to 1.80	0.08	0.12
F10008	G4000	3.00 to 3.30	0.70 to 1.00	2.10 to 1.80	0.07	0.16
F10012	G4000	3.10 to 3.60	0.60 to 0.90	1.95 to 2.40	0.07	0.12

Source: data from ASM Metals Reference Book, Second Edition, American Society for Metals, Metals Park, Ohio 44073, p.166, (1984).

Table 48. Composition Limits of Ductile Irons

Specification No.		Composition (%)				
Grade or Class	UNS	TC	Si	Mn	P	S
ASTM A395 ASME SA395 60-40-18	F32800	3.00 min	2.50 max	0.08 max		
ASTM A476 SAE AM55316 80-60-03	F34100	3.00 min	3.0 max	0.08 max		
SAE J434c D4018	F32800	3.20–4.10	1.80-3.00	0.10-1.00	0.015-0.10	0.005–0.035
MIL-1-24137 (Ships)						
Class A	F33101	3.0 min	2.50 max	0.08 max		
Class B	F43020	2.40-3.00	1.80-3.20	0.80-1.50	0.20 max	
Class C	F43021	2.70-3.10	2.00-3.00	1.90-2.50	0.15 max	

Source: Data from ASM Metals Reference Book, Second Edition, American Society for Metals, Metals Park, Ohio 44073, p.168-169, (1984).

Table 49. COMPOSITION RANGES FOR MALLEABLE IRONS

Type	Composition (%)				
	TC	Mn	Si	P	S
Ferritic					
Grade 32510	2.30-2.70	0.25-0.55	1.00-1.75	0.05 max	0.03-0.18
Grade 35018	2.00-2.45	0.25-0.55	1.00-1.35	0.05 max	0.03-0.18
Pearlitic	2.00-2.70	0.25-1.25	1.00-1.75	0.05 max	0.03-0.18

Source: Data from *ASM Metals Reference Book, Second Edition*, American Society for Metals, Metals Park, Ohio 44073, p170, (1984).

Table 50. Composition Ranges for Carbon Steels

AISI–SAE Designation	UNS Designation	Composition Range (%)	
		C	Mn
1015	G10150	0.12-0.18	0.30-0.60
1020	G10200	0.17-0.23	0.30-0.60
1022	G10220	0.17-0.23	0.70-1.00
1030	G10900	0.27-0.34	0.60-0.90
1040	G10400	0.36-0.44	0.60-0.90
1050	G10500	0.47-0.55	0.60-0.90
1060	G10600	0.55-0.66	0.60-0.90
1080	G10800	0.74-0.88	0.60-0.90
1095	G10950	0.90-1.04	0.30-0.50

Source: Data from *ASM Metals Reference Book, Second Edition*, American Society for Metals, Metals Park, Ohio 44073, p184, (1984).

Table 51. COMPOSITION RANGES FOR RESULFURIZED CARBON STEELS

AISI–SAE Designation	UNS Designation	Composition Range (%)		
		C	Mn	S
1118	G11180	0.14-0.20	1.30-1.60	0.08-0.13
1137	G11370	0.32-0.39	1.35-1.65	0.08-0.13
1141	G11410	0.37-0.45	1.35-1.65	0.08-0.13
1144	G11440	0.40-0.48	1.35-1.65	0.24-0.33

Source: Data from *ASM Metals Reference Book, Second Edition*, American Society for Metals, Metals Park, Ohio 44073, p185, (1984).

Table 52. Composition Ranges for Alloy Steels

(Sheet 1 of 3)

AISI–SAE Designation	UNS Designation	Composition Range (%)							
		C	Mn	P (max)	S (max)	Si	Cr	Ni	Mo
1330	G13300	0.28-0.33	1.60-1.90	0.035	0.040	0.15-0.30			
1340	G13400	0.38-0.43	1.60-1.90	0.035	0.040	0.15-0.30			
3140		0.38-0.43	0.70-0.90	0.040	0.040	0.20-0.35	0.55-0.75	1.10-1.40	
4037	G40370	0.35-0.40	0.70-0.90	0.035	0.040	0.15-0.30	0.20-0.30		
4042	G40420	0.40-0.45	0.70-0.90	0.035	0.040	0.15-0.30	0.20-0.30		
4130	G41300	0.28-0.33	0.40-0.60	0.035	0.040	0.15-0.30	0.80-1.10		0.15-0.25
4140	G41400	0.38-0.43	0.75-1.00	0.035	0.040	0.15-0.30	0.80-1.10		0.15-0.25
4150	G41500	0.48-0.53	0.75-1.00	0.035	0.040	0.15-0.30	0.80-1.10		0.15-0.25
4320	G43200	0.17-0.22	0.45-0.65	0.035	0.040	0.15-0.30	0.40-0.60	1.65-2.00	0.20-0.30
4340	G43400	0.38-0.43	0.60-0.80	0.035	0.040	0.15-0.30	0.70-0.90	1.65-2.00	0.20-0.30
4620	G46200	0.17-0.22	0.45-0.65	0.035	0.040	0.15-0.30	1.65-2.00		0.20-0.30
4820	G48200	0.18-0.23	0.50-0.70	0.035	0.040	0.15-0.30	3.25-3.75		0.20-0.30

(a) Contains 0.15% min vanadium.

Source: Data from *ASM Metals Reference Book, Second Edition*, American Society for Metals, Metals Park, Ohio 44073, p186-193, (1984).

Table 52. Composition Ranges for Alloy Steels
(Sheet 2 of 3)

AISI–SAE Designation	UNS Designation	Composition Range (%)							
		C	Mn	P (max)	S (max)	Si	Cr	Ni	Mo
5046	G50460	0.43-0.48	0.75-1.00	0.035	0.040	0.15-0.30	0.20-0.35		
50B46	G50461	0.44-0.49	0.75-1.00	0.035	0.040	0.15-0.30	0.20-0.35		
5060	G50600	0.56-0.64	0.75-1.00	0.035	0.040	0.15-0.30	0.40-0.60		
50B60	G50461	0.56-0.64	0.75-1.00	0.035	0.040	0.15-0.30	0.40-0.60		
5130	G51300	0.28-0.33	0.70-0.90	0.035	0.040	0.15-0.30	0.80-1.10		
5140	G51400	0.38-0.43	0.70-0.90	0.035	0.040	0.15-0.30	0.70-0.90		
5150	G51500	0.48-0.53	0.70-0.90	0.035	0.040	0.15-0.30	0.70-0.90		
5160	G51600	0.56-0.64	0.75-1.00	0.035	0.040	0.15-0.30	0.70-0.90		
51B60	G51601	0.56-0.64	0.75-1.00	0.035	0.040	0.15-0.30	0.70-0.90		
6150(a)	G61500	0.48-0.53	0.70-0.90	0.035	0.040	0.15-0.30	0.80-1.10		
81B45	G81451	0.43-0.48	0.75-1.00	0.035	0.040	0.15-0.30	0.35-0.55	0.20-0.40	0.08-0.15
8620	G86200	0.18-0.23	0.70-0.90	0.035	0.040	0.15-0.30	0.40-0.60	0.40-0.70	0.15-0.25
8630	G86300	0.28-0.33	0.70-0.90	0.035	0.040	0.15-0.30	0.40-0.60	0.40-0.70	0.15-0.25
8640	G86400	0.38-0.43	0.75-1.00	0.035	0.040	0.15-0.30	0.40-0.60	0.40-0.70	0.15-0.25

(a) Contains 0.15% min vanadium.

Source: Data from *ASM Metals Reference Book, Second Edition*, American Society for Metals, Metals Park, Ohio 44073, p186-193, (1984).

Table 52. Composition Ranges for Alloy Steels

(Sheet 3 of 3)

AISI–SAE Designation	UNS Designation	Composition Range (%)							
		C	Mn	P (max)	S (max)	Si	Cr	Ni	Mo
86B45	G86451	0.43-0.48	0.75-1.00	0.035	0.040	0.15-0.30	0.40-0.60	0.40-0.70	0.15-0.25
8650	G86500	0.48-0.53	0.75-1.00	0.035	0.040	0.15-0.30	0.40-0.60	0.40-0.70	0.15-0.25
8660	G86600	0.56-0.64	0.75-1.00	0.035	0.040	0.15-0.30	0.40-0.60	0.40-0.70	0.15-0.25
8740	G87400	0.38-0.43	0.75-1.00	0.035	0.040	0.15-0.30	0.40-0.60	0.40-0.70	0.20-0.30
9255	G92550	0.51-0.59	0.70-0.95	0.035	0.040	1.80-2.20			
9260	G92600	0.56-0.64	0.75-1.00	0.035	0.040	1.80-2.20			
9310	G93106	0.08-0.13	0.45-0.65	0.025	0.025	0.15-0.30	1.00-1.40	3.00-3.50	0.08-0.15
94B30	G94301	0.28-0.33	0.75-1.00	0.035	0.040	0.15-0.30	0.30-0.50	0.30-0.60	0.08-0.15

(a) Contains 0.15% min vanadium.

Source: Data from *ASM Metals Reference Book, Second Edition*, American Society for Metals, Metals Park, Ohio 44073, p186-193, (1984).

Table 53. COMPOSITION OF STAINLESS STEELS

(SHEET 1 OF 8)

Type	UNS Number	Composition (%)							
		C	Mn	Si	Cr	Ni	P	S	Others
Austenitic									
201	S20100	0.15	5.5–7.5	1.00	16.0–18.0	3.5–5.5	0.06	0.03	0.25 N
202	S20200	0.15	7.5-10.0	1.00	17.0–19.0	4.0–6.0	0.06	0.03	0.25 N
205	S20500	0.12–0.25	14.0–15.5	1.00	16.5–18.0	1.0–1.75	0.06	0.03	0.32–0.40 N
301	S30100	0.15	2.00	1.00	16.0–18.0	6.0–8.0	0.045	0.03	—
302	S30200	0.15	2.00	1.00	17.0–19.0	8.0–10.0	0.045	0.03	—
302B	S30215	0.15	2.00	2.0–3.0	17.0–19.0	8.0–10.0	0.045	0.03	—
303	S30300	0.15	2.00	1.00	17.0–19.0	8.0–10.0	0.20	0.15 min	0.6 Mo(c)
303Se	S30323	0.15	2.00	1.00	17.0–19.0	8.0–10.0	0.20	0.06	0.15 Se min
304	S30400	0.08	2.00	1.00	18.0–20.0	8.0–10.5	0.045	0.03	
304H	S30409	0.04-0.10	2.00	1.00	18.0–20.0	8.0–10.5	0.045	0.03	—
304L	S30403	0.03	2.00	1.00	18.0–20.0	8.0–12.0	0.045	0.03	—
304LN	—	0.03	2.00	1.00	18.0–20.0	8.0–10.5	0.045	0.03	0.10-0.15 N

Data from *ASM Metals Reference Book, Third Edition*, Michael Bauccio, Ed., ASM International, Materials Park, OH, p357-358, (1993).

Table 53. Composition of Stainless Steels

(Sheet 2 of 8)

Type	UNS Number	Composition (%)							
		C	Mn	Si	Cr	Ni	P	S	Others
S30430	S30430	0.08	2.00	1.00	17.0–19.0	8.0–10.0	0.045	0.03	3.0–4.0 Cu
304N	S30451	0.08	2.00	1.00	18.0-20.0	8.0–10.5	0.045	0.03	0.10–0.16 N
30S	S30500	0.12	2.00	1.00	17.0–19.0	10.S–13.0	0.045	0.03	—
308	S30800	0.08	2.00	1.00	19.0–21.0	10.0–12.0	0.045	0.03	—
309	S30900	0.20	2.00	1.00	22.0–24.0	12.0 15.0	0.045	0.03	—
309S	S30908	0.08	2.00	1.00	22.0–24.0	12.0–15.0	0.045	0.03	—
310	S31000	0.25	2.00	1.50	24.0–26.0	19.0–22.0	0.045	0.03	—
310S	S31008	0.08	2.00	1.50	24.0–26.0	19.0–22.0	0.045	0.03	—
314	S31400	0.25	2.00	1.5–3.0	23.0–26.0	19.0–22.0	0.045	0.03	—
316	S31600	0.08	2.00	1.00	16.0–18.0	10.0–14.0	0.045	0.03	2.0–3.0 Mo
316F	S31620	0.08	2.00	1.00	16.0–18.0	10.0–14.0	0.20	0.10 min	1.75–2.5 Mo
316H	S31609	0.04-0.10	2.00	1.00	16.0–18.0	10.0–14.0	0.045	0.03	2.0–3.0 Mo
316L	S31603	0.03	2.00	1.00	16.0-18.0	10.0–14.0	0.045	0.03	2.0–3.0 Mo
316LN	—	0.03	2.00	1.00	16.0–18.0	10.0–14.0	0.045	0.03	2.0–3.0 Mo;0.10–0.30 N
316N	S31651	0.08	2.00	1.00	16.0–18.0	10.0–14.0	0.045	0.03	2.0–3.0 Mo;0.10–0.16 N
317	S31700	0.08	2.00	1.00	18.0–20.0	11.0–15.0	0.045	0.03	3.0–4.0 Mo

Data from *ASM Metals Reference Book, Third Edition*, Michael Bauccio, Ed., ASM International, Materials Park, OH, p357-358, (1993).

Table 53. Composition of Stainless Steels
(Sheet 3 of 8)

Type	UNS Number	Composition (%)							
		C	Mn	Si	Cr	Ni	P	S	Others
317L	S31703	0.03	2.00	1.00	18.0–20.0	11.0–15.0	0.045	0.03	3.0–4.0 Mo
321	S32100	0.08	2.00	1.00	17.0–19.0	9.0–12.0	0.045	0.03	5 x %C Ti min
321H	S32109	0.04–0.10	2.00	1.00	17.0–19.0	9.0–12.0	0.045	0.03	5 x %C Ti min
329	S32900	0.10	2.00	1.00	25.0–30.0	3.0–6.0	0.045	0.03	1.0–2.0 Mo
330	N08330	0.08	2.00	0.75–1.5	17.0–20.0	34.0–37.0	0.04	0.03	—
347	S34700	0.08	2.00	1.00	17.0–19.0	9.0–13.0	0.045	0.03	10 x %C Nb + Ta(d)min
347H	S34709	0.04–0.10	2.00	1.00	17.0–19.0	9.0–13.0	0.045	0.03	10 x %C Nb + Ta min
348	S34800	0.08	2.00	1.00	17.0–19.0	9.0–13.0	0.045	0.03	0.2 Cu; 10 x %C Nb + Ta(d) min
348H	S34809	0.04–0.10	2.00	1.00	17.0–19.0	9.0–13.0	0.045	0.03	0.2 Cu; 10 x %C Nb + Ta(d) min
384	S38400	0.08	2.00	1.00	15.0–17.0	17.0–19.0	0.045	0.03	—

Data from *ASM Metals Reference Book, Third Edition*, Michael Bauccio, Ed., ASM International, Materials Park, OH, p357-358, (1993).

Table 53. Composition of Stainless Steels

(Sheet 4 of 8)

Type	UNS Number	Composition (%)							
		C	Mn	Si	Cr	Ni	P	S	Others
Ferritic									
405	S40500	0.08	1.00	1.00	11.5–14.5	—	0.04	0.03	0.10-0.30 Al
409	S40900	0.08	1.00	1.00	10.5–11.75	—	0.045	0.045	6 x %C Ti(e)min
429	S42900	0.12	1.00	1.00	14.0–16.0	—	0.04	0.03	—
430	S43000	0.12	1.00	1.00	16.0–18.0	—	0.04	0.03	—
430F	S43020	0.12	1.25	1.00	16.0–18.0	—	0.06	0.15	0.6 Mo(c)
430FSe	S43023	0.12	1.25	1.00	16.0–18.0	—	0.06	0.06	0.15 Se min
434	S43400	0.12	1.00	1.00	16.0–18.0	—	0.04	0.03	0.75–1.25 Mo
436	S43600	0.12	1.00	1.00	16.0–18.0	—	0.04	0.03	0.75–1.25 Mo;5 x %C Nb + Ta(f) min
442	S44200	0.20	1.00	1.00	18.0–23.0	—	0.04	0.03	—
446	S44600	0.20	1.50	1.00	23.0–27.0	—	0.04	0.03	0.25 N

Data from *ASM Metals Reference Book, Third Edition*, Michael Bauccio, Ed., ASM International, Materials Park, OH, p357-358, (1993).

Table 53. COMPOSITION OF STAINLESS STEELS
(SHEET 5 OF 8)

Type	UNS Number	Composition (%)							
		C	Mn	Si	Cr	Ni	P	S	Others
Martensitic									
403	S40300	0.15	1.00	0.50	11.5–13.0	—	0.04	0.03	—
410	S41000	0.15	1.00	1.00	11.5–13.0	—	0.04	0.03	
414	S41400	0.15	1.00	1.00	11.5–13.5	1.25–2.50	0.04	0.03	—
416	S41600	0.15	1.25	1.00	12.0–14.0	—	0.04	0.03	0.6 Mo(c)
416Se	S41623	0.15	1.25	1.00	12.0–14.0	—	0.06	0.06	0.15 Se min
420	S42000	0.15 min	1.00	1.00	12.0–14.0	—	0.04	0.03	—
420F	S42020	0.15 min	1.25	1.00	12.0–14.0	—	0.06	0.15 min	0.6Mo(c)
422	S42200	0.2–0.25	1.00	0.75	11.0–13.0	0.5–1.0	0.025	0.025	0.75–1.25 Mo;0.75–1.25 W; 0.15–0.3 V
431	S43100	0.20	1.00	1.00	15.0–17.0	1.25–2.50	0.04	0.03	—
440A	S44002	0.60–0.75	1.00	1.00	16.0–18.0	—	0.04	0.03	0.75 Mo

Data from *ASM Metals Reference Book, Third Edition*, Michael Bauccio, Ed., ASM International, Materials Park, OH, p357-358, (1993).

Table 53. Composition of Stainless Steels
(Sheet 6 of 8)

Type	UNS Number	Composition (%)							
		C	Mn	Si	Cr	Ni	P	S	Others
440B	S44003	0.75–0.95	1.00	1.00	16.0–18.0	—	0.04	0.03	0.75 Mo
440C	S44004	0.95–1.20	1.00	1.00	16.0–18.0	—	0.04	0.03	0.75 Mo
501	S50100	0.10 min	1.00	1.00	4.0–6.0	—	0.04	0.03	0.40–0.65Mo
501A	S50300	0.15	0.30–0.60	0.50–1.00	6.0–8.0	—	0.03	0.03	0.45–0.65Mo
501B	S50400	0.15	0.30–0.60	0.50–1.00	8.0–10.0	—	0.03	0.03	0.9–1.1 Mo
502	S50200	0.10	1.00	1.00	4.0–6.0	—	0.04	0.03	0.40–4.65Mo
503	S50300	0.15	1.00	1.00	6.0–8.0	—	0.04	0.04	0.45–0.65Mo
504	S50400	0.15	1.00	1.00	8.0–10.0	—	0.04	0.04	0.9–1.1 Mo
Precipitation–hardening									
PH 13–8Mo	S13800	0.05	0.10	0.10	12.25–13.25	7.5–8.5	0.01	0.008	2.0–2.5 Mo; 0.90–1.35 Al; 0.01 N
15–5 PH	S15500	0.07	1.00	1.00	14.0–15.5	3.5–5.5	0.04	0.03	2.5–4.5 Cu;0.15–0.45 Nb+Ta

Data from *ASM Metals Reference Book, Third Edition*, Michael Bauccio, Ed., ASM International, Materials Park, OH, p357-358, (1993).

Table 53. COMPOSITION OF STAINLESS STEELS
(SHEET 7 OF 8)

Type	UNS Number	Composition (%)							
		C	Mn	Si	Cr	Ni	P	S	Others
17–4 PH	S17400	0.07	1.00	1.00	15.5–17.5	3.0–5.0	0.04	0.03	3.0–5.0 Cu;0.15–0.45 Nb+Ta
17–7 PH	S17700	0.09	1.00	1.00	16.0–18.0	6.5–7.75	0.04	0.03	0.75–1.5 Al
414	S41400	0.15	1.00	1.00	11.5–13.5	1.25–2.50	0.04	0.03	—
416	S41600	0.15	1.25	1.00	12.0–14.0	—	0.04	0.03	0.6 Mo(c)
416Se	S41623	0.15	1.25	1.00	12.0–14.0	—	0.06	0.06	0.15 Se min
420	S42000	0.15 min	1.00	1.00	12.0–14.0	—	0.04	0.03	—
420F	S42020	0.15 min	1.25	1.00	12.0–14.0	—	0.06	0.15 min	0.6Mo(c)
422	S42200	0.2–0.25	1.00	0.75	11.0–13.0	0.5–1.0	0.025	0.025	0.75–1.25 Mo;0.75–1.25 W; 0.15–0.3 V
431	S43100	0.20	1.00	1.00	15.0–17.0	1.25–2.50	0.04	0.03	—
440A	S44002	0.60–0.75	1.00	1.00	16.0–18.0	—	0.04	0.03	0.75 Mo
440B	S44003	0.75–4.95	1.00	1.00	16.0–18.0	—	0.04	0.03	0.75 Mo
440C	S44004	0.95–1.20	1.00	1.00	16.0–18.0	—	0.04	0.03	0.75 Mo
501	S50100	0.10 min	1.00	1.00	4.0–6.0	—	0.04	0.03	0.40–0.65Mo
501A	S50300	0.15	0.30–0.60	0.50–1.00	6.0–8.0	—	0.03	0.03	0.45–0.65Mo

Data from *ASM Metals Reference Book, Third Edition*, Michael Bauccio, Ed., ASM International, Materials Park, OH, p357-358, (1993).

Table 53. COMPOSITION OF STAINLESS STEELS

(SHEET 8 OF 8)

Type	UNS Number	Composition (%)							
		C	Mn	Si	Cr	Ni	P	S	Others
501B	S50400	0.15	0.30–0.60	0.50–1.00	8.0–10.0	—	0.03	0.03	0.9–1.1 Mo
502	S50200	0.10	1.00	1.00	4.0–6.0	—	0.04	0.03	0.40–4.65Mo
503	S50300	0.15	1.00	1.00	6.0–8.0	—	0.04	0.04	0.45–0.65Mo
504	S50400	0.15	1.00	1.00	8.0–10.0	—	0.04	0.04	0.9–1.1 Mo

Data from *ASM Metals Reference Book, Third Edition*, Michael Bauccio, Ed., ASM International, Materials Park, OH, p357-358, (1993).

Table 54. COMPOSITION OF WROUGHT COPPERS AND COPPER ALLOYS*

(SHEET 1 OF 11)

UNS Number and Name	Nominal Composition (%)	Commercial Forms(a)
C10100 Oxygen-free electronic	99.99 Cu	F, R, W, T, P, S
C10200 Oxygen-free copper	99.95 Cu	F, R, W, T, P, S
C10300 Oxygen-free extra-low phosporus	99.95 Cu, 0.003 P	F, R, T, P, S
C10400, C10500, C10700 Oxygen-free, silver-bearing	99.95 Cu(b)	F, R, W, S
C10800 Oxygen-free, low phosporus	99.95 Cu, 0.009 P	F, R, T, P
CS11000 Electrolytic tough pitch copper	99.90 Cu, 0.04 O	F, R, W, T, P, S
C11100 Electrolytic tough pitch, anneal resistant	99.90 Cu, 0.04 O, 0.01 Cd	W
C11300, C11400, C11500, C11600 Silver-bearing tough pitch copper	99.90 Cu, 0.04 O, Ag(c)	F, R, W, T, S
C12000, C12100	99.9 Cu(d)	F, T, P
C12200 Phosphorus deoxidized copper, high residual phosphorus	99.90 Cu, 0.02 P	F, R, T, P
C12500, C12700, C12800, C12900, C13000 Fire-refined tough pitch with silver	99.88 Cu(e)	F, R, W, S
C14200 Phosphorus deoxidized, arsenical	99.68 Cu, 0.3 As, 0.02 P	F, R, T

Data from *ASM Metals Reference Book, Third Edition*, Michael Bauccio, Ed., ASM International, Materials Park, OH, p442–454, (1993).

Table 54. Composition of Wrought Coppers and Copper Alloys*

(Sheet 2 of 11)

UNS Number and Name	Nominal Composition (%)	Commercial Forms(a)
C19200	98.97 Cu, 1.0 Fe, 0.03 P	F, T
C14300	99.9 Cu, 0.1 Cd	F
C14310	99.8 Cu, 0.2 Cd	F
C14500 Phosphorus deoxidized, tellurium bearing	99.5 Cu, 0.50 Te, 0.008 P	F, R, W, T
C14700 Sulfur bearing	99.6 Cu, 0.40 S	R, W
C15000 Zirconium copper	99.8 Cu, 0.15 Zr	R, W
C15500	99.75 Cu, 0.06 P, 0.11 Mg, Ag(f)	F
C15710	99.8 Cu, 0.2 Al_2O_3	R, W
C15720	99.6 Cu, 0.4 Al_2O_3	F, R
C15735	99.3 Cu, 0.7 Al_2O_3	R
C15760	98.9 Cu, 1.1 Al_2O_3	F, R
C16200 Cadmium copper	99.0 Cu, 1.0 Cd	F, R, W

Data from *ASM Metals Reference Book, Third Edition*, Michael Bauccio, Ed., ASM International, Materials Park, OH, p442–454, (1993).

Table 54. COMPOSITION OF WROUGHT COPPERS AND COPPER ALLOYS*

(SHEET 3 OF 11)

UNS Number and Name	Nominal Composition (%)	Commercial Forms(a)
C16500	98.6 Cu, 0.8 Cd, 0.6 Sn	F, R, W
C17000 Beryllium copper	99.5 Cu, 1.7 Be, 0.20 Co	F, R
C17200 Beryllium copper	99.5 Cu, 1.9 Be , 0.20 Co	F, R, W, T, P, S
C17300 Beryllium copper	99.5 Cu, 1.9 Be, 0.40 Pb	R
C17500 Copper-cobalt-beryllium alloy	99.5 Cu, 2.5 Co, 0.6 Be	F, R
C18200, C18400, C18500 Chromium copper	99.5 Cu(g)	F, W, R, S, T
C18700 leaded copper	99.0 Cu, 1.0 Pb	R
C18900	98.75 Cu, 0.75 Sn, 0.3 Si, 0.20 Mn	R, W
C19000 Copper-nickel-phosphorus alloy	98.7 Cu, 1.1 Ni, 0.25 P	F, R, W
C19100 Copper-nickel-phosphorus-tellurium alloy	98.15 Cu, 1.1 Ni, 0.50 Te, 0.25 P	R, F
C19400	97.5 Cu, 2.4 Fe, 0.13 Zn, 0.03 P	F

Data from *ASM Metals Reference Book, Third Edition*, Michael Bauccio, Ed., ASM International, Materials Park, OH, p442–454, (1993).

Table 54. Composition of Wrought Coppers and Copper Alloys*

(Sheet 4 of 11)

UNS Number and Name	Nominal Composition (%)	Commercial Forms(a)
C19500	97.0 Cu, 1.5 Fe, 0.6 Sn, 0.10 P, 0.80 Co	F
C21000 Gilding, 95%	95.0 Cu, 5.0 Zn	F, W
C22000 Commercial bronze, 90%	90.0 Cu, 10.0 Zn	F, R, W, T
C22600 Jewelry bronze, 87.5%	87.5 Cu, 12.5 Zn	F, W
C23000 Red brass, 85%	85.0 Cu, 15.0 Zn	F, W, T, P
C24000 Low brass, 80%	80.0 Cu, 20.0 Zn	F, W
C26000 Cartridge brass, 70%	70.0 Cu, 30.0 Zn	F, R, W, T
C26800, C27000 Yellow brass	65.0 Cu, 35.0 Zn	F, R, W
C28000 Muntz metal	60.0 Cu, 40.0 Zn	F, R, T
C31400 Leaded commercial bronze	89.0 Cu, 1.75 Pb, 9.25 Zn	F, R
C31600 Leaded commercial bronze, nickel-bearing	89.0 Cu, 1.9 Pb, 1.0 Ni, 8.1 Zn	F, R
C33000 Low-leaded brass tube	66.0 Cu, 0.5 Pb, 33.5 Zn	T

Data from *ASM Metals Reference Book, Third Edition*, Michael Bauccio, Ed., ASM International, Materials Park, OH, p442–454, (1993).

Table 54. COMPOSITION OF WROUGHT COPPERS AND COPPER ALLOYS*

(SHEET 5 OF 11)

UNS Number and Name	Nominal Composition (%)	Commercial Forms(a)
C33200 High-leaded brass tube	66.0 Cu, 1.6 Pb, 32.4 Zn	T
C33500 Low-leaded brass	65.0 Cu, 0.5 Pb, 34.5 Zn	F
C34000 Medium-leaded brass	65.0 Cu, 1.0 Pb, 34.0 Zn	F, R, W, S
C34200 High-leaded brass	64.5 Cu, 2.0 Pb, 33.5 Zn	F, R
C34900	62.2 Cu, 0.35 Pb, 37.45 Zn	R, W
C35000 Medium-leaded brass	62.5 Cu, 1.1 Pb, 36.4 Zn	F, R
C35300 High-leaded brass	62.0 Cu, 1.8 Pb, 36.2 Zn	F, R
C35600 Extra-high-leaded brass	63.0 Cu, 2.5 Pb, 34.5 Zn	F
C36000 Free-cutting brass	61.5 Cu, 3.0 Pb, 35.5 Zn	F, R, S
C36500 to C36800 Leaded Muntz metal	60.0 Cu(h), 0.6 Pb, 39.4 Zn	F
C37000 Free-cutting Muntz metal	60.0 Cu, 1.0 Pb, 39.0 Zn	T
C37700 Forging brass	59.0 Cu, 2.0 Pb, 39.0 Zn	R, S

Data from *ASM Metals Reference Book, Third Edition*, Michael Bauccio, Ed., ASM International, Materials Park, OH, p442–454, (1993).

Table 54. COMPOSITION OF WROUGHT COPPERS AND COPPER ALLOYS*

(SHEET 6 OF 11)

UNS Number and Name	Nominal Composition (%)	Commercial Forms(a)
C38500 Architectural bronze	57.0 Cu, 3.0 Pb, 40.0 Zn	R, S
C40500	95 Cu, 1 Sn, 4 Zn	F
C40800	95 Cu, 2Sn, 3 Zn	F
C41100	91 Cu, 0.5 Sn, 8.5 Zn	F, W
C41300	90.0 Cu, 1.0 Sn, 9.0 Zn	F, R, W
C41500	91 Cu, 1.8 Sn, 7.2 Zn	F
C42200	87.5 Cu, 1.1 Sn, 11.4 Zn	F
C42500	88.5 Cu, 2.0 Sn, 9.5 Zn	F
C43000	87.0 Cu, 2.2 Sn, 10.8 Zn	F
C43400	85.0 Cu, 0.7 Sn, 14.3 Zn	F
C43500	81.0 Cu, 0.9 Sn, 18.1 Zn	F, T
C44300, C44400, C44500 Inhibited admiralty	71.0 Cu, 28.0 Zn, 1.0 Sn	F, W, T

Data from ASM *Metals Reference Book, Third Edition*, Michael Bauccio, Ed., ASM International, Materials Park, OH, p442–454, (1993).

Table 54. Composition of Wrought Coppers and Copper Alloys*

(Sheet 7 of 11)

UNS Number and Name	Nominal Composition (%)	Commercial Forms(a)
C46400 to C46700 Naval brass	60.0 Cu, 39.25 Zn, 0.75 Sn	F, R, T, S
C48200 Naval brass, medium-leaded	60.5 Cu, 0.7 Pb, 0.8 Sn, 38.0 Zn	F, R, S
C48500 Leaded naval brass	60.0 Cu, 1.75 Pb, 37.5 Zn, 0.75 Sn	F, R, S
C50500 Phosphor bronze, 1.25% E	98.75 Cu, 1.25 Sn, trace P	F, W
C51000 Phosphor bronze, 5% A	95.0 Cu, 5.0 Sn, trace P	F, R, W, T
C51100	95.6 Cu, 4.2 Sn, 0.2 P	F
C52100 Phosphor bronze, 8% C	92.0 Cu, 8.0 Sn, trace P	F, R, W
C52400 Phosphor bronze, 10% D	90.0 Cu, 10.0 Sn, trace P	F, R, W
C54400 Free-cutting phosphor bronze	88.0 Cu, 4.0 Pb, 4.0 Zn, 4.0 Sn	F, R
C60800 Aluminum bronze, 5%	95.0 Cu, 5.0 Al	T
C61000	92.0 Cu, 8.0 Al	R, W
C61300	92.65 Cu, 0.35 Sn, 7.0 Al	F, R, T, P, S

Data from *ASM Metals Reference Book, Third Edition*, Michael Bauccio, Ed., ASM International, Materials Park, OH, p442–454, (1993).

Table 54. COMPOSITION OF WROUGHT COPPERS AND COPPER ALLOYS*
(SHEET 8 OF 11)

UNS Number and Name	Nominal Composition (%)	Commercial Forms(a)
C61400 Aluminum bronze, D	91.0 Cu, 7.0 Al, 2.0 Fe	F, R, W, T, P, S
C61500	90.0 Cu, 8.0 Al, 2.0 Ni	F
C61800	89.0 Cu, 1.0 Fe, 10.0 Al	R
C61900	86.5 Cu, 4.0 Fe, 9.5 Al	F
C62300	87.0 Cu, 10.0 Al, 3.0 Fe	F, R
C62400	86.0 Cu, 3.0 Fe, 11.0 Al	F, R
C62500	82.7 Cu, 4.3 Fe, 13.0 Al	F, R
C63000	82.0 Cu, 3.0 Fe, 10.0 Al, 5.0 Ni	F, R
C63200	82.0 Cu, 4.0 Fe, 9.0 Al, 5.0 Ni	F, R
C63600	95.5 Cu, 3.5 Al, 1.0 Si	R, W
C63800	99.5 Cu, 2.8 Al, 1.8 Si, 0.40 Co	F
C64200	91.2 Cu, 7.0 Al	F, R

Data from *ASM Metals Reference Book, Third Edition*, Michael Bauccio, Ed., ASM International, Materials Park, OH, p442–454, (1993).

Table 54. COMPOSITION OF WROUGHT COPPERS AND COPPER ALLOYS*

(SHEET 9 OF 11)

UNS Number and Name	Nominal Composition (%)	Commercial Forms(a)
C65100 Low-silicon bronze, B	98.5 Cu, 1.5 Si	R, W, T
C65500 High-silicon bronze, A	97.0 Cu, 3.0 Si	F, R, W, T
C66700 Manganese brass	70.0 Cu, 28.8 Zn, 1.2 Mn	F, W
C67400	58.5 Cu, 36.5 Zn, 1.2 Al, 2.8 Mn, 1.0 Sn	F, R
C67500 Manganese bronze, A	58.5 Cu, 1.4 Fe, 39.0 Zn, 1.0 Sn, 0.1 Mn	R, S
C68700 Aluninum brass, arsenical	77.5 Cu, 20.5 Zn, 2.0 Al, 0.1 As	T
C68800	73.5 Cu, 22.7 Zn, 3.4 Al, 0.40 Co	F
C69000	73.3 Cu, 3.4 Al, 0.6 Ni, 22.7 Zn	F
C69400 Silicon red brass	81.5 Cu, 14.5 Zn, 4.0 Si	R
C70400	92.4 Cu, 1.5 Fe, 5.5 Ni, 0.6 Mn	F, T
C70600 Copper nickel, 10%	88.7 Cu, 1.3 Fe, 10.0 Ni	F, T
C71000 Copper nickel, 20%	79.0 Cu, 21.0 Ni	F, W, T

Data from ASM *Metals Reference Book, Third Edition*, Michael Bauccio, Ed., ASM International, Materials Park, OH, p442–454, (1993).

Table 54. COMPOSITION OF WROUGHT COPPERS AND COPPER ALLOYS*
(SHEET 10 OF 11)

UNS Number and Name	Nominal Composition (%)	Commercial Forms(a)
C71500 Copper nickel, 30%	70.0 Cu, 30.0 Ni	F, R, T
C71700	67.8 Cu, 0.7 Fe, 31.0 Ni, 0.5 Be	F, R, W
C72500	88.20 Cu, 9.5 Ni, 2.3 Sn	F, R, W, T
C73500	72.0 Cu, 18.0 Ni , 10.0 Zn	F, R, W, T
C74500 Nickel silver, 65-10	65.0 Cu, 25.0 Zn, 10.0 Ni	F, W
C75200 Nickel silver, 65-18	65.0 Cu, 17.0 Zn, 18.0 Ni	F, R, W
C75400 Nickel silver, 65-15	65.0 Cu, 20.0 Zn, 15.0 Ni	F
C75700 Nickel silver, 65-12	65.0 Cu, 23.0 Zn, 12.0 Ni	F, W

Data from ASM *Metals Reference Book, Third Edition*, Michael Bauccio, Ed., ASM International, Materials Park, OH, p442–454, (1993).

Table 54. COMPOSITION OF WROUGHT COPPERS AND COPPER ALLOYS*

(SHEET 11 OF 11)

UNS Number and Name	Nominal Composition (%)	Commercial Forms(a)
C76200	59.0 Cu, 29.0 Zn, 12.0 Ni	F, T
C77000 Nickel silver, 55-18	55.0 Cu, 27.0 Zn, 18.0 Ni	F, R, W
C72200	82.0 Cu, 16.0 Ni, 0.5 Cr, 0.8 Fe, 0.5 Mn	F, T
C78200 Leaded nickel silver, 65-8-2	65.0 Cu, 2.0 Pb, 25.0 Zn, 8.0 Ni	F

Data from *ASM Metals Reference Book, Third Edition*, Michael Bauccio, Ed., ASM International, Materials Park, OH, p442–454, (1993).

*

(a) F, flat products; R, rod; W, wire; T, tube; P, pipe; S, shapes.
(b) C10400, 8 oz/ton Ag, C10500, 10 oz/ton C10700, 25 oz/ton .
(c) C11300, 8 oz/ton Ag, C11400,10 oz/ton, C11500, 16 oz/ton C11600, 25 oz/ton
(d) C12000, 0.008 P; C12100, 0.008 P and 4 oz/ton Ag;
(e) C12700, 8 oz/ton Ag; C12800,10 oz/ton; C12900,16 oz/ton; C13000, 25 oz/ton.
(f) 8.30 oz/ton Ag.
(g) C18200, 0.9 Cr, C18400, 0.9 Cr; C18500, 0.7 Cr
(h) Rod, 61.0 Cu min.

Table 55. CLASSIFICATION OF COPPER AND COPPER ALLOYS

Family	Wrought Alloys UNS Numbers	Principal Alloying Element
Coppers, high copper alloys	C10000	< 8 at %
Brasses	C20000, C30000, C40000, C66400 to C69800	Zn
Phosphor bronzes	C50000	Sn
Aluminum bronzes	C60600 to C64200	Al
Silicon bronzes	C64700 to C66100	Si
Copper nickels, nickel silvers	C70000	Ni

Data from *ASM Metals Reference Book, Third Edition*, Michael Bauccio, Ed., ASM International, Materials Park, OH, p439, (1993).

Table 56. Composition Ranges for Cast Aluminum Alloys

(Sheet 1 of 3)

AA Number	Composition, (%)				
	Cu	Mg	Mn	Si	Others
201.0	4.6	0.35	0.35	—	0.7 Ag, 0.25 Ti
206.0	4.6	0.25	0.35	0.10 (max)	0.22 Ti, 0.15 Fe (max)
A206.0	4.6	0.25	0.35	0.05 (max)	0.22 Ti, 0.10 Fe (max)
208.0	4.0	—	—	3.0	—
242.0	4.0	1.5	—		2.0 Ni
295.0	4.5	—	—	0.8	—
296.0	4.5	—	—	2.5	
308.0	4.5	—	—	5.5	—
319.0	3.5	—	—	6.0	—
336.0	1.0	1.0		12.0	2.5 Ni
354.0	1.8	0.50	—	9.0	—
355.0	1.2	0.50	0.50 (max)	5.0	0.6 Fe (max), 0.35 Zn (max)
C355.0	1.2	0.50	0.10 (max)	5.0	0.20 Fe (max), 0.10 Zn (max)
356.0	0.25 (max)	0.32	0.35 (max)	7.0	0.6 Fe (max), 0.35 Zn (max)

Source: Data from *ASM Metals Reference Book, Second Edition*, American Society for Metals, Metals Park, Ohio 44073, p.303 (1984).

Table 56. COMPOSITION RANGES FOR CAST ALUMINUM ALLOYS

(SHEET 2 OF 3)

AA Number	Composition, (%)				
	Cu	Mg	Mn	Si	Others
A356.0	0.20 (max)	0.35	0.10 (max)	7.0	0.20 Fe (max), 0.10 Zn (max)
357.0	—	0.50	—	7.0	—
A357.0	—	0.6	—	7.0	0.15 Ti, 0.005 Be
359.0	—	0.6	—	9.0	—
360.0	—	0.50	—	9.5	2.0 Fe (max)
A360.0	—	0.50		9.5	1.3 Fe (max)
380.0	3.5	—	—	8.5	2.0 Fe (max)
A380.0	3.5	—	—	8.5	1.3 Fe (max)
383.0	2.5	—	—	10.5	—
384.0	3.8		—	11.2	3.0 Zn (max)
A384.0	3.8	—	—	11.2	1.0 Zn (max)
390.0	4.5	0.6	—	17.0	1.3 Zn (max)
A390.0	4.5	0.6	—	17.0	0.5 Zn (max)
413.0	—	—	—	12.0	2.0 Fe (max)

Source: Data from *ASM Metals Reference Book, Second Edition*, American Society for Metals, Metals Park, Ohio 44073, p.303 (1984).

Table 56. COMPOSITION RANGES FOR CAST ALUMINUM ALLOYS
(SHEET 3 OF 3)

AA Number	Composition, (%)				
	Cu	Mg	Mn	Si	Others
A413.0		—	—	12.0	1.3 Fe (max)
4430	0.6 (max)	—	—	5.2	—
A443.0	0.30 (max)	—	—	5.2	—
B443.0	0.15 (max)	—	—	5.2	—
C443.0	0.6 (max)	—	—	5.2	2.0 Fe (max)
514.0	—	4.0	—	—	—
518.0	—	8.0	—		
520.0	—	10.0	—	—	—
535.0	—	6.8	0.18	—	0.18 Ti
A535.0		7.0	0.18	—	—
B535.0	—	7.0	—	—	0.18 Ti
712.0	—	0.6	—	—	5.8 Zn, 0.5 Cr, 0.20 Ti
713.0	0.7	0.35	—	—	7.5 Zn, 0.7 Cu
771.0	—	0.9	—	—	7.0 Zn, 0.13 Cr, 0.15 Ti
850.0	1.0	—	—	—	6.2 Sn, 1.0 Ni

Source: Data from *ASM Metals Reference Book, Second Edition*, American Society for Metals, Metals Park, Ohio 44073, p.303 (1984).

Table 57. COMPOSITION RANGES FOR WROUGHT ALUMINUM ALLOYS
(SHEET 1 OF 3)

AA	Composition (%)							
Number	Al	Si	Cu	Mn	Mg	Cr	Zn	Other
1060	99.60 min	—	—	—	—	—	—	—
1100	99.00 min	—	0.12	—	—	—	—	—
2011	93.7	—	5.5	—	—	—	—	0.4Bi; 0.4Pb
2014	93.5	0.8	4.4	0.8	0.5	—	—	—
2024	93.5	—	4.4	0.6	1.5	—	—	—
2219	93.0	—	6.3	0.3	—	—	—	0.06Ti; 0.10V; 0.18Zr
2319	93.0	—	6.3	0.3	—	—	—	0.18Zn, 0.15Ti; 0.10V
2618	93.7	0.18	2.3	—	1.6	—	—	1.1Fe; 1.0Ni; 0.07Ti
3003	98.6	—	—	0.12	1.2	—	—	—
3004	97.8	—	—	1.2	1.0	—	—	—
3105	99.0		—	0.55	0.50	—	—	—
4032	85.0	12.2	0.9		1.0	—	—	0.9Ni
4043	94.8	5.2	—	—	—	—	—	—

Source: Data from *ASM Metals Reference Book, Second Edition*, American Society for Metals, Metals Park, Ohio 44073, p.292, (1984).

Table 57. COMPOSITION RANGES FOR WROUGHT ALUMINUM ALLOYS

(SHEET 2 OF 3)

AA	Composition (%)							
Number	Al	Si	Cu	Mn	Mg	Cr	Zn	Other
5005	99.2	—	—	—	0.8	—	—	—
5050	98.6	—	—	—	1.4	—	—	—
5052	97.2	—	—	—	2.5	0.25	—	—
5056	95.0	—	—	0.12	5.0	0.12	—	—
5083	94.7	—	—	0.7	4.4	0.15	—	—
5086	95.4	—	—	0.4	4.0	0.15	—	—
5154	96.2	—	—	—	3.5	0.25	—	—
5182	95.2	—	—	0.35	4.5	—	—	—
5252	97.5	—	—	—	2.5	—	—	—
5254	96.2	—	—	—	3.5	0.25	—	—
5356	94.6	—	—	0.12	5.0	0.12	—	0.13Ti
5454	96.3		—	0.8	2.7	0.12	—	—
5456	93.9	—	—	0.8	5.1	0.12	—	—
5457	98.7	—	—	0.3	1.0	—	—	—
5652	97.2	—	—	—	2.5	0.25	—	—
5657	99.2	—	—	—	0.8	—	—	—

Source: Data from *ASM Metals Reference Book, Second Edition*, American Society for Metals, Metals Park, Ohio 44073, p.292, (1984).

Table 57. COMPOSITION RANGES FOR WROUGHT ALUMINUM ALLOYS

(SHEET 3 OF 3)

AA	Composition (%)							
Number	Al	Si	Cu	Mn	Mg	Cr	Zn	Other
6005	98.7	0.8	—	—	0.5	—	—	—
6009	97.7	0.8	0.35	0.5	0.6	—	—	—
6010	97.3	1.0	0.35	0.5	0.8	—	—	—
6061	97.9	0.6	0.28	—	1.0	0.2	—	—
6063	98.9	0.4	—	—	0.7	—	—	—
6066	95.7	1.4	1.0	0.8	1.1	—	—	—
6070	96.8	1.4	0.28	0.7	0.8	—	—	—
6101	98.9	0.5	—	—	0.6	—	—	—
6151	98.2	0.9	—	—	0.6	0.25	—	—
6201	98.5	0.7	—	—	0.8	—	—	—
6205	98.4	0.8	—	0.1	0.5	0.1	—	0.1Zr
7049	88.2	—	1.5	—	2.5	0.15	7.6	—
7075	90.0	—	1.6	—	2.5	0.23	5.6	—

Source: Data from *ASM Metals Reference Book, Second Edition*, American Society for Metals, Metals Park, Ohio 44073, p.292, (1984).

Table 58. COMPOSITION OF TIN AND TIN ALLOYS

Grade			Composition (% max)										
ASTM B339	Designation	Class	Sn	Sb	As	Bi	Cd	Cu	Fe	Pb	Ni + Co	S	Zn
AAA	Electrolytic	Extra-high purity	99.98	0.008	0.0005	0.001	0.001	0.002	0.005	0.010	0.005	0.002	0.001
AA	Electrolytic	High purity	99.95	0.02	0.01	0.01	0.01	0.02	0.01	0.02	0.01	0.01	0.005
A	A. Straits	High purity; commercial	99.80	0.04	0.05	0.015	0.001	0.04	0.015	0.05	0.01	0.01	0.005
B	B	General purpose	99.80	—	0.05	—	—	—	—	—	—	—	—
C	C	Intermediate grade	99.65	—	—	—	—	—	—	—	—	—	—
D	D	Lower intermediate grade	99.50	—	—	—	—	—	—	—	—	—	—
E	E	Common	99.00	—	—	—	—	—	—	—	—	—	—

Data from *ASM Metals Reference Book, Third Edition*, Michael Bauccio, Ed., ASM International, Materials Park, OH, p488, (1993).

Table 59. Compositions of ACI Heat-Resistant Casting Alloys

(Sheet 1 of 2)

ACI Designation	UNS Number	ASTM specification *	Composition (%) †							
			C	Cr	Ni	Si (max)	Mn	P & S	Mb	Other
HC	J92605	A297, A608	0.50 max	26 to 30	4 max	2.00	1	0.04 max	0.5 max	
HD	J93005	A297, A608	0.50 max	26 to 30	4 to 7	2.00	1.5	0.04 max	0.5 max	
HE	J93403	A297, A608	0.20 to 0.50	26 to 30	8 to 11	2.00	2	0.04 max	0.5 max	
HF	J92603	A297, A608	0.20 to 0.40	19 to 23	9 to 12	2.00	2	0.04 max	0.5 max	
HH	J93503	A297, A608	0.20 to 0.50	24 to 28	11 to 14	2.00	2	0.04 max	0.5 max	0.2 N max
HI	J94003	A297, A567, A608	0.20 to 0.50	26 to 30	14 to 18	2.00	2	0.04 max	0.5 max	
HK	J94224	A297, A351, A567, A608	0.20 to 0.60	24 to 28	18 to 22	2.00	2	0.04 max	0.5 max	
HL	J94604	A297, A608	0.20 to 0.60	28 to 32	18 to 22	2.00	2	0.04 max	0.5 max	
HN	J94213	A297, A608	0.20 to 0.50	19 to 32	23 to 27	2.00	2	0.04 max	0.5 max	
HP	—	A297	0.35 to 0.75	24 to 28	33 to 37	2.00	2	0.04 max	0.5 max	
HT	J94605	A297,A351,A567,A608	0.35 to 0.75	13 to17	33 to37	2.50	2	0.04 max	0.5 max	
HU	—	A297, A608	0.35 to 0.75	17 to 21	37 to 41	2.50	2	0.04 max	0.5 max	

Data from *ASM Metals Reference Book, Third Edition*, Michael Bauccio, Ed., ASM International, Materials Park, OH, p384, (1993).

Table 59. Compositions of ACI Heat-Resistant Casting Alloys
(Sheet 2 of 2)

ACI Designation	UNS Number	ASTM specification *	Composition (%) †							
			C	Cr	Ni	Si (max)	Mn	P & S	Mb	Other
HW	—	A297, A608	0.35 to 0.75	10 to 14	58 to 62	2.50	2	0.04 max	0.5 max	
HX	—	A297, A608	0.35 to 0.75	15 to 19	64 to 68	2.50	2	0.04 max	0.5 max	

Data from *ASM Metals Reference Book, Third Edition*, Michael Bauccio, Ed., ASM International, Materials Park, OH, p384, (1993).

* ASTM designations are same as ACI designations.

† Rem Fe in all compositions.

Table 60. Composition of Zinc Die Casting Alloys

Alloy	Form	Composition (% max)								
		Cu	Al	Mg	Pb	Cd	Sn	Fe	Others	Zn
AG40A	Ingot	0.10	3.9 to 4.3	0.025 to 0.05	0.004	0.003	0.002	0.075	Ni 0.02, Cr 0.02, Si 0.035, Mn 0.5	rem
AC41A	Die castings	0.25 to 0.75	3.5 to 4.3	0.020 to 0.05(a)	0.005	0.004	0.003	0.100	Ni 0.02, Cr 0.02, Si 0.035, Mn 0.5	rem
Alloy 7	Ingot	0.75 to 1.25	3.9 to 4.3	0.03 to 0.06	0.004	0.003	0.002	0.075	Ni 0.02, Cr 0.02, Si 0.035, Mn 0.5	rem
ILZRO 16	Die castings	0.75 to 1.25	3.5 to 4.3	0.03 to 0.08(a)	0.005	0.004	0.003	0.100	Ni 0.02, Cr 0.02, Si 0.035, Mn 0.5	rem
	Die castings	0.25	3.5 to 4.3	0.010 to 0.02	0.0020	0.0020	0.0010	0.050	—	rem
	Die castings	1.0 to 1.5	0.01 to 0.04	—	—	—	—		Ti 0.15–0.25, Cr 0.10–0.20, Ti + Cr 0.30–0.40	rem

(a) Magnesium content may be as low as 0.015% provided that lead, cadmium and tin contents do not exceed 0.003. 0.003 and 0.001%, respectively.

Data from *ASM Metals Reference Book, Third Edition*, Michael Bauccio, Ed., ASM International, Materials Park, OH, p490, (1993).

Table 61. COMPOSITIONS OF WROUGHT SUPERALLOYS

(SHEET 1 OF 2)

Alloy	UNS number	Composition (%)										
		Cr	Ni	Co	Mo	W	Nb	Ti	Al	Fe	C	Other
Astroloy	—	15.0	56.5	15.0	5.25	—	—	3.5	4.4	<0.3	0.06	0.03 B; 0.06 Zr
D–979	N09979	15.0	45.0	—	4.0	4.0	—	3.0	1.0	27.0	0.05	0.01 B
IN 102	N06102	15.0	67.0	—	2.9	3.0	2.9	0.5	0.5	7.0	0.06	0.005 B; 0.02 Mg; 0.03 Zr
Inconel 706	N09706	16.0	41.5	—	—	—	—	1.75	0.2	37.5	0.03	2.9(Nb+Ta); 0.15 Cu max
Inconel 718	N07718	19.0	52.5	—	3.0	—	5.1	0.9	0.5	18.5	0.08 max	0.15 Cu max
Inconel 751	—	15.5	72.5	—	—	—	1.0	2.3	1.2	7.0	0.05	0.25 Cu max
Inconel X750	N07750	15.5	73.0	—	—	—	1.0	2.5	0.7	7.0	0.04	0.25 Cu max
M252	N07252	19.0	56.5	10.0	10.0	—	—	2.6	1.0	<0.75	0.15	0.005 B
Nimonic 80A	N07080	19.5	73.0	1.0	—	—	—	2.25	1.4	1.5	0.05	0.10 Cu max
Nimonic 90	N07090	19.5	55.5	18.0	—	—	—	2.4	1.4	1.5	0.06	—
Nimonic 95	—	19.5	53.5	18.0	—	—	—	2.9	2.0	5.0 max	0.15 max	+B; +Zr
Nimonic 100	—	11.0	56.0	20.0	5.0	—	—	1.5	5.0	2.0 max	0.30 max	+B; +Zr

Data from *ASM Metals Reference Book, Third Edition*, Michael Bauccio, Ed., ASM International, Materials Park, OH, p382-383, (1993).

Table 61. COMPOSITIONS OF WROUGHT SUPERALLOYS

(SHEET 2 OF 2)

Alloy	UNS number	Composition (%)										
		Cr	Ni	Co	Mo	W	Nb	Ti	Al	Fe	C	Other
Nimonic 105	—	15.0	54.0	20.0	5.0	—	—	1.2	4.7	—	0.08	0.005 B
Nimonic 115	—	15.0	55.0	15.0	4.0	—	—	4.0	5.0	1.0	0.20	0.04 Zr
Nimonic 263	—	20.0	51.0	20.0	5.9	—	—	2.1	0.45	0.7 max	0.06	—
Pyromet 860	—	13.0	44.0	4.0	6.0	—	—	3.0	1.0	28.9	0.05	0.01 B
René 41	N07041	19.0	55.0	11.0	10.0	—	—	3.1	1.5	<0.3	0.09	0.01 B
René 95	—	14.0	61.0	8.0	3.5	3.5	3.5	2.5	3.5	<0.3	0.16	0.01 B; 0.05 Zr
Udimet 500	N07500	19.0	48.0	19.0	4.0	—	—	3.0	3.0	4.0 max	0.08	0.005 B
Udimet 520	—	19.0	57.0	12.0	6.0	1.0	—	3.0	2.0	—	0.08	0.005 B
Udimet 700	—	15.09	53.0	18.5	5.0	—	—	3.4	4.3	<1.0	0.07	0.03 B
Udimet 710	—	18.0	55.0	14.8	3.0	1.5	—	5.0	2.5	—	0.07	0.01 B
Unitemp AF2–1DA	—	12.0	59.0	10.0	3.0	6.0	—	3.0	4.6	<0.5	0.35	1.5 Ta; 0.015 B; 0.1 Zr
Waspaloy	N07001	19.5	57.0	13.5	4.3	—	—	3.0	1.4	2.0 max	0.07	0.006 B; 0.09 Zr

Data from *ASM Metals Reference Book, Third Edition*, Michael Bauccio, Ed., ASM International, Materials Park, OH, p382-383, (1993).

Table 62. TYPICAL COMPOSITION OF GLASS-CERAMICS

Glass–Ceramic	Typical composition (wt %)							
	SiO_2	Li_2O	Al_2O_3	MgO	ZnO	B_2O_3	TiO_2 (nucleating agent)	P_2O_5 (nucleating agent)
Li_2O-Al_2O_3-SiO_2 system	74	4	16	—	—	—	6	—
MgO-Al_2O_3-SiO_2 system	65	—	19	9	—	—	7	—
Li_2O-MgO-SiO_2 system	73	11	—	7	—	6	—	3
Li_2O-ZnO-SiO_2 system	58	23	—	—	16	—	—	3

Source: data compiled by J.S. Park from P.C. McMillan, *Glass-Ceramics, 2nd edition*, Academic Press, (1979).

CHAPTER 3 **Phase Diagram Sources**

List of Tables

Phase Diagrams are especially useful tools for the field of materials science and engineering. In the last decade, a substantial effort has been made within the materials community to provide a comprehensive set of accurate phase equilibria information. Cooperative efforts involving academia, industry, and government have been coordinated through the professional societies, ASM International and the American Ceramic Society. As a result, the following references are available and new updates will become available on a regular basis.

Table 63. PHASE DIAGRAM SOURCES

Society	Source
American Ceramic Society	*Phase Diagrams for Ceramists,* Vols. 1-12, American Ceramic Society, Westerville, Ohio, 1964, 1969, 1975, 1981, 1983, 1987, 1989, 1989, 1992, 1994, 1995, and 1996.
ASM International	*Binary Alloy Phase Diagrams, Second Edition,* Vols. 1, 2 and 3, T.B. Massalski, et.al., ed., ASM International, Materials Park, Ohio, 1990.
ASM International	*ASM Handbook,* Vol. 3, ASM International, Materials Park, Ohio, 1992.

CHAPTER 4

Thermodynamic and Kinetic Data

List of Tables

List of Tables (Continued)

Vapor Pressure

Diffusion

Table 64. BOND STRENGTHS IN DIATOMIC MOLECULES*

(SHEET 1 OF 18)

Molecule	$kcal \cdot mol^{-1}$	
H–H	104.207	± 0.001
H–D	105.030	± 0.001
D–D	106.010	± 0.001
H–Li	56.91	± 0.01
H–Be	54	
H–B	79	± 1
H–C	80.9	
H–N	75	± 4
H–O	102.34	± 0.30
H–F	135.9	± 0.3
H–Na	48	± 5
H–Mg	47	± 12
H–Al	68	± 2
H–Si	71.4	± 1.2
H–P	82	± 7
H–S	82.3	± 2.9
H–Cl	103.1	
H–K	43.8	± 3.5
H–Ca	40.1	
H–Cr	67	± 12
H–Mn	56	± 7
H–Ni	61	± 7
H –Cu	67	± 2
H–Zn	20.5	± 0.5
H –Ga	68	± 5
H–Ge	76.8	± 0.2
H–As	65	± 3
H–Se	73	± 1

To convert **kcal** to **KJ**, multiply by **4.184**.

Source: data from: Kerr, J. A., Parsonage, M. J., and Trotman–Dickenson, A. F., in *Handbook of Chemistry and Physics, 55th ed.*, Weast, R. C., Ed., CRC Press, Cleveland, 1974, F–204.

Table 64. Bond Strengths in Diatomic Molecules*
(Sheet 2 of 18)

Molecule	kcal · mol^{-1}	
H–Br	87.4	± 0.5
H–Rb	40	± 5
H–Sr	39	± 2
H–Ag	59	± 1
H–Cd	16.5	± 0.1
H–In	59	± 2
H–Sn	63	± 1
H–Te	64	± 1
H–I	71.4	± 0.2
H–Cs	42.6	± 0.9
H–Ba	42	± 4
H–Yb	38	± 1
H–Pt	84	± 9
H–Au	75	± 3
H–Hg	9.5	
H–Ti	45	± 2
H–Pb	42	± 5
H–Bi	59	± 7
Li–Li	24.55	± 0.14
Li–O	78	± 6
Li– F	137.5	± 1
Li–Cl	111.9	± 2
Li–Br	100.2	± 2
Li–I	84.6	± 2
Be–Be	17	
Be–0	98	± 7
Be–F	136	± 2
Be–S	89	± 14

To convert **kcal** to **KJ**, multiply by **4.184**.

Source: data from: Kerr, J. A., Parsonage, M. J., and Trotman–Dickenson, A. F., in *Handbook of Chemistry and Physics, 55th ed.*, Weast, R. C., Ed., CRC Press, Cleveland, 1974, F–204.

Table 64. BOND STRENGTHS IN DIATOMIC MOLECULES*

(SHEET 3 OF 18)

Molecule	$kcal \cdot mol^{-1}$	
Be–Cl	92.8	± 2.2
Be–Au	~ 67	
B–B	~ 67	± 5
B–N	93	± 12
B–0	192.7	± 1.2
B–F	180	± 3
B–S	138.8	± 2.2
B–Cl	119	
B–Se	110	± 4
B–Br	101	± 5
B–Ru	107	± 5
B–Rh	114	± 5
B–Pd	79	± 5
B–Te	85	± 5
B–Ce	~ 100	
B–Ir	123	± 4
B–Pt	114	± 4
B–Au	82	± 4
B–Th	71	
C–C	144	± 5
C–N	184	± 1
C–0	257.26	± 0.77
C–F	128	± 5
C–Si	104	± 5
C–P	139	± 23
C–S	175	± 7
C–Cl	93	
C–Ti	~128	

To convert **kcal** to **KJ**, multiply by **4.184**.

Source: data from: Kerr, J. A., Parsonage, M. J., and Trotman–Dickenson, A. F., in *Handbook of Chemistry and Physics, 55th ed.*, Weast, R. C., Ed., CRC Press, Cleveland, 1974, F–204.

Table 64. BOND STRENGTHS IN DIATOMIC MOLECULES*

(SHEET 4 OF 18)

Molecule	kcal • mol^{-1}	
C–V	133	
C–Ge	110	± 5
C–Se	139	± 23
C–Br	67	± 5
C–Ru	152	± 3
C–Rh	139	± 2
C–I	50	± 5
C–Ce	109	± 7
C–Ir	149	± 3
C–Pt	146	± 2
C–U	111	± 7
N–N	226.8	± 1.5
N–O	150.8	± 0.2
N–F	62.6	± 0.8
N–Al	71	± 23
N–Si	105	± 9
N–P	148	± 5
N–S	~ 120	± 6
N–Cl	93	± 12
N–Ti	111	
N–As	116	± 23
N–Se	105	± 23
N–Br	67	± 5
N–Sb	72	± 12
N–I	~.38	
N–Xe	55	
N–Th	138	± 1
N–U	127	± 1

To convert **kcal** to **KJ**, multiply by **4.184**.

Source: data from: Kerr, J. A., Parsonage, M. J., and Trotman–Dickenson, A. F., in *Handbook of Chemistry and Physics, 55th ed.*, Weast, R. C., Ed., CRC Press, Cleveland, 1974, F–204.

Table 64. BOND STRENGTHS IN DIATOMIC MOLECULES*

(SHEET 5 OF 18)

Molecule	kcal • mol^{-1}	
O–O	118.86	± 0.04
O–F	56	± 9
O–Na	61	± 4
O–Mg	79	± 7
O–Al	116	± 5
O–Si	184	± 3
O–P	119.6	± 3
O–S	124.69	± 0.03
O–Cl	64.29	± 0.03
O–K	57	± 8
O–Ca	84	± 7
O–Sc	155	± 5
O–Ti	158	± 8
O–V	154	± 5
O–Cr	110	± 10
O–Mn	96	± 8
O–Fe	96	± 5
O–Co	88	± 5
O–Ni	89	± 5
O–Cu	82	± 15
O–Zn	≤ 66	
O–Ga	68	± 15
O–Ge	158.2	± 3
O–As	115	± 3
O–Se	101	
O–Br	56.2	± 0.6
O–Rb	(61)	± 20
O–Sr	93	± 6

To convert **kcal** to **KJ**, multiply by **4.184**.

Source: data from: Kerr, J. A., Parsonage, M. J., and Trotman–Dickenson, A. F., in *Handbook of Chemistry and Physics, 55th ed.*, Weast, R. C., Ed., CRC Press, Cleveland, 1974, F–204.

Table 64. BOND STRENGTHS IN DIATOMIC MOLECULES*

(SHEET 6 OF 18)

Molecule	$kcal \cdot mol^{-1}$	
O–Y	162	± 5
O–Zr	181	± 10
O–Nb	189	± 10
O–Mo	115	± 12
O–Ru	115	± 15
O–Rh	90	± 15
O–Pd	56	± 7
O–Ag	51	± 20
O–Cd	≤ 67	
O–In	≤ 77	
O–Sn	127	± 2
O–Sb	89	± 20
O–Fe	93.4	± 2
O–I	47	± 7
O–Xe	9	± 5
O–Cs	67	± 8
O–Ba	131	± 6
O–La	188	± 5
O–Ce	188	± 6
O–Pr	183.7	
O–Nd	168	± 8
O–Sm	134	± 8
O–Eu	130	± 10
O–Gd	162	± 6
O–Tb	165	± 8
O–Dy	146	± 10
O–Ho	149	± 10
O–Er	147	± 10

To convert **kcal** to **KJ**, multiply by **4.184**.

Source: data from: Kerr, J. A., Parsonage, M. J., and Trotman–Dickenson, A. F., in *Handbook of Chemistry and Physics, 55th ed.*, Weast, R. C., Ed., CRC Press, Cleveland, 1974, F–204.

Table 64. BOND STRENGTHS IN DIATOMIC MOLECULES*

(SHEET 7 OF 18)

Molecule	kcal • mol⁻¹	
O–Tm	122	± 15
O–Yb	98	± 15
O–Lu	159	± 8
O–Hf	185	± 10
O–Ta	183	± 15
O–W	156	± 6
O–Os	< 142	
O–Ir	≤ 94	
O–Pt	83	± 8
O–Pb	90.3	± 1.0
O–Bi	81.9	± 1.5
O–Th	192	± 10
O–U	182	± 8
O–Np	172	± 7
O–Pu	163	± 15
O–Cm	≤ 134	
F–F	37.5	± 2.3
F–Na	114	± 1
F–Mg	110	± 1
F–Al	159:	± 3
F–Si	116	± 12
F–P	105	± 23
F–Cl	59.9	± 0.1
F–K	118.9	± 0.6
F–Ca	125	± 5
F–Sc	141	± 3
F–Ti	136	± 8
F–Cr	104.5	± 4.7

To convert kcal to KJ, multiply by 4.184.

Source: data from: Kerr, J. A., Parsonage, M. J., and Trotman–Dickenson, A. F., in *Handbook of Chemistry and Physics, 55th ed.*, Weast, R. C., Ed., CRC Press, Cleveland, 1974, F–204.

Table 64. BOND STRENGTHS IN DIATOMIC MOLECULES*

(SHEET 8 OF 18)

Molecule	kcal • mol^{-1}	
F–Mn	101.2	± 3.5
F–Ni	89	± 4
F–Cu	88	± 9
F–Ga	138	± 4
F–Ge	116	± 5
F–Br	55.9	
F–Rb	116.1	± 1
F–Sr	129.5	± 1.6
F–Y	144	± 5
Mg–I	~.68	
Mg–Au	59	± 23
Al–Al	44	
Al–P	52	± 3
Al–S	79	
Al–Cl	119.0	± 1
Al–Br	103.1	
Al–I	88	
Al–Au	65	
Al–U	78	± 7
Si–Si	76	± 5
Si–S	148	± 3
Si–Cl	105	± 12
Si–Fe	71	± 6
Si–Co	66	± 4
Si–Ni	76	± 4
Si–Ge	72	± 5
Si–Se	127	± 4
Si–Br	82	± 12

To convert kcal to KJ, multiply by **4.184.**

Source: data from: Kerr, J. A., Parsonage, M. J., and Trotman–Dickenson, A. F., in *Handbook of Chemistry and Physics, 55th ed.*, Weast, R. C., Ed., CRC Press, Cleveland, 1974, F–204.

Table 64. BOND STRENGTHS IN DIATOMIC MOLECULES*

(SHEET 9 OF 18)

Molecule	kcal • mol^{-1}	
Si–Ru	95	± 5
Si–Rh	95	± 5
Si–Pd	75	± 4
Si–Te	121	± 9
Si–Ir	110	± 5
Si–Pt	120	± 5
Si–Au	75	± 3
P–P	117	± 3
F–Ag	84.7	± 3.9
F–Cd	73	± 5
F–In	121	± 4
F–Sn	111.5	± 3
F–Sb	105	± 23
F–I	67?	
F–Xe	11	
F–Cs	119.6	± 1
F–Ba	140.3	± 1.6
F–Nd	130	± 3
F–Sm	126.9	± 4.4
F–Eu	126.1	± 4.4
F–Gd	141.	± 46.5
F–Hg	31	± 9
F–Ti	106.4	± 4.6
F–Pb	85	± 2
F–Bi	62	
F–Pu	129	± 7
Na–Na	18.4	
Na–Cl	97.5	± 0.5

To convert **kcal** to **KJ**, multiply by **4.184**.

Source: data from: Kerr, J. A., Parsonage, M. J., and Trotman–Dickenson, A. F., in *Handbook of Chemistry and Physics, 55th ed.*, Weast, R. C., Ed., CRC Press, Cleveland, 1974, F–204.

Table 64. BOND STRENGTHS IN DIATOMIC MOLECULES*
(SHEET 10 OF 18)

Molecule	kcal • mol^{-1}	
Na–K	15.2	± 0.7
Na–Br	86.7	± 1
Na–Rb	14	± 1
Na–I	72.7	± 1
Mg–Mg	8?	
Mg–S	56?	
Mg–Cl	76	± 3
Mg–Br	75	± 23
P–S	70	
P–Ga	56	
P–W	73	± 1
P–Th	90	
S–S	101.9	± 2.5
S–Ca	75	± 5
S–Sc	114	± 3
S–Mn	72	± 4
S–Fe	78	
S–Cu	72	± 12
S–Zn	49	± 3
S–Ge	131.7	± 0.6
S–Se	91	± 5
S–Sr	75	± 5
S–Y	127	± 3
S–Cd	48	
S–In	69	± 4
S–Sn	111	± 1
S–Te	81	± 5
S–Ba	96	± 5

To convert **kcal** to **KJ**, multiply by **4.184**.

Source: data from: Kerr, J. A., Parsonage, M. J., and Trotman–Dickenson, A. F., in *Handbook of Chemistry and Physics, 55th ed.*, Weast, R. C., Ed., CRC Press, Cleveland, 1974, F–204.

Table 64. BOND STRENGTHS IN DIATOMIC MOLECULES*

(SHEET 11 OF 18)

Molecule	kcal • mol^{-1}	
S–La	137	± 3
S–Ce	137	± 3
S–Pr	122.7	
S– Nd	113	± 4
S–Eu	87	± 4
S–Gd	126	± 4
S–Ho	102	± 4
S–Lu	121	± 4
S–Au	100	± 6
S–Hg	51	
S–Pb	82.7	± 0.4
S–Bi	75.4	± 1.1
S–U	135	± 2
Cl–Cl	58.066	± 0.001
Cl–K	101.3	± 0.5
Cl–Ca	95	± 3
Cl–Sc	79	
Cl–Ti	26	± 2
Cl–Cr	87.5	± 5.8
Cl–Mn	86.2	± 2.3
Cl–Fe	84?	
Cl–Ni	89	± 5
Cl–Cu	84	± 6
Cl–Zn	54.7	± 4.7
Cl–Ga	114.5	
Cl–Ge	82?	
Cl–Br	52.3	± 0.2
Cl–Rb	100.7	± 1

To convert **kcal to KJ,** multiply by **4.184.**

Source: data from: Kerr, J. A., Parsonage, M. J., and Trotman–Dickenson, A. F., in *Handbook of Chemistry and Physics, 55th ed.*, Weast, R. C., Ed., CRC Press, Cleveland, 1974, F–204.

Table 64. BOND STRENGTHS IN DIATOMIC MOLECULES*

(SHEET 12 OF 18)

Molecule	kcal • mol^{-1}	
Cl–Sr	97	± 3
Cl–Y	82	± 23
Cl–Ag	75	± 9
Cl–Cd	49.9	
Cl–In	103.3	
CI–Sn	75?	
Cl–Sb	86	± 12
Cl–I	50.5	± 0.1
Cl–Cs	106.2	± 1
Cl–Ba	106	± 3
Cl–Au	82	± 2
Cl–Hg	24	± 2
Cl–Ti	89.0	± 0.5
Cl–Pb	72	± 7
Cl–Bi	72	± 1
Cl–Ra	82	± 18
Ar–Ar	0.2	
K–K	12.8	
K–Br	90.9	± 0.5
K–I	76.8	± 0.5
Ca–I	70	± 23
Ca–Au	18	
Sc–Sc	25.9	± 5
Ti–Ti	34	± 5
V–V	58	± 5
Cr–Cr	<37	
Cr –Cu	37	± 5
Cr–Ge	41	± 7

To convert **kcal** to **KJ**, multiply by **4.184**.

Source: data from: Kerr, J. A., Parsonage, M. J., and Trotman–Dickenson, A. F., in *Handbook of Chemistry and Physics, 55th ed.*, Weast, R. C., Ed., CRC Press, Cleveland, 1974, F–204.

Table 64. BOND STRENGTHS IN DIATOMIC MOLECULES*

(SHEET 13 OF 18)

Molecule	$kcal \cdot mol^{-1}$	
Cr–Br	78.4	± 5
Cr–I	68.6	± 5.8
Cr–Au	51.3	± 3.5
Mn–Mn	4	± 3
Mn–Se	48	± 3
Mn–Br	75.1	± 23
Mn–I	67.6	± 2.3
Mn–Au	44	± 3
Fe–Fe	24	± 5
Fe–Ge	50	± 7
Fe–Br	59	± 23
Fe–Au	45	± 4
Co–Co	40	± 6
Co–Cu	39	± 5
Co–Ge	57	± 6
Co–Au	51	± 3
Ni–Ni	55.5	± 5
Ni–Cu	48	± 5
Ni–Ge	67.3	± 4
Ni–Br	86	± 3
Ni–I	70	± 5
Ni–Au	59	± 5
Cu–Cu	46.6	± 2.2
Cu–Ge	49	± 5
Cu–Se	70	± 9
Cu–Br	79	± 6 5
Cu–Ag	41.6	± 2.2
Cu–Sn	42.3	± 4

To convert **kcal** to **KJ**, multiply by **4.184**.

Source: data from: Kerr, J. A., Parsonage, M. J., and Trotman–Dickenson, A. F., in *Handbook of Chemistry and Physics, 55th ed.*, Weast, R. C., Ed., CRC Press, Cleveland, 1974, F–204.

Table 64. Bond Strengths in Diatomic Molecules*
(Sheet 14 of 18)

Molecule	kcal • mol^{-1}	
Cu–Te	42	± 9
Cu–I	47?	
Cu–Au	55.4	± 2.2
Zn–Zn	7	
Zn–Se	33	± 3
Zn–Te	49?	
Zn–I	33	± 7
Ga–Ga	3	± 3
Ga–As	50.1	± 0.3
Ga–Br	101	± 4
Ga–Ag	4	± 3
Ga–Te	60	± 6
Ga–I	81	± 2
Ga–Au	51	± 23
Ge–Ge	65.8	± 3
Ge–Se	114	±
Ge–Br	61	± 7
Ge–Te	93	± 2
Ge–Au	70	± 23
As–As	91.7	
As–Se	23	
Se–Se	79.5	± 0.1
Se–Cd	~75	
Se–in	59	± 4
Se–Sn	95.9	± 1.4
Se–Te	64	± 2
Se–La	114	± 4
Se–Nd	92	± 4

To convert **kcal** to **KJ**, multiply by **4.184**.

Source: data from: Kerr, J. A., Parsonage, M. J., and Trotman–Dickenson, A. F., in *Handbook of Chemistry and Physics, 55th ed.*, Weast, R. C., Ed., CRC Press, Cleveland, 1974, F–204.

Table 64. Bond Strengths in Diatomic Molecules*

(Sheet 15 of 18)

Molecule	kcal • mol^{-1}	
Se–Eu	72	± 4
Se–Gd	103	± 4
Se–Ho	80	± 4
Se–Lu	100	± 4
Se–Pb	72.4	± 1
Se–Bi	67.0	± 1.5
Bi–Br	46.336	± 0.001
Br–Rb	90.4	± 1
Br–Ag	70	± 7
Br–Cd	~38	
Br–In	93	
Bi–Sn	47	± 23
Br–Sb	75	± 14
Br–I	42.8	± 0.1
Br–Cs	96.5	± 1
Br–Hg	17.3	
Br–Ti	79.8	± 0.4
Br–Pb	59	± 9
Br–Bi	63.9	± 1
Rb–Rb	12.2	
Rb–I	76.7	± 1
Sr–Au	63	± 23
Y–Y	38.3	
Y–La	48.3	
Pd–Pd	33?	
Pd–Au	34.2	± 5
Ag–Ag	41	± 2
Ag–Sn	32.5	± 5

To convert kcal to KJ, multiply by **4.184**.

Source: data from: Kerr, J. A., Parsonage, M. J., and Trotman–Dickenson, A. F., in *Handbook of Chemistry and Physics, 55th ed.*, Weast, R. C., Ed., CRC Press, Cleveland, 1974, F–204.

Table 64. BOND STRENGTHS IN DIATOMIC MOLECULES*
(SHEET 16 OF 18)

Molecule	$kcal \cdot mol^{-1}$	
Ag–Te	70	± 23
Ag–I	56	± 7
Ag–Au	48.5	± 2.2
Cd–Cd	2.7	± 0.2
Cd–I	33	± 5
In–In	23.3	± 2.5
In–Sb	36.3	± 2.5
In–Te	52	± 4
In–I	80	
Sn–Sn	46.7	± 4
Sn–Te	76	± 1
Sn–Au	58.4	± 4
Sb–Sb	71.5	± 1.5
Sb–Te	61	± 4
Sb–Bi	60	± 1
Te–Te	63.2	± 0.2
Te–La	91	± 4
Te–Nd	73	± 4
Te–Eu	58	± 4
Te–Gd	82	± 4
Te–Ho	62	± 4
Te–Lu	78	± 4
Te–Au	59	± 16
Te–Pb	60	± 3
Te–Bi	56	± 3
I–I	36.460	± 0.002
I–Cs	82.4	± 1
I–Hg	9	

To convert **kcal** to **KJ**, multiply by **4.184**.

Source: data from: Kerr, J. A., Parsonage, M. J., and Trotman–Dickenson, A. F., in *Handbook of Chemistry and Physics, 55th ed.*, Weast, R. C., Ed., CRC Press, Cleveland, 1974, F–204.

Table 64. BOND STRENGTHS IN DIATOMIC MOLECULES*

(SHEET 17 OF 18)

Molecule	$kcal \cdot mol^{-1}$	
I–Ti	65	± 2
I–Pb	47	± 9
I–Bi	52	± 1
Xe–Xe	~ 0.7	
Cs–Cs	11.3	
Ba–Au	38	± 14
La–Ld	58.6	
La–Au	80	± 5
Ce–Ce	66	± 1
Ce–Au	76	± 4
Pr–Au	74	± 5
Nd–Au	70	± 6
Au–Au	52.4	± 2.2
Au–Pb	31	± 23
Au–U	76	± 7
Hg–Hg	4.1	± 0.5
Hg–Tl	1	
Tl–Tl	15?	
Pb–Pb	24	± 5
Pb–Bi	32	± 5

To convert kcal to KJ, multiply by 4.184.

Source: data from: Kerr, J. A., Parsonage, M. J., and Trotman–Dickenson, A. F., in *Handbook of Chemistry and Physics, 55th ed.*, Weast, R. C., Ed., CRC Press, Cleveland, 1974, F–204.

Table 64. BOND STRENGTHS IN DIATOMIC MOLECULES*

(SHEET 18 OF 18)

Molecule	$kcal \cdot mol^{-1}$	
Bi–Bi	45	± 2
Po–Po	44.4	± 2.3
At–At	19	
Th–Th	<69	

To convert kcal to KJ, multiply by 4.184.

Source: data from: Kerr, J. A., Parsonage, M. J., and Trotman–Dickenson, A. F., in *Handbook of Chemistry and Physics, 55th ed.*, Weast, R. C., Ed., CRC Press, Cleveland, 1974, F–204.

* Notes for Table of Bond Strengths in Diatomic Molecules

The strength of a chemical bond, D (R–X), often known as the bond dissociation energy, is defined as the heat of the reaction: RX –> R + X. It is given by: $D(R–X) = \Delta H_f^\circ(R) + \Delta H_f^\circ(X) - \Delta H_f^\circ(RX)$. Some authors list bond strengths for 0K, but here the values for 298K are given because more thermodynamic data are available for this temperature. Bond strengths, or bond dissociation energies, are not equal to, and may differ considerable from, mean bond energies derived solely from thermochemical data on molecules and atoms.

The values in this table have usually been measured spectroscopically or by mass spectrometric analysis of hot gases effusing from a Knudsen cell.

Table 65. BOND STRENGTHS OF POLYATOMIC MOLECULES*

(SHEET 1 OF 7)

Molecule	Kcal • mol^{-1}	
	Value	Error
H–CH	102	± 2
H–CH_2	110	± 2
H–CH_3	104	± 1
H–ethynyl	128	± 5
H–vinyl	≥ 108	± 2
H–C_2H_5	98	± 1
H–propargyl	93.9	± 1.2
H–allyl	89	± 1
H–cyclopropyl	100.7	± 1
H–n–C_3H_7	98	± 1
H–i–C_3H_7	95	± 1
H–cyclobutyl	96.5	± 1
H–cyclopropycarbinyl	97.4	± 1.6
H–methdllyl	83	± 1
H–s–C_4H_9	95	± 1
H–t–C_4H_9	92	± 1.2
H–cyclopentadien–1,3–yl–5	81.2	± 1.2
H–pentadien–1,4–yi–3	80	± 1
H–OH	119	± 1
H–OCH_3	103.6	± 1
H–OC_2H_5	103.9	± 1
H–$OC(CH_3)_3$	104.7	± 1
H–OC_6H_5	88	± 5
H–O_2H	90	± 2

To convert **kcal** to **KJ**, multiply by **4.184**.

Source: data from: Kerr, J. A., Parsonage, M. J., and Trotman–Dickenson, A. F., in *Handbook of Chemistry and Physics, 55th ed.*, Weast, R. C., Ed., CRC Press, Cleveland, 1974, F–213.

Table 65. BOND STRENGTHS OF POLYATOMIC MOLECULES*

(SHEET 2 OF 7)

Molecule	Kcal • mol^{-1}	
	Value	Error
$H–O_2CCH_3$	112	± 4
$H–O_2CC_2H_3$	110	± 4
$H–O_2Cn–C_3H_7$	103	± 4
H–ONO	78.3	± 0.5
$H–ONO_2$	101.2	± 0.5
H–SH	90	± 2
H–SCH	≥ 88	
$H–SiH_3$	94	± 3
$H–Si(CH_3)_3$	90	± 3
$BH_3–BH_3$	35	
HC=CH	230	± 2
$H_2C=CH_2$	172	± 2
$H_3C–CH_3$	88	± 2
$CH_3–C(CH_3)_2CH:CH_2$	69.4	
$C_6H_5CH_2–C_2H_5$	69	± 2
$C_6H_5CH(CH_3)–CH_3$	71	
$C_6H_5CH_2–n–C_3H_7$	67	± 2
$CH_3–CH_2CN$	72.7	± 2
$CH_3–C(CH_3)_2CN$	70.2	± 2
$C_6H_5C(CH_3)(CN)–CH_3$	59.9	
NC–CN	128	± 1
$C_6H_5CH_2CO–CH_2C_6H_5$	65.4	
$C_6H_5CO–CF_3$	73.8	
$CH_3CO–COCH_3$	67.4	± 2.3

To convert kcal to KJ, multiply by **4.184**.

Source: data from: Kerr, J. A., Parsonage, M. J., and Trotman–Dickenson, A. F., in *Handbook of Chemistry and Physics, 55th ed.*, Weast, R. C., Ed., CRC Press, Cleveland, 1974, F–213.

Table 65. BOND STRENGTHS OF POLYATOMIC MOLECULES*

(SHEET 3 OF 7)

Molecule	Kcal • mol^{-1}	
	Value	Error
$C_6H_5CH_2$– COOH	68.1	
$C_6H_5CH_2$– O_2CCH_3	67	
C_6H_5CO– COC_6H_5	66.4	
$C_6H_5CH_2$– $O_2CC_6H_5$	69	
$(C_6H_5CH_2)_2CH$–COOH	59.4	
CH_2F–CH_2F	88	± 2
CF_2=CF_2	76.3	± 3
CF_3–CF_3	96.9	± 2
$C_6H_5CH_2$–NH_2	71.9	± 1
C_6H_5NH–CH_3	67.7	
$C_6H_5CH_2$–$NHCH_3$	68.7	± 1
$C_6H_5N(CH_2)$–CH_3	65.2	
$C_6H_5CH_2$–$N(CH_3)_2$	60.9	± 1
CF_3–NF_2	65	± 2.5
$CH_2 = N_2$	≤ 41.7	± 1
CH_3N:N–CH_3	52.5	
C_2H_5N:N–C_2H_5	50.0	
i –C_3H_7N:N–i –C_3H_7	47.5	
n –C_4H_9N:N–n –C_4H_9	50.0	
i –C_4H_9N:N–i –C_4H_9	49.0	

To convert **kcal** to **KJ**, multiply by **4.184**.

Source: data from: Kerr, J. A., Parsonage, M. J., and Trotman–Dickenson, A. F., in *Handbook of Chemistry and Physics, 55th ed.*, Weast, R. C., Ed., CRC Press, Cleveland, 1974, F–213.

Table 65. Bond Strengths of Polyatomic Molecules*
(Sheet 4 of 7)

Molecule	Kcal • mol^{-1}	
	Value	Error
s –C_4H_9N:N–s –C_4H_9	46.7	
t –C_4H_9N:N–t –C_4H_9	43.5	
$C_6H_5CH_2N$:N–$C_6H_5CH_2$	37.6	
CF_3N:N–CF_3	55.2	
C_2H_5–NO_2	62	
O=CO	127.2	± 0.1
CH_3–O_2SCH_3	66.8	
Allyl–O_2SCH_3	49.6	
$C_6H_5CH_2$–O_2SCH_3	52.9	
C_6H_5S–CH_3	60	
$C_6H_5CH_2$–SCH_3	53.8	
F–CH_3	103	± 3
Cl–CN	97	± 1
Cl–COC_6H_5	74	± 3
Cl–CF_3	86.1	± 0.8
Cl–CCl_2F	73	± 2
Cl–C_2F5	82.7	± 1.7
Br–CH_3	70.0	± 1.2
Br–CN	83	± 1
Br–COC_6H_5	64.2	
Br–CF_3	70.6	± 1.0
Br–CBr_3	56.2	± 1.8
Br–C_2F_5	68.7	± 1.5
Br –n –C_3F	66.5	± 2.5

To convert **kcal** to **KJ**, multiply by **4.184**.

Source: data from: Kerr, J. A., Parsonage, M. J., and Trotman–Dickenson, A. F., in *Handbook of Chemistry and Physics, 55th ed.*, Weast, R. C., Ed., CRC Press, Cleveland, 1974, F–213.

Table 65. BOND STRENGTHS OF POLYATOMIC MOLECULES*

(SHEET 5 OF 7)

Molecule	Kcal • mol^{-1}	
	Value	Error
I–CH_3	56.3	± 1
1–norbornyl	62.5	± 2.5
I–CN	73	± 1
I–CF_3	53.5	± 2
CH_3–$Ga(CH_3)_2$	59.5	
CH_3–$CdCH_3$	54.4	
CH_3–$HgCH_3$	57.5	
C_2H_5–HgC_2H_5	43.7	± 1
n –C_3H_7–Hg n –C_3H_7	47.1	
i –C_3H_7–Hg i –C_3H_7	40.7	
C_6H_5–HgC_6H_5	68	
CH_3 –$Tl(CH_3)_2$	36.4	± 0.6
CH_3–$Pb(CH_3)_3$	49.4	± 1
NH_2–NH_2	70.8	± 2
NH_2–$NHCH_3$	64.8	
NH_2 –$N(CH_3)_2$	62.7	
NH_2 –NHC_6H_5	51.1	
NO–NO_2	9.5	± 0.5
NO_2–NO_2	12.9	± 0.5
NF_2–NF_2	21	± 1
O–N_2	40	
O–NO	73	
HO–N:$CHCH_3$	49.7	
Cl–NF_2	≈ 32	

To convert **kcal** to **KJ**, multiply by **4.184**.

Source: data from: Kerr, J. A., Parsonage, M. J., and Trotman–Dickenson, A. F., in *Handbook of Chemistry and Physics, 55th ed.*, Weast, R. C., Ed., CRC Press, Cleveland, 1974, F–213.

Table 65. BOND STRENGTHS OF POLYATOMIC MOLECULES*

(SHEET 6 OF 7)

Molecule	Kcal • mol^{-1}	
	Value	Error
HO–OH	51	± 1
$CH_3O–OCH_3$	36.9	± 1
$HO–OC(CH_3)_3$	42.5	
$C_2H_5O–OC_2H_5$	37.3	± 1.2
n $-C_3H_7O–O$ n $-C_3H_7$	37.2	± 1
i $-C_3H_7O–O$ i $-C_3H_7$	37.0	± 1
s $-C_4H_9O–O$ s $-C_4H_9$	36.4	± 1
t $-C_4H_9O–O$ t $-C_4H_9$	37.4	± 1
$(CH_3)_3CCH_2O–OCH_2C(CH_3)_3$	36.4	± 1
$O–O_2CIF$	58.4	
$CH_3CO_2–O_2CCH_3$	30.4	± 2
$C_2H_5CO_2–O_2CC_2H_5$	30.4	± 2
n $-C_3H_7CO_2–O_2C$n $-C_3H_7$	30.4	± 2
O–SO	132	± 2
$F–OCF_3$	43.5	± 0.5
Cl–OH	60	± 3
O–ClO	59	± 3
Br–OH	56	± 3
I–OH	56	± 3
$ClO_3–ClO_4$	58.4	

To convert **kcal** to **KJ**, multiply by **4.184**.

Source: data from: Kerr, J. A., Parsonage, M. J., and Trotman–Dickenson, A. F., in *Handbook of Chemistry and Physics, 55th ed.*, Weast, R. C., Ed., CRC Press, Cleveland, 1974, F–213.

Table 65. BOND STRENGTHS OF POLYATOMIC MOLECULES*

(SHEET 7 OF 7)

Molecule	$Kcal \cdot mol^{-1}$	
	Value	Error
$O = PF_3$	130	± 5
$O = PCl_3$	122	± 5
$O = PBr_3$	119	± 5
$SiH_3–SiH_3$	81	± 4
$(CH_3)_3Si–Si(CH_3)_3$	80.5	

To convert **kcal** to **KJ**, multiply by **4.184**.

Source: data from: Kerr, J. A., Parsonage, M. J., and Trotman–Dickenson, A. F., in *Handbook of Chemistry and Physics, 55th ed.*, Weast, R. C., Ed., CRC Press, Cleveland, 1974, F–213.

* The values refer to a temperature of 298 K and have mostly been determined by kinetic methods. Some have been calculated from formation of the species involved according to equations:

$$D(R–X) = \Delta H_f^\circ(R^\cdot) + \Delta H_f^\circ(X^\cdot) - \Delta H_f^\circ(RX) \quad \text{or} \quad D(R–X) = 2\Delta H_f^\circ(R^\cdot) - \Delta H_f^\circ(RR)$$

Table 66. SOLUBILITY OF COPPER AND COPPER ALLOYS

Family	Wrought Alloys UNS Numbers	Principal Alloying Element	Solid Solubility at 20 °C (at. %)
Brasses	C20000, C30000, C40000, C66400 to C69800	Zn	37
Phosphor bronzes	C50000	Sn	9
Aluminum bronzes	C60600 to C64200	Al	19
Silicon bronzes	C64700 to C66100	Si	8
Copper nickels, nickel silvers	C70000	Ni	100

Data from *ASM Metals Reference Book, Third Edition*, Michael Bauccio, Ed., ASM International, Materials Park, OH, p439, (1993).

Table 67. Heat of Formation of Inorganic Oxides

(Sheet 1 of 16)

Reaction	Temperature range of validity	ΔH_0	2.303a	b	c	I
2 Ac(c) + 3/2 O_2(g) = Ac_2O_3(c)	298.16–1,000K	–446,090	–16.12	–	–	+109.89
2 Al(c) + 1/2 O_2(g) = Al_2O(g)	298.16–931.7K	–31,660	+14.97	–	–	–72.74
2 Al(l) + 1/2 O_2(g) = Al_2O(g)	931.7–2,000K	–38,670	+10.36	–	–	–51.53
Al(c) + 1/2 O_2(g) = AlO(g)	298.16–931.7K	+10,740	+5.76	–	–	–37.61
Al(l) + 1/2 O_2(g) = AlO(g)	931.7–2,000K	+8,170	+5.76	–	–	–34.85
2 Al(c) + 3/2 O_2(g) = Al_2O_3 (corundum)	298.16–931.7K	–404,080	–15.68	+2.18	+3.935	+123.64
2 Al(l) + 3/2 O_2(g) = Al_2O_3 (corundum)	931.7–2,000K	–407,950	–6.19	–0.78	+3.935	+102.37
2 Sb(c) + 3/2 O_2(g) = Sb_2O_3 (cubic)	298.16–842K	–169,450	+6.12	–6.01	–0.30	+52.21
2 Sb(c) + 3/2 O_2(g) = Sb_2O_3 (orthorhombic)	298.16–903K	–168,060	+6.12	–6.01	–0.30	+50.56
2 As(c) + 3/2 O_2(g) = As_2O_3 (orthorhombic)	298.16–542K	–154,870	+29.54	–21.33	–0.30	–8.83
2 As(c) + 3/2 O_2(g) = As_2O_3 (monoclinic)	298.16–586K	–150,760	+29.54	–21.33	–0.30	–16.95
2 As(c) + 5/2 O_2(g) = As_2O_5(c)	298.16–883K	–217,080	+12.32	–4.65	–0.50	+80.50

The ΔH_o values are given in gram calories per mole. The a, b, and *I* values listed here make it possible for one to calculate the ΔF and ΔS values by use of the following equations:

$$\Delta F_t = \Delta H_o + 2.303aT \log T + b \times 10^{-3} T^2 + c \times 10^5 T^{-1} + IT$$

$$\Delta S_t = -a - 2.303a \log T - 2b \times 10^{-3} T + c \times 10^5 T^{-2} - I$$

Source: data from *CRC Handbook of Materials Science, Vol I*, Charles T. Lynch, Ed., CRC Press, Cleveland, (1974).

Table 67. Heat of Formation of Inorganic Oxides

(Sheet 2 of 16)

Reaction	Temperature range of validity	ΔH_0	2.303a	b	c	I
$Ba(\alpha) + 1/2\ O_2(g) = BaO(c)$	298.16–648K	–134,590	–7.60	+0.87	+0.42	+45.76
$Ba(\beta) + 1/2\ O_2(g) = BaO(c)$	648–977K	–134,140	–3.34	–0.56	+0.42	+34.01
$Be(c) + 1/2\ O_2(g) = BeO(c)$	298.16–1,556K	–144,220	–1.91	–0.46	+1.24	+30.64
$Bi(c) + 1/2\ O_2(g) = BiO(c)$	298.16–544K	–50,450	–4.61	–	–	+35.51
$Bi(l) + 1/2\ O_2(g) = BiO(c)$	544–1,600K	–52,920	–4.61	–	–	+40.05
$2\ Bi(c) + 3/2\ O_2(g) = Bi_2O_3(c)$	298.16–544K	–139,000	–11.56	+2.15	–0.30	+96.52
$2\ Bi(l) + 3/2\ O_2(g) = Bi_2O_3(c)$	544–1,090K	–142,270	+2.30	–3.25	–0.30	+67.55
$2\ B(c) + 3/2\ O_2(g) = B_2O(c)$	298.16–723K	–304,690	+11.72	–7.55	+0.355	+34.25
$2\ B(c) + 3/2\ O_2(g) = B_2O_3(gl)$	298.16–723K	–298,670	+26.57	–15.90	–0.30	–10.40
$Cd(c) + 1/2\ O_2(g) = CdO(c)$	298.16–594K	–62,330	–2.05	+0.71	–0.10	+29.17
$Cd(l) + 1/2\ O_2(g) = CdO(c)$	594–1,038K	–63,240	+2.07	–0.76	–0.10	+20.14
$Ca(\alpha) + 1/2\ O_2(g) = CaO(c)$	298.16–673K	–151,850	–6.56	+1.46	+0.68	+43.93
$Ca(\beta) + 1/2\ O_2(g) = CaO(c)$	673–1,124K	–151,730	–4.14	+0.41	+0.68	+37.63

The ΔH_o values are given in gram calories per mole. The a, b, and *I* values listed here make it possible for one to calculate the ΔF and ΔS values by use of the following equations:

$$\Delta F_t = \Delta H_o + 2.303aT \log T + b \times 10^{-3} T^2 + c \times 10^5 T^{-1} + IT$$

$$\Delta S_t = -a - 2.303a \log T - 2b \times 10^{-3} T + c \times 10^5 T^{-2} - I$$

Source: data from *CRC Handbook of Materials Science, Vol I*, Charles T. Lynch, Ed., CRC Press, Cleveland, (1974).

Table 67. Heat of Formation of Inorganic Oxides

(Sheet 3 of 16)

Reaction	Temperature range of validity	ΔH_0	2.303a	b	c	I
C(graphite) + 1/2 O_2(g) = CO(g)	298.16–2,000K	–25,400	+2.05	+0.27	–1.095	–28.79
C(graphite) + O_2(g) = CO_2(g)	298.16–2,000K	–93,690	+1.63	–0.7	–0.23	–5.64
2 Ce(c) + 3/2 O_2(g) = Ce_2O_3(c)	298.16–1,048K	–435,600	–4.60	–	–	+92.84
2 Ce(l) + 3/2 O_2(g) = Ce_2O_3(c)	1,048–1,900K	–440,400	–4.60	–	–	+97.42
Ce(c) + O_2(g) = CeO_2(c)	298.16–1,048K	–245,490	–6.42	+2.34	–0.20	+67.79
Ce(l) + O_2(g) = CeO_2(c)	1,048–2,000K	–247,930	+0.71	–0.66	–0.20	+51.73
2 Cs(c) + 1/2 O_2(g) = Cs_2O(c)	298.16–301.5K	–75,900	–	–	–	+36.60
2 Cs(l) + 1/2 O_2(g) = Cs_2O(c)	301.5–763K	–76,900	–	–	–	+39.92
2 Cs(l) + 1/2 O_2(g) = Cs_2O(l)	763–963K	–75,370	–9.21	–	–	+64.47
2 Cs(g) + 1/2 O_2(g) = Cs_2O(l)	963–1,500K	–113,790	–23.03	–	–	+145.60
2 Cs(c) + 3/2 O_2(g) = Cs_2O_3(c)	298.16–301.5K	–112,690	–11.51	–	–	+110.10
2 Cs(l) + 3/2 O_2(g) = Cs_2O_3(c)	301.5–775K	–113,840	–12.66	–	–	+116.77

The ΔH_o values are given in gram calories per mole. The a, b, and *I* values listed here make it possible for one to calculate the ΔF and ΔS values by use of the following equations:

$$\Delta F_t = \Delta H_o + 2.303aT \log T + b \times 10^{-3} T^2 + c \times 10^5 T^{-1} + IT$$

$$\Delta S_t = -a - 2.303a \log T - 2b \times 10^{-3} T + c \times 10^5 T^{-2} - I$$

Source: data from *CRC Handbook of Materials Science, Vol I*, Charles T. Lynch, Ed., CRC Press, Cleveland, (1974).

Table 67. HEAT OF FORMATION OF INORGANIC OXIDES

(SHEET 4 OF 16)

Reaction	Temperature range of validity	ΔH_0	2.303a	b	c	I
$2\ Cs(l) + 3/2\ O_2(g) = Cs_2O_3(l)$	775–963K	–110,740	–26.48	–	–	+152.70
$2\ Cs(g) + 3/2\ O_2(g) = Cs_2O_3(l)$	963–1,500K	–148,680	–39.14	–	–	+229.87
$Cl_2(g) + 1/2\ O_2(g) = Cl_2O(g)$	298.16–2,000K	+17,770	–0.71	–0.12	+0.49	+16.81
$1/2\ Cl_2(g) + 1/2\ O_2(g) = ClO(g)$	298.16–1,000K	+33,000	–	–	–	0.24
$2\ Cl_2(g) + 3/2\ O_2(g) = ClO(g)$	298.16–500K	+37,740	+5.76	–	–	+21.42
$2\ Cr(c) + 3/2\ O_2(g) = Cr_2O_3(\beta)$	298.16–1,823K	–274,670	–14.07	+2.01	+0.69	+105.65
$2\ Cr(l) + 3/2\ O_2(g) = Cr_2O_3(\beta)$	1,823–2,000K	–278,030	+2.33	–0.35	+1.57	+58.29
$Cr(c) + O_2(g) = CrO_2\ (c)$	298.16–1,000K	–142,500	–	–	–	+42.00
$Cr(c) + 3/2\ O_2(g) = CrO_3(c)$	298.16–471K	–141,590	–13.82	–	–	+103.90
$Cr(c) + 3/2\ O_2(g) = Cr_2O_3(l)$	471–600K	–141,580	–32.24	–	–	+153.14
$Co(\alpha,\beta) + 1/2\ O_2(g) = CoO(c)$	298.16–1,400K	–56,910	+0.69	–	–	+16.03
$Co(\gamma) + 1/2\ O_2(g) = CoO(c)$	1,400–1,763K	–58,160	–1.15	–	–	+22.71
$2\ Cu(c) + 1/2\ O_2(g) = Cu_2O(c)$	298.16–1,357K	+10,550	–1.15	–1.10	–0.10	+21.92
$2\ Cu(l) + 1/2\ O_2(g) = Cu_2O(c)$	1,357–1,502K	–43,880	+8.47	–2.60	–0.10	–3.72

The ΔH_o values are given in gram calories per mole. The a, b, and *I* values listed here make it possible for one to calculate the ΔF and ΔS values by use of the following equations:

$$\Delta F_t = \Delta H_o + 2.303aT \log T + b \times 10^{-3} T^2 + c \times 10^5 T^{-1} + IT$$

$$\Delta S_t = -a - 2.303a \log T - 2b \times 10^{-3} T + c \times 10^5 T^{-2} - I$$

Source: data from *CRC Handbook of Materials Science, Vol I*, Charles T. Lynch, Ed., CRC Press, Cleveland, (1974).

Table 67. HEAT OF FORMATION OF INORGANIC OXIDES

(SHEET 5 OF 16)

Reaction	Temperature range of validity	ΔH_0	2.303a	b	c	I
$2\ Cu(l) + 1/2\ O_2(g) = Cu_2O(l)$	1,502–2,000K	–37,710	–12.48	+0.25	–0.10	+54.44
$Cu(c) + 1/2\ O_2(g) = CuO(c)$	298.16–1,357K	–37,740	–0.64	–1.40	–0.10	+24.87
$Cu(l) + 1/2\ O_2(g) = CuO(c)$	1,357–1,720K	–39,410	+4.17	–2.15	–0.10	+12.05
$Cu(l) + 1/2\ O_2(g) = CuO(l)$	1,720–2,000K	–41,060	–11.35	+0.25	–0.10	+59.09
$2\ Au(c) + 3/2\ O_2(g) = Au_2O_3(c)$	298.16–500K	–2,160	–10.36	–	–	+95.14
$Hf(c) + O_2(g) = HfO_2$ (monoclinic)	298.16–2,000K	–268,380	–9.74	–0.28	+1.54	+78.16
$H_2(g) + 1/2\ O_2(g) = H_2O(l)$	298.16–373.16K	–70,600	–18.26	+0.64	–0.04	+91.67
$H_2(g) + 1/2\ O_2(g) = H_2O(g)$	298.16–2,000K	–56,930	+6.75	–0.64	–0.08	–8.74
$D_2(g) + 1/2\ O_2(g) = D_2O(l)$	298.16–374.5K	–72,760	–18.10	–	–	+93.59
$D_2(g) + 1/2\ O_2(g) = D_2O(g)$	298.16–2,000K	–58,970	+5.50	–0.75	+0.085	–3.74
$0.947\ Fe(\alpha) + 1/2\ O_2(g) = Fe_{0.947}O(c)$	298.16–1,033K	–65,320	–11.26	+2.61	+0.44	+48.60
$0.947\ Fe(\alpha) + 1/2\ O_2(g) = Fe_{0.947}O(c)$	1,033–1,179K	–62,380	+4.08	–0.75	+0.235	+3.00

The ΔH_o values are given in gram calories per mole. The a, b, and I values listed here make it possible for one to calculate the ΔF and ΔS values by use of the following equations:

$$\Delta F_t = \Delta H_o + 2.303aT \log T + b \times 10^{-3} T^2 + c \times 10^5 T^{-1} + IT$$

$$\Delta S_t = -a - 2.303a \log T - 2b \times 10^{-3} T + c \times 10^5 T^{-2} - I$$

Source: data from *CRC Handbook of Materials Science, Vol I*, Charles T. Lynch, Ed., CRC Press, Cleveland, (1974).

Table 67. Heat of Formation of Inorganic Oxides

(Sheet 6 of 16)

Reaction	Temperature range of validity	ΔH_0	2.303a	b	c	I
0.947 Fc(β) + 1/2 O_2(g) = $Fe_{0.9470}$(c)	1,179–1,650K	–66,750	–8.04	+0.67	–0.10	+42.28
0.947 Fe(γ) + 1/2 O_2(g) = $Fe_{0.9470}$(l)	1,650–1,674K	–64,200	–18.72	+1.67	–0.10	+73.45
0.947 Fe(γ) + 1/2 O_2(g) = $Fe_{0.9470}$(l)	1,647–1,803K	–59,650	–6.84	+0.25	–0.10	+34.81
0.947 Fe(δ) + 1/2 O_2(g) = $Fe_{0.9470}$(l)	1,803–2,000K	–63,660	–7.48	+0.25	–0.10	+39.12
3 Fe(α) + 2 O_2(g) = Fe_3O_4(magnetite)	298.16–900K	–268,310	+5.87	–12.45	+0.245	+73.11
3 Fe(α) + 2 O_2(g) = Fe_3O_4(β)	900–1,033K	–272,300	–54.27	+11.65	+0.245	+233.52
3 Fe(β) + 2 O_2(g) = Fe_3O_4(β)	1,033–1,179K	–262,990	–5.71	+1.00	–0.40	+89.19
3 Fe(γ) + 2 O_2(g) = Fe_3O_4(β)	1,179–1,674K	–276,990	~4.05	+5.50	–0.40	+213.52
2 Fe(α) + 3/2 O_2(g) = Fe_2O_3(hematite)	298.16–950K	–200,000	–13.84	–1.45	+1.905	+108.26
2 Fe(α) + 3/2 O_2(g) = Fe_2O_3(β)	950–1,033K	–202,960	–42.64	+7.85	+0.13	+188.48
2 Fe(β) + 3/2 O_2(g) = Fe_2O_3(β)	1,033–1,050K	–196,740	–10.27	+0.75	–0.30	+92.26
2 Fe(β) + 3/2 O_2(g) = Fe_2O_3(γ)	1,050–1,179K	–193,200	–0.39	–0.13	–0.30	+59.96
2 Fe(γ) + 3/2 O_2(g) = Fe_2O_3(γ)	1,179–1,674K	–202,540	–25.95	+2.87	–0.30	+142.85

The ΔH_o values are given in gram calories per mole. The a, b, and *I* values listed here make it possible for one to calculate the ΔF and ΔS values by use of the following equations:

$$\Delta F_t = \Delta H_o + 2.303aT \log T + b \times 10^{-3} T^2 + c \times 10^5 T^{-1} + IT$$

$$\Delta S_t = -a - 2.303a \log T - 2b \times 10^{-3} T + c \times 10^5 T^{-2} - I$$

Source: data from *CRC Handbook of Materials Science, Vol I*, Charles T. Lynch, Ed., CRC Press, Cleveland, (1974).

Table 67. HEAT OF FORMATION OF INORGANIC OXIDES

(SHEET 7 OF 16)

Reaction	Temperature range of validity	ΔH_0	2.303a	b	c	I
$2 Fe(\alpha) + 3/2 O_2(g) = Fe_2O_3(\gamma)$	1,674–1,800K	–192,920	–0.85	–0.13	–0.30	+61.21
$Pb(c) + 1/2 O_2(g) = PbO$ (red)	298.16–600.5K	–52,800	–2.76	–0.80	–0.10	+32.49
$Pb(l) + 1/2 O_2(g) = PbO$ (red)	600.5–762K	–53,780	–0.51	–1.75	–0.10	+28.44
$Pb(c) + 1/2 O_2(g) = PbO$ (yellow)	298.16–600.5K	–52,040	+0.81	–2.00	–0.10	+22.13
$Pb(l) + 1/2 O_2(g) = PbO$ (yellow)	600.5–1,159K	–53,020	+3.06	–2.95	–0.10	+18.08
$I_2(c) + 5/2 O_2(g) = I_2O_5(c)$	298.16–386.8K	–42,040	+2.30	–	–	+113.71
$I_2(l) + 5/2 O_2(g) = I_2O_5(c)$	386.8–456K	–43,490	+16.12	–	–	+81.70
$I_2(g) + 5/2 O_2(g) = I_2O_5(c)$	456–500K	–58,020	–6.91	–	–	+174.79
$Ir(c) + O_2(g) = IrO_2(c)$	298.16–1,300K	–39,480	+8.17	–6.39	–0.20	+20.33
$3 Pb(c) + 2 O_2(g) = Pb_3O_4(c)$	298.16–600.5K	–174,920	+8.82	–8.20	–0.40	+72.78
$Pb(c) + O_2(g) = PbO_2(c)$	298.16–600.5K	–66,120	+0.64	–2.45	–0.20	+45.58
$2 Li(c) + 1/2 O_2(g) = Li_2O(c)$	298.16–452K	–142,220	–3.06	+5.77	–0.10	+34.19

The ΔH_o values are given in gram calories per mole. The a, b, and *I* values listed here make it possible for one to calculate the ΔF and ΔS values by use of the following equations:

$$\Delta F_t = \Delta H_o + 2.303aT \log T + b \times 10^{-3} T^2 + c \times 10^5 T^{-1} + IT$$

$$\Delta S_t = -a - 2.303a \log T - 2b \times 10^{-3} T + c \times 10^5 T^{-2} - I$$

Source: data from *CRC Handbook of Materials Science, Vol I*, Charles T. Lynch, Ed., CRC Press, Cleveland, (1974).

Table 67. HEAT OF FORMATION OF INORGANIC OXIDES

(SHEET 8 OF 16)

Reaction	Temperature range of validity	ΔH_0	2.303a	b	c	I
$Mg(c) + 1/2\ O_2(g) = MgO$ (periclase)	298.16–923K	–144,090	–1.06	+0.13	+0.25	+29.16
$Mg(l) + 1/2\ O_2(g) = MgO$ (periclase)	923–1,393K	–145,810	+1.84	–0.62	+0.64	+23.07
$Mg(g) + 1/2\ O_2(g) = MgO$ (periclase)	1,393–2,000K	–180,700	–3.75	–0.62	+0.64	+65.69
$Mn(\alpha) + 1/2\ O_2(g) = MnO(c)$	298.16–1,000K	–92,600	–4.21	+0.97	+0.155	+29.66
$Mn(\beta) + 1/2\ O_2(g) = MnO(c)$	1,000–1,374K	–91,900	+1.84	–0.39	+0.34	+12.15
$Mn(\gamma) + 1/2\ O_2(g) = Mno(c)$	1,374–1,410K	–89,810	+7.30	–0.72	+0.34	–6.05
$Mn(\delta) + 1/2\ O_2(g) = MnO(c)$	1,410–1,517K	–89,390	+8.68	–0.72	+0.34	–10.70
$Mn(l) + 1/2\ O_2(g) = MnO(c)$	1,517–2,000K	–93,350	+7.99	–0.72	+0.34	–5.90
$3\ Mn(\alpha) + 2\ O_2(g) = Mn_3O_4(\alpha)$	298.16–1,000K	–332,400	–7.41	+0.66	+0.145	+106.62
$2\ Mn(\alpha) + 3/2\ O_2(g) = Mn_2O_3(c)$	298.16–1,000K	–230,610	–5.96	–0.06	+0.945	+80.74
$Mn(\alpha) + O_2(g) = MnO_2(c)$	298.16–1,000K	–126,400	–8.61	+0.97	+1.555	+70.14
$2\ Hg(l) + 1/2\ O_2(g) = Hg_2O(c)$	298.16–629.88K	–22,400	–4.61	–	–	+43.29
$Hg(l) + 1/2\ O_2(g) = HgO$ (red)	298.16–629.88K	–21,760	+0.85	–2.47	–0.10	+24.81
$Mo(c) + O_2(g) = MoO_2(c)$	298.16–2,000K	–132,910	–3.91	–	–	+47.42

The ΔH_o values are given in gram calories per mole. The a, b, and *I* values listed here make it possible for one to calculate the ΔF and ΔS values by use of the following equations:

$$\Delta F_t = \Delta H_o + 2.303aT \log T + b \times 10^{-3} T^2 + c \times 10^5 T^{-1} + IT$$

$$\Delta S_t = -a - 2.303a \log T - 2b \times 10^{-3} T + c \times 10^5 T^{-2} - I$$

Source: data from *CRC Handbook of Materials Science, Vol I*, Charles T. Lynch, Ed., CRC Press, Cleveland, (1974).

Table 67. HEAT OF FORMATION OF INORGANIC OXIDES

(SHEET 9 OF 16)

Reaction	Temperature range of validity	ΔH_0	2.303a	b	c	I
$Mo(c) + 3/2\ O_2(g) = MoO_3(c)$	298.16–1,068K	–182,650	–8.86	–1.55	+1.54	+90.07
$Ni(\alpha) + 1/2\ O_2(g) = NiO(c)$	298.16–633K	–57,640	–4.61	+2.16	–0.10	+34.41
$Ni(\beta) + 1/2\ O_2(g) = NiO(c)$	633–1,725K	–57,460	–0.14	–0.46	–0.10	+23.27
$2\ Nb(c) + 2\ O_2(g) = Nb_2O_4(c)$	298.16–2,000K	–382,050	–9.67	–	–	+116.23
$2\ Nb(c) + 5/2\ O_2(g) = Nb_2O_5(c)$	298.16–1,785K	–458,640	–16.14	–0.56	+1.94	+157.66
$2\ Nb(c) + 5/2\ O_2(g) = Nb_2O_5(l)$	1,785–2,000K	–463,630	–66.04	+2.21	–0.50	+317.84
$N_2(g) + 1/2\ O_2(g) = N_2O(g)$	298.16–2,000K	18,650	–1.57	–0.27	+0.92	+23.47
$3/2\ O_2(g) = O_3(g)$	298.16–2,000K	+33,980	+2.03	–0.48	+0.36	+11.45
$P\ (white) + 1/2\ O_2(g) = PO(g)$	298.16–317.4K	–9,370	+2.53	–	–	–25.40
$P(l) + 1/2\ O_2(g) = PO(g)$	317.4–553K	–9,390	+3.45	–	–	–27.63
$4\ P\ (white) + 5\ O_2(g) = P_4H_{10}$ (hexagonal)	298.16–317.4K	–711,520	+95.67	–51.50	–1.00	–28.24
$2\ K(c) + 1/2\ O_2(g) = K_2O(c)$	298.16–336.4K	–86,400	–	–	–	+33.90

The ΔH_o values are given in gram calories per mole. The a, b, and *I* values listed here make it possible for one to calculate the ΔF and ΔS values by use of the following equations:

$$\Delta F_t = \Delta H_o + 2.303aT \log T + b \times 10^{-3} T^2 + c \times 10^5 T^{-1} + IT$$

$$\Delta S_t = -a - 2.303a \log T - 2b \times 10^{-3} T + c \times 10^5 T^{-2} - I$$

Source: data from *CRC Handbook of Materials Science, Vol I*, Charles T. Lynch, Ed., CRC Press, Cleveland, (1974).

Table 67. Heat of Formation of Inorganic Oxides

(Sheet 10 of 16)

Reaction	Temperature range of validity	ΔH_0	2.303a	b	c	I
$2\,K(l) + 1/2\,O_2(g) = K_2O(c)$	336.4–1,049K	–87,380	+1.15	–	–	+33.90
$2\,K(g) + 1/2\,O_2(g) = K_2O(c)$	1,049–1,500K	–133,090	–16.12	–	–	+129.64
$Ra(c) + 1/2\,O_2(g) = RaO(c)$	298.16–1,000K	–130,000	–	–	–	+23.50
$Re(c) + 3/2\,O_2(g) = ReO_3(c)$	298.16–433K	–149,090	–16.12	–	–	+110.49
$Re(c) + 3/2\,O_2(g) = ReO_3(l)$	433–1,000K	–146,750	–31.32	–	–	+145.16
$2Re(c) + 7/2\,0_2(g) = Re_2O_7(c)$	298.16–569K	–301,470	–34.64	–	–	+250.57
$2\,Re(c) + 7/2\,0_2(g) = Re_2O_7(l)$	569–635.5K	–295,810	–73.68	–	–	+348.45
$2\,Re(c) + 4\,O_2(g) = Re_2O_8(l)$	420–600K	–318,470	–87.50	–	–	+425.32
$2\,Rb(c) + 1/2\,O_2(g) = Rb_2O(c)$	298.16–312.2K	–78,900	–	–	–	+32.20
$2\,Rb(l) + 1/2\,O_2(g) = Rb_2O(c)$	312.2–750K	–79,950	–	–	–	+35.56
$Se(c) + 1/2\,O_2(g) = SeO(g)$	298.16–490K	+9,280	–3.04	+4.40	+0.30	–14.78
$Se(l) + 1/2\,O_2(g) = SeO(g)$	490–1,027K	+9,420	+8.70	–	+0.30	–44.50
$1/2\,Se_2(g) + 1/2\,O_2(g) = SeO(g)$	1,027–2,000K	–7,400	–0.37	–	+0.19	–0.80
$Si(c) + 1/2\,O_2(g) = SiO(g)$	298.16–1,683K	–21,090	+3.84	–0.16	–0.295	–33.14

The ΔH_o values are given in gram calories per mole. The a, b, and *I* values listed here make it possible for one to calculate the ΔF and ΔS values by use of the following equations:

$$\Delta F_t = \Delta H_o + 2.303aT \log T + b \times 10^{-3} T^2 + c \times 10^5 T^{-1} + IT$$

$$\Delta S_t = -a - 2.303a \log T - 2b \times 10^{-3} T + c \times 10^5 T^{-2} - I$$

Source: data from *CRC Handbook of Materials Science, Vol I*, Charles T. Lynch, Ed., CRC Press, Cleveland, (1974).

Table 67. HEAT OF FORMATION OF INORGANIC OXIDES

(SHEET 11 OF 16)

Reaction	Temperature range of validity	ΔH_0	2.303a	b	c	I
$Si(l) + 1/2\ O_2(g) = SiO(g)$	1,683–2,000K	–30,170	–7.78	–0.12	+0.25	–40.01
$Si(c) + O_2(g) = SiO_2(\alpha\text{–quartz})$	298.16–848K	–210,070	+3.98	–3.32	+0.605	+34.59
$Si(c) + O_2(g) = SiO_2(\beta\text{–quartz})$	848–1,683K	–209,920	–3.36	–0.19	–0.745	+53.44
$Si(l) + O_2(g) = SiO_2(l)$	1,883–2,000K	–228,590	–15.66	–	–	+103.97
$Si(c) + O_2(g) = SiO_2(\alpha\text{–cristobalite})$	298.16–523K	–207,330	+19.96	–9.75	–0.745	–9.78
$Si(c) + O_2(g) = SiO_2(\beta\text{–cristobalite})$	523–1,683K	–209,820	–3.34	–0.24	–0.745	+53.35
$Si(c) + 0_2(g) = SiO_2(\alpha\text{–tridymite})$	298.16–390K	–207,030	+22.29	–11.62	–0.745	–15.64
$Si(c) + O_2(g) = SiO_2(\beta\text{–tridymite})$	390–1,683K	–209,350	–1.59	–0.54	–0.745	+47.86
$2\ Ag(c) + 1/2\ O_2(g) = Ag_2O_2(c)$	298.16–1,000K	–7,740	–4.14	–	–	+27.84
$2\ Ag(c) + O_2(g) = Ag_2O_2(c)$	298.16–500K	–6,620	–3.22	–	–	+52.17
$2\ Na(c) + 1/2\ O_2(g) = Na_2O(c)$	298.16–371K	–99,820	–7.51	+5.47	–0.10	+50.43
$2\ Na(l) + 1/2\ O_2(g) = Na_2O(c)$	371–1,187K	–100,150	+4.97	–2.45	–0.10	+22.19

The ΔH_o values are given in gram calories per mole. The a, b, and *I* values listed here make it possible for one to calculate the ΔF and ΔS values by use of the following equations:

$$\Delta F_t = \Delta H_o + 2.303aT \log T + b \times 10^{-3} T^2 + c \times 10^5 T^{-1} + IT$$

$$\Delta S_t = -a - 2.303a \log T - 2b \times 10^{-3} T + c \times 10^5 T^{-2} - I$$

Source: data from *CRC Handbook of Materials Science, Vol I*, Charles T. Lynch, Ed., CRC Press, Cleveland, (1974).

Table 67. Heat of Formation of Inorganic Oxides

(Sheet 12 of 16)

Reaction	Temperature range of validity	ΔH_0	2.303a	b	c	I
$2 Na(c) + O_2(g) = Na_2O_2(c)$	298.16–371K	–122,500	–2.30	–	–	+57.51
$Sr(c) + 1/2\ O_2(g) = SrO(c)$	298.16–1,043K	–142,410	–6.79	+0.305	+0.675	+44.33
$S(rhombohedral) + 1/2\ O_2(g) = SO(g)$	298.16–368.6K	+19,250	–1.24	+2.95	+0.225	–18.84
$S(monoclmic) + 1/2\ O_2(g) = SO(g)$	368.6–392K	+19,200	–1.29	+3.31	+0.225	–18.72
$S(l\lambda,\mu) + 1/2\ O_2(g) = SO(g)$	392–718K	+20,320	+10.22	–0.17	+0.225	–50.05
$1/2\ S_2(g) + 1/2\ O_2(g) = SO(g)$	298.16–2,000K	+3,890	+0.07	–	–	–1.50
$S(rhombohedral) + O_2(g) = SO_2(g)$	298.16–368.6K	–70,980	+0.83	+2.35	+0.51	–5.85
$S(monoclinic) + O_2(g) = SO_2(g)$	368.6–392K	–71,020	+0.78	+2.71	+0.51	–5.74
$S(l\lambda,\mu) + O_2(g) = SO_2(g)$	392–718K	–69,900	+12.30	–0.77	+0.51	–37.10
$1/2\ S_2(g) + O_2(g) = SO_2(g)$	298.16–2,000K	–86,330	+2.42	–0.70	+0.31	+10.71
$S(rhombohedral) + 3/2\ O_2(g) = SO_3(c\text{–}I)$	298.16–335.4K	–111,370	–6.45	–	–	+88.32
$S(rhombohedral) + 3/2\ O_2(g) = SO_3(c\text{–}II)$	298.16–305.7K	–108,680	–11.97	–	–	+94.95
$S(rhombohedral) + 3/2\ O_2(g) = SO_3(l)$	298.16–335.4K	–107,430	–21.18	–	–	+113.76

The ΔH_o values are given in gram calories per mole. The a, b, and *I* values listed here make it possible for one to calculate the ΔF and ΔS values by use of the following equations:

$$\Delta F_t = \Delta H_o + 2.303aT \log T + b \times 10^{-3} T^2 + c \times 10^5 T^{-1} + IT$$

$$\Delta S_t = -a - 2.303a \log T - 2b \times 10^{-3} T + c \times 10^5 T^{-2} - I$$

Source: data from *CRC Handbook of Materials Science, Vol I*, Charles T. Lynch, Ed., CRC Press, Cleveland, (1974).

Table 67. Heat of Formation of Inorganic Oxides

(Sheet 13 of 16)

Reaction	Temperature range of validity	ΔH_0	2.303a	b	c	I
S(rhombohedral) + 3/2 $O_2(g) = SO_3(g)$	298.16–368.6K	–95,070	+1.43	+0.66	+1.26	+16.81
S(monoclinic) + 3/2 $O_2(g) = SO_3(g)$	368.6–392K	–95,120	+1.38	+1.02	+1.26	+16.93
S(lλ,μ) + 3/2 $O_2(g) = SO_3(g)$	392–718K	–94,010	+12.89	–2.46	+126	–14.40
1/2 $S_2(g)$ + 3/2 $O_2(g) = SO_3(g)$	298.16–1,500K	–110,420	+3.02	–2.39	+106	+33.41
2 Ta(c) + 5/2 $O_2(g) = Ta_2O_5(c)$	298.16–2,000K	–492,790	–17.18	–1.25	+2.46	+161.68
Te(c) + 1/2 $O_2(g)$ = TeO(g)	298.16–723K	+43,110	+1.91	+0.84	+0.315	–27.22
Te(l) + 1/2 $O_2(g)$ = TeO(g)	723–1,360K	+39,750	+6.08	+0.09	+0.315	–33.94
2 Tl(α) + $O_2(g) = Tl_{2r}O(c)$	298.16–505.5K	–44,110	–6.91	–	–	+42.30
2 Tl(β) + $O_2(g) = Tl_{2r}O(c)$	505.5–573K	–44,260	–6.91	–	–	+42.60
2 Tl(α) + 3/2 $O_2(g) = Tl_{2r}O_3(c)$	298.16–505.5K	–99,410	–16.12	–	–	+119.09
Th(c) + $O_2(g) = ThO_2(c)$	298.16–2,000K	–294,350	–5.25	+0.59	+0.775	+62.81
Sn(c) + 1/2 $O_2(g)$ = SnO(c)	298.16–505K	–68,600	–3.57	+1.65	–0.10	+32.59

The ΔH_o values are given in gram calories per mole. The a, b, and *I* values listed here make it possible for one to calculate the ΔF and ΔS values by use of the following equations:

$$\Delta F_t = \Delta H_o + 2.303aT \log T + b \times 10^{-3} T^2 + c \times 10^5 T^{-1} + IT$$

$$\Delta S_t = -a - 2.303a \log T - 2b \times 10^{-3} T + c \times 10^5 T^{-2} - I$$

Source: data from *CRC Handbook of Materials Science, Vol I*, Charles T. Lynch, Ed., CRC Press, Cleveland, (1974).

Table 67. HEAT OF FORMATION OF INORGANIC OXIDES

(SHEET 14 OF 16)

Reaction	Temperature range of validity	ΔH_0	2.303a	b	c	I
$Sn(l) + 1/2\ O_2(g) = SnO(c)$	505–1,300K	–69,670	+3.06	–1.50	–0.10	+18.39
$Sn(c) + O_2(g) = SnO_2(c)$	298.16–505K	–0,142	–14.00	+2.45	+2.38	+90.74
$Ti(\alpha) + 1/2\ O_2(g) = TiO(\alpha)$	298.16–1,150K	–125,010	–4.01	–0.29	+0.83	+36.28
$Ti(\alpha) + 1/2\ O_2(g) = TiO(\alpha)$	1,150–1,264K	–125,040	+1.17	–1.55	+0.83	+21.90
$2\ Ti(\alpha) + 3/2\ O_2(g) = Ti_2O_3(\alpha)$	298.16–473K	–360,660	+32.08	–23.49	–0.30	–10.66
$2\ Ti(\alpha) + 3/2\ O_2(g) = Ti_2O_3(\beta)$	473–1,150K	–369,710	–30.95	+2.62	+4.80	+162.79
$Ti(\alpha) + O_2(g) = TiO_2$ (rutile)	298.16–1,150K	–228,360	–12.80	+1.62	+1.975	+82.81
$Ti(\alpha) + O_2(g) = TiO_2$ (rutile)	1,150–2,000K	–228,380	–7.62	+0.36	+1.975	+68.43
$W(c) + O_2(g) = WO_2(c)$	298.16–1,500K	–137,180	–1.38	–	–	+45.56
$4W(c) + 11/2\ O_2(g) = W_4O_{11}(c)$	298.16–1,700K	–745,730	–32.70	–	–	+321.84
$W(c) + 3/2\ O_2(g) = WO_3(c)$	298.16–1,743K	–201,180	–2.92	–1.81	–0.30	+70.89
$W(c) + 3/2\ O_2(g) = WO_3(l)$	1,743–2,000K	–203,140	–35.74	+1.13	–0.30	+173.27
$U(\alpha) + O_2(g) = UO_2(c)$	298.16–935K	–262,880	–19.92	+3.70	+2.13	+100.54
$U(\beta) + O_2(g) = UO_2(c)$	935–1,045K	–260,660	–4.28	–0.31	+1.78	+55.50

The ΔH_o values are given in gram calories per mole. The a, b, and *I* values listed here make it possible for one to calculate the ΔF and ΔS values by use of the following equations:

$$\Delta F_t = \Delta H_o + 2.303aT \log T + b \times 10^{-3} T^2 + c \times 10^5 T^{-1} + IT$$

$$\Delta S_t = -a - 2.303a \log T - 2b \times 10^{-3} T + c \times 10^5 T^{-2} - I$$

Source: data from *CRC Handbook of Materials Science, Vol I,* Charles T. Lynch, Ed., CRC Press, Cleveland, (1974).

Table 67. HEAT OF FORMATION OF INORGANIC OXIDES

(SHEET 15 OF 16)

Reaction	Temperature range of validity	ΔH_0	2.303a	b	c	I
$U(\gamma) + O_2(g) = UO_2(c)$	1,045–1,405K	–262,830	–6.54	–0.31	+1.78	+64.41
$U(l) + O_2(g) = UO_2(l)$	1,405–1,500K	–264,790	–5.92	–	–	+63.50
$3\,U(\alpha) + 4\,O_2(g) = U_3O_8(c)$	298.16–935K	–863,370	–56.57	+10.68	+5.20	+330.19
$3\,U(\beta) + 4\,O_2(g) = U_3O_8(c)$	935–1,045K	–856,720	–9.67	–1.35	+4.15	+195.12
$3\,U(\gamma) + 4\,O_2(g) = U_3O_8(c)$	1,045–1,405K	–863,230	–16.44	–1.35	+4.15	+221.79
$3\,U(l) + 4\,O_2(g) = U_3O_8(c)$	1,405–1,500K	–869,460	–10.91	–1.35	+4.15	+208.82
$U(\alpha) + 3/2\,O_2(g) = UO_3$ (hexagonal)	298.16–935K	–294,090	–18.33	+3.49	+1.535	+114.94
$U(\beta) + 3/2\,O_2(g) = UO_3$ (hexagonal)	935–1,045K	–291,870	–2.69	–0.52	+1.185	+69.90
$U(\gamma) + 3/2\,O_2(g) = UO_3$ (hexagonal)	1,045–1,400K	–294,040	–4.95	–0.52	+1.185	+78.80
$V(c) + 1/2\,O_2(g) = VO(c)$	298.16–2,000K	–101,090	–5.39	–0.36	+0.53	+38.69
$V(c) + 1/2\,O_2(g) = VO(g)$	298.16–2,000K	+52,090	+1.80	+1.04	+0.35	–28.42
$2\,V(c) + 3/2\,O_2(g) = V_2O_3(c)$	298.16–2,000K	–299,910	–17.98	+0.37	+2.41	+118.83

The ΔH_o values are given in gram calories per mole. The a, b, and *I* values listed here make it possible for one to calculate the ΔF and ΔS values by use of the following equations:

$$\Delta F_t = \Delta H_o + 2.303aT \log T + b \times 10^{-3} T^2 + c \times 10^5 T^{-1} + IT$$

$$\Delta S_t = -a - 2.303a \log T - 2b \times 10^{-3} T + c \times 10^5 T^{-2} - I$$

Source: data from *CRC Handbook of Materials Science, Vol I*, Charles T. Lynch, Ed., CRC Press, Cleveland, (1974).

Table 67. HEAT OF FORMATION OF INORGANIC OXIDES

(SHEET 16 OF 16)

Reaction	Temperature range of validity	ΔH_0	2.303a	b	c	I
$2\,V(c) + 2\,O_2(g) = V_2O_4(\alpha)$	209.16–345K	–342,890	–11.03	+3.00	–0.40	+117.38
$2\,V(c) + 2\,O_2(g) = V_2O_4(\beta)$	345–1,818K	–345,330	–24.36	+1.30	+3.545	+155.55
$6\,V(c) + 13/2\,O_2(g) = V_6O_{13}(c)$	298.16–1,000K	–1,076,340	–95.33	–	–	+557.61
$2\,V(c) + 5/2\,O_2(g) = V_2O_5(c)$	298.16–943K	–381,960	–41.08	+5.20	+6.11	+228.50
$2\,Y(c) + 3/2\,O_2(g) = Y_2O_3(c)$	298.16–1,773K	–419,600	+2.76	–1.73	–0.30	+66.36
$Zn(c) + 1/2\,O_2(g) = ZnO(c)$	298.16–692.7K	–84,670	–6.40	+0.84	+0.99	+43.25
$Zr(\alpha) + O_2(g) = ZrO_2(\alpha)$	298.16–1,135K	–262,980	–6.10	+0.16	+1.045	+65.00
$Zr(\beta) + O_2(g) = ZrO_2(\alpha)$	1,135–1,478K	–264,190	–5.09	–0.40	+1.48	+63.58
$Zr(\beta) + O_2(g) = ZrO_2(\beta)$	1.478–2,000K	–262,290	–7.76	+0.50	–0.20	+69.50

The ΔH_o values are given in gram calories per mole. The a, b, and *I* values listed here make it possible for one to calculate the ΔF and ΔS values by use of the following equations:

$$\Delta F_t = \Delta H_o + 2.303aT \log T + b \times 10^{-3} T^2 + c \times 10^5 T^{-1} + IT$$

$$\Delta S_t = -a - 2.303a \log T - 2b \times 10^{-3} T + c \times 10^5 T^{-2} - I$$

Source: data from *CRC Handbook of Materials Science, Vol I*, Charles T. Lynch, Ed., CRC Press, Cleveland, (1974).

Table 68. PHASE CHANGE THERMODYNAMIC PROPERTIES FOR THE ELEMENTS

(SHEET 1 OF 7)

Element	Phase	Transition Temperature (K)	Heat of Transition (kcal • g mole^{-1})	Entropy of Transition (e.u.)
Ac	solid	(1090)	(2.5)	(2.3)
	liquid	(2750)	(70)	(25)
Ag	solid	1234	2.855	2.313
	liquid	2485	60.72	24.43
Al	solid	931.7	2.57	2.76
	liquid	2600	67.9	26
Am	solid	(1200)	(2.4)	(2.0)
	liquid	2733	51.7	18.9
As	solid	883	3.25	35.25
Au	solid	1336.16	3.03	2.27
	liquid	2933	74.21	25.30
B	solid	2313	(3.8)	(1.6)
	liquid	2800	75	27
Ba	solid, α	648	0.14	0.22
	solid, β	977	1.83	1.87
	liquid	1911	35.665	18.63
Be	solid	1556	2.919	1.501
	liquid	–		
Bi	solid	544.2	2.63	4.83
	liquid	1900	41.1	21.6
C	solid	–	–	–
Ca	solid, α	723	0.24	0.33
	solid, β	1123	2.2	1.96
	liquid	1755	38.6	22.0

Source: data from Weast, R. C. Ed., *Handbook of Chemistry and Physics, 69th ed.*, CRC Press, Boca Raton, Fla., 1988, D44.

Table 68. Phase Change Thermodynamic Properties for The Elements

(Sheet 2 of 7)

Element	Phase	Transition Temperature (K)	Heat of Transition (kcal • g mole^{-1})	Entropy of Transition (e.u.)
Cd	solid	594.1	1.46	2.46
	liquid	1040	23.86	22.94
Ce	solid	1048	2.1	2.0
	liquid	2800	73	26
Cl_2	gas	–	–	–
Co	solid, α	723	0.005	0.007
	solid, β	1398	0.095	0.068
	solid, γ	1766	3.7	2.1
	liquid	3370	93	28
Cr	solid	2173	3.5	1.6
	liquid	2495	72.97	29.25
Cs	solid	301.9	0.50	1.7
	liquid	963	16.32	17.0
Cu	solid	1356.2	3.11	2.29
	liquid	2868	72.8	25.4
F_2	gas	–	–	–
Fe	solid, α	1033	0.410	0.397
	solid, β	1180	0.217	0.184
	solid, γ	1673	0.15	0.084
	solid, δ	1808	3.86	2.14
	liquid	3008	84.62	28.1
Ga	solid	302.94	1.335	4.407
	liquid	2700	–	–

Source: data from Weast, R. C. Ed., *Handbook of Chemistry and Physics, 69th ed.*, CRC Press, Boca Raton, Fla., 1988, D44.

Table 68. PHASE CHANGE THERMODYNAMIC PROPERTIES FOR THE ELEMENTS

(SHEET 3 OF 7)

Element	Phase	Transition Temperature (K)	Heat of Transition (kcal • g mole^{-1})	Entropy of Transition (e.u.)
Ge	solid	1232	8.3	6.7
	liquid	2980	68	23
H_2	gas	–	–	–
Hf	solid	(2600)	(6.0)	(2.3)
Hg	liquid	629.73	13.985	22.208
In	solid	430	0.775	1.80
	liquid	2440	53.8	22.0
Ir	solid	2727	6.6	2.4
K	solid	336.4	0.5575	1.657
	liquid	1052	18.88	17.95
La	solid	1153	(2.3)	(2.0)
	liquid	3000	80	27
Li	solid	459	0.69	1.5
	liquid	1640	32.48	19.81
Mg	solid	923	2.2	2.4
	liquid	1393	31.5	22.6
Mn	solid, α	1000	0.535	0.535
	solid, β	1374	0.545	0.397
	solid, γ	1410	0.430	0.305
	solid, δ	1517	3.5	2.31
	liquid	2368	53.7	22.7
Mo	solid	2883	(5.8)	(2.0)

Source: data from Weast, R. C. Ed., *Handbook of Chemistry and Physics, 69th ed.*, CRC Press, Boca Raton, Fla., 1988, D44.

Table 68. PHASE CHANGE THERMODYNAMIC PROPERTIES FOR THE ELEMENTS

(SHEET 4 OF 7)

Element	Phase	Transition Temperature (K)	Heat of Transition (kcal • g mole^{-1})	Entropy of Transition (e.u.)
N_2	gas	–	–	–
Na	solid	371	0.63	1.7
	liquid	1187	23.4	20.1
Nb	solid	2760	(5.8)	(2.1)
Nd	solid	1297	(2.55)	(197)
	liquid	(2750)	(61)	(22)
Ni	solid α	626	0.092	0.15
	solid β	1728	4.21	2.44
	liquid	3110	90.48	29.0
Np	solid	913	(2.3)	(2.5)
	liquid	(2525)	(55)	(22)
O_2	gas	–	–	–
Os	solid	2970	(6.4)	(2.2)
P_4	solid, white	317.4	0.601	1.89
	liquid	553	11.9	21.5
Pa	solid	(18.25)	(4.0)	(2.2)
	liquid	(4500)	(115)	(26)
Pb	solid	600.6	1.141	1.900
	liquid	2023	42.5	21.0
Pd	solid	1828	4.12	2.25
	liquid	3440	89	26

Source: data from Weast, R. C. Ed., *Handbook of Chemistry and Physics, 69th ed.*, CRC Press, Boca Raton, Fla., 1988, D44.

Table 68. PHASE CHANGE THERMODYNAMIC PROPERTIES FOR THE ELEMENTS

(SHEET 5 OF 7)

Element	Phase	Transition Temperature (K)	Heat of Transition (kcal • g mole^{-1})	Entropy of Transition (e.u.)
Po	solid	525	(2.4)	(4.6)
	liquid	(1235)	(24.6)	(19.9)
Pr	solid	1205	(25)	(2.1)
	liquid	3563	–	–
Pt	solid	2042.5	5.2	25
	liquid	4100	122	29.8
Pu	solid	913	(2.26)	(2.48)
	liquid	–		
Ra	solid	1233	(2.3)	(1.9)
	liquid	(1700)	(35)	(21)
Rb	solid	312.0	0.525	1.68
	liquid	952	18.11	19.0
Re	solid	3440	(7.9)	(2.3)
Rh	solid	2240	(5.2)	(2.3)
	liquid	4150	127	30.7
Ru	solid, α	1308	0.034	0.026
	solid, β	1473	0	–
	solid, γ	1773	0.23	0.13
	solid, δ	2700	(6.1)	(2.3)
S	solid, α	368.6	0.088	0.24
	solid, β	392	0.293	0.747
	liquid	717.76	2.5	3.5
Sb	solid (α, β, γ)	903.7	4.8	5.3
	liquid	1713	46.665	27.3

Source: data from Weast, R. C. Ed., *Handbook of Chemistry and Physics, 69th ed.*, CRC Press, Boca Raton, Fla., 1988, D44.

Table 68. PHASE CHANGE THERMODYNAMIC PROPERTIES FOR THE ELEMENTS

(SHEET 6 OF 7)

Element	Phase	Transition Temperature (K)	Heat of Transition (kcal • g mole^{-1})	Entropy of Transition (e.u.)
Sc	solid	1670	(4.0)	(2.4)
	liquid	3000	80	27
Se	solid	490.6	1.25	2.55
	liquid	1000	14.27	14.27
Si	solid	1683	11.1	6.60
	liquid	2750	71	26
Sm	solid	1623	3.7	2.3
	liquid	(2800)	(70)	(25)
Sn	solid, α, β	505.1	1.69	335
	liquid	2473	(55)	(22)
Sr	solid	1043	2.2	2.1
	liquid	1657	33.61	20.28
Ta	solid	3250	7.5	2.3
Tc	solid	(2400)	(5.5)	(2.3)
	liquid	(3800)	(120)	(32)
Te	solid, α	621	0.13	0.21
	solid, β	723	4.28	5.92
	liquid	1360	11.9	8.75
Th	solid	2173	(4.6)	(2.1)
	liquid	4500	(130)	(29)
Ti	solid, α	1155	0.950	0.822
	solid, β	2000	(4.6)	(23)
	liquid	3550	(101)	(28)

Source: data from Weast, R. C. Ed., *Handbook of Chemistry and Physics, 69th ed.*, CRC Press, Boca Raton, Fla., 1988, D44.

Table 68. PHASE CHANGE THERMODYNAMIC PROPERTIES FOR THE ELEMENTS

(SHEET 7 OF 7)

Element	Phase	Transition Temperature (K)	Heat of Transition (kcal • g mole^{-1})	Entropy of Transition (e.u.)
Tl	solid, α	508.3	0.082	0.16
	solid, β	576.8	1.03	1.79
	liquid	1730	38.81	22.4
U	solid, α	938	0.665	0.709
	solid, β	1049	1.165	1.111
	solid, γ	1405	(3.0)	(2.1)
	liquid	3800	–	–
V	solid	2003	(4.0)	(2.0)
	liquid	3800	–	–
W	solid	3650	8.42	2.3
Y	solid	1750	(4.0)	(2.3)
	liquid	3500	(90)	(26)
Zn	solid	692.7	1.595	2.303
	liquid	1180	27.43	23.24
Zr	solid, α	1135	0.920	0.811
	solid, β	2125	(4.9)	(2.3)
	liquid	(3900)	(100)	(26)

Source: data from Weast, R. C. Ed., *Handbook of Chemistry and Physics, 69th ed.*, CRC Press, Boca Raton, Fla., 1988, D44.

Table 69. PHASE CHANGE THERMODYNAMIC PROPERTIES OF OXIDES (SHEET 1 OF 10)

Oxide	Phase	Transition Temperature (K)	Heat of Transition (kcal • g mole^{-1})	Entropy of Transition (e.u.)
Ac_2O_3	Solid	(2250)	(20)	(8.9)
	Liquid	–	–	–
Ag_2O	Solid	dec. 460	–	–
Ag_2O_2	Solid	dec.	–	–
Al_2O_3	Solid	2300	26	11
	Liquid	dec.	–	–
Am_2O_3	Solid	(2225)	(17)	(7.6)
	Liquid	(3400)	(85)	(25)
AmO_2	Solid	dec.	–	–
As_2O_3	Solid, α	503	4.1	8.2
	Solid, β	586	4.4	7.5
	Liquid	730	7.15	9.79
AsO_2	Solid	(1200)	(9.0)	(7.5)
	Liquid	(dec.)	–	–
As_2O_5	Solid	dec. >1100	–	–
Au_2O_3	Solid	dec.	–	–
B_2O_3	Solid	723	5.27	7.29
	Liquid	2520	(55)	(22)
Ba_2O	Solid	(880)	(5.2)	(5.9)
	Liquid	(1040)	(20)	(19)
BaO	Solid	2196	13.8	6.28
	Liquid	3000	(62)	(21)
BaO_2	Solid	723	(5.7)	(7.9)
	Liquid	dec. 1110	–	–

Source: data from Weast, R. C., Ed., *Handbook of Chemistry and Physics, 55th ed.*, CRC Press, Cleveland, 1974, D-58.

Table 69. PHASE CHANGE THERMODYNAMIC PROPERTIES OF OXIDES (SHEET 2 OF 10)

Oxide	Phase	Transition Temperature (K)	Heat of Transition (kcal • g mole^{-1})	Entropy of Transition (e.u.)
BeO	Solid	dec.	–	–
BiO	Solid	(1175)	(3.7)	(3.1)
	Liquid	(1920)	(54)	(28)
Bi_2O_3	Solid	1090	6.8	6.2
	Liquid	(dec.)	–	–
CO	Gas	–	–	–
CO_2	Gas	–	–	–
CaO	Solid	2860	(18)	(6.3)
CdO	Solid	dec.	–	–
Ce_2O_3	Solid	1960	(20)	(10)
	Liquid	(3500)	(80)	(23)
CeO_2	Solid	3000	(19)	(6.3)
CoO	Solid	2078	(12)	(5.8)
	Liquid	(2900)	(61)	(21)
Co_3O_4	Solid	dec. 1240	–	–
Cr_2O_3	Solid	2538	(25)	(10)
CrO_2	Solid	dec. 700	–	–
CrO_3	Solid	460	(6.1)	(13)
	Liquid	(1000)	(25)	(25)
Cs_2O	Solid	763	(4.58)	(6.0)
	Liquid	dec.	–	–
Cs_2O_2	Solid	867	(5.5)	(6.3)
	Liquid	dec.	–	–
Cs_2O_3	Solid	775	(7.75)	(10)
	Liquid	dec.	–	–

Source: data from Weast, R. C., Ed., *Handbook of Chemistry and Physics, 55th ed.*, CRC Press, Cleveland, 1974, D-58.

Table 69. PHASE CHANGE THERMODYNAMIC PROPERTIES OF OXIDES (SHEET 3 OF 10)

Oxide	Phase	Transition Temperature (K)	Heat of Transition (kcal • g mole^{-1})	Entropy of Transition (e.u.)
Cu_2O	Solid	1503	13.4	8.92
	Liquid	dec.	–	–
CuO	Solid	1609	(8.9)	(5.5)
	Liquid	dec.	–	–
FeO	Solid	1641	7.5	4.6
	Liquid	(2700)	(55)	(20)
Fe_3O_4	Solid, α	900	(0)	(0)
	Solid, β	dec.	–	–
Fe_2O_3	Solid, α	950	0.16	0.17
	Solid, β	1050	0	0
	Solid, γ	dec.	–	–
Ga_2O	Solid	(925)	(8.5)	(9.2)
	Liquid	(1000)	(20)	(20)
Ga_2O_3	Solid	2013	(22)	(11)
	Liquid	(2900)	(75)	(26)
GeO	Solid	983	(50)	(51)
GeO_2	Solid (α,β)	1389	10.5	7.56
	Liquid	(2625)	(61)	(23)
In_2O	Solid	(600)	(4.5)	(7.5)
	Liquid	(800)	(16)	(20)
InO	Solid	(1325)	(4.0)	(3.0)
	Liquid	(2000)	(60)	(30)
In_2O_3	Solid	(2000)	(20)	(10)
	Liquid	(3600)	(85)	(24)
Ir_2O_3	Solid	(1450)	(10)	(6.8)
	Liquid	(2250)	(50)	(22)
IrO_2	Solid	dec. 1373	–	–

Source: data from Weast, R. C., Ed., *Handbook of Chemistry and Physics, 55th ed.*, CRC Press, Cleveland, 1974, D-58.

Table 69. PHASE CHANGE THERMODYNAMIC PROPERTIES OF OXIDES (SHEET 4 OF 10)

Oxide	Phase	Transition Temperature (K)	Heat of Transition (kcal • g mole^{-1})	Entropy of Transition (e.u.)
K_2O	Solid	(980)	(6.8)	(6.9)
	Liquid	dec.	–	–
K_2O_2	Solid	763	(7.0)	(9.2)
	Liquid	(1800)	(45)	(25)
K_2O_3	Solid	703	(6.1)	(8.7)
	Liquid	(975)	(25)	(26)
KO_2	Solid	653	(4.9)	(7.5)
	Liquid	dec.	–	–
La_2O_3	Solid	2590	(18)	(7)
Li_2O	Solid	2000	(14)	(7)
	Liquid	2600	(56)	(22)
Li_2O_2	Solid	dec.470	–	–
MgO	Solid	3075	18.5	5.8
MgO_2	Solid	dec. 361	–	–
MnO	Solid	2058	13.0	6.32
	Liquid	dec.	–	–
Mn_3O_4	Solid, α	1445	4.97	3.44
	Solid, β	1863	(33)	(18)
	Liquid	(2900)	(75)	(26)
Mn_2O_3	Solid	dec. 1620	–	–
MnO_2	Solid	dec. 1120	–	–
MoO_2	Solid	(2200)	(16)	(7.3)
	Liquid	dec. 2250	–	–
MoO_3	Solid	1068	12.54	11.74
	Liquid	1530	33	22
N_2O	Gas	–	–	–

Source: data from Weast, R. C., Ed., *Handbook of Chemistry and Physics, 55th ed.*, CRC Press, Cleveland, 1974, D-58.

Table 69. PHASE CHANGE THERMODYNAMIC PROPERTIES OF OXIDES (SHEET 5 OF 10)

Oxide	Phase	Transition Temperature (K)	Heat of Transition ($kcal \cdot g\ mole^{-1}$)	Entropy of Transition (e.u.)
Na_2O	Solid	1193	(7.1)	(6.0)
	Liquid	dec.	–	–
Na_2O_2	Solid	dec. 919	–	–
NaO_2	Solid	(825)	(6.2)	(7.5)
	Liquid	(1300)	(28)	(22)
NbO	Solid	(2650)	(16)	(6.0)
NbO_2	Solid	(2275)	(16)	(7.0)
	Liquid	(3800)	(85)	(22)
Nb_2O_5	Solid	1733	(28)	(16)
	Liquid	(3200)	(80)	(25)
Nd_2O_3	Solid	2545	(22)	(8.8)
NiO	Solid	2230	(12.1)	(5.43)
	Liquid	dec.	–	–
NpO_2	Solid	(2600)	(15)	(5.7)
Np_2O_5	Solid	dec. 800–900 K	–	–
OsO_2	Solid	dec. 923	–	–
OsO_4	Solid	313.3	3.41	10.9
	Liquid	403	9.45	23.4
P_2O_3	Liquid	448.5	4.5	10
PO_2	Solid	(350)	(2.7)	(7.7)
	Liquid	(dec.)	–	–
P_2O_5	Solid	631	8.8	13.9
PaO_2	Solid	(2560)	(20)	(7.8)
Pa_2O_5	Solid	(2050)	(26)	(13)
	Liquid	(3350)	(95)	(28)

Source: data from Weast, R. C., Ed., *Handbook of Chemistry and Physics, 55th ed.*, CRC Press, Cleveland, 1974, D-58.

Table 69. PHASE CHANGE THERMODYNAMIC PROPERTIES OF OXIDES (SHEET 6 OF 10)

Oxide	Phase	Transition Temperature (K)	Heat of Transition (kcal • g mole^{-1})	Entropy of Transition (e.u.)
PbO	Solid, red	762	(0.4)	(0.5)
	Solid, yellow	1159	2.8	2.4
	Liquid	1745	51	29
Pb_2O_4	Solid	dec.	–	–
PbO_2	Solid	dec.	–	–
PdO	Solid	dec. 1150	–	–
PoO_2	Solid	(825)	(5.5)	(6.7)
	Liquid	(dec.)	–	–
Pr_2O_3	Solid	(2200)	(22)	(10)
	Liquid	(4000	(90)	(23)
PrO_2	Solid	dec. 700	–	–
PtO	Solid	dec. 780	–	–
Pt_3O_4	Solid	(dec.)	–	–
PtO_2	Solid	723	(4.6)	(6.4)
	Liquid	dec. 750	–	–
PuO	Solid	(1290)	(7.2)	(5.6)
	Liquid	(2325)	(47)	(20)
Pu_2O_3	Solid	(1880)	(16)	(8.5)
	Liquid	(3250)	(75)	(23)
PuO_2	Solid	(2400)	(15)	(6.2)
	Liquid	(3500)	(90)	(26)
RaO	Solid	(>2500)	–	–

Source: data from Weast, R. C., Ed., *Handbook of Chemistry and Physics, 55th ed.*, CRC Press, Cleveland, 1974, D-58.

Table 69. PHASE CHANGE THERMODYNAMIC PROPERTIES OF OXIDES (SHEET 7 OF 10)

Oxide	Phase	Transition Temperature (K)	Heat of Transition (kcal • g mole^{-1})	Entropy of Transition (e.u.)
Rb_2O	Solid	(910)	(5.7)	(6.3)
	Liquid	dec.	–	–
Rb_2O_2	Solid	843	(7.3)	(8.7)
	Liquid	(dec.)	–	–
Rb_2O_3	Solid	762	(7.6)	(10)
	Liquid	dec.	–	–
RbO_2	Solid	685	(4.1)	(6.0)
	Liquid	dec.	–	–
ReO_2	Solid	(1475)	(12)	(8.1)
	Liquid	(3250)	(80)	(25)
ReO_3	Solid	433	5.2	12
	Liquid	dec.	–	–
Re_2O_7	Solid	569	15.8	27.8
	Liquid	635.5	17.7	27.9
ReO_4	Solid	420	(4.2)	(10)
	Liquid	(460)	(9.3)	(20)
Rh_2O	Solid	dec. 1400	–	–
RhO	Solid	dec. 1394	–	–
Rh_2O_3	Solid	dec. 1388	–	–
RuO_2	Solid	dec. 1400	–	–
RuO4	Solid	300	(3.2)	(11)
	Liquid	dec.	–	–
SO_2	Gas	–	–	–
Sb_2O_3	Solid	928	14.74	15.88
	Liquid	1698	8.92	5.25

Source: data from Weast, R. C., Ed., *Handbook of Chemistry and Physics, 55th ed.*, CRC Press, Cleveland, 1974, D-58.

Table 69. PHASE CHANGE THERMODYNAMIC PROPERTIES OF OXIDES (SHEET 8 OF 10)

Oxide	Phase	Transition Temperature (K)	Heat of Transition (kcal • g mole^{-1})	Entropy of Transition (e.u.)
SbO_2	Solid	dec.	–	–
Sb_2O_5	Solid	dec.	–	–
Sc_2O_3	Solid	(2500)	(23)	(9.3)
SeO	Solid	(1375)	(7.6)	(5.5)
	Liquid	(2075)	(45)	(22)
SeO_2	Solid	603	(24.5)	(40.6)
SiO	Solid	(2550)	(12)	(4.7)
SiO_2	Solid, β	856	0.15	0.18
	Solid, α	1883	2.04	1.08
	Liquid	dec. 2250	–	–
Sm_2O_3	Solid	(2150)	(20)	(9.3)
	Liquid	(3800)	(80)	(21)
SnO	Solid	(1315)	(6.4)	(4.9)
	Liquid	(1800)	(60)	(33)
SnO_2	Solid	1898	(11.39)	(5.95)
	Liquid	(3200)	(75)	(23)
SrO	Solid	2703	16.7	6.2
SrO_2	Solid	dec.488	–	–
Ta_2O_5	Solid	2150	(16)	(7.4)
	Liquid	–	–	–
TcO_2	Solid	(2400)	(18)	(7.5)
	Liquid	(4000)	(105)	(26)
TcO_3	Solid	(dec. <1200)	–	–
Tc_2O_7	Solid	392.7	(11)	(28)
	Liquid	583.8	(14)	(24)

Source: data from Weast, R. C., Ed., *Handbook of Chemistry and Physics, 55th ed.*, CRC Press, Cleveland, 1974, D-58.

Table 69. PHASE CHANGE THERMODYNAMIC PROPERTIES OF OXIDES (SHEET 9 OF 10)

Oxide	Phase	Transition Temperature (K)	Heat of Transition (kcal • g $mole^{-1}$)	Entropy of Transition (e.u.)
TeO	Solid	(1020)	(7.1)	(7.0)
	Liquid	(1775)	(50)	(28)
TeO_3	Solid	1006	3.2	3.2
	Liquid	dec.	–	–
TeO_2	Solid	(2150)	(13)	(6.0)
	Liquid	(3250)	(65)	(20)
ThO_2	Solid	3225	(18)	(5.6)
TiO	Solid, α	1264	0.82	0.65
	Solid, β	dec. 2010	–	–
Ti_2O_3	Solid, α	473	0.215	0.455
	Solid, β	2400	(24)	(10)
	Liquid	3300		
Ti_3O_5	Solid, α	450	2.24	4.98
	Solid, β	(2450)	(50)	(20)
	Liquid	(3600)	(85)	(24)
TiO_2	Solid	2128	(16)	(7.5)
	Liquid	dec. 3200		
Ti_2O	Solid	573	(5.0)	(8.7)
	Liquid	773	(17)	(22)
Tl_2O_3	Solid	990	(12.4)	(13)
	Liquid	(dec.)	–	–
UO	Solid	(2750)	(14)	(5.1)
UO_2	Solid	3000	–	–
U_3O_8	Solid	dec.	–	–
UO_3	Solid	dec. 925	–	–

Source: data from Weast, R. C., Ed., *Handbook of Chemistry and Physics, 55th ed.*, CRC Press, Cleveland, 1974, D-58.

Table 69. PHASE CHANGE THERMODYNAMIC PROPERTIES OF OXIDES (SHEET 10 OF 10)

Oxide	Phase	Transition Temperature (K)	Heat of Transition (kcal • g mole^{-1})	Entropy of Transition (e.u.)
VO	Solid	(2350)	(15)	(6.4)
	Liquid	(3400)	(70)	(21)
V_2O_3	Solid	2240	(24)	(11)
	Liquid	dec. 3300	–	–
V_3O_4	Solid	(2100)	(42)	(20)
	Liquid	(dec.)	–	–
VO_2	Solid, α	345	1.02	2.96
	Solid, β	1818	13.60	7.48
	Liquid	dec. 3300	–	–
V_2O_5	Solid	943	15.56	16.50
	Liquid	(2325)	(63)	(27)
WO_2	Solid	(1543)	(11.5)	(7.45)
	Liquid	dec. 2125	–	–
WO_3	Solid	1743	(17)	(9.8)
	Liquid	(2100)	(43)	(20)
Y_2O_3	Solid	(2500)	(25)	(10)
ZnO	Solid	dec.	–	–
ZrO_2	Solid, α	1478	1.420	0.961
	Solid, β	2950	20.8	7.0

Source: data from Weast, R. C., Ed., *Handbook of Chemistry and Physics, 55th ed.*, CRC Press, Cleveland, 1974, D-58.

Table 70. MELTING POINTS OF THE ELEMENTS

(SHEET 1 OF 4)

At. No.	Element	Symbol	Melting Point (°C)
1	Hydrogen	H	-259.14
2	Helium	He	-272.2
3	Lithium	Li	180.54
4	Beryllium	Be	1278
5	Boron	B	2300
6	Carbon	C	~3550
7	Nitrogen	N	-209.86
8	Oxygen	O	-218.4
9	Fluorine	F	-219.62
10	Neon	N	-248.67
11	Sodium	Na	97.81
12	Magnesium	Mg	648.8
13	Aluminum	Al	660.37
14	Silicon	Si	1410
15	Phosphorus (White)	P	44.1
16	Sulfur	S	112.8
17	Chlorine	Cl	-100.98
18	Argon	Ar	-189.2
19	Potassium	K	63.65
20	Calcium	Ca	839
21	Scandium	Sc	1539
22	Titanium	Ti	1660
23	Vanadium	V	1890
24	Chromium	Cr	1857
25	Manganese	Mn	1244
26	Iron	Fe	1535
27	Cobalt	Co	1495

Source: data from James F. Shackelford, *Introduction to Materials Science for Engineers, Second Edition*, Macmillan Publishing Company, New York, pp.686-688, (1988).

Table 70. Melting Points of the Elements

(Sheet 2 of 4)

At. No.	Element	Symbol	Melting Point (°C)
28	Nickel	Ni	1453
29	Copper	Cu	1083.4
30	Zinc	Zn	419.58
31	Gallium	Ga	29.78
32	Germanium	Ge	937.4
33	Arsenic	As	817
34	Selenium	Se	217
35	Bromine	Br	-7.2
36	Krypton	Kr	-156.6
37	Rubidium	Rb	38.89
38	Strontium	Sr	769
39	Yttrium	Y	1523
40	Zirconium	Zr	1852
41	Niobium	Nb	2408
42	Molybdenum	Mo	2617
43	Technetium	Tc	2172
44	Ruthenium	Ru	2310
45	Rhodium	Rh	1966
46	Palladium	Pd	1552
47	Silver	Ag	961.93
48	Cadmium	Cd	320.9
49	Indium	In	156.61
50	Tin	Sn	231.9681
51	Antimony	Sb	630.74
52	Tellurium	Te	449.5
53	Iodine	I	113.5
54	Xenon	Xe	-111.9
55	Cesium (-10°)	Ce	28.4

Source: data from James F. Shackelford, *Introduction to Materials Science for Engineers, Second Edition*, Macmillan Publishing Company, New York, pp.686-688, (1988).

Table 70. MELTING POINTS OF THE ELEMENTS
(SHEET 3 OF 4)

At. No.	Element	Symbol	Melting Point (°C)
56	Barium	Ba	7.25
57	Lantium	La	920
58	Cerium	Ce	798
59	Praseodymium	Pr	931
60	Neodymium	Nd	1010
61	Promethium	Pm	~1080
62	Samarium	Sm	1072
63	Europium	Eu	822
64	Gadolinium	Gd	1311
65	Terbium	Tb	1360
66	Dysprosium	Dy	1409
67	Holmium	Ho	1470
68	Erbium	Er	1522
69	Thulium	Tm	1545
70	Ytterbium	Yb	824
71	Lutetium	Lu	1659
72	Hafnium	Hf	2227
73	Tantalum	Ta	2996
74	Tungsten	W	3410
75	Rhenium	Re	3180
76	Osmium	Os	3045
77	Iridium	Ir	2410
78	Platinum	Pt	1772
79	Gold	Au	1064.43
80	Mercury	Hg	-38.87
81	Thallium	Tl	303.5
82	Lead	Pb	327.502
83	Bismuth	Bi	271.3

Source: data from James F. Shackelford, *Introduction to Materials Science for Engineers, Second Edition*, Macmillan Publishing Company, New York, pp.686-688, (1988).

Table 70. MELTING POINTS OF THE ELEMENTS
(SHEET 4 OF 4)

At. No.	Element	Symbol	Melting Point (°C)
84	Polonium	Po	254
85	Asatine	At	302
86	Radon	Rn	-71
87	Francium	Fr	~27
88	Radium	Ra	700
89	Actinium	Ac	1050
90	Thorium	Th	1750
91	Protoactinium	Pa	<1600
92	Uranium	U	1132
93	Neptunium	Np	640
94	Plutonium	Pu	641
95	Americium	Am	994
96	Curium	Cm	1340

Source: data from James F. Shackelford, *Introduction to Materials Science for Engineers, Second Edition,* Macmillan Publishing Company, New York, pp.686-688, (1988).

Table 71. MELTING POINTS OF ELEMENTS AND INORGANIC COMPOUNDS (SHEET 1 OF 13)

Compound	Formula	Melting Point °C
Actinium[227]	Ac	1050±50
Aluminum	Al	658.5
Aluminum bromide	Al_2Br_6	87.4
Aluminum chloride	Al_2Cl_6	192.4
Aluminum iodide	Al_2I_6	190.9
Aluminum oxide	Al_2O_3	2045.0
Antimony	Sb	630
Antimony pentachloride	$SbCl_5$	4.0
Antimony tribromide	$SbBr_3$	96.8
Antimony trichloride	$SbCl_3$	73.3
Antimony trioxide	Sb_4O_6	655.0
Antimony trisulfide	Sb_4S_6	546.0
Argon	Ar	190.2
Arsenic	As	816.8
Arsenic pentafluoride	AsF_5	80.8
Arsenic tribromide	$AsBr_3$	30.0
Arsenic trichloride	$AsCl_3$	–16.0
Arsenic trifluoride	AsF_3	–6.0
Arsenic trioxide	As_4O_6	312.8
Barium	Ba	725
Barium bromide	$BaBr_2$	846.8
Barium chloride	$BaCl_2$	959.8
Barium fluoride	BaF_2	1286.8
Barium iodide	BaI_2	710.8

Source: data from: Weast, R C., Ed., *Handbook of Chemistry and Physics, 55th ed.*, CRC Press, Cleveland, (1974); and Bolz, R. E. and Tuve, G. L., Eds., Handbook of Tables for Applied Engineering Science, 2nd ed., CRC Press, Cleveland, (1973), p.479 .

Table 71. MELTING POINTS OF ELEMENTS AND INORGANIC COMPOUNDS (SHEET 2 OF 13)

Compound	Formula	Melting Point °C
Barium nitrate	$Ba(NO_3)_2$	594.8
Barium oxide	BaO	1922.8
Barium phosphate	$Ba_3(PO_4)_2$	1727
Barium sulfate	$BaSO_4$	1350
Beryllium	Be	1278
Beryllium bromide	$BeBr_2$	487.8
Beryllium chloride	$BeCl_2$	404.8
Beryllium oxide	BeO	2550.0
Bismuth	Bi	271
Bismuth trichloride	$BiCl_3$	223.8
Bismuth trifluoride	BiF_3	726.0
Bismuth trioxide	Bi_2O_3	815.8
Boron	B	2300
Boron tribromide	BBr_3	–48.8
Boron trichloride	BCl_3	–107.8
Boron trifluoride	BF_3	–128.0
Boron trioxide	B_2O_3	448.8
Bromine	Br_2	–7.2
Bromine pentafluoride	BrF_5	–61.4
Cadmium	Cd	320.8
Cadmium bromide	$CdBr_2$	567.8
Cadmium chloride	$CdCl_2$	567.8
Cadmium fluoride	CdF_2	1110
Cadmium iodide	CdI_2	386.8

Source: data from: Weast, R C., Ed., *Handbook of Chemistry and Physics, 55th ed.*, CRC Press, Cleveland, (1974); and Bolz, R. E. and Tuve, G. L., Eds., Handbook of Tables for Applied Engineering Science, 2nd ed., CRC Press, Cleveland, (1973), p.479 .

Table 71. MELTING POINTS OF ELEMENTS AND INORGANIC COMPOUNDS (SHEET 3 OF 13)

Compound	Formula	Melting Point °C
Cadmium sulfate	$CdSO_4$	1000
Calcium	Ca	851
Calcium bromide	$CaBr_2$	729.8
Calcium carbonate	$CaCO_3$	1282
Calcium chloride	$CaCl_2$	782
Calcium fluoride	CaF_2	1382
Calcium metasilicate	$CaSiO_3$	1512
Calcium nitrate	$Ca(NO_3)_2$	560.8
Calcium oxide	CaO	2707
Calcium sulfate	$CaSO_4$	1297
Carbon dioxide	CO_2	–57.6
Carbon monoxide	CO	–205
Cyanogen	C_2N_2	–27.2
Cyanogen chloride	CNCl	–5.2
Cerium	Ce	775
Cesium	Cs	28.3
Cesium chloride	CsCl	38.5
Cesium nitrate	$CsNO_3$	406.8
Chlorine	Cl_2	–103±5
Chromium	Cr	1890
Chromium (II) chloride	$CrCl_2$	814
Chromium (III) sequioxide	Cr_2O_3	2279
Chromium trioxide	CrO_3	197
Cobalt	Co	1490

Source: data from: Weast, R C., Ed., *Handbook of Chemistry and Physics, 55th ed.*, CRC Press, Cleveland, (1974); and Bolz, R. E. and Tuve, G. L., Eds., Handbook of Tables for Applied Engineering Science, 2nd ed., CRC Press, Cleveland, (1973), p.479 .

Table 71. MELTING POINTS OF ELEMENTS AND INORGANIC COMPOUNDS (SHEET 4 OF 13)

Compound	Formula	Melting Point °C
Cobalt (II) chloride	$CoCl_2$	727
Copper	Cu	1083
Copper (II) chloride	$CuCl_2$	430
Copper (I) chloride	CuCl	429
Copper(l) cyanide	$Cu_2(CN)_2$	473
Copper (I) iodide	CuI	587
Copper (II) oxide	CuO	1446
Copper (I) oxide	Cu_2O	1230
Copper (I) sulfide	Cu_2S	1129
Dysprosium	Dy	1407
Erbium	Er	1496
Europium	Eu	826
Europium trichloride	$EuCl_3$	622
Fluorine	F_2	–219.6
Gadolinium	Gd	1312
Gallium	Ga	29
Germanium	Ge	959
Gold	Au	1063
Hafnium	Hf	2214
Holmium	Ho	1461
Hydrogen	H_2	–259.25
Hydrogen bromide	HBr	–86.96
Hydrogen chloride	HCl	–114.3
Hydrogen fluoride	HF	83.11
Hydrogen iodide	HI	–50.91
Hydrogen nitrate	HNO_3	–47.2
Hydrogen oxide (water)	H_2O	0
Deuterium oxide	D_2O	3.78

Source: data from: Weast, R C., Ed., *Handbook of Chemistry and Physics, 55th ed.*, CRC Press, Cleveland, (1974); and Bolz, R. E. and Tuve, G. L., Eds., Handbook of Tables for Applied Engineering Science, 2nd ed., CRC Press, Cleveland, (1973), p.479 .

Table 71. MELTING POINTS OF ELEMENTS AND INORGANIC COMPOUNDS (SHEET 5 OF 13)

Compound	Formula	Melting Point °C
Hydrogen peroxide	H_2O_2	–0.7
Hydrogen selenate	H_2SeO_4	57.8
Hydrogen sulfate	H_2SO_4	10.4
Hydrogen sulfide	H_2S	–85.6
Hydrogen sulfide, di–	H_2S_2	–89.7
Hydrogen telluride	H_2Te	–49.0
Indium	In	156.3
Iodine	I_2	112.9
Iodine chloride (α)	ICl	17.1
Iodine chloride (β)	ICl	13.8
Iron	Fe	1530.0
Iron carbide	Fe_3C	1226.8
Iron (III) chloride	Fe_2Cl_6	303.8
Iron (II) chloride	$FeCl_2$	677
Iron (II) oxide	FeO	1380
Iron oxide	Fe_3O_4	1596
Iron pentacarbonyl	$Fe(CO)_5$	–21.2
Iron (II) sulfide	FeS	1195
Lanthanum	La	920
Lead	Pb	327.3
Leadbromide	$PbBr_2$	487.8
Lead chloride	$PbCl_2$	497.8
Lead fluoride	PbF_2	823
Lead iodide	PbI_2	412

Source: data from: Weast, R C., Ed., *Handbook of Chemistry and Physics, 55th ed.*, CRC Press, Cleveland, (1974); and Bolz, R. E. and Tuve, G. L., Eds., Handbook of Tables for Applied Engineering Science, 2nd ed., CRC Press, Cleveland, (1973), p.479 .

Table 71. MELTING POINTS OF ELEMENTS AND INORGANIC COMPOUNDS (SHEET 6 OF 13)

Compound	Formula	Melting Point °C
Lead molybdate	$PbMoO_4$	1065
Lead oxide	PbO	890
Lead sulfate	$PbSO_4$	1087
Lead sulfide	PbS	1114
Lithium	Li	178.8
Lithium bromide	$LiBr$	552
Lithium chloride	$LiCl$	614
Lithium fluoride	LiF	896
Lithium hydroxide	$LiOH$	462
Lithium iodide	LiI	440
Lithium metasilicate	Li_2SiO_3	1177
Lithium molybdate	Li_2MoO_4	705
Lithium nitrate	$LiNO_3$	250
Lithium orthosilicate	Li_4SiO_4	1249
Lithium sulfate	Li_2SO_4	857
Lithium tungstate	Li_2WO_4	742
Lutetium	Lu	1651
Magnesium	Mg	650
Magnesium bromide	$MgBr_2$	711
Magnesium chloride	$MgCl_2$	712
Magnesium fluoride	MgF_2	1221
Magnesium oxide	MgO	2642
Magnesium silicate	$MgSiO_3$	1524
Magnesium sulfate	$MgSO_4$	1327

Source: data from: Weast, R C., Ed., *Handbook of Chemistry and Physics, 55th ed.*, CRC Press, Cleveland, (1974); and Bolz, R. E. and Tuve, G. L., Eds., Handbook of Tables for Applied Engineering Science, 2nd ed., CRC Press, Cleveland, (1973), p.479 .

Table 71. MELTING POINTS OF ELEMENTS AND INORGANIC COMPOUNDS (SHEET 7 OF 13)

Compound	Formula	Melting Point °C
Manganese	Mn	1220
Manganese dichloride	$MnCl_2$	650
Manganese metasilicate	$MnSiO_3$	1274
Manganese (II) oxide	MnO	1784
Manganese oxide	Mn_3O_4	1590
Mercury	Hg	–39
Mercury bromide	$HgBr_2$	241
Mercury chloride	$HgCl_2$	276.8
Mercury iodide	HgI_2	250
Mercury sulfate	$HgSO_4$	850
Molybdenum	Mo	2622
Molybdenum dichloride	$MoCl_2$	726.8
Molybdenum hexafluoride	MoF_6	17
Molybdenum trioxide	MoO_3	795
Neodymium	Nd	1020
Neon	Ne	– 248.6
Nickel	Ni	1452
Nickel chloride	$NiCl_2$	1030
Nickel subsulfide	Ni_3S_2	790
Niobium	Nb	2496
Niobium pentachloride	$NbCl_5$	21 l
Niobium pentoxide	Nb_2O_5	1511
Nitric oxide	NO	–163.7
Nitrogen	N_2	–210

Source: data from: Weast, R C., Ed., *Handbook of Chemistry and Physics, 55th ed.*, CRC Press, Cleveland, (1974); and Bolz, R. E. and Tuve, G. L., Eds., Handbook of Tables for Applied Engineering Science, 2nd ed., CRC Press, Cleveland, (1973), p.479 .

Table 71. MELTING POINTS OF ELEMENTS AND INORGANIC COMPOUNDS (SHEET 8 OF 13)

Compound	Formula	Melting Point °C
Nitrogen tetroxide	N_2O_4	–13.2
Nitrous oxide	N_2O	–90.9
Osmium	Os	2700
Osmium tetroxide (white)	OsO_4	41.8
Osmium tetroxide (yellow)	OsO_4	55.8
Oxygen	O_2	–218.8
Palladium	Pd	1555
Phosphoric acid	H_3PO_4	42.3
Phosphoric acid. hypo–	$H_4P_2O_6$	54.8
Phosphorus acid, hypo–	H_3PO_2	17.3
Phosphorus acid, ortho–	H_3PO_3	73.8
Phosphorus oxychloride	$POCl_3$	1.0
Phosphorus pentoxide	P_4O_{10}	569.0
Phosphorus trioxide	P_4O_6	23.7
Phosphorus, yellow	P_4	44.1
Platinum	Pt	1770
Potassium	K	63.4
Potassium borate, meta–	KBO_2	947
Potassium bromide	KBr	742
Potassium carbonate	K_2CO_3	897
Potassium chloride	KCl	770
Potassium chromate	K_2CrO_4	984
Potassium cyanide	KCN	623
Potassium dichromate	$K_2Cr_2O_7$	398

Source: data from: Weast, R C., Ed., *Handbook of Chemistry and Physics, 55th ed.*, CRC Press, Cleveland, (1974); and Bolz, R. E. and Tuve, G. L., Eds., Handbook of Tables for Applied Engineering Science, 2nd ed., CRC Press, Cleveland, (1973), p.479 .

Table 71. MELTING POINTS OF ELEMENTS AND INORGANIC COMPOUNDS (SHEET 9 OF 13)

Compound	Formula	Melting Point °C
Potassium fluoride	KF	875
Potassium hydroxide	KOH	360
Potassium iodide	Kl	682
Potassium nitrate	KNO_3	338
Potassium peroxide	K_2O_2	490
Potassium phosphate	K_3PO_4	1340
Potassium pyro– phosphate	$K_4P_2O_7$	1092
Potassium sulfate	K_2SO_4	1074
Potassium thiocyanate	KSCN	179
Praseodymium	Pr	931
Rhenium	Re	3167±60
Rhenium heptoxide	Re_2O_7	296
Rhenium hexafluoride	ReF_6	19.0
Rubidium	Rb	38 .9
Rubidium bromide	RbBr	677
Rubidium chloride	RbCl	717
Rubidium fluoride	RbF	833
Rubidium iodide	Rbl	638
Rubidium nitrate	$RbNO_3$	305
Samarium	Sm	1072
Scandium	Sc	1538
Selenium	Se	217
Seleniumoxychloride	$SeOCl_3$	9.8
Silane, hexaHuoro–	Si_2F_6	–28.6
Silicon	Si	1427
Silicon dioxide (Cristobalite)	SiO_2	1723
Silicon tetrachloride	$SiCl_4$	–67.7

Source: data from: Weast, R C., Ed., *Handbook of Chemistry and Physics, 55th ed.*, CRC Press, Cleveland, (1974); and Bolz, R. E. and Tuve, G. L., Eds., Handbook of Tables for Applied Engineering Science, 2nd ed., CRC Press, Cleveland, (1973), p.479 .

Table 71. MELTING POINTS OF ELEMENTS AND INORGANIC COMPOUNDS (SHEET 10 OF 13)

Compound	Formula	Melting Point °C
Silver	Ag	961
Silver bromide	AgBr	430
Silver chloride	AgCl	455
Silver cyanide	AgCN	350
Silver iodide	AgI	557
Silver nitrate	$AgNO_3$	209
Silver sulfate	Ag_2SO_4	657
Silver sulfide	Ag_2S	841
Sodium	Na	97.8
Sodium borate, meta–	$NaBO_2$	966
Sodium bromide	NaBr	747
Sodium carbonate	Na_2CO_3	854
Sodium chlorate	$NaClO_3$	255
Sodium chloride	NaCl	800
Sodium cyanide	NaCN	562
Sodium fluoride	NaF	992
Sodium hydroxide	NaOH	322
Sodium iodide	NaI	662
Sodium molybdate	Na_2MoO_4	687
Sodium nitrate	$NaNO_3$	310
Sodium peroxide	Na_2O_2	460
Sodium phosphate, meta–	$NaPO_3$	988
Sodium pyrophosphate	$Na_4P_2O_7$	970
Sodiumsilicate,aluminum–	$NaAlSi_3O_8$	1107

Source: data from: Weast, R C., Ed., *Handbook of Chemistry and Physics, 55th ed.*, CRC Press, Cleveland, (1974); and Bolz, R. E. and Tuve, G. L., Eds., Handbook of Tables for Applied Engineering Science, 2nd ed., CRC Press, Cleveland, (1973), p.479 .

Table 71. Melting Points of Elements and Inorganic Compounds (Sheet 11 of 13)

Compound	Formula	Melting Point °C
Sodium silicate, di–	$Na_2Si_2O_5$	884
Sodium silicate, meta–	Na_2SiO_3	1087
Sodium sulfate	Na_2SO_4	884
Sodium sulfide	Na_2S	920
Sodium thiocyanate	$NaSCN$	323
Sodium tungstate	Na_2WO_4	702
Strontium	Sr	757
Strontium bromide	$SrBr_2$	643
Strontium chloride	$SrCl_2$	872
Strontium fluoride	SrF_2	1400
Strontium oxide	SrO	2430
Sulfur (monatomic)	S	119,
Sulfur dioxide	SO_2	– 73.2
Sulfur trioxide (α)	SO_3	16.8
Sulfur trioxide (β)	SO_3	32.3
Sulfur trioxide (γ)	SO_3	62.1
Tantalum	Ta	2996 ± 50
Tantalum pentachloride	$TaCl_5$	206.8
Tantalum pentoxide	Ta_2O_5	1877
Tellurium	Te	453
Terbium	Tb	1356
Thallium	Tl	302.4
Thallium bromide, mono–	TlBr	460
Thallium carbonate	Tl_2CO_3	273

Source: data from: Weast, R C., Ed., *Handbook of Chemistry and Physics, 55th ed.*, CRC Press, Cleveland, (1974); and Bolz, R. E. and Tuve, G. L., Eds., Handbook of Tables for Applied Engineering Science, 2nd ed., CRC Press, Cleveland, (1973), p.479 .

Table 71. MELTING POINTS OF ELEMENTS AND INORGANIC COMPOUNDS (SHEET 12 OF 13)

Compound	Formula	Melting Point °C
Thallium chloride, mono–	$TlCl$	427
Thallium iodide, mono–	TlI	440
Thallium nitrate	$TlNO_3$	207
Thallium sulfate	Tl_2SO_4	632
Thallium sulfide	Tl_2S	449
Thorium	Th	1845
Thorium chloride	$ThCl_4$	765
Thorium dioxide	ThO_2	2952
Thulium	Tm	1545
Tin	Sn	231.7
Tin bromide, di–	$SnBr_2$	231.8
Tin bromide, tetra–	$SnBr_4$	29.8
Tin chloride, di–	$SnCl_2$	247
Tinchloride,tetra–	$SnCl_4$	–33.3
Tin iodide, tetra–	SnI_4	143.4
Tin oxide	SnO	1042
Titanium	Ti	1800
Titanium bromide, tetra–	$TiBr_4$	38
Titanium chloride, tetra–	$TiCl_4$	–23.2
Titanium dioxide	TiO_2	1825
Titanium oxide	TiO	991
Tungsten	W	3387
Tungsten dioxide	WO_2	1270
Tungsten hexafluoride	WF_6	–0.5

Source: data from: Weast, R C., Ed., *Handbook of Chemistry and Physics, 55th ed.*, CRC Press, Cleveland, (1974); and Bolz, R. E. and Tuve, G. L., Eds., Handbook of Tables for Applied Engineering Science, 2nd ed., CRC Press, Cleveland, (1973), p.479 .

Table 71. MELTING POINTS OF ELEMENTS AND INORGANIC COMPOUNDS (SHEET 13 OF 13)

Compound	Formula	Melting Point °C
Tungsten tetrachloride	WCl_4	327
Tungsten trioxide	WO_3	1470
Uranium[235]	U	~1133
Uranium tetrachloride	UCl_4	590
Vanadium	V	1917
Vanadium dichloride	VCl_2	1027
Vanadium oxide	VO	2077
Vanadium pentoxide	V_2O_5	670
Xenon	Xe	–111.6
Ytterbium	Yb	823
Yttrium	Y	1504
Yttrium oxide	Y_2O_3	2227
Zinc	Zn	419.4
Zincchloride	$ZnCl_2$	283
Zinc oxide	ZnO	1975
Zinc sulfide	ZnS	1745
Zirconium	Zr	1857
Zirconium dichloride	$ZrCl_2$	727
Zirconium oxide	ZrO_2	2715

Source: data from: Weast, R C., Ed., *Handbook of Chemistry and Physics, 55th ed.*, CRC Press, Cleveland, (1974); and Bolz, R. E. and Tuve, G. L., Eds., Handbook of Tables for Applied Engineering Science, 2nd ed., CRC Press, Cleveland, (1973), p.479 .

Table 72. MELTING POINTS OF CERAMICS
(SHEET 1 OF 11)

Compound	(K)
AgBr	703
AgCl	728
AgF	708
AgI	831
$AgNO_3$	483
Ag_2O	573
Ag_2SO_4	933
Ag_2S	1098
$AlBr_3$	371
Al_4C_3	2000
$AlCl_3$	465
AlF_3	1564
AlI	464
AlN	>2475
Al_2O_3	2322
$Al_2(SO_4)_3$	1043
Al_2S_3	1373
BBr_3	227
B_4C	2720
BCl_3	166
BF_3	146
BN	3000
B_2O_3	723
BS_4	663

Source: data from: Lynch, Charles T., Ed., *CRC Handbook of Materials Science, Vol. 1*, CRC Press, Boca Raton, 1974, 348.

Table 72. MELTING POINTS OF CERAMICS
(SHEET 2 OF 11)

Compound	(K)
BaB_4	2543
$BaBr_2$	1123
$BaCl_2$	1235
BaF_2	1627
BaI_2	1013
$Ba(NO_3)_2$	865
BaO	2283
$BaSO_4$	1853
BaS	1473
BeB_2	>2243
$BeBr_2$	793
Be_2C	>2375
$BeCl_2$	713
BeF_2	813
BeI_2	783
Be_3N_2	2513
BeO	2725
$BeSO_4$	848
$BiBr_3$	491
$BiCl_3$	507
BiF_3	1000
BiI_3	681
B_2O_3	1098
$Bi(SO_4)_3$	678

Source: data from: Lynch, Charles T., Ed., *CRC Handbook of Materials Science, Vol. 1*, CRC Press, Boca Raton, 1974, 348.

Table 72. MELTING POINTS OF CERAMICS
(SHEET 3 OF 11)

Compound	(K)
Bi_2S_3	1020
$CaBr_2$	1003
$CaCl_2$	1055
CaF_2	1675
CaI_2	848
$Ca(NO_3)_2$	623
Ca_3N_2	1468
CaO	3183
$CaSO_4$	1723
$CdBr_2$	841
$CdCl_2$	841
CdF_2	1373
CdI_2	423
$Cd(NO_3)_2$	834
CdO	1773
$CdSO_4$	1273
CdS	2023
CeB_6	2463
$CeCl_3$	1095
CeF_2	1710
CeI_3	1025
CeO_2	>2873
CeS	2400
$Ce(SO_4)_2$	468

Source: data from: Lynch, Charles T., Ed., *CRC Handbook of Materials Science, Vol. 1*, CRC Press, Boca Raton, 1974, 348.

Table 72. MELTING POINTS OF CERAMICS
(SHEET 4 OF 11)

Compound	(K)
CrB_2	2123
$Cr3C_2$	2168
CrN	1770
Cr_2O_3	>2603
$CrSi_2$	1843
CuBr	777
CuCl	695
CuF_2	1129
CuI	878
Cu_3N	573
Cu_2O	1508
Cu_4Si	1123
Cu_2S	1400
$FeBr_2$	955
Fe_3C	2110
$FeCl_2$	945
FeF_3	>1275
Fe_2O_3	1864
$Fe_2(SO_4)_3$	753
FeS	1468
$InBr_3$	709
InCl	498
InF_3	1443
InI_3	483

Source: data from: Lynch, Charles T., Ed., *CRC Handbook of Materials Science, Vol. 1*, CRC Press, Boca Raton, 1974, 348.

Table 72. MELTING POINTS OF CERAMICS
(SHEET 5 OF 11)

Compound	(K)
In_2O_3	2183
In_2S_3	1323
KBr	1008
KCl	1043
KF	1131
KI	958
KNO_3	610
K_2O_3	703
K_2SO_4	1342
K_2S	1113
LiBr	823
LiCl	883
LiF	1119
LiI	722
$LiNO_3$	527
Li_3N	1118
Li_2O	>1975
Li_2SO_4	1132
Li_2S	1198
$MgBr_2$	984
$MgCl_2$	987
MgF_2	1535
MgI_2	<910
MgO	3098
Mg_2Si	1375
MgS	>2275
$MgSO_4$	1397
$MnCl_2$	923

Source: data from: Lynch, Charles T., Ed., *CRC Handbook of Materials Science, Vol. 1*, CRC Press, Boca Raton, 1974, 348.

Table 72. MELTING POINTS OF CERAMICS
(SHEET 6 OF 11)

Compound	(K)
MnF_2	1129
MnO	1840
MoB	2625
Mo_2C	2963
MoF_6	290
MoI_4	373
MoO_3	1068
$MoSi_2$	2553
MoS_2	1458
NaBr	1023
NaC_2	973
NaCl	1073
NaF	1267
NaI	935
$NaNO_3$	583
Na_2N	573
Na_2SO_4	1157
Na_2S	1453
NbB	>2270
NbC	3770
NbN	2323
Nb_2O_5	1764
$NbSi_2$	2203
$NiBr_2$	1236
$NiCl_3$	1274
NiF_2	1273
NiI_2	1070
NiO	2257

Source: data from: Lynch, Charles T., Ed., *CRC Handbook of Materials Science, Vol. 1*, CRC Press, Boca Raton, 1974, 348.

Table 72. MELTING POINTS OF CERAMICS
(SHEET 7 OF 11)

Compound	(K)
$NiSO_4$	1121
NiS	1070
$PbBr_2$	643
$PbCl_2$	771
PbF_2	1095
PbI_2	675
$Pb(NO_3)_2$	743
PbO	1159
$PbSO_4$	1443
PbS	1387
$PtBr_2$	523
$PtCl_2$	854
PtI_2	633
PtS_2	508
$SbBr_3$	370
$SbCl_3$	346
SbF_3	565
SbI_3	443
Sb_2O_3	928
SbS_3	820
SiC	2970
SiF_4	183
Si_3N_4	2715
SiO_2	1978

Source: data from: Lynch, Charles T., Ed., *CRC Handbook of Materials Science, Vol. 1*, CRC Press, Boca Raton, 1974, 348.

Table 72. MELTING POINTS OF CERAMICS
(SHEET 8 OF 11)

Compound	(K)
$SnBr_2$	488
$SnCl_2$	581
SnF_4	978
SnI_2	788
SnO	1353
$SnSO_4$	>635
SnS	1153
SrB_6	2508
$SrBr_2$	916
SrC_2	>1970
$SrCl_2$	1148
SrF_2	1736
SrI_2	593
$Sr(NO_3)_2$	643
SrO	2933
$SrSO_4$	1878
SrS	>2275
TaB	>2270
$TaBr_5$	538
TaC	3813
$TaCl_5$	489
TaF_5	370
Ta_2N	3360
Ta_2O_5	2100

Source: data from: Lynch, Charles T., Ed., *CRC Handbook of Materials Science, Vol. 1,* CRC Press, Boca Raton, 1974, 348.

Table 72. MELTING POINTS OF CERAMICS

(SHEET 9 OF 11)

Compound	(K)
$TaSi_2$	2670
TaS_4	>1575
$TeBr_2$	612
$TeCl_2$	448
TeO_2	1006
ThB_4	>2270
$ThBr_4$	883
ThC	2898
$ThCl_4$	1043
ThF_4	1375
ThN	2903
ThO_2	3493
ThS_2	2198
TiB_2	3253
$TiBr_4$	312
TiC	3433
$TiCl_4$	250
TiF_3	1475
TiI_2	873
TiN	3200
TiO_2	2113
$TiSi_2$	1813
UB_2	>1770
UBr_4	789

Source: data from: Lynch, Charles T., Ed., *CRC Handbook of Materials Science, Vol. 1*, CRC Press, Boca Raton, 1974, 348.

Table 72. MELTING POINTS OF CERAMICS

(SHEET 10 OF 11)

Compound	(K)
UC	2863
UCl_4	843
UF_4	1233
UI_4	779
UN	3123
UO_2	3151
USi_2	1970
US_2	>1375
VB_2	2373
VC	3600
VCl_4	245
VF_3	>1075
FI_2	1048
VN	2593
V_2O_5	947
VSi_2	2023
V_2S_3	>875
WB	3133
WC	2900
WCl_6	548
WO_3	1744
WSi_2	2320
WS_2	1523
$ZnBr_2$	667
$ZnCl_2$	548
ZnF_2	1145
ZnI_2	719
ZnO	2248

Source: data from: Lynch, Charles T., Ed., *CRC Handbook of Materials Science, Vol. 1*, CRC Press, Boca Raton, 1974, 348.

Table 72. MELTING POINTS OF CERAMICS
(SHEET 11 OF 11)

Compound	(K)
$ZnSO_4$	873
ZrB_2	3313
$ZrBr_2$	>625
ZrC	3533
$ZrCl_2$	623
ZrF_4	873
ZrI_4	772
ZrN	3250
ZrO_2	3123
$Zr(SO_4)_2$	683
ZrS_2	1823

Source: data from: Lynch, Charles T., Ed., *CRC Handbook of Materials Science, Vol. 1*, CRC Press, Boca Raton, 1974, 348.

Table 73. HEAT OF FUSION FOR ELEMENTS AND INORGANIC COMPOUNDS (SHEET 1 OF 16)

Compound	Formula	Melting point °C	Heat of fusion	
			cal/g	cal/g mole
Actinium[227]	Ac	1050±50	(11.0)	(3400)
Aluminum	Al	658.5	94.5	2550
Aluminum bromide	Al_2Br_6	87.4	10.1	5420
Aluminum chloride	Al_2Cl_6	192.4	63.6	19600
Aluminum iodide	Al_2I_6	190.9	9.8	7960
Aluminum oxide	Al_2O_3	2045.0	(256.0)	(26000)
Antimony	Sb	630	39.1	4770
Antimony pentachloride	$SbCl_5$	4.0	8.0	2400
Antimony tribromide	$SbBr_3$	96.8	9.7	3510
Antimony trichloride	$SbCl_3$	73.3	13.3	3030
Antimony trioxide	Sb_4O_6	655.0	(46.3)	(26990)
Antimony trisulfide	Sb_4S_6	546.0	33.0	11200
Argon	Ar	190.2	7.25	290
Arsenic	As	816.8	(22.0)	(6620)
Arsenic pentafluoride	AsF_5	80.8	16.5	2800
Arsenic tribromide	$AsBr_3$	30.0	8.9	2810
Arsenic trichloride	$AsCl_3$	–16.0	13.3	2420
Arsenic trifluoride	AsF_3	–6.0	18.9	2486
Arsenic trioxide	As_4O_6	312.8	22.2	8000
Barium	Ba	725	13.3	1830

For heat of fusion in **J/kg**, multiply values in **cal/g** by **4184.**
For heat of fusion in **J/mol**, multiply values in **cal/g-mol** (=cal/mol) by **4.184.**
For melting point in **K**, add **273.15** to values in °C.

Values in parentheses are of uncertain reliability.

Source: data from Weast, R C., Ed., *Handbook of Chemistry and Physics, 55th ed.*, CRC Press, Cleveland, (1974); and Bolz, R. E. and Tuve, G. L., Eds., *Handbook of Tables for Applied Engineering Science, 2nd ed.*, CRC Press, Cleveland, (1973)

Table 73. HEAT OF FUSION FOR ELEMENTS AND INORGANIC COMPOUNDS (SHEET 2 OF 16)

Compound	Formula	Melting point °C	Heat of fusion	
			cal/g	cal/g mole
Barium bromide	$BaBr_2$	846.8	21.9	6000
Barium chloride	$BaCl_2$	959.8	25.9	5370
Barium fluoride	BaF_2	1286.8	17.1	3000
Barium iodide	BaI_2	710.8	(17.3)	(6800)
Barium nitrate	$Ba(NO_3)_2$	594.8	(22.6)	(5900)
Barium oxide	BaO	1922.8	93.2	13800
Barium phosphate	$Ba_3(PO_4)_2$	1727	30.9	18600
Barium sulfate	$BaSO_4$	1350	41.6	9700
Beryllium	Be	1278	260.0	–
Beryllium bromide	$BeBr_2$	487.8	(26.6)	(4500)
Beryllium chloride	BeCl2	404.8	(30)	(3000)
Beryllium oxide	BeO	2550.0	679.7	17000
Bismuth	Bi	271	12.0	2505
Bismuth trichloride	$BiCl_3$	223.8	8.2	2600
Bismuth trifluoride	BiF_3	726.0	(23.3)	(6200)
Bismuth trioxide	Bi_2O_3	815.8	14.6	6800
Boron	B	2300	(490)	(5300)
Boron tribromide	BBr_3	–48.8	(2.9)	(700)
Boron trichloride	BCl_3	–107.8	(4.3)	(500)
Boron trifluoride	BF_3	–128.0	7.0	480

For heat of fusion in **J/kg**, multiply values in **cal/g** by **4184**.
For heat of fusion in **J/mol**, multiply values in **cal/g-mol** (=cal/mol) by **4.184**.
For melting point in **K**, add **273.15** to values in °C.

Values in parentheses are of uncertain reliability.

Source: data from Weast, R C., Ed., *Handbook of Chemistry and Physics, 55th ed.*, CRC Press, Cleveland, (1974); and Bolz, R. E. and Tuve, G. L., Eds., *Handbook of Tables for Applied Engineering Science, 2nd ed.*, CRC Press, Cleveland, (1973)

Table 73. HEAT OF FUSION FOR ELEMENTS AND INORGANIC COMPOUNDS (SHEET 3 OF 16)

Compound	Formula	Melting point °C	Heat of fusion	
			cal/g	cal/g mole
Boron trioxide	B_2O_3	448.8	78.9	5500
Bromine	Br_2	–7.2	16.1	2580
Bromine pentafluoride	BrF_5	–61.4	7.07	1355
Cadmium	Cd	320.8	12.9	1460
Cadmium bromide	$CdBr_2$	567.8	(18.4)	(5000)
Cadmium chloride	$CdCl_2$	567.8	28.8	5300
Cadmium fluoride	CdF_2	1110	(35.9)	(5400)
Cadmium iodide	CdI_2	386.8	10.0	3660
Cadmium sulfate	$CdSO_4$	1000	22.9	4790
Calcium	Ca	851	55.7	2230
Calcium bromide	$CaBr_2$	729.8	20.9	4180
Calcium carbonate	$CaCO_3$	1282	(126)	(12700)
Calcium chloride	$CaCl_2$	782	55	6100
Calcium fluoride	CaF_2	1382	52.5	4100
Calcium metasilicate	$CaSiO_3$	1512	115.4	13400
Calcium nitrate	$Ca(NO_3)_2$	560.8	31.2	5120
Calcium oxide	CaO	2707	(218.1)	(12240)
Calcium sulfate	$CaSO_4$	1297	49.2	6700
Carbon dioxide	CO_2	–57.6	43.2	1900
Carbon monoxide	CO	–205	7.13	199.7

For heat of fusion in J/kg, multiply values in **cal/g** by **4184.**
For heat of fusion in **J/mol**, multiply values in **cal/g-mol** (=cal/mol) by **4.184.**
For melting point in K, add 273.15 to values in °C.

Values in parentheses are of uncertain reliability.

Source: data from Weast, R C., Ed., *Handbook of Chemistry and Physics, 55th ed.*, CRC Press, Cleveland, (1974); and Bolz, R. E. and Tuve, G. L., Eds., *Handbook of Tables for Applied Engineering Science, 2nd ed.*, CRC Press, Cleveland, (1973)

Table 73. HEAT OF FUSION FOR ELEMENTS AND INORGANIC COMPOUNDS (SHEET 4 OF 16)

Compound	Formula	Melting point °C	Heat of fusion	
			cal/g	cal/g mole
Cyanogen	C_2N_2	–27.2	39.6	2060
Cyanogen chloride	CNCl	–5.2	36.4	2240
Cerium	Ce	775	27.2	2120
Cesium	Cs	28.3	3.7	500
Cesium chloride	CsCl	38.5	21.4	3600
Cesium nitrate	$CsNO_3$	406.8	16.6	3250
Chlorine	Cl_2	–103±5	22.8	1531
Chromium	Cr	1890	62.1	3660
Chromium (II) chloride	$CrCl_2$	814	65.9	7700
Chromium (III) sequioxide	Cr_2O_3	2279	27.6	4200
Chromium trioxide	CrO_3	197	37.7	3770
Cobalt	Co	1490	62.1	3640
Cobalt (II) chloride	$CoCl_2$	727	56.9	7390
Copper	Cu	1083	49.0	3110
Copper (II) chloride	$CuCl_2$	430	24.7	4890
Copper (I) chloride	CuCl	429	26.4	2620
Copper(I) cyanide	$Cu_2(CN)_2$	473	(30.1)	(5400)
Copper (I) iodide	CuI	587	(13.6)	(2600)
Copper (II) oxide	CuO	1446	35.4	2820
Copper (I) oxide	Cu_2O	1230	(93.6)	(13400)

For heat of fusion in J/kg, multiply values in cal/g by 4184.
For heat of fusion in J/mol, multiply values in cal/g-mol (=cal/mol) by 4.184.
For melting point in K, add 273.15 to values in °C.

Values in parentheses are of uncertain reliability.

Source: data from Weast, R C., Ed., *Handbook of Chemistry and Physics, 55th ed.*, CRC Press, Cleveland, (1974); and Bolz, R. E. and Tuve, G. L., Eds., *Handbook of Tables for Applied Engineering Science, 2nd ed.*, CRC Press, Cleveland, (1973)

Table 73. HEAT OF FUSION FOR ELEMENTS AND INORGANIC COMPOUNDS (SHEET 5 OF 16)

Compound	Formula	Melting point °C	Heat of fusion	
			cal/g	cal/g mole
Copper (I) sulfide	Cu_2S	1129	62.3	5500
Dysprosium	Dy	1407	25.2	4100
Erbium	Er	1496	24.5	4100
Europium	Eu	826	16.4	2500
Europium trichloride	$EuCl_3$	622	(20.9)	(8000)
Fluorine	F_2	–219.6	6.4	244.0
Gadolinium	Gd	1312	23.8	3700
Gallium	Ga	29	19.1	1336
Germanium	Ge	959	(114.3)	(8300)
Gold	Au	1063	(15.3)	3030
Hafnium	Hf	2214	(34.1)	(6000)
Holmium	Ho	1461	24.8	4100
Hydrogen	H_2	–259.25	13.8	28
Hydrogen bromide	HBr	–86.96	7.1	575.1
Hydrogen chloride	HCl	–114.3	13.0	476.0
Hydrogen fluoride	HF	83.11	54.7	1094
Hydrogen iodide	HI	–50.91	5.4	686.3
Hydrogen nitrate	HNO_3	–47.2	9.5	601
Hydrogen oxide (water)	H_2O	0	79.72	1436
Deuterium oxide	D_2O	3.78	75.8	1516

For heat of fusion in **J/kg**, multiply values in **cal/g** by **4184**.
For heat of fusion in **J/mol**, multiply values in **cal/g-mol** (=cal/mol) by **4.184**.
For melting point in **K**, add **273.15** to values in °C.

Values in parentheses are of uncertain reliability.

Source: data from Weast, R C., Ed., *Handbook of Chemistry and Physics, 55th ed.*, CRC Press, Cleveland, (1974); and Bolz, R. E. and Tuve, G. L., Eds., *Handbook of Tables for Applied Engineering Science, 2nd ed.*, CRC Press, Cleveland, (1973)

Table 73. HEAT OF FUSION FOR ELEMENTS AND INORGANIC COMPOUNDS (SHEET 6 OF 16)

Compound	Formula	Melting point °C	Heat of fusion cal/g	Heat of fusion cal/g mole
Hydrogen peroxide	H_2O_2	–0.7	8.58	2920
Hydrogen selenate	H_2SeO_4	57.8	23.8	3450
Hydrogen sulfate	H_2SO_4	10.4	24.0	2360
Hydrogen sulfide	H_2S	–85.6	16.8	5683
Hydrogen sulfide, di–	H_2S_2	–89.7	27.3	1805
Hydrogen telluride	H_2Te	–49.0	12.9	1670
Indium	In	156.3	6.8	781
Iodine	I_2	112.9	14.3	3650
Iodine chloride (α)	ICl	17.1	16.4	2660
Iodine chloride (β)	ICl	13.8	13.3	2270
Iron	Fe	1530.0	63.7	3560
Iron carbide	Fe_3C	1226.8	68.6	12330
Iron (III) chloride	Fe_2Cl_6	303.8	63.2	20500
Iron (II) chloride	$FeCl_2$	677	61.5	7800
Iron (II) oxide	FeO	1380	(107.2)	(7700)
Iron oxide	Fe_3O_4	1596	142.5	33000
Iron pentacarbonyl	$Fe(CO)_5$	–21.2	16.5	3250
Iron (II) sulfide	FeS	1195	56.9	5000
Lanthanum	La	920	17.4	2400
Lead	Pb	327.3	5.9	1224

For heat of fusion in J/kg, multiply values in **cal/g** by **4184.**
For heat of fusion in J/mol, multiply values in **cal/g-mol** (=cal/mol) by **4.184.**
For melting point in K, add 273.15 to values in °C.

Values in parentheses are of uncertain reliability.

Source: data from Weast, R C., Ed., *Handbook of Chemistry and Physics, 55th ed.*, CRC Press, Cleveland, (1974); and Bolz, R. E. and Tuve, G. L., Eds., *Handbook of Tables for Applied Engineering Science, 2nd ed.*, CRC Press, Cleveland, (1973)

Table 73. HEAT OF FUSION FOR ELEMENTS AND INORGANIC COMPOUNDS (SHEET 7 OF 16)

Compound	Formula	Melting point °C	Heat of fusion	
			cal/g	cal/g mole
Leadbromide	$PbBr_2$	487.8	11 7	4290
Lead chloride	$PbCl_2$	497 8	20.3	5650
Lead fluoride	PbF_2	823	7.6	1860
Lead iodide	PbI_2	412	17.9	5970
Lead molybdate	$PbMoO_4$	1065	70.8	(25800)
Lead oxide	PbO	890	12.6	2820
Lead sulfate	$PbSO_4$	1087	31.6	9600
Lead sulfide	PbS	1114	17.3	4150
Lithium	Li	178.8	158.5	1100
Lithium bromide	LiBr	552	33 4	2900
Lithium chloride	LiCl	614	75.5	3200
Lithium fluoride	LiF	896	(91.1)	(2360)
Lithium hydroxide	LiOH	462	103.3	2480
Lithium iodide	LiI	440	(10.6)	(1420)
Lithium metasilicate	Li_2SiO_3	1177	80.2	7210
Lithium molybdate	Li_2MoO_4	705	24.1	4200
Lithium nitrate	$LiNO_3$	250	87.8	6060
Lithium orthosilicate	Li_4SiO_4	1249	60.5	7430
Lithium sulfate	Li_2SO_4	857	27.6	3040
Lithium tungstate	Li_2WO_4	742	(25.6)	(6700)

For heat of fusion in **J/kg**, multiply values in **cal/g** by **4184.**
For heat of fusion in **J/mol**, multiply values in **cal/g-mol** (=cal/mol) by **4.184.**
For melting point in K, add **273.15** to values in °C.

Values in parentheses are of uncertain reliability.

Source: data from Weast, R C., Ed., *Handbook of Chemistry and Physics, 55th ed.*, CRC Press, Cleveland, (1974); and Bolz, R. E. and Tuve, G. L., Eds., *Handbook of Tables for Applied Engineering Science, 2nd ed.*, CRC Press, Cleveland, (1973)

Table 73. HEAT OF FUSION FOR ELEMENTS AND INORGANIC COMPOUNDS (SHEET 8 OF 16)

Compound	Formula	Melting point °C	Heat of fusion cal/g	Heat of fusion cal/g mole
Lutetium	Lu	1651	26.3	4600
Magnesium	Mg	650	88.9	2160
Magnesium bromide	$MgBr_2$	711	45.0	8300
Magnesium chloride	$MgCl_2$	712	82.9	8100
Magnesium fluoride	MgF_2	1221	94.7	5900
Magnesium oxide	MgO	2642	459.0	18500
Magnesium silicate	$MgSiO_3$	1524	146.4	14700
Magnesium sulfate	$MgSO_4$	1327	28.9	3500
Manganese	Mn	1220	62.7	3450
Manganese dichloride	$MnCl_2$	650	58.4	7340
Manganese metasilicate	$MnSiO_3$	1274	(62.6)	(8200)
Manganese (II) oxide	MnO	1784	183.3	13000
Manganese oxide	Mn_3O_4	1590	(170.4)	(39000)
Mercury	Hg	–39	2.7	557.2
Mercury bromide	$HgBr_2$	241	10.9	3960
Mercury chloride	$HgCl_2$	276.8	15.3	4150
Mercury iodide	HgI_2	250	9.9	4500
Mercury sulfate	$HgSO_4$	850	(4.8)	(1440)
Molybdenum	Mo	2622	(68.4)	(6600)
Molybdenum dichloride	$MoCl_2$	726.8	3.58	6000

For heat of fusion in **J/kg**, multiply values in **cal/g** by **4184.**
For heat of fusion in **J/mol**, multiply values in **cal/g-mol** (=cal/mol) by **4.184.**
For melting point in **K**, add **273.15** to values in °C.

Values in parentheses are of uncertain reliability.

Source: data from Weast, R C., Ed., *Handbook of Chemistry and Physics, 55th ed.*, CRC Press, Cleveland, (1974); and Bolz, R. E. and Tuve, G. L., Eds., *Handbook of Tables for Applied Engineering Science, 2nd ed.*, CRC Press, Cleveland, (1973)

Table 73. Heat of Fusion for Elements and Inorganic Compounds (Sheet 9 of 16)

Compound	Formula	Melting point °C	Heat of fusion	
			cal/g	cal/g mole
Molybdenum hexafluoride	MoF_6	17	11.9	2500
Molybdenum trioxide	MoO_3	795	(17.3)	(2500)
Neodymium	Nd	1020	11.8	1700
Neon	Ne	– 248.6	3.83	77.4
Nickel	Ni	1452	71.5	4200
Nickel chloride	$NiCl_2$	1030	142 5	18470
Nickel subsulfide	Ni_3S_2	790	25.8 1	5800
Niobium	Nb	2496	(68.9)	(6500)
Niobium pentachloride	$NbCl_5$	21.1	30 8	8400
Niobium pentoxide	Nb_2O_5	1511	91.0	24200
Nitric oxide	NO	–163.7	18.3	549.5
Nitrogen	N_2	–210	6.15	172.3
Nitrogen tetroxide	N_2O_4	–13.2	60.2	5540
Nitrous oxide	N_2O	–90.9	35.5	1563
Osmium	Os	2700	(36.7)	(7000)
Osmium tetroxide (white)	OsO_4	41.8	9.2	2340
Osmium tetroxide (yellow)	OsO_4	55.8	15.5	4060
Oxygen	O_2	–218.8	3.3	106.3
Palladium	Pd	1555	38.6	4120
Phosphoric acid	H_3PO_4	42.3	25.8	2520

For heat of fusion in **J/kg**, multiply values in **cal/g** by **4184**.
For heat of fusion in **J/mol**, multiply values in **cal/g-mol** (=cal/mol) by **4.184**.
For melting point in K, add **273.15** to values in °C.

Values in parentheses are of uncertain reliability.

Source: data from Weast, R C., Ed., *Handbook of Chemistry and Physics, 55th ed.*, CRC Press, Cleveland, (1974); and Bolz, R. E. and Tuve, G. L., Eds., *Handbook of Tables for Applied Engineering Science, 2nd ed.*, CRC Press, Cleveland, (1973)

Table 73. HEAT OF FUSION FOR ELEMENTS AND INORGANIC COMPOUNDS (SHEET 10 OF 16)

Compound	Formula	Melting point °C	Heat of fusion	
			cal/g	cal/g mole
Phosphoric acid. hypo–	$H_4P_2O_6$	54.8	51.2	8300
Phosphorus acid, hypo–	H_3PO_2	17.3	35.0	2310
Phosphorus acid, ortho–	H_3PO_3	73.8	37.4	3070
Phosphorus oxychloride	$POCl_3$	1.0	20.3	3110
Phosphorus pentoxide	P_4O_{10}	569.0	60.1	17080
Phosphorus trioxide	P_4O_6	23.7	15.3	3360
Phosphorus, yellow	P_4	44.1	4.8	600
Platinum	Pt	1770	24.1	4700
Potassium	K	63.4	14.6	574
Potassium borate, meta–	KBO_2	947	(69.1)	(5660)
Potassium bromide	KBr	742	42.0	5000
Potassium carbonate	K_2CO_3	897	56.4	7800
Potassium chloride	KCl	770	85.9	6410
Potassium chromate	K_2CrO_4	984	35.6	6920
Potassium cyanide	KCN	623	(53.7)	(3500)
Potassium dichromate	$K_2Cr_2O_7$	398	29.8	8770
Potassium fluoride	KF	875	111.9	6500
Potassium hydroxide	KOH	360	(35.3)	(1980)
Potassium iodide	KI	682	24.7	4100
Potassium nitrate	KNO_3	338	78.1	2840

For heat of fusion in J/kg, multiply values in cal/g by 4184.
For heat of fusion in J/mol, multiply values in cal/g-mol (=cal/mol) by 4.184.
For melting point in K, add 273.15 to values in °C.

Values in parentheses are of uncertain reliability.

Source: data from Weast, R C., Ed., *Handbook of Chemistry and Physics, 55th ed.*, CRC Press, Cleveland, (1974); and Bolz, R. E. and Tuve, G. L., Eds., *Handbook of Tables for Applied Engineering Science, 2nd ed.*, CRC Press, Cleveland, (1973)

Table 73. HEAT OF FUSION FOR ELEMENTS AND INORGANIC COMPOUNDS (SHEET 11 OF 16)

Compound	Formula	Melting point °C	Heat of fusion	
			cal/g	cal/g mole
Potassium peroxide	K_2O_2	490	55.3	6100
Potassium phosphate	K_3PO4	1340	41.9	8900
Potassium pyro– phosphate	$K_4P_2O_7$	1092	42.4	14000
Potassium sulfate	K_2SO_4	1074	46.4	8100
Potassium thiocyanate	KSCN	179	23.1	2250
Praseodymium	Pr	931	19.0	2700
Rhenium	Re	3167±60	(42.4)	(7900)
Rhenium heptoxide	Re_2O_7	296	30.1	15340
Rhenium hexafluoride	ReF_6	19.0	16.6	5000
Rubidium	Rb	38 .9	6. 1	525
Rubidium bromide	RbBr	677	22.4	3700
Rubidium chloride	RbCl	717	36.4	4400
Rubidium fluoride	RbF	833	39.5	4130
Rubidium iodide	Rbl	638	14.0	2990
Rubidium nitrate	$RbNO_3$	305	9.1	1340
Samarium	Sm	1072	17.3	2600
Scandium	Sc	1538	84.4	3800
Selenium	Se	217	15.4	1220
Seleniumoxychloride	$SeOCl_3$	9.8	6.1	1010
Silane, hexaHuoro–	Si_2F_6	–28.6	22.9	3900

For heat of fusion in **J/kg**, multiply values in **cal/g** by **4184.**
For heat of fusion in **J/mol**, multiply values in **cal/g-mol** (=cal/mol) by **4.184.**
For melting point in K, add **273.15** to values in °C.

Values in parentheses are of uncertain reliability.

Source: data from Weast, R C., Ed., *Handbook of Chemistry and Physics, 55th ed.*, CRC Press, Cleveland, (1974); and Bolz, R. E. and Tuve, G. L., Eds., *Handbook of Tables for Applied Engineering Science, 2nd ed.*, CRC Press, Cleveland, (1973)

Table 73. Heat of Fusion for Elements and Inorganic Compounds (Sheet 12 of 16)

Compound	Formula	Melting point °C	Heat of fusion cal/g	Heat of fusion cal/g mole
Silicon	Si	1427	337.0	9470
Silicon dioxide (Cristobalite)	SiO_2	1723	35.0	2100
Silicon tetrachloride	$SiCl_4$	–67.7	10.8	1845
Silver	Ag	961	25.0	2700
Silver bromide	AgBr	430	11.6	2180
Silver chloride	AgCl	455	22.0	3155
Silver cyanide	AgCN	350	20.5	2750
Silver iodide	AgI	557	9.5	2250
Silver nitrate	$AgNO_3$	209	16.2	2755
Silver sulfate	Ag_2SO_4	657	(13.7)	(4280)
Silver sulfide	Ag_2S	841	13.5	3360
Sodium	Na	97.8	27.4	630
Sodium borate, meta–	$NaBO_2$	966	134.6	8660
Sodium bromide	NaBr	747	59.7	6140
Sodium carbonate	Na_2CO_3	854	66.0	7000
Sodium chlorate	$NaClO_3$	255	49.7	5290
Sodium chloride	NaCl	800	123.5	7220
Sodium cyanide	NaCN	562	(88.9)	(4360)
Sodium fluoride	NaF	992	166.7	7000
Sodium hydroxide	NaOH	322	50.0	2000

For heat of fusion in J/kg, multiply values in **cal/g** by **4184.**
For heat of fusion in **J/mol**, multiply values in **cal/g-mol** (=cal/mol) by **4.184.**
For melting point in K, add 273.15 to values in °C.

Values in parentheses are of uncertain reliability.

Source: data from Weast, R C., Ed., *Handbook of Chemistry and Physics, 55th ed.*, CRC Press, Cleveland, (1974); and Bolz, R. E. and Tuve, G. L., Eds., *Handbook of Tables for Applied Engineering Science, 2nd ed.*, CRC Press, Cleveland, (1973)

Table 73. HEAT OF FUSION FOR ELEMENTS AND INORGANIC COMPOUNDS (SHEET 13 OF 16)

Compound	Formula	Melting point °C	Heat of fusion	
			cal/g	cal/g mole
Sodium iodide	NaI	662	35.1	5340
Sodium molybdate	Na_2MoO_4	687	17.5	3600
Sodium nitrate	$NaNO_3$	310	44.2	3760
Sodium peroxide	Na_2O_2	460	75.1	5860
Sodium phosphate, meta–	$NaPO_3$	988	(48.6)	(4960)
Sodium pyrophosphate	$Na_4P_2O_7$	970	(51.5)	(13700)
Sodiumsilicate,aluminum–	$NaAlSi_3O_8$	1107	50.1	13150
Sodium silicate, di–	$Na_2Si_2O_5$	884	46.4	8460
Sodium silicate, meta–	Na_2SiO_3	1087	84.4	10300
Sodium sulfate	Na_2SO_4	884	41.0	5830
Sodium sulfide	Na_2S	920	15.4	(1200)
Sodium thiocyanate	NaSCN	323	54.8	4450
Sodium tungstate	Na_2WO_4	702	19.6	5800
Strontium	Sr	757	25.0	2190
Strontium bromide	$SrBr_2$	643	19.3	4780
Strontium chloride	$SrCl_2$	872	26.5	4100
Strontium fluoride	SrF_2	1400	34.0	4260
Strontium oxide	SrO	2430	161.2	16700
Sulfur (monatomic)	S	119	9.2	295
Sulfur dioxide	SO_2	–73.2	32.2	2060

For heat of fusion in **J/kg**, multiply values in **cal/g** by **4184.**
For heat of fusion in **J/mol**, multiply values in **cal/g-mol** (=cal/mol) by **4.184.**
For melting point in K, add **273.15** to values in °C.

Values in parentheses are of uncertain reliability.

Source: data from Weast, R C., Ed., *Handbook of Chemistry and Physics, 55th ed.*, CRC Press, Cleveland, (1974); and Bolz, R. E. and Tuve, G. L., Eds., *Handbook of Tables for Applied Engineering Science, 2nd ed.*, CRC Press, Cleveland, (1973)

Table 73. HEAT OF FUSION FOR ELEMENTS AND INORGANIC COMPOUNDS (SHEET 14 OF 16)

Compound	Formula	Melting point °C	Heat of fusion	
			cal/g	cal/g mole
Sulfur trioxide (α)	SO_3	16.8	25.8	2060
Sulfur trioxide (β)	SO_3	32.3	36.1	2890
Sulfur trioxide (γ)	SO_3	62.1	79.0	6310
Tantalum	Ta	2996 ± 50	34.6–41.5	(7500)
Tantalum pentachloride	$TaCl_5$	206.8	25.1	9000
Tantalum pentoxide	Ta_2O_5	1877	108.6	48000
Tellurium	Te	453	25.3	3230
Terbium	Tb	1356	24.6	3900
Thallium	Tl	302.4	5.0	1030
Thallium bromide, mono–	TlBr	460	21.0	5990
Thallium carbonate	Tl_2CO_3	273	9.5	4400
Thallium chloride, mono–	TICl	427	17.7	4260
Thallium iodide, mono–	TlI	440	9.4	3125
Thallium nitrate	$TINO_3$	207	8.6	2290
Thallium sulfate	Tl_2SO_4	632	10.9	5500
Thallium sulfide	Tl_2S	449	6.8	3000
Thorium	Th	1845	(<19.8)	(<4600)
Thorium chloride	$ThCl_4$	765	61.6	22500
Thorium dioxide	ThO_2	2952	1102.0	291100
Thulium	Tm	1545	26.0	4400

For heat of fusion in J/kg, multiply values in cal/g by **4184.**
For heat of fusion in J/mol, multiply values in cal/g-mol (=cal/mol) by **4.184.**
For melting point in K, add 273.15 to values in °C.

Values in parentheses are of uncertain reliability.

Source: data from Weast, R C., Ed., *Handbook of Chemistry and Physics, 55th ed.*, CRC Press, Cleveland, (1974); and Bolz, R. E. and Tuve, G. L., Eds., *Handbook of Tables for Applied Engineering Science, 2nd ed.*, CRC Press, Cleveland, (1973)

Table 73. HEAT OF FUSION FOR ELEMENTS AND INORGANIC COMPOUNDS (SHEET 15 OF 16)

Compound	Formula	Melting point °C	Heat of fusion cal/g	Heat of fusion cal/g mole
Tin	Sn	231.7	14.4	1720
Tin bromide, di–	$SnBr_2$	231.8	(6.1)	(1720)
Tin bromide, tetra–	$SnBr_4$	29.8	6.8	3000
Tin chloride, di–	$SnCl_2$	247	16.0	3050
Tin chloride,tetra–	$SnCl_4$	–33.3	8.4	2190
Tin iodide, tetra–	SnI_4	143.4	(6.9)	(4330)
Tin oxide	SnO	1042	(46.8)	(6400)
Titanium	Ti	1800	(104.4)	(5000)
Titanium bromide, tetra–	$TiBr_4$	38	(5.6)	(2060)
Titanium chloride, tetra–	$TiCl_4$	–23.2	11.9	2240
Titanium dioxide	TiO_2	1825	(142.7)	(11400)
Titanium oxide	TiO	991	219	14000
Tungsten	W	3387	(45.8)	(8420)
Tungsten dioxide	WO_2	1270	60 1	13940
Tungsten hexafluoride	WF_6	–0.5	6.0	1800
Tungsten tetrachloride	WCl_4	327	18.4	6000
Tungsten trioxide	WO_3	1470	60 1	13940
Uranium^{235}	U	~1133	20	3700
Uranium tetrachloride	UCl_4	590	27.1	10300
Vanadium	V	1917	(70)	(4200)

For heat of fusion in J/kg, multiply values in cal/g by 4184.
For heat of fusion in J/mol, multiply values in cal/g-mol (=cal/mol) by 4.184.
For melting point in K, add 273.15 to values in °C.

Values in parentheses are of uncertain reliability.

Source: data from Weast, R C., Ed., *Handbook of Chemistry and Physics, 55th ed.*, CRC Press, Cleveland, (1974); and Bolz, R. E. and Tuve, G. L., Eds., *Handbook of Tables for Applied Engineering Science, 2nd ed.*, CRC Press, Cleveland, (1973)

Table 73. HEAT OF FUSION FOR ELEMENTS AND INORGANIC COMPOUNDS (SHEET 16 OF 16)

Compound	Formula	Melting point °C	Heat of fusion	
			cal/g	cal/g mole
Vanadium dichloride	VCl_2	1027	65.6	8000
Vanadium oxide	VO	2077	224.0	15000
Vanadium pentoxide	V_2O_5	670	85.5	15560
Xenon	Xe	–111.6	5.6	740
Ytterbium	Yb	823	12.7	2200
Yttrium	Y	1504	46.1	4100
Yttrium oxide	Y_2O_3	2227	110.7	25000
Zinc	Zn	419.4	24.4	1595
Zinc chloride	$ZnCl_2$	283	(406)	(5540)
Zinc oxide	ZnO	1975	54.9	4470
Zinc sulfide	ZnS	1745	(93.3)	(9100)
Zirconium	Zr	1857	(60)	(5500)
Zirconium dichloride	$ZrCl_2$	727	45.0	7300
Zirconium oxide	ZrO_2	2715	168.8	20800

For heat of fusion in **J/kg**, multiply values in **cal/g** by **4184.**
For heat of fusion in **J/mol**, multiply values in **cal/g-mol** (=cal/mol) by **4.184.**
For melting point in K, add **273.15** to values in °C.

Values in parentheses are of uncertain reliability.

Source: data from Weast, R C., Ed., *Handbook of Chemistry and Physics, 55th ed.*, CRC Press, Cleveland, (1974); and Bolz, R. E. and Tuve, G. L., Eds., *Handbook of Tables for Applied Engineering Science, 2nd ed.*, CRC Press, Cleveland, (1973)

Table 74. HEATS OF SUBLIMATION OF METALS AND THEIR OXIDES

Metal	kcal/mole (25°C)	kJ/mole (25°C)
Al	78	326
Cu	81	338
Fe	100	416
Mg	113	473
Metal Oxide		
FeO	122	509
MgO	145	605
α-TiO	143	597
TiO_2 (rutile)	153	639

Data from: *JANAF Thermochemical Tables, 2nd ed.*, National Standard Reference Data Series, Natl. Bur. Std. (U.S.), 37 (1971) *and* Supplement in *J. Phys. Chem. Ref. Data* 4(1), 1-175 (1975).

Table 75. KEY TO TABLES OF THERMODYNAMIC COEFFICIENTS

(SHEET 1 OF 4)

Thermodynamic calculations over a wide range of temperatures are generally made with the aid of algebraic equations representing the characteristic properties of the substances being considered. The necessary integrations and differentiations, or other mathematical manipulations, are then most easily effected. The most convenient starting point in making such calculations for a given substance is the heat capacity at constant pressure. From this quantity and a knowledge of the properties of any phase transitions, the other thermodynamic properties may be computed by the well-known equations given in standard texts on thermodynamics. Please note that the units for a, b, c, and d are cal/g mole, whereas those for A are kcal/g mole. The necessary adjustment must be made when the data are substituted into the equations. Empirical heat capacity equations are generated in the form of a power series, with the absolute temperature T as the independent variable:

$$C_p = a' + (b' \times 10^{-3})T + (c' \times 10^{-6})T_2$$

or

$$C_p = a'' + (b'' \times 10^{-3})T + \frac{d \times 10^5}{T^2}.$$

Since both forms are used in the following, let

$$C_p = a + (b \times 10^{-3})T + (c \times 10^{-6})T^2 + \frac{d \times 10^5}{T^2}.$$

Table 75. KEY TO TABLES OF THERMODYNAMIC COEFFICIENTS

(SHEET 2 OF 4)

The constants a, b, c, and d are to be determined either experimentally or by some theoretical or semi-empirical approach. The heat content, or enthalpy (H), is determined from the heat capacity by a simple integration of the range of temperatures for which the formula for c_p is valid. Thus, if 298K is taken as a reference temperature,

$$H_T - H_{298} = \int_{298}^{T} C_p dT$$

$$= a(T - 298) + \tfrac{1}{2}(b \times 10^{-3})(T_2 - 298^2) + \frac{1}{3}(c \times 10^{-6})(T^3 - 298^3) - (d \times 10^5)\left(\frac{1}{T} - \frac{1}{298}\right)$$

$$= aT + \tfrac{1}{2}(b \times 10^{-3})T^2 + \frac{1}{3}(c \times 10^{-6})T^3 - \frac{d \times 10^5}{T} - A,$$

where all the constants on the right-hand side of the equation have been incorporated in the term –A.

In general, the enthalpy is given by a sum of terms for each phase of the substance involved in the temperature range considered plus terms that represent the heats of transitions:

$$H_T - H_{298} = \Sigma \int_{T_1}^{T_2} C_p dT + \Sigma \Delta H_{tr}.$$

In a similar manner, the entropy S is obtained by performing the integration

Table 75. KEY TO TABLES OF THERMODYNAMIC COEFFICIENTS

(SHEET 3 OF 4)

$$S_T - S'_{298} = \int_{298}^{T} (C_p/T)dt$$

$$= a\ln(T/298) + (b \times 10^{-3})(T - 298) + ½(c \times 10^{-6})(T^2 - 298^2) - ½(d \times 10^5)\left(\frac{1}{T^2} - \frac{1}{298^2}\right)$$

$$= a\ln T + (b \times 10^{-3})T + ½(c \times 10^{-6})T^2 - \frac{½(d \times 10^5)}{T^2} - B'$$

or

$$S_T = 2.303\, a \log T + (b \times 10^{-3})T + ½(c \times 10^{-6})T^2 - \frac{½(d \times 10^5)}{T^2} - B$$

where

$$B = B' - S_{298}.$$

From the definition of free energy (F):

$$F = H - TS$$

the quantity

$$F_T - H_{298} = (H_T - H_{298}) - TS_T$$

Table 75. Key to Tables of Thermodynamic Coefficients

(Sheet 4 of 4)

may be written as:

$$F_T - H_{298} = -2.303aT \log T - \tfrac{1}{2}(b \times 10^{-3})T^2 - \frac{1}{6}(c \times 10^{-6})T^3 - \frac{\tfrac{1}{2}(d \times 10^5)}{T} + (B + a)T - A$$

and also the free energy function

$$\frac{F_T - H_{298}}{T} = -2.303a \log T - \tfrac{1}{2}(b \times 10^{-3})T - \frac{1}{6}(c \times 10^{-6})T^2 - \frac{\tfrac{1}{2}(d \times 10^5)}{T^2} + (B + a) - \frac{A}{T}$$

Values of these thermodynamic coefficients are given in the following tables. The first column in each table lists the material. The second column gives the phase to which the coefficients are applicable. The remaining columns list the values of the constants a, b, c, d, A, and B required in the thermodynamic equations. All values that represent estimates are enclosed in parentheses. The heat capacities at temperatures beyond the range of experimental determination were estimated by extrapolation. Where no experimental values were found, analogy with compounds of neighboring elements in the periodic table was used.

Table 76. THERMODYNAMIC COEFFICIENTS FOR SELECTED ELEMENTS*

(SHEET 1 OF 14)

Element	Phase	a –	b –	c (cal • g $mole^{-1}$)	d –	A (kcal • g $mole^{-1}$)	B (e.u.)
Ac	solid	(5.4)	(3.0)	–	–	(1.743)	(18.7)
	liquid	(8)	–	–	–	(0.295)	(31.3)
Ag	solid	5.09	1.02	–	0.36	1.488	19.21
	liquid	7.30	–	–	–	0.164	30.12
	gas	(4.97)	–	–	–	(–66.34)	(–12.52)
Al	solid	4.94	2.96	–	–	1.604	22.26
	liquid	7.0	–	–	–	0.33	30.83
Am	solid	(4.9)	(4.4)	–	–	(1.657)	(16.2)
	liquid	(8.5)	–	–	–	(0.409)	(34.5)
As	solid	5.17	2.34	–	–	1.646	21.8
Au	solid	6.14	–0.175	0.92	–	1.831	23.65
	liquid	7.00	–	–	–	–0.631	26.99
B	solid	1.54	4.40	–	–	0.655	8.67
	liquid	(6.0)	–	–	–	(–4.599)	(31.4)

Source: *data from* Weast, R. C. Ed., *Handbook of Chemistry and Physics, 69th ed.*, CRC Press, Boca Raton, Fla., 1988, D44.

Table 76. THERMODYNAMIC COEFFICIENTS FOR SELECTED ELEMENTS*

(SHEET 2 OF 14)

Element	Phase	a –	b –	c (cal • g mole^{-1})	d –	A (kcal • g mole^{-1})	B (e.u.)
Ba	solid, α	5.55	4.50	–	–	1.722	16.1
	solid, β	5.55	1.50	–	–	1.582	15.9
	liquid	(7.4)	–	–	–	(0.843)	(25.3)
	gas	(497)	–	–	–	(–39.65)	(–11.7)
Be	solid	5.07	1.21	–	–1.15	1.951	27.62
	liquid	5.27	–	–	–	–1.611	25.68
Bi	solid	5.38	2.60	–	–	1.720	17.8
	liquid	7.60	–	–	–	–0.087	25.6
	gas	(4.97)	–	–	–	(–46.19)	(–15.9)
C	solid	4.10	1.02	–	–2.10	1.972	23.484
Ca	solid, α	5.24	3.50	–	–	1.718	2095
	solid, β	6.29	1.40	–	–	1.689	26.01
	liquid	7.4	–	–	–	–0.147	30.28
	gas	(4.97)	–	–	–	(–43.015)	(–9.88)

Source: *data from* Weast, R. C. Ed., *Handbook of Chemistry and Physics, 69th ed.*, CRC Press, Boca Raton, Fla., 1988, D44.

Table 76. Thermodynamic Coefficients for Selected Elements*

(Sheet 3 of 14)

Element	Phase	a –	b –	c (cal • g mole^{-1})	d –	A (kcal • g mole^{-1})	B (e.u.)
Cd	solid	5.31	2.94	–	–	1.714	18.8
	liquid	7.10	–	–	–	0.798	26.1
	gas	(4.97)	–	–	–	(–25.28)	(–11.7)
Ce	solid	4.40	6.0	–	–	1.579	13.1
	liquid	(7.9)	–	–	–	(–0.148)	(29.1)
Cl_2	gas	8.76	0.27	–	–0.65	2.845	–2.929
Co	solid, α	4.72	4.30	–	–	1.598	21.4
	solid, β	3.30	5.86	–	–	0.974	3.1
	solid, γ	9.60	–	–	–	3.961	50.5
	liquid	8.30	–	–	–	–2.034	38.7
Cr	solid	5.35	2.36	–	–0.44	1.848	25.75
	liquid	9.40	–	–	–	1.556	50.13
	gas	(4.97)	–	–	–	(–82.47)	(–13.8)

Source: *data from* Weast, R. C. Ed., *Handbook of Chemistry and Physics, 69th ed.*, CRC Press, Boca Raton, Fla., 1988, D44.

Table 76. THERMODYNAMIC COEFFICIENTS FOR SELECTED ELEMENTS *

(SHEET 4 OF 14)

Element	Phase	a –	b –	c (cal • g mole^{-1})	d –	A (kcal • g mole^{-1})	B (e.u.)
Cs	solid	7.42	–	–	–	2.212	22.5
	liquid	8.00	–	–	–	1.887	24.1
	gas	(4.97)	–	–	–	(–17.35)	(–13.6)
Cu	solid	5.41	1.50	–	–	1.680	23.30
	liquid	7.50	–	–	–	0.024	34.05
F_2	gas	8.29	0.44	–	–0.80	2.760	–0.76
Fe	solid, α	3.37	7.10	–	0.43	1.176	14.59
	solid, β	10.40	–	–	–	4.281	55.66
	solid, γ	4.85	3.00	–	–	0.396	19.76
	solid, δ	10.30	–	–	–	4.382	55.11
	liquid	10.00	–	–	–	–0.021	50.73
Ga	solid	5.237	3.33	–	–	1.710	21.01
	liquid	(6.645)	–	–	–	(0.648)	(23.64)

Source: *data from* Weast, R. C. Ed., *Handbook of Chemistry and Physics, 69th ed.*, CRC Press, Boca Raton, Fla., 1988, D44.

Table 76. THERMODYNAMIC COEFFICIENTS FOR SELECTED ELEMENTS *

(SHEET 5 OF 14)

Element	Phase	a –	b –	c (cal • g mole^{-1})	d –	A (kcal • g mole^{-1})	B (e.u.)
Ge	solid	5.90	1.13	–	–	1.764	23.8
	liquid	(7.3)	–	–	–	(–5.668)	(25.7)
H_2	gas	6.62	0.81	–	–	2.010	6.75
Hf	solid	(6.00)	(0.52)	–	–	(1.812)	(21.2)
Hg	liquid	6.61	–	–	–	1.971	19.20
	gas	4.969	–	–	–	–13.048	–13.54
In	solid	5.81	2.50	–	–	1.844	19.97
	liquid	7.50	–	–	–	1.564	27.34
	gas	(4.97)	–	–	–	(–58.42)	(–14.46)
Ir	solid	5.56	1.42	–	–	1.721	23.4
K	solid	1.3264	19.405	–	–	1.258	–1.86
	liquid	8.8825	4.565	2.9369	–	1.923	32.55
	gas	(4.97)	–	–	–	(–19.689)	(–9.46)

Source: *data from* Weast, R. C. Ed., *Handbook of Chemistry and Physics, 69th ed.*, CRC Press, Boca Raton, Fla., 1988, D44.

Table 76. THERMODYNAMIC COEFFICIENTS FOR SELECTED ELEMENTS *

(SHEET 6 OF 14)

Element	Phase	a –	b –	c (cal • g mole^{-1})	d –	A (kcal • g mole^{-1})	B (e.u.)
La	solid	6.17	1.60	–	–	1.911	21.9
	liquid	(7.3)	–	–	–	(–0.15)	(26.0)
Li	solid	3.05	8.60	–	–	1.292	12.92
	liquid	7.0	–	–	–	1.509	32.00
	gas	(4.97)	–	–	–	(–34.30)	(–2.84)
Mg	solid	5.33	2.45	–	–0.103	1.733	23.39
	liquid	(8.0)	–	–	–	0.942	36.967
	gas	(4.97)	–	–	–	(–34.78)	(–7.60)
Mn	solid, α	6.70	3.38	–	–0.37	1.974	26.11
	solid, β	8.33	0.66	–	–	2.672	41.02
	solid, γ	10.70	–	–	–	4.760	56.84
	solid, δ	11.30	–	–	–	5.176	60.88
	liquid	11.00	–	–	–	1.221	56.38
	gas	6.26	–			–63.704	–3.13
Mo	solid	5.48	1.30	–	–	1.692	24.78

Source: *data from* Weast, R. C. Ed., *Handbook of Chemistry and Physics, 69th ed.*, CRC Press, Boca Raton, Fla., 1988, D44.

Table 76. THERMODYNAMIC COEFFICIENTS FOR SELECTED ELEMENTS*

(SHEET 7 OF 14)

Element	Phase	a –	b –	c (cal • g mole^{-1})	d –	A (kcal • g mole^{-1})	B (e.u.)
N_2	gas	6.76	0.606	0.13	–	2.044	–7.064
Na	solid	5.657	3.252	0.5785	–	1.836	20.92
	liquid	8.954	–4.577	2.540	–	1.924	36.0
		(4.97)	–	–	–	(–24.40)	(–8.7)
Nb	solid	5.66	0.96	–	–	1.730	24.24
Nd	solid	5.61	5.34	–	–	1.910	19.7
	liquid	(9.1)	–	–	–	(–0.606)	35.8
Ni	solid α	4.06	7.04	–	–	1.523	18.095
	solid β	6.00	1.80	–	–	1.619	27.16
	liquid	9.20	–	–	–	0.251	45.47
Np	solid	(5.3)	(3.4)	–	–	(1.731)	(17.9)
	liquid	(9.0)	–	–	–	(1.392)	(37.5)
O_2	gas	8.27	0.258	–	–1.877	3.007	–0.750

Source: *data from* Weast, R. C. Ed., *Handbook of Chemistry and Physics, 69th ed.*, CRC Press, Boca Raton, Fla., 1988, D44.

Table 76. THERMODYNAMIC COEFFICIENTS FOR SELECTED ELEMENTS *

(SHEET 8 OF 14)

Element	Phase	a –	b –	c (cal • g mole^{-1})	d –	A (kcal • g mole^{-1})	B (e.u.)
Os	solid	5.69	0.88	–	–	1.736	24.9
P_4	solid, white	13.62	28.72	–	–	5.338	43.8
	liquid	19.23	0.51	–	–2.98	6.035	66.7
	gas	(19.5)	(–0.4)	(1.3)	–	(–6.32)	(46.1)
Pa	solid	(5.2)	(4.0)	–	–	(1.728)	(17.3)
	liquid	(8.0)	–	–	–	(–3.823)	(28.8)
Pb	solid	5.64	2.30	–	–	1.784	17.33
	liquid	7.75	–0.73	–	–	1.362	27.11
	gas	(4.97)	–	–	–	(–45.25)	(–13.6)
Pd	solid	5.80	1.38	–	–	1.791	24.6
	liquid	(9.0)	–	–	–	(1.215)	(43.8)
Po	solid	(5.2)	(3.2)	–	–	(1.693)	(17.6)
	liquid	(9.0)	–	–	–	(0.847)	(35.2)
	gas	(4.97)	–	–	–	(–28.73)	(–13.5)

Source: *data from* Weast, R. C. Ed., *Handbook of Chemistry and Physics, 69th ed.*, CRC Press, Boca Raton, Fla., 1988, D44.

Table 76. THERMODYNAMIC COEFFICIENTS FOR SELECTED ELEMENTS *

(SHEET 9 OF 14)

Element	Phase	a –	b –	c (cal • g mole^{-1})	d –	A (kcal • g mole^{-1})	B (e.u.)
Pr	solid	(5.0)	(4.6)	–	–	(1.705)	(16.4)
	liquid	(8.0)	–	–	–	(–0.519)	(30.0)
Pt	solid	5.74	1.34	–	0.10	1.737	23.0
	liquid	(9.0)	–	–	–	(0.406)	(42.6)
Pu	solid	(5.2)	(3.6)	–	–	(1.710)	(17.7)
	liquid	(8.0)	–	–	–	(0.506)	(31.0)
Ra	solid	(5.8)	(1.2)	–	–	(1.783)	(16.4)
	liquid	(8.0)	–	–	–	(1.284)	(28.6)
	gas	(4.97)	–	–	–	(–38.87)	(–14.5)
Rb	solid	3.27	13.1	–	–	1.557	5.9
	liquid	7.85	–	–	–	1.814	26.5
	gas	(4.97)	–	–	–	(–19.04)	(–12.3)
Re	solid	(5.85)	(0.8)	–	–	(1.780)	(24.7)

Source: *data from* Weast, R. C. Ed., *Handbook of Chemistry and Physics, 69th ed.*, CRC Press, Boca Raton, Fla., 1988, D44.

Table 76. THERMODYNAMIC COEFFICIENTS FOR SELECTED ELEMENTS*

(SHEET 10 OF 14)

Element	Phase	a –	b –	c (cal • g mole^{-1})	d –	A (kcal • g mole^{-1})	B (e.u.)
Rh	solid	5.40	2.19	–	–	1.707	23.8
	liquid	(9.0)	–	–	–	(–0.923)	(44.4)
Ru	solid, α	5.25	1.50	–	–	1.632	23.5
	solid, β	7.20	–	–	–	2.867	35.5
	solid, γ	7.20	–	–	–	2.867	35.5
	solid, δ	7.50	–	–	–	3.169	37.6
S	solid, α	3.58	6.24	–	–	1.345	14.64
	solid, β	3.56	6.95	–	–	1.298	14.54
	liquid	5.4	5.0	–	–	1.576	24.02
$^{1}/_{2}$ S_2	gas	(4.25)	(0.15)	–	(–1.0)	(–2.859)	(9.57)
Sb	solid, α, β, γ	5.51	1.74	–	–	1.720	21.4
	liquid	7.50	–	–	–	1.992	28.1
1/2 Sb_2	gas	4.47	–	–	–0.11	–53.876	–21.7

Source: *data from* Weast, R. C. Ed., *Handbook of Chemistry and Physics, 69th ed.*, CRC Press, Boca Raton, Fla., 1988, D44.

Table 76. THERMODYNAMIC COEFFICIENTS FOR SELECTED ELEMENTS *

(SHEET 11 OF 14)

Element	Phase	a –	b –	c (cal • g mole^{-1})	d –	A (kcal • g mole^{-1})	B (e.u.)
Sc	solid	(5.13)	(3.0)	–	–	1.663	21.1
	liquid	(7.50)	–			(–2.563)	31.3
Se	solid	3.30	8.80	–	–	1.375	11.28
	liquid	7.0	–	–	–	0.881	27.34
Si	solid	5.70	1.02	–	–1.06	2.100	28.88
	liquid	7.4	–			7.646	33.17
Sm	solid	(6.7)	(3.4)	–	–	(2.149)	(24.2)
	liquid	(9.0)	–	–	–	(–2.296)	(33.4)
Sn	solid, α, β	4.42	6.30	–	–	1.598	14.8
	liquid	7.30	–	–	–	0.559	26.2
	gas	(4.97)	–	–	–	60.21)	(–14.3)
Sr	solid	(5.60)	(1.37)	–	–	(1.731)	(19.3)
	liquid	(7.7)	–	–	–	(0.976)	(30.4)
	gas	(4.97)	–	–	–	(37.16)	(–10.2)

Source: *data from* Weast, R. C. Ed., *Handbook of Chemistry and Physics, 69th ed.*, CRC Press, Boca Raton, Fla., 1988, D44.

Table 76. THERMODYNAMIC COEFFICIENTS FOR SELECTED ELEMENTS*

(SHEET 12 OF 14)

Element	Phase	a –	b –	c (cal • g mole^{-1})	d –	A (kcal • g mole^{-1})	B (e.u.)
Ta	solid	5.82	0.78	–	–	1.770	23.4
Tc	solid	(5.6)	(2.0)	–	–	(1.759)	(24.5)
	liquid	–	–	–	–	(3.459)	(59.4)
Te	solid, α	4.58	5.25	–	–	1.599	15.78
	solid, β	4.58	5.25	–	–	1.469	15.57
	liquid	9.0	–	–	–	–0.988	34.96
$^{1}/_{2}$ Te_2	gas	4.47	–	–	–0.10	–19.048	–6.47
Th	solid	8.2	–0.77	2.04	–	2.591	33.64
	liquid	(8.0)	–	–	–	(–7.602)	(26.84)
Ti	solid, α	5.25	2.52	–	–	1.677	23.33
	solid, β	7.50	–	–	–	1.645	35.46
	liquid	(7.8)	–	–	–	(–2.355)	(35.45)

Source: *data from* Weast, R. C. Ed., *Handbook of Chemistry and Physics, 69th ed.*, CRC Press, Boca Raton, Fla., 1988, D44.

Table 76. Thermodynamic Coefficients for Selected Elements *

(Sheet 13 of 14)

Element	Phase	a –	b –	c ($cal \cdot g\ mole^{-1}$)	d –	A ($kcal \cdot g\ mole^{-1}$)	B (e.u.)
Tl	solid, α	5.26	3.46	–	–	1.722	15.6
	solid, β	7.30	–	–	–	2.230	26.4
	liquid	7.50	–	–	–	1.315	25.9
	gas	(4.97)	–	–	–	(–41.88)	(–15.4)
U	solid, α	3.25	8.15	–	0.80	1.063	8.47
	solid, β	10.28	–	–	–	3.493	48.27
	solid, γ	9.12	–	–	–	1.110	39.09
	liquid	(8.99)	–	–	–	(–2.073)	36.01
V	solid	5.57	0.97	–	–	1.704	24.97
	liquid	(8.6)	–	–	–	1.827	44.06
W	solid	5.74	0.76	–	–	1.745	24.9
Y	solid	(5.6)	(2.2)	–	–	(1.767)	(21.6)
	liquid	(7.5)	–	–	–	2.277)	(29.6)

Source: *data from* Weast, R. C. Ed., *Handbook of Chemistry and Physics, 69th ed.*, CRC Press, Boca Raton, Fla., 1988, D44.

Table 76. THERMODYNAMIC COEFFICIENTS FOR SELECTED ELEMENTS*

(SHEET 14 OF 14)

Element	Phase	a –	b –	c (cal • g mole^{-1})	d –	A (kcal • g mole^{-1})	B (e.u.)
Zn	solid	5.35	2.40	–	–	1.702	21.25
	liquid	7.50	–	–	–	1.020	31.35
		(4.97)	–	–	–	(–29.407)	(–9.81)
Zr	solid, α	6.83	1.12	–	–0.87	2.378	30.45
	solid, β	7.27	–	–	–	1.159	31.43
	liquid	(8.0)	–	–	–	(–2.190)	(34.7)

Source: *data from* Weast, R. C. Ed., *Handbook of Chemistry and Physics, 69th ed.*, CRC Press, Boca Raton, Fla., 1988, D44.

* Please refer to Table 75, "Key to Tables of Thermodynamic Coefficients" on page 257 for an explanation of the coefficients.

Table 77. THERMODYNAMIC COEFFICIENTS FOR OXIDES

(SHEET 1 OF 23)

Oxide	Phase	a –	b (cal • g mole^{-1})	c –	d –	A (kcal • g mole^{-1})	B (e.u.)
Ac_2O_3	Solid	(20.0)	(20.4)	–	–	(6.870)	(80.9)
	Liquid	(40)	–	–	–	(–19.767)	(180.5)
Ag_2O	Solid	13.26	7.04	–	–	4.266	48.56
Ag_2O_2	Solid	(16.4)	(12.2)	–	–	(5.432)	(76.7)
Al_2O_3	Solid	26.12	4.388	–	–7.269	10.422	142.03
	Liquid	(33)	–	–	–	(– 11.655)	(174.1)
Am_2O_3	Solid	(20.0)	(15.6)	–	–	(6.657)	(81.6)
	Liquid	(38.5)	–	–	–	(–7.796)	(181.8)
AmO_2	Solid	(14.0)	(6.8)	–	–	(4.477)	(61.8)

For discussion of these coefficients, please see Table 75, Key to Tables of Thermodynamic Coefficients on page 257

Source: data from Weast, R. C., Ed., *Handbook of Chemistry and Physics, 55th ed.*, CRC Press, Cleveland, 1974, D-58.

Table 77. THERMODYNAMIC COEFFICIENTS FOR OXIDES

(SHEET 2 OF 23)

Oxide	Phase	a –	b (cal • g mole^{-1})	c –	d –	A (kcal • g mole^{-1})	B (e.u.)
As_2O_3	Solid, α	8.37	48.6	–	–	4.656	36.6
	Solid, β	8.37	48.6	–	–	0.556	28.4
	Liquid	(39)	–	–	–	(5.760)	(187.6)
	Gas	(21.5)	–	–	–	(–14.164)	(62.5)
AsO_2	Solid	(8.5)	(9.4)	–	–	(2.952)	(38.2)
	Liquid	(21)	–	–	–	(2.184)	(108.0)
As_2O_5	Solid	(31.1)	(16.4)	–	(–5.4)	(11.813)	(159.9)
Au_2O_3	Solid	(23.5)	(4.8)	–	–	(7.220)	(105.3)
B_2O_3	Solid	8.73	25.40	–	–1.31	4.171	45.04
	Liquid	30.50	–	–	–	7.822	161.59
Ba_2O	Solid	(20.0)	(2.2)	–	–	(6.061)	(91.1)
	Liquid	(22)	–	–	–	(1.769)	(96.8)
	Gas	(15)	–	–	–	(–25.51)	(29.0)

For discussion of these coefficients, please see Table 75, Key to Tables of Thermodynamic Coefficients on page 257
Source: data from Weast, R. C., Ed., *Handbook of Chemistry and Physics, 55th ed.*, CRC Press, Cleveland, 1974, D-58.

Table 77. THERMODYNAMIC COEFFICIENTS FOR OXIDES

(SHEET 3 OF 23)

Oxide	Phase	a –	b (cal • g mole^{-1})	c –	d –	A (kcal • g mole^{-1})	B (e.u.)
BaO	Solid	12.74	1.040	–	–1.984	4.510	57.2
	Liquid	(13.9)	–	–	–	(–9.341)	(57.5)
BaO_2	Solid	(13.6)	(2.0)	–	–	(4.144)	(59.6)
	Liquid	(21)	–	–	–	(3.241)	(99.0)
BeO	Solid	8.69	3.65	–	–3.13	3.803	48.99
BiO	Solid	(9.7)	(3.0)	–	–	(3.025)	(41.2)
	Liquid	(14)	–	–	–	(2.306)	(64.9)
	Gas	(8.9)	–	–	–	(–61.49)	(–1.8)
Bi_2O_3	Solid	23.27	11.05	–	–	7.429	99.7
	Liquid	(35.7)	–	–	–	(7.614)	(168.3)
CO	Gas	6.60	1.2	–	–	2.021	–9.34
CO_2	Gas	7.70	5.3	–0.83	–	2.490	–5.64
CaO	Solid	10.00	4.84	–	–1.08	3.559	49.5

For discussion of these coefficients, please see Table 75, Key to Tables of Thermodynamic Coefficients on page 257
Source: data from Weast, R. C., Ed., *Handbook of Chemistry and Physics, 55th ed.*, CRC Press, Cleveland, 1974, D-58.

Table 77. THERMODYNAMIC COEFFICIENTS FOR OXIDES

(SHEET 4 OF 23)

Oxide	Phase	a –	b (cal • g mole^{-1})	c –	d –	A (kcal • g mole^{-1})	B (e.u.)
CdO	Solid	9.65	2.08	–	–	2.970	42.5
Ce_2O_3	Solid	(–23.0)	(9.0)	–	–	(7.258)	(100.2)
	Liquid	(37)	–	–	–	(–2.591)	(178.5)
CeO_2	Solid	15.0	2.5	–	–	4.579	68.5
CoO	Solid	(9.8)	(2.2)	–	–	(3.020)	(46.0)
	Liquid	(15.5)	–	–	–	(–1.886)	(79.2)
Co_3O_4	Solid	(29.5)	(17.0)	–	–	(9.551)	(137.6)
Cr_2O_3	Solid	28.53	2.20	–	–3.736	9.857	145.9
CrO_2	Solid	(16.1)	(3.0)	–	(–3.0)	(5.946)	(82.8)
CrO_3	Solid	(18.1)	(4.0)	–	(–2.0)	(6.245)	(87.9)
	Liquid	(27)	–	–	–	(3.381)	(127.0)
	Gas	(20)	–	–	–	(–28.62)	(53.6)

For discussion of these coefficients, please see Table 75, Key to Tables of Thermodynamic Coefficients on page 257

Source: data from Weast, R. C., Ed., *Handbook of Chemistry and Physics, 55th ed.*, CRC Press, Cleveland, 1974, D-58.

Table 77. Thermodynamic Coefficients for Oxides

(Sheet 5 of 23)

Oxide	Phase	a –	b (cal • g mole^{-1})	c –	d –	A (kcal • g mole^{-1})	B (e.u.)
Cs_2O	Solid	(16.51)	(5.4)	–	–	(5.160)	(72.6)
	Liquid	(22)	–	–	–	(3.205)	(99.0)
Cs_2O_2	Solid	(21.4)	(11.4)	–	–	(6.887)	(85.3)
	Liquid	(29.5)	–	–	–	(4.125)	(123.8)
Cs_2O_3	Solid	(24.0)	(22.6)	–	–	(8.160)	(96.5)
	Liquid	(35)	–	–	–	(2.148)	(142.2)
Cu_2O	Solid	(13.4)	(8.6)	–	–	(4.378)	(96.0)
	Liquid	(21.5)	–	–	–	(3.721)	(54.9)
CuO	Solid	14.34	6.2	–	–	4.551	61.11
	Liquid	(22)	–	–	–	(–4.339)	(98.91)
FeO	Solid	9.27	4.80	–	–	(2.977)	(43.8)
	Liquid	(14.5)	–	–	–	(–3.721)	(69.2)
Fe_3O_4	Solid, α	12.38	1.62	–	–0.38	3.826	58.3
	Solid, β	(14.5)	–	–	–	(–2.399)	(66.7)

For discussion of these coefficients, please see Table 75, Key to Tables of Thermodynamic Coefficients on page 257

Source: data from Weast, R. C., Ed., *Handbook of Chemistry and Physics, 55th ed.*, CRC Press, Cleveland, 1974, D-58.

Table 77. THERMODYNAMIC COEFFICIENTS FOR OXIDES

(SHEET 6 OF 23)

Oxide	Phase	a –	b (cal • g mole^{-1})	c –	d –	A (kcal • g mole^{-1})	B (e.u.)
Fe_2O_3	Solid, α						
	Solid, β	21.88	48.20	–	–	8.666	104.0
	Solid, γ	48.00	18.6	–	–	12.652	238.3
Ga_2O	Solid	23.49	18.6	–	–3.55	9.021	119.9
	Liquid	36.00	–	–	–	11.979	187.6
	Gas	31.71	1.8	–	–	8.467	159.7
Ga_2O_3	Solid	(13.8)	–	–	–	(4.497)	(58.7)
GeO	Liquid	(21.5)	–	–	–	(–0.559)	(94.1)
	Solid	(14)	–	–	–	(–28.06)	(22.3)
	Gas	11.77	25.2	–	–	(4.630)	(54.35)
GeO_2	Solid (α,β)	(35.5)	–	–	–	(–20.66)	(173.2)
	Liquid	(10.4)	(2.6)	–	(–0.5)	(3.3)	(47.8)

For discussion of these coefficients, please see Table 75, Key to Tables of Thermodynamic Coefficients on page 257
Source: data from Weast, R. C., Ed., *Handbook of Chemistry and Physics, 55th ed.*, CRC Press, Cleveland, 1974, D-58.

Table 77. THERMODYNAMIC COEFFICIENTS FOR OXIDES

(SHEET 7 OF 23)

Oxide	Phase	a –	b (cal • g mole^{-1})	c –	d –	A (kcal • g mole^{-1})	B (e.u.)
In_2O	Solid	(14.7)	(7.8)	–	–	(4.730)	(58.1)
	Liquid	(22)	–	–	–	(3.206)	(92.6)
	Gas	(15)	–	–	–	(–18.39)	(25.8)
InO	Solid	(10.0)	(3.2)	–	–	(3.124)	(43.4)
	Liquid	(14)	–	–	–	(1.615)	(64.9)
	Gas	(9.0)	–	–	–	(–68.38)	(–3.1)
In_2O_3	Solid	(22.6)	(6.0)	–	–	(7.005)	(100.5)
	Liquid	(35)	–	–	–	(–0.195)	(172.8)
Ir_2O_3	Solid	(21.8)	(14.4)	–	–	(7.140)	(102.0)
	Liquid	(35)	–	–	–	(0.706)	(170.3)
	Gas	(20)	(10)	–	–	(–57.73)	(54.8)
IrO_2	Solid	9.17	15.20	–	–	3.410	40.9
K_2O	Solid	(15.9)	(6.4)	–	–	(5.025)	(69.5)
	Liquid	(22)	–	–	–	(1.130)	(98.3)

For discussion of these coefficients, please see Table 75, Key to Tables of Thermodynamic Coefficients on page 257

Source: data from Weast, R. C., Ed., *Handbook of Chemistry and Physics, 55th ed.*, CRC Press, Cleveland, 1974, D-58.

Table 77. Thermodynamic Coefficients for Oxides
(Sheet 8 of 23)

Oxide	Phase	a –	b (cal • g $mole^{-1}$)	c –	d –	A (kcal • g $mole^{-1}$)	B (e.u.)
K_2O_2	Solid	(20.8)	(5.4)	–	–	(6.442)	(93.1)
	Liquid	(29)	–	–	–	(4.127)	(134.2)
	Gas	(20)	–	–	–	(–57.07)	(41.7)
K_2O_3	Solid	(19.1)	(23.2)	–	–	(6.750)	(82.2)
	Liquid	(35.5)	–	–	–	(6.447)	(164.7)
	Gas	(20)	(5.0)	–	–	(–31.29)	(37.3)
KO_2	Solid	(15.0)	(12.0)	–	–	(5.006)	(61.1)
	Liquid	(24)	–	–	–	(3.424)	(105.5)
La_2O_3	Solid	28.86	3.076	–	–3.275	9.840	(130.7)
Li_2O	Solid	(11.4)	(5.4)	–	–	(3.639)	(57.5)
	Liquid	(21)	–	–	–	(–1.961)	(112.7)
Li_2O_2	Solid	(17.0)	(5.4)	–	–	(5.309)	(82.0)

For discussion of these coefficients, please see Table 75, Key to Tables of Thermodynamic Coefficients on page 257

Source: data from Weast, R. C., Ed., *Handbook of Chemistry and Physics, 55th ed.*, CRC Press, Cleveland, 1974, D-58.

Table 77. THERMODYNAMIC COEFFICIENTS FOR OXIDES

(SHEET 9 OF 23)

Oxide	Phase	a –	b (cal • g mole^{-1})	c –	d –	A (kcal • g mole^{-1})	B (e.u.)
MgO	Solid	10.86	1.197	–	–2.087	3.991	57.0
MgO_2	Solid	(12.1)	(2A)	–	–	(3.714)	(49.2)
MnO	Solid	11.11	1.94	–	–0.88	3.689	50.10
	Liquid	(13.5)	–	–	–	(–8.543)	(58.02)
Mn_3O_4	Solid, α	34.64	10.82	–	–2.20	11.312	166.3
	Solid, β	50.20	–	–	–	17.376	260.4
	Liquid	(49)	–	–	–	(–17.86)	(233.4)
Mn_2O_3	Solid	24.73	8.38	–	–3.23	8.829	118.8
MnO_2	Solid	16.60	2.44	–	–3.88	6.359	84.8
MoO_2	Solid	(16.2)	(3.0)	–	(–3.0)	(5.973)	(80.4)
	Liquid	(23)	–	–	–	(–2.463)	(118.4)

For discussion of these coefficients, please see Table 75, Key to Tables of Thermodynamic Coefficients on page 257

Source: data from Weast, R. C., Ed., *Handbook of Chemistry and Physics, 55th ed.*, CRC Press, Cleveland, 1974, D-58.

Table 77. THERMODYNAMIC COEFFICIENTS FOR OXIDES

(SHEET 10 OF 23)

Oxide	Phase	a –	b (cal • g mole^{-1})	c –	d –	A (kcal • g mole^{-1})	B (e.u.)
MoO_3	Solid	13.6	13.5	–	–	4.655	62.83
	Liquid	(28.4)	–	–	–	(0.222)	(139.88)
	Gas	(18.1)	–	–	–	(–48.54)	(42.8)
N_2O	Gas	(10.92)	2.06	–	–2.04	4.032	11.40
Na_2O	Solid	15.70	5.40	–	–	4.921	73.7
	Liquid	(22)	–	–	–	(1.494)	(105.9)
Na_2O_2	Solid	(20.2)	(3.8)	–	–	(6.192)	(93.6)
NaO_2	Solid	(16.2)	(3.6)	–	–	(4.990)	(65.7)
	Liquid	(23)	–	–	–	(3.175)	(100.9)
	Gas	(15)	–	–	–	(–35.22)	(22.0)
NbO	Solid	(9.6)	(4.4)	–	–	(3.058)	(44.0)

For discussion of these coefficients, please see Table 75, Key to Tables of Thermodynamic Coefficients on page 257
Source: data from Weast, R. C., Ed., *Handbook of Chemistry and Physics, 55th ed.*, CRC Press, Cleveland, 1974, D-58.

Table 77. THERMODYNAMIC COEFFICIENTS FOR OXIDES

(SHEET 11 OF 23)

Oxide	Phase	a –	b (cal • g mole^{-1})	c –	d –	A (kcal • g mole^{-1})	B (e.u.)
NbO_2	Solid	(17.1)	(1.6)	–	(–2.8)	(6.109)	(84.6)
	Liquid	(24)	–	–	–	(1.033)	(127.2)
Nb_2O_5	Solid	21.88	28.2	–	–	7.776	100.3
	Liquid	(44.2)	–	–	–	(–24.09)	(201.6)
Nd_2O_3	Solid	28.99	5.760	–	(–4.159)	10.295	(133.9)
NiO	Solid	13.69	0.83	–	–2.915	5.097	70.67
	Liquid	(14.3)	–	–	–	(–7.861)	(67.91)
NpO_2	Solid	(17.7)	(3.2)	–	(–2.6)	(6.292)	(84.08)
Np_2O_5	Solid	(32.4)	(12.6)	–	–	(10.22)	(145.4)
OsO_2	Solid	(11.5)	(6.0)	–	–	(3.696)	(52.8)

For discussion of these coefficients, please see Table 75, Key to Tables of Thermodynamic Coefficients on page 257
Source: data from Weast, R. C., Ed., *Handbook of Chemistry and Physics, 55th ed.*, CRC Press, Cleveland, 1974, D-58.

Table 77. THERMODYNAMIC COEFFICIENTS FOR OXIDES

(SHEET 12 OF 23)

Oxide	Phase	a –	b (cal • g mole^{-1})	c –	d –	A (kcal • g mole^{-1})	B (e.u.)
OsO_4	Solid	(16.4)	(23.1)	–	(–2.4)	(6.726)	(67.0)
	Liquid	(33)	–	–	–	(6.612)	(143.0)
	Gas	16.46	8.60	–	–4.6	(–7.644)	(25.3)
P_2O_3	Liquid	(34.5)	–	–	–	(10.287)	(162.6)
	Gas	(153)	(10)	–	–	(–1.953)	(38.0)
PO_2	Solid	(11.3)	(5.0)	–	–	(3.591)	(54.4)
	Liquid	(20)	–	–	–	(3.640)	(95.9)
P_2O_5	Solid	8.375	5.40	–	–	4.897	30.3
	Gas	36.80	–	–	–	3.284	165.6
PaO_2	Solid	(14.4)	(2.6)	–	–	(4.409)	(65.0)
Pa_2O_5	Solid	(28.4)	(11.4)	–	–	(8.975)	(127.7)
	Liquid	(48)	–	–	–	(–0.800)	(241.1)

For discussion of these coefficients, please see Table 75, Key to Tables of Thermodynamic Coefficients on page 257
Source: data from Weast, R. C., Ed., *Handbook of Chemistry and Physics, 55th ed.*, CRC Press, Cleveland, 1974, D-58.

Table 77. Thermodynamic Coefficients for Oxides

(Sheet 13 of 23)

Oxide	Phase	a –	b (cal • g mole^{-1})	c –	d –	A (kcal • g mole^{-1})	B (e.u.)
PbO	Solid, red	10.60	4.00	–	–	3.338	45.4
	Solid, yellow	9.05	6.40	–	–	2.454	36.4
	Liquid	(14.6)	–	–	–	1.788	65.7
	Gas	(8.1)	(0.4)	–	–	(–59.94)	(–11.0)
Pb_2O_4	Solid	(31.1)	(17.6)	–	–	(10.055)	(132.0)
PbO_2	Solid	12.7	7.80	–	–	4.133	56.4
PdO	Solid	3.30	14.2	–	–	1.615	(13.9)
PoO_2	Solid	(14.3)	(5.6)	–	–	(4.513)	(66.1)
	Liquid	(22)	–	–	–	(3.460)	(106.5)
Pr_2O_3	Solid	(29.0)	(4.0)	–	(–4.0)	(10.166)	(133.2)
	Liquid	(36)	–	–	–	(–6.298)	(168.3)
PrO_2	Solid	(17.6)	(3.4)	–	(–2.8)	(6.338)	(85.9)

For discussion of these coefficients, please see Table 75, Key to Tables of Thermodynamic Coefficients on page 257

Source: data from Weast, R. C., Ed., *Handbook of Chemistry and Physics, 55th ed.*, CRC Press, Cleveland, 1974, D-58.

Table 77. THERMODYNAMIC COEFFICIENTS FOR OXIDES

(SHEET 14 OF 23)

Oxide	Phase	a –	b (cal • g mole^{-1})	c –	d –	A (kcal • g mole^{-1})	B (e.u.)
PtO	Solid	(9.0)	(6.4)	–	–	(2.968)	(39.7)
Pt_3O_4	Solid	(30.8)	(17.4)	–	–	(9.957)	(139.7)
PtO_2	Solid	(11.1)	(9.6)	–	–	(3.736)	(49.6)
	Liquid	(21)	–	–	–	(3.785)	(101.5)
PuO	Solid	(12.0)	(2.4)	–	–	(3.685)	(49.1)
	Liquid	(14.5)	–	–	–	(–2.287)	(58.3)
	Gas	(8.9)	–	–	–	(–62.307)	(–5.3)
Pu_2O_3	Solid	(21.2)	(18.2)	–	–	(7.130)	(88.2)
	Liquid	(40)	–	–	–	(–5.691)	(187.2)
PuO_2	Solid	(17.1)	(3.4)	–	(–2.6)	(6.122)	(80.2)
	Liquid	(20.5)	–	–	–	(–10.62)	(92.2)
RaO	Solid	(10.5)	(2.0)	–	–	(3.220)	(43.4)

For discussion of these coefficients, please see Table 75, Key to Tables of Thermodynamic Coefficients on page 257
Source: data from Weast, R. C., Ed., *Handbook of Chemistry and Physics, 55th ed.*, CRC Press, Cleveland, 1974, D-58.

Table 77. Thermodynamic Coefficients for Oxides

(Sheet 15 of 23)

Oxide	Phase	a –	b (cal • g mole^{-1})	c –	d –	A (kcal • g mole^{-1})	B (e.u.)
Rb_2O	Solid	(15.4)	(5.8)	–	–	(4.850)	(62.5)
	Liquid	(22)	–	–	–	(2.754)	(95.9)
Rb_2O_2	Solid	(20.9)	(8.0)	–	–	(6.587)	(94.0)
	Liquid	(29)	–	–	–	(3.273)	(133.2)
Rb_2O_3	Solid	(20.5)	(13.0)	–	–	(6.690)	(88.2)
	Liquid	(34)	–	–	–	(5.603)	(157.8)
RbO_2	Solid	(13.8)	(6.4)	–	–	(4.399)	(59.0)
	Liquid	(21)	–	–	–	(3.720)	(95.7)
ReO_2	Solid	(10.8)	(9.8)	–	–	(3.656)	(49.5)
	Liquid	(24.5)	–	–	–	(1.204)	(127.0)
ReO_3	Solid	(18.0)	(5.8)	–	–	(5.625)	(84.5)
	Liquid	29	–	–	–	(4.644)	(136.8)

For discussion of these coefficients, please see Table 75, Key to Tables of Thermodynamic Coefficients on page 257

Source: data from Weast, R. C., Ed., *Handbook of Chemistry and Physics, 55th ed.*, CRC Press, Cleveland, 1974, D-58.

Table 77. Thermodynamic Coefficients for Oxides

(Sheet 16 of 23)

Oxide	Phase	a –	b (cal • g mole^{-1})	c –	d –	A (kcal • g mole^{-1})	B (e.u.)
Re_2O_7	Solid	(41.8)	(14.8)	–	(–3.0)	(14.127)	(200.3)
	Liquid	(65.7)	–	–	–	(9.203)	(314.7)
	Gas	(38.2)	–	–	–	(–25.97)	(109.3)
ReO_4	Solid	(21.4)	(10.8)	–	(–2.0)	(7.531)	(91.8)
	Liquid	(33)	–	–	–	(6.775)	(146.7)
	Gas	(16.5)	(8.6)	–	(–5.0)	(–8.118)	(30.6)
Rh_2O	Solid	15.59	6.47	–	–	4.936	(65.3)
RhO	Solid	(9.84)	(553)	–	–	(3.179)	(45.7)
Rh_2O_3	Solid	20.73	13.80	–	–	6.794	(99.2)
RuO_2	Solid	(11.4)	(6.0)	–	–	3.666	(54.2)
RuO_4	Solid	(20)	–	–	–	(5.963)	(81.5)
	Liquid	(33)	–	–	–	(6.663)	(144.9)
SO_2	Gas	11.4	1.414	–	–2.045	4.148	7.12

For discussion of these coefficients, please see Table 75, Key to Tables of Thermodynamic Coefficients on page 257
Source: data from Weast, R. C., Ed., *Handbook of Chemistry and Physics, 55th ed.*, CRC Press, Cleveland, 1974, D-58.

Table 77. THERMODYNAMIC COEFFICIENTS FOR OXIDES

(SHEET 17 OF 23)

Oxide	Phase	a –	b (cal • g mole^{-1})	c –	d –	A (kcal • g mole^{-1})	B (e.u.)
Sb_2O_3	Solid	19.10	17.1	–	–	6.455	84.5
	Liquid	(36)	–	–	–	(0.035)	(168.2)
	Gas	(20.8)	–	–	–	(–34.70)	(49.9)
SbO_2	Solid	11.30	8.1	–	–	3.725	51.6
Sb_2O_5	Solid	(22.4)	(23.6)	–	–	(7.723)	(104.8)
Sc_2O_3	Solid	23.17	5.64	–	–	7.159	1089
SeO	Solid	(9.1)	(3.8)	–	–	(2.882)	(42.0)
	Liquid	(15.5)	–	–	–	(0.490)	(77.5)
	Gas	8.20	0.50	–	–0.80	(–58.54)	(0.7)
SeO_2	Solid	(12.8)	(6.1)	–	(–0.2)	(4.150)	(59.9)
	Gas	(14.5)	–	–	–	(–20.45)	(26.4)
SiO	Solid	(7.3)	(2.4)	–	–	(2.283)	(35.8)

For discussion of these coefficients, please see Table 75, Key to Tables of Thermodynamic Coefficients on page 257

Source: data from Weast, R. C., Ed., *Handbook of Chemistry and Physics, 55th ed.*, CRC Press, Cleveland, 1974, D-58.

Table 77. THERMODYNAMIC COEFFICIENTS FOR OXIDES

(SHEET 18 OF 23)

Oxide	Phase	a –	b (cal • g mole^{-1})	c –	d –	A (kcal • g mole^{-1})	B (e.u.)
SiO_2	Solid, β	11.22	8.20	–	–2.70	4.615	57.83
	Solid, α	14.41	1.94	–	–	4.602	73.67
	Liquid	(20)	–	–	–	(9.649)	(111.08)
Sm_2O_3	Solid	(25.9)	(7.0)	–	–	(8.033)	(113.2)
	Liquid	(36)	–	–	–	(–6.431)	(166.3)
SnO	Solid	9.40	3.62	–	–	2.964	41.1
	Liquid	(14.5)	–	–	–	(0.141)	(68.1)
	Gas	(9.0)	–	–	–	(–69.76)	(–6.4)
SnO_2	Solid	17.66	2.40	–	–5.16	7.103	91.7
	Liquid	(22.5)	–	–	–	(0.304)	(117.7)
SrO	Solid	12.34	1.120	–	–1.806	4.335	58.7
SrO_2	Solid	(16.8)	(2.2)	–	(–3.0)	(6.113)	(83.3)

For discussion of these coefficients, please see Table 75, Key to Tables of Thermodynamic Coefficients on page 257
Source: data from Weast, R. C., Ed., *Handbook of Chemistry and Physics, 55th ed.*, CRC Press, Cleveland, 1974, D-58.

Table 77. THERMODYNAMIC COEFFICIENTS FOR OXIDES

(SHEET 19 OF 23)

Oxide	Phase	a –	b (cal • g mole^{-1})	c –	d –	A (kcal • g mole^{-1})	B (e.u.)
Ta_2O_5	Solid	29.2	10.0	–	–	9.151	135.2
	Liquid	(46)	–	–	–	(6.158)	(235.1)
TcO_2	Solid	(10.4)	(9.2)	–	–	(3.510)	(48.6)
	Liquid	(25)	–	–	–	(–5.946)	(132.7)
TcO_3	Solid	(19.4)	(5.2)	–	(–2.0)	(6.686)	(93.7)
Tc_2O_7	Solid	(39.1)	(18.6)	–	(–2.4)	(13.29)	(187.2)
	Liquid	(64)	–	–	–	(10.02)	(299.8)
	Gas	(25)	(28)	–	–	(–21.98)	(43.8)
TeO	Solid	(8.6)	(6.2)	–	–	(2.840)	(37.8)
	Liquid	(15.5)	–	–	–	(–0.448)	(72.3)
	Gas	(8.9)	–	–	–	(–62.16)	(–5.2)
TeO_3	Solid	13.85	6.87	–	–	4.435	63.97
	Liquid	(20)	–	–	–	(3.940)	(96.4)

For discussion of these coefficients, please see Table 75, Key to Tables of Thermodynamic Coefficients on page 257

Source: data from Weast, R. C., Ed., *Handbook of Chemistry and Physics, 55th ed.*, CRC Press, Cleveland, 1974, D-58.

Table 77. THERMODYNAMIC COEFFICIENTS FOR OXIDES

(SHEET 20 OF 23)

Oxide	Phase	a –	b (cal • g mole^{-1})	c –	d –	A (kcal • g mole^{-1})	B (e.u.)
TeO_2	Solid	(11.0)	(2.4)	–	–	(3.386)	(47.4)
	Liquid	(15)	–	–	–	(–6.561)	(66.9)
ThO_2	Solid	16.45	2.346	–	–2.124	5.721	80.03
TiO	Solid, α	10.57	3.60	–	–1.86	3.935	54.03
	Solid, β	11.85	3.00	–	–	4.108	61.71
Ti_2O_3	Solid, α	7.31	53.52	–	–	4.559	38.78
	Solid, β	34.68	1.30	–	–10.20	13.605	184.48
	Liquid	(37.5)	–	–	–	(–7.796)	(193.2)
Ti_3O_5	Solid, α	35.47	29.50	–	–	11.887	179.98
	Solid, β	41.60	8.00	–	–	10.230	202.80
	Liquid	(60)	–	–	–	(–18.701)	(306.4)

For discussion of these coefficients, please see Table 75, Key to Tables of Thermodynamic Coefficients on page 257
Source: data from Weast, R. C., Ed., *Handbook of Chemistry and Physics, 55th ed.*, CRC Press, Cleveland, 1974, D-58.

Table 77. THERMODYNAMIC COEFFICIENTS FOR OXIDES

(SHEET 21 OF 23)

Oxide	Phase	a –	b (cal • g mole^{-1})	c –	d –	A (kcal • g mole^{-1})	B (e.u.)
TiO_2	Solid	17.97	0.28	–	–4.35	6.829	92.92
	Liquid	(21.4)	–	–	–	(–2.610)	(111.08
Ti_2O	Solid	(15.8)	(6.0)	–	(–0.3)	(5.078)	(68.2)
	Liquid	(22.1)	–	–	–	(2.651)	(96.0)
	Gas	(13.7)	–	–	–	(–20.94)	(18.0)
Tl_2O_3	Solid	(23.0)	(5.0)	–	–	(7.080)	(99.0)
	Liquid	(35.5)	–	–	–	(4.604)	(167.8)
UO	Solid	(10.6)	(2.0)	–	–	(3.249)	(45.0)
UO_2	Solid	19.20	1.62	–	–3.957	7.124	93.37
U_3O_8	Solid	(65)	(7.5)	–	(–10.9)	(23.37)	(312.7)
UO_3	Solid	22.09	2.54	–	–2.973	7.969	104.72
VO	Solid	11.32	1.61	–	–1.26	3.869	56.4
	Liquid	(14.5)	–	–	–	(–8.157)	(70.9)

For discussion of these coefficients, please see Table 75, Key to Tables of Thermodynamic Coefficients on page 257

Source: data from Weast, R. C., Ed., *Handbook of Chemistry and Physics, 55th ed.*, CRC Press, Cleveland, 1974, D-58.

Table 77. THERMODYNAMIC COEFFICIENTS FOR OXIDES

(SHEET 22 OF 23)

Oxide	Phase	a –	b (cal • g mole^{-1})	c –	d –	A (kcal • g mole^{-1})	B (e.u.)
V_2O_3	Solid	29.35	4.76	–	–5.42	10.780	148.12
	Liquid	(38)	–	–	–	(–6.028)	(193.4)
V_3O_4	Solid	(36)	(30)	–	–	(12.07)	(182.1)
	Liquid	(55.6)	–	–	–	(–54.72)	(249.1)
VO_2	Solid, α	14.96	–	–	–	4.460	72.92
	Solid, β	17.85	1.70	–	–3.94	5.680	89.09
	Liquid	25.50	–	–	–	2.962	135.87
V_2O_5	Solid	46.54	–390	–	–13.22	18.136	240.2
	Liquid	45.60	–	–	–	2.122	220.1
	Gas	(40)	–	–	–	(–73.90)	(149.6)
WO_2	Solid	(17.6)	(4.2)	–	(–4.0)	(6.772)	(88.8)
	Liquid	(24)	–	–	–	(–0.112)	(121.8)

For discussion of these coefficients, please see Table 75, Key to Tables of Thermodynamic Coefficients on page 257

Source: data from Weast, R. C., Ed., *Handbook of Chemistry and Physics, 55th ed.*, CRC Press, Cleveland, 1974, D-58.

Table 77. THERMODYNAMIC COEFFICIENTS FOR OXIDES

(SHEET 23 OF 23)

Oxide	Phase	a –	b (cal • g mole^{-1})	c –	d –	A (kcal • g mole^{-1})	B (e.u.)
WO_3	Solid	17.33	7.74	–	–	5.511	81.15
	Liquid	(30)	–	–	–	(–1.162)	(152.5)
	Gas	(18)	–	–	–	(–69.36)	(40.2)
Y_2O_3	Solid	(26.0)	(8.2)	–	(–2.2)	(8.846)	(122.3)
ZnO	Solid	11.71	1.22	–	–2.18	4.277	57.88
ZrO_2	Solid, α	16.64	1.80	–	–3.36	6.168	85.21
	Solid, β	17.80	–	–	–	4.270	89.96

For discussion of these coefficients, please see Table 75, Key to Tables of Thermodynamic Coefficients on page 257

Source: data from Weast, R. C., Ed., *Handbook of Chemistry and Physics, 55th ed.*, CRC Press, Cleveland, 1974, D-58.

Table 78. ENTROPY OF THE ELEMENTS

(SHEET 1 OF 3)

Element	Phase	Entropy at 298K (e.u.)
Ac	solid	(13)
Ag	solid	10.20
Al	solid	6.769
Am	solid	(13)
As	solid	8.4
Au	solid	11.32
B	solid	1.42
Ba	solid, α	16
Be	solid	2.28
Bi	solid	13.6
C	solid	1.3609
Ca	solid, α	9.95
Cd	solid	12.3
Ce	solid	13.8
Cl_2	gas	53.286
Co	solid, α	6.8
Cr	solid	5.68
Cs	solid	19.8
Cu	solid	7.97
F_2	gas	48.58
Fe	solid, α	6.491
Ga	solid	9.82
Ge	solid	10.1
H_2	gas	31.211
Hf	solid	13.1
Hg	liquid	18.46
In	solid	13.88
Ir	solid	8.7

Source: data from Weast, R. C. Ed., *Handbook of Chemistry and Physics, 69th ed.*, CRC Press, Boca Raton, Fla., 1988, D44.

Table 78. ENTROPY OF THE ELEMENTS
(SHEET 2 OF 3)

Element	Phase	Entropy at 298K (e.u.)
K	solid	15.2
La	solid	13.7
Li	solid	6.70
Mg	solid	7.77
Mn	solid, α	7.59
Mo	solid	6.83
N_2	gas	45.767
Na	solid	12.31
Nb	solid	8.3
Nd	solid	13.9
Ni	solid, α	7.137
Np	solid	(14)
O_2	gas	49.003
Os	solid	7.8
P_4	solid, white	42.4
Pa	solid	(13.5)
Pb	solid	15.49
Pd	solid	8.9
Po	solid	13
Pr	solid	(13.5)
Pt	solid	10.0
Pu	solid	(13.0)
Ra	solid	(17)
Rb	solid	16.6
Re	solid	(8.89)
Rh	solid	7.6
Ru	solid, α	6.9
S	solid, α	7.62

Source: data from Weast, R. C. Ed., *Handbook of Chemistry and Physics, 69th ed.*, CRC Press, Boca Raton, Fla., 1988, D44.

Table 78. ENTROPY OF THE ELEMENTS
(SHEET 3 OF 3)

Element	Phase	Entropy at 298K (e.u.)
Sb	solid (α, β, γ)	10.5
Sc	solid	(9.0)
Se	solid	10.144
Si	solid	4.50
Sm	solid	(15)
Sn	solid (α, β)	12.3
Sr	solid	13.0
Ta	solid	9.9
Tc	solid	(8.0)
Te	solid, α	11.88
Th	solid	12.76
Ti	solid, α	7.334
Tl	solid, α	15.4
U	solid, α	12.03
V	solid	7.05
W	solid	8.0
Y	solid	(11)
Zn	solid	9.95
Zr	solid, α	9.29

Source: data from Weast, R. C. Ed., *Handbook of Chemistry and Physics, 69th ed.*, CRC Press, Boca Raton, Fla., 1988, D44.

Table 79. Vapor Pressure of the Elements at Very Low Pressures (Sheet 1 of 2)

Element	Melting point (°C)	Pressure (mm Hg)					
		10^{-5}	10^{-4}	10^{-3}	10^{-2}	10^{-1}	1
Ag	961	767	848	936	1047	1184	1353
Al	660	724	808	889	996	1123	1279
Au	1063	1083	1190	1316	1465	1646	1867
Ba	717	418	476	546	629	730	858
Be	1284	942	1029	1130	1246	1395	1582
Bi	271	474	536	609	698	802	934
C		2129	2288	2471	2681	2926	3214
Cd	321	148	180	220	264	321	
Co	1478	1249	1362	1494	1649	1833	2056
Cr	1900	907	992	1090	1205	1342	1504
Cu	1083	946	1035	1141	1273	1432	1628
Fe	1535	1094	1195	1310	1447	1602	1783
Hg	−38.9	−23.9	−5.5	18.0	48.0	82.0	126
In	157	667	746	840	952	1088	1260
Ir	2454	1993	2154	2340	2556	2811	3118
Mg	651	287	331	383	443	515	605
Mn	1244	717	791	878	980	1103	1251
Mo	2622	1923	2095	2295	2533		
Ni	1455	1157	1257	1371	1510	1679	1884
Os	2697	2101	2264	2451	2667	2920	3221

To convert **mm Hg** (torr) to N/m^2, divide by **133.3**

To convert **atm** to MN/m^2, divide by **0.1013**

This table lists the temperature in degrees Celsius (Centigrade) at which an element has a vapor pressure indicated by the headings of the columns.

The values given in this table are from a variety of sources that are not always in agreement; for that reason, the table should be used only as a general guide.

Source: from Dushman, S., *Scientific Foundations of Vacuum Technique*, John Wiley & Sons, New York, (1949)

Table 79. VAPOR PRESSURE OF THE ELEMENTS AT VERY LOW PRESSURES (SHEET 2 OF 2)

Element	Melting point (°C)	Pressure (mm Hg)					
		10^{-5}	10^{-4}	10^{-3}	10^{-2}	10^{-1}	1
Pb	328	483	548	625	718	832	975
Pd	1555	1156	1271	1405	1566	1759	2000
Pt	1774	1606	1744	1904	2090	2313	2582
Sb	630	466	525	595	678	779	904
Si	1410	1024	1116	1223	1343	1485	1670
Sn	232	823	922	1042	1189	1373	1609
Ta	2996	2407	2599	2820			
W	3382	2554	2767	3016	3309		
Zn	419	211	248	292	343	405	
Zr	2127	1527	1660	1816	2001	2212	2459

To convert **mm Hg** (torr) to N/m^2, divide by **133.3**

To convert **atm** to MN/m^2, divide by **0.1013**

This table lists the temperature in degrees Celsius (Centigrade) at which an element has a vapor pressure indicated by the headings of the columns.

The values given in this table are from a variety of sources that are not always in agreement; for that reason, the table should be used only as a general guide.

Source: from Dushman, S., *Scientific Foundations of Vacuum Technique*, John Wiley & Sons, New York, (1949)

Table 80. VAPOR PRESSURE OF THE ELEMENTS AT MODERATE PRESSURES (SHEET 1 OF 3)

Element	Symbol	Pressure (mm Hg)				
		1	10	100	400	760
Aluminum	Al	1540	1780	2080	2320	2467
Antimony	Sb		960	1280	1570	1750
Arsenic	As	380	440	510	580	610
Barium	Ba	860	1050	1300	1520	1640
Beryllium	Be	1520	1860	2300	2770	2970
Bismuth	Bi		1060	1280	1450	1560
Boron	B	2660	3030	3460	3810	4000
Bromine	Br	–60	–30	+9	39	59
Cadmium	Cd	393	486	610	710	765
Calcium	Ca	800	970	1200	1390	1490
Cesium	Cl		373	513	624	690
Chlorine	Cl	–123	–101	–71	–46	–34
Chromium	Cr	1610	1840	2140	2360	2480
Cobalt	Co	1910	2170	2500	2760	2870
Copper	Cu		1870	2190	2440	2600
Fluorine	F			–203	–193	–188
Gallium	Ca	1350	1570	1850	2060	2180
Germanium	Ge		2080	2440	2710	2830
Gold	Au	1880	2160	2520	2800	2940
Indium	In				1960	2080

To convert **mm Hg** (torr) to N/m^{2,} divide by **133.3**
To convert **atm** to **MN/m^2**, divide by **0.1013**

This table lists the temperature in degrees Celsius (Centigrade) at which an element has a vapor pressure indicated by the headings of the columns.

The values given in this table are from a variety of sources that are not always in agreement; for that reason, the table should be used only as a general guide.

Source: from Dushman, S., *Scientific Foundations of Vacuum Technique*, John Wiley & Sons, New York, (1949)

Table 80. VAPOR PRESSURE OF THE ELEMENTS AT MODERATE PRESSURES (SHEET 2 OF 3)

Element	Symbol	Pressure (mm Hg)				
		1	10	100	400	760
Iodine	I	40	72	115	160	185
Iridium	Ir	2830	3170	3630	3960	4130
Iron	Fe	1780	2040	2370	2620	2750
Lanthanum	La				3230	3420
Lead	Pb	970	1160	1420	1630	1740
Lithium	Li	750	890	1080	1240	1310
Magnesium	Mg	620	740	900	1040	1110
Manganese	Mn		1510	1810	2050	2100
Mercury	Hg			260	330	356.9
Molybdenum	Mo	3300	3770	4200	4580	4830
Neodymium	Nd				2870	3100
Nickel	Ni	1800	2090	2370	2620	2730
Palladium	Pd	1470	2290	2670	2950	3140
Phosphorus	P		127	199	253	283
Platinum	Pt	2600	2940	3360	3650	3830
Polonium	Po	472	587	752	890	960
Potassium	K			590	710	770
Rhodium	Rh	2530	2850	3260	3590	3760
Rubidium	Rb		390	527	640	700
Selenium	Se		429	547	640	685

To convert **mm Hg** (torr) to N/m^2, divide by **133.3**
To convert **atm** to MN/m^2, divide by **0.1013**

This table lists the temperature in degrees Celsius (Centigrade) at which an element has a vapor pressure indicated by the headings of the columns.

The values given in this table are from a variety of sources that are not always in agreement; for that reason, the table should be used only as a general guide.

Source: from Dushman, S., *Scientific Foundations of Vacuum Technique*, John Wiley & Sons, New York, (1949)

Table 80. VAPOR PRESSURE OF THE ELEMENTS AT MODERATE PRESSURES (SHEET 3 OF 3)

Element	Symbol	Pressure (mm Hg)				
		1	10	100	400	760
Silver	Ag	1310	1540	1850	2060	2210
Sodium	Na	440	546	700	830	890
Strontium	Sr	740	900	1100	1280	1380
Sulfur	S		246	333	407	445
Tellurium	Te	520	633	792	900	962
Thallium	Tl		1000	1210	1370	1470
Tin	Sn	1610	1890	2270	2580	2750
Titanium	Ti	2180	2480	2860	3100	3260
Tungsten	W	3980	4490	5160	5470	5940
Uranium	U	2450	2800	3270	3620	3800
Vanadium	V	2290	2570	2950	3220	3380
Zinc	Zn		590	730	840	907

To convert **mm Hg** (torr) to N/m^2, divide by **133.3**
To convert **atm** to **MN/m^2**, divide by **0.1013**

This table lists the temperature in degrees Celsius (Centigrade) at which an element has a vapor pressure indicated by the headings of the columns.

The values given in this table are from a variety of sources that are not always in agreement; for that reason, the table should be used only as a general guide.

Source: from Dushman, S., *Scientific Foundations of Vacuum Technique*, John Wiley & Sons, New York, (1949)

Table 81. Vapor Pressure of the Elements at High Pressures (Sheet 1 of 3)

Element	Symbol	Pressure (atm)				
		2	5	10	20	40
Aluminum	Al	2610	2850	3050	3270	3530
Antimony	Sb	1960	2490			
Arsenic	As					
Barium	Ba	1790	2030	2230		
Beryllium	Be	3240	3730	4110	4720	5610
Bismuth	Bi	1660	1850	2000	2180	
Boron	B					
Bromine	Br	78	110			
Cadmium	Cd	830	930	1030	1120	1240
Calcium	Ca	1630	1850	2020	2290	
Cesium	Cl					
Chlorine	Cl	–17	+9	30	55	97
Chromium	Cr	2630	2850	3010	3180	
Cobalt	Co	3040	3270			
Copper	Cu	2760	3010	3500	3460	3740
Fluorine	F	–180.7	–169.1	–159.6		
Gallium	Ca	2320	2560	2730		
Germanium	Ge	2970	3200	3430		
Gold	Au	3120	3490	3630	3890	
Indium	In	2230	2440	2600		
Iodine	I	216	265			
Iridium	Ir	4310	4650			
Iron	Fe	2900	3150	3360	3570	
Lanthanum	La	3620	3960	4270		

To convert **atm** to $\mathbf{MN/m^2}$ divide by **0.1013**

This table lists the temperature in degrees Celsius (Centigrade) at which an element has a vapor pressure indicated by the headings of the columns.

Source: from Loebel, R., in *Handbook of Chemistry and Physics, 55th ed.*, Weast, R. C., Ed., CRC Press, Cleveland, (1974)

Table 81. VAPOR PRESSURE OF THE ELEMENTS AT HIGH PRESSURES (SHEET 2 OF 3)

Element	Symbol	Pressure (atm)				
		2	5	10	20	40
Lead	Pb	1880	2140	2320	2620	
Lithium	Li	1420	1518			
Magnesium	Mg	1190	1330	1430	1560	
Manganese	Mn	2360	2580	2850		
Mercury	Hg	398	465	517	581	657
Molybdenum	Mo	5050	5340	5680	5980	
Neodymium	Nd	3300	3680	3990		
Nickel	Ni	2880	3120	3300	3310	
Palladium	Pd	3270	3560	3840		
Phosphorus	P	319				
Platinum	Pt	4000	4310	4570	4860	
Polonium	Po	1060	1200	1340		
Potassium	K	850	950	1110	1240	1420
Rhodium	Rh	3930	4230	4440		
Rubidium	Rb					
Selenium	Se	750	850	920	1010	1120
Silver	Ag	2360	2600	2850	3050	3300
Sodium	Na	980	1120	1230	1370	
Strontium	Sr	1480	1670	1850	2030	
Sulfur	S	493	574	640	720	
Tellurium	Te	1030	1160	1250		
Thallium	Tl	1560	1750	1900	2050	2260
Tin	Sn	2950	3270	3540	3890	
Titanium	Ti	3400	3650	3800		

To convert **atm** to **MN/m^2** divide by **0.1013**

This table lists the temperature in degrees Celsius (Centigrade) at which an element has a vapor pressure indicated by the headings of the columns.

Source: from Loebel, R., in *Handbook of Chemistry and Physics, 55th ed.*, Weast, R. C., Ed., CRC Press, Cleveland, (1974)

Table 81. VAPOR PRESSURE OF THE ELEMENTS AT HIGH PRESSURES (SHEET 3 OF 3)

Element	Symbol	Pressure (atm)				
		2	5	10	20	40
Tungsten	W	6260	6670	7250	7670	
Uranium	U	4040	4420			
Vanadium	V	3540	3800			
Zinc	Zn	970	1090	1180	1290	

To convert **atm** to MN/m^2 divide by **0.1013**

This table lists the temperature in degrees Celsius (Centigrade) at which an element has a vapor pressure indicated by the headings of the columns.

Source: from Loebel, R., in *Handbook of Chemistry and Physics, 55th ed.*, Weast, R. C., Ed., CRC Press, Cleveland, (1974)

Table 82. VAPOR PRESSURE OF ELEMENTS AND INORGANIC COMPOUNDS* (SHEET 1 OF 5)

Compound	Formula	a	b	Temperature. Range of Validity °C
Aluminum Oxide	Al_2O_3	540000	14.22	1840 to 2200 liq.
Ammonia	NH_3	31211	9.9974	–127 to –78 sol.
Ammonium Bromide	NH_4Br	90208	9.9404	250 to 400 sol.
Ammonium Chloride	NH_4Cl	83486	10.0164	100 to 400 sol.
Ammonium Cyanide	NH_4CN	41484	9.978	7 to 17 sol.
Ammonium Iodide	NH_4I	95730	10.2700	300 to 400 sol.
Ammonium Sulfhydrate	NH_4HS	46025	10.7500	6 to 40 sol.
Antimony	Sb	189000	9.051	1070 to 1325 liq.
Argon	Ar	7814.5	7.5741	–208 to –189 sol.
		6826	6.9605	–189 to –183 liq.
Arsenic	As	47100	6.692	800 to 860 liq.
		133000	10.800	440 to 815 sol.
Arsenous Oxide	As_2O_3	111350	12.127	100 to 315 sol.
		52120	6.513	315 to 490 liq.
Barium	Ba	350000	15.765	930 to 1130 liq.
Bismuth	Bi	200000	8.876	1210 to 1420 liq.
Bismuth Trichloride	$BiCl_3$	13125	2.681	91 to 213 sol.
Cadimium	Cd	10900	8.564	150 to 320.9 sol..
		99900	7.897	500 to 840 liq
Cadimium Iodide	CdI_2	122200	9.269	385 to 450 liq.
Cesium	Cs	73400	6.949	200 to 350 liq.
Cesium Chloride	CsCl	163200	8.340	986 to 1295 liq.
Calcium	Ca	370000	16.240	960 to 1110 liq.
Carbon	C	540000	9.596	3880 to 4430 liq.
Carbon Dioxide	CO_2	26179.3	9.9082	–135 to –56.7 liq.
Carbon Monooxide	CO	6354	6.976	–220 to –206 liq.
Chlorine	Cl	29293	9.950	–154 to –103 liq.
Cobalt	Co	309000	7.571	2375 liq.

Source: data compiled by J.S. Park *from CRC Handbook of Chemistry and Physics,* David R. Lide, Ed., CRC Press, Boca Raton, (1990).

Table 82. VAPOR PRESSURE OF ELEMENTS AND INORGANIC COMPOUNDS* (SHEET 2 OF 5)

Compound	Formula	a	b	Temperature. Range of Validity °C
Copper	Cu	468000	12.344	2100 to 2310 liq.
Cuprous Chloride	Cu_2Cl_2	80700	5.454	878 to 1369 liq.
Cyanogen	$(CN)_2$	32437	9.6539	–72 to –28 sol.
		23750	7.808	–32 to –6 liq.
Ferrous Chloride	$FeCl_2$	135200	8.33	700 to 390 sol.
Gold	Au	385000	9.853	2315 to 2500 liq.
Hydriodic Acid	HI	24160	8.259	–97 to –51 sol.
		21580	7.630	–50 to –34 liq.
Hydrobromic Acid	HBr	22420	8.734	–114 to –86 sol.
		17960	7.427	–86 ot –66 liq.
Hydrochloric Acid	HCl	19588	8.4430	–158 to –110 sol.
Hydrocyanic Acid	HCN	27830	7.7446	–8 to 27 liq.
Hydrofluoric Acid	HF	25180	7.370	–83 to 48 liq.
Hydrogen Peroxide	H_2O_2	48530	8.853	10 to 90 liq.
Hydrogen Sulfide	H_2S	20690	7.880	–110 to –83 sol.
Iron	Fe	309000	7.482	2220 to 2450 liq.
Krypton	Kr	10065	7.1770	–189 to –169 sol.
		9377	6.92387	–169 to –150 liq.
Lead	Pb	188500	7.827	525 to 1325 liq.
Lead Bromide	$PbBr_2$	118000	7.827	735 to 918 liq.
Lead Chloride	$PbCl_2$	141900	8.961	500 to 950 liq.
Lithium Bromide	LiBr	152700	8.068	1010 to 1265 liq.
Lithium Chloride	LiCl	155900	7.939	1045 to 1325 liq.
Lithium Fluoride	LiF	218400	8.753	1398 to 1666 liq.
Lithium Iodide	LiI	143600	8.011	940 to 1140 liq.

Source: data compiled by J.S. Park *from CRC Handbook of Chemistry and Physics*, David R. Lide, Ed., CRC Press, Boca Raton, (1990).

Table 82. Vapor Pressure of Elements and Inorganic Compounds* (Sheet 3 of 5)

Compound	Formula	a	b	Temperature. Range of Validity °C
Magnesium	Mg	260000	12.993	900 to 1070 liq.
Magnase	Mn	267000	9.300	1510 to 1900 liq.
Mercuric Bromide	$HgBr_2$	79800	10.181	111 to 235 sol.
		61250	8.284	238 to 331 lig.
Mercuric Chloride	$HgCl_2$	85030	10.888	60 to 130 sol.
		78850	10.094	130 to 270 sol.
		61020	8.409	275 to 309 liq.
Mercuric Iodide	HgI_2	82340	10.057	100 to 250 sol.
		62770	8.115	266 to 360 liq.
Mercury	Hg	73000	10.383	–80 to –38.87 sol.
		58700	7.752	400 to 1300 liq.
Molybdenum	Mo	680000	10.844	1800 to 2240 sol.
Nitrogen	N_2	6881.3	7.66558	–215 to –210 sol.
Nitrogen Dioxide	NO	16423	10.048	–200 to –161 sol.
		13040	8.440	–163.7 to –148 liq.
Nitrogen Monoxide	N_2O	23590	9.579	–144 to –90 sol.
		16440	7.535	–90.1 to –88.7 liq.
Nitrogen Pentoxide	N_2O_5	57180	12.647	–30 to 30 sol.
Nitrogen Tetroxide	N_2O_4	55160	13.400	–100 to –40 sol.
		45440	11.214	–40 to –10 sol.
		33430	8.814	–8 to 43.2 liq.
Nitrogen Trioxide	N_2O_3	39400	10.30	–25 to 0 liq.
Phosphorus (White)	P	63123	9.6511	20 to 44.1 sol.
Phosphorus (Violet)	P	108510	11.0842	380 to 590 sol.
Platinum	Pt	486000	7.786	1425 to 1765 sol.
Potassium	K	84900	7.183	260 to 760 liq.

Source: data compiled by J.S. Park *from CRC Handbook of Chemistry and Physics,* David R. Lide, Ed., CRC Press, Boca Raton, (1990).

Table 82. VAPOR PRESSURE OF ELEMENTS AND INORGANIC COMPOUNDS* (SHEET 4 OF 5)

Compound	Formula	a	b	Temperature. Range of Validity °C
Potassium Bromide	KBr	168100	8.2470	906 to 1063 liq.
		163800	7.936	1095 to 1375 liq.
Potassium Chloride	KCl	174500	8.3526	906 to 1105 liq.
		169700	8.130	1116 to 1428 liq.
Potassium Flouride	KF	207500	9.000	1278 to 1500 liq.
Potassium Hydroxide	KOH	136000	7.330	1170 to 1327 liq.
Potassium Iodide	KI	157600	8.0957	843 to 1028 liq.
		155700	7.949	1063 to 1333 liq.
Rubidium	Rb	76000	6.976	250 to 370 liq.
Rubidium Chloride	RbCl	198600	9.111	1142 to 1395 liq.
Silicon	Si	170000	5.950	1200 to 1320 sol.
Silicon Dioxide	SiO_2	506000	13.43	1860 to 2230 liq.
Silver	Ag	250000	8.762	1650 to 1950 liq.
Silver Chloride	AgCl	185500	8.179	1255 to 1442 liq.
Sodium	Na	103300	7.553	180 to 883 liq.
Sodium Bromide	NaBr	161600	7.948	1138 to 1394 liq.
Sodium Chloride	NaCl	180300	8.3297	976 to 1155 liq.
		185800	8.548	1156 to 1430 liq.
Sodium Cyanide	NaCN	155520	7.472	800 to 1360 liq.
Sodium Fluoride	NaF	218200	8.640	1562 to 1701 liq.
Sodium Hydroxide	NaOH	132000	7.030	1010 to 1042 liq.
Sodium Iodide	NaI	165100	8.371	1063 to 1307 liq.
Stannic Chloride	$SnCl_4$	46740	9.824	–52 to –38 liq.
Stronium	Sr	360000	16.056	940 to 1140 liq.

Source: data compiled by J.S. Park *from CRC Handbook of Chemistry and Physics*, David R. Lide, Ed., CRC Press, Boca Raton, (1990).

Table 82. VAPOR PRESSURE OF ELEMENTS AND INORGANIC COMPOUNDS* (SHEET 5 OF 5)

Compound	Formula	a	b	Temperature. Range of Validity °C
Sulfur Dioxide	SO_2	35827	10.5916	–95 to –75 liq.
Sulfur Trioxide	SO_3	43450	10.022	24 to 48 liq.
Thallium	Tl	120000	6.140	950 to 1200 liq.
Thallium Chloride	TlCl	105200	7.947	665 to 807 liq.
Tin	Sn	328000	9.643	1950 to 2270 liq.
Tungsten	W	897000	9.920	2230 to 2770 liq.
Zinc	Zn	133000	9.200	250 to 491.4 sol.
		118000	8.108	600 to 985 liq.

Source: data compiled by J.S. Park *from CRC Handbook of Chemistry and Physics,* David R. Lide, Ed., CRC Press, Boca Raton, (1990).

* The vapor pressure with respect to temperature may be represented by the following equation:

$$\log_{10} p = -0.05223a/T + b$$

where p is the pressure in mm of mercury of the saturated vapor at the absolute temperature T. (T = t°C + 273.1) The values obtained by the use of the equation given above are valid within the temperature ranges indicated for each of the compounds.

Table 83. VALUES OF THE ERROR FUNCTION

(SHEET 1 OF 2)

z	erf (z)
0.00	0.0000
0.01	0.0113
0.02	0.0226
0.03	0.0338
0.04	0.0451
0.05	0.0564
0.10	0.1125
0.15	0.1680
0.20	0.2227
0.25	0.2763
0.30	0.3286
0.35	0.3794
0.40	0.4284
0.45	0.4755
0.50	0.5205
0.55	0.5633
0.60	0.6039
0.65	0.6420
0.70	0.6778
0.75	0.7112
0.80	0.7421
0.85	0.7707
0.90	0.7969
0.95	0.8209
1.00	0.8427
1.10	0.8802
1.20	0.9103
1.30	0.9340

Source: from: *Handbook of Mathematical Functions*, M. Abramowitz and I. A. Stegun, eds., National Bureau of Standards, Applied Mathematics Series 55, Washington, D.C., 1972.

Table 83. VALUES OF THE ERROR FUNCTION

(SHEET 2 OF 2)

z	erf (z)
1.40	0.9523
1.50	0.9661
1.60	0.9763
1.70	0.9838
1.80	0.9891
1.90	0.9928
2.00	0.9953

Source: from: *Handbook of Mathematical Functions*, M. Abramowitz and I. A. Stegun, eds., National Bureau of Standards, Applied Mathematics Series 55, Washington, D.C., 1972.

Table 84. DIFFUSION IN METALLIC SYSTEMS*

(SHEET 1 OF 34)

Metal	Tracer	Crystalline Form	Purity (%)	Temperature Range (°C)	Activation energy, Q (kcal • mol^{-1})	Frequency factor, D_o (cm^2 • s^{-1})
Aluminum	Ag^{110}	S	99.999	371–655	27.83	0.118
	Al^{27}	S		450–650	34.0	1.71
	Au^{198}	S	99.999	423–609	27.0	0.077
	Cd^{115}	S	99.999	441–631	29.7	1.04
	Ce^{141}	P	99.995	450–630	26.60	1.9×10^{-6}
	Co^{60}	S	99.999	369–655	27.79	0.131
	Cr^{51}	S	99.999	422–654	41.74	464
	Cu^{64}	S	99.999	433–652	32.27	0.647
	Fe^{59}	S	99.99	550–636	46.0	135
	Ga^{72}	S	99.999	406–652	29.24	0.49
	Ge^{71}	S	99.999	401–653	28.98	0.481
	In^{114}	P	99.99	400–600	27.6	0.123

Source: data from Askill, J.,in *Handbook of Chemistry and Physics, 55th ed.*,Weast, R.C., Ed., CRC Press, Cleveland,1974, F61.

Table 84. DIFFUSION IN METALLIC SYSTEMS*

(SHEET 2 OF 34)

Metal	Tracer	Crystalline Form	Purity (%)	Temperature Range (°C)	Activation energy, Q (kcal • mol^{-1})	Frequency factor, D_o (cm^2 • s^{-1})
Aluminum (con't)	La^{140}	P	99.995	500–630	27.0	1.4×10^{-6}
	Mn^{54}	P	99.99	450–650	28.8	0.22
	Mo^{99}	P	99.995	400–630	13.1	1.04×10^{-9}
	Nb^{95}	P	99.95	350–480	19.65	1.66×10^{-7}
	Nd^{147}	P	99.995	450–630	25.0	4.8×10^{-7}
	Ni^{63}	P	99.99	360–630	15.7	2.9×10^{-8}
	Pd^{103}	P	99.995	400–630	20.2	1.92×10^{-7}
	Pr^{142}	P	99.995	520–630	23.87	3.58×10^{-7}
	Sb^{124}	P		448–620	29.1	0.09
	Sm^{153}	P	99.995	450–630	22.88	3.45×10^{-7}
	Sn^{113}	P		400–600	28.5	0.245
	V^{48}	P	99.995	400–630	19.6	6.05×10^{-8}

Source: data from Askill, J.,in *Handbook of Chemistry and Physics, 55th ed.,*Weast, R.C., Ed., CRC Press, Cleveland,1974, F61.

Table 84. DIFFUSION IN METALLIC SYSTEMS*

(SHEET 3 OF 34)

Metal	Tracer	Crystalline Form	Purity (%)	Temperature Range (°C)	Activation energy, Q ($kcal \cdot mol^{-1}$)	Frequency factor, D_o ($cm^2 \cdot s^{-1}$)
Aluminum (con't)	Zn^{65}	S	99.999	357–653	28.86	0.259
Beryllium	Ag^{110}	S⊥c	99.75	650–900	43.2	1.76
	Ag^{110}	S∥c	99.75	650–900	39.3	0.43
	Be^{7}	S⊥c	99.75	565–1065	37.6	0.52
	Be^{7}	S∥c	99.75	565–1065	39.4	0.62
	Fe^{59}	S	99.75	700–1076	51.6	0.67
	Ni^{63}	P		800–1250	58.0	0.2
Cadmium	Ag^{110}	S	99.99	180–300	25.4	2.21
	Cd^{115}	S	99.95	110–283	19.3	0.14
	Zn^{65}	S	99.99	180–300	19.0	0.0016

Source: data from Askill, J., in *Handbook of Chemistry and Physics, 55th ed.*, Weast, R.C., Ed., CRC Press, Cleveland, 1974, F61.

Table 84. DIFFUSION IN METALLIC SYSTEMS*

(SHEET 4 OF 34)

Metal	Tracer	Crystalline Form	Purity (%)	Temperature Range (°C)	Activation energy, Q (kcal • mol^{-1})	Frequency factor, D_o ($cm^2 \cdot s^{-1}$)
Calcium	C^{14}		99.95	550–800	29.8	3.2×10^{-5}
	Ca^{45}		99.95	500–800	38.5	8.3
	Fe^{59}		99.95	500–800	23.3	2.7×10^{-3}
	Ni^{63}		99.95	550–800	28.9	1.0×10^{-6}
	U^{235}		99.95	500–700	34.8	1.1×10^{-5}
Carbon	Ag^{110}	⊥c		750–1050	64.3	9280
	C^{14}			2000–2200	163	5
	Ni^{63}	⊥c		540–920	47.2	102
	Ni^{63}	‖c		750–1060	53.3	2.2
	Th^{228}	⊥c		1400–2200	145.4	1.33×10^{-5}
	Th^{228}	‖c		1800–2200	114.7	2.48
	U^{232}	⊥c		140~2200	115.0	6760
	U^{232}	‖c		1400 1820	129.5	385

Source: data from Askill, J.,in *Handbook of Chemistry and Physics, 55th ed.*,Weast, R.C., Ed., CRC Press, Cleveland,1974, F61.

Table 84. Diffusion in Metallic Systems*

(Sheet 5 of 34)

Metal	Tracer	Crystalline Form	Purity (%)	Temperature Range (°C)	Activation energy, Q ($kcal \cdot mol^{-1}$)	Frequency factor, D_o ($cm^2 \cdot s^{-1}$)
Chromium	C^{14}	P		120–1500	26.5	9.0×10^{-3}
	Cr^{51}	P	99.98	1030–1545	73.7	0.2
	Fe^{59}	P	99.8	980–1420	79.3	0.47
	Mo^{99}	P		1100–1420	58.0	2.7×10^{-3}
Cobalt	C^{14}	P	99.82	600–1400	34.0	0.21
	Co^{60}	P	99.9	1100–1405	67.7	0.83
	Fe^{59}	P	99.9	1104–1303	62.7	0.21
	Ni^{63}	P		1192–1297	60.2	0.10
	S^{35}	P	99.99	1150–1250	5.4	1.3
Copper	Ag^{110}	S, P		580–980	46.5	0.61
	As^{76}	P		810–1075	42.13	0.20
	Au^{193}	S, P		400–1050	42.6	0.03
	Cd^{115}	S	99.98	725–950	45.7	0.935

Source: data from Askill, J.,in *Handbook of Chemistry and Physics, 55th ed.,*Weast, R.C., Ed., CRC Press, Cleveland,1974, F61.

Table 84. DIFFUSION IN METALLIC SYSTEMS*

(SHEET 6 OF 34)

Metal	Tracer	Crystalline Form	Purity (%)	Temperature Range (°C)	Activation energy, Q ($kcal \cdot mol^{-1}$)	Frequency factor, D_o ($cm^2 \cdot s^{-1}$)
Copper (Con't)	Ce^{141}	P	99.999	766–947	27.6	2.17×10^{-3}
	Cr^{51}	S, P		800–1070	53.5	1.02
	Co^{60}	S	99.998	701–1077	54.1	1.93
	Cu^{67}	S	99.999	698–1061	50.5	0.78
	Eu^{152}	P	99.999	750–970	26.85	1.17×10^{-7}
	Fe^{59}	S. P		460–1070	52.0	1.36
	Ga^{72}			–	45.90	0.55
	Ge^{68}	S	99.998	653–1015	44.76	0.397
	Hg^{203}	P		–	44.0	0.35
	Lu^{177}	P	99.999	857–1010	26.15	4.3×10^{-9}
	Mn^{54}	S	99.99	754–950	91.4	10^{7}
	Nb^{95}	P	99.999	807–906	60.06	2.04

Source: data from Askill, J.,in *Handbook of Chemistry and Physics, 55th ed.,*Weast, R.C., Ed., CRC Press, Cleveland,1974, F61.

Table 84. DIFFUSION IN METALLIC SYSTEMS*

(SHEET 7 OF 34)

Metal	Tracer	Crystalline Form	Purity (%)	Temperature Range (°C)	Activation energy, Q (kcal • mol^{-1})	Frequency factor, D_o (cm^2 • s^{-1})
Copper (Con't)	Ni^{63}	P		620–1080	53.8	1.1
	Pd^{102}	S	99.999	807–1056	54.37	1.71
	Pm^{147}	P	99.999	720–955	27.5	3.62×10^{-8}
	Pt^{195}	P		843–997	37.5	4.8×10^{-4}
	S^{35}	S	99.999	800–1000	49.2	23
	Sb^{124}	S	99.999	600–1000	42.0	0.34
	Sn^{113}	P		680–910	45.0	0.11
	Tb^{160}	P	99.999	770–980	27.45	8.96×10^{-9}
	Tl^{204}	S	99.999	785–996	43.3	0.71
	Tm^{170}	P	99.999	705–950	24.15	7.28×10^{-9}
	Zn^{65}	P	99.999	890–1000	47.50	0.73

Source: data from Askill, J.,in *Handbook of Chemistry and Physics, 55th ed.*,Weast, R.C., Ed., CRC Press, Cleveland,1974, F61.

Table 84. DIFFUSION IN METALLIC SYSTEMS*

(SHEET 8 OF 34)

Metal	Tracer	Crystalline Form	Purity (%)	Temperature Range (°C)	Activation energy, Q ($kcal \cdot mol^{-1}$)	Frequency factor, D_o ($cm^2 \cdot s^{-1}$)
Germanium	Cd^{115}	S		750–950	102.0	1.75×10^9
	Fe^{59}	S		775–930	24.8	0.13
	Ge^{71}	S		766–928	68.5	7.8
	In^{114}	S		600–920	39.9	2.9×10^{-4}
	Sb^{124}	S		720–900	50.2	0.22
	Te^{125}	S		770–900	56.0	2.0
	Tl^{204}	S		800–930	78.4	1700
Gold	Ag^{110}	S	99.99	699–1007	40.2	0.072
	Au^{198}	S	99.97	850–1050	42.26	0.107
	Co^{60}	P	99.93	702–948	41.6	0.068
	Fe^{59}	P	99.93	701–948	41.6	0.082

Source: data from Askill, J.,in *Handbook of Chemistry and Physics, 55th ed.*,Weast, R.C., Ed., CRC Press, Cleveland,1974, F61.

Table 84. Diffusion in Metallic Systems*

(Sheet 9 of 34)

Metal	Tracer	Crystalline Form	Purity (%)	Temperature Range (°C)	Activation energy, Q ($kcal \cdot mol^{-1}$)	Frequency factor, D_o ($cm^2 \cdot s^{-1}$)
Gold (Con't)	Hg^{203}	S	99.994	600–1027	37.38	0.116
	Ni^{63}	P	99.96	880–940	46.0	0.30
	Pt^{195}	P, S	99.98	800–1060	60.9	7.6
β–Hafnium	Hf^{181}	P	97.9	1795–1995	38.7	1.2×10^{-3}
Indium	Ag^{110}	S⊥c	99.99	25–140	12.8	0.52
	Ag^{110}	S‖c	99.99	25–140	11.5	0.11
	Au^{198}	S	99.99	25–140	6.7	9×10^{-3}
	In^{114}	S⊥c	99.99	44–144	18.7	3.7
	In^{114}	S‖c	99.99	44–144	18.7	2.7
	Tl^{204}	S	99.99	49–157	15.5	0.049

Source: data from Askill, J.,in *Handbook of Chemistry and Physics, 55th ed.,*Weast, R.C., Ed., CRC Press, Cleveland,1974, F61.

Table 84. DIFFUSION IN METALLIC SYSTEMS*

(SHEET 10 OF 34)

Metal	Tracer	Crystalline Form	Purity (%)	Temperature Range (°C)	Activation energy, Q ($kcal \cdot mol^{-1}$)	Frequency factor, D_o ($cm^2 \cdot s^{-1}$)
α-Iron	Ag^{110}	P		748–888	69.0	1950
	Au^{198}	P	99.999	800–900	62.4	31
	C^{14}	P	99.98	616–844	29.3	2.2
	Co^{60}	P	99.995	638–768	62.2	7.19
	Cr^{51}	P	99.95	775–875	57.5	2.53
	Cu^{64}	P	99.9	800 1050	57.0	0.57
	Fe^{55}	P	99.92	809–889	60.3	5.4
	K^{42}	P	99.92	500–800	42.3	0.036
	Mn^{54}	P	99.97	800–900	52.5	0.35
	Mo^{99}	P		750–875	73.0	7800
	Ni^{63}	P	99.97	680–800	56.0	1.3
	P^{32}	P		860–900	55.0	2.9

Source: data from Askill, J.,in *Handbook of Chemistry and Physics, 55th ed.*,Weast, R.C., Ed., CRC Press, Cleveland,1974, F61.

Table 84. DIFFUSION IN METALLIC SYSTEMS*

(SHEET 11 OF 34)

Metal	Tracer	Crystalline Form	Purity (%)	Temperature Range (°C)	Activation energy, Q (kcal • mol^{-1})	Frequency factor, D_o (cm^2 • s^{-1})
α-Iron (Con't)	Sb^{124}	P		800–900	66.6	1100
	V^{48}	P		755–875	55.4	1.43
	W^{185}	P		755–875	55.1	0.29
γ-Iron	Be^{7}	P	99.9	1100–1350	57.6	0.1
	C^{14}	P	99.34	800–1400	34.0	0.15
	Co^{60}	P	99.98	1138–1340	72.9	1.25
	Cr^{51}	P	99.99	950–1400	69.7	10.8
	Fe^{59}	P	99.98	1171–1361	67.86	0.49
	Hf^{181}	P	99.99	1110–1360	97.3	3600
	Mn^{54}	P	99.97	920–1280	62.5	0.16
	Ni^{63}	P	99.97	930–2050	67.0	0.77

Source: data from Askill, J.,in *Handbook of Chemistry and Physics, 55th ed.*,Weast, R.C., Ed., CRC Press, Cleveland,1974, F61.

Table 84. DIFFUSION IN METALLIC SYSTEMS*

(SHEET 12 OF 34)

Metal	Tracer	Crystalline Form	Purity (%)	Temperature Range (°C)	Activation energy, Q (kcal • mol^{-1})	Frequency factor, D_o (cm^2 • s^{-1})
γ-Iron (Con't)	P^{32}	P	99.99	950–1200	43.7	0.01
	S^{35}	P		900–1250	53.0	1.7
	V^{48}	P	99.99	1120–1380	69.3	0.28
	W^{185}	P	99.5	1050–1250	90.0	1000
δ-Iron	Co^{60}	P	99.995	1428–1521	61.4	6.38
	Fe^{59}	P	99.95	1428–1492	57.5	2.01
	P^{32}	P	99.99	1370–1460	55.0	2.9
Lanthanum	Au^{198}	P	99.97	600–800	45.1	1.5
	La^{140}	P	99.97	690–850	18.1	2.2×10^{-2}
Lead	Ag^{110}	P	99.9	200–310	14.4	0.064
	Au^{198}	S	99.999	190–320	10.0	8.7×10^{-3}
	Cd^{115}	S	99.999	150–320	21.23	0.409

Source: data from Askill, J.,in *Handbook of Chemistry and Physics, 55th ed.,*Weast, R.C., Ed., CRC Press, Cleveland,1974, F61.

Table 84. DIFFUSION IN METALLIC SYSTEMS*

(SHEET 13 OF 34)

Metal	Tracer	Crystalline Form	Purity (%)	Temperature Range (°C)	Activation energy, Q (kcal • mol^{-1})	Frequency factor, D_o (cm^2 • s^{-1})
Lead (Con't)	Cu^{64}	S		150–320	14.44	0.046
	Pb^{204}	S	99.999	150–320	25.52	0.887
	Tl^{205}	P	99.999	207–322	24.33	0.511
Lithium	Ag^{110}	P	92.5	65–161	12.83	0.37
	Au^{195}	P	92.5	47–153	10.49	0.21
	Bi	P	99.95	141–177	47.3	5.3×10^{13}
	Cd^{115}	P	92.5	80–174	16.05	2.35
	Cu^{64}	P	99.98	51–120	9.22	0.47
	Ga^{72}	P	99.98	58–173	12.9	0.21
	Hg^{203}	P	99.98	58–173	14.18	1.04
	In^{114}	P	92.5	80–175	15.87	0.39

Source: data from Askill, J.,in *Handbook of Chemistry and Physics, 55th ed.,*Weast, R.C., Ed., CRC Press, Cleveland,1974, F61.

Table 84. DIFFUSION IN METALLIC SYSTEMS*

(SHEET 14 OF 34)

Metal	Tracer	Crystalline Form	Purity (%)	Temperature Range (°C)	Activation energy, Q ($kcal \cdot mol^{-1}$)	Frequency factor, D_o ($cm^2 \cdot s^{-1}$)
Lithium (Con't)	Li^{6}	P	99.98	35–178	12.60	0.14
	Na^{22}	P	92.5	52–176	12.61	0.41
	Pb^{204}	P	99.95	129–169	25.2	160
	Sb^{124}	P	99.95	141–176	41.5	1.6×10^{10}
	Sn^{113}	P	99.95	108–174	15.0	0.62
	Zn^{65}	P	92.5	60–175	12.98	0.57
Magnesium	Ag^{110}	P	99.9	476–621	28.50	0.34
	Fe^{59}	P	99.95	400–600	21.2	4×10^{-6}
	In^{114}	P	99.9	472–610	28.4	5.2×10^{-2}
	Mg^{28}	S⊥c		467–635	32.5	1.5

Source: data from Askill, J.,in *Handbook of Chemistry and Physics, 55th ed.,*Weast, R.C., Ed., CRC Press, Cleveland,1974, F61.

Table 84. DIFFUSION IN METALLIC SYSTEMS*

(SHEET 15 OF 34)

Metal	Tracer	Crystalline Form	Purity (%)	Temperature Range (°C)	Activation energy, Q ($kcal \cdot mol^{-1}$)	Frequency factor, D_o ($cm^2 \cdot s^{-1}$)
Magnesium (Con't)	Mg^{28}	S‖c		467–635	32.2	1.0
	Ni^{63}	P	99.95	400 600	22.9	1.2×10^{-5}
	U^{235}	P	99.95	500–620	27.4	1.6×10^{-5}
	Zn^{65}	P	99.9	467–620	28.6	0.41
Molybdenum	C^{14}	P	99.98	1200–1600	41.0	2.04×10^{-2}
	Co^{60}	P	99.98	1850–2350	106.7	18
	Cr^{51}	P		1000–1500	54.0	2.5 x 10–4
	Cs^{134}	S	99.99	1000–1470	28.0	8.7×10^{-11}
	K^{42}	S		800–1100	25.04	5.5×10^{-9}
	Mo^{99}	P		1850–2350	96.9	0.5
	Na^{24}	S		800–1100	21.25	2.95×10^{-9}
	Nb^{95}	P	99.98	1850–2350	108.1	14

Source: data from Askill, J.,in *Handbook of Chemistry and Physics, 55th ed.,*Weast, R.C., Ed., CRC Press, Cleveland,1974, F61.

Table 84. DIFFUSION IN METALLIC SYSTEMS*

(SHEET 16 OF 34)

Metal	Tracer	Crystalline Form	Purity (%)	Temperature Range (°C)	Activation energy, Q ($kcal \cdot mol^{-1}$)	Frequency factor, D_o ($cm^2 \cdot s^{-1}$)
Molybdenum (Con't)	P^{32}	P	99.97	2000–2200	80.5	0.19
	Re^{186}	P		1700–2100	94.7	0.097
	S^{35}	S	99.97	2220–2470	101.0	320
	Ta^{182}	P		1700–2150	83.0	3.5×10^{-4}
	U^{235}	P	99.98	1500–2000	76.4	7.6×10^{-3}
	W^{185}	P	99.98	1700–2260	110	1.7
Nickel	Au^{198}	S,P	99.999	700–1075	55.0	0.02
	Be^{7}	P	99.9	1020–1400	46.2	0.019
	C^{14}	P	99.86	600–1400	34.0	0.012
	Co^{60}	P	99.97	1149–1390	65.9	1.39

Source: data from Askill, J.,in *Handbook of Chemistry and Physics, 55th ed.*,Weast, R.C., Ed., CRC Press, Cleveland,1974, F61.

Table 84. DIFFUSION IN METALLIC SYSTEMS*

(SHEET 17 OF 34)

Metal	Tracer	Crystalline Form	Purity (%)	Temperature Range (°C)	Activation energy, Q ($kcal \cdot mol^{-1}$)	Frequency factor, D_o ($cm^2 \cdot s^{-1}$)
Nickel (Con't)	Cr^{51}	P	99.95	1100–1270	65.1	1.1
	Cu^{64}	P	99.95	1050–1360	61.7	0.57
	Fe^{59}	P		1020–1263	58.6	0.074
	Mo^{99}	P		900–1200	51.0	1.6×10^{-3}
	Ni^{63}	P	99.95	1042–1404	68.0	1.9
	Pu^{238}	P		1025–1125	51.0	0.5
	Sb^{124}	P	99.97	1020–1220	27.0	1.8×10^{-5}
	Sn^{113}	P	99.8	700–1350	58.0	0.83
	V^{48}	P	99.99	800–1300	66.5	0.87
	W^{185}	P	99.95	1100–1300	71.5	2.0

Source: data from Askill, J.,in *Handbook of Chemistry and Physics, 55th ed.*,Weast, R.C., Ed., CRC Press, Cleveland,1974, F61.

Table 84. DIFFUSION IN METALLIC SYSTEMS*

(SHEET 18 OF 34)

Metal	Tracer	Crystalline Form	Purity (%)	Temperature Range (°C)	Activation energy, Q (kcal • mol^{-1})	Frequency factor, D_o (cm^2 • s^{-1})
Niobium	C^{14}	P		800–1250	32.0	1.09×10^{-5}
	Co^{60}	P	99.85	1500–2100	70.5	0.74
	Cr^{51}	S		943–1435	83.5	0.30
	Fe^{51}	P	99.85	1400–2100	77.7	1.5
	K^{42}	S		900–1100	22.10	2.38×10^{-7}
	Nb^{95}	P, S	99.99	878–2395	96.0	1.1
	P^{32}	P	99.0	1300–1800	51.5	5.1×10^{-2}
	S^{35}	S	99.9	1100–1500	73.1	2600
	Sn^{113}	P	99.85	1850–2400	78.9	0.14
	Ta^{182}	P, S	99.997	878–2395	99.3	1.0

Source: data from Askill, J.,in *Handbook of Chemistry and Physics, 55th ed.*,Weast, R.C., Ed., CRC Press, Cleveland,1974, F61.

Table 84. DIFFUSION IN METALLIC SYSTEMS*

(SHEET 19 OF 34)

Metal	Tracer	Crystalline Form	Purity (%)	Temperature Range (°C)	Activation energy, Q ($kcal \cdot mol^{-1}$)	Frequency factor, D_o ($cm^2 \cdot s^{-1}$)
Niobium (Cont)	Ti^{44}	S		994–1492	86.9	0.099
	U^{235}	P	99.55	1500–2000	76.8	8.9×10^{-3}
	V^{48}	S	99.99	1000–1400	85.0	2.21
	W^{185}	P	99.8	1800–2200	91.7	5×10^{-4}
Palladium	Pd^{103}	S	99.999	1060–1500	63.6	0.205
Phosphorus	P^{32}	P		0–44	9.4	1.07×10^{-3}
Platinum	Co^{60}	P	99.99	900–1050	74.2	19.6
	Cu^{64}	P		1098–1375	59.5	0.074
	Pt^{195}	P	99.99	1325–1600	68.2	0.33

Source: data from Askill, J.,in *Handbook of Chemistry and Physics, 55th ed.*,Weast, R.C., Ed., CRC Press, Cleveland,1974, F61.

Table 84. DIFFUSION IN METALLIC SYSTEMS*

(SHEET 20 OF 34)

Metal	Tracer	Crystalline Form	Purity (%)	Temperature Range (°C)	Activation energy, Q (kcal • mol^{-1})	Frequency factor, D_o (cm^2 • s^{-1})
Potassium	Au^{198}	P	99.95	5.6–52.5	3.23	1.29×10^{-3}
	K^{42}	S	99.7	–52–61	9.36	0.16
	Na^{22}	P	99.7	0–62	7.45	0.058
	Rb^{86}	P	99.95	0.1–59.9	8.78	0.090
γ–Plutonium	Pu^{238}	P		190–310	16.7	2.1×10^{-5}
δ–Plutonium	Pu^{238}	P		350–440	23.8	4.5×10^{-3}
ε-Plutonium	Pu^{238}	P		500–612	18.5	2.0×10^{-2}
α-Praseodymium	Ag^{110}	P	99.93	610–730	25.4	0.14
	Au^{195}	P	99.93	650–780	19.7	4.3×10^{-2}
	Co^{60}	P	99.93	660–780	16.4	4.7×10^{-2}
	Zn^{65}	P	99.96	766–603	24.8	0.18

Source: data from Askill, J.,in *Handbook of Chemistry and Physics, 55th ed.,*Weast, R.C., Ed., CRC Press, Cleveland,1974, F61.

Table 84. DIFFUSION IN METALLIC SYSTEMS*
(SHEET 21 OF 34)

Metal	Tracer	Crystalline Form	Purity (%)	Temperature Range (°C)	Activation energy, Q (kcal • mol^{-1})	Frequency factor, D_o (cm^2 • s^{-1})
β-Praseodymium	Ag^{110}	P	99.93	800–900	21.5	3.2 x 10^{-2}
	Au^{195}	P	99.93	800–910	20.1	3.3 x 10^{-2}
	Ho^{166}	P	99.96	800–930	26.3	9.5
	In^{114}	P	99.96	800–930	28.9	9.6
	La^{140}	P	99.96	800–930	25.7	1.8
	Pr^{142}	P	99.93	800–900	29.4	8.7
	Zn^{65}	P	99.96	822–921	27.0	0.63
Selenium	Fe^{59}	P		40–100	8.88	—
	Hg^{203}	P	99.996	25–100	1.2	—
	S^{35}	S⊥c		60–90	29.9	1700
	S^{35}	S‖c		60–90	15.6	1100
	Se^{75}	P		35–140	11.7	1.4 x 10^{-4}

Source: data from Askill, J.,in *Handbook of Chemistry and Physics, 55th ed.*,Weast, R.C., Ed., CRC Press, Cleveland,1974, F61.

Table 84. DIFFUSION IN METALLIC SYSTEMS*

(SHEET 22 OF 34)

Metal	Tracer	Crystalline Form	Purity (%)	Temperature Range (°C)	Activation energy, Q ($kcal \cdot mol^{-1}$)	Frequency factor, D_o ($cm^2 \cdot s^{-1}$)
Silicon	Au^{198}	S		700–1300	47.0	2.75×10^{-3}
	C^{14}	P		1070–1400	67.2	0.33
	Cu^{64}	P		800–1100	23.0	4×10^{-2}
	Fe^{59}	S		1000–1200	20.0	6.2×10^{-3}
	Ni^{63}	P		450–800	97.5	1000
	P^{32}	S		1100–1250	41.5	–
	Sb^{124}	S		1190–1398	91.7	12.9
	Si^{31}	S	99.99999	1225–1400	110.0	1800
Silver	Au^{198}	P	99.99	718–942	48.28	0.85
	Ag^{110}	S	99.999	640–955	45.2	0.67
	Cd^{115}	S	99.99	592–937	41.69	0.44
	Co^{60}	S	99.999	700–940	48.75	1.9

Source: data from Askill, J.,in *Handbook of Chemistry and Physics, 55th ed.,*Weast, R.C., Ed., CRC Press, Cleveland,1974, F61.

Table 84. DIFFUSION IN METALLIC SYSTEMS*

(SHEET 23 OF 34)

Metal	Tracer	Crystalline Form	Purity (%)	Temperature Range (°C)	Activation energy, Q (kcal • mol^{-1})	Frequency factor, D_o (cm^2 • s^{-1})
Silver (Con't)	Cu^{64}	P	99.99	717–945	46.1	1.23
	Fe^{59}	S	99.99	720–930	49.04	2.42
	Ge^{77}	P		640–870	36.5	0.084
	Hg^{203}	P	99.99	653–948	38.1	0.079
	In^{114}	S	99.99	592–937	40.80	0.41
	Ni^{63}	S	99.99	749–950	54.8	21.9
	Pb^{210}	P		700–865	38.1	0.22
	Pd^{102}	S	99.999	736–939	56.75	9.56
	Ru^{103}	S	99.99	793–945	65.8	180
	S^{35}	S	99.999	600–900	40.0	1.65
	Sb^{124}	P	99.999	780–950	39.07	0.234
	Sn^{113}	S	99.99	592–937	39.30	0.255

Source: data from Askill, J.,in *Handbook of Chemistry and Physics, 55th ed.*,Weast, R.C., Ed., CRC Press, Cleveland,1974, F61.

Table 84. DIFFUSION IN METALLIC SYSTEMS*

(SHEET 24 OF 34)

Metal	Tracer	Crystalline Form	Purity (%)	Temperature Range (°C)	Activation energy, Q (kcal • mol^{-1})	Frequency factor, D_o (cm^2 • s^{-1})
Silver (Con't)	Te^{125}	P		770–940	38.90	0.47
	Tl^{204}	P		640–870	37.9	0.15
	Zn^{65}	S	99.99	640–925	41.7	0.54
Sodium	Au^{198}	P	99.99	1.0–77	2.21	3.34×10^{-4}
	K^{42}	P	99.99	0–91	8.43	0.08
	Na^{22}	P	99.99	0–98	10.09	0.145
	Rb^{86}	P	99.99	0–85	8.49	0.15
Tantalum	C^{14}	P		1450–2200	40.3	1.2×10^{-2}
	Fe^{59}	P		930–1240	71.4	0.505
	Mo^{99}	P		1750–2220	81.0	1.8×10^{-3}
	Nb^{95}	P, S	99.996	921–2484	98.7	0.23
	S^{35}	P	99.0	1970–2110	70.0	100
	Ta^{182}	P, S	99.996	1250–2200	98.7	1.24

Source: data from Askill, J.,in *Handbook of Chemistry and Physics, 55th ed.,*Weast, R.C., Ed., CRC Press, Cleveland,1974, F61.

Table 84. DIFFUSION IN METALLIC SYSTEMS*

(SHEET 25 OF 34)

Metal	Tracer	Crystalline Form	Purity (%)	Temperature Range (°C)	Activation energy, Q ($kcal \cdot mol^{-1}$)	Frequency factor, D_o ($cm^2 \cdot s^{-1}$)
Tellurium	Hg^{203}	P		270–440	18.7	3.14×10^{-5}
	Se^{75}	P		320–440	28.6	2.6×10^{-2}
	Tl^{204}	P		360–430	41.0	320
	Te^{127}	S⊥c	99.9999	300–400	46.7	3.91×10^{4}
	Te^{127}	S‖c	99.9999	300–400	35.5	130
α-Thallium	Ag^{110}	P⊥c	99.999	80–250	11.8	3.8×10^{-2}
	Ag^{110}	P‖c	99.999	80–250	11.2	2.7×10^{-2}
	Au^{198}	P⊥c	99.999	110–260	2.8	2.0×10^{-5}
	Au^{198}	P‖c	99.999	110–260	5.2	5.3×10^{-4}
	Tl^{204}	S⊥c	99.9	135–230	22.6	0.4
	Tl^{204}	S‖c	99.9	135–230	22.9	0.4

Source: data from Askill, J.,in *Handbook of Chemistry and Physics, 55th ed.*,Weast, R.C., Ed., CRC Press, Cleveland,1974, F61.

Table 84. DIFFUSION IN METALLIC SYSTEMS*

(SHEET 26 OF 34)

Metal	Tracer	Crystalline Form	Purity (%)	Temperature Range (°C)	Activation energy, Q (kcal • mol^{-1})	Frequency factor, D_o (cm^2 • s^{-1})
β-Thallium	Ag^{110}	P	99.999	230–310	11.9	4.2×10^{-2}
	Au^{198}	P	99.999	230–310	6.0	5.2×10^{-4}
	Tl^{204}	S	99.9	230–280	20.7	0.7
α-Thorium	Pa^{231}	P	99.85	770–910	74.7	126
	Th^{228}	P	99.85	720–880	716	395
	U^{233}	P	99.85	700–880	79.3	2210
Tin	Ag^{110}	S⊥c		135–225	18.4	0.18
	Ag^{110}	S‖c		135–225	12.3	7.1×10^{-3}
	Au^{198}	S⊥c		135–225	17.7	0.16
	Au^{198}	S‖c		135–225	11.0	5.8×10^{-3}
	Co^{60}	S,P		140–217	22.0	5.5
	In^{114}	S⊥c	99.998	181–221	25.8	34.1

Source: data from Askill, J.,in *Handbook of Chemistry and Physics, 55th ed.,*Weast, R.C., Ed., CRC Press, Cleveland,1974, F61.

Table 84. DIFFUSION IN METALLIC SYSTEMS*

(SHEET 27 OF 34)

Metal	Tracer	Crystalline Form	Purity (%)	Temperature Range (°C)	Activation energy, Q (kcal • mol^{-1})	Frequency factor, D_o (cm^2 • s^{-1})
Tin (Con't)	In^{114}	S‖c	99.998	181–221	25.6	12.2
	Sn^{113}	S⊥c	99.999	160–226	25.1	10.7
	Sn^{113}	S‖c	99.999	160–226	25.6	7.7
	Tl^{204}	P	99.999	137–216	14.7	1.2×10^{-3}
α-Titanium	Ti^{44}	P	99.99	700–850	35.9	8.6×10^{-6}
β-Titanium	Ag^{110}	P	99.95	940 1570	43.2	3×10^{-3}
	Be^{7}	P	99.96	915–1300	40.2	0.8
	C^{14}	P	99.62	1100–1600	20.0	3.02×10^{-3}
	Cr^{51}	P	99.7	950–1600	35.1	5×10^{-3}
	Co^{60}	P	99.7	900–1600	30.6	1.2×10^{-2}
	Fe^{59}	P	99.7	900–1600	31.6	7.8×10^{-3}
	Mo^{99}	P	99.7	900–1600	43.0	8.0×10^{-3}
	Mn^{54}	P	99.7	900–1600	33.7	6.1×10^{-3}

Source: data from Askill, J.,in *Handbook of Chemistry and Physics, 55th ed.,*Weast, R.C., Ed., CRC Press, Cleveland,1974, F61.

Table 84. DIFFUSION IN METALLIC SYSTEMS*

(SHEET 28 OF 34)

Metal	Tracer	Crystalline Form	Purity (%)	Temperature Range (°C)	Activation energy, Q ($kcal \cdot mol^{-1}$)	Frequency factor, D_o ($cm^2 \cdot s^{-1}$)
β-Titanium (Con't)	Nb^{95}	P	99.7	1000–1600	39.3	5.0×10^{-3}
	Ni^{63}	P	99.7	925–1600	29.6	9.2×10^{-3}
	P^{32}	P	99.7	950–1600	24.1	3.62×10^{-3}
	Sc^{46}	P	99.95	940–1590	32.4	4.0×10^{-3}
	Sn^{113}	P	99.7	950–1600	31.6	3.8×10^{-4}
	Ti^{44}	P	99.95	900–1540	31.2	3.58×10^{-4}
	U^{235}	P	99.9	900–400	29.3	5.1×10^{-4}
	V^{48}	P	99.95	900–1545	32.2	3.1×10^{-4}
	W^{185}	P	99.94	900–1250	43.9	3.6×10^{-3}
	Zr^{95}	P	98.94	920–1500	35.4	4.7×10^{-3}
Tungsten	C^{14}	P	99.51	1200–1600	53.5	8.91×10^{-3}
	Fe^{59}	P		940–1240	66.0	1.4×10^{-2}
	Mo^{99}	P		1700–2100	101.0	0.3

Source: data from Askill, J.,in *Handbook of Chemistry and Physics, 55th ed.*,Weast, R.C., Ed., CRC Press, Cleveland,1974, F61.

Table 84. DIFFUSION IN METALLIC SYSTEMS*

(SHEET 29 OF 34)

Metal	Tracer	Crystalline Form	Purity (%)	Temperature Range (°C)	Activation energy, Q (kcal • mol^{-1})	Frequency factor, D_o (cm^2 • s^{-1})
Tungsten (Con't)	Nb^{95}	P	99.99	1305–2367	137.6	3.01
	Re^{186}	S		2100–2400	141.0	19.5
	Ta^{182}	P	99.99	1305–2375	139.9	3.05
	W^{185}	P	99.99	1800–2403	140.3	1.88
α–Uranium	U^{234}	P		580–650	40.0	2×10^{-3}
β–Uranium	Co^{60}	P	99.999	692–763	27.45	1.5×10^{-2}
	U^{235}	P		690–750	44.2	2.8×10^{-3}
γ-Uranium	Au^{195}	P	99.99	785–1007	30.4	4.86×10^{-3}
	Co^{60}	P	99.99	783–989	12.57	3.51×10^{-4}
	Cr^{51}	P	99.99	797–1037	24.46	5.37×10^{-3}
	Cu^{64}	P	99.99	787–1039	24.06	1.96×10^{-3}

Source: data from Askill, J.,in *Handbook of Chemistry and Physics, 55th ed.*,Weast, R.C., Ed., CRC Press, Cleveland,1974, F61.

Table 84. DIFFUSION IN METALLIC SYSTEMS*

(SHEET 30 OF 34)

Metal	Tracer	Crystalline Form	Purity (%)	Temperature Range (°C)	Activation energy, Q ($kcal \cdot mol^{-1}$)	Frequency factor, D_o ($cm^2 \cdot s^{-1}$)
γ-Uranium (Con't)	Fe^{55}	P	99.99	787–990	12.0	2.69×10^{-4}
	Mn^{54}	P	99.99	787–939	13.88	1.81×10^{-4}
	Nb^{95}	P	99.99	791–1102	39.65	4.87×10^{-2}
	Ni^{63}	P	99.99	787–1039	15.66	5.36×10^{-4}
	U^{233}	P	99.99	800–1070	28.5	2.33×10^{-3}
	Zr^{95}	P		800–1000	16.5	3.9×10^{-4}
Vanadium	C^{14}	P	99.7	845–1130	27.3	4.9×10^{-3}
	Cr^{51}	P	99.8	960–1200	64.6	9.54×10^{-3}
	Fe^{59}	P		960–1350	71.0	0.373
	P^{32}	P	99.8	1200–1450	49.8	2.45×10^{-2}
	S^{35}	P	99.8	1320–1520	34.0	3.1×10^{-2}
	V^{48}	S,P	99.99	880–1360	73.65	0.36
	V^{48}	S,P	99.99	1360–1830	94.14	214.0

Source: data from Askill, J.,in *Handbook of Chemistry and Physics, 55th ed.,*Weast, R.C., Ed., CRC Press, Cleveland,1974, F61.

Table 84. DIFFUSION IN METALLIC SYSTEMS*

(SHEET 31 OF 34)

Metal	Tracer	Crystalline Form	Purity (%)	Temperature Range (°C)	Activation energy, Q (kcal • mol^{-1})	Frequency factor, D_o (cm^2 • s^{-1})
Yttrium	Y^{90}	S⊥c		900–1300	67.1	5.2
	Y^{90}	S‖c		900–1300	60.3	0.82
Zinc	Ag^{110}	S⊥c	99.999	271–413	27.6	0.45
	Ag^{110}	S‖c	99.999	271–413	26.0	0.32
	Au^{198}	S⊥c	99.999	315–415	29.72	0.29
	Au^{198}	S‖c	99.999	315–415	29.73	0.97
	Cd^{115}	S⊥c	99.999	225–416	20.12	0.117
	Cd^{115}	S‖c	99.999	225–416	20.54	0.114
	Cu^{64}	S⊥c	99.999	338–415	~20	~2
	Cu^{64}	S‖c	99.999	338–415	29.53	2.22

Source: data from Askill, J.,in *Handbook of Chemistry and Physics, 55th ed.*,Weast, R.C., Ed., CRC Press, Cleveland,1974, F61.

Table 84. Diffusion in Metallic Systems*

(Sheet 32 of 34)

Metal	Tracer	Crystalline Form	Purity (%)	Temperature Range (°C)	Activation energy, Q ($kcal \cdot mol^{-1}$)	Frequency factor, D_o ($cm^2 \cdot s^{-1}$)
Zinc (Con't)	Ga^{72}	S⊥c		240–403	18.15	0.018
	Ga^{72}	S‖c		240 403	18.4	0.016
	Hg^{203}	S⊥c		260–413	20.18	0.073
	Hg^{203}	S‖c		260–413	19.70	0.056
	In^{114}	S⊥c		271–413	19.60	0.14
	In^{114}	S‖c		271–413	19.10	0.062
	Sn^{113}	S⊥c		298–400	18.4	0.13
	Sn^{113}	S‖c		298–400	19.4	0.15
	Zn^{65}	S⊥c	99.999	240–418	23.0	0.18
	Zn^{65}	S‖c	99.999	240–418	21.9	0.13

Source: data from Askill, J.,in *Handbook of Chemistry and Physics, 55th ed.*,Weast, R.C., Ed., CRC Press, Cleveland,1974, F61.

Table 84. DIFFUSION IN METALLIC SYSTEMS*

(SHEET 33 OF 34)

Metal	Tracer	Crystalline Form	Purity (%)	Temperature Range (°C)	Activation energy, Q (kcal • mol^{-1})	Frequency factor, D_o (cm^2 • s^{-1})
α-Zirconium	Cr^{51}	P	99.9	700–850	18.0	1.19×10^{-8}
	Fe^{55}	P		750–840	48.0	2.5×10^{-2}
	Mo^{99}	P		600–850	24.76	6.22×10^{-8}
	Nb^{95}	P	99.99	740–857	31.5	6.6×10^{-6}
	Sn^{113}	P		300–700	22.0	1.0×10^{-8}
	Ta^{182}	P	99.6	700–800	70.0	100
	V^{48}	P	99.99	600–850	22.9	1.12×10^{-8}
	Zr^{95}	P	99.95	750–850	45.5	5.6×10^{-4}
β–Zirconium	Be^{7}	P	99.7	915–1300	31.1	8.33×10^{-2}
	C^{14}	P	96.6	1100–1600	34.2	3.57×10^{-2}
	Ce^{141}	P		880–1600	41.4	3.16
	Co^{60}	P	99.99	920–1600	21.82	3.26×10^{-3}

Source: data from Askill, J.,in *Handbook of Chemistry and Physics, 55th ed.*,Weast, R.C., Ed., CRC Press, Cleveland,1974, F61.

Table 84. DIFFUSION IN METALLIC SYSTEMS*

(SHEET 34 OF 34)

Metal	Tracer	Crystalline Form	Purity (%)	Temperature Range (°C)	Activation energy, Q ($kcal \cdot mol^{-1}$)	Frequency factor, D_o ($cm^2 \cdot s^{-1}$)
β–Zirconium (Con't)	Cr^{51}	P	99.9	700–850	18.0	1.19×10^{-8}
	Fe^{55}	P		750–840	48.0	2.5×10^{-2}
	Mo^{99}	P		900–1635	35.2	1.99×10^{-6}
	Nb^{95}	P		1230–1635	36.6	7.8×10^{-4}
	P^{32}	P	99.94	950–1200	33.3	0.33
	Sn^{113}	P		300–700	22.0	1×10^{-8}
	Ta^{182}	P	99.6	900–1200	27.0	5.5×10^{-5}
	U^{235}	P		900–1065	30.5	5.7×10^{-4}
	V^{48}	P	99.99	870–1200	45.8	7.59×10^{-3}
	V^{48}	P	99.99	1200–1400	57.7	0.32
	W^{185}	P	99.7	900–1250	55.8	0.41
	Zr^{95}	P		1100–1500	30.1	2.4×10^{-4}

Source: data from Askill, J.,in *Handbook of Chemistry and Physics, 55th ed.*,Weast, R.C., Ed., CRC Press, Cleveland,1974, F61.

* The diffusion coefficient D_T at a temperature T(K) is given by the following:

$$D_T = D_o\, e^{-Q/RT}$$

For activation energy in **KJ/mol**, multiply values in **Kcal/mol** by **4.184**.
For frequency factor in **m²/s**, multiply values in **cm²/s** by **10^{-4}**.

Abbreviations:

P= polycrystalline
S = single crystal
⊥ c = perpendicular to c direction
|| c = parallel to c direction

Table 85. DIFFUSION OF METALS INTO METALS

(SHEET 1 OF 11)

Diffusing Metal	Matrix Metal	Diffusion Temperature (°C)	Coefficient ($cm^2 \cdot hr^{-1}$)
Ag	Al	466	$6.84–8.1 \times 10^{-7}$
		500	$7.2–3.96 \times 10^{-8}$
		573	1.26×10^{-5}
	Pb	220	5.40×10^{-5}
		250	1.08×10^{-4}
		285	3.29×10^{-4}
	Sn	500	1.73×10^{-1}
Al	Cu	500	6.12×10^{-9}
		850	7.92×10^{-6}
As	Si		$0.32\ e^{-82,000/RT}$
Au	Ag	456	1.76×10^{-9}
		491	$0.92–2.38 \times 10^{-13}$
		585	3.6×10^{-8}
		601	3.96×10^{-8}
		624	$2.5–5 \times 10^{-11}$
		717	$1.04–2.25 \times 10^{-9}$
		729	1.76×10^{-9}
		767	1.15×10^{-6}

For diffusion coefficients in m^2/s, multiply values in cm^2/hr by 2.778×10^{-8}.

Source: data from Loebel, R., in *Handbook of Chemistry and Physics, 51st ed.*, Weast, R. C., Ed., Chemical Rubber, Cleveland, 1970, F-55.

Table 85. DIFFUSION OF METALS INTO METALS

(SHEET 2 OF 11)

Diffusing Metal	Matrix Metal	Diffusion Temperature (°C)	Coefficient ($cm^2 \cdot hr^{-1}$)
Au (Con't)	Ag (Con't)	847	2.30×10^{-6}
		858	3.63×10^{-8}
		861	3.92×10^{-8}
		874	3.92×10^{-8}
		916	5.40×10^{-6}
		1040	1.17×10^{-6}
		1120	2.29×10^{-5}
		1189	5.42×10^{-6}
	Au	800	1.17×10^{-8}
		900	9×10^{-8}
		1020	5.4×10^{-7}
	Bi	500	1.88×10^{-1}
	Cu	970	5.04×10^{-6}
	Hg	11	3×10^{-2}
	Pb	100	8.28×10^{-8}
		150	1.80×10^{-4}
		200	3.10×10^{-4}
		240	1.58×10^{-3}
		300	5.40×10^{-3}
		500	1.33×10^{-1}
			$0.001e^{-25,000/RT}$
	Sn	500	1.94×10^{-1}
B	Si		$10.5\ e^{-85,000/RT}$

For diffusion coefficients in **m^2/s**, multiply values in **cm^2/hr** by **2.778×10^{-8}**.

Source: data from Loebel, R., in *Handbook of Chemistry and Physics, 51st ed.*, Weast, R. C., Ed., Chemical Rubber, Cleveland, 1970, F-55.

Table 85. DIFFUSION OF METALS INTO METALS

(SHEET 3 OF 11)

Diffusing Metal	Matrix Metal	Diffusion Temperature (°C)	Coefficient ($cm^2 \cdot hr^{-1}$)
Ba	Hg	7.8	2.17×10^{-2}
Bi	Si		$1030e^{-107,000/RT}$
	Pb	220	1.73×10^{-7}
		250	1.33×10^{-6}
		285	1.58×10^{-6}
C	W	1700	1.87×10^{-3}
	Fe	930	$7.51–9.18 \times 10^{-9}$
Ca	Hg	10.2	2.25×10^{-2}
Cd	Ag	650	9.36×10^{-7}
		800	4.68×10^{-6}
		900	2.23×10^{-5}
	Hg	8.7	6.05×10^{-2}
		15	6.51×10^{-2}
		20	5.47×10^{-2}
		99.1	1.23×10^{-1}
	Pb	200	4.59×10^{-7}
		252	3.10×10^{-6}
Cd, 1 atom%	Pb	167	1.66×10^{-7}
Ce	W	1727	3.42×10^{-6}

For diffusion coefficients in m^2/s, multiply values in cm^2/hr by 2.778×10^{-8}.

Source: data from Loebel, R., in *Handbook of Chemistry and Physics, 51st ed.*, Weast, R. C., Ed., Chemical Rubber, Cleveland, 1970, F-55.

Table 85. DIFFUSION OF METALS INTO METALS

(SHEET 4 OF 11)

Diffusing Metal	Matrix Metal	Diffusion Temperature (°C)	Coefficient ($cm^2 \cdot hr^{-1}$)
Cs	Hg	7.3	1.88 x 10^{-2}
	W	27	4.32 x 10^{-3}
		227	5.40 x 10^{-4}
		427	2.88 x 10^{-2}
		540	1.44 x 10^{-1}
Cu	Al	440	1.8 x 10^{-7}
		457	2.88 x 10^{-7}
		540	5.04 x 10^{-6}
		565	4.68–5.00 x 10^{-4}
	Ag	650	1.04 x 10^{-6}
		760	1.30 x 10^{-6}
		895	3.38 x 10^{-6}
	Au	301	5.40 x 10^{-10}
		443	8.64 x 10^{-9}
		560	3.38 x 10^{-7}
		604	5.10 x 10^{-7}
		616	7.92 x 10^{-7}
		740	3.35 x 10^{-6}
	Cu	650	1.15 x 10^{-5}
		750	2.34 x 10^{-8}
		830	1 .44 x 10^{-7}

For diffusion coefficients in m^2/s, multiply values in cm^2/hr by 2.778 x 10^{-8}.

Source: data from Loebel, R., in *Handbook of Chemistry and Physics, 51st ed.*, Weast, R. C., Ed., Chemical Rubber, Cleveland, 1970, F-55.

Table 85. DIFFUSION OF METALS INTO METALS (SHEET 5 OF 11)

Diffusing Metal	Matrix Metal	Diffusion Temperature (°C)	Coefficient ($cm^2 \cdot hr^{-1}$)
Cu (Con't)	Cu (Con't)	850	9.36×10^{-7}
		950	2.30×10^{-6}
		1030	1.01×10^{-5}
	Ge	700–900	$1.01 \pm 0.1 \times 10^{-1}$
	Pt	1041	$7.83–9 \times 10^{-8}$
		1213	5.04×10^{-7}
		1401	6.12×10^{-6}
Fe	Au	753	1.94×10^{-6}
		1003	2.70×10^{-5}
			$0.0062\ e^{-20,000/RT}$
Ga	Si		$3.6\ e^{-81,000/RT}$
Ge	Al	630	3.31×10^{-1}
	Au	529	1.84×10^{-1}
		563	2.80×10^{-1}
	Ge	766–928	$7.8\ e^{-68,509/RT}$
		1060–1200•K	$87\ e^{-73,000/RT}$
Hg	Cd	156	9.36×10^{-7}
		176	2.55×10^{-6}
		202	9×10^{-6}
	Pb	177	8.34×10^{-8}
		197	2.09×10^{-5}

For diffusion coefficients in m^2/s, multiply values in cm^2/hr by 2.778×10^{-8}.

Source: data from Loebel, R., in *Handbook of Chemistry and Physics, 51st ed.*, Weast, R. C., Ed., Chemical Rubber, Cleveland, 1970, F-55.

Table 85. DIFFUSION OF METALS INTO METALS

(SHEET 6 OF 11)

Diffusing Metal	Matrix Metal	Diffusion Temperature (°C)	Coefficient ($cm^2 \cdot hr^{-1}$)
In	Ag	650	1.04×10^{-6}
		800	6.84×10^{-6}
		895	4.68×10^{-5}
			$16.5\ e^{-90,000/RT}$
K	Hg	10.5	2.21×10^{-2}
	W	207	2.05×10^{-2}
		317	3.6×10^{-1}
		507	$1.1 \times 10^{+1}$
Li	Hg	8.2	2.75×10^{-2}
Mg	Al	365	3.96×10^{-8}
		395	$1.98–2.41 \times 10^{-7}$
		420	$2.38–2.74 \times 10^{-7}$
		440	1.19×10^{-7}
		447	9.36×10^{-7}
		450	6.84×10^{-6}
		500	$3.96–7.56 \times 10^{-6}$
		577	1.58×10^{-5}
	Pb	220	4.32×10^{-7}
Mn	Cu	400	7.2×10^{-10}
		850	4.68×10^{-7}

For diffusion coefficients in m^2/s, multiply values in cm^2/hr by 2.778×10^{-8}.

Source: data from Loebel, R., in *Handbook of Chemistry and Physics, 51st ed.*, Weast, R. C., Ed., Chemical Rubber, Cleveland, 1970, F-55.

Table 85. DIFFUSION OF METALS INTO METALS

(SHEET 7 OF 11)

Diffusing Metal	Matrix Metal	Diffusion Temperature (°C)	Coefficient ($cm^2 \cdot hr^{-1}$)
Mo	W	1533	9.36×10^{-10}
		1770	4.32×10^{-9}
		2010	7.92×10^{-8}
		2260	2.81×10^{-7}
Na	W	20	2.88×10^{-2}
		227	1.80
		417	9.72
		527	1.19×10^{-1}
Ni	Au	800	2.77×10^{-6}
		1003	2.48×10^{-5}
	Cu	550	2.56×10^{-9}
		950	7.56×10^{-7}
		320	1.26×10^{-6}
	Pt	1043	1.81×10^{-8}
		1241	1.73×10^{-6}
		1401	5.40×10^{-6}
Ni, 1 atom %	Pb	285	8.34×10^{-7}
Ni, 3 atom%	Pb	252	1.25×10^{-7}
Pb	Cd	252	2.88×10^{-8}
	Pb	250	5.42×10^{-8}
		285	2.92×10^{-7}
	Sn	500	1.33×10^{-1}

For diffusion coefficients in m^2/s, multiply values in cm^2/hr by 2.778×10^{-8}.

Source: data from Loebel, R., in *Handbook of Chemistry and Physics, 51st ed.*, Weast, R. C., Ed., Chemical Rubber, Cleveland, 1970, F-55.

Table 85. DIFFUSION OF METALS INTO METALS
(SHEET 8 OF 11)

Diffusing Metal	Matrix Metal	Diffusion Temperature (°C)	Coefficient ($cm^2 \cdot hr^{-1}$)
Pb, 2 atom %	Hg	9.4	6.46×10^{-9}
		15.6	5.71×10^{-2}
		99.2	8×10^{-2}
Pd	Ag	444	4.68×10^{-9}
		571	1.33×10^{-7}
		642	4.32×10^{-7}
		917	4.32×10^{-6}
	Au	727	2.09×10^{-8}
		970	1.15×10^{-6}
	Cu	490	3.24×10^{-9}
		950	$9.0–10.44 \times 10^{-7}$
Po	Au	470	4.59×10^{-11}
	Al	20	1.08×10^{-9}
		500	1.80×10^{-7}
	Bi	150	1.80×10^{-7}
		200	1.80×10^{-6}
	Pb	150	4.59×10^{-11}
		200	4.59×10^{-9}
		310	5.41×10^{-7}
Pt	Au	740	1.69×10^{-8}
		986	$6.12–10.08 \times 10^{-7}$

For diffusion coefficients in **m^2/s**, multiply values in **cm^2/hr** by **2.778×10^{-8}**.

Source: data from Loebel, R., in *Handbook of Chemistry and Physics, 51st ed.*, Weast, R. C., Ed., Chemical Rubber, Cleveland, 1970, F-55.

Table 85. DIFFUSION OF METALS INTO METALS
(SHEET 9 OF 11)

Diffusing Metal	Matrix Metal	Diffusion Temperature (°C)	Coefficient ($cm^2 \cdot hr^{-1}$)
Pt (Con't)	Cu	490	2.01×10^{-9}
		960	$3.96–8.28 \times 10^{-7}$
	Pb	490	7.04×10^{-2}
Ra	Au	470	1.42×10^{-8}
	Pt	470	3.42×10^{-8}
Ra(β+γ)	Ag	470	1.57×10^{-8}
Rb	Hg	7.3	1.92×10^{-9}
Rh	Pb	500	1.27×10^{-1}
Sb	Ag	650	1.37×10^{-6}
		760	5.40×10^{-6}
		895	1.55×10^{-5}
			$5.6\ e^{-91,000/RT}$
Si	Al	465	1.22×10^{-6}
		510	7.2×10^{-6}
		600	3.35×10^{-5}
		667	1.44×10^{-1}
		697	3.13×10^{-1}
	Fe+C*	1400–1600	$3.24–5.4 \times 10^{-2}$
Sn	Ag	650	2.23×10^{-6}
		895	2.63×10^{-6}

For diffusion coefficients in m^2/s, multiply values in cm^2/hr by 2.778×10^{-8}.

Source: data from Loebel, R., in *Handbook of Chemistry and Physics, 51st ed.*, Weast, R. C., Ed., Chemical Rubber, Cleveland, 1970, F-55.

Table 85. DIFFUSION OF METALS INTO METALS
(SHEET 10 OF 11)

Diffusing Metal	Matrix Metal	Diffusion Temperature (°C)	Coefficient ($cm^2 \cdot hr^{-1}$)
Sn (Con't)	Cu	400	1.69×10^{-9}
		650	2.48×10^{-7}
		850	1.40×10^{-5}
	Hg	10.7	6.38×10^{-2}
	Pb	245	1.12×10^{-7}
		250	1.83×10^{-7}
		285	5.76×10^{-7}
Sr	Hg	9.4	1.96×10^{-2}
Th	Mo	1615	1.30×10^{-6}
		2000	3.60×10^{-3}
	Tl	285	8.76×10^{-7}
	W	1782	3.96×10^{-7}
		2027	4.03×10^{-6}
		2127	1.29×10^{-5}
		2227	2.45×10^{-5}
Th (β)	Pb	165	2.54×10^{-12}
		260	2.54×10^{-8}
		324	5.84×10^{-6}
Tl	Hg	11.5	3.63×10^{-2}

For diffusion coefficients in m^2/s, multiply values in cm^2/hr by 2.778×10^{-8}.

Source: data from Loebel, R., in *Handbook of Chemistry and Physics, 51st ed.*, Weast, R. C., Ed., Chemical Rubber, Cleveland, 1970, F-55.

Table 85. DIFFUSION OF METALS INTO METALS
(SHEET 11 OF 11)

Diffusing Metal	Matrix Metal	Diffusion Temperature (°C)	Coefficient ($cm^2 \cdot hr^{-1}$)
Tl (Con't)	Pb	220	1.01×10^{-7}
		250	7.92×10^{-7}
		270	3.96×10^{-7}
		285	1.12×10^{-6}
		315	2.09×10^{-6}
			$16.5\, e^{-85{,}000/RT}$
U	W	1727	4.68×10^{-8}
Y	W	1727	6.55×10^{-5}
Zn	Ag	750	1.66×10^{-5}
		850	4.37×10^{-5}
	Al	415	9×10^{-7}
		473	1.91×10^{-6}
		500	$7.2–13.68 \times 10^{-6}$
		555	1.8×10^{-5}
	Hg	11.5	9.09×10^{-2}
		15	8.72×10^{-2}
		99.2	1.20×10^{-1}
	Pb	285	5.84
Zr	W	1727	1.17×10^{-5}

For diffusion coefficients in m^2/s, multiply values in cm^2/hr by 2.778×10^{-8}.

Source: data from Loebel, R., in *Handbook of Chemistry and Physics, 51st ed.*, Weast, R. C., Ed., Chemical Rubber, Cleveland, 1970, F-55.

* Saturated FeC Alloy.

Table 86. DIFFUSION IN SEMICONDUCTORS*

(SHEET 1 OF 8)

Semiconductor	Diffusing Element	D_o ($cm^2 \cdot s^{-1}$)	ΔE (eV)	Temperature Range of Validity (°C)
Aluminum antimonide (AlSb)	Al		~1.8	
	Cu	$3.5x10^{-3}$	0.36	150–500
	Sb		~1.5	
	Zn	0.33±.15	1.93±0.04	660–860
Cadmium selenide (CdSe)	Se	$2.6x10^{-3}$	1.55	700–1800
Cadmium sulfide (CdS)	Ag	$2.5x10^{+1}$	1.2	250–500
	Cd	3.4	2.0	750–1000
	Cu	$1.5x10^{-3}$	0.76	450–750
Cadmium telluride (CdTe)	Au	$6.7x10^{+1}$	2.0	600–1000
	In	$4.1x10^{-1}$	1.6	450–1000
Calcium ferrate (III) ($CaFe_2O_4$)	Ca	30	3.7	
	Fe	0.4	3.1	
α-Calcium metasilicate ($CaSiO_3$)	Ca	$7.4x10^{+4}$	4.8	
Gallium antimonide (GaSb)	Ga	$3.2x10^{+3}$	3.15	650–700
	In	$1.2x10^{-7}$	0.53	400–650
	Sb	$3.4x10^{+4}$	3.44	650–700
		$8.7x10^{+2}$	1.13	470–570
	Sn	$2.4x10^{-5}$	0.80	320–570
	Te	$3.8x10^{-4}$	1.2	400–650

Source: from Bolz, R. E. and Tuve, G. L., Eds., *Handbook of Tables for Applied Engineering Science, 2nd ed.*, CRC Press, Cleveland, 1973, 251.

Table 86. DIFFUSION IN SEMICONDUCTORS*
(SHEET 2 OF 8)

Semiconductor	Diffusing Element	D_o $(cm^2 \cdot s^{-1})$	ΔE (eV)	Temperature Range of Validity (°C)
Gallium arsenide (GaAs)	Ag	$2.5x10^{-3}$	1.5	
		$4x10^{-4}$	0.8±0.05	500–1160
	As	$4x10^{21}$	10.2±1.2	1200–1250
	Au	10^{-3}	1.0±0.2	740–1024
	Cd	0.05±0.04	2.43±0.06	868–1149
		[b]$50x10^{-2}$	2.8[a]	
	Cu	0.03	0.52	100–600
	Ga	$1x10^{+7}$	5.60±0.32	1125–1250
	Li	0.53	1.0	250–400
	Mg	$1.4x10^{-4}$	1.89	
		$2.3x10^{-2}$	2.6	740–1024
		[b]$2.6x10^{-2}$	2.7[a]	
		[b]$6.5x10^{-1}$	2.49[a]	
		$8.5x10^{-3}$	1.7	740–1024
	S	$1.2x10^{-4}$	1.8	
		[b]$1.6x10^{-5}$	1.63[a]	
		$2.6x10^{-5}$	1.86	
		$4x10^{3}$	4.04±0.15	1000–1200
	Se	$3x10^{3}$	4.16±0.16	1000–1200
	Sn	[b]$3.8x10^{-2}$	2.7	
		$6x10^{-4}$	2.5	1069–1215

Source: from Bolz, R. E. and Tuve, G. L., Eds., *Handbook of Tables for Applied Engineering Science, 2nd ed.*, CRC Press, Cleveland, 1973, 251.

Table 86. DIFFUSION IN SEMICONDUCTORS*

(SHEET 3 OF 8)

Semiconductor	Diffusing Element	D_o ($cm^2 \cdot s^{-1}$)	ΔE (eV)	Temperature Range of Validity (°C)
Gallium arsenide (GaAs) (Con't)	Zn	[b]$2.5x10^{-1}$	3.0[a]	
		$3.0x10^{-7}$	1.0	
		$6.0x10^{-7}$	0.6	
		15±7	2.49±0.05	800
Gallium phosphide (GaP)	Zn	1.0	2.1	700–1300
Germanium (Ge)	Ag	$4.4x10^{-2}$	1.0	700–900
	As	6.3	2.4	600–850
	Au	$2.2x10^{-2}$	2.5	
	B	$1.6x10^{-9}$	4.6	600–850
	Cu	$1.9x10^{-4}$	0.18	600–850
	Fe	$1.3x10^{-1}$	1.1	750–850
	Ga	$4.0x10^{+1}$	3.1	600–850
	Ge	$8.7x10^{+1}$	3.2	750–920
	He	$6.1x10^{-3}$	0.69	750–850
	In	$3x10^{-2}$	2.4	600–850
	Li	$1.3x10^{-4}$	0.47	200–600
	Ni	$8x10^{-1}$	0.9	700–875
	P	2.5	2.5	600–850
	Pb	–	3.6	600–850
	Sb	4.0	2.4	600–850
	Sn	$1.7x10^{-2}$	1.9	600–850
	Zn	$1.0x10^{+1}$	2.5	600–850

Source: from Bolz, R. E. and Tuve, G. L., Eds., *Handbook of Tables for Applied Engineering Science, 2nd ed.*, CRC Press, Cleveland, 1973, 251.

Table 86. DIFFUSION IN SEMICONDUCTORS*

(SHEET 4 OF 8)

Semiconductor	Diffusing Element	D_o ($cm^2 \cdot s^{-1}$)	ΔE (eV)	Temperature Range of Validity (°C)
Indium antimonide (InSb)	Ag	$1.0x10^{-7}$	0.25	
	Au	[b]$7x10^{-4}$	0.32[a]	140–510
	Cd	[b]$1.0x10^{-5}$	1.1[a]	250–500
		$1.23x10^{-9}$	0.52	442–519
		1.26	1.75	
		$1.3x10^{-4}$	1.2	360–500
	Co	$2.7x10^{-11}$	0.39	
		10^{-7}	0.25	440–510
	Cu	$3.0x10^{-5}$	0.37	
		[b]$9.0x10^{-4}$	1.08[a]	
	Fe	10^{-7}	0.25	440–510
	Hb	[b]$4.0x10^{-6}$	1.17[a]	
	In	0.05	1.81	450–500
		$1.8x10^{-9}$	0.28	
	Ni	10^{-7}	0.25	440–510
	Sb	0.05	1.94	450–500
		$1.4x10^{-6}$	0.75	
	Sn	$5.5x10^{-8}$	0.75	390–512
	Te	$1.7x10^{-7}$	0.57	300–500
	Zn	0.5	1.35	360–500
		$1.6x10^{-6}$	2.3±0.3	360–500
		5.5	1.6	360–500
	(Polycrystal)	$1.7x10^{-7}$	0.85	390–512
		[b]$5.3x10^{+7}$	2.61	

Source: from Bolz, R. E. and Tuve, G. L., Eds., *Handbook of Tables for Applied Engineering Science, 2nd ed.*, CRC Press, Cleveland, 1973, 251.

Table 86. DIFFUSION IN SEMICONDUCTORS*

(SHEET 5 OF 8)

Semiconductor	Diffusing Element	D_o ($cm^2 \cdot s^{-1}$)	ΔE (eV)	Temperature Range of Validity (°C)
Indium antimonide (InSb) (Con't)	Zinc (con't) (High concentration)	$6.3x10^{+8}$	2.61	
		[b]$3.7x10^{-10}$	0.7[a]	
	(Conc. = 2.2 x 10^{20} cm^{-3})	$9.0x10^{-10}$	~0	
	(Single crystal)	$1.4x10^{-7}$	0.86	390–512
Indium arsenide (InAs)	Cd	$4.35x10^{-4}$	1.17	600–900
	Cu		0.52a	
	Ge	$3.74x10^{-6}$	1.17	600–900
	Mg	$1.98x10^{-6}$	1.17	600–900
	S	6.78	2.20	600–900
	Se	12.55	2.20	600–900
	Sn	$1.49x10^{-6}$	1.17	600–900
	Te	$3.43x10^{-5}$	1.28	600–900
	Zn	$3.11x10^{-3}$	1.17	600–900
Indium phosphide (InP)	In	$1x10^{+5}$	3.85	850–1000
	P	$7x10^{+10}$	5.65	850–1000
Iron oxide (Fe_3O_4)	Fe	5.2	2.4	
Lead metasilicate ($PbSiO_3$)	Pb	85	2.6	
Lead orthosilicate ($PbSiO_4$)	Pb	8.2	2.0	

Source: from Bolz, R. E. and Tuve, G. L., Eds., *Handbook of Tables for Applied Engineering Science, 2nd ed.*, CRC Press, Cleveland, 1973, 251.

Table 86. DIFFUSION IN SEMICONDUCTORS*
(SHEET 6 OF 8)

Semiconductor	Diffusing Element	D_o ($cm^2 \cdot s^{-1}$)	ΔE (eV)	Temperature Range of Validity (°C)
Mercury selenide (HgSe)	Sb	$6.3x10^{-5}$	0.85	540–630
Nickel aluminate ($NiAl_2O_4$)	Cr	$1.17x10^{-3}$	2.2	
	Fe	1.33	3.5	
Nickel chromate (III) ($NiCr_2O_4$)	Cr	0.74	3.1	
	Cr	$2.03x10^{-5}$	1.9	
	Fe	$1.35x10^{-3}$	2.6	
	Ni	0.85	3.2	
Selenium (Se) (amorphous)	Fe	$1.1x10^{-5}$	0.38	300–400
	Ge	$9.4x10^{-6}$	0.39	300–400
	In	$5.2x10^{-6}$	0.32	300–400
	Sb	$2.8x10^{-8}$	0.29	300–400
	Se	$7.6x10^{-10}$	0.14	300–400
	Sn	$4.8x10^{-8}$	0.39	300–400
	Te	$5.4x10^{-6}$	0.53	300–400
	Tl	$1.4x10^{-6}$	0.35	300–400
	Zn	$3.8x10^{-7}$	0.29	300–400
Silicon (Si)	Al	8.0	3.5	1100–1400
	Ag	$2x10^{-3}$	1.6	1100–1350
	As	$3.2x10^{-1}$	3.5	1100–1350
	Au	$1.1x10^{-3}$	1.1	800–1200

Source: from Bolz, R. E. and Tuve, G. L., Eds., *Handbook of Tables for Applied Engineering Science, 2nd ed.*, CRC Press, Cleveland, 1973, 251.

Table 86. DIFFUSION IN SEMICONDUCTORS*
(SHEET 7 OF 8)

Semiconductor	Diffusing Element	D_o ($cm^2 \cdot s^{-1}$)	ΔE (eV)	Temperature Range of Validity (°C)
Silicon (Si) (Con't)	B	$1.0x10^{+1}$	3.7	950–1200
	Bi	$1.04x10^{+3}$	4.6	1100–1350
	Cu	$4x10^{-1}$	1.0	800–1100
	Fe	$6.2x10^{-3}$	0.86	1000–1200
	Ga	3.6	3.5	1150–1350
	Hl	$9.4x10^{-3}$	0.47	1000–1200
	He	$1.1x10^{-1}$	0.86	1000–1200
	In	$1.65x10^{+1}$	3.9	1100–1350
	Li	$9.4x10^{-3}$	0.78	100–800
	P	$1.0x10^{+1}$	3.7	1100–1350
	Sb	5.6	3.9	1100–1350
	Tl	$1.65x10^{+1}$	3.9	1100–1350
Silicon carbide (SiC)	Al	$2.0x10^{-1}$	4.9	1800–2250
	B	$1.6x10^{+1}$	5.6	1850–2250
	Cr	$2.3x10^{-1}$	4.8	1700–1900
Sulfur (S)	S	$2.8x10^{+13}$	2.0	>100
Tin zinc oxide ($SnZn_2O_4$)	Sn	$2x10^{+5}$	4.7	
	Zn	37	3.3	
Zinc aluminate ($ZnAl_2O_4$)	Zn	$2.5x10^{+2}$	3.4	
Zinc chromate (III) ($ZnCr_2O_4$)	Cr	8.5	3.5	
	Zn	60	3.7	

Source: from Bolz, R. E. and Tuve, G. L., Eds., *Handbook of Tables for Applied Engineering Science, 2nd ed.*, CRC Press, Cleveland, 1973, 251.

Table 86. DIFFUSION IN SEMICONDUCTORS*

(SHEET 8 OF 8)

Semiconductor	Diffusing Element	D_o $(cm^2 \bullet s^{-1})$	ΔE (eV)	Temperature Range of Validity (°C)
Zinc ferrate (III) ($ZnFe_2O_4$)	Fe	$8.5x10^{+2}$	3.5	
	Zn	$8.8x10^{+2}$	3.7	
Zinc selenide (ZnSe)	Cu	$1.7x10^{-5}$	0.56	200–570
Zinc sulfide (ZnS)	Zn	$1.0x10^{+16}$	6.50	>1030
		$1.5x10^{+4}$	3.25	940–1030
		$3.0x10^{-4}$	1.52	<940

Source: from Bolz, R. E. and Tuve, G. L., Eds., *Handbook of Tables for Applied Engineering Science, 2nd ed.*, CRC Press, Cleveland, 1973, 251.

* The diffusion coefficient D at a temperature T(K) is given by the following: $D=D_o e^{-\Delta E/kT}$

For $\mathbf{D_o}$ in $\mathbf{m^2/s}$, multiply values in $\mathbf{cm^2/s}$ by $\mathbf{10^{-4}}$.

a Values obtained at the low concentration limit.

CHAPTER 5

Thermal Properties of Materials

List of Tables

Specific Heat & Heat Capacity

Thermal Conductivity

(Continued)

List of Tables (Continued)

Table 87. SPECIFIC HEAT OF THE ELEMENTS AT 25 °C
(SHEET 1 OF 4)

Element	C_p (cal • g^{-1} • K^{-1})
Aluminum	0.215
Antimony	0.049
Argon	0.124
Arsenic	0.0785
Barium	0.046
Beryllium	0.436
Bismuth	0.0296
Boron	0.245
Bromine (Br_2)	0.113
Cadmium	0.0555
Calcium	0.156
Carbon, diamond	0.124
Carbon, graphite	0.170
Cerium	0.049
Cesium	0.057
Chlorine (Cl_2)	0.114
Chromium	0.107
Cobalt	0.109
Columbium (see Niobium)	
Copper	0.092
Dysprosium	0.0414
Erbium	0.0401
Europium	0.0421
Fluorine (F_2)	0.197
Gadolinium	0.055
Gallium	0.089
Germanium	0.077
Gold	0.0308

Source: data from Weast, R. C., Ed., Handbook of Chemistry and Physics, 55th ed., CRC Press, Cleveland, 1974, D-144., Kelly, K. K., Bulletin 592, Bureau of Mines, Washington, D. C., 1961.and Hultgren, R., Orr, R L., Anderson, P. D., and Kelly, K. K., Selected Values of Thermodynamic Properties of Metals and Alloys, John Wiley & Sons, New York, (1963).

Table 87. SPECIFIC HEAT OF THE ELEMENTS AT 25 °C

(SHEET 2 OF 4)

Element	C_p ($cal \cdot g^{-1} \cdot K^{-1}$)
Hafnium	0.035
Helium	1.24
Hollnium	0.0393
Hydrogen (H_2)	3.41
Indium	0.056
Iodine (I_2)	0.102
Iridium	0.0317
Iron (α)	0.106
Krypton	0.059
Lanthanum	0.047
Lead	0.038
Lithium	0.85
Lutetium	0.037
Magnesium	0.243
Manganese, α	0.114
Manganese, β	1.119
Mercury	0.0331
Molybdenum	0.599
Neodymium	0.049
Neon	0.246
Nickel	0.106
Niobium	0.064
Nitrogen (N_2)	0.249
Osmium	0.03127
Oxygen (O_2)	0.219
Palladium	0.0584
Phosphorus, white	0.181
Phosphorus, red, triclinic	0.160

Source: data from Weast, R. C., Ed., Handbook of Chemistry and Physics, 55th ed., CRC Press, Cleveland, 1974, D-144., Kelly, K. K., Bulletin 592, Bureau of Mines, Washington, D. C., 1961.and Hultgren, R., Orr, R L., Anderson, P. D., and Kelly, K. K., Selected Values of Thermodynamic Properties of Metals and Alloys, John Wiley & Sons, New York, (1963).

Table 87. SPECIFIC HEAT OF THE ELEMENTS AT 25 °C

(SHEET 3 OF 4)

Element	C_p ($cal \cdot g^{-1} \cdot K^{-1}$)
Platinum	0.0317
Polonium	0.030
Potassium	0.180
Praseodymium	0.046
Promethium	0.0442
Protactinium	0.029
Radium	0.0288
Radon	0.0224
Rhenium	0.0329
Rhodium	0.0583
Rubidium	0.0861
Ruthenium	0.057
Samarium	0.043
Scandium	0.133
Selenium (Se_2)	0.0767
Silicon	0.168
Silver	0.0566
Sodium	0.293
Strontium	0.0719
Sulfur, yellow	0.175
Tantalum	0.0334
Technetium	0.058
Tellurium	0.0481
Terbium	0.0437
Thallium	0.0307
Thorium	0.0271
Thulium	0.0382
Tin (α)	0.0510

Source: data from Weast, R. C., Ed., Handbook of Chemistry and Physics, 55th ed., CRC Press, Cleveland, 1974, D-144., Kelly, K. K., Bulletin 592, Bureau of Mines, Washington, D. C., 1961.and Hultgren, R., Orr, R L., Anderson, P. D., and Kelly, K. K., Selected Values of Thermodynamic Properties of Metals and Alloys, John Wiley & Sons, New York, (1963).

Table 87. SPECIFIC HEAT OF THE ELEMENTS AT 25 °C

(SHEET 4 OF 4)

Element	C_p (cal • g^{-1} • K^{-1})
Tin (β)	0.0530
Titanium	0.125
Tungsten	0.0317
Uranium	0.0276
Vanadium	0.116
Xenon	0.0378
Ytterbium	0.0346
Yttrium	0.068
Zinc	0.0928
Zirconium	0.0671

Source: data from Weast, R. C., Ed., Handbook of Chemistry and Physics, 55th ed., CRC Press, Cleveland, 1974, D-144., Kelly, K. K., Bulletin 592, Bureau of Mines, Washington, D. C., 1961.and Hultgren, R., Orr, R L., Anderson, P. D., and Kelly, K. K., Selected Values of Thermodynamic Properties of Metals and Alloys, John Wiley & Sons, New York, (1963).

Table 88. HEAT CAPACITY OF CERAMICS
(SHEET 1 OF 2)

Class	Ceramic	Heat Capacity, C_p (cal/mole/K)
Borides	Chromium Diboride (CrB_2)	$9.61 + 10.72x10^{-3}T$ cal/mole at 494-1010K
	Hafnium Diboride (HfB_2)	$9.61 + 10.72x10^{-3}T$ cal/mole at 494-1010K
	Tantalum Diboride (TaB_2)	0.04 cal/g°C
	Titanium Diboride (TiB_2)	$10.93 + 7.08x10^{-3}T$ cal/mole at 420-1180 K
	Zirconium Diboride (ZrB_2)	$15.81T + 4.20x10^{-3}T - 3.52x105T^{-2}$ for 429-1171K
Carbides	Hafnium Monocarbide (HfC)	0.05 at room temp. 15 ± 0.15 at 925°C 16 ± 0.16 at 1525°C
	Silicon Carbide (SiC)	0.26 at 540°C 0.27 at 700°C 0.30 at 1000°C 0.32 at 1200°C 0.33 at 1350°C 0.35 at 1550°C
	Titanium MonoCarbide (TiC)	0.150-0.170 cal/g at 150°C 0.170-0.187 cal/g at 300°C 0.183-0.196 cal/g at 450°C 0.192-0.201 cal/g at 600°C 0.20-0.207 cal/g at 750°C 0.209 cal/g at 900°C 0.210 cal/g at 1000°C 0.211 cal/g at 1100°C

Source: data compiled by J.S. Park from No. 1 Materials Index, Peter T.B. Shaffer, Plenum Press, New York, (1964); SmithellsBrandes, ed., in association with Fulmer Research Institute Ltd. 6th ed. London, Butterworths, Boston, (1983); and Ceramic Source, American Ceramic Society (1986-1991)

Table 88. HEAT CAPACITY OF CERAMICS

(SHEET 2 OF 2)

Class	Ceramic	Heat Capacity, C_p (cal/mole/K)
Nitrides	Aluminum Nitride (AlN)	0.1961 cal/g/°C ; 0-100°C 0.2277 cal/g/°C ; 0-420°C 0.2399 cal/g/°C ; 0-598°C
	Trisilicon tetranitride (Si_3N_4)	0.17 cal/g/°C
Oxides	Cerium Dioxide (CeO_2)	$14.24T + 5.62x10^{-3}T$ 491-1140K
Silicides	Molybdenum Disilicide ($MoSi_2$)	10-14 cal/g/°C; 425-1000°C
	Tungsten Disilicide (WSi_2)	8 cal/g/°C; 425-1450°C

Source: data compiled by J.S. Park from No. 1 Materials Index, Peter T.B. Shaffer, Plenum Press, New York, (1964); SmithellsBrandes, ed., in association with Fulmer Research Institute Ltd. 6th ed. London, Butterworths, Boston, (1983); and Ceramic Source, American Ceramic Society (1986-1991)

Table 89. SPECIFIC HEAT OF POLYMERS
(SHEET 1 OF 4)

Polymer Class	Polymer Subclass	Specific heat (Btu/lb/°F)
ABS Resins; Molded, Extruded	Medium impact	0.36—0.38
	High impact	0.36—0.38
	Very high impact	0.36—0.38
	Low temperature impact	0.35—0.38
	Heat resistant	0.37—0.39
Acrylics; Cast, Molded, Extruded	**Cast Resin Sheets, Rods:**	
	General purpose, type I	0.35
	General purpose, type II	0.35
	Moldings:	
	Grades 5, 6, 8	0.35
	High impact grade	0.34
Thermoset Carbonate	Allyl diglycol carbonate	0.3
Cellulose Acetate; Molded, Extruded	ASTM Grade:	
	H6—1	0.3—0.42
	H4—1	0.3—0.42
	H2—1	0.3—0.42
	MH—1, MH—2	0.3—0.42
	MS—1, MS—2	0.3—0.42
	S2—1	0.3—0.42
Cellulose Acetate Butyrate; Molded, Extruded	ASTM Grade:	
	H4	0.3—0.4
	MH	0.3—0.4
	S2	0.3—0.4
Cellusose Acetate Propionate; Molded, Extruded	ASTM Grade:	
	1	0.3—0.4
	3	0.3—0.4
	6	0.3—0.4

Source: data compiled by J.S. Park from Charles T. Lynch, CRC Handbook of Materials Science, Vol. 3, CRC Press, Boca Raton, Florida and Engineered Materials Handbook, Vol.2, Engineering Plastics, ASM International, Metals Park, Ohio, 1988.

Table 89. Specific Heat of Polymers

(Sheet 2 of 4)

Polymer Class	Polymer Subclass	Specific heat (Btu/lb/°F)
Chlorinated polyvinyl chloride	Chlorinated polyvinyl chloride	0.3
Polycarbonate		0.3
Fluorocarbons; Molded,Extruded	Polytrifluoro chloroethylene (PTFCE)	0.22
	Polytetrafluoroethylene (PTFE)	0.25
	Fluorinated ethylene propylene(FEP)	0.28
	Polyvinylidene— fluoride (PVDF)	0.33
Epoxies; Cast, Molded, Reinforced	Standard epoxies (diglycidyl ethers of bisphenol A)	
	Cast rigid	0.4-0.5
	High strength laminate	0.21
	Filament wound composite	0.24
Nylons; Molded, Extruded	Type 6	
	General purpose	0.4
	Cast	0.4
	Type 8	0.4
	Type 11	0.58
	Type 12	0.28
Nylons; Molded, Extruded	6/6 Nylon	
	General purpose molding	0.3—0.5
	General purpose extrusion	0.3—0.5
	6/10 Nylon	
	General purpose	0.3—0.5

Source: data compiled by J.S. Park from Charles T. Lynch, CRC Handbook of Materials Science, Vol. 3, CRC Press, Boca Raton, Florida and Engineered Materials Handbook, Vol.2, Engineering Plastics, ASM International, Metals Park, Ohio, 1988.

Table 89. SPECIFIC HEAT OF POLYMERS

(SHEET 3 OF 4)

Polymer Class	Polymer Subclass	Specific heat (Btu/lb/°F)
Phenolics; Molded	**Type and filler:**	
	General: woodflour and flock	0.35—0.40
	Shock: paper, flock, or pulp	—
	High shock: chopped fabric or cord	0.30—0.35
	Very high shock: glass fiber	0.28—0.32
Phenolics: Molded	Arc resistant—mineral	0.27—0.37
	Rubber phenolic—woodflour or flock	0.33
	PVC—Acrylic Alloy	
	PVC—acrylic sheet	0.293
Polymides	Unreinforced	0.31
	Unreinforced 2nd value	0.25—0.35
	Glass reinforced	0.15—0.27
Polyacetals	Standard	0.35
	Copolymer:	
	Standard	0.35
	High flow	0.35
Polyesters: Thermosets	**Cast polyyester**	
	Rigid	0.30—0.55
	Reinforced polyester moldings	
	High strength (glass fibers)	0.25—0.35
	Sheet molding compounds, general purpose	0.20—0.25
Phenylene oxides (Noryl)	Standard	0.24
Polypropylene:	General purpose	0.45
	High impact	0.45—0.48
Polyphenylene sulfide:	Standard	0.26

Source: data compiled by J.S. Park from Charles T. Lynch, CRC Handbook of Materials Science, Vol. 3, CRC Press, Boca Raton, Florida and Engineered Materials Handbook, Vol.2, Engineering Plastics, ASM International, Metals Park, Ohio, 1988.

Table 89. SPECIFIC HEAT OF POLYMERS
(SHEET 4 OF 4)

Polymer Class	Polymer Subclass	Specific heat (Btu/lb/°F)
Polyethylenes; Molded, Extruded	Type I—lower density (0.910—0.925)	
	Melt index 0.3—3.6	0.53—0.55
	Melt index 6—26	0.53—0.55
	Melt index 200	0.53—0.55
	Type II—medium density (0.926—0.940)	
	Melt index 20	0.53—0.55
	Melt index l.0—1.9	0.53—0.55
	Type III—higher density (0.941—0.965)	
	Melt index 0.2—0.9	0.46—0.55
	Melt Melt index 0.l—12.0	0.46—0.55
	Melt index 1.5—15	0.46—0.55
Polystyrenes; Molded	Polystyrenes	
	General purpose	0.30—0.35
	Medium impact	0.30—0.35
	High impact	0.30—0.35
	Glass fiber -30% reinforced	0.256
	Styrene acrylonitrile (SAN)	0.33
Polyvinyl Chloride And Copolymers; Molded, Extruded	Vinylidene chloride	0.32
Silicones; Molded, Laminated	Woven glass fabric/ silicone laminate	0.246

Source: data compiled by J.S. Park from Charles T. Lynch, CRC Handbook of Materials Science, Vol. 3, CRC Press, Boca Raton, Florida and Engineered Materials Handbook, Vol.2, Engineering Plastics, ASM International, Metals Park, Ohio, 1988.

Table 90. SPECIFIC HEAT OF FIBERGLASS REINFORCED PLASTICS

Class	Material	Glass fiber content (wt%)	Specific heat (Btu/lb–°F)
Glass fiber reinforced thermosets	Sheet molding compound (SMC)	15 to 30	0.30 to 0.35
	Bulk molding compound(BMC)	15 to 35	0.30 to 0.35
	Preform/mat(compression molded)	25 to 50	0.30 to 0.33
	Cold press molding–polyester	20 to 30	0.30 to 0.33
	Spray–up–polyester	30 to 50	0.30 to 0.34
	Filament wound–epoxy	30 to 80	0.23 to 0.25
	Rod stock–polyester	40 to 80	0.22 to 0.25
	Molding compound–phenolic	5 to 25	0.20 to 0.30
Glass–fiber–reinforced thermoplastics	Nylon	6 to 60	0.30 to 0.35
	Polystyrene	20 to 35	0.23 to 0.35

Data from *ASM Engineering Materials Reference Book, Second Edition*, Michael Bauccio, Ed., ASM International, Materials Park, OH, p106, (1994).

Table 91. THERMAL CONDUCTIVITY OF METALS (PART 1)

(SHEET 1 OF 2)

T (K)	Aluminum	Cadmium	Chromium	Copper	Gold
1	7.8	48.7	0.401	28.7	4.4
2	15.5	89.3	0.802	57.3	8.9
3	23.2	104	1.20	85.5	13.1
4	30.8	92.0	1.60	113	17.1
5	38.1	69.0	1.99	138	20.7
6	45.1	44.2	2.38	159	23.7
7	51.5	28.0	2.77	177	26.0
8	57.3	18.0	3.14	189	27.5
9	62.2	12.2	3.50	195	28.2
10	66.1	8.87	3.85	196	28.2
11	69.0	6.91	4.18	193	27.7
12	70.8	5.56	4.49	185	26.7
13	71.5	4.67	4.78	176	25.5
14	71.3	4.01	5.04	166	24.1
15	70.2	3.55	5.27	156	22.6
16	68.4	3.16	5.48	145	20.9
18	63.5	2.62	5.81	124	17.7
20	56.5	2.26	6.01	105	15.0
25	40.0	1.79	6.07	68	10.2
30	28.5	1.56	5.58	43	7.6
35	21.0	1.41	5.03	29	6.1
40	16.0	1.32	4.30	20.5	5.2
45	12.5	1.25	3.67	15.3	4.6
50	10.0	1.20	3.17	12.2	4.2
60	6.7	1.13	2.48	8.5	3.8

Values are in watt • cm^{-1} • K^{-1}.

Note: Values in parentheses are for liquid state

These data apply only to metals of purity of at least 99.9%.

The third significant figure may not be accurate.

Source: data from Ho, C. Y., Powell, R. W., and Liley, P. E., Thermal Conductictivity of Selected Materials, NSRDS–NBS–8 and NSRD-NBS-16, Part 2 , National Standard Reference Data System–National Bureau of Standards, Part 1, 1966; Part 2, 1968.

Table 91. THERMAL CONDUCTIVITY OF METALS (PART 1)
(SHEET 2 OF 2)

T (K)	Aluminum	Cadmium	Chromium	Copper	Gold
70	5.0	1.08	2.08	6.7	3.58
80	4.0	1.06	1.82	5.7	3.52
90	3.4	1.04	1.68	5.14	3.48
100	3.0	1.03	1.58	4.83	3.45
200	2.37	0.993	1.11	4.13	3.27
273	2.36	0.975	0.948	4.01	3.18
300	2.37	0.968	0.903	3.98	3.15
400	2.4	0.947	0.873	3.92	3.12
500	2.37	0.92	0.848	3.88	3.09
600	2.32	(0.42)	0.805	3.83	3.04
700	2.26	(0.49)	0.757	3.77	2.98
800	2.2	(0.559)	0.713	3.71	2.92
900	2.13		0.678	3.64	2.85
1000	(0.93)		0.653	3.57	2.78
1100	(0.96)		0.636	3.5	2.71
1200	(0.99)		0.624	3.42	2.62
1400			0.611		

Values are in watt • cm^{-1} • K^{-1}.

Note: Values in parentheses are for liquid state

These data apply only to metals of purity of at least 99.9%.

The third significant figure may not be accurate.

Source: data from Ho, C. Y., Powell, R. W., and Liley, P. E., Thermal Conductictivity of Selected Materials, NSRDS–NBS–8 and NSRD-NBS-16, Part 2 , National Standard Reference Data System–National Bureau of Standards, Part 1, 1966; Part 2, 1968.

Table 92. THERMAL CONDUCTIVITY OF METALS (PART 2)
(SHEET 1 OF 2)

T (K)	Iron	Lead	Magnesium	Mercury	Molybdenum
1	0.75	27.7	1.30		0.146
2	1.49	42.4	2.59		0.292
3	2.24	34.0	3.88		0.438
4	2.97	22.4	5.15		0.584
5	3.71	13.8	6.39		0.730
6	4.42	8.2	7.60		0.876
7	5.13	4.9	8.75		1.02
8	5.80	3.2	9.83		1.17
9	6.45	2.3	10.8		1.31
10	7.05	1.78	11.7		1.45
11	7.62	1.46	12.5		1.60
12	8.13	1.23	13.1		1.74
13	8.58	1.07	13.6		1.88
14	8.97	0.94	14.0		2.01
15	9.30	0.84	14.3		2.15
16	9.56	0.77	14.4		2.28
18	9.88	0.66	14.3		2.53
20	9.97	0.59	13.9		2.77
25	9.36	0.507	12.0		3.25
30	8.14	0.477	9.5		3.55
35	6.81	0.462	7.4		3.62
40	5.55	0.451	5.7		3.51
45	4.50	0.442	4.57		3.26
50	3.72	0.435	3.75		3.00
60	2.65	0.424	2.74		2.60
70	2.04	0.415	2.23		2.30
80	1.68	0.407	1.95		2.09

Values are in watt • cm^{-1} • K^{-1}.

Note: Values in parentheses are for liquid state

These data apply only to metals of purity of at least 99.9%.

The third significant figure may not be accurate.

Table 92. THERMAL CONDUCTIVITY OF METALS (PART 2)
(SHEET 2 OF 2)

T (K)	Iron	Lead	Magnesium	Mercury	Molybdenum
90	1.46	0.401	1.78		1.92
100	1.32	0.396	1.69		1.79
200	0.94	0.366	1.59		1.43
273	0.835	0.355	1.57	(0.078)	1.39
300	0.803	0.352	1.56	(0.084)	1.38
400	0.694	0.338	1.53	(0.098)	1.34
500	0.613	0.325	1.51	(0.109)	1.3
600	0.547	0.312	1.49	(0.12)	1.26
700	0.487	(0.174)	1.47	(0.127)	1.22
800	0.433	(0.19)	1.46	(0.13)	1.18
900	0.38	(0.203)	1.45		1.15
1000	0.326	(0.215)	(0.84)		1.12
1100	0.297		(0.91)		1.08
1200	0.282		(0.98)		1.05
1400	0.309				0.996
1600	0.327				0.946
1800					0.907
2000					0.88
2200					0.858
2600					0.825

Values are in watt • cm^{-1} • K^{-1}.

Note: Values in parentheses are for liquid state

These data apply only to metals of purity of at least 99.9%.

The third significant figure may not be accurate.

Table 93. THERMAL CONDUCTIVITY OF METALS (PART 3)

(SHEET 1 OF 2)

T (K)	Nickel	Niobium	Platinum	Silver	Tantalum
1	0.64	0.251	2.31	39.4	0.115
2	1.27	0.501	4.60	78.3	0.230
3	1.91	0.749	6.79	115	0.345
4	2.54	0.993	8.8	147	0.459
5	3.16	1.23	10.5	172	0.571
6	3.77	1.46	11.8	187	0.681
7	4.36	1.67	12.6	193	0.788
8	4.94	1.86	12.9	190	0.891
9	5.49	2.04	12.8	181	0.989
10	6.00	2.18	12.3	168	1.08
11	6.48	2.30	11.7	154	1.16
12	6.91	2.39	10.9	139	1.24
13	7.30	2.46	10.1	124	1.30
14	7.64	2.49	9.3	109	1.36
15	7.92	2.50	8.4	96	1.40
16	8.15	2.49	7.6	85	1.44
18	8.45	2.42	6.1	66	1.47
20	8.56	2.29	4.9	51	1.47
25	8.15	1.87	3.15	29.5	1.36
30	6.95	1.45	2.28	19.3	1.16
35	5.62	1.16	1.80	13.7	0.99
40	4.63	0.97	1.51	10.5	0.87
45	3.91	0.84	1.32	8.4	0.78
50	3.36	0.76	1.18	7.0	0.72
60	2.63	0.66	1.01	5.5	0.651

Values are in watt • cm^{-1} • K^{-1}.

Note: Values in parentheses are for liquid state

These data apply only to metals of purity of at least 99.9%.

The third significant figure may not be accurate.

Table 93. THERMAL CONDUCTIVITY OF METALS (PART 3)

(SHEET 2 OF 2)

T (K)	Nickel	Niobium	Platinum	Silver	Tantalum
70	2.21	0.61	0.90	4.97	0.616
80	1.93	0.58	0.84	4.71	0.603
90	1.72	0.563	0.81	4.60	0.596
100	1.58	0.552	0.79	4.50	0.592
200	1.06	0.526	0.748	4.3	0.575
273	0.94	0.533	0.734	4.28	0.574
300	0.905	0.537	0.73	4.27	0.575
400	0.801	0.552	0.722	4.2	0.578
500	0.721	0.567	0.719	4.13	0.582
600	0.655	0.582	0.72	4.05	0.586
700	0.653	0.598	0.723	3.97	0.59
800	0.674	0.613	0.729	3.89	0.594
900	0.696	0.629	0.737	3.82	0.598
1000	0.718	0.644	0.748	3.74	0.602
1100	0.739	0.659	0.76	3.66	0.606
1200	0.761	0.675	0.775	3.58	0.610
1400	0.804	0.705	0.807		0.618
1600		0.735	0.842		0.626
1800		0.764	0.877		0.634
2000		0.791	0.913		0.640
2200		0.815			0.647
2600					0.658
3000					0.665

Values are in watt • cm^{-1} • K^{-1}.

Note: Values in parentheses are for liquid state

These data apply only to metals of purity of at least 99.9%.

The third significant figure may not be accurate.

Table 94. THERMAL CONDUCTIVITY OF METALS (PART 4)

(SHEET 1 OF 2)

T (K)	Tin	Titanium	Tungsten	Zinc	Zirconium
1		0.0144	14.4	19.0	0.111
2		0.0288	28.7	37.9	0.223
3	297	0.0432	42.6	55.5	0.333
4	181	0.0576	55.6	69.7	0.442
5	117	0.0719	67.1	77.8	0.549
6	76	0.0863	76.2	78.0	0.652
7	52	0.101	82.4	71.7	0.748
8	36	0.115	85.3	61.8	0.837
9	26	0.129	85.1	51.9	0.916
10	19.3	0.144	82.4	43.2	0.984
11	14.8	0.158	77.9	36.4	1.04
12	11.6	0.172	72.4	30.8	1.08
13	9.3	0.186	66.4	26.1	1.11
14	7.6	0.200	60.4	22.4	1.13
15	6.3	0.214	54.8	19.4	1.13
16	5.3	0.227	49.3	16.9	1.12
18	4.0	0.254	40.0	13.3	1.08
20	3.2	0.279	32.6	10.7	1.01
25	2.22	0.337	20.4	6.9	0.85
30	1.76	0.382	13.1	4.9	0.74
35	1.50	0.411	8.9	3.72	0.65
40	1.35	0.422	6.5	2.97	0.58
45	1.23	0.416	5.07	2.48	0.535
50	1.15	0.401	4.17	2.13	0.497
60	1.04	0.377	3.18	1.71	0.442

Values are in watt • cm^{-1} • K^{-1}.

Note: Values in parentheses are for liquid state

These data apply only to metals of purity of at least 99.9%.

The third significant figure may not be accurate.

Table 94. THERMAL CONDUCTIVITY OF METALS (PART 4)

(SHEET 2 OF 2)

T (K)	Tin	Titanium	Tungsten	Zinc	Zirconium
70	0.96	0.356	2.76	1.48	0.403
80	0.91	0.339	2.56	1.38	0.373
90	0.88	0.324	2.44	1.34	0.350
100	0.85	0.312	2.35	1.32	0.332
200	0.733	0.245	1.97	1.26	0.252
273	0.682	0.224	1.82	1.22	0.232
300	0.666	0.219	1.78	1.21	0.227
400	0.622	0.204	1.62	1.16	0.216
500	0.596	0.197	1.49	1.11	0.210
600	(0.323)	0.194	1.39	1.05	0.207
700	(0.343)	0.194	1.33	(0.499)	0.209
800	(0.364)	0.197	1.28	(0.557)	0.216
900	(0.384)	0.202	1.24	(0.615)	0.226
1000	(0.405)	0.207	1.21	(0.673)	0.237
1100	(0.425)	0.213	1.18	(0.73)	0.248
1200	(0.446)	0.220	1.15		0.257
1400	(0.487)	0.236	1.11		0.275
1600		0.253	1.07		0.290
1800		0.271	1.03		0.302
2000			1.00		0.313
2200			0.98		
2600			0.94		
3000			0.915		

Values are in watt • cm^{-1} • K^{-1}.

Note: Values in parentheses are for liquid state

These data apply only to metals of purity of at least 99.9%.

The third significant figure may not be accurate.

Table 95. THERMAL CONDUCTIVITY OF ALLOY CAST IRONS

Description	Description	Thermal Conductivity W/(m • K)
Abrasion–Resistant White Irons	Low–C white iron	22
	Martensitic nickel–chromium iron	30
Corrosion–Resistant Irons	High–nickel gray iron	38 to 40
	High–nickel ductile iron	13.4
Heat–Resistant Gray Irons	Medium–silicon iron	37
	High–chromium iron	20
	High–nickel iron	37 to 40
	Nickel–chromium–silicon iron	30
Heat–Resistant Ductile Iron	High–nickel ductile (20 Ni)	13

Source: Data from *ASM Metals Reference Book, Second Edition*, American Society for Metals, Metals Park, Ohio 44073, p172, (1984).

Table 96. THERMAL CONDUCTIVITY OF IRON AND IRON ALLOYS

Metal or alloy	Thermal Conductivity near room temperature (cal / cm^2 • cm • s • °C)
Pure iron	0.178
Cast iron (3.16C, 1.54Si, 0.57Mn)	0.112
Carbon steel(0.23C, 0.64Mn)	0.124
Carbon steel(1.22 C, 0.35 Mn)	0.108
Alloy steel (0.34 C, 0.55 Mn, 0.78 Cr, 3.53 Ni, 0.39 Mo, 0.05 Cu)	0.079
Type 410	0.057
Type 304	0.036
T1 tool steel	0.058

Data from *ASM Metals Reference Book, Third Edition,* Michael Bauccio, Ed., ASM International, Materials Park, OH, p156, (1993).

Table 97. THERMAL CONDUCTIVITY OF ALUMINUM AND ALUMINUM ALLOYS (SHEET 1 OF 2)

Metal or alloy	Designation	Thermal Conductivity near room temperature (cal / cm^2 • cm • s • °C)
Wrought alloys	EC(O)	0.57
	1060(O)	0.56
	1100	0.53
	2011(T3)	0.34
	2014(O)	0.46
	2024(O)	0.45
	2218(T72)	0.37
	3003(O)	0.46
	4032(O)	0.37
	5005	0.48
	5050(O)	0.46
	5052(O)	0.33
	5056(O)	0.28
	5083	0.28
	5086	0.30
	5154	0.30
	5357	0.40
	5456	0.28
	6061(O)	0.41
	6063(O)	0.52
	6101(T6)	0.52
	6151(O)	0.49
	7075(T6)	0.29
	7079(T6)	0.29
	7178	0.29

Data from *ASM Metals Reference Book, Third Edition*, Michael Bauccio, Ed., ASM International, Materials Park, OH, p156, (1993).

Table 97. Thermal Conductivity of Aluminum and Aluminum Alloys (Sheet 2 of 2)

Metal or alloy	Designation	Thermal Conductivity near room temperature (cal / cm^2 • cm • s • °C)
Casting alloys	A13	0.29
	43(F)	0.34
	108(F)	0.29
	A108	0.34
	A132(T551)	0.28
	D132(T5)	0.25
	F132	0.25
	138	0.24
	142 (T21, sand)	0.40
	195 (T4, T62)	0.33
	B195 (T4, T6)	0.31
	214	0.33
	200(T4)	0.21
	319	0.26
	355(T51, sand)	0.40
	356(T51, sand)	0.40
	360	0.35
	380	0.23
	750	0.44
	40E	0.33

Data from *ASM Metals Reference Book, Third Edition*, Michael Bauccio, Ed., ASM International, Materials Park, OH, p156, (1993).

Table 98. THERMAL CONDUCTIVITY OF COPPER AND COPPER ALLOYS (SHEET 1 OF 3)

Metal or alloy	Designation	Thermal Conductivity near room temperature (cal / cm^2 • cm • s • °C)
Wrought coppers	Pure Copper	0.941
	Electrolytic tough pitch copper (ETP)	0.934
	Deoxidized copper high residual phosphorus (DHP)	0.81
	Free–machining copper (0.5% Te)	0.88
	Free–machining copper (1% Pb)	0.92
Wrought alloys	Gilding, 95%	0.56
	Commercial bronze, 90%	0.45
	Jewelry bronze, 87.5%	0.41
	Red brass, 85%	0.38
	Low brass, 80%	0.33
	Cartridge brass, 70%	0.29
	Yellow brass	0.28
	Muntz metal	0.29
	Leaded commercial bronze	0.43
	Low–leaded brass (tube)	0.28
	Medium leaded brass	0.28
	High–leaded brass (tube)	0.28
	High–leaded brass	0.28
	Extra–high–leaded brass	0.28
	Leaded Muntz metal	0.29
	Forging brass	0.28
	Architectual bronze	0.29
	Inhibited admiralty	0.26
	Naval brass	0.28

Data from *ASM Metals Reference Book, Third Edition*, Michael Bauccio, Ed., ASM International, Materials Park, OH, p156, (1993).

Table 98. THERMAL CONDUCTIVITY OF COPPER AND COPPER ALLOYS (SHEET 2 OF 3)

Metal or alloy	Designation	Thermal Conductivity near room temperature (cal / cm^2 • cm • s • °C)
Wrought alloys (Con't)	Leaded naval brass	0.28
	Manganese bronze	0.26
	Phosphor bronze,5%	0.17
	Pbosphor bronze, 8%	0.15
	Phosphor bronze, 10%	0.12
	Phosphor bronze, 1.25%	0.49
	Free cutting phosphor bronze	0.18
	Cupro-nickel,30%	0.07
	Cupro-nickel,10%	0.095
	Nickel silver, 65–18	0.08
	Nickel silver, 55–18	0.07
	Nickel silver, 65–12	0.10
	High–silicon bronze	0.09
	Low–silicon bronze	0.14
	Aluminum bronze, 5%Al	0.198
	Aluminum bronze	0.18
	Aluminum–silicon bronze	0.108
	Aluminum bronze	0.144
	Aluminum bronze	0.091
	Beryllium copper	0.20
Casting alloys	Chromium copper (1% Cr)	0.4
	89cu–11Sn	0.121
	88Cu–6Sn–1.5Pb–4.5Zn	18% of Cu
	87Cu–8Sn–1Pb–4Zn	12% of Cu
	87Cu–10Sn–1Pb–2Zn	12% of Cu
	80Cu–10Sn–10Pb	12% of Cu
	Manganese bronze, 110 ksi	9.05% of Cu

Data from *ASM Metals Reference Book, Third Edition*, Michael Bauccio, Ed., ASM International, Materials Park, OH, p156, (1993).

Table 98. THERMAL CONDUCTIVITY OF COPPER AND COPPER ALLOYS (SHEET 3 OF 3)

Metal or alloy	Designation	Thermal Conductivity near room temperature (cal / cm^2 • cm • s • °C)
Casting alloys (Con't)	Aluminum bronze, Alloy 9A	15% of Cu
	Aluminum bronze, Alloy 9B	16% of Cu
	Aluminum bronze, Alloy 9C	18% of Cu
	Aluminum bronze, Alloy 9D	12% of Cu
	Propeller bronze	11% of Cu
	Nickel silver, 12% Ni	7% of Cu
	Nickel silver, 16% Ni	7% of Cu
	Nickel silver, 20% Ni	6% of Cu
	Nickel silver. 25% Ni	6.5% of Cu
	Silicon bronze	7% of Cu

Data from *ASM Metals Reference Book, Third Edition*, Michael Bauccio, Ed., ASM International, Materials Park, OH, p156, (1993).

Table 99. THERMAL CONDUCTIVITY OF MAGNESIUM AND MAGNESIUM ALLOYS

Metal or alloy	Designation	Thermal Conductivity near room temperature ($cal / cm^2 \cdot cm \cdot s \cdot °C$)
Pure	Magnesium (99.8%)	0.367
Casting alloys	AM100A	0.17
	AZ63A	0.18
	AZ81A(T4)	0.12
	AZ91A, B, C	0.17
	AZ92A	0.17
	HK31A (T6, sand cast)	0.22
	HZ32A	0.26
	ZH42	0.27
	ZH62A	0.26
	ZK51A	0.26
	ZE41A(T5)	0.27
	EZ33A	0.24
	EK30A	0.26
	EK41A(T5)	0.24
Wrought alloys	M1A	0.33
	AZ31B	0.23
	AZ61A	0.19
	AZ80A	0.18
	ZK60A,B(F)	0.28
	ZE10A(O)	0.33
	HM21A(O)	0.33
	HM31A	0.25

Data from *ASM Metals Reference Book, Third Edition*, Michael Bauccio, Ed., ASM International, Materials Park, OH, p156, (1993).

Table 100. THERMAL CONDUCTIVITY OF NICKEL AND NICKEL ALLOYS

Metal or alloy	Thermal Conductivity near room temperature (cal / cm^2 • cm • s • °C)
Nickel (99.95% Ni + Co)	0.22
"A" nickel	0.145
"D" nickel	0.115
Monel	0.062
"K" Monel	0.045
Inconel	0.036
Hastelloy B	0.027
Hastelloy C	0.03
Hastelloy D	0.05
Illium G	0.029
Illium R	0.031
60Ni–24Fe–16Cr	0.032
35Ni–45Fe–20Cr	0.031
Constantan	0.051

Data from *ASM Metals Reference Book, Third Edition*, Michael Bauccio, Ed., ASM International, Materials Park, OH, p156, (1993).

Table 101. Thermal Conductivity of Lead and Lead Alloys

Metal or alloy	Thermal Conductivity near room temperature (cal / cm^2 • cm • s • °C)
Corroding lead (99.73 + % Pb)	0.083
5–95 solder	0.085
20–80 solder	0.089
50-50 solder	0.111
1% antimonial lead	0.080
Hard lead (96Pb-4Sb)	0.073
Hard lead (94Pb–6Sb)	0.069
8% antimonial lead	0.065
9% antimonial lead	0.064
Lead-base babbitt (SAE 14)	0.057
Lead-base babbitt (alloy 8)	0.058

Data from *ASM Metals Reference Book, Third Edition*, Michael Bauccio, Ed., ASM International, Materials Park, OH, p156, (1993).

Table 102. THERMAL CONDUCTIVITY OF TIN, TITANIUM, ZINC AND THEIR ALLOYS

Metal or alloy	Designation	Thermal Conductivity near room temperature (cal / cm^2 • cm • s • °C)
Tin and Tin Alloys	Pure tin	0.15
	Soft solder (63Sn–37Pb)	0.12
	Tin foil (92Sn–8Zn)	0.14
Titanium and Titanium Alloys	Titanium(99.0%)	0.043
	Ti–5Al–2.5Sn	0.019
	Ti–2Fe-2Cr–2Mo	0.028
	Ti–8Mn	0.026
Zinc and Zinc Alloys	Pure zinc	0.27
	AG40A alloy	0.27
	AC41A alloy	0.26
	Commercial rolled zinc 0.08 Pb	0.257
	Commercial rolled zinc 0.06 Pb, 0.06 Cd	0.257
	Rolled zinc alloy (1 CU, 0.010 Mg)	0.25
	Zn-Cu–Ti alloy (0.8 Cu, 0.15 Ti)	0.25

Data from *ASM Metals Reference Book, Third Edition*, Michael Bauccio, Ed., ASM International, Materials Park, OH, p156, (1993).

Table 103. THERMAL CONDUCTIVITY OF PURE METALS

Metal or alloy	Thermal Conductivity near room temperature (cal / cm^2 • cm • s • °C)
Beryllium	0.35
Cadmium	0.22
Chromium	0.16
Cobalt	0.165
Germanium	0.14
Gold	0.71
Indium	0.057
Iridium	0.14
Lithium	0.17
Molybdenum	0.34
Niobium	0.13
Palladium	0.168
Platinum	0.165
Plutonium	0.020
Rhenium	0.17
Rhodium	0.21
Silicon	0.20
Silver	1.0
Sodium	0.32
Tantalum	0.130
Thallium	0.093
Thorium	0.090
Tungsten	0.397
Uranium	0.071
Vanadium	0.074
Yttrium	0.035

Data from *ASM Metals Reference Book, Third Edition*, Michael Bauccio, Ed., ASM International, Materials Park, OH, p156, (1993).

Table 104. THERMAL CONDUCTIVITY OF CERAMICS

(SHEET 1 OF 12)

Class	Ceramic	Thermal Conductivity ($cal \cdot cm^{-1} \cdot sec^{-1} \cdot K^{-1}$)
Borides	Chromium Diboride (CrB_2)	0.049-0.076 at room temp.
	Hafnium Diboride (HfB_2)	0.015 at room temp.
	Tantalum Diboride (TaB_2)	0.026 at room temp.
		0.033 at 200 °C.
	Titanium Diboride (TiB_2)	0.058-0.062 at room temp.
		0.063 at 200 °C
	Zirconium Diboride (ZrB_2)	0.055-0.058 at room temp.
		0.055-0.060 at 200 °C
Carbides	Boron Carbide (B_4C)	0.065-0.069 at room temp.
		0.198 at 425 °C
	Hafnium Monocarbide (HfC)	0.053 at room temp.
		0.15 + 1.20x10 T watts cm^{-1} K^{-1} from 1000-2000K
	Silicon Carbide (SiC)	
	(with 1 wt% Be addictive)	0.621
	(with 1 wt% B addictive)	0.406
	(with 1 wt% Al addictive)	0.143
	(with 2 wt% BN addictive)	0.263
	(with 1.6 wt% BeO addictive)	0.645 at room temp.
	(with 3.2 wt% BeO addictive)	0.645 at room temp.
		0.098-0.10 at 20°C

Source: data compiled by J.S. Park from *No. 1 Materials Index,* Peter T.B. Shaffer, Plenum Press, New York, (1964); *Smithells Metals Reference Book,* Eric A. Brandes, ed., in association with Fulmer Research Institute Ltd. 6th ed. London, Butterworths, Boston, (1983); and Ceramic Source, American Ceramic Society (1986-1991)

Table 104. THERMAL CONDUCTIVITY OF CERAMICS
(SHEET 2 OF 12)

Class	Ceramic	Thermal Conductivity (cal • cm^{-1} • sec^{-1} • K^{-1})
	(cubic, CVD)	0.289 at 127°C
		0.049-0.080 at 600°C
		0.061 at 800°C
		0.051 at 1000°C
		0.0059 at 1250°C
		0.0827 at 1327°C
		0.0032 at 1530°C
	Tantalum Monocarbide (TaC)	0.053 at room temp.
	Titanium Monocarbide (TiC)	0.041-0.074 at room temp.
		0.0135 at 1000 °C
	Trichromium Dicarbide (Cr_3C_2)	0.454
	Tungsten Monocarbide (WC)	0.201 at 20 °C
	(6% Co, 1-3μm grain size)	0.239
	(12% Co, 1-3μm grain size)	0.251
	(24% Co, 1-3μm grain size)	0.239
	(6% Co, 2-4μm grain size)	0.251
	(6% Co, 3-6μm grain size)	0.256
	Zirconium Monocarbide (ZrC)	0.049 at room temp.
		0.098 at 50°C
		0.069 at 150°C
		0.065 at 188°C

Source: data compiled by J.S. Park from *No. 1 Materials Index,* Peter T.B. Shaffer, Plenum Press, New York, (1964); *Smithells Metals Reference Book,* Eric A. Brandes, ed., in association with Fulmer Research Institute Ltd. 6th ed. London, Butterworths, Boston, (1983); and Ceramic Source, American Ceramic Society (1986-1991)

Table 104. THERMAL CONDUCTIVITY OF CERAMICS
(SHEET 3 OF 12)

Class	Ceramic	Thermal Conductivity ($cal \cdot cm^{-1} \cdot sec^{-1} \cdot K^{-1}$)
		0.061 at 288°C
		0.080 at 600°C
		0.083 at 800°C
		0.086 at 1000°C
		0.089 at 1200°C
		0.092 at 1400°C
		0.096 at 1600°C
		0.099 at 1800°C
		0.103 at 2000°C
		0.105 at 2200°C
Nitrides	Aluminum Nitride (AlN)	0.072 at 25°C
		0.060 at 200°C
		0.053 at 400°C
		0.048 at 600°C
		0.042 at 800°C
	Boron Nitride (BN) parallel to c axis	0.0687 at 300°C
		0.0646 at 700°C
		0.0637 at 1000°C
	parallel to a axis	0.0362 at 300°C
		0.0318 at 700°C
		0.0295 at 1000°C

Source: data compiled by J.S. Park from *No. 1 Materials Index,* Peter T.B. Shaffer, Plenum Press, New York, (1964); *Smithells Metals Reference Book,* Eric A. Brandes, ed., in association with Fulmer Research Institute Ltd. 6th ed. London, Butterworths, Boston, (1983); and Ceramic Source, American Ceramic Society (1986-1991)

Table 104. THERMAL CONDUCTIVITY OF CERAMICS
(SHEET 4 OF 12)

Class	Ceramic	Thermal Conductivity ($cal \cdot cm^{-1} \cdot sec^{-1} \cdot K^{-1}$)
	Titanium Mononitride (TiN)	0.069 at 25 °C
		0.057 at 127 °C
		0.040 at 200 °C
		0.027 at 650 °C
		0.020 at 1000 °C
		0.162 at 1500 °C
		0.136 at 2300 °C
	Trisilicon tetranitride (Si_3N_4) (pressureless sintered)	0.072 at room temp.
		0.022-0.072 at 127 °C
		0.041 at 200-750 °C
		0.036-0.042 at 500 °C
	(pressureless sintered)	0.038 at 1000 °C
		0.033-0.034 at 1200 °C
	Zirconium Mononitride (ZrN)	0.040 at 200 °C
		0.025 at 425 °C
		0.018 at 650 °C
		0.016 at 875 °C
		0.015 at 1100 °C
Oxides	Aluminum Oxide (Al_2O_3)	0.06 at room temp.
		0.04-0.069 at 100°C
		0.03-0.064 at 200°C
		0.037 at 315°C

Source: data compiled by J.S. Park from *No. 1 Materials Index,* Peter T.B. Shaffer, Plenum Press, New York, (1964); *Smithells Metals Reference Book*, Eric A. Brandes, ed., in association with Fulmer Research Institute Ltd. 6th ed. London, Butterworths, Boston, (1983); and Ceramic Source, American Ceramic Society (1986-1991)

Table 104. THERMAL CONDUCTIVITY OF CERAMICS
(SHEET 5 OF 12)

Class	Ceramic	Thermal Conductivity (cal • cm^{-1} • sec^{-1} • K^{-1})
		0.02-0.031 at 400°C
		0.035 at 500°C
		0.021-0.022 at 600°C
		0.015-0.017 at 800°C
		0.014-0.016 at 1000°C
		0.013-0.015 at 1200°C
		0.013 at 1400°C
		0.014 at 1600°C
		0.017 at 1800°C
	Aluminum Oxide (Al_2O_3) (single crystal)	0.103 at 20°C
		0.047 at 300°C
		0.029 at 800°C
	Beryllium Oxide (BeO)	0.038-0.47 at 20°C
		0.032-0.34 at 100°C
		0.14-0.16 at 400°C
		0.089-0.1137 at 600°C
		0.060-0.093 at 800°C
		0.043 at 1100°C
		0.041-0.054 at 1200°C
		0.038 at 1300°C

Source: data compiled by J.S. Park from *No. 1 Materials Index,* Peter T.B. Shaffer, Plenum Press, New York, (1964); *Smithells Metals Reference Book,* Eric A. Brandes, ed., in association with Fulmer Research Institute Ltd. 6th ed. London, Butterworths, Boston, (1983); and Ceramic Source, American Ceramic Society (1986-1991)

Table 104. THERMAL CONDUCTIVITY OF CERAMICS

(SHEET 6 OF 12)

Class	Ceramic	Thermal Conductivity ($cal \cdot cm^{-1} \cdot sec^{-1} \cdot K^{-1}$)
		0.036 at 1400°C
		0.034 at 1500°C
		0.033-0.039 at 1600°C
		0.033 at 1700°C
		0.036 at 1800°C
		0.036 at 1900°C
		0.036 at 2000°C
	Calcium Oxide (CaO)	0.037 at 100°C
		0.027 at 200°C
		0.022 at 400°C
		0.020 at 600°C
		0.019 at 800°C
		0.0186-0.019 at 1000°C
	Cerium Dioxide (CeO_2)	0.0229 at 400K
		0.00287 at 1400K
	Dichromium Trioxide (Cr_2O_3)	0.0239-0.0788
	Hafnium Dioxide (HfO_2)	0.0273 at 25-425°C
	Magnesium Oxide (MgO)	0.097 at room temp.
		0.078-0.082 at 100°C
		0.064-0.065 at 200°C
		0.038-0.045 at 400°C

Source: data compiled by J.S. Park from *No. 1 Materials Index,* Peter T.B. Shaffer, Plenum Press, New York, (1964); *Smithells Metals Reference Book*, Eric A. Brandes, ed., in association with Fulmer Research Institute Ltd. 6th ed. London, Butterworths, Boston, (1983); and Ceramic Source, American Ceramic Society (1986-1991)

Table 104. THERMAL CONDUCTIVITY OF CERAMICS
(SHEET 7 OF 12)

Class	Ceramic	Thermal Conductivity $(cal \cdot cm^{-1} \cdot sec^{-1} \cdot K^{-1})$
		0.0198-0.026 at 800°C
		0.016-0.020 at 1000°C
		0.0139-0.0148 at 1200°C
		0.012-0.014 at 1400°C
		0.0108-0.016 at 1600°C
		0.0096-0.0191 at 1800°C
	Nickel monoxide (NiO)	
	(0% porosity)	0.029 at 100°C
	(0% porosity)	0.024 at 200°C
	(0% porosity)	0.017 at 400°C
	(0% porosity)	0.012 at 800°C
	(0% porosity)	0.011 at 1000°C
	Silicon Dioxide (SiO_2)	0.0025 at 200°C
		0.003 at 400°C
		0.004 at 800°C
		0.005 at 1200°C
		0.006 at 1600°C
	Thorium Dioxide (ThO_2)	
	(0% porosity)	0.024 at room temp.
	(0% porosity)	0.020 at 100°C
	(0% porosity)	0.019 at 200°C
	(0% porosity)	0.014 at 400°C
	(0% porosity)	0.010 at 600°C
	(0% porosity)	0.008 at 800°C

Source: data compiled by J.S. Park from *No. 1 Materials Index,* Peter T.B. Shaffer, Plenum Press, New York, (1964); *Smithells Metals Reference Book,* Eric A. Brandes, ed., in association with Fulmer Research Institute Ltd. 6th ed. London, Butterworths, Boston, (1983); and Ceramic Source, American Ceramic Society (1986-1991)

Table 104. THERMAL CONDUCTIVITY OF CERAMICS
(SHEET 8 OF 12)

Class	Ceramic	Thermal Conductivity ($cal \cdot cm^{-1} \cdot sec^{-1} \cdot K^{-1}$)
	(0% porosity)	0.007-0.0074 at 1000°C
	(0% porosity)	0.006-0.0076 at 1200°C
	(0% porosity)	0.006 at 1400°C
	Titanium Oxide (TiO_2)	
	(0% porosity)	0.016 at 100°C
	(0% porosity)	0.012 at 200°C
	(0% porosity)	0.009 at 400°C
	(0% porosity)	0.008 at 600°C
	(0% porosity)	0.008 at 800°C
	(0% porosity)	0.008 at 1000°C
	(0% porosity)	0.008 at 1200°C
	Uranium Dioxide (UO_2)	
	(0% porosity)	0.025 at 100°C
	(0% porosity)	0.020 at 200°C
	(0% porosity)	0.015 at 400°C
	(0% porosity)	0.010 at 600°C
	(0% porosity)	0.009 at 800°C
	(0% porosity)	0.008 at 1000°C
		0.018 at 100°C
		0.012 at 400°C
		0.008 at 600°C

Source: data compiled by J.S. Park from *No. 1 Materials Index,* Peter T.B. Shaffer, Plenum Press, New York, (1964); *Smithells Metals Reference Book,* Eric A. Brandes, ed., in association with Fulmer Research Institute Ltd. 6th ed. London, Butterworths, Boston, (1983); and Ceramic Source, American Ceramic Society (1986-1991)

Table 104. THERMAL CONDUCTIVITY OF CERAMICS
(SHEET 9 OF 12)

Class	Ceramic	Thermal Conductivity ($cal \cdot cm^{-1} \cdot sec^{-1} \cdot K^{-1}$)
		0.008 at 700°C
		0.006 at 1000°C
		0.006 at 1200°C
	Zirconium Oxide (ZrO_2)	
	(stabilized, 0% porosity)	0.005 at 100°C
	(stabilized, 0% porosity)	0.005 at 200°C
	(stabilized, 0% porosity)	0.005 at 400°C
	(stabilized, 0% porosity)	0.0055 at 800°C
	(stabilized, 0% porosity)	0.006 at 1200°C
	(stabilized, 0% porosity)	0.0065 at 1400°C
	(stabilized)	0.004 at 100°C
	(stabilized)	0.0044 at 500°C
	(stabilized)	0.0048-0.0055 at 1000°C
	(stabilized)	0.0049-0.0050 at 1200°C
	(MgO stabilized)	0.0076 at room temp.
	(MgO stabilized)	0.0057 at 800°C
	(Y_2O_3 stabilized)	0.0055 at room temp.
	(Y_2O_3 stabilized)	0.0053 at 800°C
	(plasma sprayed)	0.0019-0.0031 at room temp.
	(plasma sprayed)	0.0019-0.0022 at 800°C
	(plasma sprayed and coated with Cr_2O_3)	0.0033 at room temp.
	(plasma sprayed and coated with Cr_2O_3)	0.0033 at 800°C

Source: data compiled by J.S. Park from *No. 1 Materials Index,* Peter T.B. Shaffer, Plenum Press, New York, (1964); *Smithells Metals Reference Book*, Eric A. Brandes, ed., in association with Fulmer Research Institute Ltd. 6th ed. London, Butterworths, Boston, (1983); and Ceramic Source, American Ceramic Society (1986-1991)

Table 104. THERMAL CONDUCTIVITY OF CERAMICS
(SHEET 10 OF 12)

Class	Ceramic	Thermal Conductivity ($cal \cdot cm^{-1} \cdot sec^{-1} \cdot K^{-1}$)
	(5-10% CaO stabilized)	0.0045 at 400°C
	(5-10% CaO stabilized)	0.0049 at 800°C
	(5-10% CaO stabilized)	0.0057 at 1200°C
	Cordierite ($2MgO\ 2Al_2O_3\ 5SiO_2$)	
	(ρ=2.3g/cm^3)	0.0077 at 20°C
	(ρ=2.3g/cm^3)	0.0062 at 300°C
	(ρ=2.3g/cm^3)	0.0055 at 500°C
	(ρ=2.3g/cm^3)	0.0055 at 800°C
	(ρ=2.1g/cm^3)	0.0043 at 20°C
	(ρ=2.1g/cm^3)	0.0041 at 300°C
	(ρ=2.1g/cm^3)	0.0040 at 500°C
	(ρ=2.1g/cm^3)	0.0038 at 800°C
	Mullite ($3Al_2O_3\ 2SiO_2$)	
	(0% porosity)	0.0145 at 100°C
	(0% porosity)	0.013 at 200°C
	(0% porosity)	0.011 at 400°C
	(0% porosity)	0.010 at 600°C
	(0% porosity)	0.0095 at 800°C
	(0% porosity)	0.009 at 1000°C
	(0% porosity)	0.009 at 1200°C
	(0% porosity)	0.009 at 1400°C

Source: data compiled by J.S. Park from *No. 1 Materials Index,* Peter T.B. Shaffer, Plenum Press, New York, (1964); *Smithells Metals Reference Book,* Eric A. Brandes, ed., in association with Fulmer Research Institute Ltd. 6th ed. London, Butterworths, Boston, (1983); and Ceramic Source, American Ceramic Society (1986-1991)

Table 104. THERMAL CONDUCTIVITY OF CERAMICS

(SHEET 11 OF 12)

Class	Ceramic	Thermal Conductivity $(cal \cdot cm^{-1} \cdot sec^{-1} \cdot K^{-1})$
	Sillimanite (Al_2O_3 SiO_2)	
	(0% porosity)	0.0042 at 100°C
	(0% porosity)	0.004 at 400°C
	(0% porosity)	0.0035 at 800°C
	(0% porosity)	0.0035 at 1200°C
	(0% porosity)	0.003 at 1500°C
	Spinel (Al_2O_3 MgO)	
	(0% porosity)	0.035 at 100°C
	(0% porosity)	0.031 at 200°C
	(0% porosity)	0.024 at 400°C
	(0% porosity)	0.019 at 600°C
	(0% porosity)	0.015 at 800°C
	(0% porosity)	0.013-0.0138 at 1000°C
	(0% porosity)	0.013 at 1200°C
	Zircon (SiO_2 ZrO_2)	
	(0% porosity)	0.0145 at 100°C
	(0% porosity)	0.0135 at 200°C
	(0% porosity)	0.012 at 400°C
	(0% porosity)	0.010 at 800°C
	(0% porosity)	0.0095 at 1200°C
	(0% porosity)	0.0095 at 1400°C

Source: data compiled by J.S. Park from *No. 1 Materials Index,* Peter T.B. Shaffer, Plenum Press, New York, (1964); *Smithells Metals Reference Book,* Eric A. Brandes, ed., in association with Fulmer Research Institute Ltd. 6th ed. London, Butterworths, Boston, (1983); and Ceramic Source, American Ceramic Society (1986-1991)

Table 104. THERMAL CONDUCTIVITY OF CERAMICS

(SHEET 12 OF 12)

Class	Ceramic	Thermal Conductivity ($cal \cdot cm^{-1} \cdot sec^{-1} \cdot K^{-1}$)
Silicides	Molybdenum Disilicide ($MoSi_2$)	0.129 at 150°C 0.074 at 425°C 0.053 at 540°C 0.057 at 650°C 0.046 at 875°C 0.041 at 1100°C

Source: data compiled by J.S. Park from *No. 1 Materials Index,* Peter T.B. Shaffer, Plenum Press, New York, (1964); *Smithells Metals Reference Book,* Eric A. Brandes, ed., in association with Fulmer Research Institute Ltd. 6th ed. London, Butterworths, Boston, (1983); and Ceramic Source, American Ceramic Society (1986-1991)

Table 105. THERMAL CONDUCTIVITY OF GLASSES

(SHEET 1 OF 5)

Glass	Description	Thermal Conductivity	Units	Temperature Range of Validity
SiO_2 glass		0.00329	cal/cm s K	20°C
		0.59	W/m K	80°C
		0.67	W/m K	100°C
		0.88	W/m K	150°C
		1.10	W/m K	200°C
		1.28	W/m K	250°C
		1.32	W/m K	273.1°C
		1.36	W/m K	300°C
		1.43	W/m K	350°C
		1.50	W/m K	400°C
		1.62	W/m K	500°C
		1.72	W/m K	600°C
		1.80	W/m K	700°C

Source: compiled by J.S. Park *from* O. V. Mazurin, M. V. Streltsina and T. P. Shvaiko-Shvaikovskaya, *Handbook of Glass Data*, Part A and Part B, Elsevier, New York, 1983

Table 105. THERMAL CONDUCTIVITY OF GLASSES

(SHEET 2 OF 5)

Glass	Description	Thermal Conductivity	Units	Temperature Range of Validity
SiO_2-Na_2O glass	(22% mol Na_2O)	0.70	kcal/m hr K	450°C
	(22% mol Na_2O)	0.90	kcal/m hr K	850°C
	(22% mol Na_2O)	1.20	kcal/m hr K	1050°C
	(22% mol Na_2O)	1.55	kcal/m hr K	1250°C
	(22% mol Na_2O)	2.25	kcal/m hr K	1500°C
	(25% mol Na_2O)	0.15	W/m K	35 K
	(25% mol Na_2O)	0.25	W/m K	60 K
	(25% mol Na_2O)	0.40	W/m K	80 K
	(25% mol Na_2O)	0.50	W/m K	100 K
	(25% mol Na_2O)	0.60	W/m K	140 K
	(25% mol Na_2O)	0.65	W/m K	150 K
	(25% mol Na_2O)	0.80	W/m K	190 K
	(25% mol Na_2O)	0.85	W/m K	240 K

Source: compiled by J.S. Park *from* O. V. Mazurin, M. V. Streltsina and T. P. Shvaiko-Shvaikovskaya, *Handbook of Glass Data*, Part A and Part B, Elsevier, New York, 1983

Table 105. THERMAL CONDUCTIVITY OF GLASSES

(SHEET 3 OF 5)

Glass	Description	Thermal Conductivity	Units	Temperature Range of Validity
SiO_2-Na_2O glass (Con't)	(25% mol Na_2O)	0.90	W/m K	280 K
	(25% mol Na_2O)	0.95	W/m K	300 K
	(27% mol Na_2O)	0.68	kcal/m hr K	450°C
	(27% mol Na_2O)	0.85	kcal/m hr K	850°C
	(27% mol Na_2O)	1.10	kcal/m hr K	1050°C
	(27% mol Na_2O)	1.45	kcal/m hr K	1250°C
	(27% mol Na_2O)	1.80	kcal/m hr K	1500°C
	(34.05% mol Na_2O)	0.5	kcal/m hr K	450°C
	(34.05% mol Na_2O)	0.75	kcal/m hr K	850°C
	(34.05% mol Na_2O)	0.75	kcal/m hr K	1050°C
	(34.05% mol Na_2O)	1.20	kcal/m hr K	1250°C
	(34.05% mol Na_2O)	1.5	kcal/m hr K	1500°C

Source: compiled by J.S. Park *from* O. V. Mazurin, M. V. Streltsina and T. P. Shvaiko-Shvaikovskaya, *Handbook of Glass Data*, Part A and Part B, Elsevier, New York, 1983

Table 105. THERMAL CONDUCTIVITY OF GLASSES

(SHEET 4 OF 5)

Glass	Description	Thermal Conductivity	Units	Temperature Range of Validity
SiO_2-PbO glass	(51.9% mol PbO)	0.00089	cal/cm s K	-150°C
	(51.9% mol PbO)	0.00100	cal/cm s K	-100°C
	(51.9% mol PbO)	0.00111	cal/cm s K	-50°C
	(51.9% mol PbO)	0.00123	cal/cm s K	0°C
	(51.9% mol PbO)	0.00134	cal/cm s K	50°C
	(51.9% mol PbO)	0.00146	cal/cm s K	100°C
	(49.3% mol PbO)	0.00130	cal/cm s K	40°C
	(66.2% mol PbO)	0.00112	cal/cm s K	40°C
B_2O_3 glass		0.5	mW/cm K	2K
		0.75	mW/cm K	5K
		1.5	mW/cm K	20K
B_2O_3-Na_2O glass	(3% mol Na_2O)	1.7 + 0.0054 (T-900)	W/m K	1173-1373 K
	(7% mol Na_2O)	1.5 + 0.0045 (T-900)	W/m K	1173-1373 K
	(11% mol Na_2O)	1.25 + 0.0037 (T-900)	W/m K	1173-1373 K

Source: compiled by J.S. Park *from* O. V. Mazurin, M. V. Streltsina and T. P. Shvaiko-Shvaikovskaya, *Handbook of Glass Data*, Part A and Part B, Elsevier, New York, 1983

Table 105. THERMAL CONDUCTIVITY OF GLASSES

(SHEET 5 OF 5)

Glass	Description	Thermal Conductivity	Units	Temperature Range of Validity
B_2O_3-Na_2O glass (Con't)	(14% mol Na_2O)	1.15 + 0.0020 (T-900)	W/m K	1173-1373 K
	(19% mol Na_2O)	1.0 + 0.0012 (T-900)	W/m K	1173-1373 K
	(25% mol Na_2O)	0.85 + 0.00075 (T-900)	W/m K	1173-1373 K
	(31% mol Na_2O)	0.9 + 0.00080 (T-900)	W/m K	1173-1373 K
B_2O_3-PbO glass	(27.6% mol PbO)	0.522±0.022	W/m K	30°C
	(31.9% mol PbO)	0.483±0.016	W/m K	30°C
	(36.7% mol PbO)	0.464±0.010	W/m K	30°C
	(42.1% mol PbO)	0.433±0.018	W/m K	30°C
	(48.3% mol PbO)	0.406±0.020	W/m K	30°C
	(55.5% mol PbO)	0.381±0.015	W/m K	30°C
	(64.0% mol PbO)	0.351±0.011	W/m K	30°C

Source: compiled by J.S. Park *from* O. V. Mazurin, M. V. Streltsina and T. P. Shvaiko-Shvaikovskaya, *Handbook of Glass Data*, Part A and Part B, Elsevier, New York, 1983

Table 106. THERMAL CONDUCTIVITY OF CRYOGENIC INSULATION

Class*	Cryogenic Insulation	Thermal Conductivity Range ($mW \cdot m^{-1} \cdot K^{-1}$)	Interspace Pressure (mm Hg)
2	Multilayer	0.04—0.2	10^{-4}
3	Opacified powder	0.26—0.7	10^{-4}
4	Evacuated powder	1.0—2.0	10^{-4}
5	Vacuum flask	5.0	10^{-6}
6	Gas–filled powder	1.7—7.0	760
7	Expanded foam	5.0—35	760
8	Fiber blanket	35—45	760

To convert **mm Hg** to **$N \cdot m^{-2}$** multiply by **133.32.**

Source: From Boltz, R. E. and Tuve, G. L., Eds., *Handbook of Tables for Applied Engineering Science, 2nd ed.*, CRC Press, Cleveland, 1973, 529.

* 1. Liquid and vapor shields – Very low–temperature, valuable, or dangerous liquids such as helium or fluorine are often shielded by an intermediate cryogenic liquid or vapor container that must in turn be insulated by one of the methods described below.

2. Multilayer reflecting shields – Foil or aluminized plastic alternated with paper-thin glass or plastic-fiber sheets; lowest conductivity, low density, and heat storage; good stability; minimum support structure.

3. Opacified evacuated powders - Contain metallic flakes to reduce radiation; conform to irregular shapes.

4. Evacuated dielectric powders - Very fine powders of low-conductivity adsorbent; moderate vacuum requirement; minimum fire hazard in oxygen.

5. Vacuum flasks (Dewar) - Tight shield-space with highly. reflecting walls and high vacuum; minimum heat capacity; rugged; small thickness.

6. Gas-filled powders – Same powders as Class 4 but with air or inert gas; low cost; easy application; no vacuum requirement.

7. Expanded foams – Very light foamed plastic; inexpensive; minimum weight but bulky; self supporting.

8. Porous fiber blankets – Blanket material of fine fibers, usually glass; minimum cost and easy installation but not an adequate insulation for most cryogenic applications.

Table 107. THERMAL CONDUCTIVITY OF CRYOGENIC SUPPORTS

Insulation Support	Mean Thermal Conductivity* ($W \cdot m^{-1} \cdot K^{-1}$)
Aluminum alloy	86
"K" Monel®	17
Stainbss steel	9.3
Titanhm alloy	6.1
Nylon	0.29
Teflon	0.24

Source: From Boltz, R. E. and Tuve, G. L., Eds., *Handbook of Tables for Applied Engineering Science, 2nd ed.*, CRC Press, Cleveland, 1973, 529.

*Range of Validity is 20–300 K.

Table 108. THERMAL CONDUCTIVITY OF SPECIAL CONCRETES*

Description; type of aggregate	Thermal Conductivity Btu / (hr • ft • °F)
Frost resisting; 1% $CaCl_2$; normal aggregates	1.0
Frost-resisting porous;6% air entrainment	0.85
Lightweight; with expanded shale or clay	0.25
Lightweight; with foamed slag	0.20
Cinder concrete; fine and coarse	0.25
Pulverized fuel ash	0.25
Lightweight refractory concrete with aluminous cement	0.20
Lightweight; insulating, with perlite	0.15
Lightweight; insulating, with expanded vermiculite	0.10

Source: from Bolz, R. E. and Tuve, C. L., Eds., *Handbook of Tables for Applied Engineering Science, 2nd ed.*, CRC Press, Cleveland, 1973, p.645.

* A great many varieties of aggregates have been used for concrete, dependent largely on the materials available. In general, high density concretes have high strength and high thermal conductivity, although such variables as water/cement ratio, percentage of fines, and curing conditions may result in wide differences in properties with the same materials.

Table 109. THERMAL CONDUCTIVITY OF SIC-WHISKER-REINFORCED CERAMICS

Composite	Thermal Conductivity (W/m • K)	
	at 22 °C	at 600 °C
Alumina	36 ± 5	12 ± 3
Alumina with 20 vol% SiC whiskers	32	16
SiC	95	50
Mullite with 20 vol% SiC whiskers	7.2	—

Data from *ASM Engineering Materials Reference Book, Second Edition*, Michael Bauccio, Ed., ASM International, Materials Park, OH, p173,(1994).

Table 110. THERMAL CONDUCTIVITY OF POLYMERS

(SHEET 1 OF 10)

Class	Polymer	Thermal Conductivity (ASTM C177) Btu / (hr • ft • °F)
ABS Resins; Molded, Extruded	Medium impact	0.08—0.18
	High impact	0.12—0.16
	Very high impact	0.01—0.14
	Low temperature impact	0.08—0.14
	Heat resistant	0.12—0.20
Acrylics; Cast, Molded, Extruded	Cast Resin Sheets, Rods:	
	General purpose, type I	0.12
	General purpose, type II	0.12
	Moldings:	
	Grades 5, 6, 8	0.12
	High impact grade	0.12
Thermoset Carbonate	Allyl diglycol carbonate	1.45

Source: data compiled by J.S. Park *from* Charles T. Lynch, *CRC Handbook of Materials Science, Vol. 3*, CRC Press, Boca Raton, Florida, 1975 and Engineered Materials Handbook, Vol.2, Engineering Plastics, ASM International, Metals Park, Ohio, 1988.

Table 110. THERMAL CONDUCTIVITY OF POLYMERS

(SHEET 2 OF 10)

Class	Polymer	Thermal Conductivity (ASTM C177) Btu / (hr • ft • °F)
Alkyds; Molded	Putty (encapsulating)	0.35—0.60
	Rope (general purpose)	0.35—0.60
	Granular (high speed molding)	0.35—0.60
	Glass reinforced (heavy duty parts)	0.20—0.30
Cellulose Acetate; Molded, Extruded	ASTM Grade:	
	H6—1	0.10—0.19
	H4—1	0.10—0.19
	H2—1	0.10—0.19
	MH—1, MH—2	0.10—0.19
	MS—1, MS—2	0.10—0.19
	S2—1	0.10—0.19

Source: data compiled by J.S. Park *from* Charles T. Lynch, *CRC Handbook of Materials Science, Vol. 3*, CRC Press, Boca Raton, Florida, 1975 and Engineered Materials Handbook, Vol.2, Engineering Plastics, ASM International, Metals Park, Ohio, 1988.

Table 110. THERMAL CONDUCTIVITY OF POLYMERS

(SHEET 3 OF 10)

Class	Polymer	Thermal Conductivity (ASTM C177) Btu / (hr • ft • °F)
Cellulose Acetate Butyrate; Molded, Extruded	ASTM Grade:	
	H4	0.10—0.19
	MH	0.10—0.19
	S2	0.10—0.19
Cellulose Acetate Propionate; Molded, Extruded	ASTM Grade:	
	1	0.10—0.19
	3	0.10—0.19
	6	0.10—0.19
Chlorinated Polymers	Chlorinated polyether	0.91
	Chlorinated polyvinyl chloride	0.95
Polycarbonates	Polycarbonate	0.11
	Polycarbonate (40% glass fiber reinforced)	0.13

Source: data compiled by J.S. Park *from* Charles T. Lynch, *CRC Handbook of Materials Science, Vol. 3*, CRC Press, Boca Raton, Florida, 1975 and Engineered Materials Handbook, Vol.2, Engineering Plastics, ASM International, Metals Park, Ohio, 1988.

Table 110. THERMAL CONDUCTIVITY OF POLYMERS

(SHEET 4 OF 10)

Class	Polymer	Thermal Conductivity (ASTM C177) Btu / (hr • ft • °F)
Fluorocarbons; Molded,Extruded	Polytrifluoro chloroethylene (PTFCE)	0.145
	Polytetrafluoroethylene (PTFE)	0.14
	Fluorinated ethylene propylene(FEP)	0.12
	Polyvinylidene— fluoride (PVDF)	0.14
Epoxies; Cast, Molded, Reinforced	Standard epoxies (diglycidyl ethers of bisphenol A)	
	Cast rigid	0.1—0.3
	Molded	0.1—0.5
	High strength laminate	2.35
Melamines; Molded	Filler & type	
	Cellulose electrical	0.17—0.20
	Glass fiber	0.28
Nylons; Molded, Extruded	Type 6	
	General purpose	1.2—1.69
	Glass fiber (30%) reinforced	1.69—3.27
	Cast	1.2—1.7

Source: data compiled by J.S. Park *from* Charles T. Lynch, *CRC Handbook of Materials Science, Vol. 3*, CRC Press, Boca Raton, Florida, 1975 and Engineered Materials Handbook, Vol.2, Engineering Plastics, ASM International, Metals Park, Ohio, 1988.

Table 110. THERMAL CONDUCTIVITY OF POLYMERS
(SHEET 5 OF 10)

Class	Polymer	Thermal Conductivity (ASTM C177) Btu / (hr • ft • °F)
Nylons; Molded, Extruded (Con't)	Type 11	1.5
	Type 12	1.7
	6/6 Nylon	
	General purpose molding	1.69—1.7
	Glass fiber reinforced	1.5— 3.3
	General purpose extrusion	1.7
	6/10 Nylon	
	General purpose	1.5
	Glass fiber (30%) reinforced	3.5
Phenolics; Molded	Type and filler	
	General: woodflour and flock	0.097—0.3
	Shock: paper, flock, or pulp	0.1—0.16
	High shock: chopped fabric or cord	0.097—0.170
	Very high shock: glass fiber	0.2

Source: data compiled by J.S. Park *from* Charles T. Lynch, *CRC Handbook of Materials Science, Vol. 3*, CRC Press, Boca Raton, Florida, 1975 and Engineered Materials Handbook, Vol.2, Engineering Plastics, ASM International, Metals Park, Ohio, 1988.

Table 110. THERMAL CONDUCTIVITY OF POLYMERS

(SHEET 6 OF 10)

Class	Polymer	Thermal Conductivity (ASTM C177) Btu / (hr • ft • °F)
Phenolics: Molded	Arc resistant—mineral	0.24—0.34
	Rubber phenolic—woodflour or flock	0.12
	Rubber phenolic—chopped fabric	0.05
	Rubber phenolic—asbestos	0.04
	ABS–Polycarbonate Alloy	2.46 (per ft)
PVC–Acrylic Alloy	PVC–acrylic sheet	1.01
	PVC–acrylic injection molded	0.98
Polymides	Unreinforced	6.78
	Unreinforced 2nd value	3.8
	Glass reinforced	3.59

Source: data compiled by J.S. Park *from* Charles T. Lynch, *CRC Handbook of Materials Science, Vol. 3*, CRC Press, Boca Raton, Florida, 1975 and Engineered Materials Handbook, Vol.2, Engineering Plastics, ASM International, Metals Park, Ohio, 1988.

Table 110. THERMAL CONDUCTIVITY OF POLYMERS

(SHEET 7 OF 10)

Class	Polymer	Thermal Conductivity (ASTM C177) Btu / (hr • ft • °F)
Polyacetals	Homopolymer:	
	Standard	0.13
	Copolymer:	
	Standard	0.16
	High flow	1.6
Polyester; Thermoplastic	Injection Moldings:	
	General purpose grade	0.36—0.55
Polyesters: Thermosets	Cast polyyester	
	Rigid	0.10—0.12
	Reinforced polyester moldings	
	High strength (glass fibers)	1.32—1.68
Phenylene Oxides	SE—100	1.1
	SE—1	1.5
	Glass fiber reinforced	1.15,1.1

Source: data compiled by J.S. Park *from* Charles T. Lynch, *CRC Handbook of Materials Science, Vol. 3*, CRC Press, Boca Raton, Florida, 1975 and Engineered Materials Handbook, Vol.2, Engineering Plastics, ASM International, Metals Park, Ohio, 1988.

Table 110. THERMAL CONDUCTIVITY OF POLYMERS

(SHEET 8 OF 10)

Class	Polymer	Thermal Conductivity (ASTM C177) Btu / (hr • ft • °F)
Phenylene oxides (Noryl)	Standard	1.8
	Polyarylsulfone	1.1
Polypropylene:	General purpose	1.21—1.36
	High impact	1.72
	Polyphenylene sulfide:	
	Standard	2
	40% glass reinforced	2
Polyethylenes; Molded, Extruded	Type I—lower density (0.910—0.925)	
	Melt index 0.3—3.6	0.19
	Melt index 6—26	0.19
	Melt index 200	0.19

Source: data compiled by J.S. Park *from* Charles T. Lynch, *CRC Handbook of Materials Science, Vol. 3*, CRC Press, Boca Raton, Florida, 1975 and Engineered Materials Handbook, Vol.2, Engineering Plastics, ASM International, Metals Park, Ohio, 1988.

Table 110. THERMAL CONDUCTIVITY OF POLYMERS

(SHEET 9 OF 10)

Class	Polymer	Thermal Conductivity (ASTM C177) Btu / (hr • ft • °F)
Polyethylenes; Molded, Extruded (Con't)	Type II—medium density (0.926—0.940)	
	Melt index 20	0.19
	Melt index l.0—1.9	0.19
	Type III—higher density (0.941—0.965)	
	Melt index 0.2—0.9	0.19
	Melt Melt index 0.l—12.0	0.19
	Melt index 1.5—15	0.19
	High molecular weight	0.19
Polystyrenes; Molded	Polystyrenes	
	General purpose	0.058—0.090
	Medium impact	0.024—0.090
	High impact	0.024—0.090
	Glass fiber -30% reinforced	0.117

Source: data compiled by J.S. Park *from* Charles T. Lynch, *CRC Handbook of Materials Science, Vol. 3*, CRC Press, Boca Raton, Florida, 1975 and Engineered Materials Handbook, Vol.2, Engineering Plastics, ASM International, Metals Park, Ohio, 1988.

Table 110. THERMAL CONDUCTIVITY OF POLYMERS

(SHEET 10 OF 10)

Class	Polymer	Thermal Conductivity (ASTM C177) Btu / (hr • ft • °F)
Polyvinyl Chloride And Copolymers; Molded, Extruded	Nonrigid—general	0.07—0.10
	Nonrigid—electrical	0.07—0.10
	Rigid—normal impact	0.07—0.10
	Vinylidene chloride	0.053
Silicones; Molded, Laminated	Fibrous (glass) reinforced silicones	0.18
	Granular (silica) reinforced silicones	0.25—0.5
	Woven glass fabric/ silicone laminate	0.075—0.125
Ureas; Molded	Alpha—cellulose filled (ASTM Type l)	0.17—0.244

Source: data compiled by J.S. Park *from* Charles T. Lynch, *CRC Handbook of Materials Science, Vol. 3*, CRC Press, Boca Raton, Florida, 1975 and Engineered Materials Handbook, Vol.2, Engineering Plastics, ASM International, Metals Park, Ohio, 1988.

Table 111. THERMAL CONDUCTIVITY OF FIBERGLASS REINFORCED PLASTICS

Class	Material	Glass fiber content (wt%)	Thermal conductivity (Btu • in/ft^2 • h •°F)
Glass fiber reinforced thermosets	Sheet molding compound (SMC)	15 to 30	1.3 to 1.7
	Bulk molding compound(BMC)	15 to 35	1.3 to 1.7
	Preform/mat(compression molded)	25 to 50	1.3 to 1.8
	Cold press molding–polyester	20 to 30	1.3 to 1.8
	Spray–up–polyester	30 to 50	1.2 to 1.6
	Filament wound–epoxy	30 to 80	1.92 to 2.28
	Rod stock–polyester	40 to 80	1.92 to 2.28
	Molding compound–phenolic	5 to 25	1.1 to 2.0
Glass–fiber–reinforced thermoplastic	Thermoplastic polyester	20 to 35	1.3

To convert (Btu • in/ft^2 • h •°F) to (W/m • K), multiply by 0.144

Data from *ASM Engineering Materials Reference Book, Second Edition*, Michael Bauccio, Ed., ASM International, Materials Park, OH, p106, (1994).

Table 112. THERMAL EXPANSION OF WROUGHT STAINLESS STEELS* (SHEET 1 OF 2)

Type	UNS Designation	Coefficient of Thermal Expansion (μm/m • °C)		
		0-100°C	100-315°C	0-538°C
201	S20100	15.7	17.5	18.4
202	S20200	17.5	18.4	19.2
205	S20500	—	17.9	19.1
301	S30100	17.0	17.2	18.2
302	S30200	17.2	17.8	18.4
302B	S30215	16.2	18.0	19.4
303	S30300	17.2	17.8	18.4
304	S30400	17.2	17.8	18.4
S30430	S30430	17.2	17.8	—
305	S30500	17.2	17.8	18.4
308	S30800	17.2	17.8	18.4
309	S30900	15.0	16.6	17.2
310	S31000	15.9	16.2	17.0
314	S31400	—	15.1	—
316	S31600	15.9	16.2	17.5
317	S31700	15.9	16.2	17.5
317L	S31703	16.5	—	18.1
321	S32100	16.6	17.2	18.6
330	N08330	14.4	16.0	16.7
347	S34700	16.6	17.2	18.6
384	S38400	17.2	17.8	18.4
405	S40500	10.8	11.6	12.1
409	S40900	11.7	—	—
410	S41000	9.9	11.4	11.6
414	S41400	10.4	11.0	12.1
416	S41600	9.9	11.0	11.6
420	S42000	10.3	10.8	11.7

Data from *ASM Metals Reference Book, Third Edition*, Michael Bauccio, Ed., ASM International, Materials Park, OH, p360, (1993).

Table 112. THERMAL EXPANSION OF WROUGHT STAINLESS STEELS* (SHEET 2 OF 2)

Type	UNS Designation	Coefficient of Thermal Expansion (μm/m • °C)		
		0-100°C	100-315°C	0-538°C
422	S42200	11.2	11.4	11.9
429	S42900	10.3	—	—
430	S43000	10.4	11.0	11.4
430F	S43020	10.4	11.0	11.4
431	S43100	10.2	12.1	—
434	S43400	10.4	11.0	11.4
436	S43600	9.3	—	—
440A	S44002	10.2	—	—
440C	S44004	10.2	—	—
444	S44400	10.0	10.6	11.4
446	S44600	10.4	10.8	11.2
PH 13–8 Mo	S13800	10.6	11.2	11.9
15–5 PH	S15500	10.8	11.4	—
17–4 PH	S17400	10.8	11.6	—
17–7 PH	S17700	11.0	11.6	—

Data from *ASM Metals Reference Book, Third Edition*, Michael Bauccio, Ed., ASM International, Materials Park, OH, p360, (1993).

* Annealed Condition.

Table 113. THERMAL EXPANSION OF WROUGHT TITANIUM ALLOYS

(SHEET 1 OF 2)

Class	Metal or Alloy	Coefficient of Linear Thermal Expansion (μm/m • K)						
		20-100 °C	20-205 °C	20-315 °C	20-425 °C	20-540 °C	20-650 °C	20-815 °C
Commercially Pure	99.5Ti	8.6	—	9.2	—	9.7	10.1	10.1
	99.2Ti	8.6	—	9.2	—	9.7	10.1	10.1
	99.1Ti	8.6	—	9.2	—	9.7	10.1	10.1
	99.0Ti	8.6	—	9.2	—	9.7	10.1	10.1
	99.2 Ti–0.2Pd	8.6	—	9.2	—	9.7	10.1	10.1
Alpha Alloys	Ti-5Al-2.5Sn	9.4	—	9.5	—	9.5	9.7	10.1
	Ti-5Al-2.5Sn (low O_2)	9.4	—	9.5	—	9.7	9.9	10.1
Near Alpha Alloys	Ti-8Al-1Mo-1V	8.5	—	9.0	—	10.1	10.3	—
	Ti-11Sn-1Mo-2.25Al-5.0Zr-1Mo-0.2Si	8.5	—	9.2	—	9.4	—	—
	Ti-6Al-2Sn-4Zr-2Mo	7.7	—	8.1	—	8.1	—	—
	Ti-5Al-5Sn-2Zr-2Mo-0.25Si	—	—	—	—	—	—	10.3
	Ti-6Al-2Nb-1Ta-1Mo	—	—	—	—	—	9.0	—

Data from *ASM Metals Reference Book, Third Edition*, Michael Bauccio, Ed., ASM International, Materials Park, OH, p511, (1993).

Table 113. THERMAL EXPANSION OF WROUGHT TITANIUM ALLOYS

(SHEET 2 OF 2)

Class	Metal or Alloy	Coefficient of Linear Thermal Expansion (μm/m • K)						
		20-100 °C	20-205 °C	20-315 °C	20-425 °C	20-540 °C	20-650 °C	20-815 °C
Alpha-Beta Alloys	Ti-8Mn	8.6	9.2	9.7	10.3	10.8	11.7	12.6
	Ti-3Al-2.5V	9.5	—	9.9	—	9.9	—	—
	Ti-6Al-4V	8.6	9.0	9.2	9.4	9.5	9.7	—
	Ti-6Al-4V (low O_2)	8.6	9.0	9.2	9.4	9.5	9.7	—
	Ti-6Al-6V-2Sn	9.0	—	9.4	—	9.5	—	—
	Ti-7Al-4Mo	9.0	9.2	9.4	9.7	10.1	10.4	11.2
	Ti-6Al-2Sn-4Zr-6Mo	9.0	9.2	9.4	9.5	9.5	—	—
	Ti-6Al-2Sn-2Zr-2Mo-2Cr-0.25Si	—	—	9.2	—	—	—	—
Beta Alloys	Ti-13V-11Cr-3Al	9.4	—	10.1	—	10.6	—	—
	Ti-8Mo-8V-2Fe-3Al	—	—	—	—	—	—	—
	Ti-3Al-8V-6Cr-4Mo-4Zr	—	—	—	9.68 (to 900 •F)	—	—	—

Data from *ASM Metals Reference Book, Third Edition*, Michael Bauccio, Ed., ASM International, Materials Park, OH, p511, (1993).

Table 114. Thermal Expansion of Graphite Magnesium Castings*

Fiber Type	Fiber content	Fiber orientation	Casting	Fiber Preform Method	Coefficient of Thermal Expansion (10^{-6}/K)
P75	40%	±16°	Hollow cylinder	Filament wound	1.3
	plus 9%	90°	Hollow cylinder	Filament wound	1.3
P100	40%	± 16°	Hollow cylinder	Filament wound	–0.07
P55	40%	0°	Plate	Prepreg	2.3
	30%	0° plus	Plate	Prepreg	4.5
	10%	90°	Plate	Prepreg	4.5

Data from *ASM Engineering Materials Reference Book, Second Edition*, Michael Bauccio, Ed., ASM International, Materials Park, OH, p148,(1994).

* Pitch-base fibers

Table 115. LINEAR THERMAL EXPANSION OF METALS AND ALLOYS (SHEET 1 OF 8)

Class	Metal or Alloy	Temperature (°C)	Coefficient of Thermal Expansion (μm/m • °C)
Aluminum and Aluminum Alloys	Aluminum (99.996%)	20 to 100	23.6
Wrought Alloys	EC, 1060, 1100	20 to 100	23.6
	2011, 2014	20 to 100	23.0
	2024	20 to 100	22.8
	2218	20 to 100	22.3
	3003	20 to 100	23.2
	4032	20 to 100	19.4
	5005, 5050, 5052	20 to 100	23.8
	5056	20 to 100	24.1
	5083	20 to 100	23.4
	5086	60 to 300	23.9
	5154	20 to 100	23.9
	5357	20 to 100	23.7
	5456	20 to 100	23.9
	6061, 6063	20 to 100	23.4
	6101, 6151	20 to 100	23.0
	7075	20 to 100	23.2
	7079, 7178	20 to 100	23.4
Casting Alloys	A13	20 to 100	20.4
	43 and 108	20 to 100	22.0
	A108	20 to 100	21.5
	A132	20 to 100	19.0
	D132	20 to 100	20.05
	F132	20 to 100	20.7
	138	20 to 100	21.4
	142	20 to 100	22.5

Data from *ASM Metals Reference Book, Third Edition*, Michael Bauccio, Ed., ASM International, Materials Park, OH, p154-155, (1993).

Table 115. LINEAR THERMAL EXPANSION OF METALS AND ALLOYS (SHEET 2 OF 8)

Class	Metal or Alloy	Temperature (°C)	Coefficient of Thermal Expansion (μm/m • °C)
	195	20 to 100	23.0
	B195	20 to 100	22.0
	214	20 to 100	24.0
	220	20 to 100	25.0
	319	20 to 100	21.5
	355	20 to 100	22.0
	356	20 to 100	21.5
	360	20 to 100	21.0
	750	20 to 100	23.1
	40E	21 to 93	24.7
Copper and Copper Alloys			
Wrought Coppers	Pure copper	20	16.5
	Electrolytic tough pitch copper (ETP)	20 to 100	16.8
	Deoxidized copper, high residual phosphorus (DHP)	20 to 300	17.7
	Oxygen-free copper	20 to 300	17.7
	Free machining copper, 0.5% Te or 1% Pb	20 to 300	17.7
Wrought Alloys	Gilding, 95%	20 to 300	18.1
	Commercial bronze, 90%	20 to 300	18.4
	Jewelry bronze, 87.5%	20 to 300	18.6
	Red brass, 85%	20 to 300	18.7
	Low brass, 80%	20 to 300	19.1
	Cartridge brass, 70%	20 to 300	19.9
	Yellow brass	20 to 300	20.3

Data from *ASM Metals Reference Book, Third Edition*, Michael Bauccio, Ed., ASM International, Materials Park, OH, p154-155, (1993).

Table 115. LINEAR THERMAL EXPANSION OF METALS AND ALLOYS (SHEET 3 OF 8)

Class	Metal or Alloy	Temperature (°C)	Coefficient of Thermal Expansion (µm/m • °C)
	Muntz metal	20 to 300	20.8
	Leaded commercial bronze	20 to 300	18.4
	Low-leaded brass	20 to 300	20.2
	Medium-leaded brass	20 to 300	20.3
	High-leaded brass	20 to 300	20.3
	Extra-high-leaded brass	20 to 300	20.5
	Free-cutting brass	20 to 300	20.5
	Leaded Muntz metal	20 to 300	20.8
	Forging brass	20 to 300	20.7
	Architectural bronze	20 to 300	20.9
	Inhibited admiralty	20 to 300	20.2
	Naval brass	20 to 300	21.2
	Leaded naval brass	20 to 300	21.2
	Manganese bronze (longitudinal)	20 to 300	21.2
	Manganese bronze (transverse)	20 to 300	23.4
	Phosphor bronze, 5% (longitudinal)	20 to 300	17.8
	Phosphor bronze, 5% (transverse)	20 to 300	23.4
	Phosphor bronze, 8% (longitudinal)	20 to 300	18.2
	Phosphor bronze, 8% (transverse)	20 to 300	19.4
	Phosphor bronze, 10% (longitudinal)	20 to300	18.4
	Phosphor bronze, 1.25%	20 to 300	17.8
	Free-cutting phosphor bronze	20 to 300	17.3
	Cupro-nickel, 30%	20 to 300	16.2
	Cupro-nickel, 10%	20 to 300	17.1

Data from *ASM Metals Reference Book, Third Edition*, Michael Bauccio, Ed., ASM International, Materials Park, OH, p154-155, (1993).

Table 115. Linear Thermal Expansion of Metals and Alloys (Sheet 4 of 8)

Class	Metal or Alloy	Temperature (°C)	Coefficient of Thermal Expansion (μm/m • °C)
	Nickel silver, 65-18	20 to 300	16.2
	Nickel silver, 55-18	20 to 300	16.7
	Nickel silver, 65-12	20 to 300	16.2
	High-silicon bronze (longitudinal)	20 to 300	18.0
	High-silicon bronze (transverse)	20 to 300	23.4
	Low-silicon bronze (longitudinal)	20 to 300	17.9
	Low-silicon bronze (transverse)	20 to 300	21.1
	Aluminum bronze	20 to 300	16.4
	Aluminum-silicon bronze	20 to 300	18.0
	Aluminum bronze	20 to 300	16.8
	Beryllium copper	20 to 300	17.8
Casting Alloys	88Cu-8Sn-4Zn	21 to 177	18.0
	89Cu-11Sn	20 to 300	18.4
	88Cu-6Sn-1.5Pb-4 .5Zn	21 to 260	18.5
	87Cu-8Sn-1Pb-4Zn	21 to 177	18.0
	87Cu-10Sn-1Pb-2Zn	21 to 177	18.0
	80Cu-10Sn-10Pb	21 to 204	18.5
	78Cu-7Sn-15Pb	21 to 204	18.5
	85Cu-5Sn-5Pb- 5Zn	21 to 204	18.1
	72Cu-1Sn-3Pb-24Zn	21 to 93	20.7
	67Cu-1Sn-3Pb-29Zn	21 to 93	20.2
	61Cu-1Sn-1Pb-37Zn	21 to 260	21.6
Manganese bronze	Manganese bronze, 60 ksi	21 to 204	20.5
	Manganese bronze, 65 ksi	21 to 93	21.6
	Manganese bronze, 110 ksi	21 to 260	19.8

Data from *ASM Metals Reference Book, Third Edition*, Michael Bauccio, Ed., ASM International, Materials Park, OH, p154-155, (1993).

Table 115. LINEAR THERMAL EXPANSION OF METALS AND ALLOYS (SHEET 5 OF 8)

Class	Metal or Alloy	Temperature (°C)	Coefficient of Thermal Expansion (μm/m • °C)
Iron and Iron Alloys	Pure iron	20	11.7
	Fe-C alloy 0.06% C	20 to 100	11.7
	Fe-C alloy 0.22% C	20 to 100	11.7
	Fe-C alloy 0.40% C	20 to 100	11.3
	Fe-C alloy 0.56% C	20 to 100	11.0
	Fe-C alloy 1.08% C	20 to 100	10.8
	Fe-C alloy 1.45% C	20 to 100	10.1
	Invar (36% Ni)	20	0-2
	13Mn-1.2C	20	18.0
	13Cr-0.35C	20 to 100	10.0
	12.3Cr-0.4Ni-0.09C	20 to 100	9.8
	17.7Cr-9.6Ni-0.06C	20 to 100	16.5
	18W-4Cr-1V	0 to 100	11.2
	Gray cast iron	0 to 100	10.5
	Malleable iron (pearlitic)	20 to 400	12
Lead and Lead Alloys	Corroding lead (99.73 + % Pb)	17 to 100	29.3
	5-95 solder	15 to 110	28.7
	20-80 solder	15 to 110	26.5
	50-50 solder	15 to 110	23.4
	1% antimonial lead	20 to 100	28.8
	8% antimonial lead	20 to 100	26.7
	9% antimonial lead	20 to 100	26.4
	Hard lead(96Pb-4Sb)	20 to 100	27.8
	Hard lead(94Pb-6Sb)	20 to 100	27.2
	Lead-base babbitt SAE 14	20 to 100	19.6
	Lead-base babbitt Alloy 8	20 to 100	24.0
Magnesium and Magnesium Alloys	Magnesium (99.8%)	20	25.2

Data from *ASM Metals Reference Book, Third Edition*, Michael Bauccio, Ed., ASM International, Materials Park, OH, p154-155, (1993).

Table 115. LINEAR THERMAL EXPANSION OF METALS AND ALLOYS (SHEET 6 OF 8)

Class	Metal or Alloy	Temperature (°C)	Coefficient of Thermal Expansion (μm/m • °C)
Casting alloys	AM100A	18 to 100	25.2
	AZ63A	20 to 100	26.1
	AZ91A,B,C	20 to 100	26
	AZ92A	18 to 100	25.2
	HZ32A	20 to 200	26.7
	ZH42	20 to 200	27
	ZH62A	20 to 200	27.1
	ZK51A	20	26.1
	EZ33A	20 to 100	26.1
	EK30A, EK41A	20 to 100	26.1
Wrought Alloys	M1A, A3A	20 to 100	26
	AZ31B,PE	20 to 100	26
	AZ61A, Z80A	20 to 100	26
	ZK60A, B	20 to 100	26
	HM31A	20 to 93	26.1
Nickel and Nickel Alloys	Nickel (99.95% Ni + Co)	0 to 100	13.3
	Duranickel	0 to 100	13.0
	Monel	0 to 100	14.0
	Monel (cast)	25 to 100	12.9
	Inconel	20 to 100	11.5
	Ni-o-nel	27 to 93	12.9
	Hastelloy B	0 to 100	10.0
	Hastelloy C	0 to 100	11.3
	Hastelloy D	0 to 100	11.0
	Hastelloy F	20 to 100	14.2
	Hastelloy N	21 to 204	10.4
	Hastelloy W	23 to 100	11.3
	Hastelloy X	26 to 100	13.8

Data from *ASM Metals Reference Book, Third Edition*, Michael Bauccio, Ed., ASM International, Materials Park, OH, p154-155, (1993).

Table 115. LINEAR THERMAL EXPANSION OF METALS AND ALLOYS (SHEET 7 OF 8)

Class	Metal or Alloy	Temperature (°C)	Coefficient of Thermal Expansion (μm/m • °C)
	Illium G	0 to 100	12.19
	Illium R	0 to 100	12.02
	80Ni-20Cr	20 to 1000	17.3
	60Ni-24Fc-l6Cr	20 to 1000	17.0
	35Ni-45Fe-20Cr	20 to 500	15.8
	Constantan	20 to 1000	18.8
Tin and Tin Alloys	Pure tin	0 to 100	23
	Solder (70Sn-30Pb)	15 to 110	21.6
	Solder (63Sn-37Pb)	15 to 110	24.7
Titanium and Titanium Alloys	99.9% Ti	20	8.41
	99.0% Ti	93	8.55
	Ti-5Al-2.5Sn	93	9.36
	Ti-8Mn	93	8.64
Zinc and Zinc Alloys	Pure zinc	20 to 250	39.7
	AG40A alloy	20 to 100	27.4
	AC41A alloy	20 to 100	27.4
	Commercial rolled zinc 0.08 Pb	20 to 40	32.5
	Commercial rolled zinc 0.3 Pb, 0.3 Cd	20 to 98	33.9
	Rolled zinc alloy (1 Cu, 0.010 Mg)	20 to 100	34.8
	Zn-Cu-Ti alloy (0.8 Cu, 0.15 Ti)	20 to 100	24.9
Pure Metals	Beryllium	25 to 100	11.6
	Cadmium	20	29.8
	Calcium	0 to 400	22.3
	Chromium	20	6.2

Data from *ASM Metals Reference Book, Third Edition*, Michael Bauccio, Ed., ASM International, Materials Park, OH, p154-155, (1993).

Table 115. LINEAR THERMAL EXPANSION OF METALS AND ALLOYS (SHEET 8 OF 8)

Class	Metal or Alloy	Temperature (°C)	Coefficient of Thermal Expansion (μm/m • °C)
	Cobalt	20	13.8
	Gold	20	14.2
	Iridium	20	6.8
	Lithium	20	56
	Manganese	0 to 100	22
	Palladium	20	11.76
	Platinum	20	8.9
	Rhenium	20 to 500	6.7
	Rhodium	20 to 100	8.3
	Ruthenium	20	9.1
	Silicon	0 to 1400	5
	Silver	0 to 100	19.68
	Tungsten	27	4.6
	Vanadium	23 to 100	8.3
	Zirconium	—	5.85

Data from *ASM Metals Reference Book, Third Edition*, Michael Bauccio, Ed., ASM International, Materials Park, OH, p154-155, (1993).

Table 116. THERMAL EXPANSION OF CERAMICS

(SHEET 1 OF 34)

Class	Ceramic	Thermal Expansion ($°C^{-1}$)
Borides	Chromium Diboride (CrB_2)	$4.6–11.1 \times 10^{-6}$ for 20–1000°C
	Hafnium Diboride (HfB_2)	$5.5–5.54 \times 10^{-6}$ for room temp.–1000°C
	Tantalum Diboride (TaB_2)	5.1×10^{-6} at room temp.
	Titanium Diboride (TiB_2)	$4.6–8.1 \times 10^{-6}$
	Zirconium Diboride (ZrB_2)	5.69×10^{-6} for 25–500°C
		$5.5–6.57 \times 10^{-6}$ °C for 25–1000°C
		6.98×10^{-6} for 20–1500°C
Carbides	Boron Carbide (B_4C)	4.5×10^{-6} for room temp.–800°C
		4.78×10^{-6} for 25–500°C
		5.54×10^{-6} for 25–1000°C

Source: data compiled by J.S. Park from *No. 1 Materials Index*, Peter T.B. Shaffer, Plenum Press, New York, (1964); Smithells Metals Reference Book, Eric A. Brandes, ed., in association with Fulmer Research Institute Ltd. 6th ed. London, Butterworths, Boston, (1983); and *Ceramic Source*, American Ceramic Society (1986-1991)

Table 116. THERMAL EXPANSION OF CERAMICS

(SHEET 2 OF 34)

Class	Ceramic	Thermal Expansion ($°C^{-1}$)
Carbides (Con't)	Boron Carbide (B_4C) (Con't)	6.02×10^{-6} for 25–1500°C
		6.53×10^{-6} for 25–2000°C
		7.08×10^{-6} for 25–2500°C
	Hafnium Monocarbide (HfC)	$6.27–6.59 \times 10^{-6}$ for 25–650°C
		6.25×10^{-6} for 25–1000°C
	Silicon Carbide (SiC)	4.63×10^{-6} for 25–500°C
		5.12×10^{-6} for 25–1000°C
		5.48×10^{-6} for 25–1500°C
		5.77×10^{-6} for 25–2000°C
		5.94×10^{-6} for 25–2500°C
		4.70×10^{-6} for 20–1500°C
		4.70×10^{-6} for 0–1700°C

Source: data compiled by J.S. Park from *No. 1 Materials Index*, Peter T.B. Shaffer, Plenum Press, New York, (1964); Smithells Metals Reference Book, Eric A. Brandes, ed., in association with Fulmer Research Institute Ltd. 6th ed. London, Butterworths, Boston, (1983); and *Ceramic Source*, American Ceramic Society (1986-1991)

Table 116. THERMAL EXPANSION OF CERAMICS

(SHEET 3 OF 34)

Class	Ceramic	Thermal Expansion (°C^{-1})
Carbides (Con't)	Tantalum Monocarbide (TaC)	6.29–6.32 x 10^{-6} for 25–500°C
		6.67 x 10^{-6} for 25–1000°C
		7.12 x 10^{-6} for 25–1500°C
		7.64 x 10^{-6} for 25–2000°C
		8.40 x 10^{-6} for 25–2500°C
		6.50 x 10^{-6} for 0–1000°C
		6.64 x 10^{-6} for 0–1200°C
	Titanium Monocarbide (TiC)	6.52–7.15 x 10^{-6} for 25–500°C
		7.18–7.45 x 10^{-6} for 25–750°C
		7.40–8.82 x 10^{-6} for 25–1000°C
		9.32 x 10^{-6} for 25–1250°C

Source: data compiled by J.S. Park from *No. 1 Materials Index,* Peter T.B. Shaffer, Plenum Press, New York, (1964); Smithells Metals Reference Book, Eric A. Brandes, ed., in association with Fulmer Research Institute Ltd. 6th ed. London, Butterworths, Boston, (1983); and *Ceramic Source*, American Ceramic Society (1986-1991)

Table 116. THERMAL EXPANSION OF CERAMICS

(SHEET 4 OF 34)

Class	Ceramic	Thermal Expansion ($°C^{-1}$)
Carbides (Con't)	Titanium Monocarbide (TiC) (Con't)	8.15–9.45 x 10^{-6} for 25–1500°C 8.81 x 10^{-6} for 25–2000°C 7.90 x 10^{-6} for 0–2500°C 7.08 x 10^{-6} for 0–750°C 7.85–7.86 x 10^{-6} for 0–1000°C 8.02 x 10^{-6} for 0–1275°C 8.29 x 10^{-6} for 0–1400°C 8.26 x 10^{-6} for 0–1525°C 8.40 x 10^{-6} for 0–1775°C

Source: data compiled by J.S. Park from *No. 1 Materials Index,* Peter T.B. Shaffer, Plenum Press, New York, (1964); Smithells Metals Reference Book, Eric A. Brandes, ed., in association with Fulmer Research Institute Ltd. 6th ed. London, Butterworths, Boston, (1983); and *Ceramic Source,* American Ceramic Society (1986-1991)

Table 116. THERMAL EXPANSION OF CERAMICS

(SHEET 5 OF 34)

Class	Ceramic	Thermal Expansion ($°C^{-1}$)
Carbides (Con't)	Trichromium Dicarbide (Cr_3C_2)	8.00×10^{-6} for 25–500°C
		9.95×10^{-6} for 25–500°C
		8.8×10^{-6} for 25–120°C
		10.9×10^{-6} for 150–980°C
	Tungsten Monocarbide (WC)	4.42×10^{-6} for 25–500°C
		$4.84–4.92 \times 10^{-6}$ for 25–1000°C
		$5.35–5.8 \times 10^{-6}$ for 25–1500°C
		$5.82–7.4 \times 10^{-6}$ for 25–2000°C
	Zirconium Monocarbide (ZrC)	6.10×10^{-6} for 25–500°C
		6.65×10^{-6} for 25–800°C
		6.56×10^{-6} for 25–1000°C
		7.06×10^{-6} for 25–1500°C

Source: data compiled by J.S. Park from *No. 1 Materials Index*, Peter T.B. Shaffer, Plenum Press, New York, (1964); Smithells Metals Reference Book, Eric A. Brandes, ed., in association with Fulmer Research Institute Ltd. 6th ed. London, Butterworths, Boston, (1983); and *Ceramic Source*, American Ceramic Society (1986-1991)

Table 116. THERMAL EXPANSION OF CERAMICS
(SHEET 6 OF 34)

Class	Ceramic	Thermal Expansion ($°C^{-1}$)
Carbides (Con't)	Zirconium Monocarbide (ZrC) (Con't)	7.65x 10^{-6} for 25–650°C
		6.10–6.73 x 10^{-6} for 25–650°C
		6.32x 10^{-6} for 0–750°C
		6.46–6.66x 10^{-6} for 0–1000°C
		6.68x 10^{-6} for 0–1275°C
		6.83x 10^{-6} for 0–1525°C
		6.98x 10^{-6} for 0–1775°C
		9.0x 10^{-6} for 1000–2000°C
Nitrides	Aluminum Nitride (AlN)	4.03 x 10^{-6} for 25 to 200°C
		4.84 x 10^{-6} for 25 to 500°C
		4.83 x 10^{-6} for 25 to 600°C
		5.54–5.64 x 10^{-6} for 25 to 1000°C
		6.09 x 10^{-6} for 25 to 1350°C

Source: data compiled by J.S. Park from *No. 1 Materials Index*, Peter T.B. Shaffer, Plenum Press, New York, (1964); Smithells Metals Reference Book, Eric A. Brandes, ed., in association with Fulmer Research Institute Ltd. 6th ed. London, Butterworths, Boston, (1983); and *Ceramic Source*, American Ceramic Society (1986-1991)

Table 116. THERMAL EXPANSION OF CERAMICS

(SHEET 7 OF 34)

Class	Ceramic	Thermal Expansion ($°C^{-1}$)
Nitrides (Con't)	Boron Nitride (BN)	12.2×10^{-6} for 25 to 500°C 13.3×10^{-6} for 25 to 1000°C
	parallel to c axis	10.15×10^{-6} for 25 to 350°C 8.06×10^{-6} for 25 to 700°C 7.15×10^{-6} for 25 to 1000°C
	parallel to a axis	0.59×10^{-6} for 25 to 350°C 0.89×10^{-6} for 25 to 700°C 0.77×10^{-6} for 25 to 1000°C
	Titanium Mononitride (TiN)	9.35×10^{-6}

Source: data compiled by J.S. Park from *No. 1 Materials Index,* Peter T.B. Shaffer, Plenum Press, New York, (1964); Smithells Metals Reference Book, Eric A. Brandes, ed., in association with Fulmer Research Institute Ltd. 6th ed. London, Butterworths, Boston, (1983); and *Ceramic Source*, American Ceramic Society (1986-1991)

Table 116. THERMAL EXPANSION OF CERAMICS

(SHEET 8 OF 34)

Class	Ceramic	Thermal Expansion ($°C^{-1}$)
Nitrides (Con't)	Trisilicon tetranitride (Si_3N_4)	2.11×10^{-6} for 25 to 500°C
		2.87×10^{-6} for 25 to 1000°C
		3.66×10^{-6} for 25 to 1500°C
	(hot pressed)	$3\text{–}3.9 \times 10^{-6}$ for 20 to 1000°C
	(sintered)	3.5×10^{-6} for 20 to 1000°C
	(reaction sintered)	2.9×10^{-6} for 20 to 1000°C
	(pressureless sintered)	3.7×10^{-6} for 40 to 1000°C
	Zirconium Mononitride (TiN)	6.13×10^{-6} for 20–450°C
		7.03×10^{-6} for 20–680°C

Source: data compiled by J.S. Park from *No. 1 Materials Index*, Peter T.B. Shaffer, Plenum Press, New York, (1964); Smithells Metals Reference Book, Eric A. Brandes, ed., in association with Fulmer Research Institute Ltd. 6th ed. London, Butterworths, Boston, (1983); and *Ceramic Source*, American Ceramic Society (1986-1991)

Table 116. THERMAL EXPANSION OF CERAMICS
(SHEET 9 OF 34)

Class	Ceramic	Thermal Expansion (°C^{-1})
Oxides	Aluminum Oxide (Al_2O_3)	
	parallel to c axis	1.95×10^{-6} for 0 to –273°C
		3.01×10^{-6} for 0 to –173°C
		4.39×10^{-6} for 0 to –73°C
		5.31×10^{-6} for 0 to 27°C
		6.26×10^{-6} for 0 to 127°C
		6.86×10^{-6} for 0 to 227°C
		7.31×10^{-6} for 0 to 327°C
		7.68×10^{-6} for 0 to 427°C
		7.96×10^{-6} for 0 to 527°C
		8.19×10^{-6} for 0 to 627°C
		8.38×10^{-6} for 0 to 727°C
		8.52×10^{-6} for 0 to 827°C

Source: data compiled by J.S. Park from *No. 1 Materials Index,* Peter T.B. Shaffer, Plenum Press, New York, (1964); Smithells Metals Reference Book, Eric A. Brandes, ed., in association with Fulmer Research Institute Ltd. 6th ed. London, Butterworths, Boston, (1983); and *Ceramic Source*, American Ceramic Society (1986-1991)

Table 116. THERMAL EXPANSION OF CERAMICS

(SHEET 10 OF 34)

Class	Ceramic	Thermal Expansion (°C^{-1})
Oxides (Con't)	Aluminum Oxide (Al_2O_3) parallel to c axis (Con't)	8.65 x 10^{-6} for 0 to 927°C
		8.75 x 10^{-6} for 0 to 1027°C
		8.84 x 10^{-6} for 0 to 1127°C
		8.92 x 10^{-6} for 0 to 1227°C
		8.98 x 10^{-6} for 0 to 1327°C
		9.02 x 10^{-6} for 0 to 1427°C
		9.08 x 10^{-6} for 0 to 1527°C
		9.13 x 10^{-6} for 0 to 1627°C
		9.18 x 10^{-6} for 0 to 1727°C

Source: data compiled by J.S. Park from *No. 1 Materials Index,* Peter T.B. Shaffer, Plenum Press, New York, (1964); Smithells Metals Reference Book, Eric A. Brandes, ed., in association with Fulmer Research Institute Ltd. 6th ed. London, Butterworths, Boston, (1983); and *Ceramic Source,* American Ceramic Society (1986-1991)

Table 116. THERMAL EXPANSION OF CERAMICS

(SHEET 11 OF 34)

Class	Ceramic	Thermal Expansion ($°C^{-1}$)
Oxides (Con't)	Aluminum Oxide (Al_2O_3) (single crystal) perpendicular to c axis	1.65×10^{-6} for 0 to –273°C
		2.55×10^{-6} for 0 to –173°C
		3.75×10^{-6} for 0 to –73°C
		4.78×10^{-6} for 0 to 27°C
		5.51×10^{-6} for 0 to 127°C
		6.10×10^{-6} for 0 to 227°C
		6.52×10^{-6} for 0 to 327°C
		6.88×10^{-6} for 0 to 427°C
		7.15×10^{-6} for 0 to 527°C
		7.35×10^{-6} for 0 to 627°C
		7.53×10^{-6} for 0 to 727°C
		7.67×10^{-6} for 0 to 827°C

Source: data compiled by J.S. Park from *No. 1 Materials Index*, Peter T.B. Shaffer, Plenum Press, New York, (1964); Smithells Metals Reference Book, Eric A. Brandes, ed., in association with Fulmer Research Institute Ltd. 6th ed. London, Butterworths, Boston, (1983); and *Ceramic Source*, American Ceramic Society (1986-1991)

Table 116. THERMAL EXPANSION OF CERAMICS

(SHEET 12 OF 34)

Class	Ceramic	Thermal Expansion ($°C^{-1}$)
Oxides (Con't)	Aluminum Oxide (Al_2O_3) (single crystal) (Con't) perpendicular to c axis (Con't)	7.80×10^{-6} for 0 to 927°C
		7.88×10^{-6} for 0 to 1027°C
		7.96×10^{-6} for 0 to 1127°C
		8.05×10^{-6} for 0 to 1227°C
		8.12×10^{-6} for 0 to 1327°C
		8.16×10^{-6} for 0 to 1427°C
		8.20×10^{-6} for 0 to 1527°C
		8.26×10^{-6} for 0 to 1627°C
		8.30×10^{-6} for 0 to 1727°C

Source: data compiled by J.S. Park from *No. 1 Materials Index*, Peter T.B. Shaffer, Plenum Press, New York, (1964); Smithells Metals Reference Book, Eric A. Brandes, ed., in association with Fulmer Research Institute Ltd. 6th ed. London, Butterworths, Boston, (1983); and *Ceramic Source*, American Ceramic Society (1986-1991)

Table 116. THERMAL EXPANSION OF CERAMICS
(SHEET 13 OF 34)

Class	Ceramic	Thermal Expansion ($°C^{-1}$)
Oxides (Con't)	Aluminum Oxide (Al_2O_3) (Con't) (polycrystalline)	1.89×10^{-6} for 0 to –273°C
		2.91×10^{-6} for 0 to –173°C
		4.10×10^{-6} for 0 to –73°C
		5.60×10^{-6} for 0 to 27°C
		6.03×10^{-6} for 0 to 127°C
		6.55×10^{-6} for 0 to 227°C
		6.93×10^{-6} for 0 to 327°C
		7.24×10^{-6} for 0 to 427°C
		7.50×10^{-6} for 0 to 527°C
		7.69×10^{-6} for 0 to 627°C
		7.83×10^{-6} for 0 to 727°C
		7.97×10^{-6} for 0 to 827°C

Source: data compiled by J.S. Park from *No. 1 Materials Index,* Peter T.B. Shaffer, Plenum Press, New York, (1964); Smithells Metals Reference Book, Eric A. Brandes, ed., in association with Fulmer Research Institute Ltd. 6th ed. London, Butterworths, Boston, (1983); and *Ceramic Source*, American Ceramic Society (1986-1991)

Table 116. THERMAL EXPANSION OF CERAMICS

(SHEET 14 OF 34)

Class	Ceramic	Thermal Expansion ($°C^{-1}$)
Oxides (Con't)	Aluminum Oxide (Al_2O_3) (Con't) (polycrystalline) (Con't)	8.08×10^{-6} for 0 to 927°C 8.18×10^{-6} for 0 to 1027°C 8.25×10^{-6} for 0 to 1127°C 8.32×10^{-6} for 0 to 1227°C 8.39×10^{-6} for 0 to 1327°C 8.45×10^{-6} for 0 to 1427°C 8.49×10^{-6} for 0 to 1527°C 8.53×10^{-6} for 0 to 1627°C 8.58×10^{-6} for 0 to 1727°C
Source: data compiled by J.S. Park from *No. 1 Materials Index*, Peter T.B. Shaffer, Plenum Press, New York, (1964); Smithells Metals Reference Book, Eric A. Brandes, ed., in association with Fulmer Research Institute Ltd. 6th ed. London, Butterworths, Boston, (1983); and *Ceramic Source*, American Ceramic Society (1986-1991)		

Table 116. THERMAL EXPANSION OF CERAMICS
(SHEET 15 OF 34)

Class	Ceramic	Thermal Expansion (°C^{-1})
Oxides (Con't)	Beryllium Oxide (BeO) (single crystal) parallel to c axis	6.3×10^{-6} for 28 to 252°C 6.7×10^{-6} for 28 to 474°C 7.8×10^{-6} for 28 to 749°C 8.2×10^{-6} for 28 to 872°C 8.9×10^{-6} for 28 to 1132°C
	Beryllium Oxide (BeO) (single crystal) perpendicular to c axis	7.1×10^{-6} for 28 to 252°C 7.8×10^{-6} for 28 to 474°C 8.5×10^{-6} for 28 to 749°C 9.2×10^{-6} for 28 to 872°C 9.9×10^{-6} for 28 to 1132°C

Source: data compiled by J.S. Park from *No. 1 Materials Index*, Peter T.B. Shaffer, Plenum Press, New York, (1964); Smithells Metals Reference Book, Eric A. Brandes, ed., in association with Fulmer Research Institute Ltd. 6th ed. London, Butterworths, Boston, (1983); and *Ceramic Source*, American Ceramic Society (1986-1991)

Table 116. THERMAL EXPANSION OF CERAMICS

(SHEET 16 OF 34)

Class	Ceramic	Thermal Expansion (°C^{-1})
Oxides (Con't)	Beryllium Oxide (BeO) (single crystal) average for (2a+c)/3	6.83×10^{-6} for 28 to 252°C 7.43×10^{-6} for 28 to 474°C 8.27×10^{-6} for 28 to 749°C 8.87×10^{-6} for 28 to 872°C 9.57×10^{-6} for 28 to 1132°C
	Beryllium Oxide (BeO) (polycrystalline)	2.4×10^{-6} for 25–200°C $6.3–6.4 \times 10^{-6}$ for 25–300°C 7.59×10^{-6} for 25–500°C $8.4–8.5 \times 10^{-6}$ for 25–800°C 9.03×10^{-6} for 25–1000°C 9.18×10^{-6} for 25–1250°C

Source: data compiled by J.S. Park from *No. 1 Materials Index*, Peter T.B. Shaffer, Plenum Press, New York, (1964); Smithells Metals Reference Book, Eric A. Brandes, ed., in association with Fulmer Research Institute Ltd. 6th ed. London, Butterworths, Boston, (1983); and *Ceramic Source*, American Ceramic Society (1986-1991)

Table 116. Thermal Expansion of Ceramics
(Sheet 17 of 34)

Class	Ceramic	Thermal Expansion (°C^{-1})
Oxides (Con't)	Beryllium Oxide (BeO) (polycrystalline) (Con't)	10.3 x 10^{-6} for 25–1500°C
		11.1 x 10^{-6} for 25–2000°C
		9.40 x 10^{-6} for 500–1200°C
	Cerium Dioxide (CeO_2)	8.22 x 10^{-6} for 25–500°C
		8.92 x 10^{-6} for 25–1000°C
		8.5 + 0.54T for 0–1000°C
	Dichromium Trioxide (Cr_2O_3)	8.43 x 10^{-6} for 25–500°C
		8.62 x 10^{-6} for 25–1000°C
		8.82 x 10^{-6} for 25–1500°C
		9.55 x 10^{-6} for 20–1400°C

Source: data compiled by J.S. Park from *No. 1 Materials Index,* Peter T.B. Shaffer, Plenum Press, New York, (1964); Smithells Metals Reference Book, Eric A. Brandes, ed., in association with Fulmer Research Institute Ltd. 6th ed. London, Butterworths, Boston, (1983); and *Ceramic Source,* American Ceramic Society (1986-1991)

Table 116. THERMAL EXPANSION OF CERAMICS

(SHEET 18 OF 34)

Class	Ceramic	Thermal Expansion (°C^{-1})
Oxides (Con't)	Hafnium Dioxide (HfO_2) (monoclinic single crystal)	$6.8x10^{-6}$ for 28–262°C
	parallel to a axis	$6.2x10^{-6}$ for 28–494°C
		$6.7x10^{-6}$ for 28–697°C
		$7.5x10^{-6}$ for 28–903°C
		$7.9x10^{-6}$ for 28–1098°C
	parallel to b axis	0 for 28–262°C
		$0.9x10^{-6}$ for 28–494°C
		$1.3x10^{-6}$ for 28–697°C
		$1.4x10^{-6}$ for 28–903°C
		$2.1x10^{-6}$ for 28–1098°C

Source: data compiled by J.S. Park from *No. 1 Materials Index*, Peter T.B. Shaffer, Plenum Press, New York, (1964); Smithells Metals Reference Book, Eric A. Brandes, ed., in association with Fulmer Research Institute Ltd. 6th ed. London, Butterworths, Boston, (1983); and *Ceramic Source*, American Ceramic Society (1986-1991)

Table 116. THERMAL EXPANSION OF CERAMICS
(SHEET 19 OF 34)

Class	Ceramic	Thermal Expansion (°C^{-1})
Oxides (Con't)	Hafnium Dioxide (HfO_2) (monoclinic single crystal) parallel to c axis	$11x10^{-6}$ for 28–262°C $11.4x10^{-6}$ for 28–494°C $10.8x10^{-6}$ for 28–697°C $11.9x10^{-6}$ for 28–903°C $12.1x10^{-6}$ for 28–1098°C
	Hafnium Dioxide (HfO_2) (monoclinic polycrystalline)	5.47×10^{-6} for 25–500°C 5.85×10^{-6} for 25–1000°C 5.8×10^{-6} for 25–1300°C 6.30×10^{-6} for 25–1500°C 6.45×10^{-6} for 20–1700°C

Source: data compiled by J.S. Park from *No. 1 Materials Index*, Peter T.B. Shaffer, Plenum Press, New York, (1964); Smithells Metals Reference Book, Eric A. Brandes, ed., in association with Fulmer Research Institute Ltd. 6th ed. London, Butterworths, Boston, (1983); and *Ceramic Source*, American Ceramic Society (1986-1991)

Table 116. THERMAL EXPANSION OF CERAMICS

(SHEET 20 OF 34)

Class	Ceramic	Thermal Expansion ($°C^{-1}$)
Oxides (Con't)	Hafnium Dioxide (HfO_2) (tetragonal polycrystalline)	1.31×10^{-6} for 25–1700°C 3.03×10^{-6} for 25–2000°C
	Magnesium Oxide (MgO)	12.83×10^{-6} for 25–500°C 13.63×10^{-6} for 25–1000°C 15.11×10^{-6} for 25–1500°C 15.89×10^{-6} for 25–1800°C 14.0×10^{-6} for 20–1400°C $14.2–14.9 \times 10^{-6}$ for 20–1700°C 13.3×10^{-6} for 20–1700°C 13.90×10^{-6} for 0–1000°C 14.46×10^{-6} for 0–1200°C 15.06×10^{-6} for 0–1400°C

Source: data compiled by J.S. Park from *No. 1 Materials Index*, Peter T.B. Shaffer, Plenum Press, New York, (1964); Smithells Metals Reference Book, Eric A. Brandes, ed., in association with Fulmer Research Institute Ltd. 6th ed. London, Butterworths, Boston, (1983); and *Ceramic Source*, American Ceramic Society (1986-1991)

Table 116. THERMAL EXPANSION OF CERAMICS

(SHEET 21 OF 34)

Class	Ceramic	Thermal Expansion (°C^{-1})
Oxides (Con't)	Silicon Dioxide (SiO_2)	
	α quartz	19.35 x 10^{-6} for 25–500°C
		22.2 x 10^{-6} for 25–575°C
	β quartz	27.8 x 10^{-6} for 25–575°C
		14.58 x 10^{-6} for 25–1000°C
	α tridymite	18.5 x 10^{-6} for 25–117°C
	$β_1$ tridymite	25.0 x 10^{-6} for 25–117°C
		27.5 x 10^{-6} for 25–163°C
	$β_2$ tridymite	31.9 x 10^{-6} for 25–163°C
		19.35 x 10^{-6} for 25–500°C
		10.45 x 10^{-6} for 25–1000°C

Source: data compiled by J.S. Park from *No. 1 Materials Index*, Peter T.B. Shaffer, Plenum Press, New York, (1964); Smithells Metals Reference Book, Eric A. Brandes, ed., in association with Fulmer Research Institute Ltd. 6th ed. London, Butterworths, Boston, (1983); and *Ceramic Source*, American Ceramic Society (1986-1991)

Table 116. THERMAL EXPANSION OF CERAMICS

(SHEET 22 OF 34)

Class	Ceramic	Thermal Expansion ($°C^{-1}$)
Oxides (Con't)	Silicon Dioxide (SiO_2) (Con't)	0.527×10^{-6} for 25–500°C
	Vitreous	0.564×10^{-6} for 25–1000°C
		0.5×10^{-6} for 20–1250°C
	Thorium Dioxide (ThO_2)	3.67×10^{-6} for 0 to –273°C
		5.32×10^{-6} for 0 to –173°C
		6.47×10^{-6} for 0 to –73°C
		8.10×10^{-6} for 0 to 27°C
		8.06×10^{-6} for 0 to 127°C
		8.31×10^{-6} for 0 to 227°C
		8.53×10^{-6} for 0 to 327°C
		8.71×10^{-6} for 0 to 427°C

Source: data compiled by J.S. Park from *No. 1 Materials Index*, Peter T.B. Shaffer, Plenum Press, New York, (1964); Smithells Metals Reference Book, Eric A. Brandes, ed., in association with Fulmer Research Institute Ltd. 6th ed. London, Butterworths, Boston, (1983); and *Ceramic Source*, American Ceramic Society (1986-1991)

Table 116. THERMAL EXPANSION OF CERAMICS

(SHEET 23 OF 34)

Class	Ceramic	Thermal Expansion ($°C^{-1}$)
Oxides (Con't)	Thorium Dioxide (ThO_2) (Con't)	8.87×10^{-6} for 0 to 527°C
		9.00×10^{-6} for 0 to 627°C
		9.14×10^{-6} for 0 to 727°C
		9.24×10^{-6} for 0 to 827°C
		9.34×10^{-6} for 0 to 927°C
		9.42×10^{-6} for 0 to 1027°C
		9.53×10^{-6} for 0 to 1127°C
		9.60×10^{-6} for 0 to 1227°C
		9.68×10^{-6} for 0 to 1327°C
		9.76×10^{-6} for 0 to 1427°C
		9.83×10^{-6} for 0 to 1527°C
		9.91×10^{-6} for 0 to 1627°C

Source: data compiled by J.S. Park from *No. 1 Materials Index*, Peter T.B. Shaffer, Plenum Press, New York, (1964); Smithells Metals Reference Book, Eric A. Brandes, ed., in association with Fulmer Research Institute Ltd. 6th ed. London, Butterworths, Boston, (1983); and *Ceramic Source*, American Ceramic Society (1986-1991)

Table 116. THERMAL EXPANSION OF CERAMICS
(SHEET 24 OF 34)

Class	Ceramic	Thermal Expansion ($°C^{-1}$)
Oxides (Con't)	Thorium Dioxide (ThO_2) (Con't)	9.97×10^{-6} for 0 to 1727°C
		8.63×10^{-6} for 25 to 500°C
		9.44×10^{-6} for 25 to 1000°C
		10.17×10^{-6} for 25 to 1500°C
		10.43×10^{-6} for 25 to 1700°C
		9.55×10^{-6} for 20 to 800°C
		9.55×10^{-6} for 20 to 1400°C
		7.8×10^{-6} for 27 to 223°C
		8.7×10^{-6} for 27 to 498°C
		8.9×10^{-6} for 27 to 755°C
		9.2×10^{-6} for 27 to 994°C
		9.1×10^{-6} for 27 to 1087°C

Source: data compiled by J.S. Park from *No. 1 Materials Index,* Peter T.B. Shaffer, Plenum Press, New York, (1964); Smithells Metals Reference Book, Eric A. Brandes, ed., in association with Fulmer Research Institute Ltd. 6th ed. London, Butterworths, Boston, (1983); and *Ceramic Source,* American Ceramic Society (1986-1991)

Table 116. THERMAL EXPANSION OF CERAMICS

(SHEET 25 OF 34)

Class	Ceramic	Thermal Expansion (°C^{-1})
Oxides (Con't)	Thorium Dioxide (ThO_2) (Con't)	8.96×10^{-6} for 0 to 1000°C 9.35×10^{-6} for 0 to 1200°C 9.84×10^{-6} for 0 to 1400°C
	α_l (linear expansion coefficient)	$0.6216 \times 10^{-5} + 3.541 \times 10^{-9}T - 0.1124T^{-2}$ from 298–1073K
	α_v (volume expansion coefficient)	$1.85 \times 10^{-5} + 10.96 \times 10^{-9}T - 0.3375T^{-2}$ from 298–1073K
	Titanium Oxide (TiO_2) (polycrystalline)	8.22×10^{-6} for 25–500°C 8.83×10^{-6} for 25–1000°C

Source: data compiled by J.S. Park from *No. 1 Materials Index,* Peter T.B. Shaffer, Plenum Press, New York, (1964); Smithells Metals Reference Book, Eric A. Brandes, ed., in association with Fulmer Research Institute Ltd. 6th ed. London, Butterworths, Boston, (1983); and *Ceramic Source,* American Ceramic Society (1986-1991)

Table 116. THERMAL EXPANSION OF CERAMICS

(SHEET 26 OF 34)

Class	Ceramic	Thermal Expansion ($°C^{-1}$)
Oxides (Con't)	Titanium Oxide (TiO_2) (polycrystalline) (Con't)	9.50×10^{-6} for 25–1500°C
		7.8×10^{-6} for 20–600°C
		8.98×10^{-6} for 0–1000°C
	Titanium Oxide (TiO_2) (single crystal) parallel to c axis	9.8×10^{-6} for 26 to 240°C
		10.5×10^{-6} for 26 to 455°C
		10.6×10^{-6} for 26 to 670°C
		10.5×10^{-6} for 26 to 940°C
		10.8×10^{-6} for 26 to 1110°C
	Titanium Oxide (TiO_2) (single crystal)) perpendicular to a axis	7.9×10^{-6} for 26 to 240°C
		8.2×10^{-6} for 26 to 455°C

Source: data compiled by J.S. Park from *No. 1 Materials Index*, Peter T.B. Shaffer, Plenum Press, New York, (1964); Smithells Metals Reference Book, Eric A. Brandes, ed., in association with Fulmer Research Institute Ltd. 6th ed. London, Butterworths, Boston, (1983); and *Ceramic Source*, American Ceramic Society (1986-1991)

Table 116. THERMAL EXPANSION OF CERAMICS

(SHEET 27 OF 34)

Class	Ceramic	Thermal Expansion ($°C^{-1}$)
Oxides (Con't)	Titanium Oxide (TiO_2) (single crystal)) (Con't) perpendicular to a axis (Con't)	8.1×10^{-6} for 26 to 670°C
		8.2×10^{-6} for 26 to 940°C
		8.3×10^{-6} for 26 to 1110°C
	Titanium Oxide (TiO_2) (single crystal)) average for (2a+c)/3	8.53×10^{-6} for 26 to 240°C
		8.97×10^{-6} for 26 to 455°C
		8.93×10^{-6} for 26 to 670°C
		8.97×10^{-6} for 26 to 940°C
		9.13×10^{-6} for 26 to 1110°C
	Uranium Dioxide (UO_2)	9.47×10^{-6} for 25 to 500°C
		11.19×10^{-6} for 25 to 1000°C

Source: data compiled by J.S. Park from *No. 1 Materials Index*, Peter T.B. Shaffer, Plenum Press, New York, (1964); Smithells Metals Reference Book, Eric A. Brandes, ed., in association with Fulmer Research Institute Ltd. 6th ed. London, Butterworths, Boston, (1983); and *Ceramic Source*, American Ceramic Society (1986-1991)

Table 116. THERMAL EXPANSION OF CERAMICS

(SHEET 28 OF 34)

Class	Ceramic	Thermal Expansion ($°C^{-1}$)
Oxides (Con't)	Uranium Dioxide (UO_2) (Con't)	12.19×10^{-6} for 25 to 1200°C
		11.15×10^{-6} for 25 to 1750°C
		9.18×10^{-6} for 27 to 400°C
	(heating)	9.07×10^{-6} for 27 to 400°C
		11.1×10^{-6} for 400 to 800°C
		13.0×10^{-6} for 800 to 1200°C
	(cooling)	9.28×10^{-6} for 27 to 400°C
		10.8×10^{-6} for 400 to 800°C
		10.8×10^{-6} for 400 to 800°C
		12.6×10^{-6} for 800 to 1250°C
		12.9×10^{-6} for 800 to 1200°C

Source: data compiled by J.S. Park from *No. 1 Materials Index*, Peter T.B. Shaffer, Plenum Press, New York, (1964); Smithells Metals Reference Book, Eric A. Brandes, ed., in association with Fulmer Research Institute Ltd. 6th ed. London, Butterworths, Boston, (1983); and *Ceramic Source*, American Ceramic Society (1986-1991)

Table 116. THERMAL EXPANSION OF CERAMICS

(SHEET 29 OF 34)

Class	Ceramic	Thermal Expansion ($°C^{-1}$)
Oxides (Con't)	Zirconium Oxide (ZrO_2) (monoclinic)	6.53×10^{-6} for 25 to 500°C
		7.59×10^{-6} for 25 to 1000°C
		7.72×10^{-6} for 25 to 1050°C
		8.0×10^{-6} for 25 to 1080°C
	Zirconium Oxide (ZrO_2) (tetragonal)	-21.7×10^{-6} for 25 to 1050°C
		-11.11×10^{-6} for 25 to 1500°C
		-9.53×10^{-6} for 25 to 1600°C
		4.0×10^{-6} for 0 to 500°C
		10.5×10^{-6} for 0 to 1000°C
		10.52×10^{-6} for 0 to 1000°C (MgO)
		10.6×10^{-6} for 0 to 1200°C (CaO)
		5.0×10^{-6} for 0 to 1400°C

Source: data compiled by J.S. Park from *No. 1 Materials Index*, Peter T.B. Shaffer, Plenum Press, New York, (1964); Smithells Metals Reference Book, Eric A. Brandes, ed., in association with Fulmer Research Institute Ltd. 6th ed. London, Butterworths, Boston, (1983); and *Ceramic Source*, American Ceramic Society (1986-1991)

Table 116. THERMAL EXPANSION OF CERAMICS

(SHEET 30 OF 34)

Class	Ceramic	Thermal Expansion ($°C^{-1}$)
Oxides (Con't)	Zirconium Oxide (ZrO_2) (tetragonal) (Con't)	11.0×10^{-6} for 0 to 1500°C
		$5.5–5.58 \times 10^{-6}$ for 20 to 1200°C
		7.2×10^{-6} for –10 to 1000°C
		8.64×10^{-6} for –20 to 600°C
	Zirconium Oxide (ZrO_2) (tetragonal, single crystal) parallel to a axis	8.4×10^{-6} for 27 to 264°C
		7.5×10^{-6} for 27 to 504°C
		6.8×10^{-6} for 27 to 759°C
		7.8×10^{-6} for 27 to 964°C
		8.7×10^{-6} for 27 to 1110°C
	Zirconium Oxide (ZrO_2) (tetragonal, single crystal) parallel to b axis	3×10^{-6} for 27 to 264°C
		2×10^{-6} for 27 to 504°C

Source: data compiled by J.S. Park from *No. 1 Materials Index*, Peter T.B. Shaffer, Plenum Press, New York, (1964); Smithells Metals Reference Book, Eric A. Brandes, ed., in association with Fulmer Research Institute Ltd. 6th ed. London, Butterworths, Boston, (1983); and *Ceramic Source*, American Ceramic Society (1986-1991)

Table 116. THERMAL EXPANSION OF CERAMICS

(SHEET 31 OF 34)

Class	Ceramic	Thermal Expansion ($°C^{-1}$)
Oxides (Con't)	Zirconium Oxide (ZrO_2) (tetragonal, single crystal) (Con't)	1.1×10^{-6} for 27 to 759°C
	parallel to b axis	1.5×10^{-6} for 27 to 964°C
		1.9×10^{-6} for 27 to 1110°C
	Zirconium Oxide (ZrO_2) (tetragonal, single crystal) parallel to c axis	14×10^{-6} for 27 to 264°C
		13×10^{-6} for 27 to 504°C
		11.9×10^{-6} for 27 to 759°C
		12.8×10^{-6} for 27 to 964°C
		13.6×10^{-6} for 27 to 1110°C
	Cordierite ($2MgO\ 2Al_2O_3\ 5SiO_2$)	
	(ρ=2.51g/cm^3)	2.7×10^{-6} for 25 to 1100°C
	(ρ=2.3g/cm^3)	2.3×10^{-6} for 25 to 400°C

Source: data compiled by J.S. Park from *No. 1 Materials Index*, Peter T.B. Shaffer, Plenum Press, New York, (1964); Smithells Metals Reference Book, Eric A. Brandes, ed., in association with Fulmer Research Institute Ltd. 6th ed. London, Butterworths, Boston, (1983); and *Ceramic Source*, American Ceramic Society (1986-1991)

Table 116. THERMAL EXPANSION OF CERAMICS

(SHEET 32 OF 34)

Class	Ceramic	Thermal Expansion ($°C^{-1}$)
Oxides (Con't)	(ρ=2.3g/cm^3)	3.3 x 10^{-6} for 25 to 700°C
	(ρ=2.3g/cm^3)	3.7 x 10^{-6} for 25 to 900°C
	(ρ=2.1g/cm^3)	2.2 x 10^{-6} for 25 to 400°C
	(ρ=2.1g/cm^3)	2.8 x 10^{-6} for 25 to 700°C
	(ρ=2.1g/cm^3)	2.8 x 10^{-6} for 25 to 900°C
	(ρ=1.8g/cm^3)	0.6 x 10^{-6} for 25 to 400°C
	(ρ=1.8g/cm^3)	1.5 x 10^{-6} for 25 to 700°C
	(ρ=1.8g/cm^3)	1.7 x 10^{-6} for 25 to 900°C
	(glass)	3.7–3.8 x 10^{-6} for 25 to 900°C
	Mullite ($3Al_2O_3$ $2SiO_2$)	4.5 x 10^{-6} for 20 to 1325°C
		4.63 x 10^{-6} for 25 to 500°C

Source: data compiled by J.S. Park from *No. 1 Materials Index,* Peter T.B. Shaffer, Plenum Press, New York, (1964); Smithells Metals Reference Book, Eric A. Brandes, ed., in association with Fulmer Research Institute Ltd. 6th ed. London, Butterworths, Boston, (1983); and *Ceramic Source,* American Ceramic Society (1986-1991)

Table 116. THERMAL EXPANSION OF CERAMICS

(SHEET 33 OF 34)

Class	Ceramic	Thermal Expansion ($°C^{-1}$)
Oxides (Con't)	Mullite ($3Al_2O_3$ $2SiO_2$) (Con't)	5.0×10^{-6} for 25 to 800°C
		5.13×10^{-6} for 25 to 1000°C
		5.62×10^{-6} for 20 to 1500°C
	Sillimanite (Al_2O_3 SiO_2)	6.58×10^{-6} at 20°C
	Spinel (Al_2O_3 MgO)	7.79×10^{-6} for 25 to 500°C
		8.41×10^{-6} for 25 to 1000°C
		9.17×10^{-6} for 25 to 1500°C
		9.0×10^{-6} for 20 to 1250°C
	Zircon (SiO_2 ZrO_2)	5.5×10^{-6} for 20 to 1200°C
		3.79×10^{-6} for 25 to 500°C
		4.62×10^{-6} for 25 to 1000°C
		5.63×10^{-6} for 20 to 1500°C

Source: data compiled by J.S. Park from *No. 1 Materials Index*, Peter T.B. Shaffer, Plenum Press, New York, (1964); Smithells Metals Reference Book, Eric A. Brandes, ed., in association with Fulmer Research Institute Ltd. 6th ed. London, Butterworths, Boston, (1983); and *Ceramic Source*, American Ceramic Society (1986-1991)

Table 116. THERMAL EXPANSION OF CERAMICS

(SHEET 34 OF 34)

Class	Ceramic	Thermal Expansion ($°C^{-1}$)
Silicides	Molybdenum Disilicide ($MoSi_2$)	7.79 x 10^{-6} for 25–500°C
	Molybdenum Disilicide ($MoSi_2$)	8.51 x 10^{-6} for 25–1000°C
		9.00–9.18 x 10^{-6} for 25–1500°C
		8.41 x 10^{-6} for 0–1000°C
		8.56 x 10^{-6} for 0–1400°C
	Tungsten Disilicide (WSi_2)	7.79 x 10^{-6} for 25–500°C
		8.31 x 10^{-6} for 25–1000°C
		8.21 x 10^{-6} for 0–1000°C
		8.81 x 10^{-6} for 0–1400°C

Source: data compiled by J.S. Park from *No. 1 Materials Index*, Peter T.B. Shaffer, Plenum Press, New York, (1964); Smithells Metals Reference Book, Eric A. Brandes, ed., in association with Fulmer Research Institute Ltd. 6th ed. London, Butterworths, Boston, (1983); and *Ceramic Source*, American Ceramic Society (1986-1991)

Table 117. THERMAL EXPANSION OF SIC-WHISKER-REINFORCED CERAMICS

Composite	Linear Coefficient of Thermal Expansion at 22 to 1100 °C (10^{-6}/K)
Alumina	7.8 to 8.2
Alumina with 20 vol% SiC whiskers	7.35
Alumina with 30 vol% SiC whiskers	6.70
Alumina with 60 vol% SiC whiskers	5.82
SiC	4.8
Mullite with 20 vol% SiC whiskers	5.60

Data from *ASM Engineering Materials Reference Book, Second Edition*, Michael Bauccio, Ed., ASM International, Materials Park, OH, p173,(1994).

Table 118. THERMAL EXPANSION OF GLASSES

(SHEET 1 OF 21)

Glass	Composition	Thermal Expansion	Temperature Range of Validity
SiO_2 glass	Pure	$3.50x10^{-7}/K$	–60—20°C
		$3.80x10^{-7}/K$	–40—20°C
		$4.00x10^{-7}/K$	–20—20°C
		$4.30x10^{-7}/K$	0–20°C
		$5.35x10^{-7}/K$	20–100°C
		$5.75x10^{-7}/K$	20–150°C
		$5.85x10^{-7}/K$	20–200°C
		$5.92x10^{-7}/K$	20–250°C
		$5.94x10^{-7}/K$	20–300°C
		$5.90x10^{-7}/K$	20–350°C

Source: data compiled by Jun S. Park from O. V. Mazurin, M. V. Streltsina and T. P. Shvaiko–Shvaikovskaya, *Handbook of Glass Data, Part A and Part B,* Elsevier, New York, 1983

Table 118. THERMAL EXPANSION OF GLASSES

(SHEET 2 OF 21)

Glass	Composition	Thermal Expansion	Temperature Range of Validity
SiO_2–B_2O_3 glass	(39.2% mol B_2O_3)	$47.5x10^{-7}$/K	0–100°C
	(39.2% mol B_2O_3)	$44.9x10^{-7}$/K	100–200°C
	(39.2% mol B_2O_3)	$301x10^{-7}$/K	390–410°C
	(44.2% mol B_2O_3)	$49.8x10^{-7}$/K	0–100°C
	(44.2% mol B_2O_3)	$50.8x10^{-7}$/K	100–200°C
	(44.2% mol B_2O_3)	$450x10^{-7}$/K	380–400°C
	(50.8% mol B_2O_3)	$57.6x10^{-7}$/K	0–100°C
	(50.8% mol B_2O_3)	$54.8x10^{-7}$/K	100–200°C
	(50.8% mol B_2O_3)	$579x10–^{7}$/K	350–370°C
	(58.4% mol B_2O_3)	$71.9x10^{-7}$/K	0–100°C
	(58.4% mol B_2O_3)	$70.1x10^{-7}$/K	100–200°C
	(58.4% mol B_2O_3)	$694x10^{-7}$/K	320–340°C

Source: data compiled by Jun S. Park from O. V. Mazurin, M. V. Streltsina and T. P. Shvaiko–Shvaikovskaya, *Handbook of Glass Data, Part A and Part B,* Elsevier, New York, 1983

Table 118. THERMAL EXPANSION OF GLASSES

(SHEET 3 OF 21)

Glass	Composition	Thermal Expansion	Temperature Range of Validity
SiO_2–B_2O_3 glass (Con't)	(72.7% mol B_2O_3)	$87.0x10^{-7}$/K	0–100°C
	(72.7% mol B_2O_3)	$89.7x10^{-7}$/K	100–200°C
	(72.7% mol B_2O_3)	$899x10^{-7}$/K	300–320°C
	(83.2% mol B_2O_3)	$111.4x10^{-7}$/K	0–100°C
	(83.2% mol B_2O_3)	$116.6x10^{-7}$/K	100–200°C
	(83.2% mol B_2O_3)	$970x10^{-7}$/K	280–300°C
	(88.6% mol B_2O_3)	$118.1x10^{-7}$/K	0–100°C
	(88.6% mol B_2O_3)	$126.0x10^{-7}$/K	100–200°C
	(88.6% mol B_2O_3)	$1023x10^{-7}$/K	280–300°C
	(94.0% mol B_2O_3)	$131.7x10^{-7}$/K	0–100°C
	(94.0% mol B_2O_3)	$141.9x10^{-7}$/K	100–200°C
	(94.0% mol B_2O_3)	$1200x10^{-7}$/K	270–290°C

Source: data compiled by Jun S. Park from O. V. Mazurin, M. V. Streltsina and T. P. Shvaiko–Shvaikovskaya, *Handbook of Glass Data, Part A and Part B,* Elsevier, New York, 1983

Table 118. Thermal Expansion of Glasses

(Sheet 4 of 21)

Glass	Composition	Thermal Expansion	Temperature Range of Validity
SiO_2–Al_2O_3 glass	(13.9% mol Al_2O_3, 1000°C for 115 hr)	$22.7x10^{-7}/K$	20–900°C
	(13.9% mol Al_2O_3, water quenching)	$17.2x10^{-7}/K$	20–600°C
	(17.4% mol Al_2O_3, 1000°C for 115 hr)	$28.3x10^{-7}/K$	20–800°C
	(17.4% mol Al_2O_3, water quenching)	$20.7x10^{-7}/K$	20–700°C
	(3.1% mol Al_2O_3, 1000°C for 115 hr)	$6.2x10^{-7}/K$	20–980°C
	(3.1% mol Al_2O_3, water quenching)	$6.2x10^{-7}/K$	20–980°C
	(5.4% mol Al_2O_3, 1130°C for 20 hr)	$12.2x10^{-7}/K$	20–350°C

Source: data compiled by Jun S. Park from O. V. Mazurin, M. V. Streltsina and T. P. Shvaiko–Shvaikovskaya, *Handbook of Glass Data, Part A and Part B*, Elsevier, New York, 1983

Table 118. THERMAL EXPANSION OF GLASSES

(SHEET 5 OF 21)

Glass	Composition	Thermal Expansion	Temperature Range of Validity
SiO_2–Al_2O_3 glass	(8.2% mol Al_2O_3, 1000°C for 115 hr)	$14.5x10^{-7}$/K	20–950°C
	(8.2% mol Al_2O_3, water quenching)	$8.8x10^{-7}$/K	20–800°C
SiO_2–CaO glass	(30% mol CaO)	$66±5x10^{-6}$/K	1700°C
	(35% mol CaO)	$53±5x10^{-6}$/K	1700°C
	(40% mol CaO)	$64±4x10^{-6}$/K	1700°C
	(42.5% mol CaO)	$76±4x10^{-6}$/K	1700°C
	(45% mol CaO)	$85–100±4x10^{-6}$/K	1700°C
	(47.5% mol CaO)	$76±4x10^{-6}$/K	1700°C
	(50% mol CaO)	$84–85±4x10^{-6}$/K	1700°C
	(52.5% mol CaO)	$76–107±4x10^{-6}$/K	1700°C

Source: data compiled by Jun S. Park from O. V. Mazurin, M. V. Streltsina and T. P. Shvaiko–Shvaikovskaya, *Handbook of Glass Data, Part A and Part B,* Elsevier, New York, 1983

Table 118. THERMAL EXPANSION OF GLASSES

(SHEET 6 OF 21)

Glass	Composition	Thermal Expansion	Temperature Range of Validity
SiO_2–CaO glass	(55% mol CaO)	94–95±4x10^{-6}/K	1700°C
	(57.5% mol CaO)	95±4x10^{-6}/K	1700°C
	(60% mol CaO)	103±4x10^{-6}/K	1700°C
SiO_2–PbO glass	(25.7% mol PbO)	51.45–52.23x10^{-7}/K	20–170°C
	(30.0% mol PbO)	57.68–59.08x10^{-7}/K	20–170°C
	(32.5% mol PbO)	60.62–62.31x10^{-7}/K	20–170°C
	(33.2% mol PbO)	61.58–63.33x10^{-7}/K	20–170°C
	(35.0% mol PbO)	63.99–66.17x10^{-7}/K	20–170°C
	(37.5% mol PbO)	68.75–71.44x10^{-7}/K	20–170°C
	(42.6% mol PbO)	75.16–78.58x10^{-7}/K	20–170°C
	(45.8% mol PbO)	78.85–82.60x10^{-7}/K	20–170°C

Source: data compiled by Jun S. Park from O. V. Mazurin, M. V. Streltsina and T. P. Shvaiko–Shvaikovskaya, *Handbook of Glass Data, Part A and Part B*, Elsevier, New York, 1983

Table 118. THERMAL EXPANSION OF GLASSES

(SHEET 7 OF 21)

Glass	Composition	Thermal Expansion	Temperature Range of Validity
SiO_2–PbO glass	(47.8% mol PbO)	83.03–87.03x10^{-7}/K	20–170°C
	(49.8% mol PbO)	85.57–89.82x10^{-7}/K	20–170°C
	(50% mol PbO)	723x10^{-7}/K	1100°C
	(53.8% mol PbO)	90.62–95.25x10^{-7}/K	20–170°C
	(57.5% mol PbO)	95.64–100.45x10^{-7}/K	20–170°C
	(59.0% mol PbO)	97.00–101.90x10^{-7}/K	20–170°C
	(61.0% mol PbO)	100.66–105.58x10^{-7}/K	20–170°C
	(61.75% mol PbO)	101.36–106.30x10^{-7}/K	20–170°C
	(66.7% mol PbO)	867x10^{-7}/K	1100°C
	(67.7% mol PbO)	110.38–115.48x10^{-7}/K	20–170°C

Source: data compiled by Jun S. Park from O. V. Mazurin, M. V. Streltsina and T. P. Shvaiko–Shvaikovskaya, *Handbook of Glass Data, Part A and Part B,* Elsevier, New York, 1983

Table 118. THERMAL EXPANSION OF GLASSES

(SHEET 8 OF 21)

Glass	Composition	Thermal Expansion	Temperature Range of Validity
SiO_2–Na_2O glass	(20% mol Na_2O)	$6.7x10^{-5}/K$	liquidus temp. to 1400°C
	(20% mol Na_2O, $T_g = 478°C$)	$120x10^{-7}/K$	below T_g
	(20% mol Na_2O, $T_g = 478°C$)	$315x10^{-7}/K$	above T_g
	(20.3% mol Na_2O)	$97.5x10^{-7}/K$	room temp–100°C
	(20.3% mol Na_2O)	$99.3x10^{-7}/K$	100–200°C
	(20.3% mol Na_2O)	$100.6x10^{-7}/K$	200–300°C
	(20.3% mol Na_2O)	$106.9x10^{-7}/K$	300–400°C
	(24.0% mol Na_2O)	$109.7x10^{-7}/K$	room temp–100°C
	(24.0% mol Na_2O)	$114.3x10^{-7}/K$	100–200°C
	(24.0% mol Na_2O)	$116.6x10^{-7}/K$	200–300°C
	(24.0% mol Na_2O)	$121.7x10^{-7}/K$	300–400°C

Source: data compiled by Jun S. Park from O. V. Mazurin, M. V. Streltsina and T. P. Shvaiko–Shvaikovskaya, *Handbook of Glass Data, Part A and Part B*, Elsevier, New York, 1983

Table 118. THERMAL EXPANSION OF GLASSES

(SHEET 9 OF 21)

Glass	Composition	Thermal Expansion	Temperature Range of Validity
SiO_2–Na_2O glass	(30% mol Na_2O, T_g = 455°C)	$152x10^{-7}$/K	below T_g
	(30% mol Na_2O, T_g = 455°C)	$402x10^{-7}$/K	above T_g
	(31.1% mol Na_2O)	$136.0x10^{-7}$/K	room temp–100°C
	(31.1% mol Na_2O)	$142.5x10^{-7}$/K	100–200°C
	(31.1% mol Na_2O)	$148.3x10^{-7}$/K	200–300°C
	(31.1% mol Na_2O)	$160.0x10^{-7}$/K	300–400°C
	(33% mol Na_2O, T_g = 445°C)	$165x10^{-7}$/K	below T_g
	(33% mol Na_2O, T_g = 445°C)	$465x10^{-7}$/K	above T_g
	(33.3% mol Na_2O)	$17.2x10^{-5}$/K	liquidus temp.to 1400°C

Source: data compiled by Jun S. Park from O. V. Mazurin, M. V. Streltsina and T. P. Shvaiko–Shvaikovskaya, *Handbook of Glass Data, Part A and Part B*, Elsevier, New York, 1983

Table 118. THERMAL EXPANSION OF GLASSES

(SHEET 10 OF 21)

Glass	Composition	Thermal Expansion	Temperature Range of Validity
SiO_2–Na_2O glass	(33.8% mol Na_2O)	$143.9x10^{-7}$/K	room temp–100°C
	(33.8% mol Na_2O)	$153.6x10^{-7}$/K	100–200°C
	(33.8% mol Na_2O)	$159.1x10^{-7}$/K	200–300°C
	(33.8% mol Na_2O)	$173.6x10^{-7}$/K	300–400°C
	(37.2% mol Na_2O)	$152.1x10^{-7}$/K	room temp–100°C
	(37.2% mol Na_2O)	$160.9x10^{-7}$/K	100–200°C
	(37.2% mol Na_2O)	$171.6x10^{-7}$/K	200–300°C
	(37.2% mol Na_2O)	$187.7x10^{-7}$/K	300–400°C
	(40% mol Na_2O)	$20.0x10^{-5}$/K	liquidus temp. to 1400°C
	(40% mol Na_2O, T_g = 421°C)	$179x10^{-7}$/K	below T_g
	(40% mol Na_2O, T_g = 421°C)	$500x10^{-7}$/K	above T_g

Source: data compiled by Jun S. Park from O. V. Mazurin, M. V. Streltsina and T. P. Shvaiko–Shvaikovskaya, *Handbook of Glass Data, Part A and Part B*, Elsevier, New York, 1983

Table 118. THERMAL EXPANSION OF GLASSES

(SHEET 11 OF 21)

Glass	Composition	Thermal Expansion	Temperature Range of Validity
SiO_2–Na_2O glass	(45% mol Na_2O, T_g = 417°C)	219x10^{-7}/K	below T_g
	(45% mol Na_2O, T_g = 417°C)	574x10^{-7}/K	above T_g
	(50% mol Na_2O)	23.7x10^{-5}/K	liquidus temp. to 1400°C
B_2O_3 glass		154.5–183x10^{-7}/K	0–100°C
		154.5–169x10^{-7}/K	100–200°C
		150±3–158±3x10^{-7}/K	20–200°C
B_2O_3–Na_2O glass	(0.01% mol Na_2O)	140x10^{-7}/K	–196—25°C
	(0.01% mol Na_2O)	149.3x10^{-7}/K	20–50°C
	(0.01% mol Na_2O)	149.0x10^{-7}/K	20–150°C

Source: data compiled by Jun S. Park from O. V. Mazurin, M. V. Streltsina and T. P. Shvaiko–Shvaikovskaya, *Handbook of Glass Data, Part A and Part B,* Elsevier, New York, 1983

Table 118. Thermal Expansion of Glasses

(Sheet 12 of 21)

Glass	Composition	Thermal Expansion	Temperature Range of Validity
B_2O_3–Na_2O glass	(4.4% mol Na_2O)	$94.6x10^{-7}$/K	–196—25°C
	(4.4% mol Na_2O)	$103.0x10^{-7}$/K	20–50°C
	(4.4% mol Na_2O)	$109.9x10^{-7}$/K	20–150°C
	(4.4% mol Na_2O)	$116.0x10^{-7}$/K	20–250°C
	(5% mol Na_2O, T_g = 318°C)	$115x10^{-7}$/K	below T_g
	(5% mol Na_2O, T_g = 318°C)	$1400x10^{-7}$/K	above T_g
	(8.7% mol Na_2O)	$98.8x10^{-7}$/K	20–50°C
	(8.7% mol Na_2O)	$100.5x10^{-7}$/K	20–150°C
	(8.7% mol Na_2O)	$105.3x10^{-7}$/K	20–250°C

Source: data compiled by Jun S. Park from O. V. Mazurin, M. V. Streltsina and T. P. Shvaiko–Shvaikovskaya, *Handbook of Glass Data, Part A and Part B,* Elsevier, New York, 1983

Table 118. THERMAL EXPANSION OF GLASSES

(SHEET 13 OF 21)

Glass	Composition	Thermal Expansion	Temperature Range of Validity
B_2O_3–Na_2O glass	(10% mol Na_2O, T_g = 354°C)	$77x10^{-7}$/K	below T_g
	(10% mol Na_2O, T_g = 354°C)	$1230x10^{-7}$/K	above T_g
	(11.5% mol Na_2O)	$71.5x10^{-7}$/K	–196—25°C
	(11.5% mol Na_2O)	$88.7x10^{-7}$/K	20–50°C
	(11.5% mol Na_2O)	$94.9x10^{-7}$/K	20–150°C
	(11.5% mol Na_2O)	$97.9x10^{-7}$/K	20–250°C
	(13.7% mol Na_2O)	$69.3x10^{-7}$/K	–196—25°C
	(13.7% mol Na_2O)	$87.5x10^{-7}$/K	20–50°C
	(13.7% mol Na_2O)	$92.3x10^{-7}$/K	20–150°C
	(13.7% mol Na_2O)	$90.9x10^{-7}$/K	20–250°C

Source: data compiled by Jun S. Park from O. V. Mazurin, M. V. Streltsina and T. P. Shvaiko–Shvaikovskaya, *Handbook of Glass Data, Part A and Part B,* Elsevier, New York, 1983

Table 118. THERMAL EXPANSION OF GLASSES

(SHEET 14 OF 21)

Glass	Composition	Thermal Expansion	Temperature Range of Validity
B_2O_3–Na_2O glass	(15% mol Na_2O, T_g = 407°C)	69x10^{-7}/K	below T_g
	(15% mol Na_2O, T_g = 407°C)	761x10^{-7}/K	above T_g
	(15.8% mol Na_2O)	67.4x10^{-7}/K	–196—25°C
	(15.8% mol Na_2O)	80.7x10^{-7}/K	20–50°C
	(15.8% mol Na_2O)	87.8x10^{-7}/K	20–150°C
	(15.8% mol Na_2O)	93.3x10^{-7}/K	20–250°C
	(15.8% mol Na_2O)	97.9x10^{-7}/K	20–350°C
	(16.2% mol Na_2O)	65.9x10^{-7}/K	–196—25°C
	(16.2% mol Na_2O)	86.0x10^{-7}/K	20–50°C

Source: data compiled by Jun S. Park from O. V. Mazurin, M. V. Streltsina and T. P. Shvaiko–Shvaikovskaya, *Handbook of Glass Data, Part A and Part B*, Elsevier, New York, 1983

Table 118. THERMAL EXPANSION OF GLASSES

(SHEET 15 OF 21)

Glass	Composition	Thermal Expansion	Temperature Range of Validity
B_2O_3–Na_2O glass	(16.2% mol Na_2O)	$87.7x10^{-7}$/K	20–150°C
	(16.2% mol Na_2O)	$90.9x10^{-7}$/K	20–250°C
	(16.2% mol Na_2O)	$96.9x10^{-7}$/K	20–350°C
	(17.4% mol Na_2O)	$85.6x10^{-7}$/K	20–50°C
	(17.4% mol Na_2O)	$89.1x10^{-7}$/K	20–150°C
	(17.4% mol Na_2O)	$92.4x10^{-7}$/K	20–250°C
	(17.4% mol Na_2O)	$96.3x10^{-7}$/K	20–350°C
	(18.4% mol Na_2O)	$69.1x10^{-7}$/K	–196—25°C
	(18.4% mol Na_2O)	$86.2x10^{-7}$/K	20–50°C
	(18.4% mol Na_2O)	$89.2x10^{-7}$/K	20–150°C
	(18.4% mol Na_2O)	$94.1x10^{-7}$/K	20–250°C
	(18.4% mol Na_2O)	$96.2x10^{-7}$/K	20–350°C

Source: data compiled by Jun S. Park from O. V. Mazurin, M. V. Streltsina and T. P. Shvaiko–Shvaikovskaya, *Handbook of Glass Data, Part A and Part B*, Elsevier, New York, 1983

Table 118. THERMAL EXPANSION OF GLASSES
(SHEET 16 OF 21)

Glass	Composition	Thermal Expansion	Temperature Range of Validity
B_2O_3–Na_2O glass	(19.6% mol Na_2O)	$86.8x10^{-7}$/K	20–50°C
	(19.6% mol Na_2O)	$91.2x10^{-7}$/K	20–150°C
	(19.6% mol Na_2O)	$95.3x10^{-7}$/K	20–250°C
	(19.6% mol Na_2O)	$99.6x10^{-7}$/K	20–350°C
	(20.0% mol Na_2O)	$87.6x10^{-7}$/K	20–50°C
	(20.0% mol Na_2O)	$91.6x10^{-7}$/K	20–150°C
	(20.0% mol Na_2O)	$97.6x10^{-7}$/K	20–250°C
	(20.0% mol Na_2O)	$101.3x10^{-7}$/K	20–350°C
	(20% mol Na_2O, T_g = 456°C)	$86x10^{-7}$/K	below T_g
	(20% mol Na_2O, T_g = 456°C)	$586x10^{-7}$/K	above T_g

Source: data compiled by Jun S. Park from O. V. Mazurin, M. V. Streltsina and T. P. Shvaiko–Shvaikovskaya, *Handbook of Glass Data, Part A and Part B*, Elsevier, New York, 1983

Table 118. Thermal Expansion of Glasses

(Sheet 17 of 21)

Glass	Composition	Thermal Expansion	Temperature Range of Validity
B_2O_3–Na_2O glass	(22.5% mol Na_2O)	$71.9x10^{-7}/K$	–196—25°C
	(22.5% mol Na_2O)	$90.4x10^{-7}/K$	20–50°C
	(22.5% mol Na_2O)	$94.7x10^{-7}/K$	20–150°C
	(22.5% mol Na_2O)	$98.7x10^{-7}/K$	20–250°C
	(22.5% mol Na_2O)	$104.0x10^{-7}/K$	20–350°C
	(23.6% mol Na_2O)	$90.4x10^{-7}/K$	20–50°C
	(23.6% mol Na_2O)	$96.7x10^{-7}/K$	20–150°C
	(23.6% mol Na_2O)	$101.2x10^{-7}/K$	20–250°C
	(23.6% mol Na_2O)	$106.5x10^{-7}/K$	20–350°C

Source: data compiled by Jun S. Park from O. V. Mazurin, M. V. Streltsina and T. P. Shvaiko–Shvaikovskaya, *Handbook of Glass Data, Part A and Part B*, Elsevier, New York, 1983

Table 118. THERMAL EXPANSION OF GLASSES

(SHEET 18 OF 21)

Glass	Composition	Thermal Expansion	Temperature Range of Validity
B2O3–Na2O glass	(25% mol Na_2O, T_g = 466°C)	$95x10^{-7}/K$	below T_g
	(25% mol Na_2O, T_g = 466°C)	$834x10^{-7}/K$	above T_g
	(28.9% mol Na_2O)	$81.4x10^{-7}/K$	–196—25°C
	(28.9% mol Na_2O)	$102.1x10^{-7}/K$	20–50°C
	(28.9% mol Na_2O)	$107.4x10^{-7}/K$	20–150°C
	(28.9% mol Na_2O)	$112.8x10^{-7}/K$	20–250°C
	(28.9% mol Na_2O)	$117.1x10^{-7}/K$	20–350°C
	(30% mol Na_2O, T_g = 468°C)	$128x10^{-7}/K$	below T_g
	(30% mol Na_2O, T_g = 468°C)	$1150x10^{-7}/K$	above T_g

Source: data compiled by Jun S. Park from O. V. Mazurin, M. V. Streltsina and T. P. Shvaiko–Shvaikovskaya, *Handbook of Glass Data, Part A and Part B*, Elsevier, New York, 1983

Table 118. THERMAL EXPANSION OF GLASSES
(SHEET 19 OF 21)

Glass	Composition	Thermal Expansion	Temperature Range of Validity
B_2O_3–CaO glass	(29.3% mol CaO)	54.9–56.4x10^{-7}/K	room temp. to 100°C
	(29.3% mol CaO)	60.2–60.8x10^{-7}/K	100–200°C
	(29.3% mol CaO)	63.9–65.4x10^{-7}/K	200–300°C
	(29.3% mol CaO)	71.3–71.6x10^{-7}/K	300–400°C
	(29.3% mol CaO)	76.9–77.1x10^{-7}/K	400–500°C
	(29.3% mol CaO)	80.9–86.8x10^{-7}/K	500–600°C
	(31.4% mol CaO)	57.3–58.2x10^{-7}/K	room temp. to 100°C
	(31.4% mol CaO)	63.5–65.1x10^{-7}/K	100–200°C
	(31.4% mol CaO)	67.4–68.1x10^{-7}/K	200–300°C
	(31.4% mol CaO)	76.5–76.7x10^{-7}/K	300–400°C
	(31.4% mol CaO)	79.2–81.0x10^{-7}/K	400–500°C
	(31.4% mol CaO)	83.1–88.5x10^{-7}/K	500–600°C

Source: data compiled by Jun S. Park from O. V. Mazurin, M. V. Streltsina and T. P. Shvaiko–Shvaikovskaya, *Handbook of Glass Data, Part A and Part B,* Elsevier, New York, 1983

Table 118. THERMAL EXPANSION OF GLASSES
(SHEET 20 OF 21)

Glass	Composition	Thermal Expansion	Temperature Range of Validity
B_2O_3–CaO glass	(34.9% mol CaO)	$60.1–66.2x10^{-7}/K$	room temp. to 100°C
	(34.9% mol CaO)	$67.5–67.6x10^{-7}/K$	100–200°C
	(34.9% mol CaO)	$74.7–75.2x10^{-7}/K$	200–300°C
	(34.9% mol CaO)	$77.8–78.5x10^{-7}/K$	300–400°C
	(34.9% mol CaO)	$83.8–95.0x10^{-7}/K$	400–500°C
	(34.9% mol CaO)	$91.8–92.1x10^{-7}/K$	500–600°C
	(37.1% mol CaO)	$63.1–64.0x10^{-7}/K$	room temp. to 100°C
	(37.1% mol CaO)	$68.4–70.4x10^{-7}/K$	100–200°C
	(37.1% mol CaO)	$74.6–75.8x10^{-7}/K$	200–300°C

Source: data compiled by Jun S. Park from O. V. Mazurin, M. V. Streltsina and T. P. Shvaiko–Shvaikovskaya, *Handbook of Glass Data, Part A and Part B,* Elsevier, New York, 1983

Table 118. THERMAL EXPANSION OF GLASSES

(SHEET 21 OF 21)

Glass	Composition	Thermal Expansion	Temperature Range of Validity
B_2O_3–CaO glass	(37.1% mol CaO)	81.6–82.2x10^{-7}/K	300–400°C
	(37.1% mol CaO)	86.9–87.6x10^{-7}/K	400–500°C
	(37.1% mol CaO)	93.5–95.5x10^{-7}/K	500–600°C

Source: data compiled by Jun S. Park from O. V. Mazurin, M. V. Streltsina and T. P. Shvaiko–Shvaikovskaya, *Handbook of Glass Data, Part A and Part B,* Elsevier, New York, 1983

Table 119. THERMAL EXPANSION OF POLYMERS

(SHEET 1 OF 13)

Type	Polymer	Thermal Expansion Coefficient ASTM D696 ($°F^{-1}$)
ABS Resins; Molded, Extruded	Medium impact	$3.2—4.8 \times 10^{-6}$
	High impact	$5.5—6.0 \times 10^{-6}$
	Very high impact	$5.0—6.0 \times 10^{-6}$
	Low temperature impact	$5.0—6.0 \times 10^{-6}$
	Heat resistant	$3.0—4.0 \times 10^{-6}$
Acrylics; Cast, Molded, Extruded	Cast Resin Sheets, Rods:	
	General purpose, type I	4.5×10^{-6}
	General purpose, type II	4.5×10^{-6}
	Moldings:	
	Grades 5, 6, 8	$3—4 \times 10^{-6}$
	High impact grade	$4—6 \times 10^{-6}$

Source: *data compiled by* J.S. Park from Charles T. Lynch, *CRC Handbook of Materials Science, Vol. 3*, CRC Press, Boca Raton, Florida and *Engineered Materials Handbook, Vol.2*, Engineering Plastics, ASM International, Metals Park, Ohio, 1988.

Table 119. THERMAL EXPANSION OF POLYMERS

(SHEET 2 OF 13)

Type	Polymer	Thermal Expansion Coefficient ASTM D696 ($°F^{-1}$)
Thermoset Carbonate	Allyl diglycol carbonate	6×10^{-5}
Alkyds; Molded	Putty (encapsulating)	1.3×10^{-5}
	Rope (general purpose)	1.3×10^{-5}
	Granular (high speed molding)	1.3×10^{-5}
	Glass reinforced (heavy duty parts)	1.3×10^{-5}
Cellulose Acetate; Molded, Extruded	ASTM Grade:	
	H6—1	$4.4—9.0 \times 10^{-5}$
	H4—1	$4.4—9.0 \times 10^{-5}$
	H2—1	$4.4—9.0 \times 10^{-5}$
	MH—1, MH—2	$4.4—9.0 \times 10^{-5}$
	MS—1, MS—2	$4.4—9.0 \times 10^{-5}$
	S2—1	$4.4—9.0 \times 10^{-5}$

Source: *data compiled by* J.S. Park from Charles T. Lynch, *CRC Handbook of Materials Science, Vol. 3*, CRC Press, Boca Raton, Florida and *Engineered Materials Handbook, Vol.2*, Engineering Plastics, ASM International, Metals Park, Ohio, 1988.

Table 119. THERMAL EXPANSION OF POLYMERS
(SHEET 3 OF 13)

Type	Polymer	Thermal Expansion Coefficient ASTM D696 (•F^{-1})
Cellulose Acetate Butyrate; Molded, Extruded	ASTM Grade:	
	H4	$6—9 \times 10^{-5}$
	MH	$6—9 \times 10^{-5}$
	S2	$6—9 \times 10^{-5}$
Cellusose Acetate Propionate; Molded, Extruded	ASTM Grade:	
	1	$6—9 \times 10^{-5}$
	3	$6—9 \times 10^{-5}$
	6	$6—9 \times 10^{-5}$
Chlorinated Polymers	Chlorinated polyether	6.6×10^{-6}
	Chlorinated polyvinyl chloride	4.4×10^{-6}
Polycarbonates	Polycarbonate	3.75×10^{-6}
	Polycarbonate (40% glass fiber reinforced)	$1.0—1.1 \times 10^{-6}$

Source: *data compiled by* J.S. Park from Charles T. Lynch, *CRC Handbook of Materials Science, Vol. 3*, CRC Press, Boca Raton, Florida and *Engineered Materials Handbook, Vol.2*, Engineering Plastics, ASM International, Metals Park, Ohio, 1988.

Table 119. THERMAL EXPANSION OF POLYMERS

(SHEET 4 OF 13)

Type	Polymer	Thermal Expansion Coefficient ASTM D696 ($\cdot F^{-1}$)
Diallyl Phthalates; Molded	Orlon filled	5.0×10^{-5}
	Dacron filled	5.2×10^{-5}
	Asbestos filled	4.0×10^{-5}
	Glass fiber filled	$2.2—2.6 \times 10^{-5}$
Fluorocarbons; Molded,Extruded	Polytrifluoro chloroethylene (PTFCE)	3.88×10^{-5}
	Polytetrafluoroethylene (PTFE)	55×10^{-5}
	Ceramic reinforced (PTFE)	$1.7—2.0 \times 10^{-5}$
	Fluorinated ethylene propylene(FEP)	$8.3—10.5 \times 10^{-5}$
	Polyvinylidene— fluoride (PVDF)	8.5×10^{-5}

Source: *data compiled by* J.S. Park from Charles T. Lynch, *CRC Handbook of Materials Science, Vol. 3*, CRC Press, Boca Raton, Florida and *Engineered Materials Handbook, Vol.2*, Engineering Plastics, ASM International, Metals Park, Ohio, 1988.

Table 119. THERMAL EXPANSION OF POLYMERS

(SHEET 5 OF 13)

Type	Polymer	Thermal Expansion Coefficient ASTM D696 (•F^{-1})
Epoxies; Cast, Molded, Reinforced	Standard epoxies (diglycidyl ethers of bisphenol A)	
	Cast rigid	3.3×10^{-5}
	Cast flexible	$3—5 \times 10^{-5}$
	Molded	$1—2 \times 10^{-5}$
	General purpose glass cloth laminate	$3.3—4.8 \times 10^{-6}$
	High strength laminate	$3.3—4.8 \times 10^{-6}$
	Filament wound composite	$2—6 \times 10^{-5}$
Epoxies—Molded, Extruded	High performance resins (cycloaliphatic diepoxides)	
	Molded	$1.7—2.2 \times 10^{-6}$
	Epoxy novolacs	
	Cast, rigid	$1.6—3.0 \times 10^{-6}$

Source: *data compiled by* J.S. Park from Charles T. Lynch, *CRC Handbook of Materials Science, Vol. 3*, CRC Press, Boca Raton, Florida and *Engineered Materials Handbook, Vol.2*, Engineering Plastics, ASM International, Metals Park, Ohio, 1988.

Table 119. THERMAL EXPANSION OF POLYMERS

(SHEET 6 OF 13)

Type	Polymer	Thermal Expansion Coefficient ASTM D696 (•F^{-1})
Melamines; Molded	Filler & type	
	Cellulose electrical	$1.11—2.78 \times 10^{-5}$
	Glass fiber	0.82×10^{-5}
Nylons; Molded, Extruded		
Type 6 Nylon	General purpose	4.8×10^{-5}
	Glass fiber (30%) reinforced	1.2×10^{-5}
	Cast	4.4×10^{-5}
	Type 11	5.5×10^{-5}
	Type 12	7.2×10^{-5}
6/6 Nylon	General purpose molding	$1.69—1.7 \times 10^{-5}$
	Glass fiber reinforced	$1.5–3.3 \times 10^{-5}$
	General purpose extrusion	1.7×10^{-5}

Source: *data compiled by* J.S. Park from Charles T. Lynch, *CRC Handbook of Materials Science, Vol. 3*, CRC Press, Boca Raton, Florida and *Engineered Materials Handbook, Vol.2*, Engineering Plastics, ASM International, Metals Park, Ohio, 1988.

Table 119. THERMAL EXPANSION OF POLYMERS

(SHEET 7 OF 13)

Type	Polymer	Thermal Expansion Coefficient ASTM D696 ($\cdot F^{-1}$)
6/10 Nylon	General purpose	1.5×10^{-5}
	Glass fiber (30%) reinforced	3.5×10^{-5}
Phenolics; Molded	Type and filler	
	General: woodflour and flock	1.66—2.50×10^{-5}
	Shock: paper, flock, or pulp	1.6—2.3×10^{-5}
	High shock: chopped fabric or cord	1.60—2.22×10^{-5}
	Very high shock: glass fiber	0.88×10^{-5}
Phenolics: Molded	Rubber phenolic—woodflour or flock	0.83—2.20×10^{-5}
	Rubber phenolic—chopped fabric	1.7×10^{-5}
	Rubber phenolic—asbestos	2.2×10^{-5}
ABS–Polycarbonate Alloy	ABS–Polycarbonate Alloy	6.12×10^{-5}

Source: *data compiled by* J.S. Park from Charles T. Lynch, *CRC Handbook of Materials Science, Vol. 3*, CRC Press, Boca Raton, Florida and *Engineered Materials Handbook, Vol.2*, Engineering Plastics, ASM International, Metals Park, Ohio, 1988.

Table 119. THERMAL EXPANSION OF POLYMERS

(SHEET 8 OF 13)

Type	Polymer	Thermal Expansion Coefficient ASTM D696 (•F^{-1})
PVC–Acrylic	PVC–Acrylic Alloy	
	PVC–acrylic sheet	3.5 x 10^{-5}
Polymides	Unreinforced	2.5 x 10^{-6}
	Unreinforced 2nd value	3.0—4.5 x 10^{-6}
	Glass reinforced	0.8 x 10^{-6}
Homopolymer	Standard	4.5 x 10^{-5}
	20% glass reinforced	2.0—4.5 x 10^{-5}
	22% TFE reinforced	4.5 x 10^{-5}
	Copolymer:	
	Standard	4.7 x 10^{-5}
	25% glass reinforced	2.2—4.7 x 10^{-5}
	High flow	4.7 x 10^{-5}

Source: *data compiled by* J.S. Park from Charles T. Lynch, *CRC Handbook of Materials Science, Vol. 3*, CRC Press, Boca Raton, Florida and *Engineered Materials Handbook, Vol.2*, Engineering Plastics, ASM International, Metals Park, Ohio, 1988.

Table 119. THERMAL EXPANSION OF POLYMERS

(SHEET 9 OF 13)

Type	Polymer	Thermal Expansion Coefficient ASTM D696 (•F^{-1})
Polyester; Thermoplastic	Injection Moldings:	
	General purpose grade	5.3×10^{-5}
	Glass reinforced grades	$2.7—3.3 \times 10^{-5}$
	Glass reinforced self extinguishing	3.5×10^{-5}
	General purpose grade	$4.9—13.0 \times 10^{-5}$
Polyesters: Thermosets	Cast polyyester	
	Rigid	$3.9—5.6 \times 10^{-5}$
	Reinforced polyester moldings	
	High strength (glass fibers)	$13—19 \times 10^{-6}$
Phenylene Oxides	SE—100	3.8×10^{-5}
	SE—1	3.3×10^{-5}
	Glass fiber reinforced	$1.4–2.0 \times 10^{-5}$

Source: *data compiled by* J.S. Park from Charles T. Lynch, *CRC Handbook of Materials Science, Vol. 3*, CRC Press, Boca Raton, Florida and *Engineered Materials Handbook, Vol.2*, Engineering Plastics, ASM International, Metals Park, Ohio, 1988.

Table 119. THERMAL EXPANSION OF POLYMERS

(SHEET 10 OF 13)

Type	Polymer	Thermal Expansion Coefficient ASTM D696 (•F^{-1})
Phenylene Oxides (Con't)	Phenylene oxides (Noryl)	
	Standard	3.1×10^{-5}
	Glass fiber reinforced	$1.2–1.6 \times 10^{-5}$
Polyarylsulfone	Polyarylsulfone	2.6×10^{-5}
Polypropylene	General purpose	$3.8—5.8 \times 10^{-5}$
	High impact	$4.0—5.9 \times 10^{-5}$
	Asbestos filled	$2—3 \times 10^{-5}$
	Glass reinforced	$1.6—2.4 \times 10^{-5}$
Polyphenylene sulfide	Standard	$3.0—4.9 \times 10^{-5}$
	40% glass reinforced	4×10^{-5}

Source: *data compiled by* J.S. Park from Charles T. Lynch, *CRC Handbook of Materials Science, Vol. 3*, CRC Press, Boca Raton, Florida and *Engineered Materials Handbook, Vol.2*, Engineering Plastics, ASM International, Metals Park, Ohio, 1988.

Table 119. THERMAL EXPANSION OF POLYMERS

(SHEET 11 OF 13)

Type	Polymer	Thermal Expansion Coefficient ASTM D696 (•F^{-1})
Polyethylenes; Molded, Extruded	Type I—lower density (0.910¯0.925)	
	Melt index 0.3—3.6	8.9—11.0 x 10^{-5}
	Melt index 6—26	8.9—11.0 x 10^{-5}
	Melt index 200	11 x 10^{-5}
	Type II—medium density (0.926—0.940)	
	Melt index 20	8.3—16.7 x 10^{-5}
	Melt index l.0—1.9	8.3—16.7 x 10^{-5}
	Type III—higher density (0.941—0.965)	
	Melt index 0.2—0.9	8.3—16.7 x 10^{-5}
	Melt Melt index 0.l—12.0	8.3—16.7 x 10^{-5}
	Melt index 1.5—15	8.3—16.7 x 10^{-5}

Source: *data compiled by* J.S. Park from Charles T. Lynch, *CRC Handbook of Materials Science, Vol. 3*, CRC Press, Boca Raton, Florida and *Engineered Materials Handbook, Vol.2*, Engineering Plastics, ASM International, Metals Park, Ohio, 1988.

Table 119. THERMAL EXPANSION OF POLYMERS

(SHEET 12 OF 13)

Type	Polymer	Thermal Expansion Coefficient ASTM D696 (•F^{-1})
Polystyrenes; Molded	General purpose	3.3—4.8 x 10^{-5}
	Medium impact	3.3—4.7 x 10^{-5}
	High impact	2.2—5.6 x 10^{-5}
	Glass fiber -30% reinforced	1.8 x 10^{-5}
	Styrene acrylonitrile (SAN)	3.6—3.7 x 10^{-5}
	Glass fiber (30%) reinforced SAN	1.6 x 10^{-5}
Polyvinyl Chloride And Copolymers; Molded, Extruded		
	Rigid—normal impact	2.8—3 .3 x 10^{-5}
Vinylidene chloride	Vinylidene chloride	8.78 x 10^{-5}

Source: *data compiled by* J.S. Park from Charles T. Lynch, *CRC Handbook of Materials Science, Vol. 3*, CRC Press, Boca Raton, Florida and *Engineered Materials Handbook, Vol.2*, Engineering Plastics, ASM International, Metals Park, Ohio, 1988.

Table 119. THERMAL EXPANSION OF POLYMERS
(SHEET 13 OF 13)

Type	Polymer	Thermal Expansion Coefficient ASTM D696 ($\cdot F^{-1}$)
Silicones; Molded, Laminated	Fibrous (glass) reinforced silicones	3.17—3.23 x 10^{-5}
	Granular (silica) reinforced silicones	2.5—5.0 x 10^{-5}
Ureas; Molded	Alpha—cellulose filled (ASTM Type l)	1.22—1 .50 x 10^{-5}

Source: *data compiled by* J.S. Park from Charles T. Lynch, *CRC Handbook of Materials Science, Vol. 3*, CRC Press, Boca Raton, Florida and *Engineered Materials Handbook, Vol.2*, Engineering Plastics, ASM International, Metals Park, Ohio, 1988.

Table 120. THERMAL EXPANSION COEFFICIENTS OF MATERIALS FOR INTEGRATED CIRCUITS

Material	Coefficient Range	Temperature Range (°C)
Aluminum oxide ceramic	6.0–7.0	25–300
Brass	17.7–21.2	25–300
Kanthal A	13.9–15.1	20–900
Kovar	5.0	25–300
Pyrex glass	3.2	25–300
Pyroceram (#9608)	420	25–300
Pyroceram cement (Vitreous #45)	4	0–300
Pyroceram cement (Devitrified)	2.4	25–300
Pyroceram cement (#89, #95)	8–10	—
Silicon carbide	4.8	0–1,000
Silicon nitride (α)	2.9	25–1,000
Silicon nitride (β)	2.25	25–1,000
Solder glass (Kimble CV-101)	809	0–300

Coefficient of Linear Thermal Expansion of Selected Materials per K

Note: Multiply all values by 10^{-6}.

Source: *from* Beadles, R. L., *Interconnections and Encapsulation, Integrated Silicon Device Technology, Vol. 14*, Research Triangle Institute, Research Triangle Park, N. C., 1967. in *CRC Handbook of Materials Science*, Charles T. Lynch, Ed., CRC Press, Cleveland, (1974).

Table 121. THERMAL EXPANSION OF SILICON CARBIDE SCS–2–AL

Fiber orientation	No. of plies	Coefficient of Thermal Expansion, (10^{-6}/K)
0°	6, 8, 12	6.6
90°	6, 12,40	21.3

Data from *ASM Engineering Materials Reference Book, Second Edition*, Michael Bauccio, Ed., ASM International, Materials Park, OH, p149,(1994).

Table 122. ASTM B 601 Temper Designation Codes for Copper and Copper Alloys (Sheet 1 of 2)

Class	Temper Designation	Temper Name or Material Condition
Cold-worked tempers(a)	H00	1/8 hard
	H01	1/4 hard
	H02	1/2 hard
	H03	3/4 hard
	H04	Hard
	H06	Extra hard
	H08	Spring
	H10	Extra spring
	H12	Special Spring
	H13	Ultra Spring
	H14	Super Spring
Cold-worked tempers(b)	H50	Extruded and drawn
	H52	Pierced and drawn
	H55	Light drawn, light cold rolled
	H58	Drawn general purpose
	H60	Cold heading; forming
	H63	Rivet
	H64	Screw
	H66	Bolt
	H70	Bending
	H80	Hard drawn
	H85	Medium-hard-drawn electrical wire
	H86	Hard-drawn electrical wire
	H90	As-finned

Data from *ASM Metals Reference Book, Third Edition*, Michael Bauccio, Ed., ASM International, Materials Park, OH, p439, (1993).

Table 122. ASTM B 601 Temper Designation Codes for Copper and Copper Alloys (Sheet 2 of 2)

Class	Temper Designation	Temper Name or Material Condition
Cold-worked and stress-relieved tempers	HR01	H01 and stress relieved
	HR02	H02 and stress relieved
	HR04	H04 and stress relieved
	HR08	H08 and stress relieved
	HR10	H10 and stress relieved
	HR20	As-finned
	HR50	Drawn and stress relieved
Cold-rolled and order-strengthened tempers(c)	HT04	H04 and order heat treated
	HT08	H08 and order heat treated

Data from *ASM Metals Reference Book, Third Edition*, Michael Bauccio, Ed., ASM International, Materials Park, OH, p439, (1993).

(a) Cold-worked tempers to meet standard requirements based on cold rolling or cold drawing.

(b) Cold-worked tempers to meet standard requirements based on temper names applicable to specific processes.

(c) Tempers produced by controlled amounts of cold work following by thermal treatment to produce order strengthening.

Table 123. TEMPER DESIGNATION SYSTEM FOR ALUMINUM ALLOYS

Temper	Definition
F	As fabricated
O	Annealed
H1	Strain-hardened only
H2	Strain-hardened and partially annealed
H3	Strain-hardened and stabilized (mechanical properties stabilized by low-temperature thermal treatment)
T1	Cooled from an elevated-temperature shaping process and naturally aged to a substantially stable condition
T2	Cooled from an elevated temperature shaping process, cold-worked, and naturally aged to a substantially stable condition
T3	Solution heat-treated, cold-worked, and naturally aged to a substantially stable condition
T4	Solution heat-treated and naturally aged to a substantially stable condition
T5	Cooled from an elevated-temperature shaping process and artificially aged
T6	Solution heat-treated and artificially aged
T7	Solution heat-treated and stabilized
T8	Solution heat-treated, cold-worked, and artificially aged
T9	Solution heat-treated, artificially aged, and cold-worked
T10	Cooled from an elevated temperature shaping process, cold-worked, and artificially aged

Source: data from *Metals Handbook, 9th ed., Vol. 2*, American Society for Metals, Metals Park, Ohio, 1979, 24-27.

Table 124. Tool Steel Softening After 100 Hours

Type	Original Hardness (HRC)	Hardness (HRC) after 100 h at					
		480°C	540°C	600°C	650°C	700°C	760°C
H13	60.2	48.7	46.3	29.0	22.7	20.1	13.9
	41.7	38.6	39.3	27.7	23.7	20.2	13.2
H21	49.2	48.7	47.6	37.2	27.4	19.8	16.2
	36.7	34.8	34.9	32.6	27.1	19.8	14.9
H23	40.8	40.0	40.6	40.8	38.6	33.2	26.8
	38.9	38.9	38.0	38.0	37.1	32.6	26.6
H26	61.0	60.6	60.3	47.1	38.4	26.9	21.3
	42.9	42.4	42.3	41.3	34.9	26.4	21.1

Source: Data from ASM Metals Reference Book, Second Edition, American Society for Metals, Metals Park, Ohio 44073, p.426, (1984).

See also: Mechanical Properties of Tool Steels

Table 125. THERMOPLASTIC POLYESTER SOFTENING WITH TEMPERATURE

Polymer	Flexural modulus 10^6 psi	Tensile strength 10^3 psi D638	212•F	302• F
Injection Molding Types:				
General purpose grade		7.5—8		
Glass reinforced grades		17—25	7	5.5
Glass reinforced grades	1.2—1.5		0.63	0.53

Source: *data compiled by* J.S. Park *from* Charles T. Lynch, *CRC Handbook of Materials Science, Vol. 3*, CRC Press, Boca Raton, Florida and *Engineered Materials Handbook, Vol.2*, Engineering Plastics, ASM International, Metals Park, Ohio, 1988.

See also: Mechanical Properties of Polymers

Table 126. HEAT-DEFLECTION TEMPERATURE OF CARBON- AND GLASS-REINFORCED ENGINEERING THERMOPLASTICS (SHEET 1 OF 2)

Class	Resin Type	Composition	Heat-Deflection Temperature (°C)
Amorphous	Acrylonitrile-butadiene-styrene(ABS)	30% glass fiber	105
		30% carbon fiber	105
	Nylon	30% glass fiber	140
		30% carbon fiber	145
	Polycarbonate	30% glass fiber	150
		30% carbon fiber	150
	Polyetherimide	30% glass fiber	215
		30% carbon fiber	215
	Polyphenylene oxide (PPO)	30% glass fiber	155
		30% carbon fiber	155
	Polysulfone	30% glass fiber	185
		30% carbon fiber	185
	Styrene-maleic-anhydride (SMA)	30% glass fiber	120
	Thermoplastic polyurethane	30% glass fiber	170

Data from *ASM Engineering Materials Reference Book, Second Edition*, Michael Bauccio, Ed., ASM International, Materials Park, OH, p111–112, (1994).

Table 126. Heat-Deflection Temperature of Carbon- and Glass-Reinforced Engineering Thermoplastics (Sheet 2 of 2)

Class	Resin Type	Composition	Heat-Deflection Temperature (°C)
Crystalline	Acetal	30% glass fiber	165
		20% carbon fiber	160
	Nylon 66%	30% glass fiber	255
		30% carbon fiber	257
	Polybutylene terephthalate (PBT)	30% glass fiber	210
		30% carbon fiber	210
	Polythylene terephthalate (PET)	30% glass fiber	225
	Polyphenylene sulfide (PPS)	30% glass fiber	260
		30% carbon fiber	265

Data from *ASM Engineering Materials Reference Book, Second Edition*, Michael Bauccio, Ed., ASM International, Materials Park, OH, p111–112, (1994).

CHAPTER 6

Mechanical Properties of Materials

List of Tables

Tensile Strength

(continued)

List of Tables (Continued)

Tensile Strength (continued)

Compressive Strength

List of Tables

(Continued)

Yield Strength

Flextural Strength

List of Tables (Continued)

Shear Strength

Hardness

Coefficient of Friction

Abrasion Resistance

Fatique

List of Tables

(Continued)

Impact

Fracture Toughness

Tensile Modulus

Young's Modulus

(continues)

List of Tables (Continued)

Elastic Modulus

Compression Modulus

Bulk Modulus

Sheer Modulus

Torsion Modulus

Flexural Modulus

Modulus of Rupture

List of Tables (Continued)

Poisson's Ratio

Elongation

List of Tables (Continued)

Area Reduction

Ratios

Viscosity

Surface Tension

Table 127. TENSILE STRENGTH OF TOOL STEELS

(SHEET 1 OF 2)

Type	Condition	Tensile Strength (MPa)
L2	Annealed	710
	Oil quenched from 855 •C and single tempered at:	
	205 •C	2000
	315 •C	1790
	425 •C	1550
	540 •C	1275
	650 •C	930
L6	Annealed	655
	Oil quenched from 845 •C and single tempered at:	
	315 •C	2000
	425 •C	1585
	540 •C	1345
	650 •C	965
S1	Annealed	690
	Oil quenched from 930 •C and single tempered at:	
	205 •C	2070
	315 •C	2030
	425 •C	1790
	540 •C	1680
	650 •C	1345
S5	Annealed	725
	Oil quenched from 870 •C and single tempered at:	
	205 •C	2345
	315 •C	2240
	425 •C	1895
	540 •C	1520
	650 •C	1035

Source: data from *ASM Metals Reference Book, Second Edition*, American Society for Metals, Metals Park, Ohio 44073, p241, (1984).

Table 127. TENSILE STRENGTH OF TOOL STEELS

(SHEET 2 OF 2)

Type	Condition	Tensile Strength (MPa)
S7	Annealed	640
	Fan cooled from 940 •C and single tempered at:	
	205 •C	2170
	315 •C	1965
	425 •C	1895
	540 •C	1820
	650 •C	1240

Source: data from *ASM Metals Reference Book, Second Edition*, American Society for Metals, Metals Park, Ohio 44073, p241, (1984).

Table 128. TENSILE STRENGTH OF GRAY CAST IRONS

SAE grade	Maximum Tensile Strength (MPa)
G1800	118
G2500	173
G2500a	173
G3000	207
C3500	241
G3500b	1241
G3500c	1241
G4000	276
G4000d	1276

Source: data from *ASM Metals Reference Book, Second Edition*, American Society for Metals, Metals Park, Ohio 44073, p166-167, (1984).

Table 129. TENSILE STRENGTH OF GRAY CAST IRON BARS

ASTM Class	Tensile Strength (MPa)
ASTM Class	Tensile Strength (MPa)
20	152
25	179
30	214
35	252
40	293
50	362
60	431

Source: data from *ASM Metals Reference Book, Second Edition*, American Society for Metals, Metals Park, Ohio 44073, p166-167, (1984).

Table 130. TENSILE STRENGTH OF DUCTILE IRONS

Specification Number	Grade or Class	Tensile Strength (MPa)
ASTM A395-76 ASME SA395	60-40-18	414
ASTM A476-70(d); SAE AMS5316	80-60-03	552
ASTM A536-72, MIL-1-11466B(MR)	60-40-18	414
	65-45-12	448
	80-55-06	552
	100-70-03	689
	120-90-02	827
SAE J434c	D4018	414
	D4512	448
	D5506	552
	D7003	689
MIL-I-24137(Ships)	Class A	414
	Class B	379
	Class C	345

Source: data from *ASM Metals Reference Book, Second Edition*, American Society for Metals, Metals Park, Ohio 44073, p169, (1984).

Table 131. TENSILE STRENGTH OF MALLEABLE IRON CASTINGS

Specification Number	Grade or Class	Tensile Strength (MPa)
Ferritic		
ASTM A47, A338; ANSI G48.1;		
FED QQ–I–666c	32510	345
	35018	365
ASTM A197		276
Pearlitic and Martensitic		
ASTM A220; ANSI C48.2;		
MIL–I–11444B	40010	414
	45008	448
	45006	448
	50005	483
	60004	552
	70003	586
	80002	655
	90001	724
Automotive		
ASTM A602; SAE J158	M3210	345
	M4504(a)	448
	M5003(a)	517
	M5503(b)	517
	M7002(b)	621
	M8501(b)	724

(a) Air quenched and tempered
(b) Liquid quenched and tempered

Source: data from ASM Metals Reference Book, Second Edition, American Society for Metals, Metals Park, Ohio 44073, p171, (1984).

Table 132. TENSILE STRENGTH OF AUSTENITIC STAINLESS STEELS
(SHEET 1 OF 5)

Type	Form	Condition	ASTM Specification	Tensile Strength (MPa)
Type 301(UNS S30100)	Bar,Wire,Plate,Sheet, Strip	Annealed	A167	515
Type 302 (UNS S30200)	Bar	Hot finished and annealed	A276	515
		Cold finished and annealed(a)	A276	620
		Cold finished and annealed(b)	A276	515
Type 302B (UNS S30215)	Bar	Hot finished and annealed	A276	515
		Cold finished and annealed(a)	A276	620
		Cold finished and annealed(b)	A276	515
Type 302Cu(UNS S30430)	Bar	Annealed	A493	450 to 585
Types 303 (UNS S30300) and 303Se (UNS S30323)	Bar	Annealed	A581	585
	Wire	Annealed	A581	585 to 860
		Cold worked	A581	790 to 1000

(a) Up to 13 mm thick (b) Over 13 mm thick.

Data from *ASM Metals Reference Book, Third Edition*, Michael Bauccio, Ed., ASM International, Materials Park, OH, p364-366 (1993).

Table 132. Tensile Strength of Austenitic Stainless Steels

(Sheet 2 of 5)

Type	Form	Condition	ASTM Specification	Tensile Strength (MPa)
Type 304(UNS S30400)	Bar	Hot finished and annealed	A276	515
		Cold finished and annealed(a)	A276	620
		Cold finished and annealed(b)	A276	515
Type 304L (UNS S30403)	Bar	Hot finished and annealed	A276	480
		Cold finished and annealed(a)	A276	620
		Cold finished and annealed(b)	A276	480
Types 304N (UNS S30451) and 316N(UNS S31651)	Bar	Annealed	A276	550
Type 304LN	Bar	Annealed	—	515
Type 305 (UNS S30500)	Bar	Hot finished and annealed	A276	515
		Cold finished and annealed(a)	A276	260
		Cold finished and annealed(b)	A276	515

(a) Up to 13 mm thick (b) Over 13 mm thick.

Data from *ASM Metals Reference Book, Third Edition*, Michael Bauccio, Ed., ASM International, Materials Park, OH, p364-366 (1993).

Table 132. TENSILE STRENGTH OF AUSTENITIC STAINLESS STEELS

(SHEET 3 OF 5)

Type	Form	Condition	ASTM Specification	Tensile Strength (MPa)
Types 308 (UNS S30800),321(UNS S32100),347(UNS34700) and 348 (UNS S34800)	Bar	Hot finished and annealed	A276	515
		Cold finished and annealed(a)	A276	620
		Cold finished and annealed(b)	A276	515
Type 308L	Bar	Annealed	—	550
Types 309 (UNS S30900), 309S (UNS S30908), 310 (UNS S31000) and 310S (UNS S31008)	Bar	Hot finished and annealed	A276	515
		Cold finished and annealed(a)	A276	620
		Cold finished and annealed(b)	A276	515

(a) Up to 13 mm thick (b) Over 13 mm thick.

Data from *ASM Metals Reference Book, Third Edition*, Michael Bauccio, Ed., ASM International, Materials Park, OH, p364-366 (1993).

Table 132. Tensile Strength of Austenitic Stainless Steels

(Sheet 4 of 5)

Type	Form	Condition	ASTM Specification	Tensile Strength (MPa)
Type 312 Weld metal	—		MIL–E–19933	655
Type 314 (UNS S31400)	Bar	Hot finished and annealed	A276	515
		Cold finished and annealed(a)	A276	620
		Cold finished and annealed(b)	A276	515
Type 316 (UNS S31600)	Bar	Hot finished and annealed	A276	515
		Cold finished and annealed(a)	A276	620
		Cold finished and annealed(b)	A276	515
Type 316F (UNS S31620)	Bar	Annealed	—	585
Type 316L (UNS S31603)	Bar	Hot finished and annealed	A276	480
		Cold finished and annealed(a)	A276	620
		Cold finished and annealed(b)	A276	480

(a) Up to 13 mm thick (b) Over 13 mm thick.

Data from *ASM Metals Reference Book, Third Edition*, Michael Bauccio, Ed., ASM International, Materials Park, OH, p364-366 (1993).

Table 132. TENSILE STRENGTH OF AUSTENITIC STAINLESS STEELS

(SHEET 5 OF 5)

Type	Form	Condition	ASTM Specification	Tensile Strength (MPa)
Type 316LN	Bar	Annealed	—	515
Type 317 (UNS S31700)	Bar	Hot finished and annealed	A276	515
		Cold finished and annealed(a)	A276	620
		Cold finished and annealed(b)	A276	515
Type 317L (UNS S31703)	Bar	Annealed	—	585
Type 317LM	Bar,Plate,Sheet, Strip	Annealed	—	515
Type 329 (UNS S32900)	Bar	Annealed	—	724
Type 330 (UNS N08330)	Bar	Annealed	B511	480
Type 330HC	Bar,Wire,Strip	Annealed	—	585
Types 384 (UNS S38400)	Bar	Annealed	A493	415 to 550
Types 385 (UNS38500)	Bar	Annealed	A493	415 to 550

(a) Up to 13 mm thick (b) Over 13 mm thick.

Data from *ASM Metals Reference Book, Third Edition*, Michael Bauccio, Ed., ASM International, Materials Park, OH, p364-366 (1993).

Table 133. TENSILE STRENGTH OF FERRITIC STAINLESS STEELS
(SHEET 1 OF 2)

Type	ASTM Specification	Form	Condition	Tensile Strength (MPa)
Type 405 (UNS S40500)	A580	Wire	Annealed	480
	A580		Annealed, Cold Finished	480
Type 409 (UNS S40900)	—	Bar	Annealed	450(a)
Type 429 (UNS S42900)	—	Bar	Annealed	490(a)
Type 430 (UNS S43000)	A276	Bar	Annealed, Hot Finished	480
	A276		Annealed, Cold Finished	480
Type 430F (UNS S43020)	A581	Wire	Annealed	585 to 860
Type 430Ti(UNS S43036)	—	Bar	Annealed	515(a)
Type 434 (UNS S43400)	—	Wire	Annealed	545(a)
Type 436 (UNS S43600)	—	Sheet, Strip	Annealed	530(a)

(a) Typical Values

Data from *ASM Metals Reference Book, Third Edition*, Michael Bauccio, Ed., ASM International, Materials Park, OH, p368 (1993).

Table 133. TENSILE STRENGTH OF FERRITIC STAINLESS STEELS

(SHEET 2 OF 2)

Type	ASTM Specification	Form	Condition	Tensile Strength (MPa)
Type 442 (UNS S44200)	—	Bar	Annealed	550(a)
Type 444 (UNS S44400)	A176	Plate, Sheet, Strip	Annealed	415
Type 446 (UNS S44600)	A276	Bar	Annealed, Hot Finished	480
	A276		Annealed, Cold Finished	480

(a) Typical Values

Data from *ASM Metals Reference Book, Third Edition*, Michael Bauccio, Ed., ASM International, Materials Park, OH, p368 (1993).

Table 134. Tensile Strength of Precipitation-Hardening Austenitic Stainless Steels

Type	Form	Condition	Tensile Strength (MPa)
PH 13–8 Mo (UNS S13800)	Bar, Plate, Sheet, Strip	H950	1520
		H1000	1380
15–5 PH (UNS S15500) and 17–4 PH (UNS S17400)	Bar, Plate, Sheet, Stript	H900	1310
		H925	1170
		H1025	1070
		H1075	1000
		H1100	965
		H1150	930
		H1150M	795
17–7 PH (UNS S17700)	Bar	RH950	1275
		TH1050	1170

Data from *ASM Metals Reference Book, Third Edition*, Michael Bauccio, Ed., ASM International, Materials Park, OH, p371 (1993).

Table 135. Tensile Strength of High–Nitrogen Austenitic Stainless Steels

Type	ASTM Specification	Form	Condition	Tensile Strength (MPa)
Type 201 (UNS S20100)	A276	Bar	Annealed	515
Type 202 (UNS S20200)	A276	Bar	Annealed	515
Type 205 (UNS S20500)	—	Plate	Annealed*	830
Type 304N (UNS S30451)	A276	Bar	Annealed	550
Type 304HN (UNS S30452)	—	Bar	Annealed	620
Type 316N (UNS S31651)	A276	Bar	Annealed	550

Data from *ASM Metals Reference Book, Third Edition*, Michael Bauccio, Ed., ASM International, Materials Park, OH, p367 (1993).

* Typical Values.

Table 136. TENSILE STRENGTH OF MARTENSITIC STAINLESS STEELS
(SHEET 1 OF 3)

Type	ASTM Specification	Form	Condition	Tensile Strength (MPa)
Type 403 (UNS S40300)	A276	Bar	Annealed, hot finished	485
	A276		Annealed, cold finished	485
	A276		Intermediate temper, hot finished	690
	A276		Intermediate temper, cold finished	690
	A276		Hard temper, hot finished	825
	A276		Hard temper, cold finished	825
Type 410 (UNS S41000)	A276	Bar	Annealed, hot finished	485
	A276		Annealed, cold finished	485
	A276		Intermediate temper, hot finished	690
	A276		Intermediate temper, cold finished	690
	A276		Hard temper, hot finished	825
	A276		Hard temper, cold finished	825
Type 410S (UNS S41008)	A176	Plate, Sheet, Strip	Annealed	415
Type 410Cb (UNS S41040)	A276	Bar	Annealed, hot finished	485
	A276		Annealed, cold finished	485
	A276		Intermediate temper, hot finished	860
	A276		Intermediate temper, cold finished	860

Data from *ASM Metals Reference Book, Third Edition*, Michael Bauccio, Ed., ASM International, Materials Park, OH, p369-370 (1993).

Table 136. TENSILE STRENGTH OF MARTENSITIC STAINLESS STEELS

(SHEET 2 OF 3)

Type	ASTM Specification	Form	Condition	Tensile Strength (MPa)
Type 414 (UNS S41400)	A276	Bar	Intermediate temper, hot finished	795
	A276		Intermediate temper, cold finished	795
Type 414L	—	Bar	Annealed	795
Types 416 (UNS S41600) and 416Se (UNS S41623)	A581	Wire	Annealed	585 to 860
	A581		Intermediate temper	795 to 1000
	A581		Hard temper	965 to 1210
Type 420 (UNS S42000)	—	Bar	Tempered 205 °C	1720
	A580	Wire	Annealed, cold finished	860 max
Type 422 (UNS S42200)	A565	Bar	Intermediate and hard tempers*	965
Type 431 (UNS S43100)	—	Bar	Tempered 260 °C	1370
	—		Tempered 595 °C	965

Data from *ASM Metals Reference Book, Third Edition*, Michael Bauccio, Ed., ASM International, Materials Park, OH, p369-370 (1993).

Table 136. Tensile Strength of Martensitic Stainless Steels

(Sheet 3 of 3)

Type	ASTM Specification	Form	Condition	Tensile Strength (MPa)
Type 440A (UNS S44002)	—	Bar	Annealed	725
	—		Tempered 315 °C	1790
Type 440B (UNS S44003)	—	Bar	Annealed	740
	—		Tempered 315 °C	1930
Type 440C (UNS S44004)	—	Bar	Annealed	760
	—		Tempered 315 °C	1970
Type 501 (UNS S50100)	—	Bar, Plate	Annealed	485
	—		Tempered 540 °C	1210
Type 502 (UNS S50200)	—	Bar, Plate	Annealed	485

Data from *ASM Metals Reference Book, Third Edition*, Michael Bauccio, Ed., ASM International, Materials Park, OH, p369-370 (1993).

*Heat treated for high-temperature service

Table 137. Tensile Strength of Wrought Coppers and Copper Alloys

(Sheet 1 of 11)

UNS Number and Name	Nominal Composition (%)	Commercial Forms(a)	Tensile Strength (MPa)
C10100 Oxygen-free electronic	99.99 Cu	F, R, W, T, P, S	221-455
C10200 Oxygen-free copper	99.95 Cu	F, R, W, T, P, S	221-455
C10300 Oxygen-free extra-low phosporus	99.95 Cu, 0.003 P	F, R, T, P, S	221-379
C10400, C10500, C10700 Oxygen-free, silver-bearing	99.95 Cu(e)	F, R, W, S	221-455
C10800 Oxygen-free, low phosporus	99.95 Cu, 0.009 P	F, R, T, P	221-379
CS11000 Electrolytic tough pitch copper	99.90 Cu, 0.04 O	F, R, W, T, P, S	221-455
C11100 Electrolytic tough pitch, anneal resistant	99.90 Cu, 0.04 O, 0.01 Cd	W	455
C11300, C11400, C11500, C11600 Silver-bearing tough pitch copper	99.90 Cu, 0.04 O, Ag(f)	F, R, W, T, S	221-455
C12000, C12100	99.9 Cu(g)	F, T, P	221-393
C12200 Phosphorus deoxidized copper, high residual phosphorus	99.90 Cu, 0.02 P	F, R, T, P	221-379
C12500, C12700, C12800, C12900, C13000 Fire-refined tough pitch with silver	99.88 Cu(h)	F, R, W, S	221-462
C14200 Phosphorus deoxidized, arsenical	99.68 Cu, 0.3 As, 0.02 P	F, R, T	221-379

(a) F, flat products; R, rod; W, wire; T, tube; P, pipe; S, shapes.

Data from *ASM Metals Reference Book, Third Edition*, Michael Bauccio, Ed., ASM International, Materials Park, OH, p442–454, (1993).

Table 137. TENSILE STRENGTH OF WROUGHT COPPERS AND COPPER ALLOYS

(SHEET 2 OF 11)

UNS Number and Name	Nominal Composition (%)	Commercial Forms(a)	Tensile Strength (MPa)
C19200	98.97 Cu, 1.0 Fe, 0.03 P	F, T	255-531
C14300	99.9 Cu, 0.1 Cd	F	221-400
C14310	99.8 Cu, 0.2 Cd	F	221-400
C14500 Phosphorus deoxidized, tellurium bearing	99.5 Cu, 0.50 Te, 0.008 P	F, R, W, T	221-386
C14700 Sulfur bearing	99.6 Cu, 0.40 S	R, W	221-393
C15000 Zirconium copper	99.8 Cu, 0.15 Zr	R, W	200-524
C15500	99.75 Cu, 0.06 P, 0.11 Mg, Ag(i)	F	276-552
C15710	99.8 Cu, 0.2 Al_2O_3	R, W	324-724
C15720	99.6 Cu, 0.4 Al_2O_3	F, R	462-614
C15735	99.3 Cu, 0.7 Al_2O_3	R	483-586
C15760	98.9 Cu, 1.1 Al_2O_3	F, R	483-648
C16200 Cadmium copper	99.0 Cu, 1.0 Cd	F, R, W	241-689

(a) F, flat products; R, rod; W, wire; T, tube; P, pipe; S, shapes.

Data from *ASM Metals Reference Book, Third Edition*, Michael Bauccio, Ed., ASM International, Materials Park, OH, p442–454, (1993).

Table 137. TENSILE STRENGTH OF WROUGHT COPPERS AND COPPER ALLOYS

(SHEET 3 OF 11)

UNS Number and Name	Nominal Composition (%)	Commercial Forms(a)	Tensile Strength (MPa)
C16500	98.6 Cu, 0.8 Cd, 0.6 Sn	F, R, W	276-655
C17000 Beryllium copper	99.5 Cu, 1.7 Be, 0.20 Co	F, R	483-1310
C17200 Beryllium copper	99.5 Cu, 1.9 Be , 0.20 Co	F, R, W, T, P, S	469-1462
C17300 Beryllium copper	99.5 Cu, 1.9 Be, 0.40 Pb	R	469-1479
C17500 Copper-cobalt-beryllium alloy	99.5 Cu, 2.5 Co, 0.6 Be	F, R	310-793
C18200, C18400, C18500 Chromium copper	99.5 Cu(j)	F, W, R, S, T	234-593
C18700 leaded copper	99.0 Cu, 1.0 Pb	R	221-379
C18900	98.75 Cu, 0.75 Sn, 0.3 Si, 0.20 Mn	R, W	262-655
C19000 Copper-nickel-phosphorus alloy	98.7 Cu, 1.1 Ni, 0.25 P	F, R, W	262-793
C19100 Copper-nickel-phosphorus-tellurium alloy	98.15 Cu, 1.1 Ni, 0.50 Te, 0.25 P	R, F	248-717
C19400	97.5 Cu, 2.4 Fe, 0.13 Zn, 0.03 P	F	310-524

(a) F, flat products; R, rod; W, wire; T, tube; P, pipe; S, shapes.

Data from ASM *Metals Reference Book, Third Edition*, Michael Bauccio, Ed., ASM International, Materials Park, OH, p442–454, (1993).

Table 137. TENSILE STRENGTH OF WROUGHT COPPERS AND COPPER ALLOYS
(SHEET 4 OF 11)

UNS Number and Name	Nominal Composition (%)	Commercial Forms(a)	Tensile Strength (MPa)
C19500	97.0 Cu, 1.5 Fe, 0.6 Sn, 0.10 P, 0.80 Co	F	552-669
C21000 Gilding, 95%	95.0 Cu, 5.0 Zn	F, W	234-441
C22000 Commercial bronze, 90%	90.0 Cu, 10.0 Zn	F, R, W, T	255-496
C22600 Jewelry bronze, 87.5%	87.5 Cu, 12.5 Zn	F, W	269-669
C23000 Red brass, 85%	85.0 Cu, 15.0 Zn	F, W, T, P	269-724
C24000 Low brass, 80%	80.0 Cu, 20.0 Zn	F, W	290-862
C26000 Cartridge brass, 70%	70.0 Cu, 30.0 Zn	F, R, W, T	303-896
C26800, C27000 Yellow brass	65.0 Cu, 35.0 Zn	F, R, W	317-883
C28000 Muntz metal	60.0 Cu, 40.0 Zn	F, R, T	372-510
C31400 Leaded commercial bronze	89.0 Cu, 1.75 Pb, 9.25 Zn	F, R	255-414
C31600 Leaded commercial bronze, nickel-bearing	89.0 Cu, 1.9 Pb, 1.0 Ni, 8.1 Zn	F, R	255-462
C33000 Low-leaded brass tube	66.0 Cu, 0.5 Pb, 33.5 Zn	T	324-517

(a) F, flat products; R, rod; W, wire; T, tube; P, pipe; S, shapes.

Data from *ASM Metals Reference Book, Third Edition*, Michael Bauccio, Ed., ASM International, Materials Park, OH, p442–454, (1993).

Table 137. Tensile Strength of Wrought Coppers and Copper Alloys

(Sheet 5 of 11)

UNS Number and Name	Nominal Composition (%)	Commercial Forms(a)	Tensile Strength (MPa)
C33200 High-leaded brass tube	66.0 Cu, 1.6 Pb, 32.4 Zn	T	359-517
C33500 Low-leaded brass	65.0 Cu, 0.5 Pb, 34.5 Zn	F	317-510
C34000 Medium-leaded brass	65.0 Cu, 1.0 Pb, 34.0 Zn	F, R, W, S	324-607
C34200 High-leaded brass	64.5 Cu, 2.0 Pb, 33.5 Zn	F, R	338-586
C34900	62.2 Cu, 0.35 Pb, 37.45 Zn	R, W	365-469
C35000 Medium-leaded brass	62.5 Cu, 1.1 Pb, 36.4 Zn	F, R	310-655
C35300 High-leaded brass	62.0 Cu, 1.8 Pb, 36.2 Zn	F, R	338-586
C35600 Extra-high-leaded brass	63.0 Cu, 2.5 Pb, 34.5 Zn	F	338-510
C36000 Free-cutting brass	61.5 Cu, 3.0 Pb, 35.5 Zn	F, R, S	339-469
C36500 to C36800 Leaded Muntz metal	60.0 Cu(k), 0.6 Pb, 39.4 Zn	F	372 (As hot rolled)
C37000 Free-cutting Muntz metal	60.0 Cu, 1.0 Pb, 39.0 Zn	T	372-552
C37700 Forging brass	59.0 Cu, 2.0 Pb, 39.0 Zn	R, S	359 (as extruded)

(a) F, flat products; R, rod; W, wire; T, tube; P, pipe; S, shapes.

Data from *ASM Metals Reference Book, Third Edition*, Michael Bauccio, Ed., ASM International, Materials Park, OH, p442–454, (1993).

Table 137. TENSILE STRENGTH OF WROUGHT COPPERS AND COPPER ALLOYS

(SHEET 6 OF 11)

UNS Number and Name	Nominal Composition (%)	Commercial Forms(a)	Tensile Strength (MPa)
C38500 Architectural bronze	57.0 Cu, 3.0 Pb, 40.0 Zn	R, S	414 (as extruded)
C40500	95 Cu, 1 Sn, 4 Zn	F	269-538
C40800	95 Cu, 2 Sn, 3 Zn	F	290-545
C41100	91 Cu, 0.5 Sn, 8.5 Zn	F, W	269-731
C41300	90.0 Cu, 1.0 Sn, 9.0 Zn	F, R, W	283-724
C41500	91 Cu, 1.8 Sn, 7.2 Zn	F	317-558
C42200	87.5 Cu, 1.1 Sn, 11.4 Zn	F	296-607
C42500	88.5 Cu, 2.0 Sn, 9.5 Zn	F	310-634
C43000	87.0 Cu, 2.2 Sn, 10.8 Zn	F	317-648
C43400	85.0 Cu, 0.7 Sn, 14.3 Zn	F	310-607
C43500	81.0 Cu, 0.9 Sn, 18.1 Zn	F, T	317-552
C44300, C44400, C44500 Inhibited admiralty	71.0 Cu, 28.0 Zn, 1.0 Sn	F, W, T	331-379

(a) F, flat products; R, rod; W, wire; T, tube; P, pipe; S, shapes.

Data from *ASM Metals Reference Book, Third Edition*, Michael Bauccio, Ed., ASM International, Materials Park, OH, p442–454, (1993).

Table 137. TENSILE STRENGTH OF WROUGHT COPPERS AND COPPER ALLOYS

(SHEET 7 OF 11)

UNS Number and Name	Nominal Composition (%)	Commercial Forms(a)	Tensile Strength (MPa)
C46400 to C46700 Naval brass	60.0 Cu, 39.25 Zn, 0.75 Sn	F, R, T, S	379-607
C48200 Naval brass, medium-leaded	60.5 Cu, 0.7 Pb, 0.8 Sn, 38.0 Zn	F, R, S	386-517
C48500 Leaded naval brass	60.0 Cu, 1.75 Pb, 37.5 Zn, 0.75 Sn	F, R, S	379-531
C50500 Phosphor bronze, 1.25% E	98.75 Cu, 1.25 Sn, trace P	F, W	276-545
C51000 Phosphor bronze, 5% A	95.0 Cu, 5.0 Sn, trace P	F, R, W, T	324-965
C51100	95.6 Cu, 4.2 Sn, 0.2 P	F	317-710
C52100 Phosphor bronze, 8% C	92.0 Cu, 8.0 Sn, trace P	F, R, W	379-965
C52400 Phosphor bronze, 10% D	90.0 Cu, 10.0 Sn, trace P	F, R, W	455-1014
C54400 Free-cutting phosphor bronze	88.0 Cu, 4.0 Pb, 4.0 Zn, 4.0 Sn	F, R	303-517
C60800 Aluminum bronze, 5%	95.0 Cu, 5.0 Al	T	414
C61000	92.0 Cu, 8.0 Al	R, W	483-552
C61300	92.65 Cu, 0.35 Sn, 7.0 Al	F, R, T, P, S	483-586

(a) F, flat products; R, rod; W, wire; T, tube; P, pipe; S, shapes.

Data from *ASM Metals Reference Book, Third Edition*, Michael Bauccio, Ed., ASM International, Materials Park, OH, p442–454, (1993).

Table 137. Tensile Strength of Wrought Coppers and Copper Alloys

(Sheet 8 of 11)

UNS Number and Name	Nominal Composition (%)	Commercial Forms(a)	Tensile Strength (MPa)
C61400 Aluminum bronze, D	91.0 Cu, 7.0 Al, 2.0 Fe	F, R, W, T, P, S	524-614
C61500	90.0 Cu, 8.0 Al, 2.0 Ni	F	483-1000
C61800	89.0 Cu, 1.0 Fe, 10.0 Al	R	552-586
C61900	86.5 Cu, 4.0 Fe, 9.5 Al	F	634-1048
C62300	87.0 Cu, 10.0 Al, 3.0 Fe	F, R	517-676
C62400	86.0 Cu, 3.0 Fe, 11.0 Al	F, R	621-724
C62500	82.7 Cu, 4.3 Fe, 13.0 Al	F, R	689
C63000	82.0 Cu, 3.0 Fe, 10.0 Al, 5.0 Ni	F, R	621-814
C63200	82.0 Cu, 4.0 Fe, 9.0 Al, 5.0 Ni	F, R	621-724
C63600	95.5 Cu, 3.5 Al, 1.0 Si	R, W	414-579
C63800	99.5 Cu, 2.8 Al, 1.8 Si, 0.40 Co	F	565-896
C64200	91.2 Cu, 7.0 Al	F, R	517-703

(a) F, flat products; R, rod; W, wire; T, tube; P, pipe; S, shapes.

Data from *ASM Metals Reference Book, Third Edition*, Michael Bauccio, Ed., ASM International, Materials Park, OH, p442–454, (1993).

Table 137. TENSILE STRENGTH OF WROUGHT COPPERS AND COPPER ALLOYS

(SHEET 9 OF 11)

UNS Number and Name	Nominal Composition (%)	Commercial Forms(a)	Tensile Strength (MPa)
C65100 Low-silicon bronze, B	98.5 Cu, 1.5 Si	R, W, T	276-655
C65500 High-silicon bronze, A	97.0 Cu, 3.0 Si	F, R, W, T	386-1000
C66700 Manganese brass	70.0 Cu, 28.8 Zn, 1.2 Mn	F, W	315-689
C67400	58.5 Cu, 36.5 Zn, 1.2 Al, 2.8 Mn, 1.0 Sn	F, R	483-634
C67500 Manganese bronze, A	58.5 Cu, 1.4 Fe, 39.0 Zn, 1.0 Sn, 0.1 Mn	R, S	448-579
C68700 Aluninum brass, arsenical	77.5 Cu, 20.5 Zn, 2.0 Al, 0.1 As	T	414
C68800	73.5 Cu, 22.7 Zn, 3.4 Al, 0.40 Co	F	565-889
C69000	73.3 Cu, 3.4 Al, 0.6 Ni, 22.7 Zn	F	496-896
C69400 Silicon red brass	81.5 Cu, 14.5 Zn, 4.0 Si	R	552-689
C70400	92.4 Cu, 1.5 Fe, 5.5 Ni, 0.6 Mn	F, T	262-531
C70600 Copper nickel, 10%	88.7 Cu, 1.3 Fe, 10.0 Ni	F, T	303-414
C71000 Copper nickel, 20%	79.00 Cu, 21.0 Ni	F, W, T	338-655

(a) F, flat products; R, rod; W, wire; T, tube; P, pipe; S, shapes.

Data from *ASM Metals Reference Book, Third Edition*, Michael Bauccio, Ed., ASM International, Materials Park, OH, p442–454, (1993).

Table 137. Tensile Strength of Wrought Coppers and Copper Alloys
(Sheet 10 of 11)

UNS Number and Name	Nominal Composition (%)	Commercial Forms(a)	Tensile Strength (MPa)
C71500 Copper nickel, 30%	70.0 Cu, 30.0 Ni	F, R, T	372-517
C71700	67.8 Cu, 0.7 Fe, 31.0 Ni, 0.5 Be	F, R, W	483-1379
C72500	88.20 Cu, 9.5 Ni, 2.3 Sn	F, R, W, T	379-827
C73500	72.0 Cu, 18.0 Ni , 10.0 Zn	F, R, W, T	345-758
C74500 Nickel silver, 65-10	65.0 Cu, 25.0 Zn, 10.0 Ni	F, W	338-896
C75200 Nickel silver, 65-18	65.0 Cu, 17.0 Zn, 18.0 Ni	F, R, W	386-710
C75400 Nickel silver, 65-15	65.0 Cu, 20.0 Zn, 15.0 Ni	F	365-634
C75700 Nickel silver, 65-12	65.0 Cu, 23.0 Zn, 12.0 Ni	F, W	359-641

(a) F, flat products; R, rod; W, wire; T, tube; P, pipe; S, shapes.

Data from *ASM Metals Reference Book, Third Edition*, Michael Bauccio, Ed., ASM International, Materials Park, OH, p442–454, (1993).

Table 137. TENSILE STRENGTH OF WROUGHT COPPERS AND COPPER ALLOYS
(SHEET 11 OF 11)

UNS Number and Name	Nominal Composition (%)	Commercial Forms(a)	Tensile Strength (MPa)
C76200	59.0 Cu, 29.0 Zn, 12.0 Ni	F, T	393-841
C77000 Nickel silver, 55-18	55.0 Cu, 27.0 Zn, 18.0 Ni	F, R, W	414-1000
C72200	82.0 Cu, 16.0 Ni, 0.5 Cr, 0.8 Fe, 0.5 Mn	F, T	317-483
C78200 Leaded nickel silver, 65-8-2	65.0 Cu, 2.0 Pb, 25.0 Zn, 8.0 Ni	F	365-627

(a) F, flat products; R, rod; W, wire; T, tube; P, pipe; S, shapes.

Data from *ASM Metals Reference Book, Third Edition*, Michael Bauccio, Ed., ASM International, Materials Park, OH, p442–454, (1993).

(d) Based on 100% for C360000.
(e) C10400, 8 oz/ton Ag; C10500, 10 oz/ton; C10700, 25 oz/ton .
(f) C11300, 8 oz/ton Ag; C11400,10 oz/ton; C11500, 16 oz/ton; C11600, 25 oz/ton
(g) C12000, 0.008 P; C12100, 0.008 P and 4 oz/ton Ag;
(h) C12700, 8 oz/ton Ag; C12800,10 oz/ton; C12900,16 oz/ton; C13000, 25 oz/ton.
(i) 8.30 oz/ton Ag.
(j) C18200, 0.9 Cr; C18400, 0.8 Cr; C18500, 0.7 Cr
(k) Rod, 61.0 Cu min.

Table 138. TENSILE STRENGTH OF ALUMINUM CASTING ALLOYS (SHEET 1 OF 3)

Alloy AA No.	Temper	Tensile Strength (MPa)
201.0	T4	365
	T6	485
	T7	460
206.0, A206.0	T7	435
208.0	F	145
242.0	T21	185
	T571	220
	T77	205
	T571	275
	T61	325
295.0	T4	220
	T6	250
	T62	285
296.0	T4	255
	T6	275
	T7	270
308.0	F	195
319.0	F	185
	T6	250
	F	235
	T6	280
336.0	T551	250
	T65	325
354.0	T61	380
355.0	T51	195
	T6	240
	T61	270
	T7	265

Source: data from *ASM Metals Reference Book, Second Edition*, American Society for Metals, Metals Park, Ohio 44073, (1984).

Table 138. TENSILE STRENGTH OF ALUMINUM CASTING ALLOYS (SHEET 2 OF 3)

Alloy AA No.	Temper	Tensile Strength (MPa)
355.0 (Con't)	T71	175
	T51	210
	T6	290
	T62	310
	T7	280
	T71	250
356.0	T51	175
	T6	230
	T7	235
	T71	195
	T6	265
	T7	220
357.0, A357.0	T62	360
359.0	T61	330
	T62	345
360.0	F	325
A360.0	F	320
380.0	F	330
383.0	F	310
384.0, A384.0	F	330
390.0	F	280
	T5	300
A390.0	F,T5	180
	T6	280
	T7	250
	F,T5	200
	T6	310
	T7	260

Source: data from *ASM Metals Reference Book, Second Edition*, American Society for Metals, Metals Park, Ohio 44073, (1984).

Table 138. TENSILE STRENGTH OF ALUMINUM CASTING ALLOYS (SHEET 3 OF 3)

Alloy AA No.	Temper	Tensile Strength (MPa)
413.0	F	300
A413.0	F	290
443.0	F	130
B443.0	F	159
C443.0	F	228
514.0	F	170
518.0	F	310
520.0	T4	330
535.0	F	275
712.0	F	240
713.0	T5	210
	T5	220
771.0	T6	345
850.0	T5	160

Source: data from *ASM Metals Reference Book, Second Edition*, American Society for Metals, Metals Park, Ohio 44073, (1984).

Table 139. TENSILE STRENGTH OF WROUGHT ALUMINUM ALLOYS (SHEET 1 OF 7)

Alloy	Temper	Tensile Strength (MPa)
1050	0	76
	H14	110
	H16	130
	H18	160
1060	0	69
	H12	83
	H14	97
	H16	110
	H18	130
1100	0	90
	H12	110
	H14	125
	H16	145
	H18	165
1350	0	83
	H12	97
	H14	110
	H16	125
	H19	185
2011	T3	380
	T8	405
2014	0	185
	T4	425
	T6	485
Alclad 2014	0	170
	T3	435
	T4	420
	T6	470

Source: data from *ASM Metals Reference Book, Second Edition*, American Society for Metals, Metals Park, Ohio 44073, p.299—302, (1984).

Table 139. TENSILE STRENGTH OF WROUGHT ALUMINUM ALLOYS (SHEET 2 OF 7)

Alloy	Temper	Tensile Strength (MPa)
2024	0	185
	T3	485
	T4, T351	470
	T361	495
Alclad 2024	0	180
	T	450
	T4, T351	440
	T361	460
	T81, T851	450
	T861	485
2036	T4	340
2048		455
2124	T851	490
2218	T61	405
	T71	345
	T72	330
2219	0	170
	T42	360
	T31, T351	360
	T37	395
	T62	415
	T81, T851	455
	T87	475
2618	All	440
3003	0	110
Alclad	H12	130
3003	H14	150
	H16	180
	H18	200
3004	0	180
Alclad	H32	215

Source: data from *ASM Metals Reference Book, Second Edition*, American Society for Metals, Metals Park, Ohio 44073, p.299—302, (1984).

Table 139. TENSILE STRENGTH OF WROUGHT ALUMINUM ALLOYS (SHEET 3 OF 7)

Alloy	Temper	Tensile Strength (MPa)
3004	H34	240
	H36	260
	H38	285
3105	0	115
	H12	150
	H14	170
	H16	195
	H18	215
	H25	180
4032	T6	380
4043	0	145
	H18	285
5005	0	125
	H12	140
	H14	160
	H16	180
	H18	200
	H32	140
	H34	160
	H36	180
	H38	200
5050	0	145
	H32	170
	H34	195
	H36	205
	H38	220
5052	0	195
	H32	230
	H34	260
	H36	275
	H38	290

Source: data from *ASM Metals Reference Book, Second Edition*, American Society for Metals, Metals Park, Ohio 44073, p.299—302, (1984).

Table 139. TENSILE STRENGTH OF WROUGHT ALUMINUM ALLOYS (SHEET 4 OF 7)

Alloy	Temper	Tensile Strength (MPa)
5056	0	290
	H18	435
	H38	415
5083	0	290
	H112	305
	H113	315
	H321	315
	H323, H32	325
	H343, H34	345
5086	0	260
	H32, H116, H117	290
	H34	325
	H112	270
5154	0	240
	H32	270
	H34	290
	H36	310
	H38	330
	H112	240
5182	0	275
	H32	315
	H34	340
	H19(n)	420
5252	H25	235
	H28, H38	285
5254	0	240
5254	H32	270
	H34	290
	H36	310
	H38	330
	H112	240

Source: data from *ASM Metals Reference Book, Second Edition*, American Society for Metals, Metals Park, Ohio 44073, p.299—302, (1984).

Table 139. Tensile Strength of Wrought Aluminum Alloys (Sheet 5 of 7)

Alloy	Temper	Tensile Strength (MPa)
5454	0	250
	H32	275
	H34	305
	H36	340
	H38	370
	H111	260
	H112	250
	H311	260
5456	0	310
	H111	325
	H112	310
	H321, H116	350
5457	0	130
	H25	180
	H28, H38	205
5652	0	195
	H32	230
	H34	260
	H36	275
	H38	290
5657	H25	160
	H28, H38	195
6005	T1	170
	T5	260
6009	T4	235
	T6	345
6010	T4	255

Source: data from *ASM Metals Reference Book, Second Edition*, American Society for Metals, Metals Park, Ohio 44073, p.299—302, (1984).

Table 139. TENSILE STRENGTH OF WROUGHT ALUMINUM ALLOYS (SHEET 6 OF 7)

Alloy	Temper	Tensile Strength (MPa)
6061	0	125
	T4, T451	240
	T6, T651	310
Alclad 6061	0	115
	T4, T451	230
	T6, T651	290
6063	0	90
	T1	150
	T4	170
	T5	185
	T6	240
	T83	255
	T831	205
	T832	290
6066	0	150
	T4, T451	360
	T6, T651	395
6070	0	145
	T4	315
	T6	380
6101	Hlll	97
6151	T6	220
6201	T6	330
	T81	330
6205	Tl	260
	T5	310
6262	T9	400

Source: data from *ASM Metals Reference Book, Second Edition*, American Society for Metals, Metals Park, Ohio 44073, p.299—302, (1984).

Table 139. TENSILE STRENGTH OF WROUGHT ALUMINUM ALLOYS (SHEET 7 OF 7)

Alloy	Temper	Tensile Strength (MPa)
6351	T4	250
	T6	310
6463	Tl	150
	T5	185
	T6	240
7005	0	193
	T53	393
	T6,T63,T6351	372
7050	T736	515
7075	0	230
	T6,T651	570
	T73	505
Alclad 7075	0	220
	T6,T651	525
7175	T66	595
	T736	525
7475	T61	525

Source: data from *ASM Metals Reference Book, Second Edition*, American Society for Metals, Metals Park, Ohio 44073, p.299—302, (1984).

Table 140. TENSILE STRENGTH OF COBALT-BASE SUPERALLOYS

Alloy	Temperature (°C)	Tensile Strength (MPa)
Haynes 25 (L–605) sheet	21	1010
	540	800
	650	710
	760	455
	870	325
Haynes 188, sheet	21	960
	540	740
	650	710
	760	635
	870	420
S-816, bar	21	965
	540	840
	650	765
	760	650
	870	360

Data from *ASM Metals Reference Book, Third Edition*, Michael Bauccio, Ed., ASM International, Materials Park, OH, p387, (1993).

Table 141. TENSILE STRENGTH OF NICKEL-BASE SUPERALLOYS (SHEET 1 OF 5)

Alloy	Temperature (°C)	Tensile Strength (MPa)
Astroloy, bar	21	1410
	540	1240
	650	1310
	760	1160
	870	770
D–979, bar	21	1410
	540	1300
	650	1100
	760	7
	870	345
Hastelloy X, sheet	21	785
	540	650
	650	570
	760	435
	870	255
IN–102, bar	21	960
	540	825
	650	710
	760	440
	870	215
Inconel 600, bar	21	620
	540	580
	650	450
	760	185
	870	105
Inconel 601, sheet	21	740
	540	725
	650	525
	760	290
	870	160

Data from *ASM Metals Reference Book, Third Edition*, Michael Bauccio, Ed., ASM International, Materials Park, OH, p387-389, (1993).

Table 141. TENSILE STRENGTH OF NICKEL-BASE SUPERALLOYS (SHEET 2 OF 5)

Alloy	Temperature (°C)	Tensile Strength (MPa)
Inconel 625, bar	21	855
	540	745
	650	710
	760	505
	870	285
Inconel 706, bar	21	1300
	540	1120
	650	1010
	760	690
Inconel 718, bar	21	1430
	540	1280
	650	1230
	760	950
	870	340
Inconel 718, sheet	21	1280
	540	1140
	650	1030
	760	675
Inconel X-750, bar	21	1120
	540	965
	650	825
	760	485
	870	235
M-252, bar	21	1240
	540	1230
	650	1160
	760	945
	870	510

Data from *ASM Metals Reference Book, Third Edition*, Michael Bauccio, Ed., ASM International, Materials Park, OH, p387-389, (1993).

Table 141. TENSILE STRENGTH OF NICKEL-BASE SUPERALLOYS (SHEET 3 OF 5)

Alloy	Temperature (°C)	Tensile Strength (MPa)
Nimonic 75, bar	21	750
	540	635
	650	538
	760	290
	870	145
Nimonic 80A, bar	21	1240
	540	1100
	650	1000
	760	760
	870	400
Nimonic 90, bar	21	1240
	540	1100
	650	1030
	760	825
	870	430
Nimonic 105, bar	21	1140
	540	1100
	650	1080
	760	965
	870	605
Nimonic 115, bar	21	1240
	540	1090
	650	1120
	760	1080
	870	825
Pyromet 860, bar	21	1300
	540	1250
	650	1110
	760	910

Data from *ASM Metals Reference Book, Third Edition,* Michael Bauccio, Ed., ASM International, Materials Park, OH, p387-389, (1993).

Table 141. TENSILE STRENGTH OF NICKEL-BASE SUPERALLOYS (SHEET 4 OF 5)

Alloy	Temperature (°C)	Tensile Strength (MPa)
René 41, bar	21	1420
	540	1400
	650	1340
	760	1100
	870	620
René 95, bar	21	1620
	540	1540
	650	1460
	760	1170
Udimet 500, bar	21	1310
	540	1240
	650	1210
	760	1040
	870	640
Udimet 520, bar	21	1310
	540	1240
	650	1170
	760	725
	870	515
Udimet 700, bar	21	1410
	540	1280
	650	1240
	760	1030
	870	690
Udimet 710, bar	21	1190
	540	1150
	650	1290
	760	1020
	870	705

Data from *ASM Metals Reference Book, Third Edition*, Michael Bauccio, Ed., ASM International, Materials Park, OH, p387-389, (1993).

Table 141. TENSILE STRENGTH OF NICKEL-BASE SUPERALLOYS (SHEET 5 OF 5)

Alloy	Temperature (°C)	Tensile Strength (MPa)
Unitemp AF2–1DA, bar	21	1290
	540	1340
	650	1360
	760	1150
	870	830
Waspaloy, bar	21	1280
	540	1170
	650	1120
	760	795
	870	525

Data from *ASM Metals Reference Book, Third Edition*, Michael Bauccio, Ed., ASM International, Materials Park, OH, p387-389, (1993).

Table 142. Tensile Strength of Wrought Titanium Alloys at Room Temperature (Sheet 1 of 3)

Class	Alloy	Condition	Tensile Strength (MPa)
Commercially Pure	99.5 Ti	Annealed	331
	99.2 Ti	Annealed	434
	99.1 Ti	Annealed	517
	99.0 Ti	Annealed	662
	99.2Ti-0.2Pd	Annealed	434
	Ti-0.8Ni-0.3Mo	Annealed	517
Alpha Alloys	Ti-5Al-2.5Sn	Annealed	862
	Ti-5Al-2.5Sn (low O_2)	Annealed	807
Near Alpha Alloys	Ti-8Al-1Mo-1V	Duplex Annealed	1000
	Ti-11Sn-1Mo-2.25Al-5.0Zr-1Mo-0.2Si	Duplex Annealed	1103
	Ti-6Al-2Sn-4Zr-2Mo	Duplex Annealed	979
	Ti-5Al-2Sn-2Zr-2Mo-0.25Si	975 °C (1/2h), AC + 595°C (2h), AC	1048
	Ti-6Al-2Nb-1Ta-1Mo	As rolled 2.5 cm (1 in.) plate	855
	Ti-6Al-2Sn-1.5Zr-1Mo- 0.35Bi-0.1Si	Beta forge + duplex anneal	1014

Data from *ASM Metals Reference Book, Third Edition*, Michael Bauccio, Ed., ASM International, Materials Park, OH, p512, (1993).

Table 142. Tensile Strength of Wrought Titanium Alloys at Room Temperature (Sheet 2 of 3)

Class	Alloy	Condition	Tensile Strength (MPa)
Alpha-Beta Alloys	Ti-8Mn	Annealed	945
	Ti-3Al-2.5V	Annealed	689
	Ti-6Al-4V	Annealed	993
		Solution + age	1172
	Ti-6Al-4V(low O_2)	Annealed	896
	Ti-6Al-6V-2Sn	Annealed	1069
		Solution + age	1276
	Ti-7Al-4Mo	Solution + age	1103
	Ti-6Al-2Sn-4Zr-6Mo	Solution + age	1269
	Ti-6Al-2Sn-2Zr-2Mo- 2Cr-0.25Si	Solution + age	1276
	Ti-10V-2Fe-3Al	Solution + age	1276

Data from *ASM Metals Reference Book, Third Edition*, Michael Bauccio, Ed., ASM International, Materials Park, OH, p512, (1993).

Table 142. TENSILE STRENGTH OF WROUGHT TITANIUM ALLOYS AT ROOM TEMPERATURE (SHEET 3 OF 3)

Class	Alloy	Condition	Tensile Strength (MPa)
Beta Alloys	Ti-13V-1Cr-3Al	Solution + age	1220
			1276
	Ti-8Mo-8V-2Fe-3Al	Solution + age	1310
	Ti-3Al-8V-6Cr-4Mo-4Zr	Solution + age	1448
		Annealed	883
	Ti-11.5Mo-6Zr-4.5Sn	Solution + age	1386

Data from *ASM Metals Reference Book, Third Edition*, Michael Bauccio, Ed., ASM International, Materials Park, OH, p512, (1993).

Table 143. TENSILE STRENGTH OF WROUGHT TITANIUM ALLOYS AT HIGH TEMPERATURE (SHEET 1 OF 4)

Class	Alloy	Condition	Test Temperature (°C)	Tensile Strength (MPa)
Commercially Pure	99.5 Ti	Annealed	315	152
	99.2 Ti	Annealed	315	193
	99.1 Ti	Annealed	315	234
	99.0 Ti	Annealed	315	310
	99.2Ti-0.2Pd	Annealed	315	186
	Ti-0.8Ni-0.3Mo	Annealed	205	345
	Ti-0.8Ni-0.3Mo	Annealed	315	324
Alpha Alloys	Ti-5Al-2.5Sn	Annealed	315	565
	Ti-5Al-2.5Sn (low O_2)	Annealed	-195	1241
			-255	1579
Near Alpha Alloys	Ti-8Al-1Mo-1V	Duplex Annealed	315	793
			425	738
			540	621

Data from *ASM Metals Reference Book, Third Edition*, Michael Bauccio, Ed., ASM International, Materials Park, OH, p512, (1993).

Table 143. Tensile Strength of Wrought Titanium Alloys at High Temperature (Sheet 2 of 4)

Class	Alloy	Condition	Test Temperature (°C)	Tensile Strength (MPa)
	Ti-11Sn-1Mo-2.25Al-5.0Zr-1Mo-0.2Si	Duplex Annealed	315	896
			425	827
			540	758
	Ti-6Al-2Sn-4Zr-2Mo	Duplex Annealed	315	772
			425	703
			540	648
	Ti-5Al-2Sn-2Zr-2Mo-0.25Si	975 °C (1/2h), AC + 595 °C (2h), AC	315	793
			425	779
			540	689
	Ti-6Al-2Nb-1Ta-1Mo	As rolled 2.5 cm (1 in.) plate	315	586
			425	517
			540	483
	Ti-6Al-2Sn-1.5Zr-1Mo- 0.35Bi-0.1Si	Beta forge + duplex anneal	480	724

Data from *ASM Metals Reference Book, Third Edition*, Michael Bauccio, Ed., ASM International, Materials Park, OH, p512, (1993).

Table 143. Tensile Strength of Wrought Titanium Alloys at High Temperature (Sheet 3 of 4)

Class	Alloy	Condition	Test Temperature (°C)	Tensile Strength (MPa)
Alpha-Beta Alloys	Ti-8Mn	Annealed	315	717
	Ti-3Al-2.5V	Annealed	315	483
	Ti-6Al-4V	Annealed	315	724
		Annealed	425	669
		Annealed	540	531
		Solution + age	315	862
		Solution + age	425	800
		Solution + age	540	655
	Ti-6Al-4V(low O_2)	Annealed	160	1517
	Ti-6Al-6V-2Sn	Annealed	315	931
		Solution + age	315	979
	Ti-7Al-4Mo	Solution + age	315	976
			425	848

Data from *ASM Metals Reference Book, Third Edition*, Michael Bauccio, Ed., ASM International, Materials Park, OH, p512, (1993).

Table 143. TENSILE STRENGTH OF WROUGHT TITANIUM ALLOYS AT HIGH TEMPERATURE (SHEET 4 OF 4)

Class	Alloy	Condition	Test Temperature (°C)	Tensile Strength (MPa)
Beta Alloys	Ti-6Al-2Sn-4Zr-6Mo	Solution + age	315	1020
			425	951
			540	848
	Ti-6Al-2Sn-2Zr-2Mo- 2Cr-0.25Si	Solution + age	315	979
	Ti-10V-2Fe-3Al	Solution + age	205	1117
			315	1103
	Ti-13V-1Cr-3Al	Solution + age	315	883
			425	1103
	Ti-8Mo-8V-2Fe-3Al	Solution + age	315	1131
	Ti-3Al-8V-6Cr-4Mo-4Zr	Solution + age	315	1034
			425	938
	Ti-11.5Mo-6Zr-4.5Sn	Annealed	315	724
		Solution + age	315	903

Data from *ASM Metals Reference Book, Third Edition*, Michael Bauccio, Ed., ASM International, Materials Park, OH, p512, (1993).

Table 144. Tensile Strength of Refractory Metal Alloys
(Sheet 1 of 3)

Class	Alloy	Alloying Additions (%)	Form	Condition	Temperature (°F)	Tensile Strength (ksi)
Niobium and Niobium Alloys	Pure Niobium	—	All	Recrystallized	2000	10
	Nb–1Zr	1 Zr	All	Recrystallized	2000	23
	C103(KbI–3)	10 Hf, 1 Ti 0.7 Zr	All	Recrystallized	2000	27
	SCb291	10 Ta, 10 W	Bar, Sheet	Recrystallized	2000	32
	C129	10 W, 10 Hf, 0.1 Y	Sheet	Recrystallized	2400	26
	FS85	28 Ta, 11 W, 0.8 Zr	Sheet	Recrystallized	2400	23
	SU31	17 W, 3.5 Hf, 0.12 C, 0.03 Si	Bar, Sheet	Special Thermal Processing	2400	40
Molybdenum and Molybdenum Alloys	Pure Molybdenum	—	All	Stress-relieved Annealed	1800	52
	Doped Mo	K, Si; ppm levels	Wire, Sheet	Cold Worked	3000	30
	Low C Mo	None	All	Stress-relieved Annealed	1800	50
	TZM	0.5 Ti, 0.08 Zr, 0.015 C	All	Stress-relieved Annealed	2400	45
	TZC	1.0 Ti, 0.14 Zr, 0.02 to 0.08 C	All	Stress-relieved Annealed	2400	55
	Mo–5Re	5 Re	All	Stress-relieved Annealed	3000	2
	Mo–30W	30 W	All	Stress-relieved Annealed	2000	50

Data from ASM *Metals Reference Book, Third Edition*, Michael Bauccio, Ed., ASM International, Materials Park, OH, p390, (1993).

Table 144. TENSILE STRENGTH OF REFRACTORY METAL ALLOYS
(SHEET 2 OF 3)

Class	Alloy	Alloying Additions (%)	Form	Condition	Temperature (°F)	Tensile Strength (ksi)
Tantalum Alloys	Unalloyed	None	All	Recrystallized	2400	8.5
	FS61	7.5 W(P/M)	Wire, Sheet	Cold Worked	75	165
	FS63	2.5 W, 0.15 Nb	All	Recrystallized	200	46
	TA–10W	10 W	All	Recrystallized	2400	50
	KBI–40	40 Nb	All	Recrystallized	500	42
Tungsten Alloys	Unalloyed	None	Bar, Sheet, Wire	Stress-relieved Annealed	3000	25
	Doped	K, Si, Al; ppm levels	Wire	Cold Worked	3000	94
	W–1 ThO_2	1ThO_2	Bar, Sheet, Wire	Stress-relieved Annealed	3000	37
	W–2 ThO_2	2 ThO_2	Bar, Sheet, Wire	Stress-relieved Annealed	3000	30
	W–3 ThO_2	3 ThO_2	Bar, Wire	Stress-relieved Annealed	3000	30
	W–4 ThO_2	4 ThO_2	Bar	Stress-relieved Annealed	3000	30

Data from ASM *Metals Reference Book, Third Edition*, Michael Bauccio, Ed., ASM International, Materials Park, OH, p390, (1993).

Table 144. Tensile Strength of Refractory Metal Alloys

(Sheet 3 of 3)

Class	Alloy	Alloying Additions (%)	Form	Condition	Temperature (°F)	Tensile Strength (ksi)
	W–15 Mo	15 Mo	Bar, Wire	Stress-relieved Annealed	3000	36
	W–50 Mo	50 Mo	Bar, Wire	Stress-relieved Annealed	3000	20
	W–3 Re	3 Re	Wire	Cold Worked	—	—
	W–25 Re	25 Re	Bar, Sheet, Wire	Stress-relieved Annealed	3000	33

Data from ASM *Metals Reference Book, Third Edition*, Michael Bauccio, Ed., ASM International, Materials Park, OH, p390, (1993).

Table 145. TENSILE STRENGTH OF CERAMICS
(SHEET 1 OF 4)

Type	Ceramic	Tensile Strength (psi)	Temperature
Borides	Chromium Diboride (CrB_2)	$10.6x10^4$	
	Titanium Diboride (TiB_2)	$18.4x10^3$	
	Zirconium Diboride (ZrB_2)	$28.7x10^3$	
Carbides	Boron Carbide (B_4C)	$22.5x10^3$	980°C
	Silicon Carbide (SiC)	$5\text{-}20x10^3$ psi	25°C
	(hot pressed)	$29x10^3$ psi	20°C
	(hot pressed)	$5.75\text{-}21.75\ x10^3$ psi	1400°C
	(reaction bonded)	$11.17x10^3$ psi	20°C
	Tantalum Monocarbide (TaC)	$2\text{-}42x10^3$ psi	
	Titanium Monocarbide (TiC)	$17.2x10^3$	1000°C
	Tungsten Monocarbide (WC)	$50x10^3$ psi	
	Zirconium Monocarbide (ZrC)	$16.0x10^3$	room temp.
		$11.7\text{-}14.45x10^3$	980°C
		$12.95\text{-}15.85x10^3$	1250°C
Nitrides	Boron Nitride (BN)	$0.35x10^3$	1000°C
		$0.35x10^3$	1500°C
		$1.15x10^3$	1800°C
		$2.25x10^3$	2000°C
		$6.80x10^3$	2400°C
	Trisilicon tetranitride (Si_3N_4)		
	(hot pressed)	$54.4\ x10^3$	20°C
	(hot pressed)	$21.8\ x10^3$	1400°C
	(reaction bonded)	$24.7\ x10^3$	20°C
	(reaction bonded)	$20.3\ x10^3$	1400°C

To convert **psi** to **MPa**, multiply by 145.

Source: data compiled by J.S. Park from *No. 1 Materials Index*, Peter T.B. Shaffer, Plenum Press, New York, (1964); *Smithells Metals Reference Book*, Eric A. Brandes, ed., in association with Fulmer Research Institute Ltd. 6th ed. London, Butterworths, Boston, (1983); and *Ceramic Source*, American Ceramic Society (1986-1991).

Table 145. TENSILE STRENGTH OF CERAMICS

(SHEET 2 OF 4)

Type	Ceramic	Tensile Strength (psi)	Temperature
Oxides	Aluminum Oxide (Al_2O_3)	37-37.8 x10^3	room temp.
		33.6 x10^3	300°C
		40 x10^3	500°C
		34.6 x10^3	800°C
		35 x10^3	1000°C
		33.9 x10^3	1050°C
		31.4 x10^3	1140°C
		18.5-20 x10^3	1200°C
		6.4 x10^3	1300°C
		4.3 x10^3	1400°C
		1.5 x10^3	1460°C
	Beryllium Oxide (BeO)	13.5-20 x10^3	room temp.
		11.1 x10^3	500°C
		7.0 x10^3	900°C
		5.0 x10^3	1000°C
		2.0 x10^3	1140°C
		0.6 x10^3	1300°C
	Magnesium Oxide (MgO)	14 x10^3	room temp.
		14 x10^3	200°C
		15.2 x10^3	400°C
		16 x10^3	800°C
		11.5 x10^3	1000°C
		10 x10^3	1100°C
		8 x10^3	1200°C
		6 x10^3	1300°C

To convert psi to MPa, multiply by 145.

Source: data compiled by J.S. Park from *No. 1 Materials Index*, Peter T.B. Shaffer, Plenum Press, New York, (1964); *Smithells Metals Reference Book*, Eric A. Brandes, ed., in association with Fulmer Research Institute Ltd. 6th ed. London, Butterworths, Boston, (1983); and *Ceramic Source*, American Ceramic Society (1986-1991).

Table 145. TENSILE STRENGTH OF CERAMICS

(SHEET 3 OF 4)

Type	Ceramic	Tensile Strength (psi)	Temperature
Oxides (Con't)	Thorium Dioxide (ThO_2)	14x10^3	room temp.
	Zircoium Oxide (ZrO_2)	17.9-20x10^3	room temp.
		16.8x10^3	200°C
		17.5x10^3	400°C
		20.0x10^3	500°C
		17.6x10^3	600°C
		16.0x10^3	800°C
		6.75-17.0x10^3	1000°C
		13.0-13.5x10^3	1100°C
		12.1x10^3	1200°C
		10.2x10^3	1300°C
	(MgO stabilized)	21x10^6 psi	room temp.
	Cordierite ($2MgO\ 2Al_2O_3\ 5SiO_2$)		
	(ρ=2.51g/cm^3)	7.8x10^3	25°C
	(ρ=2.1g/cm^3)	3.5x10^3	800°C
	(ρ=1.8g/cm^3)	2.5x10^3	1200°C
	Mullite ($3Al_2O_3\ 2SiO_2$)	16x10^3	25°C
	Spinel (Al_2O_3 MgO)	19.2x10^3	room temp.
		13.7x10^3	550°C
		110.8x10^3	900°C
		6.1x10^3	1150°C
		1.1x10^3	1300°C

To convert **psi** to **MPa**, multiply by **145**.

Source: data compiled by J.S. Park from *No. 1 Materials Index*, Peter T.B. Shaffer, Plenum Press, New York, (1964); *Smithells Metals Reference Book*, Eric A. Brandes, ed., in association with Fulmer Research Institute Ltd. 6th ed. London, Butterworths, Boston, (1983); and *Ceramic Source*, American Ceramic Society (1986-1991).

Table 145. TENSILE STRENGTH OF CERAMICS

(SHEET 4 OF 4)

Type	Ceramic	Tensile Strength (psi)	Temperature
Oxides (Con't)	Zircon (SiO_2 ZrO_2)	$12.7x10^3$	room temp.
		$8.7x10^3$	1050°C
		$3.6x10^3$	1200°C
Silicide	Molybdenum Disilicide ($MoSi_2$)	$40x10^3$	980°C
		$42.16x10^3$	1090°C
		$42.8x10^3$	1200°C
		$41.07x10^3$	1300°C

To convert psi to MPa, multiply by 145.

Source: data compiled by J.S. Park from *No. 1 Materials Index*, Peter T.B. Shaffer, Plenum Press, New York, (1964); *Smithells Metals Reference Book*, Eric A. Brandes, ed., in association with Fulmer Research Institute Ltd. 6th ed. London, Butterworths, Boston, (1983); and *Ceramic Source*, American Ceramic Society (1986-1991).

Table 146. TENSILE STRENGTH OF GLASS

(SHEET 1 OF 3)

Type	Glass	Tensile Strength ($Kg \cdot mm^{-2}$)	Temperature
SiO_2 glass	(48 μm diameter fiber)	49.6	
	(56 μm diameter fiber)	44.3	
	(60 μm diameter fiber)	42.3	
	(65 μm diameter fiber)	39.7	
	(74 μm diameter fiber)	36.5	
	(78 μm diameter fiber)	35.8	
	(108 μm diameter fiber)	28.8	
	(112 μm diameter fiber)	28.3	
	(1.5 mm diameter rod, 0.5 $g/mm^2 \cdot s$ stress rate)	5.84–7.08	
	(1.5 mm diameter rod, 50 $g/mm^2 \cdot s$ stress rate)	9.73±2.13	
	(1.5 mm diameter rod, 54 $g/mm^2 \cdot s$ stress rate)	8.52±2.52	
	(Corning 7940 silica glass)	5.6	100°C
	(Corning 7940 silica glass)	6.2	300°C
	(Corning 7940 silica glass)	6.6	500°C

Source: data compiled by J.S. Park from O. V. Mazurin, M. V. Streltsina and T. P. Shvaiko–Shvaikovskaya, *Handbook of Glass Data, Part A and Part B,* Elsevier, New York, 1983

Table 146. TENSILE STRENGTH OF GLASS

(SHEET 2 OF 3)

Type	Glass	Tensile Strength ($Kg \cdot mm^{-2}$)	Temperature
SiO_2 glass (Con't)	(Corning 7940 silica glass)	7.1	700°C
	(Corning 7940 silica glass)	7.6	900°C
SiO_2–Na_2O glass	(6.0μm diameter fiber, 19.5% mol Na_2O)	173±1.36	
	(8.6μm diameter fiber, 19.5% mol Na_2O)	134±1.34	
	(25.7μm diameter fiber, 19.5% mol Na_2O)	92.5±10.08	
	(5 mm diameter rod, 20% mol Na_2O)	15	
	(3.6μm diameter fiber, 25.5% mol Na_2O)	142±0.189	
	(6.3μm diameter fiber, 25.5% mol Na_2O)	127±0.259	
	(12.8μm diameter fiber, 25.5% mol Na_2O)	103±1.020	
	(5.4μm diameter fiber, 36.3% mol Na_2O)	107.6±0.308	
	(8.6μm diameter fiber, 36.3% mol Na_2O)	98.0±0.344	
	(11.4μm diameter fiber, 36.3% mol Na_2O)	91.2±1.480	

Source: data compiled by J.S. Park from O. V. Mazurin, M. V. Streltsina and T. P. Shvaiko–Shvaikovskaya, *Handbook of Glass Data, Part A and Part B,* Elsevier, New York, 1983

Table 146. TENSILE STRENGTH OF GLASS
(SHEET 3 OF 3)

Type	Glass	Tensile Strength ($Kg \cdot mm^{-2}$)	Temperature
SiO_2–PbO glass	(3.0 μm diameter fiber, 50% mol PbO)	70.8	
	(4.3 μm diameter fiber, 50% mol PbO)	64	
	(5.7 μm diameter fiber, 50% mol PbO)	66–67.2	
	(7.1 μm diameter fiber, 50% mol PbO)	62–71.3	
	(8.0 μm diameter fiber, 50% mol PbO)	64.5	
	(11.4 μm diameter fiber, 50% mol PbO)	51.9–56	
	(17.2 μm diameter fiber, 50% mol PbO)	43–51.6	
B_2O_3 glass	(10–30 μm diameter fiber)	60	
B_2O_3–Na_2O glass	(10–30 μm diameter fiber, 10% mol Na_2O)	102	
	(10–30 μm diameter fiber, 20% mol Na_2O)	137	
	(10–30 μm diameter fiber, 30% mol Na_2O)	152	

Source: data compiled by J.S. Park from O. V. Mazurin, M. V. Streltsina and T. P. Shvaiko–Shvaikovskaya, *Handbook of Glass Data, Part A and Part B*, Elsevier, New York, 1983

Table 147. Tensile Strength of Polymers

(Sheet 1 of 12)

Class	Polymer	Tensile Strength, (ASTM D638) (10^3 psi)
ABS Resins; Molded, Extruded	Medium impact	6.3—8.0
	High impact	5.0—6.0
	Very high impact	4.5—6.0
	Low temperature impact	4—6
	Heat resistant	7.0—8.0
Acrylics; Cast, Molded, Extruded	Cast Resin Sheets, Rods:	
	General purpose, type I	6—9
	General purpose, type II	8—10
	Moldings:	
	Grades 5, 6, 8	8.8—10.5
	High impact grade	5.5—8.0
Thermoset Carbonate	Allyl diglycol carbonate	5—6

To convert **psi** to **MPa**, multiply by 145.

Source: data compiled by J.S. Park from Charles T. Lynch, *CRC Handbook of Materials Science, Vol. 3*, CRC Press, Boca Raton, Florida, 1975 and *Engineered Materials Handbook, Vol.2*, Engineering Plastics, ASM International, Metals Park, Ohio, 1988.

Table 147. TENSILE STRENGTH OF POLYMERS

(SHEET 2 OF 12)

Class	Polymer	Tensile Strength, (ASTM D638) (10^3 psi)
Alkyds; Molded	Putty (encapsulating)	4—5
	Rope (general purpose)	7—8
	Granular (high speed molding)	3—4
	Glass reinforced (heavy duty parts)	5—9
Cellulose Acetate; Molded, Extruded	ASTM Grade:	(Tensile Strength at Fracture)
	H4—1	7—8
	H2—1	5.8—7.2
	MH—1, MH—2	4.8—6.3
	MS—1, MS—2	3.9—5.3
	S2—1	3.0—4.4
Cellulose Acetate Butyrate;	Molded, Extruded	(Tensile Strength at Fracture)
	ASTM Grade:	
	H4	6.9
	MH	5.0—6.0
	S2	3.0—4.0

To convert **psi** to **MPa**, multiply by 145.

Source: data compiled by J.S. Park from Charles T. Lynch, *CRC Handbook of Materials Science, Vol. 3*, CRC Press, Boca Raton, Florida, 1975 and *Engineered Materials Handbook, Vol.2*, Engineering Plastics, ASM International, Metals Park, Ohio, 1988.

Table 147. TENSILE STRENGTH OF POLYMERS

(SHEET 3 OF 12)

Class	Polymer	Tensile Strength, (ASTM D638) (10^3 psi)
Cellusose Acetate Propionate; Molded, Extruded	ASTM Grade:	
	1	5.9—6.5
	3	5.1—5.9
	6	4
Chlorinated Polymers	Chlorinated polyether	6
	Chlorinated polyvinyl chloride	7.3
Polycarbonates	Polycarbonate	9.5
	Polycarbonate (40% glass fiber reinforced)	18
Diallyl Phthalates; Molded	Orlon filled	4.5—6
	Dacron filled	4.6—6.2
	Asbestos filled	4—6.5
	Glass fiber filled	5.5—11

To convert **psi** to **MPa**, multiply by **145.**

Source: data compiled by J.S. Park from Charles T. Lynch, *CRC Handbook of Materials Science, Vol. 3*, CRC Press, Boca Raton, Florida, 1975 and *Engineered Materials Handbook, Vol.2*, Engineering Plastics, ASM International, Metals Park, Ohio, 1988.

Table 147. TENSILE STRENGTH OF POLYMERS

(SHEET 4 OF 12)

Class	Polymer	Tensile Strength, (ASTM D638) (10^3 psi)
Fluorocarbons; Molded,Extruded	Polytrifluoro chloroethylene (PTFCE)	4.6—5.7
	Polytetrafluoroethylene (PTFE)	2.5—6.5
	Ceramic reinforced (PTFE)	0.75—2.5
	Fluorinated ethylene propylene(FEP)	2.5—4.0
	Polyvinylidene— fluoride (PVDF)	5.2—8.6
Epoxies; Cast, Molded, Reinforced	Standard epoxies (diglycidyl ethers of bisphenol A)	
	Cast rigid	9.5-11.5
	Cast flexible	1.4—7.6
	Molded	8—11
	General purpose glass cloth laminate	50-58
	High strength laminate	160
	Filament wound composite	230-240 (hoop)

To convert psi to MPa, multiply by 145.

Source: data compiled by J.S. Park from Charles T. Lynch, *CRC Handbook of Materials Science, Vol. 3*, CRC Press, Boca Raton, Florida, 1975 and *Engineered Materials Handbook, Vol.2*, Engineering Plastics, ASM International, Metals Park, Ohio, 1988.

Table 147. TENSILE STRENGTH OF POLYMERS

(SHEET 5 OF 12)

Class	Polymer	Tensile Strength, (ASTM D638) (10^3 psi)
Epoxies—Molded, Extruded	High performance resins (cycloaliphatic diepoxides)	
	Cast, rigid	8—12
	Molded	5.2—5.3
	Glass cloth laminate	50—52
	Epoxy novolacs	
	Cast, rigid	9.6—12.0
	Glass cloth laminate	59.2
Melamines; Molded	Filler & type	
	Cellulose electrical	5—9
	Glass fiber	6—9
	Alpha cellulose and mineral	5—8

To convert psi to MPa, multiply by 145.

Source: data compiled by J.S. Park from Charles T. Lynch, *CRC Handbook of Materials Science, Vol. 3*, CRC Press, Boca Raton, Florida, 1975 and *Engineered Materials Handbook, Vol.2*, Engineering Plastics, ASM International, Metals Park, Ohio, 1988.

Table 147. TENSILE STRENGTH OF POLYMERS

(SHEET 6 OF 12)

Class	Polymer	Tensile Strength, (ASTM D638) (10^3 psi)
Nylons; Molded, Extruded	Type 6	
	General purpose	9.5—12.5
	Glass fiber (30%) reinforced	21—24
	Cast	12.8
	Flexible copolymers	7.5—10.0
	Type 12	7.1—8.5
	6/6 Nylon	
	General purpose molding	11.2—11.8
	Glass fiber reinforced	25—30
	Glass fiber Molybdenum disulfide filled	19—22
	General purpose extrusion	1.26–8.6
	6/10 Nylon	
	General purpose	7.1—8.5
	Glass fiber (30%) reinforced	19

To convert psi to MPa, multiply by 145.

Source: data compiled by J.S. Park from Charles T. Lynch, *CRC Handbook of Materials Science, Vol. 3*, CRC Press, Boca Raton, Florida, 1975 and *Engineered Materials Handbook, Vol.2*, Engineering Plastics, ASM International, Metals Park, Ohio, 1988.

Table 147. Tensile Strength of Polymers

(Sheet 7 of 12)

Class	Polymer	Tensile Strength, (ASTM D638) (10^3 psi)
Phenolics; Molded	Phenolics; Molded	(ASTM D651)
	Type and filler	
	General: woodflour and flock	5.0—8.5
	Shock: paper, flock, or pulp	5.0—8.5
	High shock: chopped fabric or cord	5—9
	Very high shock: glass fiber	5—10
	Arc resistant—mineral	6
	Rubber phenolic—woodflour or flock	4.5—9
	Rubber phenolic—chopped fabric	3—5
	Rubber phenolic—asbestos	4
	ABS–Polycarbonate Alloy	8.2

To convert **psi** to **MPa**, multiply by **145.**

Source: data compiled by J.S. Park from Charles T. Lynch, *CRC Handbook of Materials Science, Vol. 3*, CRC Press, Boca Raton, Florida, 1975 and *Engineered Materials Handbook, Vol.2*, Engineering Plastics, ASM International, Metals Park, Ohio, 1988.

Table 147. TENSILE STRENGTH OF POLYMERS

(SHEET 8 OF 12)

Class	Polymer	Tensile Strength, (ASTM D638) (10^3 psi)
Polyacetals	Homopolymer:	
	Standard	10
	20% glass reinforced	8.5
	22% TFE reinforced	6.9
	Copolymer:	
	Standard	8.8
	25% glass reinforced	18.5
	High flow	8.8
Polyesters: Thermosets	Cast polyyester	
	Rigid	5—15
	Flexible	1—8
Reinforced polyester moldings	High strength (glass fibers)	5—10
	Heat and chemical resistant (asbestos)	4—6
	Sheet molding compounds, general purpose	15—17

To convert **psi** to **MPa**, multiply by 145.

Source: data compiled by J.S. Park from Charles T. Lynch, *CRC Handbook of Materials Science, Vol. 3*, CRC Press, Boca Raton, Florida, 1975 and *Engineered Materials Handbook, Vol.2*, Engineering Plastics, ASM International, Metals Park, Ohio, 1988.

Table 147. TENSILE STRENGTH OF POLYMERS
(SHEET 9 OF 12)

Class	Polymer	Tensile Strength, (ASTM D638) (10^3 psi)
Polyarylsulfone	Polyarylsulfone	13
Polypropylene:	General purpose	4.5—6.0
Polyethylenes; Molded, Extruded	Type I—lower density (0.910—0.925)	(ASTM D412)
	Melt index 0.3—3.6	1.4—2.5
	Melt index 6—26	1.4—2.0
	Melt index 200	0.9—1.1
	Type II—medium density (0.926—0.940)	
	Melt index 20	2
	Melt index 1.0—1.9	2.3—2.4

To convert psi to MPa, multiply by 145.

Source: data compiled by J.S. Park from Charles T. Lynch, *CRC Handbook of Materials Science, Vol. 3*, CRC Press, Boca Raton, Florida, 1975 and *Engineered Materials Handbook, Vol.2*, Engineering Plastics, ASM International, Metals Park, Ohio, 1988.

Table 147. TENSILE STRENGTH OF POLYMERS
(SHEET 10 OF 12)

Class	Polymer	Tensile Strength, (ASTM D638) (10^3 psi)
Polyethylenes; Molded, Extruded (Con't)	Type III—higher density (0.941—0.965)	
	Melt index 0.2—0.9	4.4
	Melt Melt index 0.l—12.0	2.9—4.0
	Melt index 1.5—15	4.4
	High molecular weight	5.4
Olefin Copolymers; Molded	EEA (ethylene ethyl acrylate)	0.2
	EVA (ethylene vinyl acetate)	0.36
	Ethylene butene	0.35
Propylene—ethylene	Propylene—ethylene	0.4
	Ionomer	0.4
	Polyallomer	3—4.3

To convert **psi to MPa**, multiply by **145**.

Source: data compiled by J.S. Park from Charles T. Lynch, *CRC Handbook of Materials Science, Vol. 3*, CRC Press, Boca Raton, Florida, 1975 and *Engineered Materials Handbook, Vol.2*, Engineering Plastics, ASM International, Metals Park, Ohio, 1988.

Table 147. Tensile Strength of Polymers
(Sheet 11 of 12)

Class	Polymer	Tensile Strength, (ASTM D638) (10^3 psi)
Polystyrenes	Polystyrenes; Molded	
	General purpose	5.0—10
	Medium impact	4.0—6.0
	High impact	3.3—5.1
	Glass fiber -30% reinforced	14
	Styrene acrylonitrile (SAN)	8.3—12.0
	Glass fiber (30%) reinforced SAN	18
Polyvinyl Chloride And Copolymers;	Polyvinyl Chloride And Copolymers; Molded, Extruded	D412
	Nonrigid—general	1—3.5
	Nonrigid—electrical	2—3.2
	Rigid—normal impact	5.5—8
	Vinylidene chloride	4—40

To convert **psi** to **MPa**, multiply by **145**.

Source: data compiled by J.S. Park from Charles T. Lynch, *CRC Handbook of Materials Science, Vol. 3*, CRC Press, Boca Raton, Florida, 1975 and *Engineered Materials Handbook, Vol.2*, Engineering Plastics, ASM International, Metals Park, Ohio, 1988.

Table 147. Tensile Strength of Polymers

(Sheet 12 of 12)

Class	Polymer	Tensile Strength, (ASTM D638) (10^3 psi)
Silicones	Silicones; Molded, Laminated	(ASTM D651)
	Fibrous (glass) reinforced silicones	6.5
	Granular (silica) reinforced silicones	4—6
	Woven glass fabric/ silicone laminate	30—35
Ureas; Molded	Alpha—cellulose filled (ASTM Type l)	5—10

To convert **psi** to **MPa**, multiply by **145**.

Source: data compiled by J.S. Park from Charles T. Lynch, *CRC Handbook of Materials Science, Vol. 3*, CRC Press, Boca Raton, Florida, 1975 and *Engineered Materials Handbook, Vol.2*, Engineering Plastics, ASM International, Metals Park, Ohio, 1988.

Table 148. TENSILE STRENGTH OF FIBERGLASS REINFORCED PLASTICS

Class	Material	Glass fiber content (wt%)	Tensile strength at yield (ksi)
Glass fiber reinforced thermosets	Sheet molding compound (SMC)	15 to 30	8 to 20
	Bulk molding compound(BMC)	15 to 35	4 to 10
	Preform/mat(compression molded)	25 to 50	25 to 30
	Cold press molding–polyester	20 to 30	12 to 20
	Spray–up–polyester	30 to 50	9 to 18
	Filament wound–epoxy	30 to 80	80 to 250
	Rod stock–polyester	40 to 80	60 to 180
	Molding compound–phenolic	5 to 25	7 to 17
Glass–fiber–reinforced thermoplastics	Acetal	20 to 40	9 to 18
	Nylon	6 to 60	13 to 33
	Polycarbonate	20 to 40	12 to 25
	Polyethylene	10 to 40	6.5 to 11
	Polypropylene	20 to 40	5.5 to 10.5
	Polystyrene	20 to 35	10 to 15
	Polysulfone	20 to 40	13 to 20
	ABS(acrylonitrile butadiene styrene)	20 to 40	11 to 16
	PVC (polyvinyl chloride)	15 to 35	14 to 18
	Polyphenylene oxide(modified)	20 to 40	15 to 22
	SAN (styrene acrylonitrile)	20 to 40	13 to 18
	Thermoplastic polyester	20 to 35	14 to 19

To convert (ksi) to (MPa), multiply by 6.89

Data from *ASM Engineering Materials Reference Book, Second Edition*, Michael Bauccio, Ed., ASM International, Materials Park, OH, p106, (1994).

Table 149. TENSILE STRENGTH OF CARBON- AND GLASS-REINFORCED ENGINEERING THERMOPLASTICS (SHEET 1 OF 2)

Class	Resin Type	Composition	Tensile Strength (MPa)
Amorphous	Acrylonitrile-butadiene-styrene(ABS)	30% glass fiber	100
		30% carbon fiber	130
	Nylon	30% glass fiber	148
		30% carbon fiber	207
	Polycarbonate	30% glass fiber	128
		30% carbon fiber	165
	Polyetherimide	30% glass fiber	197
		30% carbon fiber	234
	Polyphenylene oxide (PPO)	30% glass fiber	145
		30% carbon fiber	159
	Polysulfone	30% glass fiber	124
		30% carbon fiber	159
	Styrene-maleic-anhydride (SMA)	30% glass fiber	103
	Thermoplastic polyurethane	30% glass fiber	57

Data from *ASM Engineering Materials Reference Book, Second Edition*, Michael Bauccio, Ed., ASM International, Materials Park, OH, p111–112, (1994).

Table 149. TENSILE STRENGTH OF CARBON- AND GLASS-REINFORCED ENGINEERING THERMOPLASTICS (SHEET 2 OF 2)

Class	Resin Type	Composition	Tensile Strength (MPa)
Crystalline	Acetal	30% glass fiber	134
		20% carbon fiber	81
	Nylon 66	30% glass fiber	179
		30% carbon fiber	241
	Polybutylene telphthalate (PBT)	30% glass fiber	134
		30% carbon fiber	152
	Polythylene terephthalate (PET)	30% glass fiber	159
	Polyphenylene sulfide (PPS)	30% glass fiber	138
		30% carbon fiber	186

Data from *ASM Engineering Materials Reference Book, Second Edition*, Michael Bauccio, Ed., ASM International, Materials Park, OH, p111–112, (1994).

Table 150. STRENGTH OF GRAPHITE FIBER REINFORCED METALS

Composite	Fiber content (vol%)	Strength (ksi)
Graphite(a)/lead	41	104
Graphite(b)/lead	35	72
Graphite(a)/zinc	35	110.9
Graphite(a)/magnesium	42	65

(a) Thornel 75 fiber (b) Courtaulds HM fiber

To convert **psi** to **MPa**, multiply by **145**.

Data from *ASM Engineering Materials Reference Book, Second Edition*, Michael Bauccio, Ed., ASM International, Materials Park, OH, p148,(1994).

Table 151. TENSILE STRENGTH OF GRAPHITE/MAGNESIUM CASTINGS*

Fiber Type	Fiber content	Fiber orientation	Casting	Fiber Preform Method	Tensile Strength (GPa)	Tensile Strength,90° (GPa)
P75	40%	±16°	Hollow cylinder	Filament wound	0.45	0.061
	plus 9%	90°	Hollow cylinder	Filament wound	0.45	0.061
P100	40%	± 16°	Hollow cylinder	Filament wound	0.56	0.38
P55	40%	0°	Plate	Prepreg	0.48	0.02
	30%	0° plus	Plate	Prepreg	0.28	0.010
	10%	90°	Plate	Prepreg	0.28	0.010
	20%	0° plus	Plate	Prepreg	8.45	0.24
	20%	90°	Plate	Prepreg	8.45	0.24

Data from *ASM Engineering Materials Reference Book, Second Edition*, Michael Bauccio, Ed., ASM International, Materials Park, OH, p148,(1994).

* Pitch-base fibers

Table 152. TENSILE STRENGTH OF GRAPHITE/ALUMINUM COMPOSITES

Composite	Fiber loading (vol %)	Wire diameter (mm)	Tensile Strength (MPa)
VS0054/201 Al	48 to 52	0.64 (2-strand)	1035 to 1070
GY70SE/201 Al	37 to 38	0.71(8-strand)	793 to 827

Data from *ASM Engineering Materials Reference Book, Second Edition*, Michael Bauccio, Ed., ASM International, Materials Park, OH, p148,(1994).

Table 153. TENSILE STRENGTH OF GRAPHITE/ALUMINUM COMPOSITES

Thornel Fiber	Longitudinal Tensile Strength (MPa)	Transverse Tensile Strength (MPa)
P55	517 to 621	28 to 48
P75	621 to 724	28 to 48
P100	552 to 834	28 to ~48

Data from *ASM Engineering Materials Reference Book, Second Edition*, Michael Bauccio, Ed., ASM International, Materials Park, OH, p148,(1994).

Table 154. TENSILE STRENGTH OF SILICON CARBIDE SCS–2–AL

Fiber orientation	No. of plies	Tensile Strength (MPa)
0°	6, 8, 12	1462
90°	6, 12,40	86.2
$[0°/90°/0°/90°]_s$	8	673
$[0_2\ °90°20°]_s$	8	1144
$[90_2/0°/90°]_s$	8	341.3
± 45°	8, 12, 40	309.5
$[0°\pm45°/0°]_{s+2s}$	8, 16	800.0
$[0°\pm45°/90°]_s$	8	572.3

Data from *ASM Engineering Materials Reference Book, Second Edition*, Michael Bauccio, Ed., ASM International, Materials Park, OH, p149,(1994).

Table 155. ULTIMATE TENSILE STRENGTH OF INVESTMENT CAST SILICON CARBIDE SCS–AL

Fiber orientation	Fiber vol (%)	Ultimate Tensile Strength (MPa)	Range of Measurement (%)
$0°_3/90°_6/0°_3$	33	458.5	75
$90°_3/0°_6/90°_3$	33	584.0	95
0°	34	1034.2	85

Data from *ASM Engineering Materials Reference Book, Second Edition*, Michael Bauccio, Ed., ASM International, Materials Park, OH, p149,(1994).

Table 156. ULTIMATE TENSILE STRENGTH OF SILICON CARBIDE–ALUMINUM ALLOY COMPOSITES *

Material	Fiber (vol %)	Ultimate Tensile Strength (MPa)	
		Base	Reinforced
Pure Aluminum	11	59	235
6061–T6	16	300	441
2024–T4	20	470	565

Data from *ASM Engineering Materials Reference Book, Second Edition*, Michael Bauccio, Ed., ASM International, Materials Park, OH, p149,(1994).

* Room Temperature

Table 157. TENSILE STRENGTH OF SIC-WHISKER–REINFORCED ALUMINUM ALLOY

Fiber Content (vol %)	Tensile Strength		
	(MPa)	Standard Deviation	Range of Measurement
0	297	1.8	3.5
12	359	33.6	85.6
16	374	8.0	23.0
20	383.6	15.2	38.8

Data from *ASM Engineering Materials Reference Book, Second Edition*, Michael Bauccio, Ed., ASM International, Materials Park, OH, p150,(1994).

Table 158. ULTIMATE TENSILE STRENGTH OF ALUMINUM ALLOY REINFORCED WITH SiC WHISKERS VS. TEMPERATURE

Fiber (Vol%)	Ultimate Tensile Strength (MPa) 350 °C	300 °C	250 °C
Polycrystalline alumina			
0	55	70	115
0.05	63	88	134
0.12	74	—	—
0.20	112	155	198
SiC whiskers			
0	55	70	115
0.12	124	180	226
0.16	147	—	—
0.20	184	235	284

Data from *ASM Engineering Materials Reference Book, Second Edition*, Michael Bauccio, Ed., ASM International, Materials Park, OH, p150,(1994).

Table 159. ULTIMATE TENSILE STRENGTH OF REINFORCED ALUMINUM ALLOY VS. TEMPERATURE

Fiber	Vol %	Ultimate Tensile Strength (MPa)		
		350°C	300°C	250°C
Polycrystalline alumina	0	55	70	115
	5	63	88	134
	12	74	—	—
	20	112	155	198
SiC whiskers	0	55	70	115
	12	124	180	226
	16	147	—	—
	20	184	235	284

Data from *ASM Engineering Materials Reference Book, Second Edition*, Michael Bauccio, Ed., ASM International, Materials Park, OH, p154,(1994).

Table 160. TENSILE STRENGTH OF POLYCRYSTALLINE–ALUMINA–REINFORCED ALUMINUM ALLOY

Fiber Content (vol %)	Tensile Strength		
	(MPa)	Standard Deviation	Range of Measurement
0	297	1.8	3.5
5	282	6.5	15.1
12	273	19.6	49.6
20	312	16.0	42.3

Data from *ASM Engineering Materials Reference Book, Second Edition*, Michael Bauccio, Ed., ASM International, Materials Park, OH, p154,(1994).

Table 161. TENSILE STRENGTH OF BORON/ALUMINUM COMPOSITES*

Matrix	Fiber Orientation	Tensile Strength (MPa)
Al-6061	0°	1515
	90°	138
Al-2024	0°	1550
	90°	214

Data from *ASM Engineering Materials Reference Book, Second Edition*, Michael Bauccio, Ed., ASM International, Materials Park, OH, p157,(1994).

* These samples contain 48% Avco (142 μm) boron. Longitudinal tensile specimens are 152 mm by 7.9 mm by 6 ply. Transverse tensile bars are 152 mm by 12.7 mm by 6 ply.

Table 162. COMPRESSIVE STRENGTH OF GRAY CAST IRON BARS

ASTM Class	Compressive Strength (MPa)
20	572
25	669
30	752
35	855
40	965
50	1130
60	1293

Source: data from *ASM Metals Reference Book, Second Edition*, American Society for Metals, Metals Park, Ohio 44073, p166-167, (1984).

Table 163. COMPRESSIVE STRENGTH OF CERAMICS

(SHEET 1 OF 3)

Class	Ceramic	Compressive Strength (psi)	Temperature
Boride	Titanium Diboride (TiB_2)	47-97x10^3	
Carbides	Boron Carbide (B_4C)	41.4x10^4	room temp.
	Silicon Carbide (SiC)	82-200x10^3	25°C
Nitrides	Titanium Monocarbide (TiC)	10.9-19x10^4	room temp.
	Trichromium Dicarbide (Cr_3C_2)	60x10^4	
	Zirconium Monocarbide (ZrC)	238x10^3	room temp.
	Boron Nitride (BN)		
	parallel to c axis	34.0x10^3	
	parallel to a axis	45x10^3	
	Titanium Mononitirde (TiN)	141x10^3	
	Trisilicon tetranitride (Si_3N_4)	10-100x10^3	25°C
		10-30x10^3	1000°C
Oxides	Aluminum Oxide (Al_2O_3)	427 x10^3	room temp.
		214 x10^3	400°C
		199 x10^3	600°C
		183 x10^3	800°C
		128 x10^3	1000°C
		85 x10^3	1100°C
		71 x10^3	1200°C

To convert **psi** to **MPa**, multiply by **145**.

Source: data compiled by J.S. Park from *No. 1 Materials Index*, Peter T.B. Shaffer, Plenum Press, New York, (1964); *Smithells Metals Reference Book*, Eric A. Brandes, ed., in association with Fulmer Research Institute Ltd. 6th ed. London, Butterworths, Boston, (1983); and *Ceramic Source*, American Ceramic Society (1986-1991).

Table 163. COMPRESSIVE STRENGTH OF CERAMICS
(SHEET 2 OF 3)

Class	Ceramic	Compressive Strength (psi)	Temperature
Oxides (Con't)	Aluminum Oxide (Al_2O_3) (Con't)	35.6 x10^3	1400°C
		14 x10^3	1500°C
		7 x10^3	1600°C
Oxides (Con't)	Beryllium Oxide (BeO)	114-310 x10^3	room temp.
		71 x10^3	500°C
		64 x10^3	800°C
		35.5-40 x10^3	1000°C
		28.5 x10^3	1145°C
		24 x10^3	1400°C
		17 x10^3	1500°C
		7 x10^3	1600°C
	Magnesium Oxide (MgO)	112 x10^3	room temp.
	Thorium Dioxide (ThO_2)	146-214x10^3	room temp.
		156x10^3	400°C
		85x10^3	600°C
		71x10^3	800°C
		51x10^3	1000°C
		28.5x10^3	1200°C
		5.7x10^3	1400°C
		1.5x10^3	1500°C

To convert **psi** to **MPa,** multiply by **145.**

Source: data compiled by J.S. Park from *No. 1 Materials Index,* Peter T.B. Shaffer, Plenum Press, New York, (1964); *Smithells Metals Reference Book,* Eric A. Brandes, ed., in association with Fulmer Research Institute Ltd. 6th ed. London, Butterworths, Boston, (1983); and *Ceramic Source,* American Ceramic Society (1986-1991).

Table 163. COMPRESSIVE STRENGTH OF CERAMICS
(SHEET 3 OF 3)

Class	Ceramic	Compressive Strength (psi)	Temperature
Oxides (Con't)	Zircoium Oxide (ZrO_2)	$205\text{-}300x10^3$	room temp.
		$228x10^3$	500°C
		$171x10^3$	1000°C
		$114x10^3$	1200°C
		$18.5x10^3$	1400°C
		$2.8x10^3$	1500°C
	(CaO stabilized)	$85\text{-}190x10^6$	room temp.
	Cordierite ($2MgO\ 2Al_2O_3\ 5SiO_2$)		
	(ρ=2.51g/cm^3)	$50x10^3$	25oC
	(ρ=2.3g/cm^3)	$50x10^3$	400oC
	(ρ=2.1g/cm^3)	$30x10^3$	800oC
	(ρ=1.8g/cm^3)	$18.5x10^3$	1200oC
	Mullite ($3Al_2O_3\ 2SiO_2$)	$80\text{-}190x10^3$	25°C
	Spinel (Al_2O_3 MgO)	$270x10^3$	room temp.
		$199x10^3$	500°C
		$171x10^3$	800°C
		$85.5x10^3$	1100°C
		$71x10^3$	1200°C
		$21.4x10^3$	1400°C
		$8.5x10^3$	1600°C

To convert **psi** to **MPa**, multiply by **145**.

Source: data compiled by J.S. Park from *No. 1 Materials Index,* Peter T.B. Shaffer, Plenum Press, New York, (1964); *Smithells Metals Reference Book*, Eric A. Brandes, ed., in association with Fulmer Research Institute Ltd. 6th ed. London, Butterworths, Boston, (1983); and *Ceramic Source*, American Ceramic Society (1986-1991).

Table 164. COMPRESSIVE STRENGTH OF FIBERGLASS REINFORCED PLASTIC

Class	Material	Glass fiber content (wt%)	Compressive strength (ksi)
Glass fiber reinforced thermosets	Sheet molding compound (SMC)	15 to 30	15 to 30
	Bulk molding compound(BMC)	15 to 35	20 to 30
	Preform/mat(compression molded)	25 to 50	15 to 30
	Spray–up–polyester	30 to 50	15 to 25
	Filament wound–epoxy	30 to 80	45 to 70
	Rod stock–polyester	40 to 80	30 to 70
	Molding compound–phenolic	5 to 25	14 to 35
Glass–fiber–reinforced thermoplastics	Acetal	20 to 40	11 to 17
	Nylon	6 to 60	13 to 24
	Polycarbonate	20 to 40	14 to 24
	Polyethylene	10 to 40	4 to 8
	Polypropylene	20 to 40	6 to 8
	Polystyrene	20 to 35	13.5 to 19
	Polysulfone	20 to 40	21 to 26
	ABS(acrylonitrile butadiene styrene)	20 to 40	12 to 22
	PVC (polyvinyl chloride)	15 to 35	13.4 to 16.8
	Polyphenylene oxide(modified)	20 to 40	18 to 20
	SAN (styrene acrylonitrile)	20 to 40	12 to 23
	Thermoplastic polyester	20 to 35	16 to 18

To convert (ksi) to (Mpa), multiply by 6.89

Data from *ASM Engineering Materials Reference Book, Second Edition*, Michael Bauccio, Ed., ASM International, Materials Park, OH, p106, (1994).

Table 165. ULTIMATE COMPRESSIVE STRENGTH OF INVESTMENT CAST SILICON CARBIDE SCS–AL

Fiber orientation	Fiber vol (%)	Ultimate Compressive Strength (MPa)	Compressive Modulus (GPa)
$0°_3/90°_6/0°_3$	33	1378 9	—
$90°_3/0°_6/90°_3$	33	1378.9	—
0°	34	1896.1	186.2

Data from *ASM Engineering Materials Reference Book, Second Edition*, Michael Bauccio, Ed., ASM International, Materials Park, OH, p149,(1994).

Table 166. YIELD STRENGTH OF TOOL STEELS

Type	Condition	0.2% Yield Strength (MPa)
L2	Annealed	510
	Oil quenched from 855 •C and single tempered at:	
	205 •C	1790
	315 •C	1655
	425 •C	1380
	540 •C	1170
	650 •C	760
L6	Annealed	380
	Oil quenched from 845 •C and single tempered at:	
	315 •C	1790
	425 •C	1380
	540 •C	1100
	650 •C	830
S1	Annealed	415
	Oil quenched from 930 •C and single tempered at:	
	205 •C	1895
	315 •C	1860
	425 •C	1690
	540 •C	1525
	650 •C	1240
S5	Annealed	440
	Oil quenched from 870 •C and single tempered at:	
	205 •C	1930
	315 •C	1860
	425 •C	1690
	540 •C	1380
	650 •C	1170
S7	Annealed	380
	Fan cooled from 940 •C and single tempered at:	
	205 •C	1450
	315 •C	1585
	425 •C	1410
	540 •C	1380
	650 •C	1035

Source: Data from *ASM Metals Reference Book, Second Edition*, American Society for Metals, Metals Park, Ohio 44073, p241, (1984).

Table 167. YIELD STRENGTH OF DUCTILE IRONS

Specification Number	Grade or Class	Yield Strength (MPa)
ASTM A395-76; ASME SA395	60-40-18	276
ASTM A476-70(d); SAE AMS5316	80-60-03	414
ASTM A536-72, MIL-1-11466B(MR)	60-40-18	276
	65-45-12	310
	80-55-06	379
	100-70-03	483
	120-90-02	621
SAE J434c	D4018	276
	D4512	310
	D5506	379
	D7003	483
MIL-I-24137(Ships)	Class A	310
	Class B	207
	Class C	172

Source: data from *ASM Metals Reference Book, Second Edition*, American Society for Metals, Metals Park, Ohio 44073, p169, (1984).

Table 168. Yield Strength of Malleable Iron Castings

Specification Number	Grade or Class	Yield Strength (MPa)
Ferritic		
ASTM A47, A338; ANSI G48.1;		
FED QQ–I–666c	32510	224
	35018	241
ASTM A197		207
Pearlitic and Martensitic		
ASTM A220; ANSI C48.2;		
MIL–I–11444B	40010	276
	45008	310
	45006	310
	50005	345
	60004	414
	70003	483
	80002	552
	90001	621
Automotive		
ASTM A602; SAE J158	M3210	224
	M4504(a)	310
	M5003(a)	345
	M5503(b)	379
	M7002(b)	483
	M8501(b)	586

(a) Air quenched and tempered
(b) Liquid quenched and tempered

Source: data from *ASM Metals Reference Book, Second Edition*, American Society for Metals, Metals Park, Ohio 44073, p171, (1984).

Table 169. YIELD STRENGTH OF AUSTENITIC STAINLESS STEELS

(SHEET 1 OF 5)

Type	Form	Condition	ASTM Specification	0.2% Yield Strength (MPa)
Type 301(UNS S30100)	Bar,Wire,Plate, Sheet,Strip	Annealed	A167	205
Type 302 (UNS S30200)	Bar	Hot finished and annealed	A276	205
		Cold finished and annealed(a)	A276	310
		Cold finished and annealed(b)	A276	205
Type 302B (UNS S30215)	Bar	Hot finished and annealed	A276	205
		Cold finished and annealed(a)	A276	310
		Cold finished and annealed(b)	A276	205
Type 302Cu(UNS S30430)	Bar	Annealed	A493	—
Types 303 (UNS S30300) and 303Se (UNS S30323)	Bar	Annealed	A581	240
	Wire	Annealed	A581	—
		Cold worked	A581	—

(a) Up to 13 mm thick (b) Over 13 mm thick.

Data from *ASM Metals Reference Book, Third Edition*, Michael Bauccio, Ed., ASM International, Materials Park, OH, p364-366 (1993).

Table 169. Yield Strength of Austenitic Stainless Steels

(Sheet 2 of 5)

Type	Form	Condition	ASTM Specification	0.2% Yield Strength (MPa)
Type 304(UNS S30400)	Bar	Hot finished and annealed	A276	205
		Cold finished and annealed(a)	A276	310
		Cold finished and annealed(b)	A276	205
Type 304L (UNS S30403)	Bar	Hot finished and annealed	A276	170
		Cold finished and annealed(a)	A276	310
		Cold finished and annealed(b)	A276	170
Types 304N (UNS S30451) and 316N(UNS S31651)	Bar	Annealed	A276	240
Type 304LN	Bar	Annealed	—	205
Type 305 (UNS S30500)	Bar	Hot finished and annealed	A276	205
		Cold finished and annealed(a)	A276	310
		Cold finished and annealed(b)	A276	205

(a) Up to 13 mm thick (b) Over 13 mm thick.

Data from *ASM Metals Reference Book, Third Edition*, Michael Bauccio, Ed., ASM International, Materials Park, OH, p364-366 (1993).

Table 169. YIELD STRENGTH OF AUSTENITIC STAINLESS STEELS

(SHEET 3 OF 5)

Type	Form	Condition	ASTM Specification	0.2% Yield Strength (MPa)
Types 308 (UNS S30800),321(UNS S32100),347(UNS34700) and 348 (UNS S34800)	Bar	Hot finished and annealed	A276	205
		Cold finished and annealed(a)	A276	310
		Cold finished and annealed(b)	A276	205
Type 308L	Bar	Annealed	—	207
Types 309 (UNS S30900), 309S (UNS S30908), 310 (UNS S31000) and 310S (UNS S31008)	Bar	Hot finished and annealed	A276	205
		Cold finished and annealed(a)	A276	310
		Cold finished and annealed(b)	A276	205

(a) Up to 13 mm thick (b) Over 13 mm thick.

Data from *ASM Metals Reference Book, Third Edition*, Michael Bauccio, Ed., ASM International, Materials Park, OH, p364-366 (1993).

Table 169. YIELD STRENGTH OF AUSTENITIC STAINLESS STEELS

(SHEET 4 OF 5)

Type	Form	Condition	ASTM Specification	0.2% Yield Strength (MPa)
Type 314 (UNS S31400)	Bar	Hot finished and annealed	A276	205
		Cold finished and annealed(a)	A276	310
		Cold finished and annealed(b)	A276	205
Type 316 (UNS S31600)	Bar	Hot finished and annealed	A276	205
		Cold finished and annealed(a)	A276	310
		Cold finished and annealed(b)	A276	205
Type 316F (UNS S31620)	Bar	Annealed	—	240
Type 316L (UNS S31603)	Bar	Hot finished and annealed	A276	170
		Cold finished and annealed(a)	A276	310
		Cold finished and annealed(b)	A276	170
Type 316LN	Bar	Annealed	—	205

(a) Up to 13 mm thick (b) Over 13 mm thick.

Data from *ASM Metals Reference Book, Third Edition*, Michael Bauccio, Ed., ASM International, Materials Park, OH, p364-366 (1993).

Table 169. YIELD STRENGTH OF AUSTENITIC STAINLESS STEELS

(SHEET 5 OF 5)

Type	Form	Condition	ASTM Specification	0.2% Yield Strength (MPa)
Type 317 (UNS S31700)	Bar	Hot finished and annealed	A276	205
		Cold finished and annealed(a)	A276	310
		Cold finished and annealed(b)	A276	205
Type 317L (UNS S31703)	Bar	Annealed	—	240
Type 317LM	Bar,Plate,Sheet, Strip	Annealed	—	205
Type 329 (UNS S32900)	Bar	Annealed	—	550
Type 330 (UNS N08330)	Bar	Annealed	B511	210
Type 330HC	Bar,Wire,Strip	Annealed	—	290
Types 384 (UNS S38400)	Bar	Annealed	A493	—
Types 385 (UNS38500)	Bar	Annealed	A493	—

(a) Up to 13 mm thick (b) Over 13 mm thick.

Data from *ASM Metals Reference Book, Third Edition*, Michael Bauccio, Ed., ASM International, Materials Park, OH, p364-366 (1993).

Table 170. Yield Strength of Ferritic Stainless Steels

(Sheet 1 of 2)

Type	ASTM Specification	Form	Condition	0.2% Yield Strength (MPa)
Type 405 (UNS S40500)	A580	Wire	Annealed	275
	A580		Annealed, Cold Finished	275
Type 409 (UNS S40900)	—	Bar	Annealed	240(a)
Type 429 (UNS S42900)	—	Bar	Annealed	310(a)
Type 430 (UNS S43000)	A276	Bar	Annealed, Hot Finished	275
	A276		Annealed, Cold Finished	275
Type 430Ti(UNS S43036)	—	Bar	Annealed	310(a)
Type 434 (UNS S43400)	—	Wire	Annealed	415(a)
Type 436 (UNS S43600)	—	Sheet, Strip	Annealed	365(a)

(a) Typical Values

Data from *ASM Metals Reference Book, Third Edition*, Michael Bauccio, Ed., ASM International, Materials Park, OH, p368 (1993).

Table 170. YIELD STRENGTH OF FERRITIC STAINLESS STEELS
(SHEET 2 OF 2)

Type	ASTM Specification	Form	Condition	0.2% Yield Strength (MPa)
Type 442 (UNS S44200)	—	Bar	Annealed	310(a)
Type 444 (UNS S44400)	A176	Plate, Sheet, Strip	Annealed	275
Type 446 (UNS S44600)	A276	Bar	Annealed, Hot Finished	275
	A276		Annealed, Cold Finished	275

(a) Typical Values

Data from *ASM Metals Reference Book, Third Edition*, Michael Bauccio, Ed., ASM International, Materials Park, OH, p368 (1993).

Table 171. Yield Strength of Martensitic Stainless Steels
(Sheet 1 of 3)

Type	ASTM Specification	Form	Condition	0.2% Yield Strength (MPa)
Type 403 (UNS S40300)	A276	Bar	Annealed, hot finished	275
	A276		Annealed, cold finished	275
	A276		Intermediate temper, hot finished	550
	A276		Intermediate temper, cold finished	550
	A276		Hard temper, hot finished	620
	A276		Hard temper, cold finished	620
Type 410 (UNS S41000)	A276	Bar	Annealed, hot finished	275
	A276		Annealed, cold finished	275
	A276		Intermediate temper, hot finished	550
	A276		Intermediate temper, cold finished	550
	A276		Hard temper, hot finished	620
	A276		Hard temper, cold finished	620
Type 410S (UNS S41008)	A176	Plate, Sheet, Strip	Annealed	205
Type 410Cb (UNS S41040)	A276	Bar	Annealed, hot finished	275
	A276		Annealed, cold finished	275
	A276		Intermediate temper, hot finished	690
	A276		Intermediate temper, cold finished	690

Data from *ASM Metals Reference Book, Third Edition*, Michael Bauccio, Ed., ASM International, Materials Park, OH, p369-370 (1993).

Table 171. YIELD STRENGTH OF MARTENSITIC STAINLESS STEELS
(SHEET 2 OF 3)

Type	ASTM Specification	Form	Condition	0.2% Yield Strength (MPa)
Type 414 (UNS S41400)	A276	Bar	Intermediate temper, hot finished	620
	A276		Intermediate temper, cold finished	620
Type 414L	—	Bar	Annealed	550
Type 420 (UNS S42000)	—	Bar	Tempered 205 °C	1480
Type 422 (UNS S42200)	A565	Bar	Intermediate and hard tempers for high-temperature service	760
Type 431 (UNS S43100)	—	Bar	Tempered 260 °C	1030
	—		Tempered 595 °C	795
Type 440A (UNS S44002)	—	Bar	Annealed	415
	—		Tempered 315 °C	1650
Type 440B (UNS S44003)	—	Bar	Annealed	425
	—		Tempered 315 °C	1860
Type 440C (UNS S44004)	—	Bar	Annealed	450
	—		Tempered 315 °C	1900

Data from ASM *Metals Reference Book, Third Edition*, Michael Bauccio, Ed., ASM International, Materials Park, OH, p369-370 (1993).

Table 171. YIELD STRENGTH OF MARTENSITIC STAINLESS STEELS
(SHEET 3 OF 3)

Type	ASTM Specification	Form	Condition	0.2% Yield Strength (MPa)
Type 501 (UNS S50100)	—	Bar, Plate	Annealed	205
	—		Tempered 540 °C	965
Type 502 (UNS S50200)	—	Bar, Plate	Annealed	205

Data from *ASM Metals Reference Book, Third Edition*, Michael Bauccio, Ed., ASM International, Materials Park, OH, p369-370 (1993).

Table 172. YIELD STRENGTH OF PRECIPITATION-HARDENING AUSTENITIC STAINLESS STEELS

Type	Form	Condition	0.2% Yield Strength (MPa)
PH 13–8 Mo (UNS S13800)	Bar, Plate, Sheet, Strip	H950	1410
		H1000	1310
15–5 PH (UNS S15500) and 17–4 PH (UNS S17400)	Bar, Plate, Sheet, Stript	H900	1170
		H925	1070
		H1025	1000
		H1075	860
		H1100	795
		H1150	725
		H1150M	515
17–7 PH (UNS S17700)	Bar	RH950	1030
		TH1050	965

Data from *ASM Metals Reference Book, Third Edition*, Michael Bauccio, Ed., ASM International, Materials Park, OH, p371 (1993).

Table 173. YIELD STRENGTH OF HIGH–NITROGEN AUSTENITIC STAINLESS STEELS

Type	ASTM Specification	Form	Condition	0.2% Yield Strength (MPa)
Type 201 (UNS S20100)	A276	Bar	Annealed	275
Type 202 (UNS S20200)	A276	Bar	Annealed	275
Type 205 (UNS S20500)	—	Plate	Annealed*	475
Type 304N (UNS S30451)	A276	Bar	Annealed	240
Type 304HN (UNS S30452)	—	Bar	Annealed	345
Type 316N (UNS S31651)	A276	Bar	Annealed	240

Data from *ASM Metals Reference Book, Third Edition*, Michael Bauccio, Ed., ASM International, Materials Park, OH, p367 (1993).

* Typical Values

Table 174. YIELD STRENGTH OF WROUGHT COPPERS AND COPPER ALLOYS

(SHEET 1 OF 11)

UNS Number and Name	Nominal Composition (%)	Commercial Forms(a)	Yield Strength (MPa)
C10100 Oxygen-free electronic	99.99 Cu	F, R, W, T, P, S	69-365
C10200 Oxygen-free copper	99.95 Cu	F, R, W, T, P, S	69-365
C10300 Oxygen-free extra-low phosporus	99.95 Cu, 0.003 P	F, R, T, P, S	69-345
C10400, C10500, C10700 Oxygen-free, silver-bearing	99.95 Cu(e)	F, R, W, S	69-365
C10800 Oxygen-free, low phosporus	99.95 Cu, 0.009 P	F, R, T, P	69-345
CS11000 Electrolytic tough pitch copper	99.90 Cu, 0.04 O	F, R, W, T, P, S	69-365
C11100 Electrolytic tough pitch, anneal resistant	99.90 Cu, 0.04 O, 0.01 Cd	W	—
C11300, C11400, C11500, C11600 Silver-bearing tough pitch copper	99.90 Cu, 0.04 O, Ag(f)	F, R, W, T, S	69-365
C12000, C12100	99.9 Cu(g)	F, T, P	69-365
C12200 Phosphorus deoxidized copper, high residual phosphorus	99.90 Cu, 0.02 P	F, R, T, P	69-345
C12500, C12700, C12800, C12900, C13000 Fire-refined tough pitch with silver	99.88 Cu(h)	F, R, W, S	69-365
C14200 Phosphorus deoxidized, arsenical	99.68 Cu, 0.3 As, 0.02 P	F, R, T	69-345

(a) F, flat products; R, rod; W, wire; T, tube; P, pipe; S, shapes.

Data from *ASM Metals Reference Book, Third Edition*, Michael Bauccio, Ed., ASM International, Materials Park, OH, p442–454, (1993).

Table 174. Yield Strength of Wrought Coppers and Copper Alloys
(Sheet 2 of 11)

UNS Number and Name	Nominal Composition (%)	Commercial Forms(a)	Yield Strength (MPa)
C19200	98.97 Cu, 1.0 Fe, 0.03 P	F, T	76-510
C14300	99.9 Cu, 0.1 Cd	F	76-386
C14310	99.8 Cu, 0.2 Cd	F	76-386
C14500 Phosphorus deoxidized, tellurium bearing	99.5 Cu, 0.50 Te, 0.008 P	F, R, W, T	69-352
C14700 Sulfur bearing	99.6 Cu, 0.40 S	R, W	69-379
C15000 Zirconium copper	99.8 Cu, 0.15 Zr	R, W	41-496
C15500	99.75 Cu, 0.06 P, 0.11 Mg, Ag(i)	F	124-496
C15710	99.8 Cu, 0.2 Al_2O_3	R, W	268-689
C15720	99.6 Cu, 0.4 Al_2O_3	F, R	365-586
C15735	99.3 Cu, 0.7 Al_2O_3	R	414-565
C15760	98.9 Cu, 1.1 Al_2O_3	F, R	386-552
C16200 Cadmium copper	99.0 Cu, 1.0 Cd	F, R, W	48-476

(a) F, flat products; R, rod; W, wire; T, tube; P, pipe; S, shapes.

Data from *ASM Metals Reference Book, Third Edition*, Michael Bauccio, Ed., ASM International, Materials Park, OH, p442–454, (1993).

Table 174. YIELD STRENGTH OF WROUGHT COPPERS AND COPPER ALLOYS

(SHEET 3 OF 11)

UNS Number and Name	Nominal Composition (%)	Commercial Forms(a)	Yield Strength (MPa)
C16500	98.6 Cu, 0.8 Cd, 0.6 Sn	F, R, W	97-490
C17000 Beryllium copper	99.5 Cu, 1.7 Be, 0.20 Co	F, R	221-1172
C17200 Beryllium copper	99.5 Cu, 1.9 Be , 0.20 Co	F, R, W, T, P, S	172-1344
C17300 Beryllium copper	99.5 Cu, 1.9 Be, 0.40 Pb	R	172-1255
C17500 Copper-cobalt-beryllium alloy	99.5 Cu, 2.5 Co, 0.6 Be	F, R	172-758
C18200, C18400, C18500 Chromium copper	99.5 Cu(j)	F, W, R, S, T	97-531
C18700 leaded copper	99.0 Cu, 1.0 Pb	R	69-345
C18900	98.75 Cu, 0.75 Sn, 0.3 Si, 0.20 Mn	R, W	62-359
C19000 Copper-nickel-phosphorus alloy	98.7 Cu, 1.1 Ni, 0.25 P	F, R, W	138-552
C19100 Copper-nickel-phosphorus-tellurium alloy	98.15 Cu, 1.1 Ni, 0.50 Te, 0.25 P	R, F	69-634
C19400	97.5 Cu, 2.4 Fe, 0.13 Zn, 0.03 P	F	165-503

(a) F, flat products; R, rod; W, wire; T, tube; P, pipe; S, shapes.

Data from *ASM Metals Reference Book, Third Edition*, Michael Bauccio, Ed., ASM International, Materials Park, OH, p442–454, (1993).

Table 174. YIELD STRENGTH OF WROUGHT COPPERS AND COPPER ALLOYS
(SHEET 4 OF 11)

UNS Number and Name	Nominal Composition (%)	Commercial Forms(a)	Yield Strength (MPa)
C19500	97.0 Cu, 1.5 Fe, 0.6 Sn, 0.10 P, 0.80 Co	F	448-655
C21000 Gilding, 95%	95.0 Cu, 5.0 Zn	F, W	69-400
C22000 Commercial bronze, 90%	90.0 Cu, 10.0 Zn	F, R, W, T	69-427
C22600 Jewelry bronze, 87.5%	87.5 Cu, 12.5 Zn	F, W	76-427
C23000 Red brass, 85%	85.0 Cu, 15.0 Zn	F, W, T, P	69-434
C24000 Low brass, 80%	80.0 Cu, 20.0 Zn	F, W	83-448
C26000 Cartridge brass, 70%	70.0 Cu, 30.0 Zn	F, R, W, T	76-448
C26800, C27000 Yellow brass	65.0 Cu, 35.0 Zn	F, R, W	97-427
C28000 Muntz metal	60.0 Cu, 40.0 Zn	F, R, T	145-379
C31400 Leaded commercial bronze	89.0 Cu, 1.75 Pb, 9.25 Zn	F, R	83-379
C31600 Leaded commercial bronze, nickel-bearing	89.0 Cu, 1.9 Pb, 1.0 Ni, 8.1 Zn	F, R	83-407
C33000 Low-leaded brass tube	66.0 Cu, 0.5 Pb, 33.5 Zn	T	103-414

(a) F, flat products; R, rod; W, wire; T, tube; P, pipe; S, shapes.

Data from *ASM Metals Reference Book, Third Edition*, Michael Bauccio, Ed., ASM International, Materials Park, OH, p442–454, (1993).

Table 174. YIELD STRENGTH OF WROUGHT COPPERS AND COPPER ALLOYS

(SHEET 5 OF 11)

UNS Number and Name	Nominal Composition (%)	Commercial Forms(a)	Yield Strength (MPa)
C33200 High-leaded brass tube	66.0 Cu, 1.6 Pb, 32.4 Zn	T	138-414
C33500 Low-leaded brass	65.0 Cu, 0.5 Pb, 34.5 Zn	F	97-414
C34000 Medium-leaded brass	65.0 Cu, 1.0 Pb, 34.0 Zn	F, R, W, S	103-414
C34200 High-leaded brass	64.5 Cu, 2.0 Pb, 33.5 Zn	F, R	117-427
C34900	62.2 Cu, 0.35 Pb, 37.45 Zn	R, W	110-379
C35000 Medium-leaded brass	62.5 Cu, 1.1 Pb, 36.4 Zn	F, R	90-483
C35300 High-leaded brass	62.0 Cu, 1.8 Pb, 36.2 Zn	F, R	117-427
C35600 Extra-high-leaded brass	63.0 Cu, 2.5 Pb, 34.5 Zn	F	117-414
C36000 Free-cutting brass	61.5 Cu, 3.0 Pb, 35.5 Zn	F, R, S	124-310
C36500 to C36800 Leaded Muntz metal	60.0 Cu(k), 0.6 Pb, 39.4 Zn	F	138
C37000 Free-cutting Muntz metal	60.0 Cu, 1.0 Pb, 39.0 Zn	T	138-414
C37700 Forging brass	59.0 Cu, 2.0 Pb, 39.0 Zn	R, S	138

(a) F, flat products; R, rod; W, wire; T, tube; P, pipe; S, shapes.

Data from *ASM Metals Reference Book, Third Edition*, Michael Bauccio, Ed., ASM International, Materials Park, OH, p442–454, (1993).

Table 174. YIELD STRENGTH OF WROUGHT COPPERS AND COPPER ALLOYS

(SHEET 6 OF 11)

UNS Number and Name	Nominal Composition (%)	Commercial Forms(a)	Yield Strength (MPa)
C38500 Architectural bronze	57.0 Cu, 3.0 Pb, 40.0 Zn	R, S	138
C40500	95 Cu, 1 Sn, 4 Zn	F	83-483
C40800	95 Cu, 2 Sn, 3 Zn	F	90-517
C41100	91 Cu, 0.5 Sn, 8.5 Zn	F, W	76-496
C41300	90.0 Cu, 1.0 Sn, 9.0 Zn	F, R, W	83-565
C41500	91 Cu, 1.8 Sn, 7.2 Zn	F	117-517
C42200	87.5 Cu, 1.1 Sn, 11.4 Zn	F	103-517
C42500	88.5 Cu, 2.0 Sn, 9.5 Zn	F	124-524
C43000	87.0 Cu, 2.2 Sn, 10.8 Zn	F	124-503
C43400	85.0 Cu, 0.7 Sn, 14.3 Zn	F	103-517
C43500	81.0 Cu, 0.9 Sn, 18.1 Zn	F, T	110-469
C44300, C44400, C44500 Inhibited admiralty	71.0 Cu, 28.0 Zn, 1.0 Sn	F, W, T	124-152

(a) F, flat products; R, rod; W, wire; T, tube; P, pipe; S, shapes.

Data from *ASM Metals Reference Book, Third Edition*, Michael Bauccio, Ed., ASM International, Materials Park, OH, p442–454, (1993).

Table 174. Yield Strength of Wrought Coppers and Copper Alloys

(Sheet 7 of 11)

UNS Number and Name	Nominal Composition (%)	Commercial Forms(a)	Yield Strength (MPa)
C46400 to C46700 Naval brass	60.0 Cu, 39.25 Zn, 0.75 Sn	F, R, T, S	172-455
C48200 Naval brass, medium-leaded	60.5 Cu, 0.7 Pb, 0.8 Sn, 38.0 Zn	F, R, S	172-365
C48500 Leaded naval brass	60.0 Cu, 1.75 Pb, 37.5 Zn, 0.75 Sn	F, R, S	172-365
C50500 Phosphor bronze, 1.25% E	98.75 Cu, 1.25 Sn, trace P	F, W	97-345
C51000 Phosphor bronze, 5% A	95.0 Cu, 5.0 Sn, trace P	F, R, W, T	131-552
C51100	95.6 Cu, 4.2 Sn, 0.2 P	F	345-552
C52100 Phosphor bronze, 8% C	92.0 Cu, 8.0 Sn, trace P	F, R, W	165-552
C52400 Phosphor bronze, 10% D	90.0 Cu, 10.0 Sn, trace P	F, R, W	193 (Annealed)
C54400 Free-cutting phosphor bronze	88.0 Cu, 4.0 Pb, 4.0 Zn, 4.0 Sn	F, R	131-434
C60800 Aluminum bronze, 5%	95.0 Cu, 5.0 Al	T	186
C61000	92.0 Cu, 8.0 Al	R, W	207-379
C61300	92.65 Cu, 0.35 Sn, 7.0 Al	F, R, T, P, S	207-400

(a) F, flat products; R, rod; W, wire; T, tube; P, pipe; S, shapes.

Data from *ASM Metals Reference Book, Third Edition*, Michael Bauccio, Ed., ASM International, Materials Park, OH, p442–454, (1993).

Table 174. Yield Strength of Wrought Coppers and Copper Alloys

(Sheet 8 of 11)

UNS Number and Name	Nominal Composition (%)	Commercial Forms(a)	Yield Strength (MPa)
C61400 Aluminum bronze, D	91.0 Cu, 7.0 Al, 2.0 Fe	F, R, W, T, P, S	228-414
C61500	90.0 Cu, 8.0 Al, 2.0 Ni	F	152-965
C61800	89.0 Cu, 1.0 Fe, 10.0 Al	R	269-293
C61900	86.5 Cu, 4.0 Fe, 9.5 Al	F	338-1000
C62300	87.0 Cu, 10.0 Al, 3.0 Fe	F, R	241-359
C62400	86.0 Cu, 3.0 Fe, 11.0 Al	F, R	276-359
C62500	82.7 Cu, 4.3 Fe, 13.0 Al	F, R	379
C63000	82.0 Cu, 3.0 Fe, 10.0 Al, 5.0 Ni	F, R	345-517
C63200	82.0 Cu, 4.0 Fe, 9.0 Al, 5.0 Ni	F, R	310-365
C63600	95.5 Cu, 3.5 Al, 1.0 Si	R, W	—
C63800	99.5 Cu, 2.8 Al, 1.8 Si, 0.40 Co	F	372-786
C64200	91.2 Cu, 7.0 Al	F, R	241-469

(a) F, flat products; R, rod; W, wire; T, tube; P, pipe; S, shapes.

Data from *ASM Metals Reference Book, Third Edition*, Michael Bauccio, Ed., ASM International, Materials Park, OH, p442–454, (1993).

Table 174. Yield Strength of Wrought Coppers and Copper Alloys
(Sheet 9 of 11)

UNS Number and Name	Nominal Composition (%)	Commercial Forms(a)	Yield Strength (MPa)
C65100 Low-silicon bronze, B	98.5 Cu, 1.5 Si	R, W, T	1 03-476
C65500 High-silicon bronze, A	97.0 Cu, 3.0 Si	F, R, W, T	145-483
C66700 Manganese brass	70.0 Cu, 28.8 Zn, 1.2 Mn	F, W	83-638
C67400	58.5 Cu, 36.5 Zn, 1.2 Al, 2.8 Mn, 1.0 Sn	F, R	234-379
C67500 Manganese bronze, A	58.5 Cu, 1.4 Fe, 39.0 Zn, 1.0 Sn, 0.1 Mn	R, S	207-414
C68700 Aluninum brass, arsenical	77.5 Cu, 20.5 Zn, 2.0 Al, 0.1 As	T	186
C68800	73.5 Cu, 22.7 Zn, 3.4 Al, 0.40 Co	F	379-786
C69000	73.3 Cu, 3.4 Al, 0.6 Ni, 22.7 Zn	F	345-807
C69400 Silicon red brass	81.5 Cu, 14.5 Zn, 4.0 Si	R	276-393
C70400	92.4 Cu, 1.5 Fe, 5.5 Ni, 0.6 Mn	F, T	276-524
C70600 Copper nickel, 10%	88.7 Cu, 1.3 Fe, 10.0 Ni	F, T	110-393
C71000 Copper nickel, 20%	79.00 Cu, 21.0 Ni	F, W, T	90-586

(a) F, flat products; R, rod; W, wire; T, tube; P, pipe; S, shapes.

Data from *ASM Metals Reference Book, Third Edition*, Michael Bauccio, Ed., ASM International, Materials Park, OH, p442–454, (1993).

Table 174. Yield Strength of Wrought Coppers and Copper Alloys

(Sheet 10 of 11)

UNS Number and Name	Nominal Composition (%)	Commercial Forms(a)	Yield Strength (MPa)
C71500 Copper nickel, 30%	70.0 Cu, 30.0 Ni	F, R, T	138-483
C71700	67.8 Cu, 0.7 Fe, 31.0 Ni, 0.5 Be	F, R, W	207-1241
C72500	88.20 Cu, 9.5 Ni, 2.3 Sn	F, R, W, T	152-745
C73500	72.0 Cu, 18.0 Ni , 10.0 Zn	F, R, W, T	103-579
C74500 Nickel silver, 65-10	65.0 Cu, 25.0 Zn, 10.0 Ni	F, W	124-524
C75200 Nickel silver, 65-18	65.0 Cu, 17.0 Zn, 18.0 Ni	F, R, W	172-621
C75400 Nickel silver, 65-15	65.0 Cu, 20.0 Zn, 15.0 Ni	F	124-545
C75700 Nickel silver, 65-12	65.0 Cu, 23.0 Zn, 12.0 Ni	F, W	124-545

(a) F, flat products; R, rod; W, wire; T, tube; P, pipe; S, shapes.

Data from *ASM Metals Reference Book, Third Edition*, Michael Bauccio, Ed., ASM International, Materials Park, OH, p442–454, (1993).

Table 174. YIELD STRENGTH OF WROUGHT COPPERS AND COPPER ALLOYS
(SHEET 11 OF 11)

UNS Number and Name	Nominal Composition (%)	Commercial Forms(a)	Yield Strength (MPa)
C76200	59.0 Cu, 29.0 Zn, 12.0 Ni	F, T	145-758
C77000 Nickel silver, 55-18	55.0 Cu, 27.0 Zn, 18.0 Ni	F, R, W	186-621
C72200	82.0 Cu, 16.0 Ni, 0.5 Cr, 0.8 Fe, 0.5 Mn	F, T	124-455
C78200 Leaded nickel silver, 65-8-2	65.0 Cu, 2.0 Pb, 25.0 Zn, 8.0 Ni	F	159-524

(a) F, flat products; R, rod; W, wire; T, tube; P, pipe; S, shapes.

Data from *ASM Metals Reference Book, Third Edition*, Michael Bauccio, Ed., ASM International, Materials Park, OH, p442–454, (1993).

(d) Based on 100% for C360000.
(e) C10400, 8 oz/ton Ag; C10500, 10 oz/ton; C10700, 25 oz/ton .
(f) C11300, 8 oz/ton Ag; C11400,10 oz/ton; C11500, 16 oz/ton; C11600, 25 oz/ton
(g) C12000, 0.008 P; C12100, 0.008 P and 4 oz/ton Ag;
(h) C12700, 8 oz/ton Ag; C12800,10 oz/ton; C12900,16 oz/ton; C13000, 25 oz/ton.
(i) 8.30 oz/ton Ag.
(j) C18200, 0.9 Cr; C18400, 0.8 Cr; C18500, 0.7 Cr
(k) Rod, 61.0 Cu min.

Table 175. YIELD STRENGTH OF CAST ALUMINUM ALLOYS
(SHEET 1 OF 3)

Alloy AA No.	Temper	Yield Strength (MPa)
201.0	T4	215
	T6	435
	T7	415
206.0, A206.0	T7	345
208.0	F	97
242.0	T21	125
	T571	205
	T77	160
	T571	235
	T61	290
295.0	T4	110
	T6	165
	T62	220
296.0	T4	130
	T6	180
	T7	140
308.0	F	110
319.0	F	125
	T6	165
	F	130
	T6	185
336.0	T551	195
	T65	295
354.0	T61	285
355.0	T51	160
	T6	175
	T61	240
	T7	250

Source: data from *ASM Metals Reference Book, Second Edition,* American Society for Metals, Metals Park, Ohio 44073, (1984).

Table 175. YIELD STRENGTH OF CAST ALUMINUM ALLOYS
(SHEET 2 OF 3)

Alloy AA No.	Temper	Yield Strength (MPa)
355.0 (Con't)	T71	200
	T51	165
	T6	190
	T62	280
	T7	210
	T71	215
356.0	T51	140
	T6	165
	T7	210
	T71	145
	T6	185
	T7	165
357.0, A357.0	T62	290
359.0	T61	255
	T62	290
360.0	F	170
A360.0	F	165
380.0	F	165
383.0	F	150
384.0, A384.0	F	165
390.0	F	240
	T5	260
A390.0	F,T5	180
	T6	280
	T7	250
	F,T5	200
	T6	310
	T7	260

Source: data from *ASM Metals Reference Book, Second Edition*, American Society for Metals, Metals Park, Ohio 44073, (1984).

Table 175. YIELD STRENGTH OF CAST ALUMINUM ALLOYS

(SHEET 3 OF 3)

Alloy AA No.	Temper	Yield Strength (MPa)
413.0	F	140
A413.0	F	130
443.0	F	55
B443.0	F	62
C443.0	F	110
514.0	F	85
518.0	F	190
520.0	T4	180
535.0	F	140
712.0	F	170
713.0	T5	150
	T5	150
771.0	T6	275
850.0	T5	75

Source: data from *ASM Metals Reference Book, Second Edition*, American Society for Metals, Metals Park, Ohio 44073, (1984).

Table 176. YIELD STRENGTH OF WROUGHT ALUMINUM ALLOYS (SHEET 1 OF 7)

Alloy	Temper	Yield Strength (MPa)
1050	0	28
	H14	105
	H16	125
	H18	145
1060	0	28
	H12	76
	H14	90
	H16	105
	H18	125
1100	0	34
	H12	105
	H14	115
	H16	140
	H18	150
1350	0	28
	H12	83
	H14	97
	H16	110
	H19	165
2011	T3	295
	T8	310
2014	0	97
	T4	290
	T6	415
Alclad 2014	0	69
	T3	275
	T4	255
	T6	415

Source: data from *ASM Metals Reference Book, Second Edition*, American Society for Metals, Metals Park, Ohio 44073, p.299–302, (1984)

Table 176. YIELD STRENGTH OF WROUGHT ALUMINUM ALLOYS (SHEET 2 OF 7)

Alloy	Temper	Yield Strength (MPa)
2024	0	76
	T3	345
	T4, T351	325
	T361	395
Alclad 2024	0	76
	T	310
	T4, T351	290
	T361	365
	T81, T851	415
	T861	455
2036	T4	195
2048		415
2124	T851	440
2218	T61	305
	T71	275
	T72	255
2219	0	76
	T42	185
	T31, T351	250
	T37	315
	T62	290
	T81, T851	350
	T87	395
2618	All	370
3003	0	42
Alclad	H12	125
3003	H14	145
	H16	170
	H18	185

Source: data from *ASM Metals Reference Book, Second Edition*, American Society for Metals, Metals Park, Ohio 44073, p.299–302, (1984)

Table 176. YIELD STRENGTH OF WROUGHT ALUMINUM ALLOYS (SHEET 3 OF 7)

Alloy	Temper	Yield Strength (MPa)
3004 Alclad	0	69
	H32	170
3004	H34	200
	H36	230
	H38	250
3105	0	55
	H12	130
	H14	150
	H16	170
	H18	195
	H25	160
4032	T6	315
4043	0	69
	H18	270
5005	0	41
	H12	130
	H14	150
	H16	170
	H18	195
	H32	115
	H34	140
	H36	165
	H38	185
5050	0	55
	H32	145
	H34	165
	H36	180
	H38	200

Source: data from *ASM Metals Reference Book, Second Edition*, American Society for Metals, Metals Park, Ohio 44073, p.299–302, (1984)

Table 176. YIELD STRENGTH OF WROUGHT ALUMINUM ALLOYS (SHEET 4 OF 7)

Alloy	Temper	Yield Strength (MPa)
5052	0	90
	H32	195
	H34	215
	H36	240
	H38	255
5056	0	150
	H18	405
	H38	345
5083	0	145
	H112	195
	H113	230
	H321	230
	H323, H32	250
	H343, H34	285
5086	0	115
	H32, H116, H117	205
	H34	255
	H112	130
5154	0	115
	H32	205
	H34	230
	H36	250
	H38	270
	H112	115
5182	0	140
	H32	235
	H34	285
	H19(n)	395
5252	H25	170
	H28, H38	240

Source: data from *ASM Metals Reference Book, Second Edition,* American Society for Metals, Metals Park, Ohio 44073, p.299–302, (1984)

Table 176. YIELD STRENGTH OF WROUGHT ALUMINUM ALLOYS (SHEET 5 OF 7)

Alloy	Temper	Yield Strength (MPa)
5254	0	115
	H32	205
	H34	230
	H36	250
	H38	270
	H112	115
5454	0	115
	H32	205
	H34	240
	H36	275
	H38	310
	H111	180
	H112	125
	H311	180
5456	0	160
	H111	230
	H112	165
	H321, H116	255
5457	0	48
	H25	160
	H28, H38	185
5652	0	90
	H32	195
	H34	215
	H36	240
	H38	255
5657	H25	140
	H28, H38	165
6005	T1	105
	T5	240

Source: data from *ASM Metals Reference Book, Second Edition*, American Society for Metals, Metals Park, Ohio 44073, p.299–302, (1984)

Table 176. YIELD STRENGTH OF WROUGHT ALUMINUM ALLOYS (SHEET 6 OF 7)

Alloy	Temper	Yield Strength (MPa)
6009	T4	130
	T6	325
6010	T4	170
6061	0	55
	T4, T451	145
	T6, T651	275
Alclad 6061	0	48
	T4, T451	130
	T6, T651	255
6063	0	48
	T1	90
	T4	90
	T5	145
	T6	215
	T83	240
	T831	185
	T832	270
6066	0	83
	T4, T451	205
	T6, T651	360
6070	0	69
	T4	170
	T6	350
6101	Hlll	76
6151	T6	195
6201	T6	300
	T81	310
6205	Tl	140
	T5	290

Source: data from *ASM Metals Reference Book, Second Edition*, American Society for Metals, Metals Park, Ohio 44073, p.299–302, (1984)

Table 176. YIELD STRENGTH OF WROUGHT ALUMINUM ALLOYS (SHEET 7 OF 7)

Alloy	Temper	Yield Strength (MPa)
6262	T9	380
6351	T4	150
	T6	285
6463	Tl	90
	T5	145
	T6	215
7005	0	83
	T53	345
	T6,T63,T6351	315
7049	T73	
7050	T736	455
7072	0	
	H12	
	H14	
7075	0	105
	T6,T651	505
	T73	435
Alclad 7075	0	95
	T6,T651	460
7175	T66	525
	T736	455
7475	T61	460

Source: data from *ASM Metals Reference Book, Second Edition*, American Society for Metals, Metals Park, Ohio 44073, p.299–302, (1984)

Table 177. Yield Strength of Wrought Titanium Alloys at Room Temperature (Sheet 1 of 3)

Class	Alloy	Condition	Yield Strength (MPa)
Commercially Pure	99.5 Ti	Annealed	241
	99.2 Ti	Annealed	345
	99.1 Ti	Annealed	448
	99.0 Ti	Annealed	586
	99.2Ti-0.2Pd	Annealed	345
	Ti-0.8Ni-0.3Mo	Annealed	448
Alpha Alloys	Ti-5Al-2.5Sn	Annealed	807
	Ti-5Al-2.5Sn (low O_2)	Annealed	745
Near Alpha Alloys	Ti-8Al-1Mo-1V	Duplex Annealed	951
	Ti-11Sn-1Mo-2.25Al-5.0Zr-1Mo-0.2Si	Duplex Annealed	993
	Ti-6Al-2Sn-4Zr-2Mo	Duplex Annealed	896
	Ti-5Al-2Sn-2Zr-2Mo-0.25Si	975 °C (1/2h), AC + 595 °C (2h), AC	965
	Ti-6Al-2Nb-1Ta-1Mo	As rolled 2.5 cm (1 in.) plate	758
	Ti-6Al-2Sn-1.5Zr-1Mo- 0.35Bi-0.1Si	Beta forge + duplex anneal	945

Data from *ASM Metals Reference Book, Third Edition*, Michael Bauccio, Ed., ASM International, Materials Park, OH, p512, (1993).

***Table 177.* Yield Strength of Wrought Titanium Alloys at Room Temperature (Sheet 2 of 3)**

Class	Alloy	Condition	Yield Strength (MPa)
Alpha-Beta Alloys	Ti-8Mn	Annealed	862
	Ti-3Al-2.5V	Annealed	586
	Ti-6Al-4V	Annealed	924
		Solution + age	1103
	Ti-6Al-4V(low O_2)	Annealed	827
	Ti-6Al-6V-2Sn	Annealed	1000
		Solution + age	1172
	Ti-7Al-4Mo	Solution + age	1034
	Ti-6Al-2Sn-4Zr-6Mo	Solution + age	1172
	Ti-6Al-2Sn-2Zr-2Mo- 2Cr-0.25Si	Solution + age	1138
	Ti-10V-2Fe-3Al	Solution + age	1200

Data from *ASM Metals Reference Book, Third Edition*, Michael Bauccio, Ed., ASM International, Materials Park, OH, p512, (1993).

Table 177. YIELD STRENGTH OF WROUGHT TITANIUM ALLOYS AT ROOM TEMPERATURE (SHEET 3 OF 3)

Class	Alloy	Condition	Yield Strength (MPa)
Beta Alloys	Ti-13V-1Cr-3Al	Solution + age	1172
			1207
	Ti-8Mo-8V-2Fe-3Al	Solution + age	1241
	Ti-3Al-8V-6Cr-4Mo-4Zr	Solution + age	1379
		Annealed	834
	Ti-11.5Mo-6Zr-4.5Sn	Solution + age	1317

Data from ASM *Metals Reference Book, Third Edition*, Michael Bauccio, Ed., ASM International, Materials Park, OH, p512, (1993).

Table 178. YIELD STRENGTH OF WROUGHT TITANIUM ALLOYS AT HIGH TEMPERATURE (SHEET 1 OF 4)

Class	Alloy	Condition	Test Temperature (°C)	Yield Strength (MPa)
Commercially Pure	99.5 Ti	Annealed	315	97
	99.2 Ti	Annealed	315	117
	99.1 Ti	Annealed	315	138
	99.0 Ti	Annealed	315	172
	99.2Ti-0.2Pd	Annealed	315	110
	Ti-0.8Ni-0.3Mo	Annealed	205	248
	Ti-0.8Ni-0.3Mo	Annealed	315	207
Alpha Alloys	Ti-5Al-2.5Sn	Annealed	315	448
	Ti-5Al-2.5Sn (low O_2)	Annealed	-195	1158
			-255	1420
Near Alpha Alloys	Ti-8Al-1Mo-1V	Duplex Annealed	315	621
			425	565
			540	517

Data from *ASM Metals Reference Book, Third Edition*, Michael Bauccio, Ed., ASM International, Materials Park, OH, p512, (1993).

Table 178. YIELD STRENGTH OF WROUGHT TITANIUM ALLOYS AT HIGH TEMPERATURE (SHEET 2 OF 4)

Class	Alloy	Condition	Test Temperature (°C)	Yield Strength (MPa)
	Ti-11Sn-1Mo-2.25Al-5.0Zr-1Mo-0.2Si	Duplex Annealed	315	758
			425	676
			540	586
	Ti-6Al-2Sn-4Zr-2Mo	Duplex Annealed	315	586
			425	586
			540	489
	Ti-5Al-2Sn-2Zr-2Mo-0.25Si	975 °C (1/2h), AC + 595 °C (2h), AC	315	565
			425	531
			540	503
	Ti-6Al-2Nb-1Ta-1Mo	As rolled 2.5 cm (1 in.) plate	315	462
			425	414
			540	379
	Ti-6Al-2Sn-1.5Zr-1Mo- 0.35Bi-0.1Si	Beta forge + duplex anneal	480	586

Data from *ASM Metals Reference Book, Third Edition*, Michael Bauccio, Ed., ASM International, Materials Park, OH, p512, (1993).

Table 178. YIELD STRENGTH OF WROUGHT TITANIUM ALLOYS AT HIGH TEMPERATURE (SHEET 3 OF 4)

Class	Alloy	Condition	Test Temperature (°C)	Yield Strength (MPa)
Alpha-Beta Alloys	Ti-8Mn	Annealed	315	565
	Ti-3Al-2.5V	Annealed	315	345
	Ti-6Al-4V	Annealed	315	655
		Annealed	425	572
		Annealed	540	427
		Solution + age	315	703
		Solution + age	425	621
		Solution + age	540	483
	Ti-6Al-4V(low O_2)	Annealed	160	1413
	Ti-6Al-6V-2Sn	Annealed	315	807
		Solution + age	315	896
	Ti-7Al-4Mo	Solution + age	315	745
			425	717

Data from *ASM Metals Reference Book, Third Edition*, Michael Bauccio, Ed., ASM International, Materials Park, OH, p512, (1993).

Table 178. Yield Strength of Wrought Titanium Alloys at High Temperature (Sheet 4 of 4)

Class	Alloy	Condition	Test Temperature (°C)	Yield Strength (MPa)
Beta Alloys	Ti-6Al-2Sn-4Zr-6Mo	Solution + age	315	841
			425	758
			540	655
	Ti-6Al-2Sn-2Zr-2Mo- 2Cr-0.25Si	Solution + age	315	807
	Ti-10V-2Fe-3Al	Solution + age	205	1048
			315	979
	Ti-13V-1Cr-3Al	Solution + age	315	793
			425	827
	Ti-8Mo-8V-2Fe-3Al	Solution + age	315	979
	Ti-3Al-8V-6Cr-4Mo-4Zr	Solution + age	315	896
			425	758
		Annealed	315	655
	Ti-11.5Mo-6Zr-4.5Sn	Solution + age	315	848

Data from *ASM Metals Reference Book, Third Edition*, Michael Bauccio, Ed., ASM International, Materials Park, OH, p512, (1993).

Table 179. YIELD STRENGTH OF COBALT-BASE SUPERALLOYS

Alloy	Temperature (°C)	Yield Strength (MPa)
Haynes 25 (L–605) sheet	21	460
	540	250
	650	240
	760	260
	870	240
Haynes 188, sheet	21	485
	540	305
	650	305
	760	290
	870	260
S-816, bar	21	385
	540	310
	650	305
	760	285
	870	240

Data from *ASM Metals Reference Book, Third Edition*, Michael Bauccio, Ed., ASM International, Materials Park, OH, p387, (1993).

Table 180. YIELD STRENGTH OF NICKEL-BASE SUPERALLOYS (SHEET 1 OF 5)

Alloy	Temperature (°C)	Yield Strength (MPa)
Astroloy, bar	21	1050
	540	965
	650	965
	760	910
	870	690
D–979, bar	21	1010
	540	925
	650	890
	760	655
	870	305
Hastelloy X, sheet	21	360
	540	290
	650	275
	760	260
	870	180
IN–102, bar	21	505
	540	400
	650	400
	760	385
	870	200
Inconel 600, bar	21	250
	540	195
	650	180
	760	115
	870	62
Inconel 601, sheet	21	340
	540	150
	650	180
	760	200
	870	140

Data from *ASM Metals Reference Book, Third Edition*, Michael Bauccio, Ed., ASM International, Materials Park, OH, p387-389, (1993).

Table 180. YIELD STRENGTH OF NICKEL-BASE SUPERALLOYS (SHEET 2 OF 5)

Alloy	Temperature (°C)	Yield Strength (MPa)
Inconel 625, bar	21	490
	540	405
	650	420
	760	420
	870	475
Inconel 706, bar	21	980
	540	895
	650	825
	760	675
Inconel 718, bar	21	1190
	540	1060
	650	1020
	760	740
	870	330
Inconel 718, sheet	21	1050
	540	945
	650	870
	760	625
Inconel X-750, bar	21	635
	540	580
	650	565
	760	455
	870	165
M-252, bar	21	840
	540	765
	650	745
	760	715
	870	485

Data from *ASM Metals Reference Book, Third Edition*, Michael Bauccio, Ed., ASM International, Materials Park, OH, p387-389, (1993).

Table 180. YIELD STRENGTH OF NICKEL-BASE SUPERALLOYS (SHEET 3 OF 5)

Alloy	Temperature (°C)	Yield Strength (MPa)
Nimonic 80A, bar	21	620
	540	530
	650	550
	760	505
	870	260
Nimonic 90, bar	21	805
	540	725
	650	685
	760	540
	870	260
Nimonic 105, bar	21	815
	540	775
	650	800
	760	655
	870	365
Nimonic 115, bar	21	860
	540	795
	650	815
	760	800
	870	550
Pyromet 860, bar	21	835
	540	840
	650	850
	760	835
René 41, bar	21	1060
	540	1010
	650	1000
	760	940
	870	550

Data from *ASM Metals Reference Book, Third Edition*, Michael Bauccio, Ed., ASM International, Materials Park, OH, p387-389, (1993).

Table 180. YIELD STRENGTH OF NICKEL-BASE SUPERALLOYS (SHEET 4 OF 5)

Alloy	Temperature (°C)	Yield Strength (MPa)
René 95, bar	21	1310
	540	1250
	650	1220
	760	1100
Udimet 500, bar	21	840
	540	795
	650	760
	760	730
	870	495
Udimet 520, bar	21	860
	540	825
	650	795
	760	725
	870	515
Udimet 700, bar	21	965
	540	895
	650	855
	760	825
	870	635
Udimet 710, bar	21	910
	540	850
	650	860
	760	815
	870	635
Unitemp AF2–1DA, bar	21	1050
	540	1080
	650	1080
	760	1010
	870	715

Data from *ASM Metals Reference Book, Third Edition*, Michael Bauccio, Ed., ASM International, Materials Park, OH, p387-389, (1993).

Table 180. YIELD STRENGTH OF NICKEL-BASE SUPERALLOYS (SHEET 5 OF 5)

Alloy	Temperature (°C)	Yield Strength (MPa)
Waspaloy, bar	21	795
	540	725
	650	690
	760	675
	870	515

Data from *ASM Metals Reference Book, Third Edition*, Michael Bauccio, Ed., ASM International, Materials Park, OH, p387-389, (1993).

Table 181. YIELD STRENGTH OF COMMERCIALLY PURE TIN

Temperature (°C)	Yield Strength (MPa)
Strained at 0.2 mm/m • min	
-200	36.2
-160	90.3
-120	87.6
-80	38.9
-40	20.1
0	12.5
23	11.0
Strained at 0.4 mm/m • min	
15	14.5
50	12.4
100	11.0
150	7.6
200	4.5

Data from *ASM Metals Reference Book, Third Edition*, Michael Bauccio, Ed., ASM International, Materials Park, OH, p488, (1993).

Table 182. YIELD STRENGTH OF POLYMERS

(SHEET 1 OF 3)

Class	Polymer	Yield Strength, (ASTM D638) (10^3 psi)
Chlorinated Polyether	Chlorinated Polyether	5.9
Polycarbonate	Polycarbonate	8.5
Nylons; Molded, Extruded	Type 6	
	General purpose	8.5—12.5
	Cast	12.8
	Flexible copolymers	7.5—10.0
	Type 8	3.9
	Type 11	8.5
	Type 12	5.5—6.5
	6/6 Nylon	
	General purpose molding	8.0—11.8
	Glass fiber reinforced	25
	General purpose extrusion	8.6—12.6
	6/10 Nylon	
	General purpose	7.1—8.5
	ABS–Polycarbonate Alloy	8.2
PVC–Acrylic Alloy	PVC–acrylic sheet	6.5
	PVC–acrylic injection molded	5.5
Polymides	Unreinforced	7.5
	Unreinforced 2nd value	5
	Glass reinforced	28

To convert **psi** to **MPa**, multiply by **145**.

Source: data compiled by J.S. Park from Charles T. Lynch, *CRC Handbook of Materials Science, Vol. 3*, CRC Press, Boca Raton, Florida, 1975 and *Engineered Materials Handbook, Vol.2*, Engineering Plastics, ASM International, Metals Park, Ohio, 1988.

Table 182. YIELD STRENGTH OF POLYMERS

(SHEET 2 OF 3)

Class	Polymer	Yield Strength, (ASTM D638) (10^3 psi)
Polyacetals	Homopolymer:	
	Standard	10
	Copolymer:	
	Standard	8.8
	25% glass reinforced	18.5
	High flow	8.8
Polyester; Thermoplastic	Injection Moldings:	
	General purpose grade	7.5—8
	Glass reinforced grades	17—25
	Glass reinforced self extinguishing	17
	General purpose grade	8.2
	Glass reinforced grade	14
	Asbestos—filled grade	12
Phenylene Oxides	SE—100	7.8
	SE—1	9.6
	Glass fiber reinforced	14.5—17.0
Phenylene oxides (Noryl)	Standard	10.2
	Glass fiber reinforced	17—19
	Polyarylsulfone	8—12
	Polypropylene:	
	General purpose	4.5—6.0
	High impact	2.8—4.3
	Asbestos filled	3.3—8.2
	Glass reinforced	7—11
	Flame retardant	3.6—4.2

To convert **psi** to **MPa**, multiply by **145**.

Source: data compiled by J.S. Park from Charles T. Lynch, *CRC Handbook of Materials Science, Vol. 3*, CRC Press, Boca Raton, Florida, 1975 and *Engineered Materials Handbook, Vol.2*, Engineering Plastics, ASM International, Metals Park, Ohio, 1988.

Table 182. YIELD STRENGTH OF POLYMERS

(SHEET 3 OF 3)

Class	Polymer	Yield Strength, (ASTM D638) (10^3 psi)
Polyphenylene sulfide:	Standard	9.511
	40% glass reinforced	20—21
Polystyrenes; Molded	Polystyrenes	
	General purpose	5.0—10
	Medium impact	3.7—6.0
	High impact	2.8—5.3
	Glass fiber -30% reinforced	14
Styrene acrylonitrile (SAN)	Glass fiber (30%) reinforced SAN	18

To convert **psi** to **MPa**, multiply by **145**.

Source: data compiled by J.S. Park from Charles T. Lynch, *CRC Handbook of Materials Science, Vol. 3*, CRC Press, Boca Raton, Florida, 1975 and *Engineered Materials Handbook, Vol.2*, Engineering Plastics, ASM International, Metals Park, Ohio, 1988.

Table 183. YIELD STRENGTH OF SIC-WHISKER–REINFORCED ALUMINUM ALLOY

Fiber Content (vol %)	Yield Strength (0.2%)		
	(MPa)	Standard Deviation	Range of Measurement
0	210	3.8	9.5
0.12	266.5	4.2	10.6
0.16	264.5	0.6	1.6
0.20	298	4.0	10.2

Data from *ASM Engineering Materials Reference Book, Second Edition*, Michael Bauccio, Ed., ASM International, Materials Park, OH, p150,(1994).

Table 184. YIELD STRENGTH OF REINFORCED ALUMINUM ALLOY VS. TEMPERATURE

Fiber	Vol %	Yield Strength (MPa)		
		350°C	300°C	250°C
Polycrystalline alumina	0	35	—	70
	5	54	79	112
	12	68	—	—
	20	110	154	186
SiC whiskers	0	35	—	70
	12	94	153	197
	16	120	—	—
	20	163	207	268

Data from *ASM Engineering Materials Reference Book, Second Edition*, Michael Bauccio, Ed., ASM International, Materials Park, OH, p154,(1994).

Table 185. YIELD STRENGTH OF POLYCRYSTALLINE–ALUMINA–REINFORCED ALUMINUM ALLOY

Fiber Content (vol %)	Yield Strength (0.2%)		
	(MPa)	Standard Deviation	Range of Measurement
0	210	3.8	9.5
5	232	4.2	10.4
12	251.5	14.6	38.3
20	282.5	11.3	25.2

Data from *ASM Engineering Materials Reference Book, Second Edition*, Michael Bauccio, Ed., ASM International, Materials Park, OH, p154,(1994).

Table 186. COMPRESSIVE YIELD STRENGTH OF POLYMERS

(SHEET 1 OF 2)

Class	Polymer	Compressive Yield Strength (ASTM D690 or D695) (0.1% offset, 1000 psi)
Acrylics; Cast, Molded, Extruded	Cast Resin Sheets, Rods:	
	General purpose, type I	12—14
	General purpose, type II	14—18
	Moldings:	
	Grades 5, 6, 8	14.5—17
	High impact grade	7.3—12.0
Cellulose Acetate; Molded, Extruded	ASTM Grade:	
	H4—1	6.5—10.6
	H2—1	4.3—9.6
	MH—1, MH—2	4.4—8.4
	MS—1, MS—2	3.2—7.2
	S2—1	3.15—6.1
Cellulose Acetate Butyrate; Molded, Extruded	ASTM Grade:	
	H4	8.8
	MH	5.3—7.1
	S2	2.6—4.3
Cellusose Acetate Propionate; Molded, Extruded	ASTM Grade:	
	1	6.2—7.3
	3	4.9—5.8
Fluorocarbons; Molded,Extruded	Polytrifluoro chloroethylene (PTFCE)	2
	Polytetrafluoroethylene (PTFE)	0.7—1.8
	Ceramic reinforced (PTFE)	1.4—1.8
	Fluorinated ethylene propylene(FEP)	1.6
	Polyvinylidene— fluoride (PVDF)	12.8—14.2

To convert from **psi** to **MPa**, multiply by **145**.

Source: data compiled by J.S. Park from Charles T. Lynch, *CRC Handbook of Materials Science, Vol. 3*, CRC Press, Boca Raton, Florida, 1975 and *Engineered Materials Handbook, Vol.2, Engineering Plastics*, ASM International, Metals Park, Ohio, 1988.

Table 186. COMPRESSIVE YIELD STRENGTH OF POLYMERS

(SHEET 2 OF 2)

Class	Polymer	Compressive Yield Strength (ASTM D690 or D695) (0.1% offset, 1000 psi)
Nylons; Molded, Extruded	Type 6	
	General purpose	9.7
	Glass fiber (30%) reinforced	19—20
	Cast	14
	Nylons; Molded, Extruded	
	6/6 Nylon	
	General purpose molding	4.9
	Glass fiber reinforced	20—24
	General purpose extrusion	4.9
	6/10 Nylon	
	General purpose	3.0
	Glass fiber (30%) reinforced	18
Polyacetals	Homopolymer:	
	Standard	5.2
	20% glass reinforced	5.2
	22% TFE reinforced	4.5
	Copolymer:	
	Standard	4.5
	High flow	4.5
Polypropylene:	General purpose	5.5—6.5
	High impact	4.4
	Asbestos filled	7
	Glass reinforced	6.5—7
Polyvinyl Chloride And Copolymers; Molded, Extruded	Rigid—normal impact	10—11
Vinylidene chloride	Vinylidene chloride	75—85

To convert from **psi** to **MPa**, multiply by **145**.

Source: data compiled by J.S. Park from Charles T. Lynch, *CRC Handbook of Materials Science, Vol. 3*, CRC Press, Boca Raton, Florida, 1975 and *Engineered Materials Handbook, Vol.2, Engineering Plastics*, ASM International, Metals Park, Ohio, 1988.

Table 187. FLEXURAL STRENGTH OF POLYMERS
(SHEET 1 OF 6)

Class	Polymer	Flexural Strength (ASTM D790) (10^3 psi)
ABS Resins; Molded, Extruded	Medium impact	9.9—11.8
	High impact	7.5—9.5
	Very high impact	6.0—9.8
	Low temperature impact	5—8
	Heat resistant	11.0—12.0
Acrylics; Cast, Molded, Extruded	Cast Resin Sheets, Rods:	
	General purpose, type I	12—14
	General purpose, type II	15—17
	Moldings:	
	Grades 5, 6, 8	15—16
	High impact grade	8.7—12.0
Alkyds; Molded	Putty (encapsulating)	8—11
	Rope (general purpose)	19—20
	Granular (high speed molding)	7—10
	Glass reinforced (heavy duty parts)	12—17
Cellulose Acetate; Molded, Extruded	ASTM Grade:	
	H4—1	8.1—11.15 (yield)
	H2—1	6.0—10.0 (yield)
	MH—1, MH—2	4.4—8.65 (yield)
	MS—1, MS—2	3.8—7.1 (yield)
	S2—1	3.5—5.7 (yield)
Cellulose Acetate Butyrate; Molded, Extruded	ASTM Grade:	
	H4	9 (yield)
	MH	5.6—6.7 (yield)
	S2	2.5—3.95 (yield)

To convert from **psi** to **MPa**, multiply by **145**.

Source: data compiled by J.S. Park from Charles T. Lynch, *CRC Handbook of Materials Science, Vol. 3*, CRC Press, Boca Raton, Florida, 1975 and *Engineered Materials Handbook, Vol.2, Engineering Plastics*, ASM International, Metals Park, Ohio, 1988.

Table 187. FLEXURAL STRENGTH OF POLYMERS

(SHEET 2 OF 6)

Class	Polymer	Flexural Strength (ASTM D790) (10^3 psi)
Cellusose Acetate Propionate; Molded, Extruded	ASTM Grade:	
	1	6.8—7.9 (yield)
	3	5.6—6.2 (yield)
Chlorinated Polymers	Chlorinated polyether	5 (0.1% offset)
	Chlorinated polyvinyl chloride	14.5
Polycarbonates	Polycarbonate	13.5
	Polycarbonate (40% glass fiber reinforced)	27
Diallyl Phthalates; Molded	Orlon filled	7.5—10.5
	Dacron filled	9—11.5
	Asbestos filled	8—10
	Glass fiber filled	10—18
Fluorocarbons; Molded,Extruded	Polytrifluoro chloroethylene (PTFCE)	3.5 (0.1% offset)
	Fluorinated ethylene propylene(FEP)	3 (0.1% offset)
	Polyvinylidene— fluoride (PVDF)	8.6—10.8 (0.1% offset)
Epoxies; Cast, Molded, Reinforced	Standard epoxies (diglycidyl ethers of bisphenol A)	
	Cast rigid	14—18
	Cast flexible	1.2—12.7
	Molded	19—22
	General purpose glass cloth laminate	80—90
	High strength laminate	165—177
	Filament wound composite	180—170

To convert from **psi** to **MPa**, multiply by **145**.

Source: data compiled by J.S. Park from Charles T. Lynch, *CRC Handbook of Materials Science, Vol. 3*, CRC Press, Boca Raton, Florida, 1975 and *Engineered Materials Handbook, Vol.2, Engineering Plastics*, ASM International, Metals Park, Ohio, 1988.

Table 187. FLEXURAL STRENGTH OF POLYMERS
(SHEET 3 OF 6)

Class	Polymer	Flexural Strength (ASTM D790) (10^3 psi)
Epoxies—Molded, Extruded	High performance resins (cycloaliphatic diepoxides)	
	Cast, rigid	11—16
	Molded	10—12
	Glass cloth laminate	70—72
	Epoxy novolacs	
	Cast, rigid	12—13
	Glass cloth laminate	84—89
Melamines; Molded	Filler & type	
	Unfilled	9.5—14
	Cellulose electrical	6—15
	Glass fiber	14—18
	Alpha cellulose and mineral	11—16, 8—10(mineral)
Nylons; Molded, Extruded	Type 6	
	General purpose	Unbreakable
	Glass fiber (30%) reinforced	26—34
	Cast	16.5
	Flexible copolymers	3.4—16.4
	6/6 Nylon	
	General purpose molding	Unbreakable
	Glass fiber reinforced	26—35
	Glass fiber Molybdenum disulfide filled	26—28
	6/10 Nylon	
	General purpose	8
	Glass fiber (30%) reinforced	23

To convert from **psi** to **MPa**, multiply by **145**.

Source: data compiled by J.S. Park from Charles T. Lynch, *CRC Handbook of Materials Science, Vol. 3*, CRC Press, Boca Raton, Florida, 1975 and *Engineered Materials Handbook, Vol.2, Engineering Plastics*, ASM International, Metals Park, Ohio, 1988.

Table 187. Flexural Strength of Polymers

(Sheet 4 of 6)

Class	Polymer	Flexural Strength (ASTM D790) (10^3 psi)
Phenolics; Molded	Type and filler	
	General: woodflour and flock	8.5—12
	Shock: paper, flock, or pulp	8.0—11.5
	High shock: chopped fabric or cord	8—15
	Very high shock: glass fiber	10—45
	Arc resistant—mineral	10—13
	Rubber phenolic—woodflour or flock	7—12
	Rubber phenolic—chopped fabric	7
	Rubber phenolic—asbestos	7
ABS–Polycarbonate Alloy	ABS–Polycarbonate Alloy	14.3
PVC–Acrylic Alloy	PVC–acrylic sheet	10.7
	PVC–acrylic injection molded	8.7
Polymides	Unreinforced	6.6—11
	Glass reinforced	56
Polyacetals	Homopolymer:	
	Standard	14.1
	Copolymer:	
	Standard	13
	25% glass reinforced	28
	High flow	13
Polyester; Thermoplastic	Injection Moldings:	
	General purpose grade	12.8
	Glass reinforced grades	22—24
	Glass reinforced self extinguishing	23

To convert from **psi** to **MPa**, multiply by **145**.

Source: data compiled by J.S. Park from Charles T. Lynch, *CRC Handbook of Materials Science, Vol. 3*, CRC Press, Boca Raton, Florida, 1975 and *Engineered Materials Handbook, Vol.2, Engineering Plastics*, ASM International, Metals Park, Ohio, 1988.

Table 187. FLEXURAL STRENGTH OF POLYMERS

(SHEET 5 OF 6)

Class	Polymer	Flexural Strength (ASTM D790) (10^3 psi)
Polyester; Thermoplastic (Con't)	General purpose grade	12
	Glass reinforced grade	19
	Asbestos—filled grade	19
Polyesters: Thermosets	Cast polyyester	
	Rigid	8—24
	Flexible	4—16
Reinforced polyester moldings	High strength (glass fibers)	6—26
	Heat and chemical resistant (asbestos)	10—13
	Sheet molding compounds, general purpose	26—32
Phenylene Oxides	SE—100	12.8
	SE—1	13.5
	Glass fiber reinforced	20.5—22
Phenylene oxides (Noryl)	Standard	15.4
	Glass fiber reinforced	25—28
Polyarylsulfone	Polyarylsulfone	16.1—17.2
Polypropylene:	General purpose	6—7 (yield)
	High impact	4.1 (yield)
	Asbestos filled	7.5—9 (yield)
	Glass reinforced	8—11 (yield)
Polyphenylene sulfide:	Standard	20
	40% glass reinforced	37
Polystyrenes; Molded	Polystyrenes	
	General purpose	10—15
	Glass fiber —30% reinforced	17

To convert from **psi** to **MPa**, multiply by **145**.

Source: data compiled by J.S. Park from Charles T. Lynch, *CRC Handbook of Materials Science, Vol. 3,* CRC Press, Boca Raton, Florida, 1975 and *Engineered Materials Handbook, Vol.2, Engineering Plastics,* ASM International, Metals Park, Ohio, 1988.

Table 187. FLEXURAL STRENGTH OF POLYMERS
(SHEET 6 OF 6)

Class	Polymer	Flexural Strength (ASTM D790) (10^3 psi)
Styrene acrylonitrile (SAN)	Glass fiber (30%) reinforced SAN	22
Polyvinyl Chloride And Copolymers;	Molded, Extruded:	
	Rigid—normal impact	11—16
	Vinylidene chloride	15—17
Silicones; Molded, Laminated	Fibrous (glass) reinforced silicones	16—19
	Granular (silica) reinforced silicones	6—10
	Woven glass fabric/ silicone laminate	33—47
Ureas; Molded	Alpha—cellulose filled (ASTM Type l)	8—18
	Cellulose filled (ASTM Type 2)	7.5—13
	Woodflour filled	7.5—12.0

To convert from **psi** to **MPa**, multiply by **145**.

Source: data compiled by J.S. Park from Charles T. Lynch, *CRC Handbook of Materials Science, Vol. 3*, CRC Press, Boca Raton, Florida, 1975 and *Engineered Materials Handbook, Vol.2, Engineering Plastics*, ASM International, Metals Park, Ohio, 1988.

Table 188. FLEXTURAL STRENGTH OF FIBERGLASS REINFORCED PLASTICS

Class	Material	Glass fiber content (wt%)	Flexural strength (ksi)
Glass fiber reinforced thermosets			
	Sheet molding compound (SMC)	15 to 30	18 to 30
	Bulk molding compound(BMC)	15 to 35	10 to 20
	Preform/mat(compression molded)	25 to 50	10 to 40
	Cold press molding–polyester	20 to 30	22 to 37
	Spray–up–polyester	30 to 50	16 to 28
	Filament wound–epoxy	30 to 80	100 to 270
	Rod stock–polyester	40 to 80	100 to 180
	Molding compound–phenolic	5 to 25	18 to 24
Glass–fiber–reinforced thermoplastics			
	Acetal	20 to 40	15 to 28
	Nylon	6 to 60	7 to 50
	Polycarbonate	20 to 40	17 to 30
	Polyethylene	10 to 40	7 to 12
	Polypropylene	20 to 40	7 to 11
	Polystyrene	20 to 35	10 to 17
	Polysulfone	20 to 40	21 to 27
	ABS(acrylonitrile butadiene styrene)	20 to 40	23 to 26
	PVC (polyvinyl chloride)	15 to 35	20 to 25
	Polyphenylene oxide(modified)	20 to 40	17 to 31
	SAN (styrene acrylonitrile)	20 to 40	15 to 21
	Thermoplastic polyester	20 to 35	19 to 29

To convert (ksi) to (MPa), multiply by 6.89

Data from *ASM Engineering Materials Reference Book, Second Edition*, Michael Bauccio, Ed., ASM International, Materials Park, OH, p106, (1994).

Table 189. SHEAR STRENGTH OF WROUGHT ALUMINUM ALLOYS (SHEET 1 OF 6)

Alloy AA No.	Temper	Shear Strength (MPa)
1050	0	62
	H14	69
	H16	76
	H18	83
1060	0	48
	H12	55
	H14	62
	H16	69
	H18	76
1100	0	62
	H12	69
	H14	76
	H16	83
	H18	90
1350	0	55
	H12	62
	H14	69
	H16	76
	H19	105
2011	T3	220
	T8	240
2014	0	125
	T4	260
	T6	290
Alclad 2014	0	125
	T3	255
	T4	255
	T6	285

Source: Data from *ASM Metals Reference Book, Second Edition*, American Society for Metals, Metals Park, Ohio 44073, (1984).

Table 189. SHEAR STRENGTH OF WROUGHT ALUMINUM ALLOYS (SHEET 2 OF 6)

Alloy AA No.	Temper	Shear Strength (MPa)
2024	0	125
	T3	285
	T4, T351	285
	T361	290
Alclad 2024	0	125
	T	275
	T4, T351	275
	T361	285
	T81, T851	275
	T861	290
2218	T72	205
2618	All	260
3003	0	76
Alclad	H12	83
3003	H14	97
	H16	105
	H18	110
3004	0	110
Alclad	H32	115
3004	H34	125
	H36	140
	H38	145
3105	0	83
	H12	97
	H14	105
	H16	110
	H18	115
	H25	105
4032	T6	260

Source: Data from *ASM Metals Reference Book, Second Edition*, American Society for Metals, Metals Park, Ohio 44073, (1984).

Table 189. SHEAR STRENGTH OF WROUGHT ALUMINUM ALLOYS (SHEET 3 OF 6)

Alloy AA No.	Temper	Shear Strength (MPa)
5005	0	76
	H12	97
	H14	97
	H16	105
	H18	110
	H32	97
	H34	97
	H36	105
	H38	110
5050	0	105
	H32	115
	H34	125
	H36	130
	H38	140
5052	0	125
	H32	140
	H34	145
	H36	160
	H38	165
5056	0	180
	H18	235
	H38	220
5083	0	170
5086	0	160
	H34	185
5154	0	150
	H32	150
	H34	165
	H36	180
	H38	195

Source: Data from *ASM Metals Reference Book, Second Edition*, American Society for Metals, Metals Park, Ohio 44073, (1984).

Table 189. SHEAR STRENGTH OF WROUGHT ALUMINUM ALLOYS (SHEET 4 OF 6)

Alloy AA No.	Temper	Shear Strength (MPa)
5182	0	150
5252	H25	145
	H28, H38	160
5254	0	150
5254	H32	150
	H34	165
	H36	180
	H38	195
5454	0	160
	H32	165
	H34	180
	H111	160
	H112	160
	H311	160
5456	H321, H116	205
5457	0	83
	H25	110
	H28, H38	125
5652	0	125
	H32	140
	H34	145
	H36	160
	H38	165
5657	H25	97
	H28, H38	105
6005	T5	205
6009	T4	150

Source: Data from *ASM Metals Reference Book, Second Edition*, American Society for Metals, Metals Park, Ohio 44073, (1984).

Table 189. SHEAR STRENGTH OF WROUGHT ALUMINUM ALLOYS (SHEET 5 OF 6)

Alloy AA No.	Temper	Shear Strength (MPa)
6061	0	83
	T4, T451	165
	T6, T651	205
Alclad 6061	0	76
	T4, T451	150
	T6, T651	185
6063	0	69
	T1	97
	T5	115
	T6	150
	T83	150
	T831	125
	T832	185
6066	0	97
	T4, T451	200
	T6, T651	235
6070	0	97
	T4	205
	T6	235
6151	T6	140
6205	T5	205
6262	T9	240
6351	T6	200
6463	Tl	97
	T5	115
	T6	150
7005	0	117
	T53	221
	T6,T63,T6351	214

Source: Data from *ASM Metals Reference Book, Second Edition*, American Society for Metals, Metals Park, Ohio 44073, (1984).

Table 189. SHEAR STRENGTH OF WROUGHT ALUMINUM ALLOYS (SHEET 6 OF 6)

Alloy AA No.	Temper	Shear Strength (MPa)
7072	0	55
	H12	62
	H14	69
7075	0	150
	T6,T651	330
Alclad 7075	0	150
	T6,T651	315
7175	T66	325
	T736	290
7475	T651	295
	T7351	270
	T7651	270

Source: Data from *ASM Metals Reference Book, Second Edition*, American Society for Metals, Metals Park, Ohio 44073, (1984).

Table 190. TORSION SHEAR STRENGTH OF GRAY CAST FE

ASTM Class	Torsional Shear Strength (MPa)
20	179
25	220
30	276
35	334
40	393
50	503
60	610

Source: data from *ASM Metals Reference Book, Second Edition*, American Society for Metals, Metals Park, Ohio 44073, p166-167, (1984).

Table 191. HARDNESS OF GRAY CAST IRONS

SAE grade	Hardness (HB)
G1800	187 max
G2500	170 to 229
G2500a	170 to 229
G3000	187 to 241
C3500	207 to 255
G3500b	207 to 255
G3500c	207 to 255
G4000	217 to 269
G4000d	241 to 321

Source: data from *ASM Metals Reference Book, Second Edition*, American Society for Metals, Metals Park, Ohio 44073, p166-167, (1984).

Table 192. HARDNESS OF GRAY CAST IRON BARS

ASTM Class	Hardness (HB)
20	156
25	174
30	210
35	212
40	235
50	262
60	302

Source: data from *ASM Metals Reference Book, Second Edition*, American Society for Metals, Metals Park, Ohio 44073, p166-167, (1984).

Table 193. HARDNESS OF MALLEABLE IRON CASTINGS

Specification Number	Grade or Class	Hardness (HB)
Ferritic		
ASTM A47, A338; ANSI G48.1; FED QQ–I–666c	32510	156 max
	35018	156 max
ASTM A197		156 max
Pearlitic and Martensitic		
ASTM A220; ANSI C48.2; MIL–I–11444B	40010	149–197
	45008	156–197
	45006	156–207
	50005	179–229
	60004	197–241
	70003	217–269
	80002	241–285
	90001	269–321
Automotive		
ASTM A602; SAE J158	M3210	156 max
	M4504(a)	163–217
	M5003(a)	187–241
	M5503(b)	187–241
	M7002(b)	229–269
	M8501(b)	269–302

(a) Air quenched and tempered
(b) Liquid quenched and tempered

Source: data from ASM Metals Reference Book, Second Edition, American Society for Metals, Metals Park, Ohio 44073, p171, (1984).

Table 194. HARDNESS OF DUCTILE IRONS

Specification Number	Grade or Class	Hardness (HB)
ASTM A395-76 ASME SA395	60-40-18	143-187
ASTM A476-70(d); SAE AMS5316	80-60-03	201 min
ASTM A536-72, MIL-1-11466B(MR)	60-40-18	
	65-45-12	
	80-55-06	
	100-70-03	
	120-90-02	
SAE J434c	D4018	170 max
	D4512	156-217
	D5506	187-255
	D7003	241-302
MIL-I-24137(Ships)	Class A	190 max
	Class B	190 max
	Class C	175 max

Source: data from *ASM Metals Reference Book, Second Edition*, American Society for Metals, Metals Park, Ohio 44073, p169, (1984).

Table 195. HARDNESS OF TOOL STEELS

(SHEET 1 OF 2)

Type	Condition	Hardness (HRC)
L2	Annealed	96 HRB
	Oil quenched from 855 •C and single tempered at:	
	205 •C	54
	315 •C	52
	425 •C	47
	540 •C	41
	650 •C	30
L6	Oil quenched from 845 •C and single tempered at:	
	315 •C	54
	425 •C	46
	540 •C	42
	650 •C	32
S1	Annealed	96 HRB
	Oil quenched from 930 •C and single tempered at:	
	205 •C	57.5
	315 •C	54
	425 •C	50.5
	540 •C	47.5
	650 •C	42
S5	Annealed	96 HRB
	Oil quenched from 870 •C and single tempered at:	
	205 •C	59
	315 •C	58
	425 •C	52
	540 •C	48
	650 •C	37

Source: Data from *ASM Metals Reference Book, Second Edition*, American Society for Metals, Metals Park, Ohio 44073, p241, (1984).

Table 195. HARDNESS OF TOOL STEELS
(SHEET 2 OF 2)

Type	Condition	Hardness (HRC)
S7	Annealed	95 HRB
	Fan cooled from 940 •C and single tempered at:	
	205 •C	58
	315 •C	55
	425 •C	53
	540 •C	51
	650 •C	39

Source: Data from *ASM Metals Reference Book, Second Edition*, American Society for Metals, Metals Park, Ohio 44073, p241, (1984).

Table 196. HARDNESS OF AUSTENITIC STAINLESS STEELS

Type	Form	Condition	ASTM Specification	Hardness (HRB)
Type 301(UNS S30100)	Bar,Wire,Plate,Sheet,Strip	Annealed	A167	88 max
Type 317L (UNS S31703)	Bar	Annealed	—	85max
Type 317LM	Bar,Plate,Sheet, Strip	Annealed	—	95 max

Data from *ASM Metals Reference Book, Third Edition*, Michael Bauccio, Ed., ASM International, Materials Park, OH, p364-366 (1993).

Table 197. HARDNESS OF FERRITIC STAINLESS STEELS

Type	ASTM Specification	Form	Condition	Hardness (HRB)
Type 409 (UNS S40900)	—	Bar	Annealed	75 max(a)
Type 434 (UNS S43400)	—	Wire	Annealed	90 max(a)
Type 436 (UNS S43600)	—	Sheet, Strip	Annealed	83 max(a)
Type 442 (UNS S44200)	—	Bar	Annealed	90 max(a)
Type 444 (UNS S44400)	A176	Plate, Sheet, Strip	Annealed	95 max

(a) Typical Values

Data from *ASM Metals Reference Book, Third Edition*, Michael Bauccio, Ed., ASM International, Materials Park, OH, p368 (1993).

Table 198. HARDNESS OF MARTENSITIC STAINLESS STEELS

Type	ASTM Specification	Form	Condition	Rockwell Hardness
Type 410S (UNS S41008)	A176	Plate, Sheet, Strip	Annealed	95 HRB max
Type 420 (UNS S42000)	—	Bar	Tempered 205 °C	52 HRC
Type 440A (UNS S44002)	—	Bar	Annealed	95 HRB
	—		Tempered 315 °C	51 HRC
Type 440B (UNS S44003)	—	Bar	Annealed	96 HRB
	—		Tempered 315 °C	55 HRC
Type 440C (UNS S44004)	—	Bar	Annealed	97 HRB
	—		Tempered 315 °C	57 HRC

Data from ASM *Metals Reference Book, Third Edition*, Michael Bauccio, Ed., ASM International, Materials Park, OH, p369-370 (1993).

Table 199. Hardness of Precipitation-Hardening Austenitic Stainless Steels

Type	Form	Condition	Hardness (HRC)	
			Minimum	Maximum
PH 13–8 Mo (UNS S13800)	Bar, Plate, Sheet, Strip	H950	45	—
		H1000	43	—
15–5 PH (UNS S15500) and 17–4 PH (UNS S17400)	Bar, Plate, Sheet, Stript	H900	40	48
		H925	38	47(a)
		H1025	35(a)	42(a)
		H1075	32(a)	38(a)
		H1100	31(a)	38(a)
		H1150	28(a)	36(a)
		H1150M	24(a)	34(a)
17–7 PH (UNS S17700)	Bar	RH950	41	—
		TH1050	38	—

(a) For flat rolled products, value varies with thickness.

Data from *ASM Metals Reference Book, Third Edition*, Michael Bauccio, Ed., ASM International, Materials Park, OH, p371 (1993).

Table 200. MACHINABILITY RATING OF WROUGHT COPPERS AND COPPER ALLOYS (SHEET 1 OF 9)

UNS Number and Name	Nominal Composition (%)	Commercial Forms(a)	Machinability Rating (d)
C10100 Oxygen-free electronic	99.99 Cu	F, R, W, T, P, S	20
C10200 Oxygen-free copper	99.95 Cu	F, R, W, T, P, S	20
C10300 Oxygen-free extra-low phosporus	99.95 Cu, 0.003 P	F, R, T, P, S	20
C10400, C10500, C10700 Oxygen-free, silver-bearing	99.95 Cu(e)	F, R, W, S	20
C10800 Oxygen-free, low phosporus	99.95 Cu, 0.009 P	F, R, T, P	20
CS11000 Electrolytic tough pitch copper	99.90 Cu, 0.040 O	F, R, W, T, P, S	20
C11100 Electrolytic tough pitch, anneal resistant	99.90 Cu, 0.04 O, 0.01 Cd	W	20
C11300, C11400, C11500, C11600 Silver-bearing tough pitch copper	99.90 Cu, 0.04 O, Ag(f)	F, R, W, T, S	20
C12000, C12100	99.9 Cu(g)	F, T, P	20
C12200 Phosphorus deoxidized copper, high residual phosphorus	99.90 Cu, 0.02 P	F, R, T, P	20
C12500, C12700, C12800, C12900, C13000 Fire-refined tough pitch with silver	99.88 Cu(h)	F, R, W, S	20
C14200 Phosphorus deoxidized, arsenical	99.68 Cu, 0.3 As, 0.02 P	F, R, T	20

(a) F, flat products; R, rod; W, wire; T, tube; P, pipe; S, shapes.

Data from ASM *Metals Reference Book, Third Edition*, Michael Bauccio, Ed., ASM International, Materials Park, OH, p442–454, (1993).

Table 200. MACHINABILITY RATING OF WROUGHT COPPERS AND COPPER ALLOYS (SHEET 2 OF 9)

UNS Number and Name	Nominal Composition (%)	Commercial Forms(a)	Machinability Rating (d)
C19200	98.97 Cu, 1.0 Fe, 0.03 P	F, T	20
C14300	99.9 Cu, 0.1 Cd	F	20
C14310	99.8 Cu, 0.2 Cd	F	20
C14500 Phosphorus deoxidized, tellurium bearing	99.5 Cu, 0.50 Te, 0.008 P	F, R, W, T	85
C14700 Sulfur bearing	99.6 Cu, 0.40 S	R, W	85
C15000 Zirconium copper	99.8 Cu, 0.15 Zr	R, W	20
C15500	99.75 Cu, 0.06 P, 0.11 Mg, Ag(i)	F	20
C16200 Cadmium copper	99.0 Cu, 1.0 Cd	F, R, W	20
C16500	98.6 Cu, 0.8 Cd, 0.6 Sn	F, R, W	20
C17000 Beryllium copper	99.5 Cu, 1.7 Be, 0.20 Co	F, R	20
C17200 Beryllium copper	99.5 Cu, 1.9 Be , 0.20 Co	F, R, W, T, P, S	20
C17300 Beryllium copper	99.5 Cu, 1.9 Be, 0.40 Pb	R	50

(a) F, flat products; R, rod; W, wire; T, tube; P, pipe; S, shapes.

Data from *ASM Metals Reference Book, Third Edition*, Michael Bauccio, Ed., ASM International, Materials Park, OH, p442–454, (1993).

Table 200. MACHINABILITY RATING OF WROUGHT COPPERS AND COPPER ALLOYS (SHEET 3 OF 9)

UNS Number and Name	Nominal Composition (%)	Commercial Forms(a)	Machinability Rating (d)
C18200, C18400, C18500 Chromium copper	99.5 Cu(j)	F, W, R, S, T	20
C18700 leaded copper	99.0 Cu, 1.0 Pb	R	85
C18900	98.75 Cu, 0.75 Sn, 0.3 Si, 0.20 Mn	R, W	20
C19000 Copper-nickel-phosphorus alloy	98.7 Cu, 1.1 Ni, 0.25 P	F, R, W	30
C19100 Copper-nickel-phosphorus-tellurium alloy	98.15 Cu, 1.1 Ni, 0.50 Te, 0.25 P	R, F	75
C19400	97.5 Cu, 2.4 Fe, 0.13 Zn, 0.03 P	F	20
C19500	97.0 Cu, 1.5 Fe, 0.6 Sn, 0.10 P, 0.80 Co	F	20
C21000 Gilding, 95%	95.0 Cu, 5.0 Zn	F, W	20
C22000 Commercial bronze, 90%	90.0 Cu, 10.0 Zn	F, R, W, T	20
C22600 Jewelry bronze, 87.5%	87.5 Cu, 12.5 Zn	F, W	30
C23000 Red brass, 85%	95.0 Cu, 15.0 Zn	F, W, T, P	30
C24000 Low brass, 80%	80.0 Cu, 20.0 Zn	F, W	30
C26000 Cartridge brass, 70%	70.00 Cu, 30.0 Zn	F, R, W, T	30
C26800, C27000 Yellow brass	65.0 Cu, 35.0 Zn	F, R, W	30

(a) F, flat products; R, rod; W, wire; T, tube; P, pipe; S, shapes.

Data from *ASM Metals Reference Book, Third Edition*, Michael Bauccio, Ed., ASM International, Materials Park, OH, p442–454, (1993).

Table 200. MACHINABILITY RATING OF WROUGHT COPPERS AND COPPER ALLOYS (SHEET 4 OF 9)

UNS Number and Name	Nominal Composition (%)	Commercial Forms(a)	Machinability Rating (d)
C28000 Muntz metal	60.0 Cu, 40.0 Zn	F, R, T	40
C31400 Leaded commercial bronze	89.0 Cu, 1.75 Pb, 9.25 Zn	F, R	80
C31600 Leaded commercial bronze, nickel-bearing	89.0 Cu, 1.9 Pb, 1.0 Ni, 8.1 Zn	F, R	80
C33000 Low-leaded brass tube	66.0 Cu, 0.5 Pb, 33.5 Zn	T	60
C33200 High-leaded brass tube	66.0 Cu, 1.6 Pb, 32.4 Zn	T	80
C33500 Low-leaded brass	65.0 Cu, 0.5 Pb, 34.5 Zn	F	60
C34000 Medium-leaded brass	65.0 Cu, 1.0 Pb, 34.0 Zn	F, R, W, S	70
C34200 High-leaded brass	64.5 Cu, 2.0 Pb, 33.5 Zn	F, R	90
C34900	62.2 Cu, 0.35 Pb, 37.45 Zn	R, W	50
C35000 Medium-leaded brass	62.5 Cu, 1.1 Pb, 36.4 Zn	F, R	70
C35300 High-leaded brass	62.0 Cu, 1.8 Pb, 36.2 Zn	F, R	90
C35600 Extra-high-leaded brass	63.0 Cu, 2.5 Pb, 34.5 Zn	F	100

(a) F, flat products; R, rod; W, wire; T, tube; P, pipe; S, shapes.

Data from *ASM Metals Reference Book, Third Edition*, Michael Bauccio, Ed., ASM International, Materials Park, OH, p442–454, (1993).

Table 200. MACHINABILITY RATING OF WROUGHT COPPERS AND COPPER ALLOYS (SHEET 5 OF 9)

UNS Number and Name	Nominal Composition (%)	Commercial Forms(a)	Machinability Rating (d)
C36000 Free-cutting brass	61.5 Cu, 3.0 Pb, 35.5 Zn	F, R, S	100
C36500 to C36800 Leaded Muntz metal	60.0 Cu(k), 0.6 Pb, 39.4 Zn	F	60
C37000 Free-cutting Muntz metal	60.0 Cu, 1.0 Pb, 39.0 Zn	T	70
C37700 Forging brass	59.0 Cu, 2.0 Pb, 39.0 Zn	R, S	80
C38500 Architectural bronze	57.0 Cu, 3.0 Pb, 40.0 Zn	R, S	90
C40500	95 Cu, 1 Sn, 4 Zn	F	20
C40800	95 Cu, 2 Sn, 3 Zn	F	20
C41100	91 Cu, 0.5 Sn, 8.5 Zn	F, W	20
C41300	90.0 Cu, 1.0 Sn, 9.0 Zn	F, R, W	2
C41500	91 Cu, 1.8 Sn, 7.2 Zn	F	30
C42200	87.5 Cu, 1.1 Sn, 11.4 Zn	F	30
C42500	88.5 Cu, 2.0 Sn, 9.5 Zn	F	30

(a) F, flat products; R, rod; W, wire; T, tube; P, pipe; S, shapes.

Data from *ASM Metals Reference Book, Third Edition*, Michael Bauccio, Ed., ASM International, Materials Park, OH, p442–454, (1993).

Table 200. MACHINABILITY RATING OF WROUGHT COPPERS AND COPPER ALLOYS (SHEET 6 OF 9)

UNS Number and Name	Nominal Composition (%)	Commercial Forms(a)	Machinability Rating (d)
C43000	87.0 Cu, 2.2 Sn, 10.8 Zn	F	30
C43400	85.0 Cu, 0.7 Sn, 14.3 Zn	F	30
C43500	81.0 Cu, 0.9 Sn, 18.1 Zn	F, T	30
C44300, C44400, C44500 Inhibited admiralty	71.0 Cu, 28.0 Zn, 1.0 Sn	F, W, T	30
C46400 to C46700 Naval brass	60.0 Cu, 39.25 Zn, 0.75 Sn	F, R, T, S	30
C48200 Naval brass, medium-leaded	60.5 Cu, 0.7 Pb, 0.8 Sn, 38.0 Zn	F, R, S	50
C48500 Leaded naval brass	60.0 Cu, 1.75 Pb, 37.5 Zn, 0.75 Sn	F, R, S	70
C50500 Phosphor bronze, 1.25% E	98.75 Cu, 1.25 Sn, trace P	F, W	20
C51000 Phosphor bronze, 5% A	95.0 Cu, 5.0 Sn, trace P	F, R, W, T	20
C51100	95.6 Cu, 4.2 Sn, 0.2 P	F	20
C52100 Phosphor bronze, 8% C	92.0 Cu, 8.0 Sn, trace P	F, R, W	20
C52400 Phosphor bronze, 10% D	90.0 Cu, 10.0 Sn, trace P	F, R, W	20

(a) F, flat products; R, rod; W, wire; T, tube; P, pipe; S, shapes.

Data from *ASM Metals Reference Book, Third Edition*, Michael Bauccio, Ed., ASM International, Materials Park, OH, p442–454, (1993).

Table 200. MACHINABILITY RATING OF WROUGHT COPPERS AND COPPER ALLOYS (SHEET 7 OF 9)

UNS Number and Name	Nominal Composition (%)	Commercial Forms(a)	Machinability Rating (d)
C54400 Free-cutting phosphor bronze	88.0 Cu, 4.0 Pb, 4.0 Zn, 4.0 Sn	F, R	80
C60800 Aluminum bronze, 5%	95.0 Cu, 5.0 Al	T	20
C61000	92.0 Cu, 8.0 Al	R, W	20
C61300	92.65 Cu, 0.35 Sn, 7.0 Al	F, R, T, P, S	30
C61400 Aluminum bronze, D	91.0 Cu, 7.0 Al, 2.0 Fe	F, R, W, T, P, S	20
C61500	90.0 Cu, 8.0 Al, 2.0 Ni	F	30
C61800	89.0 Cu, 1.0 Fe, 10.0 Al	R	40
C62300	87.0 Cu, 10.0 Al, 3.0 Fe	F, R	50
C62400	86.0 Cu, 3.0 Fe, 11.0 Al	F, R	50
C62500	82.7 Cu, 4.3 Fe, 13.0 Al	F, R	20
C63000	82.0 Cu, 3.0 Fe, 10.0 Al, 5.0 Ni	F, R	30
C63200	82.0 Cu, 4.0 Fe, 9.0 Al, 5.0 Ni	F, R	30
C63600	95.5 Cu, 3.5 Al, 1.0 Si	R, W	40
C64200	91.2 Cu, 7.0 Al	F, R	60

(a) F, flat products; R, rod; W, wire; T, tube; P, pipe; S, shapes.

Data from *ASM Metals Reference Book, Third Edition*, Michael Bauccio, Ed., ASM International, Materials Park, OH, p442–454, (1993).

Table 200. Machinability Rating of Wrought Coppers and Copper Alloys (Sheet 8 of 9)

UNS Number and Name	Nominal Composition (%)	Commercial Forms(a)	Machinability Rating (d)
C65100 Low-silicon bronze, B	98.5 Cu, 1.5 Si	R, W, T	30
C65500 High-silicon bronze, A	97.0 Cu, 3.0 Si	F, R, W, T	30
C66700 Manganese brass	70.0 Cu, 28.8 Zn, 1.2 Mn	F, W	30
C67400	58.5 Cu, 36.5 Zn, 1.2 Al, 2.8 Mn, 1.0 Sn	F, R	25
C67500 Manganese bronze, A	58.5 Cu, 1.4 Fe, 39.0 Zn, 1.0 Sn, 0.1 Mn	R, S	30
C68700 Aluninum brass, arsenical	77.5 Cu, 20.5 Zn, 2.0 Al, 0.1 As	T	30
C69400 Silicon red brass	81.5 Cu, 14.5 Zn, 4.0 Si	R	30
C70400	92.4 Cu, 1.5 Fe, 5.5 Ni, 0.6 Mn	F, T	20
C70600 Copper nickel, 10%	88.7 Cu, 1.3 Fe, 10.0 Ni	F, T	20
C71000 Copper nickel, 20%	79.00 Cu, 21.0 Ni	F, W, T	20
C71500 Copper nickel, 30%	70.0 Cu, 30.0 Ni	F, R, T	20
C71700	67.8 Cu, 0.7 Fe, 31.0 Ni, 0.5 Be	F, R, W	20
C72500	88.20 Cu, 9.5 Ni, 2.3 Sn	F, R, W, T	20
C73500	72.0 Cu, 18.0 Ni , 10.0 Zn	F, R, W, T	20

(a) F, flat products; R, rod; W, wire; T, tube; P, pipe; S, shapes.

Data from *ASM Metals Reference Book, Third Edition*, Michael Bauccio, Ed., ASM International, Materials Park, OH, p442–454, (1993).

Table 200. MACHINABILITY RATING OF WROUGHT COPPERS AND COPPER ALLOYS (SHEET 9 OF 9)

UNS Number and Name	Nominal Composition (%)	Commercial Forms(a)	Machinability Rating (d)
C74500 Nickel silver, 65-10	65.0 Cu, 25.0 Zn, 10.0 Ni	F, W	20
C75200 Nickel silver, 65-18	65.0 Cu, 17.0 Zn, 18.0 Ni	F, R, W	20
C75400 Nickel silver, 65-15	65.0 Cu, 20.0 Zn, 15.0 Ni	F	20
C75700 Nickel silver, 65-12	65.0 Cu, 23.0 Zn, 12.0 Ni	F, W	20
C77000 Nickel silver, 55-18	55.0 Cu, 27.0 Zn, 18.0 Ni	F, R, W	30
C78200 Leaded nickel silver, 65-8-2	65.0 Cu, 2.0 Pb, 25.0 Zn, 8.0 Ni	F	60

(a) F, flat products; R, rod; W, wire; T, tube; P, pipe; S, shapes.

Data from *ASM Metals Reference Book, Third Edition*, Michael Bauccio, Ed., ASM International, Materials Park, OH, p442–454, (1993).

(d) Based on 100% for C360000.
(e) C10400, 8 oz/ton Ag, C10500, 10 oz/ton C10700, 25 oz/ton .
(f) C11300, 8 oz/ton Ag, C11400,10 oz/ton, C11500, 16 oz/ton C11600, 25 oz/ton
(g) C12000, 0.008 P; C12100, 0.008 P and 4 oz/ton Ag;
(h) C12700, 8 oz/ton Ag; C12800,10 oz/ton; C12900,16 oz/ton; C13000, 25 oz/ton.
(i) 0.98.30 oz/ton Ag.
(j) C18200, 0.9 Cr, C18400, 0.9 Cr; C18500, 0.7 Cr
(k) Rod, 61.0 Cu min.

Table 201. HARDNESS OF WROUGHT ALUMINUM ALLOYS
(SHEET 1 OF 5)

Alloy AA No.	Temper	Hardness (BHN)
1060	0	19
	H12	23
	H14	26
	H16	30
	H18	35
1100	0	23
	H12	28
	H14	32
	H16	38
	H18	44
2011	T3	95
	T8	100
2014	0	45
	T4	105
	T6	135
2024	0	47
	T3	120
	T4, T351	120
	T361	130
2218	T61	115
	T71	105
	T72	95
3003	0	28
Alclad	H12	35
3003	H14	40
	H16	47
	H18	55

Source: data from *ASM Metals Reference Book, Second Edition*, American Society for Metals, Metals Park, Ohio 44073, (1984).

Table 201. HARDNESS OF WROUGHT ALUMINUM ALLOYS
(SHEET 2 OF 5)

Alloy AA No.	Temper	Hardness (BHN)
3004	0	45
Alclad	H32	52
3004	H34	63
	H36	70
	H38	77
4032	T6	120
5005	0	28
	H32	36
	H34	41
	H36	46
	H38	51
5050	0	36
	H32	46
	H34	53
	H36	58
	H38	63
5052	0	47
	H32	60
	H34	68
	H36	73
	H38	77
5056	0	65
	H18	105
	H38	100
5154	0	58
	H32	67
	H34	73
	H36	78
	H38	80
	H112	63

Source: data from *ASM Metals Reference Book, Second Edition*, American Society for Metals, Metals Park, Ohio 44073, (1984).

Table 201. HARDNESS OF WROUGHT ALUMINUM ALLOYS
(SHEET 3 OF 5)

Alloy AA No.	Temper	Hardness (BHN)
5182	0	58
5252	H25	68
	H28, H38	75
5254	0	58
5254	H32	67
	H34	73
	H36	78
	H38	80
	H112	63
5454	0	62
	H32	73
	H34	81
	H111	70
	H112	62
	H311	70
5456	H321, H116	90
5457	0	32
	H25	48
	H28, H38	55
5652	0	47
	H32	60
	H34	68
	H36	73
	H38	77
5657	H25	40
	H28, H38	50
6005	T5	95
6009	T4	70
6010	T4	76

Source: data from *ASM Metals Reference Book, Second Edition*, American Society for Metals, Metals Park, Ohio 44073, (1984).

Table 201. HARDNESS OF WROUGHT ALUMINUM ALLOYS
(SHEET 4 OF 5)

Alloy AA No.	Temper	Hardness (BHN)
6061	0	30
	T4, T451	65
	T6, T651	95
6063	0	25
	T1	42
	T5	60
	T6	73
	T83	82
	T831	70
	T832	95
6066	0	43
	T4, T451	90
	T6, T651	120
6070	0	35
	T4	90
	T6	120
6151	T6	71
6201	T6	90
6205	T1	65
	T5	95
6262	T9	120
6351	T6	95
6463	T1	42
	T5	60
	T6	74
7049	T73	135
7072	0	20
	H12	28
	H14	32

Source: data from *ASM Metals Reference Book, Second Edition*, American Society for Metals, Metals Park, Ohio 44073, (1984).

Table 201. HARDNESS OF WROUGHT ALUMINUM ALLOYS

(SHEET 5 OF 5)

Alloy AA No.	Temper	Hardness (BHN)
7075	0	60
	T6,T651	150
7175	T66	150
	T736	145

Source: data from *ASM Metals Reference Book, Second Edition*, American Society for Metals, Metals Park, Ohio 44073, (1984).

Table 202. HARDNESS OF WROUGHT TITANIUM ALLOYS AT ROOM TEMPERATURE (SHEET 1 OF 2)

Class	Alloy	Condition	Hardness (HRC)
Commercially Pure	99.5 Ti	Annealed	120(a)
	99.2 Ti	Annealed	200(a)
	99.1 Ti	Annealed	225(a)
	99.0 Ti	Annealed	265(a)
	99.2Ti-0.2Pd	Annealed	200(a)
Alpha Alloys	Ti-5Al-2.5Sn	Annealed	36
	Ti-5Al-2.5Sn (low O_2)	Annealed	35
Near Alpha Alloys	Ti-8Al-1Mo-1V	Duplex Annealed	35
	Ti-11Sn-1Mo-2.25Al-5.0Zr-1Mo-0.2Si	Duplex Annealed	36
	Ti-6Al-2Sn-4Zr-2Mo	Duplex Annealed	32
	Ti-6Al-2Nb-1Ta-1Mo	As rolled 2.5 cm (1 in.) plate	30
Alpha-Beta Alloys	Ti-6Al-4V	Annealed	36
		Solution + age	41

(a) Hardness, HB

Data from *ASM Metals Reference Book, Third Edition*, Michael Bauccio, Ed., ASM International, Materials Park, OH, p512, (1993).

Table 202. Hardness of Wrought Titanium Alloys at Room Temperature (Sheet 2 of 2)

Class	Alloy	Condition	Hardness (HRC)
	Ti-6Al-4V(low O_2)	Annealed	35
	Ti-6Al-6V-2Sn	Annealed	38
		Solution + age	42
	Ti-7Al-4Mo	Solution + age	38
Beta Alloys	Ti-13V-1Cr-3Al	Solution + age	40
	Ti-8Mo-8V-2Fe-3Al	Solution + age	40
	Ti-3Al-8V-6Cr-4Mo-4Zr	Solution + age	42

(a) Hardness, HB

Data from *ASM Metals Reference Book, Third Edition*, Michael Bauccio, Ed., ASM International, Materials Park, OH, p512, (1993).

Table 203. HARDNESS OF CERAMICS
(SHEET 1 OF 6)

Class	Ceramic	Hardness
Borides	Chromium Diboride (CrB_2)	micro 100g: 1800 kg/mm^2
		Vickers 50g: 1800 kg/mm^2
		Knoop 100g: 1700 kg/mm^2
	Hafnium Diboride (HfB_2)	
	(polycrystalline)	Knoop 160g : 2400kg/mm at 24 °C
	(single crystal)	Knoop 160g : 3800kg/mm at 24 °C
	Tantalum Diboride (TaB_2)	micro : 1700 kg/mm^2
		Knoop 30g: 2537 kg/mm^2
		Knoop 100g: 2615 ± 120 kg/mm^2
		Rockwell A : 89
	Titanium Diboride (TiB_2)	Vickers 50g: 3400 kg/mm^2
		Knoop 30g: 3370 kg/mm^2
		Knoop 100g: 2710-3000 kg/mm^2
		Knoop 160g: 3500 kg/mm^2
	(single crystal)	Knoop 100g: 3250±100 kg/mm^2
	Zirconium Diboride (ZrB_2)	Rockwell A: 87-89
		Vickers 50g: 2200 kg/mm^2
		Knoop 100g: 1560 kg/mm^2
		Knoop 160g: 2100 kg/mm^2
	(single crystal)	Knoop 160g: 2000 kg/mm^2
Carbides	Boron Carbide (B_4C)	Knoop 100g: 2800 kg/mm^2
		Knoop 1000g: 2230 kg/mm^2
		Vickers : 2400 kg/mm^2

Source: data compiled by J.S. Park from *No. 1 Materials Index,* Peter T.B. Shaffer, Plenum Press, New York, (1964); *Smithells Metals Reference Book,* Eric A. Brandes, ed., in association with Fulmer Research Institute Ltd. 6th ed. London, Butterworths, Boston, (1983); and *Ceramic Source,* American Ceramic Society (1986-1991).

Table 203. HARDNESS OF CERAMICS

(SHEET 2 OF 6)

Class	Ceramic	Hardness
Carbides (Con't)	Hafnium Monocarbide (HfC)	Knoop : 1790-1870 kg/mm^2
		Vickers 50g : 2533-3202 kg/mm^2
	Silicon Carbide (SiC)	Moh : 9.2
		Vickers 25g : 3000-3500 kg/mm^2
		Knoop 100g : 2500-2550 kg/mm^2
		Knoop 100g : 2960 kg/mm^2 (black)
		Knoop 100g : 2745 kg/mm^2 (green)
	(cubic, CVD)	Knoop or Vickers : 2853-4483 kg/mm^2
	Tantalum Monocarbide (TaC)	Knoop 50g: 1800-1952 kg/mm^2
		Knoop 100g: 825 kg/mm^2
		Vickers 50g: 1800 kg/mm^2
		Rockwell A: 89
		Brinell: 840
	Titanium Monocarbide (TiC)	Knoop 100g: 2470 kg/mm^2
		Knoop 1000g: 1905 kg/mm^2
		Vickers 50g: 2900-3200 kg/mm^2
		Vickers 100g: 2850-3390 kg/mm^2
		micro 20g: 3200 kg/mm^2
	(98.6% density)	Rockwell A: 88-89
	(99.5% density)	Rockwell A: 91-93.5
	(100% density)	Rockwell A: 91-93.5
	Trichromium Dicarbide (Cr_3C_2)	Knoop or Vickers : 1019-1834 kg/mm^2

Source: data compiled by J.S. Park from *No. 1 Materials Index*, Peter T.B. Shaffer, Plenum Press, New York, (1964); *Smithells Metals Reference Book*, Eric A. Brandes, ed., in association with Fulmer Research Institute Ltd. 6th ed. London, Butterworths, Boston, (1983); and *Ceramic Source*, American Ceramic Society (1986-1991).

Table 203. HARDNESS OF CERAMICS
(SHEET 3 OF 6)

Class	Ceramic	Hardness
Carbides (Con't)	Tungsten Monocarbide (WC)	Knoop 100g: 1870-1880 kg/mm^2
		Vickers 50g: 2400 kg/mm^2
		Vickers 100g: 1730 kg/mm^2
		Rockwell A: 92
	(6% Co, 1-3μm grain size)	Rockwell A: 81.4 ± 0.4
	(12% Co, 1-3μm grain size)	Rockwell A: 89.4 ± 0.5
	(24% Co, 1-3μm grain size)	Rockwell A: 86.9 ± 0.6
	(6% Co, 2-4μm grain size)	Rockwell A: 88.6 ± 0.5
	(6% Co, 3-6μm grain size)	Rockwell A: 87.3 ± 0.5
	Zirconium Monocarbide (ZrC)	Knoop : 2138 kg/mm^2
		Vickers 50g : 2600 kg/mm^2
		Vickers 100g : 2836-3840 kg/mm^2
		micro : 2090 kg/mm^2
		Rockwell A: 92.5
Nitrides	Aluminum Nitride (AlN)	Mohs: 5-5.5
		Knoop 100g: 1225-1230 kg/mm^2
	(thick film)	Rockwell 15N: 94.5
	(thin film)	Rockwell 15N: 94.0
	Boron Nitride (BN)	Mohs: 2 (hexagonal)
	Titanium Mononitirde (TiN)	Mohs: 8-10
		Knoop 30g : 2160 kg/mm^2
		Knoop 100g : 1770 kg/mm^2
	Trisilicon tetranitride (Si_3N_4)	Mohs: 9+
	(α)	Knoop or Vickers: 815-1936kg/mm^2
		Rockwell A: 99

Source: data compiled by J.S. Park from *No. 1 Materials Index*, Peter T.B. Shaffer, Plenum Press, New York, (1964); *Smithells Metals Reference Book*, Eric A. Brandes, ed., in association with Fulmer Research Institute Ltd. 6th ed. London, Butterworths, Boston, (1983); and *Ceramic Source*, American Ceramic Society (1986-1991).

Table 203. HARDNESS OF CERAMICS
(SHEET 4 OF 6)

Class	Ceramic	Hardness
Nitrides (Con't)	Zirconium Mononitride (ZrN)	Mohs: 8+
		Knoop 30g : 1983 kg/mm^2
		Knoop 100g : 1510 kg/mm^2
Oxides	Aluminum Oxide (Al_2O_3) (single crystal)	Mohs : 9
		Knoop 100g : 2000-2050 kg/mm^2
		Vickers 20g : 2600 kg/mm^2
		Vickers 50g : 2720 kg/mm^2
		R45N : 78-90
	Beryllium Oxide (BeO)	Knoop 100g : 1300 kg/mm^2
		R45N : 64-67
	Calcium Oxide (CaO)	Knoop 100g : 560 kg/mm^2
	Dichromium Trioxide (Cr_2O_3)	Knoop or Vickers : 2955 kg/mm^2
	Magnesium Oxide (MgO)	Mohs : 5.5
	Silicon Dioxide (SiO_2)	
	(parallel to optical axis)	Knoop 100g : 710 kg/mm^2
	(normal to optical axis)	Knoop 100g : 790 kg/mm^2
	(parallel to optical axis)	Vickers 500g : 1260 kg/mm^2
	(normal to optical axis)	Vickers 500g : 1103 kg/mm^2
		Vickers 500g : 1120 kg/mm^2
	(1010 face) 10 µm diagonal	Vickers 500g :1120-1230 kg/mm^2
	(1011 face) 10 µm diagonal	Vickers 500g : 1040-1130 kg/mm^2
	(polished 1010 face) 10 µm diagonal	Vickers 500g : 1300 kg/mm^2

Source: data compiled by J.S. Park from *No. 1 Materials Index,* Peter T.B. Shaffer, Plenum Press, New York, (1964); *Smithells Metals Reference Book*, Eric A. Brandes, ed., in association with Fulmer Research Institute Ltd. 6th ed. London, Butterworths, Boston, (1983); and *Ceramic Source*, American Ceramic Society (1986-1991).

Table 203. HARDNESS OF CERAMICS
(SHEET 5 OF 6)

Class	Ceramic	Hardness
Oxides (Con't)	Thorium Dioxide (ThO_2)	Mohs : 6.5 Knoop 100g : 945 kg/mm^2
	Titanium Oxide (TiO_2)	Knoop or Vickers : 713-1121 kg/mm^2
	Uranium Dioxide (UO_2)	Mohs : 6-7 Knoop 100g : 600 kg/mm^2
	Zirconium Oxide (ZrO_2)	Mohs : 6.5 Knoop 100g : 1200 kg/mm^2
	(partially stabilized)	Knoop or Vickers : 1019-1121 kg/mm^2
	(fully stabilized)	Knoop or Vickers : 1019-1529 kg/mm^2
	Cordierite ($2MgO\ 2Al_2O_3\ 5SiO_2$)	Vickers : 835.6 kg/mm^2
	(glass)	Vickers : 672.5 kg/mm^2
	Mullite ($3Al_2O_3\ 2SiO_2$)	Mohs: 7.5 Vickers : 1120 kg/mm^2 R45N: 71
	Sillimanite ($Al_2O_3\ SiO_2$)	Mohs: 6-7
	Zircon ($SiO_2\ ZrO_2$)	Mohs: 7.5
Silicides	Molybdenum Disilicide ($MoSi_2$)	Knoop 100g : 1257 kg/mm^2 Vickers 100g : 1290-1550 kg/mm^2 Micro 50g : 1200 kg/mm^2 Micro 100g : 1290 kg/mm^2

Source: data compiled by J.S. Park from *No. 1 Materials Index,* Peter T.B. Shaffer, Plenum Press, New York, (1964); *Smithells Metals Reference Book,* Eric A. Brandes, ed., in association with Fulmer Research Institute Ltd. 6th ed. London, Butterworths, Boston, (1983); and *Ceramic Source,* American Ceramic Society (1986-1991).

Table 203. HARDNESS OF CERAMICS
(SHEET 6 OF 6)

Class	Ceramic	Hardness
Silicides (Con't)	Tungsten Disilicide (WSi_2)	Knoop 100g : 1090 kg/mm^2 Vickers 100g : 1090 kg/mm^2 Vickers 10g : 1632 kg/mm^2 Micro 50g : 1260 kg/mm^2

Source: data compiled by J.S. Park from *No. 1 Materials Index,* Peter T.B. Shaffer, Plenum Press, New York, (1964); *Smithells Metals Reference Book*, Eric A. Brandes, ed., in association with Fulmer Research Institute Ltd. 6th ed. London, Butterworths, Boston, (1983); and *Ceramic Source*, American Ceramic Society (1986-1991).

Table 204. MICROHARDNESS OF GLASS
(SHEET 1 OF 2)

Class	Glass	Microhardness (Kg/mm^2)
SiO_2	SiO_2 glass	Knoop
		500–679
	SiO_2–Na_2O glass	Vickers
	(25% mol Na_2O)	423±4
	(30% mol Na_2O)	413±3
	(35% mol Na_2O)	414±4
	(40% mol Na_2O)	394±2
	(45% mol Na_2O)	378±2
	SiO_2–B_2O_3 glass	Vickers
	(60% mol B_2O_3)	328–345
	(65% mol B_2O_3)	293–297
	(70% mol B_2O_3)	251–279
	(75% mol B_2O_3)	237–269–345
	(80% mol B_2O_3)	239–271
	(85% mol B_2O_3)	239–267
	(90% mol B_2O_3)	231–257
	(95% mol B_2O_3)	227–253

Source: data compiled by J.S. Park from O. V. Mazurin, M. V. Streltsina and T. P. Shvaiko–Shvaikovskaya, *Handbook of Glass Data*, Part A and Part B, Elsevier, New York, 1983

Table 204. MICROHARDNESS OF GLASS
(SHEET 2 OF 2)

Class	Glass	Microhardness (Kg/mm^2)
B_2O_3	B_2O_3 glass	Vickers 194–205
	B_2O_3–Na_2O glass	Vickers
	(5% mol Na_2O)	276
	(10% mol Na_2O)	292
	(15% mol Na_2O)	297
	(20% mol Na_2O)	380
	(25% mol Na_2O)	460
	(30% mol Na_2O)	503

Source: data compiled by J.S. Park from O. V. Mazurin, M. V. Streltsina and T. P. Shvaiko–Shvaikovskaya, *Handbook of Glass Data*, Part A and Part B, Elsevier, New York, 1983

Table 205. Hardness of Polymers

(Sheet 1 of 7)

Class	Polymer	Hardness, (ASTM D785) (Rockwell)
ABS Resins; Molded, Extruded	Medium impact	R108—115
	High impact	R95—113
	Very high impact	R85—105
	Low temperature impact	R75—95
	Heat resistant	R107—116
Acrylics; Cast, Molded, Extruded	Cast Resin Sheets, Rods:	
	General purpose, type I	M80—90
	General purpose, type II	M96—102
	Moldings:	
	Grades 5, 6, 8	M80—103
	High impact grade	M38—45
Thermoset Carbonate	Allyl diglycol carbonate	M95—M100 (Barcol)
Alkyds; Molded	Putty (encapsulating)	60—70 (Barcol)
	Rope (general purpose)	70—75 (Barcol)
	Granular (high speed molding)	60—70 (Barcol)
	Glass reinforced (heavy duty parts)	70—80 (Barcol)
Cellulose Acetate; Molded, Extruded	ASTM Grade:	
	H4—1	R103—120
	H2—1	R89—112
	MH—1, MH—2	R74—104
	MS—1, MS—2	R54—96
	S2—1	R49—88

Source: data compiled by J.S. Park from Charles T. Lynch, CRC *Handbook of Materials Science, Vol. 3*, CRC Press, Boca Raton, Florida, 1975 and *Engineered Materials Handbook, Vol.2*, Engineering Plastics, ASM International, Metals Park, Ohio, 1988.

Table 205. HARDNESS OF POLYMERS

(SHEET 2 OF 7)

Class	Polymer	Hardness, (ASTM D785) (Rockwell)
Cellulose Acetate Butyrate; Molded, Extruded	ASTM Grade:	
	H4	R114
	MH	R80—100
	S2	R23—42
Cellusose Acetate Propionate; Molded, Extruded	ASTM Grade:	
	1	100—109
	3	92—96
	6	57
	Chlorinated Polymers	
	Chlorinated polyether	R100
	Chlorinated polyvinyl chloride	R118
Polycarbonates	Polycarbonate	M70
	Polycarbonate (40% glass fiber reinforced)	M97
Diallyl Phthalates; Molded	Orlon filled	M108
	Asbestos filled	M107
	Glass fiber filled	M108
Fluorocarbons; Molded,Extruded	Polytrifluoro chloroethylene (PTFCE)	R110—115
	Polytetrafluoroethylene (PTFE)	52D
	Ceramic reinforced (PTFE)	R35—55
	Fluorinated ethylene propylene(FEP)	57—58D
	Polyvinylidene— fluoride (PVDF)	109—110R

Source: data compiled by J.S. Park from Charles T. Lynch, CRC *Handbook of Materials Science, Vol. 3*, CRC Press, Boca Raton, Florida, 1975 and *Engineered Materials Handbook, Vol.2*, Engineering Plastics, ASM International, Metals Park, Ohio, 1988.

Table 205. HARDNESS OF POLYMERS

(SHEET 3 OF 7)

Class	Polymer	Hardness, (ASTM D785) (Rockwell)
Epoxies; Cast, Molded, Reinforced	Standard epoxies (diglycidyl ethers of bisphenol A)	
	Cast rigid	106M
	Cast flexible	50-100M
	Molded	75-80 (Barcol)
	General purpose glass cloth laminate	115—117M
	High strength laminate	70—72 (Barcol)
	Filament wound composite	98-120M
Epoxies—Molded, Extruded	High performance resins	
	(cycloaliphatic diepoxides)	
	Cast, rigid	107—112
	Molded	94—96D
	Glass cloth laminate	75—80
Melamines; Molded	Filler & type	
	Unfilled	E110
	Cellulose electrical	M115—125
Nylons; Molded, Extruded	Type 6	
	General purpose	R118—R120
	Glass fiber (30%) reinforced	R93—121
	Cast	R116
	Flexible copolymers	R72—Rll9
	Type 11	Rl00—R108
	Type 12	R106

Source: data compiled by J.S. Park from Charles T. Lynch, CRC *Handbook of Materials Science, Vol. 3*, CRC Press, Boca Raton, Florida, 1975 and *Engineered Materials Handbook, Vol.2*, Engineering Plastics, ASM International, Metals Park, Ohio, 1988.

Table 205. HARDNESS OF POLYMERS
(SHEET 4 OF 7)

Class	Polymer	Hardness, (ASTM D785) (Rockwell)
Nylons; Molded, Extruded	6/6 Nylon	
	General purpose molding	R118—120, R108
	Glass fiber reinforced	E60—E80
	Glass fiber Molybdenum disulfide filled	M95—100
	General purpose extrusion	R118—108
	6/10 Nylon	
	General purpose	R111
	Glass fiber (30%) reinforced	E40—50
Phenolics; Molded	Type and filler	
	General: woodflour and flock	E85—100
	Shock: paper, flock, or pulp	E85—95
	High shock: chopped fabric or cord	E80—90
	Very high shock: glass fiber	E50—70
	Arc resistant—mineral	M105—115
	Rubber phenolic—woodflour or flock	M40—90
	Rubber phenolic—chopped fabric	M57
	Rubber phenolic—asbestos	M50
ABS–Polycarbonate Alloy	ABS–Polycarbonate Alloy	R118
PVC–Acrylic Alloy	PVC–Acrylic Alloy	
	PVC–acrylic sheet	R105
	PVC–acrylic injection molded	R104
Polymide	Glass reinforced	114E

Source: data compiled by J.S. Park from Charles T. Lynch, CRC *Handbook of Materials Science, Vol. 3,* CRC Press, Boca Raton, Florida, 1975 and *Engineered Materials Handbook, Vol.2,* Engineering Plastics, ASM International, Metals Park, Ohio, 1988.

Table 205. HARDNESS OF POLYMERS
(SHEET 5 OF 7)

Class	Polymer	Hardness, (ASTM D785) (Rockwell)
Polyacetals	Homopolymer:	
	Standard	M94
	20% glass reinforced	M90
	22% TFE reinforced	M78
	Copolymer:	
	Standard	M80
	25% glass reinforced	M79
	High flow	M80
Polyester; Thermoplastic	Injection Moldings:	
	General purpose grade	R117
	Glass reinforced grades	R118—M90
	Glass reinforced self extinguishing	R119
	General purpose grade	R117
	Glass reinforced grade	R117—M85
	Asbestos—filled grade	M85
Polyesters: Thermosets	Cast polyyester	
	Rigid	35—50 (Barcol)
	Flexible	6—40 (Barcol)
	Reinforced polyester moldings	
	High strength (glass fibers)	60—80 (Barcol)
	Heat and chemical resistant (asbestos)	40—70 (Barcol)
	Sheet molding compounds, general purpose	45—60 (Barcol)
Phenylene Oxides	SE—100	R115
	SE—1	R119
	Glass fiber reinforced	L106, L108
Phenylene oxides (Noryl)	Standard	R120
	Glass fiber reinforced	M84

Source: data compiled by J.S. Park from Charles T. Lynch, CRC *Handbook of Materials Science, Vol. 3,* CRC Press, Boca Raton, Florida, 1975 and *Engineered Materials Handbook, Vol.2,* Engineering Plastics, ASM International, Metals Park, Ohio, 1988.

Table 205. HARDNESS OF POLYMERS
(SHEET 6 OF 7)

Class	Polymer	Hardness, (ASTM D785) (Rockwell)
Polyarylsulfone	Polyarylsulfone	M85—110
Polypropylene:	General purpose	R80—R100
	High impact	R28—95
	Asbestos filled	R90—R110
	Glass reinforced	R90—R115
	Flame retardant	R60—R105
Polyphenylene sulfide:	Standard	R120—124
	40% glass reinforced	R123
Polyethylenes; Molded, Extruded	Type I—lower density (0.910—0.925)	
	Melt index 0.3—3.6	C73, D50—52 (Shore)
	Melt index 6—26	C73, D47—53 (Shore)
	Melt index 200	D45 (Shore)
	Type II—medium density (0.926—0.940)	
	Melt index 20	D55 (Shore)
	Melt index l.0—1.9	D55—D56 (Shore)
	Type III—higher density (0.941—0.965)	
	Melt index 0.2—0.9	D68—70 (Shore)
	Melt Melt index 0.l—12.0	D60—70 (Shore)
	Melt index 1.5—15	D68—70 (Shore)
	High molecular weight	D60—65 (Shore)
Olefin Copolymers; Molded	EEA (ethylene ethyl acrylate)	D35 (Shore)
	EVA (ethylene vinyl acetate)	D36 (Shore)
	Ethylene butene	D65 (Shore)
	Propylene—ethylene ionomer	D60 (Shore)

Source: data compiled by J.S. Park from Charles T. Lynch, CRC *Handbook of Materials Science, Vol. 3,* CRC Press, Boca Raton, Florida, 1975 and *Engineered Materials Handbook, Vol.2,* Engineering Plastics, ASM International, Metals Park, Ohio, 1988.

Table 205. HARDNESS OF POLYMERS
(SHEET 7 OF 7)

Class	Polymer	Hardness, (ASTM D785) (Rockwell)
Polystyrenes; Molded	Polystyrenes	
	General purpose	M72
	Medium impact	M47—65
	High impact	M3—43
	Glass fiber -30% reinforced	M85—95
	Styrene acrylonitrile (SAN)	M75—85
	Glass fiber (30%) reinforced SAN	M90—123
Polyvinyl Chloride And Copolymers; Molded, Extruded	Rigid—normal impact	R110—120
	Vinylidene chloride	M50—65
	Polyvinyl Chloride And Copolymers;	(ASTM D676)
	Molded, Extruded:	
	Nonrigid—general	A50—100 (Shore)
	Nonrigid—electrical	A78—100 (Shore)
	Rigid—normal impact	D70—85 (Shore)
	Vinylidene chloride	>A95 (Shore)
Silicones; Molded, Laminated	Fibrous (glass) reinforced silicones	M87
	Granular (silica) reinforced silicones	M71—95
	Woven glass fabric/ silicone laminate	75 (Barcol)
Ureas; Molded	Alpha—cellulose filled (ASTM Type l)	E94—97, M116—120
	Woodflour filled	M116—120

Source: data compiled by J.S. Park from Charles T. Lynch, CRC *Handbook of Materials Science, Vol. 3,* CRC Press, Boca Raton, Florida, 1975 and *Engineered Materials Handbook, Vol.2,* Engineering Plastics, ASM International, Metals Park, Ohio, 1988.

Table 206. HARDNESS OF Si_3N_4 AND Al_2O_3 COMPOSITES

Matrix	Dispersed Phase	Knoop Hardness (GPa)
Si_3N_4+ 6 wt % Y_2O_3	None	13.4 ± 0.3
Si_3N_4+ 6 wt % Y_2O_3	TiC	15.21 ± 0.3
	(Ti, W) C	14.06 ± 0.3
	WC	14.4 ± 0.4
	TaC	12.6 ± 0.2
	HfC	14.1 ± 0.4
	SiC	13.6 ± 0.2
Al_2O_3	TiC	17.2 ± 0.2

Containing 30 Vol % of Metal Carbide Dispersoid (2 μm average particle diameter)

Data from *ASM Engineering Materials Reference Book, Second Edition*, Michael Bauccio, Ed., ASM International, Materials Park, OH, p169,(1994).

Table 207. COEFFICIENT OF STATIC FRICTION FOR POLYMERS

Class	Polymer	Coefficient of Static Friction (Against Self) (Dimensionless)
ABS–Polycarbonate Alloy	ABS–Polycarbonate Alloy	0.2
Polycarbonates	Polycarbonate	0.52
Nylons; Molded, Extruded	Type 6	
	Cast	0.32 (dynamic)
	6/6 Nylon	
	General purpose molding	0.04—0.13
Polyacetals	Homopolymer:	
	Standard	0.1—0.3 (against steel)
	20% glass reinforced	0.1—0.3 (against steel)
	22% TFE reinforced	0.05—0.15 (against steel)
	Copolymer:	
	Standard	0.15 (against steel)
	25% glass reinforced	0.15 (against steel)
	High flow	0.15 (against steel)
Polyester; Thermoplastic		(ASTM D1894)
	Injection Moldings:	
	General purpose grade	0.17
	Glass reinforced grades	0.16
	Glass reinforced self extinguishing	0.16
Polyester; Thermoplastic	Injection Moldings:	
	General purpose grade	0.13 (against steel)
	Glass reinforced grades	0.14 (against steel)
	Glass reinforced self extinguishing	0.14 (against steel)
Phenylene oxides (Noryl)	Standard	0.67
Polyarylsulfone	Polyarylsulfone	0.1—0.3

Source: data compiled by J.S. Park from Charles T. Lynch, CRC *Handbook of Materials Science, Vol. 3*, CRC Press, Boca Raton, Florida, 1975 and *Engineered Materials Handbook, Vol.2*, Engineering Plastics, ASM International, Metals Park, Ohio, 1988.

Table 208. ABRASION RESISTANCE OF POLYMERS

(SHEET 1 OF 2)

Class	Polymer	Abrasion Resistance (Taber, CS—17 wheel, ASTM D1044) (mg / 1000 cycles)
Fluorocarbons; Molded,Extruded	Polytrifluoro chloroethylene (PTFCE)	0.008 (g/cycle)
	Polyvinylidene— fluoride (PVDF)	0.0006—0.0012 (g/cycle)
Polycarbonates	Polycarbonate	10
	Polycarbonate (40% glass fiber reinforced)	40
	Nylons; Molded, Extruded	
	Type 6	
	General purpose	5
	Cast	2.7
Nylons; Molded, Extruded	6/6 Nylon	
	General purpose molding	3—8
	General purpose extrusion	3—5
PVC–Acrylic Alloy	PVC–acrylic sheet	0.073 (CS—10 wheel)
	PVC–acrylic injection molded	0.0058 (CS—10 wheel)
Polymides	Unreinforced	0.08
	Unreinforced 2nd value	0.004
	Glass reinforced	20
Polyacetals	Homopolymer:	
	Standard	14—20
	20% glass reinforced	33
	22% TFE reinforced	9
	Copolymer:	
	Standard	14
	25% glass reinforced	40
	High flow	14

Source: data compiled by J.S. Park from Charles T. Lynch, *CRC Handbook of Materials Science, Vol. 3,* CRC Press, Boca Raton, Florida, 1975 and *Engineered Materials Handbook, Vol.2, Engineering Plastics,* ASM International, Metals Park, Ohio, 1988.

Table 208. ABRASION RESISTANCE OF POLYMERS
(SHEET 2 OF 2)

Class	Polymer	Abrasion Resistance (Taber, CS—17 wheel, ASTM D1044) (mg / 1000 cycles)
Polyester; Thermoplastic	Injection Moldings:	
	General purpose grade	6.5
	Glass reinforced grades	9—50
	Glass reinforced self extinguishing	11
Phenylene Oxides	SE—100	100
	SE—1	20
	Glass fiber reinforced	35
Phenylene oxides (Noryl)	Standard	20
Polyarylsulfone	Polyarylsulfone	40
Polystyrenes; Molded	Glass fiber -30% reinforced	164

Source: data compiled by J.S. Park from Charles T. Lynch, *CRC Handbook of Materials Science, Vol. 3*, CRC Press, Boca Raton, Florida, 1975 and *Engineered Materials Handbook, Vol.2, Engineering Plastics*, ASM International, Metals Park, Ohio, 1988.

Table 209. FATIGUE STRENGTH OF WROUGHT ALUMINUM ALLOYS (SHEET 1 OF 4)

Alloy AA No.	Temper	Fatigue Strength (MPa)
1060	0	21
	H12	28
	H14	34
	H16	45
	H18	45
1100	0	34
	H12	41
	H14	48
	H16	62
	H18	62
1350	H19	48
2011	T3	125
	T8	125
2014	0	90
	T4	140
	T6	125
2024	0	90
	T3	140
	T4, T351	140
	T361	125
2036	T4	125
2048		220
2219	T62	105
	T81, T851	105
	T87	105
2618	All	125
3003	0	48
Alclad	H12	55

Source: data from *ASM Metals Reference Book, Second Edition*, American Society for Metals, Metals Park, Ohio 44073, (1984).

Table 209. FATIGUE STRENGTH OF WROUGHT ALUMINUM ALLOYS (SHEET 2 OF 4)

Alloy AA No.	Temper	Fatigue Strength (MPa)
3003	H14	62
	H16	69
	H18	69
3004	0	97
Alclad	H32	105
3004	H34	105
	H36	110
	H38	110
4032	T6	110
5050	0	83
	H32	90
	H34	90
	H36	97
	H38	97
5052	0	110
	H32	115
	H34	125
	H36	130
	H38	140
5056	0	140
	H18	150
	H38	150
5083	H321	160
5154	0	115
	H32	125
	H34	130
	H36	140
	H38	145
	H112	115
5182	0	140
5254	0	115

Source: data from *ASM Metals Reference Book, Second Edition*, American Society for Metals, Metals Park, Ohio 44073, (1984).

Table 209. FATIGUE STRENGTH OF WROUGHT ALUMINUM ALLOYS (SHEET 3 OF 4)

Alloy AA No.	Temper	Fatigue Strength (MPa)
5254	H32	125
	H34	130
	H36	140
	H38	145
	H112	115
5652	0	110
	H32	115
	H34	125
	H36	130
	H38	140
6005	T1	97
	T5	97
6009	T4	115
6010	T4	115
6061	0	62
	T4, T451	97
	T6, T651	97
6063	0	55
	T1	62
	T5	69
	T6	69
6066	T6, T651	110
6070	0	62
	T4	90
	T6	97
6205	T5	105
6262	T9	90
6351	T6	90

Source: data from *ASM Metals Reference Book, Second Edition*, American Society for Metals, Metals Park, Ohio 44073, (1984).

Table 209. FATIGUE STRENGTH OF WROUGHT ALUMINUM ALLOYS (SHEET 4 OF 4)

Alloy AA No.	Temper	Fatigue Strength (MPa)
6463	T1	69
	T5	69
	T6	69
7005	T53	140
	T6,T63,T6351	125
7049	T73	295
7050	T736	240
7075	T6,T651	160
7175	T66	160
	T736	160
7475	T7351	220

Source: data from *ASM Metals Reference Book, Second Edition*, American Society for Metals, Metals Park, Ohio 44073, (1984).

Table 210. REVERSED BENDING FATIGUE LIMIT OF GRAY CAST IRON BARS

ASTM Class	Reversed Bending Fatigue Limit (MPa)
20	69
25	79
30	97
35	110
40	128
50	148
60	169

Source: data from ASM *Metals Reference Book, Second Edition*, American Society for Metals, Metals Park, Ohio 44073, p166-167, (1984).

Table 211. IMPACT ENERGY OF TOOL STEELS
(SHEET 1 OF 2)

Type	Condition	Impact Energy (J)
L2	Oil quenched from 855 °C and single tempered at:	
	205 °C	28(a)
	315 °C	19(a)
	425 °C	26(a)
	540 °C	39(a)
	650 °C	125(a)
L6	Annealed	93 HRB
	Oil quenched from 845 °C and single tempered at:	
	315 °C	12(a)
	425 °C	18(a)
	540 °C	23(a)
	650 °C	81(a)
S1	Oil quenched from 930 °C and single tempered at:	
	205 °C	249(b)
	315 °C	233(b)
	425 °C	203(b)
	540 °C	230(b)
	650 °C	
S5	Oil quenched from 870 °C and single tempered at:	
	205 °C	206(b)
	315 °C	232(b)
	425 °C	243(b)
	540 °C	188(b)
	650 °C	

(a) Charpy V-notch.

(b) Charpy unnotched.

Source: Data from *ASM Metals Reference Book, Second Edition*, American Society for Metals, Metals Park, Ohio 44073, p241, (1984).

Table 211. IMPACT ENERGY OF TOOL STEELS

(SHEET 2 OF 2)

Type	Condition	Impact Energy (J)
S7	Fan cooled from 940 °C and single tempered at:	
	205 °C	244(b)
	315 °C	309(b)
	425 °C	243(b)
	540 °C	324(b)
	650 °C	358(b)

(a) Charpy V-notch.

(b) Charpy unnotched.

Source: Data from *ASM Metals Reference Book, Second Edition*, American Society for Metals, Metals Park, Ohio 44073, p241, (1984).

Table 212. IMPACT STRENGTH OF WROUGHT TITANIUM ALLOYS AT ROOM TEMPERATURE

Class	Alloy	Condition	Charpy Impact Strength (J)
Commercially Pure	99.2Ti	Annealed	43
	99.1Ti	Annealed	38
	99.0 Ti	Annealed	20
	99.2Ti-0.2Pd	Annealed	43
Alpha Alloys	Ti-5Al-2.5Sn	Annealed	26
	Ti-5Al-2.5Sn (low O_2)	Annealed	27
Near alpha alloys	Ti-8Al-1Mo-1V	Duplex Annealed	32
	Ti-6Al-2Nb-1Ta-1Mo	As rolled 2.5 cm (1 in.) plate	31
Alpha-Beta Alloys	Ti-6Al-4V	Annealed	19
	Ti-6Al-4V(low O_2)	Annealed	24
	Ti-6Al-6V-2Sn	Annealed	18
	Ti-7Al-4Mo	Solution + age	18
Beta Alloys	Ti-13V-1Cr-3Al	Solution + age	11
	Ti-3Al-8V-6Cr-4Mo-4Zr	Solution + age	10

Data from *ASM Metals Reference Book, Third Edition*, Michael Bauccio, Ed., ASM International, Materials Park, OH, p512, (1993).

Table 213. IMPACT STRENGTH OF POLYMERS
(SHEET 1 OF 7)

Class	Polymer	Impact Strength (Izod notched, ASTM D256) (ft—lb / in.)
ABS Resins; Molded, Extruded	Medium impact	2.0—4.0
	High impact	3.0—5.0
	Very high impact	5.0—7.5
	Low temperature impact	6—10
	Heat resistant	2.0—4.0
Acrylics; Cast, Molded, Extruded	Cast Resin Sheets, Rods:	
	General purpose, type I	0.4
	General purpose, type II	0.4
	Moldings:	
	Grades 5, 6, 8	0.2—0.4
	High impact grade	0.8—2.3
Thermoset Carbonate	Allyl diglycol carbonate	0.2—0.4
Alkyds; Molded	Putty (encapsulating)	0.25—0.35
	Rope (general purpose)	2.2
	Granular (high speed molding)	0.30—0.35
	Glass reinforced (heavy duty parts)	8—12
Cellulose Acetate Butyrate; Molded, Extruded	ASTM Grade:	
	H4	3
	MH	4.4—6.9
	S2	7.5—10.0

To convert **ft—lb / in.** to **N•m/m**, multiply by **53.38**

Source: data compiled by J.S. Park from Charles T. Lynch, *CRC Handbook of Materials Science, Vol. 3,* CRC Press, Boca Raton, Florida, 1975 and *Engineered Materials Handbook, Vol.2,* Engineering Plastics, ASM International, Metals Park, Ohio, 1988.

Table 213. IMPACT STRENGTH OF POLYMERS
(SHEET 2 OF 7)

Class	Polymer	Impact Strength (Izod notched, ASTM D256) (ft—lb / in.)
Cellusose Acetate Propionate; Molded, Extruded	ASTM Grade:	
	1	1.7—2.7
	3	3.5—5.6
	6	9.4
Chlorinated Polymers	Chlorinated polyether	0.4 (D758)
	Chlorinated polyvinyl chloride	6.3
Polycarbonate	Polycarbonate	12—16
Diallyl Phthalates; Molded	Orlon filled	0.5—1.2
	Dacron filled	1.7—5.0
	Asbestos filled	0.30—0.50
	Glass fiber filled	0.5—15.0
Fluorocarbons; Molded,Extruded	Polytrifluoro chloroethylene (PTFCE)	3.50—3.62
	Polytetrafluoroethylene (PTFE)	2.0—4.0
	Fluorinated ethylene propylene(FEP)	No break
	Polyvinylidene— fluoride (PVDF)	3.0—10.3
Epoxies; Cast, Molded, Reinforced	Standard epoxies (diglycidyl ethers of bisphenol A)	
	Cast rigid	0.2—0.5
	Cast flexible	0.3—0.2

To convert **ft—lb / in.** to **N•m/m**, multiply by **53.38**

Source: data compiled by J.S. Park from Charles T. Lynch, *CRC Handbook of Materials Science, Vol. 3,* CRC Press, Boca Raton, Florida, 1975 and *Engineered Materials Handbook, Vol.2*, Engineering Plastics, ASM International, Metals Park, Ohio, 1988.

Table 213. IMPACT STRENGTH OF POLYMERS
(SHEET 3 OF 7)

Class	Polymer	Impact Strength (Izod notched, ASTM D256) (ft—lb / in.)
Epoxies; Cast, Molded, Reinforced (Con't)	Molded	0.4—0.5
	General purpose glass cloth laminate	12—15
	High strength laminate	60—61
Epoxies—Molded, Extruded	High performance resins	
	(cycloaliphatic diepoxides)	
	Cast, rigid	0.5
	Molded	0.3—0.5
Epoxy novolacs	Cast, rigid	13—17
Melamines; Molded	Filler & type	
	Cellulose electrical	0.27—0.36
	Glass fiber	0.5—12.0
	Alpha cellulose and mineral	0.30—0.35, 0.2(mineral)
Nylons; Molded, Extruded	Type 6	
	General purpose	0.6—1.2
	Glass fiber (30%) reinforced	2.2—3.4
	Cast	1.2
	Flexible copolymers	1.5—19
	Type 8	>16
	Type 11	3.3—3.6
	Type 12	1.2—4.2
	6/6 Nylon	(ASTM D638)
	General purpose molding	0.55—1.0,2.0
	Glass fiber reinforced	2.5—3.4
	General purpose extrusion	1.3

To convert **ft—lb / in.** to **N•m/m**, multiply by **53.38**

Source: data compiled by J.S. Park from Charles T. Lynch, *CRC Handbook of Materials Science, Vol. 3,* CRC Press, Boca Raton, Florida, 1975 and *Engineered Materials Handbook, Vol.2,* Engineering Plastics, ASM International, Metals Park, Ohio, 1988.

Table 213. IMPACT STRENGTH OF POLYMERS
(SHEET 4 OF 7)

Class	Polymer	Impact Strength (Izod notched, ASTM D256) (ft—lb / in.)
Nylons; Molded, Extruded (Con't)	6/10 Nylon	
	General purpose	0.6–1.6
	Glass fiber (30%) reinforced	3.4
Phenolics; Molded	Type and filler	
	General: woodflour and flock	0.24—0.50
	Shock: paper, flock, or pulp	0.4—1.0
	High shock: chopped fabric or cord	0.6—8.0
	Very high shock: glass fiber	10—33
	Arc resistant—mineral	0.30—0.45
	Rubber phenolic—woodflour or flock	0.34—1.0
	Rubber phenolic—chopped fabric	2.0—2.3
	Rubber phenolic—asbestos	0.3—0.4
ABS–Polycarbonate Alloy	ABS–Polycarbonate Alloy	10 (ASTM D638)
PVC–Acrylic Alloy	PVC–acrylic sheet	15
	PVC–acrylic injection molded	15
Polymides	Unreinforced	0.5
	Unreinforced 2nd value	0.5
	Glass reinforced	17
Polyacetals		(ASTM D638)
	Homopolymer:	
	Standard	1.4
	20% glass reinforced	0.8
	22% TFE reinforced	0.7

To convert **ft—lb / in.** to **N•m/m**, multiply by **53.38**

Source: data compiled by J.S. Park from Charles T. Lynch, *CRC Handbook of Materials Science, Vol. 3*, CRC Press, Boca Raton, Florida, 1975 and *Engineered Materials Handbook, Vol.2*, Engineering Plastics, ASM International, Metals Park, Ohio, 1988.

Table 213. IMPACT STRENGTH OF POLYMERS

(SHEET 5 OF 7)

Class	Polymer	Impact Strength (Izod notched, ASTM D256) (ft—lb / in.)
Polyacetals (Con't)	Copolymer:	
	Standard	1.3
	25% glass reinforced	1.8
	High flow	1
Polyester; Thermoplastic	Injection Moldings:	
	General purpose grade	1.0—1.2
	Glass reinforced grades	1.3—2.2
	Glass reinforced self extinguishing	1.8
	General purpose grade	1
	Glass reinforced grade	1
	Asbestos—filled grade	0.5
Polyesters: Thermosets	Cast polyyester	
	Rigid	0.18—0.40
	Flexible	4
	Reinforced polyester moldings	
	High strength (glass fibers)	1—10
	Heat and chemical resistsnt (asbestos)	0.45—1.0
	Sheet molding compounds, general purpose	5—15
Phenylene Oxides		(ASTM D638)
	SE—100	5
	SE—1	5
	Glass fiber reinforced	2.3
Phenylene oxides (Noryl)	Standard	1.2—1.3
	Glass fiber reinforced	1.8—2.0

To convert **ft—lb / in.** to **N•m/m**, multiply by **53.38**

Source: data compiled by J.S. Park from Charles T. Lynch, *CRC Handbook of Materials Science, Vol. 3,* CRC Press, Boca Raton, Florida, 1975 and *Engineered Materials Handbook, Vol.2*, Engineering Plastics, ASM International, Metals Park, Ohio, 1988.

Table 213. IMPACT STRENGTH OF POLYMERS

(SHEET 6 OF 7)

Class	Polymer	Impact Strength (Izod notched, ASTM D256) (ft—lb / in.)
Polyarylsulfone	Polyarylsulfone	1.6—5.0
Polypropylene	General purpose	0.4—2.2
	High impact	1.5—12
	Asbestos filled	0.5—1.5
	Glass reinforced	0.5—2
	Flame retardant	2.2
Polyphenylene sulfide	Standard	0.3
	40% glass reinforced	1.09
Polyethylenes; Molded, Extruded	Type III—higher density (0.941—0.965)	
	Melt index 0.2—0.9	4.0—14
	Melt Melt index 0.l—12.0	0.4—6.0
	Melt index 1.5—15	1.2—2.5
	High molecular weight	>20
Olefin Copolymers; Molded	Ethylene butene	0.4
	Propylene—ethylene	1.1
	Ionomer	9—14
	Polyallomer	1.5
Polystyrenes; Molded	Polystyrenes	(ASTM D638)
	General purpose	0.2—0.4
	Medium impact	0.5—1.2
	High impact	0.8—1.8
	Glass fiber —30% reinforced	2.5
	Styrene acrylonitrile (SAN)	0.29—0.54
	Glass fiber (30%) reinforced SAN	1.35—3.0

To convert **ft—lb / in.** to **N•m/m**, multiply by **53.38**

Source: data compiled by J.S. Park from Charles T. Lynch, *CRC Handbook of Materials Science, Vol. 3,* CRC Press, Boca Raton, Florida, 1975 and *Engineered Materials Handbook, Vol.2*, Engineering Plastics, ASM International, Metals Park, Ohio, 1988.

Table 213. IMPACT STRENGTH OF POLYMERS
(SHEET 7 OF 7)

Class	Polymer	Impact Strength (Izod notched, ASTM D256) (ft—lb / in.)
Polyvinyl Chloride And Copolymers; Molded, Extruded	Nonrigid—general	Variable
	Nonrigid—electrical	Variable
	Rigid—normal impact	0.5—10
	Vinylidene chloride	2—8
Silicones; Molded, Laminated	Fibrous (glass) reinforced silicones	10
	Granular (silica) reinforced silicones	0.34
	Woven glass fabric/ silicone laminate	10—25
Ureas; Molded	Alpha—cellulose filled (ASTM Type l)	0.20—0.35
	Cellulose filled (ASTM Type 2)	0.20—0.275
	Woodflour filled	0.25—0.35

To convert **ft—lb / in.** to **N•m/m**, multiply by **53.38**

Source: data compiled by J.S. Park from Charles T. Lynch, *CRC Handbook of Materials Science, Vol. 3,* CRC Press, Boca Raton, Florida, 1975 and *Engineered Materials Handbook, Vol.2*, Engineering Plastics, ASM International, Metals Park, Ohio, 1988.

Table 214. IMPACT STRENGTH OF FIBERGLASS REINFORCED PLASTICS

Class	Material	Glass fiber content (wt%)	Izod Impact strength (ft • lb/in. of notch)
Glass fiber reinforced thermosets	Sheet molding compound (SMC)	15 to 30	8 to 22
	Bulk molding compound(BMC)	15 to 35	2 to 10
	Preform/mat(compression molded)	25 to 50	10 to 20
	Cold press molding–polyester	20 to 30	9 to 12
	Spray–up–polyester	30 to 50	4 to 12
	Filament wound–epoxy	30 to 80	40 to 60
	Rod stock–polyester	40 to 80	45 to 60
	Molding compound–phenolic	5 to 25	1 to 8
Glass–fiber–reinforced thermoplastics	Acetal	20 to 40	0.8 to 2.8
	Nylon	6 to 60	0.8 to 4.5
	Polycarbonate	20 to 40	1.5 to 3.5
	Polyethylene	10 to 40	1.2 to 4.0
	Polypropylene	20 to 40	1 to 4
	Polystyrene	20 to 35	0.4 to 4.5
	Polysulfone	20 to 40	1.3 to 2.5
	ABS(acrylonitrile butadiene styrene)	20 to 40	1 to 2.4
	PVC (polyvinyl chloride)	15 to 35	0.8 to 1.6
	Polyphenylene oxide(modified)	20 to 40	1.6 to 2.2
	SAN (styrene acrylonitrile)	20 to 40	0.4 to 2.4
	Thermoplastic polyester	20 to 35	1.0 to 2.7

To convert (ft • lb/in. of notch) to (J/cm of notch), multiply by 0.534

Data from *ASM Engineering Materials Reference Book, Second Edition,* Michael Bauccio, Ed., ASM International, Materials Park, OH, p106, (1994).

Table 215. IMPACT STRENGTH OF CARBON- AND GLASS-REINFORCED ENGINEERING THERMOPLASTICS (SHEET 1 OF 2)

Class	Resin Type	Composition	Impact Strength, Notched/Unnotched (J/cm)
Amorphous	Acrylonitrile-butadiene-styrene(ABS)	30% glass fiber 30% carbon fiber	0.75/3.5 0.59/2.4
	Nylon	30% glass fiber 30% carbon fiber	0.64/3.7 0.64/4.3
	Polycarbonate	30% glass fiber 30% carbon fiber	2.0/9.34 0.96/5.34
	Polyetherimide	30% glass fiber 30% carbon fiber	0.75/5.60 0.75/6.67
	Polyphenylene oxide (PPO)	30% glass fiber 30% carbon fiber	1.2/5.1 0.53/3.0
	Polysulfone	30% glass fiber 30% carbon fiber	0.96/7.5 0.64/3.5
	Styrene-maleic-anhydride (SMA)	30% glass fiber	0.59/2.4
	Thermoplastic polyurethane	30% glass fiber	5.1/15

Data from *ASM Engineering Materials Reference Book, Second Edition*, Michael Bauccio, Ed., ASM International, Materials Park, OH, p111–112, (1994).

Table 215. Impact Strength of Carbon- and Glass-Reinforced Engineering Thermoplastics (Sheet 2 of 2)

Class	Resin Type	Composition	Impact Strength, Notched/Unnotched (J/cm)
Crystalline	Acetal	30% glass fiber	0.96/4.8
		20% carbon fiber	0.53/1.6
	Nylon 66	30% glass fiber	1.5/11
		30% carbon fiber	0.80/6.4
	Polybutylene telphthalate (PBT)	30% glass fiber	1.4/9.1
		30% carbon fiber	0.64/3.5
	Polythylene terephthalate (PET)	30% glass fiber	1.0/—
	Polyphenylene sulfide (PPS)	30% glass fiber	0.75/4.5
		30% carbon fiber	0.59/2.9

Data from *ASM Engineering Materials Reference Book, Second Edition*, Michael Bauccio, Ed., ASM International, Materials Park, OH, p111–112, (1994).

Table 216. Fracture Toughness of Si_3N_4 and Al_2O_3 Composites

Matrix	Dispersed Phase	Fracture Toughness (K_{Ic}), (MPa $\sqrt{m}$)
Si_3N_4+ 6 wt % Y_2O_3	None	4.8 ± 0.3
Si_3N_4+ 6 wt % Y_2O_3	TiC	4.4 ± 0.5
	(Ti, W) C	3.5 ± 0.3
	WC	5.2 ± 0.4
	TaC	4.6 ± 0.4
	HfC	3.6 ± 0.2
	SiC	3.65 ± 0.5
Al_2O_3	TiC	3.2 ± 0.4

Containing 30 Vol % of Metal Carbide Dispersoid (2 μm average particle diameter)

Data from *ASM Engineering Materials Reference Book, Second Edition*, Michael Bauccio, Ed., ASM International, Materials Park, OH, p169,(1994).

Table 217. TENSILE MODULUS OF GRAY CAST IRONS

ASTM Class	Tensile Modulus (GPa)
20	66 to 97
25	79 to 102
30	90 to 113
35	100 to 119
40	110 to 138
50	130 to 157
60	141 to 162

Source: data from *ASM Metals Reference Book, Second Edition,* American Society for Metals, Metals Park, Ohio 44073, p166-167, (1984).

Table 218. TENSION MODULUS OF TREATED DUCTILE IRONS

Treatment	Tension Modulus (MPa)
60-40-18	169
65-45-12	168
80-55-06	168
120 90-02	164

Source: data from *ASM Metals Reference Book, Second Edition,* American Society for Metals, Metals Park, Ohio 44073, p169-170, (1984).

Table 219. TENSILE MODULUS OF FIBERGLASS REINFORCED PLASTICS

Class	Material	Glass fiber content (wt%)	Tensile modulus (10^5 psi)
Glass fiber reinforced thermosets	Sheet molding compound (SMC)	15 to 30	16 to 25
	Bulk molding compound(BMC)	15 to 35	16 to 25
	Preform/mat(compression molded)	25 to 50	9 to 20
	Spray–up–polyester	30 to 50	8 to l8
	Filament wound–epoxy	30 to 80	40 to 90
	Rod stock–polyester	40 to 80	40 to 60
	Molding compound–phenolic	5 to 25	26 to 29
Glass–fiber–reinforced thermoplastics	Acetal	20 to 40	8 to 15
	Nylon	6 to 60	2 to 20
	Polycarbonate	20 to 40	7.5 to 17
	Polyethylene	10 to 40	4 to 9
	Polypropylene	20 to 40	4.5 to 9
	Polystyrene	20 to 35	8.4 to 12.1
	Polysulfone	20 to 40	15
	ABS(acrylonitrile butadiene styrene)	20 to 40	6 to 10
	PVC (polyvinyl chloride)	15 to 35	10 to 18
	Polyphenylene oxide(modified)	20 to 40	9.5 to 15
	SAN (styrene acrylonitrile)	20 to 40	9 to 18.5
	Thermoplastic polyester	20 to 35	13 to 15.5

To convert from **psi** to **MPa**, multiply by **145**.

Data from *ASM Engineering Materials Reference Book, Second Edition*, Michael Bauccio, Ed., ASM International, Materials Park, OH, p106, (1994).

Table 220. TENSILE MODULUS OF GRAPHITE/ALUMINUM COMPOSITES

Composite	Fiber loading (vol %)	Wire diameter (mm)	Tensile Modulus (GPa)
VS0054/201 Al	48 to 52	0.64 (2-strand)	345
GY70SE/201 Al	37 to 38	0.71(8-strand)	207

Data from *ASM Engineering Materials Reference Book, Second Edition*, Michael Bauccio, Ed., ASM International, Materials Park, OH, p148,(1994).

Table 221. TENSILE MODULUS OF INVESTMENT CAST SILICON CARBIDE SCS–AL

Fiber orientation	Fiber vol (%)	Tensile Modulus (GPa)	Range of Measurement (%)
$0°_3/90°_6/0°_3$	33	122.0	107
$90°_3/0°_6/90°_3$	33	124.8	110
0°	34	172.4	100

Data from *ASM Engineering Materials Reference Book, Second Edition*, Michael Bauccio, Ed., ASM International, Materials Park, OH, p149,(1994).

Table 222. Tensile Modulus of Silicon Carbide SCS–2–Al

Fiber orientation	No. of plies	Tensile Modulus (GPa)
0°	6, 8, 12	204.1
90°	6, 12,40	118.0
$[0°/90°/0°/90°]_s$	8	136.5
$[0_2\ °99°20°]_s$	8	180.0
$[90_2/0°/90°]_s$	8	96.5
± 45°	8, 12, 40	94.5
$[0°±45°/0°]_{s+2s}$	8, 16	146.2
$[0°±45°/90°]_s$	8	127.0

Data from *ASM Engineering Materials Reference Book, Second Edition*, Michael Bauccio, Ed., ASM International, Materials Park, OH, p149,(1994).

Table 223. YOUNG'S MODULUS OF CERAMICS
(SHEET 1 OF 7)

Class	Ceramic	Young's Modulus (psi)	Temperature
Borides	Chromium Diboride (CrB_2)	$30.6x10^6$	
	Tantalum Diboride (TaB_2)	$37 x10^6$	
	Titanium Diboride (TiB_2)	$53.2x10^6$	
	(6.0 μm grain size, ρ=4.46g/cm^3)	$81.6x10^6$	
	(3.5 μm grain size, ρ=4.37g/cm^3, 0.8wt% Ni)	$75.0x10^6$	
	(6.0 μm grain size, ρ=4.56g/cm^3, 0.16wt% Ni)	$77.9x10^6$	
	(12.0 μm grain size, ρ=4.66g/cm^3, 9.6wt% Ni)	$6.29x10^6$	
	Zirconium Diboride (ZrB_2)	$49.8\text{-}63.8x10^6$	
	(22.4% density,foam)	$3.305x10^6$	
Carbides	Boron Carbide (B_4C)	$42\text{-}65.2x10^6$	room temp.
	Hafnium Monocarbide (HfC) (ρ = 11.94 g/cm^3)	$61.55x10^6$	room temp.
	Silicon Carbide (SiC)		
	(pressureless sintered)	$43.9x10^6$	room temp.
	(hot pressed)	$63.8x10^6$	room temp.
	(self bonded)	$59.5x10^6$	room temp.
	(cubic, CVD)	$60.2\text{-}63.9x10^6$	room temp.

To convert from **psi** to **MPa**, multiply by **145**.

Source: data compiled by J.S. Park from No. *1 Materials Index*, Peter T.B. Shaffer, Plenum Press, New York, (1964); *Smithells Metals Reference Book*, Eric A. Brandes, ed., in association with Fulmer Research Institute Ltd. 6th ed. London, Butterworths, Boston, (1983); and *Ceramic Source*, American Ceramic Society (1986-1991)

Table 223. YOUNG'S MODULUS OF CERAMICS
(SHEET 2 OF 7)

Class	Ceramic	Young's Modulus (psi)	Temperature
Carbides (Con't)	(ρ = 3.128 g/cm^3)	58.2x10^6	room temp.
	(ρ = 3.120 g/cm^3)	59.52x10^6	room temp.
	(hot pressed)	62.4-65.3x10^6	20°C
	(sintered)	54.38-60.9x10^6	20°C
	(reaction sintered)	50.75-54.38x10^6	20°C
		55x10^6	400°C
		53x10^6	800°C
		51x10^6	1200°C
	(hot pressed)	55.1x10^6	1400°C
	(sintered)	43.5-58.0x10^6	1400°C
	(reaction sintered)	29-46.4x10^6	1400°C
	Tantalum Monocarbide (TaC)	41.3-91.3x10^6	room temp.
	Titanium Monocarbide (TiC)	63.715x10^6	room temp.
		45-55x10^6	1000°C
	Trichromium Dicarbide (Cr_3C_2)	54.1x10^6	
	Tungsten Monocarbide (WC)	96.91-103.5x10^6	room temp.
	Zirconium Monocarbide (ZrC)	28.3-69.6x10^6	room temp.

To convert from **psi** to **MPa**, multiply by **145.**

Source: data compiled by J.S. Park from No. *1 Materials Index*, Peter T.B. Shaffer, Plenum Press, New York, (1964); *Smithells Metals Reference Book*, Eric A. Brandes, ed., in association with Fulmer Research Institute Ltd. 6th ed. London, Butterworths, Boston, (1983); and *Ceramic Source*, American Ceramic Society (1986-1991)

Table 223. YOUNG'S MODULUS OF CERAMICS
(SHEET 3 OF 7)

Class	Ceramic	Young's Modulus (psi)	Temperature
Nitrides	Aluminum Nitride (AlN)	$50x10^6$	25°C
		$46x10^6$	1000°C
		$40x10^6$	1400°C
	Boron Nitride (BN)		
	parallel to c axis	$4.91x10^6$	23°C
		$3.47x10^6$	300°C
		$0.51x10^6$	700°C
	parallel to a axis	$12.46x10^6$	23°C
		$8.79x10^6$	300°C
		$1.54x10^6$	700°C
		$1.65x10^6$	1000°C
	Titanium Mononitride (TiN)	$11.47\text{-}36.3x10^6$	
	Trisilicon tetranitride (Si_3N_4)		
	(hot pressed)	$36.25\text{-}47.13x10^6$	20°C
	(sintered)	$28.28\text{-}45.68x10^6$	20°C
	(reaction sintered)	$14.5\text{-}31.9x10^6$	20°C
	(hot pressed)	$25.38\text{-}36.25x10^6$	1400°C
	(reaction sintered)	$17.4\text{-}29.0x10^6$	1400°C
Oxides	Aluminum Oxide (Al_2O_3)	$50\text{-}59.3x10^6$	room temp.
		$50\text{-}57.275\ x10^6$	500°C
		$51.2\ x10^6$	800°C
		$45.5\text{-}50\ x10^6$	1000°C

To convert from **psi** to **MPa**, multiply by **145**.

Source: data compiled by J.S. Park from No. *1 Materials Index*, Peter T.B. Shaffer, Plenum Press, New York, (1964); *Smithells Metals Reference Book*, Eric A. Brandes, ed., in association with Fulmer Research Institute Ltd. 6th ed. London, Butterworths, Boston, (1983); and *Ceramic Source*, American Ceramic Society (1986-1991)

Table 223. YOUNG'S MODULUS OF CERAMICS
(SHEET 4 OF 7)

Class	Ceramic	Young's Modulus (psi)	Temperature
Oxides (Con't)	Aluminum Oxide (Al_2O_3) (Con't)	39.8-53.65 x10^6	1200°C
		32 x10^6	1250°C
		32.7 x10^6	1400°C
		25.6 x10^6	1500°C
	Beryllium Oxide (BeO)	42.8-45.5x10^6	room temp.
		40 x10^6	800°C
		33 x10^6	1000°C
		20 x10^6	1145°C
	Cerium Dioxide (CeO_2)	24.9x10^6	
	Dichromium Trioxide (Cr_2O_3)	>14.9x10^6	
	Hafnium Dioxide (HfO_2)	8.2x10^6	
	Magnesium Oxide (MgO)	30.5-36.3x10^6	room temp.
		29.5 x10^6	600°C
		21 x10^6	1000°C
		10 x10^6	1200°C
		4 x10^6	1300°C
	($\rho = 3.506$ g/cm^3)	42.74x10^6	room temp.

To convert from **psi** to **MPa**, multiply by **145**.

Source: data compiled by J.S. Park from No. *1 Materials Index*, Peter T.B. Shaffer, Plenum Press, New York, (1964); *Smithells Metals Reference Book*, Eric A. Brandes, ed., in association with Fulmer Research Institute Ltd. 6th ed. London, Butterworths, Boston, (1983); and *Ceramic Source*, American Ceramic Society (1986-1991)

Table 223. YOUNG'S MODULUS OF CERAMICS
(SHEET 5 OF 7)

Class	Ceramic	Young's Modulus (psi)	Temperature
Oxides (Con't)	Thorium Dioxide (ThO_2)	$17.9\text{-}34.87x10^6$	room temp.
		$18\text{-}18.5x10^6$	800°C
		$17.1x10^6$	1000°C
		$12.8x10^6$	1200°C
	Titanium Oxide (TiO_2)	$41x10^6$	
	Uranium Dioxide (UO_2)	$21x10^6$	0-1000°C
		$25x10^6$	20°C
	(ρ=10.37 g/cm^3)	$27.98x10^6$	room temp.
	Zirconium Oxide (ZrO_2)		
	(partially stabilized)	$29.7x10^6$	room temp.
	(fully stabilized)	$14.1\text{-}30.0x10^6$	room temp.
	(plasma sprayed)	$6.96x10^6$	room temp.
		$24.8\text{-}27x10^6$	room temp.
		$36x10^6$	20°C
		$2x10^6$	500°C
		$18.9x10^6$	800°C
		$18.5\text{-}25x10^6$	1000°C
		$3.05x10^6$	1100°C
		$17.1\text{-}18.0x10^6$	1200°C
		$14.2x10^6$	1400°C
		$12.8x10^6$	1500°C

To convert from **psi** to **MPa**, multiply by **145**.

Source: data compiled by J.S. Park from No. *1 Materials Index*, Peter T.B. Shaffer, Plenum Press, New York, (1964); *Smithells Metals Reference Book*, Eric A. Brandes, ed., in association with Fulmer Research Institute Ltd. 6th ed. London, Butterworths, Boston, (1983); and *Ceramic Source*, American Ceramic Society (1986-1991)

Table 223. YOUNG'S MODULUS OF CERAMICS
(SHEET 6 OF 7)

Class	Ceramic	Young's Modulus (psi)	Temperature
Oxides (Con't)	Uranium Dioxide (UO_2) (Con't)		
	(stabilized, ρ=5.634 g/cm^3)	19.96x10^6	room temp.
	Cordierite ($2MgO\ 2Al_2O_3\ 5SiO_2$)	20.16x10^6	
	(glass)	13.92x10^6	
	Mullite ($3Al_2O_3\ 2SiO_2$)		
	(ρ=2.779 g/cm^3)	20.75x10^6	room temp.
	(ρ=2.77 g/cm^3)	18.42x10^6	25°C
	(ρ=2.77 g/cm^3)	18.89x10^6	400°C
	(ρ=2.77 g/cm^3)	14.79x10^6	800°C
	(ρ=2.77 g/cm^3)	4.00x10^6	1200°C
	(full density)	33.35x10^6	room temp.
	Spinel ($Al_2O_3\ MgO$)	34.5x10^6	room temp.
		34.4x10^6	200°C
		34.5x10^6	400°C
		34x10^6	600°C
		32.9x10^6	800°C
		30.4x10^6	1000°C
		25.0x10^6	1200°C
		20.1x10^6	1300°C
	(ρ=3.510 g/cm^3)	38.23x10^6	room temp.

To convert from **psi** to **MPa**, multiply by **145**.

Source: data compiled by J.S. Park from No. *1 Materials Index*, Peter T.B. Shaffer, Plenum Press, New York, (1964); *Smithells Metals Reference Book*, Eric A. Brandes, ed., in association with Fulmer Research Institute Ltd. 6th ed. London, Butterworths, Boston, (1983); and *Ceramic Source*, American Ceramic Society (1986-1991)

Table 223. YOUNG'S MODULUS OF CERAMICS
(SHEET 7 OF 7)

Class	Ceramic	Young's Modulus (psi)	Temperature
Oxides (Con't)	Zircon (SiO_2 ZrO_2)	$24x10^6$	room temp.
Silicide	Molybdenum Disilicide ($MoSi_2$)	39.3-$56.36x10^6$	room temp.

To convert from **psi** to **MPa**, multiply by **145**.

Source: data compiled by J.S. Park from No. *1 Materials Index*, Peter T.B. Shaffer, Plenum Press, New York, (1964); *Smithells Metals Reference Book*, Eric A. Brandes, ed., in association with Fulmer Research Institute Ltd. 6th ed. London, Butterworths, Boston, (1983); and *Ceramic Source*, American Ceramic Society (1986-1991)

Table 224. YOUNG'S MODULUS OF GLASS
(SHEET 1 OF 2)

Class	Glass	Young's Modulus (GPa)	Temperature
SiO_2 glass		72.76–74.15	20°C
		79.87	998°C (annealing point)
		80.80	1096°C (straining point)
	SiO_2–Na_2O glass		
	(15% mol Na_2O)	64.4	room temp.
	(20% mol Na_2O)	62.0	room temp.
	(25% mol Na_2O)	56.9	–196°C
	(25% mol Na_2O)	61.4	room temp.
	(25% mol Na_2O)	53.9	200–250°C
	(30% mol Na_2O)	60.5	room temp.
	(33% mol Na_2O)	54.9	–196°C
	(33% mol Na_2O)	60.3	room temp.
	(33% mol Na_2O)	51.0	200–250°C
	(35% mol Na_2O)	60.2	room temp.
	(40% mol Na_2O)	51.9	–196°C
	(40% mol Na_2O)	46.1	200–250°C
	SiO_2–PbO glass		
	(24.6% mol PbO)	47.1	
	(30.0% mol PbO)	50.1	
	(35.7% mol PbO)	46.3	
	(38.4% mol PbO)	52.8	
	(45.0% mol PbO)	51.7	
	(50.0% mol PbO)	44.1	
	(55.0% mol PbO)	49.3	
	(60.0% mol PbO)	43.6	
	(65.0% mol PbO)	41.2	

Source: data compiled by J.S. Park from O. V. Mazurin, M. V. Streltsina and T. P. Shvaiko–Shvaikovskaya, *Handbook of Glass Data, Part A and Part B*, Elsevier, New York, 1983.

Table 224. YOUNG'S MODULUS OF GLASS

(SHEET 2 OF 2)

Class	Glass	Young's Modulus (GPa)	Temperature
SiO_2–B_2O_3 glass	(60% mol B_2O_3)	23.3	
	(65% mol B_2O_3)	22.5	
	(70% mol B_2O_3)	23.5	
	(75% mol B_2O_3)	24.1	
	(80% mol B_2O_3)	22.8	
	(85% mol B_2O_3)	21.2	
	(90% mol B_2O_3)	20.9	
	(95% mol B_2O_3)	21.2	
B_2O_3 glass	B_2O_3 glass	17.2–17.7	room temp.
	B_2O_3–Na_2O glass		
	(10% mol Na_2O)	31.4	15°C
	(20% mol Na_2O)	43.2	15°C
	(25% mol Na_2O)	53.7	15°C
	(33.3% mol Na_2O)	59.4	15°C
	(37% mol Na_2O)	57.1	15°C

Source: data compiled by J.S. Park from O. V. Mazurin, M. V. Streltsina and T. P. Shvaiko–Shvaikovskaya, *Handbook of Glass Data, Part A and Part B*, Elsevier, New York, 1983.

Table 225. ELASTIC MODULUS OF WROUGHT STAINLESS STEELS*

(SHEET 1 OF 2)

Type	UNS Designation	Elastic Modulus (GPa)
201	S20100	197
205	S20500	197
301	S30100	193
302	S30200	193
302B	S30215	193
303	S30300	193
304	S30400	193
S30430	S30430	193
304N	S30451	196
305	S30500	193
308	S30800	193
309	S30900	200
310	S31000	200
314	S31400	200
316	S31600	193
316N	S31651	196
317	S31700	193
317L	S31703	200
321	S32100	193
330	N08330	196
347	S34700	193
384	S38400	193
405	S40500	200
410	S41000	200
414	S41400	200
416	S41600	200
420	S42000	200
429	S42900	200

Data from *ASM Metals Reference Book, Third Edition*, Michael Bauccio, Ed., ASM International, Materials Park, OH, p360, (1993).

Table 225. ELASTIC MODULUS OF WROUGHT STAINLESS STEELS*

(SHEET 2 OF 2)

Type	UNS Designation	Elastic Modulus (GPa)
430	S43000	200
430F	S43020	200
431	S43100	200
434	S43400	200
436	S43600	200
440A	S44002	200
440C	S44004	200
444	S44400	200
446	S44600	200
PH 13–8 Mo	S13800	203
15–5 PH	S15500	196
17–4 PH	S17400	196
17–7 PH	S17700	204

Data from *ASM Metals Reference Book, Third Edition*, Michael Bauccio, Ed., ASM International, Materials Park, OH, p360, (1993).

* Annealed Condition.

Table 226. Modulus of Elasticity of Wrought Titanium Alloys

Class	Metal or Alloy	Modulus of Elasticity (GPa)
Commercially Pure	99.5 Ti	102.7
	99.2 Ti	102.7
	99.1 Ti	103.4
	99.0Ti	104.1
	99.2 Ti–0.2Pd	102.7
Alpha Alloys	Ti-5Al-2.5Sn	110.3
	Ti-5Al-2.5Sn (low O_2)	110.3
Near Alpha Alloys	Ti-8Al-1Mo-1V	124.1
	Ti-11Sn-1Mo-2.25Al-5.0Zr-1Mo-0.2Si	113.8
	Ti-6Al-2Sn-4Zr-2Mo	113.8
	Ti-5Al-5Sn-2Zr-2Mo-0.25Si	113.8
	Ti-6Al-2Nb-1Ta-1Mo	113.8
Alpha-Beta Alloys	Ti-8Mn	113.1
	Ti-3Al-2.5V	106.9
	Ti-6Al-4V	113.8
	Ti-6Al-4V (low O_2)	113.8
	Ti-6Al-6V-2Sn	110.3
	Ti-7Al-4Mo	113.8
	Ti-6Al-2Sn-4Zr-6Mo	113.8
	Ti-6Al-2Sn-2Zr-2Mo-2Cr-0.25Si	122.0
	Ti-10V-2Fe-3Al	111.7
Beta Alloys	Ti-13V-11Cr-3Al	101.4
	Ti-8Mo-8V-2Fe-3Al	106.9
	Ti-3Al-8V-6Cr-4Mo-4Zr	105.5
	Ti-11.5Mo-6Zr-4.5Sn	103.4

Data from *ASM Metals Reference Book, Third Edition*, Michael Bauccio, Ed., ASM International, Materials Park, OH, p511, (1993).

Table 227. MODULUS OF ELASTICITY IN TENSION FOR POLYMERS

(SHEET 1 OF 6)

Class	Polymer	Modulus of Elasticity in Tension, (ASTM D638) (10^5 psi)
ABS Resins; Molded, Extruded	Medium impact	3.3—4.0
	High impact	2.6—3.2
	Very high impact	2.0—3.1
	Low temperature impact	2.0—3.1
	Heat resistant	3.5—4.2
Acrylics; Cast, Molded, Extruded	Cast Resin Sheets, Rods:	
	General purpose, type I	3.5—4.5
	General purpose, type II	4.0—5.0
	Moldings:	
	Grades 5, 6, 8	3.5—5.0
	High impact grade	2.3—3.3
Chlorinated Polymers	Chlorinated polyether	1.5
	Chlorinated polyvinyl chloride	3.7

To convert **psi** to **MPa**, multiply by **145**.

Source: data compiled by J.S. Park from Charles T. Lynch, *CRC Handbook of Materials Science, Vol. 3*, CRC Press, Boca Raton, Florida, 1975 and *Engineered Materials Handbook, Vol.2*, Engineering Plastics, ASM International, Metals Park, Ohio, 1988.

Table 227. MODULUS OF ELASTICITY IN TENSION FOR POLYMERS

(SHEET 2 OF 6)

Class	Polymer	Modulus of Elasticity in Tension, (ASTM D638) (10^5 psi)
Polycarbonates	Polycarbonate	3.45
	Polycarbonate (40% glass fiber reinforced)	17
Diallyl Phthalates; Molded	Orlon filled	6
	Asbestos filled	12
Fluorocarbons; Molded,Extruded	Polytrifluoro chloroethylene (PTFCE)	1.9—3.0
	Polytetrafluoroethylene (PTFE)	0.38—0.65
	Ceramic reinforced (PTFE)	1.5—2.0
	Fluorinated ethylene propylene(FEP)	0.5—0.7
	Polyvinylidene— fluoride (PVDF)	1.7—2
Epoxies; Cast, Molded, Reinforced	Standard epoxies (diglycidyl ethers of bisphenol A)	
	Cast rigid	4.5
	Cast flexible	0.5—2.5

To convert **psi** to **MPa**, multiply by **145**.

Source: data compiled by J.S. Park from Charles T. Lynch, *CRC Handbook of Materials Science, Vol. 3*, CRC Press, Boca Raton, Florida, 1975 and *Engineered Materials Handbook, Vol.2*, Engineering Plastics, ASM International, Metals Park, Ohio, 1988.

Table 227. MODULUS OF ELASTICITY IN TENSION FOR POLYMERS

(SHEET 3 OF 6)

Class	Polymer	Modulus of Elasticity in Tension, (ASTM D638) (10^5 psi)
Epoxies; Cast, Molded, Reinforced (Con't)	Molded:	
	General purpose glass cloth laminate	33—36
	High strength laminate	57—58
	Filament wound composite	72—64
	High performance resins (cycloaliphatic diepoxides)	
	Cast, rigid	4—5
	Molded	
	Glass cloth laminate	32—33
	Epoxy novolacs	
	Cast, rigid	4.8—5.0
	Glass cloth laminate	27.5
Melamines; Molded	Unfilled	
	Cellulose electrical	10—11

To convert **psi** to **MPa**, multiply by **145**.

Source: data compiled by J.S. Park from Charles T. Lynch, *CRC Handbook of Materials Science, Vol. 3*, CRC Press, Boca Raton, Florida, 1975 and *Engineered Materials Handbook, Vol.2*, Engineering Plastics, ASM International, Metals Park, Ohio, 1988.

Table 227. Modulus of Elasticity in Tension for Polymers

(Sheet 4 of 6)

Class	Polymer	Modulus of Elasticity in Tension, (ASTM D638) (10^5 psi)
Phenolics; Molded	Type and filler	
	General: woodflour and flock	8—13
	Shock: paper, flock, or pulp	8—12
	High shock: chopped fabric or cord	9—14
	Very high shock: glass fiber	30—33
	Arc resistant—mineral	10—30
	Rubber phenolic—woodflour or flock	4—6
	Rubber phenolic—chopped fabric	3.5—6
	Rubber phenolic—asbestos	5—9
Polyesters: Thermosets	Cast polyyester	
	Rigid	1.5—6.5
	Flexible	0.001—0.10

To convert **psi** to **MPa**, multiply by **145**.

Source: data compiled by J.S. Park from Charles T. Lynch, *CRC Handbook of Materials Science, Vol. 3*, CRC Press, Boca Raton, Florida, 1975 and *Engineered Materials Handbook, Vol.2*, Engineering Plastics, ASM International, Metals Park, Ohio, 1988.

Table 227. Modulus of Elasticity in Tension for Polymers

(Sheet 5 of 6)

Class	Polymer	Modulus of Elasticity in Tension, (ASTM D638) (10^5 psi)
Polyesters: Thermosets (Con't)	Reinforced polyester moldings	
	High strength (glass fibers)	16—20
	Heat and chemical resistsnt (asbestos)	12—15
	Sheet molding compounds, general purpose	15—20
Polyethylenes; Molded, Extruded	Type I—lower density (0.910—0.925)	
	Melt index 0.3—3.6	0.21—0.27
	Melt index 6—26	0.20—0.24
Polystyrenes; Molded	Polystyrenes	D638
	General purpose	4.6—5.0
	Medium impact	2.6—4.7
	High impact	1.50—3.80
	Glass fiber -30% reinforced	12.1

To convert **psi** to **MPa**, multiply by **145**.

Source: data compiled by J.S. Park from Charles T. Lynch, *CRC Handbook of Materials Science, Vol. 3*, CRC Press, Boca Raton, Florida, 1975 and *Engineered Materials Handbook, Vol.2*, Engineering Plastics, ASM International, Metals Park, Ohio, 1988.

Table 227. MODULUS OF ELASTICITY IN TENSION FOR POLYMERS

(SHEET 6 OF 6)

Class	Polymer	Modulus of Elasticity in Tension, (ASTM D638) (10^5 psi)
SAN	Styrene acrylonitrile (SAN)	4.0—5.2
	Glass fiber (30%) reinforced SAN	17.5
Polyvinyl Chloride And Copolymers;		ASTM D412
	Molded, Extruded	
	Nonrigid—general	0.004—0.03
	Nonrigid—electrical	0.01—0.03
	Rigid—normal impact	3 5—4.0
	Vinylidene chloride	0.7—2.0
Silicones; Molded, Laminated		ASTM D651
	Woven glass fabric/ silicone laminate	28
Ureas; Molded	Alpha—cellulose filled (ASTM Type l)	13—16
	Woodflour filled	11—14

To convert **psi** to **MPa**, multiply by **145**.

Source: data compiled by J.S. Park from Charles T. Lynch, *CRC Handbook of Materials Science, Vol. 3*, CRC Press, Boca Raton, Florida, 1975 and *Engineered Materials Handbook, Vol.2*, Engineering Plastics, ASM International, Metals Park, Ohio, 1988.

Table 228. Modulus of Elasticity of 55MSI Graphite/6061 Aluminum Composites

Material	Reinforcement content (vol %)	Fiber orientation	Modulus of Elasticity (GPa)
55MSI graphite/6061 aluminum composites	34	0°	182.2±6.6
55MSI graphite/6061 aluminum composites	34	90°	33

Data from *ASM Engineering Materials Reference Book, Second Edition*, Michael Bauccio, Ed., ASM International, Materials Park, OH, p148,(1994).

Table 229. Modulus of Elasticity of Graphite/Magnesium Castings*

Fiber Type	Fiber content	Fiber orientation	Casting	Fiber Preform Method	Modulus of Elasticity, 0° (GPa)	Modulus of Elasticity,90° (GPa)
P75	40%	±16°	Hollow cylinder	Filament wound	179	86
	plus 9%	90°		Filament wound		
P100	40%	± 16°	Hollow cylinder	Filament wound	228	30
P55	40%	0°	Plate	Prepreg	159	21
	30%	0° plus	Plate	Prepreg	83	34
	10%	90°				
	20%	0° plus	Plate	Prepreg	90	90
	20%	90°				

Data from *ASM Engineering Materials Reference Book, Second Edition*, Michael Bauccio, Ed., ASM International, Materials Park, OH, p148,(1994).

* Pitch-base fibers

Table 230. MODULUS OF ELASTICITY OF GRAPHITE/ALUMINUM COMPOSITES

Thornel Fiber	Longitudinal Modulus of Elasticity (GPa)	Transverse Modulus of Elasticity (GPa)
P55	207 to 221	28 to 41
P75	276 to 296	28 to 41
P100	379 to 414	28 to 41

Data from *ASM Engineering Materials Reference Book, Second Edition*, Michael Bauccio, Ed., ASM International, Materials Park, OH, p148,(1994).

Table 231. MODULUS OF ELASTICITY OF GRAPHITE FIBER REINFORCED METALS

Composite	Fiber content (vol%)	Modulus of Elasticity (10^6psi)
Graphite(a)/lead	41	29.0
Graphite(b)/lead	35	17.4
Graphite(a)/zinc	35	16.9
Graphite(a)/magnesium	42	26.6

(a) Thornel 75 fiber (b) Courtaulds HM fiber

To convert from **psi** to **MPa,** multiply by **145.**

Data from *ASM Engineering Materials Reference Book, Second Edition*, Michael Bauccio, Ed., ASM International, Materials Park, OH, p148,(1994).

Table 232. Modulus of Elasticity of SiC-Whisker–Reinforced Aluminum Alloy

Fiber Content (vol %)	Modulus of Elasticity		
	(GPa)	Standard Deviation	Range of Measurement
0	71.9	4.5	13
12	95.3	1.6	6
16	90.0	3.7	9
20	111.0	5.0	13

Data from *ASM Engineering Materials Reference Book, Second Edition*, Michael Bauccio, Ed., ASM International, Materials Park, OH, p150,(1994).

Table 233. MODULUS OF ELASTICITY OF POLYCRYSTALLINE–ALUMINA–REINFORCED ALUMINUM ALLOY

Fiber Content (vol %)	Modulus of Elasticity (GPa)	Standard Deviation	Range of Measurement
0	71.9	4.5	13
5	78.4	2.3	6
12	83.0	7.8	21
20	95.2	2.7	7

Data from *ASM Engineering Materials Reference Book, Second Edition*, Michael Bauccio, Ed., ASM International, Materials Park, OH, p154,(1994).

Table 234. MODULUS OF ELASTICITY OF BORON/ALUMINUM COMPOSITES*

Matrix	Fiber Orientation	Modulus of Elasticity (GPa)
Al-6061	0°	207
	90°	138
Al-2024	0°	207
	90°	145

Data from *ASM Engineering Materials Reference Book, Second Edition*, Michael Bauccio, Ed., ASM International, Materials Park, OH, p157,(1994).

* These samples contain 48% Avco (142 μm) boron. Longitudinal tensile specimens are 152 mm by 7.9 mm by 6 ply. Transverse tensile bars are 152 mm by 12.7 mm by 6 ply.

Table 235. Compression Modulus of Treated Ductile Irons

Treatment	Compression Modulus (MPa)
60-40-18	164
65-45-12	163
80-55-06	165
120 90-02	164

Source: data from ASM *Metals Reference Book, Second Edition*, American Society for Metals, Metals Park, Ohio 44073, p169-170, (1984).

Table 236. Modulus of Elasticity in Compression for Polymers

Polymer	Modulus of Elasticity in Compression, (ASTM D638) (10^5 psi)
Fluorocarbons; Molded,Extruded	
Polytrifluoro chloroethylene (PTFCE)	1.8
Polytetrafluoroethylene (PTFE)	0 70—0.90
Ceramic reinforced (PTFE)	1.5—2.0
Fluorinated ethylene propylene(FEP)	0.6—0.8
Polyvinylidene— fluoride (PVDF)	1.7—2

To convert from **psi** to **MPa**, multiply by **145.**

Source: data compiled by J.S. Park from Charles T. Lynch, CRC *Handbook of Materials Science, Vol. 3*, CRC Press, Boca Raton, Florida, 1975 and *Engineered Materials Handbook, Vol.2*, Engineering Plastics, ASM International, Metals Park, Ohio, 1988.

Table 237. BULK MODULUS OF GLASS

Glass	Bulk Modulus (GPa)	Temperature
SiO_2 glass	31.01-37.62	
SiO_2-Na_2O glass		
(15% mol Na_2O)	33.8	room temp.
(20% mol Na_2O)	34.8	room temp.
(25% mol Na_2O)	36.5	room temp.
(30% mol Na_2O)	38.2	room temp.
(33% mol Na_2O)	40.1	room temp.
(35% mol Na_2O)	39.8	room temp.
SiO_2-PbO glass		
(24.6% mol PbO)	33.9	
(30.0% mol PbO)	25.6	
(35.7% mol PbO)	31.1	
(38.4% mol PbO)	25.1	
(45.0% mol PbO)	30.6	
(50.0% mol PbO)	30.5	
(55.0% mol PbO)	29.5	
(60.0% mol PbO)	33.1	
(65.0% mol PbO)	31.6	
B_2O_3-Na_2O glass		
(10% mol Na_2O)	23.2	15°C
(20% mol Na_2O)	33.6	15°C
(25% mol Na_2O)	39.2	15°C
(33.3% mol Na_2O)	44.4	15°C
(37% mol Na_2O)	42.1	15°C

Source: data compiled by J.S. Park from O. V. Mazurin, M. V. Streltsina and T. P. Shvaiko-Shvaikovskaya, *Handbook of Glass Data, Part A and Part B*, Elsevier, New York, 1983

Table 238. SHEAR MODULUS OF GLASS

(SHEET 1 OF 2)

Class	Glass	Shear Modulus (GPa)	Temperature
SiO_2 glass	SiO_2 glass	31.38	20°C
		33.57	998°C (annealing point)
		34.15	1096°C (straining point)
SiO_2–Na_2O glass	(5% mol Na_2O)	27.2	–100°C
	(5% mol Na_2O)	27.4	0°C
	(5% mol Na_2O)	27.6	80°C
	(5% mol Na_2O)	27.2	160°C
	(7.5% mol Na_2O)	26.9	–100—160°C
	(15% mol Na_2O)	27.2	room temp.
	(18% mol Na_2O)	25.8	–100°C
	(18% mol Na_2O)	25.0	0°C
	(18% mol Na_2O)	24.8	80°C
	(18% mol Na_2O)	24.2	160°C
	(20% mol Na_2O)	25.8	room temp.
	(25% mol Na_2O)	25.2	room temp.
	(30% mol Na_2O)	24.5	room temp.
	(33% mol Na_2O)	24.2	room temp.
	(35% mol Na_2O)	24.1	room temp.

Source: *data compiled by* J.S. Park *from* O. V. Mazurin, M. V. Streltsina and T. P. Shvaiko–Shvaikovskaya, *Handbook of Glass Data, Part A and Part B,* Elsevier, New York, 1983.

Table 238. SHEAR MODULUS OF GLASS
(SHEET 2 OF 2)

Class	Glass	Shear Modulus (GPa)	Temperature
SiO_2–PbO glass	(24.6% mol PbO)	20.4	
	(30.0% mol PbO)	21.4	
	(35.7% mol PbO)	18.5	
	(38.4% mol PbO)	23.0	
	(45.0% mol PbO)	21.2	
	(50.0% mol PbO)	17.5	
	(55.0% mol PbO)	20.2	
	(60.0% mol PbO)	17.0	
	(65.0% mol PbO)	16.1	
B_2O_3 glass		6.55	room temp.
		6.29	250°C
		6.07	260°C
		5.78	270°C
		5.49	280°C
		5.15	290°C
		4.75	300°C
B_2O_3–Na_2O glass	(10% mol Na_2O)	12.3	15°C
	(20% mol Na_2O)	16.8	15°C
	(25% mol Na_2O)	21.1	15°C
	(33.3% mol Na_2O)	23.2	15°C
	(37% mol Na_2O)	22.4	15°C

Source: *data compiled by* J.S. Park *from* O. V. Mazurin, M. V. Streltsina and T. P. Shvaiko–Shvaikovskaya, *Handbook of Glass Data, Part A and Part B*, Elsevier, New York, 1983.

Table 239. Torsional Modulus of Gray Cast Irons

ASTM Class	Torsional Modulus (GPa)
20	27 to 39
25	32 to 41
30	36 to 45
35	40 to 48
40	44 to 54
50	50 to 55
60	54 to 59

Source: data from *ASM Metals Reference Book, Second Edition*, American Society for Metals, Metals Park, Ohio 44073, p166-167, (1984).

Table 240. Torsion Modulus of Treated Ductile Irons

Treatment	Torsion Modulus (MPa)
60-40-18	63
65-45-12	64
80-55-06	62
120 90-02	63.4

Source: data from *ASM Metals Reference Book, Second Edition*, American Society for Metals, Metals Park, Ohio 44073, p169-170, (1984).

Table 241. Modulus of Elasticity in Flexure for Polymers (Sheet 1 of 13)

Polymer Class	Polymer	Modulus of Elasticity in Flexure (ASTM D790) (10^5 psi)
ABS Resins; Molded, Extruded	Medium impact	3.5—4.0
	High impact	2.5—3.2
	Very high impact	2.0—3.2
	Low temperature impact	2.0—3.2
	Heat resistant	3.5—4.2
Acrylics; Cast, Molded, Extruded	Cast Resin Sheets, Rods:	
	General purpose, type I	3.5—4.5
	General purpose, type II	4.0—5.0
	Moldings:	
	Grades 5, 6, 8	3.5—5.0
	High impact grade	2.7—3.6

To convert from **psi** to **MPa**, multiply by **145**.

Source: data compiled by J.S. Park from Charles T. Lynch, *CRC Handbook of Materials Science, Vol. 3*, CRC Press, Boca Raton, Florida, 1975 and *Engineered Materials Handbook, Vol.2*, Engineering Plastics, ASM International, Metals Park, Ohio, 1988.

Table 241. MODULUS OF ELASTICITY IN FLEXURE FOR POLYMERS (SHEET 2 OF 13)

Polymer Class	Polymer	Modulus of Elasticity in Flexure (ASTM D790) (10^5 psi)
Thermoset Carbonate	Allyl diglycol carbonate	2.5—3.3
Alkyds; Molded	Rope (general purpose)	22—27
	Granular (high speed molding)	22—27
	Glass reinforced (heavy duty parts)	22—28
Cellulose Acetate; Molded, Extruded	ASTM Grade:	(ASTM D747)
	H4—1	2.0—2.55
	H2—1	1.50—2.35
	MH—1, MH—2	1.50—2.15
	MS—1, MS—2	1.25—1.90
	S2—1	1.05—1.65

To convert from **psi** to **MPa**, multiply by **145**.

Source: data compiled by J.S. Park from Charles T. Lynch, *CRC Handbook of Materials Science, Vol. 3*, CRC Press, Boca Raton, Florida, 1975 and *Engineered Materials Handbook, Vol.2*, Engineering Plastics, ASM International, Metals Park, Ohio, 1988.

Table 241. MODULUS OF ELASTICITY IN FLEXURE FOR POLYMERS (SHEET 3 OF 13)

Polymer Class	Polymer	Modulus of Elasticity in Flexure (ASTM D790) (10^5 psi)
Cellulose Acetate Butyrate; Molded, Extruded		(ASTM D747)
	ASTM Grade:	
	H4	1.8
	MH	1.20—1.40
	S2	0.70—0.90
Cellusose Acetate Propionate; Molded, Extruded	ASTM Grade:	
	1	1.7—1.8
	3	1.45—1.55
	6	1.1
Chlorinated Polymers	Chlorinated polyether	1.3 (0.1% offset)
	Chlorinated polyvinyl chloride	3.85
Polycarbonates	Polycarbonate	3.4
	Polycarbonate (40% glass fiber reinforced)	12

To convert from **psi** to **MPa**, multiply by **145**.

Source: data compiled by J.S. Park from Charles T. Lynch, *CRC Handbook of Materials Science, Vol. 3*, CRC Press, Boca Raton, Florida, 1975 and *Engineered Materials Handbook, Vol.2*, Engineering Plastics, ASM International, Metals Park, Ohio, 1988.

Table 241. MODULUS OF ELASTICITY IN FLEXURE FOR POLYMERS (SHEET 4 OF 13)

Polymer Class	Polymer	Modulus of Elasticity in Flexure (ASTM D790) (10^5 psi)
Fluorocarbons; Molded,Extruded	Polytrifluoro chloroethylene (PTFCE)	2.0—2.5
	Polytetrafluoroethylene (PTFE)	0.6—1.1
	Ceramic reinforced (PTFE)	4.64
	Fluorinated ethylene propylene(FEP)	0.8
	Polyvinylidene— fluoride (PVDF)	1.75—2.0
Epoxies; Cast, Molded, Reinforced	Standard epoxies (diglycidyl ethers of bisphenol A)	
	Cast rigid	4.5—5.4
	Cast flexible	0.36—3.9
	Molded	15—25
	General purpose glass cloth laminate	36—39
	High strength laminate	53—55
	Filament wound composite	69—75

To convert from **psi** to **MPa**, multiply by **145**.

Source: data compiled by J.S. Park from Charles T. Lynch, *CRC Handbook of Materials Science, Vol. 3*, CRC Press, Boca Raton, Florida, 1975 and *Engineered Materials Handbook, Vol.2*, Engineering Plastics, ASM International, Metals Park, Ohio, 1988.

Table 241. Modulus of Elasticity in Flexure for Polymers (Sheet 5 of 13)

Polymer Class	Polymer	Modulus of Elasticity in Flexure (ASTM D790) (10^5 psi)
Epoxies—Molded, Extruded	High performance resins (cycloaliphatic diepoxides)	
	Cast, rigid	4—5
	Glass cloth laminate	28—31
	Epoxy novolacs	
	Cast, rigid	4.4—4.8
	Glass cloth laminate	32—35
Melamines; Molded	Filler & type	
	Unfilled	10—13
	Cellulose electrical	10—13
	Glass fiber	24

To convert from **psi** to **MPa**, multiply by **145**.

Source: data compiled by J.S. Park from Charles T. Lynch, *CRC Handbook of Materials Science, Vol. 3*, CRC Press, Boca Raton, Florida, 1975 and *Engineered Materials Handbook, Vol.2*, Engineering Plastics, ASM International, Metals Park, Ohio, 1988.

Table 241. MODULUS OF ELASTICITY IN FLEXURE FOR POLYMERS (SHEET 6 OF 13)

Polymer Class	Polymer	Modulus of Elasticity in Flexure (ASTM D790) (10^5 psi)
Nylons; Molded, Extruded	Type 6	
	General purpose	1.4—3.9
	Glass fiber (30%) reinforced	1.0—1.4
	Cast	5.05
	Flexible copolymers	0.92—3.2
	Type 8	0.4
	Type 11	1.51
	6/6 Nylon	
	General purpose molding	1.75–4.5
	Glass fiber reinforced	10—18
	Glass fiber Molybdenum disulfide filled	11—13
	General purpose extrusion	1.75—4.1

To convert from **psi** to **MPa**, multiply by **145**.

Source: data compiled by J.S. Park from Charles T. Lynch, *CRC Handbook of Materials Science, Vol. 3*, CRC Press, Boca Raton, Florida, 1975 and *Engineered Materials Handbook, Vol.2*, Engineering Plastics, ASM International, Metals Park, Ohio, 1988.

Table 241. Modulus of Elasticity in Flexure for Polymers (Sheet 7 of 13)

Polymer Class	Polymer	Modulus of Elasticity in Flexure (ASTM D790) (10^5 psi)
Nylons; Molded, Extruded (Con't)	6/10 Nylon	
	General purpose	1.6–2.8
	Glass fiber (30%) reinforced	8.5
Phenolics; Molded	Type and filler	
	General: woodflour and flock	8—12
	Shock: paper, flock, or pulp	8—12
	High shock: chopped fabric or cord	9—13
	Very high shock: glass fiber	30—33
	Arc resistant—mineral	10—30
	Rubber phenolic—woodflour or flock	4—6
	Rubber phenolic—chopped fabric	3.5
	Rubber phenolic—asbestos	5
ABS–Polycarbonate Alloy	ABS–Polycarbonate Alloy	4

To convert from **psi** to **MPa**, multiply by **145**.

Source: data compiled by J.S. Park from Charles T. Lynch, *CRC Handbook of Materials Science, Vol. 3*, CRC Press, Boca Raton, Florida, 1975 and *Engineered Materials Handbook, Vol.2*, Engineering Plastics, ASM International, Metals Park, Ohio, 1988.

Table 241. Modulus of Elasticity in Flexure for Polymers (Sheet 8 of 13)

Polymer Class	Polymer	Modulus of Elasticity in Flexure (ASTM D790) (10^5 psi)
PVC–Acrylic Alloy	PVC–acrylic sheet	4
	PVC–acrylic injection molded	3
Polymides	Unreinforced	7
	Unreinforced 2nd value	5
	Glass reinforced	38.4
Polyacetals	Homopolymer:	
	Standard	4.1
	20% glass reinforced	8.8
	22% TFE reinforced	4
	Copolymer:	
	Standard	3.75
	25% glass reinforced	11
	High flow	3.75

To convert from **psi** to **MPa**, multiply by **145**.

Source: data compiled by J.S. Park from Charles T. Lynch, *CRC Handbook of Materials Science, Vol. 3*, CRC Press, Boca Raton, Florida, 1975 and *Engineered Materials Handbook, Vol.2*, Engineering Plastics, ASM International, Metals Park, Ohio, 1988.

Table 241. Modulus of Elasticity in Flexure for Polymers (Sheet 9 of 13)

Polymer Class	Polymer	Modulus of Elasticity in Flexure (ASTM D790) (10^5 psi)
Polyester; Thermoplastic	Injection Moldings:	
	General purpose grade	3.4
	Glass reinforced grades	12—15
	Glass reinforced self extinguishing	12
	General purpose grade	33
	Glass reinforced grade	87
	Asbestos—filled grade	90
Polyesters: Thermosets	Cast polyyester	
	Rigid	1—9
	Flexible	0.001—0.39
	Reinforced polyester moldings	
	High strength (glass fibers)	15—25
	Sheet molding compounds, general purpose	15—18

To convert from **psi** to **MPa**, multiply by **145**.

Source: data compiled by J.S. Park from Charles T. Lynch, *CRC Handbook of Materials Science, Vol. 3*, CRC Press, Boca Raton, Florida, 1975 and *Engineered Materials Handbook, Vol.2*, Engineering Plastics, ASM International, Metals Park, Ohio, 1988.

Table 241. MODULUS OF ELASTICITY IN FLEXURE FOR POLYMERS (SHEET 10 OF 13)

Polymer Class	Polymer	Modulus of Elasticity in Flexure (ASTM D790) (10^5 psi)
Phenylene Oxides	SE—100	3.6
	SE—1	3.6
	Glass fiber reinforced	7.4—10.4
Phenylene oxides (Noryl)	Standard	3.9
	Glass fiber reinforced	12, 15.5
	Polyarylsulfone	4
Polypropylene:	General purpose	1.7—2.5
	High impact	1.0—2.0
	Asbestos filled	3.4—6.5
	Glass reinforced	4—8.2
	Flame retardant	1.9—6.1

To convert from **psi** to **MPa**, multiply by **145**.

Source: data compiled by J.S. Park from Charles T. Lynch, *CRC Handbook of Materials Science, Vol. 3*, CRC Press, Boca Raton, Florida, 1975 and *Engineered Materials Handbook, Vol.2*, Engineering Plastics, ASM International, Metals Park, Ohio, 1988.

Table 241. Modulus of Elasticity in Flexure for Polymers (Sheet 11 of 13)

Polymer Class	Polymer	Modulus of Elasticity in Flexure (ASTM D790) (10^5 psi)
Polyphenylene sulfide:	Standard	5.5—6.0
	40% glass reinforced	17—22
Polyethylenes; Molded, Extruded		(ASTM D747)
	Type I—lower density (0.910—0.925)	
	Melt index 0.3—3.6	0.13—0.27
	Melt index 6—26	0.12—0.3
	Melt index 200	0.1
	Type II—medium density (0.926—0.940)	
	Melt index 20	0.35—0.5
	Melt index 1.0—1.9	0.35—0.5

To convert from **psi** to **MPa**, multiply by **145**.

Source: data compiled by J.S. Park from Charles T. Lynch, *CRC Handbook of Materials Science, Vol. 3*, CRC Press, Boca Raton, Florida, 1975 and *Engineered Materials Handbook, Vol.2*, Engineering Plastics, ASM International, Metals Park, Ohio, 1988.

Table 241. Modulus of Elasticity in Flexure for Polymers (Sheet 12 of 13)

Polymer Class	Polymer	Modulus of Elasticity in Flexure (ASTM D790) (10^5 psi)
Polyethylenes; Molded, Extruded (Con't)	Type III—higher density (0.941—0.965)	
	Melt index 0.2—0.9	1.3—1.5
	Melt Melt index 0.1—12.0	0.9—0.25
	Melt index 1.5—15	1.5
	High molecular weight	0.75
Olefin Copolymers; Molded	Ethylene butene	165 (psi)
	Propylene—ethylene	140 (psi)
	Polyallomer	0.7—1.3
Polystyrenes; Molded	Polystyrenes:	
	General purpose	4—5
	Medium impact	3.5—5.0
	High impact	2.3—4.0
	Glass fiber -30% reinforced	12

To convert from **psi** to **MPa**, multiply by **145**.

Source: data compiled by J.S. Park from Charles T. Lynch, *CRC Handbook of Materials Science, Vol. 3*, CRC Press, Boca Raton, Florida, 1975 and *Engineered Materials Handbook, Vol.2*, Engineering Plastics, ASM International, Metals Park, Ohio, 1988.

Table 241. Modulus of Elasticity in Flexure for Polymers (Sheet 13 of 13)

Polymer Class	Polymer	Modulus of Elasticity in Flexure (ASTM D790) (10^5 psi)
Styrene acrylonitrile (SAN):	Glass fiber (30%) reinforced SAN	14.5
Polyvinyl Chloride And Copolymers; Molded, Extruded	Rigid—normal impact	3.8—5.4
Silicones; Molded, Laminated	Fibrous (glass) reinforced silicones	25
	Granular (silica) reinforced silicones	14—17
	Woven glass fabric/ silicone laminate	26—32

To convert from **psi** to **MPa**, multiply by **145**.

Source: data compiled by J.S. Park from Charles T. Lynch, *CRC Handbook of Materials Science, Vol. 3*, CRC Press, Boca Raton, Florida, 1975 and *Engineered Materials Handbook, Vol.2*, Engineering Plastics, ASM International, Metals Park, Ohio, 1988.

Table 242. FLEXURAL MODULUS OF FIBERGLASS REINFORCED PLASTICS

Class	Material	Glass fiber content (wt%)	Flexural modulus (10^5 psi)
Glass fiber reinforced thermosets	Sheet molding compound (SMC)	15 to 30	14 to 20
	Bulk molding compound(BMC)	15 to 35	14 to 20
	Preform/mat(compression molded)	25 to 50	13 to 18
	Cold press molding–polyester	20 to 30	13 to 19
	Spray–up–polyester	30 to 50	10 to 12
	Filament wound–epoxy	30 to 80	50 to 70
	Rod stock–polyester	40 to 80	40 to 60
	Molding compound–phenolic	5 to 25	30
Glass–fiber–reinforced thermoplastics	Acetal	20 to 40	8 to 13
	Nylon	6 to 60	2 to 28
	Polycarbonate	20 to 40	7.5 to 15
	Polyethylene	10 to 40	2.1 to 6
	Polypropylene	20 to 40	3.5 to 8.2
	Polystyrene	20 to 35	8 to 12
	Polysulfone	20 to 40	8 to 15
	ABS(acrylonitrile butadiene styrene)	20 to 40	9.2 to 15
	PVC (polyvinyl chloride)	15 to 35	9 to 16
	Polyphenylene oxide(modified)	20 to 40	8 to 15
	SAN (styrene acrylonitrile)	20 to 40	8.0 to 18
	Thermoplastic polyester	20 to 35	8.7 to 15

To convert from **psi** to **MPa**, multiply by **145**.

Data from *ASM Engineering Materials Reference Book, Second Edition*, Michael Bauccio, Ed., ASM International, Materials Park, OH, p106, (1994).

Table 243. FLEXURAL MODULUS OF CARBON- AND GLASS-REINFORCED ENGINEERING THERMOPLASTICS (SHEET 1 OF 2)

Class	Resin Type	Composition	Flexural Modulus (GPa)
Amorphous	Acrylonitrile-butadiene-styrene(ABS)	30% glass fiber	7.6
		30% carbon fiber	12.4
	Nylon	30% glass fiber	7.9
		30% carbon fiber	15.2
	Polycarbonate	30% glass fiber	8.3
		30% carbon fiber	13.1
	Polyetherimide	30% glass fiber	8.6
		30% carbon fiber	17.2
	Polyphenylene oxide (PPO)	30% glass fiber	9.0
		30% carbon fiber	11.7
	Polysulfone	30% glass fiber	8.3
		30% carbon fiber	14.5
	Styrene-maleic-anhydride (SMA)	30% glass fiber	9.0
	Thermoplastic polyurethane	30% glass fiber	1.3

Data from *ASM Engineering Materials Reference Book, Second Edition*, Michael Bauccio, Ed., ASM International, Materials Park, OH, p111–112, (1994).

Table 243. FLEXURAL MODULUS OF CARBON- AND GLASS-REINFORCED ENGINEERING THERMOPLASTICS (SHEET 2 OF 2)

Class	Resin Type	Composition	Flexural Modulus (GPa)
Crystalline	Acetal	30% glass fiber	9.7
		20% carbon fiber	9.3
	Nylon 66	30% glass fiber	9.0
		30% carbon fiber	20.0
	Polybutylene telphthalate (PBT)	30% glass fiber	9.7
		30% carbon fiber	15.9
	Polythylene terephthalate (PET)	30% glass fiber	9.0
	Polyphenylene sulfide (PPS)	30% glass fiber	11.0
		30% carbon fiber	16.9

Data from *ASM Engineering Materials Reference Book, Second Edition*, Michael Bauccio, Ed., ASM International, Materials Park, OH, p111–112, (1994).

Table 244. MODULUS OF RUPTURE FOR CERAMICS

(SHEET 1 OF 10)

Class	Ceramic	Modulus of Rupture (psi)	Temperature
Borides	Titanium Diboride (TiB_2)	$19x10^3$	
	(98% dense)	$5.37x10^3$	
	(6.0 µm grain size, ρ=4.46g/cm^3)	$6.2x10^3$	
	(3.5 µm grain size, ρ=4.37g/cm^3, 0.8wt% Ni)	$5.7x10^3$	
	(6.0 µm grain size, ρ=4.56g/cm^3, 0.16wt% Ni)	$11.0x10^3$	
	(12.0 µm grain size, ρ=4.66g/cm^3, 9.6wt% Ni)	$6.29x10^3$	
Carbides	Hafnium Monocarbide (HfC)		
	(ρ = 11.9 g/cm^3)	$34.67x10^3$	room temp.
	(ρ = 11.9 g/cm^3)	$12.64x10^3$	2000 °C
	(ρ = 11.9 g/cm^3)	$4.78x10^3$	2200 °C

To convert from **psi** to **MPa**, multiply by **145**.

Source: data compiled by J.S. Park from *No. 1 Materials Index*, Peter T.B. Shaffer, Plenum Press, New York, (1964); *Smithells Metals Reference Book*, Eric A. Brandes, ed., in association with Fulmer Research Institute Ltd. 6th ed. London, Butterworths, Boston, (1983); and *Ceramic Source*, American Ceramic Society (1986-1991).

Table 244. MODULUS OF RUPTURE FOR CERAMICS

(SHEET 2 OF 10)

Class	Ceramic	Modulus of Rupture (psi)	Temperature
Carbides (Con't)	Silicon Carbide (SiC)	$27x10^3$	room temp.
		$25x10^3$	1300 °C
		$11x10^3$	1400 °C
		$15x10^3$	1800 °C
	(with 1 wt% Be addictive)	$58x10^3$	
	(with 1wt% B addictive)	$42x10^3$	
	(with 1wt% Al addictive)	$136x10^3$	
	Titanium Monocarbide (TiC)		
	(ρ = 4.85 g/cm^3)	$32.67x10^3$	room temp.
	(ρ = 4.85 g/cm^3)	$13.6x10^3$	2000°C

To convert from **psi** to **MPa**, multiply by **145**.

Source: data compiled by J.S. Park from *No. 1 Materials Index*, Peter T.B. Shaffer, Plenum Press, New York, (1964); *Smithells Metals Reference Book*, Eric A. Brandes, ed., in association with Fulmer Research Institute Ltd. 6th ed. London, Butterworths, Boston, (1983); and *Ceramic Source*, American Ceramic Society (1986-1991).

Table 244. MODULUS OF RUPTURE FOR CERAMICS
(SHEET 3 OF 10)

Class	Ceramic	Modulus of Rupture (psi)	Temperature
Carbides (Con't)	Tungsten Monocarbide (WC)	$55.65\text{-}84 \times 10^3$	room temp.
Carbides (Con't)	Zirconium Monocarbide (ZrC)	$16.6\text{-}22.5 \times 10^3$	room temp.
		8.3×10^3	1250 °C
		5.14×10^3	1750 °C
		2.5×10^3	2000 °C
Nitrides	Aluminum Nitride (AlN) (hot pressed)	38.5×10^3	25°C
		27×10^3	1000°C
		18.1×10^3	1400°C

To convert from **psi** to **MPa**, multiply by **145**.

Source: data compiled by J.S. Park from *No. 1 Materials Index*, Peter T.B. Shaffer, Plenum Press, New York, (1964); *Smithells Metals Reference Book*, Eric A. Brandes, ed., in association with Fulmer Research Institute Ltd. 6th ed. London, Butterworths, Boston, (1983); and *Ceramic Source*, American Ceramic Society (1986-1991).

Table 244. MODULUS OF RUPTURE FOR CERAMICS

(SHEET 4 OF 10)

Class	Ceramic	Modulus of Rupture (psi)	Temperature
Nitrides (Con't)	Boron Nitride (BN)		
	parallel to c axis	$7.28\text{-}13.2x10^3$	25 °C
		$7.03x10^3$	300 °C
		$1.90x10^3$	700 °C
		$1.08x10^3$	1000 °C
		$1.25x10^3$	1500 °C
		$1.50x10^3$	1800 °C
		$2.45x10^3$	2000 °C

To convert from **psi** to **MPa**, multiply by **145**.

Source: data compiled by J.S. Park from *No. 1 Materials Index*, Peter T.B. Shaffer, Plenum Press, New York, (1964); *Smithells Metals Reference Book*, Eric A. Brandes, ed., in association with Fulmer Research Institute Ltd. 6th ed. London, Butterworths, Boston, (1983); and *Ceramic Source*, American Ceramic Society (1986-1991).

Table 244. MODULUS OF RUPTURE FOR CERAMICS
(SHEET 5 OF 10)

Class	Ceramic	Modulus of Rupture (psi)	Temperature
Nitrides (Con't)	parallel to a axis	$15.88x10^3$	25 °C
		$15.14x10^3$	300 °C
		$3.84x10^3$	700 °C
		$2.18x10^3$	1000 °C
	Titanium Mononitride (TiN)	$34x10^3$	
	(10wt% AlO and 10wt% AlN)	$13.34x10^3$	
	(30wt% AlO and 10wt% AlN)	$23.93x10^3$	
	(30wt% AlO and 30wt% AlN)	$33.25x10^3$	

To convert from **psi** to **MPa**, multiply by **145**.

Source: data compiled by J.S. Park from *No. 1 Materials Index*, Peter T.B. Shaffer, Plenum Press, New York, (1964); *Smithells Metals Reference Book*, Eric A. Brandes, ed., in association with Fulmer Research Institute Ltd. 6th ed. London, Butterworths, Boston, (1983); and *Ceramic Source*, American Ceramic Society (1986-1991).

Table 244. Modulus of Rupture for Ceramics

(Sheet 6 of 10)

Class	Ceramic	Modulus of Rupture (psi)	Temperature
Nitrides (Con't)	Trisilicon Tetranitride (Si_3N_4)		
	(hot pressed)	$65.3\text{-}159.5 \times 10^3$	20°C
	(sintered)	$39.9\text{-}121.8 \times 10^3$	20°C
	(reaction sintered)	$7.25\text{-}43.5 \times 10^3$	20°C
Oxides	Aluminum Oxide (Al_2O_3)		
	(single crystal)	131×10^3	
		60×10^3	room temp.
	(80% dense, 3μm grain size)	56×10^3	20 °C
	(80% dense, 3μm grain size)	62×10^3	600 °C
	(80% dense, 3μm grain size)	58×10^3	900 °C
	(80% dense, 3μm grain size)	42×10^3	1100 °C

To convert from **psi** to **MPa**, multiply by **145**.

Source: data compiled by J.S. Park from *No. 1 Materials Index*, Peter T.B. Shaffer, Plenum Press, New York, (1964); *Smithells Metals Reference Book*, Eric A. Brandes, ed., in association with Fulmer Research Institute Ltd. 6th ed. London, Butterworths, Boston, (1983); and *Ceramic Source*, American Ceramic Society (1986-1991).

Table 244. MODULUS OF RUPTURE FOR CERAMICS

(SHEET 7 OF 10)

Class	Ceramic	Modulus of Rupture (psi)	Temperature
Oxides (Con't)	Aluminum Oxide (Al_2O_3) (Con't)		
	(80% dense, 20μm grain size)	$30x10^3$	20 °C
	(80% dense, 20μm grain size)	$28x10^3$	600 °C
	(80% dense, 20μm grain size)	$31x10^3$	900 °C
	(80% dense, 20μm grain size)	$30x10^3$	1100 °C
	(zirconia toughened alumina, 15 vol% ZrO_2)	$137x10^3$	
	(zirconia toughened alumina, 25 vol% ZrO_2)	$139x10^3$	
	(zirconia toughened alumina, 50 vol% ZrO_2)	$145x10^3$	

To convert from **psi** to **MPa**, multiply by **145**.

Source: data compiled by J.S. Park from *No. 1 Materials Index*, Peter T.B. Shaffer, Plenum Press, New York, (1964); *Smithells Metals Reference Book*, Eric A. Brandes, ed., in association with Fulmer Research Institute Ltd. 6th ed. London, Butterworths, Boston, (1983); and *Ceramic Source*, American Ceramic Society (1986-1991).

Table 244. MODULUS OF RUPTURE FOR CERAMICS

(SHEET 8 OF 10)

Class	Ceramic	Modulus of Rupture (psi)	Temperature
Oxides (Con't)	Beryllium Oxide (BeO)	$24\text{-}29 \times 10^3$	room temp.
	Dichromium Trioxide (Cr_2O_3)	$>38 \times 10^3$	
	Hafnium Dioxide (HfO_2)	10×10^3	
	Titanium Oxide (TiO_2)	$10\text{-}14.9 \times 10^3$	room temp.
	Zirconium Oxide (ZrO_2)		
	(5-10 CaO stabilized)	$20\text{-}35 \times 10^3$	room temp.
	(MgO stabilized)	30×10^3	room temp.
	(hot pressed yittria doped zirconia)	222×10^3	
	(sintered yittria doped zirconia)	148×10^3	

To convert from **psi** to **MPa**, multiply by **145**.

Source: data compiled by J.S. Park from *No. 1 Materials Index*, Peter T.B. Shaffer, Plenum Press, New York, (1964); *Smithells Metals Reference Book*, Eric A. Brandes, ed., in association with Fulmer Research Institute Ltd. 6th ed. London, Butterworths, Boston, (1983); and *Ceramic Source*, American Ceramic Society (1986-1991).

Table 244. MODULUS OF RUPTURE FOR CERAMICS

(SHEET 9 OF 10)

Class	Ceramic	Modulus of Rupture (psi)	Temperature
Oxides (Con't)	Cordierite ($2MgO\ 2Al_2O_3\ 5SiO_2$)		
	(ρ=2.51g/cm^3)	16x10^3	25°C
	(ρ=2.3g/cm^3)	15x10^3	400°C
	(ρ=2.1g/cm^3)	8x10^3	800°C
	(ρ=1.8g/cm^3)	3.4x10^3	1200°C
	Mullite ($3Al_2O_3\ 2SiO_2$)	6-27x10^3	25°C
	(ρ=2.77g/cm^3)	8.5x10^3	25°C
	(ρ=2.77g/cm^3)	13.5x10^3	400°C
	(ρ=2.77g/cm^3)	16.7x10^3	800°C
	(ρ=2.77g/cm^3)	11.5x10^3	1200°C

To convert from **psi** to **MPa**, multiply by **145**.

Source: data compiled by J.S. Park from *No. 1 Materials Index*, Peter T.B. Shaffer, Plenum Press, New York, (1964); *Smithells Metals Reference Book*, Eric A. Brandes, ed., in association with Fulmer Research Institute Ltd. 6th ed. London, Butterworths, Boston, (1983); and *Ceramic Source*, American Ceramic Society (1986-1991).

Table 244. Modulus of Rupture for Ceramics

(Sheet 10 of 10)

Class	Ceramic	Modulus of Rupture (psi)	Temperature
Silicide	Molybdenum Disilicide ($MoSi_2$) ($\rho = 5.57\ g/cm^3$)	$18.57x10^3$	room temp.
	(sintered)	$50.7x10^3$	room temp.
	(sintered)	$67.25x10^3$	980°C
	(sintered)	$86.00x10^3$	1090°C
	(hot pressed)	$36\text{-}57x10^3$	room temp.
	(hot pressed)	$72.00x10^3$	1090°C
	(hot pressed)	$55.00x10^3$	1200°C

To convert from **psi** to **MPa**, multiply by **145**.

Source: data compiled by J.S. Park from *No. 1 Materials Index*, Peter T.B. Shaffer, Plenum Press, New York, (1964); *Smithells Metals Reference Book*, Eric A. Brandes, ed., in association with Fulmer Research Institute Ltd. 6th ed. London, Butterworths, Boston, (1983); and *Ceramic Source*, American Ceramic Society (1986-1991).

Table 245. Rupture Strength of Refractory Metal Alloys

(Sheet 1 of 2)

Class	Alloy	Alloying Additions (%)	Form	Condition	Temperature (°F)	10-h rupture (ksi)
Niobium and Niobium Alloys	Pure Niobium	—	All	Recrystallized	2000	5.4
	Nb–1Zr	1 Zr	All	Recrystallized	2000	14
	SCb291	10 Ta, 10 W	Bar, Sheet	Recrystallized	2000	9
	C129	10 W, 10 Hf, 0.1 Y	Sheet	Recrystallized	2400	15
	FS85	28 Ta, 11 W, 0.8 Zr	Sheet	Recrystallized	2400	12
	SU31	17 W, 3.5 Hf, 0.12 C, 0.03 Si	Bar, Sheet	Special Thermal Processing	2400	22
Molybdenum and Molybdenum Alloys	Pure Molybdenum	—	All	Stress-relieved Annealed	1800	25
	Low C Mo	None	All	Stress-relieved Annealed	1800	24
	TZM	0.5 Ti, 0.08 Zr, 0.015 C	All	Stress-relieved Annealed	2400	23
	TZC	1.0 Ti, 0.14 Zr, 0.02 to 0.08 C	All	Stress-relieved Annealed	2400	28
	Mo–5Re	5 Re	All	Stress-relieved Annealed	3000	1
	Mo–30W	30 W	All	Stress-relieved Annealed	2000	20

To convert (ksi) to (MPa), multiply by 6.89

Data from *ASM Engineering Materials Reference Book, Second Edition*, Michael Bauccio, Ed., ASM International, Materials Park, OH, p106, (1994).

Table 245. Rupture Strength of Refractory Metal Alloys

(Sheet 2 of 2)

Class	Alloy	Alloying Additions (%)	Form	Condition	Temperature (°F)	10-h rupture (ksi)
Tantalum Alloys	Unalloyed	None	All	Recrystallized	2400	2.5
	TA–10W	10 W	All	Recrystallized	2400	20
Tungsten Alloys	Unalloyed	None	Bar, Sheet, Wire	Stress-relieved Annealed	3000	6.8
	W–2 ThO_2	2 ThO_2	Bar, Sheet, Wire	Stress-relieved Annealed	3000	18
	W–3 ThO_2	3 ThO_2	Bar, Wire	Stress-relieved Annealed	3000	18
	W–4 ThO_2	4 ThO_2	Bar	Stress-relieved Annealed	3000	18
	W–15 Mo	15 Mo	Bar, Wire	Stress-relieved Annealed	3000	12
	W–50 Mo	50 Mo	Bar, Wire	Stress-relieved Annealed	3000	12
	W–25 Re	25 Re	Bar, Sheet, Wire	Stress-relieved Annealed	3000	10

To convert (ksi) to (MPa), multiply by 6.89

Data from *ASM Engineering Materials Reference Book, Second Edition*, Michael Bauccio, Ed., ASM International, Materials Park, OH, p106, (1994).

Table 246. Rupture Strength of Superalloys (Sheet 1 of 3)

Alloy*	Temperature (°C)	Stress Rupture	
		100 h (MPa)	1000 h (MPa)
Incoloy 800	650	220	145
	760	115	69
	870	45	33
Incoloy 801	650	250	—
	730	145	—
	815	62	—
Incoloy 802	650	240	170
	760	145	105
	870	97	62
Inconel 600	815	55	39
	870	37	24
Inconel 601(a)	540	—	400
	870	48	30
	980	23	14
Inconel 617(b)	815	140	97
	925	62	—
	980	41	—
Inconel 625(a)	650	440	370
	815	130	93
	870	72	48
Inconel 718(c)	540	—	951
	595	860	760
	650	690	585
Inconel 751(d)	815	200	125
	870	120	69

Data from *ASM Metals Reference Book, Third Edition*, Michael Bauccio, Ed., ASM International, Materials Park, OH, p391, (1993).

Table 246. RUPTURE STRENGTH OF SUPERALLOYS

(SHEET 2 OF 3)

Alloy*	Temperature (°C)	Stress Rupture 100 h (MPa)	Stress Rupture 1000 h (MPa)
Inconel X–750(e)	540	—	827
	870	83	45
	925	58	21
N–155, bar(f)	650	360	295
	730	195	150
	870	97	66
N–155(g)	650	380	290
N–155, sheet(f)	980	39	20
Nimonic 75(h)	815	38	24
	870	23	15
	925	14	10
	980	—	7.6
Nimonic 80A(j)	540	—	825
	815	185	115
	870	105	—
Nimonic 90(j)	815	240	155
	870	150	69
	925	69	—
Nimonic 105(k)	815	325	225
	870	210	135
Nimonic 115(m)	815	425	315
	870	315	205
	925	205	130

Data from *ASM Metals Reference Book, Third Edition*, Michael Bauccio, Ed., ASM International, Materials Park, OH, p391, (1993).

Table 246. RUPTURE STRENGTH OF SUPERALLOYS

(SHEET 3 OF 3)

Alloy*	Temperature (°C)	Stress Rupture 100 h (MPa)	Stress Rupture 1000 h (MPa)
Nimonic 263(n)	815	170	105
	870	93	46
	925	45	—

Data from *ASM Metals Reference Book, Third Edition*, Michael Bauccio, Ed., ASM International, Materials Park, OH, p391, (1993).

* (a) Solution treat 1150 °C.

(b) Solution treat 1175 °C.

(c) Heat treat to 980 °C plus 720 °C hold for 8 h, furnace cool to 620 °C hold for 8 h.

(d) 730 °C hold for 2h.

(e) Heat treat to 1150 °C plus 840 °C hold for 24h, plus 705 °C hold for 20h.

(f) Solution treated and aged.

(g) Stress-relieved forging.

(h) Heat treat to 1050 °C hold for 1 h.

(j) Heat treat to 1080 °C hold for 8 h, plus 700 °C hold for 16 h.

(k) Heat treat to 1150 °C hold for 4 h, plus 1050 °C hold for 16 h, plus 850 °C hold for 16 h.

(m) Heat treat to 1190 °C hold for 1.5 h, plus 1100 °C hold for 6 h.

(n) Heat treat to 1150 °C hold for 2 h, water quench, plus 800 °C hold for 8 h.

Table 247. Modulus of Rupture for Si_3N_4 and Al_2O_3 Composites

Matrix	Dispersed Phase	Modulus of Rupture (MPa)		
		RT	1000 °C	1200 °C
Si_3N_4+ 6 wt % Y_2O_3	None	110.9 ± 1.6	88.3 ± 3.5	49.2 ± 5.0
Si_3N_4+ 6 wt % Y_2O_3	TiC	80.6 ± 5.9	120.4 ± 12.2	64.4 ± 2.9
	(Ti, W) C	75.5 ± 3.2	86 ± 0	52.9 ± 0.5
	WC	89.1 ± 31.8	136.4 ± 1.6	55.7 ± 0.5
	TaC	86.2 ± 7.3	124.5 ± 16.0	43.2 ± 2.0
	HfC	86 ± 0.8	—	68.6 ± 0.5
	SiC	97.6 ± 8.5	94.0 ± 4.9	52.3 ± 3.2
Al_2O_3	TiC	72.2 ± 13.0	69.4 ± 4.3	57.0 ± 4.1

Containing 30 Vol % of Metal Carbide Dispersoid (2 μm average particle diameter)

Data from *ASM Engineering Materials Reference Book, Second Edition*, Michael Bauccio, Ed., ASM International, Materials Park, OH, p169,(1994).

Table 248. POISSON'S RATIO OF WROUGHT TITANIUM ALLOYS

Class	Metal or Alloy	Poisson's Ratio
Commercially Pure	99.5 Ti	0.34
	99.2 Ti	0.34
	99.1 Ti	0.34
	99.0Ti	0.34
	99.2 Ti–0.2Pd	0.34
Near Alpha Alloys	Ti-8Al-1Mo-1V	0.32
	Ti-5Al-5Sn-2Zr-2Mo-0.25Si	0.326
Alpha-Beta Alloys	Ti-6Al-4V	0.342
	Ti-6Al-4V (low O_2)	0.342
	Ti-6Al-2Sn-2Zr-2Mo-2Cr-0.25Si	0.327
Beta Alloys	Ti-13V-11Cr-3Al	0.304

Data from *ASM Metals Reference Book, Third Edition*, Michael Bauccio, Ed., ASM International, Materials Park, OH, p511, (1993).

Table 249. POISSON'S RATIO FOR CERAMICS
(SHEET 1 OF 2)

Class	Ceramic	Poisson's Ratio
Borides	Titanium Diboride (TiB_2)	0.09-0.28
	(6.0 μm grain size, ρ=4.46g/cm^3)	0.10
	(3.5 μm grain size, ρ=4.37g/cm^3, 0.8wt% Ni)	0.12
	(6.0 μm grain size, ρ=4.56g/cm^3, 0.16wt% Ni)	0.11
	(12.0 μm grain size, ρ=4.66g/cm^3, 9.6wt% Ni)	0.15
	Zirconium Diboride (ZrB_2)	0.144
Carbides	Boron Carbide (B_4C)	0.207
	Hafnium Monocarbide (HfC)	0.166
	Silicon Carbide (SiC) (ρ = 3.128 g/cm^3)	0.183-0.192 at room temp.
	Tantalum Monocarbide (TaC)	0.1719 -0.24
	Titanium Monocarbide (TiC)	0.187-189
	Tungsten Monocarbide (WC)	0.24
	Zirconium Monocarbide (ZrC) (ρ = 6.118 g/cm^3)	0.257
Nitrides	Trisilicon tetranitride (Si_3N_4)	0.24
	(presureless sintered)	0.22-0.27
Oxides	Aluminum Oxide (Al_2O_3)	0.21-0.27
	Beryllium Oxide (BeO)	0.26-0.34
	Cerium Dioxide (CeO_2)	0.27-0.31
	Magnesium Oxide (MgO) (ρ = 3.506 g/cm^3)	0.163 at room temp.
	Thorium Dioxide (ThO_2) (ρ=9.722 g/cm^3)	0.275

Source: data compiled by J.S. Park from *No. 1 Materials Index,* Peter T.B. Shaffer, Plenum Press, New York, (1964); *Smithells Metals Reference Book,* Eric A. Brandes, ed., in association with Fulmer Research Institute Ltd. 6th ed. London, Butterworths, Boston, (1983); and *Ceramic Source,* American Ceramic Society (1986-1991)

Table 249. POISSON'S RATIO FOR CERAMICS

(SHEET 2 OF 2)

Class	Ceramic	Poisson's Ratio
Oxides (Con't)	Titanium Oxide (TiO_2)	0.28
	Uranium Dioxide (UO_2)	
	(ρ=10.37 g/cm^3)	0.302
	Zirconium Oxide (ZrO_2)	0.324-0.337 at room temp.
	(partially stabilized)	0.23
	(fully stabilized)	0.23-0.32
	(plasma sprayed)	0.25
	Cordierite ($2MgO\ 2Al_2O_3\ 5SiO_2$)	
	(ρ=2.3g/cm^3)	0.21
	(ρ=2.1g/cm^3)	0.17
	(glass)	0.26
	Mullite ($3Al_2O_3\ 2SiO_2$)	
	(ρ=2.779 g/cm^3)	0.238
	Spinel (Al_2O_3 MgO)	
	(ρ=3.510 g/cm^3)	0.294
Silicide	Molybdenum Disilicide ($MoSi_2$)	0.158-0.172

Source: data compiled by J.S. Park from *No. 1 Materials Index,* Peter T.B. Shaffer, Plenum Press, New York, (1964); *Smithells Metals Reference Book*, Eric A. Brandes, ed., in association with Fulmer Research Institute Ltd. 6th ed. London, Butterworths, Boston, (1983); and *Ceramic Source,* American Ceramic Society (1986-1991)

Table 250. POISSON'S RATIO OF GLASS
(SHEET 1 OF 2)

Class	Composition	Poisson's Ratio	Temperature
SiO_2 glass		0.166–0.177	room temp.
SiO_2–Na_2O glass	(15% mol Na_2O)	0.183	room temp.
	(20% mol Na_2O)	0.203	room temp.
	(25% mol Na_2O)	0.219	room temp.
	(30% mol Na_2O)	0.236	room temp.
	(33% mol Na_2O)	0.249	room temp.
	(35% mol Na_2O)	0.248	room temp.
SiO_2–PbO glass	(24.6% mol PbO)	0.249	
	(30.0% mol PbO)	0.174	
	(35.7% mol PbO)	0.252	
	(38.4% mol PbO)	0.150	
	(45.0% mol PbO)	0.219	
	(50.0% mol PbO)	0.259	
	(55.0% mol PbO)	0.222	
	(60.0% mol PbO)	0.281	
	(65.0% mol PbO)	0.283	
B_2O_3 glass		0.288–0.309	room temp.

Source: data compiled by J.S. Park from O. V. Mazurin, M. V. Streltsina and T. P. Shvaiko–Shvaikovskaya, *Handbook of Glass Data, Part A and Part B*, Elsevier, New York, 1983

Table 250. POISSON'S RATIO OF GLASS
(SHEET 2 OF 2)

Class	Composition	Poisson's Ratio	Temperature
B_2O_3–Na_2O glass	(5.5% mol Na_2O)	0.279	
	(10% mol Na_2O)	0.2740	15°C
	(15.4% mol Na_2O)	0.271	
	(20% mol Na_2O)	0.2860	15°C
	(22.8% mol Na_2O)	0.272	
	(25% mol Na_2O)	0.2713	15°C
	(29.8% mol Na_2O)	0.274	
	(33.3% mol Na_2O)	0.2771	15°C
	(37% mol Na_2O)	0.2739	15°C
	(37.25% mol Na_2O)	0.292	

Source: data compiled by J.S. Park from O. V. Mazurin, M. V. Streltsina and T. P. Shvaiko–Shvaikovskaya, *Handbook of Glass Data, Part A and Part B*, Elsevier, New York, 1983

Table 251. POISSON'S RATIO OF SILICON CARBIDE SCS–2–AL

Fiber orientation	No. of plies	Poisson's Ratio
0°	6, 8, 12	0.268
90°	6, 12,40	0.124
± 45°	8, 12, 40	0.395

Data from *ASM Engineering Materials Reference Book, Second Edition*, Michael Bauccio, Ed., ASM International, Materials Park, OH, p149,(1994).

Table 252. COMPRESSION POISSON'S RATIO OF TREATED DUCTILE IRONS

Treatment	Compression Poisson's Ratio
60-40-18	0.26
65-45-12	0.31
80-55-06	0.31
120 90-02	0.27

Source: data from ASM *Metals Reference Book, Second Edition*, American Society for Metals, Metals Park, Ohio 44073, p169-170, (1984).

Table 253. TORSION POISSON'S RATIO OF TREATED DUCTILE IRONS

Treatment	Torsion Poisson's Ratio
60-40-18	0.29
65-45-12	0.29
80-55-06	0.31
120 90-02	0.28

Source: data from *ASM Metals Reference Book, Second Edition*, American Society for Metals, Metals Park, Ohio 44073, p169-170, (1984).

Table 254. ELONGATION OF TOOL STEELS

Type	Condition	Elongation (%)
L2	Annealed	25
	Oil quenched from 855 •C and single tempered at:	
	205 •C	5
	315 •C	10
	425 •C	12
	540 •C	15
	650 •C	25
L6	Annealed	25
	Oil quenched from 845 •C and single tempered at:	
	315 •C	4
	425 •C	8
	540 •C	12
	650 •C	20
S1	Annealed	24
	Oil quenched from 930 •C and single tempered at:	
	205 •C	
	315 •C	4
	425 •C	5
	540 •C	9
	650 •C	12
S5	Annealed	25
	Oil quenched from 870 •C and single tempered at:	
	205 •C	5
	315 •C	7
	425 •C	9
	540 •C	10
	650 •C	15
S7	Annealed	25
	Fan cooled from 940 •C and single tempered at:	
	205 •C	7
	315 •C	9
	425 •C	10
	540 •C	10
	650 •C	14

Source: Data from *ASM Metals Reference Book, Second Edition*, American Society for Metals, Metals Park, Ohio 44073, p241, (1984).

Table 255. ELONGATION OF DUCTILE IRONS

Specification Number	Grade or Class	Elongation (%)
ASTM A395-76 ASME SA395	60-40-18	18
ASTM A476-70(d); SAE AMS5316	80-60-03	3
ASTM A536-72, MIL-1-11466B(MR)	60-40-18	18
	65-45-12	12
	80-55-06	6
	100-70-03	3
	120-90-02	2
SAE J434c	D4018	18
	D4512	12
	D5506	6
	D7003	3
MIL-I-24137(Ships)	Class A	15
	Class B	7
	Class C	20

Source: data from ASM *Metals Reference Book, Second Edition*, American Society for Metals, Metals Park, Ohio 44073, p169, (1984).

Table 256. ELONGATION OF MALLEABLE IRON CASTINGS

Specification Number	Grade or Class	Elongation (%)
Ferritic		
ASTM A47, A338; ANSI G48.1;		
FED QQ-I-666c	32510	10
	35018	18
ASTM A197		5
Pearlitic and Martensitic		
ASTM A220; ANSI C48.2;		
MIL-I-11444B	40010	10
	45008	8
	45006	6
	50005	5
	60004	4
	70003	3
	80002	2
	90001	1
Automotive		
ASTM A602; SAE J158	M3210	10
	M4504(a)	4
	M5003(a)	3
	M5503(b)	3
	M7002(b)	2
	M8501(b)	1

(a) Air quenched and tempered
(b) Liquid quenched and tempered

Source: data from *ASM Metals Reference Book, Second Edition*, American Society for Metals, Metals Park, Ohio 44073, p171, (1984).

Table 257. ELONGATION OF FERRITIC STAINLESS STEELS

(SHEET 1 OF 2)

Type	ASTM Specification	Form	Condition	Elongation (%)
Type 405 (UNS S40500)	A580	Wire	Annealed	20
	A580		Annealed, Cold Finished	16
Type 409 (UNS S40900)	—	Bar	Annealed	25(a)
Type 429 (UNS S42900)	—	Bar	Annealed	30(a)
Type 430 (UNS S43000)	A276	Bar	Annealed, Hot Finished	20
	A276		Annealed, Cold Finished	16
Type 430Ti(UNS S43036)	—	Bar	Annealed	30(a)
Type 434 (UNS S43400)	—	Wire	Annealed	33(a)
Type 436 (UNS S43600)	—	Sheet, Strip	Annealed	23(a)

(a) Typical Values

Data from *ASM Metals Reference Book, Third Edition*, Michael Bauccio, Ed., ASM International, Materials Park, OH, p368 (1993).

Table 257. ELONGATION OF FERRITIC STAINLESS STEELS

(SHEET 2 OF 2)

Type	ASTM Specification	Form	Condition	Elongation (%)
Type 442 (UNS S44200)	—	Bar	Annealed	20(a)
Type 444 (UNS S44400)	A176	Plate, Sheet, Strip	Annealed	20
Type 446 (UNS S44600)	A276	Bar	Annealed, Hot Finished	20
	A276		Annealed, Cold Finished	16

(a) Typical Values

Data from *ASM Metals Reference Book, Third Edition*, Michael Bauccio, Ed., ASM International, Materials Park, OH, p368 (1993).

Table 258. ELONGATION OF MARTENSITIC STAINLESS STEELS
(SHEET 1 OF 3)

Type	ASTM Specification	Form	Condition	Elongation (%)
Type 403 (UNS S40300)	A276	Bar	Annealed, hot finished	20
	A276		Annealed, cold finished	16
	A276		Intermediate temper, hot finished	15
	A276		Intermediate temper, cold finished	12
	A276		Hard temper, hot finished	12
	A276		Hard temper, cold finished	12
Type 410 (UNS S41000)	A276	Bar	Annealed, hot finished	20
	A276		Annealed, cold finished	16
	A276		Intermediate temper, hot finished	15
	A276		Intermediate temper, cold finished	12
	A276		Hard temper, hot finished	12
	A276		Hard temper, cold finished	12
Type 410S (UNS S41008)	A176	Plate, Sheet, Strip	Annealed	22
Type 410Cb (UNS S41040)	A276	Bar	Annealed, hot finished	13
	A276		Annealed, cold finished	12
	A276		Intermediate temper, hot finished	13
	A276		Intermediate temper, cold finished	12

Data from *ASM Metals Reference Book, Third Edition*, Michael Bauccio, Ed., ASM International, Materials Park, OH, p369-370 (1993).

Table 258. ELONGATION OF MARTENSITIC STAINLESS STEELS

(SHEET 2 OF 3)

Type	ASTM Specification	Form	Condition	Elongation (%)
Type 414 (UNS S41400)	A276	Bar	Intermediate temper, hot finished	15
	A276		Intermediate temper, cold finished	15
Type 414L	—	Bar	Annealed	20
Type 420 (UNS S42000)	—	Bar	Tempered 205 °C	8
Type 422 (UNS S42200)	A565	Bar	Intermediate and hard tempers for high-temperature service	13
Type 431 (UNS S43100)	—	Bar	Tempered 260 °C	16
	—		Tempered 595 °C	19
Type 440A (UNS S44002)	—	Bar	Annealed	20
	—		Tempered 315 °C	5
Type 440B (UNS S44003)	—	Bar	Annealed	18
	—		Tempered 315 °C	3
Type 440C (UNS S44004)	—	Bar	Annealed	14
	—		Tempered 315 °C	2

Data from *ASM Metals Reference Book, Third Edition*, Michael Bauccio, Ed., ASM International, Materials Park, OH, p369-370 (1993).

Table 258. ELONGATION OF MARTENSITIC STAINLESS STEELS

(SHEET 3 OF 3)

Type	ASTM Specification	Form	Condition	Elongation (%)
Type 501 (UNS S50100)	—	Bar, Plate	Annealed	28
	—		Tempered 540 °C	15
Type 502 (UNS S50200)	—	Bar, Plate	Annealed	30

Data from *ASM Metals Reference Book, Third Edition*, Michael Bauccio, Ed., ASM International, Materials Park, OH, p369-370 (1993).

Table 259. Elongation of Precipitation-Hardening Austenitic Stainless Steels

Type	Form	Condition	Elongation (%)
PH 13–8 Mo (UNS S13800)	Bar, Plate, Sheet, Strip	H950	6-10
		H1000	6-10
15–5 PH (UNS S15500) and 17–4 PH (UNS S17400)	Bar, Plate, Sheet, Stript	H900	10(a)
		H925	10(a)
		H1025	12(a)
		H1075	13(a)
		H1100	14(a)
		H1150	16(a)
		H1150M	18(a)
17–7 PH (UNS S17700)	Bar	RH950	6
		TH1050	6

(a) For flat rolled products, value varies with thickness.

Data from *ASM Metals Reference Book, Third Edition*, Michael Bauccio, Ed., ASM International, Materials Park, OH, p371 (1993).

Table 260. Elongation of High–Nitrogen Austenitic Stainless Steels

Type	ASTM Specification	Form	Condition	Elongation (%)
Type 201 (UNS S20100)	A276	Bar	Annealed	40
Type 202 (UNS S20200)	A276	Bar	Annealed	40
Type 205 (UNS S20500)	—	Plate	Annealed*	58
Type 304N (UNS S30451)	A276	Bar	Annealed	30
Type 304HN (UNS S30452)	—	Bar	Annealed	30
Type 316N (UNS S31651)	A276	Bar	Annealed	30

Data from *ASM Metals Reference Book, Third Edition*, Michael Bauccio, Ed., ASM International, Materials Park, OH, p367 (1993).

* Typical values

Table 261. TOTAL ELONGATION OF CAST ALUMINUM ALLOYS (SHEET 1 OF 3)

Alloy AA No.	Temper	Elongation (in 2 in.) (%)
201.0	T4	20
	T6	7
	T7	4.5
206.0, A206.0	T7	11.7
208.0	F	2.5
242.0	T21	1.0
	T571	0.5
	T77	2.0
	T571	1.0
	T61	0.5
295.0	T4	8.5
	T6	5.0
	T62	2.0
296.0	T4	9.0
	T6	5.0
	T7	4.5
308.0	F	2.0
319.0	F	2.0
	T6	2.0
	F	2.5
	T6	3.0
336.0	T551	0.5
	T65	0.5
354.0	T61	6.0
355.0	T51	1.5
	T6	3.0
	T61	1.0
	T7	0.5

Source: data from *ASM Metals Reference Book, Second Edition*, American Society for Metals, Metals Park, Ohio 44073, (1984).

Table 261. TOTAL ELONGATION OF CAST ALUMINUM ALLOYS (SHEET 2 OF 3)

Alloy AA No.	Temper	Elongation (in 2 in.) (%)
355.0 (Con't)	T71	1.5
	T51	2.0
	T6	4.0
	T62	1.5
	T7	2.0
	T71	3.0
356.0	T51	2.0
	T6	3.5
	T7	2.0
	T71	3.5
	T6	5.0
	T7	6.0
357.0, A357.0	T62	8.0
359.0	T61	6.0
	T62	5.5
360.0	F	3.0
A360.0	F	5.0
380.0	F	3.0
383.0	F	3.5
384.0, A384.0	F	2.5
390.0	F	1.0
	T5	1.0
A390.0	F,T5	<1.0
	T6	<1.0
	T7	<1.0
	F,T5	1.0
	T6	<1.0
	T7	<1.0

Source: data from *ASM Metals Reference Book, Second Edition*, American Society for Metals, Metals Park, Ohio 44073, (1984).

Table 261. TOTAL ELONGATION OF CAST ALUMINUM ALLOYS (SHEET 3 OF 3)

Alloy AA No.	Temper	Elongation (in 2 in.) (%)
413.0	F	2.5
A413.0	F	3.5
443.0	F	8.0
B443.0	F	10.0
C443.0	F	9.0
514.0	F	9.0
518.0	F	5.0—8.0
520.0	T4	16
535.0	F	13
712.0	F	5.0
713.0	T5	3.0
	T5	4.0
771.0	T6	9.0
850.0	T5	10.0

Source: data from *ASM Metals Reference Book, Second Edition*, American Society for Metals, Metals Park, Ohio 44073, (1984).

Table 262. ELONGATION OF WROUGHT COPPERS AND COPPER ALLOYS

(SHEET 1 OF 11)

UNS Number and Name	Nominal Composition (%)	Commercial Forms(a)	Elongation in 2 In (%)
C10100 Oxygen-free electronic	99.99 Cu	F, R, W, T, P, S	55(4)
C10200 Oxygen-free copper	99.95 Cu	F, R, W, T, P, S	55(4)
C10300 Oxygen-free extra-low phosporus	99.95 Cu, 0.003 P	F, R, T, P, S	50(6)
C10400, C10500, C10700 Oxygen-free, silver-bearing	99.95 Cu(e)	F, R, W, S	55(4)
C10800 Oxygen-free, low phosporus	99.95 Cu, 0.009 P	F, R, T, P	50(4)
CS11000 Electrolytic tough pitch copper	99.90 Cu, 0.04 O	F, R, W, T, P, S	55(4)
C11100 Electrolytic tough pitch, anneal resistant	99.90 Cu, 0.04 O, 0.01 Cd	W	(60)
C11300, C11400, C11500, C11600 Silver-bearing tough pitch copper	99.90 Cu, 0.04 O, Ag(f)	F, R, W, T, S	55(4)
C12000, C12100	99.9 Cu(g)	F, T, P	55(4)
C12200 Phosphorus deoxidized copper, high residual phosphorus	99.90 Cu, 0.02 P	F, R, T, P	45(8)
C12500, C12700, C12800, C12900, C13000 Fire-refined tough pitch with silver	99.88 Cu(h)	F, R, W, S	55(4)
C14200 Phosphorus deoxidized, arsenical	99.68 Cu, 0.3 As, 0.02 P	F, R, T	45(8)

(a) F, flat products; R, rod; W, wire; T, tube; P, pipe; S, shapes.

Data from *ASM Metals Reference Book, Third Edition*, Michael Bauccio, Ed., ASM International, Materials Park, OH, p442–454, (1993).

Table 262. ELONGATION OF WROUGHT COPPERS AND COPPER ALLOYS

(SHEET 2 OF 11)

UNS Number and Name	Nominal Composition (%)	Commercial Forms(a)	Elongation in 2 In (%)
C19200	98.97 Cu, 1.0 Fe, 0.03 P	F, T	40
C14300	99.9 Cu, 0.1 Cd	F	42(1)
C14310	99.8 Cu, 0.2 Cd	F	42(1)
C14500 Phosphorus deoxidized, tellurium bearing	99.5 Cu, 0.50 Te, 0.008 P	F, R, W, T	50(3)
C14700 Sulfur bearing	99.6 Cu, 0.40 S	R, W	52(8)
C15000 Zirconium copper	99.8 Cu, 0.15 Zr	R, W	54(1.5)
C15500	99.75 Cu, 0.06 P, 0.11 Mg, Ag(i)	F	40(3)
C15710	99.8 Cu, 0.2 Al_2O_3	R, W	20(10)
C15720	99.6 Cu, 0.4 Al_2O_3	F, R	20(3.5)
C15735	99.3 Cu, 0.7 Al_2O_3	R	16(10)
C15760	98.9 Cu, 1.1 Al_2O_3	F, R	20(8)
C16200 Cadmium copper	99.0 Cu, 1.0 Cd	F, R, W	57(1)

(a) F, flat products; R, rod; W, wire; T, tube; P, pipe; S, shapes.

Data from ASM *Metals Reference Book, Third Edition*, Michael Bauccio, Ed., ASM International, Materials Park, OH, p442–454, (1993).

Table 262. ELONGATION OF WROUGHT COPPERS AND COPPER ALLOYS

(SHEET 3 OF 11)

UNS Number and Name	Nominal Composition (%)	Commercial Forms(a)	Elongation in 2 In (%)
C16500	98.6 Cu, 0.8 Cd, 0.6 Sn	F, R, W	53(1.5)
C17000 Beryllium copper	99.5 Cu, 1.7 Be, 0.20 Co	F, R	45(3)
C17200 Beryllium copper	99.5 Cu, 1.9 Be , 0.20 Co	F, R, W, T, P, S	48(1)
C17300 Beryllium copper	99.5 Cu, 1.9 Be, 0.40 Pb	R	48(3)
C17500 Copper-cobalt-beryllium alloy	99.5 Cu, 2.5 Co, 0.6 Be	F, R	28(5)
C18200, C18400, C18500 Chromium copper	99.5 Cu(j)	F, W, R, S, T	40(5)
C18700 leaded copper	99.0 Cu, 1.0 Pb	R	45(8)
C18900	98.75 Cu, 0.75 Sn, 0.3 Si, 0.20 Mn	R, W	48(14)
C19000 Copper-nickel-phosphorus alloy	98.7 Cu, 1.1 Ni, 0.25 P	F, R, W	50
C19100 Copper-nickel-phosphorus-tellurium alloy	98.15 Cu, 1.1 Ni, 0.50 Te, 0.25 P	R, F	27(6)
C19400	97.5 Cu, 2.4 Fe, 0.13 Zn, 0.03 P	F	32

(a) F, flat products; R, rod; W, wire; T, tube; P, pipe; S, shapes.

Data from *ASM Metals Reference Book, Third Edition*, Michael Bauccio, Ed., ASM International, Materials Park, OH, p442–454, (1993).

Table 262. Elongation of Wrought Coppers and Copper Alloys

(Sheet 4 of 11)

UNS Number and Name	Nominal Composition (%)	Commercial Forms(a)	Elongation in 2 In (%)
C19500	97.0 Cu, 1.5 Fe, 0.6 Sn, 0.10 P, 0.80 Co	F	15
C21000 Gilding, 95%	95.0 Cu, 5.0 Zn	F, W	45(4)
C22000 Commercial bronze, 90%	90.0 Cu, 10.0 Zn	F, R, W, T	50(3)
C22600 Jewelry bronze, 87.5%	87.5 Cu, 12.5 Zn	F, W	46(3)
C23000 Red brass, 85%	85.0 Cu, 15.0 Zn	F, W, T, P	55(3)
C24000 Low brass, 80%	80.0 Cu, 20.0 Zn	F, W	55(3)
C26000 Cartridge brass, 70%	70.0 Cu, 30.0 Zn	F, R, W, T	66(3)
C26800, C27000 Yellow brass	65.0 Cu, 35.0 Zn	F, R, W	65(3)
C28000 Muntz metal	60.0 Cu, 40.0 Zn	F, R, T	52(10)
C31400 Leaded commercial bronze	89.0 Cu, 1.75 Pb, 9.25 Zn	F, R	45(10)
C31600 Leaded commercial bronze, nickel-bearing	89.0 Cu, 1.9 Pb, 1.0 Ni, 8.1 Zn	F, R	45(12)
C33000 Low-leaded brass tube	66.0 Cu, 0.5 Pb, 33.5 Zn	T	60(7)

(a) F, flat products; R, rod; W, wire; T, tube; P, pipe; S, shapes.

Data from *ASM Metals Reference Book, Third Edition*, Michael Bauccio, Ed., ASM International, Materials Park, OH, p442–454, (1993).

Table 262. ELONGATION OF WROUGHT COPPERS AND COPPER ALLOYS

(SHEET 5 OF 11)

UNS Number and Name	Nominal Composition (%)	Commercial Forms(a)	Elongation in 2 In (%)
C33200 High-leaded brass tube	66.0 Cu, 1.6 Pb, 32.4 Zn	T	50(7)
C33500 Low-leaded brass	65.0 Cu, 0.5 Pb, 34.5 Zn	F	65(8)
C34000 Medium-leaded brass	65.0 Cu, 1.0 Pb, 34.0 Zn	F, R, W, S	60(7)
C34200 High-leaded brass	64.5 Cu, 2.0 Pb, 33.5 Zn	F, R	52(5)
C34900	62.2 Cu, 0.35 Pb, 37.45 Zn	R, W	72(18)
C35000 Medium-leaded brass	62.5 Cu, 1.1 Pb, 36.4 Zn	F, R	66(1)
C35300 High-leaded brass	62.0 Cu, 1.8 Pb, 36.2 Zn	F, R	52(5)
C35600 Extra-high-leaded brass	63.0 Cu, 2.5 Pb, 34.5 Zn	F	50(7)
C36000 Free-cutting brass	61.5 Cu, 3.0 Pb, 35.5 Zn	F, R, S	53(18)
C36500 to C36800 Leaded Muntz metal	60.0 Cu(k), 0.6 Pb, 39.4 Zn	F	45
C37000 Free-cutting Muntz metal	60.0 Cu, 1.0 Pb, 39.0 Zn	T	40(6)
C37700 Forging brass	59.0 Cu, 2.0 Pb, 39.0 Zn	R, S	45

(a) F, flat products; R, rod; W, wire; T, tube; P, pipe; S, shapes.

Data from *ASM Metals Reference Book, Third Edition*, Michael Bauccio, Ed., ASM International, Materials Park, OH, p442–454, (1993).

Table 262. Elongation of Wrought Coppers and Copper Alloys
(Sheet 6 of 11)

UNS Number and Name	Nominal Composition (%)	Commercial Forms(a)	Elongation in 2 In (%)
C38500 Architectural bronze	57.0 Cu, 3.0 Pb, 40.0 Zn	R, S	30
C40500	95 Cu, 1 Sn, 4 Zn	F	49(3)
C40800	95 Cu, 2 Sn, 3 Zn	F	43(3)
C41100	91 Cu, 0.5 Sn, 8.5 Zn	F, W	13
C41300	90.0 Cu, 1.0 Sn, 9.0 Zn	F, R, W	45
C41500	91 Cu, 1.8 Sn, 7.2 Zn	F	44
C42200	87.5 Cu, 1.1 Sn, 11.4 Zn	F	46
C42500	88.5 Cu, 2.0 Sn, 9.5 Zn	F	49
C43000	87.0 Cu, 2.2 Sn, 10.8 Zn	F	55(3)
C43400	85.0 Cu, 0.7 Sn, 14.3 Zn	F	49(3)
C43500	81.0 Cu, 0.9 Sn, 18.1 Zn	F, T	46(7)
C44300, C44400, C44500 Inhibited admiralty	71.0 Cu, 28.0 Zn, 1.0 Sn	F, W, T	65(0)

(a) F, flat products; R, rod; W, wire; T, tube; P, pipe; S, shapes.

Data from *ASM Metals Reference Book, Third Edition*, Michael Bauccio, Ed., ASM International, Materials Park, OH, p442–454, (1993).

Table 262. ELONGATION OF WROUGHT COPPERS AND COPPER ALLOYS
(SHEET 7 OF 11)

UNS Number and Name	Nominal Composition (%)	Commercial Forms(a)	Elongation in 2 In (%)
C46400 to C46700 Naval brass	60.0 Cu, 39.25 Zn, 0.75 Sn	F, R, T, S	50(17)
C48200 Naval brass, medium-leaded	60.5 Cu, 0.7 Pb, 0.8 Sn, 38.0 Zn	F, R, S	43(15)
C48500 Leaded naval brass	60.0 Cu, 1.75 Pb, 37.5 Zn, 0.75 Sn	F, R, S	40(15)
C50500 Phosphor bronze, 1.25% E	98.75 Cu, 1.25 Sn, trace P	F, W	48(4)
C51000 Phosphor bronze, 5% A	95.0 Cu, 5.0 Sn, trace P	F, R, W, T	64
C51100	95.6 Cu, 4.2 Sn, 0.2 P	F	48
C52100 Phosphor bronze, 8% C	92.0 Cu, 8.0 Sn, trace P	F, R, W	70
C52400 Phosphor bronze, 10% D	90.0 Cu, 10.0 Sn, trace P	F, R, W	70(3)
C54400 Free-cutting phosphor bronze	88.0 Cu, 4.0 Pb, 4.0 Zn, 4.0 Sn	F, R	50(15)
C60800 Aluminum bronze, 5%	95.0 Cu, 5.0 Al	T	55
C61000	92.0 Cu, 8.0 Al	R, W	65(25)
C61300	92.65 Cu, 0.35 Sn, 7.0 Al	F, R, T, P, S	42(35)

(a) F, flat products; R, rod; W, wire; T, tube; P, pipe; S, shapes.

Data from *ASM Metals Reference Book, Third Edition*, Michael Bauccio, Ed., ASM International, Materials Park, OH, p442–454, (1993).

Table 262. Elongation of Wrought Coppers and Copper Alloys
(Sheet 8 of 11)

UNS Number and Name	Nominal Composition (%)	Commercial Forms(a)	Elongation in 2 In (%)
C61400 Aluminum bronze, D	91.0 Cu, 7.0 Al, 2.0 Fe	F, R, W, T, P, S	45(32)
C61500	90.0 Cu, 8.0 Al, 2.0 Ni	F	55(1)
C61800	89.0 Cu, 1.0 Fe, 10.0 Al	R	28(23)
C61900	86.5 Cu, 4.0 Fe, 9.5 Al	F	30(1)
C62300	87.0 Cu, 10.0 Al, 3.0 Fe	F, R	35(22)
C62400	86.0 Cu, 3.0 Fe, 11.0 Al	F, R	18(14)
C62500	82.7 Cu, 4.3 Fe, 13.0 Al	F, R	1
C63000	82.0 Cu, 3.0 Fe, 10.0 Al, 5.0 Ni	F, R	20(15)
C63200	82.0 Cu, 4.0 Fe, 9.0 Al, 5.0 Ni	F, R	25(20)
C63600	95.5 Cu, 3.5 Al, 1.0 Si	R, W	64(29)
C63800	99.5 Cu, 2.8 Al, 1.8 Si, 0.40 Co	F	36(4)
C64200	91.2 Cu, 7.0 Al	F, R	32(22)

(a) F, flat products; R, rod; W, wire; T, tube; P, pipe; S, shapes.

Data from *ASM Metals Reference Book, Third Edition*, Michael Bauccio, Ed., ASM International, Materials Park, OH, p442–454, (1993).

Table 262. Elongation of Wrought Coppers and Copper Alloys
(Sheet 9 of 11)

UNS Number and Name	Nominal Composition (%)	Commercial Forms(a)	Elongation in 2 In (%)
C65100 Low-silicon bronze, B	98.5 Cu, 1.5 Si	R, W, T	55(11)
C65500 High-silicon bronze, A	97.0 Cu, 3.0 Si	F, R, W, T	63(3)
C66700 Manganese brass	70.0 Cu, 28.8 Zn, 1.2 Mn	F, W	60
C67400	58.5 Cu, 36.5 Zn, 1.2 Al, 2.8 Mn, 1.0 Sn	F, R	28(20)
C67500 Manganese bronze, A	58.5 Cu, 1.4 Fe, 39.0 Zn, 1.0 Sn, 0.1 Mn	R, S	33(19)
C68700 Aluninum brass, arsenical	77.5 Cu, 20.5 Zn, 2.0 Al, 0.1 As	T	55
C68800	73.5 Cu, 22.7 Zn, 3.4 Al, 0.40 Co	F	36
C69000	73.3 Cu, 3.4 Al, 0.6 Ni, 22.7 Zn	F	40
C69400 Silicon red brass	81.5 Cu, 14.5 Zn, 4.0 Si	R	25(20)
C70400	92.4 Cu, 1.5 Fe, 5.5 Ni, 0.6 Mn	F, T	46
C70600 Copper nickel, 10%	88.7 Cu, 1.3 Fe, 10.0 Ni	F, T	42(10)
C71000 Copper nickel, 20%	79.00 Cu, 21.0 Ni	F, W, T	40(3)

(a) F, flat products; R, rod; W, wire; T, tube; P, pipe; S, shapes.

Data from *ASM Metals Reference Book, Third Edition*, Michael Bauccio, Ed., ASM International, Materials Park, OH, p442–454, (1993).

Table 262. ELONGATION OF WROUGHT COPPERS AND COPPER ALLOYS

(SHEET 10 OF 11)

UNS Number and Name	Nominal Composition (%)	Commercial Forms(a)	Elongation in 2 In (%)
C71500 Copper nickel, 30%	70.0 Cu, 30.0 Ni	F, R, T	45(15)
C71700	67.8 Cu, 0.7 Fe, 31.0 Ni, 0.5 Be	F, R, W	40(4)
C72500	88.20 Cu, 9.5 Ni, 2.3 Sn	F, R, W, T	35(1)
C73500	72.0 Cu, 18.0 Ni , 10.0 Zn	F, R, W, T	37(1)
C74500 Nickel silver, 65-10	65.0 Cu, 25.0 Zn, 10.0 Ni	F, W	50(1)
C75200 Nickel silver, 65-18	65.0 Cu, 17.0 Zn, 18.0 Ni	F, R, W	45(3)
C75400 Nickel silver, 65-15	65.0 Cu, 20.0 Zn, 15.0 Ni	F	43
C75700 Nickel silver, 65-12	65.0 Cu, 23.0 Zn, 12.0 Ni	F, W	48

(a) F, flat products; R, rod; W, wire; T, tube; P, pipe; S, shapes.

Data from *ASM Metals Reference Book, Third Edition*, Michael Bauccio, Ed., ASM International, Materials Park, OH, p442–454, (1993).

Table 262. Elongation of Wrought Coppers and Copper Alloys

(Sheet 11 of 11)

UNS Number and Name	Nominal Composition (%)	Commercial Forms(a)	Elongation in 2 In (%)
C76200	59.0 Cu, 29.0 Zn, 12.0 Ni	F, T	50(1)
C77000 Nickel silver, 55-18	55.0 Cu, 27.0 Zn, 18.0 Ni	F, R, W	40
C72200	82.0 Cu, 16.0 Ni, 0.5 Cr, 0.8 Fe, 0.5 Mn	F, T	46(6)
C78200 Leaded nickel silver, 65-8-2	65.0 Cu, 2.0 Pb, 25.0 Zn, 8.0 Ni	F	40(3)

(a) F, flat products; R, rod; W, wire; T, tube; P, pipe; S, shapes.

Data from *ASM Metals Reference Book, Third Edition*, Michael Bauccio, Ed., ASM International, Materials Park, OH, p442–454, (1993).

(d) Based on 100% for C360000.
(e) C10400, 8 oz/ton Ag; C10500, 10 oz/ton; C10700, 25 oz/ton .
(f) C11300, 8 oz/ton Ag; C11400,10 oz/ton; C11500, 16 oz/ton; C11600, 25 oz/ton
(g) C12000, 0.008 P; C12100, 0.008 P and 4 oz/ton Ag;
(h) C12700, 8 oz/ton Ag; C12800,10 oz/ton; C12900,16 oz/ton; C13000, 25 oz/ton.
(i) 8.30 oz/ton Ag.
(j) C18200, 0.9 Cr; C18400, 0.8 Cr; C18500, 0.7 Cr
(k) Rod, 61.0 Cu min.

Table 263. ELONGATION OF COMMERCIALLY PURE TIN

Temperature (°C)	Elongation in 25mm (%)
Strained at 0.2 mm/m • min	
-200	6
-160	15
-120	60
-80	89
-40	86
0	64
23	57
Strained at 0.4 mm/m • min	
15	75
50	85
100	55
150	55
200	45

Data from *ASM Metals Reference Book, Third Edition*, Michael Bauccio, Ed., ASM International, Materials Park, OH, p488, (1993).

Table 264. ELONGATION OF COBALT-BASE SUPERALLOYS

Alloy	Temperature (°C)	Elongation (%)
Haynes 25 (L–605) sheet	21	64
	540	59
	650	35
	760	12
	870	30
Haynes 188, sheet	21	56
	540	70
	650	61
	760	43
	870	73
S-816, bar	21	30
	540	27
	650	25
	760	21
	870	16

Data from *ASM Metals Reference Book, Third Edition*, Michael Bauccio, Ed., ASM International, Materials Park, OH, p387, (1993).

Table 265. ELONGATION OF NICKEL-BASE SUPERALLOYS (SHEET 1 OF 5)

Alloy	Temperature (°C)	Elongation (%)
Astroloy, bar	21	16
	540	16
	650	18
	760	21
	870	25
D–979, bar	21	15
	540	15
	650	21
	760	17
	870	18
Hastelloy X, sheet	21	43
	540	45
	650	37
	760	37
	870	50
IN–102, bar	21	47
	540	48
	650	64
	760	110
	870	110
Inconel 600, bar	21	47
	540	47
	650	39
	760	46
	870	80
Inconel 601, sheet	21	45
	540	38
	650	45
	760	73
	870	92

Data from *ASM Metals Reference Book, Third Edition*, Michael Bauccio, Ed., ASM International, Materials Park, OH, p387-389, (1993).

Table 265. ELONGATION OF NICKEL-BASE SUPERALLOYS (SHEET 2 OF 5)

Alloy	Temperature (°C)	Elongation (%)
Inconel 625, bar	21	50
	540	50
	650	35
	760	42
	870	125
Inconel 706, bar	21	19
	540	19
	650	21
	760	32
Inconel 718, bar	21	21
	540	18
	650	19
	760	25
	870	88
Inconel 718, sheet	21	22
	540	26
	650	15
	760	8
Inconel X-750, bar	21	24
	540	22
	650	9
	760	9
	870	47
M-252, bar	21	16
	540	15
	650	11
	760	10
	870	18

Data from *ASM Metals Reference Book, Third Edition*, Michael Bauccio, Ed., ASM International, Materials Park, OH, p387-389, (1993).

Table 265. ELONGATION OF NICKEL-BASE SUPERALLOYS (SHEET 3 OF 5)

Alloy	Temperature (°C)	Elongation (%)
Nimonic 75, bar	21	41
	540	41
	650	42
	760	70
	870	68
Nimonic 80A, bar	21	24
	540	24
	650	18
	760	20
	870	34
Nimonic 90, bar	21	23
	540	23
	650	20
	760	10
	870	16
Nimonic 105, bar	21	12
	540	18
	650	24
	760	22
	870	25
Nimonic 115, bar	21	25
	540	26
	650	25
	760	22
	870	18
Pyromet 860, bar	21	22
	540	15
	650	17
	760	18

Data from *ASM Metals Reference Book, Third Edition*, Michael Bauccio, Ed., ASM International, Materials Park, OH, p387-389, (1993).

Table 265. ELONGATION OF NICKEL-BASE SUPERALLOYS (SHEET 4 OF 5)

Alloy	Temperature (°C)	Elongation (%)
René 41, bar	21	14
	540	14
	650	14
	760	11
	870	19
René 95, bar	21	15
	540	12
	650	14
	760	15
Udimet 500, bar	21	32
	540	28
	650	28
	760	39
	870	20
Udimet 520, bar	21	21
	540	20
	650	17
	760	15
	870	20
Udimet 700, bar	21	17
	540	16
	650	16
	760	20
	870	27
Udimet 710, bar	21	7
	540	10
	650	15
	760	25
	870	29

Data from *ASM Metals Reference Book, Third Edition*, Michael Bauccio, Ed., ASM International, Materials Park, OH, p387-389, (1993).

Table 265. ELONGATION OF NICKEL-BASE SUPERALLOYS (SHEET 5 OF 5)

Alloy	Temperature (°C)	Elongation (%)
Unitemp AF2–1DA, bar	21	10
	540	13
	650	13
	760	8
	870	8
Waspaloy, bar	21	25
	540	23
	650	34
	760	28
	870	35

Data from *ASM Metals Reference Book, Third Edition*, Michael Bauccio, Ed., ASM International, Materials Park, OH, p387-389, (1993).

Table 266. Ductility of Refractory Metal Alloys

(Sheet 1 of 3)

Class	Alloy	Alloying Additions (%)	Form	Condition	Low Temperature Ductility*
Niobium and Niobium Alloys	Pure Niobium	—	All	Recrystallized	A
	Nb–1Zr	1 Zr	All	Recrystallized	A
	C103(KbI–3)	10 Hf, 1 Ti 0.7 Zr	All	Recrystallized	A
	SCb291	10 Ta, 10 W	Bar, Sheet	Recrystallized	A
	C129	10 W, 10 Hf, 0.1 Y	Sheet	Recrystallized	A
	FS85	28 Ta, 11 W, 0.8 Zr	Sheet	Recrystallized	A
	SU31	17 W, 3.5 Hf, 0.12 C, 0.03 Si	Bar, Sheet	Special Thermal Processing	C
Molybdenum and Molybdenum Alloys	Pure Molybdenum	—	All	Stress-relieved Annealed	B–C
	Doped Mo	K, Si; ppm levels	Wire, Sheet	Cold Worked	B
	Low C Mo	None	All	Stress-relieved Annealed	B
	TZM	0.5 Ti, 0.08 Zr, 0.015 C	All	Stress-relieved Annealed	B–C

* A excellent cryogenic ductility;
B excellent room-temperature ductility;
C may have marginal ductility at room temperature;
D normally brittle at room temperature.

Data from *ASM Metals Reference Book, Third Edition*, Michael Bauccio, Ed., ASM International, Materials Park, OH, p390, (1993).

Table 266. Ductility of Refractory Metal Alloys

(Sheet 2 of 3)

Class	Alloy	Alloying Additions (%)	Form	Condition	Low Temperature Ductility*
	TZC	1.0 Ti, 0.14 Zr, 0.02 to 0.08 C	All	Stress-relieved Annealed	B–C
	Mo–5Re	5 Re	All	Stress-relieved Annealed	B
	Mo–30W	30 W	All	Stress-relieved Annealed	B–C
Tantalum Alloys	Unalloyed	None	All	Recrystallized	A
	FS61	7.5 W(P/M)	Wire, Sheet	Cold Worked	A
	FS63	2.5 W, 0.15 Nb	All	Recrystallized	A
	TA–10W	10 W	All	Recrystallized	A
	KBI–40	40 Nb	All	Recrystallized	A
Tungsten Alloys	Unalloyed	None	Bar, Sheet, Wire	Stress-relieved Annealed	D
	Doped	K, Si, Al; ppm levels	Wire	Cold Worked	C

* A excellent cryogenic ductility;
B excellent room-temperature ductility;
C may have marginal ductility at room temperature;
D normally brittle at room temperature.

Data from *ASM Metals Reference Book, Third Edition*, Michael Bauccio, Ed., ASM International, Materials Park, OH, p390, (1993).

Table 266. Ductility of Refractory Metal Alloys

(Sheet 3 of 3)

Class	Alloy	Alloying Additions (%)	Form	Condition	Low Temperature Ductility*
	W–1 ThO_2	1ThO_2	Bar, Sheet, Wire	Stress-relieved Annealed	D
	W–2 ThO_2	2 ThO_2	Bar, Sheet, Wire	Stress-relieved Annealed	D
	W–3 ThO_2	3 ThO_2	Bar, Wire	Stress-relieved Annealed	D
	W–4 ThO_2	4 ThO_2	Bar	Stress-relieved Annealed	D
	W–15 Mo	15 Mo	Bar, Wire	Stress-relieved Annealed	D
	W–50 Mo	50 Mo	Bar, Wire	Stress-relieved Annealed	D
	W–3 Re	3 Re	Wire	Cold Worked	C
	W–25 Re	25 Re	Bar, Sheet, Wire	Stress-relieved Annealed	B

* A excellent cryogenic ductility;
B excellent room-temperature ductility;
C may have marginal ductility at room temperature;
D normally brittle at room temperature.

Data from *ASM Metals Reference Book, Third Edition*, Michael Bauccio, Ed., ASM International, Materials Park, OH, p390, (1993).

Table 267. Elongation of Wrought Titanium Alloys at Room Temperature (Sheet 1 of 3)

Class	Alloy	Condition	Elongation (%)
Commercially Pure	99.5 Ti	Annealed	30
	99.2 Ti	Annealed	28
	99.1 Ti	Annealed	25
	99.0 Ti	Annealed	20
	99.2Ti-0.2Pd	Annealed	28
	Ti-0.8Ni-0.3Mo	Annealed	25
Alpha Alloys	Ti-5Al-2.5Sn	Annealed	16
	Ti-5Al-2.5Sn (low O_2)	Annealed	16
Near Alpha Alloys	Ti-8Al-1Mo-1V	Duplex Annealed	15
	Ti-11Sn-1Mo-2.25Al-5.0Zr-1Mo-0.2Si	Duplex Annealed	15
	Ti-6Al-2Sn-4Zr-2Mo	Duplex Annealed	15
	Ti-5Al-2Sn-2Zr-2Mo-0.25Si	975 °C (1/2h), AC + 595 °C (2h), AC	13

Data from *ASM Metals Reference Book, Third Edition*, Michael Bauccio, Ed., ASM International, Materials Park, OH, p512, (1993).

Table 267. Elongation of Wrought Titanium Alloys at Room Temperature (Sheet 2 of 3)

Class	Alloy	Condition	Elongation (%)
Alpha-Beta Alloys	Ti-6Al-2Nb-1Ta-1Mo	As rolled 2.5 cm (1 in.) plate	13
	Ti-6Al-2Sn-1.5Zr-1Mo- 0.35Bi-0.1Si	Beta forge + duplex anneal	11
	Ti-8Mn	Annealed	15
	Ti-3Al-2.5V	Annealed	20
	Ti-6Al-4V	Annealed	14
		Solution + age	10
	Ti-6Al-4V(low O_2)	Annealed	15
	Ti-6Al-6V-2Sn	Annealed	14
		Solution + age	10
	Ti-7Al-4Mo	Solution + age	16
	Ti-6Al-2Sn-4Zr-6Mo	Solution + age	10
	Ti-6Al-2Sn-2Zr-2Mo- 2Cr-0.25Si	Solution + age	11
	Ti-10V-2Fe-3Al	Solution + age	10

Data from *ASM Metals Reference Book, Third Edition*, Michael Bauccio, Ed., ASM International, Materials Park, OH, p512, (1993).

Table 267. ELONGATION OF WROUGHT TITANIUM ALLOYS AT ROOM TEMPERATURE (SHEET 3 OF 3)

Class	Alloy	Condition	Elongation (%)
Beta Alloys	Ti-13V-1Cr-3Al	Solution + age	8
	Ti-8Mo-8V-2Fe-3Al	Solution + age	8
	Ti-3Al-8V-6Cr-4Mo-4Zr	Solution + age	7
		Annealed	15
	Ti-11.5Mo-6Zr-4.5Sn	Solution + age	11

Data from *ASM Metals Reference Book, Third Edition*, Michael Bauccio, Ed., ASM International, Materials Park, OH, p512, (1993).

Table 268. ELONGATION OF WROUGHT TITANIUM ALLOYS AT HIGH TEMPERATURE (SHEET 1 OF 4)

Class	Alloy	Condition	Test Temperature (°C)	Elongation (%)
Commercially Pure	99.5 Ti	Annealed	315	32
	99.2 Ti	Annealed	315	35
	99.1 Ti	Annealed	315	34
	99.0 Ti	Annealed	315	25
	99.2Ti-0.2Pd	Annealed	315	37
	Ti-0.8Ni-0.3Mo	Annealed	205	37
	Ti-0.8Ni-0.3Mo	Annealed	315	32
Alpha Alloys	Ti-5Al-2.5Sn	Annealed	315	18
	Ti-5Al-2.5Sn (low O_2)	Annealed	-195	16
			-255	15
Near Alpha Alloys	Ti-8Al-1Mo-1V	Duplex Annealed	315	20
			425	20
			540	25

Data from *ASM Metals Reference Book, Third Edition*, Michael Bauccio, Ed., ASM International, Materials Park, OH, p512, (1993).

Table 268. ELONGATION OF WROUGHT TITANIUM ALLOYS AT HIGH TEMPERATURE (SHEET 2 OF 4)

Class	Alloy	Condition	Test Temperature (°C)	Elongation (%)
	Ti-11Sn-1Mo-2.25Al-5.0Zr-1Mo-0.2Si	Duplex Annealed	315	20
			425	22
			540	24
	Ti-6Al-2Sn-4Zr-2Mo	Duplex Annealed	315	16
			425	21
			540	26
	Ti-5Al-2Sn-2Zr-2Mo-0.25Si	975 °C (1/2h), AC + 595 °C (2h), AC	315	15
			425	17
			540	19
	Ti-6Al-2Nb-1Ta-1Mo	As rolled 2.5 cm (1 in.) plate	315	20
			425	20
			540	20
	Ti-6Al-2Sn-1.5Zr-1Mo- 0.35Bi-0.1Si	Beta forge + duplex anneal	480	15

Data from *ASM Metals Reference Book, Third Edition*, Michael Bauccio, Ed., ASM International, Materials Park, OH, p512, (1993).

Table 268. ELONGATION OF WROUGHT TITANIUM ALLOYS AT HIGH TEMPERATURE (SHEET 3 OF 4)

Class	Alloy	Condition	Test Temperature (°C)	Elongation (%)
Alpha-Beta Alloys	Ti-8Mn	Annealed	315	18
	Ti-3Al-2.5V	Annealed	315	25
	Ti-6Al-4V	Annealed	315	14
		Annealed	425	18
		Annealed	540	35
		Solution + age	315	10
		Solution + age	425	12
		Solution + age	540	22
	Ti-6Al-4V(low O_2)	Annealed	160	14
	Ti-6Al-6V-2Sn	Annealed	315	18
		Solution + age	315	12
	Ti-7Al-4Mo	Solution + age	315	18
			425	20

Data from *ASM Metals Reference Book, Third Edition*, Michael Bauccio, Ed., ASM International, Materials Park, OH, p512, (1993).

Table 268. ELONGATION OF WROUGHT TITANIUM ALLOYS AT HIGH TEMPERATURE (SHEET 4 OF 4)

Class	Alloy	Condition	Test Temperature (°C)	Elongation (%)
Beta Alloys	Ti-6Al-2Sn-4Zr-6Mo	Solution + age	315	18
			425	19
			540	19
	Ti-6Al-2Sn-2Zr-2Mo- 2Cr-0.25Si	Solution + age	315	14
	Ti-10V-2Fe-3Al	Solution + age	205	13
			315	13
	Ti-13V-1Cr-3Al	Solution + age	315	19
			425	12
	Ti-8Mo-8V-2Fe-3Al	Solution + age	315	15
	Ti-3Al-8V-6Cr-4Mo-4Zr	Solution + age	315	20
			425	17
		Annealed	315	22
	Ti-11.5Mo-6Zr-4.5Sn	Solution + age	315	16

Data from ASM *Metals Reference Book, Third Edition*, Michael Bauccio, Ed., ASM International, Materials Park, OH, p512, (1993).

Table 269. TOTAL ELONGATION OF POLYMERS

(SHEET 1 OF 10)

Class	Polymer	Elongation (in 2 in.), (ASTM D638) (%)
ABS Resins; Molded, Extruded	Medium impact	5—20
	High impact	5—50
	Very high impact	20—50
	Low temperature impact	30—200
	Heat resistant	20
Acrylics; Cast, Molded, Extruded	Cast Resin Sheets, Rods:	
	General purpose, type I	2—7
	General purpose, type II	2—7
	Moldings:	
	Grades 5, 6, 8	3—5
	High impact grade	>25
	Chlorinated Polymers	
	Chlorinated polyether	130

Source: data compiled by J.S. Park from Charles T. Lynch, CRC *Handbook of Materials Science, Vol. 3*, CRC Press, Boca Raton, Florida, 1975 and *Engineered Materials Handbook, Vol.2, Engineering Plastics*, ASM International, Metals Park, Ohio, 1988.

Table 269. TOTAL ELONGATION OF POLYMERS

(SHEET 2 OF 10)

Class	Polymer	Elongation (in 2 in.), (ASTM D638) (%)
Polycarbonates	Polycarbonate	110
	Polycarbonate (40% glass fiber reinforced)	0—5
Fluorocarbons; Molded,Extruded	Polytrifluoro chloroethylene (PTFCE)	125—175
	Polytetrafluoroethylene (PTFE)	250—350
	Ceramic reinforced (PTFE)	10—200
	Fluorinated ethylene propylene(FEP)	250—330
	Polyvinylidene— fluoride (PVDF)	200—300
Epoxies; Cast, Molded, Reinforced	Standard epoxies (diglycidyl ethers of bisphenol A)	
	Cast rigid	4.4
	Cast flexible	1.5-60

Source: data compiled by J.S. Park from Charles T. Lynch, CRC *Handbook of Materials Science, Vol. 3*, CRC Press, Boca Raton, Florida, 1975 and *Engineered Materials Handbook, Vol.2, Engineering Plastics*, ASM International, Metals Park, Ohio, 1988.

Table 269. TOTAL ELONGATION OF POLYMERS

(SHEET 3 OF 10)

Class	Polymer	Elongation (in 2 in.), (ASTM D638) (%)
Epoxies—Molded, Extruded	High performance resins (cycloaliphatic diepoxides)	
	Cast, rigid	2—5
	Epoxy novolacs	
	Glass cloth laminate	2.2—4.8
Melamines; Molded	Cellulose electrical	0.6
Nylons; Molded, Extruded	Type 6	
	General purpose	30—100
	Glass fiber (30%) reinforced	2.2—3.6
	Cast	20
	Flexible copolymers	200—320
	Type 8	400
	Type 11	100—120
	Type 12	120—350

Source: data compiled by J.S. Park from Charles T. Lynch, CRC *Handbook of Materials Science, Vol. 3*, CRC Press, Boca Raton, Florida, 1975 and *Engineered Materials Handbook, Vol.2, Engineering Plastics*, ASM International, Metals Park, Ohio, 1988.

Table 269. Total Elongation of Polymers

(Sheet 4 of 10)

Class	Polymer	Elongation (in 2 in.), (ASTM D638) (%)
Nylons; Molded, Extruded (Con't)	6/6 Nylon	
	General purpose molding	15—300
	Glass fiber reinforced	1.8—2.2
	Glass fiber Molybdenum disulfide filled	3
	General purpose extrusion	90—240
	6/10 Nylon	
	General purpose	85—220
	Glass fiber (30%) reinforced	1.9
Phenolics; Molded	Type and filler	
	General: woodflour and flock	0.4—0.8
	High shock: chopped fabric or cord	0.37—0.57
	Very high shock: glass fiber	0.2
	Rubber phenolic—woodflour or flock	0.75—2.25
ABS–Polycarbonate Alloy	ABS–Polycarbonate Alloy	110

Source: data compiled by J.S. Park from Charles T. Lynch, CRC *Handbook of Materials Science, Vol. 3*, CRC Press, Boca Raton, Florida, 1975 and *Engineered Materials Handbook, Vol.2, Engineering Plastics*, ASM International, Metals Park, Ohio, 1988.

Table 269. TOTAL ELONGATION OF POLYMERS

(SHEET 5 OF 10)

Class	Polymer	Elongation (in 2 in.), (ASTM D638) (%)
PVC–Acrylic Alloy	PVC–acrylic sheet	>100
	PVC–acrylic injection molded	150
Polymides	Unreinforced	<1
	Unreinforced 2nd value	1.2
	Glass reinforced	<1
Polyacetals	Homopolymer:	
	Standard	25
	20% glass reinforced	7
	22% TFE reinforced	12
	Copolymer:	
	Standard	60—75
	25% glass reinforced	3
	High flow	40

Source: data compiled by J.S. Park from Charles T. Lynch, CRC *Handbook of Materials Science, Vol. 3*, CRC Press, Boca Raton, Florida, 1975 and *Engineered Materials Handbook, Vol.2, Engineering Plastics*, ASM International, Metals Park, Ohio, 1988.

Table 269. TOTAL ELONGATION OF POLYMERS

(SHEET 6 OF 10)

Class	Polymer	Elongation (in 2 in.), (ASTM D638) (%)
Polyester; Thermoplastic	Injection Moldings:	
	General purpose grade	300
	Glass reinforced grades	1—5
	Glass reinforced self extinguishing	5
	General purpose grade	250
	Glass reinforced grade	<5
	Asbestos—filled grade	<5
Polyesters: Thermosets	Cast polyyester	
	Rigid	1.7—2.6
	Flexible	25—300
Reinforced polyester moldings	High strength (glass fibers)	0.3—0.5
Phenylene Oxides	SE—100	50
	SE—1	60
	Glass fiber reinforced	4—6

Source: data compiled by J.S. Park from Charles T. Lynch, CRC *Handbook of Materials Science, Vol. 3*, CRC Press, Boca Raton, Florida, 1975 and *Engineered Materials Handbook, Vol.2, Engineering Plastics*, ASM International, Metals Park, Ohio, 1988.

Table 269. TOTAL ELONGATION OF POLYMERS

(SHEET 7 OF 10)

Class	Polymer	Elongation (in 2 in.), (ASTM D638) (%)
Phenylene oxides (Noryl)	Standard	50—100
Polyarylsulfone	Polyarylsulfone	15—40
Polypropylene	General purpose	100—600
	High impact	30—>200
Polypropylene (Con't)	Asbestos filled	3—20
	Glass reinforced	2—4
	Flame retardant	3—15
Polyphenylene sulfide	Standard	3
	40% glass reinforced	3—9
Polyethylenes; Molded, Extruded	Type I—lower density (0.910—0.925)	(ASTM D412)
	Melt index 0.3—3.6	500—725
	Melt index 6—26	125—675
	Melt index 200	80—100

Source: data compiled by J.S. Park from Charles T. Lynch, CRC *Handbook of Materials Science, Vol. 3*, CRC Press, Boca Raton, Florida, 1975 and *Engineered Materials Handbook, Vol.2, Engineering Plastics*, ASM International, Metals Park, Ohio, 1988.

Table 269. TOTAL ELONGATION OF POLYMERS
(SHEET 8 OF 10)

Class	Polymer	Elongation (in 2 in.), (ASTM D638) (%)
Polyethylenes; Molded, Extruded (Con't)	Type II—medium density (0.926—0.940)	
	Melt index 20	200
	Melt index 1.0—1.9	200—425
	Type III—higher density (0.941—0.965)	
	Melt index 0.2—0.9	700—1,000
	Melt Melt index 0.1—12.0	50—1,000
	Melt index 1.5—15	100—700
	High molecular weight	400
Olefin Copolymers; Molded	EEA (ethylene ethyl acrylate)	650
	EVA (ethylene vinyl acetate)	650
	Ethylene butene	20
	Ionomer	450
	Polyallomer	300—400

Source: data compiled by J.S. Park from Charles T. Lynch, CRC *Handbook of Materials Science, Vol. 3*, CRC Press, Boca Raton, Florida, 1975 and *Engineered Materials Handbook, Vol.2, Engineering Plastics*, ASM International, Metals Park, Ohio, 1988.

Table 269. TOTAL ELONGATION OF POLYMERS

(SHEET 9 OF 10)

Class	Polymer	Elongation (in 2 in.), (ASTM D638) (%)
Polystyrenes; Molded	General purpose	1.0—2.3
	Medium impact	3.0—40
	Glass fiber -30% reinforced	1.1
	Styrene acrylonitrile (SAN)	0.5—4.5
	Glass fiber (30%) reinforced SAN	1.4—1.6
Polyvinyl Chloride And Copolymers; Molded, Extruded	Nonrigid—general	200—450
	Nonrigid—electrical	220—360
	Rigid—normal impact	1—10
	Vinylidene chloride	15—30

Source: data compiled by J.S. Park from Charles T. Lynch, CRC *Handbook of Materials Science, Vol. 3*, CRC Press, Boca Raton, Florida, 1975 and *Engineered Materials Handbook, Vol.2, Engineering Plastics*, ASM International, Metals Park, Ohio, 1988.

Table 269. TOTAL ELONGATION OF POLYMERS

(SHEET 10 OF 10)

Class	Polymer	Elongation (in 2 in.), (ASTM D638) (%)
Silicones; Molded, Laminated		(ASTM D651)
	Fibrous (glass) reinforced silicones	<3
	Granular (silica) reinforced silicones	<3
Ureas; Molded	Alpha—cellulose filled (ASTM Type l)	1

Source: data compiled by J.S. Park from Charles T. Lynch, CRC *Handbook of Materials Science, Vol. 3*, CRC Press, Boca Raton, Florida, 1975 and *Engineered Materials Handbook, Vol.2, Engineering Plastics*, ASM International, Metals Park, Ohio, 1988.

Table 270. Elongation at Yield for Polymers

(Sheet 1 of 3)

Class	Polymer	Elongation at Yield, (ASTM D638) (%)
Chlorinated polyether	Chlorinated polyether	15
Polycarbonates	Polycarbonate	5
Nylons; Molded, Extruded	Type 6	
	Cast	5
	Type 12	5.8
	6/6 Nylon:	
	General purpose molding	5—25
	General purpose extrusion	5—30
	6/10 Nylon:	
	General purpose	5—30

Source: data compiled by J.S. Park from Charles T. Lynch, *CRC Handbook of Materials Science, Vol. 3*, CRC Press, Boca Raton, Florida, 1975 and *Engineered Materials Handbook, Vol.2, Engineering Plastics*, ASM International, Metals Park, Ohio, 1988.

Table 270. ELONGATION AT YIELD FOR POLYMERS

(SHEET 2 OF 3)

Class	Polymer	Elongation at Yield, (ASTM D638) (%)
Polyacetals	Homopolymer:	
	Standard	12
	Copolymer:	
	Standard	12
	25% glass reinforced	3
	High flow	12
Phenylene oxides (Noryl)	Standard	5.6
	Glass fiber reinforced	1.6—2
Polyarylsulfone	Polyarylsulfone	6.5—13
Polypropylene:	General purpose	9—15
	High impact	7—13
	Asbestos filled	5
Polyphenylene sulfide:	Standard	1.6
	40% glass reinforced	1.25

Source: data compiled by J.S. Park from Charles T. Lynch, *CRC Handbook of Materials Science, Vol. 3*, CRC Press, Boca Raton, Florida, 1975 and *Engineered Materials Handbook, Vol.2, Engineering Plastics*, ASM International, Metals Park, Ohio, 1988.

Table 270. ELONGATION AT YIELD FOR POLYMERS

(SHEET 3 OF 3)

Class	Polymer	Elongation at Yield, (ASTM D638) (%)
Polystyrenes; Molded	General purpose	1.0—2.3
	Medium impact	1.2—3.0
	High impact	1.5—2.0
	Glass fiber -30% reinforced	1.1
	Glass fiber (30%) reinforced SAN	1.4—1.6

Source: data compiled by J.S. Park from Charles T. Lynch, *CRC Handbook of Materials Science, Vol. 3*, CRC Press, Boca Raton, Florida, 1975 and *Engineered Materials Handbook, Vol.2, Engineering Plastics*, ASM International, Metals Park, Ohio, 1988.

Table 271. ULTIMATE TENSILE ELONGATION OF FIBERGLASS REINFORCED PLASTICS

Class	Material	Glass fiber content (wt%)	Ultimate tensile elongation (%)
Glass fiber reinforced thermosets	Sheet molding compound (SMC)	15 to 30	0.3 to 1.5
	Bulk molding compound(BMC)	15 to 35	0.3 to 5
	Preform/mat(compression molded)	25 to 50	1 to 2
	Cold press molding–polyester	20 to 30	1 to 2
	Spray–up–polyester	30 to 50	1.0 to 1.2
	Filament wound–epoxy	30 to 80	1.6 to 2.8
	Rod stock–polyester	40 to 80	1.6 to 2.5
	Molding compound–phenolic	5 to 25	0.25 to 0.6
Glass–fiber–reinforced thermoplastics	Acetal	20 to 40	2
	Nylon	6 to 60	2 to 10
	Polycarbonate	20 to 40	2
	Polyethylene	10 to 40	1.5 to 3.5
	Polypropylene	20 to 40	1 to 3
	Polystyrene	20 to 35	1.0 to 1.4
	Polysulfone	20 to 40	2 to 3
	ABS(acrylonitrile butadiene styrene)	20 to 40	3 to 3.4
	PVC (polyvinyl chloride)	15 to 35	2 to 4
	Polyphenylene oxide(modified)	20 to 40	1.7 to 5
	SAN (styrene acrylonitrile)	20 to 40	1.1 to 1.6
	Thermoplastic polyester	20 to 35	1 to 5

To convert (ksi) to (Mpa), multiply by 6.89

Data from *ASM Engineering Materials Reference Book, Second Edition*, Michael Bauccio, Ed., ASM International, Materials Park, OH, p106, (1994).

Table 272. TOTAL STRAIN OF SILICON CARBIDE SCS–2–AL

Fiber orientation	No. of plies	Total Strain
0°	6, 8, 12	0.89
90°	6, 12,40	0.08
$[0°/90°/0°/90°]_s$	8	0.90
$[0_2\ °99°20°]_s$	8	0.92
$[90_2/90°/90°]_s$	8	1.01
± 45°	8, 12, 40	10.6
$[0°±45°/0°]_{s+2s}$	8, 16	0.86
$[0°±45°/90°]_s$	8	1.0

Data from *ASM Engineering Materials Reference Book, Second Edition*, Michael Bauccio, Ed., ASM International, Materials Park, OH, p149,(1994).

Table 273. AREA REDUCTION OF TOOL STEELS

(SHEET 1 OF 2)

Type	Condition	Area Reduction (%)
L2	Annealed	50
	Oil quenched from 855 •C and single tempered at:	
	205 •C	15
	315 •C	30
	425 •C	35
	540 •C	45
	650 •C	55
L6	Annealed	55
	Oil quenched from 845 •C and single tempered at:	
	315 •C	9
	425 •C	20
	540 •C	30
	650 •C	48
S1	Annealed	52
	Oil quenched from 930 •C and single tempered at:	
	205 •C	
	315 •C	12
	425 •C	17
	540 •C	23
	650 •C	37
S5	Annealed	50
	Oil quenched from 870 •C and single tempered at:	
	205 •C	20
	315 •C	24
	425 •C	28
	540 •C	30
	650 •C	40

Area Reduction in 50 mm or 2 in.

Source: data from *ASM Metals Reference Book, Second Edition*, American Society for Metals, Metals Park, Ohio 44073, p241, (1984).

Table 273. AREA REDUCTION OF TOOL STEELS
(SHEET 2 OF 2)

Type	Condition	Area Reduction (%)
S7	Annealed	55
	Fan cooled from 940 •C and single tempered at:	
	205 •C	20
	315 •C	25
	425 •C	29
	540 •C	33
	650 •C	45

Area Reduction in 50 mm or 2 in.

Source: data from *ASM Metals Reference Book, Second Edition*, American Society for Metals, Metals Park, Ohio 44073, p241, (1984).

Table 274. Reduction in Area of Austenitic Stainless Steels
(Sheet 1 of 4)

Type	Form	Condition	ASTM Specification	Reduction in Area (%)
Type 302 (UNS S30200)	Bar	Hot finished and annealed	A276	50
		Cold finished and annealed(a)	A276	40
		Cold finished and annealed(b)	A276	40
Type 302B (UNS S30215)	Bar	Hot finished and annealed	A276	50
		Cold finished and annealed(a)	A276	40
		Cold finished and annealed(b)	A276	40
Types 303 (UNS S30300) and 303Se (UNS S30323)	Bar	Annealed	A581	55
Type 304(UNS S30400)	Bar	Hot finished and annealed	A276	50
		Cold finished and annealed(a)	A276	40
		Cold finished and annealed(b)	A276	40
Type 304L (UNS S30403)	Bar	Hot finished and annealed	A276	50
		Cold finished and annealed(a)	A276	40
		Cold finished and annealed(b)	A276	40

(a) Up to 13 mm thick (b) Over 13 mm thick.

Data from *ASM Metals Reference Book, Third Edition*, Michael Bauccio, Ed., ASM International, Materials Park, OH, p364-366 (1993).

Table 274. REDUCTION IN AREA OF AUSTENITIC STAINLESS STEELS

(SHEET 2 OF 4)

Type	Form	Condition	ASTM Specification	Reduction in Area (%)
Type 305 (UNS S30500)	Bar	Hot finished and annealed	A276	50
		Cold finished and annealed(a)	A276	40
		Cold finished and annealed(b)	A276	40
Types 308 (UNS S30800),321(UNS S32100),347(UNS34700) and 348 (UNS S34800)	Bar	Hot finished and annealed	A276	50
		Cold finished and annealed(a)	A276	40
		Cold finished and annealed(b)	A276	40
Type 308L	Bar	Annealed	—	70
Types 309 (UNS S30900), 309S (UNS S30908), 310 (UNS S31000) and 310S (UNS S31008)	Bar	Hot finished and annealed	A276	50
		Cold finished and annealed(a)	A276	40
		Cold finished and annealed(b)	A276	40

(a) Up to 13 mm thick (b) Over 13 mm thick.

Data from *ASM Metals Reference Book, Third Edition*, Michael Bauccio, Ed., ASM International, Materials Park, OH, p364-366 (1993).

Table 274. REDUCTION IN AREA OF AUSTENITIC STAINLESS STEELS

(SHEET 3 OF 4)

Type	Form	Condition	ASTM Specification	Reduction in Area (%)
Type 314 (UNS S31400)	Bar	Hot finished and annealed	A276	50
		Cold finished and annealed(a)	A276	40
		Cold finished and annealed(b)	A276	40
Type 316 (UNS S31600)	Bar	Hot finished and annealed	A276	50
		Cold finished and annealed(a)	A276	40
		Cold finished and annealed(b)	A276	40
Type 316F (UNS S31620)	Bar	Annealed	—	55
Type 316L (UNS S31603)	Bar	Hot finished and annealed	A276	50
		Cold finished and annealed(a)	A276	40
		Cold finished and annealed(b)	A276	40
Type 316LN	Bar	Annealed	—	70

(a) Up to 13 mm thick (b) Over 13 mm thick.

Data from *ASM Metals Reference Book, Third Edition*, Michael Bauccio, Ed., ASM International, Materials Park, OH, p364-366 (1993).

Table 274. Reduction in Area of Austenitic Stainless Steels

(Sheet 4 of 4)

Type	Form	Condition	ASTM Specification	Reduction in Area (%)
Type 317 (UNS S31700)	Bar	Hot finished and annealed	A276	50
		Cold finished and annealed(a)	A276	40
		Cold finished and annealed(b)	A276	40
Type 317L (UNS S31703)	Bar	Annealed	—	65
Type 317LM	Bar,Plate,Sheet, Strip	Annealed	—	50
Type 329 (UNS S32900)	Bar	Annealed	—	50
Type 330HC	Bar,Wire,Strip	Annealed	—	65

(a) Up to 13 mm thick (b) Over 13 mm thick.

Data from *ASM Metals Reference Book, Third Edition*, Michael Bauccio, Ed., ASM International, Materials Park, OH, p364-366 (1993).

Table 275. Reduction in Area of Ferritic Stainless Steels

Type	ASTM Specification	Form	Condition	(%)
Type 405 (UNS S40500)	A580	Wire	Annealed	45
	A580		Annealed, Cold Finished	45
Type 429 (UNS S42900)	—	Bar	Annealed	65(a)
Type 430 (UNS S43000)	A276	Bar	Annealed, Hot Finished	45
	A276		Annealed, Cold Finished	45
Type 430Ti(UNS S43036)	—	Bar	Annealed	65(a)
Type 434 (UNS S43400)	—	Wire	Annealed	78(a)
Type 442 (UNS S44200)	—	Bar	Annealed	40(a)
Type 446 (UNS S44600)	A276	Bar	Annealed, Hot Finished	45
	A276		Annealed, Cold Finished	45

(a) Typical Values

Data from *ASM Metals Reference Book, Third Edition*, Michael Bauccio, Ed., ASM International, Materials Park, OH, p368 (1993).

Table 276. Reduction in Area of High–Nitrogen Austenitic Stainless Steels

Type	ASTM Specification	Form	Condition	Reduction in Area (%)
Type 201 (UNS S20100)	A276	Bar	Annealed	45
Type 205 (UNS S20500)	—	Plate	Annealed*	62
Type 304HN (UNS S30452)	—	Bar	Annealed	50

Data from *ASM Metals Reference Book, Third Edition*, Michael Bauccio, Ed., ASM International, Materials Park, OH, p367 (1993).

* Typical values

Table 277. Reduction in Area of Precipitation-Hardening Austenitic Stainless Steels

Type	Form	Condition	Reduction in Area (%)
PH 13–8 Mo (UNS S13800)	Bar, Plate, Sheet, Strip	H950	45
		H1000	45
15–5 PH (UNS S15500) and 17–4 PH (UNS S17400)	Bar, Plate, Sheet, Stript	H900	35(a)
		H925	38(a)
		H1025	45(a)
		H1075	45(a)
		H1100	45(a)
		H1150	50(a)
		H1150M	55(a)
17–7 PH (UNS S17700)	Bar	RH950	10
		TH1050	25

(a) For flat rolled products, value generally lower and varies with thickness.

Data from *ASM Metals Reference Book, Third Edition*, Michael Bauccio, Ed., ASM International, Materials Park, OH, p371 (1993).

Table 278. Reduction in Area of Martensitic Stainless Steels (Sheet 1 of 2)

Type	ASTM Specification	Form	Condition	Reduction in Area (%)
Type 403 (UNS S40300)	A276	Bar	Annealed, hot finished	45
	A276		Annealed, cold finished	45
	A276		Intermediate temper, hot finished	45
	A276		Intermediate temper, cold finished	40
	A276		Hard temper, hot finished	40
	A276		Hard temper, cold finished	40
Type 410 (UNS S41000)	A276	Bar	Annealed, hot finished	45
	A276		Annealed, cold finished	45
	A276		Intermediate temper, hot finished	45
	A276		Intermediate temper, cold finished	40
	A276		Hard temper, hot finished	40
	A276		Hard temper, cold finished	40
Type 410Cb (UNS S41040)	A276	Bar	Annealed, hot finished	45
	A276		Annealed, cold finished	35
	A276		Intermediate temper, hot finished	45
	A276		Intermediate temper, cold finished	35

Data from *ASM Metals Reference Book, Third Edition*, Michael Bauccio, Ed., ASM International, Materials Park, OH, p369-370 (1993).

Table 278. REDUCTION IN AREA OF MARTENSITIC STAINLESS STEELS
(SHEET 2 OF 2)

Type	ASTM Specification	Form	Condition	Reduction in Area (%)
Type 414 (UNS S41400)	A276	Bar	Intermediate temper, hot finished	45
	A276		Intermediate temper, cold finished	45
Type 414L	—	Bar	Annealed	60
Type 420 (UNS S42000)	—	Bar	Tempered 205 °C	25
Type 422 (UNS S42200)	A565	Bar	Intermediate and hard tempers for high-temperature service	30
Type 431 (UNS S43100)	—	Bar	Tempered 260 °C	55
	—		Tempered 595 °C	57
	—		Tempered 315 °C	20
	—		Tempered 315 °C	15
	—		Tempered 315 °C	10
Type 501 (UNS S50100)	—	Bar, Plate	Annealed	65
	—		Tempered 540 °C	50
Type 502 (UNS S50200)	—	Bar, Plate	Annealed	70

Data from *ASM Metals Reference Book, Third Edition*, Michael Bauccio, Ed., ASM International, Materials Park, OH, p369-370 (1993).

Table 279. Reduction in Area of Commercially Pure Tin

Temperature (°C)	Reduction in Area (%)
Strained at 0.2 mm/m • min	
-200	6
-160	10
-120	97
-80	100
-40	100
0	100
23	100

Data from *ASM Metals Reference Book, Third Edition*, Michael Bauccio, Ed., ASM International, Materials Park, OH, p488, (1993).

Table 280. Area Reduction of Wrought Titanium Alloys at Room Temperature (Sheet 1 of 2)

Class	Alloy	Condition	Reduction in Area (%)
Commercially Pure	99.5 Ti	Annealed	55
	99.2 Ti	Annealed	50
	99.1 Ti	Annealed	45
	99.0 Ti	Annealed	40
	99.2Ti-0.2Pd	Annealed	50
	Ti-0.8Ni-0.3Mo	Annealed	42
Alpha Alloys	Ti-5Al-2.5Sn	Annealed	40
Near Alpha Alloys	Ti-8Al-1Mo-1V	Duplex Annealed	28
	Ti-11Sn-1Mo-2.25Al-5.0Zr-1Mo-0.2Si	Duplex Annealed	35
	Ti-6Al-2Sn-4Zr-2Mo	Duplex Annealed	35
	Ti-6Al-2Nb-1Ta-1Mo	As rolled 2.5 cm (1 in.) plate	34
Alpha-Beta Alloys	Ti-8Mn	Annealed	32
	Ti-6Al-4V	Annealed	30
		Solution + age	25

Data from *ASM Metals Reference Book, Third Edition*, Michael Bauccio, Ed., ASM International, Materials Park, OH, p512, (1993).

Table 280. AREA REDUCTION OF WROUGHT TITANIUM ALLOYS AT ROOM TEMPERATURE (SHEET 2 OF 2)

Class	Alloy	Condition	Reduction in Area (%)
	Ti-6Al-4V(low O_2)	Annealed	35
	Ti-6Al-6V-2Sn	Annealed	30
		Solution + age	20
	Ti-7Al-4Mo	Solution + age	22
	Ti-6Al-2Sn-4Zr-6Mo	Solution + age	23
	Ti-6Al-2Sn-2Zr-2Mo- 2Cr-0.25Si	Solution + age	33
	Ti-10V-2Fe-3Al	Solution + age	19

Data from *ASM Metals Reference Book, Third Edition*, Michael Bauccio, Ed., ASM International, Materials Park, OH, p512, (1993).

Table 281. AREA REDUCTION OF WROUGHT TITANIUM ALLOYS AT HIGH TEMPERATURE (SHEET 1 OF 3)

Class	Alloy	Condition	Test Temperature (°C)	Reduction in Area (%)
Commercially Pure	99.5 Ti	Annealed	315	80
	99.2 Ti	Annealed	315	75
	99.1 Ti	Annealed	315	75
	99.0 Ti	Annealed	315	70
	99.2Ti-0.2Pd	Annealed	315	75
Alpha Alloys	Ti-5Al-2.5Sn	Annealed	315	45
Near Alpha Alloys	Ti-8Al-1Mo-1V	Duplex Annealed	315	38
			425	44
			540	55
	Ti-11Sn-1Mo-2.25Al-5.0Zr-1Mo-0.2Si	Duplex Annealed	315	44
			425	48
			540	50

Data from *ASM Metals Reference Book, Third Edition*, Michael Bauccio, Ed., ASM International, Materials Park, OH, p512, (1993).

Table 281. Area Reduction of Wrought Titanium Alloys at High Temperature (Sheet 2 of 3)

Class	Alloy	Condition	Test Temperature (°C)	Reduction in Area (%)
Alpha-Beta Alloys	Ti-6Al-2Sn-4Zr-2Mo	Duplex Annealed	315	42
			425	55
			540	60
	Ti-6Al-4V	Annealed	315	35
		Annealed	425	40
		Annealed	540	50
		Solution + age	315	28
		Solution + age	425	35
		Solution + age	540	45
	Ti-6Al-6V-2Sn	Annealed	315	42
		Solution + age	315	28
	Ti-7Al-4Mo	Solution + age	315	50
			425	55

Data from *ASM Metals Reference Book, Third Edition*, Michael Bauccio, Ed., ASM International, Materials Park, OH, p512, (1993).

Table 281. AREA REDUCTION OF WROUGHT TITANIUM ALLOYS AT HIGH TEMPERATURE (SHEET 3 OF 3)

Class	Alloy	Condition	Test Temperature (°C)	Reduction in Area (%)
	Ti-6Al-2Sn-4Zr-6Mo	Solution + age	315	55
			425	67
			540	70
	Ti-6Al-2Sn-2Zr-2Mo- 2Cr-0.25Si	Solution + age	315	27
	Ti-10V-2Fe-3Al	Solution + age	205	33
			315	42

Data from ASM *Metals Reference Book, Third Edition*, Michael Bauccio, Ed., ASM International, Materials Park, OH, p512, (1993).

Table 282. STRENGTH DENSITY RATIO OF GRAPHITE FIBER REINFORCED METALS

Composite	Fiber content (vol%)	Strength / Density (10^6in)
Graphite(a)/lead	41	0.385
Graphite(b)/lead	35	0.260
Graphite(a)/zinc	35	0.580
Graphite(a)/magnesium	42	1.016

(a) Thornel 75 fiber (b) Courtaulds HM fiber

Data from *ASM Engineering Materials Reference Book, Second Edition*, Michael Bauccio, Ed., ASM International, Materials Park, OH, p148,(1994).

Table 283. MODULUS DENSITY RATIO OF GRAPHITE FIBER REINFORCED METALS

Composite	Fiber content (vol%)	Modulus/ Density (10^6in)
Graphite(a)/lead	41	107.0
Graphite(b)/lead	35	62.3
Graphite(a)/zinc	35	88.5
Graphite(a)/magnesium	42	393.7

(a) Thornel 75 fiber (b) Courtaulds HM fiber

Data from *ASM Engineering Materials Reference Book, Second Edition*, Michael Bauccio, Ed., ASM International, Materials Park, OH, p148,(1994).

Table 284. VISCOSITY OF GLASSES
(SHEET 1 OF 15)

Glass	Composition	Viscosity	Temperature °C
SiO_2 glass		12.6-14.4 logP	1100
		11.4-12.8 logP	1200
		10.4-11.83 logP	1300
		9.43-10.65 logP	1400
		8.54-9.52 logP	1500
		7.8-8.53 logP	1600
		7.1-7.65 logP	1700
		6.43-6.9 logP	1800
		5.88-6.2 logP	1900
		5.2-5.4 logP	2000
SiO_2-Na_2O glass	(21.7 % mol Na_2O)	4.28 logP	900
	(21.7 % mol Na_2O)	3.66 logP	1000
	(21.7 % mol Na_2O)	3.17 logP	1100
	(21.7 % mol Na_2O)	2.76 logP	1200
	(21.7 % mol Na_2O)	2.40 logP	1300
	(21.7 % mol Na_2O)	2.08 logP	1400
	(23.8 % mol Na_2O)	3.88 logP	900
	(23.8 % mol Na_2O)	3.28 logP	1000
	(23.8 % mol Na_2O)	2.82 logP	1100
	(23.8 % mol Na_2O)	2.44 logP	1200
	(23.8 % mol Na_2O)	2.10 logP	1300
	(23.8 % mol Na_2O)	1.84 logP	1400
	(27.7 % mol Na_2O)	4.33 logP	800
	(27.7 % mol Na_2O)	3.71 logP	900

Source: data compiled by J. S. Park from O. V. Mazurin, M. V. Streltsina and T. P. Shvaiko-Shvaikovskaya, Handbook of Glass Data, Part A and Part B, Elsevier, New York, 1983

Table 284. VISCOSITY OF GLASSES
(SHEET 2 OF 15)

Glass	Composition	Viscosity	Temperature °C
SiO_2-Na_2O glass	(27.7 % mol Na_2O)	3.16 logP	1000
	(27.7 % mol Na_2O)	2.69 logP	1100
	(27.7 % mol Na_2O)	2.31 logP	1200
	(27.7 % mol Na_2O)	1.98 logP	1300
	(27.7 % mol Na_2O)	1.65 logP	1400
	(31.7 % mol Na_2O)	4.17 logP	800
	(31.7 % mol Na_2O)	3.45 logP	900
	(31.7 % mol Na_2O)	2.92 logP	1000
	(31.7 % mol Na_2O)	2.48 logP	1100
	(31.7 % mol Na_2O)	2.12 logP	1200
	(31.7 % mol Na_2O)	1.83 logP	1300
	(31.7 % mol Na_2O)	1.59 logP	1400
	(33.7 % mol Na_2O)	4.06 logP	800
	(33.7 % mol Na_2O)	3.39 logP	900
	(33.7 % mol Na_2O)	2.66 logP	1000
	(33.7 % mol Na_2O)	2.20 logP	1100
	(33.7 % mol Na_2O)	1.81 logP	1200
	(33.7 % mol Na_2O)	1.52 logP	1300
	(36.3 % mol Na_2O)	4.13 logP	800
	(36.3 % mol Na_2O)	3.40 logP	900
	(36.3 % mol Na_2O)	2.86 logP	1000
	(36.3 % mol Na_2O)	2.42 logP	1100
	(36.3 % mol Na_2O)	2.06 logP	1200

Source: data compiled by J. S. Park from O. V. Mazurin, M. V. Streltsina and T. P. Shvaiko-Shvaikovskaya, Handbook of Glass Data, Part A and Part B, Elsevier, New York, 1983

Table 284. VISCOSITY OF GLASSES
(SHEET 3 OF 15)

Glass	Composition	Viscosity	Temperature °C
SiO_2-Na_2O glass	(36.3 % mol Na_2O)	1.76 logP	1300
	(36.3 % mol Na_2O)	1.51 logP	1400
	(38.9 % mol Na_2O)	3.91 logP	800
	(38.9 % mol Na_2O)	3.20 logP	900
	(38.9 % mol Na_2O)	2.63 logP	1000
	(38.9 % mol Na_2O)	2.18 logP	1100
	(38.9 % mol Na_2O)	1.78 logP	1200
	(38.9 % mol Na_2O)	1.47 logP	1300
	(41.9 % mol Na_2O)	3.56 logP	800
	(41.9 % mol Na_2O)	2.83 logP	900
	(41.9 % mol Na_2O)	2.29 logP	1000
	(41.9 % mol Na_2O)	1.85 logP	1100
	(41.9 % mol Na_2O)	1.50 logP	1200
	(44.0 % mol Na_2O)	3.65 logP	800
	(44.0 % mol Na_2O)	2.81 logP	900
	(44.0 % mol Na_2O)	2.24 logP	1000
	(44.0 % mol Na_2O)	1.80 logP	1100
	(44.0 % mol Na_2O)	1.43 logP	1200
SiO_2-CaO glass	(30.5% mol CaO)	13.6 P	1700
	(30.5% mol CaO)	10.4 P	1750
	(30.5% mol CaO)	8.5 P	1800
	(34.5% mol CaO)	10.0 P	1650
	(34.5% mol CaO)	7.8 P	1700

Source: data compiled by J. S. Park from O. V. Mazurin, M. V. Streltsina and T. P. Shvaiko-Shvaikovskaya, Handbook of Glass Data, Part A and Part B, Elsevier, New York, 1983

Table 284. VISCOSITY OF GLASSES
(SHEET 4 OF 15)

Glass	Composition	Viscosity	Temperature °C
SiO_2-CaO glass	(34.5% mol CaO)	6.05 P	1750
	(34.5% mol CaO)	4.5 P	1800
	(41.6% mol CaO)	9.35 P	1500
	(41.6% mol CaO)	6.48 P	1550
	(41.6% mol CaO)	4.68 P	1600
	(41.6% mol CaO)	3.57 P	1650
	(41.6% mol CaO)	2.75 P	1700
	(41.6% mol CaO)	2.16 P	1750
	(41.6% mol CaO)	1.8 P	1800
	(48.7% mol CaO)	4.35 P	1500
	(48.7% mol CaO)	3.17 P	1550
	(48.7% mol CaO)	2.41 P	1600
	(48.7% mol CaO)	1.90 P	1650
	(48.7% mol CaO)	1.50 P	1700
	(48.7% mol CaO)	1.20 P	1750
	(48.7% mol CaO)	0.99 P	1800
	(52.7% mol CaO)	3.03 P	1500
	(52.7% mol CaO)	2.20 P	1550
	(52.7% mol CaO)	1.66 P	1600
	(52.7% mol CaO)	1.28 P	1650
	(52.7% mol CaO)	1.01 P	1700
	(52.7% mol CaO)	0.83 P	1750
	(52.7% mol CaO)	0.72 P	1800
	(54.7% mol CaO)	2.57 P	1500
	(54.7% mol CaO)	1.39 P	1550
	(54.7% mol CaO)	1.40 P	1600
	(54.7% mol CaO)	1.10 P	1650
	(54.7% mol CaO)	0.90 P	1700

Source: data compiled by J. S. Park from O. V. Mazurin, M. V. Streltsina and T. P. Shvaiko-Shvaikovskaya, Handbook of Glass Data, Part A and Part B, Elsevier, New York, 1983

Table 284. VISCOSITY OF GLASSES
(SHEET 5 OF 15)

Glass	Composition	Viscosity	Temperature °C
SiO_2-CaO glass	(54.7% mol CaO)	0.75 P	1750
	(54.7% mol CaO)	0.66 P	1800
	(57.7% mol CaO)	1.13 P	1600
	(57.7% mol CaO)	0.90 P	1650
	(57.7% mol CaO)	0.74 P	1700
	(57.7% mol CaO)	0.62 P	1750
	(57.7% mol CaO)	0.54 P	1800
SiO_2-PbO glass	(35% mol PbO)	7380 P	840
	(35% mol PbO)	1920 P	900
	(35% mol PbO)	620 P	960
	(35% mol PbO)	302 P	1020
	(35% mol PbO)	164 P	1080
	(35% mol PbO)	100.0 P	1140
	(35% mol PbO)	62.0 P	1200
	(35% mol PbO)	38.2 P	1260
	(35% mol PbO)	25.0 P	1320
	(35% mol PbO)	16.2 P	1380
	(35% mol PbO)	11.8 P	1440
	(40% mol PbO)	2970 P	780
	(40% mol PbO)	830 P	840
	(40% mol PbO)	329 P	900
	(40% mol PbO)	164 P	960
	(40% mol PbO)	91.0 P	1020
	(40% mol PbO)	51.8 P	1080
	(40% mol PbO)	31.8 P	1140
	(40% mol PbO)	20.4 P	1200

Source: data compiled by J. S. Park from O. V. Mazurin, M. V. Streltsina and T. P. Shvaiko-Shvaikovskaya, Handbook of Glass Data, Part A and Part B, Elsevier, New York, 1983

Table 284. VISCOSITY OF GLASSES
(SHEET 6 OF 15)

Glass	Composition	Viscosity	Temperature °C
SiO_2-PbO glass	(40% mol PbO)	13.5 P	1260
	(40% mol PbO)	10.2 P	1320
	(46% mol PbO)	2260 P	720
	(46% mol PbO)	494 P	780
	(46% mol PbO)	166 P	840
	(46% mol PbO)	85.0 P	900
	(46% mol PbO)	47.4 P	960
	(46% mol PbO)	29.4 P	1020
	(46% mol PbO)	18.6 P	1080
	(46% mol PbO)	12.7 P	1140
	(46% mol PbO)	8.8 P	1200
	(46% mol PbO)	6.3 P	1260
	(46% mol PbO)	5.2 P	1320
	(46% mol PbO)	4.9 P	1380
	(50% mol PbO)	21200 P	600
	(50% mol PbO)	1600 P	660
	(50% mol PbO)	292 P	720
	(50% mol PbO)	105 P	780
	(50% mol PbO)	43.8 P	840
	(50% mol PbO)	22.5 P	900
	(50% mol PbO)	13.9 P	960
	(50% mol PbO)	8.8 P	1020
	(50% mol PbO)	6.0 P	1080
	(50% mol PbO)	4.3 P	1140
	(50% mol PbO)	2.9 P	1200
	(55% mol PbO)	51.0 P	720
	(55% mol PbO)	22.4 P	780
	(55% mol PbO)	12.6 P	840

Source: data compiled by J. S. Park from O. V. Mazurin, M. V. Streltsina and T. P. Shvaiko-Shvaikovskaya, Handbook of Glass Data, Part A and Part B, Elsevier, New York, 1983

Table 284. VISCOSITY OF GLASSES
(SHEET 7 OF 15)

Glass	Composition	Viscosity	Temperature °C
SiO_2-PbO glass	(55% mol PbO)	7.10 P	900
	(55% mol PbO)	4.44 P	960
	(55% mol PbO)	3.00 P	1020
	(55% mol PbO)	2.06 P	1080
	(55% mol PbO)	1.40 P	1140
	(55% mol PbO)	0.98 P	1200
	(60% mol PbO)	37.6 P	660
	(60% mol PbO)	12.4 P	720
	(60% mol PbO)	5.8 P	780
	(60% mol PbO)	3.2 P	840
	(60% mol PbO)	2.2 P	900
	(60% mol PbO)	1.5 P	960
	(60% mol PbO)	1.00 P	1020
	(60% mol PbO)	0.7 P	1080
	(64% mol PbO)	5.2 P	720
	(64% mol PbO)	2.5 P	780
	(64% mol PbO)	1.23 P	840
	(64% mol PbO)	1.00 P	900
	(64% mol PbO)	0.70 P	960
	(64% mol PbO)	0.50 P	1020
	(64% mol PbO)	0.30 P	1080
	(66.7% mol PbO)	1.60 P	780
	(66.7% mol PbO)	1.00 P	840
	(66.7% mol PbO)	0.70 P	900
	(66.7% mol PbO)	0.50 P	960
	(66.7% mol PbO)	0.35 P	1020
	(70% mol PbO)	1.80 P	720
	(70% mol PbO)	1.17 P	780
	(70% mol PbO)	0.80 P	840

Source: data compiled by J. S. Park from O. V. Mazurin, M. V. Streltsina and T. P. Shvaiko-Shvaikovskaya, Handbook of Glass Data, Part A and Part B, Elsevier, New York, 1983

Table 284. VISCOSITY OF GLASSES
(SHEET 8 OF 15)

Glass	Composition	Viscosity	Temperature °C
SiO_2-PbO glass	(70% mol PbO)	0.40 P	900
	(70% mol PbO)	0.20 P	960
SiO_2-Al_2O_3 glass	(37.1% mol Al_2O_3)	5.8 P	1850
	(37.1% mol Al_2O_3)	4.1 P	1900
	(37.1% mol Al_2O_3)	3.1 P	1950
	(37.1% mol Al_2O_3)	2.5 P	2000
	(37.1% mol Al_2O_3)	2.2 P	2050
	(37.1% mol Al_2O_3)	1.9 P	2100
	(46.9% mol Al_2O_3)	3.3 P	1850
	(46.9% mol Al_2O_3)	2.4 P	1900
	(46.9% mol Al_2O_3)	1.8 P	1950
	(46.9% mol Al_2O_3)	1.5 P	2000
	(46.9% mol Al_2O_3)	1.3 P	2050
	(46.9% mol Al_2O_3)	1.2 P	2100
	(70.2% mol Al_2O_3)	0.9 P	1950
	(70.2% mol Al_2O_3)	0.8 P	2000
	(70.2% mol Al_2O_3)	0.7 P	2050
	(70.2% mol Al_2O_3)	0.6 P	2100
SiO_2-B_2O_3 glass	(6.2% mol B_2O_3)	33.0 kP	1763
	(6.2% mol B_2O_3)	26.6 kP	1783
	(6.2% mol B_2O_3)	16.9 kP	1815
	(6.2% mol B_2O_3)	13.1 kP	1840
	(10.1% mol B_2O_3)	13.3 kP	1727
	(10.1% mol B_2O_3)	11.2 kP	1730

Source: data compiled by J. S. Park from O. V. Mazurin, M. V. Streltsina and T. P. Shvaiko-Shvaikovskaya, Handbook of Glass Data, Part A and Part B, Elsevier, New York, 1983

Table 284. VISCOSITY OF GLASSES
(SHEET 9 OF 15)

Glass	Composition	Viscosity	Temperature °C
SiO_2-B_2O_3 glass	(10.1% mol B_2O_3)	10.9 kP	1736
	(10.1% mol B_2O_3)	11.4 kP	1738
	(10.1% mol B_2O_3)	11.0 kP	1740
	(10.1% mol B_2O_3)	9.07 kP	1757
	(10.1% mol B_2O_3)	8.57 kP	1768
	(10.1% mol B_2O_3)	7.78 kP	1775
	(10.1% mol B_2O_3)	6.54 kP	1778
	(10.1% mol B_2O_3)	5.83 kP	1792
	(14.5% mol B_2O_3)	3.51 kP	1691
	(14.5% mol B_2O_3)	3.37 kP	1693
	(14.5% mol B_2O_3)	2.63 kP	1720
	(14.5% mol B_2O_3)	2.45 kP	1725
	(14.5% mol B_2O_3)	1.92 kP	1752
	(14.5% mol B_2O_3)	1.85 kP	1757
	(14.5% mol B_2O_3)	1.47 kP	1778
	(14.5% mol B_2O_3)	1.45 kP	1783
	(14.5% mol B_2O_3)	1.17 kP	1797
	(14.5% mol B_2O_3)	1.14 kP	1800
	(14.5% mol B_2O_3)	1.12 kP	1802
	(14.5% mol B_2O_3)	1.00 kP	1812
	(14.5% mol B_2O_3)	0.97 kP	1816
	(25.2% mol B_2O_3)	127.0 kP	1303
	(25.2% mol B_2O_3)	89.8 kP	1329

Source: data compiled by J. S. Park from O. V. Mazurin, M. V. Streltsina and T. P. Shvaiko-Shvaikovskaya, Handbook of Glass Data, Part A and Part B, Elsevier, New York, 1983

Table 284. VISCOSITY OF GLASSES
(SHEET 10 OF 15)

Glass	Composition	Viscosity	Temperature °C
SiO_2-B_2O_3 glass	(25.2% mol B_2O_3)	67.4 kP	1355
	(25.2% mol B_2O_3)	44.5 kP	1376
	(25.2% mol B_2O_3)	32.0 kP	1418
	(25.2% mol B_2O_3)	21.9 kP	1444
	(42.4% mol B_2O_3)	-2.37+9823/T log P	1100-1460
	(53.1% mol B_2O_3)	-1.96+8239/T log P	1380-1530
	(62.4% mol B_2O_3)	-1.99+7687/T log P	1280-1460
	(71.9% mol B_2O_3)	-1.24+5740/T log P	1130-1410
	(75.4 % mol B_2O_3)	119000 P	530
	(75.4 % mol B_2O_3)	15230 P	630
	(75.4 % mol B_2O_3)	3400 P	800
	(79.7 % mol B_2O_3)	49500 P	530
	(79.7 % mol B_2O_3)	9300 P	630
	(79.7 % mol B_2O_3)	1400 P	800
	(81.9% mol B_2O_3)	11.61-14.06 log P	243-306
	(82.5% mol B_2O_3)	0.90+4576/T log P	1050-1360
	(86.3 % mol B_2O_3)	17000 P	530
	(86.3 % mol B_2O_3)	4000 P	630
	(86.3 % mol B_2O_3)	425 P	800
	(90.0% mol B_2O_3)	0.42+3434/T log P	1030-1360

Source: data compiled by J. S. Park from O. V. Mazurin, M. V. Streltsina and T. P. Shvaiko-Shvaikovskaya, Handbook of Glass Data, Part A and Part B, Elsevier, New York, 1983

Table 284. VISCOSITY OF GLASSES
(SHEET 11 OF 15)

Glass	Composition	Viscosity	Temperature °C
SiO_2-B_2O_3 glass	(90.4 % mol B_2O_3)	15300 P	530
	(90.4 % mol B_2O_3)	4400 P	630
	(90.4 % mol B_2O_3)	565 P	800
	(93.1 % mol B_2O_3)	7150 P	530
	(93.1 % mol B_2O_3)	2200 P	630
	(93.1 % mol B_2O_3)	420 P	800
	(93.91% mol B_2O_3)	0.68+3655/T log P	1070-1350
	(97.7 % mol B_2O_3)	6900 P	530
	(97.7 % mol B_2O_3)	2730 P	630
	(97.7 % mol B_2O_3)	410 P	800
B_2O_3 glass		9.799 log P	325
		8.602 log P	350
		7.602 log P	375
		6.415 log P	411
		5.484 log P	450
		4.611 log P	500
		4.029 log P	550
		3.561 log P	600
		2.959 log P	700
		2.549 log P	800
		2.245 log P	900
		2.000 log P	1000
		1.785 log P	1100
		1.603 log P	1200
		1.462 log P	1300
		1.335 log P	1400

Source: data compiled by J. S. Park from O. V. Mazurin, M. V. Streltsina and T. P. Shvaiko-Shvaikovskaya, Handbook of Glass Data, Part A and Part B, Elsevier, New York, 1983

Table 284. VISCOSITY OF GLASSES
(SHEET 12 OF 15)

Glass	Composition	Viscosity	Temperature °C
B_2O_3 glass		4.65 P	1829
		3.87 P	1863
B_2O_3-CaO glass	(32.0 % mol CaO)	12.51 log P	646.5
	(32.0 % mol CaO)	12.02 log P	654.8
	(32.0 % mol CaO)	10.64 log P	674.8
	(32.0 % mol CaO)	9.17 log P	697.2
	(34.0 % mol CaO)	11.32 log P	656.1
	(34.0 % mol CaO)	10.68 log P	667.1
	(34.0 % mol CaO)	9.88 log P	681.3
	(34.0 % mol CaO)	10.51 log P	671.3
	(34.0 % mol CaO)	11.60 log P	653.6
	(34.0 % mol CaO)	10.48 log P	668.9
	(34.0 % mol CaO)	9.09 log P	691.5
	(34.0 % mol CaO)	11.37 log P	657.2
	(55.0 % mol CaO)	12.92 log P	650
	(55.0 % mol CaO)	9.84 log P	700
	(55.0 % mol CaO)	7.32 log P	750
	(55.0 % mol CaO)	5.38 log P	800
	(55.0 % mol CaO)	2.60 log P	900
	(55.0 % mol CaO)	1.96 log P	950
	(55.0 % mol CaO)	1.38 log P	1000
	(55.0 % mol CaO)	0.96 log P	1050
	(55.0 % mol CaO)	0.74 log P	1100
B_2O_3-Na_2O glass	(5% mol Na_2O)	7.83×10^{14} P	285
	(5% mol Na_2O)	5.86×10^{13} P	300
	(5% mol Na_2O)	1.99×10^{13} P	309
	(9.9% mol Na_2O)	3.371 log P	630
	(9.9% mol Na_2O)	3.095 log P	650
	(9.9% mol Na_2O)	2.586 log P	700

Source: data compiled by J. S. Park from O. V. Mazurin, M. V. Streltsina and T. P. Shvaiko-Shvaikovskaya, Handbook of Glass Data, Part A and Part B, Elsevier, New York, 1983

Table 284. VISCOSITY OF GLASSES
(SHEET 13 OF 15)

Glass	Composition	Viscosity	Temperature °C
B_2O_3-Na_2O glass	(9.9% mol Na_2O)	2.181 log P	750
	(9.9% mol Na_2O)	1.884 log P	800
	(9.9% mol Na_2O)	1.647 log P	850
	(9.9% mol Na_2O)	1.569 log P	870
	(10% mol Na_2O)	1.28×10^{15} P	328
	(10% mol Na_2O)	1.41×10^{14} P	340
	(10% mol Na_2O)	2.06×10^{13} P	351
	(12.8% mol Na_2O)	3.566 log P	630
	(12.8% mol Na_2O)	3.257 log P	650
	(12.8% mol Na_2O)	2.695 log P	700
	(12.8% mol Na_2O)	2.252 log P	750
	(12.8% mol Na_2O)	1.923 log P	800
	(12.8% mol Na_2O)	1.661 log P	850
	(12.8% mol Na_2O)	1.574 log P	870
	(15% mol Na_2O)	1.44×10^{15} P	381
	(15% mol Na_2O)	1.65×10^{14} P	394
	(15% mol Na_2O)	2.75×10^{13} P	405
	(15.1% mol Na_2O)	3.825 log P	630
	(15.1% mol Na_2O)	3.457 log P	650
	(15.1% mol Na_2O)	2.818 log P	700
	(15.1% mol Na_2O)	2.319 log P	750
	(15.1% mol Na_2O)	1.942 log P	800
	(15.1% mol Na_2O)	1.652 log P	850
	(15.1% mol Na_2O)	1.560 log P	870

Source: data compiled by J. S. Park from O. V. Mazurin, M. V. Streltsina and T. P. Shvaiko-Shvaikovskaya, Handbook of Glass Data, Part A and Part B, Elsevier, New York, 1983

Table 284. VISCOSITY OF GLASSES

(SHEET 14 OF 15)

Glass	Composition	Viscosity	Temperature °C
B_2O_3-Na_2O glass	(17.5% mol Na_2O)	4.050 log P	630
	(17.5% mol Na_2O)	3.623 log P	650
	(17.5% mol Na_2O)	2.881 log P	700
	(17.5% mol Na_2O)	2.332 log P	750
	(17.5% mol Na_2O)	1.931 log P	800
	(17.5% mol Na_2O)	1.633 log P	850
	(17.5% mol Na_2O)	1.545 log P	870
	(19.7% mol Na_2O)	4.110 log P	630
	(19.7% mol Na_2O)	3.712 log P	650
	(19.7% mol Na_2O)	2.945 log P	700
	(19.7% mol Na_2O)	2.324 log P	750
	(19.7% mol Na_2O)	1.875 log P	800
	(19.7% mol Na_2O)	1.540 log P	850
	(19.7% mol Na_2O)	1.435 log P	870
	(20% mol Na_2O)	5.19×10^{15} P	435
	(20% mol Na_2O)	1.31×10^{14} P	445
	(20% mol Na_2O)	1.57×10^{13} P	457
	(21.9% mol Na_2O)	4.185 log P	630
	(21.9% mol Na_2O)	3.746 log P	650
	(21.9% mol Na_2O)	2.951 log P	700
	(21.9% mol Na_2O)	2.324 log P	750
	(21.9% mol Na_2O)	1.810 log P	800
	(21.9% mol Na_2O)	1.506 log P	850
	(21.9% mol Na_2O)	1.392 log P	870

Source: data compiled by J. S. Park from O. V. Mazurin, M. V. Streltsina and T. P. Shvaiko-Shvaikovskaya, Handbook of Glass Data, Part A and Part B, Elsevier, New York, 1983

Table 284. VISCOSITY OF GLASSES
(SHEET 15 OF 15)

Glass	Composition	Viscosity	Temperature °C
B_2O_3-Na_2O glass	(24.0% mol Na_2O)	4.050 log P	630
	(24.0% mol Na_2O)	3.598 log P	650
	(24.0% mol Na_2O)	2.824 log P	700
	(24.0% mol Na_2O)	2.228 log P	750
	(24.0% mol Na_2O)	1.782 log P	800
	(24.0% mol Na_2O)	1.455 log P	850
	(24.0% mol Na_2O)	1.344 log P	870
	(25% mol Na_2O)	$6.67x10^{14}$ P	445
	(25% mol Na_2O)	$1.29x10^{14}$ P	455
	(25% mol Na_2O)	$1.31x10^{13}$ P	466
	(26.4% mol Na_2O)	3.865 log P	630
	(26.4% mol Na_2O)	3.448 log P	650
	(26.4% mol Na_2O)	2.679 log P	700
	(26.4% mol Na_2O)	2.086 log P	750
	(26.4% mol Na_2O)	1.684 log P	800
	(26.4% mol Na_2O)	1.395 log P	850
	(26.4% mol Na_2O)	1.300 log P	870
	(30% mol Na_2O)	$2.12x10^{15}$ P	448
	(30% mol Na_2O)	$8.06x10^{14}$ P	457
	(30% mol Na_2O)	$1.02x10^{13}$ P	467

Source: data compiled by J. S. Park from O. V. Mazurin, M. V. Streltsina and T. P. Shvaiko-Shvaikovskaya, Handbook of Glass Data, Part A and Part B, Elsevier, New York, 1983

Table 285. INTERNAL FRICTION OF SiO_2 GLASS

Glass	Internal Friction	Temperature	Frequency
SiO_2 glass	$4\text{-}80 \times 10^{-7}$	100°C	(1.6 MHz)
	$2\text{-}60 \times 10^{-7}$	200°C	(1.6 MHz)
	$2.5\text{-}30 \times 10^{-7}$	300°C	(1.6 MHz)
	$3.5\text{-}9 \times 10^{-7}$	400°C	(1.6 MHz)
	$4.5\text{-}5 \times 10^{-7}$	500°C	(1.6 MHz)
	$5.5\text{-}9 \times 10^{-7}$	600°C	(1.6 MHz)
	$8\text{-}15 \times 10^{-7}$	700°C	(1.6 MHz)
	$10.5\text{-}50 \times 10^{-7}$	800°C	(1.6 MHz)
	$13.5\text{-}95 \times 10^{-7}$	900°C	(1.6 MHz)
	$15\text{-}150 \times 10^{-7}$	1000°C	(1.6 MHz)

Source: *data compiled by* J.S. Park *from* O. V. Mazurin, M. V. Streltsina and T. P. Shvaiko-Shvaikovskaya, *Handbook of Glass Data, Part A and Part B*, Elsevier, New York, 1983

Table 286. SURFACE TENSION OF ELEMENTS AT MELTING (SHEET 1 OF 6)

Element	Purity (wt. %)	σ_{mp} (dyn/cm)	Atmosphere
Ag	99.7	863±25	Ar
	99.99	(785)	vac.
	99.99	860±20	Ar
	99.99	865	vac.
	99.99	(825)	Ar
	99.99	866	He
	99.999	(828)	vac.
	99.999	873	He
	spect. pure	921	
	spect. pure	918	
Au		(754)	vac.
	99.999	1130	He
	99.999	(731)	vac.
B	99.8	1060±50	vac.
Ba	99.5	276	
Bi	99.9	380±10	Ar
	99.98	378	vac., Ar, H_2
	99.98	380±10	Ar
	99.99	376	vac.
	99.999	380±3	Ar
	99.99995	375	
Ca	p.a.	360	
Cd	99.9	(550±10)	Ar
		(525±30)	H_2
	99.9999	590±5	—

Values in parentheses are less certain.

Source: *data from* Lang,G.,in *Handbook of Chemistry and Physics, 55th ed.*, Weast, R.C., Ed., CRC Press, Cleveland, 1974, F-23.

Table 286. SURFACE TENSION OF ELEMENTS AT MELTING (SHEET 2 OF 6)

Element	Purity (wt. %)	σ_{mp} (dyn/cm)	Atmosphere
Co	99.99	(1520)	H_2, He
	99.9983	1880	vac.
Cr	99.9997	1700±50	Ar
Cs	99.995	68.6	He
Cu	99.9	(11802±40)	Ar
	99.9	(1127)	vac.
	99.98	(1085)	vac.
	99.98	1270	vac.
	99.997	1352	vac.
	99.997	1355	He, H_2
	99.997	1358	Ar
	99.99999	1300	vac.
Fe	99.69	1760±20	He, H_2
	99.85	(1619)	vac.
	99.93	(1510)	vac.
	99.93	1860±40	He
	99.985	(1560)	
	99.99	(1384)	vac.
	99.99	(1650)	He, H_2
	99.99	(1700)	vac.
	99.9992	1773	He, H_2
	99.9998	1880	vac.
Fr	99.9998	718	vac., Al_2O_3
		650	vac.
		632±5	N_2, He
Hf	97.5±2.5	1630	vac.

Values in parentheses are less certain.

Source: *data from* Lang,G.,in *Handbook of Chemistry and Physics, 55th ed.*, Weast, R.C., Ed., CRC Press, Cleveland, 1974, F-23.

Table 286. SURFACE TENSION OF ELEMENTS AT MELTING
(SHEET 3 OF 6)

Element	Purity (wt. %)	σ_{mp} (dyn/cm)	Atmosphere
In	99.95	559	H_2
	99.995	556.0	Ar, He
Ir	99.9980	2250	vac.
K	99.895	101	Ar
	99.895	110.3± 1	—
	99.895	117	vac.
	99.936	(79.2)	He
	99.936	95 ±9.5	—
	99.97±0.64	111.35	He
	99.986	116.95	Ar
Mg	99.5	583	—
	99.91	(525±10)	Ar
Mn	99.9985	1100 ± 50	Ar
Mo	99.7	2080	vac.
	99.98	2049	vac.
	99.98	2130	vac.
	99.9996	2250	vac.
Na	99.96	210.12	Ar
	99.982	187.4	He
	99.995	191	Ar
	99.995	200.2 ±0.6	—
	99.995	202	vac.
Nb, Cb	99.99	2020	vac.
	99.9986	1900	vac.
Nd		688	Ar

Values in parentheses are less certain.

Source: *data from* Lang,G.,in *Handbook of Chemistry and Physics, 55th ed.*, Weast, R.C., Ed., CRC Press, Cleveland, 1974, F-23.

Table 286. SURFACE TENSION OF ELEMENTS AT MELTING (SHEET 4 OF 6)

Element	Purity (wt. %)	σ_{mp} (dyn/cm)	Atmosphere
Ni	99.7	1725	vac.
	99.999	1770±13	vac.
	99.999	1728±10	vac.
	99.999	1822±8	vac.
	99.999	(1670)	vac.
	99.999	1760	vac.
	99.999	(1687)	vac.
	99.99975	(1977)	He
	–	1809±20	H_2, He,
Os	99.9998	2500	vac.
Pb	99.9	(410±5)	Ar
	99.98	450	He
	99.98	451	vac.
	99.998	480	H_2
	99.999	470	Ar
	99.9995	470	
Pd	—	1470	vac.
	99.998	1500	vac.
	99.998	1460	He
Pt	—	1869	CO_2
	99.84	(1740±20)	vac.
	99.9980	1865	vac.
Rb	—	(77±5)	vac.
	—	99.8	Ar
	99.92	91 17	Ar
	99.997	85.7	He

Values in parentheses are less certain.

Source: *data from* Lang,G.,in *Handbook of Chemistry and Physics, 55th ed.*, Weast, R.C., Ed., CRC Press, Cleveland, 1974, F-23.

Table 286. SURFACE TENSION OF ELEMENTS AT MELTING
(SHEET 5 OF 6)

Element	Purity (wt. %)	σ_{mp} (dyn/cm)	Atmosphere
Re	99.4	2610	vac.
	99.9999	2700	vac.
Ru	99.9980	2250	vac.
Rh	—	1940	vac.
	99.9975	2000	vac.
S	—	60.9	vac.
Sb	99.15	395±20	Ar
	99.5	383	H_2, N_2
	99.99	395±20	Ar
Sn	99.89	543.7	—
	99.89	562	vac.
	99.9	(526±10)	Ar
	99.96	552	vac.
	99.96	552	Ar
	99.99	537	vac.
	99.99	530	He
	99.998	566	H_2
	99.998	610	vac.
	99.999	590	vac.
	99.999	555.8±1.9	—
Sr	99.5	303	
Ta	99.9	(1884)	vac.
	99.9983	2150	vac.
	—	2360	vac.
	—	2030	vac.
	—	1910	vac.

Values in parentheses are less certain.

Source: *data from* Lang,G.,in *Handbook of Chemistry and Physics, 55th ed.*, Weast, R.C., Ed., CRC Press, Cleveland, 1974, F-23.

Table 286. SURFACE TENSION OF ELEMENTS AT MELTING (SHEET 6 OF 6)

Element	Purity (wt. %)	σ_{mp} (dyn/cm)	Atmosphere
Te	99.4	186±2	Ar
	—	178	—
Ti	98.7	1510	vac.
	99.69	1402	vac.
	99.92	1390	Ar
	99.92	1460	vac.
	99.9991	1650	vac.
Tl	—	464.5	Ar
	99.999	467	—
U	99.94	(1294)	vac.
	—	1500±75	—
	—	1550	Ar
V	99.9977	1950	vac.
	—	(1760)	vac.
W	99.8	2220	vac.
	99.9	(2000)	vac.
	99.9999	2500	vac.
	—	2310	vac.
Zn	99.9	750 ±20	Ar
	99.99	757.0±5	vac.
	99.999	761.0	vac.
	99.9999	767.5	vac.
Zr	—	1400	Ar
	99.5	1411±70	vac.
	99.7	(1533)	vac.
	99.9998	1480	vac.

Values in parentheses are less certain.

Source: *data from* Lang,G.,in *Handbook of Chemistry and Physics, 55th ed.*, Weast, R.C., Ed., CRC Press, Cleveland, 1974, F-23.

Table 287. SURFACE TENSION OF LIQUID ELEMENTS*

(SHEET 1 OF 15)

Element	Purity (wt. %)	σ_t (dyn/cm)	Temperature •C	Atmosphere
Ag	99.7	$\sigma = (863+25) - 0.33\ (t-t_{mp})$		Ar
	99.96	893	1000	H_2
		862	1150	
		849	1250	
		908	1000	vac.
	99.72	840	950	vac.
	99.99	890	1000	Ar, H_2
		916	1000	H_2
		$\sigma = 865-0.14\ (t-t_{mp})$	vac.	
		$\sigma = 825-0.05\ (T-993)$ *	Ar	
		$\sigma = 866-0.15\ (t-t_{mp})$	He	
	99.995	907	1000	H_2
		894	1100	
		876	1200	
	99.999	905±10	980	Ar
		890±10	1108	
		725	1600	He
		$\sigma = 873-0.15\ (t-t_{mp})$		
	spect. pure	$\sigma = 1136-0.174\ T$	(valid 1300 to 2200 K)	
		$\sigma = 918-0.149\ (t-t_{mp})$		

* T in Kelvin (t in °C). Values in parentheses are less certain.

Source: *data from* Lang,G.,in *Handbook of Chemistry and Physics, 55th ed.*, Weast, R.C., Ed., CRC Press, Cleveland, 1974, F-23.

Table 287. Surface Tension of Liquid Elements*

(Sheet 2 of 15)

Element	Purity (wt. %)	σ_t (dyn/cm)	Temperature •C	Atmosphere
Au	99.999	1130±10	1108	Ar
		1070	1200	He
		1020	1300	
Ba	–	224	720	Ar
	99.5	σ = 351–0.075 T	(valid 1410 to 1880 K) *	
Be	99.98	1100	1500	vac.
Bi	99.9	362	350	Ar
		350	700	vac.
	99.90	343	800	H_2
		328	1000	
		(382)	450	vac.
	99.98	380	450	–
		379	300	vac.
	99.999	σ = 380–0.142 $(t–t_{mp})$	(valid MP to 555•C)	Ar
	99.99995	σ = 423–0.088 T	(valid 1352 to 1555 K) *	
Ca	–	337	850	Ar
	p.a.	σ = 472–0.100 T	(valid 1445 to 1655 K) *	
Cd	99.9	604	390	Ar
Co		1836	1550	Ar
	99.99	1800	1520	vac., Al_2O_3
	99.99	(1630)	1520	He,Al_2O_3
	99.99	(1640)	1520	He, BeO

* T in Kelvin (t in °C). Values in parentheses are less certain.

Source: *data from* Lang,G.,in *Handbook of Chemistry and Physics, 55th ed.*, Weast, R.C., Ed., CRC Press, Cleveland, 1974, F-23.

Table 287. SURFACE TENSION OF LIQUID ELEMENTS*

(SHEET 3 OF 15)

Element	Purity (wt. %)	σ_t (dyn/cm)	Temperature •C	Atmosphere
Co (Con't)	99.99	(1560)	1520	He, MgO
	99.99	1780	1520	H, Al_2O_3
	99.99	(1620)	1520	He
	99.99	(1590)	1520	H_2
		1870	1500	vac.
		1815	1600	vac.
	99.99	1812	1600	vac., Al_2O_3
	99.99	1845	1550	H_2, He
	99.99	1780	1550	
Cr	–	1590±50	1950	vac.
	99.9997			Ar
Cs		68.4	62	Ar
		67.5	62	Ar
		62.9	146	
	99.95	69.5	39	Ar
		42.8	494	
		34.6	642	
	99.995	$\sigma = 68.6 – 0.047 (t – t_{mp})$	(valid 52 to 1100•C)	He
Cu		1269±20	1120	Ar
		1285±10	1120	Ar
	99.9	1220	1100	Ar
		1370	1150	vac.
		(1130)	1183	Ar

* T in Kelvin (t in °C). Values in parentheses are less certain.

Source: *data from* Lang,G.,in *Handbook of Chemistry and Physics, 55th ed.*, Weast, R.C., Ed., CRC Press, Cleveland, 1974, F-23.

Table 287. SURFACE TENSION OF LIQUID ELEMENTS*
(SHEET 4 OF 15)

Element	Purity (wt. %)	σ_t (dyn/cm)	Temperature •C	Atmosphere
Cu (Con't)	99.99	$\sigma = 73.74 - 1.791 \cdot 10^{-2} (t - t_{mp}) - 9.610 \cdot 10^{-5} (t - t_{mp})^2 + 6.629 \cdot 10^{-8} (t - t_{mp})^3$	(valid 71 to 1011•C)	Ar
	99.98	1301	1100	H_2
		1295	1165	
		1287	1255	
		1285	1120	vac.
		1298	1440	
	99.98	1230	1600	Ar
	99.99	1290	1250	He
	99.99	1300	1250	H_2
	99.997			He, H_2
	99.997	$\sigma = 1352 - 0.17 (t - t_{mp})$		vac.
	99.997	$\sigma = 1358 - 0.20 (t - t_{mp})$		Ar
		1285±10	1120	Ar, He
	99.99	1320	1100	Ar, H_2
	99.99	1265	1550	H_2, He
	99.999	1341	1100	N_2
		1338	1150	
		1335	1200	

* T in Kelvin (t in °C). Values in parentheses are less certain.

Source: *data from* Lang,G.,in *Handbook of Chemistry and Physics, 55th ed.*, Weast, R.C., Ed., CRC Press, Cleveland, 1974, F-23.

Table 287. SURFACE TENSION OF LIQUID ELEMENTS*
(SHEET 5 OF 15)

Element	Purity (wt. %)	σ_t (dyn/cm)	Temperature •C	Atmosphere
Cu (Con't)	99.99999	1268±60	1130	vac.
Fe	Armco	1795	1550	Ar, N_2
		1754	1550	vac.
	99.69	(1727)	1550	He, Al_2O_3
		(1734)	1550	H_2, Al_2O_3
		$\sigma = 1760–0.35 (t–t_{mp})$		He, H_2
	99.94	(1710)	1560	vac., Al_2O_3
	99.97	1830±6	1550	vac., BeO
	99.985	1788	1550	Ar
	99.987	(1730)	1550	vac.
	99.99	(1610)	1650	He
		(1430)	1650	He
		(1400)	1650	H_2
		1865	1550	vac., He
		(1430)	1650	He
		(1400)	1650	H
		(1640)	1650	
	99.9992	$\sigma = 773+0.65\ t$	(valid 1550 to 1780•C)	He, H_2
Fr	—	58.4	100	
				Ar
	—	718	350	vac.
	—	559	1500	He, Al_2O_3
	99.9998	$\sigma = 718–0.101 (t–t_{mp})$		vac., Al_2O_3

* T in Kelvin (t in °C). Values in parentheses are less certain.

Source: *data from* Lang,G.,in *Handbook of Chemistry and Physics, 55th ed.*, Weast, R.C., Ed., CRC Press, Cleveland, 1974, F-23.

Table 287. SURFACE TENSION OF LIQUID ELEMENTS*
(SHEET 6 OF 15)

Element	Purity (wt. %)	σ_t (dyn/cm)	Temperature •C	Atmosphere
Fr (Con't)	99.9998 (Con't)	530	1200	vac.
		650	1000	vac.
Hg		(437)	20	
		(350.5)	21	
		476	25	
		472	25	
		(464)	25	
		(516)	25	
		(435)	25	
		488	25	
		(498)	25	
		476	25	
		484±1.5	25	
		484.9±1.8	25	
		449.7	103	
		387.1	350	
		(410)	16	air
		(435.5)	20	air
		(454.7)	20	Ar
		(542)	20	H
		473	19	H_2
		476	25	H_2
		472	20	vac
		(402)	20	vac
		(432)	20	vac
		(436)	20	vac.

* T in Kelvin (t in °C). Values in parentheses are less certain.

Source: *data from* Lang,G.,in *Handbook of Chemistry and Physics, 55th ed.*, Weast, R.C., Ed., CRC Press, Cleveland, 1974, F-23.

Table 287. SURFACE TENSION OF LIQUID ELEMENTS*

(SHEET 7 OF 15)

Element	Purity (wt. %)	σ_t (dyn/cm)	Temperature •C	Atmosphere
Hg (Con't)		480	20	vac.
		(420)	20	vac.
		(410)	20	vac.
		(455)	20	vac.
		(465.2)	20	vac.
		485.5±1.0	20	vac.
		(468)	22	vac.
		473	25	vac.
		$\sigma = 489.5 - 0.20\,t$		
	99.9			
		487	–10	
		487.3	16.5	
		(500±15)	20	
		484.6±1.3	20	
		482.5 ± 3.0	20	
		484.9±0.3	21.5	
		(465)	22	
		482.8±9.7	23–25	
		483.5±1.0	25	
		485.1	25	
		485.4±1.2	25	
		480	25	
		$\sigma = 468.7 - 1.61 \cdot 10^{-1} t - 1.815 \cdot 10^{-2} t^2$		
		$\sigma = 485.5 - 0.149\,t - 2.84 \cdot 10^{-4} t^2$		

* T in Kelvin (t in °C). Values in parentheses are less certain.

Source: *data from* Lang,G.,in *Handbook of Chemistry and Physics, 55th ed.*, Weast, R.C., Ed., CRC Press, Cleveland, 1974, F-23.

Table 287. SURFACE TENSION OF LIQUID ELEMENTS*
(SHEET 8 OF 15)

Element	Purity (wt. %)	σ_t (dyn/cm)	Temperature •C	Atmosphere
Hg (Con't)	99.99	475	20	
In	99.95	515	600	H_2
		540	623	
	99.995	592	185	vac
		514	600	H_2
		541	300	
	99.999	556	200	Ar
		535	400	
		527.8	550	
	99.9994	539	350	vac.
	99.9999	$\sigma = 568.0–0.04 t–7.08 \cdot 10^{-5} t^2$		
K	99.895	$\sigma = 117– 0.66 (t–t_{mp})$		vac.
		112	87	Ar
		80	457	
		64.8	677	
	99.986	$\sigma = 116.95–6.742 \cdot 10^{-2} (t–t_{mp}) – 3.836 \cdot 10^{-5} (t–t_{mp})^2 + 3.707 \cdot 10^{-8}(t–t_{mp})^3$	(valid 77 to 983•C)	Ar
	99.936	$(\sigma = 76.8–70.3 \cdot 10^{-4} (t–400))$	(valid 600 to 1126•C)	He

* T in Kelvin (t in °C). Values in parentheses are less certain.

Source: *data from* Lang,G.,in *Handbook of Chemistry and Physics, 55th ed.*, Weast, R.C., Ed., CRC Press, Cleveland, 1974, F-23.

Table 287. SURFACE TENSION OF LIQUID ELEMENTS*

(SHEET 9 OF 15)

Element	Purity (wt. %)	σ_t (dyn/cm)	Temperature •C	Atmosphere
K (Con't)	99.97±0.64	σ = 115.51–0.0653 t	(valid 70 to 713•C)	He
Li	99.95	397.5	180	Ar
		380	300	
		351.5	500	
	99.98	386	287	Ar
		275	922	
		253	1077	
Mg	99.5	σ = 721–0.149 T *	(valid 1125 to 1326•K)	
	99.8	552	670	N_2
		542	700	
		528	740	
	99.9	550±15	700	Ar
Mn	99.94	1030	1550	vac.
		1010	1550	
Na	99.982	σ = 144–0.108 (t–500)	(valid 400 to 1125•C)	He
	99.995	198	123	vac.
		198.5	129	
		190	140	
		σ = 202–0.092(t–t_{mp})	(valid 100 to 1000°C)	vac

* T in Kelvin (t in °C). Values in parentheses are less certain.

Source: *data from* Lang,G.,in *Handbook of Chemistry and Physics, 55th ed.*, Weast, R.C., Ed., CRC Press, Cleveland, 1974, F-23.

Table 287. SURFACE TENSION OF LIQUID ELEMENTS*
(SHEET 10 OF 15)

Element	Purity (wt. %)	σ_t (dyn/cm)	Temperature •C	Atmosphere
Na (Con't)	99.96	$\sigma = 210.12 - 8.105 \cdot 10^{-2} (t - t_{mp}) - 8.064 \cdot 10^{-5} (t - t_{mp})^2 + 3.380 \cdot 10^{-8} (t - t_{mp})^3$	(valid 141 to 992•C)	Ar
	p.a.	144	617	Ar
		130	764	
		120.4	855	
Nd		674	1186	Ar
Ni	99.7	(1615)	1470	He
		(1570)	1470	H_2
		1735	1470	vac.
		1725	1475	vac.
		(1934)	1550	Ar
	99.99	(1490)	1470	He
		(1500)	1470	He, BeO
		(1530)	1470	He, MgO
		(1530)	1470	H_2
		(1600)	1520	H_2,Al_2O_3
		(1650)	1530	H_2
		1700	1470	H_2, He
		1720	1500	vac.
		1705	1640	vac.

* T in Kelvin (t in °C). Values in parentheses are less certain.

Source: *data from* Lang,G.,in *Handbook of Chemistry and Physics, 55th ed.*, Weast, R.C., Ed., CRC Press, Cleveland, 1974, F-23.

Table 287. SURFACE TENSION OF LIQUID ELEMENTS*

(SHEET 11 OF 15)

Element	Purity (wt. %)	σ_t (dyn/cm)	Temperature •C	Atmosphere
Ni (Con't)	99.99 (Con't)	1740	1520	vac., Al_2O_3
		1770	1520	He,Ar,Al_2O_3
		1780	1550	vac., Al_2O_3
		1810	1560	vac., Al_2O_3
	99.999	1745 σ = 1770–0.39 (t–1550)	1500	He
	99.99975	σ = 1665 + 0.215 t	(valid 1475 to 1650•C)	He
P(white)		69.7	50	
		64.95	68.7	
Pb	99.9	388	1000	H_2
		445	350	Ar
	99.98	448	340	H, N_2
		442	390	
		439	440	
		452	360	air
		442	340	vac.
		435	400	
		440	425	
		450	350–450	
	99.998	428	700	vac.
		474	623	H_2
		455	362	vac.

* T in Kelvin (t in °C). Values in parentheses are less certain.

Source: *data from* Lang,G.,in *Handbook of Chemistry and Physics, 55th ed.*, Weast, R.C., Ed., CRC Press, Cleveland, 1974, F-23.

Table 287. SURFACE TENSION OF LIQUID ELEMENTS*
(SHEET 12 OF 15)

Element	Purity (wt. %)	σ_t (dyn/cm)	Temperature •C	Atmosphere
Pb (Con't)	99.999	456	390	He
		310	1600	
		$\sigma = 470–0.164 (t–t_{mp})$	(valid mp to 535•C)	Ar
	99.9994	438	450	vac
	99.9995	$\sigma = 538–0.114 T$ *	(valid 1440 to 1970•K)	
Pt	99.999	(1699±20)	1800	Ar
Rb		84	52	Ar
		55	477	
		46.8	632	
	99.92	$\sigma = 91.17– 9.189\ 10^{-2} (t–t_{mp}) + 7.228 \cdot 10^{-5} (t–t_{mp})^2 – 3.830 \cdot 10^{-8} (t–t_{mp})^3$	(valid 1104 to 1006•C)	Ar
	99.997	$\sigma = 85.7– 0.054 (t–t_{mp})$	(valid 53 to 1115•C)	He
S	–	51.1	250	vac.
Sb		349	640	H_2
		349	700	
		368	750	

* T in Kelvin (t in °C). Values in parentheses are less certain.

Source: *data from* Lang,G.,in *Handbook of Chemistry and Physics, 55th ed.*, Weast, R.C., Ed., CRC Press, Cleveland, 1974, F-23.

Table 287. SURFACE TENSION OF LIQUID ELEMENTS*

(SHEET 13 OF 15)

Element	Purity (wt. %)	σ_t (dyn/cm)	Temperature •C	Atmosphere
Sb (Con't)		361	900	
		342	974	
		348	1100	
		367.9	640	vac.
		364.9	762	
	99.5	384	675	H_2, N_2
		380	800	
	99.995	350.2	650	Ar
		347.6	700	
		345.0	800	
	99.999	359	800	N_2
		351	1000	
		345	1100	
		320	1600	He
Se	–	88.0±5	230–250	Ar
Si		725	1450	He
		720	1550	vac.
	99.99	750	1550	vac.
	99.9999	825	1500	Ar
Sn	99.9	600	290	vac.
	99.93	549	250	vac.
		539	400	
		526	600	
	99.96	470	1000	vac.
		$\sigma = 552 - 0.167 (t - t_{mp})$	(valid MP to 500•C)	Ar

* T in Kelvin (t in °C). Values in parentheses are less certain.

Source: *data from* Lang,G.,in *Handbook of Chemistry and Physics, 55th ed.*, Weast, R.C., Ed., CRC Press, Cleveland, 1974, F-23.

Table 287. SURFACE TENSION OF LIQUID ELEMENTS*
(SHEET 14 OF 15)

Element	Purity (wt. %)	σ_t (dyn/cm)	Temperature •C	Atmosphere
Sn (Con't)	99.965	508	740	H_2
		489.5	950	
		479.5	1115	
	99.89	554	300	vac.
	99.99	524	500	vac.
		508	600	
		543	489	H_2
		528	572	
		503	692	
		536	250	
		530	450	
		545	250	
		530	600	H_2, He
	99.998	559	623	H_2
		500	800	vac.
		538	300	–
		546	290	–
	99.999	(520)	290	H_2
		(524)	290	vac.
		$\sigma = 566.84 - 4.76 \cdot 10^{-2}\, t$		
	99.9994	537	350	vac.
	99.9999	552.7	246	H_2

* T in Kelvin (t in °C). Values in parentheses are less certain.

Source: *data from* Lang,G.,in *Handbook of Chemistry and Physics, 55th ed.*, Weast, R.C., Ed., CRC Press, Cleveland, 1974, F-23.

Table 287. SURFACE TENSION OF LIQUID ELEMENTS*
(SHEET 15 OF 15)

Element	Purity (wt. %)	σ_t (dyn/cm)	Temperature •C	Atmosphere
Sr		288	775	Ar
		282	830	
		282	893	
	99.5	$\sigma = 392–0.085$ T	(valid 1152 to 1602 K)	
Te	99.4	178±1.5	460	vac.
	–	(162)	475	vac.
		$\sigma = 178–0.024$ $(t–t_{mp})$		
Ti	99.0	1576	1680	vac.
	99.99999	1588	1680	vac.
Tl	–	450	450	vac.
	99.999	450	450	vac.
		($\sigma = 536 –$ 0.119 T)*	(valid 1270 to 1695•K)	

* T in Kelvin (t in °C). Values in parentheses are less certain.

Source: *data from* Lang,G.,in *Handbook of Chemistry and Physics, 55th ed.*, Weast, R.C., Ed., CRC Press, Cleveland, 1974, F-23.

* The data are a compilation of several studies and measurements were obtained from the "sessile drop", "maximum bubble pressure" ,and the "pendant drop" methods. The accuracy varies with both method and the study.

CHAPTER 7

Electrical Properties of Materials

List of Tables

List of Tables (Continued)

Arc Resistance

Table 288. ELECTRICAL CONDUCTIVITY OF METALS
(SHEET 1 OF 7)

Class	Metal or Alloy	Electrical Conductivity (%IACS)
Aluminum and Aluminum Alloys	Aluminum (99.996%)	64.95
	EC(O, H19)	62
	5052 (O, H38)	35
	5056 (H38)	27
	6101 (T6)	56
Copper and Copper Alloys: Wrought Copper	Pure copper	103.06
	Electolytic (ETP)	101
	Oxygen–free copper (OF)	101
	Free–machining copper 0.5% Te	95
	Free–machining copper 1.0% Pb	98
Wrought Alloys	Cartridge brass, 70%	28
	Yellow brass	27
	Leaded commercial bronze	42
	Phosphor bronze,1.25%	48
	Nickel silver, 55–18	5.5
	Low–silicon bronze(B)	12
	Beryllium copper	22 to 30
Copper and copper Alloys: Casting Alloys	Chromium copper (1% Cr)	80 to 90
	88Cu–8Sn–4Zn	11
	87Cu–10Sn–1Pb–2Zn	11
Electrical Contact Materials: Copper Alloys	0.04 oxide	100
	1.25 Sn + P	48
	5 Sn+P	18
	8 Sn+P	13
	15 Zn	37
	20 Zn	32

Data from *ASM Metals Reference Book, Third Edition*, Michael Bauccio, Ed., ASM International, Materials Park, OH, p157-158, (1993).

Table 288. ELECTRICAL CONDUCTIVITY OF METALS
(SHEET 2 OF 7)

Class	Metal or Alloy	Electrical Conductivity (%IACS)
	35 Zn	27
	2 Be+Ni or Co	17 to 21
Electrical Contact Materials: Silver and Silver Alloys	Fine silver	106
	92.5 Ag–7.5Cu	85
	90Ag–10Cu	85
	72Ag–28Cu	87
	72Ag–26Cu–2Ni	60
	85Ag–15Cd	35
	97Ag–3Pt	50
	97Ag–3Pd	60
	90Ag–10Pd	30
	90Ag–10Au	40
	60Ag–40Pd	8
	70Ag–30Pd	12
Electrical Contact Materials: Platinum and Platinum Alloys	Platinum	16
	95Pt–5Ir	9
	90Pt–10Ir	7
	85Pt–15Ir	6
	80Pt–20Ir	5.6
	75Pt–25Ir	5.5
	70Pt–30Ir	5
	65Pt–35Ir	5
	95Pt–5Ru	5.5
	90Pt–10Ru	4
	89Pt–11Ru	4
	86Pt–14Ru	3.5
	96Pt–4W	5

Data from *ASM Metals Reference Book, Third Edition*, Michael Bauccio, Ed., ASM International, Materials Park, OH, p157-158, (1993).

Table 288. ELECTRICAL CONDUCTIVITY OF METALS
(SHEET 3 OF 7)

Class	Metal or Alloy	Electrical Conductivity (%IACS)
Electrical Contact Materials: Palladium and Palladium Alloys		
	Palladium	16
	95.5Pd–4.5Ru	7
	90Pd–10Ru	6.5
	70Pd–30Ag	4.3
	60Pd 40Ag	4.0
	50Pd–50Ag	5.5
	72Pd–26Ag–2Ni	4
	60Pd–40Cu	5
	45Pd–30Ag–20Au–5Pt	4.5
	35Pd–30Ag–14Cu–l0Pt–l0Au–1Zn	5
Electrical Contact Materials: Gold and Gold Alloys		
	Gold	75
	90Au–10Cu	16
	75Au–25Ag	16
	72.5Au–14Cu–8.5Pt–4Ag–1Zn	10
	69Au–25Ag–6Pt	11
	41.7Au–32.5Cu–18.8Ni–7Zn	4.5
Electrical Heating Alloys: Ni–Cr and Ni–Cr–Fe Alloys		
	78.5Ni–20Cr–1.5Si (80–20)	1.6
	73.5Ni–20Cr–5Al–1.5Si	1.2
	68Ni–20Cr–8.5Fe–2Si	1.5
	60Ni–16Cr–22.5Fe–1.5Si	1.5
	35Ni–20Cr–43.5Fe–1.5Si	1.7
Electrical Heating Alloys: Fe–Cr–Al Alloys		
	72Fe–23Cr–5Al	1.3
	55Fe–37.5Cr–75Al	1.2
Pure Metals	Molybdenum	34
	Platinum	16
	Tantalum	13.9
	Tungsten	30

Data from *ASM Metals Reference Book, Third Edition*, Michael Bauccio, Ed., ASM International, Materials Park, OH, p157-158, (1993).

Table 288. ELECTRICAL CONDUCTIVITY OF METALS

(SHEET 4 OF 7)

Class	Metal or Alloy	Electrical Conductivity (%IACS)
Nonmetallic Heating Element Materials	Silicon carbide, SiC	1 to 1.7
	Molybdenum disilicide, $MoSi_2$	4.5
Instrument and Control Alloys: Cu–Ni Alloys	98Cu–2Ni	35
	94Cu–6Ni	17
	89Cu–11Ni	11
	78Cu–22Ni	5.7
	55Cu–45Ni (constantan)	3.5
Instrument and Control Alloys: Cu–Mn–Ni Alloys	87Cu–13Mn(manganin)	3.5
	83Cu–13Mn 4Ni(manganin)	3.5
	85Cu–10Mn–4Ni (shunt manganin)	45
	70Cu–20Ni–10Mn	3.6
	67Cu–5Ni–27Mn	1.8
Instrument and Control Alloys: Ni–Base Alloys	99.8 Ni	23
	71Ni–29Fe	9
	80Ni–20Cr	1.5
	75Ni–20Cr–3Al+Cu or Fe	1.3
	76Ni–17Cr–4Si–3Mn	1.3
	60Ni–16Cr–24Fe	1.5
	35Ni–20Cr–45Fe	1.7
Instrument and Control Alloys: Fe–Cr–Al alloy	72Fe–23Cr–5Al–0.5Co	1.3
Instrument and Control Alloys: Pure Metals	Iron(99.99%)	17.75

Data from *ASM Metals Reference Book, Third Edition*, Michael Bauccio, Ed., ASM International, Materials Park, OH, p157-158, (1993).

Table 288. ELECTRICAL CONDUCTIVITY OF METALS
(SHEET 5 OF 7)

Class	Metal or Alloy	Electrical Conductivity (%IACS)
Thermostat Metals	75Fe–22Ni–3Cr	3
	72Mn–18Cu–10Ni	1.5
	67Ni–30Cu–1.4Fe–1Mn	3.5
	75Fe–22Ni–3Cr	12
	66.5Fe–22Ni–8.5Cr	3.3
Permanent Magnet Materials: Steels	Carbon Steel (0.65%)	9.5
	Carbon Steel (1% C)	8
	Chromium Steel (3.5% Cr)	6.1
	Tungsten Steel (6% W)	6
	Cobalt Steel (17% Co)	6.3
	Cobalt Steel (36% Co)	6.5
Permanent Magnet Materials: Intermediate Alloys	Cunico	7.5
	Cunife	9.5
	Comol	3.6
Permanent Magnet Materials: Alnico Alloys	Alnico I	3.3
	Alnico II	3.3
	Alnico III	3.3
	Alnico IV	3.3
	Alnico V	3.5
	Alnico VI	3.5
Magnetically Soft Materials: Electrical Steel Sheet	M–50	9.5
	M–43	6 to 9
	M–36	5.5 to 7.5
	M–27	3.5 to 5.5
	M–22	3.5 to 5
	M–19	3.5 to 5
	M–17	3 to 3.5
	M–15	3 to 3.5

Data from *ASM Metals Reference Book, Third Edition,* Michael Bauccio, Ed., ASM International, Materials Park, OH, p157-158, (1993).

Table 288. ELECTRICAL CONDUCTIVITY OF METALS

(SHEET 6 OF 7)

Class	Metal or Alloy	Electrical Conductivity (%IACS)
	M–14	3 to 3.5
	M–7	3 to 3.5
	M–6	3 to 3.5
	M–5	3 to 3.5
Moderately High–Permeability Materials	Thermenol	0.5
	16 Alfenol	0.7
	Sinimax	2
	Monimax	2.5
	Supermalloy	3
	4–79 Moly Pemalloy, Hymu 80	3
	Mumetal	3
	1040 alloy	3
	High Permalloy 49, A–L 4750, Armco 48	3.6
	45 Permalloy	3.6
High–Permeability Materials	Supermendur	4.5
	2V Pamendur	4.5
	35% Co, 1% Cr	9
	Ingot iron	17.5
	0.5% Si Steel	6
	1.75% Si Steel	4.6
	3.0% Si Steel	3.6
	Grain–oriented 3.0% Si Steel	3.5
	Grain–oriented 50% Ni iron	3.6
	50% Ni iron	3.5
Relay Steels and Alloys After Annealing		
Low–carbon Iron and Steel	Low–carbon iron	17.5
	1010 Steel	14.5

Data from *ASM Metals Reference Book, Third Edition*, Michael Bauccio, Ed., ASM International, Materials Park, OH, p157-158, (1993).

Table 288. ELECTRICAL CONDUCTIVITY OF METALS
(SHEET 7 OF 7)

Class	Metal or Alloy	Electrical Conductivity (%IACS)
Silicon Steels	1% Si	7.5
	2.5% Si	4
	3% Si	3.5
	3% Si, grain–oriented	3.5
	4% Si	3
Stainless Steels	Type 410	3
	Type 416	3
	Type 430	3
	Type 443	3
	Type 446	3
Nickel Irons	50% Ni	3.5
	78% Ni	11
	77% Ni (Cu, Cr)	3
	79% Ni (Mo)	3
Stainless and Heat Resisting Alloys	Type 302	3
	Type 309	2.5
	Type 316	2.5
	Type 317	2.5
	Type 347	2.5
	Type 403	3
	Type 405	3
	Type 501	4.5
	HH	2.5
	HK	2
	HT	1.7

Data from *ASM Metals Reference Book, Third Edition*, Michael Bauccio, Ed., ASM International, Materials Park, OH, p157-158, (1993).

Table 289. ELECTRICAL RESISTIVITY OF METALS

(SHEET 1 OF 7)

Class	Metal or Alloy	Electrical Resistivity (μΩ • cm)
Aluminum and Aluminum Alloys	Aluminum (99.996%)	2.65
	EC(O, H19)	2.8
	5052 (O, H38)	4.93
	5056 (H38)	6.4
	6101 (T6)	3.1
Copper and Copper Alloys: Wrought Copper	Pure copper	1.67
	Electolytic (ETP)	1.71
	Oxygen–free copper (OF)	1.71
	Free–machining copper 0.5% Te	1.82
	Free–machining copper 1.0% Pb	1.76
Wrought Alloys	Cartridge brass, 70%	6.2
	Yellow brass	6.4
	Leaded commercial bronze	4.1
	Phosphor bronze, 1.25%	3.6
	Nickel silver, 55–18	31
	Low–silicon bronze(B)	14.3
	Beryllium copper	5.7 to 7.8
Copper and Copper Alloys: Casting Alloys	Chromium copper (1% Cr)	2.10
	88Cu–8Sn–4Zn	15
	87Cu–10Sn–1Pb–2Zn	15
Electrical Contact Materials: Copper Alloys	0.04 oxide	1.72
	1.25 Sn + P	3.6
	5 Sn+P	11
	8 Sn+P	13
	15 Zn	4.7
	20 Zn	5.4

Data from *ASM Metals Reference Book, Third Edition*, Michael Bauccio, Ed., ASM International, Materials Park, OH, p157-158, (1993).

Table 289. ELECTRICAL RESISTIVITY OF METALS
(SHEET 2 OF 7)

Class	Metal or Alloy	Electrical Resistivity (μΩ • cm)
	35 Zn	6.4
	2 Be+Ni or Co	9.6 to 11.5
Electrical Contact Materials: Silver and Silver Alloys	Fine silver	1.59
	92.5 Ag–7.5Cu	2
	90Ag–10Cu	2
	72Ag–28Cu	2
	72Ag–26Cu–2Ni	2.9
	85Ag–15Cd	4.93
	97Ag–3Pt	3.5
	97Ag–3Pd	2.9
	90Ag–10Pd	5.3
	90Ag–10Au	4.2
	60Ag–40Pd	23
	70Ag–30Pd	14.3
Electrical Contact Materials: Platinum and Platinum Alloys	Platinum	10.6
	95Pt–5Ir	19
	90Pt–10Ir	25
	85Pt–15Ir	28.5
	80Pt–20Ir	31
	75Pt–25Ir	33
	70Pt–30Ir	35
	65Pt–35Ir	36
	95Pt–5Ru	31.5
	90Pt–10Ru	43
	89Pt–11Ru	43
	86Pt–14Ru	46
	96Pt–4W	36

Data from *ASM Metals Reference Book, Third Edition*, Michael Bauccio, Ed., ASM International, Materials Park, OH, p157-158, (1993).

Table 289. ELECTRICAL RESISTIVITY OF METALS

(SHEET 3 OF 7)

Class	Metal or Alloy	Electrical Resistivity (μΩ • cm)
Electrical Contact Materials: Palladium and Palladium Alloys	Palladium	10.8
	95.5Pd–4.5Ru	24.2
	90Pd–10Ru	27
	70Pd–30Ag	40
	60Pd–40Ag	43
	50Pd–50Ag	31.5
	72Pd–26Ag–2Ni	43
	60Pd–40Cu	35
	45Pd–30Ag–20Au–5Pt	39
	35Pd–30Ag–14Cu–l0Pt–10Au–1Zn	35
Electrical Contact Materials: Gold and Gold Alloys	Gold	2.35
	90Au–10Cu	10.8
	75Au–25Ag	10.8
	72.5Au–14Cu–8.5Pt–4Ag–1Zn	17
	69Au–25Ag–6Pt	15
	41.7Au–32.5Cu–18.8Ni–7Zn	39
Electrical Heating Alloys: Ni–Cr and Ni–Cr–Fe Alloys	78.5Ni–20Cr–1.5Si (80–20)	108.05
	73.5Ni–20Cr–5Al–1.5Si	137.97
	68Ni–20Cr–8.5Fe–2Si	116.36
	60Ni–16Cr–22.5Fe–1.5Si	112.20
	35Ni–20Cr–43.5Fe–1.5Si	101.4
Electrical Heating Alloys: Fe–Cr–Al Alloys	72Fe–23Cr–5Al	138.8
	55Fe–37.5Cr–7.5Al	166.23
Pure Metals	Molybdenum	5.2
	Platinum	10.64
	Tantalum	12.45
	Tungsten	5.65

Data from *ASM Metals Reference Book, Third Edition*, Michael Bauccio, Ed., ASM International, Materials Park, OH, p157-158, (1993).

Table 289. ELECTRICAL RESISTIVITY OF METALS

(SHEET 4 OF 7)

Class	Metal or Alloy	Electrical Resistivity (μΩ • cm)
Nonmetallic Heating Element Materials	Silicon carbide, SiC	100 to 200
	Molybdenum disilicide, $MoSi_2$	37.24
	Graphite	910.1
Instrument and Control Alloys: Cu–Ni Alloys	98Cu–2Ni	4.99
	94Cu–6Ni	9.93
	89Cu–11Ni	14.96
	78Cu–22Ni	29.92
	55Cu–45Ni (constantan)	49.87
Instrument and Control Alloys: Cu–Mn–Ni Alloys	87Cu–13Mn(manganin)	48.21
	83Cu–13Mn–4Ni(manganin)	48.21
	85Cu–10Mn–4Ni (shunt manganin)	38.23
	70Cu–20Ni–10Mn	48.88
	67Cu–5Ni–27Mn	99.74
Instrument and Control Alloys: Ni–base Alloys	99.8 Ni	7.98
	71Ni–29Fe	19.95
	80Ni–20Cr	112.2
	75Ni–20Cr–3Al+Cu or Fe	132.98
	76Ni–17Cr–4Si–3Mn	132.98
	60Ni–16Cr–24Fe	112.2
	35Ni–20Cr–45Fe	101.4
Instrument and Control Alloys: Fe–Cr–Al alloy	72Fe–23Cr–5Al–0.5Co	135.48
Instrument and Control Alloys: Pure Metals	Iron(99.99%)	9.71

Data from *ASM Metals Reference Book, Third Edition*, Michael Bauccio, Ed., ASM International, Materials Park, OH, p157-158, (1993).

Table 289. ELECTRICAL RESISTIVITY OF METALS
(SHEET 5 OF 7)

Class	Metal or Alloy	Electrical Resistivity (μΩ • cm)
Thermostat Metals	75Fe–22Ni–3Cr	78.13
	72Mn–18Cu–10Ni	112.2
	67Ni–30Cu–1.4Fe–1Mn	56.52
	75Fe–22Ni–3Cr	15.79
	66.5Fe–22Ni–8.5Cr	58.18
Permanent Magnet Materials: Steels	Carbon Steel (0.65%)	18
	Carbon Steel (1% C)	20
	Chromium Steel (3.5% Cr)	29
	Tungsten Steel (6% W)	30
	Cobalt Steel (17% Co)	28
	Cobalt Steel (36% Co)	27
Permanent Magnet Materials: Intermediate Alloys	Cunico	24
	Cunife	18
	Comol	45
Permanent Magnet Materials: Alnico Alloys	Alnico I	75
	Alnico II	65
	Alnico III	60
	Alnico IV	75
	Alnico V	47
	Alnico VI	50
Magnetically Soft Materials: Electrical Steel Sheet	M–50	18
	M–43	20 to 28
	M–36	24 to 33
	M–27	32 to 47
	M–22	41 to 52
	M–19	41 to 56
	M–17	45 to 58
	M–15	45 to 69

Data from *ASM Metals Reference Book, Third Edition*, Michael Bauccio, Ed., ASM International, Materials Park, OH, p157-158, (1993).

Table 289. ELECTRICAL RESISTIVITY OF METALS
(SHEET 6 OF 7)

Class	Metal or Alloy	Electrical Resistivity (μΩ • cm)
Moderately High–Permeability Materials	M–14	58 to 69
	M–7	45 to 52
	M–6	45 to 52
	M–5	45 to 52
	Thermenol	162
	16 Alfenol	153
	Sinimax	90
	Monimax	80
High–Permeability Materials	Supermalloy	65
	4–79 Moly Pemalloy, Hymu 80	58
	Mumetal	60
	1040 alloy	56
	High Permalloy 49, A–L 4750, Armco 48	48
	45 Permalloy	45
	Supermendur	40
	2V Pamendur	40
	35% Co, 1% Cr	20
	Ingot iron	10
	0.5% Si Steel	28
	1.75% Si Steel	37
	3.0% Si Steel	47
	Grain–oriented 3.0% Si Steel	50
	Grain–oriented 50% Ni iron	45
	50% Ni iron	50
Relay Steels and Alloys After Annealing		
Low–Carbon Iron and Steel	Low–carbon iron	10
	1010 Steel	12

Data from *ASM Metals Reference Book, Third Edition*, Michael Bauccio, Ed., ASM International, Materials Park, OH, p157-158, (1993).

Table 289. ELECTRICAL RESISTIVITY OF METALS
(SHEET 7 OF 7)

Class	Metal or Alloy	Electrical Resistivity (μΩ • cm)
Silicon Steels	1% Si	23
	2.5% Si	41
	3% Si	48
	3% Si, grain–oriented	48
	4% Si	59
Stainless Steels	Type 410	57
	Type 416	57
	Type 430	60
	Type 443	68
	Type 446	61
Nickel Irons	50% Ni	48
	78% Ni	16
	77% Ni (Cu, Cr)	60
	79% Ni (Mo)	58
Stainless and Heat Resisting Alloys	Type 302	72
	Type 309	78
	Type 316	74
	Type 317	74
	Type 347	73
	Type 403	57
	Type 405	60
	Type 501	40
	HH	80
	HK	90
	HT	100

Data from *ASM Metals Reference Book, Third Edition*, Michael Bauccio, Ed., ASM International, Materials Park, OH, p157-158, (1993).

Table 290. ELECTRICAL RESISTIVITY OF ALLOY CAST IRONS

Class	Description	Electrical Resistivity (mΩ • m)
Abrasion–Resistant White Irons	Low–C white iron	0.53
	Martensitic nickel–chromium iron	0.80
Corrosion–Resistant Irons	High– Silicon iron	0.50
	High–nickel gray iron	1.0[a]
	High–nickel ductile iron	1.0[a]
	Heat–Resistant Gray Irons	
Heat–Resistant Gray Irons	Medium–silicon iron	
	High–chromium iron	
	High–nickel iron	1.4 to 1.7
	Nickel–chromium–silicon iron	1.5 to 1.7
	High–aluminum iron	2.4
Heat–Resistant Ductile Irons	Medium–silicon ductile iron	0.58 to 0.87
	High–nickel ductile (20 Ni)	1.02
	High–nickel ductile (23 Ni)	1.0[a]

[a] Estimated.

Source: data from ASM Metals Reference Book, Second Edition, American Society for Metals, Metals Park, Ohio 44073, (1984).

Table 291. RESISTIVITY OF CERAMICS

(SHEET 1 OF 6)

Class	Ceramic	Resistivity (Ω–cm)	Temperature Range of Validity
Borides	Chromium Diboride (CrB_2)	$21x10^6$	
	Hafnium Diboride (HfB_2)	10–12 x 10^6	room temp.
	Tantalum Diboride (TaB_2)	68 $x10^6$	
	Titanium Diboride (TiB_2) (polycrystalline)		
	(85% dense)	$26.5–28.4x10^6$	room temp.
	(85% dense)	$9.0x10^6$	room temp.
	(100% dense, extrapolated values)	$8.7–14.1x10^6$	room temp.
		$3.7x10^6$	liquid air temp.
	Titanium Diboride (TiB_2) (monocrystalline)		
	(crystal length 5 cm, 39 deg. and 59 deg. orientation with respect to growth axis)	$6.6±0.2x10^6$	room temp.
	(crystal length 1.5 cm, 16.5 deg. and 90 deg. orientation with respect to growth axis)	$6.7±0.2x10^6$	room temp.
	Zirconium Diboride (ZrB_2)	$9.2x10^6$	20 °C
		$1.8x10^6$	liquid air temperature
Carbides	Boron Carbide (B_4C)	0.3–0.8	

Source: data compiled by J.S. Park from No. 1 Materials Index, Peter T.B. Shaffer, Plenum Press, New York, (1964); Smithells Metals Reference Book, Eric A. Brandes, ed., in association with Fulmer Research Institute Ltd. 6th ed. London, Butterworths, Boston, (1983); and Ceramic Source, American Ceramic Society (1986–1991).

Table 291. RESISTIVITY OF CERAMICS

(SHEET 2 OF 6)

Class	Ceramic	Resistivity (Ω–cm)	Temperature Range of Validity
Carbides (Con't)	Hafnium Monocarbide (HfC)	$41x10^6$	4.2K
		$41x10^6$	80K
		$45x10^6$	160K
		$49x10^6$	240K
		$60x10^6$	300K
		$(30 + 0.0628T) x10^6$	300–2000K
	Silicon Carbide (SiC)	$10^2 - 10^{12}$	20°C
	(with 1 wt% Be additive)	$3x10^{13}$	
	(with 1 wt% B additive)	$2x10^4$	
	(with 1 wt% Al additive)	0.8	
	(with 1.6 wt% BeO additive)	$>10^{13}$	
	(with 3.2 wt% BeO additive)	$4x10^{13}$	
	(with 2.0 wt% BN additive)	$1x10^{11}$	
	Tantalum Monocarbide (TaC)		
	(80% dense)	$8x10^6$	4.2K
	(80% dense)	$10x10^6$	80K
	(80% dense)	$15x10^6$	160K
	(80% dense)	$20x10^6$	240K
	(80% dense)	$25x10^6$	300K
	Titanium Monocarbide (TiC)	0.3–0.8	

Source: data compiled by J.S. Park from No. 1 Materials Index, Peter T.B. Shaffer, Plenum Press, New York, (1964); Smithells Metals Reference Book, Eric A. Brandes, ed., in association with Fulmer Research Institute Ltd. 6th ed. London, Butterworths, Boston, (1983); and Ceramic Source, American Ceramic Society (1986–1991).

Table 291. RESISTIVITY OF CERAMICS

(SHEET 3 OF 6)

Class	Ceramic	Resistivity (Ω–cm)	Temperature Range of Validity
Carbides (Con't)	Zirconium Monocarbide (ZrC)	$41x10^6$	4.2K
		$45x10^6$	80K
		$47x10^6$	160K
		$53x10^6$	240K
		$61–64x10^6$	300K
		$97x10^6$	773K
		$137x10^6$	1273K
Nitrides	Aluminum Nitride (AlN)	$2x10^{11}–10^{13}$	room temp.
	Boron Nitride (BN)	$1.7x10^{13}$	25°C
		$2.3x10^{10}$	480°C
		$3.1x10^4$	1000°C
	(20% humidity)	$1.0x10^{12}$	25°C
	(50% humidity)	$7.0x10^{10}$	25°C
	(90% humidity)	$5.0x10^9$	25°C
	Titanium Mononitirde (TiN)	$11.07–130x10^6$	room temp.
		$340x10^6$	melting temp.
		$8.13x10^6$	liquid air
	Trisilicon tetranitride (Si_3N_4)	$>10^{13}$	
	Zirconium Mononitride (TiN)	$11.52–160x10^6$	room temp.
		$320x10^6$	melting temp.
		$3.97x10^6$	liquid air

Source: data compiled by J.S. Park from No. 1 Materials Index, Peter T.B. Shaffer, Plenum Press, New York, (1964); Smithells Metals Reference Book, Eric A. Brandes, ed., in association with Fulmer Research Institute Ltd. 6th ed. London, Butterworths, Boston, (1983); and Ceramic Source, American Ceramic Society (1986–1991).

Table 291. RESISTIVITY OF CERAMICS

(SHEET 4 OF 6)

Class	Ceramic	Resistivity (Ω–cm)	Temperature Range of Validity
Oxides	Aluminum Oxide (Al_2O_3)	>10x10^{14}	25°C
		2x10^{13}	100°C
		1x10^{13}	300°C
		6.3x10^{10}	500°C
		5.0x10^{8}	700°C
		2x10^{6}	1000°C
	Beryllium Oxide (BeO)	>10^{17}	25°C
		>10^{15}	300°C
		1–5x10^{15}	500°C
		1.5–2x10^{15}	700°C
		4–7x10^{15}	1000°C
	Magnesium Oxide (MgO)	1.3x10^{15}	27°C
		0.2–1x10^{8}	1000°C
		4x10^{2}	1727°C
	Silicon Dioxide (SiO_2)	10^{18}	room temp.
	Zirconium Oxide (ZrO_2)		
	(stabilized)	2300	700°C
	(stabilized)	77	1200°C
	(stabilized)	9.4	1300°C
	(stabilized)	1.6	1700°C
	(stabilized)	0.59	2000°C
	(stabilized)	0.37	2200°C

Source: data compiled by J.S. Park from No. 1 Materials Index, Peter T.B. Shaffer, Plenum Press, New York, (1964); Smithells Metals Reference Book, Eric A. Brandes, ed., in association with Fulmer Research Institute Ltd. 6th ed. London, Butterworths, Boston, (1983); and Ceramic Source, American Ceramic Society (1986–1991).

Table 291. RESISTIVITY OF CERAMICS

(SHEET 5 OF 6)

Class	Ceramic	Resistivity (Ω–cm)	Temperature Range of Validity
Oxides (Con't)	Cordierite ($2MgO\ 2Al_2O_3\ 5SiO_2$)		
	(ρ=2.3g/cm^3)	$1x10^{14}$	25°C
	(ρ=2.3g/cm^3)	$2.5x10^{11}$	100°C
	(ρ=2.3g/cm^3)	$3.3x10^{7}$	300°C
	(ρ=2.3g/cm^3)	$7.7x10^{5}$	500°C
	(ρ=2.3g/cm^3)	$8.0x10^{4}$	700°C
	(ρ=2.3g/cm^3)	$1.9x10^{4}$	900°C
	(ρ=2.1g/cm^3)	$>1x10^{14}$	25°C
	(ρ=2.1g/cm^3)	$3.0x10^{13}$	100°C
	(ρ=2.1g/cm^3)	$2.0x10^{10}$	300°C
	(ρ=2.1g/cm^3)	$9.0x10^{7}$	500°C
	(ρ=2.1g/cm^3)	$3.0x10^{6}$	700°C
	(ρ=2.1g/cm^3)	$3.5x10^{5}$	900°C
	(ρ=1.8g/cm^3)	$1.0x10^{14}$	25°C
	(ρ=1.8g/cm^3)	$1.0x10^{13}$	100°C
	(ρ=1.8g/cm^3)	$3.0x10^{9}$	300°C
	(ρ=1.8g/cm^3)	$4.9x10^{7}$	500°C
	(ρ=1.8g/cm^3)	$4.7x10^{6}$	700°C
	(ρ=1.8g/cm^3)	$7.0x10^{5}$	900°C
	Mullite ($3Al_2O_3\ 2SiO_2$)	$>10^{14}$	25°C
		10^{10}	300°C
		10^{8}	500°C

Source: data compiled by J.S. Park from No. 1 Materials Index, Peter T.B. Shaffer, Plenum Press, New York, (1964); Smithells Metals Reference Book, Eric A. Brandes, ed., in association with Fulmer Research Institute Ltd. 6th ed. London, Butterworths, Boston, (1983); and Ceramic Source, American Ceramic Society (1986–1991).

Table 291. RESISTIVITY OF CERAMICS

(SHEET 6 OF 6)

Class	Ceramic	Resistivity (Ω–cm)	Temperature Range of Validity
Silicides	Molybdenum Disilicide ($MoSi_2$)	$21.5x10^6$	22°C
		$18.9x10^6$	–80°C
		$75–80x10^6$	1600°C
	Tungsten Disilicide (WSi_2)	$33.4–54.9x10^6$	

Source: data compiled by J.S. Park from No. 1 Materials Index, Peter T.B. Shaffer, Plenum Press, New York, (1964); Smithells Metals Reference Book, Eric A. Brandes, ed., in association with Fulmer Research Institute Ltd. 6th ed. London, Butterworths, Boston, (1983); and Ceramic Source, American Ceramic Society (1986–1991).

Table 292. VOLUME RESISTIVITY OF GLASS

(SHEET 1 OF 13)

Glass	Description	Resistivity	Temperature (°C)
SiO_2 glass		11.0–13.6 log Ω cm	250°C
		$3.16 \times 10^8 - 6.3 \times 10^{10}$ Ω cm	400°C
		6.3×10^7 Ω cm	500°C
		1.0×10^7 Ω cm	600°C
		3.6×10^6 Ω cm	700°C
		1.6×10^6 Ω cm	800°C
		8.0×10^5 Ω cm	900°C
		4.6×10^5 Ω cm	1000°C
		2.9×10^5 Ω cm	1100°C
		2.0×10^5 Ω cm	1200°C
		1.4×10^5 Ω cm	1300°C
		1.0×10^5 Ω cm	1400°C
		7.9×10^4 Ω cm	1500°C
	(0.5 atm Ar pressure)	4.6×10^4 Ω cm	1500°C
	(0.5 atm Ar pressure)	2.5×10^4 Ω cm	1600°C
	(0.5 atm Ar pressure)	1.0×10^4 Ω cm	1700°C
	(0.5 atm Ar pressure)	3.0×10^3 Ω cm	1800°C
	(0.5 atm Ar pressure)	1.0×10^3 Ω cm	1900°C
	(0.5 atm Ar pressure)	5.0×10^2 Ω cm	2000°C
	(0.5 atm Ar pressure)	2.0×10^2 Ω cm	2100°C
SiO_2–Na_2O glass	(5% mol Na_2O)	10.45–11.71 log Ω cm	150°C
	(5% mol Na_2O)	7.63 log Ω cm	250°C
	(5% mol Na_2O)	7.33–8.25 log Ω cm	300°C
	(5% mol Na_2O)	6.37 log Ω cm	350°C

Source: data compiled by J. S. Park from O. V. Mazurin, M. V. Streltsina and T. P. Shvaiko-Shvaikovskaya, Handbook of Glass Data, Part A and Part B, Elsevier, New York, 1983

Table 292. VOLUME RESISTIVITY OF GLASS
(SHEET 2 OF 13)

Glass	Description	Resistivity	Temperature (°C)
SiO_2–Na_2O glass (Con't)	(7.5% mol Na_2O)	7.59 log Ω cm	150°C
	(7.5% mol Na_2O)	5.30 log Ω cm	300°C
	(7.8% mol Na_2O)	$7.8x10^9$ Ω cm	100°C
	(10% mol Na_2O)	7.35 log Ω cm	150°C
	(10% mol Na_2O)	6.14 log Ω cm	250°C
	(10% mol Na_2O)	5.18 log Ω cm	300°C
	(10% mol Na_2O)	4.96 log Ω cm	350°C
	(10% mol Na_2O)	1.03 log Ω cm	1500°C
	(10% mol Na_2O)	0.92 log Ω cm	1600°C
	(13% mol Na_2O)	6.90–6.96 log Ω cm	150°C
	(13% mol Na_2O)	4.77–4.79 log Ω cm	300°C
	(15% mol Na_2O)	5.44 log Ω cm	250°C
	(15% mol Na_2O)	4.32 log Ω cm	350°C
	(15% mol Na_2O)	0.61 log Ω cm	1400°C
	(15% mol Na_2O)	0.56 log Ω cm	1500°C
	(15.1% mol Na_2O)	$1.4x10^8$ Ω cm	100°C
	(19.9% mol Na_2O)	1.68 log Ω cm	600°C
	(19.9% mol Na_2O)	1.34 log Ω cm	700°C
	(19.9% mol Na_2O)	0.96 log Ω cm	800°C
	(19.9% mol Na_2O)	0.76 log Ω cm	900°C
	(19.9% mol Na_2O)	0.61 log Ω cm	1000°C
	(19.9% mol Na_2O)	0.48 log Ω cm	1100°C
	(19.9% mol Na_2O)	0.38 log Ω cm	1200°C
	(19.9% mol Na_2O)	0.30 log Ω cm	1300°C

Source: data compiled by J. S. Park from O. V. Mazurin, M. V. Streltsina and T. P. Shvaiko-Shvaikovskaya, Handbook of Glass Data, Part A and Part B, Elsevier, New York, 1983

Table 292. VOLUME RESISTIVITY OF GLASS
(SHEET 3 OF 13)

Glass	Description	Resistivity	Temperature (°C)
SiO_2–Na_2O glass (Con't)	(20% mol Na_2O)	6.45–6.80 log Ω cm	150°C
	(20% mol Na_2O)	4.85 log Ω cm	250°C
	(20% mol Na_2O)	4.36–4.64 log Ω cm	300°C
	(20% mol Na_2O)	3.80 log Ω cm	350°C
	(24.8% mol Na_2O)	0.52 log Ω cm	900°C
	(24.8% mol Na_2O)	0.38 log Ω cm	1000°C
	(24.8% mol Na_2O)	0.26 log Ω cm	1100°C
	(24.8% mol Na_2O)	0.17 log Ω cm	1200°C
	(25% mol Na_2O)	6.05 log Ω cm	150°C
	(25% mol Na_2O)	4.50 log Ω cm	250°C
	(25% mol Na_2O)	4.03 log Ω cm	300°C
	(25% mol Na_2O)	3.52 log Ω cm	350°C
	(27% mol Na_2O)	5.87 log Ω cm	150°C
	(27% mol Na_2O)	3.94 log Ω cm	300°C
	(29.7% mol Na_2O)	1.31 log Ω cm	550°C
	(29.7% mol Na_2O)	1.16 log Ω cm	600°C
	(29.7% mol Na_2O)	0.78 log Ω cm	700°C
	(29.7% mol Na_2O)	0.52 log Ω cm	800°C
	(29.7% mol Na_2O)	0.34 log Ω cm	900°C
	(29.7% mol Na_2O)	0.20 log Ω cm	1000°C
	(29.7% mol Na_2O)	0.08 log Ω cm	1100°C
	(29.7% mol Na_2O)	–0.02 log Ω cm	1200°C
	(29.7% mol Na_2O)	–0.10 log Ω cm	1300°C
	(29.7% mol Na_2O)	–0.16 log Ω cm	1400°C

Source: data compiled by J. S. Park from O. V. Mazurin, M. V. Streltsina and T. P. Shvaiko-Shvaikovskaya, Handbook of Glass Data, Part A and Part B, Elsevier, New York, 1983

Table 292. VOLUME RESISTIVITY OF GLASS
(SHEET 4 OF 13)

Glass	Description	Resistivity	Temperature (°C)
SiO_2–Na_2O glass (Con't)	(30% mol Na_2O)	5.48–5.75 log Ω cm	150°C
	(30% mol Na_2O)	4.42 log Ω cm	250°C
	(30% mol Na_2O)	3.64–3.78 log Ω cm	300°C
	(30% mol Na_2O)	3.46 log Ω cm	350°C
	(30.2% mol Na_2O)	$3.8x10^6$ Ω cm	100°C
	(33.3% mol Na_2O)	5.06 log Ω cm	150°C
	(33.3% mol Na_2O)	3.34 log Ω cm	300°C
	(34.7% mol Na_2O)	0.12 log Ω cm	900°C
	(34.7% mol Na_2O)	0.00 log Ω cm	1000°C
	(34.7% mol Na_2O)	–0.11 log Ω cm	1100°C
	(34.7% mol Na_2O)	–0.20 log Ω cm	1200°C
	(34.7% mol Na_2O)	–0.27 log Ω cm	1300°C
	(34.7% mol Na_2O)	–0.33 log Ω cm	1400°C
	(35% mol Na_2O)	3.85 log Ω cm	250°C
	(35% mol Na_2O)	2.92 log Ω cm	350°C
	(36% mol Na_2O)	4.89 log Ω cm	150°C
	(36% mol Na_2O)	3.22 log Ω cm	300°C
	(39.5% mol Na_2O)	0.91 log Ω cm	550°C
	(39.5% mol Na_2O)	0.67 log Ω cm	600°C
	(39.5% mol Na_2O)	0.33 log Ω cm	700°C
	(39.5% mol Na_2O)	0.13 log Ω cm	800°C

Source: data compiled by J. S. Park from O. V. Mazurin, M. V. Streltsina and T. P. Shvaiko-Shvaikovskaya, Handbook of Glass Data, Part A and Part B, Elsevier, New York, 1983

Table 292. VOLUME RESISTIVITY OF GLASS
(SHEET 5 OF 13)

Glass	Description	Resistivity	Temperature (°C)
SiO_2–Na_2O glass (Con't)	(39.5% mol Na_2O)	0.00 log Ω cm	900°C
	(39.5% mol Na_2O)	–0.13 log Ω cm	1000°C
	(39.5% mol Na_2O)	–0.24 log Ω cm	1100°C
	(39.5% mol Na_2O)	–0.32 log Ω cm	1200°C
	(39.5% mol Na_2O)	–0.39 log Ω cm	1300°C
	(39.5% mol Na_2O)	–0.45 log Ω cm	1400°C
	(40% mol Na_2O)	4.58 log Ω cm	150°C
	(40% mol Na_2O)	3.59 log Ω cm	250°C
	(40% mol Na_2O)	2.97 log Ω cm	300°C
	(40% mol Na_2O)	2.66 log Ω cm	350°C
	(44.2% mol Na_2O)	$1.4x10^5$ Ω cm	100°C
	(44.5% mol Na_2O)	–0.38 log Ω cm	1100°C
	(44.5% mol Na_2O)	–0.46 log Ω cm	1200°C
	(44.5% mol Na_2O)	–0.52 log Ω cm	1300°C
	(45% mol Na_2O)	4.33 log Ω cm	150°C
	(45% mol Na_2O)	3.30 log Ω cm	250°C
	(45% mol Na_2O)	2.69 log Ω cm	300°C
	(45% mol Na_2O)	2.35 log Ω cm	350°C
	(48% mol Na_2O)	4.09 log Ω cm	150°C
	(48% mol Na_2O)	2.58 log Ω cm	300°C
	(49.3% mol Na_2O)	–0.47 log Ω cm	1100°C
	(49.3% mol Na_2O)	–0.56 log Ω cm	1200°C
	(49.3% mol Na_2O)	–0.61 log Ω cm	1300°C

Source: data compiled by J. S. Park from O. V. Mazurin, M. V. Streltsina and T. P. Shvaiko-Shvaikovskaya, Handbook of Glass Data, Part A and Part B, Elsevier, New York, 1983

Table 292. VOLUME RESISTIVITY OF GLASS
(SHEET 6 OF 13)

Glass	Description	Resistivity	Temperature (°C)
SiO_2–Na_2O glass (Con't)	(57.5% mol Na_2O)	–0.52 log Ω cm	1100°C
	(57.5% mol Na_2O)	–0.61 log Ω cm	1200°C
	(57.5% mol Na_2O)	–0.67 log Ω cm	1300°C
SiO_2–PbO glass	(30% mol PbO)	12.94 log Ω cm	200°C
	(30% mol PbO)	10.44 log Ω cm	300°C
	(33.8% mol PbO)	16.14 log Ω cm	66°C
	(33.8% mol PbO)	13.68 log Ω cm	135°C
	(35% mol PbO)	12.10 log Ω cm	200°C
	(35% mol PbO)	9.89 log Ω cm	300°C
	(38.5% mol PbO)	4.40 log Ω cm	700°C
	(38.5% mol PbO)	3.20 log Ω cm	800°C
	(38.5% mol PbO)	2.47 log Ω cm	900°C
	(38.5% mol PbO)	1.94 log Ω cm	1000°C
	(38.5% mol PbO)	1.56 log Ω cm	1100°C
	(38.5% mol PbO)	1.26 log Ω cm	1200°C
	(38.5% mol PbO)	1.04 log Ω cm	1300°C
	(40% mol PbO)	11.54 log Ω cm	200°C
	(40% mol PbO)	9.48 log Ω cm	300°C
	(40.2% mol PbO)	14.85 log Ω cm	78°C
	(40.2% mol PbO)	11.70 log Ω cm	175°C
	(44.7% mol PbO)	2.38 log Ω cm	800°C
	(44.7% mol PbO)	1.82 log Ω cm	900°C
	(44.7% mol PbO)	1.40 log Ω cm	1000°C
	(44.7% mol PbO)	1.15 log Ω cm	1100°C
	(44.7% mol PbO)	0.98 log Ω cm	1200°C
	(44.7% mol PbO)	0.82 log Ω cm	1300°C

Source: data compiled by J. S. Park from O. V. Mazurin, M. V. Streltsina and T. P. Shvaiko-Shvaikovskaya, Handbook of Glass Data, Part A and Part B, Elsevier, New York, 1983

Table 292. VOLUME RESISTIVITY OF GLASS
(SHEET 7 OF 13)

Glass	Description	Resistivity	Temperature (°C)
SiO_2–PbO glass (Con't)	(47.3% mol PbO)	14.48 log Ω cm	79°C
	(47.3% mol PbO)	11.74 log Ω cm	149°C
	(50% mol PbO)	10.69 log Ω cm	200°C
	(50% mol PbO)	8.80–9.2 log Ω cm	300°C
	(50.0% mol PbO)	1.90 log Ω cm	800°C
	(50.0% mol PbO)	1.36 log Ω cm	900°C
	(50.0% mol PbO)	1.02 log Ω cm	1000°C
	(50.0% mol PbO)	0.80 log Ω cm	1100°C
	(50.0% mol PbO)	0.60 log Ω cm	1200°C
	(51.4% mol PbO)	14.52 log Ω cm	65°C
	(51.4% mol PbO)	11.59 log Ω cm	139°C
	(51.6% mol PbO)	1.62 log Ω cm	800°C
	(51.6% mol PbO)	1.20 log Ω cm	900°C
	(51.6% mol PbO)	0.92 log Ω cm	1000°C
	(51.6% mol PbO)	0.70 log Ω cm	1100°C
	(51.6% mol PbO)	0.54 log Ω cm	1200°C
	(57.1% mol PbO)	13.70 log Ω cm	77°C
	(57.1% mol PbO)	10.14 log Ω cm	172°C
	(60% mol PbO)	10.04 log Ω cm	200°C
	(60% mol PbO)	8.11 log Ω cm	300°C
	(60% mol PbO)	1.72 log Ω cm	650°C
	(60% mol PbO)	1.74 log Ω cm	700°C
	(60% mol PbO)	1.07 log Ω cm	800°C
	(60% mol PbO)	0.76 log Ω cm	900°C
	(60% mol PbO)	0.40 log Ω cm	1000°C

Source: data compiled by J. S. Park from O. V. Mazurin, M. V. Streltsina and T. P. Shvaiko-Shvaikovskaya, Handbook of Glass Data, Part A and Part B, Elsevier, New York, 1983

Table 292. VOLUME RESISTIVITY OF GLASS
(SHEET 8 OF 13)

Glass	Description	Resistivity	Temperature (°C)
SiO_2–PbO glass (Con't)	(63.2% mol PbO)	14.29 log Ω cm	57°C
	(63.2% mol PbO)	10.34 log Ω cm	159°C
	(65% mol PbO)	9.76 log Ω cm	200°C
	(65% mol PbO)	7.81 log Ω cm	300°C
	(66.7% mol PbO)	1.32 log Ω cm	700°C
	(66.7% mol PbO)	0.82 log Ω cm	800°C
	(66.7% mol PbO)	0.50 log Ω cm	900°C
	(66.7% mol PbO)	0.26 log Ω cm	1000°C
SiO_2–CaO glass	(33.6% mol CaO)	0.97 log Ω cm	1500°C
	(33.6% mol CaO)	0.93–0.94 log Ω cm	1560°C
	(33.6% mol CaO)	0.79–0.80 log Ω cm	1600°C
	(41.3% mol CaO)	0.82 log Ω cm	1519°C
	(41.3% mol CaO)	0.76 log Ω cm	1550°C
	(41.3% mol CaO)	0.67–0.68 log Ω cm	1600°C
	(45.4% mol CaO)	0.65 log Ω cm	1550°C
	(45.4% mol CaO)	0.58–0.59 log Ω cm	1585°C
	(45.4% mol CaO)	0.52 log Ω cm	1622°C
	(50% mol CaO)	12.2 log Ω cm	300°C
	(50% mol CaO)	8.70 log Ω cm	400°C
	(51.4% mol CaO)	0.48–0.49 log Ω cm	1500°C
	(51.4% mol CaO)	0.47 log Ω cm	1560°C
	(51.4% mol CaO)	0.38 log Ω cm	1618°C
	(55.2% mol CaO)	0.51–0.53 log Ω cm	1499°C
	(55.2% mol CaO)	0.42–0.43 log Ω cm	1550°C
	(55.2% mol CaO)	0.34 log Ω cm	1600°C

Source: data compiled by J. S. Park from O. V. Mazurin, M. V. Streltsina and T. P. Shvaiko-Shvaikovskaya, Handbook of Glass Data, Part A and Part B, Elsevier, New York, 1983

Table 292. VOLUME RESISTIVITY OF GLASS
(SHEET 9 OF 13)

Glass	Description	Resistivity	Temperature (°C)
SiO_2–B_2O_3 glass	(2.74% wt B_2O_3)	5.30 log Ω cm	900°C
	(2.74% wt B_2O_3)	4.72 log Ω cm	1100°C
	(2.74% wt B_2O_3)	4.40 log Ω cm	1300°C
	(2.74% wt B_2O_3)	4.02 log Ω cm	1500°C
	(2.74% wt B_2O_3)	3.76 log Ω cm	1700°C
	(2.74% wt B_2O_3)	3.56 log Ω cm	1900°C
	(5.48% wt B_2O_3)	5.64 log Ω cm	900°C
	(5.48% wt B_2O_3)	5.16 log Ω cm	1100°C
	(5.48% wt B_2O_3)	4.56 log Ω cm	1300°C
	(5.48% wt B_2O_3)	4.30 log Ω cm	1500°C
	(5.48% wt B_2O_3)	4.10 log Ω cm	1700°C
	(5.48% wt B_2O_3)	3.94 log Ω cm	1900°C
	(10.75% wt B_2O_3)	5.74 log Ω cm	900°C
	(10.75% wt B_2O_3)	5.08 log Ω cm	1100°C
	(10.75% wt B_2O_3)	4.69 log Ω cm	1300°C
	(10.75% wt B_2O_3)	4.40 log Ω cm	1500°C
	(10.75% wt B_2O_3)	4.16 log Ω cm	1700°C
	(10.75% wt B_2O_3)	3.98 log Ω cm	1900°C
	(19.37% wt B_2O_3)	5.65 log Ω cm	900°C
	(19.37% wt B_2O_3)	4.82 log Ω cm	1100°C
	(19.37% wt B_2O_3)	4.48 log Ω cm	1300°C
	(19.37% wt B_2O_3)	4.22 log Ω cm	1500°C
	(19.37% wt B_2O_3)	4.00 log Ω cm	1700°C
	(19.37% wt B_2O_3)	3.84 log Ω cm	1900°C

Source: data compiled by J. S. Park from O. V. Mazurin, M. V. Streltsina and T. P. Shvaiko-Shvaikovskaya, Handbook of Glass Data, Part A and Part B, Elsevier, New York, 1983

Table 292. VOLUME RESISTIVITY OF GLASS
(SHEET 10 OF 13)

Glass	Description	Resistivity	Temperature (°C)
SiO_2–Al_2O_3 glass	(2.83% wt Al_2O_3)	5.74 log Ω cm	700°C
	(2.83% wt Al_2O_3)	4.82 log Ω cm	900°C
	(2.83% wt Al_2O_3)	4.29 log Ω cm	1100°C
	(2.83% wt Al_2O_3)	3.94 log Ω cm	1300°C
	(2.83% wt Al_2O_3)	3.67 log Ω cm	1500°C
	(2.83% wt Al_2O_3)	3.46 log Ω cm	1700°C
	(2.83% wt Al_2O_3)	3.28 log Ω cm	1900°C
	(5.51% wt Al_2O_3)	5.34 log Ω cm	700°C
	(5.51% wt Al_2O_3)	4.65 log Ω cm	900°C
	(5.51% wt Al_2O_3)	4.15 log Ω cm	1100°C
	(5.51% wt Al_2O_3)	3.76 log Ω cm	1300°C
	(5.51% wt Al_2O_3)	3.56 log Ω cm	1500°C
	(5.51% wt Al_2O_3)	3.36 log Ω cm	1700°C
	(5.51% wt Al_2O_3)	3.20 log Ω cm	1900°C
	(10.86% wt Al_2O_3)	5.38 log Ω cm	700°C
	(10.86% wt Al_2O_3)	4.54 log Ω cm	900°C
	(10.86% wt Al_2O_3)	4.02 log Ω cm	1100°C
	(10.86% wt Al_2O_3)	3.74 log Ω cm	1300°C
	(10.86% wt Al_2O_3)	3.52 log Ω cm	1500°C
	(10.86% wt Al_2O_3)	3.34 log Ω cm	1700°C
	(10.86% wt Al_2O_3)	3.20 log Ω cm	1900°C
B_2O_3 glass		7.6 log Ω cm	560°C
		7.3 log Ω cm	600°C
		6.9 log Ω cm	640°C

Source: data compiled by J. S. Park from O. V. Mazurin, M. V. Streltsina and T. P. Shvaiko-Shvaikovskaya, Handbook of Glass Data, Part A and Part B, Elsevier, New York, 1983

Table 292. VOLUME RESISTIVITY OF GLASS
(SHEET 11 OF 13)

Glass	Description	Resistivity	Temperature (°C)
B_2O_3 glass (Con't)		6.6 log Ω cm	680°C
		6.2 log Ω cm	730°C
		5.8 log Ω cm	780°C
		5.5 log Ω cm	840°C
B_2O_3–Na_2O glass	(3.63% mol Na_2O)	2.70 log Ω cm	800°C
	(3.63% mol Na_2O)	2.30 log Ω cm	900°C
	(3.63% mol Na_2O)	2.00 log Ω cm	1000°C
	(10% mol Na_2O)	14.20 log Ω cm	40°C
	(10% mol Na_2O)	13.21 log Ω cm	60°C
	(10% mol Na_2O)	12.40 log Ω cm	80°C
	(10% mol Na_2O)	11.61 log Ω cm	100°C
	(12.1% mol Na_2O)	2.43 log Ω cm	700°C
	(12.1% mol Na_2O)	1.89 log Ω cm	800°C
	(12.1% mol Na_2O)	1.48 log Ω cm	900°C
	(16% mol Na_2O)	15.89 log Ω cm	40°C
	(16% mol Na_2O)	15.08 log Ω cm	60°C
	(16% mol Na_2O)	14.32 log Ω cm	80°C
	(16% mol Na_2O)	13.58 log Ω cm	100°C
	(17.3% mol Na_2O)	1.39 log Ω cm	850°C
	(17.3% mol Na_2O)	1.18 log Ω cm	900°C
	(17.3% mol Na_2O)	0.89 log Ω cm	1000°C
	(20% mol Na_2O)	13.86 log Ω cm	40°C
	(20% mol Na_2O)	12.91 log Ω cm	60°C
	(20% mol Na_2O)	12.05 log Ω cm	80°C
	(20% mol Na_2O)	11.28 log Ω cm	100°C

Source: data compiled by J. S. Park from O. V. Mazurin, M. V. Streltsina and T. P. Shvaiko-Shvaikovskaya, Handbook of Glass Data, Part A and Part B, Elsevier, New York, 1983

Table 292. VOLUME RESISTIVITY OF GLASS

(SHEET 12 OF 13)

Glass	Description	Resistivity	Temperature (°C)
B_2O_3–Na_2O glass (Con't)	(21.9% mol Na_2O)	1.29 log Ω cm	800°C
	(21.9% mol Na_2O)	0.94 log Ω cm	900°C
	(21.9% mol Na_2O)	0.65 log Ω cm	1000°C
	(27.5% mol Na_2O)	1.00 log Ω cm	800°C
	(27.5% mol Na_2O)	0.70 log Ω cm	900°C
	(30% mol Na_2O)	11.90 log Ω cm	40°C
	(30% mol Na_2O)	10.14 log Ω cm	60°C
	(30% mol Na_2O)	9.43 log Ω cm	80°C
	(30% mol Na_2O)	8.82 log Ω cm	100°C
	(32.8% mol Na_2O)	1.02 log Ω cm	700°C
	(32.8% mol Na_2O)	0.60 log Ω cm	800°C
	(32.8% mol Na_2O)	0.40 log Ω cm	900°C
	(40% mol Na_2O)	10.48 log Ω cm	40°C
	(40% mol Na_2O)	9.73 log Ω cm	60°C
	(40% mol Na_2O)	9.08 log Ω cm	80°C
	(40% mol Na_2O)	8.46 log Ω cm	100°C
B_2O_3–CaO glass	(33.3% mol CaO)	14.40 log Ω cm	150°C
	(33.3% mol CaO)	13.92 log Ω cm	200°C
	(33.3% mol CaO)	13.50 log Ω cm	250°C
	(33.3% mol CaO)	13.16 log Ω cm	300°C
	(33.3% mol CaO)	3.10 log Ω cm	850°C
	(33.3% mol CaO)	2.25 log Ω cm	950°C
	(33.3% mol CaO)	1.52 log Ω cm	1050°C
	(33.3% mol CaO)	1.10 log Ω cm	1150°C
	(33.3% mol CaO)	0.85 log Ω cm	1250°C

Source: data compiled by J. S. Park from O. V. Mazurin, M. V. Streltsina and T. P. Shvaiko-Shvaikovskaya, Handbook of Glass Data, Part A and Part B, Elsevier, New York, 1983

Table 292. VOLUME RESISTIVITY OF GLASS
(SHEET 13 OF 13)

Glass	Description	Resistivity	Temperature (°C)
B_2O_3–CaO glass (Con't)	(40.0% mol CaO)	2.97 log Ω cm	850°C
	(40.0% mol CaO)	2.06 log Ω cm	950°C
	(40.0% mol CaO)	1.40 log Ω cm	1050°C
	(40.0% mol CaO)	0.98 log Ω cm	1150°C
	(40.0% mol CaO)	0.75 log Ω cm	1250°C
	(55.4% mol CaO)	6.13 log Ω cm	750°C
	(55.4% mol CaO)	3.86 log Ω cm	850°C
	(55.4% mol CaO)	2.46 log Ω cm	950°C
	(55.4% mol CaO)	1.70 log Ω cm	1050°C
	(55.4% mol CaO)	1.22 log Ω cm	1150°C

Source: data compiled by J. S. Park from O. V. Mazurin, M. V. Streltsina and T. P. Shvaiko-Shvaikovskaya, Handbook of Glass Data, Part A and Part B, Elsevier, New York, 1983

Table 293. VOLUME RESISTIVITY OF POLYMERS

(SHEET 1 OF 8)

Polymer	Type	Volume Resistivity, (ASTM D257) (Ω • cm)
ABS Resins; Molded, Extruded	Medium impact	2—4 x 10^{15}
	High impact	1—4 x 10^{15}
	Very high impact	1—4 x 10^{15}
	Low temperature impact	1—4 x 10^{15}
	Heat resistant	1—5 x 10^{15}
Acrylics; Cast, Molded, Extruded	Cast Resin Sheets, Rods:	
	General purpose, type I	>10^{15}
	General purpose, type II	>10^{15}
	Moldings:	
	Grades 5, 6, 8	>10^{14}
	High impact grade	2.0 x 10^{16}
Thermoset Carbonate	Allyl diglycol carbonate	4 x 10^{14}
Alkyds; Molded	Putty (encapsulating)	10^{14}
	Rope (general purpose)	10^{14}
	Granular (high speed molding)	10^{14} — 10^{15}
	Glass reinforced (heavy duty parts)	10^{14}
Cellulose Acetate; Molded, Extruded	ASTM Grade:	
	H6—1	10^{10}—10^{13}
	H4—1	10^{10}—10^{13}
	H2—1	10^{10}—10^{13}

Source: data compiled by J.S. Park from Charles T. Lynch, CRC Handbook of Materials Science, Vol. 3, CRC Press, Boca Raton, Florida, 1975 and Engineered Materials Handbook, Vol.2, Engineering Plastics, ASM International, Metals Park, Ohio, 1988.

Table 293. VOLUME RESISTIVITY OF POLYMERS

(SHEET 2 OF 8)

Polymer	Type	Volume Resistivity, (ASTM D257) (Ω • cm)
Cellulose Acetate; Molded, Extruded (Con't)	MH—1, MH—2	10^{10}—10^{13}
	MS—1, MS—2	10^{10}—10^{13}
	S2—1	10^{10}—10^{13}
Cellulose Acetate Butyrate; Molded, Extruded	ASTM Grade:	
	H4	10^{11}—10^{14}
	MH	10^{11}—10^{14}
	S2	10^{11}—10^{14}
Cellusose Acetate Propionate; Molded, Extruded	ASTM Grade:	
	1	10^{11}—10^{14}
	3	10^{11}—10^{14}
	6	10^{11}—10^{14}
Chlorinated Polymers	Chlorinated polyether	1.5×10^{16}
	Chlorinated polyvinyl chloride	1×10^{15}—2×10^{16}
Polycarbonates	Polycarbonate	2.1×10^{16}
	Polycarbonate (40% glass fiber reinforced)	1.4×10^{15}
Diallyl Phthalates; Molded	Orlon filled	6×10^{4}—6×10^{6}
	Dacron filled	10^{2}—2.5×10^{4}
	Asbestos filled	10^{2}—5×10^{3}
	Glass fiber filled	10^{4}—5×10^{4}

Source: data compiled by J.S. Park from Charles T. Lynch, CRC Handbook of Materials Science, Vol. 3, CRC Press, Boca Raton, Florida, 1975 and Engineered Materials Handbook, Vol.2, Engineering Plastics, ASM International, Metals Park, Ohio, 1988.

Table 293. VOLUME RESISTIVITY OF POLYMERS (SHEET 3 OF 8)

Polymer	Type	Volume Resistivity, (ASTM D257) (Ω • cm)
Fluorocarbons; Molded,Extruded	Polytetrifluoro chloroethylene (PTFCE)	10^{18}
	Polytetrafluoroethylene (PTFE)	$>10^{18}$
	Ceramic reinforced (PTFE)	10^{15}
	Fluorinated ethylene propylene(FEP)	$>2 \times 10^{18}$
	Polyvinylidene— fluoride (PVDF)	5×10^{14}
Epoxies; Cast, Molded, Reinforced	Standard epoxies (diglycidyl ethers of bisphenol A)	
	Cast rigid	6.1×10^{15}
	Cast flexible	9.1×10^{5}—6.7×10^{9}
	Molded	1—5×10^{15}
	High strength laminate	6.6×10^{7}—10^{9}
Epoxies—Molded, Extruded	High performance resins (cycloaliphatic diepoxides)	
	Cast, rigid	2.10×10^{14}
	Molded	1.4—5.5×10^{14}
Epoxy novolacs	Cast, rigid	$>10^{16}$
Melamines; Molded	Filler & type	
	Cellulose electrical	10^{12}—10^{13}
	Glass fiber	1—7×10^{11}
	Alpha cellulose and mineral	10^{12}

Source: data compiled by J.S. Park from Charles T. Lynch, CRC Handbook of Materials Science, Vol. 3, CRC Press, Boca Raton, Florida, 1975 and Engineered Materials Handbook, Vol.2, Engineering Plastics, ASM International, Metals Park, Ohio, 1988.

Table 293. VOLUME RESISTIVITY OF POLYMERS
(SHEET 4 OF 8)

Polymer	Type	Volume Resistivity, (ASTM D257) (Ω • cm)
Nylons; Molded, Extruded	**Type 6**	
	General purpose	4.5×10^{13}
	Glass fiber (30%) reinforced	2.8×10^{14}—1.5×10^{15}
	Cast	2.6×10^{14}
	Type 8	1.5×10^{11}
	Type 11	2×10^{13}
	Type 12	10^{14}—10^{15}
	6/6 Nylon	
	General purpose molding	10^{14}—10^{15}
	Glass fiber reinforced	2.6—5.5×10^{15}
	General purpose extrusion	10^{15}
	6/10 Nylon	
	General purpose	10^{15}
Phenolics; Molded	Type and filler	
	General: woodflour and flock	10^{9}—10^{13}
	Shock: paper, flock, or pulp	1—50×10^{11}
	High shock: chopped fabric or cord	$>10^{10}$
	Very high shock: glass fiber	10^{10} — 10^{11}
	Arc resistant—mineral	10^{10} — 10^{12}
	Rubber phenolic—woodflour or flock	10^{8}—10^{11}
	Rubber phenolic—chopped fabric	10^{11}
	Rubber phenolic—asbestos	10^{11}

Source: data compiled by J.S. Park from Charles T. Lynch, CRC Handbook of Materials Science, Vol. 3, CRC Press, Boca Raton, Florida, 1975 and Engineered Materials Handbook, Vol.2, Engineering Plastics, ASM International, Metals Park, Ohio, 1988.

Table 293. VOLUME RESISTIVITY OF POLYMERS

(SHEET 5 OF 8)

Polymer	Type	Volume Resistivity, (ASTM D257) (Ω • cm)
Phenolics; Molded (Con't)	ABS—Polycarbonate Alloy	2.2×10^{16}
	PVC—Acrylic Alloy	
	PVC—acrylic Sheet	$1—5 \times 10^{13}$
	PVC—acrylic injection molded	5×10^{15}
Polymides	Unreinforced	4×10^{15}
	Glass reinforced	9.2×10^{15}
Polyacetals	Homopolymer:	
	Standard	1×10^{15}
	20% glass reinforced	5×10^{14}
	Copolymer:	
	Standard	1×10^{14}
	25% glass reinforced	1.2×10^{14}
	High flow	1.0×10^{14}
Polyester; Thermoplastic	Injection Moldings:	
	General purpose grade	$1—4 \times 10^{16}$
	Glass reinforced grades	$3.2—3.3 \times 10^{16}$
	Glass reinforced self extinguishing	3.4×10^{16}
	General purpose grade	2×10^{15}
	Asbestos—filled grade	3×10^{14}
Polyesters: Thermosets	Cast polyyester	
	Rigid	10^{13}
	Flexible	10^{12}

Source: data compiled by J.S. Park from Charles T. Lynch, CRC Handbook of Materials Science, Vol. 3, CRC Press, Boca Raton, Florida, 1975 and Engineered Materials Handbook, Vol.2, Engineering Plastics, ASM International, Metals Park, Ohio, 1988.

Table 293. VOLUME RESISTIVITY OF POLYMERS

(SHEET 6 OF 8)

Polymer	Type	Volume Resistivity, (ASTM D257) (Ω • cm)
Polyesters: Thermosets (Con't)	Reinforced polyester moldings	
	High strength (glass fibers)	1×10^{12} —1×10^{13}
	Heat and chemical resistant (asbestos)	1×10^{12} —1×10^{13}
	Sheet molding compounds, general purpose	6.4×10^{15} —2.2×10^{16}
	Phenylene Oxides	
	SE—100	10^{17}
	SE—1	10^{17}
	Glass fiber reinforced	10^{17}
Phenylene oxides (Noryl)	Standard	5×10^{16}
	Glass fiber reinforced	10^{17}
	Polyarylsulfone	3.2—7.71×10^{16}
Polypropylene	General purpose	$>10^{17}$
	High impact	10^{17}
	Asbestos filled	1.5×10^{15}
	Glass reinforced	1.7×10^{16}
	Flame retardant	4×10^{16}—10^{17}
Polyphenylene sulfide	40% glass reinforced	4.5×10^{14}

Source: data compiled by J.S. Park from Charles T. Lynch, CRC Handbook of Materials Science, Vol. 3, CRC Press, Boca Raton, Florida, 1975 and Engineered Materials Handbook, Vol.2, Engineering Plastics, ASM International, Metals Park, Ohio, 1988.

Table 293. VOLUME RESISTIVITY OF POLYMERS (SHEET 7 OF 8)

Polymer	Type	Volume Resistivity, (ASTM D257) (Ω • cm)
Polyethylenes; Molded, Extruded	Type I—lower density (0.910—0.925)	
	Melt index 0.3—3.6	10^{17}—10^{19}
	Melt index 6—26	10^{17}—10^{19}
	Melt index 200	10^{17}—10^{19}
	Type II—medium density (0.926—0.940)	
	Melt index 20	$>10^{15}$
	Melt index l.0—1.9	$>10^{15}$
	Type III—higher density (0.941—0.965)	
	Melt index 0.2—0.9	$>10^{15}$
	Melt Melt index 0.l—12.0	$>10^{15}$
	Melt index 1.5—15	$>10^{15}$
	High molecular weight	$>10^{15}$
Olefin Copolymers; Molded	EEA (ethylene ethyl acrylate)	2.4×10^{15}
	EVA (ethylene vinyl acetate)	0.15×10^{15}
	Ionomer	10×10^{15}
	Polyallomer	$>10^{16}$
Polystyrenes; Molded	Polystyrenes	
	General purpose	$>10^{16}$
	Medium impact	$>10^{16}$
	High impact	$>10^{16}$

Source: data compiled by J.S. Park from Charles T. Lynch, CRC Handbook of Materials Science, Vol. 3, CRC Press, Boca Raton, Florida, 1975 and Engineered Materials Handbook, Vol.2, Engineering Plastics, ASM International, Metals Park, Ohio, 1988.

Table 293. VOLUME RESISTIVITY OF POLYMERS

(SHEET 8 OF 8)

Polymer	Type	Volume Resistivity, (ASTM D257) ($\Omega \cdot cm$)
Polystyrenes; Molded (Con't)	Glass fiber -30% reinforced	3.6×10^{16}
	Styrene acrylonitrile (SAN)	$>10^{16}$
	Glass fiber (30%) reinforced SAN	4.4×10^{16}
Polyvinyl Chloride And Copolymers; Molded, Extruded	Nonrigid—general	$1—700 \times 10^{12}$
	Nonrigid—electrical	$4—300 \times 10^{11}$
	Rigid—normal impact	$10^{14}—10^{16}$
	Vinylidene chloride	$10^{14}—10^{16}$
	Silicones; Molded, Laminated	(dry)
	Fibrous (glass) reinforced silicones	9×10^{14}
	Granular (silica) reinforced silicones	5×10^{14}
	Woven glass fabric/ silicone laminate	$2—5 \times 10^{14}$
Ureas; Molded	Alpha—cellulose filled (ASTM Type l)	$0.5—5 \times 10^{11}$
	Cellulose filled (ASTM Type 2)	$5—8 \times 10^{10}$

Source: data compiled by J.S. Park from Charles T. Lynch, CRC Handbook of Materials Science, Vol. 3, CRC Press, Boca Raton, Florida, 1975 and Engineered Materials Handbook, Vol.2, Engineering Plastics, ASM International, Metals Park, Ohio, 1988.

Table 294. CRITICAL TEMPERATURE OF SUPERCONDUCTIVE ELEMENTS (SHEET 1 OF 2)

Element	T_c(K)
Al	1.175
Be	0.026
Cd	0.518-0.52
Ga	1.0833
Ga (β)	5.90-6.2
Ga (γ)	7.62
Ga (δ)	7.85
Hg (α)	4.154
Hg (β)	3.949
In	3.405
Ir	0.11-0.14
La (α)	4.88
La (β)	6.00
Mo	0.916
Nb	9.25
Os	0.655
Pa	1.4
Pb	7.23
Re	1.697
Ru	0.493
Sb	2.6-2.7[a]
Sn	3.721
Ta	4.47
Tc	7.73-7.78
Th	1.39

[a] Metastable.

Source: data from Roberts, B. W., Properties of Selected Superconductive Materials - 1974 Supplement, NBS Technical Note 825, National Bureau of Standards, U.S. Government Printing Office, Washington,D.C., 1974, 10.

Table 294. CRITICAL TEMPERATURE OF SUPERCONDUCTIVE ELEMENTS (SHEET 2 OF 2)

Element	T_c(K)
Ti	0.39
Ti	2.332-2.39
V	5.43-5.31
W	0.0154
Zn	0.875
Zr	0.53
Zr (ω)	0.65

[a] Metastable.

Source: data from Roberts, B. W., Properties of Selected Superconductive Materials - 1974 Supplement, NBS Technical Note 825, National Bureau of Standards, U.S. Government Printing Office, Washington,D.C., 1974, 10.

Table 295. DISSIPATION FACTOR FOR POLYMERS

(SHEET 1 OF 8)

Class	Polymer	Dissipation Factor (ASTM D150)	
		60 Hz	10^6 Hz
ABS Resins; Molded, Extruded	Medium impact	0.003—0.006	0.008—0.009
	High impact	0.005—0.007	0.007—0.015
	Very high impact	0.005—0.010	0.008—0.016
	Low temperature impact	0.005—0.01	0.008—0.016
	Heat resistant	0.030—0.040	0.005—0.015
Acrylics; Cast, Molded, Extruded	Cast Resin Sheets, Rods:		
	General purpose, type I	0.05—0.06	0.02—0.03
	General purpose, type II	0.05—0.06	0.02—0.03
	Moldings:		
	Grades 5, 6, 8	0.04—0.06	0.02—0.03
	High impact grade	0.03—0.04	0.01—0.02
Thermoset Carbonate	Allyl diglycol carbonate	0.03—0.04	0.1—0.2
Alkyds; Molded	Putty (encapsulating)	0.030—0.045	0.016—0.020
	Rope (general purpose)	0.019	0.023
	Granular (high speed molding)	0.030—0.040	0.017—0.020
	Glass reinforced (heavy duty parts)	0.02—0.03	0.015—0.022
Cellulose Acetate; Molded, Extruded	ASTM Grade:		
	H4—1	0.01—0.06	0.01—0.10
	H2—1	0.01—0.06	0.01—0.10
	MH—1, MH—2	0.01—0.06	0.01—0.10
	MS—1, MS—2	0.01—0.06	0.01—0.10
	S2—1	0.01—0.06	0.01—0.10

Source: data compiled by J.S. Park from Charles T. Lynch, CRC Handbook of Materials Science, Vol. 3, CRC Press, Boca Raton, Florida, 1975 and Engineered Materials Handbook, Vol.2, Engineering Plastics, ASM International, Metals Park, Ohio, 1988.

Table 295. DISSIPATION FACTOR FOR POLYMERS

(SHEET 2 OF 8)

Class	Polymer	Dissipation Factor (ASTM D150)	
		60 Hz	10^6 Hz
Cellulose Acetate Butyrate; Molded, Extruded	ASTM Grade:		
	H4	0.01—0.04	0.02—0.05
	MH	0.01—0.04	0.02—0.05
	S2	0.01—0.04	0.02—0.05
Cellulose Acetate Propionate; Molded, Extruded	ASTM Grade:		
	1	0.01—0.04	0.02—0.05
	3	0.01—0.04	0.02—0.05
	6	0.01—0.04	0.02—0.05
	Chlorinated Polymers		
	Chlorinated polyether	0.011	0.011
	Chlorinated polyvinyl chloride	0.0189—0.0208	0.02
Polycarbonates	Polycarbonate	0.0009	0.01
	Polycarbonate (40% glass fiber reinforced)	0.006	0.007
Diallyl Phthalates; Molded	Orlon filled	0.023—0.015 (Dry)	0.045—0.040 (Wet)
	Dacron filled	0.004—0.016 (Dry)	0.009—0.017 (Wet)
	Asbestos filled	0.05—0.03 (Dry)	0.154—0.050 (Wet)
	Glass fiber filled	0.004—0.015 (Dry)	0.012—0.020 (Wet)

Source: data compiled by J.S. Park from Charles T. Lynch, CRC Handbook of Materials Science, Vol. 3, CRC Press, Boca Raton, Florida, 1975 and Engineered Materials Handbook, Vol.2, Engineering Plastics, ASM International, Metals Park, Ohio, 1988.

Table 295. DISSIPATION FACTOR FOR POLYMERS

(SHEET 3 OF 8)

Class	Polymer	Dissipation Factor (ASTM D150)	
		60 Hz	10^6 Hz
Fluorocarbons; Molded,Extruded	Polytrifluoro chloroethylene (PTFCE)	0.02	0.007—0.010
	Polytetrafluoroethylene (PTFE)	0.0002	0.0002
	Ceramic reinforced (PTFE)	0.0005–0.0015	0.0005–0.0015
	Fluorinated ethylene propylene(FEP)	0.0003	0.0003
	Polyvinylidene— fluoride (PVDF)	0.05	0.184
Epoxies; Cast, Molded, Reinforced	Standard epoxies (diglycidyl ethers of bisphenol A)		
	Cast rigid	0.0074	0.032
	Cast flexible	0.0048-0.0380	0.0369-0.0622
	Molded	0.011-0.018	0.013—0.020
	General purpose glass cloth laminate	0.004-0.006	0.024—0.026
	High strength laminate	—	0.010-0.017
Epoxies—Molded, Extruded	High performance resins		
	(cycloaliphatic diepoxides)		
	Cast, rigid	0.0055—0.0074	0.029—0.028
	Molded	0.0071—0.025	—
	Glass cloth laminate	—	0.0158
	Epoxy novolacs		
	Cast, rigid	0.001—0.007	—

Source: data compiled by J.S. Park from Charles T. Lynch, CRC Handbook of Materials Science, Vol. 3, CRC Press, Boca Raton, Florida, 1975 and Engineered Materials Handbook, Vol.2, Engineering Plastics, ASM International, Metals Park, Ohio, 1988.

Table 295. DISSIPATION FACTOR FOR POLYMERS

(SHEET 4 OF 8)

Class	Polymer	Dissipation Factor (ASTM D150)	
		60 Hz	10^6 Hz
Melamines; Molded	Filler & type		
	Unfilled	0.048—0.162	0.031—0.040
	Cellulose electrical	0.026—0.192	0.032—0.12
	Glass fiber	0.14—0.23	0.020—0.03
	Alpha cellulose	—	0.028
	Mineral	—	0.030
Nylons; Molded, Extruded	**Type 6**		
	General purpose	0.06—0.014	0.03—0.04
	Glass fiber (30%) reinforced	0.022—0.008	0.019—0.015
	Cast	0.015	0.05
	Flexible copolymers	0.007—0.010	0.010—0.015
	Type 8	0.19	0.08
	Type 11	0.03	0.02
	Type 12	0.04 (10^3 Hz)	
	6/6 Nylon		
	General purpose molding	0.014—0.04	0.04
	Glass fiber reinforced	0.009—0.018	0.017—0.018
	6/10 Nylon		
	General purpose	0.04	
Phenolics; Molded	Type and filler		
	General: woodflour and flock	0.05—0.30	0.03—0.07
	Shock: paper, flock, or pulp	0.08—0.35	0.03—0.07
	High shock: chopped fabric or cord	0.08—0.45	0.03—0.09
	Very high shock: glass fiber	0.02—0.03	0.02

Source: data compiled by J.S. Park from Charles T. Lynch, CRC Handbook of Materials Science, Vol. 3, CRC Press, Boca Raton, Florida, 1975 and Engineered Materials Handbook, Vol.2, Engineering Plastics, ASM International, Metals Park, Ohio, 1988.

Table 295. DISSIPATION FACTOR FOR POLYMERS

(SHEET 5 OF 8)

Class	Polymer	Dissipation Factor (ASTM D150)	
		60 Hz	10^6 Hz
Phenolics: Molded	Arc resistant—mineral	0.13—0.16	0.1
	Rubber phenolic—woodflour or flock	0.15—0.60	0.1—0.2
	Rubber phenolic—chopped fabric	0.5	0.09
	Rubber phenolic—asbestos	0.15	0.13
	ABS–Polycarbonate Alloy	0.0026	0.0059
PVC–Acrylic Alloy	PVC–acrylic sheet	0.076	0.094
	PVC–acrylic injection molded	0.037	0.031
Polyimides	Unreinforced	0.003	0.011
	Glass reinforced	0.0034	0.0055
Polyacetals	Homopolymer:		
	Standard	0.0048	0.0048
	20% glass reinforced	0.0047	0.0036
	Copolymer:		
	Standard	0.001 (100 Hz)	0.006
	25% glass reinforced	0.003 (100 Hz)	0.006
	High flow	0.001 (100 Hz)	0.006
Polyester; Thermoplastic	Injection Moldings:		
	General purpose grade	0.002 (10^3 Hz)	
	Glass reinforced grades	0.002—0.003 (10^3 Hz)	
	Glass reinforced self extinguishing	0.002 (10^3 Hz)	
	General purpose grade	0.023 (10^3 Hz)	
	Asbestos—filled grade	0.015 (10^3 Hz)	

Source: data compiled by J.S. Park from Charles T. Lynch, CRC Handbook of Materials Science, Vol. 3, CRC Press, Boca Raton, Florida, 1975 and Engineered Materials Handbook, Vol.2, Engineering Plastics, ASM International, Metals Park, Ohio, 1988.

Table 295. DISSIPATION FACTOR FOR POLYMERS

(SHEET 6 OF 8)

Class	Polymer	Dissipation Factor (ASTM D150)	
		60 Hz	10^6 Hz
Polyesters: Thermosets	Cast polyyester		
	Rigid	0.003—0.04	0.006—0.04
	Flexible	0.01—0.18	0.02—0.06
Reinforced polyester moldings	Sheet molding compounds, general purpose	0.0087—0.04	0.0086—0.022
Phenylene Oxides	SE—100	0.0007	0.0024
	SE—1	0.0007	0.0024
	Glass fiber reinforced	0.0009	0.0015
	Phenylene oxides (Noryl)		
	Standard	0.0008	0.0034
	Glass fiber reinforced	0.0019	0.0049
	Polyarylsulfone	0.0017—0.003	0.0056—0.012
Polypropylene	General purpose	0.0005–0.0007	0.0002–0.0003
	High impact	<0.0016	0.0002—0.0003
	Asbestos filled	0.007	0.002
	Glass reinforced	0.002	0.003
	Flame retardant	0.0007–0.017	0.0006–0.003
Polyphenylene sulfide	Standard	—	0.0007
	40% glass reinforced	—	0.0014—0.0041
Polyethylenes; Molded, Extruded	Type I—lower density (0.910—0.925)		
	Melt index 0.3—3.6	<0.0005	
	Melt index 6—26	<0.0005	
	Melt index 200	<0.0005	

Source: data compiled by J.S. Park from Charles T. Lynch, CRC Handbook of Materials Science, Vol. 3, CRC Press, Boca Raton, Florida, 1975 and Engineered Materials Handbook, Vol.2, Engineering Plastics, ASM International, Metals Park, Ohio, 1988.

Table 295. DISSIPATION FACTOR FOR POLYMERS

(SHEET 7 OF 8)

Class	Polymer	Dissipation Factor (ASTM D150)	
		60 Hz	10^6 Hz
Polyethylenes; Molded, Extruded (Con't)	Type II—medium density (0.926—0.940)		
	Melt index 20	<0.0005	
	Melt index l.0—1.9	<0.0005	
	Type III—higher density (0.941—0.965)		
	Melt index 0.2—0.9	<0.0005	
	Melt Melt index 0.l—12.0	<0.0005	
	Melt index 1.5—15	<0.0005	
	High molecular weight	<0.0005	
Olefin Copolymers; Molded	EEA (ethylene ethyl acrylate)	0.001	
	EVA (ethylene vinyl acetate)	0.003	
	Ionomer	0.003	
	Polyallomer	>0.0005	
Polystyrenes; Molded	Polystyrenes		
	General purpose	0.0001–0.0003	0.0001–0.0005
	Medium impact	0.0004–0.002	0.0004–0.002
	High impact	0.0004–0.002	0.0004–0.002
	Glass fiber -30% reinforced	0.005	0.002
	Styrene acrylonitrile (SAN)	>0.006	0.007–0.010
	Glass fiber (30%) reinforced SAN	0.005	0.009

Source: data compiled by J.S. Park from Charles T. Lynch, CRC Handbook of Materials Science, Vol. 3, CRC Press, Boca Raton, Florida, 1975 and Engineered Materials Handbook, Vol.2, Engineering Plastics, ASM International, Metals Park, Ohio, 1988.

Table 295. DISSIPATION FACTOR FOR POLYMERS

(SHEET 8 OF 8)

Class	Polymer	Dissipation Factor (ASTM D150)	
		60 Hz	10^6 Hz
Polyvinyl Chloride and Copolymers;	Molded, Extruded		
	Nonrigid—general	0.05—0.15	
	Nonrigid—electrical	0.08—0.11	
	Rigid—normal impact	0.020—0.03	
	Vinylidene chloride	0.03—0.15	
Silicones; Molded, Laminated	Fibrous (glass) reinforced silicones	0.01	0.004
	Granular (silica) reinforced silicones	0.002—0.004	0.001—0.004
	Woven glass fabric/ silicone laminate	0.02	0.002
Ureas; Molded	Alpha—cellulose filled (ASTM Type l)	0.035—0.043	0.028—0.032
	Cellulose filled (ASTM Type 2)	0.042—0.044	0.027—0.029
	Woodflour filled	0.035—0.040	0.028—0.032

Source: data compiled by J.S. Park from Charles T. Lynch, CRC Handbook of Materials Science, Vol. 3, CRC Press, Boca Raton, Florida, 1975 and Engineered Materials Handbook, Vol.2, Engineering Plastics, ASM International, Metals Park, Ohio, 1988.

Table 296. DIELECTRIC STRENGTH OF POLYMERS

(SHEET 1 OF 12)

Class	Polymer	Dielectric Strength (Short Time, ASTM D149) (V / mil)
ABS Resins; Molded, Extruded	Medium impact	385
	High impact	350—440
	Very high impact	300—375
	Low temperature impact	300—415
	Heat resistant	360—400
Acrylics; Cast, Molded, Extruded	Cast Resin Sheets, Rods:	
	General purpose, type I	450—530
	General purpose, type II	450—500
	Moldings:	
	Grades 5, 6, 8	400
	High impact grade	400—500

Source: data compiled by J.S. Park from Charles T. Lynch, *CRC Handbook of Materials Science, Vol. 3*, CRC Press, Boca Raton, Florida, 1975 and *Engineered Materials Handbook, Vol.2, Engineering Plastics*, ASM International, Metals Park, Ohio, 1988.

Table 296. DIELECTRIC STRENGTH OF POLYMERS
(SHEET 2 OF 12)

Class	Polymer	Dielectric Strength (Short Time, ASTM D149) (V / mil)
Cellulose Acetate; Molded, Extruded	ASTM Grade:	
	H6—1	250—600
	H4—1	250—600
	H2—1	250—600
	MH—1, MH—2	250—600
	MS—1, MS—2	250—600
	S2—1	250—600
Cellulose Acetate Butyrate; Molded, Extruded	ASTM Grade:	
	H4	250—400
	MH	250—400
	S2	250—400
Cellusose Acetate Propionate; Molded, Extruded	ASTM Grade:	
	1	300—450
	3	300—450
	6	300—450

Source: data compiled by J.S. Park from Charles T. Lynch, *CRC Handbook of Materials Science, Vol. 3*, CRC Press, Boca Raton, Florida, 1975 and *Engineered Materials Handbook, Vol.2, Engineering Plastics*, ASM International, Metals Park, Ohio, 1988.

Table 296. Dielectric Strength of Polymers
(Sheet 3 of 12)

Class	Polymer	Dielectric Strength (Short Time, ASTM D149) (V / mil)
Chlorinated Polymers	Chlorinated polyether	400
	Chlorinated polyvinyl chloride	1,250—1,550
Polycarbonates	Polycarbonate	400
	Polycarbonate (40% glass fiber reinforced)	475
Diallyl Phthalates; Molded	Orlon filled	400 (dry) 375 (wet)
	Dacron filled	376—400 (dry) 360—391 (wet)
	Asbestos filled	350—450 (dry) 300—400 (wet)
	Glass fiber filled	350—430 (dry) 300—420 (wet)

Source: data compiled by J.S. Park from Charles T. Lynch, *CRC Handbook of Materials Science, Vol. 3*, CRC Press, Boca Raton, Florida, 1975 and *Engineered Materials Handbook, Vol.2, Engineering Plastics*, ASM International, Metals Park, Ohio, 1988.

Table 296. DIELECTRIC STRENGTH OF POLYMERS

(SHEET 4 OF 12)

Class	Polymer	Dielectric Strength (Short Time, ASTM D149) (V / mil)
Fluorocarbons; Molded,Extruded	Polytrifluoro chloroethylene (PTFCE)	530—600
	Polytetrafluoroethylene (PTFE)	1000—2000
	Ceramic reinforced (PTFE)	300—400
	Fluorinated ethylene propylene(FEP)	2100
	Polyvinylidene— fluoride (PVDF)	260
Epoxies—Molded, Extruded	High performance resins (cycloaliphatic diepoxides)	
	Molded	280—400 (step)
	Epoxy novolacs	
	Cast, rigid	444
Melamines; Molded	Filler & type	
	Cellulose electrical	350—400
	Glass fiber	250 —300
	Alpha cellulose and mineral	375

Source: data compiled by J.S. Park from Charles T. Lynch, *CRC Handbook of Materials Science, Vol. 3*, CRC Press, Boca Raton, Florida, 1975 and *Engineered Materials Handbook, Vol.2, Engineering Plastics*, ASM International, Metals Park, Ohio, 1988.

Table 296. DIELECTRIC STRENGTH OF POLYMERS

(SHEET 5 OF 12)

Class	Polymer	Dielectric Strength (Short Time, ASTM D149) (V / mil)
Nylons; Molded, Extruded	Type 6	
	General purpose	385—400
	Glass fiber (30%) reinforced	400—450
	Cast	380
	Flexible copolymers	440
	Type 8	340
	Type 11	425
	Type 12	840
	6/6 Nylon	
	General purpose molding	385
	Glass fiber reinforced	400—480
	Glass fiber Molybdenum disulfide filled	300—400
	General purpose extrusion	
	6/10 Nylon	
	General purpose	470

Source: data compiled by J.S. Park from Charles T. Lynch, *CRC Handbook of Materials Science, Vol. 3*, CRC Press, Boca Raton, Florida, 1975 and *Engineered Materials Handbook, Vol.2, Engineering Plastics*, ASM International, Metals Park, Ohio, 1988.

Table 296. DIELECTRIC STRENGTH OF POLYMERS

(SHEET 6 OF 12)

Class	Polymer	Dielectric Strength (Short Time, ASTM D149) (V / mil)
Phenolics; Molded	Type and filler	
	General: woodflour and flock	200—425
	Shock: paper, flock, or pulp	250—350
	High shock: chopped fabric or cord	200—350
	Very high shock: glass fiber	375—425
	Phenolics: Molded	
Phenolics: Molded	Arc resistant—mineral	350—425
	Rubber phenolic—woodflour or flock	250—375
	Rubber phenolic—chopped fabric	250
	Rubber phenolic—asbestos	350
ABS–Polycarbonate Alloy	ABS–Polycarbonate Alloy	500
	PVC–Acrylic Alloy	
PVC–Acrylic Alloy	PVC–acrylic sheet	>429
	PVC–acrylic injection molded	400

Source: data compiled by J.S. Park from Charles T. Lynch, *CRC Handbook of Materials Science, Vol. 3*, CRC Press, Boca Raton, Florida, 1975 and *Engineered Materials Handbook, Vol.2, Engineering Plastics*, ASM International, Metals Park, Ohio, 1988.

Table 296. DIELECTRIC STRENGTH OF POLYMERS

(SHEET 7 OF 12)

Class	Polymer	Dielectric Strength (Short Time, ASTM D149) (V / mil)
Polyimides	Unreinforced 2nd value	310
	Glass reinforced	300
Polyacetals	Homopolymer:	
	Standard	500
	20% glass reinforced	500
	Copolymer:	
	Standard	500
	25% glass reinforced	580
	High flow	500
Polyester; Thermoplastic	Injection Moldings:	
	General purpose grade	590
	Glass reinforced grades	560—750
	Glass reinforced self extinguishing	750

Source: data compiled by J.S. Park from Charles T. Lynch, *CRC Handbook of Materials Science, Vol. 3*, CRC Press, Boca Raton, Florida, 1975 and *Engineered Materials Handbook, Vol.2, Engineering Plastics*, ASM International, Metals Park, Ohio, 1988.

Table 296. DIELECTRIC STRENGTH OF POLYMERS

(SHEET 8 OF 12)

Class	Polymer	Dielectric Strength (Short Time, ASTM D149) (V / mil)
Polyester; Thermoplastic (Con't)	General purpose grade	420—540
	Glass reinforced grade	—
	Asbestos—filled grade	580
Polyesters: Thermosets	Cast polyyester	
	Rigid	300—400
	Flexible	300—400
	Reinforced polyester moldings	
	High strength (glass fibers)	200—400
	Heat and chemical resistant (asbestos)	350
	Sheet molding compounds, general purpose	400—440
Phenylene Oxides	SE—100	400 (1/8 in.)
	SE—1	500 (1/8 in.)
	Glass fiber reinforced	1,020 (1/32 in.)

Source: data compiled by J.S. Park from Charles T. Lynch, *CRC Handbook of Materials Science, Vol. 3*, CRC Press, Boca Raton, Florida, 1975 and *Engineered Materials Handbook, Vol.2, Engineering Plastics*, ASM International, Metals Park, Ohio, 1988.

Table 296. DIELECTRIC STRENGTH OF POLYMERS

(SHEET 9 OF 12)

Class	Polymer	Dielectric Strength (Short Time, ASTM D149) (V / mil)
Phenylene oxides (Noryl)	Standard	425
	Glass fiber reinforced	480
Polyarylsulfone	Polyarylsulfone	350—383
Polypropylene	General purpose	650 (125 mil)
	High impact	450—650
	Asbestos filled	450
	Glass reinforced	317—475
	Flame retardant	485—700
Polyphenylene sulfide	Standard	450—595
	40% glass reinforced	490

Source: data compiled by J.S. Park from Charles T. Lynch, *CRC Handbook of Materials Science, Vol. 3*, CRC Press, Boca Raton, Florida, 1975 and *Engineered Materials Handbook, Vol.2, Engineering Plastics*, ASM International, Metals Park, Ohio, 1988.

Table 296. DIELECTRIC STRENGTH OF POLYMERS

(SHEET 10 OF 12)

Class	Polymer	Dielectric Strength (Short Time, ASTM D149) (V / mil)
Polyethylenes; Molded, Extruded	Type I—lower density (0.910—0.925)	
	Melt index 0.3—3.6	480
	Melt index 6—26	480
	Melt index 200	480
	Type II—medium density (0.926—0.940)	
	Melt index 20	480
	Melt index l.0—1.9	480
	Type III—higher density (0.941—0.965)	
	Melt index 0.2—0.9	480
	Melt Melt index 0.l—12.0	480
	Melt index 1.5—15	480
	High molecular weight	480

Source: data compiled by J.S. Park from Charles T. Lynch, *CRC Handbook of Materials Science, Vol. 3*, CRC Press, Boca Raton, Florida, 1975 and *Engineered Materials Handbook, Vol.2, Engineering Plastics*, ASM International, Metals Park, Ohio, 1988.

Table 296. Dielectric Strength of Polymers

(Sheet 11 of 12)

Class	Polymer	Dielectric Strength (Short Time, ASTM D149) (V / mil)
Olefin Copolymers; Molded	EEA (ethylene ethyl acrylate)	550
	EVA (ethylene vinyl acetate)	525
	Ionomer	1000
	Polyallomer	500—650
Polystyrenes; Molded	Polystyrenes	
	General purpose	>500
	Medium impact	>425
	High impact	300—650
	Glass fiber -30% reinforced	396
	Styrene acrylonitrile (SAN)	400—500
	Glass fiber (30%) reinforced SAN	515
Polyvinyl Chloride And Copolymers; Molded, Extruded	Nonrigid—electrical	24—500
	Rigid—normal impact	725—1,400

Source: data compiled by J.S. Park from Charles T. Lynch, *CRC Handbook of Materials Science, Vol. 3*, CRC Press, Boca Raton, Florida, 1975 and *Engineered Materials Handbook, Vol.2, Engineering Plastics*, ASM International, Metals Park, Ohio, 1988.

Table 296. DIELECTRIC STRENGTH OF POLYMERS

(SHEET 12 OF 12)

Class	Polymer	Dielectric Strength (Short Time, ASTM D149) (V / mil)
Silicones; Molded, Laminated	Fibrous (glass) reinforced silicones	280 (in oil)
	Granular (silica) reinforced silicones	380 (in oil)
	Woven glass fabric/ silicone laminate	725
Ureas; Molded	Ureas; Molded	
	Alpha—cellulose filled (ASTM Type l)	300—400
	Cellulose filled (ASTM Type 2)	340—370
	Woodflour filled	300—400

Source: data compiled by J.S. Park from Charles T. Lynch, *CRC Handbook of Materials Science, Vol. 3*, CRC Press, Boca Raton, Florida, 1975 and *Engineered Materials Handbook, Vol.2, Engineering Plastics*, ASM International, Metals Park, Ohio, 1988.

Table 297. Step Dielectric Strength of Polymers

(Sheet 1 of 3)

Class	Polymer	Dielectric Strength, Step by Step ASTM D149 (V/mil)	
		(dry)	(wet)
Thermoset Carbonate	Thermoset Carbonate Allyl diglycol carbonate	290	
Alkyds; Molded	Putty (encapsulating)	300—350	
	Rope (general purpose)	290	
	Granular (high speed molding)	300—350	
	Glass reinforced (heavy duty parts)	300—350	
Diallyl Phthalates; Molded	Orlon filled	350	325
	Dacron filled	350—410	350—361
	Asbestos filled	300—400	250—350
	Glass fiber filled	300—420	275—420

Source: data compiled by J.S. Park from Charles T. Lynch, *CRC Handbook of Materials Science, Vol. 3*, CRC Press, Boca Raton, Florida, 1975 and *Engineered Materials Handbook, Vol.2, Engineering Plastics*, ASM International, Metals Park, Ohio, 1988.

Table 297. Step Dielectric Strength of Polymers

(Sheet 2 of 3)

Class	Polymer	Dielectric Strength, Step by Step ASTM D149 (V/mil) (dry)	(wet)
Epoxies; Cast, Molded, Reinforced	Standard epoxies (diglycidyl ethers of bisphenol A)		
	Cast rigid	>400	
	Cast flexible	400—410	
	Molded	360—400	
	General purpose glass cloth laminate	450—550	
	High strength laminate	650-750	
	High performance resins (cycloaliphatic diepoxides)		
	Molded	280—400	

Source: data compiled by J.S. Park from Charles T. Lynch, *CRC Handbook of Materials Science, Vol. 3*, CRC Press, Boca Raton, Florida, 1975 and *Engineered Materials Handbook, Vol.2, Engineering Plastics*, ASM International, Metals Park, Ohio, 1988.

Table 297. Step Dielectric Strength of Polymers

(Sheet 3 of 3)

Class	Polymer	Dielectric Strength, Step by Step ASTM D149 (V/mil)	
		(dry)	(wet)
Polyesters: Thermosets	Cast polyyester		
	Rigid	300—400	
	Flexible	300—400	
	Reinforced polyester moldings		
	High strength (glass fibers)	200—400	
	Heat and chemical resistsnt (asbestos)	350	
	Sheet molding compounds, general purpose	400—440	

Source: data compiled by J.S. Park from Charles T. Lynch, *CRC Handbook of Materials Science, Vol. 3*, CRC Press, Boca Raton, Florida, 1975 and *Engineered Materials Handbook, Vol.2, Engineering Plastics*, ASM International, Metals Park, Ohio, 1988.

Table 298. DIELECTRIC CONSTANT OF POLYMERS
(SHEET 1 OF 14)

Polymer	Type	Dielectric Constant (ASTM D150)	
		60 Hz	10^6 Hz
ABS Resins; Molded, Extruded	Medium impact	2.8—3.2	2.75—3.0
	High impact	2.8—3.2	2.7—3.0
	Very high impact	2.8—3.5	2.4—3.0
	Low temperature impact	2.5—3.5	2.4—3.0
	Heat resistant	2.7—3.5	2.8—3.2
Acrylics; Cast, Molded, Extruded	Cast Resin Sheets, Rods:		
	General purpose, type I	3.5—4.5	2.7—3.2
	General purpose, type II	3.5—4.5	2.7—3.2
	Moldings:		
	Grades 5, 6, 8	3.5—3.9	2.7—2.9
	High impact grade	3.5—3.9	2.5—3.0
Thermoset Carbonate	Allyl diglycol carbonate	4.4	3.5—3.8

Source: data compiled by J.S. Park from Charles T. Lynch, *CRC Handbook of Materials Science, Vol. 3*, CRC Press, Boca Raton, Florida, 1975 and *Engineered Materials Handbook, Vol.2, Engineering Plastics*, ASM International, Metals Park, Ohio, 1988.

Table 298. DIELECTRIC CONSTANT OF POLYMERS
(SHEET 2 OF 14)

Polymer	Type	Dielectric Constant (ASTM D150)	
		60 Hz	10^6 Hz
Alkyds; Molded	Putty (encapsulating)	5.4—5.9	4.5—4.7
	Rope (general purpose)	7.4	6.8
	Granular (high speed molding)	5.7—6.3	4.8—5.1
	Glass reinforced (heavy duty parts)	5.2—6.0	4.5—5.0
Cellulose Acetate; Molded, Extruded	ASTM Grade:		
	H6—1	3.5—7.5	3.2—7.0
	H4—1	3.5—7.5	3.2—7.0
	H2—1	3.5—7.5	3.2—7.0
	MH—1, MH—2	3.5—7.5	3.2—7.0
	MS—1, MS—2	3.5—7.5	3.2—7.0
	S2—1	3.5—7.5	3.2—7.0

Source: data compiled by J.S. Park from Charles T. Lynch, *CRC Handbook of Materials Science, Vol. 3*, CRC Press, Boca Raton, Florida, 1975 and *Engineered Materials Handbook, Vol.2, Engineering Plastics*, ASM International, Metals Park, Ohio, 1988.

Table 298. DIELECTRIC CONSTANT OF POLYMERS

(SHEET 3 OF 14)

Polymer	Type	Dielectric Constant (ASTM D150)	
		60 Hz	10^6 Hz
Cellulose Acetate Butyrate; Molded, Extruded	ASTM Grade:		
	H4	3.5—6.4	3.2—6.2
	MH	3.5—6.4	3.2—6.2
	S2	3.5—64	3.2—6.2
Cellusose Acetate Propionate; Molded, Extruded	ASTM Grade:		
	1	3.7—4.0	3.4—3.7
	3	3.7—4.0	3.4—3.7
	6	3.7—4.0	3.7—3.4
Chlorinated Polymers	Chlorinated polyether	3.1	2.92
	Chlorinated polyvinyl chloride	3.08	3.2—3.6
Polycarbonates	Polycarbonate	3.17	2.96
	Polycarbonate (40% glass fiber reinforced)	3.8	3.58

Source: data compiled by J.S. Park from Charles T. Lynch, *CRC Handbook of Materials Science, Vol. 3*, CRC Press, Boca Raton, Florida, 1975 and *Engineered Materials Handbook, Vol.2, Engineering Plastics*, ASM International, Metals Park, Ohio, 1988.

Table 298. DIELECTRIC CONSTANT OF POLYMERS

(SHEET 4 OF 14)

Polymer	Type	Dielectric Constant (ASTM D150)	
		60 Hz	10^6 Hz
Diallyl Phthalates; Molded	Orlon filled	3.9(Dry), 3.3(Wet)	4.1(D), 3.4(W)
	Dacron filled	3.7–3.8(D), 3.5–3.6(W)	3.9(D), 3.7(W)
	Asbestos filled	5.2(D), 4.5(W)	6.5 (D), 4.8(W)
	Glass fiber filled	4.1–4.5(D), 3.5–4.5(W)	4.6 (D), 4.4(W)
Fluorocarbons; Molded,Extruded	Polytrifluoro chloroethylene (PTFCE)	2.6—2.7	
	Polytetrafluoroethylene (PTFE) (0.01 in thickness)	2.1	
	Ceramic reinforced (PTFE)	2.9—3.6	
	Fluorinated ethylene propylene(FEP) (0.01 in thickness)	2.1	
	Polyvinylidene— fluoride (PVDF) (0.125 in thickness)	10	

Source: data compiled by J.S. Park from Charles T. Lynch, *CRC Handbook of Materials Science, Vol. 3*, CRC Press, Boca Raton, Florida, 1975 and *Engineered Materials Handbook, Vol.2, Engineering Plastics*, ASM International, Metals Park, Ohio, 1988.

Table 298. DIELECTRIC CONSTANT OF POLYMERS

(SHEET 5 OF 14)

Polymer	Type	Dielectric Constant (ASTM D150) 60 Hz	Dielectric Constant (ASTM D150) 10^6 Hz
Epoxies; Cast, Molded, Reinforced	Standard epoxies (diglycidyl ethers of bisphenol A)		
	Cast rigid	4.02	3.42
	Cast flexible	4.43-4.79	2.78-3.52
	Molded	4.4-5.4	4.1-4.6
	General purpose glass cloth laminate	5.3-5.4	4.7-4.8
	High strength laminate	—	4.8-5.2
Epoxies; Molded, Extruded	High performance resins (cycloaliphatic diepoxides)		
	Cast, rigid	3.96—4.02	3.53—3.58
	Molded	4.7—5.7	4.3—4.8
	Glass cloth laminate	—	5.1
Epoxy novolacs	Cast, rigid	3.34—3.39	—
	Glass cloth laminate	4.41—4.43	—

Source: data compiled by J.S. Park from Charles T. Lynch, *CRC Handbook of Materials Science, Vol. 3*, CRC Press, Boca Raton, Florida, 1975 and *Engineered Materials Handbook, Vol.2, Engineering Plastics*, ASM International, Metals Park, Ohio, 1988.

Table 298. DIELECTRIC CONSTANT OF POLYMERS
(SHEET 6 OF 14)

Polymer	Type	Dielectric Constant (ASTM D150)	
		60 Hz	10^6 Hz
Melamines; Molded	Filler & type		
	Unfilled	7.9—11.0	6.3—7.3
	Cellulose electrical	6.2—7.7	5.2—6.0
	Glass fiber	7.0—11.1	6.0—7.9
Alpha cellulose	Alpha cellulose	—	6.4—8.1
	Mineral	—	5.6
Nylons; Molded, Extruded	Type 6		
	General purpose	4.0—5.3	3.6—3.8
	Glass fiber (30%) reinforced	4.6—5.6	3.9—5.4
	Cast	4	3.3
	Flexible copolymers	3.2—4.0	3.0—3.6

Source: data compiled by J.S. Park from Charles T. Lynch, *CRC Handbook of Materials Science, Vol. 3*, CRC Press, Boca Raton, Florida, 1975 and *Engineered Materials Handbook, Vol.2, Engineering Plastics*, ASM International, Metals Park, Ohio, 1988.

Table 298. DIELECTRIC CONSTANT OF POLYMERS

(SHEET 7 OF 14)

Polymer	Type	Dielectric Constant (ASTM D150) 60 Hz	10^6 Hz
Nylons; Molded, Extruded (Con't)	Type 8	9.3	4
	Type 11	3.3 (10^3 Hz)	—
	Type 12	3.6 (10^3 Hz)	—
	6/6 Nylon		
	General purpose molding	4	3.6
	Glass fiber reinforced	40—44	3.5—4.1
	6/10 Nylon		
	General purpose	3.9	3.5
Phenolics; Molded	Type and filler		
	General: woodflour and flock	5.0—9.0	4.0—7.0
	Shock: paper, flock, or pulp	5.6—11.0	4.5—7.0
	High shock: chopped fabric or cord	6.5—15.0	4.5—7.0
	Very high shock: glass fiber	7.1—7.2	4.6—6.6

Source: data compiled by J.S. Park from Charles T. Lynch, *CRC Handbook of Materials Science, Vol. 3*, CRC Press, Boca Raton, Florida, 1975 and *Engineered Materials Handbook, Vol.2, Engineering Plastics*, ASM International, Metals Park, Ohio, 1988.

Table 298. DIELECTRIC CONSTANT OF POLYMERS

(SHEET 8 OF 14)

Polymer	Type	Dielectric Constant (ASTM D150)	
		60 Hz	10^6 Hz
Phenolics: Molded	Arc resistant—mineral	7.4	5
	Rubber phenolic—woodflour or flock	9—16	5
	Rubber phenolic—chopped fabric	15	5
	Rubber phenolic—asbestos	15	5
ABS–Polycarbonate Alloy	ABS–Polycarbonate Alloy	2.74	2.69
PVC–Acrylic Alloy	PVC–Acrylic Alloy		
	PVC–acrylic sheet	3.86	3.44
	PVC–acrylic injection molded	4	3.4
Polyimides	Unreinforced	4.12	3.96
	Glass reinforced	4.84	4.74

Source: data compiled by J.S. Park from Charles T. Lynch, *CRC Handbook of Materials Science, Vol. 3*, CRC Press, Boca Raton, Florida, 1975 and *Engineered Materials Handbook, Vol.2, Engineering Plastics*, ASM International, Metals Park, Ohio, 1988.

Table 298. Dielectric Constant of Polymers
(Sheet 9 of 14)

Polymer	Type	Dielectric Constant (ASTM D150)	
		60 Hz	10^6 Hz
Polyacetals	Homopolymer:		
	Standard	3.7	3.7
	20% glass reinforced	4	4.0
	Copolymer:		
	Standard	3.7 (100 Hz)	3.7
	25% glass reinforced	3.9 (100 Hz)	3.9
	High flow	3.7 (100 Hz)	3.7
Polyester; Thermoplastic	Injection Moldings:		
	General purpose grade	3.1—3.3	—
	Glass reinforced grades	3.7—4.2	—
	Glass reinforced self extinguishing	3.7—3.8	—
	General purpose grade	3.16	—
	Asbestos—filled grade	3.5—4.2	—

Source: data compiled by J.S. Park from Charles T. Lynch, *CRC Handbook of Materials Science, Vol. 3*, CRC Press, Boca Raton, Florida, 1975 and *Engineered Materials Handbook, Vol.2, Engineering Plastics*, ASM International, Metals Park, Ohio, 1988.

Table 298. DIELECTRIC CONSTANT OF POLYMERS

(SHEET 10 OF 14)

Polymer	Type	Dielectric Constant (ASTM D150)	
		60 Hz	10^6 Hz
Polyesters: Thermosets	Cast polyyester		
	Rigid	2.8—4.4	2.8—4.4
	Flexible	3.18—7.0	3.7—6.1
	Reinforced polyester moldings		
	Sheet molding compounds, general purpose	4.62—5.0	4.55—4.75
Phenylene Oxides	SE—100	2.65	2.64
	SE—1	2.69	2.68
	Glass fiber reinforced	2.93	2.92
Phenylene oxides (Noryl)	Standard	3.06—3.15	3.03—3.10
	Glass fiber reinforced	3.55	3.41
Polyarylsulfone	Polyarylsulfone	3.51—3.94	3.54—3.7

Source: data compiled by J.S. Park from Charles T. Lynch, *CRC Handbook of Materials Science, Vol. 3*, CRC Press, Boca Raton, Florida, 1975 and *Engineered Materials Handbook, Vol.2, Engineering Plastics*, ASM International, Metals Park, Ohio, 1988.

Table 298. DIELECTRIC CONSTANT OF POLYMERS

(SHEET 11 OF 14)

Polymer	Type	Dielectric Constant (ASTM D150)	
		60 Hz	10^6 Hz
Polypropylene	General purpose	2.20—2.28	2.23—2.24
	High impact	2.20—2.28	2.23—2.27
	Asbestos filled	2.75	2.6—3.17
	Glass reinforced	2.3—2.5	2—2.25
	Flame retardant	2.46—2.79	2.45—2.70
Polyphenylene sulfide	Standard	—	3.22—3.8
	40% glass reinforced	—	3.88
Polyethylenes; Molded, Extruded	Type I—lower density (0.910—0.925)		
	Melt index 0.3—3.6	2.3	—
	Melt index 6—26	2.3	—
	Melt index 200	2.3	—

Source: data compiled by J.S. Park from Charles T. Lynch, *CRC Handbook of Materials Science, Vol. 3*, CRC Press, Boca Raton, Florida, 1975 and *Engineered Materials Handbook, Vol.2, Engineering Plastics*, ASM International, Metals Park, Ohio, 1988.

Table 298. DIELECTRIC CONSTANT OF POLYMERS

(SHEET 12 OF 14)

Polymer	Type	Dielectric Constant (ASTM D150) 60 Hz	Dielectric Constant (ASTM D150) 10^6 Hz
Polyethylenes; Molded, Extruded (Con't)	Type II—medium density (0.926—0.940)		
	Melt index 20	2.3	—
	Melt index 1.0—1.9	2.3	—
	Type III—higher density (0.941—0.965)		
	Melt index 0.2—0.9	2.3	—
	Melt Melt index 0.1—12.0	2.3	—
	Melt index 1.5—15	2.3	—
	High molecular weight	2.3	—
Olefin Copolymers; Molded	EEA (ethylene ethyl acrylate)	2.8	
	EVA (ethylene vinyl acetate)	3.16	
	Ionomer	2.4	
	Polyallomer	2.3	

Source: data compiled by J.S. Park from Charles T. Lynch, *CRC Handbook of Materials Science, Vol. 3*, CRC Press, Boca Raton, Florida, 1975 and *Engineered Materials Handbook, Vol.2, Engineering Plastics*, ASM International, Metals Park, Ohio, 1988.

Table 298. DIELECTRIC CONSTANT OF POLYMERS

(SHEET 13 OF 14)

Polymer	Type	Dielectric Constant (ASTM D150)	
		60 Hz	10^6 Hz
Polystyrenes; Molded	Polystyrenes		
	General purpose	2.45—2.65	2.45—2.65
	Medium impact	2.45—4.75	2.4—3.8
	High impact	2.45—4.75	2.5—4.0
	Glass fiber -30% reinforced	3.1	3
	Styrene acrylonitrile (SAN)	2.6—3.4	2.6—3.02
	Glass fiber (30%) reinforced SAN	3.5	3.4—3.6
Polyvinyl Chloride And Copolymers;	Molded, Extruded		
	Nonrigid—general	5.5—9.1	
	Nonrigid—electrical	6.0—8.0	
	Rigid—normal impact	2.3—3.7	
	Vinylidene chloride	3—5	

Source: data compiled by J.S. Park from Charles T. Lynch, *CRC Handbook of Materials Science, Vol. 3*, CRC Press, Boca Raton, Florida, 1975 and *Engineered Materials Handbook, Vol.2, Engineering Plastics*, ASM International, Metals Park, Ohio, 1988.

Table 298. DIELECTRIC CONSTANT OF POLYMERS

(SHEET 14 OF 14)

Polymer	Type	Dielectric Constant (ASTM D150)	
		60 Hz	10^6 Hz
Silicones; Molded, Laminated	Fibrous (glass) reinforced silicones	4.34	4.28
	Granular (silica) reinforced silicones	4.1—4.5	3.4 —4.3
	Woven glass fabric/ silicone laminate	3.9—4.2	3.8—397
Ureas; Molded	Alpha—cellulose filled (ASTM Type l)	7.0—9.5	6.4—6.9
	Cellulose filled (ASTM Type 2)	7.2—7.3	6.4—6.5
	Woodflour filled	7.0—9.5	6.4—6.9

Source: data compiled by J.S. Park from Charles T. Lynch, *CRC Handbook of Materials Science, Vol. 3*, CRC Press, Boca Raton, Florida, 1975 and *Engineered Materials Handbook, Vol.2, Engineering Plastics*, ASM International, Metals Park, Ohio, 1988.

Table 299. DIELECTRIC BREAKDOWN OF POLYMERS

Polymer	Type	Dielectric Breakdown, Short Time (kV)	
		(dry)	(wet)
Diallyl Phthalates; Molded	Orlon filled	65—75	60—65
	Dacron filled	65	60
	Asbestos filled	55—80	55
	Glass fiber filled	63—70	45—65

Source: data compiled by J.S. Park from Charles T. Lynch, *CRC Handbook of Materials Science, Vol. 3,* CRC Press, Boca Raton, Florida, 1975 and *Engineered Materials Handbook, Vol.2, Engineering Plastics,* ASM International, Metals Park, Ohio, 1988.

Table 300. Dielectric Breakdown of Polymers

Polymer	Type	Dielectric Breakdown, Step by Step (kV)	
		(dry)	(wet)
Diallyl Phthalates; Molded	Orlon filled	55—60	46—60
	Dacron filled	60	55
	Asbestos filled	38—70	39—60
	Glass fiber filled	55—65	45—65

Source: data compiled by J.S. Park from Charles T. Lynch, *CRC Handbook of Materials Science, Vol. 3*, CRC Press, Boca Raton, Florida, 1975 and *Engineered Materials Handbook, Vol.2, Engineering Plastics*, ASM International, Metals Park, Ohio, 1988.

Table 301. TANGENT LOSS IN GLASS
(SHEET 1 OF 5)

Glass	Composition	Frequency (Hz)	Tangent Loss (tan δ)	Temperature
SiO_2 glass	Pure	100 Hz	0.00002	25°C
		100 Hz	0.00052	200°C
		100 Hz	0.080	300°C
		100 Hz	1.0	400°C
		1 kHz	0.00002	25°C
		1 kHz	0.00012	200°C
		1 kHz	0.0072	300°C
		1 kHz	0.2	400°C
		10 kHz	0.00002	25°C
		10 kHz	0.00004	200°C
		10 kHz	0.00072	300°C
		10 kHz	0.022	400°C
		9.4 GHz	1.5×10^{-4}	20°C
		9.4 GHz	1.8×10^{-4}	200°C
		9.4 GHz	2.0×10^{-4}	400°C
		9.4 GHz	2.9×10^{-4}	600°C
		9.4 GHz	4.8×10^{-4}	800°C
		9.4 GHz	11×10^{-4}	1000°C
		9.4 GHz	25×10^{-4}	1200°C
		9.4 GHz	46×10^{-4}	1400°C
SiO_2-Na_2O glass	16% mol Na_2O	4.5×10^8 Hz	0.0058	20°C
	19.5% mol Na_2O	1kHz	0.144	room temp.
	19.5% mol Na_2O	3 kHz	0.0984	room temp.
	19.5% mol Na_2O	5 kHz	0.0832	room temp.
	19.5% mol Na_2O	10 kHz	0.0656	room temp.

Source: data compiled by J.S. Park from O. V. Mazurin, M. V. Streltsina and T. P. Shvaiko-Shvaikovskaya, *Handbook of Glass Data, Part A and Part B*, Elsevier, New York, 1983

Table 301. TANGENT LOSS IN GLASS
(SHEET 2 OF 5)

Glass	Composition	Frequency (Hz)	Tangent Loss (tan δ)	Temperature
SiO_2-Na_2O glass (Con't)	19.5% mol Na_2O	30 kHz	0.0492	room temp.
	19.5% mol Na_2O	50 kHz	0.0428	room temp.
	19.5% mol Na_2O	100 kHz	0.0364	room temp.
	19.5% mol Na_2O	300 kHz	0.0295	room temp.
	20% mol Na_2O	$4.5x10^8$ Hz	0.0073	20°C
	22.2% mol Na_2O	$4.5x10^8$ Hz	0.0081	20°C
	24.4% mol Na_2O	1kHz	0.2207	room temp.
	24.4% mol Na_2O	3 kHz	0.1455	room temp.
	24.4% mol Na_2O	5 kHz	0.1194	room temp.
	24.4% mol Na_2O	10 kHz	0.0916	room temp.
	24.4% mol Na_2O	30 kHz	0.0652	room temp.
	24.4% mol Na_2O	50 kHz	0.0563	room temp.
	24.4% mol Na_2O	100 kHz	0.0456	room temp.
	24.4% mol Na_2O	300 kHz	0.0369	room temp.
	28.6% mol Na_2O	$4.5x10^8$ Hz	0.0102	20°C
	29.4% mol Na_2O	1kHz	0.4923	room temp.
	29.4% mol Na_2O	3 kHz	0.3027	room temp.
	29.4% mol Na_2O	5 kHz	0.2426	room temp.
	29.4% mol Na_2O	10 kHz	0.1764	room temp.
	29.4% mol Na_2O	30 kHz	0.1172	room temp.
	29.4% mol Na_2O	50 kHz	0.0972	room temp.
	29.4% mol Na_2O	100 kHz	0.0758	room temp.
	29.4% mol Na_2O	300 kHz	0.0568	room temp.

Source: data compiled by J.S. Park from O. V. Mazurin, M. V. Streltsina and T. P. Shvaiko-Shvaikovskaya, *Handbook of Glass Data, Part A and Part B*, Elsevier, New York, 1983

Table 301. TANGENT LOSS IN GLASS
(SHEET 3 OF 5)

Glass	Composition	Frequency (Hz)	Tangent Loss (tan δ)	Temperature
SiO_2-Na_2O glass (Con't)	34.3% mol Na_2O	1kHz	0.10324	room temp.
	34.3% mol Na_2O	3 kHz	0.6520	room temp.
	34.3% mol Na_2O	5 kHz	0.5280	room temp.
	34.3% mol Na_2O	10 kHz	0.3752	room temp.
	34.3% mol Na_2O	30 kHz	0.2314	room temp.
	34.3% mol Na_2O	50 kHz	0.1864	room temp.
	34.3% mol Na_2O	100 kHz	0.1388	room temp.
	34.3% mol Na_2O	300 kHz	0.0936	room temp.
	36% mol Na_2O	$4.5x10^8$ Hz	0.0162	20°C
	39.3% mol Na_2O	10 kHz	0.6338	room temp.
	39.3% mol Na_2O	30 kHz	0.3835	room temp.
	39.3% mol Na_2O	50 kHz	0.3032	room temp.
	39.3% mol Na_2O	100 kHz	0.2144	room temp.
	39.3% mol Na_2O	300 kHz	0.1402	room temp.
SiO_2-PbO glass	40% mol PbO	32 GHz	0.015	-150°C
	40% mol PbO	32 GHz	0.018	-100°C
	40% mol PbO	32 GHz	0.020	-50°C
	40% mol PbO	32 GHz	0.022	0°C
	40% mol PbO	32 GHz	0.024	50°C
	40% mol PbO	100 GHz	0.005	room temp.
	40% mol PbO	1000 GHz	0.050	room temp.
SiO_2-B_2O_3 glass	46.3% mol B_2O_3	10 GHz	0.0014	

Source: data compiled by J.S. Park from O. V. Mazurin, M. V. Streltsina and T. P. Shvaiko-Shvaikovskaya, *Handbook of Glass Data, Part A and Part B*, Elsevier, New York, 1983

Table 301. TANGENT LOSS IN GLASS

(SHEET 4 OF 5)

Glass	Composition	Frequency (Hz)	Tangent Loss (tan δ)	Temperature
SiO_2-Al_2O_3 glass	0.5% mol Al_2O_3	50 K	0.0025	50 K
	0.5% mol Al_2O_3	100 K	0.0021	100 K
	0.5% mol Al_2O_3	150 K	0.0026	150 K
B_2O_3 glass	B_2O_3 glass	1 MHz	0.0004	100°C
		1 MHz	0.0005	200°C
		1 MHz	0.0009	300°C
		32 kHz	0.00005	50K
		32 kHz	0.00011	100K
		32 kHz	0.0007	150K
		32 kHz	0.0010	200K
		32 kHz	0.0008	250K
		32 kHz	0.0003	300K
B_2O_3-Na_2O glass	8% mol Na_2O	1MHz	0.0025	room temp.
	10% mol Na_2O	1MHz	0.0022	room temp.
	10% mol Na_2O	1 kHz	0.0003	134.5°C
	10% mol Na_2O	1 kHz	0.0009	214°C
	10% mol Na_2O	1 kHz	0.0038	277°C
	10% mol Na_2O	1 kHz	0.0066	298°C
	12.5% mol Na_2O	1 kHz	0.0005	134.5°C
	12.5% mol Na_2O	1 kHz	0.0022	214°C
	12.5% mol Na_2O	1 kHz	0.0100	277°C
	12.5% mol Na_2O	1 kHz	0.0170	298°C

Source: data compiled by J.S. Park from O. V. Mazurin, M. V. Streltsina and T. P. Shvaiko-Shvaikovskaya, *Handbook of Glass Data, Part A and Part B*, Elsevier, New York, 1983

Table 301. TANGENT LOSS IN GLASS
(SHEET 5 OF 5)

Glass	Composition	Frequency (Hz)	Tangent Loss (tan δ)	Temperature
B_2O_3-Na_2O glass (Con't)	15% mol Na_2O	1 kHz	0.0015	134.5°C
	15% mol Na_2O	1 kHz	0.0064	214°C
	15% mol Na_2O	1 kHz	0.0296	277°C
	15% mol Na_2O	1 kHz	0.0477	298°C
	16% mol Na_2O	1MHz	0.0031	room temp.
	20% mol Na_2O	1 kHz	0.0009	16°C
	20% mol Na_2O	1 kHz	0.0026	90.5°C
	20% mol Na_2O	1 kHz	0.0149	157°C
	20% mol Na_2O	1 kHz	0.0890	219°C
	20% mol Na_2O	1 kHz	0.2480	274°C
	25% mol Na_2O	1 kHz	0.0022	16°C
	25% mol Na_2O	1MHz	0.0063	room temp.
	25% mol Na_2O	1 kHz	0.0150	90.5°C
	25% mol Na_2O	1 kHz	0.1080	157°C
	28% mol Na_2O	1MHz	0.0081	room temp.
B_2O_3-CaO glass	33.3% mol CaO	2 MHz	0.001	25°C
	33.3% mol CaO	2 MHz	0.002	100°C
	33.3% mol CaO	2 MHz	0.0025	200°C
	33.3% mol CaO	2 MHz	0.0035	300°C
	33.3% mol CaO	2 MHz	0.0045	400°C
	33.3% mol CaO	2 MHz	0.0055	500°C
	33.3% mol CaO	2 MHz	0.007	550°C

Source: data compiled by J.S. Park from O. V. Mazurin, M. V. Streltsina and T. P. Shvaiko-Shvaikovskaya, *Handbook of Glass Data, Part A and Part B*, Elsevier, New York, 1983

Table 302. ELECTRICAL PERMITTIVITY OF GLASS

(SHEET 1 OF 6)

Glass	Composition	Frequency (Hz)	Electrical Permittivity	Temperature (°C)
SiO_2 glass	Pure	100 Hz	4.0	25
		100 Hz	4.0	200
		100 Hz	4.0	300
		100 Hz	5.5	400
		1 kHz	4.0	25
		1 kHz	4.0	200
		1 kHz	4.0	300
		1 kHz	4.1	400
		10 kHz	4.0	25
		10 kHz	4.0	200
		10 kHz	4.0	300
		10 kHz	4.0	400
		9.4 GHz	3.81	20
		9.4 GHz	3.83	200
		9.4 GHz	3.84	400
		9.4 GHz	3.86	600
		9.4 GHz	3.88	800
		9.4 GHz	3.91	1000
		9.4 GHz	3.93	1200
		9.4 GHz	3.96	1400
		10 GHz	3.82	20
		10 GHz	3.82	220
		10 GHz	3.91	888
		10 GHz	3.98	1170
		10 GHz	4.05	1335
		10 GHz	4.07	1420
		10 GHz	4.09	1480
		10 GHz	4.11	1526

Source: data compiled by J.S. Park from O. V. Mazurin, M. V. Streltsina and T. P. Shvaiko-Shvaikovskaya, *Handbook of Glass Data, Part A and Part B*, Elsevier, New York, 1983

Table 302. ELECTRICAL PERMITTIVITY OF GLASS
(SHEET 2 OF 6)

Glass	Composition	Frequency (Hz)	Electrical Permittivity	Temperature (°C)
SiO_2 glass (Con't)	Pure (Con't)	10 GHz	4.12	1584
		10 GHz	4.15	1602
		10 GHz	4.12	1647
		10 GHz	4.04	1764
		10 GHz	4.05	1764
SiO_2–Na_2O glass	(16% mol Na_2O)	4.5x10^8 Hz	6.01	20
	(19.5% mol Na_2O)	1kHz	9.40	room temp.
	(19.5% mol Na_2O)	3 kHz	8.97	room temp.
	(19.5% mol Na_2O)	5 kHz	8.56	room temp.
	(19.5% mol Na_2O)	10 kHz	8.26	room temp.
	(19.5% mol Na_2O)	30 kHz	8.00	room temp.
	(19.5% mol Na_2O)	50 kHz	7.88	room temp.
	(19.5% mol Na_2O)	100 kHz	7.74	room temp.
	(19.5% mol Na_2O)	300 kHz	7.62	room temp.
	(20% mol Na_2O)	4.5x10^8 Hz	6.48	20
	(22.2% mol Na_2O)	4.5x10^8 Hz	6.85	20
	(24.4% mol Na_2O)	1kHz	11.62	room temp.
	(24.4% mol Na_2O)	3 kHz	10.61	room temp.
	(24.4% mol Na_2O)	5 kHz	10.21	room temp.
	(24.4% mol Na_2O)	10 kHz	9.74	room temp.
	(24.4% mol Na_2O)	30 kHz	9.30	room temp.
	(24.4% mol Na_2O)	50 kHz	9.14	room temp.
	(24.4% mol Na_2O)	100 kHz	8.91	room temp.
	(24.4% mol Na_2O)	300 kHz	8.75	room temp.
	(28.6% mol Na_2O)	4.5x10^8 Hz	7.62	20

Source: data compiled by J.S. Park from O. V. Mazurin, M. V. Streltsina and T. P. Shvaiko-Shvaikovskaya, *Handbook of Glass Data, Part A and Part B*, Elsevier, New York, 1983

Table 302. ELECTRICAL PERMITTIVITY OF GLASS
(SHEET 3 OF 6)

Glass	Composition	Frequency (Hz)	Electrical Permittivity	Temperature (°C)
SiO_2–Na_2O glass (Con't)	(29.4% mol Na_2O)	1kHz	17.52	room temp.
	(29.4% mol Na_2O)	3 kHz	14.23	room temp.
	(29.4% mol Na_2O)	5 kHz	13.19	room temp.
	(29.4% mol Na_2O)	10 kHz	12.08	room temp.
	(29.4% mol Na_2O)	30 kHz	11.21	room temp.
	(29.4% mol Na_2O)	50 kHz	10.86	room temp.
	(29.4% mol Na_2O)	100 kHz	10.47	room temp.
	(29.4% mol Na_2O)	300 kHz	10.15	room temp.
	(34.3% mol Na_2O)	1kHz	38.61	room temp.
	(34.3% mol Na_2O)	3 kHz	21.30	room temp.
	(34.3% mol Na_2O)	5 kHz	18.13	room temp.
	(34.3% mol Na_2O)	10 kHz	15.22	room temp.
	(34.3% mol Na_2O)	30 kHz	13.28	room temp.
	(34.3% mol Na_2O)	50 kHz	12.57	room temp.
	(34.3% mol Na_2O)	100 kHz	11.78	room temp.
	(34.3% mol Na_2O)	300 kHz	11.14	room temp.
	(36% mol Na_2O)	$4.5x10^8$ Hz	9.40	20
	(39.3% mol Na_2O)	10 kHz	22.08	room temp.
	(39.3% mol Na_2O)	30 kHz	16.56	room temp.
	(39.3% mol Na_2O)	50 kHz	15.06	room temp.
	(39.3% mol Na_2O)	100 kHz	13.55	room temp.
	(39.3% mol Na_2O)	300 kHz	12.43	room temp.

Source: data compiled by J.S. Park from O. V. Mazurin, M. V. Streltsina and T. P. Shvaiko-Shvaikovskaya, *Handbook of Glass Data, Part A and Part B*, Elsevier, New York, 1983

Table 302. ELECTRICAL PERMITTIVITY OF GLASS
(SHEET 4 OF 6)

Glass	Composition	Frequency (Hz)	Electrical Permittivity	Temperature (°C)
SiO_2–PbO glass	(40% mol PbO)	32 GHz	4.25	–150
	(40% mol PbO)	32 GHz	4.30	–100
	(40% mol PbO)	32 GHz	4.40	–50
	(40% mol PbO)	32 GHz	4.45	0
	(40% mol PbO)	32 GHz	5.00	50
SiO_2–Al_2O_3 glass	(46.3% mol B_2O_3)	10 GHz	3.55	
B_2O_3 glass	Pure	1 kHz	3.17	500
		1 kHz	3.21	550
		1 kHz	3.27	580
		3 kHz	3.15	500
		3 kHz	3.17	550
		3 kHz	3.18	580
		3 kHz	3.21	620
		3 kHz	3.25	650
		10 kHz	3.13	500
		10 kHz	3.14	550
		10 kHz	3.145	580
		10 kHz	3.15	620
		10 kHz	3.15	650
		10 kHz	3.16	700
		50 kHz	3.10	500
		50 kHz	3.12	550
		50 kHz	3.115	580
		50 kHz	3.05	620
		50 kHz	3.10	650
		50 kHz	3.09	700
		50 kHz	3.06	750
		50 kHz	3.04	800

Source: data compiled by J.S. Park from O. V. Mazurin, M. V. Streltsina and T. P. Shvaiko-Shvaikovskaya, *Handbook of Glass Data, Part A and Part B*, Elsevier, New York, 1983

Table 302. ELECTRICAL PERMITTIVITY OF GLASS

(SHEET 5 OF 6)

Glass	Composition	Frequency (Hz)	Electrical Permittivity	Temperature (°C)
B_2O_3–Na_2O glass	(4.08% mol Na_2O)	56.8 MHz	3.72	room temp.
	(7.35% mol Na_2O)	56.8 MHz	4.20	room temp.
	(14.15% mol Na_2O)	56.8 MHz	4.94	room temp.
	(17.31% mol Na_2O)	56.8 MHz	5.27	room temp.
	(24.77% mol Na_2O)	56.8 MHz	6.24	room temp.
	(31.98% mol Na_2O)	56.8 MHz	7.03	room temp.
	(10% mol Na_2O)	1 kHz	5.00	73
	(10% mol Na_2O)	1 kHz	5.05	134.5
	(10% mol Na_2O)	1 kHz	5.15	214
	(10% mol Na_2O)	1 kHz	5.45	277
	(10% mol Na_2O)	1 kHz	5.60	298
	(12.5% mol Na_2O)	1 kHz	5.45	73
	(12.5% mol Na_2O)	1 kHz	5.60	134.5
	(12.5% mol Na_2O)	1 kHz	5.75	214
	(12.5% mol Na_2O)	1 kHz	6.30	277
	(12.5% mol Na_2O)	1 kHz	6.65	298
	(15% mol Na_2O)	1 kHz	5.80	73
	(15% mol Na_2O)	1 kHz	6.00	134.5
	(15% mol Na_2O)	1 kHz	6.50	214
	(15% mol Na_2O)	1 kHz	7.80	277
	(15% mol Na_2O)	1 kHz	8.60	298
	(20% mol Na_2O)	1 kHz	6.15	16
	(20% mol Na_2O)	1 kHz	6.43	90.5
	(20% mol Na_2O)	1 kHz	7.45	157

Source: data compiled by J.S. Park from O. V. Mazurin, M. V. Streltsina and T. P. Shvaiko-Shvaikovskaya, *Handbook of Glass Data, Part A and Part B*, Elsevier, New York, 1983

Table 302. ELECTRICAL PERMITTIVITY OF GLASS

(SHEET 6 OF 6)

Glass	Composition	Frequency (Hz)	Electrical Permittivity	Temperature (°C)
B_2O_3–Na_2O glass (Con't)	(20% mol Na_2O)	1 kHz	11.85	219
	(20% mol Na_2O)	1 kHz	31.00	274
	(25% mol Na_2O)	1 kHz	7.50	16
	(25% mol Na_2O)	1 kHz	8.90	90.5
	(25% mol Na_2O)	1 kHz	17.30	157

Source: data compiled by J.S. Park from O. V. Mazurin, M. V. Streltsina and T. P. Shvaiko-Shvaikovskaya, *Handbook of Glass Data, Part A and Part B*, Elsevier, New York, 1983

Table 303. ARC RESISTANCE OF POLYMERS

(SHEET 1 OF 8)

Polymer	Type	Arc Resistance, (ASTM D495) (seconds)
Acrylics; Cast, Molded, Extruded	Cast Resin Sheets, Rods:	
	General purpose, type I	No track
	General purpose, type II	No track
	Moldings:	
	Grades 5, 6, 8	No track
	High impact grade	No track
Thermoset Carbonate	Allyl diglycol carbonate	185
Alkyds; Molded	Putty (encapsulating)	180
	Rope (general purpose)	180
	Granular (high speed molding)	180
	Glass reinforced (heavy duty parts)	180
Polycarbonates	Polycarbonate	120 (tungsten electrode)
	Polycarbonate (40% glass fiber reinforced)	120 (tungsten electrode)

Source: data compiled by J.S. Park from Charles T. Lynch, *CRC Handbook of Materials Science, Vol. 3*, CRC Press, Boca Raton, Florida, 1975 and *Engineered Materials Handbook, Vol.2, Engineering Plastics*, ASM International, Metals Park, Ohio, 1988.

Table 303. ARC RESISTANCE OF POLYMERS

(SHEET 2 OF 8)

Polymer	Type	Arc Resistance, (ASTM D495) (seconds)
Diallyl Phthalates; Molded	Orlon filled	85—115
	Dacron filled	105—125
	Asbestos filled	125—140
	Glass fiber filled	125—140
Fluorocarbons; Molded,Extruded	Polytrifluoro chloroethylene (PTFCE)	>360
	Polytetrafluoroethylene (PTFE)	>200
	Ceramic reinforced (PTFE)	
	Fluorinated ethylene propylene(FEP)	>165
	Polyvinylidene— fluoride (PVDF)	50—60
Epoxies; Cast, Molded, Reinforced	Standard epoxies (diglycidyl ethers of bisphenol A)	
	Cast rigid	100
	Cast flexible	75—98
	Molded	135—190
	General purpose glass cloth laminate	130—180

Source: data compiled by J.S. Park from Charles T. Lynch, *CRC Handbook of Materials Science, Vol. 3*, CRC Press, Boca Raton, Florida, 1975 and *Engineered Materials Handbook, Vol.2, Engineering Plastics*, ASM International, Metals Park, Ohio, 1988.

Table 303. ARC RESISTANCE OF POLYMERS

(SHEET 3 OF 8)

Polymer	Type	Arc Resistance, (ASTM D495) (seconds)
Epoxies—Molded, Extruded	High performance resins (cycloaliphatic diepoxides)	
	Molded	180—185
Epoxy novolacs	Cast, rigid	120
Melamines; Molded	Filler & type	
	Unfilled	100—145
	Cellulose electrical	70—135
	Glass fiber	180—186
	Alpha cellulose and mineral	125
Nylons; Molded, Extruded	Type 6	
	Glass fiber (30%) reinforced	92—81
	6/6 Nylon	
	General purpose molding	120
	Glass fiber reinforced	100—148
	Glass fiber Molybdenum disulfide filled	135
	General purpose extrusion	120

Source: data compiled by J.S. Park from Charles T. Lynch, *CRC Handbook of Materials Science, Vol. 3*, CRC Press, Boca Raton, Florida, 1975 and *Engineered Materials Handbook, Vol.2, Engineering Plastics*, ASM International, Metals Park, Ohio, 1988.

Table 303. ARC RESISTANCE OF POLYMERS

(SHEET 4 OF 8)

Polymer	Type	Arc Resistance, (ASTM D495) (seconds)
Nylons; Molded, Extruded (Con't)	6/10 Nylon	
	General purpose	120
Phenolics; Molded	Type and filler	
	General: woodflour and flock	5—60
	Shock: paper, flock, or pulp	5—60
	High shock: chopped fabric or cord	5—60
	Very high shock: glass fiber	60
Phenolics: Molded	Arc resistant—mineral	180
	Rubber phenolic—woodflour or flock	7—20
	Rubber phenolic—chopped fabric	10—20
	Rubber phenolic—asbestos	5—20
ABS–Polycarbonate Alloy	ABS–Polycarbonate Alloy	96
PVC–Acrylic Alloy	PVC–acrylic sheet	80
	PVC–acrylic injection molded	25

Source: data compiled by J.S. Park from Charles T. Lynch, *CRC Handbook of Materials Science, Vol. 3*, CRC Press, Boca Raton, Florida, 1975 and *Engineered Materials Handbook, Vol.2, Engineering Plastics*, ASM International, Metals Park, Ohio, 1988.

Table 303. ARC RESISTANCE OF POLYMERS
(SHEET 5 OF 8)

Polymer	Type	Arc Resistance, (ASTM D495) (seconds)
Polyimides	Unreinforced	152
	Glass reinforced	50—180
Polyacetals	Homopolymer:	
	Standard	129
	20% glass reinforced	188
	Copolymer:	
	Standard	240
	25% glass reinforced	136
	High flow	240
Polyester; Thermoplastic	Injection Moldings:	
	General purpose grade	190
	Glass reinforced grades	130
	Glass reinforced self extinguishing	80
	General purpose grade	125
	Asbestos—filled grade	108

Source: data compiled by J.S. Park from Charles T. Lynch, *CRC Handbook of Materials Science, Vol. 3*, CRC Press, Boca Raton, Florida, 1975 and *Engineered Materials Handbook, Vol.2, Engineering Plastics*, ASM International, Metals Park, Ohio, 1988.

Table 303. Arc Resistance of Polymers
(Sheet 6 of 8)

Polymer	Type	Arc Resistance, (ASTM D495) (seconds)
Polyesters: Thermosets	Cast polyyester	
	Rigid	115—135
	Flexible	125—145
	Reinforced polyester moldings	
	High strength (glass fibers)	130—170
	Sheet molding compounds, general purpose	130—180
Phenylene Oxides	SE—100	75
	SE—1	75
	Glass fiber reinforced	120
Phenylene oxides (Noryl)	Standard	122
	Glass fiber reinforced	114
Polyarylsulfone	Polyarylsulfone	67—81
Polypropylene	General purpose	125—136
	High impact	123—140

Source: data compiled by J.S. Park from Charles T. Lynch, *CRC Handbook of Materials Science, Vol. 3*, CRC Press, Boca Raton, Florida, 1975 and *Engineered Materials Handbook, Vol.2, Engineering Plastics*, ASM International, Metals Park, Ohio, 1988.

Table 303. ARC RESISTANCE OF POLYMERS
(SHEET 7 OF 8)

Polymer	Type	Arc Resistance, (ASTM D495) (seconds)
Polypropylene (Con't)	Asbestos filled	121—125
	Glass reinforced	73—77
	Flame retardant	15—40
Polyphenylene sulfide	40% glass reinforced	34
Polystyrenes	Molded	
	General purpose	60—135
	Medium impact	20—135
	High impact	20—100
	Glass fiber -30% reinforced	28
	Styrene acrylonitrile (SAN)	100—150
	Glass fiber (30%) reinforced SAN	65

Source: data compiled by J.S. Park from Charles T. Lynch, *CRC Handbook of Materials Science, Vol. 3*, CRC Press, Boca Raton, Florida, 1975 and *Engineered Materials Handbook, Vol.2, Engineering Plastics*, ASM International, Metals Park, Ohio, 1988.

Table 303. ARC RESISTANCE OF POLYMERS
(SHEET 8 OF 8)

Polymer	Type	Arc Resistance, (ASTM D495) (seconds)
Silicones; Molded, Laminated	Fibrous (glass) reinforced silicones	240
	Granular (silica) reinforced silicones	250—310
	Woven glass fabric/ silicone laminate	225—250
Ureas; Molded	Alpha—cellulose filled (ASTM Type l)	100—135
	Cellulose filled (ASTM Type 2)	85—110
	Woodflour filled	80—110

Source: data compiled by J.S. Park from Charles T. Lynch, *CRC Handbook of Materials Science, Vol. 3*, CRC Press, Boca Raton, Florida, 1975 and *Engineered Materials Handbook, Vol.2, Engineering Plastics*, ASM International, Metals Park, Ohio, 1988.

CHAPTER 8

Optical Properties of Materials

List of Tables

Transparency & Transmission

Refractive Index

Dispersion

Table 304. TRANSMISSION RANGE OF OPTICAL MATERIALS (SHEET 1 OF 2)

Material & Crystal Structure	Transmission Region (μm, at 298 K)
Alumina (Sapphire, Single Crystal)	0.15 – 6.5
Ammonium Dihydrogen Phosphate (ADP, Single Crystal)	0.13 – 1.7
Arsenic Trisulfade (Glass)	0.6 – 13
Barium Fluoride (Single Crystal)	0.25 – 15
Cadmium Sulfide (Bulk and Hexagonal Single Crystal)	0.5 – 16
Cadmium Telluride (Hot Pressed Polycrystalline)	0.9 – 16
Calcium Carbonate (Calcite, Single Crystal)	0.2 – 5.5
Calcium Fluoride (Single Crystal)	0.13 – 12
Cesium Bromide (Single Crystal)	0.3 – 55
Cesium Iodide (Single Crystal)	0.25 – 80
Cuprous Chloride (Single Crystal)	0.4 – 19
Gallium Arsenide (Intrinsic Single Crystal)	1.0 – 15
Germanium (Intrinsic Single Crystal)	1.8 – 23
Indium Arsenide (Single Crystal)	3.8 – 7.0
Lead Sulfide (Single Crystal)	3.0 – 7.0
Lithium Fluoride (Single Crystal)	0.12 – 9.0
Lithium Niobate (Single Crystal)	0.33 – 5.2
Magnesium Fluoride (Film)	0.2 – 5.0
Magnesium Fluoride (Single Crystal)	0.1 – 9.7
Magnesium Oxide (Single Crystal)	0.25 – 8.5
Potassium Bromide (Single Crystal)	0.25 – 35
Potassium Iodide (Single Crystal)	0.25 – 45
Selenium (Amorphous)	1.0 – 20
Silica (High Purity Crystalline)	0.12 – 4.5
Silica (High Purity Fused)	0.12 – 4.5
Silicon (Single Crystal)	1.2 – 15
Silver Bromide (Single Crystal)	0.45 – 35

External transmittance ≥ 10% with 2.0 mm thickness.

Source: Data compiled by J.S. Park.

Table 304. TRANSMISSION RANGE OF OPTICAL MATERIALS (SHEET 2 OF 2)

Material & Crystal Structure	Transmission Region (μm, at 298 K)
Silver Chloride (Single Crystal)	0.4 – 2.8
Sodium Fluoride (Single Crystal)	0.19 – 15
Strontium Titanate (Single Crystal)	0.39 – 6.8
Tellurium (Polycrystalline Film)	3.5 – 8.0
Tellurium (Single Crystal)	3.5 – 8.0
Thallium Bromoiodide (KRS–5, Mixed Crystal)	0.6 – 40
Thallium Chloribromide (KRS–6, Mixed Crystal)	0.21 – 35
Titanium Dioxide (Rutile, Single Crystal)	0.43 – 6.2
Zinc Selenide (Single Crystal, Cubic)	~0.5 – 22
Zinc Sulfide (Single Crystal, Cubic)	~0.6 – 15.6

External transmittance ≥ 10% with 2.0 mm thickness.

Source: Data compiled by J.S. Park.

Table 305. TRANSPARENCY OF POLYMERS

(SHEET 1 OF 7)

Polymer	Type	Transparency (visible light) (ASTM D791) (%)
Acrylics; Cast, Molded, Extruded	Cast Resin Sheets, Rods:	(0.125 in.)
	General purpose, type I	91—92
	General purpose, type II	91—92
	Moldings:	
	Grades 5, 6, 8	>92
	High impact grade	90
Thermoset Carbonate	Allyl diglycol carbonate	89—92
Alkyds; Molded	Putty (encapsulating)	Opaque
	Rope (general purpose)	Opaque
	Granular (high speed molding)	Opaque
	Glass reinforced (heavy duty parts)	Opaque

Source: data compiled by J.S. Park from Charles T. Lynch, *CRC Handbook of Materials Science, Vol. 3*, CRC Press, Boca Raton, Florida, 1975 and *Engineered Materials Handbook, Vol.2, Engineering Plastics*, ASM International, Metals Park, Ohio, 1988.

Table 305. TRANSPARENCY OF POLYMERS

(SHEET 2 OF 7)

Polymer	Type	Transparency (visible light) (ASTM D791) (%)
Cellulose Acetate; Molded, Extruded	ASTM Grade:	
	H6—1	75—90
	H4—1	75—90
	H2—1	80—90
	MH—1, MH—2	80—90
	MS—1, MS—2	80—90
	S2—1	80—95
Cellulose Acetate Butyrate; Molded, Extruded	ASTM Grade:	
	H4	75—92
	MH	80—92
	S2	85—95
Cellusose Acetate Propionate; Molded, Extruded	ASTM Grade:	
	1	80—92
	3	80—92
	6	80—92

Source: data compiled by J.S. Park from Charles T. Lynch, *CRC Handbook of Materials Science, Vol. 3*, CRC Press, Boca Raton, Florida, 1975 and *Engineered Materials Handbook, Vol.2, Engineering Plastics*, ASM International, Metals Park, Ohio, 1988.

Table 305. TRANSPARENCY OF POLYMERS

(SHEET 3 OF 7)

Polymer	Type	Transparency (visible light) (ASTM D791) (%)
Chlorinated Polymers	Chlorinated polyether	Opaque
	Chlorinated polyvinyl chloride	Opaque
Polycarbonates	Polycarbonate	75—85
	Polycarbonate (40% glass fiber reinforced)	Translucent
Fluorocarbons; Molded,Extruded	Polytrifluoro chloroethylene (PTFCE)	80—92
Epoxies; Cast, Molded, Reinforced	Standard epoxies (diglycidyl ethers of bisphenol A)	
	Cast rigid	
	Cast flexible	90
	Molded	85
	General purpose glass cloth laminate	Opaque
	High strength laminate	Opaque
	Filament wound composite	Opaque

Source: data compiled by J.S. Park from Charles T. Lynch, *CRC Handbook of Materials Science, Vol. 3*, CRC Press, Boca Raton, Florida, 1975 and *Engineered Materials Handbook, Vol.2, Engineering Plastics*, ASM International, Metals Park, Ohio, 1988.

Table 305. Transparency of Polymers

(Sheet 4 of 7)

Polymer	Type	Transparency (visible light) (ASTM D791) (%)
Epoxies—Molded, Extruded	High performance resins (cycloaliphatic diepoxides)	
	Cast, rigid	
	Molded	Opaque
	Glass cloth laminate	Opaque
Epoxy novolacs	Glass cloth laminate	Opaque
Melamines; Molded	Filler & type	
	Unfilled	Good
	Cellulose electrical	Opaque
Nylons; Molded, Extruded	6/6 Nylon	
	General purpose molding	Translucent
	Glass fiber reinforced	Opaque
	Glass fiber Molybdenum disulfide filled	Opaque
	General purpose extrusion	Opaque

Source: data compiled by J.S. Park from Charles T. Lynch, *CRC Handbook of Materials Science, Vol. 3*, CRC Press, Boca Raton, Florida, 1975 and *Engineered Materials Handbook, Vol.2, Engineering Plastics*, ASM International, Metals Park, Ohio, 1988.

Table 305. TRANSPARENCY OF POLYMERS

(SHEET 5 OF 7)

Polymer	Type	Transparency (visible light) (ASTM D791) (%)
Nylons; Molded, Extruded (Con't)	6/10 Nylon	
	General purpose	Opaque
	Glass fiber (30%) reinforced	Opaque
ABS–Polycarbonate Alloy	ABS–Polycarbonate Alloy	Opaque
PVC–Acrylic Alloy	PVC–acrylic sheet	Opaque
	PVC–acrylic injection molded	Opaque
Poliymides	Unreinforced	Opaque
	Unreinforced 2nd value	Opaque
	Glass reinforced	Opaque
Polyesters: Thermosets	Reinforced polyester moldings	
	High strength (glass fibers)	Opaque
	Heat and chemical resistsnt (asbestos)	Opaque
	Sheet molding compounds, general purpose	Opaque

Source: data compiled by J.S. Park from Charles T. Lynch, *CRC Handbook of Materials Science, Vol. 3*, CRC Press, Boca Raton, Florida, 1975 and *Engineered Materials Handbook, Vol.2, Engineering Plastics*, ASM International, Metals Park, Ohio, 1988.

Table 305. TRANSPARENCY OF POLYMERS

(SHEET 6 OF 7)

Polymer	Type	Transparency (visible light) (ASTM D791) (%)
Phenylene Oxides	SE—100	Opaque
	SE—1	Opaque
	Glass fiber reinforced	Opaque
Phenylene oxides (Noryl)	Glass fiber reinforced	Opaque
Polypropylene	General purpose	Translucent—opaque
	High impact	Translucent—opaque
	Asbestos filled	Opaque
	Glass reinforced	Opaque
	Flame retardant	Opaque
Polyphenylene sulfide	Standard	Opaque
	40% glass reinforced	Opaque

Source: data compiled by J.S. Park from Charles T. Lynch, *CRC Handbook of Materials Science, Vol. 3*, CRC Press, Boca Raton, Florida, 1975 and *Engineered Materials Handbook, Vol.2, Engineering Plastics*, ASM International, Metals Park, Ohio, 1988.

Table 305. Transparency of Polymers

(Sheet 7 of 7)

Polymer	Type	Transparency (visible light) (ASTM D791) (%)
Polystyrenes; Molded	General purpose	Transparent
	Medium impact	Opaque
	High impact	Opaque
	Glass fiber -30% reinforced	Opaque
Styrene acrylonitrile (SAN)	Styrene acrylonitrile (SAN)	Transparent
	Glass fiber (30%) reinforced SAN	Opaque
Silicones; Molded, Laminated	Fibrous (glass) reinforced silicones	Opaque
	Granular (silica) reinforced silicones	Opaque
	Woven glass fabric/ silicone laminate	Opaque
Ureas; Molded	Alpha—cellulose filled (ASTM Type 1)	21.8
	Cellulose filled (ASTM Type 2)	Opaque
	Woodflour filled	Opaque

Source: data compiled by J.S. Park from Charles T. Lynch, *CRC Handbook of Materials Science, Vol. 3*, CRC Press, Boca Raton, Florida, 1975 and *Engineered Materials Handbook, Vol.2, Engineering Plastics*, ASM International, Metals Park, Ohio, 1988.

Table 306. Refractive Index of Polymers
(Sheet 1 of 5)

Polymer	Type	Refractive index, (ASTM D542) (n_D)
Acrylics; Cast, Molded, Extruded	Cast Resin Sheets, Rods:	
	General purpose, type I	1.485—1.500
	General purpose, type II	1.485—1.495
	Moldings:	
	Grades 5, 6, 8	1.489—1.493
	High impact grade	1.49
Thermoset Carbonate	Allyl diglycol carbonate	1.5
Cellulose Acetate; Molded, Extruded	ASTM Grade:	
	H6—1	1.46—1.50
	H4—1	1.46—1.50
	H2—1	1.46—1.50
	MH—1, MH—2	1.46—1.50
	MS—1, MS—2	1.46—1.50
	S2—1	1.46—1.50

Source: data compiled by J.S. Park from Charles T. Lynch, *CRC Handbook of Materials Science, Vol.* 3, CRC Press, Boca Raton, Florida, 1975 and *Engineered Materials Handbook, Vol.2, Engineering Plastics*, ASM International, Metals Park, Ohio, 1988.

Table 306. REFRACTIVE INDEX OF POLYMERS

(SHEET 2 OF 5)

Polymer	Type	Refractive index, (ASTM D542) (n_D)
Cellulose Acetate Butyrate; Molded, Extruded	ASTM Grade:	(D543)
	H4	1.46—1.49
	MH	1.46—1.49
	S2	1.46—1.49
Cellusose Acetate Propionate; Molded, Extruded	ASTM Grade:	
	1	1.46—1.49
	3	1.46—1.49
	6	1.46—1.49
	Polycarbonate	1.586
Fluorocarbons; Molded,Extruded	Polytrifluoro chloroethylene (PTFCE)	1.43
	Polytetrafluoroethylene (PTFE)	1.35
	Fluorinated ethylene propylene(FEP)	1.34
	Polyvinylidene— fluoride (PVDF)	1.42

Source: data compiled by J.S. Park from Charles T. Lynch, *CRC Handbook of Materials Science, Vol.* 3, CRC Press, Boca Raton, Florida, 1975 and *Engineered Materials Handbook, Vol.2, Engineering Plastics*, ASM International, Metals Park, Ohio, 1988.

Table 306. Refractive Index of Polymers

(Sheet 3 of 5)

Polymer	Type	Refractive index, (ASTM D542) (n_D)
Epoxies; Cast, Molded, Reinforced	Standard epoxies (diglycidyl ethers of bisphenol A)	
	Cast rigid	
	Cast flexible	1.61
	Molded	1.61
Polyacetals	Homopolymer:	
	Standard	Opaque
	20% glass reinforced	Opaque
	22% TFE reinforced	Opaque
	Copolymer:	
	Standard	Opaque
	25% glass reinforced	Opaque
	High flow	Opaque
Polyesters: Thermosets	Cast polyyester	
	Rigid	1.53—1.58
	Flexible	1.50—1.57

Source: data compiled by J.S. Park from Charles T. Lynch, *CRC Handbook of Materials Science, Vol.* 3, CRC Press, Boca Raton, Florida, 1975 and *Engineered Materials Handbook, Vol.2, Engineering Plastics*, ASM International, Metals Park, Ohio, 1988.

Table 306. Refractive Index of Polymers

(Sheet 4 of 5)

Polymer	Type	Refractive index, (ASTM D542) (n_D)
Phenylene oxides (Noryl)	Standard	1.63
Polyarylsulfone	Polyarylsulfone	1.651
Polyethylenes; Molded, Extruded	Type I—lower density (0.910—0.925)	
	Melt index 0.3—3.6	1.51
	Melt index 6—26	1.51
	Melt index 200	1.51
	Type II—medium density (0.926—0.940)	
	Melt index 20	1.51
	Melt index 1.0—1.9	1.51
	Type III—higher density (0.941—0.965)	
	Melt index 0.2—0.9	1.54
	Melt index 0.1—12.0	1.54
	Melt index 1.5—15	1.54

Source: data compiled by J.S. Park from Charles T. Lynch, *CRC Handbook of Materials Science, Vol.* 3, CRC Press, Boca Raton, Florida, 1975 and *Engineered Materials Handbook, Vol.2, Engineering Plastics*, ASM International, Metals Park, Ohio, 1988.

Table 306. REFRACTIVE INDEX OF POLYMERS
(SHEET 5 OF 5)

Polymer	Type	Refractive index, (ASTM D542) (n_D)
Polystyrenes; Molded	Polystyrenes	
	General purpose	1.6
	Medium impact	Opaque
	High impact	Opaque
	Glass fiber -30% reinforced	Opaque
	Styrene acrylonitrile (SAN)	1.565—1.569
	Glass fiber (30%) reinforced SAN	Opaque
Polyvinyl Chloride And Copolymers; Molded, Extruded	Vinylidene chloride	1.60—1.63

Source: data compiled by J.S. Park from Charles T. Lynch, *CRC Handbook of Materials Science, Vol.* 3, CRC Press, Boca Raton, Florida, 1975 and *Engineered Materials Handbook, Vol.2, Engineering Plastics*, ASM International, Metals Park, Ohio, 1988.

Table 307. Dispersion of Optical Materials
(Sheet 1 of 13)

Material	Dispersion Equation at 298 K
Alumina (Sapphire, Single Crystal)	$n^2-1=\sum_{i=1}^{3}\frac{A_i\lambda^2}{\lambda^2-\lambda_i^2}$ (λ in μm) where i = 1: $\lambda_i 2$ = 0.00377588, A_i = 1.023798 i = 2: $\lambda_i 2$ = 0.0122544, A_i = 1.058264 i = 3: $\lambda_i 2$ = 321.3616, A_i = 5.280792 (λ in mm)

Source: data compiled by J.S. Park from Charles T. Lynch, *CRC Handbook of Materials Science, Vol. 3*, CRC Press, Boca Raton, Florida, 1975 and *Engineered Materials Handbook, Vol.2, Engineering Plastics*, ASM International, Metals Park, Ohio, 1988.

Table 307. DISPERSION OF OPTICAL MATERIALS

(SHEET 2 OF 13)

Material	Dispersion Equation at 298 K
ArsenicTrisulfide (Glass)	$n^2-1=\sum_{i=1}^{5}\frac{K_i\lambda^2}{\lambda^2-\lambda_i^2}$ (λ in μm) where i · $\lambda_i 2$ · K_i 1 · 0.0225 · 1.8983678 2 · 0.0625 · 1.9222979 3 · 0.1225 · 0.8765134 4 · 0.2025 · 0.1188704 5 · 0.705 · 0.9569903 (λ in μm)

Source: data compiled by J.S. Park from Charles T. Lynch, *CRC Handbook of Materials Science, Vol. 3*, CRC Press, Boca Raton, Florida, 1975 and *Engineered Materials Handbook, Vol.2, Engineering Plastics*, ASM International, Metals Park, Ohio, 1988.

Table 307. Dispersion of Optical Materials

(Sheet 3 of 13)

Material	Dispersion Equation at 298 K
Barium Fluoride (Single Crystal)	$n^2-1=\sum_{i=1}^{3}\frac{A_i\lambda^2}{\lambda^2-\lambda_i^2}$ (λ in μm) where i — λ_i — A_i 1 — 0.057789 — 0.643356 2 — 0.10968 — 0.50676 3 — 46.3864 — 3.8261 (λ in μm)

Source: data compiled by J.S. Park from Charles T. Lynch, *CRC Handbook of Materials Science, Vol. 3*, CRC Press, Boca Raton, Florida, 1975 and *Engineered Materials Handbook, Vol.2, Engineering Plastics*, ASM International, Metals Park, Ohio, 1988.

Table 307. Dispersion of Optical Materials
(Sheet 4 of 13)

Material	Dispersion Equation at 298 K
Cadmium Sulfide (Bulk and Hexagonal Single Crystal)	$n_o^2=5.235+\frac{1.891\text{x}10^7}{\lambda^2-1.651\text{x}10^7}$ for ordinary ray, and $n_e^2=5.239+\frac{2.076\text{x}10^7}{\lambda^2-1.651\text{x}10^7}$ for extraordinary ray. (λ in μm)

Source: data compiled by J.S. Park from Charles T. Lynch, *CRC Handbook of Materials Science, Vol. 3*, CRC Press, Boca Raton, Florida, 1975 and *Engineered Materials Handbook, Vol.2, Engineering Plastics*, ASM International, Metals Park, Ohio, 1988.

Table 307. DISPERSION OF OPTICAL MATERIALS

(SHEET 5 OF 13)

Material	Dispersion Equation at 298 K
Calcium Fluoride (Single Crystal)	$n^2-1=\sum_{i=1}^{3}\frac{A_i\lambda^2}{\lambda^2-\lambda_i^2}$ (λ in μm) i / A_i / λ_i 1 / 0.5675888 / 0.050263605 2 / 0.4710914 / 0.1003909 3 / 3.8484723 / 34.64904
Cesium Bromide (Single Crystal)	$n^2=5.640752-3.338\times10^{-6}\lambda^2+\frac{0.0018612}{\lambda^2}+\frac{41110.49}{\lambda^2-14390.4}+\frac{0.0290764}{\lambda^2-0.024964}$ (λ in μm)

Source: data compiled by J.S. Park from Charles T. Lynch, *CRC Handbook of Materials Science, Vol. 3*, CRC Press, Boca Raton, Florida, 1975 and *Engineered Materials Handbook, Vol.2, Engineering Plastics*, ASM International, Metals Park, Ohio, 1988.

Table 307. DISPERSION OF OPTICAL MATERIALS

(SHEET 6 OF 13)

Material	Dispersion Equation at 298 K
Cesium Iodide (Single Crystal)	$n^2\text{-}1 = \sum_{i=1}^{5} \frac{K_i \lambda^2}{\lambda^2 - \lambda_i^2}$ (λ in μm) where i / $\lambda_i 2$ / K_i 1 / 0.00052701 / 0.3461725 2 / 0.02149156 / 1.0080886 3 / 0.28551800 / 0.02149156 4 / 0.39743178 / 0.044944 5 / 3.3605359 / 25921 (λ in mm)

Source: data compiled by J.S. Park from Charles T. Lynch, *CRC Handbook of Materials Science, Vol. 3*, CRC Press, Boca Raton, Florida, 1975 and *Engineered Materials Handbook, Vol.2, Engineering Plastics*, ASM International, Metals Park, Ohio, 1988.

Table 307. DISPERSION OF OPTICAL MATERIALS

(SHEET 7 OF 13)

Material	Dispersion Equation at 298 K
Germanium (Intrinsic Single Crystal)	$n = A + B\lambda + C\lambda^2 + D\lambda^2 + E\lambda^4$ where A=3.99931 B=0.391707 C=0.163492 D=–0.0000060 E=0.000000053 for $2.0\mu m \leq \lambda \leq 13.5\ \mu m$
Lithium Fluoride (Single Crystal)	$n = A + BL + CL^2 + D\lambda^2 + E\lambda^4$ where A=1.38761 B=0.001796 C=–0.000041 D=–0.0023045 E=–0.00000557 for $0.5\mu m \leq \lambda \leq 6.0\ \mu m$

Source: data compiled by J.S. Park from Charles T. Lynch, *CRC Handbook of Materials Science, Vol. 3*, CRC Press, Boca Raton, Florida, 1975 and *Engineered Materials Handbook, Vol.2, Engineering Plastics*, ASM International, Metals Park, Ohio, 1988.

Table 307. Dispersion of Optical Materials

(Sheet 8 of 13)

Material	Dispersion Equation at 298 K
Magnesium Fluoride (Single Crystal)	$n_o = 1.36957 + \frac{0.0035821}{\lambda - 0.14925}$ for ordinary wavelengths, and $n_e = 1.38100 + \frac{0.0037415}{\lambda - 0.14947}$ for wavelengths within $0.4\mu m \le \lambda \le 0.7\ \mu m$
Magnesium Oxide (Single Crystal)	$n^2 = 2.956362 - 0.1062387\ \lambda^2 - 2.04968 \times 10^{-5} \lambda^4 - \frac{0.0219577}{\lambda^2 - 0.01428322}$

Source: data compiled by J.S. Park from Charles T. Lynch, *CRC Handbook of Materials Science, Vol. 3*, CRC Press, Boca Raton, Florida, 1975 and *Engineered Materials Handbook, Vol.2, Engineering Plastics*, ASM International, Metals Park, Ohio, 1988.

Table 307. DISPERSION OF OPTICAL MATERIALS

(SHEET 9 OF 13)

Material	Dispersion Equation at 298 K
Potassium Bromide (Single Crystal)	$n^2 = 2.3618102 - 0.00058072\ \lambda^2 + \frac{0.02305269}{\lambda^2 - 0.02425381}$ for $0.4\mu m \leq \lambda \leq 0.7\ \mu m$
Potassium Chloride (Single Crystal)	$n^2 = 2.174967 + \frac{0.08344206}{\lambda^2 - 0.0119082} + \frac{0.00698382}{\lambda^2 - 0.025555} - 0.000513495\,\lambda^2 - 0.06167587\ \lambda^4$ for ultraviolet wavelengths $n^2 = 3.866619 + \frac{0.08344206}{\lambda^2 - 0.0119082} - \frac{0.00698382}{\lambda^2 - 0.025555} - \frac{5569.715}{\lambda^2 - 3292.472}$ for the visible light

Source: data compiled by J.S. Park from Charles T. Lynch, *CRC Handbook of Materials Science, Vol. 3*, CRC Press, Boca Raton, Florida, 1975 and *Engineered Materials Handbook, Vol.2, Engineering Plastics*, ASM International, Metals Park, Ohio, 1988.

Table 307. DISPERSION OF OPTICAL MATERIALS

(SHEET 10 OF 13)

Material	Dispersion Equation at 298 K
Silica (High Purity Fused)	$n^2 = 2.978645 + \frac{0.008777808}{\lambda^2 - 0.010609} + \frac{84.06224}{\lambda^2 - 96.0000}$
Silicon (Single Crystal)	$n = 3.41696 + 0.138497L + 0.013924L^2 - 0.0000209\lambda^2 + 0.000000148\lambda^4$ where $L = (\lambda^2 - 0.028)^{-1}$
Silver Bromide (Single Crystal)	$\frac{n^2 - 1}{n^2 + 2} = 0.48484 + \frac{0.10279\lambda^2}{\lambda^2 - 0.0900} - 0.004796\lambda^2$ for $0.54\mu m \le \lambda \le 0.65\ \mu m$

Source: data compiled by J.S. Park from Charles T. Lynch, *CRC Handbook of Materials Science, Vol. 3*, CRC Press, Boca Raton, Florida, 1975 and *Engineered Materials Handbook, Vol.2, Engineering Plastics*, ASM International, Metals Park, Ohio, 1988.

Table 307. Dispersion of Optical Materials
(Sheet 11 of 13)

Material	Dispersion Equation at 298 K
Silver Chloride (Single Crystal)	$n = 4.00804 - 0.00085111\lambda^2 - 0.00000019762\lambda^4 + 0.079086/(\lambda^2 - 0.04584)$
Strontium Titanate (Single Crystal)	$n = A + BL + CL^2 + D\lambda^2 + E\lambda^4$ where A=2.28355 B=0.035906 C=0.001666 D=–0.0061355 E=–0.00001502 for $1.0\ \mu m \le \lambda \le 5.3\ \mu m$

Source: data compiled by J.S. Park from Charles T. Lynch, *CRC Handbook of Materials Science, Vol. 3*, CRC Press, Boca Raton, Florida, 1975 and *Engineered Materials Handbook, Vol.2, Engineering Plastics*, ASM International, Metals Park, Ohio, 1988.

Table 307. Dispersion of Optical Materials

(Sheet 12 of 13)

Material	Dispersion Equation at 298 K
Thallium Bromoiodide (KRS-5, Mixed Crystal)	$n^2 - 1 = \sum_{i=1}^{5} \frac{K_i \lambda^2}{\lambda^2 - \lambda_i^2}$ where i — $\lambda_i 2$ — K_i 1 — 0.0225 — 1.8293958 2 — 0.0625 — 1.6675593 3 — 0.1225 — 1.1210424 4 — 0.2025 — 0.4513366 5 — 27089.737 — 12.380234 (λ in μm)

Source: data compiled by J.S. Park from Charles T. Lynch, *CRC Handbook of Materials Science, Vol. 3*, CRC Press, Boca Raton, Florida, 1975 and *Engineered Materials Handbook, Vol.2, Engineering Plastics*, ASM International, Metals Park, Ohio, 1988.

Table 307. DISPERSION OF OPTICAL MATERIALS

(SHEET 13 OF 13)

Material	Dispersion Equation at 298 K
Titanium Dioxide (Rutile, Single Crystal)	$n_o^2 = 5.913 + \frac{2.441 x 10^7}{\lambda^2 - 0.803 x 10^7}$ for ordinary wavelengths, and $n_e^2 = 7.197 + \frac{3.322 x 10^7}{\lambda^2 - 0.843 x 10^7}$ for extraordinary wavelengths. (λ in Å)
Zinc Sulfide (Single Crystal, Cubic)	$n = \frac{5.164 + 1.208 x 10^7}{12 - 0.732 \, x 10^7}$ (λ in Å)

Source: data compiled by J.S. Park from Charles T. Lynch, *CRC Handbook of Materials Science, Vol. 3*, CRC Press, Boca Raton, Florida, 1975 and *Engineered Materials Handbook, Vol.2, Engineering Plastics*, ASM International, Metals Park, Ohio, 1988.

CHAPTER 9

Chemical Properties of Materials

List of Tables

Absorption

EMF Potentials and Galvanic Series

Corrosion

List of Tables (Continued)

Corrosion (con't)

Flammability

Table 308. WATER ABSORPTION OF POLYMERS

(SHEET 1 OF 12)

Polymer	Type	Water Absorption in 24 hr, ASTM D570) (%)
ABS Resins; Molded, Extruded	Medium impact	0.2—0.4
	High impact	0.2—0.45
	Very high impact	0.2—0.45
	Low temperature impact	0.2—0.45
	Heat resistant	0.2—0.4
Acrylics; Cast, Molded, Extruded	Cast Resin Sheets, Rods:	
	General purpose, type I	0.3—0.4
	General purpose, type II	0.2—0.4
	Moldings:	
	Grades 5, 6, 8	0.3—0.4
	High impact grade	0.2—0.4
Thermoset Carbonate	Allyl diglycol carbonate	0.2

Source: data compiled by J.S. Park from Charles T. Lynch, *CRC Handbook of Materials Science, Vol. 3*, CRC Press, Boca Raton, Florida, 1975 and *Engineered Materials Handbook, Vol.2*, Engineering Plastics, ASM International, Metals Park, Ohio, 1988.

Table 308. WATER ABSORPTION OF POLYMERS

(SHEET 2 OF 12)

Polymer	Type	Water Absorption in 24 hr, ASTM D570) (%)
Alkyds; Molded	Putty (encapsulating)	0.10—0.15
	Rope (general purpose)	0.05—0.08
	Granular (high speed molding)	0.08—0.12
	Glass reinforced (heavy duty parts)	0.007—0.10
Cellulose Acetate; Molded, Extruded	ASTM Grade:	
	H4—1	1.7—2.7
	H2—1	1.7—2.7
	MH—1, MH—2	1.8—4.0
	MS—1, MS—2	2.1—4.0
	S2—1	2.3—4.0
Cellulose Acetate Butyrate; Molded, Extruded	ASTM Grade:	
	H4	2
	MH	1.3—1.6
	S2	0.9—1.3

Source: data compiled by J.S. Park from Charles T. Lynch, *CRC Handbook of Materials Science, Vol. 3*, CRC Press, Boca Raton, Florida, 1975 and *Engineered Materials Handbook, Vol.2*, Engineering Plastics, ASM International, Metals Park, Ohio, 1988.

Table 308. WATER ABSORPTION OF POLYMERS

(SHEET 3 OF 12)

Polymer	Type	Water Absorption in 24 hr, ASTM D570) (%)
Cellusose Acetate Propionate; Molded, Extruded	ASTM Grade:	
	1	1.6—2.0
	3	1.3—1.8
	6	1.6
Chlorinated Polymers	Chlorinated polyether	0.01
	Chlorinated polyvinyl chloride	0.11
Polycarbonates	Polycarbonate	0.15
	Polycarbonate (40% glass fiber reinforced)	0.08
Diallyl Phthalates; Molded		(122 •F, 48 hr), %
	Orlon filled	0.2—0.5
	Dacron filled	0.2—0.5
	Asbestos filled	0.4—0.7
	Glass fiber filled	0.2—0.4

Source: data compiled by J.S. Park from Charles T. Lynch, *CRC Handbook of Materials Science, Vol. 3*, CRC Press, Boca Raton, Florida, 1975 and *Engineered Materials Handbook, Vol.2*, Engineering Plastics, ASM International, Metals Park, Ohio, 1988.

Table 308. WATER ABSORPTION OF POLYMERS

(SHEET 4 OF 12)

Polymer	Type	Water Absorption in 24 hr, ASTM D570) (%)
Fluorocarbons; Molded,Extruded	Polytrifluoro chloroethylene (PTFCE)	0
	Polytetrafluoroethylene (PTFE)	0.01
	Ceramic reinforced (PTFE)	>0.2
	Fluorinated ethylene propylene(FEP)	<0.01
	Polyvinylidene— fluoride (PVDF)	0.03—0.06
Epoxies; Cast, Molded, Reinforced	Standard epoxies (diglycidyl ethers of bisphenol A)	
	Cast rigid	0.1—0.2
	Cast flexible	0.4—0.1
	Molded	0.3—0.8
	General purpose glass cloth laminate	0.05—0.07
	High strength laminate	0.05
	Filament wound composite	0.05—0.07

Source: data compiled by J.S. Park from Charles T. Lynch, *CRC Handbook of Materials Science, Vol. 3*, CRC Press, Boca Raton, Florida, 1975 and *Engineered Materials Handbook, Vol.2*, Engineering Plastics, ASM International, Metals Park, Ohio, 1988.

Table 308. Water Absorption of Polymers

(Sheet 5 of 12)

Polymer	Type	Water Absorption in 24 hr, ASTM D570) (%)
Epoxies—Molded, Extruded	High performance resins (cycloaliphatic diepoxides)	
	Molded	0.11—0.2
	Glass cloth laminate	0.04—0.06
Epoxy novolacs	Cast, rigid	0.1—0.7
Melamines; Molded	Filler & type	
	Unfilled	0.2—0.5
	Cellulose electrical	0.27—0.80
	Glass fiber	0.09—0.60
	Alpha cellulose and mineral	0.3—0.5
Nylons; Molded, Extruded	Type 6	
	General purpose	1.3—1.9
	Glass fiber (30%) reinforced	0.9—1.2
	Cast	0.6
	Flexible copolymers	0.8—1.4

Source: data compiled by J.S. Park from Charles T. Lynch, *CRC Handbook of Materials Science, Vol. 3*, CRC Press, Boca Raton, Florida, 1975 and *Engineered Materials Handbook, Vol.2*, Engineering Plastics, ASM International, Metals Park, Ohio, 1988.

Table 308. WATER ABSORPTION OF POLYMERS

(SHEET 6 OF 12)

Polymer	Type	Water Absorption in 24 hr, ASTM D570) (%)
Nylons; Molded, Extruded (Con't)	Type 8	9.5
	Type 11	0.4
	Type 12	0.25
	6/6 Nylon	
	General purpose molding	1.5
	Glass fiber reinforced	0.8—0.9
	Glass fiber Molybdenum disulfide filled	0.5—0.7
	General purpose extrusion	1.5
	6/10 Nylon	
	General purpose	0.4
	Glass fiber (30%) reinforced	0.2

Source: data compiled by J.S. Park from Charles T. Lynch, *CRC Handbook of Materials Science, Vol. 3*, CRC Press, Boca Raton, Florida, 1975 and *Engineered Materials Handbook, Vol.2*, Engineering Plastics, ASM International, Metals Park, Ohio, 1988.

Table 308. Water Absorption of Polymers

(Sheet 7 of 12)

Polymer	Type	Water Absorption in 24 hr, ASTM D570) (%)
Phenolics; Molded	Type and filler	
	General: woodflour and flock	0.3—0.8
	Shock: paper, flock, or pulp	0.4—1.5
	High shock: chopped fabric or cord	0.4—1.75
	Very high shock: glass fiber	0.1—1.0
Phenolics; Molded (Con't)	Arc resistant—mineral	0.5—0.7
	Rubber phenolic—woodflour or flock	0.5—2.0
	Rubber phenolic—chopped fabric	0.5—2.0
	Rubber phenolic—asbestos	0.10—0.50
ABS–Polycarbonate Alloy	ABS–Polycarbonate Alloy	0.21
PVC–Acrylic Alloy	PVC–acrylic sheet	0.06
	PVC–acrylic injection molded	0.13

Source: data compiled by J.S. Park from Charles T. Lynch, *CRC Handbook of Materials Science, Vol. 3*, CRC Press, Boca Raton, Florida, 1975 and *Engineered Materials Handbook, Vol.2*, Engineering Plastics, ASM International, Metals Park, Ohio, 1988.

Table 308. WATER ABSORPTION OF POLYMERS

(SHEET 8 OF 12)

Polymer	Type	Water Absorption in 24 hr, ASTM D570) (%)
Polyimides	Unreinforced	0.47
	Unreinforced 2nd value	0.24—0.40
	Glass reinforced	0.2
Polyacetals	Homopolymer:	
	Standard	0.25
	20% glass reinforced	0.25
	22% TFE reinforced	0.2
Polyacetals (Con't)	Copolymer:	
	Standard	0.22
	25% glass reinforced	0.29
	High flow	0.22
Polyester; Thermoplastic	Injection Moldings:	
	General purpose grade	0.08
	Glass reinforced grades	0.06—0.07
	Glass reinforced self extinguishing	0.07

Source: data compiled by J.S. Park from Charles T. Lynch, *CRC Handbook of Materials Science, Vol. 3*, CRC Press, Boca Raton, Florida, 1975 and *Engineered Materials Handbook, Vol.2*, Engineering Plastics, ASM International, Metals Park, Ohio, 1988.

Table 308. WATER ABSORPTION OF POLYMERS

(SHEET 9 OF 12)

Polymer	Type	Water Absorption in 24 hr, ASTM D570) (%)
Polyester; Thermoplastic (Con't)	General purpose grade	0.09
	Glass reinforced grade	0.07
	Asbestos—filled grade	0.1
Polyesters: Thermosets	Cast polyyester	
	Rigid	0.20—0.60
	Flexible	0.12—2.5
Polyesters: Thermosets (Con't)	Reinforced polyester moldings	
	High strength (glass fibers)	0.5—0.75
	Heat and chemical resistsnt (asbestos)	0.25—0.50
	Sheet molding compounds, general purpose	0.15—0.25
Phenylene Oxides	SE—100	0.07
	SE—1	0.07
	Glass fiber reinforced	0.06

Source: data compiled by J.S. Park from Charles T. Lynch, *CRC Handbook of Materials Science, Vol. 3*, CRC Press, Boca Raton, Florida, 1975 and *Engineered Materials Handbook, Vol.2*, Engineering Plastics, ASM International, Metals Park, Ohio, 1988.

Table 308. Water Absorption of Polymers
(Sheet 10 of 12)

Polymer	Type	Water Absorption in 24 hr, ASTM D570) (%)
Phenylene oxides (Noryl)	Standard	0.22
	Glass fiber reinforced	0.22, 0.18
Polyarylsulfone	Polyarylsulfone	0.4
Polypropylene	General purpose	<0.01—0.03
	High impact	<0.01—0.02
Polypropylene (Con't)	Asbestos filled	0.02—0.04
	Glass reinforced	0.02—0.05
	Flame retardant	0.02—0.03
Polyethylenes; Molded, Extruded	Type I—lower density (0.910—0.925)	
	Melt index 0.3—3.6	<0.01
	Melt index 6—26	<0.01
	Melt index 200	<0.01

Source: data compiled by J.S. Park from Charles T. Lynch, *CRC Handbook of Materials Science, Vol. 3*, CRC Press, Boca Raton, Florida, 1975 and *Engineered Materials Handbook, Vol.2*, Engineering Plastics, ASM International, Metals Park, Ohio, 1988.

Table 308. WATER ABSORPTION OF POLYMERS

(SHEET 11 OF 12)

Polymer	Type	Water Absorption in 24 hr, ASTM D570) (%)
Polyethylenes; Molded, Extruded (Con't)	Type II—medium density (0.926—0.940)	
	Melt index 20	<0.01
	Melt index 1.0—1.9	<0.01
	Type III—higher density (0.941—0.965)	
	Melt index 0.2—0.9	<0.01
	Melt Melt index 0.1—12.0	<0.01
	Melt index 1.5—15	<0.01
	High molecular weight	<0.01
Polystyrenes; Molded	General purpose	0.30—0.2
	Medium impact	0.03—0.09
	High impact	0.05—0.22
	Glass fiber –30% reinforced	0.07
Styrene acrylonitrile (SAN)	Styrene acrylonitrile (SAN)	0.20—0.35
	Glass fiber (30%) reinforced SAN	0.15

Source: data compiled by J.S. Park from Charles T. Lynch, *CRC Handbook of Materials Science, Vol. 3*, CRC Press, Boca Raton, Florida, 1975 and *Engineered Materials Handbook, Vol.2*, Engineering Plastics, ASM International, Metals Park, Ohio, 1988.

Table 308. WATER ABSORPTION OF POLYMERS

(SHEET 12 OF 12)

Polymer	Type	Water Absorption in 24 hr, ASTM D570) (%)
Polyvinyl Chloride And Copolymers;	Molded, Extruded	(ASTM D635)
	Nonrigid—general	0.2—1.0
	Nonrigid—electrical	0.40—0.75
	Rigid—normal impact	0.03—0.40
	Vinylidene chloride	>0.1
Silicones; Molded, Laminated	Fibrous (glass) reinforced silicones	0.1—0.15
	Granular (silica) reinforced silicones	0.08—0.1
	Woven glass fabric/ silicone laminate	0.03—0.05
Ureas; Molded	Ureas; Molded Alpha—cellulose filled (ASTM Type l)	0.4—0.8

Source: data compiled by J.S. Park from Charles T. Lynch, *CRC Handbook of Materials Science, Vol. 3*, CRC Press, Boca Raton, Florida, 1975 and *Engineered Materials Handbook, Vol.2*, Engineering Plastics, ASM International, Metals Park, Ohio, 1988.

Table 309. STANDARD ELECTROMOTIVE FORCE POTENTIALS
(SHEET 1 OF 18)

Reaction	Reduction Potential E°, (V)
$F_2 + 2H^+ + 2\,e^- = 2\,HF$	3.053
$F_2 + 2\,e^- = 2F^-$	2.866
$H_2N_2O_2 + 2H^+ + 2\,e^- = N_2 + 2H_2O$	2.65
$O(g) + 2H^+ + 2\,e^- = H_2O$	2.421
$FeO_4^{2-} + 8H^+ + 3\,e^- = Fe^{3+} + 4H_2O$	2.20
$F_2O + 2H^+ + 4\,e^- = H_2O + 2F^-$	2.153
$S_2O_8^{2-} + 2H^+ + 2\,e^- = 2HSO_4^-$	2.123
$O_3 + 2H^+ + 4\,e^- = O_2 + H_2O$	2.076
$OH + e^- = OH^-$	2.02
$S_2O_8^{2-} + 2\,e^- = 2SO_4^{2-}$	2.010
$Ag^{2+} + e^- = Ag^+$	1.980
$Co^{3+} + e^- = Co^{2+}$ (2 mol /l H_2SO_4)	1.83
$H_2O_2 + 2H^+ + 2\,e^- = 2\,H_2O$	1.776
$N_2O + 2H^+ + 2\,e^- = N_2 + H_2O$	1.766
$CeOH^{3+} + H^+ + e^- = Ce^{3+} + H_2O$	1.715
$Au^+ + e^- = Au$	1.692
$PbO_2 + SO_4^{2-} + 4H^+ + 2\,e^- = PbSO_4 + 2H_2O$	1.6913
$MnO_4^- + 4H^+ + 3\,e^- = MnO_2 + 2\,H_2O$	1.679
$NiO_2 + 4H^+ + 2\,e^- = Ni^{2+} + 2\,H_2O$	1.678
$HClO_2 + 2H^+ + 2\,e^- = HClO + H_2O$	1.645
$HClO_2 + 3H^+ + 3\,e^- = 1/2Cl_2 + 2\,H_2O$	1.628
$HClO + H^+ + e^- = 1/2Cl_2 + H_2O$	1.611

Source: data compiled by J.S. Park from Petr Vanysek, *Handbook of Physics and Chemistry, 69th Edition*, CRC Press, Boca Raton, Florida, (1988).

Table 309. STANDARD ELECTROMOTIVE FORCE POTENTIALS (SHEET 2 OF 18)

Reaction	Reduction Potential E°, (V)
$Ce^{4+} + e^- = Ce^{3+}$	1.61
$H_5IO_6 + H^+ + 2\ e^- = IO_3^- + 3\ H_2O$	1.601
$HBrO + H^+ + e^- = 1/2\ Br_2\ (l) + H_2O$	1.596
$Bi_2O_4 + 4\ H^+ + 2\ e^- = 2\ BiO^+ + 2\ H_2O$	1.593
$2\ NO + 2\ H^+ + 2\ e^- = N_2O + H_2O$	1.591
$HBrO + H^+ + e^- = 1/2\ Br_2\ (aq) + H_2O$	1.574
$HClO_2 + 3\ H^+ + 4\ e^- = Cl^- + H_2O$	1.570
$Mn^{3+} + e^- = Mn^{2+}$	1.5415
$MnO_4^- + 8\ H^+ + 5\ e^- = Mn^{2+} + 4\ H_2O$	1.507
$Au^{3+} + 3\ e^- = Au$	1.498
$HO_2 + H^+ + e^- = H_2O_2$	1.495
$HClO + H^+ + 2\ e^- = Cl^- + H_2O$	1.482
$BrO_3^- + 6\ H^+ + 5\ e^- = 1/2\ Br_2 + 3\ H_2O$	1.482
$ClO_3^- + 6H^+ + 5\ e^- = 1/2\ Cl_2 + 3\ H_2O$	1.47
$PbO_2 + 4\ H^+ + 2\ e^- = Pb^{2+} + 2\ H_2O$	1.455
$ClO_3^- + 6\ H^+ + 6\ e^- = Cl^- + H_2O$	1.451
$Au(OH)_3 + 3\ H^+ + 3\ e^- = Au^- + 3\ H_2O$	1.45
$2\ HIO + 2\ H^+ + 2\ e^- = I_2 + 2H_2O$	1.439
$BrO_3^- + 6\ H^+ + 6\ e^- = Br^- + 3\ H_2O$	1.423
$2\ NH_3OH^+ + H^+ + 2\ e^- = N_2H_5+ + 2\ H_2O$	1.42
$Au^{3+} + 2\ e^- = Au^+$	1.401
$ClO_4^- + 8\ H^+ + 7\ e^- = 1/2\ Cl_2 + 4\ H_2O$	1.39

Source: data compiled by J.S. Park from Petr Vanysek, *Handbook of Physics and Chemistry, 69th Edition*, CRC Press, Boca Raton, Florida, (1988).

Table 309. STANDARD ELECTROMOTIVE FORCE POTENTIALS
(SHEET 3 OF 18)

Reaction	Reduction Potential E°, (V)
$ClO_4^- + 8 H^+ + 8 e^- = Cl^- + 4 H_2O$	1.389
$Cl_2(g) + 2 e^- = 2Cl^-$	1.35827
$HCrO_4^- + 7 H^+ + 3 e^- = Cr^{3+} + 4 H_2O$	1.350
$HBrO + H^+ + 2 e^- = Br^- + H_2O$	1.331
$PuO_2(OH)_2 + H^+ + 3 e^- = Pu(OH)_4$	1.325
$2 HNO_2 + 4 H^+ + 4 e^- = NO_2 + 3 H_2O$	1.297
$[PdCl_6]^{2-} + 2 e^- = [PdCl_4]^{2-} + 2 Cl^-$	1.288
$ClO_2 + H^+ + e^- = HClO_2$	1.277
$N_2H_5^+ + 3 H^+ + 2 e^- = 2 NH_4^+$	1.275
$Tl^{3+} + 2 e^- = Tl^+$	1.252
$O_3 + H_2O + 2 e^- = O_2 + 2 OH^-$	1.24
$Cr_2O_7^{2-} + 14 H^+ + 3 e^- = 2 Cr^{3+} + 7 H_2O$	1.232
$O_2 + 4 H^+ + 4 e^- = 2 H_2O$	1.229
$MnO_2 + 4 H^+ + 2 e^- = Mn^{2+} + 2 H_2O$	1.224
$ClO_3^- + 3 H^+ + 2 e^- = HClO_2 + H_2O$	1.214
$2 IO_3^- + 12 H^+ + 10 e^- = I_2 + 6 H_2O$	1.195
$ClO_4^- + 2 H^+ + 2 e^- = ClO_3^- + H_2O$	1.189
$Ir^{3+} + 3 e^- = Ir$	1.156
$ClO_3^- + 2 H^+ + e^- = ClO_2 + H_2O$	1.152
$SeO_4^{2-} + 4 H^+ + 2 e^- = H_2SeO_3 + H_2O$	1.151
$[Fe(pheneathroline)_3]^{3+} + e^- = [Fe(phen)_3]^{2+}$	1.147
$RuO_2 + 4 H^+ + 2 e^- = Ru^{2+} + 2 H_2O$	1.120

Source: data compiled by J.S. Park from Petr Vanysek, *Handbook of Physics and Chemistry, 69th Edition*, CRC Press, Boca Raton, Florida, (1988).

Table 309. STANDARD ELECTROMOTIVE FORCE POTENTIALS
(SHEET 4 OF 18)

Reaction	Reduction Potential E°, (V)
$Pt^{2-} + 2\,e^- = Pt$	1.118
$Pu^{5+} + e^- = Pu^{4+}$	1.099
$Br_2(aq) + 2\,e^- = 2\,Br^-$	1.0873
$IO_3^- + 6\,H^+ + 6\,e^- = I^- + 3\,H_2O$	1.085
$Br_2(l) + 2\,e^- = 2\,Br^-$	1.066
$N_2O_4 + 2\,H^+ + 2\,e^- = 2\,HNO_2$	1.065
$PuO_2(OH)_2 + H^+ + e^- = PuO_2OH + H_2O$	1.062
$[Fe(phen)_3]^{3+} + e^- = [Fe(phen)_3]^{2+}$ (1 mol/l H_2SO_4)	1.06
$N_2O_4 + 4\,H^+ + 4\,e^- = 2\,NO + 2\,H_2O$	1.035
$H_6TeO_6 + 2\,H^+ + 2\,e^- = TeO_2 + 4\,H_2O$	1.02
$Pu^{4+} + e^- = Pu^{3+}$	1.006
$AuCl_4^- + 3\,e^- = Au + 4\,Cl^-$	1.002
$V(OH)_4^+ + 2\,H^+ + e^- = VO^{2+} + 3\,H_2O$	1.00
$RuO_4 + e^- = RuO_4^-$	1.00
$VO_2+ + 2\,H^+ + e^- = VO^{2+} + H_2O$	0.991
$HIO + H^+ + 2\,e^- = I^- + H_2O$	0.987
$HNO_2 + H^+ + e^- = NO + H_2O$	0.983
$AuBr_2^- + e^- = Au + 2\,Br^-$	0.959
$NO_3^- + 4\,H^+ + 3\,e^- = NO + 2\,H_2O$	0.957
$ClO_2(aq) + e^- = ClO_2^-$	0.954
$Pd^{2+} + 2\,e^- = Pd$	0.951
$NO_3^- + 3\,H^+ + 2\,e^- = HNO + H_2O$	0.934

Source: data compiled by J.S. Park from Petr Vanysek, *Handbook of Physics and Chemistry, 69th Edition*, CRC Press, Boca Raton, Florida, (1988).

Table 309. STANDARD ELECTROMOTIVE FORCE POTENTIALS

(SHEET 5 OF 18)

Reaction	Reduction Potential E°, (V)
$2\ Hg^{2+} + 2\ e^- = Hg_2^{2+}$	0.920
$HO_2^- + H_2O + 2\ e^- = 3\ OH^-$	0.878
$N_2O_4 + 2\ e^- = 2\ NO_2^-$	0.867
$[IrCl_6]^{2-} + e^- = [IrCl_6]^{3-}$	0.8665
$2\ HNO_2 + 4\ H^+ + 4\ e^- = H_2N_2O_2 + H_2O$	0.86
$SiO_2(quartz) + 4\ H^+ + 4\ e^- = Si + 2\ H_2O$	0.857
$AuBr_4^- + 3\ e^- = Au + 4\ Br^-$	0.854
$Hg^{2+} + 2\ e^- = Hg$	0.851
$OsO_4 + 8\ H^+ + 8\ e^- = Os + 4\ H_2O$	0.85
$ClO^- + H_2O + 2\ e^- = Cl^- + 2\ OH^-$	0.841
$2\ NO_3^- + 4\ H^+ + 2\ e^- = N_2O_4 + 2\ H_2O$	0.803
$Ag^+ + e^- = Ag$	0.7996
$Hg_2^{2+} + 2\ e^- = Hg$	0.7973
$TcO_4^- + 4\ H^+ + 3\ e^- = TcO_2 + 2\ H_2O$	0.782
$AgF + e^- = Ag + F^-$	0.779
$Fe^{3+} + e^- = Fe^{2+}$	0.771
$[IrCl_6]^{3-} + 3e = Ir + 6\ Cl^-$	0.77
$(CNS)_2 + 2\ e^- = 2\ CNS^-$	0.77
$ReO_4^- + 2\ H^+ + e^- = ReO_3 + H_2O$	0.768
$BrO^- + H_2O + 2\ e^- = Br^- + 2\ OH^-$	0.761
$2\ NO + H_2O + 2\ e^- = N_2O + 2\ OH^-$	0.76
$ClO_2^- + 2\ H_2O + 4\ e^- = Cl^- + 4\ OH^-$	0.76

Source: data compiled by J.S. Park from Petr Vanysek, *Handbook of Physics and Chemistry, 69th Edition*, CRC Press, Boca Raton, Florida, (1988).

Table 309. STANDARD ELECTROMOTIVE FORCE POTENTIALS
(SHEET 6 OF 18)

Reaction	Reduction Potential E°, (V)
$Rh^{3+} + 3\,e^- = Rh$	0.758
$[PtCl_4]^{2-} + 2\,e^- = Pt + 4\,Cl^-$	0.755
$Ag_2O_3 + H_2O + 2\,e^- = 2\,AgO + 2\,OH^-$	0.739
$H_3IO_6 + 2\,e^- = IO_3^- + 3\,OH^-$	0.7
p–benzoquinone + $2\,H^+ + 2\,e^-$ = hydroquinone	0.6992
$O_2 + 2\,H^+ + 2\,e^- = H_2O_2$	0.695
$[PtCl_6]^{2-} + 2\,e^- = [PtCl_4]^{2-} + 2\,Cl^-$	0.68
Sb_2O_5(senarmontite) $+ 4\,H^+ + 4\,e^- = Sb_2O_3 + 2\,H_2O$	0.671
$ClO_2^- + H_2O + 2\,e^- = ClO^- + 2\,OH^-$	0.66
$Ag_2SO_4 + 2\,e^- = 2\,Ag + SO_4^{2-}$	0.654
Sb_2O_5(valentinite) $+ 4\,H^+ + 4\,e^- = Sb_2O_3 + 2\,H_2O$	0.649
$Ag(ac) + e^- = Ag + (ac)^-$	0.643
$Hg_2HPO_4 + 2\,e^- = 2\,Hg + HPO_4^{2-}$	0.6359
$ClO_3^- + 3\,H_2O + 6\,e^- = Cl^- + 6\,OH^-$	0.62
$Hg_2SO_4 + 2\,e^- = 2\,Hg + SO_4^{2-}$	0.6125
$UO_2^+ + 4\,H^+ + e^- = U^{4+} + 2\,H_2O$	0.612
$BrO_3^- + 3\,H_2O + 6\,e^- = Br^- + 6\,OH^-$	0.61
$2\,AgO + H_2O + 2\,e^- = Ag_2O + 2OH^-$	0.607
$MnO_4^{2-} + 2\,H_2O + 2\,e^- = MnO_2 + 4\,OH^-$	0.60
$Rh^+ + e^- = Rh$	0.600
$Rh^{2+} + 2\,e^- = Rh$	0.600
$MnO_4^- + 2\,H_2O + 3\,e^- = MnO_2 + 4\,OH^-$	0.595

Source: data compiled by J.S. Park from Petr Vanysek, *Handbook of Physics and Chemistry, 69th Edition*, CRC Press, Boca Raton, Florida, (1988).

Table 309. STANDARD ELECTROMOTIVE FORCE POTENTIALS
(SHEET 7 OF 18)

Reaction	Reduction Potential E°, (V)
$TeO_2 + 4\ H^+ + 4\ e^- = Te + 2\ H_2O$	0.593
$[PdCl_4]^{2-} + 2\ e^- = Pd + 4\ Cl^-$	0.591
$RuO_4^- + e^- = RuO_4^{2-}$	0.59
$Sb_2O_5 + 6\ H^+ + 4\ e^- = 2\ SbO^+ + 3\ H_2O$	0.581
$Te^{4+} + 4\ e^- = Te$	0.568
$AgNO_2 + e^- = Ag + NO_2^-$	0.564
$S_2O_6^{2-} + 4\ H^+ + 2\ e^- = 2\ H_2SO_3$	0.564
$H_3AsO_4 + 2\ H^+ + 2\ e^- = HAsO_2 + 2\ H_2O$	0.560
$MnO_4^- + e^- = MnO_4^{2-}$	0.558
$AgBrO_3 + e^- = Ag + BrO_3^-$	0.546
$I_3^- + 2\ e^- = 3\ I^-$	0.536
$I_2 + 2\ e^- = 2\ I^-$	0.5355
$Cu^+ + e^- = Cu$	0.521
$Hg_2(ac)_2 + 2\ e^- = 2\ Hg + 2\ (ac)^-$	0.51163
$ReO_4^- + 4\ H^+ + 3\ e^- = ReO_2 + 2\ H_2O$	0.510
$NiO_2 + 2\ H_2O + 2\ e^- = Ni(OH)_2 + 2\ OH^-$	0.490
$IO^- + H_2O + 2\ e^- = I^- + 2\ OH^-$	0.485
$TeO_4^- + 8\ H^+ + 7\ e^- = Te + 4\ H_2O$	0.472
$Ag_2CO_3 + 2\ e^- = 2\ Ag + CO_3^{2-}$	0.47
$Ag_2WO_4 + 2\ e^- = 2\ Ag + WO_4^{2-}$	0.4660
$Ag_2C_2O_4 + 2\ e^- = 2\ Ag + C_2O_4^{2-}$	0.4647
$Ag_2MoO_4 + 2\ e^- = 2\ Ag + MoO_4^{2-}$	0.4573

Source: data compiled by J.S. Park from Petr Vanysek, *Handbook of Physics and Chemistry, 69th Edition*, CRC Press, Boca Raton, Florida, (1988).

Table 309. STANDARD ELECTROMOTIVE FORCE POTENTIALS
(SHEET 8 OF 18)

Reaction	Reduction Potential E°, (V)
$Ru^{2+} + 2\ e^- = Ru$	0.455
$H_2SO_3 + 4\ H^+ + 4\ e^- = S + 3\ H_2O$	0.449
$Ag_2CrO_4 + 2\ e^- = 2\ Ag + CrO_4^{\ 2-}$	0.4470
$[RhCl_6]^{3-} + 3\ e^- = Rh + 6\ Cl^-$	0.431
$AgOCN + e^- = Ag + OCN^-$	0.41
$O_2 + H_2O + 4\ e^- = 4\ OH^-$	0.401
$Tc^{2+} + 2\ e^- = Tc$	0.400
$(ferricinium)^+ + e^- = ferrocene$	0.400
$(CN)_2 + 2\ H^+ + 2\ e^- = 2\ HCN$	0.373
$ReO_4^{\ -} + 8\ H^+ + 7\ e^- = Re + 4\ H_2O$	0.368
$Ag_2SeO_3 + 2\ e^- = 2\ Ag + SeO_3^{\ 2-}$	0.3629
$ClO_4^{\ -} + H_2O + 2\ e^- = ClO_3^{\ -} + 2OH^-$	0.36
$[Fe(CN)_6]^{3-} + e^- = [Fe(CN)_6]^{4-}$	0.358
$AgIO_3 + e^- = Ag + IO_3^{\ -}$	0.354
$Cu^{2+} + 2\ e^- = Cu$	0.3419
$VO^{2+} + 2\ H^+ + e^- = V^{3+} + H_2O$	0.337
Calomel electrode, 0.1 mol/l KCl	0.3337
$2\ HCNO + 2\ H^+ + 2\ e^- = (CN)_2 + 2\ H_2O$	0.330
$ClO_3^{\ -} + H_2O + 2\ e^- = ClO_2^{\ -} + 2\ OH^-$	0.33
$UO_2^{\ 2+} + 4\ H^+ + 2\ e^- = U^{4+} + 2\ H_2O$	0.327
$BiO^+ + 2\ H^+ + 3\ e^- = Bi + H_2O$	0.320
$Re^{3+} + 3\ e^- = Re$	0.300

Source: data compiled by J.S. Park from Petr Vanysek, *Handbook of Physics and Chemistry, 69th Edition*, CRC Press, Boca Raton, Florida, (1988).

Table 309. STANDARD ELECTROMOTIVE FORCE POTENTIALS
(SHEET 9 OF 18)

Reaction	Reduction Potential E°, (V)
Calomel electrode, 1 mol/l KCl (NCE)	0.2801
Calomel electrode, molal KCl	0.2800
$Hg_2Cl_2 + 2\,e^- = 2\,Hg^+ + 2\,Cl^-$	0.26808
$IO_3^- + 3\,H_2O + 6\,e^- = I^- + 6\,OH^-$	0.26
$ReO_2 + 4\,H^+ + 4\,e^- = Re + 2\,H_2O$	0.2513
$Ru^{3+} + e^- = Ru^{2+}$	0.2487
$HAsO_2 + 3\,H^+ + 3\,e^- = As + 2\,H_2O$	0.248
$PbO_2 + H_2O + 2\,e^- = PbO + 2\,OH^-$	0.247
Calomel electrode, saturated KCl	0.2412
$Ge^{2+} + 2\,e^- = Ge$	0.24
Calomel electrode, saturated NaCl (SSCE)	0.2360
$As_2O_3 + 6\,H^+ + 6\,e^- = 2\,As + 3\,H_2O$	0.234
$AgCl + e^- = Ag + Cl^-$	0.22233
$SbO^+ + 2\,H^+ + 3\,e^- = Sb + H_2O$	0.212
$SO_4^{2-} + 4\,H^+ + 2\,e^- = H_2SO_3 + H_2O$	0.172
$Co(OH)_3 + e^- = Co(OH)_2 + OH^-$	0.17
$Bi(Cl)_4^- + 3\,e^- = Bi + 4\,Cl^-$	0.16
$BiOCl + 2\,H^+ + 3\,e^- = Bi + Cl^- + H_2O$	0.1583
$Cu^{2+} + e^- = Cu^+$	0.153
$Sb_2O_3 + 6\,H^+ + 6\,e^- = 2\,Sb + 3\,H_2O$	0.152
$Sn^{4+} + 2\,e^- = Sn^{2+}$	0.151
$2\,NO_2^- + 3\,H_2O + 4\,e^- = N_2O + 6\,OH^-$	0.15
$Mn(OH)_3 + e^- = Mn(OH)_2 + OH^-$	0.15
$IO_3^- + 2\,H_2O + 4\,e^- = IO^- + 4\,OH^-$	0.15

Source: data compiled by J.S. Park from Petr Vanysek, *Handbook of Physics and Chemistry, 69th Edition*, CRC Press, Boca Raton, Florida, (1988).

Table 309. STANDARD ELECTROMOTIVE FORCE POTENTIALS

(SHEET 10 OF 18)

Reaction	Reduction Potential E°, (V)
$Ag_4[Fe(CN)_6] + 4\ e^- = 4\ Ag + [Fe(CN)_6]^{4-}$	0.1478
$Np^{4+} + e^- = Np^{3+}$	0.147
$S + 2\ H^+ + 2\ e^- = H_2S(aq)$	0.142
$Pt(OH)_2 + 2\ e^- = Pt + 2\ OH^-$	0.14
$Hg_2Br_2 + 2\ e^- = 2\ Hg + 2\ Br^-$	0.13923
$Ge^{4+} + 4\ e^- = Ge$	0.124
$Hg_2O + H_2O + 2\ e^- = 2\ Hg + 2\ OH^-$	0.123
$[Co(NH_3)_6]^{3+} + e^- = [Co(NH_3)_6]^{2+}$	0.108
$2\ NO + 2\ e^- = N_2O_2{}^{2-}$	0.10
$Ir_2O_3 + 3\ H_2O + 6\ e^- = 2\ Ir + 6\ OH^-$	0.098
$HgO + H_2O + 2\ e^- = Hg + 2\ OH^-$	0.0977
$N_2 + 2\ H_2O + 6\ H^+ + 6\ e^- = 2\ NH_4OH$	0.092
$AgSCN + e^- = Ag + SCN^-$	0.08951
$S_4O_6{}^{2-} + 2\ e^- = 2\ S_2O_3{}^{2-}$	0.08
$AgBr + e^- = Ag + Br^-$	0.07133
$Pd(OH)_2 + 2\ e^- = Pd + 2\ OH^-$	0.07
$UO_2{}^{2+} + e^- = UO_2{}^+$	0.062
$SeO_4{}^{2-} + H_2O + 2\ e^- = SeO_3{}^{2-} + 2\ OH^-$	0.05
$Tl_2O_3 + 3\ H_2O + 4\ e^- = 2\ Tl^{2+} + 6\ OH^-$	0.02
$NO_3{}^- + H_2O + 2\ e^- = NO_2{}^- + 2\ OH^-$	0.01
$Ge^{4+} + 2\ e^- = Ge^{2+}$	0.00
$CuI_2{}^- + e^- = Cu + 2\ I^-$	0.00

Source: data compiled by J.S. Park from Petr Vanysek, *Handbook of Physics and Chemistry, 69th Edition*, CRC Press, Boca Raton, Florida, (1988).

Table 309. STANDARD ELECTROMOTIVE FORCE POTENTIALS
(SHEET 11 OF 18)

Reaction	Reduction Potential E°, (V)
$2\,H^+ + 2\,e^- = H_2$	0.00000
$AgCN + e^- = Ag + CN^-$	–0.017
$2\,WO_3 + 2\,H^+ + 2\,e^- = W_2O_5 + H_2O$	–0.029
$W_2O_5 + 2\,H^+ + 2\,e^- = 2\,WO_2 + H_2O$	–0.031
$D^+ + e^- = 1/2\,D_2$	–0.034
$Ag_2S + 2\,H^+ + 2\,e^- = 2\,Ag + H_2S$	–0.0366
$Fe^{3+} + 3\,e^- = Fe$	–0.037
$Hg_2I_2 + 2\,e^- = 2\,Hg + 2\,I^-$	–0.0405
$2\,D^+ + 2\,e^- = D_2$	–0.044
$Tl(OH)_3 + 2\,e^- = TlOH + 2\,OH^-$	–0.05
$TiOH^{3+} + H^+ + e^- = Ti^{3+} + H_2O$	–0.055
$2\,H_2SO_3 + H^+ + 2\,e^- = HS_2O_4^- + 2\,H_2O$	–0.056
$P(white) + 3\,H^+ + 3\,e^- = PH_3(g)$	–0.063
$O_2^- + H_2O + 2\,e^- = HO_2^- + OH^-$	–0.076
$2\,Cu(OH)_2 + 2\,e^- = Cu_2O + 2\,OH^- + H_2O$	–0.080
$WO_3 + 6\,H^+ + 6\,e^- = W + 3\,H_2O$	–0.090
$P(red) + 3\,H^+ + 3\,e^- = PH_3(g)$	–0.111
$GeO_2 + 2\,H^+ + 2\,e^- = GeO + H_2O$	–0.118
$WO_2 + 4\,H^+ + 4\,e^- = W + 2\,H_2O$	–0.119
$Pb^{2+} + 2\,e^- = Pb(Hg)$	–0.1205
$Pb^{2+} + 2\,e^- = Pb$	–0.1262
$CrO_4^{2-} + 4\,H_2O + 3\,e^- = Cr(OH)_3 + 5\,OH^-$	–0.13

Source: data compiled by J.S. Park from Petr Vanysek, *Handbook of Physics and Chemistry, 69th Edition*, CRC Press, Boca Raton, Florida, (1988).

Table 309. STANDARD ELECTROMOTIVE FORCE POTENTIALS

(SHEET 12 OF 18)

Reaction	Reduction Potential E°, (V)
$Sn^{2+} + 2\ e^- = Sn$	–0.1375
$In^+ + e^- = In$	–0.14
$O_2 + 2\ H_2O + 2\ e^- = H_2O_2 + 2\ OH^-$	–0.146
$AgI + e^- = Ag + I^-$	–0.15224
$2\ NO_2^- + 2\ H_2O + 4\ e^- = N_2O_2^{2-} + 4\ OH^-$	–0.18
$H_2GeO_3 + 4\ H^+ + 4\ e^- = Ge + 3\ H_2O$	–0.182
$CO_2 + 2\ H^+ + 2\ e^- = HCOOH$	–0.199
$Mo^{3+} + 3\ e^- = Mo$	–0.200
$2\ SO_2^{2-} + 4\ H^+ + 2\ e^- = S_2O_6^{2-} + H_2O$	–0.22
$Cu(OH)_2 + 2\ e^- = Cu + 2\ OH^-$	–0.222
$CdSO_4 + 2\ e^- = Cd + SO_4^{2-}$	–0.246
$V(OH)_4^+ + 4\ H^+ + 5\ e^- = V + 4\ H_2O$	–0.254
$V^{3+} + e^- = V^{2+}$	–0.255
$Ni^{2+} + 2\ e^- = Ni$	–0.257
$PbCl_2 + 2\ e^- = Pb + 2\ Cl^-$	–0.2675
$H_3PO_4 + 2\ H^+ + 2\ e^- = H_3PO_3 + H_2O$	–0.276
$Co^{2+} + 2\ e^- = Co$	–0.28
$PbBr_2 + 2\ e^- = Pb + 2\ Br^-$	–0.284
$Tl^+ + e^- = Tl(Hg)$	–0.3338
$Tl^+ + e^- = Tl$	–0.336
$In^{3+} + 3\ e^- = In$	–0.3382
$TlOH + e^- = Tl + OH^-$	–0.34

Source: data compiled by J.S. Park from Petr Vanysek, *Handbook of Physics and Chemistry, 69th Edition*, CRC Press, Boca Raton, Florida, (1988).

Table 309. STANDARD ELECTROMOTIVE FORCE POTENTIALS

(SHEET 13 OF 18)

Reaction	Reduction Potential E°, (V)
$PbF_2 + 2\ e^- = Pb + 2\ F^-$	–0.3444
$PbSO_4 + 2\ e^- = Pb(Hg) + SO_4^{2-}$	–0.3505
$Cd^{2+} + 2\ e^- = Cd(Hg)$	–0.3521
$PbSO_4 + 2\ e^- = Pb + SO_4^{2-}$	–0.3588
$Cu_2O + H_2O + 2e^- = 2\ Cu + 2\ OH^-$	–0.360
$Eu^{3+} + e^- = Eu^{2+}$	–0.36
$PbI_2 + 2\ e^- = Pb + 2\ I^-$	–0.365
$SeO_3^{2-} + 3\ H_2O + 4\ e^- = Se + 6\ OH^-$	–0.366
$Ti^{3+} + e^- = Ti^{2+}$	–0.368
$Se + 2\ H^+ + 2\ e^- = H_2Se(aq)$	–0.399
$In^{2+} + e^- = In^+$	–0.40
$Cd^{2+} + e^- = Cd$	–0.4030
$Cr^{3+} + e^- = Cr^{2+}$	–0.407
$2\ S + 2\ e^- = S_2^{2-}$	–0.42836
$Tl_2SO_4 + 2\ e^- = Tl + SO_4^{2-}$	–0.4360
$In^{3+} + 2\ e^- = In^+$	–0.443
$Fe^{2+} + 2\ e^- = Fe$	–0.447
$H_3PO_3 + 3\ H^+ + 3\ e^- = P + 3\ H_2O$	–0.454
$Bi_2O_3 + 3\ H_2O + 6\ e^- = 2\ Bi + 6\ OH^-$	–0.46
$NO_2^- + H_2O + e^- = NO + 2\ OH^-$	–0.46
$PbHPO_4 + 2\ e^- = Pb + HPO_4^{2-}$	–0.465
$S + 2\ e^- = S^{2-}$	–0.47627

Source: data compiled by J.S. Park from Petr Vanysek, *Handbook of Physics and Chemistry, 69th Edition*, CRC Press, Boca Raton, Florida, (1988).

Table 309. STANDARD ELECTROMOTIVE FORCE POTENTIALS (SHEET 14 OF 18)

Reaction	Reduction Potential E°, (V)
$S + H_2O + 2\,e^- = HS^- + OH^-$	–0.478
$In^{3+} + e^- = In^{2+}$	–0.49
$H_3PO_3 + 2\,H^+ + 2\,e^- = H_3PO_2 + H_2O$	–0.499
$TiO_2 + 4\,H^+ + 2\,e^- = Ti^{2+} + 2\,H_2O$	–0.502
$H_3PO_2 + H^+ + e^- = P + 2\,H_2O$	–0.508
$Sb + 3\,H^+ + 3\,e^- = SbH_3$	–0.510
$HPbO_2^- + H_2O + 2\,e^- = Pb + 3\,OH^-$	–0.537
$TlCl + e^- = Tl + Cl^-$	–0.5568
$Ga^{3+} + 3\,e^- = Ga$	–0.560
$Fe(OH)_3 + e^- = Fe(OH)_2 + OH^-$	–0.56
$TeO_3^{2-} + 3\,H_2O + 4\,e^- = Te + 6OH^-$	–0.57
$2\,SO_3^- + 3\,H_2O + 4\,e^- = S_2O_3^- + 6\,OH^-$	–0.571
$PbO + H_2O + 2\,e^- = Pb + 2\,OH^-$	–0.580
$ReO_2^- + 4\,H_2O + 7\,e^- = Re + 8\,OH^-$	–0.584
$SbO_3^- + H_2O + 2\,e^- = SbO_2^- + 2\,OH^-$	–0.59
$U^{4+} + e^- = U^{3+}$	–0.607
$As + 3\,H^+ + 3\,e^- = AsH_3$	–0.608
$Nb_2O_5 + 10\,H^+ + 3\,e^- = 2\,Nb + 5\,H_2O$	–0.644
$TlBr + e^- = Tl + Br^-$	–0.658
$SbO_2^- + 2\,H_2O + 3\,e^- = Sb + 4\,OH^-$	–0.66
$AsO_2^- + 2\,H_2O + 3\,e^- = As + 4\,OH^-$	–0.68
$Ag_2S + 2\,e^- = 2\,Ag + S^{2-}$	–0.691

Source: data compiled by J.S. Park from Petr Vanysek, *Handbook of Physics and Chemistry, 69th Edition*, CRC Press, Boca Raton, Florida, (1988).

Table 309. STANDARD ELECTROMOTIVE FORCE POTENTIALS
(SHEET 15 OF 18)

Reaction	Reduction Potential E°, (V)
$AsO_4^{3-} + 2 H_2O + 2 e^- = AsO_2^- + 4 OH^-$	–0.71
$Ni(OH)_2 + 2 e^- = Ni + 2 OH^-$	–0.72
$Co(OH)_2 + 2 e^- = Co + 2 OH^-$	–0.73
$H_2SeO_3 + 4 H^+ + 4 e^- = Se + 3 H_2O$	–0.74
$Cr^{3+} + 3 e^- = Cr$	–0.744
$Ta_2O_5 + 10 H^+ + 4 e^- = 2 Ta + 5 H_2O$	–0.75
$TlI + e^- = Tl + I^-$	–0.752
$Zn^{2+} + 2 e^- = Zn$	–0.7618
$Zn^{2+} + 2 e^- = Zn(Hg)$	–0.7628
$Te + 2 H^+ + 2 e^- = H_2Te$	–0.793
$ZnSO_4 7H_2O + 2 e^- = Zn(Hg) + SO_4^{2-}$ (Sat'd $ZnSO_4$)	–0.7993
$Cd(OH)_2 + 2 e^- = Cd(Hg) + 2 OH^-$	–0.809
$2 H_2O + 2 e^- = H_2 + 2 OH^-$	–0.8277
$2 NO_3^- + 2 H_2O + 2 e^- = N_2O_4 + 4 OH^-$	–0.85
$H_3BO_3 + 3 H^+ + 3 e^- = B + 3 H_2O$	–0.8698
$P + 3 H_2O + 3 e^- = PH_3(g) + 3 OH^-$	–0.87
$HSnO_2^- + H_2O + 3 e^- = Sn + 3 OH^-$	–0.909
$Cr^{2+} + 2 e^- = Cr$	–0.913
$Se + 2 e^- = Se^{2-}$	–0.924
$SO_4^{2-} + H_2O + 2 e^- = SO_3^{2-} + 2 OH^-$	–0.93
$Sn(OH)_6^{2-} + 2 e^- = HSnO_2^- + 3 OH^- + H_2O$	–0.93
$NpO_2 + H_2O + H^+ + e^- = Np(OH)_3$	–0.962

Source: data compiled by J.S. Park from Petr Vanysek, *Handbook of Physics and Chemistry, 69th Edition*, CRC Press, Boca Raton, Florida, (1988).

Table 309. Standard Electromotive Force Potentials
(Sheet 16 of 18)

Reaction	Reduction Potential E°, (V)
$PO_4^{3-} + 2\,H_2O + 2\,e^- = HPO_3^{2-} + 3\,OH^-$	–1.05
$Nb^{3+} + 3\,e^- = Nb$	–1.099
$2\,SO_3^{2-} + 2\,H_2O + 2\,e^- = S_2O_4^{2-} + 4\,OH^-$	–1.12
$Te + 2\,e^- = Te^{2-}$	–1.143
$V^{2+} + 2\,e^- = V$	–1.175
$Mn^{2+} + 2\,e^- = Mn$	–1.185
$CrO_2^- + 2\,H_2O + 3\,e^- = Cr + 4\,OH^-$	–1.2
$ZnO_2^- + 2\,H_2O + 2\,e^- = Zn + 4\,OH^-$	–1.215
$H_2GaO_3^- + H_2O + 3\,e^- = Ga + 4\,OH^-$	–1.219
$H_2BO_3^- + 5\,H_2O + 8\,e^- = BH_4^- + 8\,OH^-$	–1.24
$SiF_6^{2-} + 4\,e^- = Si + 6\,F^-$	–1.24
$Ce^{3+} + 3\,e^- = Ce(Hg)$	–1.4373
$UO_2^{2+} + 4\,H^+ + 6\,e^- = U + 2\,H_2O$	–1.444
$Cr(OH)_3 + 3\,e^- = Cr + 3\,OH^-$	–1.48
$HfO_2 + 4\,H^+ + 4\,e^- = Hf + 2\,H_2O$	–1.505
$ZrO_2 + 4\,H^+ + 4\,e^- = Zr + 2\,H_2O$	–1.553
$Mn(OH)_2 + 2\,e^- = Mn + 2\,OH^-$	–1.56
$Ba^{2+} + 2\,e^- = Ba(Hg)$	–1.570
$Ti^{2+} + 2\,e^- = Ti$	–1.63
$HPO_3^{2-} + 2\,H_2O + 2\,e^- = H_2PO_2^- + 3\,OH^-$	–1.65
$Al^{3+} + 3\,e^- = Al$	–1.662
$SiO_3^- + H_2O + 4\,e^- = Si + 6\,OH^-$	–1.697

Source: data compiled by J.S. Park from Petr Vanysek, *Handbook of Physics and Chemistry, 69th Edition*, CRC Press, Boca Raton, Florida, (1988).

Table 309. STANDARD ELECTROMOTIVE FORCE POTENTIALS
(SHEET 17 OF 18)

Reaction	Reduction Potential E°, (V)
$HPO_3^{2-} + 2\ H_2O + 3\ e^- = P + 5\ OH^-$	–1.71
$HfO^{2+} + 2\ H^+ + 4\ e^- = Hf + H_2O$	–1.724
$ThO_2 + 4\ H^+ + 4\ e^- = Th + 2\ H_2O$	–1.789
$H_2BO_3^- + H_2O + 3\ e^- = B + 4\ OH^-$	–1.79
$Sr^{2+} + 2\ e^- = Sr(Hg)$	–1.793
$U^{3+} + 3\ e^- = U$	–1.798
$H_2PO_2^- + e^- = P + 2\ OH^-$	–1.82
$Be^{2+} + 2\ e^- = Be$	–1.847
$Np^{3+} + 3\ e^- = Np$	–1.856
$Th^{4+} + 4\ e^- = Th$	–1.899
$Pu^{3+} + 3\ e^- = Pu$	–2.031
$AlF_6^{3-} + 3\ e^- = Al + 6\ F^-$	–2.069
$Sc^{3+} + 3\ e^- = Sc$	–2.077
$H_2 + 2\ e^- = 2\ H^-$	–2.23
$H_2AlO_3^- + H_2O + 3\ e^- = Al + 4\ OH^-$	–2.33
$ZrO(OH)_2 + H_2O + 4\ e^- = Zr + 4\ OH^-$	–2.36
$Mg^{2+} + 2\ e^- = Mg$	–2.372
$Y^{3+} + 3\ e^- = Y$	–2.372
$Eu^{3+} + 3\ e^- = Eu$	–2.407
$Nd^{3+} + 3\ e^- = Nd$	–2.431
$Th(OH)_4 + 4\ e^- = Th + 4\ OH^-$	–2.48
$Ce^{3+} + 3\ e^- = Ce$	–2.483

Source: data compiled by J.S. Park from Petr Vanysek, *Handbook of Physics and Chemistry, 69th Edition*, CRC Press, Boca Raton, Florida, (1988).

Table 309. STANDARD ELECTROMOTIVE FORCE POTENTIALS

(SHEET 18 OF 18)

Reaction	Reduction Potential E°, (V)
$HfO(OH)_2 + H_2O + 4\ e^- = Hf + 4\ OH^-$	–2.50
$La^{3+} + 3\ e^- = La$	–2.522
$Be_2O_3^{2-} + 3\ H_2O + 4\ e^- = 2\ Be + 6\ OH^-$	–2.63
$Mg(OH)_2 + 2\ e^- = Mg + 2\ OH^-$	–2.690
$Mg^+ + e^- = Mg$	–2.70
$Na^+ + e^- = Na$	–2.71
$Ca^{2+} + 2\ e^- = Ca$	–2.868
$Sr(OH)_2 + 2\ e^- = Sr + 2\ OH^-$	–2.88
$Sr^{2+} + 2\ e^- = Sr$	–2.89
$La(OH)_3 + 3\ e^- = La + 3\ OH^-$	–2.90
$Ba^{2+} + 2\ e^- = Ba$	–2.912
$Cs^+ + e^- = Cs$	–2.92
$K^+ + e^- = K$	–2.931
$Rb^+ + e^- = Rb$	–2.98
$Ba(OH)_3 + 2\ e^- = Ba + 2\ OH^-$	–2.99
$Ca(OH)_3 + 2\ e^- = Ca + 2\ OH^-$	–3.02
$Li^+ + e^- = Li$	–3.0401
$3\ N_2 + 2\ H^+ + 2\ e^- = 2\ NH_3$	–3.09
$Eu^{2+} + 2\ e^- = Eu$	–3.395
$Ca^+ + e^- = Ca$	–3.80
$Sr^+ + e^- = Sr$	–4.10

Source: data compiled by J.S. Park from Petr Vanysek, *Handbook of Physics and Chemistry, 69th Edition*, CRC Press, Boca Raton, Florida, (1988).

Table 310. GALVANIC SERIES OF METALS

Metal	Potential, volts (V)
	Anodic or Corroded End
Lithium	-3.04
Rubidium	-2.93
Potassium	-2.92
Barium	-2.90
Strontium	-2.89
Calcium	-2.8
Sodium	-2.71
Magnesium	-2.37
Beryllium	-1.7
Aluminum	-1.7
Manganese	-1.04
Zinc	-0.76
Chromium	-0.6
Cadmium	-0.4
Titanium	-0.33
Cobalt	-0.28
Nickel	-0.23
Tin	-0.14
Lead	-0.126
Hydrogen	0.00
Copper	0.52
Silver	0.80
Mercury	0.85
Palladium	1.0
Platinum	1.2
Gold	1.5
	Cathodic or Noble Metal End

Source: data compiled by J.S. Park from R. E. Bolz and G. L. Tuve, Eds., *CRC Handbook of Tables for Applied Engineering Science, 2nd edition*, CRC Press, Inc., Boca Ranton, Florida, (1973).

Table 311. GALVANIC SERIES OF METALS IN SEA WATER

(SHEET 1 OF 2)

	Metal
Active End (-)	Magnesium Magnesium Alloys Zinc Galvanized Steel
	Aluminum 1100
	Aluminum 6053 Alcad
	Cadmium
	Aluminum 2024 (4.5 Cu, 1.5 Mg, 0.6 Mn)
	Mild Steel Wrought Iron Cast Iron
	13% Chromium Stainless Steel Type 410 (Active) 18-8 Stainless Steel Type 304 (Active) 18-12-3 Stainless Steel Type 316 (Active)
	Lead-Tin Solders Lead Tin
	Muntz Metal Manganese Bronze Naval Brass
	Nickel (Active) 76 Ni-16 Cr-7 Fe alloy (Active)
	60 Ni-30 Mo-6 Fe-1 Mn

Source: data compiled by J.Park from Standard Guide for Development and Use of a Galvanic Series for Predicting Galvanic Corrosion Performance, G 82, *Annual Book of ASTM Standards*, American Society for Testing and Materials, (1989).

Table 311. GALVANIC SERIES OF METALS IN SEA WATER

(SHEET 2 OF 2)

	Metal
	Yellow Brass Admirality Brass Aluminum Brass Red Brass Copper Silicon Bronze
	70:30 Cupro Nickel G-Bronze M-Bronze Silver Solder Nickel (Passive) 76 Ni-16 Cr-7 Fe Alloy (Passive) 67 Ni-33 Cu Alloy (Monel)
	13% Chromium Stainless Steel Type 410 (Passive) Titanium
	18-8 Stainless Steel Type 304 (Passive) 18-12-3 Stainless Steel Type 316 (Passive) Silver
Noble or Passive End (+)	Graphite Gold Platinum

Source: data compiled by J.Park from Standard Guide for Development and Use of a Galvanic Series for Predicting Galvanic Corrosion Performance, G 82, *Annual Book of ASTM Standards*, American Society for Testing and Materials, (1989).

Table 312. Corrosion Rate of Metals in Acidic Solutions

Metal	Corrosive Environment		
	Sulfuric, 5% (Non-oxidizing)	**Acetic, 5% (Non-oxidizing)**	**Nitric, 5% (Oxidizing)**
Aluminum	8-100	0.5-5	15-80
Copper alloys	2-50*	2-15*	150-1500
Gold	<0.1	<0.1	<0.1
Iron	15-400*	10-400	1000-10000
Lead	0-2	10-150*	100-6000
Molybdenum	0-0.2	<0.1	high
Nickel alloys	2-35*	2-10*	0.1-1500
Platinum	<0.1	<0.1	<0.1
Silicon iron	0-5	0-0.2	0-20
Silver	0-1	<0.1	high
Stainless steel	0-100**	0-0.5	0-2
Tantalum	<0.1	<0.1	<0.1
Tin	2-500*	2-500*	100-400
Titanium	10-100	<0.1	0.1-1
Zinc	high	600-800	high
Zirconium	<0.5	<0.1	<0.1

* Aeration leads to the higher rates in the range.

** Aeration leads to passivity, scarcity of dissolved air to activity.

Corrosion Rate Ranges Expressed in Mils Penetration per Year (1 Mil = 0.001 in)

Note: The corrosion-rate ranges for the solutions are based on temperature up to 212 °F.

Source: data compiled by J.S. Park from R. E. Bolz and G. L. Tuve, CRC Handbook of Tables for Applied Engineering Science, 2nd edition, CRC Press, Inc., Boca Raton, Florida, (1973).

Table 313. Corrosion Rate of Metals in Neutral and Alkaline Solutions

Metal	Corrosive Environment		
	Sodium Hydroxide, 5%	Fresh Water	Sea Water
Aluminum	13000	0.1	1-50
Copper alloys	2-5	0-1	0.2-15*
Gold	<0.1	<0.1	<0.1
Iron	0-0.2	0.1-10*	0.1-10*
Lead	5-500*	0.1-2	0.2-15
Molybdenum	<0.1	<0.1	<0.1
Nickel alloys	0-0.2	0-0.2	0-1
Platinum	<0.1	<0.1	<0.1
Silicon iron	0-10	0-0.2	0-3
Silver	<0.1	<0.1	<0.1
Stainless steel	0-0.2	0-0.2	0-200**
Tantalum	<1	<0.1	<0.1
Tin	5-20	0-0.5	0.1
Titanium	<0.2	<0.1	<0.1
Zinc	15-200	0.5-10	0.5-10*
Zirconium	<0.1	<0.1	<0.1

* Aeration leads to the higher rates in the range.

** Aeration leads to passivity, scarcity of dissolved air to activity.

Corrosion Rate Ranges Expressed in Mils Penetration per Year (1 Mil = 0.001 in)

Note: The corrosion-rate ranges for the solutions are based on temperature up to 212 °F.

Source: data compiled by J.S. Park from R. E. Bolz and G. L. Tuve, *CRC Handbook of Tables for Applied Engineering Science, 2nd edition*, CRC Press, Inc., Boca Raton, Florida, (1973).

Table 314. CORROSION RATE OF METALS IN AIR

Metal	Normal Outdoor Air (Urban Exposure)
Aluminum	0-0.5
Copper alloys	0-0.2
Gold	<0.1
Iron	1-8
Lead	0-0.2
Molybdenum	<0.1
Nickel alloys	0-0.2
Platinum	<0.1
Silicon iron	0-0.2
Silver	<0.1
Stainless steel	0-0.2
Tantalum	<0.1
Tin	0-0.2
Titanium	<0.1
Zinc	0-0.5
Zirconium	<0.1

Corrosion Rate Ranges Expressed in Mils Penetration per Year (1 Mil = 0.001 in)

Source: data compiled by J.S. Park from R. E. Bolz and G. L. Tuve, *CRC Handbook of Tables for Applied Engineering Science, 2nd edition*, CRC Press, Inc., Boca Raton, Florida, (1973).

Table 315. CORROSION RATES OF 1020 STEEL AT 70°F *

(SHEET 1 OF 8)

Corrosive Medium	Corrosion Rate* in 10% Corrosive Medium (ipy)	Corrosion Rate** in 100% Corrosive Medium (ipy)
Acetaldehyde	<0.05	<0.002
Acetic Acid (Aerated)	>0.05	>0.05
Acetic Acid (Air Free)	>0.05	>0.05
Acetic Anhydride	—	>0.05
Acetoacetic Acid	>0.05	>0.05
Acetone	<0.05	<0.002
Acetylene	—	<0.002
Acrolein	<0.02	<0.02
Acrylonitril	—	<0.002
Alcohol (Ethyl)	<0.02	<0.002
Alcohol (Methyl)	<0.02	<0.002
Alcohol (Allyl)	—	<0.002
Alcohol (Amyl)	—	<0.02
Alcohol (Benzyl)	—	<0.002
Alcohol (Butyl)	—	<0.002
Alcohol (Cetyl)	—	<0.02
Alcohol (Isopropyl)	—	<0.002
Allylamine	<0.02 (30%)	<0.02
Allyl Chloride	—	<0.002
Allyl Sulfide	—	<0.02
Aluminum Acetate	>0.05	—
Aluminum Chloride	>0.05	<0.002
Aluminum Fluoride	<0.02	—
Aluminum Fluosilicate	—	>0.05
Aluminum Formate	<0.05	>0.05
Aluminum Hydroxide	<0.02	—
Aluminum Nitrate	>0.05	—
Aluminum Potassium Sulfate	>0.05	—

* 10% corrosive medium in 90% water. (Other % corrosive medium in parentheses.)
** Water-free, dry or maximum concentration of corrosive medium.
Source: data compiled by J.S. Park from Earl R. Parker, *Materials Data Book for Engineers and Scientists*, McGraw-Hill Book Company, New York, 1967.

Table 315. CORROSION RATES OF 1020 STEEL AT 70°F *

(SHEET 2 OF 8)

Corrosive Medium	Corrosion Rate* in 10% Corrosive Medium (ipy)	Corrosion Rate** in 100% Corrosive Medium (ipy)
Aluminum Sulfate	>0.05	—
Ammonia	<0.002	<0.002
Ammonium Acetate	—	<0.002
Ammonium Bicarbonate	<0.02	<0.002
Ammonium Bromide	>0.05	>0.05
Ammonium Carbonate	<0.02	<0.002
Ammonium Chloride	<0.05	<0.02
Ammonium Citrate	>0.05	<0.002
Ammonium Nitrate	<0.002	<0.02
Ammonium Sulfate	<0.02	—
Ammonium Sulfite	>0.05	—
Ammonium Thiocyanate	<0.02	—
Amyl Acetate	<0.002	<0.02
Amyl Chloride	>0.05	<0.02
Aniline	—	<0.002
Aniline Hydro-chloride	>0.05	>0.05
Anthracine	—	<0.02
Antimony Trichloride	>0.05	<0.05
Barium Carbonate	<0.02	<0.02
Barium Chloride	<0.02	<0.002
Barium Hydroxide	—	<0.02
Barium Nitrate	<0.02	<0.02
Barium Oxide	—	<0.002
Barium Peroxide	<0.05	<0.002
Benzaldehyde	>0.05	<0.002
Benzene	—	<0.02
Benzoic Acid	>0.05	>0.05
Boric Acid	<0.05	—

* 10% corrosive medium in 90% water. (Other % corrosive medium in parentheses.)
** Water-free, dry or maximum concentration of corrosive medium.
Source: data compiled by J.S. Park from Earl R. Parker, *Materials Data Book for Engineers and Scientists*, McGraw-Hill Book Company, New York, 1967.

Table 315. Corrosion Rates of 1020 Steel at 70°F *

(Sheet 3 of 8)

Corrosive Medium	Corrosion Rate* in 10% Corrosive Medium (ipy)	Corrosion Rate** in 100% Corrosive Medium (ipy)
Bromic Acid	>0.05	>0.05
Bromine (Dry)	—	<0.05
Bromine (Wet)	—	>0.05
Butyric Acid	<0.05	>0.05
Cadmium Chloride	>0.05	<0.002
Cadmium Sulfate	<0.02	<0.02
Calcium Acetate	<0.02	<0.05
Calcium Bicarbonate	<0.02	<0.02
Calcium Bromide	—	<0.05
Calcium Chlorate	<0.002	<0.02
Calcium Chloride	<0.002	<0.002
Calcium Hydroxide	<0.02	<0.02
Calcium Hypochlorite	<0.05	<0.02
Carbon Dioxide	—	<0.002
Carbon Monoxide	—	<0.002
Carbon Tetrachloride	—	<0.002
Carbon Acid (Air Free)	<0.02	<0.02
Chloroacetic Acid	>0.05	>0.05
Chlorine Gas	>0.05	<0.02
Chlorine Liquid	—	<0.02
Chloroform (Dry)	—	<0.002
Chromic Acid	>0.05	<0.002
Chromic Hydroxide	—	<0.02
Chromic Sulfates	>0.05	>0.05
Citric Acid	>0.05	<0.002
Copper Nitrate	>0.05	—
Copper Sulfate	>0.05	—
Diethylene Glycol	<0.002 (60%)	<0.002

* 10% corrosive medium in 90% water. (Other % corrosive medium in parentheses.)
** Water-free, dry or maximum concentration of corrosive medium.
Source: data compiled by J.S. Park from Earl R. Parker, *Materials Data Book for Engineers and Scientists*, McGraw-Hill Book Company, New York, 1967.

Table 315. CORROSION RATES OF 1020 STEEL AT 70°F *

(SHEET 4 OF 8)

Corrosive Medium	Corrosion Rate* in 10% Corrosive Medium (ipy)	Corrosion Rate** in 100% Corrosive Medium (ipy)
Ethyl Chloride	>0.05 (90%)	<0.002
Ethylene Glycol	<0.02	<0.002
Ethylene Oxide	—	<0.002
Fatty Acids	—	>0.05
Ferric Chloride	>0.05	<0.02
Ferric Nitrate	>0.05	—
Ferrous Chloride	>0.05	—
Ferrous Sulfate	>0.05	—
Fluorine	—	<0.002
Formaldehyde	<0.05 (40%)	<0.002
Formic Acid	>0.05	>0.05
Furfural	<0.02 (30%)	<0.02
Hydrazine	>0.05	>0.05
Hydrobromic Acid	>0.05	<0.02
Hydro-chloric Acid (Areated)	>0.05	—
Hydro-chloric Acid (Air Free)	>0.05	—
Hydrocyanic Acid	—	<0.002
Hydrofluoric Acid (Areated)	>0.05	<0.02
Hydrofluoric Acid (Air Free)	>0.05	<0.05
Hydrogen Chloride	>0.05 (90%)	<0.002
Hydrogen Fluoride	—	<0.002
Hydrogen Iodide	<0.05 (1%)	<0.02
Hydrogen Peroxide	>0.05 (20%)	—
Hydrogen Sulfide	<0.02	<0.02
Lactic Acid	>0.05	>0.05
Lead Acetate	>0.05 (20%)	<0.002
Lead Chromate	—	<0.02
Lead Nitrate	>0.05	<0.02

* 10% corrosive medium in 90% water. (Other % corrosive medium in parentheses.)
** Water-free, dry or maximum concentration of corrosive medium.
Source: data compiled by J.S. Park from Earl R. Parker, *Materials Data Book for Engineers and Scientists*, McGraw-Hill Book Company, New York, 1967.

Table 315. CORROSION RATES OF 1020 STEEL AT 70°F*

(SHEET 5 OF 8)

Corrosive Medium	Corrosion Rate* in 10% Corrosive Medium (ipy)	Corrosion Rate** in 100% Corrosive Medium (ipy)
Lead Sulfate	—	<0.02
Lithium Chloride	<0.02 (30%)	<0.002
Lithium Hydroxide	<0.02	<0.002
Magnesium Chloride	<0.02	<0.002
Magnesium Hydroxide	<0.02	<0.002
Magnesium Sulfate	<0.02	<0.02
Maleic Acid	>0.05	<0.002
Malic Acid	>0.05	—
Maganous Chloride	>0.05 (40%)	—
Mercuric Chloride	>0.05	—
Mercurous Nitrate	—	<0.02
Methallyl-amine	<0.02	<0.02
Methanol	<0.02	<0.002
Methyl Ethyl Ketone	<0.02	<0.002
Methyl Isobutyl Ketone	<0.02	<0.02
Methylamine	<0.02	<0.02
Methylene Chloride	—	<0.02
Monochloro-acetic Acid	>0.05	<0.002
Monorthanol-amine	<0.02	<0.02
Monoethal-amine	<0.02	<0.02
Monoethyl-amine	<0.02	<0.02
Monosodium Phosphate	>0.05	—
Nickel Chloride	>0.05	—
Nickel Nitrate	<0.02	—
Nickel Sulfate	>0.05	—
Nitric Acid	>0.05	>0.05
Nitric Acid (Red Fuming)	—	<0.05
Nitric + Hydrochloric Acid	—	>0.05

* 10% corrosive medium in 90% water. (Other % corrosive medium in parentheses.)
** Water-free, dry or maximum concentration of corrosive medium.
Source: data compiled by J.S. Park from Earl R. Parker, *Materials Data Book for Engineers and Scientists*, McGraw-Hill Book Company, New York, 1967.

Table 315. CORROSION RATES OF 1020 STEEL AT 70°F *

(SHEET 6 OF 8)

Corrosive Medium	Corrosion Rate* in 10% Corrosive Medium (ipy)	Corrosion Rate** in 100% Corrosive Medium (ipy)
Nitric + Hydrofluoric Acid	—	>0.05
Nitric + Sulfuric Acid	—	>0.05
Nitrobenzene	—	<0.002
Nitrocelluolose	—	<0.02
Nitroglycerine	—	<0.05
Nitrotolune	—	<0.02
Nitrous Acid	—	>0.05
Oleic Acid	—	<0.02
Oxalic Acid	>0.05	>0.05
Phenol	—	<0.002
Phosphoric Acid (Areated)	>0.05	>0.05
Phosphoric Acid (Air Free)	>0.05	>0.05
Picric Acid	>0.05	>0.05
Potassium Bicarbonate	<0.02	<0.002
Potassium Bromide	<0.05	>0.05
Potassium Carbonate	<0.02	<0.02
Potassium Chlorate	<0.02	<0.002
Potassium Chromate	<0.02	—
Potassium Cyanide	<0.02	<0.002
Potassium Dichromate	<0.02	—
Potassium Ferricyanide	<0.02	<0.02
Potassium Ferrocyanide	>0.05	—
Potassium Hydroxide	<0.02	<0.002
Potassium Hypochlorite	>0.05	<0.002
Potassium Iodide	<0.02	<0.02
Potassium Nitrate	<0.02	<0.002
Potassium Nitrite	<0.02	<0.02
Potassium Permanganate	<0.02	<0.002

* 10% corrosive medium in 90% water. (Other % corrosive medium in parentheses.)
** Water-free, dry or maximum concentration of corrosive medium.
Source: data compiled by J.S. Park from Earl R. Parker, *Materials Data Book for Engineers and Scientists*, McGraw-Hill Book Company, New York, 1967.

Table 315. CORROSION RATES OF 1020 STEEL AT 70°F *

(SHEET 7 OF 8)

Corrosive Medium	Corrosion Rate* in 10% Corrosive Medium (ipy)	Corrosion Rate** in 100% Corrosive Medium (ipy)
Potassium Silicate	<0.02	<0.02
Propionic Acid	>0.05	<0.02
Pyridine	<0.02	<0.02
Quinine Sulfate	>0.05	>0.05
Salicylic Acid	—	>0.05
Silicon Tetrachloride (Dry)	—	<0.002
Silicon Tetrachloride (Wet)	—	>0.05
Silver Bromide	>0.05	>0.05
Silver Chloride	>0.05	>0.05
Silver Nitrate	>0.05	—
Sodium Acetate	<0.02	<0.002
Sodium Bicarbonate	<0.02	<0.05
Sodium Bisulfate	>0.05	<0.002
Sodium Bromide	<0.02	<0.02
Sodium Carbonate	<0.002	<0.02
Sodium Chloride	<0.02	<0.002
Sodium Chromate	<0.02	<0.02
Sodium Hydroxide	<0.002	<0.02
Sodium Hypochlorite	>0.05	>0.05
Sodium Metasilicate	<0.02	<0.002
Sodium Nitrate	<0.02	<0.02
Sodium Nitrite	<0.02	<0.002
Sodium Phosphate	<0.02	<0.02
Sodium Silicate	<0.02	<0.02
Sodium Sulfate	<0.02	<0.02
Sodium Sulfide	<0.05	<0.02
Sodium Sulfite	<0.02	—
Stannic Chloride	>0.05	<0.002

* 10% corrosive medium in 90% water. (Other % corrosive medium in parentheses.)
** Water-free, dry or maximum concentration of corrosive medium.
Source: data compiled by J.S. Park from Earl R. Parker, *Materials Data Book for Engineers and Scientists*, McGraw-Hill Book Company, New York, 1967.

Table 315. CORROSION RATES OF 1020 STEEL AT 70°F *

(SHEET 8 OF 8)

Corrosive Medium	Corrosion Rate* in 10% Corrosive Medium (ipy)	Corrosion Rate** in 100% Corrosive Medium (ipy)
Stannous Chloride	>0.05	<0.02
Strontium Nitrate	>0.05	>0.05
Succinic Acid	<0.02	<0.02
Sulfur Dioxide	>0.05	<0.002
Sulfur Trioxide	—	<0.02
Sulfuric Acid (Areated)	>0.05	<0.02
Sulfuric Acid (Air Free)	>0.05	<0.02
Sulfuric Acid (Fuming)	—	<0.02
Sulfurous Acid	<0.05	>0.05
Tannic Acid	>0.05	<0.002
Tartaric Acid	>0.05	<0.05
Tetraphosphoric Acid	>0.05	>0.05
Trichloroacetic Acid	>0.05	>0.05
Trichloroethylene	—	<0.002
Urea	<0.05	—
Zinc Chloride	>0.05	<0.002
Zinc Sulfate	>0.05	—

* 10% corrosive medium in 90% water. (Other % corrosive medium in parentheses.)
** Water-free, dry or maximum concentration of corrosive medium.
Source: data compiled by J.S. Park from Earl R. Parker, *Materials Data Book for Engineers and Scientists*, McGraw-Hill Book Company, New York, 1967.

* <0.002 means that corrosion rate is likely to be less than 0.002 inch per year (Excellent).

<0.02 means that corrosion rate is likely to be less than about 0.02 inch per year (Good).

<0.05 means that corrosion rate is likely to be less than about 0.05 inch per year (Fair).

>0.05 means that corrosion rate is likely to be more than 0.05 inch per year (Poor).

Table 316. CORROSION RATES OF GREY CAST IRON AT 70°F *
(SHEET 1 OF 9)

Corrosive Medium	Corrosion Rate* in 10% Corrosive Medium (ipy)	Corrosion Rate** in 100% Corrosive Medium (ipy)
Acetaldehyde	<0.05	<0.002
Acetic Acid (Aerated)	>0.05	>0.05
Acetic Acid (Air Free)	>0.05	>0.05
Acetic Anhydride	—	>0.05
Acetoacetic Acid	>0.05	>0.05
Acetone	—	<0.002
Acetylene	—	<0.002
Acrolein	—	<0.02
Acrylonitril	—	<0.002
Alcohol (Ethyl)	<0.02	<0.02
Alcohol (Methyl)	<0.02	<0.002
Alcohol (Allyl)	—	<0.02
Alcohol (Amyl)	—	<0.02
Alcohol (Butyl)	—	<0.002
Alcohol (Isopropyl)	—	<0.02
Allylamine	—	<0.02
Allyl Chloride	—	<0.02
Allyl Sulfide	—	<0.02
Aluminum Acetate	>0.05	—
Aluminum Chloride	>0.05	>0.05
Aluminum Fluoride	<0.02	—
Aluminum Fluosilicate	—	>0.05
Aluminum Hydroxide	<0.02	—
Aluminum Nitrate	>0.05	—

* 10% corrosive medium in 90% water. (Other % corrosive medium in parentheses.)
** Water-free, dry or maximum concentration of corrosive medium.

Source: data compiled by J.S. Park from Earl R. Parker, *Materials Data Book for Engineers and Scientists*, McGraw-Hill Book Company, New York, 1967.

Table 316. Corrosion Rates of Grey Cast Iron at 70°F *

(Sheet 2 of 9)

Corrosive Medium	Corrosion Rate* in 10% Corrosive Medium (ipy)	Corrosion Rate** in 100% Corrosive Medium (ipy)
Aluminum Potassium Sulfate	>0.05	—
Aluminum Sulfate	>0.05	—
Ammonia	<0.002	<0.002
Ammonium Acetate	—	<0.02
Ammonium Bicarbonate	<0.02	<0.02
Ammonium Bromide	>0.05	>0.05
Ammonium Carbonate	<0.02	<0.02
Ammonium Chloride	>0.05	—
Ammonium Citrate	>0.05	—
Ammonium Nitrate	<0.02	<0.05
Ammonium Sulfate	<0.05	<0.02
Ammonium Sulfite	>0.05	—
Ammonium Thiocyanate	<0.02	—
Amyl Acetate	—	<0.02
Amyl Chloride	—	<0.02
Aniline	—	<0.002
Aniline Hydrochloride	>0.05	>0.05
Anthracine	—	<0.02
Antimony Trichloride	>0.05	—
Barium Carbonate	<0.02	<0.02
Barium Chloride	>0.05	<0.02
Barium Hydroxide	—	<0.02
Benzaldehyde	>0.05	>0.05
Benzene	—	<0.02

* 10% corrosive medium in 90% water. (Other % corrosive medium in parentheses.)
** Water-free, dry or maximum concentration of corrosive medium.

Source: data compiled by J.S. Park from Earl R. Parker, *Materials Data Book for Engineers and Scientists*, McGraw-Hill Book Company, New York, 1967.

Table 316. CORROSION RATES OF GREY CAST IRON AT 70°F *
(SHEET 3 OF 9)

Corrosive Medium	Corrosion Rate* in 10% Corrosive Medium (ipy)	Corrosion Rate** in 100% Corrosive Medium (ipy)
Benzoic Acid	>0.05	>0.05
Boric Acid	>0.05	—
Bromic Acid	>0.05	>0.05
Bromine (Dry)	—	>0.05
Bromine (Wet)	—	>0.05
Butyric Acid	>0.05	—
Cadmium Chloride	>0.05	—
Cadmium Sulfate	<0.02	<0.02
Calcium Acetate	<0.05	<0.05
Calcium Bicarbonate	—	<0.02
Calcium Bromide	—	<0.05
Calcium Chlorate	<0.02	<0.02
Calcium Chloride	<0.02	<0.002
Calcium Hydroxide	<0.02	<0.02
Calcium Hypochlorite	<0.05	<0.02
Carbon Dioxide	—	<0.002
Carbon Monoxide	—	<0.002
Carbon Tetrachloride	—	<0.05
Carbon Acid (Air Free)	—	<0.05
Chloroacetic Acid	>0.05	>0.05
Chlorine Gas	>0.05	<0.02
Chloroform (Dry)	—	<0.002
Chromic Acid	<0.05	<0.02
Citric Acid	>0.05	—

* 10% corrosive medium in 90% water. (Other % corrosive medium in parentheses.)
** Water-free, dry or maximum concentration of corrosive medium.

Source: data compiled by J.S. Park from Earl R. Parker, *Materials Data Book for Engineers and Scientists*, McGraw-Hill Book Company, New York, 1967.

Table 316. CORROSION RATES OF GREY CAST IRON AT 70°F *

(SHEET 4 OF 9)

Corrosive Medium	Corrosion Rate* in 10% Corrosive Medium (ipy)	Corrosion Rate** in 100% Corrosive Medium (ipy)
Copper Nitrate	>0.05	—
Copper Sulfate	>0.05	—
Ethylene Glycol	—	<0.02
Ethylene Oxide	—	<0.02
Fatty Acids	—	>0.05
Ferric Chloride	>0.05	—
Ferric Nitrate	>0.05	—
Ferrous Chloride	>0.05	—
Ferrous Sulfate	>0.05	—
Fluorine	—	>0.05
Formaldehyde	<0.05 (40%)	<0.02
Formic Acid	>0.05	>0.05
Furfural	—	<0.02
Hydrazine	>0.05	—
Hydrobromic Acid	>0.05	<0.02
Hydrochloric Acid (Areated)	>0.05	—
Hydrochloric Acid (Air Free)	>0.05	—
Hydrocyanic Acid	—	<0.02
Hydrofluoric Acid (Areated)	>0.05	>0.05
Hydrofluoric Acid (Air Free)	>0.05	>0.05
Hydrogen Chloride	>0.05 (90%)	<0.02
Hydrogen Iodide	>0.05	<0.02
Hydrogen Peroxide	>0.05 (20%)	—
Hydrogen Sulfide	<0.02	<0.02

* 10% corrosive medium in 90% water. (Other % corrosive medium in parentheses.)
** Water-free, dry or maximum concentration of corrosive medium.

Source: data compiled by J.S. Park from Earl R. Parker, *Materials Data Book for Engineers and Scientists*, McGraw-Hill Book Company, New York, 1967.

Table 316. CORROSION RATES OF GREY CAST IRON AT 70°F *

(SHEET 5 OF 9)

Corrosive Medium	Corrosion Rate* in 10% Corrosive Medium (ipy)	Corrosion Rate** in 100% Corrosive Medium (ipy)
Lactic Acid	>0.05	>0.05
Lead Acetate	>0.05	—
Lead Chromate	—	<0.02
Lead Nitrate	>0.05	<0.02
Lead Sulfate	—	<0.02
Lithium Chloride	<0.02 (30%)	<0.002
Lithium Hydroxide	<0.02	—
Magnesium Chloride	<0.02	<0.02
Magnesium Hydroxide	<0.02	—
Magnesium Sulfate	>0.05	<0.02
Maleic Acid	>0.05	—
Malic Acid	>0.05	—
Maganous Chloride	>0.05 (40%)	—
Mercuric Chloride	>0.05	—
Methallylamine	—	<0.02
Methanol	<0.02	<0.002
Methyl Ethyl Ketone	<0.02	<0.002
Methyl Isobutyl Ketone	<0.02	<0.02
Methylamine	<0.02	<0.02
Methylene Chloride	—	<0.02
Monochloroacetic Acid	>0.05	>0.05
Monorthanolamine	—	<0.02
Monoethylamine	<0.02	<0.02
Monosodium Phosphate	>0.05	—

* 10% corrosive medium in 90% water. (Other % corrosive medium in parentheses.)
** Water-free, dry or maximum concentration of corrosive medium.

Source: data compiled by J.S. Park from Earl R. Parker, *Materials Data Book for Engineers and Scientists*, McGraw-Hill Book Company, New York, 1967.

Table 316. CORROSION RATES OF GREY CAST IRON AT 70°F *

(SHEET 6 OF 9)

Corrosive Medium	Corrosion Rate* in 10% Corrosive Medium (ipy)	Corrosion Rate** in 100% Corrosive Medium (ipy)
Nickel Chloride	>0.05	—
Nickel Nitrate	<0.02	—
Nickel Sulfate	>0.05	—
Nitric Acid	>0.05	>0.05
Nitric Acid (Red Fuming)	—	>0.05
Nitric + Hydrochloric Acid	—	>0.05
Nitric + Hydrofluoric Acid	—	>0.05
Nitric + Sulfuric Acid	—	>0.05
Nitrobenzene	—	<0.02
Nitrocelluolose	—	<0.02
Nitroglycerine	—	<0.05
Nitrotolune	—	<0.02
Oleic Acid	—	<0.02
Oxalic Acid	>0.05	>0.05
Phenol	—	<0.02
Phosphoric Acid (Areated)	>0.05	>0.05
Phosphoric Acid (Air Free)	>0.05	>0.05
Picric Acid	>0.05	>0.05
Potassium Bicarbonate	<0.02	—
Potassium Bromide	<0.05	>0.05
Potassium Carbonate	<0.02	<0.02
Potassium Chromate	<0.02	—
Potassium Cyanide	>0.05	<0.02
Potassium Dichromate	<0.02	—

* 10% corrosive medium in 90% water. (Other % corrosive medium in parentheses.)
** Water-free, dry or maximum concentration of corrosive medium.

Source: data compiled by J.S. Park from Earl R. Parker, *Materials Data Book for Engineers and Scientists*, McGraw-Hill Book Company, New York, 1967.

Table 316. CORROSION RATES OF GREY CAST IRON AT 70°F *

(SHEET 7 OF 9)

Corrosive Medium	Corrosion Rate* in 10% Corrosive Medium (ipy)	Corrosion Rate** in 100% Corrosive Medium (ipy)
Potassium Ferricyanide	<0.02	<0.02
Potassium Ferrocyanide	>0.05	—
Potassium Hydroxide	<0.02	<0.02
Potassium Hypochlorite	>0.05	—
Potassium Nitrate	<0.02	<0.02
Potassium Nitrite	<0.02	<0.02
Potassium Permanganate	<0.02	<0.02
Potassium Silicate	<0.02	<0.02
Propionic Acid	>0.05	—
Pyridine	<0.02	<0.02
Quinine Sulfate	>0.05	>0.05
Salicylic Acid	—	>0.05
Silicon Tetrachloride (Dry)	—	<0.002
Silicon Tetrachloride (Wet)	—	>0.05
Silver Bromide	>0.05	>0.05
Silver Chloride	>0.05	>0.05
Silver Nitrate	>0.05	—
Sodium Acetate	—	<0.002
Sodium Bicarbonate	<0.02	<0.05
Sodium Bisulfate	>0.05	—
Sodium Bromide	—	<0.05
Sodium Carbonate	<0.002	<0.02
Sodium Chloride	<0.02	<0.02
Sodium Chromate	<0.02	<0.02

* 10% corrosive medium in 90% water. (Other % corrosive medium in parentheses.)
** Water-free, dry or maximum concentration of corrosive medium.

Source: data compiled by J.S. Park from Earl R. Parker, *Materials Data Book for Engineers and Scientists*, McGraw-Hill Book Company, New York, 1967.

Table 316. CORROSION RATES OF GREY CAST IRON AT 70°F *

(SHEET 8 OF 9)

Corrosive Medium	Corrosion Rate* in 10% Corrosive Medium (ipy)	Corrosion Rate** in 100% Corrosive Medium (ipy)
Sodium Hydroxide	<0.02	—
Sodium Hypochlorite	>0.05	—
Sodium Metasilicate	<0.02	<0.02
Sodium Nitrate	<0.02	<0.02
Sodium Nitrite	<0.02	—
Sodium Phosphate	<0.02	<0.02
Sodium Silicate	<0.02	<0.02
Sodium Sulfate	<0.02	<0.02
Sodium Sulfide	<0.05	<0.02
Sodium Sulfite	>0.05	—
Stannic Chloride	>0.05	—
Stannous Chloride	>0.05	<0.02
Strontium Nitrate	>0.05	>0.05
Succinic Acid	<0.02	<0.02
Sulfur Dioxide	—	<0.02
Sulfur Trioxide	—	<0.02
Sulfuric Acid (Areated)	>0.05	<0.02
Sulfuric Acid (Air Free)	>0.05	<0.02
Sulfuric Acid (Fuming)	—	<0.02
Sulfurous Acid	—	>0.05

* 10% corrosive medium in 90% water. (Other % corrosive medium in parentheses.)
** Water-free, dry or maximum concentration of corrosive medium.

Source: data compiled by J.S. Park from Earl R. Parker, *Materials Data Book for Engineers and Scientists*, McGraw-Hill Book Company, New York, 1967.

Table 316. CORROSION RATES OF GREY CAST IRON AT 70°F *

(SHEET 9 OF 9)

Corrosive Medium	Corrosion Rate* in 10% Corrosive Medium (ipy)	Corrosion Rate** in 100% Corrosive Medium (ipy)
Tannic Acid	—	<0.02
Tartaric Acid	>0.05	>0.05
Tetraphosphoric Acid	>0.05	>0.05
Trichloroacetic Acid	>0.05	>0.05
Trichloroethylene	—	<0.02
Zinc Chloride	>0.05	<0.02
Zinc Sulfate	>0.05	—

* 10% corrosive medium in 90% water. (Other % corrosive medium in parentheses.)
** Water-free, dry or maximum concentration of corrosive medium.

Source: data compiled by J.S. Park from Earl R. Parker, *Materials Data Book for Engineers and Scientists*, McGraw-Hill Book Company, New York, 1967.

* <0.002 means that corrosion rate is likely to be less than 0.002 inch per year (Excellent).
<0.02 means that corrosion rate is likely to be less than about 0.02 inch per year (Good).
<0.05 means that corrosion rate is likely to be less than about 0.05 inch per year (Fair).
>0.05 means that corrosion rate is likely to be more than 0.05 inch per year (Poor).

Table 317. CORROSION RATES OF NI–RESIST CAST IRON * AT 70°F (SHEET 1 OF 8)

Corrosive Medium	Corrosion Rate* in 10% Corrosive Medium (ipy)	Corrosion Rate** in 100% Corrosive Medium (ipy)
Acetaldehyde	—	<0.002
Acetic Acid (Aerated)	<0.02	>0.05
Acetic Acid (Air Free)	<0.02	>0.05
Acetic Anhydride	—	<0.02
Acetone	—	<0.002
Acetylene	—	<0.002
Acrolein	—	<0.02
Acrylonitril	—	<0.002
Alcohol (Ethyl)	<0.02	<0.02
Alcohol (Methyl)	<0.02	<0.002
Alcohol (Allyl)	—	<0.02
Alcohol (Amyl)	—	<0.02
Alcohol (Isopropyl)	—	<0.02
Allylamine	—	<0.02
Allyl Sulfide	—	<0.02
Aluminum Acetate	—	<0.02
Aluminum Chloride	>0.05	>0.05
Aluminum Hydroxide	<0.02	—
Aluminum Potassium Sulfate	>0.05	—
Aluminum Sulfate	<0.02	—
Ammonia	<0.002	<0.002
Ammonium Acetate	<0.002	<0.002
Ammonium Bicarbonate	<0.02	<0.02
Ammonium Carbonate	<0.02	<0.02

* 10% corrosive medium in 90% water. (Other % corrosive medium in parentheses.)
** Water-free, dry or maximum concentration of corrosive medium.

Source: data compiled by J.S. Park from Earl R. Parker, *Materials Data Book for Engineers and Scientists*, McGraw-Hill Book Company, New York, 1967.

Table 317. Corrosion Rates of Ni–Resist Cast Iron * at 70°F (Sheet 2 of 8)

Corrosive Medium	Corrosion Rate* in 10% Corrosive Medium (ipy)	Corrosion Rate** in 100% Corrosive Medium (ipy)
Ammonium Chloride	<0.02	—
Ammonium Citrate	>0.05	—
Ammonium Nitrate	<0.02	—
Ammonium Sulfate	>0.05	<0.02
Ammonium Sulfite	>0.05	—
Ammonium Thiocyanate	<0.02	—
Amyl Acetate	—	<0.002
Aniline	<0.02	<0.02
Aniline Hydrochloride	>0.05	>0.05
Anthracine	—	<0.02
Antimony Trichloride	>0.05	—
Barium Carbonate	<0.02	<0.02
Barium Chloride	<0.02	—
Benzaldehyde	<0.02	<0.002
Benzene	—	<0.02
Benzoic Acid	—	<0.02
Boric Acid	<0.002	<0.02
Bromine (Dry)	—	<0.02
Bromine (Wet)	—	>0.05
Butyric Acid	>0.05	>0.05
Cadmium Chloride	>0.05	—
Calcium Chlorate	<0.05	<0.02
Calcium Chloride	<0.02	—
Calcium Hydroxide	<0.02	—

* 10% corrosive medium in 90% water. (Other % corrosive medium in parentheses.)
** Water-free, dry or maximum concentration of corrosive medium.

Source: data compiled by J.S. Park from Earl R. Parker, *Materials Data Book for Engineers and Scientists*, McGraw-Hill Book Company, New York, 1967.

Table 317. CORROSION RATES OF NI–RESIST CAST IRON * AT 70°F (SHEET 3 OF 8)

Corrosive Medium	Corrosion Rate* in 10% Corrosive Medium (ipy)	Corrosion Rate** in 100% Corrosive Medium (ipy)
Calcium Hypochlorite	<0.02	—
Carbon Dioxide	—	<0.002
Carbon Monoxide	—	<0.002
Carbon Tetrachloride	—	<0.02
Carbon Acid (Air Free)	—	<0.002
Chloroacetic Acid	>0.05	>0.05
Chlorine Gas	>0.05	<0.02
Chromic Acid	<0.05	<0.02
Chromic Hydroxide	—	<0.02
Citric Acid	>0.05	>0.05
Copper Nitrate	>0.05	—
Copper Sulfate	>0.05	—
Ethylene Glycol	—	<0.02
Fatty Acids	—	<0.02
Ferric Chloride	>0.05	—
Ferrous Chloride	>0.05	—
Formaldehyde	<0.05 (40%)	—
Formic Acid	>0.05	>0.05
Furfural	<0.02 (30%)	<0.02
Hydrobromic Acid	—	>0.05
Hydrochloric Acid (Areated)	>0.05	—
Hydrochloric Acid (Air Free)	<0.05	—
Hydrocyanic Acid	—	<0.02
Hydrofluoric Acid (Areated)	<0.002	<0.02

* 10% corrosive medium in 90% water. (Other % corrosive medium in parentheses.)
** Water-free, dry or maximum concentration of corrosive medium.

Source: data compiled by J.S. Park from Earl R. Parker, *Materials Data Book for Engineers and Scientists*, McGraw-Hill Book Company, New York, 1967.

Table 317. CORROSION RATES OF NI–RESIST CAST IRON * AT 70°F (SHEET 4 OF 8)

Corrosive Medium	Corrosion Rate* in 10% Corrosive Medium (ipy)	Corrosion Rate** in 100% Corrosive Medium (ipy)
Hydrofluoric Acid (Air Free)	<0.002	<0.02
Hydrogen Chloride	—	<0.002
Hydrogen Fluoride	—	<0.02
Hydrogen Iodide	—	<0.02
Hydrogen Sulfide	<0.02	<0.02
Lactic Acid	>0.05	>0.05
Lead Chromate	—	<0.02
Lead Sulfate	—	<0.02
Lithium Chloride	<0.002 (30%)	—
Lithium Hydroxide	<0.02	—
Magnesium Chloride	<0.02	<0.02
Magnesium Hydroxide	<0.02	<0.02
Magnesium Sulfate	<0.02	<0.02
Maleic Acid	>0.05	—
Maganous Chloride	<0.05 (40%)	—
Mercuric Chloride	>0.05	—
Methallylamine	<0.02	<0.02
Methanol	<0.02	<0.002
Methyl Ethyl Ketone	<0.02	<0.002
Methyl Isobutyl Ketone	<0.02	<0.02
Methylamine	<0.02	<0.02
Methylene Chloride	—	<0.02
Monochloroacetic Acid	—	<0.05
Monorthanolamine	—	<0.02

* 10% corrosive medium in 90% water. (Other % corrosive medium in parentheses.)
** Water-free, dry or maximum concentration of corrosive medium.

Source: data compiled by J.S. Park from Earl R. Parker, *Materials Data Book for Engineers and Scientists*, McGraw-Hill Book Company, New York, 1967.

Table 317. CORROSION RATES OF NI–RESIST CAST IRON * AT 70°F (SHEET 5 OF 8)

Corrosive Medium	Corrosion Rate* in 10% Corrosive Medium (ipy)	Corrosion Rate** in 100% Corrosive Medium (ipy)
Monoethalamine	<0.02	<0.02
Monoethylamine	<0.02	<0.02
Monosodium Phosphate	>0.05	—
Nickel Chloride	>0.05	—
Nickel Nitrate	<0.02	—
Nitric Acid	>0.05	>0.05
Nitric Acid (Red Fuming)	—	>0.05
Nitric + Hydrochloric Acid	—	>0.05
Nitric + Hydrofluoric Acid	—	>0.05
Nitric + Sulfuric Acid	—	>0.05
Nitrobenzene	—	<0.02
Nitrocelluolose	—	<0.02
Nitroglycerine	—	<0.02
Nitrotolune	—	<0.02
Oleic Acid	—	<0.002
Oxalic Acid	>0.05	<0.02
Phenol	—	<0.02
Phosphoric Acid (Areated)	>0.05	>0.05
Phosphoric Acid (Air Free)	>0.05	>0.05
Picric Acid	—	>0.05
Potassium Bicarbonate	<0.02	—
Potassium Bromide	<0.02	<0.02
Potassium Carbonate	<0.02	<0.02
Potassium Chlorate	<0.02	—

* 10% corrosive medium in 90% water. (Other % corrosive medium in parentheses.)
** Water-free, dry or maximum concentration of corrosive medium.

Source: data compiled by J.S. Park from Earl R. Parker, *Materials Data Book for Engineers and Scientists*, McGraw-Hill Book Company, New York, 1967.

Table 317. CORROSION RATES OF NI–RESIST CAST IRON *
AT 70°F (SHEET 6 OF 8)

Corrosive Medium	Corrosion Rate* in 10% Corrosive Medium (ipy)	Corrosion Rate** in 100% Corrosive Medium (ipy)
Potassium Chromate	<0.02	—
Potassium Cyanide	<0.02	—
Potassium Dichromate	<0.02	<0.02
Potassium Ferricyanide	<0.02	<0.02
Potassium Ferrocyanide	<0.02	—
Potassium Hydroxide	<0.02	—
Potassium Hypochlorite	>0.05	—
Potassium Iodide	<0.02	—
Potassium Nitrate	<0.02	—
Potassium Nitrite	<0.02	<0.02
Potassium Permanganate	<0.02	—
Potassium Silicate	<0.02	<0.02
Pyridine	<0.02	<0.02
Quinine Sulfate	<0.02	<0.02
Salicylic Acid	—	<0.02
Silicon Tetrachloride (Dry)	—	<0.002
Silicon Tetrachloride (Wet)	—	>0.05
Silver Bromide	>0.05	>0.05
Sodium Acetate	<0.02	—
Sodium Bicarbonate	<0.02	<0.02
Sodium Bisulfate	<0.002	<0.002
Sodium Bromide	<0.02	<0.02
Sodium Carbonate	<0.002	<0.02
Sodium Chloride	<0.02	<0.02

* 10% corrosive medium in 90% water. (Other % corrosive medium in parentheses.)
** Water-free, dry or maximum concentration of corrosive medium.

Source: data compiled by J.S. Park from Earl R. Parker, *Materials Data Book for Engineers and Scientists*, McGraw-Hill Book Company, New York, 1967.

Table 317. CORROSION RATES OF NI–RESIST CAST IRON * AT 70°F (SHEET 7 OF 8)

Corrosive Medium	Corrosion Rate* in 10% Corrosive Medium (ipy)	Corrosion Rate** in 100% Corrosive Medium (ipy)
Sodium Chromate	<0.02	<0.02
Sodium Hydroxide	<0.002	<0.02
Sodium Hypochlorite	>0.05	—
Sodium Metasilicate	<0.002	<0.02
Sodium Nitrate	<0.02	<0.02
Sodium Nitrite	<0.02	—
Sodium Phosphate	<0.02	<0.02
Sodium Silicate	<0.02	<0.02
Sodium Sulfate	<0.02	<0.02
Sodium Sulfite	<0.02	—
Stannic Chloride	>0.05	—
Stannous Chloride	>0.05	<0.02
Strontium Nitrate	<0.02	—
Succinic Acid	<0.02	<0.02
Sulfur Dioxide	—	<0.02
Sulfur Trioxide	—	<0.02
Sulfuric Acid (Areated)	<0.02	<0.02
Sulfuric Acid (Air Free)	<0.02	<0.02
Sulfuric Acid (Fuming)	—	<0.05
Sulfurous Acid	<0.05	>0.05
Tartaric Acid	<0.02	—
Tetraphosphoric Acid	>0.05	<0.05
Trichloroacetic Acid	>0.05	>0.05

* 10% corrosive medium in 90% water. (Other % corrosive medium in parentheses.)
** Water-free, dry or maximum concentration of corrosive medium.

Source: data compiled by J.S. Park from Earl R. Parker, *Materials Data Book for Engineers and Scientists*, McGraw-Hill Book Company, New York, 1967.

Table 317. CORROSION RATES OF NI–RESIST CAST IRON * AT 70°F (SHEET 8 OF 8)

Corrosive Medium	Corrosion Rate* in 10% Corrosive Medium (ipy)	Corrosion Rate** in 100% Corrosive Medium (ipy)
Trichloroethylene	—	<0.02
Zinc Chloride	<0.02	<0.02
Zinc Sulfate	<0.02	—

* 10% corrosive medium in 90% water. (Other % corrosive medium in parentheses.)
** Water-free, dry or maximum concentration of corrosive medium.

Source: data compiled by J.S. Park from Earl R. Parker, *Materials Data Book for Engineers and Scientists*, McGraw-Hill Book Company, New York, 1967.

* <0.002 means that corrosion rate is likely to be less than 0.002 inch per year (Excellent).
<0.02 means that corrosion rate is likely to be less than about 0.02 inch per year (Good).
<0.05 means that corrosion rate is likely to be less than about 0.05 inch per year (Fair).
>0.05 means that corrosion rate is likely to be more than 0.05 inch per year (Poor).

Table 318. CORROSION RATES OF 12% CR STEEL AT 70° *

(SHEET 1 OF 9)

Corrosive Medium	Corrosion Rate* in 10% Corrosive Medium (ipy)	Corrosion Rate** in 100% Corrosive Medium (ipy)
Acetaldehyde	—	<0.002
Acetic Acid (Aerated)	<0.02	>0.05
Acetic Acid (Air Free)	<0.02	>0.05
Acetic Anhydride	—	<0.05
Acetone	<0.02	<0.002
Acetylene	—	<0.002
Acrolein	<0.02	<0.02
Acrylonitril	—	<0.002
Alcohol (Ethyl)	<0.02	<0.02
Alcohol (Methyl)	<0.02	<0.02
Alcohol (Allyl)	—	<0.02
Alcohol (Amyl)	—	<0.02
Alcohol (Benzyl)	—	<0.02
Alcohol (Butyl)	—	<0.002
Alcohol (Cetyl)	—	<0.02
Alcohol (Isopropyl)	—	<0.02
Allylamine	—	<0.02
Allyl Chloride	—	<0.02
Allyl Sulfide	—	<0.02
Aluminum Acetate	<0.02	<0.02
Aluminum Chloride	>0.05	<0.002
Aluminum Fluoride	>0.05	>0.05
Aluminum Fluosilicate	—	<0.02
Aluminum Formate	<0.02	<0.02

* 10% corrosive medium in 90% water. (Other % corrosive medium in parentheses.)
** Water-free, dry or maximum concentration of corrosive medium.

Source: data compiled by J.S. Park from Earl R. Parker, *Materials Data Book for Engineers and Scientists*, McGraw-Hill Book Company, New York, 1967.

Table 318. Corrosion Rates of 12% Cr Steel at 70° *

(Sheet 2 of 9)

Corrosive Medium	Corrosion Rate* in 10% Corrosive Medium (ipy)	Corrosion Rate** in 100% Corrosive Medium (ipy)
Aluminum Hydroxide	<0.02	—
Aluminum Nitrate	<0.02	<0.02
Aluminum Potassium Sulfate	>0.05	<0.05
Aluminum Sulfate	>0.05	>0.05
Ammonia	<0.002	<0.002
Ammonium Acetate	<0.002	<0.002
Ammonium Bicarbonate	<0.02	—
Ammonium Bromide	<0.05	>0.05
Ammonium Carbonate	<0.02	<0.02
Ammonium Chloride	<0.05	>0.05
Ammonium Nitrate	<0.02	<0.02
Ammonium Sulfate	>0.05	—
Ammonium Sulfite	>0.05	—
Amyl Acetate	—	<0.002
Amyl Chloride	—	<0.05
Aniline	<0.02	<0.02
Aniline Hydrochloride	>0.05	>0.05
Anthracine	—	<0.02
Antimony Trichloride	>0.05	>0.05
Barium Carbonate	<0.02	<0.02
Barium Chloride	<0.05	—
Barium Hydroxide	—	<0.02
Barium Oxide	—	<0.02
Barium Peroxide	>0.05	—

* 10% corrosive medium in 90% water. (Other % corrosive medium in parentheses.)
** Water-free, dry or maximum concentration of corrosive medium.

Source: data compiled by J.S. Park from Earl R. Parker, *Materials Data Book for Engineers and Scientists*, McGraw-Hill Book Company, New York, 1967.

Table 318. CORROSION RATES OF 12% CR STEEL AT 70° *

(SHEET 3 OF 9)

Corrosive Medium	Corrosion Rate* in 10% Corrosive Medium (ipy)	Corrosion Rate** in 100% Corrosive Medium (ipy)
Benzaldehyde	—	<0.02
Benzene	<0.02	<0.02
Benzoic Acid	<0.02	<0.02
Boric Acid	<0.02	<0.02
Bromic Acid	>0.05	>0.05
Bromine (Dry)	—	>0.05
Bromine (Wet)	—	>0.05
Butyric Acid	<0.05	—
Cadmium Chloride	>0.05	—
Calcium Acetate	<0.02	<0.02
Calcium Bicarbonate	—	<0.02
Calcium Bromide	<0.02	<0.02
Calcium Chlorate	<0.02	—
Calcium Chloride	<0.02	—
Calcium Hydroxide	<0.02	<0.02
Calcium Hypochlorite	>0.05	>0.05
Carbon Dioxide	—	<0.002
Carbon Monoxide	—	<0.002
Carbon Tetrachloride	>0.05	<0.02
Carbon Acid (Air Free)	—	<0.002
Chloroacetic Acid	>0.05	>0.05
Chlorine Gas	>0.05	<0.05
Chloroform (Dry)	—	<0.002
Chromic Acid	>0.05	<0.02

* 10% corrosive medium in 90% water. (Other % corrosive medium in parentheses.)
** Water-free, dry or maximum concentration of corrosive medium.

Source: data compiled by J.S. Park from Earl R. Parker, *Materials Data Book for Engineers and Scientists*, McGraw-Hill Book Company, New York, 1967.

Table 318. CORROSION RATES OF 12% CR STEEL AT 70° *
(SHEET 4 OF 9)

Corrosive Medium	Corrosion Rate* in 10% Corrosive Medium (ipy)	Corrosion Rate** in 100% Corrosive Medium (ipy)
Chromic Hydroxide	—	<0.02
Chromic Sulfates	>0.05	>0.05
Citric Acid	<0.05	—
Copper Nitrate	<0.02	—
Copper Sulfate	<0.02	—
Ethyl Chloride	>0.05 (90%)	<0.002
Ethylene Glycol	—	<0.02
Ethylene Oxide	—	<0.02
Fatty Acids	—	<0.02
Ferric Chloride	>0.05	—
Ferric Nitrate	<0.02	—
Ferrous Chloride	>0.05	—
Ferrous Sulfate	<0.02	—
Fluorine	—	>0.05
Formaldehyde	<0.02	<0.02
Formic Acid	<0.05	<0.02
Furfural	<0.02 (80%)	—
Hydrobromic Acid	>0.05	—
Hydrochloric Acid (Areated)	>0.05	—
Hydrochloric Acid (Air Free)	>0.05	—
Hydrocyanic Acid	—	>0.05
Hydrofluoric Acid (Air Free)	>0.05	>0.05
Hydrogen Chloride	>0.05 (90%)	>0.05
Hydrogen Fluoride	—	<0.02

* 10% corrosive medium in 90% water. (Other % corrosive medium in parentheses.)
** Water-free, dry or maximum concentration of corrosive medium.

Source: data compiled by J.S. Park from Earl R. Parker, *Materials Data Book for Engineers and Scientists*, McGraw-Hill Book Company, New York, 1967.

Table 318. CORROSION RATES OF 12% CR STEEL AT 70° *

(SHEET 5 OF 9)

Corrosive Medium	Corrosion Rate* in 10% Corrosive Medium (ipy)	Corrosion Rate** in 100% Corrosive Medium (ipy)
Hydrogen Iodide	<0.05	>0.05
Hydrogen Peroxide	<0.02 (20%)	<0.02
Hydrogen Sulfide	<0.02	<0.02
Lactic Acid	>0.05	—
Lead Acetate	<0.02	<0.02
Lead Chromate	—	<0.02
Lead Nitrate	<0.02	—
Lead Sulfate	—	<0.02
Lithium Hydroxide	<0.02	—
Magnesium Chloride	<0.05	—
Magnesium Hydroxide	<0.02	<0.02
Magnesium Sulfate	>0.05	<0.05
Maleic Acid	—	<0.05
Malic Acid	<0.02	—
Mercuric Chloride	>0.05	>0.05
Mercurous Nitrate	<0.02	<0.02
Methallylamine	<0.02	<0.02
Methanol	<0.02	<0.002
Methyl Ethyl Ketone	<0.02	<0.002
Methyl Isobutyl Ketone	<0.02	<0.02
Methylamine	<0.02	<0.02
Methylene Chloride	—	<0.02
Monochloroacetic Acid	>0.05	>0.05
Monorthanolamine	<0.02	—

* 10% corrosive medium in 90% water. (Other % corrosive medium in parentheses.)
** Water-free, dry or maximum concentration of corrosive medium.

Source: data compiled by J.S. Park from Earl R. Parker, *Materials Data Book for Engineers and Scientists*, McGraw-Hill Book Company, New York, 1967.

Table 318. CORROSION RATES OF 12% CR STEEL AT 70° *

(SHEET 6 OF 9)

Corrosive Medium	Corrosion Rate* in 10% Corrosive Medium (ipy)	Corrosion Rate** in 100% Corrosive Medium (ipy)
Monoethalamine	<0.02	<0.02
Monoethylamine	<0.02	<0.02
Monosodium Phosphate	>0.05	—
Nickel Chloride	>0.05	—
Nickel Nitrate	<0.02	—
Nitric Acid	<0.02	>0.05
Nitric Acid (Red Fuming)	—	<0.002
Nitric + Hydrochloric Acid	—	>0.05
Nitric + Hydrofluoric Acid	—	>0.05
Nitric + Sulfuric Acid	—	>0.05
Nitrobenzene	—	<0.02
Nitrocelluolose	—	<0.02
Nitroglycerine	—	<0.02
Nitrotolune	—	<0.02
Nitrous Acid	<0.05	—
Oleic Acid	<0.02	<0.02
Oxalic Acid	>0.05	>0.05
Phenol	—	<0.02
Phosphoric Acid (Areated)	<0.02	—
Phosphoric Acid (Air Free)	>0.05	>0.05
Picric Acid	<0.02	<0.02
Potassium Bicarbonate	<0.02	—
Potassium Bromide	<0.02	<0.002
Potassium Carbonate	<0.02	<0.02

* 10% corrosive medium in 90% water. (Other % corrosive medium in parentheses.)
** Water-free, dry or maximum concentration of corrosive medium.

Source: data compiled by J.S. Park from Earl R. Parker, *Materials Data Book for Engineers and Scientists*, McGraw-Hill Book Company, New York, 1967.

Table 318. CORROSION RATES OF 12% CR STEEL AT 70° *
(SHEET 7 OF 9)

Corrosive Medium	Corrosion Rate* in 10% Corrosive Medium (ipy)	Corrosion Rate** in 100% Corrosive Medium (ipy)
Potassium Chlorate	<0.02	<0.02
Potassium Chromate	<0.02	<0.02
Potassium Cyanide	<0.02	<0.02
Potassium Dichromate	<0.02	<0.02
Potassium Ferricyanide	<0.02	—
Potassium Ferrocyanide	>0.05	—
Potassium Hydroxide	<0.02	<0.002
Potassium Hypochlorite	>0.05	—
Potassium Iodide	>0.05	—
Potassium Nitrate	<0.02	<0.02
Potassium Nitrite	<0.02	<0.02
Potassium Permanganate	<0.002	<0.02
Potassium Silicate	<0.02	<0.02
Pyridine	<0.02	<0.02
Salicylic Acid	—	<0.02
Silicon Tetrachloride (Dry)	—	<0.002
Silicon Tetrachloride (Wet)	—	>0.05
Silver Bromide	>0.05	>0.05
Silver Chloride	>0.05	>0.05
Silver Nitrate	<0.02	
Sodium Acetate	<0.02	<0.02
Sodium Bicarbonate	<0.02	—
Sodium Bisulfate	<0.002	>0.05
Sodium Bromide	<0.05	—

* 10% corrosive medium in 90% water. (Other % corrosive medium in parentheses.)
** Water-free, dry or maximum concentration of corrosive medium.

Source: data compiled by J.S. Park from Earl R. Parker, *Materials Data Book for Engineers and Scientists*, McGraw-Hill Book Company, New York, 1967.

Table 318. CORROSION RATES OF 12% CR STEEL AT 70° *

(SHEET 8 OF 9)

Corrosive Medium	Corrosion Rate* in 10% Corrosive Medium (ipy)	Corrosion Rate** in 100% Corrosive Medium (ipy)
Sodium Carbonate	<0.02	<0.02
Sodium Chloride	<0.02	—
Sodium Chromate	<0.02	<0.02
Sodium Hydroxide	<0.002	—
Sodium Hypochlorite	>0.05	>0.05
Sodium Metasilicate	<0.002	<0.002
Sodium Nitrate	<0.02	<0.02
Sodium Nitrite	<0.02	<0.002
Sodium Phosphate	<0.02	<0.02
Sodium Silicate	<0.02	<0.02
Sodium Sulfate	<0.05	>0.05
Sodium Sulfide	>0.05	<0.02
Sodium Sulfite	<0.02	—
Stannic Chloride	>0.05	—
Stannous Chloride	>0.05	—
Strontium Nitrate	<0.02	—
Succinic Acid	<0.02	<0.02
Sulfur Dioxide	>0.05	<0.02
Sulfur Trioxide	—	<0.02
Sulfuric Acid (Areated)	<0.05	>0.05
Sulfuric Acid (Air Free)	>0.05	<0.05
Sulfuric Acid (Fuming)	—	<0.002
Sulfurous Acid	>0.05	>0.05
Tannic Acid	<0.02	<0.02

* 10% corrosive medium in 90% water. (Other % corrosive medium in parentheses.)
** Water-free, dry or maximum concentration of corrosive medium.

Source: data compiled by J.S. Park from Earl R. Parker, *Materials Data Book for Engineers and Scientists*, McGraw-Hill Book Company, New York, 1967.

Table 318. CORROSION RATES OF 12% CR STEEL AT 70° *

(SHEET 9 OF 9)

Corrosive Medium	Corrosion Rate* in 10% Corrosive Medium (ipy)	Corrosion Rate** in 100% Corrosive Medium (ipy)
Tartaric Acid	<0.02	—
Tetraphosphoric Acid	>0.05	>0.05
Trichloroacetic Acid	>0.05	>0.05
Trichloroethylene	—	<0.02
Urea	<0.02	—
Zinc Chloride	—	>0.05
Zinc Sulfate	<0.05	—

* 10% corrosive medium in 90% water. (Other % corrosive medium in parentheses.)
** Water-free, dry or maximum concentration of corrosive medium.

Source: data compiled by J.S. Park from Earl R. Parker, *Materials Data Book for Engineers and Scientists*, McGraw-Hill Book Company, New York, 1967.

* <0.002 means that corrosion rate is likely to be less than 0.002 inch per year (Excellent).
<0.02 means that corrosion rate is likely to be less than about 0.02 inch per year (Good).
<0.05 means that corrosion rate is likely to be less than about 0.05 inch per year (Fair).
>0.05 means that corrosion rate is likely to be more than 0.05 inch per year (Poor).

Table 319. CORROSION RATES OF 17% CR STEEL AT 70°F *

(SHEET 1 OF 9)

Corrosive Medium	Corrosion Rate* in 10% Corrosive Medium (ipy)	Corrosion Rate** in 100% Corrosive Medium (ipy)
Acetaldehyde	—	<0.002
Acetic Acid (Aerated)	<0.002	<0.002
Acetic Acid (Air Free)	<0.02	<0.05
Acetic Anhydride	—	<0.05
Acetoacetic Acid	<0.02	<0.02
Acetone	<0.02	<0.002
Acetylene	—	<0.002
Acrolein	<0.02	<0.02
Acrylonitril	—	<0.002
Alcohol (Ethyl)	<0.02	<0.02
Alcohol (Methyl)	<0.02	<0.02
Alcohol (Allyl)	—	<0.02
Alcohol (Amyl)	—	<0.02
Alcohol (Benzyl)	—	<0.02
Alcohol (Butyl)	—	<0.002
Alcohol (Cetyl)	—	<0.02
Alcohol (Isopropyl)	—	<0.02
Allylamine	—	<0.02
Allyl Chloride	—	<0.02
Allyl Sulfide	—	<0.02
Aluminum Chlorate	<0.002	—
Aluminum Chloride	>0.05	<0.002
Aluminum Fluoride	>0.05	>0.05
Aluminum Fluosilicate	—	<0.02

* 10% corrosive medium in 90% water. (Other % corrosive medium in parentheses.)
** Water-free, dry or maximum concentration of corrosive medium.

Source: data compiled by J.S. Park from Earl R. Parker, *Materials Data Book for Engineers and Scientists*, McGraw-Hill Book Company, New York, 1967.

Table 319. CORROSION RATES OF 17% CR STEEL AT 70°F *

(SHEET 2 OF 9)

Corrosive Medium	Corrosion Rate* in 10% Corrosive Medium (ipy)	Corrosion Rate** in 100% Corrosive Medium (ipy)
Aluminum Formate	<0.02	<0.02
Aluminum Hydroxide	<0.02	<0.02
Aluminum Nitrate	<0.02	<0.02
Aluminum Potassium Sulfate	<0.05	>0.05
Aluminum Sulfate	—	>0.05
Ammonia	<0.002	<0.002
Ammonium Acetate	<0.002	<0.002
Ammonium Bicarbonate	<0.02	—
Ammonium Bromide	<0.05	—
Ammonium Carbonate	<0.02	<0.02
Ammonium Chloride	<0.05	>0.05
Ammonium Citrate	<0.02	—
Ammonium Nitrate	<0.002	<0.02
Ammonium Sulfate	<0.05	—
Ammonium Sulfite	>0.05	—
Ammonium Thiocyanate	<0.02	—
Amyl Acetate	—	<0.02
Amyl Chloride	—	<0.05
Aniline	<0.02	<0.02
Aniline Hydrochloride	>0.05	>0.05
Anthracine	—	<0.02
Antimony Trichloride	>0.05	>0.05
Barium Carbonate	<0.02	<0.02
Barium Chloride	<0.02	<0.02

* 10% corrosive medium in 90% water. (Other % corrosive medium in parentheses.)
** Water-free, dry or maximum concentration of corrosive medium.

Source: data compiled by J.S. Park from Earl R. Parker, *Materials Data Book for Engineers and Scientists*, McGraw-Hill Book Company, New York, 1967.

Table 319. CORROSION RATES OF 17% CR STEEL AT 70°F *

(SHEET 3 OF 9)

Corrosive Medium	Corrosion Rate* in 10% Corrosive Medium (ipy)	Corrosion Rate** in 100% Corrosive Medium (ipy)
Barium Hydroxide	—	<0.02
Barium Nitrate	<0.02	—
Barium Oxide	—	<0.02
Benzaldehyde	—	<0.02
Benzene	<0.02	<0.02
Benzoic Acid	<0.02	<0.02
Boric Acid	<0.02	<0.02
Bromic Acid	>0.05	>0.05
Bromine (Dry)	—	>0.05
Bromine (Wet)	—	>0.05
Butyric Acid	<0.05	<0.05
Cadmium Chloride	>0.05	—
Cadmium Sulfate	<0.002	—
Calcium Acetate	<0.02	<0.02
Calcium Bicarbonate	—	<0.02
Calcium Bromide	<0.02	<0.02
Calcium Chlorate	<0.02	—
Calcium Chloride	<0.05	<0.02
Calcium Hydroxide	<0.02	<0.02
Calcium Hypochlorite	>0.05	>0.05
Carbon Dioxide	—	<0.002
Carbon Monoxide	—	<0.002
Carbon Tetrachloride	<0.002	<0.002
Carbon Acid (Air Free)	—	<0.002

* 10% corrosive medium in 90% water. (Other % corrosive medium in parentheses.)
** Water-free, dry or maximum concentration of corrosive medium.

Source: data compiled by J.S. Park from Earl R. Parker, *Materials Data Book for Engineers and Scientists*, McGraw-Hill Book Company, New York, 1967.

Table 319. CORROSION RATES OF 17% CR STEEL AT 70°F *

(SHEET 4 OF 9)

Corrosive Medium	Corrosion Rate* in 10% Corrosive Medium (ipy)	Corrosion Rate** in 100% Corrosive Medium (ipy)
Chloroacetic Acid	>0.05	>0.05
Chlorine Gas	>0.05	<0.05
Chloroform (Dry)	—	<0.02
Chromic Acid	<0.02	—
Chromic Hydroxide	—	<0.02
Chromic Sulfates	>0.05	>0.05
Citric Acid	<0.02	—
Copper Nitrate	<0.02	—
Copper Sulfate	<0.02	—
Diethylene Glycol	—	<0.002
Ethyl Chloride	>0.05 (90%)	<0.002
Ethylene Glycol	—	<0.02
Ethylene Oxide	—	<0.02
Fatty Acids	—	<0.02
Ferric Chloride	>0.05	—
Ferric Nitrate	<0.02	—
Ferrous Chloride	>0.05	—
Ferrous Sulfate	<0.02	—
Fluorine	—	<0.002
Formaldehyde	<0.002	<0.002
Formic Acid	<0.05	<0.05
Furfural	<0.002 (30%)	—
Hydrobromic Acid	>0.05	—
Hydrochloric Acid (Areated)	>0.05	—

* 10% corrosive medium in 90% water. (Other % corrosive medium in parentheses.)
** Water-free, dry or maximum concentration of corrosive medium.

Source: data compiled by J.S. Park from Earl R. Parker, *Materials Data Book for Engineers and Scientists*, McGraw-Hill Book Company, New York, 1967.

Table 319. CORROSION RATES OF 17% CR STEEL AT 70°F *

(SHEET 5 OF 9)

Corrosive Medium	Corrosion Rate* in 10% Corrosive Medium (ipy)	Corrosion Rate** in 100% Corrosive Medium (ipy)
Hydrochloric Acid (Air Free)	>0.05	—
Hydrocyanic Acid	—	<0.05
Hydrofluoric Acid (Air Free)	>0.05	>0.05
Hydrogen Chloride	>0.05 (90%)	>0.05
Hydrogen Fluoride	—	<0.02
Hydrogen Iodide	—	>0.05
Hydrogen Peroxide	<0.02 (20%)	<0.02
Hydrogen Sulfide	<0.02	<0.05
Lactic Acid	>0.05	—
Lead Acetate	<0.02	<0.02
Lead Chromate	—	<0.02
Lead Nitrate	<0.02	—
Lead Sulfate	—	<0.02
Lithium Hydroxide	<0.02	—
Magnesium Chloride	<0.05	—
Magnesium Hydroxide	<0.02	<0.02
Magnesium Sulfate	<0.002	<0.02
Maleic Acid	<0.02	<0.02
Malic Acid	<0.02	—
Mercuric Chloride	>0.05	>0.05
Mercurous Nitrate	<0.02	—
Methallylamine	<0.02	<0.02
Methanol	<0.02	<0.002
Methyl Ethyl Ketone	<0.02	<0.002

* 10% corrosive medium in 90% water. (Other % corrosive medium in parentheses.)
** Water-free, dry or maximum concentration of corrosive medium.

Source: data compiled by J.S. Park from Earl R. Parker, *Materials Data Book for Engineers and Scientists*, McGraw-Hill Book Company, New York, 1967.

Table 319. CORROSION RATES OF 17% CR STEEL AT 70°F *

(SHEET 6 OF 9)

Corrosive Medium	Corrosion Rate* in 10% Corrosive Medium (ipy)	Corrosion Rate** in 100% Corrosive Medium (ipy)
Methyl Isobutyl Ketone	<0.02	<0.02
Methylamine	<0.02	<0.02
Methylene Chloride	—	<0.02
Monochloroacetic Acid	>0.05	>0.05
Monorthanolamine	<0.002	—
Monoethalamine	<0.02	<0.02
Monoethylamine	<0.02	<0.02
Monosodium Phosphate	>0.05	—
Nickel Chloride	>0.05	—
Nickel Nitrate	<0.02	—
Nitric Acid	<0.02	<0.05
Nitric Acid (Red Fuming)	—	<0.002
Nitric + Hydrochloric Acid	—	>0.05
Nitric + Hydrofluoric Acid	—	>0.05
Nitric + Sulfuric Acid	—	>0.05
Nitrobenzene	—	<0.02
Nitrocelluolose	—	<0.02
Nitroglycerine	—	<0.02
Nitrotolune	—	<0.02
Nitrous Acid	<0.02	—
Oleic Acid	<0.02	<0.02
Oxalic Acid	>0.05	>0.05
Phenol	—	<0.02
Phosphoric Acid (Areated)	<0.02	—

* 10% corrosive medium in 90% water. (Other % corrosive medium in parentheses.)
** Water-free, dry or maximum concentration of corrosive medium.

Source: data compiled by J.S. Park from Earl R. Parker, *Materials Data Book for Engineers and Scientists*, McGraw-Hill Book Company, New York, 1967.

Table 319. CORROSION RATES OF 17% CR STEEL AT 70°F *

(SHEET 7 OF 9)

Corrosive Medium	Corrosion Rate* in 10% Corrosive Medium (ipy)	Corrosion Rate** in 100% Corrosive Medium (ipy)
Phosphoric Acid (Air Free)	>0.05	>0.05
Picric Acid	<0.02	<0.02
Potassium Bicarbonate	<0.02	<0.02
Potassium Bromide	<0.02	<0.02
Potassium Carbonate	<0.02	<0.02
Potassium Chlorate	<0.02	<0.02
Potassium Chromate	<0.02	<0.02
Potassium Cyanide	<0.02	<0.02
Potassium Dichromate	<0.02	<0.02
Potassium Ferricyanide	<0.02	<0.02
Potassium Ferrocyanide	<0.02	—
Potassium Hydroxide	<0.02	<0.002
Potassium Hypochlorite	>0.05	—
Potassium Iodide	>0.05	—
Potassium Nitrate	<0.02	<0.02
Potassium Nitrite	<0.02	<0.02
Potassium Permanganate	<0.02	<0.02
Potassium Silicate	<0.02	<0.02
Pyridine	<0.02	<0.02
Quinine Sulfate	<0.02	<0.02
Salicylic Acid	—	<0.02
Silicon Tetrachloride (Dry)	—	<0.002
Silicon Tetrachloride (Wet)	—	>0.05
Silver Bromide	>0.05	>0.05

* 10% corrosive medium in 90% water. (Other % corrosive medium in parentheses.)
** Water-free, dry or maximum concentration of corrosive medium.

Source: data compiled by J.S. Park from Earl R. Parker, *Materials Data Book for Engineers and Scientists*, McGraw-Hill Book Company, New York, 1967.

Table 319. CORROSION RATES OF 17% CR STEEL AT 70°F *

(SHEET 8 OF 9)

Corrosive Medium	Corrosion Rate* in 10% Corrosive Medium (ipy)	Corrosion Rate** in 100% Corrosive Medium (ipy)
Silver Chloride	>0.05	>0.05
Silver Nitrate	<0.02	—
Sodium Acetate	<0.02	<0.02
Sodium Bicarbonate	<0.02	<0.02
Sodium Bisulfate	<0.002	—
Sodium Bromide	<0.05	—
Sodium Carbonate	<0.02	<0.02
Sodium Chloride	<0.02	—
Sodium Chromate	<0.02	<0.02
Sodium Hydroxide	<0.002	—
Sodium Hypochlorite	>0.05	>0.05
Sodium Metasilicate	<0.002	<0.002
Sodium Nitrate	<0.02	<0.002
Sodium Nitrite	<0.02	—
Sodium Phosphate	<0.02	<0.02
Sodium Silicate	<0.02	<0.02
Sodium Sulfate	<0.05	>0.05
Sodium Sulfide	>0.05	>0.05
Sodium Sulfite	<0.02	—
Stannic Chloride	>0.05	—
Stannous Chloride	>0.05	<0.05
Strontium Nitrate	<0.02	<0.02
Succinic Acid	<0.02	<0.02
Sulfur Dioxide	>0.05	<0.02

* 10% corrosive medium in 90% water. (Other % corrosive medium in parentheses.)
** Water-free, dry or maximum concentration of corrosive medium.

Source: data compiled by J.S. Park from Earl R. Parker, *Materials Data Book for Engineers and Scientists*, McGraw-Hill Book Company, New York, 1967.

Table 319. CORROSION RATES OF 17% CR STEEL AT 70°F *

(SHEET 9 OF 9)

Corrosive Medium	Corrosion Rate* in 10% Corrosive Medium (ipy)	Corrosion Rate** in 100% Corrosive Medium (ipy)
Sulfur Trioxide	—	<0.02
Sulfuric Acid (Areated)	<0.05	>0.05
Sulfuric Acid (Air Free)	>0.05	<0.05
Sulfuric Acid (Fuming)	—	<0.002
Sulfurous Acid	>0.05	>0.05
Tannic Acid	<0.02	<0.02
Tartaric Acid	<0.02	—
Tetraphosphoric Acid	>0.05	>0.05
Trichloroacetic Acid	>0.05	>0.05
Trichloroethylene	—	<0.02
Urea	<0.02	—
Zinc Chloride	—	>0.05
Zinc Sulfate	<0.05	—

* 10% corrosive medium in 90% water. (Other % corrosive medium in parentheses.)
** Water-free, dry or maximum concentration of corrosive medium.

Source: data compiled by J.S. Park from Earl R. Parker, *Materials Data Book for Engineers and Scientists*, McGraw-Hill Book Company, New York, 1967.

* <0.002 means that corrosion rate is likely to be less than 0.002 inch per year (Excellent).
<0.02 means that corrosion rate is likely to be less than about 0.02 inch per year (Good).
<0.05 means that corrosion rate is likely to be less than about 0.05 inch per year (Fair).
>0.05 means that corrosion rate is likely to be more than 0.05 inch per year (Poor).

Table 320. CORROSION RATES OF 14% SI IRON AT 70°F *

(SHEET 1 OF 9)

Corrosive Medium	Corrosion Rate* in 10% Corrosive Medium (ipy)	Corrosion Rate** in 100% Corrosive Medium (ipy)
Acetaldehyde	<0.002	<0.002
Acetic Acid (Aerated)	<0.002	<0.002
Acetic Acid (Air Free)	<0.002	<0.002
Acetic Anhydride	<0.002	<0.002
Acetoacetic Acid	<0.02	<0.02
Acetone	<0.002	<0.002
Acetylene	—	<0.002
Acrolein	<0.02	<0.02
Acrylonitril	—	<0.002
Alcohol (Ethyl)	<0.002	<0.002
Alcohol (Methyl)	<0.002	<0.002
Alcohol (Allyl)	<0.02	<0.02
Alcohol (Amyl)	—	<0.02
Alcohol (Benzyl)	—	<0.02
Alcohol (Butyl)	—	<0.002
Alcohol (Cetyl)	—	<0.02
Alcohol (Isopropyl)	—	<0.02
Allylamine	<0.002 (30%)	<0.02
Allyl Chloride	—	<0.002
Allyl Sulfide	—	<0.02
Aluminum Acetate	<0.02	<0.002
Aluminum Chlorate	<0.02	<0.002
Aluminum Chloride	<0.002	<0.02
Aluminum Fluoride	>0.05	>0.05

* 10% corrosive medium in 90% water. (Other % corrosive medium in parentheses.)
** Water-free, dry or maximum concentration of corrosive medium.

Source: data compiled by J.S. Park from Earl R. Parker, *Materials Data Book for Engineers and Scientists*, McGraw-Hill Book Company, New York, 1967.

Table 320. CORROSION RATES OF 14% SI IRON AT 70°F *

(SHEET 2 OF 9)

Corrosive Medium	Corrosion Rate* in 10% Corrosive Medium (ipy)	Corrosion Rate** in 100% Corrosive Medium (ipy)
Aluminum Fluosilicate	—	<0.02
Aluminum Formate	<0.02	<0.02
Aluminum Hydroxide	<0.02	—
Aluminum Potassium Sulfate	—	<0.002
Aluminum Sulfate	<0.002	<0.02
Ammonia	<0.02	<0.02
Ammonium Acetate	<0.002	<0.02
Ammonium Bicarbonate	<0.002	<0.002
Ammonium Bromide	<0.002	—
Ammonium Carbonate	<0.002	<0.02
Ammonium Chloride	<0.002	<0.02
Ammonium Formate	<0.02	<0.02
Ammonium Nitrate	<0.002	—
Ammonium Sulfate	<0.002	<0.002
Ammonium Sulfite	<0.02	—
Ammonium Thiocyanate	<0.02	—
Amyl Acetate	<0.002	<0.002
Amyl Chloride	<0.02	<0.02
Aniline	<0.002	<0.002
Aniline Hydrochloride	<0.02	<0.02
Anthracine	—	<0.02
Antimony Trichloride	<0.002	—
Barium Carbonate	<0.02	<0.02
Barium Chloride	<0.02	—

* 10% corrosive medium in 90% water. (Other % corrosive medium in parentheses.)
** Water-free, dry or maximum concentration of corrosive medium.

Source: data compiled by J.S. Park from Earl R. Parker, *Materials Data Book for Engineers and Scientists*, McGraw-Hill Book Company, New York, 1967.

Table 320. CORROSION RATES OF 14% SI IRON AT 70°F *
(SHEET 3 OF 9)

Corrosive Medium	Corrosion Rate* in 10% Corrosive Medium (ipy)	Corrosion Rate** in 100% Corrosive Medium (ipy)
Barium Hydroxide	—	<0.02
Barium Nitrate	<0.02	<0.02
Barium Oxide	—	<0.02
Barium Peroxide	<0.02	—
Benzaldehyde	<0.02	<0.02
Benzene	<0.002	<0.002
Benzoic Acid	<0.02	<0.02
Boric Acid	<0.02	<0.02
Bromine (Dry)	—	>0.05
Bromine (Wet)	—	>0.05
Butyric Acid	<0.002	<0.002
Cadmium Chloride	<0.02	—
Cadmium Sulfate	<0.002	—
Calcium Acetate	<0.02	<0.02
Calcium Bicarbonate	—	<0.02
Calcium Bromide	—	<0.02
Calcium Chlorate	<0.02	—
Calcium Chloride	<0.002	<0.02
Calcium Hydroxide	<0.02	—
Calcium Hypochlorite	<0.02	<0.05
Carbon Dioxide	—	<0.002
Carbon Monoxide	—	<0.002
Carbon Tetrachloride	<0.002	<0.002
Carbon Acid (Air Free)	<0.02	<0.002

* 10% corrosive medium in 90% water. (Other % corrosive medium in parentheses.)
** Water-free, dry or maximum concentration of corrosive medium.

Source: data compiled by J.S. Park from Earl R. Parker, *Materials Data Book for Engineers and Scientists*, McGraw-Hill Book Company, New York, 1967.

Table 320. CORROSION RATES OF 14% SI IRON AT 70°F *

(SHEET 4 OF 9)

Corrosive Medium	Corrosion Rate* in 10% Corrosive Medium (ipy)	Corrosion Rate** in 100% Corrosive Medium (ipy)
Chloroacetic Acid	>0.05	>0.05
Chlorine Gas	—	<0.02
Chromic Acid	<0.002	<0.02
Chromic Hydroxide	—	<0.02
Chromic Sulfates	<0.002	<0.02
Citric Acid	<0.002	<0.002
Copper Nitrate	<0.002	—
Copper Sulfate	<0.002	—
Diethylene Glycol	—	<0.002
Ethyl Chloride	—	<0.002
Ethylene Glycol	<0.02	<0.02
Ethylene Oxide	—	<0.02
Fatty Acids	—	<0.002
Ferric Chloride	>0.05	—
Ferric Nitrate	<0.02	—
Ferrous Chloride	>0.05	—
Ferrous Sulfate	<0.02	—
Fluorine	—	>0.05
Formaldehyde	<0.002	<0.002
Formic Acid	<0.002	<0.002
Furfural	<0.02 (20%)	<0.02
Hydrobromic Acid	>0.05	>0.05
Hydrochloric Acid (Areated)	<0.02	—
Hydrochloric Acid (Air Free)	<0.02	—

* 10% corrosive medium in 90% water. (Other % corrosive medium in parentheses.)
** Water-free, dry or maximum concentration of corrosive medium.

Source: data compiled by J.S. Park from Earl R. Parker, *Materials Data Book for Engineers and Scientists*, McGraw-Hill Book Company, New York, 1967.

Table 320. Corrosion Rates of 14% Si Iron at 70°F *

(Sheet 5 of 9)

Corrosive Medium	Corrosion Rate* in 10% Corrosive Medium (ipy)	Corrosion Rate** in 100% Corrosive Medium (ipy)
Hydrocyanic Acid	—	<0.02
Hydrofluoric Acid (Areated)	>0.05	>0.05
Hydrofluoric Acid (Air Free)	>0.05	>0.05
Hydrogen Chloride	<0.02 (90%)	<0.02
Hydrogen Iodide	>0.05	<0.02
Hydrogen Peroxide	<0.02 (20%)	<0.02
Hydrogen Sulfide	—	<0.02
Lactic Acid	<0.002	<0.02
Lead Acetate	<0.02	<0.05
Lead Chromate	—	<0.02
Lead Nitrate	<0.002	<0.002
Lead Sulfate	—	<0.02
Lithium Chloride	<0.02 (30%)	<0.02
Lithium Hydroxide	>0.05	—
Magnesium Chloride	<0.002	>0.05
Magnesium Hydroxide	<0.02	—
Magnesium Sulfate	<0.002	<0.002
Maleic Acid	<0.02	<0.02
Mercuric Chloride	<0.02	<0.02
Mercurous Nitrate	<0.02	<0.002
Methallylamine	<0.02	<0.002
Methanol	<0.002	<0.002
Methyl Ethyl Ketone	<0.02	<0.002
Methyl Isobutyl Ketone	<0.02	<0.02

* 10% corrosive medium in 90% water. (Other % corrosive medium in parentheses.)
** Water-free, dry or maximum concentration of corrosive medium.

Source: data compiled by J.S. Park from Earl R. Parker, *Materials Data Book for Engineers and Scientists*, McGraw-Hill Book Company, New York, 1967.

Table 320. CORROSION RATES OF 14% SI IRON AT 70°F *

(SHEET 6 OF 9)

Corrosive Medium	Corrosion Rate* in 10% Corrosive Medium (ipy)	Corrosion Rate** in 100% Corrosive Medium (ipy)
Methylamine	<0.02	<0.02
Methylene Chloride	—	<0.02
Monochloroacetic Acid	<0.02	<0.02
Monoethalamine	<0.02	<0.02
Monoethylamine	<0.02	<0.02
Monosodium Phosphate	<0.02	—
Nickel Chloride	<0.02	—
Nickel Nitrate	<0.002	—
Nickel Sulfate	<0.002	—
Nitric Acid	<0.002	<0.002
Nitric Acid (Red Fuming)	—	<0.002
Nitric + Hydrochloric Acid	—	<0.05
Nitric + Hydrofluoric Acid	—	>0.05
Nitric + Sulfuric Acid	<0.02	<0.02
Nitrobenzene	—	<0.002
Nitrocelluolose	—	<0.02
Nitroglycerine	—	<0.05
Nitrotolune	—	<0.02
Nitrous Acid	<0.002	<0.002
Oleic Acid	<0.002	<0.002
Oxalic Acid	<0.02	<0.02
Phenol	—	<0.002
Phosphoric Acid (Areated)	<0.002	<0.002
Phosphoric Acid (Air Free)	<0.02	<0.02

* 10% corrosive medium in 90% water. (Other % corrosive medium in parentheses.)
** Water-free, dry or maximum concentration of corrosive medium.

Source: data compiled by J.S. Park from Earl R. Parker, *Materials Data Book for Engineers and Scientists*, McGraw-Hill Book Company, New York, 1967.

Table 320. CORROSION RATES OF 14% SI IRON AT 70°F *

(SHEET 7 OF 9)

Corrosive Medium	Corrosion Rate* in 10% Corrosive Medium (ipy)	Corrosion Rate** in 100% Corrosive Medium (ipy)
Picric Acid	<0.02	<0.02
Potassium Bicarbonate	<0.02	—
Potassium Bromide	<0.02	<0.02
Potassium Carbonate	<0.02	<0.02
Potassium Chlorate	<0.02	<0.02
Potassium Chromate	<0.02	—
Potassium Cyanide	<0.02	<0.02
Potassium Dichromate	<0.002	—
Potassium Ferricyanide	<0.02	—
Potassium Ferrocyanide	<0.02	—
Potassium Hydroxide	>0.05	>0.05
Potassium Hypochlorite	<0.002	<0.002
Potassium Iodide	<0.02	<0.02
Potassium Nitrate	<0.002	<0.002
Potassium Nitrite	<0.02	<0.02
Potassium Permanganate	<0.02	—
Potassium Silicate	<0.02	<0.02
Propionic Acid	<0.02	<0.02
Pyridine	<0.02	<0.02
Quinine Sulfate	<0.02	<0.02
Salicylic Acid	—	<0.02
Silicon Tetrachloride (Dry)	—	<0.002
Silicon Tetrachloride (Wet)	—	<0.002
Silver Bromide	—	<0.02

* 10% corrosive medium in 90% water. (Other % corrosive medium in parentheses.)
** Water-free, dry or maximum concentration of corrosive medium.

Source: data compiled by J.S. Park from Earl R. Parker, *Materials Data Book for Engineers and Scientists*, McGraw-Hill Book Company, New York, 1967.

Table 320. CORROSION RATES OF 14% SI IRON AT 70°F *

(SHEET 8 OF 9)

Corrosive Medium	Corrosion Rate* in 10% Corrosive Medium (ipy)	Corrosion Rate** in 100% Corrosive Medium (ipy)
Silver Chloride	—	<0.02
Silver Nitrate	<0.002	—
Sodium Acetate	<0.002	<0.02
Sodium Bicarbonate	<0.002	—
Sodium Bisulfate	<0.002	<0.002
Sodium Bromide	<0.05	—
Sodium Carbonate	<0.02	<0.02
Sodium Chloride	<0.02	—
Sodium Chromate	<0.02	<0.02
Sodium Hydroxide	>0.05	—
Sodium Metasilicate	<0.02	<0.02
Sodium Nitrate	<0.002	<0.002
Sodium Nitrite	<0.02	—
Sodium Phosphate	<0.02	<0.02
Sodium Silicate	<0.02	<0.02
Sodium Sulfate	<0.002	<0.002
Sodium Sulfide	<0.02	<0.02
Sodium Sulfite	<0.002	—
Stannic Chloride	>0.05	—
Stannous Chloride	<0.002	—
Strontium Nitrate	<0.02	<0.02
Succinic Acid	<0.02	—
Sulfur Dioxide	—	>0.05
Sulfur Trioxide	—	>0.05

* 10% corrosive medium in 90% water. (Other % corrosive medium in parentheses.)
** Water-free, dry or maximum concentration of corrosive medium.

Source: data compiled by J.S. Park from Earl R. Parker, *Materials Data Book for Engineers and Scientists*, McGraw-Hill Book Company, New York, 1967.

Table 320. CORROSION RATES OF 14% SI IRON AT 70°F *

(SHEET 9 OF 9)

Corrosive Medium	Corrosion Rate* in 10% Corrosive Medium (ipy)	Corrosion Rate** in 100% Corrosive Medium (ipy)
Sulfuric Acid (Areated)	<0.002	<0.02
Sulfuric Acid (Fuming)	—	<0.02
Sulfurous Acid	<0.02	<0.02
Tannic Acid	<0.002	<0.002
Tartaric Acid	<0.02	<0.02
Tetraphosphoric Acid	—	<0.05
Trichloroacetic Acid	<0.002	<0.002
Trichloroethylene	—	<0.002
Urea	<0.02	—
Zinc Sulfate	<0.002	—

* 10% corrosive medium in 90% water. (Other % corrosive medium in parentheses.)
** Water-free, dry or maximum concentration of corrosive medium.

Source: data compiled by J.S. Park from Earl R. Parker, *Materials Data Book for Engineers and Scientists*, McGraw-Hill Book Company, New York, 1967.

* <0.002 means that corrosion rate is likely to be less than 0.002 inch per year (Excellent).
<0.02 means that corrosion rate is likely to be less than about 0.02 inch per year (Good).
<0.05 means that corrosion rate is likely to be less than about 0.05 inch per year (Fair).
>0.05 means that corrosion rate is likely to be more than 0.05 inch per year (Poor).

Table 321. CORROSION RATES OF STAINLESS STEEL 301 * AT 70°F (SHEET 1 OF 9)

Corrosive Medium	Corrosion Rate* in 10% Corrosive Medium (ipy)	Corrosion Rate** in 100% Corrosive Medium (ipy)
Acetaldehyde	—	<0.002
Acetic Acid (Aerated)	<0.002	<0.002
Acetic Acid (Air Free)	<0.02	<0.002
Acetic Anhydride	—	<0.02
Acetoacetic Acid	<0.02	<0.02
Acetone	<0.02	<0.002
Acetylene	—	<0.002
Acrolein	<0.02	<0.002
Acrylonitril	—	<0.002
Alcohol (Ethyl)	<0.02	<0.02
Alcohol (Methyl)	<0.02	<0.02
Alcohol (Allyl)	—	<0.02
Alcohol (Amyl)	—	<0.02
Alcohol (Benzyl)	—	<0.02
Alcohol (Butyl)	—	<0.002
Alcohol (Cetyl)	—	<0.02
Alcohol (Isopropyl)	—	<0.02
Allylamine	<0.002 (30%)	<0.02
Allyl Chloride	—	<0.02
Allyl Sulfide	—	<0.02
Aluminum Acetate	<0.02	<0.02
Aluminum Chlorate	<0.002	—
Aluminum Chloride	>0.05	<0.002
Aluminum Fluoride	>0.05	>0.05

* 10% corrosive medium in 90% water. (Other % corrosive medium in parentheses.)
** Water-free, dry or maximum concentration of corrosive medium.

Source: data compiled by J.S. Park from Earl R. Parker, *Materials Data Book for Engineers and Scientists*, McGraw-Hill Book Company, New York, 1967.

Table 321. CORROSION RATES OF STAINLESS STEEL 301 * AT 70°F (SHEET 2 OF 9)

Corrosive Medium	Corrosion Rate* in 10% Corrosive Medium (ipy)	Corrosion Rate** in 100% Corrosive Medium (ipy)
Aluminum Fluosilicate	—	<0.02
Aluminum Formate	<0.02	<0.02
Aluminum Hydroxide	<0.02	<0.02
Aluminum Nitrate	<0.02	<0.02
Aluminum Potassium Sulfate	<0.02	<0.02
Aluminum Sulfate	<0.02	<0.02
Ammonia	<0.002	<0.002
Ammonium Acetate	<0.002	<0.002
Ammonium Bicarbonate	<0.02	<0.05
Ammonium Bromide	<0.05	<0.05
Ammonium Carbonate	<0.02	<0.02
Ammonium Chloride	<0.02	>0.05
Ammonium Citrate	<0.02	—
Ammonium Formate	<0.02	<0.02
Ammonium Nitrate	<0.002	<0.002
Ammonium Sulfate	<0.05	—
Ammonium Sulfite	<0.05	<0.05
Ammonium Thiocyanate	<0.02	—
Amyl Acetate	<0.002	<0.002
Amyl Chloride	>0.05	<0.002
Aniline	<0.02	<0.02
Aniline Hydrochloride	>0.05	>0.05
Anthracine	—	<0.02
Antimony Trichloride	>0.05	>0.05

* 10% corrosive medium in 90% water. (Other % corrosive medium in parentheses.)
** Water-free, dry or maximum concentration of corrosive medium.

Source: data compiled by J.S. Park from Earl R. Parker, *Materials Data Book for Engineers and Scientists*, McGraw-Hill Book Company, New York, 1967.

Table 321. CORROSION RATES OF STAINLESS STEEL 301 * AT 70°F (SHEET 3 OF 9)

Corrosive Medium	Corrosion Rate* in 10% Corrosive Medium (ipy)	Corrosion Rate** in 100% Corrosive Medium (ipy)
Barium Carbonate	<0.02	<0.02
Barium Chloride	<0.02	<0.05
Barium Hydroxide	—	<0.02
Barium Nitrate	<0.02	<0.02
Barium Oxide	—	<0.02
Barium Peroxide	<0.02	—
Benzaldehyde	<0.02	<0.02
Benzene	<0.02	<0.02
Benzoic Acid	<0.02	<0.02
Boric Acid	<0.002	<0.02
Bromic Acid	>0.05	—
Bromine (Dry)	—	>0.05
Bromine (Wet)	—	>0.05
Butyric Acid	<0.02	<0.02
Cadmium Chloride	<0.02	—
Cadmium Sulfate	<0.002	—
Calcium Acetate	<0.02	<0.02
Calcium Bicarbonate	—	<0.02
Calcium Bromide	<0.02	<0.02
Calcium Chlorate	<0.02	—
Calcium Chloride	<0.02	<0.02
Calcium Hydroxide	<0.02	—
Calcium Hypochlorite	<0.05	—
Carbon Dioxide	—	<0.002

* 10% corrosive medium in 90% water. (Other % corrosive medium in parentheses.)
** Water-free, dry or maximum concentration of corrosive medium.

Source: data compiled by J.S. Park from Earl R. Parker, *Materials Data Book for Engineers and Scientists*, McGraw-Hill Book Company, New York, 1967.

Table 321. CORROSION RATES OF STAINLESS STEEL 301 * AT 70°F (SHEET 4 OF 9)

Corrosive Medium	Corrosion Rate* in 10% Corrosive Medium (ipy)	Corrosion Rate** in 100% Corrosive Medium (ipy)
Carbon Monoxide	—	<0.002
Carbon Tetrachloride	>0.05	<0.02
Carbon Acid (Air Free)	<0.02	<0.02
Chloroacetic Acid	>0.05	—
Chlorine Gas	—	<0.002
Chloroform (Dry)	—	<0.002
Chromic Acid	<0.02	—
Chromic Hydroxide	—	<0.02
Chromic Sulfates	<0.02	<0.05
Citric Acid	<0.02	<0.02
Copper Nitrate	<0.02	—
Copper Sulfate	<0.02	—
Diethylene Glycol	—	<0.002
Ethyl Chloride	>0.05 (90%)	<0.002
Ethylene Glycol	—	<0.02
Ethylene Oxide	—	<0.02
Fatty Acids	—	<0.02
Ferric Chloride	>0.05	—
Ferric Nitrate	<0.02	—
Ferrous Chloride	>0.05	—
Ferrous Sulfate	<0.02	—
Fluorine	—	<0.002
Formaldehyde	<0.002 (20%)	<0.002
Formic Acid	<0.02	<0.02

* 10% corrosive medium in 90% water. (Other % corrosive medium in parentheses.)
** Water-free, dry or maximum concentration of corrosive medium.

Source: data compiled by J.S. Park from Earl R. Parker, *Materials Data Book for Engineers and Scientists*, McGraw-Hill Book Company, New York, 1967.

Table 321. CORROSION RATES OF STAINLESS STEEL 301 * AT 70°F (SHEET 5 OF 9)

Corrosive Medium	Corrosion Rate* in 10% Corrosive Medium (ipy)	Corrosion Rate** in 100% Corrosive Medium (ipy)
Furfural	<0.002 (30%)	<0.02
Hydrazine	<0.002	—
Hydrobromic Acid	>0.05	>0.05
Hydrochloric Acid (Areated)	>0.05	—
Hydrochloric Acid (Air Free)	>0.05	—
Hydrocyanic Acid	—	<0.02
Hydrofluoric Acid (Areated)	<0.002	<0.02
Hydrofluoric Acid (Air Free)	>0.05	>0.05
Hydrogen Chloride	>0.05 (90%)	<0.002
Hydrogen Fluoride	—	<0.002
Hydrogen Iodide	<0.02 (1%)	<0.02
Hydrogen Peroxide	<0.02 (20%)	<0.02
Hydrogen Sulfide	>0.05	<0.05
Lactic Acid	<0.02	<0.02
Lead Acetate	<0.02	<0.02
Lead Chromate	—	<0.02
Lead Nitrate	<0.02	<0.02
Lead Sulfate	—	<0.02
Lithium Chloride	<0.002 (30%)	<0.002
Lithium Hydroxide	<0.02	—
Magnesium Chloride	<0.05	—
Magnesium Hydroxide	<0.02	<0.02
Magnesium Sulfate	<0.002	<0.02
Maleic Acid	<0.02	<0.02

* 10% corrosive medium in 90% water. (Other % corrosive medium in parentheses.)
** Water-free, dry or maximum concentration of corrosive medium.

Source: data compiled by J.S. Park from Earl R. Parker, *Materials Data Book for Engineers and Scientists*, McGraw-Hill Book Company, New York, 1967.

Table 321. CORROSION RATES OF STAINLESS STEEL 301* AT 70°F (SHEET 6 OF 9)

Corrosive Medium	Corrosion Rate* in 10% Corrosive Medium (ipy)	Corrosion Rate** in 100% Corrosive Medium (ipy)
Malic Acid	<0.002	<0.002
Maganous Chloride	<0.02 (40%)	—
Mercuric Chloride	>0.05	>0.05
Mercurous Nitrate	<0.02	<0.02
Methallylamine	<0.02	<0.02
Methanol	<0.02	<0.002
Methyl Ethyl Ketone	<0.02	<0.002
Methyl Isobutyl Ketone	<0.02	<0.02
Methylamine	<0.02	<0.02
Methylene Chloride	<0.02	<0.02
Monochloroacetic Acid	<0.05	<0.02
Monorthanolamine	<0.002	<0.02
Monoethalamine	<0.02	<0.02
Monoethylamine	<0.02	<0.02
Monosodium Phosphate	<0.02	—
Nickel Chloride	>0.05	—
Nickel Nitrate	<0.02	—
Nickel Sulfate	<0.002	—
Nitric Acid	<0.002	<0.002
Nitric Acid (Red Fuming)	—	<0.002
Nitric + Hydrochloric Acid	—	>0.05
Nitric + Hydrofluoric Acid	—	>0.05
Nitric + Sulfuric Acid	—	>0.05
Nitrobenzene	—	<0.02

* 10% corrosive medium in 90% water. (Other % corrosive medium in parentheses.)
** Water-free, dry or maximum concentration of corrosive medium.

Source: data compiled by J.S. Park from Earl R. Parker, *Materials Data Book for Engineers and Scientists*, McGraw-Hill Book Company, New York, 1967.

Table 321. CORROSION RATES OF STAINLESS STEEL 301 [*] AT 70°F (SHEET 7 OF 9)

Corrosive Medium	Corrosion Rate* in 10% Corrosive Medium (ipy)	Corrosion Rate** in 100% Corrosive Medium (ipy)
Nitrocellulose	—	<0.02
Nitroglycerine	—	<0.02
Nitrotolune	—	<0.02
Nitrous Acid	<0.02	<0.02
Oleic Acid	<0.02	<0.02
Oxalic Acid	<0.02	>0.05
Phenol	—	<0.02
Phosphoric Acid (Areated)	<0.02	>0.05
Phosphoric Acid (Air Free)	<0.02	—
Picric Acid	<0.02	<0.02
Potassium Bicarbonate	<0.02	<0.02
Potassium Bromide	<0.02	<0.05
Potassium Carbonate	<0.02	<0.02
Potassium Chlorate	<0.02	<0.02
Potassium Chromate	<0.02	<0.02
Potassium Cyanide	<0.02	<0.02
Potassium Dichromate	<0.002	<0.02
Potassium Ferricyanide	<0.02	<0.02
Potassium Ferrocyanide	<0.02	—
Potassium Hydroxide	<0.02	<0.002
Potassium Hypochlorite	>0.05	—
Potassium Iodide	<0.02	<0.02
Potassium Nitrate	<0.02	<0.02
Potassium Nitrite	<0.02	<0.02

* 10% corrosive medium in 90% water. (Other % corrosive medium in parentheses.)
** Water-free, dry or maximum concentration of corrosive medium.

Source: data compiled by J.S. Park from Earl R. Parker, *Materials Data Book for Engineers and Scientists*, McGraw-Hill Book Company, New York, 1967.

Table 321. CORROSION RATES OF STAINLESS STEEL 301 * AT 70°F (SHEET 8 OF 9)

Corrosive Medium	Corrosion Rate* in 10% Corrosive Medium (ipy)	Corrosion Rate** in 100% Corrosive Medium (ipy)
Potassium Permanganate	<0.02	<0.02
Potassium Silicate	<0.02	<0.02
Pyridine	<0.02	<0.02
Quinine Sulfate	<0.02	<0.02
Salicylic Acid	—	<0.02
Silicon Tetrachloride (Dry)	—	<0.002
Silicon Tetrachloride (Wet)	—	>0.05
Silver Bromide	>0.05	<0.05
Silver Chloride	>0.05	>0.05
Silver Nitrate	<0.02	—
Sodium Acetate	<0.02	<0.02
Sodium Bicarbonate	<0.02	—
Sodium Bisulfate	<0.002	>0.05
Sodium Bromide	<0.05	—
Sodium Carbonate	<0.02	<0.02
Sodium Chloride	<0.02	—
Sodium Chromate	<0.02	<0.02
Sodium Hydroxide	<0.002	—
Sodium Hypochlorite	>0.05	>0.05
Sodium Metasilicate	<0.002	<0.002
Sodium Nitrate	<0.002	<0.02
Sodium Nitrite	<0.02	<0.02
Sodium Phosphate	<0.02	<0.02
Sodium Silicate	<0.02	<0.02

* 10% corrosive medium in 90% water. (Other % corrosive medium in parentheses.)
** Water-free, dry or maximum concentration of corrosive medium.

Source: data compiled by J.S. Park from Earl R. Parker, *Materials Data Book for Engineers and Scientists*, McGraw-Hill Book Company, New York, 1967.

Table 321. CORROSION RATES OF STAINLESS STEEL 301 * AT 70°F (SHEET 9 OF 9)

Corrosive Medium	Corrosion Rate* in 10% Corrosive Medium (ipy)	Corrosion Rate** in 100% Corrosive Medium (ipy)
Sodium Sulfate	<0.02	<0.002
Sodium Sulfide	<0.02	>0.05
Sodium Sulfite	<0.002	—
Stannic Chloride	>0.05	—
Stannous Chloride	>0.05	<0.05
Strontium Nitrate	<0.02	<0.02
Succinic Acid	<0.02	<0.02
Sulfur Dioxide	>0.05	<0.02
Sulfur Trioxide	—	<0.02
Sulfuric Acid (Areated)	>0.05	<0.02
Sulfuric Acid (Air Free)	>0.05	<0.05
Sulfuric Acid (Fuming)	—	<0.02
Sulfurous Acid	<0.02	>0.05
Tannic Acid	<0.02	<0.02
Tartaric Acid	<0.002	—
Tetraphosphoric Acid	—	<0.02
Trichloroacetic Acid	>0.05	>0.05
Trichloroethylene	—	<0.02
Urea	<0.02	—
Zinc Sulfate	<0.002	—

* 10% corrosive medium in 90% water. (Other % corrosive medium in parentheses.)
** Water-free, dry or maximum concentration of corrosive medium.

Source: data compiled by J.S. Park from Earl R. Parker, *Materials Data Book for Engineers and Scientists*, McGraw-Hill Book Company, New York, 1967.

*
<0.002 means that corrosion rate is likely to be less than 0.002 inch per year (Excellent).
<0.02 means that corrosion rate is likely to be less than about 0.02 inch per year (Good).
<0.05 means that corrosion rate is likely to be less than about 0.05 inch per year (Fair).
>0.05 means that corrosion rate is likely to be more than 0.05 inch per year (Poor).

Table 322. CORROSION RATES OF STAINLESS STEEL 316* AT 70°F (SHEET 1 OF 9)

Corrosive Medium	Corrosion Rate* in 10% Corrosive Medium (ipy)	Corrosion Rate** in 100% Corrosive Medium (ipy)
Acetaldehyde	—	<0.002
Acetic Acid (Aerated)	<0.002	<0.002
Acetic Acid (Air Free)	<0.002	<0.02
Acetic Anhydride	—	<0.02
Acetoacetic Acid	<0.02	<0.02
Acetone	<0.02	<0.002
Acetylene	—	<0.002
Acrolein	<0.02	<0.02
Acrylonitril	—	<0.002
Alcohol (Ethyl)	<0.002	<0.002
Alcohol (Methyl)	<0.002	<0.002
Alcohol (Allyl)	—	<0.02
Alcohol (Amyl)	—	<0.02
Alcohol (Benzyl)	—	<0.02
Alcohol (Butyl)	—	<0.002
Alcohol (Cetyl)	—	<0.02
Alcohol (Isopropyl)	—	<0.02
Allylamine	<0.002 (30%)	<0.02
Allyl Chloride	—	<0.002
Allyl Sulfide	—	<0.02
Aluminum Acetate	<0.02	<0.02
Aluminum Chloride	<0.05	—
Aluminum Fluoride	—	<0.05
Aluminum Fluosilicate	—	<0.02

* 10% corrosive medium in 90% water. (Other % corrosive medium in parentheses.)
** Water-free, dry or maximum concentration of corrosive medium.

Source: data compiled by J.S. Park from Earl R. Parker, *Materials Data Book for Engineers and Scientists*, McGraw-Hill Book Company, New York, 1967.

Table 322. CORROSION RATES OF STAINLESS STEEL 316 * AT 70°F (SHEET 2 OF 9)

Corrosive Medium	Corrosion Rate* in 10% Corrosive Medium (ipy)	Corrosion Rate** in 100% Corrosive Medium (ipy)
Aluminum Formate	<0.02	<0.02
Aluminum Hydroxide	<0.02	<0.02
Aluminum Nitrate	<0.02	<0.02
Aluminum Potassium Sulfate	<0.02	—
Aluminum Sulfate	<0.02	<0.02
Ammonia	<0.002	<0.002
Ammonium Acetate	<0.002	<0.002
Ammonium Bicarbonate	<0.02	<0.02
Ammonium Bromide	<0.02	—
Ammonium Carbonate	<0.02	<0.02
Ammonium Chloride	<0.02	—
Ammonium Citrate	<0.02	—
Ammonium Formate	<0.02	<0.02
Ammonium Nitrate	<0.002	<0.002
Ammonium Sulfate	<0.02	—
Ammonium Sulfite	<0.02	<0.02
Ammonium Thiocyanate	<0.02	—
Amyl Acetate	<0.002	<0.002
Amyl Chloride	—	<0.002
Aniline	<0.02	<0.02
Aniline Hydrochloride	>0.05	>0.05
Anthracine	—	<0.02
Antimony Trichloride	>0.05	—
Barium Carbonate	<0.02	<0.02

* 10% corrosive medium in 90% water. (Other % corrosive medium in parentheses.)
** Water-free, dry or maximum concentration of corrosive medium.

Source: data compiled by J.S. Park from Earl R. Parker, *Materials Data Book for Engineers and Scientists*, McGraw-Hill Book Company, New York, 1967.

Table 322. CORROSION RATES OF STAINLESS STEEL 316 * AT 70°F (SHEET 3 OF 9)

Corrosive Medium	Corrosion Rate* in 10% Corrosive Medium (ipy)	Corrosion Rate** in 100% Corrosive Medium (ipy)
Barium Chloride	<0.02	<0.02
Barium Hydroxide	—	<0.02
Barium Nitrate	<0.02	<0.02
Barium Oxide	—	<0.02
Barium Peroxide	<0.02	—
Benzaldehyde	—	<0.02
Benzene	<0.02	<0.02
Benzoic Acid	<0.02	<0.02
Boric Acid	<0.002	<0.02
Bromic Acid	>0.05	—
Bromine (Dry)	—	>0.05
Bromine (Wet)	—	>0.05
Butyric Acid	<0.02	<0.02
Cadmium Chloride	<0.02	—
Cadmium Sulfate	<0.002	—
Calcium Acetate	<0.02	<0.02
Calcium Bicarbonate	—	<0.02
Calcium Bromide	<0.02	<0.02
Calcium Chlorate	<0.02	—
Calcium Chloride	<0.02	<0.002
Calcium Hydroxide	<0.02	—
Calcium Hypochlorite	<0.05	—
Carbon Dioxide	—	<0.002
Carbon Monoxide	—	<0.002

* 10% corrosive medium in 90% water. (Other % corrosive medium in parentheses.)
** Water-free, dry or maximum concentration of corrosive medium.

Source: data compiled by J.S. Park from Earl R. Parker, *Materials Data Book for Engineers and Scientists*, McGraw-Hill Book Company, New York, 1967.

Table 322. CORROSION RATES OF STAINLESS STEEL 316 * AT 70°F (SHEET 4 OF 9)

Corrosive Medium	Corrosion Rate* in 10% Corrosive Medium (ipy)	Corrosion Rate** in 100% Corrosive Medium (ipy)
Carbon Tetrachloride	<0.02	<0.02
Carbon Acid (Air Free)	<0.02	<0.02
Chloroacetic Acid	>0.05	—
Chlorine Gas	—	<0.02
Chloroform (Dry)	—	<0.002
Chromic Acid	<0.02	—
Chromic Hydroxide	—	<0.02
Chromic Sulfates	<0.02	—
Citric Acid	<0.02	<0.02
Copper Nitrate	<0.002	—
Copper Sulfate	<0.02	—
Diethylene Glycol	—	<0.002
Ethyl Chloride	—	<0.002
Ethylene Glycol	—	<0.02
Ethylene Oxide	—	<0.02
Fatty Acids	—	<0.002
Ferric Chloride	>0.05	—
Ferric Nitrate	<0.02	—
Ferrous Chloride	>0.05	—
Ferrous Sulfate	<0.02	—
Fluorine	—	<0.002
Formaldehyde	<0.02	<0.002
Formic Acid	<0.002	<0.002
Furfural	<0.002	<0.02

* 10% corrosive medium in 90% water. (Other % corrosive medium in parentheses.)
** Water-free, dry or maximum concentration of corrosive medium.

Source: data compiled by J.S. Park from Earl R. Parker, *Materials Data Book for Engineers and Scientists*, McGraw-Hill Book Company, New York, 1967.

Table 322. CORROSION RATES OF STAINLESS STEEL 316 * AT 70°F (SHEET 5 OF 9)

Corrosive Medium	Corrosion Rate* in 10% Corrosive Medium (ipy)	Corrosion Rate** in 100% Corrosive Medium (ipy)
Hydrazine	<0.002	—
Hydrobromic Acid	>0.05	—
Hydrochloric Acid (Areated)	>0.05	—
Hydrochloric Acid (Air Free)	>0.05	—
Hydrocyanic Acid	—	<0.02
Hydrofluoric Acid (Areated)	<0.002	<0.02
Hydrofluoric Acid (Air Free)	>0.05	<0.02
Hydrogen Chloride	—	<0.002
Hydrogen Fluoride	—	<0.002
Hydrogen Iodide	—	<0.02
Hydrogen Peroxide	<0.02 (20%)	<0.02
Hydrogen Sulfide	<0.002	<0.02
Lactic Acid	<0.02	<0.02
Lead Acetate	<0.02	<0.02
Lead Chromate	—	<0.02
Lead Nitrate	<0.02	<0.02
Lead Sulfate	—	<0.02
Lithium Chloride	<0.002 (30%)	<0.002
Lithium Hydroxide	<0.02	—
Magnesium Chloride	<0.02	—
Magnesium Hydroxide	<0.02	<0.02
Magnesium Sulfate	<0.002	<0.02
Maleic Acid	<0.02	<0.02
Malic Acid	<0.002	<0.002

* 10% corrosive medium in 90% water. (Other % corrosive medium in parentheses.)
** Water-free, dry or maximum concentration of corrosive medium.

Source: data compiled by J.S. Park from Earl R. Parker, *Materials Data Book for Engineers and Scientists*, McGraw-Hill Book Company, New York, 1967.

Table 322. CORROSION RATES OF STAINLESS STEEL 316 * AT 70°F (SHEET 6 OF 9)

Corrosive Medium	Corrosion Rate* in 10% Corrosive Medium (ipy)	Corrosion Rate** in 100% Corrosive Medium (ipy)
Maganous Chloride	<0.02 (40%)	—
Mercuric Chloride	>0.05	—
Mercurous Nitrate	<0.02	<0.02
Methallylamine	<0.02	<0.02
Methanol	<0.02	<0.002
Methyl Ethyl Ketone	<0.02	<0.002
Methyl Isobutyl Ketone	<0.02	<0.02
Methylamine	<0.02	<0.02
Methylene Chloride	<0.02	<0.02
Monochloroacetic Acid	<0.05	<0.02
Monorthanolamine	<0.02	<0.02
Monoethalamine	<0.02	<0.02
Monoethylamine	<0.02	<0.02
Monosodium Phosphate	<0.02	—
Nickel Chloride	>0.05	—
Nickel Nitrate	<0.02	—
Nickel Sulfate	<0.02	—
Nitric Acid	<0.002	<0.002
Nitric Acid (Red Fuming)	—	<0.002
Nitric + Hydrochloric Acid	—	>0.05
Nitric + Hydrofluoric Acid	—	>0.05
Nitric + Sulfuric Acid	—	>0.05
Nitrobenzene	—	<0.02
Nitrocelluolose	—	<0.02

* 10% corrosive medium in 90% water. (Other % corrosive medium in parentheses.)
** Water-free, dry or maximum concentration of corrosive medium.

Source: data compiled by J.S. Park from Earl R. Parker, *Materials Data Book for Engineers and Scientists*, McGraw-Hill Book Company, New York, 1967.

Table 322. CORROSION RATES OF STAINLESS STEEL 316 * AT 70°F (SHEET 7 OF 9)

Corrosive Medium	Corrosion Rate* in 10% Corrosive Medium (ipy)	Corrosion Rate** in 100% Corrosive Medium (ipy)
Nitroglycerine	—	<0.02
Nitrotolune	—	<0.02
Nitrous Acid	<0.02	<0.02
Oleic Acid	<0.02	<0.02
Oxalic Acid	<0.02	>0.05
Phenol	—	<0.02
Phosphoric Acid (Areated)	<0.002	<0.02
Phosphoric Acid (Air Free)	<0.02	—
Picric Acid	<0.02	<0.02
Potassium Bicarbonate	<0.02	<0.02
Potassium Bromide	<0.02	—
Potassium Carbonate	<0.02	<0.02
Potassium Chlorate	<0.02	<0.02
Potassium Chromate	<0.02	<0.02
Potassium Cyanide	<0.02	<0.02
Potassium Dichromate	<0.002	<0.02
Potassium Ferricyanide	<0.02	<0.02
Potassium Ferrocyanide	<0.02	—
Potassium Hydroxide	<0.02	—
Potassium Hypochlorite	<0.05	<0.02
Potassium Iodide	<0.02	<0.02
Potassium Nitrate	<0.02	—
Potassium Nitrite	<0.02	<0.02
Potassium Permanganate	<0.02	—

* 10% corrosive medium in 90% water. (Other % corrosive medium in parentheses.)
** Water-free, dry or maximum concentration of corrosive medium.

Source: data compiled by J.S. Park from Earl R. Parker, *Materials Data Book for Engineers and Scientists*, McGraw-Hill Book Company, New York, 1967.

Table 322. CORROSION RATES OF STAINLESS STEEL 316 * AT 70°F (SHEET 8 OF 9)

Corrosive Medium	Corrosion Rate* in 10% Corrosive Medium (ipy)	Corrosion Rate** in 100% Corrosive Medium (ipy)
Potassium Silicate	<0.02	<0.02
Propionic Acid	—	<0.02
Pyridine	<0.02	<0.02
Quinine Sulfate	<0.02	<0.02
Salicylic Acid	—	<0.02
Silicon Tetrachloride (Dry)	—	<0.002
Silver Bromide	>0.05	—
Silver Chloride	>0.05	—
Silver Nitrate	<0.002	<0.02
Sodium Acetate	<0.02	<0.02
Sodium Bicarbonate	<0.02	—
Sodium Bisulfate	<0.002	—
Sodium Bromide	<0.05	—
Sodium Carbonate	<0.02	<0.02
Sodium Chloride	<0.02	—
Sodium Chromate	<0.02	<0.02
Sodium Hydroxide	<0.002	—
Sodium Hypochlorite	>0.05	>0.05
Sodium Metasilicate	<0.002	<0.002
Sodium Nitrate	<0.002	<0.02
Sodium Nitrite	<0.02	—
Sodium Phosphate	<0.02	<0.02
Sodium Silicate	<0.02	<0.02
Sodium Sulfate	<0.002	<0.002

* 10% corrosive medium in 90% water. (Other % corrosive medium in parentheses.)
** Water-free, dry or maximum concentration of corrosive medium.

Source: data compiled by J.S. Park from Earl R. Parker, *Materials Data Book for Engineers and Scientists*, McGraw-Hill Book Company, New York, 1967.

Table 322. CORROSION RATES OF STAINLESS STEEL 316 * AT 70°F (SHEET 9 OF 9)

Corrosive Medium	Corrosion Rate* in 10% Corrosive Medium (ipy)	Corrosion Rate** in 100% Corrosive Medium (ipy)
Sodium Sulfide	>0.05	—
Sodium Sulfite	<0.002	—
Stannic Chloride	>0.05	—
Stannous Chloride	<0.02	—
Strontium Nitrate	<0.02	<0.02
Succinic Acid	<0.02	<0.02
Sulfur Dioxide	<0.002	<0.02
Sulfur Trioxide	—	<0.02
Sulfuric Acid (Areated)	<0.002	<0.02
Sulfuric Acid (Air Free)	<0.05	<0.02
Sulfuric Acid (Fuming)	—	<0.02
Sulfurous Acid	<0.02	<0.002
Tannic Acid	<0.02	<0.02
Tartaric Acid	<0.02	—
Tetraphosphoric Acid	—	<0.02
Trichloroacetic Acid	>0.05	>0.05
Trichloroethylene	—	<0.02
Urea	<0.02	—
Zinc Sulfate	<0.02	—

* 10% corrosive medium in 90% water. (Other % corrosive medium in parentheses.)
** Water-free, dry or maximum concentration of corrosive medium.

Source: data compiled by J.S. Park from Earl R. Parker, *Materials Data Book for Engineers and Scientists*, McGraw-Hill Book Company, New York, 1967.

* <0.002 means that corrosion rate is likely to be less than 0.002 inch per year (Excellent).
<0.02 means that corrosion rate is likely to be less than about 0.02 inch per year (Good).
<0.05 means that corrosion rate is likely to be less than about 0.05 inch per year (Fair).
>0.05 means that corrosion rate is likely to be more than 0.05 inch per year (Poor).

Table 323. CORROSION RATES OF ALUMINUM AT 70°F *

(SHEET 1 OF 9)

Corrosive Medium	Corrosion Rate* in 10% Corrosive Medium (ipy)	Corrosion Rate** in 100% Corrosive Medium (ipy)
Acetaldehyde	<0.02	<0.002
Acetic Acid (Aerated)	<0.02	<0.002
Acetic Acid (Air Free)	<0.002	<0.002
Acetic Anhydride	—	<0.002
Acetoacetic Acid	<0.02	<0.02
Acetone	<0.02	<0.002
Acetylene	—	<0.002
Acrolein	<0.02	<0.02
Acrylonitril	—	<0.002
Alcohol (Ethyl)	<0.02	<0.02
Alcohol (Methyl)	—	<0.02
Alcohol (Allyl)	—	<0.02
Alcohol (Amyl)	—	<0.002
Alcohol (Benzyl)	—	<0.02
Alcohol (Butyl)	<0.002	<0.002
Alcohol (Cetyl)	—	<0.02
Alcohol (Isopropyl)	—	<0.02
Allyl Chloride	—	>0.05
Allyl Sulfide	—	<0.02
Aluminum Acetate	<0.002	<0.002
Aluminum Chloride	>0.05	<0.02
Aluminum Fluoride	<0.002	—
Aluminum Formate	<0.02	<0.02
Aluminum Hydroxide	<0.02	—

* 10% corrosive medium in 90% water. (Other % corrosive medium in parentheses.)
** Water-free, dry or maximum concentration of corrosive medium.

Source: data compiled by J.S. Park from Earl R. Parker, *Materials Data Book for Engineers and Scientists*, McGraw-Hill Book Company, New York, 1967.

Table 323. CORROSION RATES OF ALUMINUM AT 70°F *

(SHEET 2 OF 9)

Corrosive Medium	Corrosion Rate* in 10% Corrosive Medium (ipy)	Corrosion Rate** in 100% Corrosive Medium (ipy)
Aluminum Nitrate	<0.02	<0.02
Aluminum Potassium Sulfate	<0.02	<0.02
Aluminum Sulfate	<0.002	>0.05
Ammonia	<0.002	<0.002
Ammonium Acetate	<0.002	<0.002
Ammonium Bicarbonate	<0.02	<0.02
Ammonium Bromide	>0.05	—
Ammonium Carbonate	<0.02	<0.02
Ammonium Chloride	>0.05	<0.02
Ammonium Citrate	<0.02	<0.02
Ammonium Formate	<0.02	—
Ammonium Nitrate	<0.02	<0.02
Ammonium Sulfate	>0.05	<0.02
Amyl Acetate	—	<0.002
Amyl Chloride	—	<0.02
Aniline	—	<0.02
Aniline Hydrochloride	>0.05	>0.05
Anthracine	—	<0.02
Antimony Trichloride	>0.05	<0.02
Barium Carbonate	—	>0.05
Barium Chloride	<0.02	>0.05
Barium Hydroxide	>0.05	>0.05
Barium Nitrate	<0.02	—
Barium Peroxide	>0.05	—

* 10% corrosive medium in 90% water. (Other % corrosive medium in parentheses.)
** Water-free, dry or maximum concentration of corrosive medium.

Source: data compiled by J.S. Park from Earl R. Parker, *Materials Data Book for Engineers and Scientists*, McGraw-Hill Book Company, New York, 1967.

Table 323. CORROSION RATES OF ALUMINUM AT 70°F *

(SHEET 3 OF 9)

Corrosive Medium	Corrosion Rate* in 10% Corrosive Medium (ipy)	Corrosion Rate** in 100% Corrosive Medium (ipy)
Benzaldehyde	<0.02	<0.002
Benzene	<0.02	<0.02
Benzoic Acid	<0.02	<0.02
Boric Acid	<0.05	<0.02
Bromic Acid	>0.05	—
Bromine (Dry)	—	<0.02
Bromine (Wet)	—	>0.05
Butyric Acid	<0.02	<0.002
Cadmium Chloride	>0.05	—
Cadmium Sulfate	<0.02	—
Calcium Acetate	—	<0.05
Calcium Bicarbonate	—	<0.02
Calcium Bromide	<0.05	<0.05
Calcium Chlorate	<0.02	—
Calcium Chloride	<0.002	>0.05
Calcium Hydroxide	>0.05	>0.05
Calcium Hypochlorite	>0.05	—
Carbon Dioxide	—	<0.002
Carbon Monoxide	—	<0.002
Carbon Tetrachloride	—	<0.02
Carbon Acid (Air Free)	<0.02	<0.002
Chloroacetic Acid	>0.05	>0.05
Chlorine Gas	—	<0.02
Chloroform (Dry)	—	<0.02

* 10% corrosive medium in 90% water. (Other % corrosive medium in parentheses.)
** Water-free, dry or maximum concentration of corrosive medium.

Source: data compiled by J.S. Park from Earl R. Parker, *Materials Data Book for Engineers and Scientists*, McGraw-Hill Book Company, New York, 1967.

Table 323. CORROSION RATES OF ALUMINUM AT 70°F *

(SHEET 4 OF 9)

Corrosive Medium	Corrosion Rate* in 10% Corrosive Medium (ipy)	Corrosion Rate** in 100% Corrosive Medium (ipy)
Chromic Acid	>0.05	>0.05
Chromic Hydroxide	—	<0.02
Chromic Sulfates	—	<0.05
Citric Acid	<0.02	<0.02
Copper Nitrate	>0.05	—
Copper Sulfate	>0.05	>0.05
Diethylene Glycol	—	<0.02
Ethyl Chloride	—	<0.002
Ethylene Glycol	<0.002	<0.002
Ethylene Oxide	—	<0.002
Fatty Acids	—	<0.002
Ferric Chloride	>0.05	>0.05
Ferric Nitrate	>0.05	—
Ferrous Chloride	>0.05	—
Ferrous Sulfate	<0.002	—
Fluorine	—	>0.05
Formaldehyde	<0.02	<0.002
Formic Acid	<0.02	<0.02
Furfural	—	<0.02
Hydrazine	—	<0.002
Hydrobromic Acid	>0.05	>0.05
Hydrochloric Acid (Areated)	>0.05	—
Hydrochloric Acid (Air Free)	>0.05	—
Hydrocyanic Acid	<0.02	<0.002

* 10% corrosive medium in 90% water. (Other % corrosive medium in parentheses.)
** Water-free, dry or maximum concentration of corrosive medium.

Source: data compiled by J.S. Park from Earl R. Parker, *Materials Data Book for Engineers and Scientists*, McGraw-Hill Book Company, New York, 1967.

Table 323. CORROSION RATES OF ALUMINUM AT 70°F *

(SHEET 5 OF 9)

Corrosive Medium	Corrosion Rate* in 10% Corrosive Medium (ipy)	Corrosion Rate** in 100% Corrosive Medium (ipy)
Hydrofluoric Acid (Areated)	>0.05	—
Hydrofluoric Acid (Air Free)	>0.05	—
Hydrogen Chloride	—	>0.05
Hydrogen Fluoride	—	<0.02
Hydrogen Iodide	—	>0.05
Hydrogen Peroxide	<0.002	<0.002
Hydrogen Sulfide	—	<0.002
Lactic Acid	<0.02	<0.02
Lead Acetate	—	>0.05
Lead Chromate	>0.05	—
Lead Nitrate	>0.05	—
Lead Sulfate	>0.05	—
Lithium Chloride	<0.05	—
Lithium Hydroxide	>0.05	>0.05
Magnesium Chloride	>0.05	—
Magnesium Hydroxide	>0.05	>0.05
Magnesium Sulfate	<0.02	<0.02
Maleic Acid	<0.02	—
Malic Acid	<0.02	<0.002
Mercuric Chloride	>0.05	—
Mercurous Nitrate	>0.05	>0.05
Mercury	—	>0.05
Methallylamine	—	<0.02
Methanol	—	<0.02

* 10% corrosive medium in 90% water. (Other % corrosive medium in parentheses.)
** Water-free, dry or maximum concentration of corrosive medium.

Source: data compiled by J.S. Park from Earl R. Parker, *Materials Data Book for Engineers and Scientists*, McGraw-Hill Book Company, New York, 1967.

Table 323. CORROSION RATES OF ALUMINUM AT 70°F *

(SHEET 6 OF 9)

Corrosive Medium	Corrosion Rate* in 10% Corrosive Medium (ipy)	Corrosion Rate** in 100% Corrosive Medium (ipy)
Methyl Ethyl Ketone	<0.02	<0.002
Methyl Isobutyl Ketone	<0.02	<0.002
Methylamine	<0.02	<0.02
Methylene Chloride	>0.05	<0.002
Monochloroacetic Acid	>0.05	>0.05
Monorthanolamine	—	<0.02
Monoethalamine	<0.02	<0.02
Monoethylamine	<0.02	<0.02
Monosodium Phosphate	>0.05	—
Nickel Chloride	>0.05	>0.05
Nickel Nitrate	>0.05	—
Nickel Sulfate	>0.05	>0.05
Nitric Acid	>0.05	<0.02
Nitric Acid (Red Fuming)	—	<0.002
Nitric + Hydrochloric Acid	—	>0.05
Nitric + Sulfuric Acid	>0.05	>0.05
Nitrobenzene	—	<0.02
Nitrocelluolose	—	<0.002
Nitroglycerine	—	<0.002
Nitrotolune	—	<0.02
Nitrous Acid	<0.05	—
Oleic Acid	—	<0.002
Oxalic Acid	<0.02	<0.02
Phenol	—	<0.002

* 10% corrosive medium in 90% water. (Other % corrosive medium in parentheses.)
** Water-free, dry or maximum concentration of corrosive medium.

Source: data compiled by J.S. Park from Earl R. Parker, *Materials Data Book for Engineers and Scientists*, McGraw-Hill Book Company, New York, 1967.

Table 323. CORROSION RATES OF ALUMINUM AT 70°F *
(SHEET 7 OF 9)

Corrosive Medium	Corrosion Rate* in 10% Corrosive Medium (ipy)	Corrosion Rate** in 100% Corrosive Medium (ipy)
Phosphoric Acid (Areated)	>0.05	<0.02
Phosphoric Acid (Air Free)	>0.05	>0.05
Picric Acid	>0.05	<0.02
Potassium Bicarbonate	>0.05	<0.02
Potassium Bromide	<0.02	—
Potassium Carbonate	>0.05	>0.05
Potassium Chlorate	<0.02	<0.02
Potassium Chromate	<0.02	<0.02
Potassium Cyanide	>0.05	—
Potassium Dichromate	<0.002	<0.02
Potassium Ferricyanide	<0.02	—
Potassium Ferrocyanide	<0.002	<0.02
Potassium Hydroxide	>0.05	—
Potassium Hypochlorite	>0.05	—
Potassium Iodide	<0.02	—
Potassium Nitrate	<0.002	<0.02
Potassium Nitrite	<0.02	<0.02
Potassium Permanganate	<0.02	<0.02
Potassium Silicate	>0.05	<0.02
Propionic Acid	<0.02	<0.02
Pyridine	<0.02	<0.02
Salicylic Acid	>0.05	<0.02
Silicon Tetrachloride (Dry)	—	<0.02
Silicon Tetrachloride (Wet)	—	>0.05

* 10% corrosive medium in 90% water. (Other % corrosive medium in parentheses.)
** Water-free, dry or maximum concentration of corrosive medium.

Source: data compiled by J.S. Park from Earl R. Parker, *Materials Data Book for Engineers and Scientists*, McGraw-Hill Book Company, New York, 1967.

Table 323. CORROSION RATES OF ALUMINUM AT 70°F *

(SHEET 8 OF 9)

Corrosive Medium	Corrosion Rate* in 10% Corrosive Medium (ipy)	Corrosion Rate** in 100% Corrosive Medium (ipy)
Silver Bromide	>0.05	—
Silver Chloride	>0.05	—
Silver Nitrate	>0.05	—
Sodium Acetate	<0.02	<0.002
Sodium Bicarbonate	>0.05	<0.02
Sodium Bisulfate	>0.05	—
Sodium Bromide	<0.05	—
Sodium Carbonate	>0.05	—
Sodium Chloride	<0.05	—
Sodium Chromate	<0.02	<0.02
Sodium Hydroxide	>0.05	—
Sodium Hypochlorite	>0.05	>0.05
Sodium Metasilicate	>0.05	<0.02
Sodium Nitrate	<0.002	<0.02
Sodium Nitrite	<0.02	—
Sodium Phosphate	>0.05	—
Sodium Silicate	>0.05	<0.002
Sodium Sulfate	<0.002	—
Sodium Sulfide	>0.05	>0.05
Sodium Sulfite	<0.02	—
Stannic Chloride	>0.05	—
Stannous Chloride	>0.05	—
Strontium Nitrate	<0.02	<0.02
Succinic Acid	<0.02	<0.02

* 10% corrosive medium in 90% water. (Other % corrosive medium in parentheses.)
** Water-free, dry or maximum concentration of corrosive medium.

Source: data compiled by J.S. Park from Earl R. Parker, *Materials Data Book for Engineers and Scientists*, McGraw-Hill Book Company, New York, 1967.

Table 323. Corrosion Rates of Aluminum at 70°F *

(Sheet 9 of 9)

Corrosive Medium	Corrosion Rate* in 10% Corrosive Medium (ipy)	Corrosion Rate** in 100% Corrosive Medium (ipy)
Sulfur Dioxide	>0.05	<0.02
Sulfur Trioxide	—	<0.02
Sulfuric Acid (Areated)	>0.05	>0.05
Sulfuric Acid (Air Free)	>0.05	>0.05
Sulfuric Acid (Fuming)	—	<0.02
Sulfurous Acid	<0.02	<0.02
Tannic Acid	<0.02	>0.05
Tartaric Acid	<0.02	—
Tetraphosphoric Acid	>0.05	>0.05
Trichloroacetic Acid	>0.05	>0.05
Trichloroethylene	—	<0.002
Urea	<0.02	<0.02
Zinc Chloride	>0.05	—
Zinc Sulfate	<0.05	—

* 10% corrosive medium in 90% water. (Other % corrosive medium in parentheses.)
** Water-free, dry or maximum concentration of corrosive medium.

Source: data compiled by J.S. Park from Earl R. Parker, *Materials Data Book for Engineers and Scientists*, McGraw-Hill Book Company, New York, 1967.

* <0.002 means that corrosion rate is likely to be less than 0.002 inch per year (Excellent).
<0.02 means that corrosion rate is likely to be less than about 0.02 inch per year (Good).
<0.05 means that corrosion rate is likely to be less than about 0.05 inch per year (Fair).
>0.05 means that corrosion rate is likely to be more than 0.05 inch per year (Poor).

Table 324. Corrosion Resistance of Wrought Coppers and Copper Alloys
(Sheet 1 of 10)

UNS Number and Name	Nominal Composition (%)	Commercial Forms(a)	Corrosion Resistance (b)
C10100 Oxygen-free electronic	99.99 Cu	F, R, W, T, P, S	G-E
C10200 Oxygen-free copper	99.95 Cu	F, R, W, T, P, S	G-E
C10300 Oxygen-free extra-low phosporus	99.95 Cu, 0.003 P	F, R, T, P, S	G-E
C10400, C10500, C10700 Oxygen-free, silver-bearing	99.95 Cu(e)	F, R, W, S	G-E
C10800 Oxygen-free, low phosporus	99.95 Cu, 0.009 P	F, R, T, P	G-E
CS11000 Electrolytic tough pitch copper	99.90 Cu, 0.04 O	F, R, W, T, P, S	G-E
C11100 Electrolytic tough pitch, anneal resistant	99.90 Cu, 0.04 O, 0.01 Cd	W	G-E
C11300, C11400, C11500, C11600 Silver-bearing tough pitch copper	99.90 Cu, 0.04 O, Ag(f)	F, R, W, T, S	G-E
C12000, C12100	99.9 Cu(g)	F, T, P	G-E
C12200 Phosphorus deoxidized copper, high residual phosphorus	99.90 Cu, 0.02 P	F, R, T, P	G-E
C12500, C12700, C12800, C12900, C13000 Fire-refined tough pitch with silver	99.88 Cu(h)	F, R, W, S	G-E
C14200 Phosphorus deoxidized, arsenical	99.68 Cu, 0.3 As, 0.02 P	F, R, T	G-E

(a) F, flat products; R, rod; W, wire; T, tube; P, pipe; S, shapes.
(b) E, excellent; G, good; F, fair.

Data from *ASM Metals Reference Book, Third Edition*, Michael Bauccio, Ed., ASM International, Materials Park, OH, p442–454, (1993).

Table 324. **Corrosion Resistance of Wrought Coppers and Copper Alloys**

(Sheet 2 of 10)

UNS Number and Name	Nominal Composition (%)	Commercial Forms(a)	Corrosion Resistance (b)
C19200	98.97 Cu, 1.0 Fe, 0.03 P	F, T	G-E
C14300	99.9 Cu, 0.1 Cd	F	G-E
C14310	99.8 Cu, 0.2 Cd	F	G-E
C14500 Phosphorus deoxidized, tellurium bearing	99.5 Cu, 0.50 Te, 0.008 P	F, R, W, T	G-E
C14700 Sulfur bearing	99.6 Cu, 0.40 S	R, W	G-E
C15000 Zirconium copper	99.8 Cu, 0.15 Zr	R, W	G-E
C15500	99.75 Cu, 0.06 P, 0.11 Mg, Ag(i)	F	G-E
C16200 Cadmium copper	99.0 Cu, 1.0 Cd	F, R, W	G-E
C16500	98.6 Cu, 0.8 Cd, 0.6 Sn	F, R, W	G-E
C17000 Beryllium copper	99.5 Cu, 1.7 Be, 0.20 Co	F, R	G-E
C17200 Beryllium copper	99.5 Cu, 1.9 Be , 0.20 Co	F, R, W, T, P, S	G-E
C17300 Beryllium copper	99.5 Cu, 1.9 Be, 0.40 Pb	R	G-E

(a) F, flat products; R, rod; W, wire; T, tube; P, pipe; S, shapes.
(b) E, excellent; G, good; F, fair.

Data from *ASM Metals Reference Book, Third Edition*, Michael Bauccio, Ed., ASM International, Materials Park, OH, p442–454, (1993).

Table 324. CORROSION RESISTANCE OF WROUGHT COPPERS AND COPPER ALLOYS

(SHEET 3 OF 10)

UNS Number and Name	Nominal Composition (%)	Commercial Forms(a)	Corrosion Resistance (b)
C17500 Copper-cobalt-beryllium alloy	99.5 Cu, 2.5 Co, 0.6 Be	F, R	G-E
C18200, C18400, C18500 Chromium copper	99.5 Cu(j)	F, W, R, S, T	G-E
C18700 leaded copper	99.0 Cu, 1.0 Pb	R	G-E
C18900	98.75 Cu, 0.75 Sn, 0.3 Si, 0.20 Mn	R, W	G-E
C19000 Copper-nickel-phosphorus alloy	98.7 Cu, 1.1 Ni, 0.25 P	F, R, W	G-E
C19100 Copper-nickel-phosphorus-tellurium alloy	98.15 Cu, 1.1 Ni, 0.50 Te, 0.25 P	R, F	G-E
C19400	97.5 Cu, 2.4 Fe, 0.13 Zn, 0.03 P	F	G-E
C19500	97.0 Cu, 1.5 Fe, 0.6 Sn, 0.10 P, 0.80 Co	F	G-E
C21000 Gilding, 95%	95.0 Cu, 5.0 Zn	F, W	G-E
C22000 Commercial bronze, 90%	90.0 Cu, 10.0 Zn	F, R, W, T	G-E
C22600 Jewelry bronze, 87.5%	87.5 Cu, 12.5 Zn	F, W	G-E

(a) F, flat products; R, rod; W, wire; T, tube; P, pipe; S, shapes.
(b) E, excellent; G, good; F, fair.

Data from *ASM Metals Reference Book, Third Edition*, Michael Bauccio, Ed., ASM International, Materials Park, OH, p442–454, (1993).

Table 324. Corrosion Resistance of Wrought Coppers and Copper Alloys (Sheet 4 of 10)

UNS Number and Name	Nominal Composition (%)	Commercial Forms(a)	Corrosion Resistance (b)
C23000 Red brass, 85%	85.0 Cu, 15.0 Zn	F, W, T, P	G-E
C24000 Low brass, 80%	80.0 Cu, 20.0 Zn	F, W	F-E
C26000 Cartridge brass, 70%	70.0 Cu, 30.0 Zn	F, R, W, T	F-E
C26800, C27000 Yellow brass	65.0 Cu, 35.0 Zn	F, R, W	F-E
C28000 Muntz metal	60.0 Cu, 40.0 Zn	F, R, T	F-E
C31400 Leaded commercial bronze	89.0 Cu, 1.75 Pb, 9.25 Zn	F, R	G-E
C31600 Leaded commercial bronze, nickel-bearing	89.0 Cu, 1.9 Pb, 1.0 Ni, 8.1 Zn	F, R	G-E
C33000 Low-leaded brass tube	66.0 Cu, 0.5 Pb, 33.5 Zn	T	F-E
C33200 High-leaded brass tube	66.0 Cu, 1.6 Pb, 32.4 Zn	T	F-E
C33500 Low-leaded brass	65.0 Cu, 0.5 Pb, 34.5 Zn	F	F-E
C34000 Medium-leaded brass	65.0 Cu, 1.0 Pb, 34.0 Zn	F, R, W, S	F-E
C34200 High-leaded brass	64.5 Cu, 2.0 Pb, 33.5 Zn	F, R	F-E

(a) F, flat products; R, rod; W, wire; T, tube; P, pipe; S, shapes.
(b) E, excellent; G, good; F, fair.

Data from *ASM Metals Reference Book, Third Edition*, Michael Bauccio, Ed., ASM International, Materials Park, OH, p442–454, (1993).

Table 324. CORROSION RESISTANCE OF WROUGHT COPPERS AND COPPER ALLOYS

(SHEET 5 OF 10)

UNS Number and Name	Nominal Composition (%)	Commercial Forms(a)	Corrosion Resistance (b)
C34900	62.2 Cu, 0.35 Pb, 37.45 Zn	R, W	F-E
C35000 Medium-leaded brass	62.5 Cu, 1.1 Pb, 36.4 Zn	F, R	F-E
C35300 High-leaded brass	62.0 Cu, 1.8 Pb, 36.2 Zn	F, R	F-E
C35600 Extra-high-leaded brass	63.0 Cu, 2.5 Pb, 34.5 Zn	F	F-E
C36000 Free-cutting brass	61.5 Cu, 3.0 Pb, 35.5 Zn	F, R, S	F-E
C36500 to C36800 Leaded Muntz metal	60.0 Cu(k), 0.6 Pb, 39.4 Zn	F	F-E
C37000 Free-cutting Muntz metal	60.0 Cu, 1.0 Pb, 39.0 Zn	T	F-E
C37700 Forging brass	59.0 Cu, 2.0 Pb, 39.0 Zn	R, S	F-E
C38500 Architectural bronze	57.0 Cu, 3.0 Pb, 40.0 Zn	R, S	F-E
C40500	95 Cu, 1 Sn, 4 Zn	F	G-E
C40800	95 Cu, 2 Sn, 3 Zn	F	G-E
C41100	91 Cu, 0.5 Sn, 8.5 Zn	F, W	G-E

(a) F, flat products; R, rod; W, wire; T, tube; P, pipe; S, shapes.
(b) E, excellent; G, good; F, fair.

Data from *ASM Metals Reference Book, Third Edition*, Michael Bauccio, Ed., ASM International, Materials Park, OH, p442–454, (1993).

Table 324. CORROSION RESISTANCE OF WROUGHT COPPERS AND COPPER ALLOYS
(SHEET 6 OF 10)

UNS Number and Name	Nominal Composition (%)	Commercial Forms(a)	Corrosion Resistance (b)
C41300	90.0 Cu, 1.0 Sn, 9.0 Zn	F, R, W	G-E
C41500	91 Cu, 1.8 Sn, 7.2 Zn	F	G-E
C42200	87.5 Cu, 1.1 Sn, 11.4 Zn	F	G-E
C42500	88.5 Cu, 2.0 Sn, 9.5 Zn	F	G-E
C43000	87.0 Cu, 2.2 Sn, 10.8 Zn	F	G-E
C43400	85.0 Cu, 0.7 Sn, 14.3 Zn	F	G-E
C43500	81.0 Cu, 0.9 Sn, 18.1 Zn	F, T	G-E
C44300, C44400, C44500 Inhibited admiralty	71.0 Cu, 28.0 Zn, 1.0 Sn	F, W, T	G-E
C46400 to C46700 Naval brass	60.0 Cu, 39.25 Zn, 0.75 Sn	F, R, T, S	F-E
C48200 Naval brass, medium-leaded	60.5 Cu, 0.7 Pb, 0.8 Sn, 38.0 Zn	F, R, S	F-E
C48500 Leaded naval brass	60.0 Cu, 1.75 Pb, 37.5 Zn, 0.75 Sn	F, R, S	F-E
C50500 Phosphor bronze, 1.25% E	98.75 Cu, 1.25 Sn, trace P	F, W	G-E

(a) F, flat products; R, rod; W, wire; T, tube; P, pipe; S, shapes.
(b) E, excellent; G, good; F, fair.

Data from *ASM Metals Reference Book, Third Edition*, Michael Bauccio, Ed., ASM International, Materials Park, OH, p442–454, (1993).

Table 324. Corrosion Resistance of Wrought Coppers and Copper Alloys

(Sheet 7 of 10)

UNS Number and Name	Nominal Composition (%)	Commercial Forms(a)	Corrosion Resistance (b)
C51000 Phosphor bronze, 5% A	95.0 Cu, 5.0 Sn, trace P	F, R, W, T	G-E
C51100	95.6 Cu, 4.2 Sn, 0.2 P	F	G-E
C52100 Phosphor bronze, 8% C	92.0 Cu, 8.0 Sn, trace P	F, R, W	G-E
C52400 Phosphor bronze, 10% D	90.0 Cu, 10.0 Sn, trace P	F, R, W	G-E
C54400 Free-cutting phosphor bronze	88.0 Cu, 4.0 Pb, 4.0 Zn, 4.0 Sn	F, R	G-E
C60800 Aluminum bronze, 5%	95.0 Cu, 5.0 Al	T	G-E
C61000	92.0 Cu, 8.0 Al	R, W	G-E
C61300	92.65 Cu, 0.35 Sn, 7.0 Al	F, R, T, P, S	G-E
C61400 Aluminum bronze, D	91.0 Cu, 7.0 Al, 2.0 Fe	F, R, W, T, P, S	G-E
C61500	90.0 Cu, 8.0 Al, 2.0 Ni	F	G-E
C61800	89.0 Cu, 1.0 Fe, 10.0 Al	R	G-E
C61900	86.5 Cu, 4.0 Fe, 9.5 Al	F	G-E

(a) F, flat products; R, rod; W, wire; T, tube; P, pipe; S, shapes.
(b) E, excellent; G, good; F, fair.

Data from *ASM Metals Reference Book, Third Edition*, Michael Bauccio, Ed., ASM International, Materials Park, OH, p442–454, (1993).

Table 324. Corrosion Resistance of Wrought Coppers and Copper Alloys
(Sheet 8 of 10)

UNS Number and Name	Nominal Composition (%)	Commercial Forms(a)	Corrosion Resistance (b)
C62300	87.0 Cu, 10.0 Al, 3.0 Fe	F, R	G-E
C62400	86.0 Cu, 3.0 Fe, 11.0 Al	F, R	G-E
C62500	82.7 Cu, 4.3 Fe, 13.0 Al	F, R	G-E
C63000	82.0 Cu, 3.0 Fe, 10.0 Al, 5.0 Ni	F, R	G-E
C63200	82.0 Cu, 4.0 Fe, 9.0 Al, 5.0 Ni	F, R	G-E
C63600	95.5 Cu, 3.5 Al, 1.0 Si	R, W	G-E
C63800	99.5 Cu, 2.8 Al, 1.8 Si, 0.40 Co	F	G-E
C64200	91.2 Cu, 7.0 Al	F, R	G-E
C65100 Low-silicon bronze, B	98.5 Cu, 1.5 Si	R, W, T	G-E
C65500 High-silicon bronze, A	97.0 Cu, 3.0 Si	F, R, W, T	G-E
C66700 Manganese brass	70.0 Cu, 28.8 Zn, 1.2 Mn	F, W	G-E
C67400	58.5 Cu, 36.5 Zn, 1.2 Al, 2.8 Mn, 1.0 Sn	F, R	F-E

(a) F, flat products; R, rod; W, wire; T, tube; P, pipe; S, shapes.
(b) E, excellent; G, good; F, fair.

Data from *ASM Metals Reference Book, Third Edition*, Michael Bauccio, Ed., ASM International, Materials Park, OH, p442–454, (1993).

Table 324. Corrosion Resistance of Wrought Coppers and Copper Alloys

(Sheet 9 of 10)

UNS Number and Name	Nominal Composition (%)	Commercial Forms(a)	Corrosion Resistance (b)
C67500 Manganese bronze, A	58.5 Cu, 1.4 Fe, 39.0 Zn, 1.0 Sn, 0.1 Mn	R, S	F-E
C68700 Aluninum brass, arsenical	77.5 Cu, 20.5 Zn, 2.0 Al, 0.1 As	T	G-E
C68800	73.5 Cu, 22.7 Zn, 3.4 Al, 0.40 Co	F	G-E
C69000	73.3 Cu, 3.4 Al, 0.6 Ni, 22.7 Zn	F	G-E
C69400 Silicon red brass	81.5 Cu, 14.5 Zn, 4.0 Si	R	G-E
C70400	92.4 Cu, 1.5 Fe, 5.5 Ni, 0.6 Mn	F, T	G-E
C70600 Copper nickel, 10%	88.7 Cu, 1.3 Fe, 10.0 Ni	F, T	E
C71000 Copper nickel, 20%	79.00 Cu, 21.0 Ni	F, W, T	E
C71500 Copper nickel, 30%	70.0 Cu, 30.0 Ni	F, R, T	E
C71700	67.8 Cu, 0.7 Fe, 31.0 Ni, 0.5 Be	F, R, W	G-E
C72500	88.20 Cu, 9.5 Ni, 2.3 Sn	F, R, W, T	E
C73500	72.0 Cu, 18.0 Ni , 10.0 Zn	F, R, W, T	E

(a) F, flat products; R, rod; W, wire; T, tube; P, pipe; S, shapes.
(b) E, excellent; G, good; F, fair.

Data from *ASM Metals Reference Book, Third Edition*, Michael Bauccio, Ed., ASM International, Materials Park, OH, p442–454, (1993).

Table 324. Corrosion Resistance of Wrought Coppers and Copper Alloys
(Sheet 10 of 10)

UNS Number and Name	Nominal Composition (%)	Commercial Forms(a)	Corrosion Resistance (b)
C74500 Nickel silver, 65-10	65.0 Cu, 25.0 Zn, 10.0 Ni	F, W	E
C75200 Nickel silver, 65-18	65.0 Cu, 17.0 Zn, 18.0 Ni	F, R, W	E
C75400 Nickel silver, 65-15	65.0 Cu, 20.0 Zn, 15.0 Ni	F	E
C75700 Nickel silver, 65-12	65.0 Cu, 23.0 Zn, 12.0 Ni	F, W	E
C76200	59.0 Cu, 29.0 Zn, 12.0 Ni	F, T	G-E
C77000 Nickel silver, 55-18	55.0 Cu, 27.0 Zn, 18.0 Ni	F, R, W	E
C72200	82.0 Cu, 16.0 Ni, 0.5 Cr, 0.8 Fe, 0.5 Mn	F, T	G-E
C78200 Leaded nickel silver, 65-8-2	65.0 Cu, 2.0 Pb, 25.0 Zn, 8.0 Ni	F	E

(a) F, flat products; R, rod; W, wire; T, tube; P, pipe; S, shapes.
(b) E, excellent; G, good; F, fair.

Data from *ASM Metals Reference Book, Third Edition*, Michael Bauccio, Ed., ASM International, Materials Park, OH, p442–454, (1993).

Table 325. CORROSION RATES OF 70-30 BRASS AT 70°F *

(SHEET 1 OF 9)

Corrosive Medium	Corrosion Rate* in 10% Corrosive Medium (ipy)	Corrosion Rate** in 100% Corrosive Medium (ipy)
Acetaldehyde	<0.02	<0.002
Acetic Acid (Aerated)	>0.05	>0.05
Acetic Acid (Air Free)	>0.05	>0.05
Acetic Anhydride	—	>0.05
Acetone	<0.002	<0.002
Acetylene	—	<0.002
Acrolein	<0.02	<0.02
Acrylonitril	—	<0.002
Alcohol (Ethyl)	<0.002	<0.002
Alcohol (Methyl)	<0.02	<0.02
Alcohol (Allyl)	—	<0.02
Alcohol (Benzyl)	—	<0.02
Alcohol (Butyl)	—	<0.002
Alcohol (Isopropyl)	—	<0.02
Allylamine	—	>0.05
Allyl Chloride	—	<0.02
Allyl Sulfide	—	>0.05
Aluminum Acetate	—	<0.02
Aluminum Chloride	>0.05	>0.05
Aluminum Fluoride	>0.05	—
Aluminum Fluosilicate	—	<0.02
Aluminum Hydroxide	<0.02	—
Aluminum Potassium Sulfate	>0.05	>0.05
Aluminum Sulfate	<0.02	<0.05

* 10% corrosive medium in 90% water. (Other % corrosive medium in parentheses.)
** Water-free, dry or maximum concentration of corrosive medium.

Source: data compiled by J.S. Park from Earl R. Parker, *Materials Data Book for Engineers and Scientists*, McGraw-Hill Book Company, New York, 1967.

Table 325. CORROSION RATES OF 70-30 BRASS AT 70°F *

(SHEET 2 OF 9)

Corrosive Medium	Corrosion Rate* in 10% Corrosive Medium (ipy)	Corrosion Rate** in 100% Corrosive Medium (ipy)
Ammonia	>0.05	<0.002
Ammonium Acetate	—	>0.05
Ammonium Bicarbonate	>0.05	—
Ammonium Bromide	>0.05	—
Ammonium Carbonate	>0.05	—
Ammonium Chloride	>0.05	>0.05
Ammonium Citrate	>0.05	—
Ammonium Nitrate	>0.05	>0.05
Ammonium Sulfate	>0.05	<0.02
Ammonium Sulfite	>0.05	>0.05
Ammonium Thiocyanate	>0.05	—
Amyl Acetate	<0.02	<0.02
Amyl Chloride	—	<0.02
Aniline	—	>0.05
Aniline Hydrochloride	>0.05	—
Anthracine	—	<0.02
Antimony Trichloride	>0.05	—
Barium Carbonate	<0.02	<0.02
Barium Chloride	>0.05	<0.02
Barium Hydroxide	>0.05	—
Barium Nitrate	>0.05	—
Barium Peroxide	>0.05	—
Benzaldehyde	>0.05	<0.02
Benzene	<0.02	<0.02

* 10% corrosive medium in 90% water. (Other % corrosive medium in parentheses.)
** Water-free, dry or maximum concentration of corrosive medium.

Source: data compiled by J.S. Park from Earl R. Parker, *Materials Data Book for Engineers and Scientists*, McGraw-Hill Book Company, New York, 1967.

Table 325. CORROSION RATES OF 70-30 BRASS AT 70°F *

(SHEET 3 OF 9)

Corrosive Medium	Corrosion Rate* in 10% Corrosive Medium (ipy)	Corrosion Rate** in 100% Corrosive Medium (ipy)
Benzoic Acid	<0.02	<0.02
Boric Acid	<0.02	<0.02
Bromic Acid	>0.05	>0.05
Bromine (Dry)	—	<0.02
Bromine (Wet)	—	>0.05
Butyric Acid	<0.05	—
Cadmium Chloride	>0.05	—
Cadmium Sulfate	<0.02	—
Calcium Acetate	<0.02	<0.02
Calcium Bicarbonate	—	<0.02
Calcium Bromide	<0.02	<0.02
Calcium Chlorate	>0.05	<0.02
Calcium Chloride	<0.02	<0.02
Calcium Hydroxide	<0.02	—
Calcium Hypochlorite	<0.02	—
Carbon Dioxide	—	<0.002
Carbon Monoxide	—	<0.002
Carbon Tetrachloride	—	<0.05
Carbon Acid (Air Free)	—	>0.05
Chloroacetic Acid	>0.05	>0.05
Chlorine Gas	—	>0.05
Chloroform (Dry)	—	<0.02
Chromic Acid	>0.05	>0.05
Chromic Hydroxide	—	<0.02

* 10% corrosive medium in 90% water. (Other % corrosive medium in parentheses.)
** Water-free, dry or maximum concentration of corrosive medium.

Source: data compiled by J.S. Park from Earl R. Parker, *Materials Data Book for Engineers and Scientists*, McGraw-Hill Book Company, New York, 1967.

Table 325. CORROSION RATES OF 70-30 BRASS AT 70°F *

(SHEET 4 OF 9)

Corrosive Medium	Corrosion Rate* in 10% Corrosive Medium (ipy)	Corrosion Rate** in 100% Corrosive Medium (ipy)
Chromic Sulfates	<0.02	—
Citric Acid	>0.05	<0.02
Copper Nitrate	>0.05	>0.05
Copper Sulfate	>0.05	>0.05
Diethylene Glycol	—	<0.002
Ethyl Chloride	—	<0.002
Ethylene Glycol	—	<0.02
Ethylene Oxide	—	>0.05
Fatty Acids	—	<0.05
Ferric Chloride	>0.05	<0.02
Ferric Nitrate	>0.05	—
Ferrous Chloride	>0.05	—
Ferrous Sulfate	>0.05	<0.05
Fluorine	—	<0.02
Formaldehyde	<0.002	<0.02
Formic Acid	<0.05	<0.02
Furfural	<0.02	<0.02
Hydrazine	>0.05	—
Hydrobromic Acid	>0.05	>0.05
Hydrochloric Acid (Areated)	>0.05	—
Hydrochloric Acid (Air Free)	>0.05	—
Hydrocyanic Acid	>0.05	<0.02
Hydrofluoric Acid (Areated)	>0.05	—
Hydrofluoric Acid (Air Free)	>0.05	<0.02

* 10% corrosive medium in 90% water. (Other % corrosive medium in parentheses.)
** Water-free, dry or maximum concentration of corrosive medium.

Source: data compiled by J.S. Park from Earl R. Parker, *Materials Data Book for Engineers and Scientists*, McGraw-Hill Book Company, New York, 1967.

Table 325. CORROSION RATES OF 70-30 BRASS AT 70°F *

(SHEET 5 OF 9)

Corrosive Medium	Corrosion Rate* in 10% Corrosive Medium (ipy)	Corrosion Rate** in 100% Corrosive Medium (ipy)
Hydrogen Chloride	—	<0.02
Hydrogen Fluoride	—	<0.02
Hydrogen Iodide	—	>0.05
Hydrogen Peroxide	>0.05	>0.05
Hydrogen Sulfide	<0.02	<0.02
Lactic Acid	<0.05	<0.05
Lead Acetate		<0.05
Lead Chromate	—	<0.02
Lead Sulfate	—	<0.02
Lithium Chloride	<0.02 (30%)	—
Lithium Hydroxide	>0.05	—
Magnesium Chloride	<0.02	—
Magnesium Hydroxide	<0.02	<0.02
Magnesium Sulfate	<0.02	<0.02
Maleic Acid	<0.02	—
Mercuric Chloride	>0.05	>0.05
Mercurous Nitrate	>0.05	>0.05
Mercury	—	>0.05
Methallylamine	—	>0.05
Methanol	<0.02	<0.02
Methyl Ethyl Ketone	<0.02	<0.002
Methyl Isobutyl Ketone	<0.02	<0.02
Methylamine	—	>0.05
Methylene Chloride	—	<0.002

* 10% corrosive medium in 90% water. (Other % corrosive medium in parentheses.)
** Water-free, dry or maximum concentration of corrosive medium.

Source: data compiled by J.S. Park from Earl R. Parker, *Materials Data Book for Engineers and Scientists*, McGraw-Hill Book Company, New York, 1967.

Table 325. CORROSION RATES OF 70-30 BRASS AT 70°F *

(SHEET 6 OF 9)

Corrosive Medium	Corrosion Rate* in 10% Corrosive Medium (ipy)	Corrosion Rate** in 100% Corrosive Medium (ipy)
Monochloroacetic Acid	>0.05	>0.05
Monorthanolamine	—	>0.05
Monoethalamine	—	>0.05
Monoethylamine	—	>0.05
Monosodium Phosphate	<0.02	—
Nickel Chloride	>0.05	—
Nickel Nitrate	<0.05	—
Nickel Sulfate	<0.05	<0.02
Nitric Acid	>0.05	>0.05
Nitric Acid (Red Fuming)	—	>0.05
Nitric + Hydrochloric Acid	—	>0.05
Nitric + Sulfuric Acid	>0.05	>0.05
Nitrobenzene	—	<0.02
Nitrocelluolose	—	<0.02
Nitroglycerine	—	<0.02
Nitrotolune	—	<0.02
Nitrous Acid	—	>0.05
Oleic Acid	>0.05	<0.02
Oxalic Acid	<0.02	<0.05
Phenol	—	<0.002
Phosphoric Acid (Areated)	>0.05	>0.05
Phosphoric Acid (Air Free)	<0.02	>0.05
Picric Acid	>0.05	>0.05
Potassium Bicarbonate	<0.02	<0.02

* 10% corrosive medium in 90% water. (Other % corrosive medium in parentheses.)
** Water-free, dry or maximum concentration of corrosive medium.

Source: data compiled by J.S. Park from Earl R. Parker, *Materials Data Book for Engineers and Scientists*, McGraw-Hill Book Company, New York, 1967.

Table 325. CORROSION RATES OF 70-30 BRASS AT 70°F *

(SHEET 7 OF 9)

Corrosive Medium	Corrosion Rate* in 10% Corrosive Medium (ipy)	Corrosion Rate** in 100% Corrosive Medium (ipy)
Potassium Bromide	<0.02	<0.02
Potassium Carbonate	<0.02	<0.02
Potassium Chlorate	<0.02	<0.05
Potassium Chromate	<0.02	<0.02
Potassium Cyanide	>0.05	>0.05
Potassium Dichromate	<0.02	—
Potassium Ferricyanide	<0.02	—
Potassium Ferrocyanide	<0.02	—
Potassium Hydroxide	<0.02	—
Potassium Hypochlorite	>0.05	—
Potassium Nitrate	<0.02	<0.02
Potassium Nitrite	<0.02	<0.02
Potassium Permanganate	<0.02	—
Potassium Silicate	<0.02	<0.02
Propionic Acid	<0.02	—
Pyridine	<0.02	<0.02
Quinine Sulfate	<0.02	<0.02
Silicon Tetrachloride (Dry)	—	<0.002
Silicon Tetrachloride (Wet)	—	>0.05
Silver Bromide	>0.05	—
Silver Chloride	>0.05	—
Silver Nitrate	>0.05	—
Sodium Acetate	<0.02	—
Sodium Bicarbonate	<0.02	—

* 10% corrosive medium in 90% water. (Other % corrosive medium in parentheses.)
** Water-free, dry or maximum concentration of corrosive medium.

Source: data compiled by J.S. Park from Earl R. Parker, *Materials Data Book for Engineers and Scientists*, McGraw-Hill Book Company, New York, 1967.

Table 325. CORROSION RATES OF 70-30 BRASS AT 70°F *

(SHEET 8 OF 9)

Corrosive Medium	Corrosion Rate* in 10% Corrosive Medium (ipy)	Corrosion Rate** in 100% Corrosive Medium (ipy)
Sodium Bisulfate	>0.05	<0.05
Sodium Bromide	<0.05	—
Sodium Carbonate	>0.05	—
Sodium Chloride	<0.05	—
Sodium Chromate	<0.02	<0.02
Sodium Hydroxide	>0.05	—
Sodium Hypochlorite	>0.05	>0.05
Sodium Metasilicate	<0.02	<0.02
Sodium Nitrate	<0.05	<0.05
Sodium Nitrite	<0.02	—
Sodium Phosphate	<0.02	<0.02
Sodium Silicate	<0.02	<0.02
Sodium Sulfate	<0.02	>0.05
Sodium Sulfide	<0.05	>0.05
Sodium Sulfite	>0.05	>0.05
Stannic Chloride	>0.05	—
Stannous Chloride	>0.05	—
Strontium Nitrate	<0.02	<0.02
Succinic Acid	<0.02	<0.02
Sulfur Dioxide	>0.05	<0.05
Sulfur Trioxide	—	<0.02
Sulfuric Acid (Areated)	>0.05	>0.05
Sulfuric Acid (Air Free)	<0.05	—
Sulfuric Acid (Fuming)	—	>0.05

* 10% corrosive medium in 90% water. (Other % corrosive medium in parentheses.)
** Water-free, dry or maximum concentration of corrosive medium.

Source: data compiled by J.S. Park from Earl R. Parker, *Materials Data Book for Engineers and Scientists*, McGraw-Hill Book Company, New York, 1967.

Table 325. CORROSION RATES OF 70-30 BRASS AT 70°F *

(SHEET 9 OF 9)

Corrosive Medium	Corrosion Rate* in 10% Corrosive Medium (ipy)	Corrosion Rate** in 100% Corrosive Medium (ipy)
Sulfurous Acid	<0.02	>0.05
Tannic Acid	—	<0.05
Tartaric Acid	<0.05	—
Tetraphosphoric Acid	>0.05	<0.05
Trichloroacetic Acid	>0.05	>0.05
Trichloroethylene	—	<0.02
Urea	<0.02	—
Zinc Chloride	>0.05	—
Zinc Sulfate	<0.05	<0.02

* 10% corrosive medium in 90% water. (Other % corrosive medium in parentheses.)
** Water-free, dry or maximum concentration of corrosive medium.

Source: data compiled by J.S. Park from Earl R. Parker, *Materials Data Book for Engineers and Scientists*, McGraw-Hill Book Company, New York, 1967.

* <0.002 means that corrosion rate is likely to be less than 0.002 inch per year (Excellent).
<0.02 means that corrosion rate is likely to be less than about 0.02 inch per year (Good).
<0.05 means that corrosion rate is likely to be less than about 0.05 inch per year (Fair).
>0.05 means that corrosion rate is likely to be more than 0.05 inch per year (Poor).

Table 326. CORROSION RATES OF COPPER, SN-BRAZE, AL-BRAZE AT 70°F * (SHEET 1 OF 9)

Corrosive Medium	Corrosion Rate* in 10% Corrosive Medium (ipy)	Corrosion Rate** in 100% Corrosive Medium (ipy)
Acetaldehyde	<0.002	<0.002
Acetic Acid (Aerated)	>0.05	<0.02
Acetic Acid (Air Free)	<0.002	<0.002
Acetic Anhydride	—	<0.02
Acetone	<0.002	<0.002
Acetylene	—	<0.002
Acrolein	<0.02	<0.02
Acrylonitril	—	<0.002
Alcohol (Ethyl)	<0.002	<0.002
Alcohol (Methyl)	<0.02	<0.02
Alcohol (Allyl)	—	<0.02
Alcohol (Amyl)	—	<0.002
Alcohol (Benzyl)	—	<0.02
Alcohol (Butyl)	—	<0.002
Alcohol (Cetyl)	—	<0.02
Alcohol (Isopropyl)	—	<0.02
Allylamine	—	>0.05
Allyl Chloride	—	<0.02
Allyl Sulfide	—	>0.05
Aluminum Acetate	<0.02	<0.02
Aluminum Chloride	<0.02	<0.02
Aluminum Fluoride	<0.02	—
Aluminum Fluosilicate	—	<0.02
Aluminum Formate	—	<0.02

* 10% corrosive medium in 90% water. (Other % corrosive medium in parentheses.)
** Water-free, dry or maximum concentration of corrosive medium.

Source: data compiled by J.S. Park from Earl R. Parker, *Materials Data Book for Engineers and Scientists*, McGraw-Hill Book Company, New York, 1967.

Table 326. CORROSION RATES OF COPPER, SN-BRAZE, AL-BRAZE AT 70°F * (SHEET 2 OF 9)

Corrosive Medium	Corrosion Rate* in 10% Corrosive Medium (ipy)	Corrosion Rate** in 100% Corrosive Medium (ipy)
Aluminum Hydroxide	<0.02	—
Aluminum Potassium Sulfate	<0.02	<0.02
Aluminum Sulfate	<0.02	<0.002
Ammonia	>0.05	<0.002
Ammonium Acetate	—	>0.05
Ammonium Bicarbonate	>0.05	—
Ammonium Bromide	>0.05	—
Ammonium Carbonate	>0.05	—
Ammonium Chloride	>0.05	>0.05
Ammonium Citrate	>0.05	—
Ammonium Nitrate	>0.05	>0.05
Ammonium Sulfate	<0.05	<0.02
Ammonium Sulfite	>0.05	>0.05
Ammonium Thiocyanate	>0.05	—
Amyl Acetate	<0.02	<0.02
Amyl Chloride	<0.02	<0.002
Aniline	—	>0.05
Aniline Hydrochloride	>0.05	—
Anthracine	—	<0.02
Antimony Trichloride	>0.05	<0.05
Barium Carbonate	<0.02	<0.02
Barium Chloride	<0.02	<0.02
Barium Hydroxide	>0.05	—
Barium Nitrate	>0.05	—

* 10% corrosive medium in 90% water. (Other % corrosive medium in parentheses.)
** Water-free, dry or maximum concentration of corrosive medium.

Source: data compiled by J.S. Park from Earl R. Parker, *Materials Data Book for Engineers and Scientists*, McGraw-Hill Book Company, New York, 1967.

Table 326. CORROSION RATES OF COPPER, SN-BRAZE, AL-BRAZE AT 70°F * (SHEET 3 OF 9)

Corrosive Medium	Corrosion Rate* in 10% Corrosive Medium (ipy)	Corrosion Rate** in 100% Corrosive Medium (ipy)
Barium Peroxide	>0.05	—
Benzaldehyde	>0.05	<0.02
Benzene	<0.002	<0.02
Benzoic Acid	<0.02	<0.02
Boric Acid	<0.02	<0.02
Bromic Acid	>0.05	>0.05
Bromine (Dry)	—	<0.02
Bromine (Wet)	—	>0.05
Butyric Acid	<0.05	<0.02
Cadmium Chloride	<0.02	—
Cadmium Sulfate	<0.02	—
Calcium Acetate	<0.02	<0.02
Calcium Bicarbonate	—	<0.02
Calcium Bromide	<0.02	<0.02
Calcium Chlorate	<0.02	—
Calcium Chloride	<0.002	<0.02
Calcium Hydroxide	<0.02	—
Calcium Hypochlorite	<0.02	—
Carbon Dioxide	—	<0.002
Carbon Monoxide	—	<0.002
Carbon Tetrachloride	—	<0.002
Carbon Acid (Air Free)	<0.02	<0.02
Chloroacetic Acid	>0.05	>0.05
Chlorine Gas	—	<0.02

* 10% corrosive medium in 90% water. (Other % corrosive medium in parentheses.)
** Water-free, dry or maximum concentration of corrosive medium.

Source: data compiled by J.S. Park from Earl R. Parker, *Materials Data Book for Engineers and Scientists*, McGraw-Hill Book Company, New York, 1967.

Table 326. CORROSION RATES OF COPPER, SN-BRAZE, AL-BRAZE AT 70°F* (SHEET 4 OF 9)

Corrosive Medium	Corrosion Rate* in 10% Corrosive Medium (ipy)	Corrosion Rate** in 100% Corrosive Medium (ipy)
Chloroform (Dry)	—	<0.002
Chromic Acid	>0.05	—
Chromic Hydroxide	—	<0.02
Chromic Sulfates	<0.02	<0.05
Citric Acid	<0.05	<0.02
Copper Nitrate	>0.05	>0.05
Copper Sulfate	>0.05	>0.05
Diethylene Glycol	—	<0.002
Ethyl Chloride	<0.02	<0.002
Ethylene Glycol	<0.02	<0.02
Ethylene Oxide	—	>0.05
Fatty Acids	—	<0.05
Ferric Chloride	>0.05	<0.02
Ferric Nitrate	>0.05	—
Ferrous Chloride	<0.02	<0.02
Ferrous Sulfate	<0.02	<0.02
Fluorine	—	<0.002
Formaldehyde	<0.002	<0.002
Formic Acid	<0.02	<0.02
Furfural	<0.02	<0.02
Hydrazine	>0.05	—
Hydrobromic Acid	>0.05	<0.02
Hydrochloric Acid (Areated)	>0.05	—
Hydrochloric Acid (Air Free)	>0.05	—

* 10% corrosive medium in 90% water. (Other % corrosive medium in parentheses.)
** Water-free, dry or maximum concentration of corrosive medium.

Source: data compiled by J.S. Park from Earl R. Parker, *Materials Data Book for Engineers and Scientists*, McGraw-Hill Book Company, New York, 1967.

Table 326. CORROSION RATES OF COPPER, SN-BRAZE, AL-BRAZE AT 70°F* (SHEET 5 OF 9)

Corrosive Medium	Corrosion Rate* in 10% Corrosive Medium (ipy)	Corrosion Rate** in 100% Corrosive Medium (ipy)
Hydrocyanic Acid	>0.05	<0.02
Hydrofluoric Acid (Areated)	<0.02	<0.02
Hydrofluoric Acid (Air Free)	<0.02	<0.02
Hydrogen Chloride	—	<0.02
Hydrogen Fluoride	—	<0.02
Hydrogen Iodide	—	<0.02
Hydrogen Peroxide	>0.05	>0.05
Hydrogen Sulfide	<0.02	<0.02
Lactic Acid	<0.002	<0.02
Lead Acetate	<0.05	—
Lead Chromate	—	<0.02
Lead Sulfate	—	<0.02
Lithium Chloride	<0.02 (30%)	—
Lithium Hydroxide	>0.05	—
Magnesium Chloride	<0.02	<0.02
Magnesium Hydroxide	<0.02	<0.02
Magnesium Sulfate	<0.002	<0.02
Maleic Acid	<0.02	<0.02
Mercuric Chloride	>0.05	>0.05
Mercurous Nitrate	>0.05	>0.05
Mercury	—	>0.05
Methallylamine	—	>0.05
Methanol	<0.02	<0.02
Methyl Ethyl Ketone	<0.02	<0.002

* 10% corrosive medium in 90% water. (Other % corrosive medium in parentheses.)
** Water-free, dry or maximum concentration of corrosive medium.

Source: data compiled by J.S. Park from Earl R. Parker, *Materials Data Book for Engineers and Scientists*, McGraw-Hill Book Company, New York, 1967.

Table 326. CORROSION RATES OF COPPER, SN-BRAZE, AL-BRAZE AT 70°F * (SHEET 6 OF 9)

Corrosive Medium	Corrosion Rate* in 10% Corrosive Medium (ipy)	Corrosion Rate** in 100% Corrosive Medium (ipy)
Methyl Isobutyl Ketone	<0.02	<0.02
Methylamine	—	>0.05
Methylene Chloride	<0.02	<0.002
Monochloroacetic Acid	>0.05	>0.05
Monorthanolamine	—	>0.05
Monoethalamine	—	>0.05
Monoethylamine	—	>0.05
Monosodium Phosphate	<0.02	—
Nickel Chloride	>0.05	—
Nickel Nitrate	<0.05	—
Nickel Sulfate	<0.02	<0.02
Nitric Acid	>0.05	>0.05
Nitric Acid (Red Fuming)	—	>0.05
Nitric + Hydrochloric Acid	—	>0.05
Nitric + Hydrofluoric Acid	—	>0.05
Nitric + Sulfuric Acid	>0.05	>0.05
Nitrobenzene	—	<0.02
Nitroglycerine	—	<0.02
Nitrotolune	—	<0.02
Nitrous Acid	—	>0.05
Oleic Acid	—	<0.002
Oxalic Acid	<0.02	<0.05
Phenol	—	<0.002
Phosphoric Acid (Areated)	>0.05	>0.05

* 10% corrosive medium in 90% water. (Other % corrosive medium in parentheses.)
** Water-free, dry or maximum concentration of corrosive medium.

Source: data compiled by J.S. Park from Earl R. Parker, *Materials Data Book for Engineers and Scientists*, McGraw-Hill Book Company, New York, 1967.

Table 326. CORROSION RATES OF COPPER, SN-BRAZE, AL-BRAZE AT 70°F * (SHEET 7 OF 9)

Corrosive Medium	Corrosion Rate* in 10% Corrosive Medium (ipy)	Corrosion Rate** in 100% Corrosive Medium (ipy)
Phosphoric Acid (Air Free)	<0.02	—
Picric Acid	>0.05	>0.05
Potassium Bicarbonate	<0.02	<0.02
Potassium Bromide	<0.02	<0.02
Potassium Carbonate	<0.02	<0.02
Potassium Chlorate	<0.02	<0.05
Potassium Chromate	<0.02	—
Potassium Cyanide	>0.05	>0.05
Potassium Dichromate	<0.02	—
Potassium Ferricyanide	<0.02	<0.02
Potassium Ferrocyanide	<0.02	—
Potassium Hydroxide	<0.02	—
Potassium Hypochlorite	<0.02	—
Potassium Iodide	<0.02	<0.02
Potassium Nitrate	<0.02	<0.002
Potassium Nitrite	<0.02	<0.02
Potassium Permanganate	<0.02	<0.02
Potassium Silicate	<0.02	<0.02
Propionic Acid	<0.02	<0.02
Pyridine	<0.02	<0.02
Quinine Sulfate	<0.02	<0.02
Salicylic Acid	—	<0.02
Silicon Tetrachloride (Dry)	—	<0.002
Silicon Tetrachloride (Wet)	—	>0.05

* 10% corrosive medium in 90% water. (Other % corrosive medium in parentheses.)
** Water-free, dry or maximum concentration of corrosive medium.

Source: data compiled by J.S. Park from Earl R. Parker, *Materials Data Book for Engineers and Scientists*, McGraw-Hill Book Company, New York, 1967.

Table 326. CORROSION RATES OF COPPER, SN-BRAZE, AL-BRAZE AT 70°F * (SHEET 8 OF 9)

Corrosive Medium	Corrosion Rate* in 10% Corrosive Medium (ipy)	Corrosion Rate** in 100% Corrosive Medium (ipy)
Silver Bromide	>0.05	—
Silver Chloride	>0.05	<0.02
Silver Nitrate	>0.05	—
Sodium Acetate	<0.02	<0.02
Sodium Bicarbonate	<0.02	<0.02
Sodium Bisulfate	—	<0.02
Sodium Bromide	<0.02	<0.05
Sodium Carbonate	<0.02	—
Sodium Chloride	<0.02	—
Sodium Chromate	<0.02	<0.02
Sodium Hydroxide	<0.002	—
Sodium Hypochlorite	>0.05	—
Sodium Metasilicate	<0.02	<0.02
Sodium Nitrate	<0.02	<0.05
Sodium Nitrite	<0.02	—
Sodium Phosphate	<0.02	<0.02
Sodium Silicate	<0.02	<0.02
Sodium Sulfate	<0.02	<0.02
Sodium Sulfide	>0.05	>0.05
Sodium Sulfite	<0.02	<0.05
Stannic Chloride	>0.05	—
Stannous Chloride	>0.05	—
Strontium Nitrate	<0.02	<0.02
Succinic Acid	<0.02	<0.02

* 10% corrosive medium in 90% water. (Other % corrosive medium in parentheses.)
** Water-free, dry or maximum concentration of corrosive medium.

Source: data compiled by J.S. Park from Earl R. Parker, *Materials Data Book for Engineers and Scientists*, McGraw-Hill Book Company, New York, 1967.

Table 326. CORROSION RATES OF COPPER, SN-BRAZE, AL-BRAZE AT 70°F * (SHEET 9 OF 9)

Corrosive Medium	Corrosion Rate* in 10% Corrosive Medium (ipy)	Corrosion Rate** in 100% Corrosive Medium (ipy)
Sulfur Dioxide	<0.02	<0.02
Sulfur Trioxide	—	<0.02
Sulfuric Acid (Areated)	>0.05	>0.05
Sulfuric Acid (Air Free)	<0.02	—
Sulfuric Acid (Fuming)	—	>0.05
Sulfurous Acid	<0.02	<0.05
Tannic Acid	<0.02	<0.02
Tartaric Acid	<0.02	<0.02
Tetraphosphoric Acid	—	<0.05
Trichloroacetic Acid	>0.05	>0.05
Trichloroethylene	—	<0.002
Urea	<0.02	—
Zinc Chloride	<0.02	—
Zinc Sulfate	<0.02	<0.02

* 10% corrosive medium in 90% water. (Other % corrosive medium in parentheses.)
** Water-free, dry or maximum concentration of corrosive medium.

Source: data compiled by J.S. Park from Earl R. Parker, *Materials Data Book for Engineers and Scientists*, McGraw-Hill Book Company, New York, 1967.

* <0.002 means that corrosion rate is likely to be less than 0.002 inch per year (Excellent).
<0.02 means that corrosion rate is likely to be less than about 0.02 inch per year (Good).
<0.05 means that corrosion rate is likely to be less than about 0.05 inch per year (Fair).
>0.05 means that corrosion rate is likely to be more than 0.05 inch per year (Poor).

Table 327. CORROSION RATES OF SILICON BRONZE AT 70°F*

(SHEET 1 OF 9)

Corrosive Medium	Corrosion Rate* in 10% Corrosive Medium (ipy)	Corrosion Rate** in 100% Corrosive Medium (ipy)
Acetaldehyde	<0.02	<0.002
Acetic Acid (Aerated)	>0.05	>0.05
Acetic Acid (Air Free)	>0.05	<0.02
Acetic Anhydride	—	<0.02
Acetone	<0.002	<0.002
Acetylene	—	<0.002
Acrolein	<0.02	<0.02
Acrylonitril	—	<0.002
Alcohol (Ethyl)	<0.002	<0.002
Alcohol (Methyl)	<0.02	<0.02
Alcohol (Allyl)	—	<0.02
Alcohol (Amyl)	—	<0.02
Alcohol (Benzyl)	—	<0.02
Alcohol (Butyl)	—	<0.002
Alcohol (Isopropyl)	—	<0.02
Allylamine	—	>0.05
Allyl Chloride	—	<0.02
Allyl Sulfide	—	>0.05
Aluminum Acetate	<0.02	<0.02
Aluminum Chloride	<0.02	<0.02
Aluminum Fluoride	<0.02	—
Aluminum Fluosilicate	—	<0.02
Aluminum Formate	<0.02	<0.02
Aluminum Hydroxide	<0.02	<0.02

* 10% corrosive medium in 90% water. (Other % corrosive medium in parentheses.)
** Water-free, dry or maximum concentration of corrosive medium.

Source: data compiled by J.S. Park from Earl R. Parker, *Materials Data Book for Engineers and Scientists*, McGraw-Hill Book Company, New York, 1967.

Table 327. CORROSION RATES OF SILICON BRONZE AT 70°F *

(SHEET 2 OF 9)

Corrosive Medium	Corrosion Rate* in 10% Corrosive Medium (ipy)	Corrosion Rate** in 100% Corrosive Medium (ipy)
Aluminum Potassium Sulfate	<0.02	<0.02
Aluminum Sulfate	<0.02	<0.02
Ammonia	>0.05	<0.002
Ammonium Acetate	—	>0.05
Ammonium Bicarbonate	>0.05	—
Ammonium Bromide	>0.05	—
Ammonium Carbonate	>0.05	<0.02
Ammonium Chloride	>0.05	>0.05
Ammonium Citrate	>0.05	—
Ammonium Nitrate	>0.05	>0.05
Ammonium Sulfate	<0.02	<0.02
Ammonium Sulfite	>0.05	>0.05
Ammonium Thiocyanate	>0.05	—
Amyl Acetate	<0.02	<0.02
Amyl Chloride	—	<0.002
Aniline Hydrochloride	>0.05	—
Anthracine	—	<0.02
Antimony Trichloride	>0.05	—
Barium Carbonate	<0.02	<0.02
Barium Chloride	<0.02	<0.02
Barium Hydroxide	>0.05	—
Barium Nitrate	>0.05	—
Barium Peroxide	>0.05	—
Benzaldehyde	>0.05	<0.02

* 10% corrosive medium in 90% water. (Other % corrosive medium in parentheses.)
** Water-free, dry or maximum concentration of corrosive medium.

Source: data compiled by J.S. Park from Earl R. Parker, *Materials Data Book for Engineers and Scientists*, McGraw-Hill Book Company, New York, 1967.

Table 327. CORROSION RATES OF SILICON BRONZE AT 70°F *

(SHEET 3 OF 9)

Corrosive Medium	Corrosion Rate* in 10% Corrosive Medium (ipy)	Corrosion Rate** in 100% Corrosive Medium (ipy)
Benzene	<0.02	<0.02
Benzoic Acid	<0.02	<0.02
Boric Acid	<0.02	<0.02
Bromic Acid	>0.05	>0.05
Bromine (Dry)	—	<0.02
Bromine (Wet)	—	>0.05
Butyric Acid	<0.02	<0.02
Cadmium Chloride	<0.02	—
Cadmium Sulfate	<0.02	—
Calcium Acetate	<0.02	<0.02
Calcium Bicarbonate	—	<0.02
Calcium Bromide	<0.02	<0.02
Calcium Chlorate	<0.02	—
Calcium Chloride	<0.02	<0.02
Calcium Hydroxide	<0.02	—
Calcium Hypochlorite	<0.02	—
Carbon Dioxide	—	<0.002
Carbon Monoxide	—	<0.002
Carbon Tetrachloride	—	<0.002
Carbon Acid (Air Free)	<0.02	<0.02
Chloroacetic Acid	—	<0.05
Chlorine Gas	—	<0.02
Chloroform (Dry)	—	<0.02
Chromic Acid	>0.05	—

* 10% corrosive medium in 90% water. (Other % corrosive medium in parentheses.)
** Water-free, dry or maximum concentration of corrosive medium.

Source: data compiled by J.S. Park from Earl R. Parker, *Materials Data Book for Engineers and Scientists*, McGraw-Hill Book Company, New York, 1967.

Table 327. CORROSION RATES OF SILICON BRONZE AT 70°F *
(SHEET 4 OF 9)

Corrosive Medium	Corrosion Rate* in 10% Corrosive Medium (ipy)	Corrosion Rate** in 100% Corrosive Medium (ipy)
Chromic Hydroxide	—	<0.02
Chromic Sulfates	<0.02	—
Citric Acid	<0.05	<0.02
Copper Nitrate	>0.05	<0.05
Copper Sulfate	<0.02	>0.05
Diethylene Glycol	—	<0.002
Ethyl Chloride	—	<0.002
Ethylene Glycol	—	<0.02
Ethylene Oxide	—	>0.05
Fatty Acids	—	<0.05
Ferric Chloride	>0.05	<0.02
Ferric Nitrate	>0.05	—
Ferrous Chloride	<0.05	<0.02
Ferrous Sulfate	<0.02	<0.02
Fluorine	—	>0.05
Formaldehyde	<0.002	<0.02
Formic Acid	<0.02	<0.02
Furfural	<0.02	<0.02
Hydrazine	>0.05	—
Hydrobromic Acid	<0.02	<0.02
Hydrochloric Acid (Areated)	>0.05	—
Hydrochloric Acid (Air Free)	<0.02	—
Hydrocyanic Acid	>0.05	<0.02
Hydrofluoric Acid (Areated)	>0.05	—

* 10% corrosive medium in 90% water. (Other % corrosive medium in parentheses.)
** Water-free, dry or maximum concentration of corrosive medium.

Source: data compiled by J.S. Park from Earl R. Parker, *Materials Data Book for Engineers and Scientists*, McGraw-Hill Book Company, New York, 1967.

Table 327. CORROSION RATES OF SILICON BRONZE AT 70°F *

(SHEET 5 OF 9)

Corrosive Medium	Corrosion Rate* in 10% Corrosive Medium (ipy)	Corrosion Rate** in 100% Corrosive Medium (ipy)
Hydrofluoric Acid (Air Free)	<0.02	<0.02
Hydrogen Chloride	—	<0.02
Hydrogen Fluoride	—	<0.02
Hydrogen Iodide	—	<0.02
Hydrogen Peroxide	>0.05	>0.05
Hydrogen Sulfide	<0.02	<0.02
Lactic Acid	<0.05	<0.02
Lead Acetate	—	<0.02
Lead Chromate	—	<0.02
Lead Sulfate	—	<0.02
Lithium Chloride	<0.02 (30%)	—
Lithium Hydroxide	>0.05	—
Magnesium Chloride	<0.02	<0.02
Magnesium Hydroxide	<0.02	<0.02
Magnesium Sulfate	<0.002	<0.02
Maleic Acid	<0.02	—
Mercuric Chloride	>0.05	>0.05
Mercurous Nitrate	>0.05	—
Mercury	—	>0.05
Methallylamine	—	>0.05
Methanol	<0.02	<0.02
Methyl Ethyl Ketone	<0.02	<0.002
Methyl Isobutyl Ketone	<0.02	<0.02
Methylamine	—	>0.05

* 10% corrosive medium in 90% water. (Other % corrosive medium in parentheses.)
** Water-free, dry or maximum concentration of corrosive medium.

Source: data compiled by J.S. Park from Earl R. Parker, *Materials Data Book for Engineers and Scientists*, McGraw-Hill Book Company, New York, 1967.

Table 327. CORROSION RATES OF SILICON BRONZE AT 70°F *

(SHEET 6 OF 9)

Corrosive Medium	Corrosion Rate* in 10% Corrosive Medium (ipy)	Corrosion Rate** in 100% Corrosive Medium (ipy)
Methylene Chloride	<0.02	<0.02
Monochloroacetic Acid	>0.05	>0.05
Monorthanolamine	—	>0.05
Monoethalamine	—	>0.05
Monoethylamine	—	>0.05
Monosodium Phosphate	<0.02	—
Nickel Chloride	>0.05	<0.02
Nickel Nitrate	<0.05	—
Nickel Sulfate	<0.02	<0.02
Nitric Acid	>0.05	>0.05
Nitric Acid (Red Fuming)	—	>0.05
Nitric + Hydrochloric Acid	—	>0.05
Nitric + Sulfuric Acid	>0.05	>0.05
Nitrobenzene	—	<0.02
Nitrocelluolose	—	<0.02
Nitroglycerine	—	<0.02
Nitrotolune	—	<0.02
Nitrous Acid	—	>0.05
Oleic Acid	—	<0.02
Oxalic Acid	<0.02	<0.02
Phenol	—	<0.002
Phosphoric Acid (Areated)	>0.05	>0.05
Phosphoric Acid (Air Free)	<0.02	—
Picric Acid	>0.05	>0.05

* 10% corrosive medium in 90% water. (Other % corrosive medium in parentheses.)
** Water-free, dry or maximum concentration of corrosive medium.

Source: data compiled by J.S. Park from Earl R. Parker, *Materials Data Book for Engineers and Scientists*, McGraw-Hill Book Company, New York, 1967.

Table 327. CORROSION RATES OF SILICON BRONZE AT 70°F *

(SHEET 7 OF 9)

Corrosive Medium	Corrosion Rate* in 10% Corrosive Medium (ipy)	Corrosion Rate** in 100% Corrosive Medium (ipy)
Potassium Bicarbonate	<0.02	<0.02
Potassium Bromide	<0.02	<0.02
Potassium Carbonate	<0.02	<0.02
Potassium Chlorate	<0.02	<0.05
Potassium Chromate	<0.02	<0.02
Potassium Cyanide	>0.05	>0.05
Potassium Dichromate	<0.02	—
Potassium Ferricyanide	<0.02	—
Potassium Ferrocyanide	<0.02	—
Potassium Hydroxide	<0.02	>0.05
Potassium Hypochlorite	>0.05	—
Potassium Iodide	<0.02	<0.02
Potassium Nitrate	<0.02	<0.02
Potassium Nitrite	<0.02	<0.02
Potassium Permanganate	<0.02	<0.02
Potassium Silicate	<0.02	<0.02
Propionic Acid	<0.02	—
Pyridine	<0.02	<0.02
Quinine Sulfate	<0.02	<0.02
Salicylic Acid	—	<0.02
Silicon Tetrachloride (Dry)	—	<0.002
Silicon Tetrachloride (Wet)	—	>0.05
Silver Bromide	>0.05	—
Silver Chloride	>0.05	—

* 10% corrosive medium in 90% water. (Other % corrosive medium in parentheses.)
** Water-free, dry or maximum concentration of corrosive medium.

Source: data compiled by J.S. Park from Earl R. Parker, *Materials Data Book for Engineers and Scientists*, McGraw-Hill Book Company, New York, 1967.

Table 327. CORROSION RATES OF SILICON BRONZE AT 70°F *
(SHEET 8 OF 9)

Corrosive Medium	Corrosion Rate* in 10% Corrosive Medium (ipy)	Corrosion Rate** in 100% Corrosive Medium (ipy)
Silver Nitrate	>0.05	—
Sodium Acetate	<0.02	—
Sodium Bicarbonate	<0.02	—
Sodium Bisulfate	<0.02	<0.02
Sodium Bromide	<0.02	—
Sodium Carbonate	<0.02	<0.02
Sodium Chloride	<0.02	—
Sodium Chromate	<0.02	<0.02
Sodium Hydroxide	<0.02	—
Sodium Hypochlorite	<0.02	>0.05
Sodium Metasilicate	<0.02	<0.02
Sodium Nitrate	<0.02	<0.02
Sodium Nitrite	<0.02	—
Sodium Phosphate	<0.02	<0.02
Sodium Silicate	<0.02	<0.02
Sodium Sulfate	<0.02	<0.02
Sodium Sulfide	>0.05	>0.05
Sodium Sulfite	<0.02	<0.02
Stannic Chloride	>0.05	>0.05
Stannous Chloride	<0.02	<0.02
Strontium Nitrate	<0.02	<0.02
Succinic Acid	<0.02	<0.02
Sulfur Dioxide	—	<0.02
Sulfur Trioxide	—	<0.02

* 10% corrosive medium in 90% water. (Other % corrosive medium in parentheses.)
** Water-free, dry or maximum concentration of corrosive medium.

Source: data compiled by J.S. Park from Earl R. Parker, *Materials Data Book for Engineers and Scientists*, McGraw-Hill Book Company, New York, 1967.

Table 327. CORROSION RATES OF SILICON BRONZE AT 70°F [*]

(SHEET 9 OF 9)

Corrosive Medium	Corrosion Rate* in 10% Corrosive Medium (ipy)	Corrosion Rate** in 100% Corrosive Medium (ipy)
Sulfuric Acid (Areated)	>0.05	>0.05
Sulfuric Acid (Air Free)	<0.02	—
Sulfuric Acid (Fuming)	—	>0.05
Sulfurous Acid	<0.02	<0.02
Tannic Acid	<0.02	<0.02
Tartaric Acid	<0.05	<0.02
Tetraphosphoric Acid	>0.05	<0.05
Trichloroacetic Acid	—	<0.05
Trichloroethylene	—	<0.02
Urea	<0.02	—
Zinc Chloride	<0.02	>0.05
Zinc Sulfate	<0.02	<0.02

* 10% corrosive medium in 90% water. (Other % corrosive medium in parentheses.)
** Water-free, dry or maximum concentration of corrosive medium.

Source: data compiled by J.S. Park from Earl R. Parker, *Materials Data Book for Engineers and Scientists,* McGraw-Hill Book Company, New York, 1967.

* <0.002 means that corrosion rate is likely to be less than 0.002 inch per year (Excellent).
<0.02 means that corrosion rate is likely to be less than about 0.02 inch per year (Good).
<0.05 means that corrosion rate is likely to be less than about 0.05 inch per year (Fair).
>0.05 means that corrosion rate is likely to be more than 0.05 inch per year (Poor).

Table 328. CORROSION RATES OF HASTELLOY AT 70°F *

(SHEET 1 OF 9)

Corrosive Medium	Corrosion Rate* in 10% Corrosive Medium (ipy)	Corrosion Rate** in 100% Corrosive Medium (ipy)
Acetaldehyde	—	<0.002
Acetic Acid (Aerated)	<0.002	<0.002
Acetic Acid (Air Free)	<0.002	<0.002
Acetic Anhydride	<0.002	<0.002
Acetoacetic Acid	<0.02	<0.02
Acetone	<0.002	<0.002
Acetylene	—	<0.002
Acrolein	—	<0.02
Acrylonitril	—	<0.002
Alcohol (Ethyl)	<0.002	<0.002
Alcohol (Methyl)	<0.002	<0.002
Alcohol (Allyl)	—	<0.02
Alcohol (Benzyl)	<0.02	<0.02
Alcohol (Isopropyl)	—	<0.02
Allyl Chloride	—	<0.02
Aluminum Acetate	<0.02	<0.02
Aluminum Chlorate	<0.02	<0.02
Aluminum Chloride	<0.002	<0.002
Aluminum Fluoride	<0.02	—
Aluminum Fluosilicate	—	<0.02
Aluminum Formate	<0.02	<0.02
Aluminum Hydroxide	<0.02	—
Aluminum Nitrate	<0.02	—
Aluminum Potassium Sulfate	<0.02	—

* 10% corrosive medium in 90% water. (Other % corrosive medium in parentheses.)
** Water-free, dry or maximum concentration of corrosive medium.

Source: data compiled by J.S. Park from Earl R. Parker, *Materials Data Book for Engineers and Scientists*, McGraw-Hill Book Company, New York, 1967.

Table 328. CORROSION RATES OF HASTELLOY AT 70°F *

(SHEET 2 OF 9)

Corrosive Medium	Corrosion Rate* in 10% Corrosive Medium (ipy)	Corrosion Rate** in 100% Corrosive Medium (ipy)
Aluminum Sulfate	<0.002	<0.02
Ammonia	<0.002	<0.002
Ammonium Acetate	<0.002	<0.002
Ammonium Bromide	<0.02	—
Ammonium Carbonate	>0.05	—
Ammonium Chloride	<0.002	<0.02
Ammonium Citrate	<0.02	—
Ammonium Formate	<0.002	—
Ammonium Nitrate	<0.02	—
Ammonium Sulfate	<0.02	<0.02
Amyl Acetate	<0.002	<0.002
Amyl Chloride	—	<0.02
Aniline	—	<0.02
Aniline Hydrochloride	<0.02	<0.05
Anthracine	—	<0.02
Antimony Trichloride	>0.05	<0.002
Barium Carbonate	—	<0.02
Barium Chloride	<0.02	<0.02
Barium Hydroxide	<0.02	<0.02
Barium Nitrate	<0.02	<0.02
Barium Oxide	—	<0.02
Benzaldehyde	—	<0.02
Benzene	<0.02	<0.02
Benzoic Acid	<0.002	—

* 10% corrosive medium in 90% water. (Other % corrosive medium in parentheses.)
** Water-free, dry or maximum concentration of corrosive medium.

Source: data compiled by J.S. Park from Earl R. Parker, *Materials Data Book for Engineers and Scientists*, McGraw-Hill Book Company, New York, 1967.

Table 328. CORROSION RATES OF HASTELLOY AT 70°F *

(SHEET 3 OF 9)

Corrosive Medium	Corrosion Rate* in 10% Corrosive Medium (ipy)	Corrosion Rate** in 100% Corrosive Medium (ipy)
Boric Acid	<0.002	<0.002
Bromine (Dry)	—	<0.002
Bromine (Wet)	—	<0.002
Butyric Acid	<0.002	<0.002
Cadmium Chloride	<0.02	—
Cadmium Sulfate	<0.002	—
Calcium Acetate	<0.02	<0.02
Calcium Bicarbonate	—	<0.02
Calcium Bromide	<0.02	<0.02
Calcium Chlorate	<0.02	<0.02
Calcium Chloride	<0.002	<0.002
Calcium Hydroxide	<0.002	—
Calcium Hypochlorite	<0.02	<0.02
Carbon Dioxide	—	<0.002
Carbon Monoxide	—	<0.002
Carbon Tetrachloride	<0.002	<0.002
Carbon Acid (Air Free)	<0.002	<0.002
Chloroacetic Acid	<0.02	<0.002
Chlorine Gas	—	<0.02
Chloroform (Dry)	—	<0.02
Chromic Acid	<0.02	<0.02
Chromic Hydroxide	—	<0.02
Chromic Sulfates	<0.02	<0.02
Citric Acid	<0.002	<0.002

* 10% corrosive medium in 90% water. (Other % corrosive medium in parentheses.)
** Water-free, dry or maximum concentration of corrosive medium.

Source: data compiled by J.S. Park from Earl R. Parker, *Materials Data Book for Engineers and Scientists*, McGraw-Hill Book Company, New York, 1967.

Table 328. CORROSION RATES OF HASTELLOY AT 70°F *

(SHEET 4 OF 9)

Corrosive Medium	Corrosion Rate* in 10% Corrosive Medium (ipy)	Corrosion Rate** in 100% Corrosive Medium (ipy)
Copper Nitrate	<0.02	<0.02
Copper Sulfate	<0.002	<0.002
Diethylene Glycol	—	<0.02
Ethyl Chloride	—	<0.02
Ethylene Oxide	—	<0.002
Fatty Acids	—	<0.002
Ferric Chloride	<0.002	<0.02
Ferric Nitrate	<0.002	—
Ferrous Chloride	<0.02	<0.02
Ferrous Sulfate	<0.02	<0.02
Fluorine	—	<0.02
Formaldehyde	<0.02	<0.02
Formic Acid	<0.002	<0.002
Furfural	<0.02	<0.02
Hydrazine	—	<0.002
Hydrobromic Acid	<0.02	—
Hydrochloric Acid (Areated)	<0.02	—
Hydrochloric Acid (Air Free)	<0.02	—
Hydrocyanic Acid	—	<0.02
Hydrofluoric Acid (Areated)	<0.02	<0.02
Hydrofluoric Acid (Air Free)	<0.02	<0.05
Hydrogen Chloride	—	<0.002
Hydrogen Fluoride	—	<0.02
Hydrogen Iodide	—	<0.02

* 10% corrosive medium in 90% water. (Other % corrosive medium in parentheses.)
** Water-free, dry or maximum concentration of corrosive medium.

Source: data compiled by J.S. Park from Earl R. Parker, *Materials Data Book for Engineers and Scientists*, McGraw-Hill Book Company, New York, 1967.

Table 328. CORROSION RATES OF HASTELLOY AT 70°F *

(SHEET 5 OF 9)

Corrosive Medium	Corrosion Rate* in 10% Corrosive Medium (ipy)	Corrosion Rate** in 100% Corrosive Medium (ipy)
Hydrogen Peroxide	<0.002	<0.002
Hydrogen Sulfide	—	<0.002
Lactic Acid	<0.02	<0.02
Lead Acetate	<0.02	>0.05
Lead Chromate	—	<0.02
Lead Nitrate	—	<0.02
Lead Sulfate	—	<0.02
Lithium Chloride	<0.002 (30%)	—
Lithium Hydroxide	<0.02	<0.02
Magnesium Chloride	<0.002	<0.002
Magnesium Hydroxide	<0.02	—
Magnesium Sulfate	<0.002	<0.002
Maleic Acid	<0.002	<0.02
Maganous Chloride	<0.02	—
Mercuric Chloride	<0.02	—
Mercurous Nitrate	<0.02	<0.02
Mercury	—	<0.02
Methallylamine	—	<0.02
Methanol	<0.002	<0.02
Methyl Ethyl Ketone	<0.02	<0.002
Methyl Isobutyl Ketone	<0.02	<0.002
Methylene Chloride	<0.02	—
Monochloroacetic Acid	—	<0.002
Monosodium Phosphate	<0.02	—

* 10% corrosive medium in 90% water. (Other % corrosive medium in parentheses.)
** Water-free, dry or maximum concentration of corrosive medium.

Source: data compiled by J.S. Park from Earl R. Parker, *Materials Data Book for Engineers and Scientists*, McGraw-Hill Book Company, New York, 1967.

Table 328. CORROSION RATES OF HASTELLOY AT 70°F *
(SHEET 6 OF 9)

Corrosive Medium	Corrosion Rate* in 10% Corrosive Medium (ipy)	Corrosion Rate** in 100% Corrosive Medium (ipy)
Nickel Chloride	<0.002	<0.002
Nickel Nitrate	<0.02	<0.02
Nickel Sulfate	<0.02	<0.02
Nitric Acid	<0.002	—
Nitric Acid (Red Fuming)	—	<0.02
Nitric + Hydrochloric Acid	—	>0.05
Nitric + Hydrofluoric Acid	—	<0.05
Nitrobenzene	—	<0.02
Oleic Acid	—	<0.02
Oxalic Acid	<0.02	<0.02
Phenol	—	<0.002
Phosphoric Acid (Areated)	<0.002	<0.002
Phosphoric Acid (Air Free)	<0.002	<0.002
Picric Acid	<0.02	<0.02
Potassium Bicarbonate	<0.02	—
Potassium Bromide	<0.002	<0.02
Potassium Carbonate	<0.02	<0.02
Potassium Chlorate	<0.02	—
Potassium Chromate	<0.002	—
Potassium Cyanide	<0.02	—
Potassium Dichromate	<0.02	—
Potassium Ferricyanide	<0.02	—
Potassium Ferrocyanide	<0.02	—
Potassium Hydroxide	<0.02	—

* 10% corrosive medium in 90% water. (Other % corrosive medium in parentheses.)
** Water-free, dry or maximum concentration of corrosive medium.

Source: data compiled by J.S. Park from Earl R. Parker, *Materials Data Book for Engineers and Scientists*, McGraw-Hill Book Company, New York, 1967.

Table 328. CORROSION RATES OF HASTELLOY AT 70°F *
(SHEET 7 OF 9)

Corrosive Medium	Corrosion Rate* in 10% Corrosive Medium (ipy)	Corrosion Rate** in 100% Corrosive Medium (ipy)
Potassium Hypochlorite	<0.02	<0.02
Potassium Iodide	<0.02	<0.02
Potassium Nitrate	<0.02	—
Potassium Nitrite	<0.02	<0.02
Potassium Permanganate	<0.002	<0.002
Potassium Silicate	<0.02	<0.02
Pyridine	<0.02	<0.02
Quinine Sulfate	<0.02	<0.02
Salicylic Acid	—	<0.02
Silicon Tetrachloride (Dry)	—	<0.02
Silicon Tetrachloride (Wet)	—	<0.02
Silver Bromide	<0.002	—
Silver Chloride	<0.02	—
Silver Nitrate	<0.002	—
Sodium Acetate	<0.02	—
Sodium Bicarbonate	<0.02	—
Sodium Bisulfate	<0.02	<0.02
Sodium Bromide	<0.02	—
Sodium Carbonate	<0.02	<0.02
Sodium Chloride	<0.02	—
Sodium Chromate	<0.02	<0.02
Sodium Hydroxide	<0.002	<0.002
Sodium Hypochlorite	<0.002	<0.05
Sodium Metasilicate	<0.002	<0.002

* 10% corrosive medium in 90% water. (Other % corrosive medium in parentheses.)
** Water-free, dry or maximum concentration of corrosive medium.

Source: data compiled by J.S. Park from Earl R. Parker, *Materials Data Book for Engineers and Scientists*, McGraw-Hill Book Company, New York, 1967.

Table 328. CORROSION RATES OF HASTELLOY AT 70°F *

(SHEET 8 OF 9)

Corrosive Medium	Corrosion Rate* in 10% Corrosive Medium (ipy)	Corrosion Rate** in 100% Corrosive Medium (ipy)
Sodium Nitrate	<0.02	—
Sodium Nitrite	<0.02	—
Sodium Phosphate	<0.02	<0.02
Sodium Silicate	<0.02	<0.02
Sodium Sulfate	<0.02	<0.002
Sodium Sulfide	<0.02	—
Sodium Sulfite	<0.02	—
Stannic Chloride	<0.02	<0.02
Stannous Chloride	<0.02	<0.02
Strontium Nitrate	<0.02	<0.02
Succinic Acid	<0.02	—
Sulfur Dioxide	<0.002	<0.02
Sulfur Trioxide	—	<0.02
Sulfuric Acid (Areated)	<0.002	<0.02
Sulfuric Acid (Air Free)	<0.002	<0.02
Sulfuric Acid (Fuming)	—	<0.002
Sulfurous Acid	<0.02	<0.02
Tannic Acid	<0.02	—
Tartaric Acid	<0.02	<0.02
Tetraphosphoric Acid	—	<0.02
Trichloroacetic Acid	<0.02	<0.02
Trichloroethylene	—	<0.002
Urea	<0.02	—

* 10% corrosive medium in 90% water. (Other % corrosive medium in parentheses.)
** Water-free, dry or maximum concentration of corrosive medium.

Source: data compiled by J.S. Park from Earl R. Parker, *Materials Data Book for Engineers and Scientists*, McGraw-Hill Book Company, New York, 1967.

Table 328. CORROSION RATES OF HASTELLOY AT 70°F *

(SHEET 9 OF 9)

Corrosive Medium	Corrosion Rate* in 10% Corrosive Medium (ipy)	Corrosion Rate** in 100% Corrosive Medium (ipy)
Zinc Chloride	<0.02	<0.02
Zinc Sulfate	<0.02	—

* 10% corrosive medium in 90% water. (Other % corrosive medium in parentheses.)
** Water-free, dry or maximum concentration of corrosive medium.

Source: data compiled by J.S. Park from Earl R. Parker, *Materials Data Book for Engineers and Scientists*, McGraw-Hill Book Company, New York, 1967.

* <0.002 means that corrosion rate is likely to be less than 0.002 inch per year (Excellent).
<0.02 means that corrosion rate is likely to be less than about 0.02 inch per year (Good).
<0.05 means that corrosion rate is likely to be less than about 0.05 inch per year (Fair).
>0.05 means that corrosion rate is likely to be more than 0.05 inch per year (Poor).

Table 329. CORROSION RATES OF INCONEL AT 70°F *

(SHEET 1 OF 8)

Corrosive Medium	Corrosion Rate* in 10% Corrosive Medium (ipy)	Corrosion Rate** in 100% Corrosive Medium (ipy)
Acetaldehyde	—	<0.002
Acetic Acid (Aerated)	<0.02	<0.02
Acetic Acid (Air Free)	<0.02	<0.02
Acetic Anhydride	—	<0.02
Acetone	<0.002	<0.002
Acetylene	—	<0.002
Acrolein	—	<0.02
Acrylonitril	—	<0.002
Alcohol (Ethyl)	<0.002	<0.002
Alcohol (Methyl)	<0.002	<0.002
Alcohol (Allyl)	—	<0.02
Alcohol (Benzyl)	—	<0.02
Alcohol (Butyl)	—	<0.002
Alcohol (Cetyl)	—	<0.02
Alcohol (Isopropyl)	—	<0.02
Allyl Chloride	—	<0.02
Aluminum Acetate	<0.02	—
Aluminum Chlorate	<0.02	<0.02
Aluminum Chloride	>0.05	—
Aluminum Fluosilicate	—	<0.02
Aluminum Formate	<0.02	<0.02
Aluminum Nitrate	<0.02	—
Aluminum Sulfate	<0.02	—
Ammonia	<0.002	<0.002

* 10% corrosive medium in 90% water. (Other % corrosive medium in parentheses.)
** Water-free, dry or maximum concentration of corrosive medium.

Source: data compiled by J.S. Park from Earl R. Parker, *Materials Data Book for Engineers and Scientists*, McGraw-Hill Book Company, New York, 1967.

Table 329. CORROSION RATES OF INCONEL AT 70°F *

(SHEET 2 OF 8)

Corrosive Medium	Corrosion Rate* in 10% Corrosive Medium (ipy)	Corrosion Rate** in 100% Corrosive Medium (ipy)
Ammonium Acetate	<0.002	<0.002
Ammonium Carbonate	>0.05	<0.02
Ammonium Chloride	<0.02	<0.02
Ammonium Citrate	<0.02	<0.02
Ammonium Formate	<0.02	<0.02
Ammonium Sulfate	<0.02	—
Ammonium Sulfite	>0.05	—
Amyl Acetate	—	<0.02
Aniline Hydrochloride	>0.05	—
Anthracine	—	<0.02
Barium Chloride	<0.02	<0.02
Barium Hydroxide	<0.02	<0.02
Barium Nitrate	<0.02	<0.02
Barium Oxide	—	<0.02
Benzaldehyde	—	<0.02
Benzene	<0.002	<0.02
Benzoic Acid	<0.02	—
Boric Acid	<0.02	<0.02
Bromic Acid	>0.05	>0.05
Bromine (Dry)	—	<0.002
Bromine (Wet)	—	>0.05
Butyric Acid	<0.05	<0.05
Cadmium Sulfate	<0.002	—
Calcium Acetate	<0.02	<0.02

* 10% corrosive medium in 90% water. (Other % corrosive medium in parentheses.)
** Water-free, dry or maximum concentration of corrosive medium.

Source: data compiled by J.S. Park from Earl R. Parker, *Materials Data Book for Engineers and Scientists*, McGraw-Hill Book Company, New York, 1967.

Table 329. CORROSION RATES OF INCONEL AT 70°F *

(SHEET 3 OF 8)

Corrosive Medium	Corrosion Rate* in 10% Corrosive Medium (ipy)	Corrosion Rate** in 100% Corrosive Medium (ipy)
Calcium Bicarbonate	—	<0.02
Calcium Bromide	<0.02	<0.02
Calcium Chlorate	<0.02	—
Calcium Chloride	<0.002	<0.02
Calcium Hydroxide	<0.02	<0.02
Calcium Hypochlorite	>0.05	—
Carbon Dioxide	—	<0.002
Carbon Monoxide	—	<0.002
Carbon Tetrachloride	<0.002	<0.002
Carbon Acid (Air Free)	<0.02	<0.002
Chloroacetic Acid	—	<0.05
Chlorine Gas	—	<0.02
Chloroform (Dry)	—	<0.002
Chromic Acid	<0.02	—
Chromic Hydroxide	—	<0.02
Citric Acid	<0.02	<0.02
Copper Nitrate	>0.05	—
Copper Sulfate	<0.02	—
Diethylene Glycol	—	<0.02
Ethyl Chloride	—	<0.002
Ethylene Glycol	—	<0.02
Ethylene Oxide	—	<0.02
Fatty Acids	—	<0.02
Ferric Chloride	<0.05	>0.05

* 10% corrosive medium in 90% water. (Other % corrosive medium in parentheses.)
** Water-free, dry or maximum concentration of corrosive medium.

Source: data compiled by J.S. Park from Earl R. Parker, *Materials Data Book for Engineers and Scientists*, McGraw-Hill Book Company, New York, 1967.

Table 329. CORROSION RATES OF INCONEL AT 70°F *
(SHEET 4 OF 8)

Corrosive Medium	Corrosion Rate* in 10% Corrosive Medium (ipy)	Corrosion Rate** in 100% Corrosive Medium (ipy)
Ferric Nitrate	>0.05	—
Ferrous Chloride	>0.05	—
Ferrous Sulfate	<0.02	—
Fluorine	—	<0.002
Formaldehyde	<0.002	<0.02
Formic Acid	<0.02	<0.02
Furfural	<0.02	<0.02
Hydrazine	—	<0.002
Hydrochloric Acid (Areated)	>0.05	—
Hydrochloric Acid (Air Free)	>0.05	—
Hydrocyanic Acid	—	<0.02
Hydrofluoric Acid (Areated)	<0.02	<0.02
Hydrofluoric Acid (Air Free)	<0.02	<0.02
Hydrogen Chloride	—	<0.002
Hydrogen Fluoride	—	<0.02
Hydrogen Peroxide	<0.02	<0.02
Hydrogen Sulfide	<0.02	<0.02
Lactic Acid	<0.02	—
Lead Acetate	<0.02	—
Lead Chromate	—	<0.02
Lead Nitrate	—	<0.02
Lead Sulfate	—	<0.02
Lithium Chloride	<0.002 (30%)	—
Lithium Hydroxide	<0.02	<0.02

* 10% corrosive medium in 90% water. (Other % corrosive medium in parentheses.)
** Water-free, dry or maximum concentration of corrosive medium.

Source: data compiled by J.S. Park from Earl R. Parker, *Materials Data Book for Engineers and Scientists*, McGraw-Hill Book Company, New York, 1967.

Table 329. CORROSION RATES OF INCONEL AT 70°F *

(SHEET 5 OF 8)

Corrosive Medium	Corrosion Rate* in 10% Corrosive Medium (ipy)	Corrosion Rate** in 100% Corrosive Medium (ipy)
Magnesium Chloride	<0.002	<0.02
Magnesium Sulfate	<0.02	<0.02
Maleic Acid	<0.02	—
Malic Acid	<0.002	<0.02
Mercuric Chloride	>0.05	—
Mercury	—	<0.02
Methallylamine	—	<0.02
Methanol	<0.002	<0.002
Methyl Ethyl Ketone	<0.02	<0.002
Methyl Isobutyl Ketone	<0.02	<0.02
Methylene Chloride	—	<0.02
Monochloroacetic Acid	<0.02	<0.02
Monorthanolamine	—	<0.02
Monosodium Phosphate	<0.02	—
Nickel Chloride	—	<0.02
Nickel Nitrate	>0.05	<0.02
Nickel Sulfate	<0.02	<0.02
Nitric Acid	<0.02	—
Nitric Acid (Red Fuming)	—	<0.02
Nitric + Hydrochloric Acid	—	>0.05
Nitric + Sulfuric Acid	>0.05	>0.05
Nitrobenzene	—	<0.02
Nitrocelluolose	—	<0.02
Nitroglycerine	—	<0.02

* 10% corrosive medium in 90% water. (Other % corrosive medium in parentheses.)
** Water-free, dry or maximum concentration of corrosive medium.

Source: data compiled by J.S. Park from Earl R. Parker, *Materials Data Book for Engineers and Scientists*, McGraw-Hill Book Company, New York, 1967.

Table 329. CORROSION RATES OF INCONEL AT 70°F *

(SHEET 6 OF 8)

Corrosive Medium	Corrosion Rate* in 10% Corrosive Medium (ipy)	Corrosion Rate** in 100% Corrosive Medium (ipy)
Nitrotolune	—	<0.02
Oleic Acid	—	<0.002
Oxalic Acid	<0.02	<0.02
Phenol	—	<0.002
Phosphoric Acid (Areated)	<0.02	>0.05
Phosphoric Acid (Air Free)	<0.02	—
Picric Acid	—	<0.02
Potassium Bicarbonate	<0.02	—
Potassium Bromide	<0.02	<0.02
Potassium Carbonate	<0.02	<0.02
Potassium Chlorate	<0.05	—
Potassium Chromate	<0.002	—
Potassium Cyanide	<0.02	<0.02
Potassium Dichromate	<0.02	—
Potassium Ferrocyanide	<0.02	—
Potassium Hydroxide	<0.02	—
Potassium Hypochlorite	<0.05	—
Potassium Iodide	<0.02	<0.02
Potassium Nitrate	<0.02	—
Potassium Nitrite	<0.02	<0.02
Potassium Permanganate	<0.02	—
Potassium Silicate	<0.02	<0.02
Pyridine	<0.02	<0.02
Quinine Sulfate	<0.02	<0.02

* 10% corrosive medium in 90% water. (Other % corrosive medium in parentheses.)
** Water-free, dry or maximum concentration of corrosive medium.

Source: data compiled by J.S. Park from Earl R. Parker, *Materials Data Book for Engineers and Scientists*, McGraw-Hill Book Company, New York, 1967.

Table 329. CORROSION RATES OF INCONEL AT 70°F *
(SHEET 7 OF 8)

Corrosive Medium	Corrosion Rate* in 10% Corrosive Medium (ipy)	Corrosion Rate** in 100% Corrosive Medium (ipy)
Salicylic Acid	—	<0.02
Silicon Tetrachloride (Dry)	—	<0.002
Silver Nitrate	<0.02	—
Sodium Acetate	<0.02	<0.02
Sodium Bicarbonate	<0.02	—
Sodium Bisulfate	<0.02	<0.02
Sodium Bromide	<0.02	—
Sodium Carbonate	<0.02	<0.02
Sodium Chloride	<0.002	—
Sodium Chromate	<0.02	<0.02
Sodium Hydroxide	<0.002	<0.002
Sodium Hypochlorite	>0.05	—
Sodium Metasilicate	<0.002	<0.002
Sodium Nitrate	<0.002	—
Sodium Nitrite	<0.02	<0.02
Sodium Phosphate	<0.02	<0.02
Sodium Silicate	<0.02	<0.02
Sodium Sulfate	<0.02	<0.02
Sodium Sulfide	<0.02	—
Sodium Sulfite	<0.02	<0.02
Stannic Chloride	>0.05	—
Stannous Chloride	>0.05	<0.02
Strontium Nitrate	<0.02	<0.02
Succinic Acid	<0.02	<0.02

* 10% corrosive medium in 90% water. (Other % corrosive medium in parentheses.)
** Water-free, dry or maximum concentration of corrosive medium.

Source: data compiled by J.S. Park from Earl R. Parker, *Materials Data Book for Engineers and Scientists*, McGraw-Hill Book Company, New York, 1967.

Table 329. CORROSION RATES OF INCONEL AT 70°F *

(SHEET 8 OF 8)

Corrosive Medium	Corrosion Rate* in 10% Corrosive Medium (ipy)	Corrosion Rate** in 100% Corrosive Medium (ipy)
Sulfur Dioxide	<0.02	<0.02
Sulfur Trioxide	—	<0.02
Sulfuric Acid (Areated)	>0.05	>0.05
Sulfuric Acid (Air Free)	<0.05	—
Sulfuric Acid (Fuming)	—	<0.02
Sulfurous Acid	<0.05	<0.02
Tannic Acid	—	<0.02
Tartaric Acid	<0.02	—
Tetraphosphoric Acid	—	<0.02
Trichloroethylene	—	<0.02
Urea	<0.02	—
Zinc Chloride	—	<0.02
Zinc Sulfate	<0.002	—

* 10% corrosive medium in 90% water. (Other % corrosive medium in parentheses.)
** Water-free, dry or maximum concentration of corrosive medium.

Source: data compiled by J.S. Park from Earl R. Parker, *Materials Data Book for Engineers and Scientists*, McGraw-Hill Book Company, New York, 1967.

* <0.002 means that corrosion rate is likely to be less than 0.002 inch per year (Excellent).
<0.02 means that corrosion rate is likely to be less than about 0.02 inch per year (Good).
<0.05 means that corrosion rate is likely to be less than about 0.05 inch per year (Fair).
>0.05 means that corrosion rate is likely to be more than 0.05 inch per year (Poor).

Table 330. CORROSION RATES OF NICKEL AT 70°F *

(SHEET 1 OF 9)

Corrosive Medium	Corrosion Rate* in 10% Corrosive Medium (ipy)	Corrosion Rate** in 100% Corrosive Medium (ipy)
Acetaldehyde	<0.002	<0.002
Acetic Acid (Aerated)	<0.05	>0.05
Acetic Acid (Air Free)	<0.02	<0.02
Acetic Anhydride	—	<0.02
Acetoacetic Acid	<0.02	<0.02
Acetone	<0.002	<0.002
Acetylene	—	<0.002
Acrolein	—	<0.02
Acrylonitril	—	<0.002
Alcohol (Ethyl)	<0.002	<0.002
Alcohol (Methyl)	<0.002	<0.002
Alcohol (Allyl)	—	<0.02
Alcohol (Benzyl)	—	<0.02
Alcohol (Butyl)	—	<0.002
Alcohol (Cetyl)	—	<0.02
Alcohol (Isopropyl)	—	<0.02
Allyl Chloride	—	<0.02
Aluminum Acetate	<0.02	—
Aluminum Chlorate	<0.02	<0.02
Aluminum Chloride	<0.05	<0.02
Aluminum Fluoride	<0.02	—
Aluminum Fluosilicate	—	<0.02
Aluminum Formate	<0.02	<0.02
Aluminum Hydroxide	<0.02	—

* 10% corrosive medium in 90% water. (Other % corrosive medium in parentheses.)
** Water-free, dry or maximum concentration of corrosive medium.

Source: data compiled by J.S. Park from Earl R. Parker, *Materials Data Book for Engineers and Scientists*, McGraw-Hill Book Company, New York, 1967.

Table 330. CORROSION RATES OF NICKEL AT 70°F *

(SHEET 2 OF 9)

Corrosive Medium	Corrosion Rate* in 10% Corrosive Medium (ipy)	Corrosion Rate** in 100% Corrosive Medium (ipy)
Aluminum Nitrate	<0.02	—
Aluminum Potassium Sulfate	<0.02	—
Aluminum Sulfate	<0.02	<0.02
Ammonia	>0.05	<0.002
Ammonium Acetate	<0.002	<0.002
Ammonium Bromide	<0.02	—
Ammonium Carbonate	>0.05	<0.02
Ammonium Chloride	<0.02	<0.02
Ammonium Citrate	<0.02	—
Ammonium Formate	<0.02	—
Ammonium Nitrate	<0.02	<0.02
Ammonium Sulfate	<0.02	<0.02
Ammonium Sulfite	>0.05	—
Ammonium Thiocyanate	<0.02	<0.02
Amyl Acetate	—	<0.02
Amyl Chloride	<0.02	<0.02
Aniline	<0.02	<0.02
Aniline Hydrochloride	<0.05	—
Anthracine	—	<0.02
Antimony Trichloride	>0.05	<0.02
Barium Carbonate	<0.02	<0.02
Barium Chloride	<0.02	<0.02
Barium Hydroxide	<0.002	<0.02
Barium Nitrate	<0.02	—

* 10% corrosive medium in 90% water. (Other % corrosive medium in parentheses.)
** Water-free, dry or maximum concentration of corrosive medium.

Source: data compiled by J.S. Park from Earl R. Parker, *Materials Data Book for Engineers and Scientists*, McGraw-Hill Book Company, New York, 1967.

Table 330. CORROSION RATES OF NICKEL AT 70°F *

(SHEET 3 OF 9)

Corrosive Medium	Corrosion Rate* in 10% Corrosive Medium (ipy)	Corrosion Rate** in 100% Corrosive Medium (ipy)
Barium Peroxide	<0.02	—
Benzaldehyde	—	<0.02
Benzene	<0.002	<0.02
Benzoic Acid	<0.02	<0.02
Boric Acid	<0.02	<0.02
Bromic Acid	>0.05	>0.05
Bromine (Dry)	—	<0.002
Bromine (Wet)	—	>0.05
Butyric Acid	<0.05	<0.05
Cadmium Chloride	<0.02	—
Cadmium Sulfate	<0.002	—
Calcium Acetate	<0.02	<0.02
Calcium Bicarbonate	—	<0.02
Calcium Bromide	<0.02	<0.02
Calcium Chlorate	<0.02	—
Calcium Chloride	<0.002	<0.02
Calcium Hydroxide	<0.02	<0.02
Calcium Hypochlorite	>0.05	—
Carbon Dioxide	—	<0.002
Carbon Monoxide	—	<0.002
Carbon Tetrachloride	<0.02	<0.002
Carbon Acid (Air Free)	<0.02	<0.02
Chloroacetic Acid	—	<0.02
Chlorine Gas	—	<0.002

* 10% corrosive medium in 90% water. (Other % corrosive medium in parentheses.)
** Water-free, dry or maximum concentration of corrosive medium.

Source: data compiled by J.S. Park from Earl R. Parker, *Materials Data Book for Engineers and Scientists*, McGraw-Hill Book Company, New York, 1967.

Table 330. CORROSION RATES OF NICKEL AT 70°F *

(SHEET 4 OF 9)

Corrosive Medium	Corrosion Rate* in 10% Corrosive Medium (ipy)	Corrosion Rate** in 100% Corrosive Medium (ipy)
Chloroform (Dry)	—	<0.002
Chromic Acid	>0.05	—
Chromic Hydroxide	—	<0.02
Citric Acid	<0.02	<0.02
Copper Nitrate	>0.05	—
Copper Sulfate	<0.02	—
Diethylene Glycol	—	<0.02
Ethyl Chloride	—	<0.002
Ethylene Glycol	—	<0.02
Ethylene Oxide	—	<0.02
Fatty Acids	—	<0.02
Ferric Chloride	>0.05	—
Ferric Nitrate	>0.05	—
Ferrous Chloride	<0.05	—
Ferrous Sulfate	>0.05	<0.02
Fluorine	—	<0.002
Formaldehyde	<0.002	<0.002
Formic Acid	<0.02	<0.02
Furfural	<0.02	<0.02
Hydrazine	—	<0.002
Hydrobromic Acid	>0.05	<0.02
Hydrochloric Acid (Areated)	>0.05	—
Hydrochloric Acid (Air Free)	>0.05	—
Hydrocyanic Acid	—	<0.02

* 10% corrosive medium in 90% water. (Other % corrosive medium in parentheses.)
** Water-free, dry or maximum concentration of corrosive medium.

Source: data compiled by J.S. Park from Earl R. Parker, *Materials Data Book for Engineers and Scientists*, McGraw-Hill Book Company, New York, 1967.

Table 330. CORROSION RATES OF NICKEL AT 70°F *

(SHEET 5 OF 9)

Corrosive Medium	Corrosion Rate* in 10% Corrosive Medium (ipy)	Corrosion Rate** in 100% Corrosive Medium (ipy)
Hydrofluoric Acid (Areated)	<0.02	<0.02
Hydrofluoric Acid (Air Free)	<0.02	<0.02
Hydrogen Chloride	—	<0.002
Hydrogen Fluoride	—	<0.002
Hydrogen Iodide	—	<0.02
Hydrogen Peroxide	<0.02	<0.02
Hydrogen Sulfide	—	<0.02
Lactic Acid	<0.02	—
Lead Acetate	<0.02	—
Lead Chromate	—	<0.02
Lead Nitrate	<0.02	<0.02
Lead Sulfate	<0.02	<0.02
Lithium Chloride	<0.002 (30%)	—
Lithium Hydroxide	<0.02	<0.02
Magnesium Chloride	<0.002	<0.02
Magnesium Hydroxide	—	<0.02
Magnesium Sulfate	<0.02	<0.02
Maleic Acid	<0.02	—
Malic Acid	<0.02	<0.02
Mercuric Chloride	<0.05	—
Mercury	—	<0.02
Methallylamine	—	<0.02
Methanol	<0.002	<0.002
Methyl Ethyl Ketone	<0.02	<0.002

* 10% corrosive medium in 90% water. (Other % corrosive medium in parentheses.)
** Water-free, dry or maximum concentration of corrosive medium.

Source: data compiled by J.S. Park from Earl R. Parker, *Materials Data Book for Engineers and Scientists*, McGraw-Hill Book Company, New York, 1967.

Table 330. CORROSION RATES OF NICKEL AT 70°F *

(SHEET 6 OF 9)

Corrosive Medium	Corrosion Rate* in 10% Corrosive Medium (ipy)	Corrosion Rate** in 100% Corrosive Medium (ipy)
Methyl Isobutyl Ketone	<0.02	<0.02
Methylene Chloride	—	<0.02
Monochloroacetic Acid	<0.02	<0.02
Monorthanolamine	—	<0.02
Monosodium Phosphate	<0.02	—
Nickel Nitrate	>0.05	<0.02
Nickel Sulfate	<0.02	—
Nitric Acid	>0.05	>0.05
Nitric Acid (Red Fuming)	—	>0.05
Nitric + Hydrochloric Acid	—	>0.05
Nitric + Sulfuric Acid	>0.05	>0.05
Nitrobenzene	—	<0.02
Nitrocelluolose	—	<0.02
Nitrotolune	—	<0.02
Nitrous Acid	>0.05	>0.05
Oleic Acid	—	<0.002
Oxalic Acid	<0.02	<0.05
Phenol	—	<0.002
Phosphoric Acid (Areated)	<0.05	>0.05
Phosphoric Acid (Air Free)	<0.02	—
Picric Acid	>0.05	<0.02
Potassium Bicarbonate	<0.02	—
Potassium Bromide	<0.02	<0.02
Potassium Carbonate	<0.02	<0.02

* 10% corrosive medium in 90% water. (Other % corrosive medium in parentheses.)
** Water-free, dry or maximum concentration of corrosive medium.

Source: data compiled by J.S. Park from Earl R. Parker, *Materials Data Book for Engineers and Scientists*, McGraw-Hill Book Company, New York, 1967.

Table 330. Corrosion Rates of Nickel at 70°F *

(Sheet 7 of 9)

Corrosive Medium	Corrosion Rate* in 10% Corrosive Medium (ipy)	Corrosion Rate** in 100% Corrosive Medium (ipy)
Potassium Chlorate	<0.02	—
Potassium Chromate	<0.002	—
Potassium Cyanide	<0.02	<0.02
Potassium Dichromate	<0.02	—
Potassium Ferricyanide	<0.02	—
Potassium Ferrocyanide	<0.02	—
Potassium Hydroxide	<0.002	—
Potassium Hypochlorite	<0.05	—
Potassium Iodide	<0.02	<0.02
Potassium Nitrate	<0.02	<0.02
Potassium Nitrite	<0.02	<0.02
Potassium Permanganate	<0.02	—
Potassium Silicate	<0.02	<0.02
Propionic Acid	<0.02	—
Pyridine	<0.02	<0.02
Quinine Sulfate	<0.02	<0.02
Salicylic Acid	<0.02	<0.02
Silicon Tetrachloride (Dry)	—	<0.002
Silicon Tetrachloride (Wet)	—	>0.05
Silver Bromide	—	<0.02
Silver Nitrate	>0.05	—
Sodium Acetate	<0.02	<0.02
Sodium Bicarbonate	<0.02	—
Sodium Bisulfate	<0.02	<0.02

* 10% corrosive medium in 90% water. (Other % corrosive medium in parentheses.)
** Water-free, dry or maximum concentration of corrosive medium.

Source: data compiled by J.S. Park from Earl R. Parker, *Materials Data Book for Engineers and Scientists*, McGraw-Hill Book Company, New York, 1967.

Table 330. CORROSION RATES OF NICKEL AT 70°F *

(SHEET 8 OF 9)

Corrosive Medium	Corrosion Rate* in 10% Corrosive Medium (ipy)	Corrosion Rate** in 100% Corrosive Medium (ipy)
Sodium Bromide	<0.02	—
Sodium Carbonate	<0.02	<0.02
Sodium Chloride	<0.002	—
Sodium Chromate	<0.02	<0.02
Sodium Hydroxide	<0.002	<0.002
Sodium Hypochlorite	>0.05	—
Sodium Metasilicate	<0.002	<0.002
Sodium Nitrate	<0.02	<0.02
Sodium Nitrite	<0.02	<0.02
Sodium Phosphate	<0.02	<0.02
Sodium Silicate	<0.02	<0.02
Sodium Sulfate	<0.02	<0.02
Sodium Sulfide	<0.02	—
Sodium Sulfite	<0.02	—
Stannic Chloride	>0.05	—
Stannous Chloride	<0.05	<0.02
Strontium Nitrate	<0.02	<0.02
Succinic Acid	<0.02	<0.02
Sulfur Dioxide	>0.05	<0.02
Sulfur Trioxide	—	<0.02
Sulfuric Acid (Areated)	<0.05	>0.05
Sulfuric Acid (Air Free)	<0.02	>0.05
Sulfuric Acid (Fuming)	—	>0.05
Sulfurous Acid	<0.05	>0.05

* 10% corrosive medium in 90% water. (Other % corrosive medium in parentheses.)
** Water-free, dry or maximum concentration of corrosive medium.

Source: data compiled by J.S. Park from Earl R. Parker, *Materials Data Book for Engineers and Scientists*, McGraw-Hill Book Company, New York, 1967.

Table 330. CORROSION RATES OF NICKEL AT 70°F *

(SHEET 9 OF 9)

Corrosive Medium	Corrosion Rate* in 10% Corrosive Medium (ipy)	Corrosion Rate** in 100% Corrosive Medium (ipy)
Tannic Acid	—	<0.02
Tartaric Acid	<0.02	—
Tetraphosphoric Acid	—	>0.05
Trichloroacetic Acid	—	<0.02
Trichloroethylene	—	<0.002
Urea	<0.02	—
Zinc Chloride	<0.02	<0.02
Zinc Sulfate	<0.02	—

* 10% corrosive medium in 90% water. (Other % corrosive medium in parentheses.)
** Water-free, dry or maximum concentration of corrosive medium.

Source: data compiled by J.S. Park from Earl R. Parker, *Materials Data Book for Engineers and Scientists*, McGraw-Hill Book Company, New York, 1967.

* <0.002 means that corrosion rate is likely to be less than 0.002 inch per year (Excellent).
<0.02 means that corrosion rate is likely to be less than about 0.02 inch per year (Good).
<0.05 means that corrosion rate is likely to be less than about 0.05 inch per year (Fair).
>0.05 means that corrosion rate is likely to be more than 0.05 inch per year (Poor).

Table 331. CORROSION RATES OF MONEL AT 70°F *

(SHEET 1 OF 9)

Corrosive Medium	Corrosion Rate* in 10% Corrosive Medium (ipy)	Corrosion Rate** in 100% Corrosive Medium (ipy)
Acetaldehyde	<0.002	<0.002
Acetic Acid (Aerated)	<0.02	<0.02
Acetic Acid (Air Free)	<0.02	<0.02
Acetic Anhydride	—	<0.02
Acetoacetic Acid	<0.02	<0.02
Acetone	<0.002	<0.002
Acetylene	—	<0.002
Acrolein	—	<0.02
Acrylonitril	—	<0.002
Alcohol (Ethyl)	<0.002	<0.002
Alcohol (Methyl)	<0.002	<0.002
Alcohol (Allyl)	—	<0.02
Alcohol (Benzyl)	—	<0.02
Alcohol (Butyl)	—	<0.002
Alcohol (Cetyl)	—	<0.02
Alcohol (Isopropyl)	—	<0.02
Allyl Chloride	—	<0.02
Aluminum Acetate	<0.02	—
Aluminum Chlorate	<0.02	<0.02
Aluminum Chloride	<0.02	—
Aluminum Fluoride	<0.002	—
Aluminum Fluosilicate	—	<0.02
Aluminum Formate	<0.02	<0.02
Aluminum Hydroxide	<0.02	—

* 10% corrosive medium in 90% water. (Other % corrosive medium in parentheses.)
** Water-free, dry or maximum concentration of corrosive medium.

Source: data compiled by J.S. Park from Earl R. Parker, *Materials Data Book for Engineers and Scientists*, McGraw-Hill Book Company, New York, 1967.

Table 331. Corrosion Rates of Monel at 70°F *

(Sheet 2 of 9)

Corrosive Medium	Corrosion Rate* in 10% Corrosive Medium (ipy)	Corrosion Rate** in 100% Corrosive Medium (ipy)
Aluminum Nitrate	<0.02	—
Aluminum Potassium Sulfate	<0.02	—
Aluminum Sulfate	<0.02	<0.02
Ammonia	>0.05	<0.002
Ammonium Acetate	<0.002	<0.002
Ammonium Bromide	<0.02	—
Ammonium Carbonate	<0.02	<0.02
Ammonium Chloride	<0.02	<0.02
Ammonium Citrate	<0.02	—
Ammonium Formate	<0.02	—
Ammonium Nitrate	>0.05	<0.02
Ammonium Sulfate	<0.02	<0.02
Ammonium Sulfite	>0.05	—
Ammonium Thiocyanate	<0.02	<0.02
Amyl Acetate	<0.02	<0.02
Amyl Chloride	<0.02	<0.02
Aniline	<0.02	<0.02
Aniline Hydrochloride	>0.05	—
Anthracine	—	<0.02
Antimony Trichloride	>0.05	—
Barium Carbonate	<0.02	<0.02
Barium Chloride	<0.02	<0.02
Barium Hydroxide	<0.02	<0.02
Barium Oxide	—	<0.02

* 10% corrosive medium in 90% water. (Other % corrosive medium in parentheses.)
** Water-free, dry or maximum concentration of corrosive medium.

Source: data compiled by J.S. Park from Earl R. Parker, *Materials Data Book for Engineers and Scientists*, McGraw-Hill Book Company, New York, 1967.

Table 331. CORROSION RATES OF MONEL AT 70°F *

(SHEET 3 OF 9)

Corrosive Medium	Corrosion Rate* in 10% Corrosive Medium (ipy)	Corrosion Rate** in 100% Corrosive Medium (ipy)
Barium Peroxide	<0.02	—
Benzaldehyde	—	<0.02
Benzene	<0.002	<0.02
Benzoic Acid	<0.02	<0.02
Boric Acid	<0.02	<0.02
Bromic Acid	>0.05	>0.05
Bromine (Dry)	—	<0.002
Bromine (Wet)	—	>0.05
Butyric Acid	<0.05	<0.02
Cadmium Chloride	<0.02	—
Cadmium Sulfate	<0.002	—
Calcium Acetate	<0.02	<0.02
Calcium Bicarbonate	—	<0.02
Calcium Bromide	<0.02	<0.02
Calcium Chlorate	<0.02	—
Calcium Chloride	<0.002	<0.02
Calcium Hydroxide	<0.02	<0.02
Calcium Hypochlorite	>0.05	—
Carbon Dioxide	—	<0.002
Carbon Monoxide	—	<0.002
Carbon Tetrachloride	<0.02	<0.002
Carbon Acid (Air Free)	<0.02	<0.05
Chloroacetic Acid	<0.02	<0.05
Chlorine Gas	—	<0.02

* 10% corrosive medium in 90% water. (Other % corrosive medium in parentheses.)
** Water-free, dry or maximum concentration of corrosive medium.

Source: data compiled by J.S. Park from Earl R. Parker, *Materials Data Book for Engineers and Scientists*, McGraw-Hill Book Company, New York, 1967.

Table 331. CORROSION RATES OF MONEL AT 70°F *

(SHEET 4 OF 9)

Corrosive Medium	Corrosion Rate* in 10% Corrosive Medium (ipy)	Corrosion Rate** in 100% Corrosive Medium (ipy)
Chlorine Liquid	—	<0.02
Chloroform (Dry)	—	<0.002
Chromic Acid	>0.05	—
Chromic Hydroxide	—	<0.02
Chromic Sulfates	—	<0.05
Citric Acid	<0.02	<0.02
Copper Nitrate	>0.05	—
Copper Sulfate	<0.02	—
Diethylene Glycol	—	<0.02
Ethyl Chloride	<0.02	<0.02
Ethylene Glycol	—	<0.02
Ethylene Oxide	—	<0.02
Fatty Acids	—	<0.02
Ferric Chloride	>0.05	>0.05
Ferric Nitrate	>0.05	—
Ferrous Chloride	>0.05	—
Ferrous Sulfate	—	<0.02
Fluorine	—	<0.002
Formaldehyde	<0.002	<0.002
Formic Acid	<0.02	—
Furfural	<0.02	<0.02
Hydrazine	—	>0.05
Hydrobromic Acid	>0.05	—
Hydrochloric Acid (Areated)	>0.05	—

* 10% corrosive medium in 90% water. (Other % corrosive medium in parentheses.)
** Water-free, dry or maximum concentration of corrosive medium.

Source: data compiled by J.S. Park from Earl R. Parker, *Materials Data Book for Engineers and Scientists*, McGraw-Hill Book Company, New York, 1967.

Table 331. CORROSION RATES OF MONEL AT 70°F *

(SHEET 5 OF 9)

Corrosive Medium	Corrosion Rate* in 10% Corrosive Medium (ipy)	Corrosion Rate** in 100% Corrosive Medium (ipy)
Hydrochloric Acid (Air Free)	>0.05	—
Hydrocyanic Acid	>0.05	<0.02
Hydrofluoric Acid (Areated)	<0.02	<0.02
Hydrofluoric Acid (Air Free)	<0.02	<0.02
Hydrogen Chloride	—	<0.002
Hydrogen Fluoride	—	<0.02
Hydrogen Iodide	<0.02	—
Hydrogen Peroxide	<0.02	<0.002
Hydrogen Sulfide	—	<0.02
Lactic Acid	>0.05	—
Lead Acetate	<0.02	<0.02
Lead Chromate	—	<0.02
Lead Nitrate	—	<0.02
Lead Sulfate	—	<0.02
Lithium Chloride	<0.002 (30%)	<0.002
Lithium Hydroxide	<0.02	<0.02
Magnesium Chloride	<0.002	<0.02
Magnesium Hydroxide	<0.02	<0.02
Magnesium Sulfate	<0.02	<0.02
Maleic Acid	<0.05	—
Malic Acid	<0.02	—
Mercuric Chloride	>0.05	—
Mercurous Nitrate	<0.02	—
Mercury	—	<0.02

* 10% corrosive medium in 90% water. (Other % corrosive medium in parentheses.)
** Water-free, dry or maximum concentration of corrosive medium.

Source: data compiled by J.S. Park from Earl R. Parker, *Materials Data Book for Engineers and Scientists*, McGraw-Hill Book Company, New York, 1967.

Table 331. CORROSION RATES OF MONEL AT 70°F *

(SHEET 6 OF 9)

Corrosive Medium	Corrosion Rate* in 10% Corrosive Medium (ipy)	Corrosion Rate** in 100% Corrosive Medium (ipy)
Methallylamine	—	<0.05
Methanol	<0.002	<0.002
Methyl Ethyl Ketone	<0.02	<0.002
Methyl Isobutyl Ketone	<0.02	<0.02
Methylene Chloride	—	<0.002
Monochloroacetic Acid	—	<0.05
Monorthanolamine	—	<0.02
Monosodium Phosphate	<0.02	—
Nickel Chloride	<0.02	<0.02
Nickel Nitrate	>0.05	<0.02
Nickel Sulfate	—	<0.02
Nitric Acid	>0.05	>0.05
Nitric Acid (Red Fuming)	—	>0.05
Nitric + Hydrochloric Acid	—	>0.05
Nitric + Sulfuric Acid	>0.05	>0.05
Nitrobenzene	—	<0.02
Nitrocelluolose	—	<0.002
Nitroglycerine	—	<0.02
Nitrotolune	—	<0.02
Nitrous Acid	—	>0.05
Oleic Acid	—	<0.002
Oxalic Acid	<0.02	<0.02
Phenol	<0.002	<0.002
Phosphoric Acid (Areated)	<0.05	—

* 10% corrosive medium in 90% water. (Other % corrosive medium in parentheses.)
** Water-free, dry or maximum concentration of corrosive medium.

Source: data compiled by J.S. Park from Earl R. Parker, *Materials Data Book for Engineers and Scientists*, McGraw-Hill Book Company, New York, 1967.

Table 331. Corrosion Rates of Monel at 70°F *

(Sheet 7 of 9)

Corrosive Medium	Corrosion Rate* in 10% Corrosive Medium (ipy)	Corrosion Rate** in 100% Corrosive Medium (ipy)
Phosphoric Acid (Air Free)	<0.02	—
Picric Acid	<0.05	>0.05
Potassium Bicarbonate	<0.02	—
Potassium Bromide	<0.02	<0.02
Potassium Carbonate	<0.02	<0.02
Potassium Chlorate	<0.05	—
Potassium Chromate	<0.02	—
Potassium Cyanide	<0.02	<0.02
Potassium Dichromate	<0.02	—
Potassium Ferricyanide	<0.02	—
Potassium Ferrocyanide	<0.02	—
Potassium Hydroxide	<0.002	—
Potassium Hypochlorite	<0.05	—
Potassium Iodide	<0.02	<0.02
Potassium Nitrate	<0.02	<0.02
Potassium Nitrite	<0.02	<0.02
Potassium Permanganate	<0.05	—
Potassium Silicate	<0.02	<0.02
Propionic Acid	<0.02	<0.02
Pyridine	<0.02	<0.02
Quinine Sulfate	<0.02	<0.02
Salicylic Acid	<0.02	<0.02
Silicon Tetrachloride (Dry)	—	<0.002
Silicon Tetrachloride (Wet)	—	>0.05

* 10% corrosive medium in 90% water. (Other % corrosive medium in parentheses.)
** Water-free, dry or maximum concentration of corrosive medium.

Source: data compiled by J.S. Park from Earl R. Parker, *Materials Data Book for Engineers and Scientists*, McGraw-Hill Book Company, New York, 1967.

Table 331. Corrosion Rates of Monel at 70°F *

(Sheet 8 of 9)

Corrosive Medium	Corrosion Rate* in 10% Corrosive Medium (ipy)	Corrosion Rate** in 100% Corrosive Medium (ipy)
Silver Bromide	—	<0.02
Silver Nitrate	>0.05	—
Sodium Acetate	<0.05	<0.02
Sodium Bicarbonate	<0.02	—
Sodium Bisulfate	<0.02	<0.02
Sodium Bromide	<0.02	—
Sodium Carbonate	<0.02	<0.02
Sodium Chloride	<0.002	—
Sodium Chromate	<0.02	<0.02
Sodium Hydroxide	<0.002	<0.002
Sodium Hypochlorite	>0.05	<0.02
Sodium Metasilicate	<0.002	<0.002
Sodium Nitrate	<0.02	<0.02
Sodium Nitrite	<0.02	<0.002
Sodium Phosphate	<0.02	<0.02
Sodium Silicate	<0.02	<0.02
Sodium Sulfate	<0.02	<0.02
Sodium Sulfide	<0.02	—
Sodium Sulfite	<0.02	<0.02
Stannic Chloride	>0.05	—
Stannous Chloride	>0.05	<0.02
Strontium Nitrate	<0.02	<0.02
Succinic Acid	<0.02	<0.02
Sulfur Dioxide	>0.05	<0.02

* 10% corrosive medium in 90% water. (Other % corrosive medium in parentheses.)
** Water-free, dry or maximum concentration of corrosive medium.

Source: data compiled by J.S. Park from Earl R. Parker, *Materials Data Book for Engineers and Scientists*, McGraw-Hill Book Company, New York, 1967.

Table 331. CORROSION RATES OF MONEL AT 70°F *

(SHEET 9 OF 9)

Corrosive Medium	Corrosion Rate* in 10% Corrosive Medium (ipy)	Corrosion Rate** in 100% Corrosive Medium (ipy)
Sulfur Trioxide	—	<0.02
Sulfuric Acid (Areated)	<0.05	>0.05
Sulfuric Acid (Air Free)	<0.002	>0.05
Sulfuric Acid (Fuming)	—	>0.05
Sulfurous Acid	>0.05	>0.05
Tannic Acid	<0.02	<0.02
Tartaric Acid	<0.02	—
Tetraphosphoric Acid	—	<0.05
Trichloroacetic Acid	—	>0.05
Trichloroethylene	—	<0.002
Urea	<0.02	—
Zinc Chloride	<0.02	<0.02
Zinc Sulfate	<0.02	—

* 10% corrosive medium in 90% water. (Other % corrosive medium in parentheses.)
** Water-free, dry or maximum concentration of corrosive medium.

Source: data compiled by J.S. Park from Earl R. Parker, *Materials Data Book for Engineers and Scientists*, McGraw-Hill Book Company, New York, 1967.

* <0.002 means that corrosion rate is likely to be less than 0.002 inch per year (Excellent).
<0.02 means that corrosion rate is likely to be less than about 0.02 inch per year (Good).
<0.05 means that corrosion rate is likely to be less than about 0.05 inch per year (Fair).
>0.05 means that corrosion rate is likely to be more than 0.05 inch per year (Poor).

Table 332. CORROSION RATES OF LEAD AT 70°F *

(SHEET 1 OF 8)

Corrosive Medium	Corrosion Rate* in 10% Corrosive Medium (ipy)	Corrosion Rate** in 100% Corrosive Medium (ipy)
Acetaldehyde	<0.02	<0.002
Acetic Acid (Aerated)	>0.05	<0.05
Acetic Acid (Air Free)	>0.05	<0.02
Acetic Anhydride	—	<0.002
Acetoacetic Acid	—	<0.02
Acetone	<0.002	<0.02
Acetylene	—	<0.002
Acrolein	<0.02	—
Acrylonitril	—	<0.002
Alcohol (Ethyl)	<0.002	<0.002
Alcohol (Methyl)	<0.02	<0.02
Alcohol (Allyl)	—	<0.02
Alcohol (Benzyl)	—	<0.02
Alcohol (Cetyl)	—	<0.02
Alcohol (Isopropyl)	—	<0.002
Allyl Chloride	—	<0.05
Allyl Sulfide	—	>0.05
Aluminum Acetate	<0.002	<0.002
Aluminum Chlorate	<0.02	<0.02
Aluminum Chloride	>0.05	—
Aluminum Fluoride	<0.02	—
Aluminum Fluosilicate	—	<0.02
Aluminum Formate	—	<0.02
Aluminum Hydroxide	<0.02	—

* 10% corrosive medium in 90% water. (Other % corrosive medium in parentheses.)
** Water-free, dry or maximum concentration of corrosive medium.

Source: data compiled by J.S. Park from Earl R. Parker, *Materials Data Book for Engineers and Scientists*, McGraw-Hill Book Company, New York, 1967.

Table 332. CORROSION RATES OF LEAD AT 70°F *

(SHEET 2 OF 8)

Corrosive Medium	Corrosion Rate* in 10% Corrosive Medium (ipy)	Corrosion Rate** in 100% Corrosive Medium (ipy)
Aluminum Nitrate	<0.02	—
Aluminum Potassium Sulfate	<0.002	<0.02
Aluminum Sulfate	<0.02	—
Ammonia	<0.02	<0.02
Ammonium Bicarbonate	<0.02	—
Ammonium Bromide	>0.05	—
Ammonium Carbonate	<0.02	—
Ammonium Chloride	>0.05	<0.02
Ammonium Nitrate	>0.05	
Ammonium Sulfate	<0.02	<0.02
Amyl Acetate	—	<0.02
Amyl Chloride	—	>0.05
Aniline	—	>0.05
Aniline Hydrochloride	>0.05	—
Anthracine	—	<0.02
Antimony Trichloride	<0.02	<0.002
Barium Carbonate	—	>0.05
Barium Chloride	<0.02	—
Barium Hydroxide	>0.05	>0.05
Barium Nitrate	<0.02	—
Barium Peroxide	>0.05	—
Benzaldehyde	>0.05	>0.05
Benzene	<0.02	<0.02
Benzoic Acid	>0.05	>0.05

* 10% corrosive medium in 90% water. (Other % corrosive medium in parentheses.)
** Water-free, dry or maximum concentration of corrosive medium.

Source: data compiled by J.S. Park from Earl R. Parker, *Materials Data Book for Engineers and Scientists*, McGraw-Hill Book Company, New York, 1967.

Table 332. CORROSION RATES OF LEAD AT 70°F *

(SHEET 3 OF 8)

Corrosive Medium	Corrosion Rate* in 10% Corrosive Medium (ipy)	Corrosion Rate** in 100% Corrosive Medium (ipy)
Boric Acid	<0.02	<0.02
Bromic Acid	<0.02	<0.02
Bromine (Dry)	—	<0.002
Bromine (Wet)	—	>0.05
Butyric Acid	>0.05	>0.05
Cadmium Sulfate	<0.002	—
Calcium Acetate	<0.02	<0.02
Calcium Bicarbonate	—	<0.05
Calcium Bromide	<0.02	<0.02
Calcium Chlorate	<0.02	—
Calcium Chloride	>0.05	—
Calcium Hydroxide	>0.05	—
Calcium Hypochlorite	<0.05	<0.002
Carbon Dioxide	—	<0.002
Carbon Monoxide	—	<0.002
Carbon Tetrachloride	—	<0.002
Carbon Acid (Air Free)	—	>0.05
Chloroacetic Acid	>0.05	>0.05
Chlorine Gas	—	<0.02
Chlorine Liquid	—	<0.02
Chloroform (Dry)	—	<0.02
Chromic Acid	<0.02	—
Chromic Hydroxide	—	<0.02
Chromic Sulfates	<0.02	<0.02

* 10% corrosive medium in 90% water. (Other % corrosive medium in parentheses.)
** Water-free, dry or maximum concentration of corrosive medium.

Source: data compiled by J.S. Park from Earl R. Parker, *Materials Data Book for Engineers and Scientists*, McGraw-Hill Book Company, New York, 1967.

Table 332. Corrosion Rates of Lead at 70°F *

(Sheet 4 of 8)

Corrosive Medium	Corrosion Rate* in 10% Corrosive Medium (ipy)	Corrosion Rate** in 100% Corrosive Medium (ipy)
Citric Acid	<0.02	>0.05
Copper Sulfate	<0.02	<0.02
Diethylene Glycol	—	<0.02
Ethyl Chloride	—	<0.02
Ethylene Glycol	—	<0.05
Ethylene Oxide	—	<0.02
Fatty Acids	—	>0.05
Ferric Chloride	>0.05	—
Ferric Nitrate	<0.002	<0.002
Ferrous Chloride	>0.05	—
Ferrous Sulfate	<0.02	—
Fluorine	—	<0.02
Formaldehyde	<0.02	<0.02
Formic Acid	>0.05	>0.05
Furfural	—	<0.02
Hydrazine	>0.05	>0.05
Hydrobromic Acid	>0.05	—
Hydrochloric Acid (Areated)	<0.02	—
Hydrochloric Acid (Air Free)	<0.02	—
Hydrocyanic Acid	>0.05	<0.02
Hydrofluoric Acid (Areated)	>0.05	—
Hydrofluoric Acid (Air Free)	<0.002	>0.05
Hydrogen Chloride	—	<0.02
Hydrogen Fluoride	—	>0.05

* 10% corrosive medium in 90% water. (Other % corrosive medium in parentheses.)
** Water-free, dry or maximum concentration of corrosive medium.

Source: data compiled by J.S. Park from Earl R. Parker, *Materials Data Book for Engineers and Scientists*, McGraw-Hill Book Company, New York, 1967.

Table 332. CORROSION RATES OF LEAD AT 70°F *

(SHEET 5 OF 8)

Corrosive Medium	Corrosion Rate* in 10% Corrosive Medium (ipy)	Corrosion Rate** in 100% Corrosive Medium (ipy)
Hydrogen Peroxide	>0.05	<0.002
Hydrogen Sulfide	—	<0.02
Lactic Acid	>0.05	>0.05
Lead Chromate	—	<0.02
Lead Nitrate	—	<0.02
Lead Sulfate	—	<0.02
Lithium Chloride	<0.02	<0.02
Lithium Hydroxide	>0.05	—
Magnesium Chloride	>0.05	>0.05
Magnesium Hydroxide	>0.05	—
Magnesium Sulfate	<0.02	—
Mercuric Chloride	<0.05	—
Mercurous Nitrate	—	>0.05
Mercury	—	>0.05
Methanol	<0.02	<0.02
Methyl Ethyl Ketone	<0.02	<0.002
Methyl Isobutyl Ketone	<0.02	<0.002
Methylene Chloride	—	<0.02
Monochloroacetic Acid	>0.05	>0.05
Monosodium Phosphate	<0.02	—
Nickel Chloride	—	<0.02
Nickel Nitrate	—	<0.02
Nickel Sulfate	<0.02	<0.02
Nitric Acid	>0.05	>0.05

* 10% corrosive medium in 90% water. (Other % corrosive medium in parentheses.)
** Water-free, dry or maximum concentration of corrosive medium.

Source: data compiled by J.S. Park from Earl R. Parker, *Materials Data Book for Engineers and Scientists*, McGraw-Hill Book Company, New York, 1967.

Table 332. CORROSION RATES OF LEAD AT 70°F *

(SHEET 6 OF 8)

Corrosive Medium	Corrosion Rate* in 10% Corrosive Medium (ipy)	Corrosion Rate** in 100% Corrosive Medium (ipy)
Nitric + Hydrochloric Acid	—	>0.05
Nitric + Sulfuric Acid	>0.05	>0.05
Nitrobenzene	—	<0.02
Nitrocelluolose	—	<0.002
Nitroglycerine	—	<0.05
Nitrotolune	—	<0.02
Nitrous Acid	—	>0.05
Oleic Acid	—	>0.05
Oxalic Acid	>0.05	>0.05
Phenol	—	<0.02
Phosphoric Acid (Areated)	<0.02	<0.02
Phosphoric Acid (Air Free)	<0.002	<0.02
Picric Acid	>0.05	<0.02
Potassium Bicarbonate	>0.05	—
Potassium Bromide	<0.02	<0.02
Potassium Carbonate	>0.05	>0.05
Potassium Chlorate	<0.02	—
Potassium Chromate	<0.02	—
Potassium Cyanide	>0.05	—
Potassium Dichromate	<0.02	—
Potassium Ferricyanide	<0.02	—
Potassium Ferrocyanide	<0.02	—
Potassium Hydroxide	>0.05	>0.05
Potassium Hypochlorite	<0.02	—

* 10% corrosive medium in 90% water. (Other % corrosive medium in parentheses.)
** Water-free, dry or maximum concentration of corrosive medium.

Source: data compiled by J.S. Park from Earl R. Parker, *Materials Data Book for Engineers and Scientists*, McGraw-Hill Book Company, New York, 1967.

Table 332. CORROSION RATES OF LEAD AT 70°F *

(SHEET 7 OF 8)

Corrosive Medium	Corrosion Rate* in 10% Corrosive Medium (ipy)	Corrosion Rate** in 100% Corrosive Medium (ipy)
Potassium Iodide	>0.05	—
Potassium Nitrate	<0.02	—
Potassium Nitrite	<0.02	<0.02
Potassium Permanganate	<0.05	>0.05
Propionic Acid	>0.05	—
Pyridine	<0.02	<0.02
Salicylic Acid	—	<0.02
Silicon Tetrachloride (Dry)	—	<0.02
Silver Nitrate	>0.05	—
Sodium Acetate	—	<0.02
Sodium Bicarbonate	<0.02	—
Sodium Bisulfate	<0.02	—
Sodium Carbonate	<0.02	—
Sodium Chloride	<0.02	—
Sodium Chromate	<0.02	<0.02
Sodium Hydroxide	<0.02	—
Sodium Hypochlorite	>0.05	>0.05
Sodium Nitrate	>0.05	—
Sodium Nitrite	<0.02	—
Sodium Phosphate	<0.02	<0.02
Sodium Silicate	>0.05	—
Sodium Sulfate	<0.02	<0.02
Sodium Sulfide	<0.002	<0.002
Sodium Sulfite	<0.02	<0.02

* 10% corrosive medium in 90% water. (Other % corrosive medium in parentheses.)
** Water-free, dry or maximum concentration of corrosive medium.

Source: data compiled by J.S. Park from Earl R. Parker, *Materials Data Book for Engineers and Scientists*, McGraw-Hill Book Company, New York, 1967.

Table 332. CORROSION RATES OF LEAD AT 70°F *

(SHEET 8 OF 8)

Corrosive Medium	Corrosion Rate* in 10% Corrosive Medium (ipy)	Corrosion Rate** in 100% Corrosive Medium (ipy)
Stannic Chloride	>0.05	—
Stannous Chloride	>0.05	—
Succinic Acid	<0.02	<0.02
Sulfur Dioxide	—	<0.02
Sulfur Trioxide	—	<0.02
Sulfuric Acid (Areated)	<0.002	>0.05
Sulfuric Acid (Air Free)	<0.002	>0.05
Sulfuric Acid (Fuming)	—	>0.05
Sulfurous Acid	<0.02	<0.02
Tannic Acid	>0.05	>0.05
Tartaric Acid	<0.02	>0.05
Tetraphosphoric Acid	>0.05	>0.05
Trichloroacetic Acid	>0.05	>0.05
Trichloroethylene	—	>0.05
Zinc Chloride	<0.02	<0.02
Zinc Sulfate	<0.02	—

* 10% corrosive medium in 90% water. (Other % corrosive medium in parentheses.)
** Water-free, dry or maximum concentration of corrosive medium.

Source: data compiled by J.S. Park from Earl R. Parker, *Materials Data Book for Engineers and Scientists*, McGraw-Hill Book Company, New York, 1967.

* <0.002 means that corrosion rate is likely to be less than 0.002 inch per year (Excellent).
<0.02 means that corrosion rate is likely to be less than about 0.02 inch per year (Good).
<0.05 means that corrosion rate is likely to be less than about 0.05 inch per year (Fair).
>0.05 means that corrosion rate is likely to be more than 0.05 inch per year (Poor).

Table 333. CORROSION RATES OF TITANIUM AT 70°F *

(SHEET 1 OF 5)

Corrosive Medium	Corrosion Rate* in 10% Corrosive Medium (ipy)	Corrosion Rate** in 100% Corrosive Medium (ipy)
Acetaldehyde	—	<0.002
Acetic Acid (Aerated)	<0.002	<0.002
Acetic Acid (Air Free)	<0.002	<0.002
Acetic Anhydride	—	<0.002
Acetone	<0.002	<0.002
Acetylene	—	<0.002
Acrolein	—	<0.02
Acrylonitril	—	<0.002
Alcohol (Ethyl)	<0.002	<0.002
Alcohol (Allyl)	—	<0.002
Alcohol (Amyl)	—	<0.002
Alcohol (Benzyl)	—	<0.002
Alcohol (Butyl)	—	<0.002
Alcohol (Cetyl)	—	<0.002
Aluminum Acetate	—	<0.002
Aluminum Chlorate	<0.002	—
Aluminum Chloride	>0.05	—
Aluminum Formate	—	<0.002
Aluminum Hydroxide	<0.002	<0.002
Aluminum Nitrate	<0.002	<0.002
Aluminum Potassium Sulfate	—	<0.002
Aluminum Sulfate	<0.002	—
Ammonia	<0.002	<0.002
Ammonium Chloride	<0.002	—

* 10% corrosive medium in 90% water. (Other % corrosive medium in parentheses.)
** Water-free, dry or maximum concentration of corrosive medium.

Source: data compiled by J.S. Park from Earl R. Parker, *Materials Data Book for Engineers and Scientists*, McGraw-Hill Book Company, New York, 1967.

Table 333. CORROSION RATES OF TITANIUM AT 70°F *

(SHEET 2 OF 5)

Corrosive Medium	Corrosion Rate* in 10% Corrosive Medium (ipy)	Corrosion Rate** in 100% Corrosive Medium (ipy)
Ammonium Citrate	<0.002	<0.002
Ammonium Formate	<0.002	<0.002
Ammonium Nitrate	<0.05	—
Ammonium Sulfate	<0.002	—
Amyl Acetate	—	<0.002
Aniline Hydrochloride	<0.002	—
Anthracine	—	<0.002
Barium Chloride	<0.002	—
Benzene	<0.002	<0.002
Benzoic Acid	<0.002	<0.002
Boric Acid	<0.002	—
Bromine (Dry)	—	>0.05
Bromine (Wet)	—	>0.05
Butyric Acid	<0.002	<0.002
Calcium Acetate	<0.002	<0.002
Calcium Bicarbonate	—	<0.002
Calcium Bromide	—	<0.05
Calcium Chlorate	—	<0.002
Calcium Chloride	<0.002	—
Calcium Hypochlorite	<0.002	—
Carbon Dioxide	—	<0.002
Carbon Monoxide	—	<0.002
Carbon Tetrachloride	—	<0.002
Chloroacetic Acid	—	<0.002

* 10% corrosive medium in 90% water. (Other % corrosive medium in parentheses.)
** Water-free, dry or maximum concentration of corrosive medium.

Source: data compiled by J.S. Park from Earl R. Parker, *Materials Data Book for Engineers and Scientists*, McGraw-Hill Book Company, New York, 1967.

Table 333. CORROSION RATES OF TITANIUM AT 70°F *

(SHEET 3 OF 5)

Corrosive Medium	Corrosion Rate* in 10% Corrosive Medium (ipy)	Corrosion Rate** in 100% Corrosive Medium (ipy)
Chlorine Gas	—	>0.05
Chromic Acid	<0.002	—
Citric Acid	<0.002	—
Diethylene Glycol	—	<0.002
Ethyl Chloride	—	<0.002
Ethylene Oxide	—	<0.002
Fatty Acids	—	<0.002
Ferric Chloride	<0.002	—
Ferric Nitrate	<0.002	—
Ferrous Chloride	<0.002	—
Ferrous Sulfate	<0.002	—
Formaldehyde	<0.002	<0.002
Formic Acid	<0.02	<0.02
Furfural	—	<0.002
Hydrochloric Acid (Areated)	<0.02	—
Hydrochloric Acid (Air Free)	<0.02	—
Hydrofluoric Acid (Areated)	>0.05	—
Hydrofluoric Acid (Air Free)	>0.05	>0.05
Hydrogen Fluoride	—	<0.002
Hydrogen Peroxide	<0.002	>0.05
Hydrogen Sulfide	—	<0.002
Lactic Acid	<0.002	<0.002
Lead Acetate	<0.002	—
Magnesium Chloride	<0.002	<0.002

* 10% corrosive medium in 90% water. (Other % corrosive medium in parentheses.)
** Water-free, dry or maximum concentration of corrosive medium.

Source: data compiled by J.S. Park from Earl R. Parker, *Materials Data Book for Engineers and Scientists*, McGraw-Hill Book Company, New York, 1967.

Table 333. CORROSION RATES OF TITANIUM AT 70°F *

(SHEET 4 OF 5)

Corrosive Medium	Corrosion Rate* in 10% Corrosive Medium (ipy)	Corrosion Rate** in 100% Corrosive Medium (ipy)
Malic Acid	—	<0.002
Maganous Chloride	<0.002	—
Mercuric Chloride	<0.002	—
Methyl Ethyl Ketone	<0.002	<0.002
Methyl Isobutyl Ketone	<0.002	<0.002
Monochloroacetic Acid	—	<0.002
Nickel Chloride	<0.02	—
Nitric Acid	<0.002	—
Nitric Acid (Red Fuming)	—	<0.002
Nitric + Hydrochloric Acid	—	<0.02
Nitric + Hydrofluoric Acid	—	>0.05
Oleic Acid	—	<0.002
Oxalic Acid	<0.02	—
Phosphoric Acid (Areated)	<0.02	>0.05
Phosphoric Acid (Air Free)	—	>0.05
Potassium Bromide	<0.002	—
Potassium Carbonate	<0.002	—
Potassium Chlorate	<0.002	—
Potassium Cyanide	—	>0.05
Potassium Dichromate	<0.002	—
Potassium Hydroxide	<0.002	—
Potassium Hypochlorite	<0.002	—
Potassium Iodide	<0.002	<0.002
Potassium Nitrate	<0.002	—

* 10% corrosive medium in 90% water. (Other % corrosive medium in parentheses.)
** Water-free, dry or maximum concentration of corrosive medium.

Source: data compiled by J.S. Park from Earl R. Parker, *Materials Data Book for Engineers and Scientists*, McGraw-Hill Book Company, New York, 1967.

Table 333. CORROSION RATES OF TITANIUM AT 70°F *

(SHEET 5 OF 5)

Corrosive Medium	Corrosion Rate* in 10% Corrosive Medium (ipy)	Corrosion Rate** in 100% Corrosive Medium (ipy)
Potassium Nitrite	<0.002	<0.002
Propionic Acid	—	>0.05
Quinine Sulfate	—	<0.002
Silver Bromide	—	<0.002
Silver Chloride	<0.002	—
Sodium Chloride	<0.002	—
Sodium Hydroxide	<0.002	—
Sodium Hypochlorite	<0.002	<0.002
Sodium Nitrite	<0.002	—
Sodium Sulfide	<0.002	—
Stannic Chloride	<0.002	—
Succinic Acid	<0.002	<0.002
Sulfuric Acid (Areated)	<0.02	>0.05
Sulfuric Acid (Air Free)	—	>0.05
Sulfurous Acid	<0.002	<0.002
Tannic Acid	<0.002	<0.002
Tartaric Acid	<0.002	<0.002
Trichloroacetic Acid	<0.002	>0.05
Trichloroethylene	—	<0.002
Zinc Chloride	<0.002	—

* 10% corrosive medium in 90% water. (Other % corrosive medium in parentheses.)
** Water-free, dry or maximum concentration of corrosive medium.

Source: data compiled by J.S. Park from Earl R. Parker, *Materials Data Book for Engineers and Scientists*, McGraw-Hill Book Company, New York, 1967.

* <0.002 means that corrosion rate is likely to be less than 0.002 inch per year (Excellent).
<0.02 means that corrosion rate is likely to be less than about 0.02 inch per year (Good).
<0.05 means that corrosion rate is likely to be less than about 0.05 inch per year (Fair).
>0.05 means that corrosion rate is likely to be more than 0.05 inch per year (Poor).

Table 334. Corrosion Rates of ACI Heat–Resistant Castings Alloys in Air

Alloy	Oxidation Rate in Air (mils/yr)		
	(870 °C)	(980 °C)	(1090 °C)
HC	10	50	50
HD	10–	50–	50–
HE	5–	25–	35–
HF	5–	50+	100
HH	5–	25–	50
HI	5–	10+	35–
HK	10–	10–	35–
HL	10+	25–	35
HN	5	10+	50–
HP	25–	25	50
HT	5–	10+	50
HU	5–	10–	35–
HW	5–	10–	35
HX	5–	10–	35–

Based on 100–h tests.

To convert **mils/yr** to **μm/yr** multiply by 25

Data from *ASM Metals Reference Book, Third Edition*, Michael Bauccio, Ed., ASM International, Materials Park, OH, p392, (1993).

Table 335. Corrosion Rates for ACI Heat–Resistant Castings Alloys in Flue Gas

Alloy	Corrosion rate (mils/yr)			
	flue gas sulfur content 0.12 g/m^3		flue gas sulfur content 2.3 g/m^3	
	Oxidizing	Reducing	Oxidizing	Reducing
HC	25–	25+	25	25–
HD	25–	25–	25–	25–
HE	25–	25–	25–	25–
HF	50+	100+	50+	250–
HH	25–	25	25	25–
HI	25–	25–	25–	25–
HK	25–	25–	25–	25–
HL	25–	25–	25–	25–
HN	25–	25–	25	25
HP	25–	25–	25–	25–
HT	25	25–	25	100
HU	25–	25–	25–	25
HW	25	25–	50–	250
HX	25–	25–	25–	25–

Basd on 100–h tests.

To convert **mils/yr** to **μm/yr** multiply by **25**

Data from *ASM Metals Reference Book, Third Edition,* Michael Bauccio, Ed., ASM International, Materials Park, OH, p392, (1993).

Table 336. FLAMMABILITY OF POLYMERS

(SHEET 1 OF 11)

Polymer	Type	Flammability, (ASTM D635) (ipm)
ABS Resins; Molded, Extruded	Medium impact	1.0—1.6
	High impact	1.3—1.5
	Very high impact	1.3—1.5
	Low temperature impact	1.0—1.5
	Heat resistant	1.3—2.0
Acrylics; Cast, Molded, Extruded		(0.125 in.)
	Cast Resin Sheets, Rods:	
	General purpose, type I	0.5—2.2
	General purpose, type II	0.5—1.8
	Moldings:	
	Grades 5, 6, 8	0.9—1.2
	High impact grade	0.8—1.2
Thermoset Carbonate	Allyl diglycol carbonate	0.35

Source: data compiled by J.S. Park from Charles T. Lynch, *CRC Handbook of Materials Science, Vol. 3*, CRC Press, Boca Raton, Florida, 1975 and *Engineered Materials Handbook, Vol.2*, Engineering Plastics, ASM International, Metals Park, Ohio, 1988.

Table 336. FLAMMABILITY OF POLYMERS

(SHEET 2 OF 11)

Polymer	Type	Flammability, (ASTM D635) (ipm)
Alkyds; Molded	Putty (encapsulating)	Nonburning
	Rope (general purpose)	Self extinguishing
	Granular (high speed molding)	Self extinguishing
	Glass reinforced (heavy duty parts)	Nonburning
Cellulose Acetate; Molded, Extruded	ASTM Grade:	
	H6—1	0.5—2.0
	H4—1	0.5—2.0
	H2—1	0.5—2.0
	MH—1, MH—2	0.5—2.0
	MS—1, MS—2	0.5—2.0
	S2—1	0.5—2.0
Cellulose Acetate Butyrate; Molded, Extruded	ASTM Grade:	
	H4	0.5—1.5
	MH	0.5—1.5
	S2	0.5—1.5

Source: data compiled by J.S. Park from Charles T. Lynch, *CRC Handbook of Materials Science, Vol. 3*, CRC Press, Boca Raton, Florida, 1975 and *Engineered Materials Handbook, Vol.2*, Engineering Plastics, ASM International, Metals Park, Ohio, 1988.

Table 336. FLAMMABILITY OF POLYMERS

(SHEET 3 OF 11)

Polymer	Type	Flammability, (ASTM D635) (ipm)
Cellusose Acetate Propionate; Molded, Extruded	ASTM Grade:	
	1	0.5—1.5
	3	0.5—1.5
	6	0.5—1.5
Chlorinated Polymers	Chlorinated polyether	Self extinguishing
	Chlorinated polyvinyl chloride	Nonburning
Polycarbonates	Polycarbonate	Self extinguishing
	Polycarbonate (40% glass fiber reinforced)	Self extinguishing
Diallyl Phthalates; Molded		ignition time (s)
	Orlon filled	68 s
	Dacron filled	84—90 s
	Asbestos filled	70 s
	Glass fiber filled	70—400 s
Fluorocarbons; Molded,Extruded	Polytrifluoro chloroethylene (PTFCE)	Noninflammable
	Polytetrafluoroethylene (PTFE)	Noninflammable

Source: data compiled by J.S. Park from Charles T. Lynch, *CRC Handbook of Materials Science, Vol. 3*, CRC Press, Boca Raton, Florida, 1975 and *Engineered Materials Handbook, Vol.2*, Engineering Plastics, ASM International, Metals Park, Ohio, 1988.

Table 336. FLAMMABILITY OF POLYMERS

(SHEET 4 OF 11)

Polymer	Type	Flammability, (ASTM D635) (ipm)
Fluorocarbons; Molded,Extruded (Con't)	Ceramic reinforced (PTFE)	Noninflammable
	Fluorinated ethylene propylene(FEP)	Noninflammable
	Polyvinylidene— fluoride (PVDF)	Self extinguishing
Epoxies; Cast, Molded, Reinforced	Standard epoxies (diglycidyl ethers of bisphenol A)	
	Cast rigid	0.3-0.34
	Cast flexible	-
	Molded	Self extinguishing
	General purpose glass cloth laminate	Slow burn to Self extinguishing
	High strength laminate	Self extinguishing
	Filament wound composite	Self extinguishing
Epoxies—Molded, Extruded	High performance resins (cycloaliphatic diepoxides)	
	Cast, rigid	Self extinguishing
	Molded	Self extinguishing
	Glass cloth laminate	Self extinguishing

Source: data compiled by J.S. Park from Charles T. Lynch, *CRC Handbook of Materials Science, Vol. 3*, CRC Press, Boca Raton, Florida, 1975 and *Engineered Materials Handbook, Vol.2*, Engineering Plastics, ASM International, Metals Park, Ohio, 1988.

Table 336. FLAMMABILITY OF POLYMERS

(SHEET 5 OF 11)

Polymer	Type	Flammability, (ASTM D635) (ipm)
Melamines; Molded	Filler & type	
	Unfilled	Self extinguishing
	Cellulose electrical	Self extinguishing
	Glass fiber	Self extinguishing
	Alpha cellulose and mineral	Self extinguishing
Nylons; Molded, Extruded	Type 6	
	General purpose	Self extinguishing
	Glass fiber (30%) reinforced	Slow burn
	Cast	Self extinguishing
	Flexible copolymers	Slow burn, 0.6
	Type 8	Self extinguishing
	Type 11	Self extinguishing

Source: data compiled by J.S. Park from Charles T. Lynch, *CRC Handbook of Materials Science, Vol. 3*, CRC Press, Boca Raton, Florida, 1975 and *Engineered Materials Handbook, Vol.2*, Engineering Plastics, ASM International, Metals Park, Ohio, 1988.

Table 336. FLAMMABILITY OF POLYMERS

(SHEET 6 OF 11)

Polymer	Type	Flammability, (ASTM D635) (ipm)
Nylons; Molded, Extruded (Con't)	6/6 Nylon	
	General purpose molding	Self extinguishing
	Glass fiber reinforced	Slow burn
	Glass fiber Molybdenum disulfide filled	Slow burn
	General purpose extrusion	Self extinguishing
	6/10 Nylon	
	General purpose	Self extinguishing
	Glass fiber (30%) reinforced	Slow burn
Phenolics; Molded	Type and filler	
	General: woodflour and flock	Self extinguishing
	Shock: paper, flock, or pulp	Self extinguishing
	High shock: chopped fabric or cord	Self extinguishing
	Very high shock: glass fiber	Self extinguishing

Source: data compiled by J.S. Park from Charles T. Lynch, *CRC Handbook of Materials Science, Vol. 3*, CRC Press, Boca Raton, Florida, 1975 and *Engineered Materials Handbook, Vol.2*, Engineering Plastics, ASM International, Metals Park, Ohio, 1988.

Table 336. FLAMMABILITY OF POLYMERS

(SHEET 7 OF 11)

Polymer	Type	Flammability, (ASTM D635) (ipm)
Phenolics; Molded (Con't)	Arc resistant—mineral	Self extinguishing
	Rubber phenolic—woodflour or flock	Self extinguishing
	Rubber phenolic—chopped fabric	Self extinguishing
	Rubber phenolic—asbestos	Self extinguishing
ABS–Polycarbonate Alloy		0.9
PVC–Acrylic Alloy	PVC–acrylic sheet	Nonburning
	PVC–acrylic injection molded	Nonburning
	Polyimides	
	Unreinforced	IBM Class A
	Unreinforced 2nd value	IBM Class A
	Glass reinforced	UL SE—0
Polyacetals	Homopolymer:	
	Standard	1.1
	20% glass reinforced	0.8
	22% TFE reinforced	0.8

Source: data compiled by J.S. Park from Charles T. Lynch, *CRC Handbook of Materials Science, Vol. 3*, CRC Press, Boca Raton, Florida, 1975 and *Engineered Materials Handbook, Vol.2*, Engineering Plastics, ASM International, Metals Park, Ohio, 1988.

Table 336. FLAMMABILITY OF POLYMERS

(SHEET 8 OF 11)

Polymer	Type	Flammability, (ASTM D635) (ipm)
Polyacetals (Con't)	Copolymer:	
	Standard	1.1
	25% glass reinforced	1
	High flow	1.1
Polyester; Thermoplastic	Injection Moldings:	
	General purpose grade	Slow burn
	Glass reinforced grades	Slow burn
	Glass reinforced self extinguishing	Self extinguishing
	General purpose grade	Slow burn
	Glass reinforced grade	Slow burn
Polyesters: Thermosets	Cast polyyester	
	Rigid	0.87 to self extinguishing
	Flexible	Slow burn to self extinguishing

Source: data compiled by J.S. Park from Charles T. Lynch, *CRC Handbook of Materials Science, Vol. 3*, CRC Press, Boca Raton, Florida, 1975 and *Engineered Materials Handbook, Vol.2*, Engineering Plastics, ASM International, Metals Park, Ohio, 1988.

Table 336. FLAMMABILITY OF POLYMERS

(SHEET 9 OF 11)

Polymer	Type	Flammability, (ASTM D635) (ipm)
Reinforced polyester moldings	High strength (glass fibers)	Self extinguishing
	Heat and chemical resistant (asbestos)	Self extinguishing
	Sheet molding compounds, general purpose	Self extinguishing
Phenylene Oxides	SE—100	Self extinguishing
	SE—1	Self extinguishing
	Glass fiber reinforced	Self extinguishing
Phenylene oxides (Noryl)	Standard	Self extinguishing
	Glass fiber reinforced	Self extinguishing
Polyarylsulfone		Self extinguishing
Polypropylene:	General purpose	0.7—1
	High impact	1
	Asbestos filled	1
	Glass reinforced	1
	Flame retardant	Self extinguishing

Source: data compiled by J.S. Park from Charles T. Lynch, *CRC Handbook of Materials Science, Vol. 3*, CRC Press, Boca Raton, Florida, 1975 and *Engineered Materials Handbook, Vol.2*, Engineering Plastics, ASM International, Metals Park, Ohio, 1988.

Table 336. FLAMMABILITY OF POLYMERS
(SHEET 10 OF 11)

Polymer	Type	Flammability, (ASTM D635) (ipm)
Polyphenylene sulfide:	Standard	Non—burning
	40% glass reinforced	Non—burning
Polyethylenes; Molded, Extruded	Type I—lower density (0.910—0.925)	
	Melt index 0.3—3.6	1
	Melt index 6—26	1
	Melt index 200	1
	Type II—medium density (0.926—0.940)	
	Melt index 20	1
	Melt index 1.0—1.9	1
	Type III—higher density (0.941—0.965)	
	Melt index 0.2—0.9	1
	Melt index 0.1—12.0	1
	Melt index 1.5—15	1
	High molecular weight	1

Source: data compiled by J.S. Park from Charles T. Lynch, *CRC Handbook of Materials Science, Vol. 3*, CRC Press, Boca Raton, Florida, 1975 and *Engineered Materials Handbook, Vol.2*, Engineering Plastics, ASM International, Metals Park, Ohio, 1988.

Table 336. FLAMMABILITY OF POLYMERS
(SHEET 11 OF 11)

Polymer	Type	Flammability, (ASTM D635) (ipm)
Polystyrenes; Molded	Polystyrenes	
	General purpose	1.0—1.5
	Medium impact	0.5—2.0
	High impact	0.5—1.5
	Styrene acrylonitrile (SAN)	0.8
Polyvinyl Chloride And Copolymers; Molded, Extruded	Nonrigid—general	Self extinguishing
	Nonrigid—electrical	Self extinguishing
	Rigid—normal impact	Self extinguishing
	Vinylidene chloride	Self extinguishing
Silicones; Molded, Laminated	Fibrous (glass) reinforced silicones	Nonburning
	Granular (silica) reinforced silicones	Nonburning
	Woven glass fabric/ silicone laminate	0.12
Ureas; Molded	Alpha—cellulose filled (ASTM Type l)	Self extinguishing
	Cellulose filled (ASTM Type 2)	Self extinguishing
	Woodflour filled	Self extinguishing

Source: data compiled by J.S. Park from Charles T. Lynch, *CRC Handbook of Materials Science, Vol. 3*, CRC Press, Boca Raton, Florida, 1975 and *Engineered Materials Handbook, Vol.2*, Engineering Plastics, ASM International, Metals Park, Ohio, 1988.

Table 337. FLAMMABILITY OF FIBERGLASS REINFORCED PLASTICS

Class	Material	Glass fiber content (wt%)	Flammability (UL94)
Glass fiber reinforced thermosets	Sheet molding compound (SMC)	15 to 30	5V
	Bulk molding compound(BMC)	15 to 35	5V
	Preform/mat(compression molded)	25 to 50	V–0
	Cold press molding–polyester	20 to 30	V–0
	Spray–up–polyester	30 to 50	V–0
	Filament wound–epoxy	30 to 80	V–0
	Rod stock–polyester	40 to 80	V–0
	Molding compound–phenolic	5 to 25	V–0
Glass–fiber–reinforced thermoplastics	Acetal	20 to 40	HB
	Nylon	6 to 60	V–0
	Polycarbonate	20 to 40	V–0
	Polyethylene	10 to 40	V–0
	Polypropylene	20 to 40	V–0
	Polystyrene	20 to 35	V–0
	Polysulfone	20 to 40	V–0
	ABS(acrylonitrile butadiene styrene)	20 to 40	V–0
	PVC (polyvinyl chloride)	15 to 35	V–0
	Polyphenylene oxide(modified)	20 to 40	V–0
	SAN (styrene acrylonitrile)	20 to 40	V–0
	Thermoplastic polyester	20 to 35	V–0

Data from *ASM Engineering Materials Reference Book, Second Edition*, Michael Bauccio, Ed., ASM International, Materials Park, OH, p106, (1994).

CHAPTER 10

Selecting Structural Properties

List of Tables

Table 338. SELECTING ATOMIC RADII OF THE ELEMENTS*

(SHEET 1 OF 3)

Atomic Number	Symbol	Atomic Radius (nm)
1	H	0.046
8	O	0.060
7	N	0.071
6	C	0.077
5	B	0.097
16	S	0.106
17	Cl	0.107
15	P	0.109
25	Mn	0.112
4	Be	0.114
34	Se	0.116
14	Si	0.117
35	Br	0.119
32	Ge	0.122
26	Fe	0.124
24	Cr	0.125
27	Co	0.125
28	Ni	0.125
33	As	0.125
29	Cu	0.128
23	V	0.132
30	Zn	0.133
44	Ru	0.134
45	Rh	0.134
31	Ga	0.135
76	Os	0.135
77	Ir	0.135
42	Mo	0.136

Source: After a tabulation by R. A. Flinn and P. K. Trojan, *Engineering Materials and Their Applications*, Houghton Mifflin Company, Boston, 1975.

Table 338. SELECTING ATOMIC RADII OF THE ELEMENTS*

(SHEET 2 OF 3)

Atomic Number	Symbol	Atomic Radius (nm)
53	I	0.136
46	Pd	0.137
74	W	0.137
75	Re	0.138
78	Pt	0.138
92	U	0.138
84	Po	0.140
13	Al	0.143
41	Nb	0.143
52	Te	0.143
47	Ag	0.144
79	Au	0.144
22	Ti	0.147
73	Ta	0.147
48	Cd	0.150
80	Hg	0.150
3	Li	0.152
49	In	0.157
40	Zr	0.158
50	Sn	0.158
72	Hf	0.159
10	Ne	0.160
12	Mg	0.160
21	Sc	0.160
51	Sb	0.161
81	Tl	0.171
71	Lu	0.173
69	Tm	0.174

Source: After a tabulation by R. A. Flinn and P. K. Trojan, *Engineering Materials and Their Applications*, Houghton Mifflin Company, Boston, 1975.

Table 338. SELECTING ATOMIC RADII OF THE ELEMENTS*

(SHEET 3 OF 3)

Atomic Number	Symbol	Atomic Radius (nm)
68	Er	0.175
82	Pb	0.175
67	Ho	0.176
65	Tb	0.177
66	Dy	0.177
64	Gd	0.180
90	Th	0.180
39	Y	0.181
62	Sm	0.181
58	Ce	0.182
60	Nd	0.182
83	Bi	0.182
59	Pr	0.183
11	Na	0.186
57	La	0.187
18	Ar	0.192
70	Yb	0.193
20	Ca	0.197
36	Kr	0.197
63	Eu	0.204
38	Sr	0.215
56	Ba	0.217
54	Xe	0.218
19	K	0.231
37	Rb	0.251
55	Cs	0.265

Source: After a tabulation by R. A. Flinn and P. K. Trojan, *Engineering Materials and Their Applications*, Houghton Mifflin Company, Boston, 1975.

* The ionic radii are based on the calculations of V. M. Goldschmidt, who assigned radii based on known interatomic distances in various ionic crystals.

Table 339. SELECTING IONIC RADII OF THE ELEMENTS*

(SHEET 1 OF 5)

Ion	Ionic Radius (nm)
N^{5+}	0.01–0.2
C^{4+}	<0.02
B^{3+}	0.02
P^{5+}	0.03–0.04
Cr^{6+}	0.03–0.04
Se^{6+}	0.03–0.04
S^{6+}	0.034
Si^{4+}	0.039
V^{5+}	0.04
As^{5+}	~0.04
Ge^{4+}	0.044
Pd^{2+}	0.050
Mn^{4+}	0.052
Pt^{2+}	0.052
Be^{2+}	0.054
Pt^{4+}	0.055
Al^{3+}	0.057
V^{4+}	0.061
Ga^{3+}	0.062
At^{7+}	0.062
Ti^{4+}	0.064
Cr^{3+}	0.064
V^{3+}	0.065
Co^{3+}	0.065

Source: After a tabulation by R. A. Flinn and P. K. Trojan, *Engineering Materials and Their Applications*, Houghton Mifflin Company, Boston, 1975.

Table 339. SELECTING IONIC RADII OF THE ELEMENTS*

(SHEET 2 OF 5)

Ion	Ionic Radius (nm)
Mo^{6+}	0.065
Ru^{4+}	0.065
Rh^{4+}	0.065
W^{6+}	0.065
Ir^{4+}	0.066
Fe^{2+}	0.067
Os^{4+}	0.067
Po^{6+}	0.067
Mo^{4+}	0.068
Rh^{3+}	0.068
Ta^{5+}	0.068
W^{4+}	0.068
Ti^{3+}	0.069
As^{3+}	0.069
Nb^{5+}	0.069
Mn^{3+}	0.070
Re^{4+}	0.072
Nb^{4+}	0.074
Sn^{4+}	0.074
Ti^{2+}	0.076
Li^{+}	0.078
Mg^{2+}	0.078
Ni^{2+}	0.078
Co^{2+}	0.082

Source: After a tabulation by R. A. Flinn and P. K. Trojan, *Engineering Materials and Their Applications*, Houghton Mifflin Company, Boston, 1975.

Table 339. SELECTING IONIC RADII OF THE ELEMENTS*

(SHEET 3 OF 5)

Ion	Ionic Radius (nm)
Sc^{2+}	0.083
Zn^{2+}	0.083
Hf^{4+}	0.084
Pb^{4+}	0.084
Fe^{2+}	0.087
Zr^{4+}	0.087
Te^{4+}	0.089
Tb^{4+}	0.089
Sb^{3+}	0.090
Mn^{2+}	0.091
In^{3+}	0.091
I^{5+}	0.094
Cu^{+}	0.096
Na^{+}	0.098
Lu^{3+}	0.099
Pr^{4+}	0.100
Yb^{3+}	0.100
Ce^{4+}	0.102
Cd^{2+}	0.103
Er^{3+}	0.104
Tm^{3+}	0.104
Ho^{3+}	0.105
U^{4+}	0.105
Ca^{2+}	0.106

Source: After a tabulation by R. A. Flinn and P. K. Trojan, *Engineering Materials and Their Applications*, Houghton Mifflin Company, Boston, 1975.

Table 339. SELECTING IONIC RADII OF THE ELEMENTS*
(SHEET 4 OF 5)

Ion	Ionic Radius (nm)
Y^{3+}	0.106
Pm^{3+}	0.106
Tl^{3+}	0.106
Dy^{3+}	0.107
Tb^{3+}	0.109
Th^{4+}	0.110
Gd^{3+}	0.111
Hg^{2+}	0.112
Ag^{+}	0.113
Sm^{3+}	0.113
Eu^{3+}	0.113
Nd^{3+}	0.115
Pr^{3+}	0.116
Ce^{3+}	0.118
Ac^{3+}	0.118
Bi^{3+}	0.120
La^{3+}	0.122
Sr^{2+}	0.127
Ba^{2+}	0.13
O^{2-}	0.132
Pb^{2+}	0.132
F^{-}	0.133
K^{+}	0.133
Au^{+}	0.137

Source: After a tabulation by R. A. Flinn and P. K. Trojan, *Engineering Materials and Their Applications*, Houghton Mifflin Company, Boston, 1975.

Table 339. SELECTING IONIC RADII OF THE ELEMENTS*
(SHEET 5 OF 5)

Ion	Ionic Radius (nm)
Rb^{+}	0.149
Tl^{+}	0.149
Ra^{+}	0.152
H^{-}	0.154
Cs^{+}	0.165
S^{2-}	0.174
Fr^{+}	0.180
Cl^{-}	0.181
Se^{2-}	0.191
Br^{-}	0.196
Si^{4-}	0.198
Te^{2-}	0.211
Sn^{4-}	0.215
Pb^{4-}	0.215
I^{-}	0.220

Source: After a tabulation by R. A. Flinn and P. K. Trojan, *Engineering Materials and Their Applications*, Houghton Mifflin Company, Boston, 1975.

* The ionic radii are based on the calculations of V. M. Goldschmidt, who assigned radii based on known interatomic distances in various ionic crystals.

Table 340. SELECTING BOND LENGTHS BETWEEN ELEMENTS

(SHEET 1 OF 2)

Elements	Compound	Bond length (Å)		
O-H	H_2O_2	0.960	±	0.005
O-H	OD	0.9699		
N-H	HNCS	1.013	±	0.005
N-N	N_3H	1.02	±	0.01
O-H	$[OH]^+$	1.0289		
N-H	$[NH_4]^+$	1.034	±	0.003
N-H	NH	1.038		
N-H	ND	1.041		
N=O	$[NO]^+$	1.0619		
N-N	$[N_2]^+$	1.116		
N-N	N_2O	1.126	±	0.002
N=O	N_2O	1.186	±	0.002
N-O	NO_2	1.188	±	0.005
B-O	BO	1.2049		
B-H	Hydrides	1.21	±	.02
O-O	$[O_2]^+$	1.227		
N-O	NO_2Cl	1.24	±	0.01
O-O	$[O_2]^-$	1.26	±	0.2
B-F	BF	1.262		
B-F	BF_3	1.29	±	0.01
S-D	SD_2	1.345		
S-D	SD	1.3473		
N-F	NF_3	1.36	±	0.02
B-O	$B(OH)_3$	1.362	±	0.005 (av)

To convert Å to **nm**, multiply by 10^{-1}

Source: from Kennard, O., in *Handbook of Chemistry and Physics, 69th ed.*, Weast, R. C., Ed., CRC Press, Boca Raton, Fla., 1988, F-167.

Table 340. SELECTING BOND LENGTHS BETWEEN ELEMENTS
(SHEET 2 OF 2)

Elements	Compound	Bond length (Å)		
B-H bridge	Hydrides	1.39	±	.02
B-N	$(BClNH)_3$	1.42	±	.01
P-H	$[PH_4]^+$	1.42	±	0.02
P-D	PD	1.429		
S-O	SO_2	1.4321		
S-O	$SOCl_2$	1.45	±	0.02
O-O	H_2O_2	1.48	±	0.01
Si-H	SiH_4	1.480	±	0.005
O-O	$[O_2]^{--}$	1.49	±	0.02
P-N	PN	1.4910		
Si-O	$[SiO]^+$	1.504		
Si-F	SiF_4	1.561	±	0.003 (av)
N-Si	SiN	1.572		
S-F	SOF_2	1.585	±	0.005
B-Cl	BCl	1.715		
B-Cl	BCl_3	1.72	±	0.01
B-B	B_2H_6	1.770	±	0.013
N-Cl	NO_2Cl	1.79	±	0.02
P-S	$PSBr_3$ (Cl_3,F_3)	1.86	±	0.02
B-Br	BBr_3	1.87	±	0.02
B-Br	BBF	1.88		
Si-Cl	$SiCl_4$	2.03	±	1.01 (av)
S-S	S_2Cl_2	2.04	±	0.01
Si-Br	$SiBr_4$	2.17	±	1.01
S-Br	$SOBr_2$	2.27	±	0.02
Si-Si	$[Si_2Cl_2]$	2.30	±	0.02

To convert **Å** to **nm**, multiply by 10^{-1}

Source: from Kennard, O., in *Handbook of Chemistry and Physics, 69th ed.*, Weast, R. C., Ed., CRC Press, Boca Raton, Fla., 1988, F-167.

Table 341. SELECTING BOND ANGLES BETWEEN ELEMENTS

Bond	Compound	Bond angle (•)		
F-S-F	SOF_2	92.8	±	1
Br-S-Br	$SOBr_2$	96	±	2
O-O-H	H_2O_2	100	±	2
F-N-F	NF_3	102.5	±	1.5
H-N-N'	N_3H	112.65	±	0.5
O-S-O	SO_2	119.54		
O-B-O	$B(OH)_3$	119.7		
Br-B-Br	BBr_3	120	±	6
Cl- B-Cl	BCl_3	120	±	3
F-B-F	BF_3	120		
B-N-B	$(BClNH)_3$	121		
H-B-H	B_2H_6	121.5	±	7.5
O-N-O	NO_2Cl	126	±	2
H-N-C	HNCS	130.25	±	0.25
O-N-O	NO_2	134.1	±	0.25

Source: from Kennard, O., in *Handbook of Chemistry and Physics, 69th ed.*, Weast, R. C., Ed., CRC Press, Boca Raton, Fla., 1988, F-167.

Table 342. SELECTING DENSITY OF THE ELEMENTS
(SHEET 1 OF 3)

Element	At. No.	Sym.	Solid Density (Mg/m^3)
Lithium	3	Li	0.533
Potassium	19	K	0.862
Sodium	11	Na	0.966
Calcium	20	Ca	1.53
Rubidium	37	Rb	1.53
Magnesium	12	Mg	1.74
Phosphorus (White)	15	P	1.82
Beryllium	4	Be	1.85
Cesium	55	Ce	1.91
Sulfur	16	S	2.09
Carbon	6	C	2.27
Silicon	14	Si	2.33
Boron	5	B	2.47
Strontium	38	Sr	2.58
Aluminum	13	Al	2.7
Scandium	21	Sc	2.99
Barium	56	Ba	3.59
Yttrium	39	Y	4.48
Titanium	22	Ti	4.51
Selenium	34	Se	4.81
Iodine	53	I	4.95
Europium	63	Eu	5.25
Germanium	32	Ge	5.32
Arsenic	33	As	5.78
Gallium	31	Ga	5.91
Vanadium	23	V	6.09
Lanthanum	57	La	6.17

Source: data from James F. Shackelford, *Introduction to Materials Science for Engineers, Second Edition*, Macmillian Publishing Company, New York, pp.686-688, (1988).

Table 342. SELECTING DENSITY OF THE ELEMENTS
(SHEET 2 OF 3)

Element	At. No.	Sym.	Solid Density (Mg/m^3)
Tellurium	52	Te	6.25
Zirconium	40	Zr	6.51
Antimony	51	Sb	6.69
Cerium	58	Ce	6.77
Praseodymium	59	Pr	6.78
Ytterbium	70	Yb	6.97
Neodymium	60	Nd	7.00
Zinc	30	Zn	7.13
Chromium	24	Cr	7.19
Indium	49	In	7.29
Tin	50	Sn	7.29
Manganese	25	Mn	7.47
Samarium	62	Sm	7.54
Iron	26	Fe	7.87
Gadolinium	64	Gd	7.87
Terbium	65	Tb	8.27
Dysprosium	66	Dy	8.53
Niobium	41	Nb	8.58
Cadmium	48	Cd	8.65
Cobalt	27	Co	8.8
Holmium	67	Ho	8.80
Nickel	28	Ni	8.91
Copper	29	Cu	8.93
Erbium	68	Er	9.04
Polonium	84	Po	9.2
Thulium	69	Tm	9.33
Bismuth	83	Bi	9.80
Lutetium	71	Lu	9.84

Source: data from James F. Shackelford, *Introduction to Materials Science for Engineers, Second Edition*, Macmillian Publishing Company, New York, pp.686-688, (1988).

Table 342. SELECTING DENSITY OF THE ELEMENTS
(SHEET 3 OF 3)

Element	At. No.	Sym.	Solid Density (Mg/m^3)
Molybdenum	42	Mo	10.22
Silver	47	Ag	10.50
Lead	82	Pb	11.34
Technetium	43	Tc	11.5
Thorium	90	Th	11.72
Thallium	81	Tl	11.87
Palladium	46	Pd	12.00
Ruthenium	44	Ru	12.36
Rhodium	45	Rh	12.42
Hafnium	72	Hf	13.28
Tantalum	73	Ta	16.67
Uranium	92	U	19.05
Tungsten	74	W	19.25
Gold	79	Au	19.28
Plutonium	94	Pu	19.81
Rhenium	75	Re	21.02
Platinum	78	Pt	21.44
Iridium	77	Ir	22.55
Osmium	76	Os	22.58

Source: data from James F. Shackelford, *Introduction to Materials Science for Engineers, Second Edition*, Macmillian Publishing Company, New York, pp.686-688, (1988).

CHAPTER 11 Selecting Thermodynamic and Kinetic Properties

List of Tables

Table 343. SELECTING BOND STRENGTHS IN DIATOMIC MOLECULES* (SHEET 1 OF 18)

Molecule	kcal • mol^{-1}	Range
Ar–Ar	0.2	
N–I	~.38	
Mg–I	~.68	
Xe–Xe	~ 0.7	
Hg–Tl	1	
Cd–Cd	2.7	± 0.2
Ga–Ga	3	± 3
Mn–Mn	4	± 3
Ga–Ag	4	± 3
Hg–Hg	4.1	± 0.5
Zn–Zn	7	
Mg–Mg	8?	
O–Xe	9	± 5
I–Hg	9	
H–Hg	9.5	
F–Xe	11	
Cs–Cs	11.3	
Rb–Rb	12.2	
K–K	12.8	
Na–Rb	14	± 1
Tl–Tl	15?	
Na–K	15.2	± 0.7
H–Cd	16.5	± 0.1
Be–Be	17	
Br–Hg	17.3	
Ca–Au	18	
Na–Na	18.4	
At–At	19	

To convert **kcal** to **KJ**, multiply by **4.184**.

Source: *from* Kerr, J. A., Parsonage, M. J., and Trotman–Dickenson, A. F., in *Handbook of Chemistry and Physics, 55th ed.*, Weast, R. C., Ed., CRC Press, Cleveland, 1974, F-204.

Table 343. SELECTING BOND STRENGTHS IN DIATOMIC MOLECULES* (SHEET 2 OF 18)

Molecule	kcal • mol^{-1}	Range
H–Zn	20.5	± 0.5
As–Se	23	
In–In	23.3	± 2.5
Cl–Hg	24	± 2
Fe–Fe	24	± 5
Pb–Pb	24	± 5
Li–Li	24.55	± 0.14
Sc–Sc	25.9	± 5
Cl–Ti	26	± 2
F–Hg	31	± 9
Au–Pb	31	± 23
Pb–Bi	32	± 5
Ag–Sn	32.5	± 5
Zn–Se	33	± 3
Zn–I	33	± 7
Cd–I	33	± 5
Pd–Pd	33?	
Ti–Ti	34	± 5
Pd–Au	34.2	± 5
In–Sb	36.3	± 2.5
I–I	36.460	± 0.002
Cr–Cu	37	± 5
Cr–Cr	<37	
F–F	37.5	± 2.3
H–Yb	38	± 1
Br–Cd	~38	
Ba–Au	38	± 14
Y–Y	38.3	

To convert kcal to KJ, multiply by 4.184.

Source: *from* Kerr, J. A., Parsonage, M. J., and Trotman–Dickenson, A. F., in *Handbook of Chemistry and Physics, 55th ed.*, Weast, R. C., Ed., CRC Press, Cleveland, 1974, F-204.

Table 343. SELECTING BOND STRENGTHS IN DIATOMIC MOLECULES* (SHEET 3 OF 18)

Molecule	kcal • mol^{-1}	Range
H–Sr	39	± 2
Co–Cu	39	± 5
H–Rb	40	± 5
Co–Co	40	± 6
H–Ca	40.1	
Cr–Ge	41	± 7
Ag–Ag	41	± 2
Cu–Ag	41.6	± 2.2
H–Ba	42	± 4
H–Pb	42	± 5
Cu–Te	42	± 9
Cu–Sn	42.3	± 4
H–Cs	42.6	± 0.9
Br–I	42.8	± 0.1
H–K	43.8	± 3.5
Al–Al	44	
Mn–Au	44	± 3
Po–Po	44.4	± 2.3
H–Ti	45	± 2
Fe–Au	45	± 4
Bi–Bi	45	± 2
Bi–Br	46.336	± 0.001
Cu–Cu	46.6	± 2.2
Sn–Sn	46.7	± 4
H–Mg	47	± 12
O–I	47	± 7
Bi–Sn	47	± 23
I–Pb	47	± 9

To convert **kcal** to **KJ**, multiply by **4.184**.

Source: *from* Kerr, J. A., Parsonage, M. J., and Trotman–Dickenson, A. F., in *Handbook of Chemistry and Physics, 55th ed.*, Weast, R. C., Ed., CRC Press, Cleveland, 1974, F-204.

Table 343. SELECTING BOND STRENGTHS IN DIATOMIC MOLECULES* (SHEET 4 OF 18)

Molecule	$kcal \cdot mol^{-1}$	Range
Cu–I	47?	
H–Na	48	± 5
S–Cd	48	
Mn–Se	48	± 3
Ni–Cu	48	± 5
Y–La	48.3	
Ag–Au	48.5	± 2.2
S–Zn	49	± 3
Cu–Ge	49	± 5
Zn–Te	49?	
Cl–Cd	49.9	
C–I	50	± 5
Fe–Ge	50	± 7
Ga–As	50.1	± 0.3
Cl–I	50.5	± 0.1
O–Ag	51	± 20
S–Hg	51	
Co–Au	51	± 3
Ga–Au	51	± 23
Cr–Au	51.3	± 3.5
Al–P	52	± 3
In–Te	52	± 4
I–Bi	52	± 1
Cl–Br	52.3	± 0.2
Au–Au	52.4	± 2.2
H–Be	54	
Cl–Zn	54.7	± 4.7
N–Xe	55	

To convert kcal to KJ, multiply by 4.184.

Source: *from* Kerr, J. A., Parsonage, M. J., and Trotman–Dickenson, A. F., in *Handbook of Chemistry and Physics, 55th ed.*, Weast, R. C., Ed., CRC Press, Cleveland, 1974, F-204.

Table 343. SELECTING BOND STRENGTHS IN DIATOMIC MOLECULES* (SHEET 5 OF 18)

Molecule	kcal • mol^{-1}	Range
Cu–Au	55.4	± 2.2
Ni–Ni	55.5	± 5
F–Br	55.9	
H–Mn	56	± 7
O–F	56	± 9
O–Pd	56	± 7
P–Ga	56	
Ag–I	56	± 7
Te–Bi	56	± 3
Mg–S	56?	
O–Br	56.2	± 0.6
H–Li	56.91	± 0.01
O–K	57	± 8
Co–Ge	57	± 6
V–V	58	± 5
Te–Eu	58	± 4
Cl–Cl	58.066	± 0.001
Sn–Au	58.4	± 4
La–Ld	58.6	
H–Ag	59	± 1
H–In	59	± 2
H–Bi	59	± 7
Mg–Au	59	± 23
Fe–Br	59	± 23
Ni–Au	59	± 5
Se–in	59	± 4
Br–Pb	59	± 9
Te–Au	59	± 16

To convert kcal to KJ, multiply by **4.184.**

Source: *from* Kerr, J. A., Parsonage, M. J., and Trotman–Dickenson, A. F., in *Handbook of Chemistry and Physics, 55th ed.*, Weast, R. C., Ed., CRC Press, Cleveland, 1974, F-204.

Table 343. SELECTING BOND STRENGTHS IN DIATOMIC MOLECULES* (SHEET 6 OF 18)

Molecule	kcal • mol^{-1}	Range
F–Cl	59.9	± 0.1
Ga–Te	60	± 6
Sb–Bi	60	± 1
Te–Pb	60	± 3
O–Rb	(61)	± 20
H–Ni	61	± 7
O–Na	61	± 4
Ge–Br	61	± 7
Sb–Te	61	± 4
F–Bi	62	
Te–Ho	62	± 4
N–F	62.6	± 0.8
H–Sn	63	± 1
Sr–Au	63	± 23
Te–Te	63.2	± 0.2
Br–Bi	63.9	± 1
H–Te	64	± 1
Se–Te	64	± 2
O–Cl	64.29	± 0.03
H–As	65	± 3
Al–Au	65	
I–Ti	65	± 2
Ge–Ge	65.8	± 3
Si–Co	66	± 4
Ce–Ce	66	± 1
O–Zn	≤ 66	
O–Cd	≤ 67	
B–B	~ 67	± 5

To convert **kcal** to **KJ**, multiply by **4.184**.

Source: *from* Kerr, J. A., Parsonage, M. J., and Trotman–Dickenson, A. F., in *Handbook of Chemistry and Physics, 55th ed.*, Weast, R. C., Ed., CRC Press, Cleveland, 1974, F-204.

Table 343. SELECTING BOND STRENGTHS IN DIATOMIC MOLECULES* (SHEET 7 OF 18)

Molecule	kcal • mol^{-1}	Range
Be–Au	~ 67	
H–Cr	67	± 12
H –Cu	67	± 2
C–Br	67	± 5
N–Br	67	± 5
O–Cs	67	± 8
Se–Bi	67.0	± 1.5
F–I	67?	
Ni–Ge	67.3	± 4
Mn–I	67.6	± 2.3
H–Al	68	± 2
H –Ga	68	± 5
O–Ga	68	± 15
Cr–I	68.6	± 5.8
S–In	69	± 4
Th–Th	<69	
P–S	70	
Ca–I	70	± 23
Ni–I	70	± 5
Cu–Se	70	± 9
Ge–Au	70	± 23
Br–Ag	70	± 7
Ag–Te	70	± 23
Nd–Au	70	± 6
B–Th	71	
N–Al	71	± 23
Si–Fe	71	± 6
H–Si	71.4	± 1.2

To convert **kcal** to **KJ**, multiply by **4.184**.

Source: *from* Kerr, J. A., Parsonage, M. J., and Trotman–Dickenson, A. F., in *Handbook of Chemistry and Physics, 55th ed.*, Weast, R. C., Ed., CRC Press, Cleveland, 1974, F-204.

Table 343. SELECTING BOND STRENGTHS IN DIATOMIC MOLECULES * (SHEET 8 OF 18)

Molecule	kcal • mol^{-1}	Range
H–I	71.4	± 0.2
Sb–Sb	71.5	± 1.5
N–Sb	72	± 12
Si–Ge	72	± 5
S–Mn	72	± 4
S–Cu	72	± 12
Cl–Pb	72	± 7
Cl–Bi	72	± 1
Se–Eu	72	± 4
Se–Pb	72.4	± 1
Na–I	72.7	± 1
H–Se	73	± 1
F–Cd	73	± 5
P–W	73	± 1
Te–Nd	73	± 4
Pr–Au	74	± 5
H–N	75	± 4
H–Au	75	± 3
Si–Pd	75	± 4
Si–Au	75	± 3
Mg–Br	75	± 23
S–Ca	75	± 5
S–Sr	75	± 5
Cl–Ag	75	± 9
Se–Cd	~75	
Br–Sb	75	± 14
CI–Sn	75?	
Mn–Br	75.1	± 23

To convert **kcal** to **KJ**, multiply by **4.184**.

Source: *from* Kerr, J. A., Parsonage, M. J., and Trotman–Dickenson, A. F., in *Handbook of Chemistry and Physics, 55th ed.*, Weast, R. C., Ed., CRC Press, Cleveland, 1974, F-204.

Table 343. SELECTING BOND STRENGTHS IN DIATOMIC MOLECULES * (SHEET 9 OF 18)

Molecule	kcal • mol^{-1}	Range
S–Bi	75.4	± 1.1
Si–Si	76	± 5
Si–Ni	76	± 4
Mg–Cl	76	± 3
Sn–Te	76	± 1
Ce–Au	76	± 4
Au–U	76	± 7
Rb–I	76.7	± 1
H–Ge	76.8	± 0.2
K–I	76.8	± 0.5
O–In	≤ 77	
Li–O	78	± 6
Al–U	78	± 7
S–Fe	78	
Te–Lu	78	± 4
Cr–Br	78.4	± 5
H–B	79	± 1
B–Pd	79	± 5
O–Mg	79	± 7
Al–S	79	
Cl–Sc	79	
Cu–Br	79	± 6 5
Se–Se	79.5	± 0.1
Br–Ti	79.8	± 0.4
Se–Ho	80	± 4
In–I	80	
La–Au	80	± 5
H–C	80.9	

To convert **kcal** to **KJ**, multiply by **4.184**.

Source: *from* Kerr, J. A., Parsonage, M. J., and Trotman–Dickenson, A. F., in *Handbook of Chemistry and Physics, 55th ed.*, Weast, R. C., Ed., CRC Press, Cleveland, 1974, F-204.

Table 343. SELECTING BOND STRENGTHS IN DIATOMIC MOLECULES* (SHEET 10 OF 18)

Molecule	kcal • mol^{-1}	Range
S–Te	81	± 5
Ga–I	81	± 2
O–Bi	81.9	± 1.5
H–P	82	± 7
B–Au	82	± 4
O–Cu	82	± 15
Si–Br	82	± 12
Cl–Y	82	± 23
Cl–Au	82	± 2
Cl–Ra	82	± 18
Te–Gd	82	± 4
Cl–Ge	82?	
H–S	82.3	± 2.9
I–Cs	82.4	± 1
S–Pb	82.7	± 0.4
O–Pt	83	± 8
H–Pt	84	± 9
O–Ca	84	± 7
Cl–Cu	84	± 6
Cl–Fe	84?	
Li–I	84.6	± 2
F–Ag	84.7	± 3.9
B–Te	85	± 5
F–Pb	85	± 2
Cl–Sb	86	± 12
Ni–Br	86	± 3
Cl–Mn	86.2	± 2.3
Na–Br	86.7	± 1

To convert **kcal** to **KJ**, multiply by **4.184**.

Source: *from* Kerr, J. A., Parsonage, M. J., and Trotman–Dickenson, A. F., in *Handbook of Chemistry and Physics, 55th ed.*, Weast, R. C., Ed., CRC Press, Cleveland, 1974, F-204.

Table 343. SELECTING BOND STRENGTHS IN DIATOMIC MOLECULES * (SHEET 11 OF 18)

Molecule	kcal • mol^{-1}	Range
S–Eu	87	± 4
H–Br	87.4	± 0.5
Cl–Cr	87.5	± 5.8
O–Co	88	± 5
F–Cu	88	± 9
Al–I	88	
Be–S	89	± 14
O–Ni	89	± 5
O–Sb	89	± 20
F–Ni	89	± 4
Cl–Ni	89	± 5
Cl–Ti	89.0	± 0.5
O–Rh	90	± 15
P–Th	90	
O–Pb	90.3	± 1.0
Br–Rb	90.4	± 1
K–Br	90.9	± 0.5
S–Se	91	± 5
Te–La	91	± 4
As–As	91.7	
Se–Nd	92	± 4
Be–Cl	92.8	± 2.2
B–N	93	± 12
C–Cl	93	
N–Cl	93	± 12
O–Sr	93	± 6
Ge–Te	93	± 2
Br–In	93	

To convert **kcal** to **KJ**, multiply by **4.184**.

Source: *from* Kerr, J. A., Parsonage, M. J., and Trotman–Dickenson, A. F., in *Handbook of Chemistry and Physics, 55th ed.*, Weast, R. C., Ed., CRC Press, Cleveland, 1974, F-204.

Table 343. SELECTING BOND STRENGTHS IN DIATOMIC MOLECULES* (SHEET 12 OF 18)

Molecule	kcal • mol^{-1}	Range
O–Fe	93.4	± 2
O–Ir	≤ 94	
Si–Ru	95	± 5
Si–Rh	95	± 5
Cl–Ca	95	± 3
Se–Sn	95.9	± 1.4
O–Mn	96	± 8
O–Fe	96	± 5
S–Ba	96	± 5
Br–Cs	96.5	± 1
Cl–Sr	97	± 3
Na–Cl	97.5	± 0.5
Be–O	98	± 7
O–Yb	98	± 15
S–Au	100	± 6
Se–Lu	100	± 4
B–Ce	~ 100	
Li–Br	100.2	± 2
Cl–Rb	100.7	± 1
B–Br	101	± 5
O–Se	101	
Ga–Br	101	± 4
F–Mn	101.2	± 3.5
Cl–K	101.3	± 0.5
S–S	101.9	± 2.5
S–Ho	102	± 4
H–O	102.34	± 0.30
Se–Gd	103	± 4

To convert **kcal** to **KJ**, multiply by **4.184**.

Source: *from* Kerr, J. A., Parsonage, M. J., and Trotman–Dickenson, A. F., in *Handbook of Chemistry and Physics, 55th ed.*, Weast, R. C., Ed., CRC Press, Cleveland, 1974, F-204.

Table 343. SELECTING BOND STRENGTHS IN DIATOMIC MOLECULES* (SHEET 13 OF 18)

Molecule	kcal • mol^{-1}	Range
H–Cl	103.1	
Al–Br	103.1	
Cl–In	103.3	
C–Si	104	± 5
H–H	104.207	± 0.001
F–Cr	104.5	± 4.7
N–Si	105	± 9
N–Se	105	± 23
F–P	105	± 23
Si–Cl	105	± 12
F–Sb	105	± 23
H–D	105.030	± 0.001
Cl–Ba	106	± 3
D–D	106.010	± 0.001
Cl–Cs	106.2	± 1
F–Ti	106.4	± 4.6
B–Ru	107	± 5
C–Ce	109	± 7
B–Se	110	± 4
C–Ge	110	± 5
O–Cr	110	± 10
F–Mg	110	± 1
Si–Ir	110	± 5
C–U	111	± 7
N–Ti	111	
S–Sn	111	± 1
F–Sn	111.5	± 3
Li–Cl	111.9	± 2

To convert **kcal** to **KJ**, multiply by **4.184**.

Source: *from* Kerr, J. A., Parsonage, M. J., and Trotman–Dickenson, A. F., in *Handbook of Chemistry and Physics, 55th ed.*, Weast, R. C., Ed., CRC Press, Cleveland, 1974, F-204.

Table 343. SELECTING BOND STRENGTHS IN DIATOMIC MOLECULES * (SHEET 14 OF 18)

Molecule	kcal • mol^{-1}	Range
S– Nd	113	± 4
B–Rh	114	± 5
B–Pt	114	± 4
F–Na	114	± 1
S–Sc	114	± 3
Ge–Se	114	±
Se–La	114	± 4
Cl–Ga	114.5	
O–As	115	± 3
O–Mo	115	± 12
O–Ru	115	± 15
N–As	116	± 23
O–Al	116	± 5
F–Si	116	± 12
F–Ge	116	± 5
F–Rb	116.1	± 1
P–P	117	± 3
O–O	118.86	± 0.04
F–K	118.9	± 0.6
B–Cl	119	
Al–Cl	119.0	± 1
O–P	119.6	± 3
F–Cs	119.6	± 1
N–S	~ 120	± 6
Si–Pt	120	± 5
Si–Te	121	± 9
F–In	121	± 4
S–Lu	121	± 4

To convert **kcal** to **KJ**, multiply by **4.184**.

Source: *from* Kerr, J. A., Parsonage, M. J., and Trotman–Dickenson, A. F., in *Handbook of Chemistry and Physics, 55th ed.*, Weast, R. C., Ed., CRC Press, Cleveland, 1974, F-204.

Table 343. SELECTING BOND STRENGTHS IN DIATOMIC MOLECULES* (SHEET 15 OF 18)

Molecule	$kcal \cdot mol^{-1}$	Range
O–Tm	122	± 15
S–Pr	122.7	
B–Ir	123	± 4
O–S	124.69	± 0.03
F–Ca	125	± 5
S–Gd	126	± 4
F–Eu	126.1	± 4.4
F–Sm	126.9	± 4.4
N–U	127	± 1
O–Sn	127	± 2
Si–Se	127	± 4
S–Y	127	± 3
C–F	128	± 5
C–Ti	~128	
F–Pu	129	± 7
F–Sr	129.5	± 1.6
O–Eu	130	± 10
F–Nd	130	± 3
O–Ba	131	± 6
S–Ge	131.7	± 0.6
C–V	133	
O–Sm	134	± 8
O–Cm	≤ 134	
S–U	135	± 2
H–F	135.9	± 0.3
Be–F	136	± 2
F–Ti	136	± 8
S–La	137	± 3

To convert **kcal** to **KJ**, multiply by **4.184**.

Source: *from* Kerr, J. A., Parsonage, M. J., and Trotman–Dickenson, A. F., in *Handbook of Chemistry and Physics, 55th ed.*, Weast, R. C., Ed., CRC Press, Cleveland, 1974, F-204.

Table 343. SELECTING BOND STRENGTHS IN DIATOMIC MOLECULES * (SHEET 16 OF 18)

Molecule	kcal • mol^{-1}	Range
S–Ce	137	± 3
Li– F	137.5	± 1
N–Th	138	± 1
F–Ga	138	± 4
B–S	138.8	± 2.2
C–P	139	± 23
C–Se	139	± 23
C–Rh	139	± 2
F–Ba	140.3	± 1.6
F–Sc	141	± 3
F–Gd	141.	± 46.5
O–Os	< 142	
C–C	144	± 5
F–Y	144	± 5
C–Pt	146	± 2
O–Dy	146	± 10
O–Er	147	± 10
N–P	148	± 5
Si–S	148	± 3
C–Ir	149	± 3
O–Ho	149	± 10
N–O	150.8	± 0.2
C–Ru	152	± 3
O–V	154	± 5
O–Sc	155	± 5
O–W	156	± 6
O–Ti	158	± 8
O–Ge	158.2	± 3

To convert kcal to KJ, multiply by 4.184.

Source: *from* Kerr, J. A., Parsonage, M. J., and Trotman–Dickenson, A. F., in *Handbook of Chemistry and Physics, 55th ed.*, Weast, R. C., Ed., CRC Press, Cleveland, 1974, F-204.

Table 343. SELECTING BOND STRENGTHS IN DIATOMIC MOLECULES* (SHEET 17 OF 18)

Molecule	kcal • mol^{-1}	Range
O–Lu	159	± 8
F–Al	159	± 3
O–Y	162	± 5
O–Gd	162	± 6
O–Pu	163	± 15
O–Tb	165	± 8
O–Nd	168	± 8
O–Np	172	± 7
C–S	175	± 7
B–F	180	± 3
O–Zr	181	± 10
O–U	182	± 8
O–Ta	183	± 15
O–Pr	183.7	
C–N	184	± 1
O–Si	184	± 3
O–Hf	185	± 10
O–La	188	± 5
O–Ce	188	± 6
O–Nb	189	± 10

To convert **kcal** to **KJ**, multiply by **4.184**.

Source: *from* Kerr, J. A., Parsonage, M. J., and Trotman–Dickenson, A. F., in *Handbook of Chemistry and Physics, 55th ed.*, Weast, R. C., Ed., CRC Press, Cleveland, 1974, F-204.

Table 343. SELECTING BOND STRENGTHS IN DIATOMIC MOLECULES* (SHEET 18 OF 18)

Molecule	$kcal \cdot mol^{-1}$	Range
O–Th	192	± 10
B–O	192.7	± 1.2
N–N	226.8	± 1.5
C–O	257.26	± 0.77

To convert **kcal** to **KJ**, multiply by **4.184**.

Source: *from* Kerr, J. A., Parsonage, M. J., and Trotman–Dickenson, A. F., in *Handbook of Chemistry and Physics, 55th ed.*, Weast, R. C., Ed., CRC Press, Cleveland, 1974, F-204.

* The strength of a chemical bond, Δ(R - X), often known as the bond dissociation energy, is defined as the heat of the reaction: RX -> R + X. It is given by: $\Delta(R - X) = \Delta H_f^{\circ}(R) + \Delta H_f^{\circ}(X) - \Delta H_f^{\circ}(RX)$. Some authors list bond strengths for 0 K, but here the values for 298K are given because more thermodynamic data are available for this temperature. Bond strengths, or bond dissociation energies, are not equal to, and may differ considerable from, mean bond energies derived solely from thermochemical data on molecules and atoms.

The values in this table have usually been measured spectroscopically or by mass spectrometric analysis of hot gases effusing from a Knudsen cell.

Table 344. SELECTING BOND STRENGTHS OF POLYATOMIC MOLECULES* (SHEET 1 OF 6)

Bond	Strength $Kcal \cdot mol^{-1}$	
$NO - NO_2$	9.5	± 0.5
$NO_2 - NO_2$	12.9	± 0.5
$NF_2 - NF_2$	21	± 1
$CH_3CO_2 - O_2CCH_3$	30.4	± 2
$C_2H_5CO_2 - O_2CC_2H_5$	30.4	± 2
n -$C_3H_7CO_2 - O_2C$n –C_3H_7	30.4	± 2
$Cl - NF_2$	32	
BH_3–BH_3	35	
CH_3 –$Tl(CH_3)_2$	36.4	± 0.6
s -C_4H_9O – O s –C_4H_9	36.4	± 1
$(CH_3)_3CCH_2O - OCH_2C(CH_3)_3$	36.4	± 1
$CH_3O - OCH_3$	36.9	± 1
i –C_3H_7O – O i –C_3H_7	37.0	± 1
n –C_3H_7O – O n –C_3H_7	37.2	± 1
$C_2H_5O - OC_2H_5$	37.3	± 1.2
t –C_4H_9O – O t –C_4H_9	37.4	± 1
$C_6H_5CH_2N$:N–$C_6H_5CH_2$	37.6	
$O - N_2$	40	
i –C_3H_7 – Hg i –C_3H_7	40.7	
$CH_2 = N_2$	41.7	± 1
$HO - OC(CH_3)_3$	42.5	
t –C_4H_9N:N–t –C_4H_9	43.5	
$F - OCF_3$	43.5	± 0.5
$C_2H_5 - HgC_2H_5$	43.7	± 1

To convert **kcal** to **KJ**, multiply by **4.184**.

Source: data from: Kerr, J. A., Parsonage, M. J., and Trotman–Dickenson, A. F., in *Handbook of Chemistry and Physics, 55th ed.*, Weast, R. C., Ed., CRC Press, Cleveland, 1974, F–213.

Table 344. SELECTING BOND STRENGTHS OF POLYATOMIC MOLECULES* (SHEET 2 OF 6)

Bond	Strength Kcal • mol^{-1}	
s $-C_4H_9N:N-$s $-C_4H_9$	46.7	
n $-C_3H_7-Hg$ n $-C_3H_7$	47.1	
i $-C_3H_7N:N-$i $-C_3H_7$	47.5	
i $-C_4H_9N:N-$i $-C_4H_9$	49.0	
$CH_3-Pb(CH_3)_3$	49.4	± 1
Allyl–O_2SCH_3	49.6	
HO – N:$CHCH_3$	49.7	
$C_2H_5N:N-C_2H_5$	50.0	
n $-C_4H_9N:N-$n $-C_4H_9$	50.0	
HO – OH	51	± 1
$NH_2-NHC_6H_5$	51.1	
$CH_3N:N-CH_3$	52.5	
$C_6H_5CH_2-O_2SCH_3$	52.9	
$I-CF_3$	53.5	± 2
$C_6H_5CH_2-SCH_3$	53.8	
CH_3-CdCH_3	54.4	
$CF_3N:N-CF_3$	55.2	
Br – OH	56	± 3
I – OH	56	± 3
$Br-CBr_3$	56.2	± 1.8
$I-CH_3$	56.3	± 1
CH_3-HgCH_3	57.5	
$O-O_2ClF$	58.4	
ClO_3-ClO_4	58.4	

To convert kcal to KJ, multiply by 4.184.

Source: data from: Kerr, J. A., Parsonage, M. J., and Trotman–Dickenson, A. F., in *Handbook of Chemistry and Physics, 55th ed.*, Weast, R. C., Ed., CRC Press, Cleveland, 1974, F–213.

Table 344. SELECTING BOND STRENGTHS OF POLYATOMIC MOLECULES* (SHEET 3 OF 6)

Bond	Strength Kcal • mol^{-1}	
O – ClO	59	± 3
$(C_6H_5CH_2)_2CH$–COOH	59.4	
$CH_3 – Ga(CH_3)_2$	59.5	
$C_6H_5C(CH_3)(CN) – CH_3$	59.9	
$C_6H_5S – CH_3$	60	
Cl – OH	60	± 3
$C_6H_5CH_2 – N(CH_3)_2$	60.9	± 1
$C_2H_5 – NO_2$	62	
1–norbornyl	62.5	± 2.5
$NH_2 – N(CH_3)_2$	62.7	
$Br – COC_6H_5$	64.2	
$NH_2 – NHCH_3$	64.8	
$CF_3 – NF_2$	65	± 2.5
$C_6H_5N(CH_2) – CH_3$	65.2	
$C_6H_5CH_2CO – CH_2C_6H_5$	65.4	
$C_6H_5CO – COC_6H_5$	66.4	
Br –n –C_3F	66.5	± 2.5
$CH_3 – O_2SCH_3$	66.8	
$C_6H_5CH_2$–n –C_3H_7	67	± 2
$C_6H_5CH_2 – O_2CCH_3$	67	
$CH_3CO – COCH_3$	67.4	± 2.3
$C_6H_5NH–CH_3$	67.7	
$C_6H_5 – HgC_6H_5$	68	
$C_6H_5CH_2 – COOH$	68.1	

To convert **kcal** to **KJ**, multiply by **4.184**.

Source: data from: Kerr, J. A., Parsonage, M. J., and Trotman–Dickenson, A. F., in *Handbook of Chemistry and Physics, 55th ed.*, Weast, R. C., Ed., CRC Press, Cleveland, 1974, F–213.

Table 344. SELECTING BOND STRENGTHS OF POLYATOMIC MOLECULES* (SHEET 4 OF 6)

Bond	Strength Kcal • mol^{-1}	
$C_6H_5CH_2 - NHCH_3$	68.7	± 1
$Br - C_2F_5$	68.7	± 1.5
$C_6H_5CH_2–C_2H_5$	69	± 2
$C_6H_5CH_2 - O_2CC_6H_5$	69	
$CH_3–C(CH_3)_2CH{:}CH_2$	69.4	
$Br - CH_3$	70.0	± 1.2
$CH_3–C(CH_3)_2CN$	70.2	± 2
$Br - CF_3$	70.6	± 1.0
$NH_2 - NH_2$	70.8	± 2
$C_6H_5CH(CH_3) - CH_3$	71	
$C_6H_5CH_2 –NH_2$	71.9	± 1
$CH_3–CH_2CN$	72.7	± 2
$Cl - CCl_2F$	73	± 2
I – CN	73	± 1
O – NO	73	
$C_6H_5CO - CF_3$	73.8	
$Cl - COC_6H_5$	74	± 3
$CF_2 = CF_2$	76.3	± 3
H–ONO	78.3	± 0.5
H–pentadien–1,4–yi–3	80	± 1
$(CH_3)_3Si - Si(CH_3)_3$	80.5	
$SiH_3 - SiH_3$	81	± 4
H–cyclopentadien–1,3–yl–5	81.2	± 1.2
$Cl - C_2F5$	82.7	± 1.7

To convert **kcal** to **KJ**, multiply by **4.184**.

Source: data from: Kerr, J. A., Parsonage, M. J., and Trotman–Dickenson, A. F., in *Handbook of Chemistry and Physics, 55th ed.*, Weast, R. C., Ed., CRC Press, Cleveland, 1974, F–213.

Table 344. SELECTING BOND STRENGTHS OF POLYATOMIC MOLECULES* (SHEET 5 OF 6)

Bond	Strength Kcal • mol^{-1}	
H–methdllyl	83	± 1
Br – CN	83	± 1
Cl – CF_3	86.1	± 0.8
H–OC_6H_5	88	± 5
H_3C–CH_3	88	± 2
CH_2F – CH_2F	88	± 2
H–SCH?	≥ 88	
H–allyl	89	± 1
H–O_2H	90	± 2
H–SH	90	± 2
H–$Si(CH_3)_3$	90	± 3
H–t–C_4H_9	92	± 1.2
H–propargyl	93.9	± 1.2
H–SiH_3	94	± 3
H–i–C_3H_7	95	± 1
H–s–C_4H_9	95	± 1
H–cyclobutyl	96.5	± 1
CF_3 – CF_3	96.9	± 2
Cl – CN	97	± 1
H–cyclopropycarbinyl	97.4	± 1.6
H–C_2H_5	98	± 1
H–n–C_3H_7	98	± 1
H–cyclopropyl	100.7	± 1
H–ONO_2	101.2	± 0.5

To convert **kcal** to **KJ**, multiply by **4.184**.

Source: data from: Kerr, J. A., Parsonage, M. J., and Trotman–Dickenson, A. F., in *Handbook of Chemistry and Physics, 55th ed.*, Weast, R. C., Ed., CRC Press, Cleveland, 1974, F–213.

Table 344. SELECTING BOND STRENGTHS OF POLYATOMIC MOLECULES* (SHEET 6 OF 6)

Bond	Strength Kcal • mol^{-1}	
H–CH	102	± 2
H–O_2Cn–C_3H_7	103	± 4
F – CH_3	103	± 3
H–OCH_3	103.6	± 1
H–OC_2H_5	103.9	± 1
H–CH_3	104	± 1
H–OC$(CH_3)_3$	104.7	± 1
H–vinyl	≥ 108	± 2
H–CH_2	110	± 2
H–$O_2CC_2H_3$	110	± 4
H–O_2CCH_3	112	± 4
H–OH	119	± 1
O = PBr_3	119	± 5
O = PCl_3	122	± 5
O=CO	127.2	± 0.1
H–ethynyl	128	± 5
NC–CN	128	± 1
O = PF_3	130	± 5
O – SO	132	± 2
H_2C=CH_2	172	± 2
HC=CH	230	± 2

To convert **kcal** to **KJ**, multiply by **4.184**.

Source: data from: Kerr, J. A., Parsonage, M. J., and Trotman–Dickenson, A. F., in *Handbook of Chemistry and Physics, 55th ed.*, Weast, R. C., Ed., CRC Press, Cleveland, 1974, F–213.

* The values refer to a temperature of 298 K and have mostly been determined by kinetic methods. Some have been calculated from formation of the species involved according to equations:

$D(R–X) = \Delta H_f^\circ (R^\cdot) + \Delta H_f^\circ (X^\cdot) - \Delta H_f^\circ (RX)$ or $D(R–X) = 2\Delta H_f^\circ (R\cdot) - \Delta H_f^\circ (RR)$

Table 345. SELECTING HEAT OF FORMATION OF INORGANIC OXIDES (SHEET 1 OF 9)

Reaction	Temperature Range of Validity	ΔH_0
$6\,V(c) + 13/2\,O_2(g) = V_6O_{13}(c)$	298.16–1,000K	–1,076,340
$3\,U(l) + 4\,O_2(g) = U_3O_8(c)$	1,405–1,500K	–869,460
$3\,U(\alpha) + 4\,O_2(g) = U_3O_8(c)$	298.16–935K	–863,370
$3\,U(\gamma) + 4\,O_2(g) = U_3O_8(c)$	1,045–1,405K	–863,230
$3\,U(\beta) + 4\,O_2(g) = U_3O_8(c)$	935–1,045K	–856,720
$4W(c) + 11/2\,O_2(g) = W_4O_{11}(c)$	298.16–1,700K	–745,730
$4\,P$ (white) + $5\,O_2(g) = P_4H_{10}$ (hexagonal)	298.16–317.4K	–711,520
$2\,Ta(c) + 5/2\,O_2(g) = Ta_2O_5(c)$	298.16–2,000K	–492,790
$2\,Nb(c) + 5/2\,O_2(g) = Nb_2O_5(l)$	1,785–2,000K	–463,630
$2\,Nb(c) + 5/2\,O_2(g) = Nb_2O_5(c)$	298.16–1,785K	–458,640
$2\,Ac(c) + 3/2\,O_2(g) = Ac_2O_3(c)$	298.16–1,000K	–446,090
$2\,Ce(l) + 3/2\,O_2(g) = Ce_2O_3(c)$	1,048–1,900K	–440,400
$2\,Ce(c) + 3/2\,O_2(g) = Ce_2O_3(c)$	298.16–1,048K	–435,600
$2\,Y(c) + 3/2\,O_2(g) = Y_2O_3(c)$	298.16–1,773K	–419,600
$2\,Al(l) + 3/2\,O_2(g) = Al_2O_3$ (corundum)	931.7–2,000K	–407,950
$2\,Al(c) + 3/2\,O_2(g) = Al_2O_3$ (corundum)	298.16–931.7K	–404,080
$2\,Nb(c) + 2\,O_2(g) = Nb_2O_4(c)$	298.16–2,000K	–382,050
$2\,V(c) + 5/2\,O_2(g) = V_2O_5(c)$	298.16–943K	–381,960
$2\,Ti(\alpha) + 3/2\,O_2(g) = Ti_2O_3(\beta)$	473–1,150K	–369,710
$2\,Ti(\alpha) + 3/2\,O_2(g) = Ti_2O_3(\alpha)$	298.16–473K	–360,660
$2\,V(c) + 2\,O_2(g) = V_2O_4(\beta)$	345–1,818K	–345,330
$2\,V(c) + 2\,O_2(g) = V_2O_4(\alpha)$	209.16–345K	–342,890
$3\,Mn(\alpha) + 2\,O_2(g) = Mn_3O_4(\alpha)$	298.16–1,000K	–332,400
$2\,B(c) + 3/2\,O_2(g) = B_2O(c)$	298.16–723K	–304,690

The ΔH_o values are given in **gram calories per mole.**

Source: data from *CRC Handbook of Materials Science, Vol II*, Charles T. Lynch, Ed., CRC Press, Cleveland, (1974).

Table 345. SELECTING HEAT OF FORMATION OF INORGANIC OXIDES (SHEET 2 OF 9)

Reaction	Temperature Range of Validity	ΔH_0
$2\ Re(c) + 7/2\ 0_7(g) = Re_2O_7(c)$	298.16–569K	–301,470
$2\ V(c) + 3/2\ O_2(g) = V_2O_3(c)$	298.16–2,000K	–299,910
$2\ B(c) + 3/2\ O_2(g) = B_2O_3(gl)$	298.16–723K	–298,670
$2\ Re(c) + 7/2\ 0_7(g) = Re_2O_7(l)$	569–635.5K	–295,810
$Th(c) + O_2(g) = ThO_2(c)$	298.16–2,000K	–294,350
$U(\alpha) + 3/2\ O_2(g) = UO_3$ (hexagonal)	298.16–935K	–294,090
$U(\gamma) + 3/2\ O_2(g) = UO_3$ (hexagonal)	1,045–1,400K	–294,040
$U(\beta) + 3/2\ O_2(g) = UO_3$ (hexagonal)	935–1,045K	–291,870
$2\ Cr(l) + 3/2\ O_2(g) = Cr_2O_3(\beta)$	1,823–2,000K	–278,030
$3\ Fe(\gamma) + 2\ O_2(g) = Fe_3O_4(\beta)$	1,179–1,674K	–276,990
$2\ Cr(c) + 3/2\ O_2(g) = Cr_2O_3(\beta)$	298.16–1,823K	–274,670
$3\ Fe(\alpha) + 2\ O_2(g) = Fe_3O_4(\beta)$	900–1,033K	–272,300
$Hf(c) + O_2(g) = HfO_2$ (monoclinic)	298.16–2,000K	–268,380
$3\ Fe(\alpha) + 2\ O_2(g) = Fe_3O_4$(magnetite)	298.16–900K	–268,310
$U(l) + O_2(g) = UO_2(l)$	1,405–1,500K	–264,790
$Zr(\beta) + O_2(g) = ZrO_2(\alpha)$	1,135–1,478K	–264,190
$3\ Fe(\beta) + 2\ O_2(g) = Fe_3O_4(\beta)$	1,033–1,179K	–262,990
$Zr(\alpha) + O_2(g) = ZrO_2(\alpha)$	298.16–1,135K	–262,980
$U(\alpha) + O_2(g) = UO_2(c)$	298.16–935K	–262,880
$U(\gamma) + O_2(g) = UO_2(c)$	1,045–1,405K	–262,830
$Zr(\beta) + O_2(g) = ZrO_2(\beta)$	1.478–2,000K	–262,290
$U(\beta) + O_2(g) = UO_2(c)$	935–1,045K	–260,660
$Ce(l) + O_2(g) = CeO_2(c)$	1,048–2,000K	–247,930
$Ce(c) + O_2(g) = CeO_2(c)$	298.16–1,048K	–245,490

The ΔH_o values are given in gram calories per mole.

Source: data from *CRC Handbook of Materials Science, Vol II*, Charles T. Lynch, Ed., CRC Press, Cleveland, (1974).

Table 345. SELECTING HEAT OF FORMATION OF INORGANIC OXIDES (SHEET 3 OF 9)

Reaction	Temperature Range of Validity	ΔH_0
$2\ Mn(\alpha) + 3/2\ O_2(g) = Mn_2O_3(c)$	298.16–1,000K	–230,610
$Si(l) + O_2(g) = SiO_2(l)$	1,883–2,000K	–228,590
$Ti(\alpha) + O_2(g) = TiO_2$ (rutile)	1,150–2,000K	–228,380
$Ti(\alpha) + O_2(g) = TiO_2$ (rutile)	298.16–1,150K	–228,360
$2\ As(c) + 5/2\ O_2(g) = As_2O_5(c)$	298.16–883K	–217,080
$Si(c) + O_2(g) = SiO_2(\alpha\text{–quartz})$	298.16–848K	–210,070
$Si(c) + O_2(g) = SiO_2(\beta\text{–quartz})$	848–1,683K	–209,920
$Si(c) + O_2(g) = SiO_2(\beta\text{–cristobalite})$	523–1,683K	–209,820
$Si(c) + O_2(g) = SiO_2(\beta\text{–tridymite})$	390–1,683K	–209,350
$Si(c) + O_2(g) = SiO_2(\alpha\text{–cristobalite})$	298.16–523K	–207,330
$Si(c) + 0_2(g\ ) = SiO_2(\alpha\text{–tridymite})$	298.16–390K	–207,030
$W(c) + 3/2\ O_2(g) = WO_3(l)$	1,743–2,000K	–203,140
$2\ Fe(\alpha) + 3/2\ O_2(g) = Fe_2O_3(\beta)$	950–1,033K	–202,960
$2\ Fe(\gamma) + 3/2\ O_2(g) = Fe_2O_3(\gamma)$	1,179–1,674K	–202,540
$W(c) + 3/2\ O_2(g) = WO_3(c)$	298.16–1,743K	–201,180
$2\ Fe(\alpha) + 3/2\ O_2(g) = Fe_2O_3$(hematite)	298.16–950K	–200,000
$2\ Fe(\beta) + 3/2\ O_2(g) = Fe_2O_3(\beta)$	1,033–1,050K	–196,740
$2\ Fe(\beta) + 3/2\ O_2(g) = Fe_2O_3(\gamma)$	1,050–1,179K	–193,200
$2\ Fe(\alpha) + 3/2\ O_2(g) = Fe_2O_3(\gamma)$	1,674–1,800K	–192,920
$Mo(c) + 3/2\ O_2(g) = MoO_3(c)$	298.16–1,068K	–182,650
$Mg(g) + 1/2\ O_2(g) = MgO$ (periclase)	1,393–2,000K	–180,700
$3\ Pb(c) + 2\ O_2(g) = Pb_3O_4(c)$	298.16–600.5K	–174,920
$2\ Sb(c) + 3/2\ O_2(g) = Sb_2O_3$ (cubic)	298.16–842K	–169,450
$2\ Sb(c) + 3/2\ O_2(g) = Sb_2O_3$ (orthorhombic)	298.16–903K	–168,060

The ΔH_o values are given in **gram calories per mole.**

Source: data from *CRC Handbook of Materials Science, Vol II*, Charles T. Lynch, Ed., CRC Press, Cleveland, (1974).

Table 345. SELECTING HEAT OF FORMATION OF INORGANIC OXIDES (SHEET 4 OF 9)

Reaction	Temperature Range of Validity	ΔH_0
$2\ As(c) + 3/2\ O_2(g) = As_2O_3$ (orthorhombic)	298.16–542K	–154,870
$Ca(\alpha) + 1/2\ O_2(g) = CaO(c)$	298.16–673K	–151,850
$Ca(\beta) + 1/2\ O_2(g) = CaO(c)$	673–1,124K	–151,730
$2\ As(c) + 3/2\ O_2(g) = As_2O_3$ (monoclinic)	298.16–586K	–150,760
$Re(c) + 3/2\ O_2(g) = ReO_3(c)$	298.16–433K	–149,090
$2\ Cs(g) + 3/2\ O_2(g) = Cs_2O_3(l)$	963–1,500K	–148,680
$Re(c) + 3/2\ O_2(g) = ReO_3(l)$	433–1,000K	–146,750
$Mg(l) + 1/2\ O_2(g) = MgO$ (periclase)	923–1,393K	–145,810
$Be(c) + 1/2\ O_2(g) = BeO(c)$	298.16–1,556K	–144,220
$Mg(c) + 1/2\ O_2(g) = MgO$ (periclase)	298.16–923K	–144,090
$Cr(c) + O_2(g) = CrO_2\ (c)$	298.16–1,000K	–142,500
$Sr(c) + 1/2\ O_2(g) = SrO(c)$	298.16–1,043K	–142,410
$2\ Bi(l) + 3/2\ O_2(g) = Bi_2O_3(c)$	544–1,090K	–142,270
$2\ Li(c) + 1/2\ O_2(g) = Li_2O(c)$	298.16–452K	–142,220
$Cr(c) + 3/2\ O_2(g) = CrO_3(c)$	298.16–471K	–141,590
$Cr(c) + 3/2\ O_2(g) = Cr_2O_3(l)$	471–600K	–141,580
$2\ Bi(c) + 3/2\ O_2(g) = Bi_2O_3(c)$	298.16–544K	–139,000
$W(c) + O_2(g) = WO_2(c)$	298.16–1,500K	–137,180
$Ba(\alpha) + 1/2\ O_2(g) = BaO(c)$	298.16–648K	–134,590
$Ba(\beta) + 1/2\ O_2(g) = BaO(c)$	648–977K	–134,140
$2\ K(g) + 1/2\ O_2(g) = K_2O(c)$	1,049–1,500K	–133,090
$Mo(c) + O_2(g) = MoO_2(c)$	298.16–2,000K	–132,910
$Ra(c) + 1/2\ O_2(g) = RaO(c)$	298.16–1,000K	–130,000
$Mn(\alpha) + O_2(g) = MnO_2(c)$	298.16–1,000K	–126,400

The ΔH_o values are given in **gram calories per mole.**

Source: data from *CRC Handbook of Materials Science, Vol II*, Charles T. Lynch, Ed., CRC Press, Cleveland, (1974).

Table 345. SELECTING HEAT OF FORMATION OF INORGANIC OXIDES (SHEET 5 OF 9)

Reaction	Temperature Range of Validity	ΔH_0
$Ti(\alpha) + 1/2\ O_2(g) = TiO(\alpha)$	1,150–1,264K	–125,040
$Ti(\alpha) + 1/2\ O_2(g) = TiO(\alpha)$	298.16–1,150K	–125,010
$2\ Na(c) + O_2(g) = Na_2O_2(c)$	298.16–371K	–122,500
$2\ Cs(l) + 3/2\ O_2(g) = Cs_2O_3(c)$	301.5–775K	–113,840
$2\ Cs(g) + 1/2\ O_2(g) = Cs_2O(l)$	963–1,500K	–113,790
$2\ Cs(c) + 3/2\ O_2(g) = Cs_2O_3(c)$	298.16–301.5K	–112,690
$S(rhombohedral) + 3/2\ O_2(g) = SO_3(c\text{–}I)$	298.16–335.4K	–111,370
$2\ Cs(l) + 3/2\ O_2(g) = Cs_2O_3(l)$	775–963K	–110,740
$1/2\ S_2(g) + 3/2\ O_2(g) = SO_3(g)$	298.16–1,500K	–110,420
$S(rhombohedral) + 3/2\ O_2(g) = SO_3(c\text{–}II)$	298.16–305.7K	–108,680
$S(rhombohedral) + 3/2\ O_2(g) = SO_3(l)$	298.16–335.4K	–107,430
$V(c) + 1/2\ O_2(g) = VO(c)$	298.16–2,000K	–101,090
$2\ Na(l) + 1/2\ O_2(g) = Na_2O(c)$	371–1,187K	–100,150
$2\ Na(c) + 1/2\ O_2(g) = Na_2O(c)$	298.16–371K	–99,820
$2\ Tl(\alpha) + 3/2\ O_2(g) = Tl_2O_3(c)$	298.16–505.5K	–99,410
$S(monoclinic) + 3/2\ O_2(g) = SO_3(g)$	368.6–392K	–95,120
$S(rhombohedral) + 3/2\ O_2(g) = SO_3(g)$	298.16–368.6K	–95,070
$S(l\lambda,\mu) + 3/2\ O_2(g) = SO_3(g)$	392–718K	–94,010
$C(graphite) + O_2(g) = CO_2(g)$	298.16–2,000K	–93,690
$Mn(l) + 1/2\ O_2(g) = MnO(c)$	1,517–2,000K	–93,350
$Mn(\alpha) + 1/2\ O_2(g) = MnO(c)$	298.16–1,000K	–92,600
$Mn(\beta) + 1/2\ O_2(g) = MnO(c)$	1,000–1,374K	–91,900
$Mn(\gamma) + 1/2\ O_2(g) = Mno(c)$	1,374–1,410K	–89,810
$Mn(\delta) + 1/2\ O_2(g) = MnO(c)$	1,410–1,517K	–89,390

The ΔH_o values are given in **gram calories per mole.**

Source: data from *CRC Handbook of Materials Science, Vol II*, Charles T. Lynch, Ed., CRC Press, Cleveland, (1974).

Table 345. SELECTING HEAT OF FORMATION OF INORGANIC OXIDES (SHEET 6 OF 9)

Reaction	Temperature Range of Validity	ΔH_0
$2\,K(l) + 1/2\,O_2(g) = K_2O(c)$	336.4–1,049K	–87,380
$2\,K(c) + 1/2\,O_2(g) = K_2O(c)$	298.16–336.4K	–86,400
$1/2\,S_2(g) + O_2(g) = SO_2(g)$	298.16–2,000K	–86,330
$Zn(c) + 1/2\,O_2(g) = ZnO(c)$	298.16–692.7K	–84,670
$2\,Rb(l) + 1/2\,O_2(g) = Rb_2O(c)$	312.2–750K	–79,950
$2\,Rb(c) + 1/2\,O_2(g) = Rb_2O(c)$	298.16–312.2K	–78,900
$2\,Cs(l) + 1/2\,O_2(g) = Cs_2O(c)$	301.5–763K	–76,900
$2\,Cs(c) + 1/2\,O_2(g) = Cs_2O(c)$	298.16–301.5K	–75,900
$2\,Cs(l) + 1/2\,O_2(g) = Cs_2O(l)$	763–963K	–75,370
$D_2(g) + 1/2\,O_2(g) = D_2O(l)$	298.16–374.5K	–72,760
$S(monoclinic) + O_2(g) = SO_2(g)$	368.6–392K	–71,020
$S(rhombohedral) + O_2(g) = SO_2(g)$	298.16–368.6K	–70,980
$H_2(g) + 1/2\,O_2(g) = H_2O(l)$	298.16–373.16K	–70,600
$S(1\lambda,\mu) + O_2(g) = SO_2(g)$	392–718K	–69,900
$Sn(l) + 1/2\,O_2(g) = SnO(c)$	505–1,300K	–69,670
$Sn(c) + 1/2\,O_2(g) = SnO(c)$	298.16–505K	–68,600
$0.947\,Fc(\beta) + 1/2\,O_2(g) = Fe_{0.947}0(c)$	1,179–1,650K	–66,750
$Pb(c) + O_2(g) = PbO_2(c)$	298.16–600.5K	–66,120
$0.947\,Fe(\alpha) + 1/2\,O_2(g) = Fe_{0.947}0(c)$	298.16–1,033K	–65,320
$0.947\,Fe(\gamma) + 1/2\,O_2(g) = Fe_{0.947}0(l)$	1,650–1,674K	–64,200
$0.947\,Fe(\delta) + 1/2\,O_2(g) = Fe_{0.947}0(l)$	1,803–2,000K	–63,660
$Cd(l) + 1/2\,O_2(g) = CdO(c)$	594–1,038K	–63,240
$0.947\,Fe(\alpha) + 1/2\,O_2(g) = Fe_{0.947}0(c)$	1,033–1,179K	–62,380
$Cd(c) + 1/2\,O_2(g) = CdO(c)$	298.16–594K	–62,330

The ΔH_o values are given in **gram calories per mole.**

Source: data from *CRC Handbook of Materials Science, Vol II*, Charles T. Lynch, Ed., CRC Press, Cleveland, (1974).

Table 345. SELECTING HEAT OF FORMATION OF INORGANIC OXIDES (SHEET 7 OF 9)

Reaction	Temperature Range of Validity	ΔH_0
$0.947\ Fe(\gamma) + 1/2\ O_2(g) = Fe_{0.9470}(l)$	1,647–1,803K	–59,650
$D_2(g) + 1/2\ O_2(g) = D_2O(g)$	298.16–2,000K	–58,970
$Co(\gamma) + 1/2\ O_2(g) = CoO(c)$	1,400–1,763K	–58,160
$I_2(g) + 5/2\ O_2(g) = I_2O_5(c)$	456–500K	–58,020
$Ni(\alpha) + 1/2\ O_2(g) = NiO(c)$	298.16–633K	–57,640
$Ni(\beta) + 1/2\ O_2(g) = NiO(c)$	633–1,725K	–57,460
$H_2(g) + 1/2\ O_2(g) = H_2O(g)$	298.16–2,000K	–56,930
$Co(\alpha,\beta) + 1/2\ O_2(g) = CoO(c)$	298.16–1,400K	–56,910
$Pb(l) + 1/2\ O_2(g) = PbO$ (red)	600.5–762K	–53,780
$Pb(l) + 1/2\ O_2(g) = PbO$ (yellow)	600.5–1,159K	–53,020
$Bi(l) + 1/2\ O_2(g) = BiO(c)$	544–1,600K	–52,920
$Pb(c) + 1/2\ O_2(g) = PbO$ (red)	298.16–600.5K	–52,800
$Pb(c) + 1/2\ O_2(g) = PbO$ (yellow)	298.16–600.5K	–52,040
$Bi(c) + 1/2\ O_2(g) = BiO(c)$	298.16–544K	–50,450
$2\ Tl(\beta) + O_2(g) = Tl_2O(c)$	505.5–573K	–44,260
$2\ Tl(\alpha) + O_2(g) = Tl_2O(c)$	298.16–505.5K	–44,110
$2\ Cu(l) + 1/2\ O_2(g) = Cu_2O(c)$	1,357–1,502K	–43,880
$I_2(l) + 5/2\ O_2(g) = I_2O_5(c)$	386.8–456K	–43,490
$I_2(c) + 5/2\ O_2(g) = I_2O_5(c)$	298.16–386.8K	–42,040
$Cu(l) + 1/2\ O_2(g) = CuO(l)$	1,720–2,000K	–41,060
$Ir(c) + O_2(g) = IrO_2(c)$	298.16–1,300K	–39,480
$Cu(l) + 1/2\ O_2(g) = CuO(c)$	1,357–1,720K	–39,410
$2\ Al(l) + 1/2\ O_2(g) = Al_2O(g)$	931.7–2,000K	–38,670
$Cu(c) + 1/2\ O_2(g) = CuO(c)$	298.16–1,357K	–37,740

The ΔH_o values are given in **gram calories per mole.**

Source: data from *CRC Handbook of Materials Science, Vol II*, Charles T. Lynch, Ed., CRC Press, Cleveland, (1974).

Table 345. SELECTING HEAT OF FORMATION OF INORGANIC OXIDES (SHEET 8 OF 9)

Reaction	Temperature Range of Validity	ΔH_0
$2\ Cu(l) + 1/2\ O_2(g) = Cu_2O(l)$	1,502–2,000K	–37,710
$2\ Al(c) + 1/2\ O_2(g) = Al_2O(g)$	298.16–931.7K	–31,660
$Si(l) + 1/2\ O_2(g) = SiO(g)$	1,683–2,000K	–30,170
$C(graphite) + 1/2\ O_2(g) = CO(g)$	298.16–2,000K	–25,400
$2\ Hg(l) + 1/2\ O_2(g) = Hg_2O(c)$	298.16–629.88K	–22,400
$Hg(l) + 1/2\ O_2(g) = HgO\ (red)$	298.16–629.88K	–21,760
$Si(c) + 1/2\ O_2(g) = SiO(g)$	298.16–1,683K	–21,090
$P(l) + 1/2\ O_2(g) = PO(g)$	317.4–553K	–9,390
$P\ (white) + 1/2\ O_2(g) = PO(g)$	298.16–317.4K	–9,370
$2\ Ag(c) + 1/2\ O_2(g) = Ag_2O_2(c)$	298.16–1,000K	–7,740
$1/2\ Se_2(g) + 1/2\ O_2(g) = SeO(g)$	1,027–2,000K	–7,400
$2\ Ag(c) + O_2(g) = Ag_2O_2(c)$	298.16–500K	–6,620
$2\ Au(c) + 3/2\ O_2(g) = Au_2O_3(c)$	298.16–500K	–2,160
$1/2\ S_2\ (g) + 1/2\ O_2(g) = SO(g)$	298.16–2,000K	+3,890
$Al(l) + 1/2\ O_2(g) = AlO(g)$	931.7–2,000K	+8,170
$Se(c) + 1/2\ O_2(g) = SeO(g)$	298.16–490K	+9,280
$Se(l) + 1/2\ O_2(g) = SeO(g)$	490–1,027K	+9,420
$2\ Cu(c) + 1/2\ O_2(g) = Cu_2O(c)$	298.16–1,357K	+10,550
$Al(c) + 1/2\ O_2(g) = AlO(g)$	298.16–931.7K	+10,740
$Cl_2(g) + 1/2\ O_2(g) = Cl_2O(g)$	298.16–2,000K	+17,770
$S(monoclmic) + 1/2\ O_2(g) = SO(g)$	368.6–392K	+19,200
$S(rhombohedral) + 1/2\ O_2(g) = SO(g)$	298.16–368.6K	+19,250
$S(l\lambda,\mu\ ) + 1/2\ O_2(g) = SO(g)$	392–718K	+20,320
$1/2\ Cl_2(g) + 1/2\ O_2(g) = ClO(g)$	298.16–1,000K	+33,000

The ΔH_o values are given in **gram calories per mole.**

Source: data from *CRC Handbook of Materials Science, Vol II*, Charles T. Lynch, Ed., CRC Press, Cleveland, (1974).

Table 345. SELECTING HEAT OF FORMATION OF INORGANIC OXIDES (SHEET 9 OF 9)

Reaction	Temperature Range of Validity	ΔH_0
$3/2\ O_2(g) = O_3(g)$	298.16–2,000K	+33,980
$2\ Cl_2(g) + 3/2\ O_2(g) = ClO(g)$	298.16–500K	+37,740
$Te(l) + 1/2\ O_2(g) = TeO(g)$	723–1,360K	+39,750
$Te(c) + 1/2\ O_2(g) = TeO(g)$	298.16–723K	+43,110
$V(c) + 1/2\ O_2(g) = VO(g)$	298.16–2,000K	+52,090

The ΔH_o values are given in **gram calories per mole.**

Source: data from *CRC Handbook of Materials Science, Vol II*, Charles T. Lynch, Ed., CRC Press, Cleveland, (1974).

Table 346. SELECTING SPECIFIC HEAT OF ELEMENTS

(SHEET 1 OF 4)

Element	C_p at 25 °C ($cal \cdot g^{-1} \cdot K^{-1}$)
Radon	0.0224
Thorium	0.0271
Uranium	0.0276
Radium	0.0288
Protactinium	0.029
Bismuth	0.0296
Polonium	0.030
Thallium	0.0307
Gold	0.0308
Osmium	0.03127
Iridium	0.0317
Platinum	0.0317
Tungsten	0.0317
Rhenium	0.0329
Mercury	0.0331
Tantalum	0.0334
Ytterbium	0.0346
Hafnium	0.035
Lutetium	0.037
Xenon	0.0378
Lead	0.038
Thulium	0.0382
Hollnium	0.0393
Erbium	0.0401

See also: Thermodynamic Coefficients of the Elements.

Source: data from Weast, R. C., Ed., *Handbook of Chemistry and Physics, 55th ed.*, CRC Press, Cleveland, 1974, D-144., Kelly, K. K., *Bulletin 592*, Bureau of Mines, Washington, D. C., 1961.and Hultgren, R., Orr, R L., Anderson, P. D., and Kelly, K. K., *Selected Values of Thermodynamic Properties of Metals and Alloys*, John Wiley & Sons, New York, (1963).

Table 346. SELECTING SPECIFIC HEAT OF ELEMENTS

(SHEET 2 OF 4)

Element	C_p at 25 °C ($cal \cdot g^{-1} \cdot K^{-1}$)
Dysprosium	0.0414
Europium	0.0421
Samarium	0.043
Terbium	0.0437
Promethium	0.0442
Barium	0.046
Praseodymium	0.046
Lanthanum	0.047
Tellurium	0.0481
Antimony	0.049
Cerium	0.049
Neodymium	0.049
Tin (α)	0.0510
Tin (β)	0.0530
Gadolinium	0.055
Cadmium	0.0555
Indium	0.056
Silver	0.0566
Cesium	0.057
Ruthenium	0.057
Technetium	0.058
Rhodium	0.0583
Palladium	0.0584
Krypton	0.059
Niobium	0.064
Zirconium	0.0671
Yttrium	0.068
Strontium	0.0719

See also: Thermodynamic Coefficients of the Elements.

Source: data from Weast, R. C., Ed., *Handbook of Chemistry and Physics, 55th ed.*, CRC Press, Cleveland, 1974, D-144., Kelly, K. K., *Bulletin 592*, Bureau of Mines, Washington, D. C., 1961.and Hultgren, R., Orr, R L., Anderson, P. D., and Kelly, K. K., *Selected Values of Thermodynamic Properties of Metals and Alloys*, John Wiley & Sons, New York, (1963).

Table 346. SELECTING SPECIFIC HEAT OF ELEMENTS
(SHEET 3 OF 4)

Element	C_p at 25 °C ($cal \cdot g^{-1} \cdot K^{-1}$)
Selenium (Se_2)	0.0767
Germanium	0.077
Arsenic	0.0785
Rubidium	0.0861
Gallium	0.089
Copper	0.092
Zinc	0.0928
Iodine (I_2)	0.102
Iron (α)	0.106
Nickel	0.106
Chromium	0.107
Cobalt	0.109
Bromine (Br_2)	0.113
Chlorine (Cl_2)	0.114
Manganese, a	0.114
Vanadium	0.116
Argon	0.124
Carbon, diamond	0.124
Titanium	0.125
Scandium	0.133
Calcium	0.156
Phosphorus, red, triclinic	0.160
Silicon	0.168
Carbon, graphite	0.170

See also: Thermodynamic Coefficients of the Elements.

Source: data from Weast, R. C., Ed., *Handbook of Chemistry and Physics, 55th ed.*, CRC Press, Cleveland, 1974, D-144., Kelly, K. K., *Bulletin 592*, Bureau of Mines, Washington, D. C., 1961.and Hultgren, R., Orr, R L., Anderson, P. D., and Kelly, K. K., *Selected Values of Thermodynamic Properties of Metals and Alloys*, John Wiley & Sons, New York, (1963).

Table 346. SELECTING SPECIFIC HEAT OF ELEMENTS
(SHEET 4 OF 4)

Element	C_p at 25 °C ($cal \cdot g^{-1} \cdot K^{-1}$)
Sulfur, yellow	0.175
Potassium	0.180
Phosphorus, white	0.181
Fluorine (F_2)	0.197
Aluminum	0.215
Oxygen (O_2)	0.219
Magnesium	0.243
Boron	0.245
Neon	0.246
Nitrogen (N_2)	0.249
Sodium	0.293
Beryllium	0.436
Molybdenum	0.599
Lithium	0.85
Manganese (β)	1.119
Helium	1.24
Hydrogen (H_2)	3.41

See also: Thermodynamic Coefficients of the Elements.

Source: data from Weast, R. C., Ed., *Handbook of Chemistry and Physics, 55th ed.*, CRC Press, Cleveland, 1974, D-144., Kelly, K. K., *Bulletin 592*, Bureau of Mines, Washington, D. C., 1961.and Hultgren, R., Orr, R L., Anderson, P. D., and Kelly, K. K., *Selected Values of Thermodynamic Properties of Metals and Alloys*, John Wiley & Sons, New York, (1963).

Table 347. SELECTING SPECIFIC HEAT OF POLYMERS
(SHEET 1 OF 3)

Polymer	Specific Heat (Btu/lb/•F)
Polymide: Glass reinforced	0.15—0.27
Reinforced polyester moldings: Sheet molding compounds, general purpose	0.20—0.25
Standard Epoxies: High strength laminate	0.21
Polytrifluoro chloroethylene (PTFCE)	0.22
Silicone: Woven glass fabric/ silicone laminate	0.246
Phenylene oxides (Noryl): Standard	0.24
Standard Epoxies: Filament wound composite	0.24
Polystyrenes; Molded: Glass fiber -30% reinforced	0.256
Polytetrafluoroethylene (PTFE)	0.25
Polymide: Unreinforced	0.25—0.35
Reinforced polyester moldings: High strength (glass fibers)	0.25—0.35
Polyphenylene sulfide: Standard	0.26
Phenolics; Molded; General: Arc resistant—mineral filled	0.27—0.37
Fluorinated ethylene propylene(FEP)	0.28
Nylon, Type 6: Type 12	0.28
Phenolics; Molded; General: Very high shock: glass fiber filled	0.28—0.32
PVC–acrylic sheet	0.293
Phenolics; Molded; General: High shock: chopped fabric or cord filled	0.30—0.35
Polystyrenes; Molded: General purpose	0.30—0.35
Polystyrenes; Molded: High impact	0.30—0.35
Polystyrenes; Molded: Medium impact	0.30—0.35
Polyesters: Thermoset Cast; Rigid	0.30—0.55
Vinylidene chloride	0.32
Polyvinylidene— fluoride (PVDF)	0.33
Rubber phenolic—woodflour or flock	0.33
Styrene acrylonitrile (SAN)	0.33
Acrylic Moldings: High impact grade	0.34
Acrylic Moldings: Grades 5, 6, 8	0.35

Source: data compiled by J.S. Park from Charles T. Lynch, *CRC Handbook of Materials Science, Vol. 3*, CRC Press, Boca Raton, Florida and *Engineered Materials Handbook, Vol.2*, Engineering Plastics, ASM International, Metals Park, Ohio, 1988.

Table 347. SELECTING SPECIFIC HEAT OF POLYMERS

(SHEET 2 OF 3)

Polymer	Specific Heat (Btu/lb/•F)
Acrylics; Cast Resin Sheets, Rods: General purpose, type I	0.35
Acrylics; Cast Resin Sheets, Rods: General purpose, type II	0.35
Polyacetal Copolymer: High flow	0.35
Polyacetal Copolymer: Standard	0.35
Polyacetal: Standard	0.35
ABS Resins; Molded, Extruded; Low temperature impact	0.35—0.38
Phenolics; Molded; General: woodflour and flock filled	0.35—0.40
ABS Resins; Molded, Extruded; High impact	0.36—0.38
ABS Resins; Molded, Extruded; Medium impact	0.36—0.38
ABS Resins; Molded, Extruded; Very high impact	0.36—0.38
ABS Resins; Molded, Extruded; Heat resistant	0.37—0.39
Chlorinated polyvinyl chloride	0.3
Polycarbonate	0.3
Thermoset Carbonate: Allyl diglycol carbonate	0.3
Cellulose Acetate; Molded, Extruded; ASTM Grade: H2—1	0.3—0.42
Cellulose Acetate; Molded, Extruded; ASTM Grade: H4—1	0.3—0.42
Cellulose Acetate; Molded, Extruded; ASTM Grade: H6—1	0.3—0.42
Cellulose Acetate; Molded, Extruded; ASTM Grade: MH—1, MH—2	0.3—0.42
Cellulose Acetate; Molded, Extruded; ASTM Grade: MS—1, MS—2	0.3—0.42
Cellulose Acetate; Molded, Extruded; ASTM Grade: S2—1	0.3—0.42
Cellulose Acetate Butyrate; Molded, Extruded; ASTM Grade: H4	0.3—0.4
Cellulose Acetate Butyrate; Molded, Extruded; ASTM Grade: MH	0.3—0.4
Cellulose Acetate Butyrate; Molded, Extruded; ASTM Grade: S2	0.3—0.4
Cellusose Acetate Propionate; Molded, Extruded; ASTM Grade: 1	0.3—0.4
Cellusose Acetate Propionate; Molded, Extruded; ASTM Grade: 3	0.3—0.4
Cellusose Acetate Propionate; Molded, Extruded; ASTM Grade: 6	0.3—0.4
6/10 Nylon: General purpose	0.3—0.5
6/6 Nylon: General purpose extrusion	0.3—0.5

Source: data compiled by J.S. Park from Charles T. Lynch, *CRC Handbook of Materials Science, Vol. 3,* CRC Press, Boca Raton, Florida and *Engineered Materials Handbook, Vol.2,* Engineering Plastics, ASM International, Metals Park, Ohio, 1988.

Table 347. SELECTING SPECIFIC HEAT OF POLYMERS
(SHEET 3 OF 3)

Polymer	Specific Heat (Btu/lb/•F)
6/6 Nylon: General purpose molding	0.3—0.5
Standard Epoxies: Cast rigid	0.4-0.5
Polypropylene: General purpose	0.45
Polypropylene: High impact	0.45—0.48
Polyethylenes; Molded, Extruded; Type III: Melt index 0.2—0.9	0.46—0.55
Polyethylenes; Molded, Extruded; Type III: Melt index 0.l—12.0	0.46—0.55
Polyethylenes; Molded, Extruded; Type III: Melt index 1.5—15	0.46—0.55
Nylon, Type 6: Cast	0.4
Nylon, Type 6: General purpose	0.4
Nylon, Type 6: Type 8	0.4
Polyethylenes; Molded, Extruded; Type I: Melt index 0.3—3.6	0.53—0.55
Polyethylenes; Molded, Extruded; Type I: Melt index 200	0.53—0.55
Polyethylenes; Molded, Extruded; Type I: Melt index 6—26	0.53—0.55
Polyethylenes; Molded, Extruded; Type II: Melt index 20	0.53—0.55
Polyethylenes; Molded, Extruded; Type II: Melt index l.0—1.9	0.53—0.55
Nylon, Type 6: Type 11	0.58

Source: data compiled by J.S. Park from Charles T. Lynch, *CRC Handbook of Materials Science, Vol. 3*, CRC Press, Boca Raton, Florida and *Engineered Materials Handbook, Vol.2*, Engineering Plastics, ASM International, Metals Park, Ohio, 1988.

Table 348. SELECTING MELTING POINTS OF THE ELEMENTS

(SHEET 1 OF 4)

At. No.	Element	Sym.	Melting Point (°C)
2	Helium	He	-272.2
1	Hydrogen	H	-259.14
10	Neon	N	-248.67
9	Fluorine	F	-219.62
8	Oxygen	O	-218.4
7	Nitrogen	N	-209.86
18	Argon	Ar	-189.2
36	Krypton	Kr	-156.6
54	Xenon	Xe	-111.9
17	Chlorine	Cl	-100.98
86	Radon	Rn	-71
80	Mercury	Hg	-38.87
35	Bromine	Br	-7.2
56	Barium	Ba	7.25
87	Francium	Fr	~27
55	Cesium	Ce	28.4
31	Gallium	Ga	29.78
37	Rubidium	Rb	38.89
15	Phosphorus (White)	P	44.1
19	Potassium	K	63.65
11	Sodium	Na	97.81
16	Sulfur	S	112.8
53	Iodine	I	113.5
49	Indium	In	156.61
3	Lithium	Li	180.54
34	Selenium	Se	217
50	Tin	Sn	231.9681

Source: data from James F. Shackelford, *Introduction to Materials Science for Engineers, Second Edition*, Macmillian Publishing Company, New York, pp.686-688, (1988).

Table 348. SELECTING MELTING POINTS OF THE ELEMENTS
(SHEET 2 OF 4)

At. No.	Element	Sym.	Melting Point (°C)
84	Polonium	Po	254
83	Bismuth	Bi	271.3
85	Asatine	At	302
81	Thallium	Tl	303.5
48	Cadmium	Cd	320.9
82	Lead	Pb	327.502
30	Zinc	Zn	419.58
52	Tellurium	Te	449.5
51	Antimony	Sb	630.74
93	Neptunium	Np	640
94	Plutonium	Pu	641
12	Magnesium	Mg	648.8
13	Aluminum	Al	660.37
88	Radium	Ra	700
38	Strontium	Sr	769
58	Cerium	Ce	798
33	Arsenic	As	817
63	Europium	Eu	822
70	Ytterbium	Yb	824
20	Calcium	Ca	839
57	Lanthanum	La	920
59	Praseodymium	Pr	931
32	Germanium	Ge	937.4
47	Silver	Ag	961.93
95	Americium	Am	994
60	Neodymium	Nd	1010
89	Actinium	Ac	1050
79	Gold	Au	1064.43

Source: data from James F. Shackelford, *Introduction to Materials Science for Engineers, Second Edition*, Macmillian Publishing Company, New York, pp.686-688, (1988).

Table 348. SELECTING MELTING POINTS OF THE ELEMENTS

(SHEET 3 OF 4)

At. No.	Element	Sym.	Melting Point (°C)
62	Samarium	Sm	1072
61	Promethium	Pm	~1080
29	Copper	Cu	1083.4
92	Uranium	U	1132
25	Manganese	Mn	1244
4	Beryllium	Be	1278
64	Gadolinium	Gd	1311
96	Curium	Cm	1340
65	Terbium	Tb	1360
66	Dysprosium	Dy	1409
14	Silicon	Si	1410
28	Nickel	Ni	1453
67	Holmium	Ho	1470
27	Cobalt	Co	1495
68	Erbium	Er	1522
39	Yttrium	Y	1523
26	Iron	Fe	1535
21	Scandium	Sc	1539
69	Thulium	Tm	1545
46	Palladium	Pd	1552
91	Protoactinium	Pa	<1600
71	Lutetium	Lu	1659
22	Titanium	Ti	1660
90	Thorium	Th	1750
78	Platinum	Pt	1772
40	Zirconium	Zr	1852
24	Chromium	Cr	1857
23	Vanadium	V	1890

Source: data from James F. Shackelford, *Introduction to Materials Science for Engineers, Second Edition*, Macmillian Publishing Company, New York, pp.686-688, (1988).

Table 348. SELECTING MELTING POINTS OF THE ELEMENTS
(SHEET 4 OF 4)

At. No.	Element	Sym.	Melting Point (°C)
45	Rhodium	Rh	1966
43	Technetium	Tc	2172
72	Hafnium	Hf	2227
5	Boron	B	2300
44	Ruthenium	Ru	2310
41	Niobium	Nb	2408
77	Iridium	Ir	2410
42	Molybdenum	Mo	2617
73	Tantalum	Ta	2996
76	Osmium	Os	3045
75	Rhenium	Re	3180
74	Tungsten	W	3410
6	Carbon	C	~3550

Source: data from James F. Shackelford, *Introduction to Materials Science for Engineers, Second Edition*, Macmillian Publishing Company, New York, pp.686-688, (1988).

Table 349. SELECTING MELTING POINTS OF ELEMENTS AND INORGANIC COMPOUNDS (SHEET 1 OF 12)

Compound	Formula	Melting Point •C
Hydrogen	H_2	–259.25
Neon	Ne	–248.6
Fluorine	F_2	–219.6
Oxygen	O_2	–218.8
Nitrogen	N_2	–210
Carbon monoxide	CO	–205
Nitric oxide	NO	–163.7
Boron trifluoride	BF_3	–128.0
Hydrogen chloride	HCl	–114.3
Xenon	Xe	–111.6
Boron trichloride	BCl_3	–107.8
Chlorine	Cl_2	–103±5
Nitrous oxide	N_2O	–90.9
Hydrogen sulfide, di–	H_2S_2	–89.7
Hydrogen bromide	HBr	–86.96
Hydrogen sulfide	H_2S	–85.6
Sulfur dioxide	SO_2	–73.2
Silicon tetrachloride	$SiCl_4$	–67.7
Bromine pentafluoride	BrF_5	–61.4
Carbon dioxide	CO_2	–57.6
Hydrogen iodide	HI	–50.91
Hydrogen telluride	H_2Te	–49.0
Boron tribromide	BBr_3	–48.8
Hydrogen nitrate	HNO_3	–47.2

Source: data from: Weast, R C., Ed., *Handbook of Chemistry and Physics, 55th ed.*, CRC Press, Cleveland, (1974); and Bolz, R. E. and Tuve, G. L., Eds., *Handbook of Tables for Applied Engineering Science, 2nd ed.*, CRC Press, Cleveland, (1973), p.479 .

Table 349. SELECTING MELTING POINTS OF ELEMENTS AND INORGANIC COMPOUNDS (SHEET 2 OF 12)

Compound	Formula	Melting Point •C
Mercury	Hg	–39
Tin chloride,tetra–	$SnCl_4$	–33.3
Silane, hexafluoro–	Si_2F_6	–28.6
Cyanogen	C_2N_2	–27.2
Titanium chloride, tetra–	$TiCl_4$	–23.2
Iron pentacarbonyl	$Fe(CO)_5$	–21.2
Arsenic trichloride	$AsCl_3$	–16.0
Nitrogen tetroxide	N_2O_4	–13.2
Bromine	Br_2	–7.2
Arsenic trifluoride	AsF_3	–6.0
Cyanogen chloride	CNCl	–5.2
Hydrogen peroxide	H_2O_2	–0.7
Tungsten hexafluoride	WF_6	–0.5
Hydrogen oxide (water)	H_2O	0
Phosphorus oxychloride	$POCl_3$	1.0
Deuterium oxide	D_2O	3.78
Antimony pentachloride	$SbCl_5$	4.0
Seleniumoxychloride	$SeOCl_3$	9.8
Hydrogen sulfate	H_2SO_4	10.4
lodine chloride (β)	ICl	13.8
Sulfur trioxide (α)	SO_3	16.8
Molybdenum hexafluoride	MoF_6	17
lodine chloride (α)	ICl	17.1
Phosphorus acid, hypo–	H_3PO_2	17.3

Source: data from: Weast, R C., Ed., *Handbook of Chemistry and Physics, 55th ed.*, CRC Press, Cleveland, (1974); and Bolz, R. E. and Tuve, G. L., Eds., *Handbook of Tables for Applied Engineering Science, 2nd ed.*, CRC Press, Cleveland, (1973), p.479 .

Table 349. SELECTING MELTING POINTS OF ELEMENTS AND INORGANIC COMPOUNDS (SHEET 3 OF 12)

Compound	Formula	Melting Point •C
Rhenium hexafluoride	ReF_6	19.0
Niobium pentachloride	$NbCl_5$	21.1
Phosphorus trioxide	P_4O_6	23.7
Cesium	Cs	28.3
Gallium	Ga	29
Tin bromide, tetra–	$SnBr_4$	29.8
Arsenic tribromide	$AsBr_3$	30.0
Sulfur trioxide (β)	SO_3	32.3
Titanium bromide, tetra–	$TiBr_4$	38
Cesium chloride	CsCl	38.5
Rubidium	Rb	38.9
Osmium tetroxide (white)	OsO_4	41.8
Phosphoric acid	H_3PO_4	42.3
Phosphorus, yellow	P_4	44.1
Phosphoric acid. hypo–	$H_4P_2O_6$	54.8
Osmium tetroxide (yellow)	OsO_4	55.8
Hydrogen selenate	H_2SeO_4	57.8
Sulfur trioxide (γ)	SO_3	62.1
Potassium	K	63.4
Antimony trichloride	$SbCl_3$	73.3
Phosphorus acid, ortho–	H_3PO_3	73.8
Arsenic pentafluoride	AsF_5	80.8
Hydrogen fluoride	HF	83.11
Aluminum bromide	Al_2Br_6	87.4

Source: data from: Weast, R C., Ed., *Handbook of Chemistry and Physics, 55th ed.*, CRC Press, Cleveland, (1974); and Bolz, R. E. and Tuve, G. L., Eds., *Handbook of Tables for Applied Engineering Science, 2nd ed.*, CRC Press, Cleveland, (1973), p.479 .

Table 349. SELECTING MELTING POINTS OF ELEMENTS AND INORGANIC COMPOUNDS (SHEET 4 OF 12)

Compound	Formula	Melting Point •C
Antimony tribromide	$SbBr_3$	96.8
Sodium	Na	97.8
Iodine	I_2	112.9
Sulfur (monatomic)	S	119
Tin iodide, tetra–	SnI_4	143.4
Indium	In	156.3
Lithium	Li	178.8
Potassium thiocyanate	KSCN	179
Argon	Ar	190.2
Aluminum iodide	Al_2I_6	190.9
Aluminum chloride	Al_2Cl_6	192.4
Chromium trioxide	CrO_3	197
Tantalum pentachloride	$TaCl_5$	206.8
Thallium nitrate	$TINO_3$	207
Silver nitrate	$AgNO_3$	209
Selenium	Se	217
Bismuth trichloride	$BiCl_3$	223.8
Tin	Sn	231.7
Tin bromide, di–	$SnBr_2$	231.8
Mercury bromide	$HgBr_2$	241
Tin chloride, di–	$SnCl_2$	247
Lithium nitrate	$LiNO_3$	250
Mercury iodide	HgI_2	250
Sodium chlorate	$NaClO_3$	255
Bismuth	Bi	271
Thallium carbonate	Tl_2CO_3	273
Mercury chloride	$HgCl_2$	276.8
Zincchloride	$ZnCl_2$	283

Source: data from: Weast, R C., Ed., *Handbook of Chemistry and Physics, 55th ed.*, CRC Press, Cleveland, (1974); and Bolz, R. E. and Tuve, G. L., Eds., *Handbook of Tables for Applied Engineering Science, 2nd ed.*, CRC Press, Cleveland, (1973), p.479 .

Table 349. SELECTING MELTING POINTS OF ELEMENTS AND INORGANIC COMPOUNDS (SHEET 5 OF 12)

Compound	Formula	Melting Point •C
Rhenium heptoxide	Re_2O_7	296
Thallium	Tl	302.4
Iron (III) chloride	Fe_2Cl_6	303.8
Rubidium nitrate	$RbNO_3$	305
Sodium nitrate	$NaNO_3$	310
Arsenic trioxide	As_4O_6	312.8
Cadmium	Cd	320.8
Sodium hydroxide	NaOH	322
Sodium thiocyanate	NaSCN	323
Tungsten tetrachloride	WCl_4	327
Lead	Pb	327.3
Potassium nitrate	KNO_3	338
Silver cyanide	AgCN	350
Potassium hydroxide	KOH	360
Cadmium iodide	CdI_2	386.8
Potassium dichromate	$K_2Cr_2O_7$	398
Beryllium chloride	$BeCl_2$	404.8
Cesium nitrate	$CsNO_3$	406.8
Lead iodide	PbI_2	412
Zinc	Zn	419.4
Thallium chloride, mono–	TICl	427
Copper (I) chloride	CuCl	429
Copper (II) chloride	$CuCl_2$	430
Silver bromide	AgBr	430
Lithium iodide	LiI	440
Thallium iodide, mono–	TlI	440
Boron trioxide	B_2O_3	448.8
Thallium sulfide	Tl_2S	449

Source: data from: Weast, R C., Ed., *Handbook of Chemistry and Physics, 55th ed.*, CRC Press, Cleveland, (1974); and Bolz, R. E. and Tuve, G. L., Eds., *Handbook of Tables for Applied Engineering Science, 2nd ed.*, CRC Press, Cleveland, (1973), p.479 .

Table 349. SELECTING MELTING POINTS OF ELEMENTS AND INORGANIC COMPOUNDS (SHEET 6 OF 12)

Compound	Formula	Melting Point •C
Tellurium	Te	453
Silver chloride	AgCl	455
Sodium peroxide	Na_2O_2	460
Thallium bromide, mono–	TlBr	460
Lithium hydroxide	LiOH	462
Copper(l) cyanide	$Cu_2(CN)_2$	473
Beryllium bromide	$BeBr_2$	487.8
Lead bromide	$PbBr_2$	487.8
Potassium peroxide	K_2O_2	490
Antimony trisulfide	Sb_4S_6	546.0
Lithium bromide	LiBr	552
Silver iodide	Agl	557
Calcium nitrate	$Ca(NO_3)_2$	560.8
Sodium cyanide	NaCN	562
Cadmium bromide	$CdBr_2$	567.8
Cadmium chloride	$CdCl_2$	567.8
Phosphorus pentoxide	P_4O_{10}	569.0
Copper (I) iodide	CuI	587
Uranium tetrachloride	UCl_4	590
Barium nitrate	$Ba(NO_3)_2$	594.8
Lithium chloride	LiCl	614
Europium trichloride	$EuCl_3$	622
Potassium cyanide	KCN	623
Antimony	Sb	630
Thallium sulfate	Tl_2SO_4	632
Rubidium iodide	RbI	638
Strontium bromide	$SrBr_2$	643
Magnesium	Mg	650

Source: data from: Weast, R C., Ed., *Handbook of Chemistry and Physics, 55th ed.*, CRC Press, Cleveland, (1974); and Bolz, R. E. and Tuve, G. L., Eds., *Handbook of Tables for Applied Engineering Science, 2nd ed.*, CRC Press, Cleveland, (1973), p.479 .

Table 349. SELECTING MELTING POINTS OF ELEMENTS AND INORGANIC COMPOUNDS (SHEET 7 OF 12)

Compound	Formula	Melting Point •C
Manganese dichloride	$MnCl_2$	650
Antimony trioxide	Sb_4O_6	655.0
Silver sulfate	Ag_2SO_4	657
Aluminum	Al	658.5
Sodium iodide	NaI	662
Vanadium pentoxide	V_2O_5	670
Iron (II) chloride	$FeCl_2$	677
Rubidium bromide	RbBr	677
Potassium iodide	KI	682
Sodium molybdate	Na_2MoO_4	687
Sodium tungstate	Na_2WO_4	702
Lithium molybdate	Li_2MoO_4	705
Barium iodide	BaI_2	710.8
Magnesium bromide	$MgBr_2$	711
Magnesium chloride	$MgCl_2$	712
Rubidium chloride	RbCl	717
Barium	Ba	725
Bismuth trifluoride	BiF_3	726.0
Molybdenum dichloride	$MoCl_2$	726.8
Cobalt (II) chloride	$CoCl_2$	727
Zirconium dichloride	$ZrCl_2$	727
Calcium bromide	$CaBr_2$	729.8
Lithium tungstate	Li_2WO_4	742
Potassium bromide	KBr	742
Sodium bromide	NaBr	747
Strontium	Sr	757
Thorium chloride	$ThCl_4$	765
Potassium chloride	KCl	770

Source: data from: Weast, R C., Ed., *Handbook of Chemistry and Physics, 55th ed.*, CRC Press, Cleveland, (1974); and Bolz, R. E. and Tuve, G. L., Eds., *Handbook of Tables for Applied Engineering Science, 2nd ed.*, CRC Press, Cleveland, (1973), p.479 .

Table 349. SELECTING MELTING POINTS OF ELEMENTS AND INORGANIC COMPOUNDS (SHEET 8 OF 12)

Compound	Formula	Melting Point •C
Cerium	Ce	775
Calcium chloride	$CaCl_2$	782
Nickel subsulfide	Ni_3S_2	790
Molybdenum trioxide	MoO_3	795
Sodium chloride	NaCl	800
Chromium (II) chloride	$CrCl_2$	814
Bismuth trioxide	Bi_2O_3	815.8
Arsenic	As	816.8
Lead fluoride	PbF_2	823
Ytterbium	Yb	823
Europium	Eu	826
Rubidium fluoride	RbF	833
Silver sulfide	Ag_2S	841
Barium bromide	$BaBr_2$	846.8
Mercury sulfate	$HgSO_4$	850
Calcium	Ca	851
Sodium carbonate	Na_2CO_3	854
Lithium sulfate	Li_2SO_4	857
Strontium chloride	$SrCl_2$	872
Potassium fluoride	KF	875
Sodium silicate, di–	$Na_2Si_2O_5$	884
Sodium sulfate	Na_2SO_4	884
Lead oxide	PbO	890
Lithium fluoride	LiF	896
Potassium carbonate	K_2CO_3	897
Lanthanum	La	920
Sodium sulfide	Na_2S	920
Praseodymium	Pr	931

Source: data from: Weast, R C., Ed., *Handbook of Chemistry and Physics, 55th ed.*, CRC Press, Cleveland, (1974); and Bolz, R. E. and Tuve, G. L., Eds., *Handbook of Tables for Applied Engineering Science, 2nd ed.*, CRC Press, Cleveland, (1973), p.479 .

Table 349. SELECTING MELTING POINTS OF ELEMENTS AND INORGANIC COMPOUNDS (SHEET 9 OF 12)

Compound	Formula	Melting Point •C
Potassium borate, meta–	KBO_2	947
Germanium	Ge	959
Barium chloride	$BaCl_2$	959.8
Silver	Ag	961
Sodium borate, meta–	$NaBO_2$	966
Sodium pyrophosphate	$Na_4P_2O_7$	970
Potassium chromate	K_2CrO_4	984
Sodium phosphate, meta–	$NaPO_3$	988
Titanium oxide	TiO	991
Sodium fluoride	NaF	992
Cadmium sulfate	$CdSO_4$	1000
Neodymium	Nd	1020
Vanadium dichloride	VCl_2	1027
Nickel chloride	$NiCl_2$	1030
Tin oxide	SnO	1042
Actinium227	Ac	1050±50
Gold	Au	1063
Lead molybdate	$PbMoO_4$	1065
Samarium	Sm	1072
Potassium sulfate	K_2SO_4	1074
Copper	Cu	1083
Lead sulfate	$PbSO_4$	1087
Sodium silicate, meta–	Na_2SiO_3	1087
Potassium pyro–phosphate	$K_4P_2O_7$	1092
Sodiumsilicate,aluminum–	$NaAlSi_3O_8$	1107
Cadmium fluoride	CdF_2	1110
Lead sulfide	PbS	1114
Copper (I) sulfide	Cu_2S	1129

Source: data from: Weast, R C., Ed., *Handbook of Chemistry and Physics, 55th ed.*, CRC Press, Cleveland, (1974); and Bolz, R. E. and Tuve, G. L., Eds., *Handbook of Tables for Applied Engineering Science, 2nd ed.*, CRC Press, Cleveland, (1973), p.479 .

Table 349. SELECTING MELTING POINTS OF ELEMENTS AND INORGANIC COMPOUNDS (SHEET 10 OF 12)

Compound	Formula	Melting Point •C
Uranium^{235}	U	~1133
Lithium metasilicate	Li_2SiO_3	1177
Iron (II) sulfide	FeS	1195
Manganese	Mn	1220
Magnesium fluoride	MgF_2	1221
Iron carbide	Fe_3C	1226.8
Copper (I) oxide	Cu_2O	1230
Lithium orthosilicate	Li_4SiO_4	1249
Tungsten dioxide	WO_2	1270
Manganese metasilicate	$MnSiO_3$	1274
Beryllium	Be	1278
Calcium carbonate	$CaCO_3$	1282
Barium fluoride	BaF_2	1286.8
Calcium sulfate	$CaSO_4$	1297
Gadolinium	Gd	1312
Magnesium sulfate	$MgSO_4$	1327
Potassium phosphate	K_3PO_4	1340
Barium sulfate	$BaSO_4$	1350
Terbium	Tb	1356
Iron (II) oxide	FeO	1380
Calcium fluoride	CaF_2	1382
Strontium fluoride	SrF_2	1400
Dysprosium	Dy	1407
Silicon	Si	1427
Copper (II) oxide	CuO	1446
Nickel	Ni	1452
Holmium	Ho	1461

Source: data from: Weast, R C., Ed., *Handbook of Chemistry and Physics, 55th ed.*, CRC Press, Cleveland, (1974); and Bolz, R. E. and Tuve, G. L., Eds., *Handbook of Tables for Applied Engineering Science, 2nd ed.*, CRC Press, Cleveland, (1973), p.479 .

Table 349. SELECTING MELTING POINTS OF ELEMENTS AND INORGANIC COMPOUNDS (SHEET 11 OF 12)

Compound	Formula	Melting Point •C
Tungsten trioxide	WO_3	1470
Cobalt	Co	1490
Erbium	Er	1496
Yttrium	Y	1504
Niobium pentoxide	Nb_2O_5	1511
Calcium metasilicate	$CaSiO_3$	1512
Magnesium silicate	$MgSiO_3$	1524
Iron	Fe	1530.0
Scandium	Sc	1538
Thulium	Tm	1545
Palladium	Pd	1555
Manganese oxide	Mn_3O_4	1590
Iron oxide	Fe_3O_4	1596
Lutetium	Lu	1651
Barium phosphate	$Ba_3(PO_4)_2$	1727
Zinc sulfide	ZnS	1745
Platinum	Pt	1770
Manganese (II) oxide	MnO	1784
Titanium	Ti	1800
Titanium dioxide	TiO_2	1825
Thorium	Th	1845
Zirconium	Zr	1857
Tantalum pentoxide	Ta_2O_5	1877
Chromium	Cr	1890
Vanadium	V	1917
Barium oxide	BaO	1922.8
Zinc oxide	ZnO	1975
Aluminum oxide	Al_2O_3	2045.0

Source: data from: Weast, R C., Ed., *Handbook of Chemistry and Physics, 55th ed.*, CRC Press, Cleveland, (1974); and Bolz, R. E. and Tuve, G. L., Eds., *Handbook of Tables for Applied Engineering Science, 2nd ed.*, CRC Press, Cleveland, (1973), p.479 .

Table 349. SELECTING MELTING POINTS OF ELEMENTS AND INORGANIC COMPOUNDS (SHEET 12 OF 12)

Compound	Formula	Melting Point •C
Vanadium oxide	VO	2077
Hafnium	Hf	2214
Yttrium oxide	Y_2O_3	2227
Chromium (III) sequioxide	Cr_2O_3	2279
Boron	B	2300
Strontium oxide	SrO	2430
Niobium	Nb	2496
Beryllium oxide	BeO	2550.0
Molybdenum	Mo	2622
Magnesium oxide	MgO	2642
Osmium	Os	2700
Calcium oxide	CaO	2707
Zirconium oxide	ZrO_2	2715
Thorium dioxide	ThO_2	2952
Tantalum	Ta	2996 ± 50
Rhenium	Re	3167±60
Tungsten	W	3387

Source: data from: Weast, R C., Ed., *Handbook of Chemistry and Physics, 55th ed.*, CRC Press, Cleveland, (1974); and Bolz, R. E. and Tuve, G. L., Eds., *Handbook of Tables for Applied Engineering Science, 2nd ed.*, CRC Press, Cleveland, (1973), p.479 .

Table 350. SELECTING MELTING POINTS OF CERAMICS
(SHEET 1 OF 11)

Compound	(K)
TaC	3813
NbC	3770
VC	3600
ZrC	3533
ThO_2	3493
TiC	3433
Ta_2N	3360
ZrB_2	3313
TiB_2	3253
ZrN	3250
TiN	3200
CaO	3183
UO_2	3151
WB	3133
ZrO_2	3123
UN	3123
MgO	3098
BN	3000
SiC	2970
Mo_2C	2963
SrO	2933
ThN	2903
WC	2900
ThC	2898
CeO_2	>2873
UC	2863
BeO	2725
B_4C	2720

Source: data from Lynch, Charles T., Ed., *CRC Handbook of Materials Science, Vol. 1*, CRC Press, Boca Raton, 1974, 348.

Table 350. SELECTING MELTING POINTS OF CERAMICS (SHEET 2 OF 11)

Compound	(K)
Si_3N_4	2715
$TaSi_2$	2670
MoB	2625
Cr_2O_3	>2603
VN	2593
$MoSi_2$	2553
BaB_4	2543
Be_3N_2	2513
SrB_6	2508
AlN	>2475
CeB_6	2463
CeS	2400
Be_2C	>2375
VB_2	2373
NbN	2323
Al_2O_3	2322
WSi_2	2320
BaO	2283
SrS	>2275
MgS	>2275
ThB_4	>2270
TaB	>2270
NbB	>2270
NiO	2257
ZnO	2248
BeB_2	>2243
$NbSi_2$	2203
ThS_2	2198

Source: data from Lynch, Charles T., Ed., *CRC Handbook of Materials Science, Vol. 1*, CRC Press, Boca Raton, 1974, 348.

Table 350. SELECTING MELTING POINTS OF CERAMICS
(SHEET 3 OF 11)

Compound	(K)
In_2O_3	2183
Cr_3C_2	2168
CrB_2	2123
TiO_2	2113
Fe_3C	2110
Ta_2O_5	2100
VSi_2	2023
CdS	2023
Al_4C_3	2000
SiO_2	1978
Li_2O	>1975
USi_2	1970
SrC_2	>1970
$SrSO_4$	1878
Fe_2O_3	1864
$BaSO_4$	1853
$CrSi_2$	1843
MnO	1840
ZrS_2	1823
$TiSi_2$	1813
CdO	1773
UB_2	>1770
CrN	1770
Nb_2O_5	1764

Source: data from Lynch, Charles T., Ed., *CRC Handbook of Materials Science, Vol. 1*, CRC Press, Boca Raton, 1974, 348.

Table 350. SELECTING MELTING POINTS OF CERAMICS (SHEET 4 OF 11)

Compound	(K)
WO_3	1744
SrF_2	1736
$CaSO_4$	1723
CeF_2	1710
CaF_2	1675
BaF_2	1627
TaS_4	>1575
AlF_3	1564
MgF_2	1535
WS_2	1523
Cu_2O	1508
TiF_3	1475
BaS	1473
FeS	1468
Ca_3N_2	1468
MoS_2	1458
Na_2S	1453
$PbSO_4$	1443
InF_3	1443
Cu_2S	1400
$MgSO_4$	1397
PbS	1387
US_2	>1375
ThF_4	1375

Source: data from Lynch, Charles T., Ed., *CRC Handbook of Materials Science, Vol. 1*, CRC Press, Boca Raton, 1974, 348.

Table 350. SELECTING MELTING POINTS OF CERAMICS
(SHEET 5 OF 11)

Compound	(K)
Mg_2Si	1375
CdF_2	1373
Al_2S_3	1373
SnO	1353
K_2SO_4	1342
In_2S_3	1323
FeF_3	>1275
$NiCl_3$	1274
NiF_2	1273
$CdSO_4$	1273
NaF	1267
$NiBr_2$	1236
$BaCl_2$	1235
UF_4	1233
Li_2S	1198
PbO	1159
Na_2SO_4	1157
SnS	1153
$SrCl_2$	1148
ZnF_2	1145
Li_2SO_4	1132
KF	1131
MnF_2	1129
CuF_2	1129

Source: data from Lynch, Charles T., Ed., *CRC Handbook of Materials Science, Vol. 1*, CRC Press, Boca Raton, 1974, 348.

Table 350. SELECTING MELTING POINTS OF CERAMICS
(SHEET 6 OF 11)

Compound	(K)
Cu_4Si	1123
$BaBr_2$	1123
$NiSO_4$	1121
LiF	1119
Li_3N	1118
K_2S	1113
B_2O_3	1098
Ag_2S	1098
PbF_2	1095
$CeCl_3$	1095
VF_3	>1075
NaCl	1073
NiS	1070
NiI_2	1070
MoO_3	1068
$CaCl_2$	1055
FI_2	1048
$ThCl_4$	1043
KCl	1043
$Al_2(SO_4)_3$	1043
CeI_3	1025
NaBr	1023
Bi_2S_3	1020
BaI_2	1013

Source: data from Lynch, Charles T., Ed., *CRC Handbook of Materials Science, Vol. 1*, CRC Press, Boca Raton, 1974, 348.

Table 350. SELECTING MELTING POINTS OF CERAMICS

(SHEET 7 OF 11)

Compound	(K)
KBr	1008
TeO_2	1006
$CaBr_2$	1003
BiF_3	1000
$MgCl_2$	987
$MgBr_2$	984
SnF_4	978
NaC_2	973
KI	958
$FeBr_2$	955
V_2O_5	947
$FeCl_2$	945
NaI	935
Ag_2SO_4	933
Sb_2O_3	928
$MnCl_2$	923
$SrBr_2$	916
MgI_2	<910
$ThBr_4$	883
LiCl	883
CuI	878
V_2S_3	>875
ZrF_4	873
$ZnSO_4$	873

Source: data from Lynch, Charles T., Ed., *CRC Handbook of Materials Science, Vol. 1*, CRC Press, Boca Raton, 1974, 348.

Table 350. SELECTING MELTING POINTS OF CERAMICS
(SHEET 8 OF 11)

Compound	(K)
TiI_2	873
$Ba(NO_3)_2$	865
$PtCl_2$	854
CaI_2	848
$BeSO_4$	848
UCl_4	843
$CdCl_2$	841
$CdBr_2$	841
$Cd(NO_3)_2$	834
AgI	831
LiBr	823
SbS_3	820
BeF_2	813
$BeBr_2$	793
UBr_4	789
SnI_2	788
BeI_2	783
UI_4	779
CuBr	777
ZrI_4	772
$PbCl_2$	771
$Fe_2(SO_4)_3$	753
$Pb(NO_3)_2$	743
AgCl	728

Source: data from Lynch, Charles T., Ed., *CRC Handbook of Materials Science, Vol. 1*, CRC Press, Boca Raton, 1974, 348.

Table 350. SELECTING MELTING POINTS OF CERAMICS

(SHEET 9 OF 11)

Compound	(K)
B_2O_3	723
LiI	722
ZnI_2	719
$BeCl_2$	713
$InBr_3$	709
AgF	708
K_2O_3	703
AgBr	703
CuCl	695
$Zr(SO_4)_2$	683
BiI_3	681
$Bi(SO_4)_3$	678
PbI_2	675
$ZnBr_2$	667
BS_4	663
$Sr(NO_3)_2$	643
$PbBr_2$	643
$SnSO_4$	>635
PtI_2	633
$ZrBr_2$	>625
$ZrCl_2$	623
$Ca(NO_3)_2$	623
$TeBr_2$	612
KNO_3	610

Source: data from Lynch, Charles T., Ed., *CRC Handbook of Materials Science, Vol. 1*, CRC Press, Boca Raton, 1974, 348.

Table 350. SELECTING MELTING POINTS OF CERAMICS
(SHEET 10 OF 11)

Compound	(K)
SrI_2	593
$NaNO_3$	583
$SnCl_2$	581
Na_2N	573
Cu_3N	573
Ag_2O	573
SbF_3	565
$ZnCl_2$	548
WCl_6	548
$TaBr_5$	538
$LiNO_3$	527
$PtBr_2$	523
PtS_2	508
$BiCl_3$	507
InCl	498
$BiBr_3$	491
$TaCl_5$	489
$SnBr_2$	488
InI_3	483
$AgNO_3$	483
$Ce(SO_4)_2$	468
$AlCl_3$	465
AlI	464
$TeCl_2$	448

Source: data from Lynch, Charles T., Ed., *CRC Handbook of Materials Science, Vol. 1*, CRC Press, Boca Raton, 1974, 348.

Table 350. SELECTING MELTING POINTS OF CERAMICS
(SHEET 11 OF 11)

Compound	(K)
SbI_3	443
CdI_2	423
MoI_4	373
$AlBr_3$	371
TaF_5	370
$SbBr_3$	370
$SbCl_3$	346
$TiBr_4$	312
MoF_6	290
$TiCl_4$	250
VCl_4	245
BBr_3	227
SiF_4	183
BCl_3	166
BF_3	146

Source: data from Lynch, Charles T., Ed., *CRC Handbook of Materials Science, Vol. 1*, CRC Press, Boca Raton, 1974, 348.

Table 351. SELECTING HEAT OF FUSION FOR ELEMENTS AND INORGANIC COMPOUNDS* (SHEET 1 OF 15)

Compound	Formula	Melting point •C	Heat of fusion	
			cal/g	cal/g mole
Hydrogen	H_2	–259.25	13.8	28
Neon	Ne	–248.6	3.83	77.4
Oxygen	O_2	–218.8	3.3	106.3
Nitrogen	N_2	–210	6.15	172.3
Carbon monoxide	CO	–205	7.13	199.7
Fluorine	F_2	–219.6	6.4	244.0
Argon	Ar	190.2	7.25	290
Sulfur (monatomic)	S	119	9.2	295
Hydrogen chloride	HCl	–114.3	13.0	476.0
Boron trifluoride	BF_3	–128.0	7.0	480
Boron trichloride	BCl_3	–107.8	(4.3)	(500)
Cesium	Cs	28.3	3.7	500
Rubidium	Rb	38.9	6.1	525
Nitric oxide	NO	–163.7	18.3	549.5
Mercury	Hg	–39	2.7	557.2
Potassium	K	63.4	14.6	574
Hydrogen bromide	HBr	–86.96	7.1	575.1
Phosphorus, yellow	P_4	44.1	4.8	600
Hydrogen nitrate	HNO_3	–47.2	9.5	601
Sodium	Na	97.8	27.4	630
Hydrogen iodide	HI	–50.91	5.4	686.3
Boron tribromide	BBr_3	–48.8	(2.9)	(700)
Xenon	Xe	–111.6	5.6	740
Indium	In	156.3	6.8	781

For heat of fusion in **J/kg**, multiply values in **cal/g** by **4184**.
For heat of fusion in **J/mol**, multiply values in **cal/g-mol** (=**cal/mol**) by **4.184**.
For melting point in **K**, add **273.15** to values in °C.

Source: data from Weast, R C., Ed., *Handbook of Chemistry and Physics, 55th ed.*, CRC Press, Cleveland, (1974); and Bolz, R. E. and Tuve, G. L., Eds., *Handbook of Tables for Applied Engineering Science, 2nd ed.*, CRC Press, Cleveland, (1973)

Table 351. SELECTING HEAT OF FUSION FOR ELEMENTS AND INORGANIC COMPOUNDS* (SHEET 2 OF 15)

Compound	Formula	Melting point •C	Heat of fusion	
			cal/g	cal/g mole
Seleniumoxychloride	$SeOCl_3$	9.8	6.1	1010
Thallium	Tl	302.4	5.0	1030
Hydrogen fluoride	HF	83.11	54.7	1094
Lithium	Li	178.8	158.5	1100
Sodium sulfide	Na_2S	920	15.4	(1200)
Selenium	Se	217	15.4	1220
Lead	Pb	327.3	5.9	1224
Gallium	Ga	29	19.1	1336
Rubidium nitrate	$RbNO_3$	305	9.1	1340
Bromine pentafluoride	BrF_5	–61.4	7.07	1355
Lithium iodide	LiI	440	(10.6)	(1420)
Hydrogen oxide (water)	H_2O	0	79.72	1436
Mercury sulfate	$HgSO_4$	850	(4.8)	(1440)
Cadmium	Cd	320.8	12.9	1460
Deuterium oxide	D_2O	3.78	75.8	1516
Chlorine	Cl_2	–103+5	22.8	1531
Nitrous oxide	N_2O	–90.9	35.5	1563
Zinc	Zn	419.4	24.4	1595
Hydrogen telluride	H_2Te	–49.0	12.9	1670
Neodymium	Nd	1020	11.8	1700
Tin	Sn	231.7	14.4	1720
Tin bromide, di–	$SnBr_2$	231.8	(6.1)	(1720)
Tungsten hexafluoride	WF_6	–0.5	6.0	1800
Hydrogen sulfide, di–	H_2S_2	–89.7	27.3	1805

For heat of fusion in **J/kg**, multiply values in **cal/g** by **4184.**
For heat of fusion in **J/mol**, multiply values in **cal/g-mol** (**=cal/mol**) by **4.184.**
For melting point in **K**, add **273.15** to values in **°C.**

Source: data from Weast, R C., Ed., *Handbook of Chemistry and Physics, 55th ed.*, CRC Press, Cleveland, (1974); and Bolz, R. E. and Tuve, G. L., Eds., *Handbook of Tables for Applied Engineering Science, 2nd ed.*, CRC Press, Cleveland, (1973)

Table 351. SELECTING HEAT OF FUSION FOR ELEMENTS AND INORGANIC COMPOUNDS* (SHEET 3 OF 15)

Compound	Formula	Melting point •C	Heat of fusion cal/g	Heat of fusion cal/g mole
Barium	Ba	725	13.3	1830
Silicon tetrachloride	$SiCl_4$	–67.7	10.8	1845
Lead fluoride	PbF_2	823	7.6	1860
Carbon dioxide	CO_2	–57.6	43.2	1900
Potassium hydroxide	KOH	360	(35.3)	(1980)
Sodium hydroxide	NaOH	322	50.0	2000
Cyanogen	C_2N_2	–27.2	39.6	2060
Sulfur dioxide	SO_2	–73.2	32.2	2060
Sulfur trioxide (α)	SO_3	16.8	25.8	2060
Titanium bromide, tetra–	$TiBr_4$	38	(5.6)	(2060)
Silicon dioxide (Cristobalite)	SiO_2	1723	35.0	2100
Cerium	Ce	775	27.2	2120
Magnesium	Mg	650	88.9	2160
Silver bromide	AgBr	430	11.6	2180
Strontium	Sr	757	25.0	2190
Tinchloride,tetra–	$SnCl_4$	–33.3	8.4	2190
Ytterbium	Yb	823	12.7	2200
Calcium	Ca	851	55.7	2230
Cyanogen chloride	CNCl	–5.2	36.4	2240
Titanium chloride, tetra–	$TiCl_4$	–23.2	11.9	2240
Potassium thiocyanate	KSCN	179	23.1	2250
Silver iodide	AgI	557	9.5	2250
lodine chloride (β)	ICl	13.8	13.3	2270
Thallium nitrate	$TINO_3$	207	8.6	2290

For heat of fusion in **J/kg**, multiply values in **cal/g** by **4184**.
For heat of fusion in **J/mol**, multiply values in **cal/g-mol** (=**cal/mol**) by **4.184**.
For melting point in **K**, add **273.15** to values in °C.

Source: data from Weast, R C., Ed., *Handbook of Chemistry and Physics, 55th ed.*, CRC Press, Cleveland, (1974); and Bolz, R. E. and Tuve, G. L., Eds., *Handbook of Tables for Applied Engineering Science, 2nd ed.*, CRC Press, Cleveland, (1973)

Table 351. SELECTING HEAT OF FUSION FOR ELEMENTS AND INORGANIC COMPOUNDS* (SHEET 4 OF 15)

Compound	Formula	Melting point •C	Heat of fusion	
			cal/g	cal/g mole
Phosphorus acid, hypo–	H_3PO_2	17.3	35.0	2310
Osmium tetroxide (white)	OsO_4	41.8	9.2	2340
Hydrogen sulfate	H_2SO_4	10.4	24.0	2360
Lithium fluoride	LiF	896	(91.1)	(2360)
Antimony pentachloride	$SbCl_5$	4.0	8.0	2400
Lanthanum	La	920	17.4	2400
Arsenic trichloride	$AsCl_3$	–16.0	13.3	2420
Lithium hydroxide	LiOH	462	103.3	2480
Arsenic trifluoride	AsF_3	–6.0	18.9	2486
Europium	Eu	826	16.4	2500
Molybdenum hexafluoride	MoF_6	17	11.9	2500
Molybdenum trioxide	MoO_3	795	(17.3)	(2500)
Bismuth	Bi	271	12.0	2505
Phosphoric acid	H_3PO_4	42.3	25.8	2520
Aluminum	Al	658.5	94.5	2550
Bromine	Br_2	–7.2	16.1	2580
Bismuth trichloride	$BiCl_3$	223.8	8.2	2600
Copper (I) iodide	CuI	587	(13.6)	(2600)
Samarium	Sm	1072	17.3	2600
Copper (I) chloride	CuCl	429	26.4	2620
Iodine chloride (α)	ICl	17.1	16.4	2660
Praseodymium	Pr	931	19.0	2700
Silver	Ag	961	25.0	2700
Silver cyanide	AgCN	350	20.5	2750

For heat of fusion in **J/kg**, multiply values in **cal/g** by **4184.**
For heat of fusion in **J/mol**, multiply values in **cal/g-mol** (=**cal/mol**) by **4.184.**
For melting point in **K**, add **273.15** to values in °C.

Source: data from Weast, R C., Ed., *Handbook of Chemistry and Physics, 55th ed.*, CRC Press, Cleveland, (1974); and Bolz, R. E. and Tuve, G. L., Eds., *Handbook of Tables for Applied Engineering Science, 2nd ed.*, CRC Press, Cleveland, (1973)

Table 351. SELECTING HEAT OF FUSION FOR ELEMENTS AND INORGANIC COMPOUNDS* (SHEET 5 OF 15)

Compound	Formula	Melting point •C	Heat of fusion	
			cal/g	cal/g mole
Silver nitrate	$AgNO_3$	209	16.2	2755
Arsenic pentafluoride	AsF_5	80.8	16.5	2800
Arsenic tribromide	$AsBr_3$	30.0	8.9	2810
Copper (II) oxide	CuO	1446	35.4	2820
Lead oxide	PbO	890	12.6	2820
Potassium nitrate	KNO_3	338	78.1	2840
Sulfur trioxide (β)	SO_3	32.3	36.1	2890
Lithium bromide	LiBr	552	33.4	2900
Hydrogen peroxide	H_2O_2	–0.7	8.58	2920
Rubidium iodide	Rbl	638	14.0	2990
Barium fluoride	BaF_2	1286.8	17.1	3000
Beryllium chloride	BeCl2	404.8	(30)	(3000)
Thallium sulfide	Tl_2S	449	6.8	3000
Tin bromide, tetra–	$SnBr_4$	29.8	6.8	3000
Antimony trichloride	$SbCl_3$	73.3	13.3	3030
Gold	Au	1063	(15.3)	3030
Lithium sulfate	Li_2SO_4	857	27.6	3040
Tin chloride, di–	$SnCl_2$	247	16.0	3050
Phosphorus acid, ortho–	H_3PO_3	73.8	37.4	3070
Copper	Cu	1083	49.0	3110

For heat of fusion in **J/kg**, multiply values in **cal/g** by **4184**.
For heat of fusion in **J/mol**, multiply values in **cal/g-mol** (=**cal/mol**) by **4.184**.
For melting point in **K**, add **273.15** to values in °C.

Source: data from Weast, R C., Ed., *Handbook of Chemistry and Physics, 55th ed.*, CRC Press, Cleveland, (1974); and Bolz, R. E. and Tuve, G. L., Eds., *Handbook of Tables for Applied Engineering Science, 2nd ed.*, CRC Press, Cleveland, (1973)

Table 351. SELECTING HEAT OF FUSION FOR ELEMENTS AND INORGANIC COMPOUNDS* (SHEET 6 OF 15)

Compound	Formula	Melting point •C	Heat of fusion	
			cal/g	cal/g mole
Phosphorus oxychloride	$POCl_3$	1.0	20.3	3110
Thallium iodide, mono–	TlI	440	9.4	3125
Silver chloride	AgCl	455	22.0	3155
Lithium chloride	LiCl	614	75.5	3200
Tellurium	Te	453	25.3	3230
Cesium nitrate	$CsNO_3$	406.8	16.6	3250
Iron pentacarbonyl	$Fe(CO)_5$	–21.2	16.5	3250
Phosphorus trioxide	P_4O_6	23.7	15.3	3360
Silver sulfide	Ag_2S	841	13.5	3360
Actinium[227]	Ac	1050±50	(11.0)	(3400)
Hydrogen selenate	H_2SeO_4	57.8	23.8	3450
Manganese	Mn	1220	62.7	3450
Magnesium sulfate	$MgSO_4$	1327	28.9	3500
Potassium cyanide	KCN	623	(53.7)	(3500)
Antimony tribromide	$SbBr_3$	96.8	9.7	3510
Iron	Fe	1530.0	63.7	3560
Cesium chloride	CsCl	38.5	21.4	3600
Sodium molybdate	Na_2MoO_4	687	17.5	3600
Cobalt	Co	1490	62.1	3640
Iodine	I_2	112.9	14.3	3650
Cadmium iodide	CdI_2	386.8	10.0	3660
Chromium	Cr	1890	62.1	3660
Gadolinium	Gd	1312	23.8	3700
Rubidium bromide	RbBr	677	22.4	3700

For heat of fusion in **J/kg**, multiply values in **cal/g** by **4184**.
For heat of fusion in **J/mol**, multiply values in **cal/g-mol** (=**cal/mol**) by **4.184**.
For melting point in K, add **273.15** to values in °C.

Source: data from Weast, R C., Ed., *Handbook of Chemistry and Physics, 55th ed.*, CRC Press, Cleveland, (1974); and Bolz, R. E. and Tuve, G. L., Eds., *Handbook of Tables for Applied Engineering Science, 2nd ed.*, CRC Press, Cleveland, (1973)

Table 351. SELECTING HEAT OF FUSION FOR ELEMENTS AND INORGANIC COMPOUNDS* (SHEET 7 OF 15)

Compound	Formula	Melting point •C	Heat of fusion	
			cal/g	cal/g mole
Uranium[235]	U	~1133	20	3700
Sodium nitrate	$NaNO_3$	310	44.2	3760
Chromium trioxide	CrO_3	197	37.7	3770
Scandium	Sc	1538	84.4	3800
Silane, hexaHuoro–	Si_2F_6	–28.6	22.9	3900
Terbium	Tb	1356	24.6	3900
Mercury bromide	$HgBr_2$	241	10.9	3960
Osmium tetroxide (yellow)	OsO_4	55.8	15.5	4060
Calcium fluoride	CaF_2	1382	52.5	4100
Dysprosium	Dy	1407	25.2	4100
Erbium	Er	1496	24.5	4100
Holmium	Ho	1461	24.8	4100
Potassium iodide	Kl	682	24.7	4100
Strontium chloride	$SrCl_2$	872	26.5	4100
Yttrium	Y	1504	46.1	4100
Palladium	Pd	1555	38.6	4120
Rubidium fluoride	RbF	833	39.5	4130
Lead sulfide	PbS	1114	17.3	4150
Mercury chloride	$HgCl_2$	276.8	15.3	4150
Calcium bromide	$CaBr_2$	729.8	20.9	4180
Chromium (III) sequioxide	Cr_2O_3	2279	27.6	4200
Lithium molybdate	Li_2MoO_4	705	24.1	4200
Nickel	Ni	1452	71.5	4200
Vanadium	V	1917	(70)	(4200)

For heat of fusion in **J/kg**, multiply values in **cal/g** by **4184**.
For heat of fusion in **J/mol**, multiply values in **cal/g-mol** (=cal/mol) by **4.184**.
For melting point in K, add **273.15** to values in °C.

Source: data from Weast, R C., Ed., *Handbook of Chemistry and Physics, 55th ed.*, CRC Press, Cleveland, (1974); and Bolz, R. E. and Tuve, G. L., Eds., *Handbook of Tables for Applied Engineering Science, 2nd ed.*, CRC Press, Cleveland, (1973)

Table 351. SELECTING HEAT OF FUSION FOR ELEMENTS AND INORGANIC COMPOUNDS* (SHEET 8 OF 15)

Compound	Formula	Melting point •C	Heat of fusion	
			cal/g	cal/g mole
Strontium fluoride	SrF_2	1400	34.0	4260
Thallium chloride, mono–	TlCl	427	17.7	4260
Silver sulfate	Ag_2SO_4	657	(13.7)	(4280)
Leadbromide	$PbBr_2$	487.8	11.7	4290
Tin iodide, tetra–	SnI_4	143.4	(6.9)	(4330)
Sodium cyanide	NaCN	562	(88.9)	(4360)
Rubidium chloride	RbCl	717	36.4	4400
Thallium carbonate	Tl_2CO_3	273	9.5	4400
Thulium	Tm	1545	26.0	4400
Sodium thiocyanate	NaSCN	323	54.8	4450
Zinc oxide	ZnO	1975	54.9	4470
Beryllium bromide	$BeBr_2$	487.8	(26.6)	(4500)
Mercury iodide	HgI_2	250	9.9	4500
Thorium	Th	1845	(<19.8)	(<4600)
Lutetium	Lu	1651	26.3	4600
Platinum	Pt	1770	24.1	4700
Antimony	Sb	630	39.1	4770
Strontium bromide	$SrBr_2$	643	19.3	4780
Cadmium sulfate	$CdSO_4$	1000	22.9	4790
Copper (II) chloride	$CuCl_2$	430	24.7	4890
Sodium phosphate, meta–	$NaPO_3$	988	(48.6)	(4960)
Cadmium bromide	$CdBr_2$	567.8	(18.4)	(5000)
Iron (II) sulfide	FeS	1195	56.9	5000
Potassium bromide	KBr	742	42.0	5000

For heat of fusion in **J/kg**, multiply values in **cal/g** by **4184.**
For heat of fusion in **J/mol**, multiply values in **cal/g-mol** (=**cal/mol**) by **4.184.**
For melting point in K, add **273.15** to values in °C.

Source: data from Weast, R C., Ed., *Handbook of Chemistry and Physics, 55th ed.*, CRC Press, Cleveland, (1974); and Bolz, R. E. and Tuve, G. L., Eds., *Handbook of Tables for Applied Engineering Science, 2nd ed.*, CRC Press, Cleveland, (1973)

Table 351. SELECTING HEAT OF FUSION FOR ELEMENTS AND INORGANIC COMPOUNDS* (SHEET 9 OF 15)

Compound	Formula	Melting point •C	Heat of fusion	
			cal/g	cal/g mole
Rhenium hexafluoride	ReF_6	19.0	16.6	5000
Titanium	Ti	1800	(104.4)	(5000)
Calcium nitrate	$Ca(NO_3)_2$	560.8	31.2	5120
Sodium chlorate	$NaClO_3$	255	49.7	5290
Boron	B	2300	(490)	(5300)
Cadmium chloride	$CdCl_2$	567.8	28.8	5300
Sodium iodide	NaI	662	35.1	5340
Barium chloride	$BaCl_2$	959.8	25.9	5370
Cadmium fluoride	CdF_2	1110	(35.9)	(5400)
Copper(l) cyanide	$Cu_2(CN)_2$	473	(30.1)	(5400)
Aluminum bromide	Al_2Br_6	87.4	10.1	5420
Boron trioxide	B_2O_3	448.8	78.9	5500
Copper (I) sulfide	Cu_2S	1129	62.3	5500
Thallium sulfate	Tl_2SO_4	632	10.9	5500
Zirconium	Zr	1857	(60)	(5500)
Nitrogen tetroxide	N_2O_4	–13.2	60.2	5540
Zinc chloride	$ZnCl_2$	283	(406)	(5540)
Lead chloride	$PbCl_2$	497.8	20.3	5650
Potassium borate, meta–	KBO_2	947	(69.1)	(5660)
Hydrogen sulfide	H_2S	–85.6	16.8	5683

For heat of fusion in **J/kg**, multiply values in **cal/g** by **4184**.
For heat of fusion in **J/mol**, multiply values in **cal/g-mol** (=cal/mol) by **4.184**.
For melting point in **K**, add **273.15** to values in **°C**.

Source: data from Weast, R C., Ed., *Handbook of Chemistry and Physics, 55th ed.*, CRC Press, Cleveland, (1974); and Bolz, R. E. and Tuve, G. L., Eds., *Handbook of Tables for Applied Engineering Science, 2nd ed.*, CRC Press, Cleveland, (1973)

Table 351. SELECTING HEAT OF FUSION FOR ELEMENTS AND INORGANIC COMPOUNDS* (SHEET 10 OF 15)

Compound	Formula	Melting point •C	Heat of fusion	
			cal/g	cal/g mole
Nickel subsulfide	Ni_3S_2	790	25.8	5800
Sodium tungstate	Na_2WO_4	702	19.6	5800
Sodium sulfate	Na_2SO_4	884	41.0	5830
Sodium peroxide	Na_2O_2	460	75.1	5860
Barium nitrate	$Ba(NO_3)_2$	594.8	(22.6)	(5900)
Magnesium fluoride	MgF_2	1221	94.7	5900
Lead iodide	PbI_2	412	17.9	5970
Thallium bromide, mono–	TlBr	460	21.0	5990
Barium bromide	$BaBr_2$	846.8	21.9	6000
Hafnium	Hf	2214	(34.1)	(6000)
Molybdenum dichloride	$MoCl_2$	726.8	3.58	6000
Tungsten tetrachloride	WCl_4	327	18.4	6000
Lithium nitrate	$LiNO_3$	250	87.8	6060
Calcium chloride	$CaCl_2$	782	55	6100
Potassium peroxide	K_2O_2	490	55.3	6100
Sodium bromide	NaBr	747	59.7	6140
Bismuth trifluoride	BiF_3	726.0	(23.3)	(6200)
Sulfur trioxide (γ)	SO_3	62.1	79.0	6310
Tin oxide	SnO	1042	(46.8)	(6400)
Potassium chloride	KCl	770	85.9	6410

For heat of fusion in **J/kg**, multiply values in **cal/g** by **4184.**
For heat of fusion in **J/mol**, multiply values in **cal/g-mol** (=**cal/mol**) by **4.184.**
For melting point in K, add **273.15** to values in °C.

Source: data from Weast, R C., Ed., *Handbook of Chemistry and Physics, 55th ed.*, CRC Press, Cleveland, (1974); and Bolz, R. E. and Tuve, G. L., Eds., *Handbook of Tables for Applied Engineering Science, 2nd ed.*, CRC Press, Cleveland, (1973)

Table 351. SELECTING HEAT OF FUSION FOR ELEMENTS AND INORGANIC COMPOUNDS* (SHEET 11 OF 15)

Compound	Formula	Melting point •C	Heat of fusion	
			cal/g	cal/g mole
Niobium	Nb	2496	(68.9)	(6500)
Potassium fluoride	KF	875	111.9	6500
Molybdenum	Mo	2622	(68.4)	(6600)
Arsenic	As	816.8	(22.0)	(6620)
Calcium sulfate	$CaSO_4$	1297	49.2	6700
Lithium tungstate	Li_2WO_4	742	(25.6)	(6700)
Barium iodide	BaI_2	710.8	(17.3)	(6800)
Bismuth trioxide	Bi_2O_3	815.8	14.6	6800
Potassium chromate	K_2CrO_4	984	35.6	6920
Osmium	Os	2700	(36.7)	(7000)
Sodium carbonate	Na_2CO_3	854	66.0	7000
Sodium fluoride	NaF	992	166.7	7000
Lithium metasilicate	Li_2SiO_3	1177	80.2	7210
Sodium chloride	NaCl	800	123.5	7220
Zirconium dichloride	$ZrCl_2$	727	45.0	7300
Manganese dichloride	$MnCl_2$	650	58.4	7340
Cobalt (II) chloride	$CoCl_2$	727	56.9	7390
Lithium orthosilicate	Li_4SiO_4	1249	60.5	7430
Tantalum	Ta	2996 ± 50	34.6–41.5	(7500)
Chromium (II) chloride	$CrCl_2$	814	65.9	7700

For heat of fusion in **J/kg**, multiply values in **cal/g** by **4184**.
For heat of fusion in **J/mol**, multiply values in **cal/g-mol** (=cal/mol) by **4.184**.
For melting point in **K**, add **273.15** to values in °C.

Source: data from Weast, R C., Ed., *Handbook of Chemistry and Physics, 55th ed.*, CRC Press, Cleveland, (1974); and Bolz, R. E. and Tuve, G. L., Eds., *Handbook of Tables for Applied Engineering Science, 2nd ed.*, CRC Press, Cleveland, (1973)

Table 351. SELECTING HEAT OF FUSION FOR ELEMENTS AND INORGANIC COMPOUNDS* (SHEET 12 OF 15)

Compound	Formula	Melting point •C	Heat of fusion	
			cal/g	cal/g mole
Iron (II) oxide	FeO	1380	(107.2)	(7700)
Iron (II) chloride	$FeCl_2$	677	61.5	7800
Potassium carbonate	K_2CO_3	897	56.4	7800
Rhenium	Re	3167±60	(42.4)	(7900)
Aluminum iodide	Al_2I_6	190.9	9.8	7960
Arsenic trioxide	As_4O_6	312.8	22.2	8000
Europium trichloride	$EuCl_3$	622	(20.9)	(8000)
Vanadium dichloride	VCl_2	1027	65.6	8000
Magnesium chloride	$MgCl_2$	712	82.9	8100
Potassium sulfate	K_2SO_4	1074	46.4	8100
Manganese metasilicate	$MnSiO_3$	1274	(62.6)	(8200)
Germanium	Ge	959	(114.3)	(8300)
Magnesium bromide	$MgBr_2$	711	45.0	8300
Phosphoric acid. hypo–	$H_4P_2O_6$	54.8	51.2	8300
Niobium pentachloride	$NbCl_5$	21.1	30.8	8400
Tungsten	W	3387	(45.8)	(8420)
Sodium silicate, di–	$Na_2Si_2O_5$	884	46.4	8460
Sodium borate, meta–	$NaBO_2$	966	134.6	8660
Potassium dichromate	$K_2Cr_2O_7$	398	29.8	8770
Potassium phosphate	K_3PO4	1340	41.9	8900

For heat of fusion in **J/kg**, multiply values in **cal/g** by **4184.**
For heat of fusion in **J/mol**, multiply values in **cal/g-mol** (**=cal/mol**) by **4.184.**
For melting point in K, add **273.15** to values in °C.

Source: data from Weast, R C., Ed., *Handbook of Chemistry and Physics, 55th ed.*, CRC Press, Cleveland, (1974); and Bolz, R. E. and Tuve, G. L., Eds., *Handbook of Tables for Applied Engineering Science, 2nd ed.*, CRC Press, Cleveland, (1973)

Table 351. SELECTING HEAT OF FUSION FOR ELEMENTS AND INORGANIC COMPOUNDS* (SHEET 13 OF 15)

Compound	Formula	Melting point •C	Heat of fusion	
			cal/g	cal/g mole
Tantalum pentachloride	$TaCl_5$	206.8	25.1	9000
Zinc sulfide	ZnS	1745	(93.3)	(9100)
Silicon	Si	1427	337.0	9470
Lead sulfate	$PbSO_4$	1087	31.6	9600
Barium sulfate	$BaSO_4$	1350	41.6	9700
Sodium silicate, meta–	Na_2SiO_3	1087	84.4	10300
Uranium tetrachloride	UCl_4	590	27.1	10300
Antimony trisulfide	Sb_4S_6	546.0	33.0	11200
Titanium dioxide	TiO_2	1825	(142.7)	(11400)
Calcium oxide	CaO	2707	(218.1)	(12240)
Iron carbide	Fe_3C	1226.8	68.6	12330
Calcium carbonate	$CaCO_3$	1282	(126)	(12700)
Manganese (II) oxide	MnO	1784	183.3	13000
Sodiumsilicate, aluminum–	$NaAlSi_3O_8$	1107	50.1	13150
Calcium metasilicate	$CaSiO_3$	1512	115.4	13400
Copper (I) oxide	Cu_2O	1230	(93.6)	(13400)
Sodium pyrophosphate	$Na_4P_2O_7$	970	(51.5)	(13700)
Barium oxide	BaO	1922.8	93.2	13800
Tungsten dioxide	WO_2	1270	60.1	13940
Tungsten trioxide	WO_3	1470	60.1	13940

For heat of fusion in **J/kg**, multiply values in **cal/g** by **4184**.
For heat of fusion in **J/mol**, multiply values in **cal/g-mol** (**=cal/mol**) by **4.184**.
For melting point in **K**, add **273.15** to values in °C.

Source: data from Weast, R C., Ed., *Handbook of Chemistry and Physics, 55th ed.*, CRC Press, Cleveland, (1974); and Bolz, R. E. and Tuve, G. L., Eds., *Handbook of Tables for Applied Engineering Science, 2nd ed.*, CRC Press, Cleveland, (1973)

Table 351. SELECTING HEAT OF FUSION FOR ELEMENTS AND INORGANIC COMPOUNDS* (SHEET 14 OF 15)

Compound	Formula	Melting point •C	Heat of fusion	
			cal/g	cal/g mole
Potassium pyro–phosphate	$K_4P_2O_7$	1092	42.4	14000
Titanium oxide	TiO	991	219	14000
Magnesium silicate	$MgSiO_3$	1524	146.4	14700
Vanadium oxide	VO	2077	224.0	15000
Rhenium heptoxide	Re_2O_7	296	30.1	15340
Vanadium pentoxide	V_2O_5	670	85.5	15560
Strontium oxide	SrO	2430	161.2	16700
Beryllium oxide	BeO	2550.0	679.7	17000
Phosphorus pentoxide	P_4O_{10}	569.0	60.1	17080
Nickel chloride	$NiCl_2$	1030	142.5	18470
Magnesium oxide	MgO	2642	459.0	18500
Barium phosphate	$Ba_3(PO_4)_2$	1727	30.9	18600
Aluminum chloride	Al_2Cl_6	192.4	63.6	19600
Iron (III) chloride	Fe_2Cl_6	303.8	63.2	20500
Zirconium oxide	ZrO_2	2715	168.8	20800
Thorium chloride	$ThCl_4$	765	61.6	22500
Niobium pentoxide	Nb_2O_5	1511	91.0	24200
Yttrium oxide	Y_2O_3	2227	110.7	25000
Lead molybdate	$PbMoO_4$	1065	70.8	(25800)
Aluminum oxide	Al_2O_3	2045.0	(256.0)	(26000)

For heat of fusion in **J/kg**, multiply values in **cal/g** by **4184**.
For heat of fusion in **J/mol**, multiply values in **cal/g-mol** (**=cal/mol**) by **4.184**.
For melting point in **K**, add **273.15** to values in **°C**.

Source: data from Weast, R C., Ed., *Handbook of Chemistry and Physics, 55th ed.*, CRC Press, Cleveland, (1974); and Bolz, R. E. and Tuve, G. L., Eds., *Handbook of Tables for Applied Engineering Science, 2nd ed.*, CRC Press, Cleveland, (1973)

Table 351. Selecting Heat of Fusion For Elements and Inorganic Compounds* (Sheet 15 of 15)

Compound	Formula	Melting point •C	Heat of fusion	
			cal/g	cal/g mole
Antimony trioxide	Sb_4O_6	655.0	(46.3)	(26990)
Iron oxide	Fe_3O_4	1596	142.5	33000
Manganese oxide	Mn_3O_4	1590	(170.4)	(39000)
Tantalum pentoxide	Ta_2O_5	1877	108.6	48000
Thorium dioxide	ThO_2	2952	1102.0	291100

For heat of fusion in **J/kg**, multiply values in **cal/g** by **4184**.
For heat of fusion in **J/mol**, multiply values in **cal/g-mol** (=**cal/mol**) by **4.184**.
For melting point in **K**, add **273.15** to values in °C.

Source: data from Weast, R C., Ed., *Handbook of Chemistry and Physics, 55th ed.*, CRC Press, Cleveland, (1974); and Bolz, R. E. and Tuve, G. L., Eds., *Handbook of Tables for Applied Engineering Science, 2nd ed.*, CRC Press, Cleveland, (1973)

* Values in parentheses are of uncertain reliability.

Table 352. SELECTING ENTROPY OF THE ELEMENTS

(SHEET 1 OF 3)

Element	Phase	Entropy at 298K (e.u.)
C	solid	1.3609
B	solid	1.42
Be	solid	2.28
Si	solid	4.50
Cr	solid	5.68
Fe	solid, α	6.491
Li	solid	6.70
Al	solid	6.769
Co	solid, α	6.8
Mo	solid	6.83
Ru	solid, α	6.9
V	solid	7.05
Ni	solid, α	7.137
Ti	solid, α	7.334
Mn	solid, α	7.59
Rh	solid	7.6
S	solid, α	7.62
Mg	solid	7.77
Os	solid	7.8
Cu	solid	7.97
Tc	solid	8.0
W	solid	8.0
Nb	solid	8.3
As	solid	8.4
Ir	solid	8.7
Re	solid	8.89
Pd	solid	8.9
Sc	solid	9.0

Source: data from Weast, R. C. Ed., *Handbook of Chemistry and Physics, 69th ed.*, CRC Press, Boca Raton, Fla., 1988, D44.

Table 352. SELECTING ENTROPY OF THE ELEMENTS

(SHEET 2 OF 3)

Element	Phase	Entropy at 298K (e.u.)
Zr	solid, α	9.29
Ga	solid	9.82
Ca	solid, α	9.95
Zn	solid	9.95
Pt	solid	10.0
Ge	solid	10.1
Se	solid	10.144
Ag	solid	10.20
Sb	solid (α, β, γ)	10.5
Y	solid	11
Au	solid	11.32
Te	solid, α	11.88
U	solid, α	12.03
Cd	solid	12.3
Sn	solid (α, β)	12.3
Na	solid	12.31
Th	solid	12.76
Ac	solid	13
Am	solid	13
Po	solid	13
Pu	solid	13.0
Sr	solid	13.0
Hf	solid	13.1
Pa	solid	13.5
Pr	solid	13.5
Bi	solid	13.6
La	solid	13.7
Ce	solid	13.8

Source: data from Weast, R. C. Ed., *Handbook of Chemistry and Physics, 69th ed.*, CRC Press, Boca Raton, Fla., 1988, D44.

Table 352. SELECTING ENTROPY OF THE ELEMENTS (SHEET 3 OF 3)

Element	Phase	Entropy at 298K (e.u.)
In	solid	13.88
Nd	solid	13.9
Np	solid	14
Sm	solid	15
K	solid	15.2
Tl	solid, α	15.4
Pb	solid	15.49
Ba	solid, α	16
Rb	solid	16.6
Ra	solid	17
Hg	liquid	18.46
Cs	solid	19.8
H_2	gas	31.211
P_4	solid, white	42.4
N_2	gas	45.767
F_2	gas	48.58
O_2	gas	49.003
Cl_2	gas	53.286
Ta	solid	99

Source: data from Weast, R. C. Ed., *Handbook of Chemistry and Physics, 69th ed.*, CRC Press, Boca Raton, Fla., 1988, D44.

Table 353. Selecting Diffusion Activation Energy in Metallic Systems*

(Sheet 1 of 33)

Metal	Tracer	Crystal Form	Purity %	Temperature Range °C	Activation Energy, Q kcal • mol^{-1}	Frequency Factor, D_o cm^2 • s^{-1}
Selenium	Hg^{203}	P	99.996	25–100	1.2	—
Zinc	Cu64	S⊥c	99.999	338–415	2.0	2.0
Sodium	Au^{198}	P	99.99	1.0–77	2.21	3.34×10^{-4}
α-Thallium	Au^{198}	P⊥c	99.999	110–260	2.8	2.0×10^{-5}
Potassium	Au^{198}	P	99.95	5.6–52.5	3.23	1.29×10^{-3}
α-Thallium	Au^{198}	P‖c	99.999	110–260	5.2	5.3×10^{-4}
Cobalt	S^{35}	P	99.99	1150–1250	5.4	1.3
β-Thallium	Au^{198}	P	99.999	230–310	6.0	5.2×10^{-4}
Indium	Au^{198}	S	99.99	25–140	6.7	9×10^{-3}
Potassium	Na^{22}	P	99.7	0–62	7.45	0.058
Sodium	K^{42}	P	99.99	0–91	8.43	0.08
Sodium	Rb^{86}	P	99.99	0–85	8.49	0.15

Source: data from Askill, J.,in *Handbook of Chemistry and Physics, 55th ed.*,Weast, R.C., Ed., CRC Press, Cleveland,1974, F61.

Table 353. SELECTING DIFFUSION ACTIVATION ENERGY IN METALLIC SYSTEMS*

(SHEET 2 OF 33)

Metal	Tracer	Crystal Form	Purity %	Temperature Range °C	Activation Energy, Q kcal • mol^{-1}	Frequency Factor, D_o $cm^2 \cdot s^{-1}$
Potassium	Rb^{86}	P	99.95	0.1–59.9	8.78	0.090
Selenium	Fe^{59}	P		40–100	8.88	—
Lithium	Cu64	P	99.98	51–120	9.22	0.47
Potassium	K^{42}	S	99.7	–52–61	9.36	0.16
Phosphorus	P^{32}	P		0–44	9.4	1.07×10^{-3}
Lead	Au^{198}	S	99.999	190–320	10.0	8.7×10^{-3}
Sodium	Na^{22}	P	99.99	0–98	10.09	0.145
Lithium	Au^{195}	P	92.5	47–153	10.49	0.21
Tin	Au^{198}	S‖c		135–225	11.0	5.8×10^{-3}
α-Thallium	Ag^{110}	P‖c	99.999	80–250	11.2	2.7×10^{-2}
Indium	Ag^{110}	S‖c	99.99	25–140	11.5	0.11
Selenium	Se^{75}	P		35–140	11.7	1.4×10^{-4}

Source: data from Askill, J.,in *Handbook of Chemistry and Physics, 55th ed.*,Weast, R.C., Ed., CRC Press, Cleveland,1974, F61.

Table 353. Selecting Diffusion Activation Energy in Metallic Systems*

(Sheet 3 of 33)

Metal	Tracer	Crystal Form	Purity %	Temperature Range °C	Activation Energy, Q kcal • mol^{-1}	Frequency Factor, D_o cm^2 • s^{-1}
α-Thallium	Ag^{110}	P⊥c	99.999	80–250	11.8	3.8 x 10^{-2}
β-Thallium	Ag^{110}	P	99.999	230–310	11.9	4.2 x 10^{-2}
γ-Uranium	Fe^{55}	P	99.99	787–990	12.0	2.69 x 10^{-4}
Tin	Ag^{110}	S‖c		135–225	12.3	7.1 x 10^{-3}
γ-Uranium	Co^{60}	P	99.99	783–989	12.57	3.51 x 10^{-4}
Lithium	Li^{6}	P	99.98	35–178	12.60	0.14
Lithium	Na^{22}	P	92.5	52–176	12.61	0.41
Indium	Ag^{110}	S⊥c	99.99	25–140	12.8	0.52
Lithium	Ag^{110}	P	92.5	65–161	12.83	0.37
Lithium	Ga^{72}	P	99.98	58–173	12.9	0.21
Lithium	Zn^{65}	P	92.5	60–175	12.98	0.57
Aluminum	Mo^{99}	P	99.995	400–630	13.1	1.04 x 10^{-9}

Source: data from Askill, J.,in *Handbook of Chemistry and Physics, 55th ed.*,Weast, R.C., Ed., CRC Press, Cleveland,1974, F61.

Table 353. Selecting Diffusion Activation Energy in Metallic Systems*

(Sheet 4 of 33)

Metal	Tracer	Crystal Form	Purity %	Temperature Range °C	Activation Energy, Q kcal • mol^{-1}	Frequency Factor, D_o $cm^2 \cdot s^{-1}$
g-Uranium	Mn^{54}	P	99.99	787–939	13.88	1.81×10^{-4}
Lithium	Hg^{203}	P	99.98	58–173	14.18	1.04
Lead	Ag^{110}	P	99.9	200–310	14.4	0.064
Lead	Cu^{64}	S		150–320	14.44	0.046
Tin	Tl^{204}	P	99.999	137–216	14.7	1.2×10^{-3}
Lithium	Sn^{113}	P	99.95	108–174	15.0	0.62
Indium	Tl^{204}	S	99.99	49–157	15.5	0.049
Selenium	S^{35}	S‖c		60–90	15.6	1100
g-Uranium	Ni^{63}	P	99.99	787–1039	15.66	5.36×10^{-4}
Aluminum	Ni^{63}	P	99.99	360–630	15.7	2.9×10^{-8}
Lithium	In^{114}	P	92.5	80–175	15.87	0.39
Lithium	Cd^{115}	P	92.5	80–174	16.05	2.35

Source: data from Askill, J.,in *Handbook of Chemistry and Physics, 55th ed.*,Weast, R.C., Ed., CRC Press, Cleveland,1974, F61.

Table 353. SELECTING DIFFUSION ACTIVATION ENERGY IN METALLIC SYSTEMS*

(SHEET 5 OF 33)

Metal	Tracer	Crystal Form	Purity %	Temperature Range °C	Activation Energy, Q kcal • mol^{-1}	Frequency Factor, D_o $cm^2 \cdot s^{-1}$
a-Praseodymium	Co^{60}	P	99.93	660–780	16.4	4.7×10^{-2}
γ-Uranium	Zr^{95}	P		800–1000	16.5	3.9×10^{-4}
γ–Plutonium	Pu^{238}	P		190–310	16.7	2.1×10^{-5}
Tin	Au^{198}	S⊥c		135–225	17.7	0.16
a-Zirconium	Cr^{51}	P	99.9	700–850	18.0	1.19×10^{-8}
β–Zirconium	Cr^{51}	P	99.9	700–850	18.0	1.19×10^{-8}
Lanthanum	La^{140}	P	99.97	690–850	18.1	2.2×10^{-2}
Zinc	Ga^{72}	S⊥c		240–403	18.15	0.018
Tin	Ag^{110}	S⊥c		135–225	18.4	0.18
Zinc	Ga^{72}	S‖c		240–403	18.4	0.016
Zinc	Sn^{113}	S⊥c		298–400	18.4	0.13
ε-Plutonium	Pu^{238}	P		500–612	18.5	2.0×10^{-2}

Source: data from Askill, J.,in *Handbook of Chemistry and Physics, 55th ed.*,Weast, R.C., Ed., CRC Press, Cleveland,1974, F61.

Table 353. SELECTING DIFFUSION ACTIVATION ENERGY IN METALLIC SYSTEMS*

(SHEET 6 OF 33)

Metal	Tracer	Crystal Form	Purity %	Temperature Range °C	Activation Energy, Q $kcal \cdot mol^{-1}$	Frequency Factor, D_o $cm^2 \cdot s^{-1}$
Indium	In^{114}	S⊥c	99.99	44–144	18.7	3.7
Indium	In^{114}	S‖c	99.99	44–144	18.7	2.7
Tellurium	Hg^{203}	P		270–440	18.7	3.14×10^{-5}
Cadmium	Zn^{65}	S	99.99	180–300	19.0	0.0016
Zinc	In^{114}	S‖c		271–413	19.10	0.062
Cadmium	Cd^{115}	S	99.95	110–283	19.3	0.14
Zinc	Sn^{113}	S‖c		298–400	19.4	0.15
Aluminum	V^{48}	P	99.995	400–630	19.6	6.05×10^{-8}
Zinc	In^{114}	S⊥c		271–413	19.60	0.14
Aluminum	Nb^{95}	P	99.95	350–480	19.65	1.66×10^{-7}
α-Praseodymium	Au^{195}	P	99.93	650–780	19.7	4.3×10^{-2}
Zinc	Hg^{203}	S‖c		260–413	19.70	0.056

Source: data from Askill, J.,in *Handbook of Chemistry and Physics, 55th ed.*,Weast, R.C., Ed., CRC Press, Cleveland,1974, F61.

Table 353. SELECTING DIFFUSION ACTIVATION ENERGY IN METALLIC SYSTEMS*

(SHEET 7 OF 33)

Metal	Tracer	Crystal Form	Purity %	Temperature Range °C	Activation Energy, Q kcal • mol^{-1}	Frequency Factor, D_o cm^2 • s^{-1}
Silicon	Fe^{59}	S		1000–1200	20.0	6.2 x 10^{-3}
β-Titanium	C^{14}	P	99.62	1100–1600	20.0	3.02 x 10^{-3}
β-Praseodymium	Au^{195}	P	99.93	800–910	20.1	3.3 x 10^{-2}
Zinc	Cd^{115}	S⊥c	99.999	225–416	20.12	0.117
Zinc	Hg^{203}	S⊥c		260–413	20.18	0.073
Aluminum	$Pd^{10}3$	P	99.995	400–630	20.2	1.92 x 10^{-7}
Zinc	Cd^{115}	S‖c	99.999	225–416	20.54	0.114
β-Thallium	Tl^{204}	S	99.9	230–280	20.7	0.7
Magnesium	Fe^{59}	P	99.95	400–600	21.2	4 x 10^{-6}
Lead	Cd^{115}	S	99.999	150–320	21.23	0.409
Molybdenum	Na^{24}	S		800–1100	21.25	2.95 x 10^{-9}
β-Praseodymium	Ag^{110}	P	99.93	800–900	21.5	3.2 x 10^{-2}

Source: data from Askill, J.,in *Handbook of Chemistry and Physics, 55th ed.*,Weast, R.C., Ed., CRC Press, Cleveland,1974, F61.

Table 353. SELECTING DIFFUSION ACTIVATION ENERGY IN METALLIC SYSTEMS*

(SHEET 8 OF 33)

Metal	Tracer	Crystal Form	Purity %	Temperature Range °C	Activation Energy, Q kcal • mol^{-1}	Frequency Factor, D_o $cm^2 \cdot s^{-1}$
b–Zirconium	Co^{60}	P	99.99	920–1600	21.82	3.26×10^{-3}
Zinc	Zn^{65}	S‖c	99.999	240–418	21.9	0.13
Tin	Co^{60}	S,P		140–217	22.0	5.5
α-Zirconium	Sn^{113}	P		300–700	22.0	1.0×10^{-8}
b–Zirconium	Sn^{113}	P		300–700	22.0	1×10^{-8}
Niobium	K^{42}	S		900 1100	22.10	2.38×10^{-7}
α-Thallium	Tl^{204}	S⊥c	99.9	135–230	22.6	0.4
Aluminum	Sm^{153}	P	99.995	450–630	22.88	3.45×10^{-7}
Magnesium	Ni^{63}	P	99.95	400 600	22.9	1.2×10^{-5}
α-Thallium	Tl^{204}	S‖c	99.9	135–230	22.9	0.4
α-Zirconium	V^{48}	P	99.99	600–850	22.9	1.12×10^{-8}
Silicon	Cu^{64}	P		800–1100	23.0	4×10^{-2}

Source: data from Askill, J.,in *Handbook of Chemistry and Physics, 55th ed.*,Weast, R.C., Ed., CRC Press, Cleveland,1974, F61.

Table 353. SELECTING DIFFUSION ACTIVATION ENERGY IN METALLIC SYSTEMS*

(SHEET 9 OF 33)

Metal	Tracer	Crystal Form	Purity %	Temperature Range °C	Activation Energy, Q kcal • mol^{-1}	Frequency Factor, D_o cm^2 • s^{-1}
Zinc	Zn^{65}	S⊥c	99.999	240–418	23.0	0.18
Calcium	Fe^{59}		99.95	500–800	23.3	2.7 x 10^{-3}
δ–Plutonium	Pu^{238}	P		350–440	23.8	4.5 x 10^{-3}
Aluminum	Pr^{142}	P	99.995	520–630	23.87	3.58 x 10^{-7}
g-Uranium	Cu^{64}	P	99.99	787–1039	24.06	1.96 x 10^{-3}
β-Titanium	P^{32}	P	99.7	950–1600	24.1	3.62x10^{-3}
Copper	Tm^{170}	P	99.999	705–950	24.15	7.28 x 10^{-9}
Lead	Tl^{205}	P	99.999	207–322	24.33	0.511
g-Uranium	Cr^{51}	P	99.99	797–1037	24.46	5.37 X 10^{-3}
α-Zirconium	Mo^{99}	P		600–850	24.76	6.22 x 10^{-8}
Germanium	Fe^{59}	S		775–930	24.8	0.13
α-Praseodymium	Zn^{65}	P	99.96	766–603	24.8	0.18

Source: data from Askill, J.,in *Handbook of Chemistry and Physics, 55th ed.*,Weast, R.C., Ed., CRC Press, Cleveland,1974, F61.

Table 353. SELECTING DIFFUSION ACTIVATION ENERGY IN METALLIC SYSTEMS*

(SHEET 10 OF 33)

Metal	Tracer	Crystal Form	Purity %	Temperature Range °C	Activation Energy, Q kcal • mol^{-1}	Frequency Factor, D_o cm^2 • s^{-1}
Aluminum	Nd^{147}	P	99.995	450–630	25.0	4.8×10^{-7}
Molybdenum	K^{42}	S		800–1100	25.04	5.5×10^{-9}
Tin	Sn^{113}	S⊥c	99.999	160–226	25.1	10.7
Lithium	Pb^{204}	P	99.95	129–169	25.2	160
Cadmium	Ag^{110}	S	99.99	180–300	25.4	2.21
α-Praseodymium	Ag^{110}	P	99.93	610 730	25.4	0.14
Lead	Pb^{204}	S	99.999	150–320	25.52	0.887
Tin	In^{114}	S‖c	99.998	181–221	25.6	12.2
Tin	Sn^{113}	S‖c	99.999	160–226	25.6	7.7
β-Praseodymium	La^{140}	P	99.96	800–930	25.7	1.8
Tin	In^{114}	S⊥c	99.998	181–221	25.8	34.1
Zinc	Ag^{110}	S‖c	99.999	271–413	26.0	0.32

Source: data from Askill, J.,in *Handbook of Chemistry and Physics, 55th ed.*,Weast, R.C., Ed., CRC Press, Cleveland,1974, F61.

Table 353. Selecting Diffusion Activation Energy in Metallic Systems*

(Sheet 11 of 33)

Metal	Tracer	Crystal Form	Purity %	Temperature Range °C	Activation Energy, Q $kcal \cdot mol^{-1}$	Frequency Factor, D_o $cm^2 \cdot s^{-1}$
Copper	Lu^{177}	P	99.999	857–1010	26.15	4.3×10^{-9}
β-Praseodymium	Ho^{166}	P	99.96	800–930	26.3	9.5
Chromium	C^{14}	P		120~1500	26.5	9.0×10^{-3}
Aluminum	Ce^{141}	P	99.995	450–630	26.60	1.9×10^{-6}
Copper	Eu^{152}	P	99.999	750–970	26.85	1.17×10^{-7}
Aluminum	Au^{198}	S	99.999	423–609	27.0	0.077
Aluminum	La^{140}	P	99.995	500–630	27.0	1.4×10^{-6}
Nickel	Sb^{124}	P	99.97	1020–1220	27.0	1.8×10^{-5}
b-Praseodymium	Zn^{65}	P	99.96	822–921	27.0	0.63
β–Zirconium	Ta^{182}	P	99.6	900–1200	27.0	5.5×10^{-5}
Vanadium	C^{14}	P	99.7	845–1130	27.3	4.9×10^{-3}
Magnesium	U^{235}	P	99.95	500–620	27.4	1.6×10^{-5}

Source: data from Askill, J.,in *Handbook of Chemistry and Physics, 55th ed.*,Weast, R.C., Ed., CRC Press, Cleveland,1974, F61.

Table 353. SELECTING DIFFUSION ACTIVATION ENERGY IN METALLIC SYSTEMS*

(SHEET 12 OF 33)

Metal	Tracer	Crystal Form	Purity %	Temperature Range °C	Activation Energy, Q $kcal \cdot mol^{-1}$	Frequency Factor, D_o $cm^2 \cdot s^{-1}$
Copper	Tb^{160}	P	99.999	770–980	27.45	8.96×10^{-9}
β–Uranium	Co^{60}	P	99.999	692–763	27.45	1.5×10^{-2}
Copper	Pm^{147}	P	99.999	720–955	27.5	3.62×10^{-8}
Aluminum	In^{114}	P	99.99	400–600	27.6	0.123
Copper	Ce^{141}	P	99.999	766–947	27.6	2.17×10^{-3}
Zinc	Ag^{110}	S⊥c	99.999	271–413	27.6	0.45
Aluminum	Co^{60}	S	99.999	369–655	27.79	0.131
Aluminum	Ag^{110}	S	99.999	371–655	27.83	0.118
Molybdenum	Cs^{134}	S	99.99	1000–1470	28.0	8.7×10^{-11}
Magnesium	In^{114}	P	99.9	472–610	28.4	5.2×10^{-2}
Aluminum	Sn^{113}	P		400–600	28.5	0.245
γ-Uranium	U^{233}	P	99.99	800–1070	28.5	2.33×10^{-3}

Source: data from Askill, J.,in *Handbook of Chemistry and Physics, 55th ed.*,Weast, R.C., Ed., CRC Press, Cleveland,1974, F61.

Table 353. SELECTING DIFFUSION ACTIVATION ENERGY IN METALLIC SYSTEMS*

(SHEET 13 OF 33)

Metal	Tracer	Crystal Form	Purity %	Temperature Range °C	Activation Energy, Q kcal • mol^{-1}	Frequency Factor, D_o cm^2 • s^{-1}
Magnesium	Ag^{110}	P	99.9	476–621	28.50	0.34
Magnesium	Zn^{65}	P	99.9	467–620	28.6	0.41
Tellurium	Se^{75}	P		320–440	28.6	2.6×10^{-2}
Aluminum	Mn^{54}	P	99.99	450–650	28.8	0.22
Aluminum	Zn^{65}	S	99.999	357–653	28.86	0.259
Calcium	Ni^{63}		99.95	550–800	28.9	1.0×10^{-6}
β-Praseodymium	In^{114}	P	99.96	800–930	28.9	9.6
Aluminum	Ge^{71}	S	99.999	401–653	28.98	0.481
Aluminum	Sb^{124}	P		448–620	29.1	0.09
Aluminum	Ga^{72}	S	99.999	406–652	29.24	0.49
α-Iron	C^{14}	P	99.98	616–844	29.3	2.2
β-Titanium	U^{235}	P	99.9	900–400	29.3	5.1×10^{-4}

Source: data from Askill, J.,in *Handbook of Chemistry and Physics, 55th ed.*,Weast, R.C., Ed., CRC Press, Cleveland,1974, F61.

Table 353. SELECTING DIFFUSION ACTIVATION ENERGY IN METALLIC SYSTEMS*

(SHEET 14 OF 33)

Metal	Tracer	Crystal Form	Purity %	Temperature Range °C	Activation Energy, Q kcal • mol^{-1}	Frequency Factor, D_o cm^2 • s^{-1}
b-Praseodymium	$Pr^{14}2$	P	99.93	800–900	29.4	8.7
Zinc	Cu64	S‖c	99.999	338–415	29.53	2.22
β-Titanium	Ni^{63}	P	99.7	925–1600	29.6	9.2×10^{-3}
Aluminum	Cd^{115}	S	99.999	441–631	29.7	1.04
Zinc	Au^{198}	S⊥c	99.999	315–415	29.72	0.29
Zinc	Au^{198}	S‖c	99.999	315–415	29.73	0.97
Calcium	C^{14}		99.95	550–800	29.8	3.2×10^{-5}
Selenium	S^{35}	S⊥c		60–90	29.9	1700
b–Zirconium	Zr^{95}	P		1100–1500	30.1	2.4×10^{-4}
γ-Uranium	Au^{195}	P	99.99	785–1007	30.4	4.86×10^{-3}
β–Zirconium	U^{235}	P		900–1065	30.5	5.7×10^{-4}
β-Titanium	Co^{60}	P	99.7	900–1600	30.6	1.2×10^{-2}

Source: data from Askill, J.,in *Handbook of Chemistry and Physics, 55th ed.*,Weast, R.C., Ed., CRC Press, Cleveland,1974, F61.

Table 353. SELECTING DIFFUSION ACTIVATION ENERGY IN METALLIC SYSTEMS*

(SHEET 15 OF 33)

Metal	Tracer	Crystal Form	Purity %	Temperature Range °C	Activation Energy, Q kcal • mol^{-1}	Frequency Factor, D_o cm^2 • s^{-1}
b–Zirconium	Be^7	P	99.7	915–1300	31.1	8.33 x 10^{-2}
β-Titanium	Ti^{44}	P	99.95	900–1540	31.2	3.58 x 10^{-4}
α-Zirconium	Nb^{95}	P	99.99	740–857	31.5	6.6 x 10^{-6}
β-Titanium	Fe^{59}	P	99.7	900–1600	31.6	7.8 x 10^{-3}
b-Titanium	Sn^{113}	P	99.7	950–1600	31.6	3.8 x 10^{-4}
Niobium	C^{14}	P		800–1250	32.0	1.09 x 10^{-5}
Magnesium	Mg^{28}	S‖c		467–635	32.2	1.0
β-Titanium	V^{48}	P	99.95	900–1545	32.2	3.1 x 10^{-4}
Aluminum	Cu^{64}	S	99.999	433–652	32.27	0.647
β-Titanium	Sc^{46}	P	99.95	940–1590	32.4	4.0 x 10^{-3}
Magnesium	Mg^{28}	S⊥c		467–635	32.5	1.5
β–Zirconium	P^{32}	P	99.94	950–1200	33.3	0.33

Source: data from Askill, J.,in *Handbook of Chemistry and Physics, 55th ed.*,Weast, R.C., Ed., CRC Press, Cleveland,1974, F61.

Table 353. SELECTING DIFFUSION ACTIVATION ENERGY IN METALLIC SYSTEMS*

(SHEET 16 OF 33)

Metal	Tracer	Crystal Form	Purity %	Temperature Range °C	Activation Energy, Q $kcal \cdot mol^{-1}$	Frequency Factor, D_o $cm^2 \cdot s^{-1}$
b-Titanium	Mn^{54}	P	99.7	900–1600	33.7	6.1×10^{-3}
Aluminum	Al^{27}	S		450–650	34.0	1.71
Cobalt	C^{14}	P	99.82	600–1400	34.0	0.21
γ-Iron	C^{14}	P	99.34	800–1400	34.0	0.15
Nickel	C^{14}	P	99.86	600–1400	34.0	0.012
Vanadium	S^{35}	P	99.8	1320 1520	34.0	3.1×10^{-2}
β–Zirconium	C^{14}	P	96.6	1100–1600	34.2	3.57×10^{-2}
Calcium	U^{235}		99.95	500–700	34.8	1.1×10^{-5}
b-Titanium	Cr^{51}	P	99.7	950–1600	35.1	5×10^{-3}
β–Zirconium	Mo^{99}	P		900–1635	35.2	1.99×10^{-6}
β-Titanium	Zr^{95}	P	98.94	920–1500	35.4	4.7×10^{-3}
Tellurium	Te^{127}	S‖c	99.9999	300–400	35.5	130

Source: data from Askill, J.,in *Handbook of Chemistry and Physics, 55th ed.*,Weast, R.C., Ed., CRC Press, Cleveland,1974, F61.

Table 353. SELECTING DIFFUSION ACTIVATION ENERGY IN METALLIC SYSTEMS*

(SHEET 17 OF 33)

Metal	Tracer	Crystal Form	Purity %	Temperature Range °C	Activation Energy, Q kcal • mol^{-1}	Frequency Factor, D_o cm^2 • s^{-1}
a-Titanium	Ti^{44}	P	99.99	700–850	35.9	8.6 x 10^{-6}
Silver	Ge^{77}	P		640–870	36.5	0.084
β–Zirconium	Nb^{95}	P		1230–1635	36.6	7.8 x 10^{-4}
Gold	Hg^{203}	S	99.994	600–1027	37.38	0.116
Copper	Pt^{195}	P		843–997	37.5	4.8 x 10^{-4}
Beryllium	Be^{7}	S⊥c	99.75	565–1065	37.6	0.52
Silver	Tl^{204}	P		640–870	37.9	0.15
Silver	$Hg^{20}3$	P	99.99	653–948	38.1	0.079
Silver	Pb^{210}	P		700–865	38.1	0.22
Calcium	Ca^{45}		99.95	500–800	38.5	8.3
β–Hafnium	Hf^{181}	P	97.9	1795–1995	38.7	1.2 x10^{-3}
Silver	Te^{125}	P		770–940	38.90	0.47

Source: data from Askill, J.,in *Handbook of Chemistry and Physics, 55th ed.*,Weast, R.C., Ed., CRC Press, Cleveland,1974, F61.

Table 353. SELECTING DIFFUSION ACTIVATION ENERGY IN METALLIC SYSTEMS*

(SHEET 18 OF 33)

Metal	Tracer	Crystal Form	Purity %	Temperature Range °C	Activation Energy, Q kcal • mol^{-1}	Frequency Factor, D_o cm^2 • s^{-1}
Silver	Sb^{124}	P	99.999	780–950	39.07	0.234
Beryllium	Ag^{110}	S‖c	99.75	650–900	39.3	0.43
β-Titanium	Nb^{95}	P	99.7	1000–1600	39.3	5.0×10^{-3}
Silver	Sn^{113}	S	99.99	592–937	39.30	0.255
Beryllium	Be^{7}	S‖c	99.75	565–1065	39.4	0.62
γ-Uranium	Nb^{95}	P	99.99	791–1102	39.65	4.87×10^{-2}
Germanium	In^{114}	S		600–920	39.9	2.9×10^{-4}
Silver	S^{35}	S	99.999	600–900	40.0	1.65
a–Uranium	U^{234}	P		580–650	40.0	2×10^{-3}
Gold	Ag^{110}	S	99.99	699–1007	40.2	0.072
β-Titanium	Be^{7}	P	99.96	915–1300	40.2	0.8
Tantalum	C^{14}	P		1450–2200	40.3	1.2×10^{-2}

Source: data from Askill, J.,in *Handbook of Chemistry and Physics, 55th ed.*,Weast, R.C., Ed., CRC Press, Cleveland,1974, F61.

Table 353. SELECTING DIFFUSION ACTIVATION ENERGY IN METALLIC SYSTEMS*

(SHEET 19 OF 33)

Metal	Tracer	Crystal Form	Purity %	Temperature Range °C	Activation Energy, Q kcal • mol^{-1}	Frequency Factor, D_o cm^2 • s^{-1}
Silver	In^{114}	S	99.99	592–937	40.80	0.41
Molybdenum	C^{14}	P	99.98	1200–1600	41.0	2.04×10^{-2}
Tellurium	Tl^{204}	P		360–430	41.0	320
β–Zirconium	Ce^{141}	P		880–1600	41.4	3.16
Lithium	Sb^{124}	P	99.95	141–176	41.5	1.6×10^{10}
Silicon	P^{32}	S		1100–1250	41.5	–
Gold	Co^{60}	P	99.93	702–948	41.6	0.068
Gold	Fe^{59}	P	99.93	701–948	41.6	0.082
Silver	Cd^{115}	S	99.99	592–937	41.69	0.44
Silver	Zn^{65}	S	99.99	640–925	41.7	0.54
Aluminum	Cr^{51}	S	99.999	422–654	41.74	464
Copper	Sb^{124}	S	99.999	600–1000	42.0	0.34

Source: data from Askill, J.,in *Handbook of Chemistry and Physics, 55th ed.*,Weast, R.C., Ed., CRC Press, Cleveland,1974, F61.

Table 353. SELECTING DIFFUSION ACTIVATION ENERGY IN METALLIC SYSTEMS*

(SHEET 20 OF 33)

Metal	Tracer	Crystal Form	Purity %	Temperature Range °C	Activation Energy, Q kcal • mol^{-1}	Frequency Factor, D_o cm^2 • s^{-1}
Copper	As^{76}	P		810–1075	42.13	0.20
Gold	Au^{198}	S	99.97	850–1050	42.26	0.107
α-Iron	K^{42}	P	99.92	500–800	42.3	0.036
Copper	Au^{193}	S, P		400–1050	42.6	0.03
b-Titanium	Mo^{99}	P	99.7	900–1600	43.0	8.0 x 10^{-3}
Beryllium	Ag^{110}	S⊥c	99.75	650–900	43.2	1.76
β-Titanium	Ag^{110}	P	99.95	940 1570	43.2	3 x 10^{-3}
Copper	Tl^{204}	S	99.999	785–996	43.3	0.71
g-Iron	P^{32}	P	99.99	950–1200	43.7	0.01
β-Titanium	W^{185}	P	99.94	900–1250	43.9	3.6 x 10^{-3}
Copper	Hg^{203}	P		–	44.0	0.35
β–Uranium	U^{235}	P		690–750	44.2	2.8 x10^{-3}

Source: data from Askill, J.,in *Handbook of Chemistry and Physics, 55th ed.*,Weast, R.C., Ed., CRC Press, Cleveland,1974, F61.

Table 353. SELECTING DIFFUSION ACTIVATION ENERGY IN METALLIC SYSTEMS*

(SHEET 21 OF 33)

Metal	Tracer	Crystal Form	Purity %	Temperature Range °C	Activation Energy, Q kcal • mol^{-1}	Frequency Factor, D_o $cm^2 \cdot s^{-1}$
Copper	Ge^{68}	S	99.998	653–1015	44.76	0.397
Copper	Sn^{113}	P		680–910	45.0	0.11
Lanthanum	Au^{198}	P	99.97	600–800	45.1	1.5
Silver	Ag^{110}	S	99.999	640–955	45.2	0.67
a-Zirconium	Zr^{95}	P	99.95	750–850	45.5	5.6×10^{-4}
Copper	Cd^{115}	S	99.98	725–950	45.7	0.935
β–Zirconium	V^{48}	P	99.99	870–1200	45.8	7.59×10^{-3}
Copper	Ga^{72}			–	45.90	0.55
Aluminum	Fe^{59}	S	99.99	550–636	46.0	135
Gold	Ni^{63}	P	99.96	880–940	46.0	0.30
Silver	Cu^{64}	P	99.99	717–945	46.1	1.23
Nickel	Be^{7}	P	99.9	1020–1400	46.2	0.019

Source: data from Askill, J.,in *Handbook of Chemistry and Physics, 55th ed.*,Weast, R.C., Ed., CRC Press, Cleveland,1974, F61.

Table 353. SELECTING DIFFUSION ACTIVATION ENERGY IN METALLIC SYSTEMS*

(SHEET 22 OF 33)

Metal	Tracer	Crystal Form	Purity %	Temperature Range °C	Activation Energy, Q kcal • mol^{-1}	Frequency Factor, D_o cm^2 • s^{-1}
Copper	Ag^{110}	S, P		580–980	46.5	0.61
Tellurium	Te^{127}	S⊥c	99.9999	300–400	46.7	3.91×10^{4}
Silicon	Au^{198}	S		700–1300	47.0	2.75×10^{-3}
Carbon	Ni^{63}	⊥c		540–920	47.2	102
Lithium	Bi	P	99.95	141–177	47.3	5.3×10^{13}
Copper	Zn^{65}	P	99.999	890–1000	47.50	0.73
α-Zirconium	Fe^{55}	P		750–840	48.0	2.5×10^{-2}
β–Zirconium	Fe^{55}	P		750–840	48.0	2.5×10^{-2}
Silver	Au^{198}	P	99.99	718–942	48.28	0.85
Silver	Co^{60}	S	99.999	700–940	48.75	1.9
Silver	Fe^{59}	S	99.99	720–930	49.04	2.42
Copper	S^{35}	S	99.999	800–1000	49.2	23

Source: data from Askill, J.,in *Handbook of Chemistry and Physics, 55th ed.*,Weast, R.C., Ed., CRC Press, Cleveland,1974, F61.

Table 353. SELECTING DIFFUSION ACTIVATION ENERGY IN METALLIC SYSTEMS*

(SHEET 23 OF 33)

Metal	Tracer	Crystal Form	Purity %	Temperature Range °C	Activation Energy, Q kcal • mol^{-1}	Frequency Factor, D_o cm^2 • s^{-1}
Vanadium	P^{32}	P	99.8	1200–1450	49.8	2.45 x 10^{-2}
Germanium	Sb^{124}	S		720–900	50.2	0.22
Copper	Cu^{67}	S	99.999	698–1061	50.5	0.78
Nickel	Mo^{99}	P		900–1200	51.0	1.6 x 10^{-3}
Nickel	Pu^{238}	P		1025–1125	51.0	0.5
Niobium	P^{32}	P	99.0	1300–1800	51.5	5.1 x 10^{-2}
Beryllium	Fe^{59}	S	99.75	700–1076	51.6	0.67
Copper	Fe^{59}	S. P		460–1070	52.0	1.36
a-Iron	Mn^{54}	P	99.97	800–900	52.5	0.35
γ-Iron	S^{35}	P		900–1250	53.0	1.7
Carbon	Ni^{63}	‖c		750–1060	53.3	2.2
Copper	Cr^{51}	S, P		800–1070	53.5	1.02

Source: data from Askill, J.,in *Handbook of Chemistry and Physics, 55th ed.*,Weast, R.C., Ed., CRC Press, Cleveland,1974, F61.

Table 353. SELECTING DIFFUSION ACTIVATION ENERGY IN METALLIC SYSTEMS*

(SHEET 24 OF 33)

Metal	Tracer	Crystal Form	Purity %	Temperature Range °C	Activation Energy, Q kcal • mol^{-1}	Frequency Factor, D_o $cm^2 \cdot s^{-1}$
Tungsten	C^{14}	P	99.51	1200–1600	53.5	8.91×10^{-3}
Copper	Ni^{63}	P		620–1080	53.8	1.1
Molybdenum	Cr^{51}	P		1000–1500	54.0	2.5 x 10–4
Copper	Co^{60}	S	99.998	701–1077	54.1	1.93
Copper	Pd^{102}	S	99.999	807–1056	54.37	1.71
Silver	Ni^{63}	S	99.99	749–950	54.8	21.9
α-Iron	P^{32}	P		860–900	55.0	2.9
δ-Iron	P^{32}	P	99.99	1370–1460	55.0	2.9
Nickel	Au^{198}	S,P	99.999	700–1075	55.0	0.02
α-Iron	W^{185}	P		755–875	55.1	0.29
α-Iron	V^{48}	P		755–875	55.4	1.43
β–Zirconium	W^{185}	P	99.7	900–1250	55.8	0.41

Source: data from Askill, J.,in *Handbook of Chemistry and Physics, 55th ed.*,Weast, R.C., Ed., CRC Press, Cleveland,1974, F61.

Table 353. SELECTING DIFFUSION ACTIVATION ENERGY IN METALLIC SYSTEMS*

(SHEET 25 OF 33)

Metal	Tracer	Crystal Form	Purity %	Temperature Range °C	Activation Energy, Q kcal • mol^{-1}	Frequency Factor, D_o $cm^2 \cdot s^{-1}$
Germanium	Te^{125}	S		770–900	56.0	2.0
α-Iron	Ni^{63}	P	99.97	680–800	56.0	1.3
Silver	Pd^{102}	S	99.999	736–939	56.75	9.56
α-Iron	Cu^{64}	P	99.9	800 1050	57.0	0.57
a-Iron	Cr^{51}	P	99.95	775–875	57.5	2.53
δ-Iron	Fe^{59}	P	99.95	1428–1492	57.5	2.01
γ-Iron	Be^{7}	P	99.9	1100–1350	57.6	0.1
β–Zirconium	V^{48}	P	99.99	1200–1400	57.7	0.32
Beryllium	Ni^{63}	P		800–1250	58.0	0.2
Chromium	Mo^{99}	P		1100–1420	58.0	2.7×10^{-3}
Nickel	Sn^{113}	P	99.8	700–1350	58.0	0.83
Nickel	Fe^{59}	P		1020–1263	58.6	0.074

Source: data from Askill, J.,in *Handbook of Chemistry and Physics, 55th ed.*,Weast, R.C., Ed., CRC Press, Cleveland,1974, F61.

Table 353. SELECTING DIFFUSION ACTIVATION ENERGY IN METALLIC SYSTEMS*

(SHEET 26 OF 33)

Metal	Tracer	Crystal Form	Purity %	Temperature Range °C	Activation Energy, Q $kcal \cdot mol^{-1}$	Frequency Factor, D_o $cm^2 \cdot s^{-1}$
Platinum	Cu^{64}	P		1098–1375	59.5	0.074
Copper	Nb^{95}	P	99.999	807–906	60.06	2.04
Cobalt	Ni^{63}	P		1192–1297	60.2	0.10
α-Iron	Fe^{55}	P	99.92	809–889	60.3	5.4
Yttrium	Y^{90}	S‖c		900–1300	60.3	0.82
Gold	Pt^{195}	P, S	99.98	800–1060	60.9	7.6
δ-Iron	Co^{60}	P	99.995	1428–1521	61.4	6.38
Nickel	Cu^{64}	P	99.95	1050–1360	61.7	0.57
a-Iron	Co^{60}	P	99.995	638–768	62.2	7.19
α-Iron	Au^{198}	P	99.999	800–900	62.4	31
γ-Iron	Mn^{54}	P	99.97	920–1280	62.5	0.16
Cobalt	Fe^{59}	P	99.9	1104–1303	62.7	0.21

Source: data from Askill, J.,in *Handbook of Chemistry and Physics, 55th ed.*,Weast, R.C., Ed., CRC Press, Cleveland,1974, F61.

Table 353. SELECTING DIFFUSION ACTIVATION ENERGY IN METALLIC SYSTEMS*

(SHEET 27 OF 33)

Metal	Tracer	Crystal Form	Purity %	Temperature Range °C	Activation Energy, Q $kcal \cdot mol^{-1}$	Frequency Factor, D_o $cm^2 \cdot s^{-1}$
Palladium	Pd^{103}	S	99.999	1060–1500	63.6	0.205
Carbon	Ag^{110}	⊥c		750–1050	64.3	9280
Vanadium	Cr^{51}	P	99.8	960–1200	64.6	9.54×10^{-3}
Nickel	Cr^{51}	P	99.95	1100–1270	65.1	1.1
Silver	Ru^{103}	S	99.99	793–945	65.8	180
Nickel	Co^{60}	P	99.97	1149–1390	65.9	1.39
Tungsten	Fe^{59}	P		940–1240	66.0	1.4×10^{-2}
Nickel	V^{48}	P	99.99	800–1300	66.5	0.87
a-Iron	Sb^{124}	P		800–900	66.6	1100
γ-Iron	Ni^{63}	P	99.97	930–2050	67.0	0.77
Yttrium	Y^{90}	S⊥c		900–1300	67.1	5.2
Silicon	C^{14}	P		1070–1400	67.2	0.33

Source: data from Askill, J.,in *Handbook of Chemistry and Physics, 55th ed.*,Weast, R.C., Ed., CRC Press, Cleveland,1974, F61.

Table 353. SELECTING DIFFUSION ACTIVATION ENERGY IN METALLIC SYSTEMS*

(SHEET 28 OF 33)

Metal	Tracer	Crystal Form	Purity %	Temperature Range °C	Activation Energy, Q $kcal \cdot mol^{-1}$	Frequency Factor, D_o $cm^2 \cdot s^{-1}$
Cobalt	Co^{60}	P	99.9	1100–1405	67.7	0.83
γ-Iron	Fe^{59}	P	99.98	1171–1361	67.86	0.49
Nickel	Ni^{63}	P	99.95	1042–1404	68.0	1.9
Platinum	Pt^{195}	P	99.99	1325–1600	68.2	0.33
Germanium	Ge^{71}	S		766–928	68.5	7.8
α-Iron	Ag^{110}	P		748–888	69.0	1950
γ-Iron	V^{48}	P	9999	1120–1380	69.3	0.28
γ-Iron	Cr^{51}	P	99.99	950–1400	69.7	10.8
Tantalum	S^{35}	P	99.0	1970–2110	70.0	100
α-Zirconium	Ta^{182}	P	99.6	700–800	70.0	100
Niobium	Co^{60}	P	99.85	1500–2100	70.5	0.74
Vanadium	Fe^{59}	P		960–1350	71.0	0.373

Source: data from Askill, J.,in *Handbook of Chemistry and Physics, 55th ed.*,Weast, R.C., Ed., CRC Press, Cleveland,1974, F61.

Table 353. SELECTING DIFFUSION ACTIVATION ENERGY IN METALLIC SYSTEMS*

(SHEET 29 OF 33)

Metal	Tracer	Crystal Form	Purity %	Temperature Range °C	Activation Energy, Q kcal • mol^{-1}	Frequency Factor, D_o cm^2 • s^{-1}
Tantalum	Fe^{59}	P		930–1240	71.4	0.505
Nickel	W^{185}	P	99.95	1100–1300	71.5	2.0
γ-Iron	Co^{60}	P	99.98	1138–1340	72.9	1.25
α-Iron	Mo^{99}	P		750–875	73.0	7800
Niobium	S^{35}	S	99.9	1100–1500	73.1	2600
Vanadium	V^{48}	S,P	99.99	880–1360	73.65	0.36
Chromium	Cr^{51}	P	99.98	1030–1545	73.7	0.2
Platinum	Co^{60}	P	99.99	900–1050	74.2	19.6
a-Thorium	Pa^{231}	P	99.85	770–910	74.7	126
Molybdenum	U^{235}	P	99.98	1500–2000	76.4	7.6×10^{-3}
Niobium	U^{235}	P	99.55	1500–2000	76.8	8.9×10^{-3}
Niobium	Fe^{51}	P	99.85	1400–2100	77.7	1.5

Source: data from Askill, J.,in *Handbook of Chemistry and Physics, 55th ed.*,Weast, R.C., Ed., CRC Press, Cleveland,1974, F61.

Table 353. SELECTING DIFFUSION ACTIVATION ENERGY IN METALLIC SYSTEMS*

(SHEET 30 OF 33)

Metal	Tracer	Crystal Form	Purity %	Temperature Range °C	Activation Energy, Q $kcal \cdot mol^{-1}$	Frequency Factor, D_o $cm^2 \cdot s^{-1}$
Germanium	Tl^{204}	S		800–930	78.4	1700
Niobium	Sn^{113}	P	99.85	1850–2400	78.9	0.14
Chromium	Fe^{59}	P	99.8	980–1420	79.3	0.47
α-Thorium	U^{233}	P	99.85	700–880	79.3	2210
Molybdenum	P^{32}	P	99.97	2000–2200	80.5	0.19
Tantalum	Mo^{99}	P		1750–2220	81.0	1.8×10^{-3}
Molybdenum	Ta^{182}	P		1700–2150	83.0	3.5×10^{-4}
Niobium	Cr^{51}	S		943–1435	83.5	0.30
Niobium	V^{48}	S	99.99	1000–1400	85.0	2.21
Niobium	Ti^{44}	S		994–1492	86.9	0.099
γ-Iron	W^{185}	P	99.5	1050–1250	90.0	1000
Copper	Mn^{54}	S	99.99	754–950	91.4	10^7

Source: data from Askill, J.,in *Handbook of Chemistry and Physics, 55th ed.*,Weast, R.C., Ed., CRC Press, Cleveland,1974, F61.

Table 353. Selecting Diffusion Activation Energy in Metallic Systems*

(Sheet 31 of 33)

Metal	Tracer	Crystal Form	Purity %	Temperature Range °C	Activation Energy, Q $kcal \cdot mol^{-1}$	Frequency Factor, D_o $cm^2 \cdot s^{-1}$
Niobium	W^{185}	P	99.8	1800–2200	91.7	5×10^{-4}
Silicon	Sb^{124}	S		1190–1398	91.7	12.9
Vanadium	V^{48}	S,P	99.99	1360–1830	94.14	214.0
Molybdenum	Re^{186}	P		1700–2100	94.7	0.097
Niobium	Nb^{95}	P, S	99.99	878–2395	96.0	1.1
Molybdenum	Mo^{99}	P		1850–2350	96.9	0.5
Silicon	Ni^{63}	P		450–800	97.5	1000
γ-Iron	Hf^{181}	P	99.99	1110–1360	97.3	3600
Tantalum	Nb^{95}	P, S	99.996	921–2484	98.7	0.23
Tantalum	Ta^{182}	P, S	99.996	1250–2200	98.7	1.24
Niobium	Ta^{182}	P, S	99.997	878–2395	99.3	1.0
Molybdenum	S^{35}	S	99.97	2220–2470	101.0	320

Source: data from Askill, J.,in *Handbook of Chemistry and Physics, 55th ed.*,Weast, R.C., Ed., CRC Press, Cleveland,1974, F61.

Table 353. Selecting Diffusion Activation Energy in Metallic Systems*

(Sheet 32 of 33)

Metal	Tracer	Crystal Form	Purity %	Temperature Range °C	Activation Energy, Q kcal • mol^{-1}	Frequency Factor, D_o $cm^2 \cdot s^{-1}$
Tungsten	Mo^{99}	P		1700–2100	101.0	0.3
Germanium	Cd^{115}	S		750–950	102.0	1.75 x 10^9
Molybdenum	Co^{60}	P	99.98	1850–2350	106.7	18
Molybdenum	Nb^{95}	P	99.98	1850–2350	108.1	14
Molybdenum	W^{185}	P	99.98	1700–2260	110	1.7
Silicon	Si^{31}	S	99.99999	1225–1400	110.0	1800
Carbon	Th^{228}	‖c		1800–2200	114.7	2.48
Carbon	U^{232}	⊥c		140~2200	115.0	6760
Carbon	U^{232}	‖c		1400 1820	129.5	385
Tungsten	Nb^{95}	P	99.99	1305–2367	137.6	3.01
Tungsten	Ta^{182}	P	99.99	1305–2375	139.9	3.05
Tungsten	W^{185}	P	99.99	1800–2403	140.3	1.88

Source: data from Askill, J.,in *Handbook of Chemistry and Physics, 55th ed.*,Weast, R.C., Ed., CRC Press, Cleveland,1974, F61.

Table 353. SELECTING DIFFUSION ACTIVATION ENERGY IN METALLIC SYSTEMS*

(SHEET 33 OF 33)

Metal	Tracer	Crystal Form	Purity %	Temperature Range °C	Activation Energy, Q kcal • mol^{-1}	Frequency Factor, D_o cm^2 • s^{-1}
Tungsten	Re^{186}	S		2100–2400	141.0	19.5
Carbon	Th^{228}	⊥c		1400–2200	145.4	1.33 x 10^{-5}
Carbon	C^{14}			2000–2200	163	5
α-Thorium	Th^{228}	P	99.85	720–880	716	395

Source: data from Askill, J.,in *Handbook of Chemistry and Physics, 55th ed.*,Weast, R.C., Ed., CRC Press, Cleveland,1974, F61.

* The diffusion coefficient DT at a temperature T(K) is given by the following:

$D_T = D_o e^{-Q/RT}$

Abbreviations:

P= polycrystalline
S = single crystal
⊥ c = perpendicular to c direction
|| c = parallel to c direction

CHAPTER 12

Selecting Thermal Properties

List of Tables

Thermal Conductivity

Thermal Expansion

Table 354. SELECTING THERMAL CONDUCTIVITY OF METALS*
(SHEET 1 OF 29)

Metal	Temperature (K)	Thermal Conductivity ($watt \cdot cm^{-1} \cdot K^{-1}$)
Titanium	1	0.0144
Titanium	2	0.0288
Titanium	3	0.0432
Titanium	4	0.0576
Titanium	5	0.0719
Titanium	6	0.0863
Titanium	7	0.101
Zirconium	1	0.111
Titanium	8	0.115
Tantalum	1	0.115
Titanium	9	0.129
Titanium	10	0.144
Molybdenum	1	0.146
Titanium	11	0.158
Titanium	12	0.172
Titanium	13	0.186
Titanium	600	0.194
Titanium	700	0.194
Titanium	500	0.197
Titanium	800	0.197
Titanium	14	0.2
Titanium	900	0.202
Titanium	400	0.204
Zirconium	600	0.207
Titanium	1000	0.207
Zirconium	700	0.209
Zirconium	500	0.21
Titanium	1100	0.213

Source: data from Ho, C. Y., Powell, R. W., and Liley, P. E., *Thermal Conductictivity of Selected Materials,* NSRDS–NBS–8 and NSRD-NBS-16, Part 2 , National Standard Reference Data System–National Bureau of Standards, Part 1, 1966; Part 2, 1968.

Table 354. SELECTING THERMAL CONDUCTIVITY OF METALS*

(SHEET 2 OF 29)

Metal	Temperature (K)	Thermal Conductivity (watt • cm^{-1} • K^{-1})
Titanium	15	0.214
Zirconium	400	0.216
Zirconium	800	0.216
Titanium	300	0.219
Titanium	1200	0.22
Zirconium	2	0.223
Titanium	273	0.224
Zirconium	900	0.226
Zirconium	300	0.227
Titanium	16	0.227
Tantalum	2	0.23
Zirconium	273	0.232
Titanium	1400	0.236
Zirconium	1000	0.237
Titanium	200	0.245
Zirconium	1100	0.248
Niobium	1	0.251
Zirconium	200	0.252
Titanium	1600	0.253
Titanium	18	0.254
Zirconium	1200	0.257
Titanium	1800	0.271
Zirconium	1400	0.275
Titanium	20	0.279
Iron	1200	0.282
Zirconium	1600	0.29
Molybdenum	2	0.292
Iron	1100	0.297

Source: data from Ho, C. Y., Powell, R. W., and Liley, P. E., *Thermal Conductictivity of Selected Materials*, NSRDS–NBS–8 and NSRD-NBS-16, Part 2 , National Standard Reference Data System–National Bureau of Standards, Part 1, 1966; Part 2, 1968.

Table 354. SELECTING THERMAL CONDUCTIVITY OF METALS*

(SHEET 3 OF 29)

Metal	Temperature (K)	Thermal Conductivity (watt • cm^{-1} • K^{-1})
Zirconium	1800	0.302
Iron	1400	0.309
Titanium	100	0.312
Lead	600	0.312
Zirconium	2000	0.313
Titanium	90	0.324
Lead	500	0.325
Iron	1000	0.326
Iron	1600	0.327
Zirconium	100	0.332
Zirconium	3	0.333
Titanium	25	0.337
Lead	400	0.338
Titanium	80	0.339
Tantalum	3	0.345
Zirconium	90	0.35
Lead	300	0.352
Lead	273	0.355
Titanium	70	0.356
Lead	200	0.366
Zirconium	80	0.373
Titanium	60	0.377
Iron	900	0.38
Titanium	30	0.382
Lead	100	0.396
Titanium	50	0.401
Lead	90	0.401
Chromium	1	0.401

Source: data from Ho, C. Y., Powell, R. W., and Liley, P. E., *Thermal Conductictivity of Selected Materials*, NSRDS–NBS–8 and NSRD-NBS-16, Part 2 , National Standard Reference Data System–National Bureau of Standards, Part 1, 1966; Part 2, 1968.

Table 354. SELECTING THERMAL CONDUCTIVITY OF METALS*

(SHEET 4 OF 29)

Metal	Temperature (K)	Thermal Conductivity (watt • cm^{-1} • K^{-1})
Zirconium	70	0.403
Lead	80	0.407
Titanium	35	0.411
Lead	70	0.415
Titanium	45	0.416
Titanium	40	0.422
Lead	60	0.424
Iron	800	0.433
Lead	50	0.435
Molybdenum	3	0.438
Zirconium	4	0.442
Zirconium	60	0.442
Lead	45	0.442
Lead	40	0.451
Tantalum	4	0.459
Lead	35	0.462
Lead	30	0.477
Iron	700	0.487
Zirconium	50	0.497
Niobium	2	0.501
Lead	25	0.507
Niobium	200	0.526
Niobium	273	0.533
Zirconium	45	0.535
Niobium	300	0.537
Iron	600	0.547
Zirconium	5	0.549
Niobium	100	0.552

Source: data from Ho, C. Y., Powell, R. W., and Liley, P. E., *Thermal Conductictivity of Selected Materials*, NSRDS–NBS–8 and NSRD-NBS-16, Part 2 , National Standard Reference Data System–National Bureau of Standards, Part 1, 1966; Part 2, 1968.

Table 354. SELECTING THERMAL CONDUCTIVITY OF METALS*
(SHEET 5 OF 29)

Metal	Temperature (K)	Thermal Conductivity (watt • cm^{-1} • K^{-1})
Niobium	400	0.552
Niobium	90	0.563
Niobium	500	0.567
Tantalum	5	0.571
Tantalum	273	0.574
Tantalum	200	0.575
Tantalum	300	0.575
Tantalum	400	0.578
Zirconium	40	0.58
Niobium	80	0.58
Tantalum	500	0.582
Niobium	600	0.582
Molybdenum	4	0.584
Tantalum	600	0.586
Tantalum	700	0.59
Lead	20	0.59
Tantalum	100	0.592
Tantalum	800	0.594
Tin	500	0.596
Tantalum	90	0.596
Tantalum	900	0.598
Niobium	700	0.598
Tantalum	1000	0.602
Tantalum	80	0.603
Tantalum	1100	0.606
Tantalum	1200	0.61
Niobium	70	0.61
Chromium	1400	0.611

Source: data from Ho, C. Y., Powell, R. W., and Liley, P. E., *Thermal Conductictivity of Selected Materials*, NSRDS–NBS–8 and NSRD-NBS-16, Part 2 , National Standard Reference Data System–National Bureau of Standards, Part 1, 1966; Part 2, 1968.

Table 354. SELECTING THERMAL CONDUCTIVITY OF METALS*

(SHEET 6 OF 29)

Metal	Temperature (K)	Thermal Conductivity (watt • cm^{-1} • K^{-1})
Niobium	800	0.613
Iron	500	0.613
Tantalum	70	0.616
Tantalum	1400	0.618
Tin	400	0.622
Chromium	1200	0.624
Tantalum	1600	0.626
Niobium	900	0.629
Tantalum	1800	0.634
Chromium	1100	0.636
Tantalum	2000	0.64
Nickel	1	0.64
Niobium	1000	0.644
Tantalum	2200	0.647
Zirconium	35	0.65
Tantalum	60	0.651
Zirconium	6	0.652
Nickel	700	0.653
Chromium	1000	0.653
Nickel	600	0.655
Tantalum	2600	0.658
Niobium	1100	0.659
Niobium	60	0.66
Lead	18	0.66
Tantalum	3000	0.665
Tin	300	0.666
Nickel	800	0.674
Niobium	1200	0.675

Source: data from Ho, C. Y., Powell, R. W., and Liley, P. E., *Thermal Conductictivity of Selected Materials*, NSRDS–NBS–8 and NSRD-NBS-16, Part 2 , National Standard Reference Data System–National Bureau of Standards, Part 1, 1966; Part 2, 1968.

Table 354. SELECTING THERMAL CONDUCTIVITY OF METALS*

(SHEET 7 OF 29)

Metal	Temperature (K)	Thermal Conductivity (watt • cm^{-1} • K^{-1})
Chromium	900	0.678
Tantalum	6	0.681
Tin	273	0.682
Iron	400	0.694
Nickel	900	0.696
Niobium	1400	0.705
Chromium	800	0.713
Nickel	1000	0.718
Platinum	500	0.719
Tantalum	50	0.72
Platinum	600	0.72
Nickel	500	0.721
Platinum	400	0.722
Platinum	700	0.723
Platinum	800	0.729
Platinum	300	0.73
Molybdenum	5	0.73
Tin	200	0.733
Platinum	273	0.734
Niobium	1600	0.735
Platinum	900	0.737
Nickel	1100	0.739
Zirconium	30	0.74
Zirconium	7	0.748
Platinum	200	0.748
Platinum	1000	0.748
Niobium	3	0.749
Iron	1	0.75

Source: data from Ho, C. Y., Powell, R. W., and Liley, P. E., *Thermal Conductictivity of Selected Materials*, NSRDS–NBS–8 and NSRD-NBS-16, Part 2 , National Standard Reference Data System–National Bureau of Standards, Part 1, 1966; Part 2, 1968.

Table 354. SELECTING THERMAL CONDUCTIVITY OF METALS*
(SHEET 8 OF 29)

Metal	Temperature (K)	Thermal Conductivity (watt • cm^{-1} • K^{-1})
Chromium	700	0.757
Platinum	1100	0.76
Niobium	50	0.76
Nickel	1200	0.761
Niobium	1800	0.764
Lead	16	0.77
Platinum	1200	0.775
Tantalum	45	0.78
Tantalum	7	0.788
Platinum	100	0.79
Niobium	2000	0.791
Nickel	400	0.801
Chromium	2	0.802
Iron	300	0.803
Nickel	1400	0.804
Chromium	600	0.805
Platinum	1400	0.807
Platinum	90	0.81
Niobium	2200	0.815
Molybdenum	2600	0.825
Iron	273	0.835
Zirconium	8	0.837
Platinum	80	0.84
Niobium	45	0.84
Lead	15	0.84
Platinum	1600	0.842
Chromium	500	0.848
Zirconium	25	0.85

Source: data from Ho, C. Y., Powell, R. W., and Liley, P. E., *Thermal Conductictivity of Selected Materials*, NSRDS–NBS–8 and NSRD-NBS-16, Part 2 , National Standard Reference Data System–National Bureau of Standards, Part 1, 1966; Part 2, 1968.

Table 354. SELECTING THERMAL CONDUCTIVITY OF METALS*

(SHEET 9 OF 29)

Metal	Temperature (K)	Thermal Conductivity (watt • cm^{-1} • K^{-1})
Tin	100	0.85
Molybdenum	2200	0.858
Tantalum	40	0.87
Chromium	400	0.873
Molybdenum	6	0.876
Platinum	1800	0.877
Tin	90	0.88
Molybdenum	2000	0.88
Tantalum	8	0.891
Platinum	70	0.9
Chromium	300	0.903
Nickel	300	0.905
Molybdenum	1800	0.907
Tin	80	0.91
Platinum	2000	0.913
Tungsten	3000	0.915
Zirconium	9	0.916
Cadmium	500	0.92
Tungsten	2600	0.94
Nickel	273	0.94
Lead	14	0.94
Iron	200	0.94
Molybdenum	1600	0.946
Cadmium	400	0.947
Chromium	273	0.948
Tin	70	0.96
Cadmium	300	0.968
Niobium	40	0.97

Source: data from Ho, C. Y., Powell, R. W., and Liley, P. E., *Thermal Conductictivity of Selected Materials*, NSRDS–NBS–8 and NSRD-NBS-16, Part 2 , National Standard Reference Data System–National Bureau of Standards, Part 1, 1966; Part 2, 1968.

Table 354. SELECTING THERMAL CONDUCTIVITY OF METALS*

(SHEET 10 OF 29)

Metal	Temperature (K)	Thermal Conductivity (watt • cm^{-1} • K^{-1})
Cadmium	273	0.975
Tungsten	2200	0.98
Zirconium	10	0.984
Tantalum	9	0.989
Tantalum	35	0.99
Niobium	4	0.993
Cadmium	200	0.993
Molybdenum	1400	0.996
Tungsten	2000	1
Zirconium	20	1.01
Platinum	60	1.01
Molybdenum	7	1.02
Tungsten	1800	1.03
Cadmium	100	1.03
Zirconium	11	1.04
Tin	60	1.04
Cadmium	90	1.04
Zinc	600	1.05
Molybdenum	1200	1.05
Nickel	200	1.06
Cadmium	80	1.06
Tungsten	1600	1.07
Lead	13	1.07
Zirconium	12	1.08
Zirconium	18	1.08
Tantalum	10	1.08
Molybdenum	1100	1.08
Cadmium	70	1.08

Source: data from Ho, C. Y., Powell, R. W., and Liley, P. E., *Thermal Conductictivity of Selected Materials*, NSRDS–NBS–8 and NSRD-NBS-16, Part 2 , National Standard Reference Data System–National Bureau of Standards, Part 1, 1966; Part 2, 1968.

Table 354. SELECTING THERMAL CONDUCTIVITY OF METALS*

(SHEET 11 OF 29)

Metal	Temperature (K)	Thermal Conductivity (watt • cm^{-1} • K^{-1})
Zirconium	13	1.11
Zinc	500	1.11
Tungsten	1400	1.11
Chromium	200	1.11
Zirconium	16	1.12
Molybdenum	1000	1.12
Zirconium	14	1.13
Zirconium	15	1.13
Cadmium	60	1.13
Tungsten	1200	1.15
Tin	50	1.15
Molybdenum	900	1.15
Zinc	400	1.16
Tantalum	11	1.16
Tantalum	30	1.16
Niobium	35	1.16
Molybdenum	8	1.17
Tungsten	1100	1.18
Platinum	50	1.18
Molybdenum	800	1.18
Chromium	3	1.2
Cadmium	50	1.2
Zinc	300	1.21
Tungsten	1000	1.21
Zinc	273	1.22
Molybdenum	700	1.22
Tin	45	1.23
Niobium	5	1.23

Source: data from Ho, C. Y., Powell, R. W., and Liley, P. E., *Thermal Conductictivity of Selected Materials*, NSRDS–NBS–8 and NSRD-NBS-16, Part 2 , National Standard Reference Data System–National Bureau of Standards, Part 1, 1966; Part 2, 1968.

Table 354. SELECTING THERMAL CONDUCTIVITY OF METALS*

(SHEET 12 OF 29)

Metal	Temperature (K)	Thermal Conductivity (watt • cm^{-1} • K^{-1})
Lead	12	1.23
Tungsten	900	1.24
Tantalum	12	1.24
Cadmium	45	1.25
Zinc	200	1.26
Molybdenum	600	1.26
Nickel	2	1.27
Tungsten	800	1.28
Tantalum	13	1.3
Molybdenum	500	1.3
Magnesium	1	1.3
Molybdenum	9	1.31
Zinc	100	1.32
Platinum	45	1.32
Iron	100	1.32
Cadmium	40	1.32
Tungsten	700	1.33
Zinc	90	1.34
Molybdenum	400	1.34
Tin	40	1.35
Tantalum	14	1.36
Tantalum	25	1.36
Zinc	80	1.38
Molybdenum	300	1.38
Tungsten	600	1.39
Molybdenum	273	1.39
Tantalum	15	1.4
Cadmium	35	1.41

Source: data from Ho, C. Y., Powell, R. W., and Liley, P. E., *Thermal Conductictivity of Selected Materials*, NSRDS–NBS–8 and NSRD-NBS-16, Part 2 , National Standard Reference Data System–National Bureau of Standards, Part 1, 1966; Part 2, 1968.

Table 354. SELECTING THERMAL CONDUCTIVITY OF METALS*
(SHEET 13 OF 29)

Metal	Temperature (K)	Thermal Conductivity (watt • cm^{-1} • K^{-1})
Molybdenum	200	1.43
Tantalum	16	1.44
Niobium	30	1.45
Molybdenum	10	1.45
Magnesium	900	1.45
Niobium	6	1.46
Magnesium	800	1.46
Lead	11	1.46
Iron	90	1.46
Tantalum	18	1.47
Tantalum	20	1.47
Magnesium	700	1.47
Zinc	70	1.48
Tungsten	500	1.49
Magnesium	600	1.49
Iron	2	1.49
Tin	35	1.5
Platinum	40	1.51
Magnesium	500	1.51
Magnesium	400	1.53
Magnesium	300	1.56
Cadmium	30	1.56
Magnesium	273	1.57
Nickel	100	1.58
Chromium	100	1.58
Magnesium	200	1.59
Molybdenum	11	1.6
Chromium	4	1.6

Source: data from Ho, C. Y., Powell, R. W., and Liley, P. E., *Thermal Conductictivity of Selected Materials*, NSRDS–NBS–8 and NSRD-NBS-16, Part 2 , National Standard Reference Data System–National Bureau of Standards, Part 1, 1966; Part 2, 1968.

Table 354. SELECTING THERMAL CONDUCTIVITY OF METALS*

(SHEET 14 OF 29)

Metal	Temperature (K)	Thermal Conductivity (watt • cm^{-1} • K^{-1})
Tungsten	400	1.62
Niobium	7	1.67
Iron	80	1.68
Chromium	90	1.68
Magnesium	100	1.69
Zinc	60	1.71
Nickel	90	1.72
Molybdenum	12	1.74
Tin	30	1.76
Tungsten	300	1.78
Magnesium	90	1.78
Lead	10	1.78
Molybdenum	100	1.79
Cadmium	25	1.79
Platinum	35	1.8
Tungsten	273	1.82
Chromium	80	1.82
Niobium	8	1.86
Niobium	25	1.87
Molybdenum	13	1.88
Nickel	3	1.91
Molybdenum	90	1.92
Nickel	80	1.93
Magnesium	80	1.95
Tungsten	200	1.97
Chromium	5	1.99
Molybdenum	14	2.01
Niobium	9	2.04

Source: data from Ho, C. Y., Powell, R. W., and Liley, P. E., *Thermal Conductictivity of Selected Materials,* NSRDS–NBS–8 and NSRD-NBS-16, Part 2 , National Standard Reference Data System–National Bureau of Standards, Part 1, 1966; Part 2, 1968.

Table 354. SELECTING THERMAL CONDUCTIVITY OF METALS*
(SHEET 15 OF 29)

Metal	Temperature (K)	Thermal Conductivity (watt • cm^{-1} • K^{-1})
Iron	70	2.04
Chromium	70	2.08
Molybdenum	80	2.09
Zinc	50	2.13
Aluminum	900	2.13
Molybdenum	15	2.15
Niobium	10	2.18
Aluminum	800	2.2
Nickel	70	2.21
Tin	25	2.22
Magnesium	70	2.23
Iron	3	2.24
Cadmium	20	2.26
Aluminum	700	2.26
Platinum	30	2.28
Molybdenum	16	2.28
Niobium	20	2.29
Niobium	11	2.3
Molybdenum	70	2.3
Lead	9	2.3
Platinum	1	2.31
Aluminum	600	2.32
Tungsten	100	2.35
Aluminum	273	2.36
Aluminum	200	2.37
Aluminum	300	2.37
Aluminum	500	2.37
Chromium	6	2.38

Source: data from Ho, C. Y., Powell, R. W., and Liley, P. E., *Thermal Conductictivity of Selected Materials*, NSRDS–NBS–8 and NSRD-NBS-16, Part 2 , National Standard Reference Data System–National Bureau of Standards, Part 1, 1966; Part 2, 1968.

Table 354. SELECTING THERMAL CONDUCTIVITY OF METALS*

(SHEET 16 OF 29)

Metal	Temperature (K)	Thermal Conductivity (watt • cm^{-1} • K^{-1})
Niobium	12	2.39
Aluminum	400	2.4
Niobium	18	2.42
Tungsten	90	2.44
Niobium	13	2.46
Zinc	45	2.48
Chromium	60	2.48
Niobium	14	2.49
Niobium	16	2.49
Niobium	15	2.5
Molybdenum	18	2.53
Nickel	4	2.54
Tungsten	80	2.56
Magnesium	2	2.59
Molybdenum	60	2.6
Gold	1200	2.62
Cadmium	18	2.62
Nickel	60	2.63
Iron	60	2.65
Gold	1100	2.71
Magnesium	60	2.74
Tungsten	70	2.76
Molybdenum	20	2.77
Chromium	7	2.77
Gold	1000	2.78
Gold	900	2.85
Gold	800	2.92
Zinc	40	2.97

Source: data from Ho, C. Y., Powell, R. W., and Liley, P. E., *Thermal Conductictivity of Selected Materials*, NSRDS–NBS–8 and NSRD-NBS-16, Part 2 , National Standard Reference Data System–National Bureau of Standards, Part 1, 1966; Part 2, 1968.

Table 354. SELECTING THERMAL CONDUCTIVITY OF METALS*

(SHEET 17 OF 29)

Metal	Temperature (K)	Thermal Conductivity (watt • cm^{-1} • K^{-1})
Iron	4	2.97
Gold	700	2.98
Molybdenum	50	3
Aluminum	100	3.0
Gold	600	3.04
Gold	500	3.09
Gold	400	3.12
Chromium	8	3.14
Platinum	25	3.15
Gold	300	3.15
Nickel	5	3.16
Cadmium	16	3.16
Chromium	50	3.17
Tungsten	60	3.18
Gold	273	3.18
Tin	20	3.2
Lead	8	3.2
Molybdenum	25	3.25
Molybdenum	45	3.26
Gold	200	3.27
Nickel	50	3.36
Aluminum	90	3.4
Copper	1200	3.42
Gold	100	3.45
Gold	90	3.48
Copper	1100	3.5
Chromium	9	3.5
Molybdenum	40	3.51

Source: data from Ho, C. Y., Powell, R. W., and Liley, P. E., *Thermal Conductictivity of Selected Materials*, NSRDS–NBS–8 and NSRD-NBS-16, Part 2 , National Standard Reference Data System–National Bureau of Standards, Part 1, 1966; Part 2, 1968.

Table 354. SELECTING THERMAL CONDUCTIVITY OF METALS*

(SHEET 18 OF 29)

Metal	Temperature (K)	Thermal Conductivity (watt • cm^{-1} • K^{-1})
Gold	80	3.52
Molybdenum	30	3.55
Cadmium	15	3.55
Copper	1000	3.57
Silver	1200	3.58
Gold	70	3.58
Molybdenum	35	3.62
Copper	900	3.64
Silver	1100	3.66
Chromium	45	3.67
Iron	5	3.71
Copper	800	3.71
Zinc	35	3.72
Iron	50	3.72
Silver	1000	3.74
Magnesium	50	3.75
Nickel	6	3.77
Copper	700	3.77
Gold	60	3.8
Silver	900	3.82
Copper	600	3.83
Chromium	10	3.85
Magnesium	3	3.88
Copper	500	3.88
Silver	800	3.89
Nickel	45	3.91
Copper	400	3.92
Silver	700	3.97

Source: data from Ho, C. Y., Powell, R. W., and Liley, P. E., *Thermal Conductictivity of Selected Materials*, NSRDS–NBS–8 and NSRD-NBS-16, Part 2 , National Standard Reference Data System–National Bureau of Standards, Part 1, 1966; Part 2, 1968.

Table 354. SELECTING THERMAL CONDUCTIVITY OF METALS*
(SHEET 19 OF 29)

Metal	Temperature (K)	Thermal Conductivity (watt • cm^{-1} • K^{-1})
Copper	300	3.98
Tin	18	4
Aluminum	80	4.0
Copper	273	4.01
Cadmium	14	4.01
Silver	600	4.05
Silver	500	4.13
Copper	200	4.13
Tungsten	50	4.17
Chromium	11	4.18
Silver	400	4.2
Gold	50	4.2
Silver	300	4.27
Silver	273	4.28
Silver	200	4.3
Chromium	40	4.3
Nickel	7	4.36
Gold	1	4.4
Iron	6	4.42
Chromium	12	4.49
Silver	100	4.5
Iron	45	4.5
Magnesium	45	4.57
Silver	90	4.6
Platinum	2	4.6
Gold	45	4.6
Nickel	40	4.63
Cadmium	13	4.67

Source: data from Ho, C. Y., Powell, R. W., and Liley, P. E., *Thermal Conductictivity of Selected Materials,* NSRDS–NBS–8 and NSRD-NBS-16, Part 2 , National Standard Reference Data System–National Bureau of Standards, Part 1, 1966; Part 2, 1968.

Table 354. SELECTING THERMAL CONDUCTIVITY OF METALS*

(SHEET 20 OF 29)

Metal	Temperature (K)	Thermal Conductivity (watt • cm^{-1} • K^{-1})
Silver	80	4.71
Chromium	13	4.78
Copper	100	4.83
Zinc	30	4.9
Platinum	20	4.9
Lead	7	4.9
Nickel	8	4.94
Silver	70	4.97
Aluminum	70	5.0
Chromium	35	5.03
Chromium	14	5.04
Tungsten	45	5.07
Iron	7	5.13
Copper	90	5.14
Magnesium	4	5.15
Gold	40	5.2
Chromium	15	5.27
Tin	16	5.3
Chromium	16	5.48
Nickel	9	5.49
Silver	60	5.5
Iron	40	5.55
Cadmium	12	5.56
Chromium	30	5.58
Nickel	35	5.62
Magnesium	40	5.7
Copper	80	5.7
Iron	8	5.8

Source: data from Ho, C. Y., Powell, R. W., and Liley, P. E., *Thermal Conductictivity of Selected Materials*, NSRDS–NBS–8 and NSRD-NBS-16, Part 2 , National Standard Reference Data System–National Bureau of Standards, Part 1, 1966; Part 2, 1968.

Table 354. SELECTING THERMAL CONDUCTIVITY OF METALS*

(SHEET 21 OF 29)

Metal	Temperature (K)	Thermal Conductivity (watt • cm^{-1} • K^{-1})
Chromium	18	5.81
Nickel	10	6
Chromium	20	6.01
Chromium	25	6.07
Platinum	18	6.1
Gold	35	6.1
Tin	15	6.3
Magnesium	5	6.39
Iron	9	6.45
Nickel	11	6.48
Tungsten	40	6.5
Copper	70	6.7
Aluminum	60	6.7
Platinum	3	6.79
Iron	35	6.81
Zinc	25	6.9
Nickel	12	6.91
Cadmium	11	6.91
Nickel	30	6.95
Silver	50	7
Iron	10	7.05
Nickel	13	7.3
Magnesium	35	7.4
Tin	14	7.6
Platinum	16	7.6
Magnesium	6	7.6
Gold	30	7.6
Iron	11	7.62

Source: data from Ho, C. Y., Powell, R. W., and Liley, P. E., *Thermal Conductictivity of Selected Materials*, NSRDS–NBS–8 and NSRD-NBS-16, Part 2 , National Standard Reference Data System–National Bureau of Standards, Part 1, 1966; Part 2, 1968.

Table 354. SELECTING THERMAL CONDUCTIVITY OF METALS*
(SHEET 22 OF 29)

Metal	Temperature (K)	Thermal Conductivity (watt • cm^{-1} • K^{-1})
Nickel	14	7.64
Aluminum	1	7.8
Nickel	15	7.92
Iron	12	8.13
Iron	30	8.14
Nickel	16	8.15
Nickel	25	8.15
Lead	6	8.2
Silver	45	8.4
Platinum	15	8.4
Nickel	18	8.45
Copper	60	8.5
Nickel	20	8.56
Iron	13	8.58
Magnesium	7	8.75
Platinum	4	8.8
Cadmium	10	8.87
Tungsten	35	8.9
Gold	2	8.9
Iron	14	8.97
Tin	13	9.3
Platinum	14	9.3
Iron	15	9.3
Iron	25	9.36
Magnesium	30	9.5
Iron	16	9.56
Magnesium	8	9.83
Iron	18	9.88

Source: data from Ho, C. Y., Powell, R. W., and Liley, P. E., *Thermal Conductictivity of Selected Materials,* NSRDS–NBS–8 and NSRD-NBS-16, Part 2 , National Standard Reference Data System–National Bureau of Standards, Part 1, 1966; Part 2, 1968.

Table 354. SELECTING THERMAL CONDUCTIVITY OF METALS*

(SHEET 23 OF 29)

Metal	Temperature (K)	Thermal Conductivity (watt • cm^{-1} • K^{-1})
Iron	20	9.97
Aluminum	50	10.0
Platinum	13	10.1
Gold	25	10.2
Silver	40	10.5
Platinum	5	10.5
Zinc	20	10.7
Magnesium	9	10.8
Platinum	12	10.9
Tin	12	11.6
Platinum	11	11.7
Magnesium	10	11.7
Platinum	6	11.8
Magnesium	25	12
Copper	50	12.2
Cadmium	9	12.2
Platinum	10	12.3
Magnesium	11	12.5
Aluminum	45	12.5
Platinum	7	12.6
Platinum	9	12.8
Platinum	8	12.9
Tungsten	30	13.1
Magnesium	12	13.1
Gold	3	13.1
Zinc	18	13.3
Magnesium	13	13.6
Silver	35	13.7

Source: data from Ho, C. Y., Powell, R. W., and Liley, P. E., *Thermal Conductictivity of Selected Materials*, NSRDS–NBS–8 and NSRD-NBS-16, Part 2 , National Standard Reference Data System–National Bureau of Standards, Part 1, 1966; Part 2, 1968.

Table 354. SELECTING THERMAL CONDUCTIVITY OF METALS*

(SHEET 24 OF 29)

Metal	Temperature (K)	Thermal Conductivity (watt • cm^{-1} • K^{-1})
Lead	5	13.8
Magnesium	20	13.9
Magnesium	14	14
Magnesium	15	14.3
Magnesium	18	14.3
Tungsten	1	14.4
Magnesium	16	14.4
Tin	11	14.8
Gold	20	15
Copper	45	15.3
Aluminum	2	15.5
Aluminum	40	16.0
Zinc	16	16.9
Gold	4	17.1
Gold	18	17.7
Cadmium	8	18
Zinc	1	19
Tin	10	19.3
Silver	30	19.3
Zinc	15	19.4
Tungsten	25	20.4
Copper	40	20.5
Gold	5	20.7
Gold	16	20.9
Aluminum	35	21.0
Zinc	14	22.4
Lead	4	22.4
Gold	15	22.6

Source: data from Ho, C. Y., Powell, R. W., and Liley, P. E., *Thermal Conductictivity of Selected Materials,* NSRDS–NBS–8 and NSRD-NBS-16, Part 2 , National Standard Reference Data System–National Bureau of Standards, Part 1, 1966; Part 2, 1968.

Table 354. SELECTING THERMAL CONDUCTIVITY OF METALS*

(SHEET 25 OF 29)

Metal	Temperature (K)	Thermal Conductivity (watt • cm^{-1} • K^{-1})
Aluminum	3	23.2
Gold	6	23.7
Gold	14	24.1
Gold	13	25.5
Tin	9	26
Gold	7	26
Zinc	13	26.1
Gold	12	26.7
Gold	8	27.5
Lead	1	27.7
Gold	11	27.7
Cadmium	7	28
Gold	9	28.2
Gold	10	28.2
Aluminum	30	28.5
Tungsten	2	28.7
Copper	1	28.7
Copper	35	29
Silver	25	29.5
Zinc	12	30.8
Aluminum	4	30.8
Tungsten	20	32.6
Lead	3	34
Tin	8	36
Zinc	11	36.4
Zinc	2	37.9
Aluminum	5	38.1
Silver	1	39.4

Source: data from Ho, C. Y., Powell, R. W., and Liley, P. E., *Thermal Conductictivity of Selected Materials*, NSRDS–NBS–8 and NSRD-NBS-16, Part 2 , National Standard Reference Data System–National Bureau of Standards, Part 1, 1966; Part 2, 1968.

Table 354. SELECTING THERMAL CONDUCTIVITY OF METALS*

(SHEET 26 OF 29)

Metal	Temperature (K)	Thermal Conductivity (watt • cm^{-1} • K^{-1})
Tungsten	18	40
Aluminum	25	40.0
Lead	2	42.4
Tungsten	3	42.6
Copper	30	43
Zinc	10	43.2
Cadmium	6	44.2
Aluminum	6	45.1
Cadmium	1	48.7
Tungsten	16	49.3
Silver	20	51
Aluminum	7	51.5
Zinc	9	51.9
Tin	7	52
Tungsten	15	54.8
Zinc	3	55.5
Tungsten	4	55.6
Aluminum	20	56.5
Copper	2	57.3
Aluminum	8	57.3
Tungsten	14	60.4
Zinc	8	61.8
Aluminum	9	62.2
Aluminum	18	63.5
Silver	18	66
Aluminum	10	66.1
Tungsten	13	66.4
Tungsten	5	67.1

Source: data from Ho, C. Y., Powell, R. W., and Liley, P. E., *Thermal Conductictivity of Selected Materials*, NSRDS–NBS–8 and NSRD-NBS-16, Part 2 , National Standard Reference Data System–National Bureau of Standards, Part 1, 1966; Part 2, 1968.

Table 354. SELECTING THERMAL CONDUCTIVITY OF METALS*

(SHEET 27 OF 29)

Metal	Temperature (K)	Thermal Conductivity (watt • cm^{-1} • K^{-1})
Copper	25	68
Aluminum	16	68.4
Cadmium	5	69
Aluminum	11	69.0
Zinc	4	69.7
Aluminum	15	70.2
Aluminum	12	70.8
Aluminum	14	71.3
Aluminum	13	71.5
Zinc	7	71.7
Tungsten	12	72.4
Tin	6	76
Tungsten	6	76.2
Zinc	5	77.8
Tungsten	11	77.9
Zinc	6	78
Silver	2	78.3
Tungsten	7	82.4
Tungsten	10	82.4
Silver	16	85
Tungsten	9	85.1
Tungsten	8	85.3
Copper	3	85.5
Cadmium	2	89.3
Cadmium	4	92
Silver	15	96
Cadmium	3	104
Copper	20	105

Source: data from Ho, C. Y., Powell, R. W., and Liley, P. E., *Thermal Conductictivity of Selected Materials*, NSRDS–NBS–8 and NSRD-NBS-16, Part 2 , National Standard Reference Data System–National Bureau of Standards, Part 1, 1966; Part 2, 1968.

Table 354. SELECTING THERMAL CONDUCTIVITY OF METALS*

(SHEET 28 OF 29)

Metal	Temperature (K)	Thermal Conductivity (watt • cm^{-1} • K^{-1})
Silver	14	109
Copper	4	113
Silver	3	115
Tin	5	117
Silver	13	124
Copper	18	124
Copper	5	138
Silver	12	139
Copper	16	145
Silver	4	147
Silver	11	154
Copper	15	156
Copper	6	159
Copper	14	166
Silver	10	168
Silver	5	172
Copper	13	176
Copper	7	177
Tin	4	181
Silver	9	181
Copper	12	185
Silver	6	187
Copper	8	189
Silver	8	190

Source: data from Ho, C. Y., Powell, R. W., and Liley, P. E., *Thermal Conductictivity of Selected Materials,* NSRDS–NBS–8 and NSRD-NBS-16, Part 2 , National Standard Reference Data System–National Bureau of Standards, Part 1, 1966; Part 2, 1968.

Table 354. SELECTING THERMAL CONDUCTIVITY OF METALS*

(SHEET 29 OF 29)

Metal	Temperature (K)	Thermal Conductivity (watt • cm^{-1} • K^{-1})
Silver	7	193
Copper	11	193
Copper	9	195
Copper	10	196
Tin	3	297

Source: data from Ho, C. Y., Powell, R. W., and Liley, P. E., *Thermal Conductictivity of Selected Materials,* NSRDS–NBS–8 and NSRD-NBS-16, Part 2 , National Standard Reference Data System–National Bureau of Standards, Part 1, 1966; Part 2, 1968.

* These data apply only to metals of purity of at least 99.9%.
The third significant figure may not be accurate.

Table 355. SELECTING THERMAL CONDUCTIVITY OF METALS AT TEMPERATURE* (SHEET 1 OF 30)

Temperature (K)	Metal	Thermal Conductivity (watt • cm^{-1} • K^{-1})
1	Titanium	0.0144
	Zirconium	0.111
	Tantalum	0.115
	Molybdenum	0.146
	Niobium	0.251
	Chromium	0.401
	Nickel	0.64
	Iron	0.75
	Magnesium	1.3
	Platinum	2.31
	Gold	4.4
	Aluminum	7.8
	Tungsten	14.4
	Zinc	19
	Lead	27.7
	Copper	28.7
	Silver	39.4
	Cadmium	48.7
2	Titanium	0.0288
	Zirconium	0.223
	Tantalum	0.23
	Molybdenum	0.292
	Niobium	0.501
	Chromium	0.802
	Nickel	1.27
	Iron	1.49

Source: data from Ho, C. Y., Powell, R. W., and Liley, P. E., *Thermal Conductictivity of Selected Materials*, NSRDS–NBS–8 and NSRD-NBS-16, Part 2 , National Standard Reference Data System–National Bureau of Standards, Part 1, 1966; Part 2, 1968.

Table 355. SELECTING THERMAL CONDUCTIVITY OF METALS AT TEMPERATURE* (SHEET 2 OF 30)

Temperature (K)	Metal	Thermal Conductivity (watt • cm^{-1} • K^{-1})
	Magnesium	2.59
	Platinum	4.6
	Gold	8.9
	Aluminum	15.5
	Tungsten	28.7
	Zinc	37.9
	Lead	42.4
	Copper	57.3
	Silver	78.3
	Cadmium	89.3
3	Titanium	0.0432
	Zirconium	0.333
	Tantalum	0.345
	Molybdenum	0.438
	Niobium	0.749
	Chromium	1.2
	Nickel	1.91
	Iron	2.24
	Magnesium	3.88
	Platinum	6.79
	Gold	13.1
	Aluminum	23.2
	Lead	34
	Tungsten	42.6
	Zinc	55.5
	Copper	85.5
	Cadmium	104
	Silver	115
	Tin	297

Source: data from Ho, C. Y., Powell, R. W., and Liley, P. E., *Thermal Conductictivity of Selected Materials*, NSRDS–NBS–8 and NSRD-NBS-16, Part 2 , National Standard Reference Data System–National Bureau of Standards, Part 1, 1966; Part 2, 1968.

Table 355. SELECTING THERMAL CONDUCTIVITY OF METALS AT TEMPERATURE* (SHEET 3 OF 30)

Temperature (K)	Metal	Thermal Conductivity (watt • cm^{-1} • K^{-1})
4	Titanium	0.0576
	Zirconium	0.442
	Tantalum	0.459
	Molybdenum	0.584
	Niobium	0.993
	Chromium	1.6
	Nickel	2.54
	Iron	2.97
	Magnesium	5.15
	Platinum	8.8
	Gold	17.1
	Lead	22.4
	Aluminum	30.8
	Tungsten	55.6
	Zinc	69.7
	Cadmium	92
	Copper	113
	Silver	147
	Tin	181
5	Titanium	0.0719
	Zirconium	0.549
	Tantalum	0.571
	Molybdenum	0.73
	Niobium	1.23
	Chromium	1.99
	Nickel	3.16
	Iron	3.71

Source: data from Ho, C. Y., Powell, R. W., and Liley, P. E., *Thermal Conductictivity of Selected Materials*, NSRDS–NBS–8 and NSRD-NBS-16, Part 2 , National Standard Reference Data System–National Bureau of Standards, Part 1, 1966; Part 2, 1968.

Table 355. SELECTING THERMAL CONDUCTIVITY OF METALS AT TEMPERATURE* (SHEET 4 OF 30)

Temperature (K)	Metal	Thermal Conductivity (watt • cm^{-1} • K^{-1})
	Magnesium	6.39
	Platinum	10.5
	Lead	13.8
	Gold	20.7
	Aluminum	38.1
	Tungsten	67.1
	Cadmium	69
	Zinc	77.8
	Tin	117
	Copper	138
	Silver	172
6	Titanium	0.0863
	Zirconium	0.652
	Tantalum	0.681
	Molybdenum	0.876
	Niobium	1.46
	Chromium	2.38
	Nickel	3.77
	Iron	4.42
	Magnesium	7.6
	Lead	8.2
	Platinum	11.8
	Gold	23.7
	Cadmium	44.2
	Aluminum	45.1
	Tin	76
	Tungsten	76.2

Source: data from Ho, C. Y., Powell, R. W., and Liley, P. E., *Thermal Conductictivity of Selected Materials*, NSRDS–NBS–8 and NSRD-NBS-16, Part 2 , National Standard Reference Data System–National Bureau of Standards, Part 1, 1966; Part 2, 1968.

Table 355. SELECTING THERMAL CONDUCTIVITY OF METALS AT TEMPERATURE* (SHEET 5 OF 30)

Temperature (K)	Metal	Thermal Conductivity (watt • cm^{-1} • K^{-1})
	Zinc	78
	Copper	159
	Silver	187
7	Titanium	0.101
	Zirconium	0.748
	Tantalum	0.788
	Molybdenum	1.02
	Niobium	1.67
	Chromium	2.77
	Nickel	4.36
	Lead	4.9
	Iron	5.13
	Magnesium	8.75
	Platinum	12.6
	Gold	26
	Cadmium	28
	Aluminum	51.5
	Tin	52
	Zinc	71.7
	Tungsten	82.4
	Copper	177
	Silver	193
8	Titanium	0.115
	Zirconium	0.837
	Tantalum	0.891
	Molybdenum	1.17

Source: data from Ho, C. Y., Powell, R. W., and Liley, P. E., *Thermal Conductictivity of Selected Materials*, NSRDS–NBS–8 and NSRD-NBS-16, Part 2 , National Standard Reference Data System–National Bureau of Standards, Part 1, 1966; Part 2, 1968.

Table 355. SELECTING THERMAL CONDUCTIVITY OF METALS AT TEMPERATURE* (SHEET 6 OF 30)

Temperature (K)	Metal	Thermal Conductivity (watt • cm^{-1} • K^{-1})
	Niobium	1.86
	Chromium	3.14
	Lead	3.2
	Nickel	4.94
	Iron	5.8
	Magnesium	9.83
	Platinum	12.9
	Cadmium	18
	Gold	27.5
	Tin	36
	Aluminum	57.3
	Zinc	61.8
	Tungsten	85.3
	Copper	189
	Silver	190
9	Titanium	0.129
	Zirconium	0.916
	Tantalum	0.989
	Molybdenum	1.31
	Niobium	2.04
	Lead	2.3
	Chromium	3.5
	Nickel	5.49
	Iron	6.45
	Magnesium	10.8
	Cadmium	12.2
	Platinum	12.8

Source: data from Ho, C. Y., Powell, R. W., and Liley, P. E., *Thermal Conductictivity of Selected Materials*, NSRDS–NBS–8 and NSRD-NBS-16, Part 2 , National Standard Reference Data System–National Bureau of Standards, Part 1, 1966; Part 2, 1968.

Table 355. SELECTING THERMAL CONDUCTIVITY OF METALS AT TEMPERATURE* (SHEET 7 OF 30)

Temperature (K)	Metal	Thermal Conductivity (watt • cm^{-1} • K^{-1})
	Tin	26
	Gold	28.2
	Zinc	51.9
	Aluminum	62.2
	Tungsten	85.1
	Silver	181
	Copper	195
10	Titanium	0.144
	Zirconium	0.984
	Tantalum	1.08
	Molybdenum	1.45
	Lead	1.78
	Niobium	2.18
	Chromium	3.85
	Nickel	6
	Iron	7.05
	Cadmium	8.87
	Magnesium	11.7
	Platinum	12.3
	Tin	19.3
	Gold	28.2
	Zinc	43.2
	Aluminum	66.1
	Tungsten	82.4
	Silver	168
	Copper	196

Source: data from Ho, C. Y., Powell, R. W., and Liley, P. E., *Thermal Conductictivity of Selected Materials*, NSRDS–NBS–8 and NSRD-NBS-16, Part 2 , National Standard Reference Data System–National Bureau of Standards, Part 1, 1966; Part 2, 1968.

Table 355. SELECTING THERMAL CONDUCTIVITY OF METALS AT TEMPERATURE* (SHEET 8 OF 30)

Temperature (K)	Metal	Thermal Conductivity (watt • cm^{-1} • K^{-1})
11	Titanium	0.158
	Zirconium	1.04
	Tantalum	1.16
	Lead	1.46
	Molybdenum	1.6
	Niobium	2.3
	Chromium	4.18
	Nickel	6.48
	Cadmium	6.91
	Iron	7.62
	Platinum	11.7
	Magnesium	12.5
	Tin	14.8
	Gold	27.7
	Zinc	36.4
	Aluminum	69
	Tungsten	77.9
	Silver	154
	Copper	193
12	Titanium	0.172
	Zirconium	1.08
	Lead	1.23
	Tantalum	1.24
	Molybdenum	1.74
	Niobium	2.39
	Chromium	4.49
	Cadmium	5.56

Source: data from Ho, C. Y., Powell, R. W., and Liley, P. E., *Thermal Conductictivity of Selected Materials*, NSRDS–NBS–8 and NSRD-NBS-16, Part 2 , National Standard Reference Data System–National Bureau of Standards, Part 1, 1966; Part 2, 1968.

Table 355. SELECTING THERMAL CONDUCTIVITY OF METALS AT TEMPERATURE* (SHEET 9 OF 30)

Temperature (K)	Metal	Thermal Conductivity (watt • cm^{-1} • K^{-1})
	Nickel	6.91
	Iron	8.13
	Platinum	10.9
	Tin	11.6
	Magnesium	13.1
	Gold	26.7
	Zinc	30.8
	Aluminum	70.8
	Tungsten	72.4
	Silver	139
	Copper	185
13	Titanium	0.186
	Lead	1.07
	Zirconium	1.11
	Tantalum	1.3
	Molybdenum	1.88
	Niobium	2.46
	Cadmium	4.67
	Chromium	4.78
	Nickel	7.3
	Iron	8.58
	Tin	9.3
	Platinum	10.1
	Magnesium	13.6
	Gold	25.5
	Zinc	26.1
	Tungsten	66.4

Source: data from Ho, C. Y., Powell, R. W., and Liley, P. E., *Thermal Conductictivity of Selected Materials*, NSRDS–NBS–8 and NSRD-NBS-16, Part 2 , National Standard Reference Data System–National Bureau of Standards, Part 1, 1966; Part 2, 1968.

Table 355. SELECTING THERMAL CONDUCTIVITY OF METALS AT TEMPERATURE* (SHEET 10 OF 30)

Temperature (K)	Metal	Thermal Conductivity (watt • cm^{-1} • K^{-1})
	Aluminum	71.5
	Silver	124
	Copper	176
14	Titanium	0.2
	Lead	0.94
	Zirconium	1.13
	Tantalum	1.36
	Molybdenum	2.01
	Niobium	2.49
	Cadmium	4.01
	Chromium	5.04
	Tin	7.6
	Nickel	7.64
	Iron	8.97
	Platinum	9.3
	Magnesium	14
	Zinc	22.4
	Gold	24.1
	Tungsten	60.4
	Aluminum	71.3
	Silver	109
	Copper	166
15	Titanium	0.214
	Lead	0.84
	Zirconium	1.13
	Tantalum	1.4

Source: data from Ho, C. Y., Powell, R. W., and Liley, P. E., *Thermal Conductictivity of Selected Materials*, NSRDS–NBS–8 and NSRD-NBS-16, Part 2 , National Standard Reference Data System–National Bureau of Standards, Part 1, 1966; Part 2, 1968.

Table 355. SELECTING THERMAL CONDUCTIVITY OF METALS AT TEMPERATURE* (SHEET 11 OF 30)

Temperature (K)	Metal	Thermal Conductivity ($watt \cdot cm^{-1} \cdot K^{-1}$)
	Molybdenum	2.15
	Niobium	2.5
	Cadmium	3.55
	Chromium	5.27
	Tin	6.3
	Nickel	7.92
	Platinum	8.4
	Iron	9.3
	Magnesium	14.3
	Zinc	19.4
	Gold	22.6
	Tungsten	54.8
	Aluminum	70.2
	Silver	96
	Copper	156
16	Titanium	0.227
	Lead	0.77
	Zirconium	1.12
	Tantalum	1.44
	Molybdenum	2.28
	Niobium	2.49
	Cadmium	3.16
	Tin	5.3
	Chromium	5.48
	Platinum	7.6
	Nickel	8.15
	Iron	9.56

Source: data from Ho, C. Y., Powell, R. W., and Liley, P. E., *Thermal Conductictivity of Selected Materials*, NSRDS–NBS–8 and NSRD-NBS-16, Part 2 , National Standard Reference Data System–National Bureau of Standards, Part 1, 1966; Part 2, 1968.

Table 355. SELECTING THERMAL CONDUCTIVITY OF METALS AT TEMPERATURE* (SHEET 12 OF 30)

Temperature (K)	Metal	Thermal Conductivity (watt • cm^{-1} • K^{-1})
	Magnesium	14.4
	Zinc	16.9
	Gold	20.9
	Tungsten	49.3
	Aluminum	68.4
	Silver	85
	Copper	145
18	Titanium	0.254
	Lead	0.66
	Zirconium	1.08
	Tantalum	1.47
	Niobium	2.42
	Molybdenum	2.53
	Cadmium	2.62
	Tin	4
	Chromium	5.81
	Platinum	6.1
	Nickel	8.45
	Iron	9.88
	Zinc	13.3
	Magnesium	14.3
	Gold	17.7
	Tungsten	40
	Aluminum	63.5
	Silver	66
	Copper	124

Source: data from Ho, C. Y., Powell, R. W., and Liley, P. E., *Thermal Conductictivity of Selected Materials*, NSRDS–NBS–8 and NSRD-NBS-16, Part 2 , National Standard Reference Data System–National Bureau of Standards, Part 1, 1966; Part 2, 1968.

Table 355. SELECTING THERMAL CONDUCTIVITY OF METALS AT TEMPERATURE* (SHEET 13 OF 30)

Temperature (K)	Metal	Thermal Conductivity (watt • cm^{-1} • K^{-1})
20	Titanium	0.279
	Lead	0.59
	Zirconium	1.01
	Tantalum	1.47
	Cadmium	2.26
	Niobium	2.29
	Molybdenum	2.77
	Tin	3.2
	Platinum	4.9
	Chromium	6.01
	Nickel	8.56
	Iron	9.97
	Zinc	10.7
	Magnesium	13.9
	Gold	15
	Tungsten	32.6
	Silver	51
	Aluminum	56.5
	Copper	105
25	Titanium	0.337
	Lead	0.507
	Zirconium	0.85
	Tantalum	1.36
	Cadmium	1.79
	Niobium	1.87
	Tin	2.22
	Platinum	3.15

Source: data from Ho, C. Y., Powell, R. W., and Liley, P. E., *Thermal Conductictivity of Selected Materials*, NSRDS–NBS–8 and NSRD-NBS-16, Part 2 , National Standard Reference Data System–National Bureau of Standards, Part 1, 1966; Part 2, 1968.

Table 355. SELECTING THERMAL CONDUCTIVITY OF METALS AT TEMPERATURE* (SHEET 14 OF 30)

Temperature (K)	Metal	Thermal Conductivity ($watt \cdot cm^{-1} \cdot K^{-1}$)
	Molybdenum	3.25
	Chromium	6.07
	Zinc	6.9
	Nickel	8.15
	Iron	9.36
	Gold	10.2
	Magnesium	12
	Tungsten	20.4
	Silver	29.5
	Aluminum	40
	Copper	68
30	Titanium	0.382
	Lead	0.477
	Zirconium	0.74
	Tantalum	1.16
	Niobium	1.45
	Cadmium	1.56
	Tin	1.76
	Platinum	2.28
	Molybdenum	3.55
	Zinc	4.9
	Chromium	5.58
	Nickel	6.95
	Gold	7.6
	Iron	8.14
	Magnesium	9.5
	Tungsten	13.1

Source: data from Ho, C. Y., Powell, R. W., and Liley, P. E., *Thermal Conductictivity of Selected Materials*, NSRDS–NBS–8 and NSRD-NBS-16, Part 2 , National Standard Reference Data System–National Bureau of Standards, Part 1, 1966; Part 2, 1968.

Table 355. SELECTING THERMAL CONDUCTIVITY OF METALS AT TEMPERATURE* (SHEET 15 OF 30)

Temperature (K)	Metal	Thermal Conductivity (watt • cm^{-1} • K^{-1})
	Silver	19.3
	Aluminum	28.5
	Copper	43
35	Titanium	0.411
	Lead	0.462
	Zirconium	0.65
	Tantalum	0.99
	Niobium	1.16
	Cadmium	1.41
	Tin	1.5
	Platinum	1.8
	Molybdenum	3.62
	Zinc	3.72
	Chromium	5.03
	Nickel	5.62
	Gold	6.1
	Iron	6.81
	Magnesium	7.4
	Tungsten	8.9
	Silver	13.7
	Aluminum	21
	Copper	29
40	Titanium	0.422
	Lead	0.451
	Zirconium	0.58
	Tantalum	0.87

Source: data from Ho, C. Y., Powell, R. W., and Liley, P. E., *Thermal Conductictivity of Selected Materials*, NSRDS–NBS–8 and NSRD-NBS-16, Part 2 , National Standard Reference Data System–National Bureau of Standards, Part 1, 1966; Part 2, 1968.

Table 355. SELECTING THERMAL CONDUCTIVITY OF METALS AT TEMPERATURE* (SHEET 16 OF 30)

Temperature (K)	Metal	Thermal Conductivity (watt • cm^{-1} • K^{-1})
	Niobium	0.97
	Cadmium	1.32
	Tin	1.35
	Platinum	1.51
	Zinc	2.97
	Molybdenum	3.51
	Chromium	4.3
	Nickel	4.63
	Gold	5.2
	Iron	5.55
	Magnesium	5.7
	Tungsten	6.5
	Silver	10.5
	Aluminum	16
	Copper	20.5
45	Titanium	0.416
	Lead	0.442
	Zirconium	0.535
	Tantalum	0.78
	Niobium	0.84
	Tin	1.23
	Cadmium	1.25
	Platinum	1.32
	Zinc	2.48
	Molybdenum	3.26
	Chromium	3.67
	Nickel	3.91

Source: data from Ho, C. Y., Powell, R. W., and Liley, P. E., *Thermal Conductictivity of Selected Materials*, NSRDS–NBS–8 and NSRD-NBS-16, Part 2 , National Standard Reference Data System–National Bureau of Standards, Part 1, 1966; Part 2, 1968.

Table 355. SELECTING THERMAL CONDUCTIVITY OF METALS AT TEMPERATURE* (SHEET 17 OF 30)

Temperature (K)	Metal	Thermal Conductivity (watt • cm^{-1} • K^{-1})
	Iron	4.5
	Magnesium	4.57
	Gold	4.6
	Tungsten	5.07
	Silver	8.4
	Aluminum	12.5
	Copper	15.3
50	Titanium	0.401
	Lead	0.435
	Zirconium	0.497
	Tantalum	0.72
	Niobium	0.76
	Tin	1.15
	Platinum	1.18
	Cadmium	1.2
	Zinc	2.13
	Molybdenum	3
	Chromium	3.17
	Nickel	3.36
	Iron	3.72
	Magnesium	3.75
	Tungsten	4.17
	Gold	4.2
	Silver	7
	Aluminum	10
	Copper	12.2

Source: data from Ho, C. Y., Powell, R. W., and Liley, P. E., *Thermal Conductictivity of Selected Materials*, NSRDS–NBS–8 and NSRD-NBS-16, Part 2 , National Standard Reference Data System–National Bureau of Standards, Part 1, 1966; Part 2, 1968.

Table 355. SELECTING THERMAL CONDUCTIVITY OF METALS AT TEMPERATURE* (SHEET 18 OF 30)

Temperature (K)	Metal	Thermal Conductivity (watt • cm^{-1} • K^{-1})
60	Titanium	0.377
	Lead	0.424
	Zirconium	0.442
	Tantalum	0.651
	Niobium	0.66
	Platinum	1.01
	Tin	1.04
	Cadmium	1.13
	Zinc	1.71
	Chromium	2.48
	Molybdenum	2.6
	Nickel	2.63
	Iron	2.65
	Magnesium	2.74
	Tungsten	3.18
	Gold	3.8
	Silver	5.5
	Aluminum	6.7
	Copper	8.5
70	Titanium	0.356
	Zirconium	0.403
	Lead	0.415
	Niobium	0.61
	Tantalum	0.616
	Platinum	0.9
	Tin	0.96
	Cadmium	1.08

Source: data from Ho, C. Y., Powell, R. W., and Liley, P. E., *Thermal Conductictivity of Selected Materials*, NSRDS–NBS–8 and NSRD-NBS-16, Part 2 , National Standard Reference Data System–National Bureau of Standards, Part 1, 1966; Part 2, 1968.

Table 355. SELECTING THERMAL CONDUCTIVITY OF METALS AT TEMPERATURE* (SHEET 19 OF 30)

Temperature (K)	Metal	Thermal Conductivity (watt • cm^{-1} • K^{-1})
	Zinc	1.48
	Iron	2.04
	Chromium	2.08
	Nickel	2.21
	Magnesium	2.23
	Molybdenum	2.3
	Tungsten	2.76
	Gold	3.58
	Silver	4.97
	Aluminum	5
	Copper	6.7
80	Titanium	0.339
	Zirconium	0.373
	Lead	0.407
	Niobium	0.58
	Tantalum	0.603
	Platinum	0.84
	Tin	0.91
	Cadmium	1.06
	Zinc	1.38
	Iron	1.68
	Chromium	1.82
	Nickel	1.93
	Magnesium	1.95
	Molybdenum	2.09
	Tungsten	2.56
	Gold	3.52

Source: data from Ho, C. Y., Powell, R. W., and Liley, P. E., *Thermal Conductictivity of Selected Materials*, NSRDS–NBS–8 and NSRD-NBS-16, Part 2 , National Standard Reference Data System–National Bureau of Standards, Part 1, 1966; Part 2, 1968.

Table 355. SELECTING THERMAL CONDUCTIVITY OF METALS AT TEMPERATURE* (SHEET 20 OF 30)

Temperature (K)	Metal	Thermal Conductivity (watt • cm^{-1} • K^{-1})
	Aluminum	4
	Silver	4.71
	Copper	5.7
90	Titanium	0.324
	Zirconium	0.35
	Lead	0.401
	Niobium	0.563
	Tantalum	0.596
	Platinum	0.81
	Tin	0.88
	Cadmium	1.04
	Zinc	1.34
	Iron	1.46
	Chromium	1.68
	Nickel	1.72
	Magnesium	1.78
	Molybdenum	1.92
	Tungsten	2.44
	Aluminum	3.4
	Gold	3.48
	Silver	4.6
	Copper	5.14
100	Titanium	0.312
	Zirconium	0.332
	Lead	0.396
	Niobium	0.552

Source: data from Ho, C. Y., Powell, R. W., and Liley, P. E., *Thermal Conductictivity of Selected Materials*, NSRDS–NBS–8 and NSRD-NBS-16, Part 2 , National Standard Reference Data System–National Bureau of Standards, Part 1, 1966; Part 2, 1968.

Table 355. SELECTING THERMAL CONDUCTIVITY OF METALS AT TEMPERATURE* (SHEET 21 OF 30)

Temperature (K)	Metal	Thermal Conductivity (watt • cm^{-1} • K^{-1})
	Tantalum	0.592
	Platinum	0.79
	Tin	0.85
	Cadmium	1.03
	Zinc	1.32
	Iron	1.32
	Nickel	1.58
	Chromium	1.58
	Magnesium	1.69
	Molybdenum	1.79
	Tungsten	2.35
	Aluminum	3
	Gold	3.45
	Silver	4.5
	Copper	4.83
200	Titanium	0.245
	Zirconium	0.252
	Lead	0.366
	Niobium	0.526
	Tantalum	0.575
	Tin	0.733
	Platinum	0.748
	Iron	0.94
	Cadmium	0.993
	Nickel	1.06
	Chromium	1.11
	Zinc	1.26

Source: data from Ho, C. Y., Powell, R. W., and Liley, P. E., *Thermal Conductictivity of Selected Materials*, NSRDS–NBS–8 and NSRD-NBS-16, Part 2 , National Standard Reference Data System–National Bureau of Standards, Part 1, 1966; Part 2, 1968.

Table 355. SELECTING THERMAL CONDUCTIVITY OF METALS AT TEMPERATURE* (SHEET 22 OF 30)

Temperature (K)	Metal	Thermal Conductivity (watt • cm^{-1} • K^{-1})
	Molybdenum	1.43
	Magnesium	1.59
	Tungsten	1.97
	Aluminum	2.37
	Gold	3.27
	Copper	4.13
	Silver	4.3
273	Titanium	0.224
	Zirconium	0.232
	Lead	0.355
	Niobium	0.533
	Tantalum	0.574
	Tin	0.682
	Platinum	0.734
	Iron	0.835
	Nickel	0.94
	Chromium	0.948
	Cadmium	0.975
	Zinc	1.22
	Molybdenum	1.39
	Magnesium	1.57
	Tungsten	1.82
	Aluminum	2.36
	Gold	3.18
	Copper	4.01
	Silver	4.28

Source: data from Ho, C. Y., Powell, R. W., and Liley, P. E., *Thermal Conductictivity of Selected Materials*, NSRDS–NBS–8 and NSRD-NBS-16, Part 2 , National Standard Reference Data System–National Bureau of Standards, Part 1, 1966; Part 2, 1968.

Table 355. SELECTING THERMAL CONDUCTIVITY OF METALS AT TEMPERATURE* (SHEET 23 OF 30)

Temperature (K)	Metal	Thermal Conductivity (watt • cm^{-1} • K^{-1})
300	Titanium	0.219
	Zirconium	0.227
	Lead	0.352
	Niobium	0.537
	Tantalum	0.575
	Tin	0.666
	Platinum	0.73
	Iron	0.803
	Chromium	0.903
	Nickel	0.905
	Cadmium	0.968
	Zinc	1.21
	Molybdenum	1.38
	Magnesium	1.56
	Tungsten	1.78
	Aluminum	2.37
	Gold	3.15
	Copper	3.98
	Silver	4.27
400	Titanium	0.204
	Zirconium	0.216
	Lead	0.338
	Niobium	0.552
	Tantalum	0.578
	Tin	0.622
	Iron	0.694
	Platinum	0.722

Source: data from Ho, C. Y., Powell, R. W., and Liley, P. E., *Thermal Conductictivity of Selected Materials*, NSRDS–NBS–8 and NSRD-NBS-16, Part 2 , National Standard Reference Data System–National Bureau of Standards, Part 1, 1966; Part 2, 1968.

Table 355. SELECTING THERMAL CONDUCTIVITY OF METALS AT TEMPERATURE* (SHEET 24 OF 30)

Temperature (K)	Metal	Thermal Conductivity (watt • cm^{-1} • K^{-1})
	Nickel	0.801
	Chromium	0.873
	Cadmium	0.947
	Zinc	1.16
	Molybdenum	1.34
	Magnesium	1.53
	Tungsten	1.62
	Aluminum	2.4
	Gold	3.12
	Copper	3.92
	Silver	4.2
500	Titanium	0.197
	Zirconium	0.21
	Lead	0.325
	Niobium	0.567
	Tantalum	0.582
	Tin	0.596
	Iron	0.613
	Platinum	0.719
	Nickel	0.721
	Chromium	0.848
	Cadmium	0.92
	Zinc	1.11
	Molybdenum	1.3
	Tungsten	1.49
	Magnesium	1.51
	Aluminum	2.37

Source: data from Ho, C. Y., Powell, R. W., and Liley, P. E., *Thermal Conductictivity of Selected Materials*, NSRDS–NBS–8 and NSRD-NBS-16, Part 2 , National Standard Reference Data System–National Bureau of Standards, Part 1, 1966; Part 2, 1968.

Table 355. SELECTING THERMAL CONDUCTIVITY OF METALS AT TEMPERATURE* (SHEET 25 OF 30)

Temperature (K)	Metal	Thermal Conductivity (watt • cm^{-1} • K^{-1})
	Gold	3.09
	Copper	3.88
	Silver	4.13
700	Titanium	0.194
	Zirconium	0.207
	Lead	0.312
	Iron	0.547
	Niobium	0.582
	Tantalum	0.586
	Nickel	0.655
	Platinum	0.72
	Chromium	0.805
	Zinc	1.05
	Molybdenum	1.26
	Tungsten	1.39
	Magnesium	1.49
	Aluminum	2.32
	Gold	3.04
	Copper	3.83
	Silver	4.05
700	Titanium	0.194
	Zirconium	0.209
	Iron	0.487
	Tantalum	0.59
	Niobium	0.598
	Nickel	0.653
	Platinum	0.723
	Chromium	0.757

Source: data from Ho, C. Y., Powell, R. W., and Liley, P. E., *Thermal Conductictivity of Selected Materials*, NSRDS–NBS–8 and NSRD-NBS-16, Part 2 , National Standard Reference Data System–National Bureau of Standards, Part 1, 1966; Part 2, 1968.

Table 355. SELECTING THERMAL CONDUCTIVITY OF METALS AT TEMPERATURE* (SHEET 26 OF 30)

Temperature (K)	Metal	Thermal Conductivity (watt • cm^{-1} • K^{-1})
	Molybdenum	1.22
	Tungsten	1.33
	Magnesium	1.47
	Aluminum	2.26
	Gold	2.98
	Copper	3.77
	Silver	3.97
800	Titanium	0.197
	Zirconium	0.216
	Iron	0.433
	Tantalum	0.594
	Niobium	0.613
	Nickel	0.674
	Chromium	0.713
	Platinum	0.729
	Molybdenum	1.18
	Tungsten	1.28
	Magnesium	1.46
	Aluminum	2.2
	Gold	2.92
	Copper	3.71
	Silver	3.89
900	Titanium	0.202
	Zirconium	0.226
	Iron	0.38
	Tantalum	0.598

Source: data from Ho, C. Y., Powell, R. W., and Liley, P. E., *Thermal Conductictivity of Selected Materials*, NSRDS–NBS–8 and NSRD-NBS-16, Part 2 , National Standard Reference Data System–National Bureau of Standards, Part 1, 1966; Part 2, 1968.

Table 355. SELECTING THERMAL CONDUCTIVITY OF METALS AT TEMPERATURE* (SHEET 27 OF 30)

Temperature (K)	Metal	Thermal Conductivity (watt • cm^{-1} • K^{-1})
	Niobium	0.629
	Chromium	0.678
	Nickel	0.696
	Platinum	0.737
	Molybdenum	1.15
	Tungsten	1.24
	Magnesium	1.45
	Aluminum	2.13
	Gold	2.85
	Copper	3.64
	Silver	3.82
1000	Titanium	0.207
	Zirconium	0.237
	Iron	0.326
	Tantalum	0.602
	Niobium	0.644
	Chromium	0.653
	Nickel	0.718
	Platinum	0.748
	Molybdenum	1.12
	Tungsten	1.21
	Gold	2.78
	Copper	3.57
	Silver	3.74
1100	Titanium	0.213
	Zirconium	0.248
	Iron	0.297
	Tantalum	0.606

Source: data from Ho, C. Y., Powell, R. W., and Liley, P. E., *Thermal Conductictivity of Selected Materials*, NSRDS–NBS–8 and NSRD-NBS-16, Part 2 , National Standard Reference Data System–National Bureau of Standards, Part 1, 1966; Part 2, 1968.

Table 355. SELECTING THERMAL CONDUCTIVITY OF METALS AT TEMPERATURE* (SHEET 28 OF 30)

Temperature (K)	Metal	Thermal Conductivity (watt • cm^{-1} • K^{-1})
	Chromium	0.636
	Niobium	0.659
	Nickel	0.739
	Platinum	0.76
	Molybdenum	1.08
	Tungsten	1.18
	Gold	2.71
	Copper	3.5
	Silver	3.66
1200	Titanium	0.22
	Zirconium	0.257
	Iron	0.282
	Tantalum	0.61
	Chromium	0.624
	Niobium	0.675
	Nickel	0.761
	Platinum	0.775
	Molybdenum	1.05
	Tungsten	1.15
	Gold	2.62
	Copper	3.42
	Silver	3.58
1400	Titanium	0.236
	Zirconium	0.275
	Iron	0.309
	Chromium	0.611

Source: data from Ho, C. Y., Powell, R. W., and Liley, P. E., *Thermal Conductictivity of Selected Materials*, NSRDS–NBS–8 and NSRD-NBS-16, Part 2 , National Standard Reference Data System–National Bureau of Standards, Part 1, 1966; Part 2, 1968.

Table 355. SELECTING THERMAL CONDUCTIVITY OF METALS AT TEMPERATURE* (SHEET 29 OF 30)

Temperature (K)	Metal	Thermal Conductivity (watt • cm^{-1} • K^{-1})
	Tantalum	0.618
	Niobium	0.705
	Nickel	0.804
	Platinum	0.807
	Molybdenum	0.996
	Tungsten	1.11
1600	Titanium	0.253
	Zirconium	0.29
	Iron	0.327
	Tantalum	0.626
	Niobium	0.735
	Platinum	0.842
	Molybdenum	0.946
	Tungsten	1.07
1800	Titanium	0.271
	Zirconium	0.302
	Tantalum	0.634
	Niobium	0.764
	Platinum	0.877
	Molybdenum	0.907
	Tungsten	1.03
2000	Zirconium	0.313
	Tantalum	0.64
	Niobium	0.791
	Molybdenum	0.88
	Platinum	0.913
	Tungsten	1

Source: data from Ho, C. Y., Powell, R. W., and Liley, P. E., *Thermal Conductictivity of Selected Materials*, NSRDS–NBS–8 and NSRD-NBS-16, Part 2 , National Standard Reference Data System–National Bureau of Standards, Part 1, 1966; Part 2, 1968.

Table 355. SELECTING THERMAL CONDUCTIVITY OF METALS AT TEMPERATURE* (SHEET 30 OF 30)

Temperature (K)	Metal	Thermal Conductivity (watt • cm^{-1} • K^{-1})
2200	Tantalum	0.647
	Niobium	0.815
	Molybdenum	0.858
	Tungsten	0.98
2600	Tantalum	0.658
	Molybdenum	0.825
	Tungsten	0.94
3000	Tantalum	0.665
	Tungsten	0.915

Source: data from Ho, C. Y., Powell, R. W., and Liley, P. E., *Thermal Conductictivity of Selected Materials,* NSRDS–NBS–8 and NSRD-NBS-16, Part 2 , National Standard Reference Data System–National Bureau of Standards, Part 1, 1966; Part 2, 1968.

* These data apply only to metals of purity of at least 99.9%. The third significant figure may not be accurate.

Table 356. SELECTING THERMAL CONDUCTIVITY OF ALLOY CAST IRONS

Description	Thermal Conductivity W/(m • K)
Heat–Resistant High–Nickel Ductile Iron (20 Ni)	13
Corrosion–Resistant High–Nickel Ductile Iron	13.4
Heat–Resistant Gray High–Chromium Iron	20
Abrasion–Resistant Low–C White Iron	22†
Heat–Resistant Gray Nickel–Chromium–Silicon Iron	30
Abrasion–Resistant Martensitic Nickel–Chromium White Iron	30†
Heat–Resistant Gray Medium–Silicon Iron	37
Heat–Resistant Gray High–Nickel Iron	37 to 40
Corrosion–Resistant High–Nickel Gray Iron	38 to 40

† Estimated.

Source: Data from ASM *Metals Reference Book, Second Edition*, American Society for Metals, Metals Park, Ohio 44073, p172, (1984).

Table 357. SELECTING THERMAL CONDUCTIVITY OF CERAMICS (SHEET 1 OF 12)

Ceramic	Thermal Conductivity ($cal \cdot cm^{-1} \cdot sec^{-1} \cdot K^{-1}$)
Zirconium Oxide (ZrO_2) (plasma sprayed)	0.0019-0.0022 at 800°C
Zirconium Oxide (ZrO_2) (plasma sprayed)	0.0019-0.0031 at room temp.
Silicon Dioxide (SiO_2)	0.0025 at 200°C
Cerium Dioxide (CeO_2)	0.00287 at 1400K
Silicon Dioxide (SiO_2)	0.003 at 400°C
Sillimanite (Al_2O_3 SiO_2) (0% porosity)	0.003 at 1500°C
Silicon Carbide (SiC) (cubic, CVD)	0.0032 at 1530°C
Zirconium Oxide (ZrO_2) (plasma sprayed and coated with Cr_2O_3)	0.0033 at 800°C
Zirconium Oxide (ZrO_2) (plasma sprayed and coated with Cr_2O_3)	0.0033 at room temp.
Sillimanite (Al_2O_3 SiO_2) (0% porosity)	0.0035 at 800°C
Sillimanite (Al_2O_3 SiO_2) (0% porosity)	0.0035 at 1200°C
Cordierite (2MgO $2Al_2O_3$ $5SiO_2$) (ρ=2.1g/cm^3)	0.0038 at 800°C
Zirconium Oxide (ZrO_2) (stabilized)	0.004 at 100°C
Sillimanite (Al_2O_3 SiO_2) (0% porosity)	0.004 at 400°C
Silicon Dioxide (SiO_2)	0.004 at 800°C
Cordierite (2MgO $2Al_2O_3$ $5SiO_2$) (ρ=2.1g/cm^3)	0.0040 at 500°C
Cordierite (2MgO $2Al_2O_3$ $5SiO_2$) (ρ=2.1g/cm^3)	0.0041 at 300°C

Source: data compiled by J.S. Park from No. *1 Materials Index*, Peter T.B. Shaffer, Plenum Press, New York, (1964); *Smithells Metals Reference Book*, Eric A. Brandes, ed., in association with Fulmer Research Institute Ltd. 6th ed. London, Butterworths, Boston, (1983); and *Ceramic Source*, American Ceramic Society (1986-1991)

Table 357. SELECTING THERMAL CONDUCTIVITY OF CERAMICS
(SHEET 2 OF 12)

Ceramic	Thermal Conductivity ($cal \cdot cm^{-1} \cdot sec^{-1} \cdot K^{-1}$)
Sillimanite (Al_2O_3 SiO_2) (0% porosity)	0.0042 at 100°C
Cordierite (2MgO $2Al_2O_3$ $5SiO_2$) (ρ=2.1g/cm^3)	0.0043 at 20°C
Zirconium Oxide (ZrO_2) (stabilized)	0.0044 at 500°C
Zirconium Oxide (ZrO_2) (5-10% CaO stabilized)	0.0045 at 400°C
Zirconium Oxide (ZrO_2) (stabilized)	0.0048-0.0055 at 1000°C
Zirconium Oxide (ZrO_2) (stabilized)	0.0049-0.0050 at 1200°C
Zirconium Oxide (ZrO_2) (5-10% CaO stabilized)	0.0049 at 800°C
Zirconium Oxide (ZrO_2) (stabilized, 0% porosity)	0.005 at 100°C
Zirconium Oxide (ZrO_2) (stabilized, 0% porosity)	0.005 at 200°C
Zirconium Oxide (ZrO_2) (stabilized, 0% porosity)	0.005 at 400°C
Silicon Dioxide (SiO_2)	0.005 at 1200°C
Zirconium Oxide (ZrO_2) (Y_2O_3 stabilized)	0.0053 at 800°C
Cordierite (2MgO $2Al_2O_3$ $5SiO_2$) (ρ=2.3g/cm^3)	0.0055 at 500°C
Zirconium Oxide (ZrO_2) (stabilized, 0% porosity)	0.0055 at 800°C
Cordierite (2MgO $2Al_2O_3$ $5SiO_2$) (ρ=2.3g/cm^3)	0.0055 at 800°C
Zirconium Oxide (ZrO_2) (Y_2O_3 stabilized)	0.0055 at room temp.
Zirconium Oxide (ZrO_2) (MgO stabilized)	0.0057 at 800°C
Zirconium Oxide (ZrO_2) (5-10% CaO stabilized)	0.0057 at 1200°C
Silicon Carbide (SiC) (cubic, CVD)	0.0059 at 1250°C
Thorium Dioxide (ThO_2) (0% porosity)	0.006-0.0076 at 1200°C

Source: data compiled by J.S. Park from No. *1 Materials Index*, Peter T.B. Shaffer, Plenum Press, New York, (1964); *Smithells Metals Reference Book*, Eric A. Brandes, ed., in association with Fulmer Research Institute Ltd. 6th ed. London, Butterworths, Boston, (1983); and *Ceramic Source*, American Ceramic Society (1986-1991)

Table 357. SELECTING THERMAL CONDUCTIVITY OF CERAMICS (SHEET 3 OF 12)

Ceramic	Thermal Conductivity (cal • cm^{-1} • sec^{-1} • K^{-1})
Uranium Dioxide (UO_2)	0.006 at 1000°C
Uranium Dioxide (UO_2)	0.006 at 1200°C
Zirconium Oxide (ZrO_2) (stabilized, 0% porosity)	0.006 at 1200°C
Thorium Dioxide (ThO_2) (0% porosity)	0.006 at 1400°C
Silicon Dioxide (SiO_2)	0.006 at 1600°C
Cordierite (2MgO $2Al_2O_3$ $5SiO_2$) (ρ=2.3g/cm^3)	0.0062 at 300°C
Zirconium Oxide (ZrO_2) (stabilized, 0% porosity)	0.0065 at 1400°C
Thorium Dioxide (ThO_2) (0% porosity)	0.007-0.0074 at 1000°C
Zirconium Oxide (ZrO_2) (MgO stabilized)	0.0076 at room temp.
Cordierite (2MgO $2Al_2O_3$ $5SiO_2$) (ρ=2.3g/cm^3)	0.0077 at 20°C
Titanium Oxide (TiO_2) (0% porosity)	0.008 at 600°C
Uranium Dioxide (UO_2)	0.008 at 600°C
Uranium Dioxide (UO_2)	0.008 at 700°C
Thorium Dioxide (ThO_2) (0% porosity)	0.008 at 800°C
Titanium Oxide (TiO_2) (0% porosity)	0.008 at 800°C
Titanium Oxide (TiO_2) (0% porosity)	0.008 at 1000°C
Uranium Dioxide (UO_2) (0% porosity)	0.008 at 1000°C
Titanium Oxide (TiO_2) (0% porosity)	0.008 at 1200°C
Titanium Oxide (TiO_2) (0% porosity)	0.009 at 400°C
Uranium Dioxide (UO_2) (0% porosity)	0.009 at 800°C

Source: data compiled by J.S. Park from No. *1 Materials Index*, Peter T.B. Shaffer, Plenum Press, New York, (1964); *Smithells Metals Reference Book*, Eric A. Brandes, ed., in association with Fulmer Research Institute Ltd. 6th ed. London, Butterworths, Boston, (1983); and *Ceramic Source*, American Ceramic Society (1986-1991)

Table 357. SELECTING THERMAL CONDUCTIVITY OF CERAMICS
(SHEET 4 OF 12)

Ceramic	Thermal Conductivity ($cal \cdot cm^{-1} \cdot sec^{-1} \cdot K^{-1}$)
Mullite ($3Al_2O_3$ $2SiO_2$) (0% porosity)	0.009 at 1000°C
Mullite ($3Al_2O_3$ $2SiO_2$) (0% porosity)	0.009 at 1200°C
Mullite ($3Al_2O_3$ $2SiO_2$) (0% porosity)	0.009 at 1400°C
Mullite ($3Al_2O_3$ $2SiO_2$) (0% porosity)	0.0095 at 800°C
Zircon (SiO_2 ZrO_2) (0% porosity)	0.0095 at 1200°C
Zircon (SiO_2 ZrO_2) (0% porosity)	0.0095 at 1400°C
Magnesium Oxide (MgO)	0.0096-0.0191 at 1800°C
Thorium Dioxide (ThO_2) (0% porosity)	0.010 at 600°C
Uranium Dioxide (UO_2) (0% porosity)	0.010 at 600°C
Mullite ($3Al_2O_3$ $2SiO_2$) (0% porosity)	0.010 at 600°C
Zircon (SiO_2 ZrO_2) (0% porosity)	0.010 at 800°C
Magnesium Oxide (MgO)	0.0108-0.016 at 1600°C
Mullite ($3Al_2O_3$ $2SiO_2$) (0% porosity)	0.011 at 400°C
Nickel monoxide (NiO) (0% porosity)	0.011 at 1000°C
Magnesium Oxide (MgO)	0.012-0.014 at 1400°C
Titanium Oxide (TiO_2) (0% porosity)	0.012 at 200°C
Uranium Dioxide (UO_2)	0.012 at 400°C
Zircon (SiO_2 ZrO_2) (0% porosity)	0.012 at 400°C
Nickel monoxide (NiO) (0% porosity)	0.012 at 800°C
Spinel (Al_2O_3 MgO) (0% porosity)	0.013-0.0138 at 1000°C

Source: data compiled by J.S. Park from No. *1 Materials Index*, Peter T.B. Shaffer, Plenum Press, New York, (1964); *Smithells Metals Reference Book*, Eric A. Brandes, ed., in association with Fulmer Research Institute Ltd. 6th ed. London, Butterworths, Boston, (1983); and *Ceramic Source*, American Ceramic Society (1986-1991)

Table 357. SELECTING THERMAL CONDUCTIVITY OF CERAMICS (SHEET 5 OF 12)

Ceramic	Thermal Conductivity ($cal \cdot cm^{-1} \cdot sec^{-1} \cdot K^{-1}$)
Aluminum Oxide (Al_2O_3)	0.013-0.015 at 1200°C
Mullite ($3Al_2O_3$ $2SiO_2$) (0% porosity)	0.013 at 200°C
Spinel (Al_2O_3 MgO) (0% porosity)	0.013 at 1200°C
Aluminum Oxide (Al_2O_3)	0.013 at 1400°C
Zircon (SiO_2 ZrO_2) (0% porosity)	0.0135 at 200°C
Titanium Monocarbide (TiC)	0.0135 at 1000 °C
Magnesium Oxide (MgO)	0.0139-0.0148 at 1200°C
Aluminum Oxide (Al_2O_3)	0.014-0.016 at 1000°C
Thorium Dioxide (ThO_2) (0% porosity)	0.014 at 400°C
Aluminum Oxide (Al_2O_3)	0.014 at 1600°C
Mullite ($3Al_2O_3$ $2SiO_2$) (0% porosity)	0.0145 at 100°C
Zircon (SiO_2 ZrO_2) (0% porosity)	0.0145 at 100°C
Aluminum Oxide (Al_2O_3)	0.015-0.017 at 800°C
Uranium Dioxide (UO_2) (0% porosity)	0.015 at 400°C
Spinel (Al_2O_3 MgO) (0% porosity)	0.015 at 800°C
Zirconium Mononitride (TiN)	0.015 at 1100 °C
Hafnium Diboride (HfB_2)	0.015 at room temp.
Magnesium Oxide (MgO)	0.016-0.020 at 1000°C
Titanium Oxide (TiO_2) (0% porosity)	0.016 at 100°C
Zirconium Mononitride (TiN)	0.016 at 875 °C

Source: data compiled by J.S. Park from No. *1 Materials Index*, Peter T.B. Shaffer, Plenum Press, New York, (1964); *Smithells Metals Reference Book*, Eric A. Brandes, ed., in association with Fulmer Research Institute Ltd. 6th ed. London, Butterworths, Boston, (1983); and *Ceramic Source*, American Ceramic Society (1986-1991)

Table 357. SELECTING THERMAL CONDUCTIVITY OF CERAMICS
(SHEET 6 OF 12)

Ceramic	Thermal Conductivity ($cal \cdot cm^{-1} \cdot sec^{-1} \cdot K^{-1}$)
Nickel monoxide (NiO) (0% porosity)	0.017 at 400°C
Aluminum Oxide (Al_2O_3)	0.017 at 1800°C
Uranium Dioxide (UO_2)	0.018 at 100°C
Zirconium Mononitride (TiN)	0.018 at 650 °C
Calcium Oxide (CaO)	0.0186-0.019 at 1000°C
Thorium Dioxide (ThO_2) (0% porosity)	0.019 at 200°C
Spinel (Al_2O_3 MgO) (0% porosity)	0.019 at 600°C
Calcium Oxide (CaO)	0.019 at 800°C
Magnesium Oxide (MgO)	0.0198-0.026 at 800°C
Aluminum Oxide (Al_2O_3)	0.02-0.031 at 400°C
Thorium Dioxide (ThO_2) (0% porosity)	0.020 at 100°C
Uranium Dioxide (UO_2) (0% porosity)	0.020 at 200°C
Calcium Oxide (CaO)	0.020 at 600°C
Titanium Mononitride (TiN)	0.020 at 1000 °C
Aluminum Oxide (Al_2O_3)	0.021-0.022 at 600°C
Trisilicon tetranitride (Si_3N_4) (pressureless sintered)	0.022-0.072 at 127 °C
Calcium Oxide (CaO)	0.022 at 400°C
Cerium Dioxide (CeO_2)	0.0229 at 400K
Dichromium Trioxide (Cr_2O_3)	0.0239-0.0788
Nickel monoxide (NiO) (0% porosity)	0.024 at 200°C

Source: data compiled by J.S. Park from No. *1 Materials Index*, Peter T.B. Shaffer, Plenum Press, New York, (1964); *Smithells Metals Reference Book*, Eric A. Brandes, ed., in association with Fulmer Research Institute Ltd. 6th ed. London, Butterworths, Boston, (1983); and *Ceramic Source*, American Ceramic Society (1986-1991)

Table 357. SELECTING THERMAL CONDUCTIVITY OF CERAMICS
(SHEET 7 OF 12)

Ceramic	Thermal Conductivity ($cal \cdot cm^{-1} \cdot sec^{-1} \cdot K^{-1}$)
Spinel (Al_2O_3 MgO) (0% porosity)	0.024 at 400°C
Thorium Dioxide (ThO_2) (0% porosity)	0.024 at room temp.
Uranium Dioxide (UO_2) (0% porosity)	0.025 at 100°C
Zirconium Mononitride (TiN)	0.025 at 425 °C
Tantalum Diboride (TaB_2)	0.026 at room temp.
Calcium Oxide (CaO)	0.027 at 200°C
Titanium Mononitride (TiN)	0.027 at 650 °C
Hafnium Dioxide (HfO_2)	0.0273 at 25-425°C
Nickel monoxide (NiO) (0% porosity)	0.029 at 100°C
Aluminum Oxide (Al_2O_3) (single crystal)	0.029 at 800°C
Boron Nitride (BN) parallel to a axis	0.0295 at 1000°C
Aluminum Oxide (Al_2O_3)	0.03-0.064 at 200°C
Spinel (Al_2O_3 MgO) (0% porosity)	0.031 at 200°C
Boron Nitride (BN) parallel to a axis	0.0318 at 700°C
Beryllium Oxide (BeO)	0.032-0.34 at 100°C
Trisilicon tetranitride (Si_3N_4) (pressureless sintered)	0.033-0.034 at 1200 °C
Beryllium Oxide (BeO)	0.033-0.039 at 1600°C
Tantalum Diboride (TaB_2)	0.033 at 200 °C.
Beryllium Oxide (BeO)	0.033 at 1700°C
Beryllium Oxide (BeO)	0.034 at 1500°C

Source: data compiled by J.S. Park from No. *1 Materials Index*, Peter T.B. Shaffer, Plenum Press, New York, (1964); *Smithells Metals Reference Book*, Eric A. Brandes, ed., in association with Fulmer Research Institute Ltd. 6th ed. London, Butterworths, Boston, (1983); and *Ceramic Source*, American Ceramic Society (1986-1991)

Table 357. SELECTING THERMAL CONDUCTIVITY OF CERAMICS
(SHEET 8 OF 12)

Ceramic	Thermal Conductivity (cal • cm^{-1} • sec^{-1} • K^{-1})
Spinel (Al_2O_3 MgO) (0% porosity)	0.035 at 100°C
Aluminum Oxide (Al_2O_3)	0.035 at 500°C
Trisilicon tetranitride (Si_3N_4) (pressureless sintered)	0.036-0.042 at 500 °C
Beryllium Oxide (BeO)	0.036 at 1400°C
Beryllium Oxide (BeO)	0.036 at 1800°C
Beryllium Oxide (BeO)	0.036 at 1900°C
Beryllium Oxide (BeO)	0.036 at 2000°C
Boron Nitride (BN) parallel to a axis	0.0362 at 300°C
Calcium Oxide (CaO)	0.037 at 100°C
Aluminum Oxide (Al_2O_3)	0.037 at 315°C
Magnesium Oxide (MgO)	0.038-0.045 at 400°C
Beryllium Oxide (BeO)	0.038-0.47 at 20°C
Trisilicon tetranitride (Si_3N_4) (pressureless sintered)	0.038 at 1000 °C
Beryllium Oxide (BeO)	0.038 at 1300°C
Aluminum Oxide (Al_2O_3)	0.04-0.069 at 100°C
Titanium Mononitride (TiN)	0.040 at 200 °C
Zirconium Mononitride (TiN)	0.040 at 200 °C
Beryllium Oxide (BeO)	0.041-0.054 at 1200°C
Titanium Monocarbide (TiC)	0.041-0.074 at room temp.
Trisilicon tetranitride (Si_3N_4) (pressureless sintered)	0.041 at 200-750 °C

Source: data compiled by J.S. Park from No. *1 Materials Index*, Peter T.B. Shaffer, Plenum Press, New York, (1964); *Smithells Metals Reference Book*, Eric A. Brandes, ed., in association with Fulmer Research Institute Ltd. 6th ed. London, Butterworths, Boston, (1983); and *Ceramic Source*, American Ceramic Society (1986-1991)

Table 357. SELECTING THERMAL CONDUCTIVITY OF CERAMICS

(SHEET 9 OF 12)

Ceramic	Thermal Conductivity ($cal \cdot cm^{-1} \cdot sec^{-1} \cdot K^{-1}$)
Molybdenum Disilicide ($MoSi_2$)	0.041 at 1100°C
Aluminum Nitride (AlN)	0.042 at 800°C
Beryllium Oxide (BeO)	0.043 at 1100°C
Molybdenum Disilicide ($MoSi_2$)	0.046 at 875°C
Aluminum Oxide (Al_2O_3) (single crystal)	0.047 at 300°C
Aluminum Nitride (AlN)	0.048 at 600°C
Chromium Diboride (CrB_2)	0.049-0.076 at room temp.
Silicon Carbide (SiC) (cubic, CVD)	0.049-0.080 at 600°C
Zirconium Monocarbide (ZrC)	0.049 at room temp.
Silicon Carbide (SiC) (cubic, CVD)	0.051 at 1000°C
Aluminum Nitride (AlN)	0.053 at 400°C
Molybdenum Disilicide ($MoSi_2$)	0.053 at 540°C
Hafnium Monocarbide (HfC)	0.053 at room temp.
Tantalum Monocarbide (TaC)	0.053 at room temp.
Zirconium Diboride (ZrB_2)	0.055-0.058 at room temp.
Zirconium Diboride (ZrB_2)	0.055-0.060 at 200 °C
Titanium Mononitride (TiN)	0.057 at 127 °C
Molybdenum Disilicide ($MoSi_2$)	0.057 at 650°C
Titanium Diboride (TiB_2)	0.058-0.062 at room temp.
Aluminum Oxide (Al_2O_3)	0.06 at room temp.
Beryllium Oxide (BeO)	0.060-0.093 at 800°C
Aluminum Nitride (AlN)	0.060 at 200°C
Zirconium Monocarbide (ZrC)	0.061 at 288°C
Silicon Carbide (SiC) (cubic, CVD)	0.061 at 800°C

Source: data compiled by J.S. Park from No. *1 Materials Index*, Peter T.B. Shaffer, Plenum Press, New York, (1964); *Smithells Metals Reference Book*, Eric A. Brandes, ed., in association with Fulmer Research Institute Ltd. 6th ed. London, Butterworths, Boston, (1983); and *Ceramic Source*, American Ceramic Society (1986-1991)

Table 357. SELECTING THERMAL CONDUCTIVITY OF CERAMICS
(SHEET 10 OF 12)

Ceramic	Thermal Conductivity ($cal \cdot cm^{-1} \cdot sec^{-1} \cdot K^{-1}$)
Titanium Diboride (TiB_2)	0.063 at 200 °C
Boron Nitride (BN) parallel to c axis	0.0637 at 1000°C
Magnesium Oxide (MgO)	0.064-0.065 at 200°C
Boron Nitride (BN) parallel to c axis	0.0646 at 700°C
Boron Carbide (B_4C)	0.065-0.069 at room temp.
Zirconium Monocarbide (ZrC)	0.065 at 188°C
Boron Nitride (BN) parallel to c axis	0.0687 at 300°C
Titanium Mononitride (TiN)	0.069 at 25 °C
Zirconium Monocarbide (ZrC)	0.069 at 150°C
Aluminum Nitride (AlN)	0.072 at 25°C
Trisilicon tetranitride (Si_3N_4) (pressureless sintered)	0.072 at room temp.
Molybdenum Disilicide ($MoSi_2$)	0.074 at 425°C
Magnesium Oxide (MgO)	0.078-0.082 at 100°C
Zirconium Monocarbide (ZrC)	0.080 at 600°C
Silicon Carbide (SiC) (cubic, CVD)	0.0827 at 1327°C
Zirconium Monocarbide (ZrC)	0.083 at 800°C
Zirconium Monocarbide (ZrC)	0.086 at 1000°C
Beryllium Oxide (BeO)	0.089-0.1137 at 600°C
Zirconium Monocarbide (ZrC)	0.089 at 1200°C
Zirconium Monocarbide (ZrC)	0.092 at 1400°C

Source: data compiled by J.S. Park from No. *1 Materials Index*, Peter T.B. Shaffer, Plenum Press, New York, (1964); *Smithells Metals Reference Book*, Eric A. Brandes, ed., in association with Fulmer Research Institute Ltd. 6th ed. London, Butterworths, Boston, (1983); and *Ceramic Source*, American Ceramic Society (1986-1991)

Table 357. SELECTING THERMAL CONDUCTIVITY OF CERAMICS
(SHEET 11 OF 12)

Ceramic	Thermal Conductivity ($cal \cdot cm^{-1} \cdot sec^{-1} \cdot K^{-1}$)
Zirconium Monocarbide (ZrC)	0.096 at 1600°C
Magnesium Oxide (MgO)	0.097 at room temp.
Silicon Carbide (SiC)	0.098-0.10 at 20°C
Zirconium Monocarbide (ZrC)	0.098 at 50°C
Zirconium Monocarbide (ZrC)	0.099 at 1800°C
Aluminum Oxide (Al_2O_3) (single crystal)	0.103 at 20°C
Zirconium Monocarbide (ZrC)	0.103 at 2000°C
Zirconium Monocarbide (ZrC)	0.105 at 2200°C
Molybdenum Disilicide ($MoSi_2$)	0.129 at 150°C
Titanium Mononitride (TiN)	0.136 at 2300 °C
Beryllium Oxide (BeO)	0.14-0.16 at 400°C
Silicon Carbide (SiC) (with 1 wt% Al additive)	0.143
Titanium Mononitride (TiN)	0.162 at 1500 °C
Boron Carbide (B_4C)	0.198 at 425 °C
Tungsten Monocarbide (WC)	0.201 at 20 °C
Tungsten Monocarbide (WC) (6% Co, 1-3μm grain size)	0.239
Tungsten Monocarbide (WC) (24% Co, 1-3μm grain size)	0.239
Tungsten Monocarbide (WC) (12% Co, 1-3μm grain size)	0.251
Tungsten Monocarbide (WC) (6% Co, 2-4μm grain size)	0.251
Tungsten Monocarbide (WC) (6% Co, 3-6μm grain size)	0.256

Source: data compiled by J.S. Park from No. *1 Materials Index*, Peter T.B. Shaffer, Plenum Press, New York, (1964); *Smithells Metals Reference Book*, Eric A. Brandes, ed., in association with Fulmer Research Institute Ltd. 6th ed. London, Butterworths, Boston, (1983); and *Ceramic Source*, American Ceramic Society (1986-1991)

Table 357. SELECTING THERMAL CONDUCTIVITY OF CERAMICS
(SHEET 12 OF 12)

Ceramic	Thermal Conductivity ($cal \cdot cm^{-1} \cdot sec^{-1} \cdot K^{-1}$)
Silicon Carbide (SiC) (with 2 wt% BN additive)	0.263
Silicon Carbide (SiC) (cubic, CVD)	0.289 at 127°C
Silicon Carbide (SiC) (with 1 wt% B additive)	0.406
Trichromium Dicarbide (Cr_3C_2)	0.454
Silicon Carbide (SiC) (with 1 wt% Be additive)	0.621
Silicon Carbide (SiC) (with 1.6 wt% BeO additive)	0.645 at room temp.
Silicon Carbide (SiC) (with 3.2 wt% BeO additive)	0.645 at room temp.

Source: data compiled by J.S. Park from No. *1 Materials Index*, Peter T.B. Shaffer, Plenum Press, New York, (1964); *Smithells Metals Reference Book*, Eric A. Brandes, ed., in association with Fulmer Research Institute Ltd. 6th ed. London, Butterworths, Boston, (1983); and *Ceramic Source*, American Ceramic Society (1986-1991)

Table 358. SELECTING THERMAL CONDUCTIVITY OF CERAMICS AT TEMPERATURE

(SHEET 1 OF 19)

Temperature (°C)	Ceramic	Thermal Conductivity ($cal \cdot cm^{-1} \cdot sec^{-1} \cdot K^{-1}$)
20	Zirconium Oxide (ZrO_2) (plasma sprayed)	0.0019-0.0031
20	Zirconium Oxide (ZrO_2) (plasma sprayed and coated with Cr_2O_3)	0.0033
20	Cordierite ($2MgO\ 2Al_2O_3\ 5SiO_2$) (ρ=2.1g/cm^3)	0.0043
20	Zirconium Oxide (ZrO_2) (Y_2O_3 stabilized)	0.0055
20	Zirconium Oxide (ZrO_2) (MgO stabilized)	0.0076
20	Cordierite ($2MgO\ 2Al_2O_3\ 5SiO_2$) (ρ=2.3g/cm^3)	0.0077
20	Hafnium Diboride (HfB_2)	0.015
20	Thorium Dioxide (ThO_2) (0% porosity)	0.024
20	Tantalum Diboride (TaB_2)	0.026
20	Beryllium Oxide (BeO)	0.038-0.47
20	Titanium Monocarbide (TiC)	0.041-0.074

Source: data compiled by J.S. Park from *No. 1 Materials Index*, Peter T.B. Shaffer, Plenum Press, New York, (1964); *Smithells Metals Reference Book*, Eric A. Brandes, ed., in association with Fulmer Research Institute Ltd. 6th ed. London, Butterworths, Boston, (1983); and *Ceramic Source*, American Ceramic Society (1986-1991)

Table 358. SELECTING THERMAL CONDUCTIVITY OF CERAMICS AT TEMPERATURE

(SHEET 2 OF 19)

Temperature (°C)	Ceramic	Thermal Conductivity (cal • cm^{-1} • sec^{-1} • K^{-1})
20	Zirconium Monocarbide (ZrC)	0.049
20	Chromium Diboride (CrB_2)	0.049-0.076
20	Hafnium Monocarbide (HfC)	0.053
20	Tantalum Monocarbide (TaC)	0.053
20	Zirconium Diboride (ZrB_2)	0.055-0.058
20	Titanium Diboride (TiB_2)	0.058-0.062
20	Aluminum Oxide (Al_2O_3)	0.06
20	Boron Carbide (B_4C)	0.065-0.069
20	Trisilicon tetranitride (Si_3N_4) (pressureless sintered)	0.072
20	Magnesium Oxide (MgO)	0.097
20	Silicon Carbide (SiC)	0.098-0.10
20	Aluminum Oxide (Al_2O_3) (single crystal)	0.103

Source: data compiled by J.S. Park from *No. 1 Materials Index*, Peter T.B. Shaffer, Plenum Press, New York, (1964); *Smithells Metals Reference Book*, Eric A. Brandes, ed., in association with Fulmer Research Institute Ltd. 6th ed. London, Butterworths, Boston, (1983); and *Ceramic Source*, American Ceramic Society (1986-1991)

Table 358. SELECTING THERMAL CONDUCTIVITY OF CERAMICS AT TEMPERATURE

(SHEET 3 OF 19)

Temperature (°C)	Ceramic	Thermal Conductivity ($cal \cdot cm^{-1} \cdot sec^{-1} \cdot K^{-1}$)
20	Tungsten Monocarbide (WC)	0.201
20	Silicon Carbide (SiC) (with 1.6 wt% BeO additive)	0.645
20	Silicon Carbide (SiC) (with 3.2 wt% BeO additive)	0.645
25-425	Hafnium Dioxide (HfO_2)	0.0273
25	Titanium Mononitride (TiN)	0.069
25	Aluminum Nitride (AlN)	0.072
50	Zirconium Monocarbide (ZrC)	0.098
100	Zirconium Oxide (ZrO_2) (stabilized)	0.004
100	Sillimanite (Al_2O_3 SiO_2) (0% porosity)	0.0042
100	Zirconium Oxide (ZrO_2) (stabilized, 0% porosity)	0.005
100	Mullite ($3Al_2O_3$ $2SiO_2$) (0% porosity)	0.0145

Source: data compiled by J.S. Park from *No. 1 Materials Index*, Peter T.B. Shaffer, Plenum Press, New York, (1964); *Smithells Metals Reference Book*, Eric A. Brandes, ed., in association with Fulmer Research Institute Ltd. 6th ed. London, Butterworths, Boston, (1983); and *Ceramic Source*, American Ceramic Society (1986-1991)

Table 358. SELECTING THERMAL CONDUCTIVITY OF CERAMICS AT TEMPERATURE

(SHEET 4 OF 19)

Temperature (°C)	Ceramic	Thermal Conductivity ($cal \cdot cm^{-1} \cdot sec^{-1} \cdot K^{-1}$)
100	Zircon (SiO_2 ZrO_2) (0% porosity)	0.0145
100	Titanium Oxide (TiO_2) (0% porosity)	0.016
100	Uranium Dioxide (UO_2)	0.018
100	Thorium Dioxide (ThO_2) (0% porosity)	0.020
100	Uranium Dioxide (UO_2) (0% porosity)	0.025
100	Nickel monoxide (NiO) (0% porosity)	0.029
100	Beryllium Oxide (BeO)	0.032-0.34
100	Spinel (Al_2O_3 MgO) (0% porosity)	0.035
100	Calcium Oxide (CaO)	0.037
100	Aluminum Oxide (Al_2O_3)	0.04-0.069
100	Magnesium Oxide (MgO)	0.078-0.082

Source: data compiled by J.S. Park from *No. 1 Materials Index*, Peter T.B. Shaffer, Plenum Press, New York, (1964); *Smithells Metals Reference Book*, Eric A. Brandes, ed., in association with Fulmer Research Institute Ltd. 6th ed. London, Butterworths, Boston, (1983); and *Ceramic Source*, American Ceramic Society (1986-1991)

Table 358. SELECTING THERMAL CONDUCTIVITY OF CERAMICS AT TEMPERATURE

(SHEET 5 OF 19)

Temperature (°C)	Ceramic	Thermal Conductivity ($cal \cdot cm^{-1} \cdot sec^{-1} \cdot K^{-1}$)
127	Trisilicon tetranitride (Si_3N_4) (pressureless sintered)	0.022-0.072
127	Titanium Mononitride (TiN)	0.057
127	Silicon Carbide (SiC) (cubic, CVD)	0.289
150	Zirconium Monocarbide (ZrC)	0.069
150	Molybdenum Disilicide ($MoSi_2$)	0.129
188	Zirconium Monocarbide (ZrC)	0.065
200	Silicon Dioxide (SiO_2)	0.0025
200	Zirconium Oxide (ZrO_2) (stabilized, 0% porosity)	0.005
200	Titanium Oxide (TiO_2) (0% porosity)	0.012
200	Mullite ($3Al_2O_3$ $2SiO_2$) (0% porosity)	0.013

Source: data compiled by J.S. Park from *No. 1 Materials Index*, Peter T.B. Shaffer, Plenum Press, New York, (1964); *Smithells Metals Reference Book*, Eric A. Brandes, ed., in association with Fulmer Research Institute Ltd. 6th ed. London, Butterworths, Boston, (1983); and *Ceramic Source*, American Ceramic Society (1986-1991)

Table 358. SELECTING THERMAL CONDUCTIVITY OF CERAMICS AT TEMPERATURE

(SHEET 6 OF 19)

Temperature (˚C)	Ceramic	Thermal Conductivity ($cal \cdot cm^{-1} \cdot sec^{-1} \cdot K^{-1}$)
200	Zircon (SiO_2 ZrO_2) (0% porosity)	0.0135
200	Thorium Dioxide (ThO_2) (0% porosity)	0.019
200	Uranium Dioxide (UO_2) (0% porosity)	0.020
200	Nickel monoxide (NiO) (0% porosity)	0.024
200	Calcium Oxide (CaO)	0.027
200	Aluminum Oxide (Al_2O_3)	0.03-0.064
200	Spinel (Al_2O_3 MgO) (0% porosity)	0.031
200.	Tantalum Diboride (TaB_2)	0.033
200	Titanium Mononitride (TiN)	0.040
200	Zirconium Mononitride (TiN)	0.040
200-750	Trisilicon tetranitride (Si_3N_4) (pressureless sintered)	0.041
200	Zirconium Diboride (ZrB_2)	0.055-0.060

Source: data compiled by J.S. Park from *No. 1 Materials Index*, Peter T.B. Shaffer, Plenum Press, New York, (1964); *Smithells Metals Reference Book*, Eric A. Brandes, ed., in association with Fulmer Research Institute Ltd. 6th ed. London, Butterworths, Boston, (1983); and *Ceramic Source*, American Ceramic Society (1986-1991)

Table 358. SELECTING THERMAL CONDUCTIVITY OF CERAMICS AT TEMPERATURE

(SHEET 7 OF 19)

Temperature (°C)	Ceramic	Thermal Conductivity ($cal \cdot cm^{-1} \cdot sec^{-1} \cdot K^{-1}$)
200	Aluminum Nitride (AlN)	0.060
200	Titanium Diboride (TiB_2)	0.063
200	Magnesium Oxide (MgO)	0.064-0.065
288	Zirconium Monocarbide (ZrC)	0.061
300	Cordierite ($2MgO\ 2Al_2O_3\ 5SiO_2$) (ρ=2.1g/cm^3)	0.0041
300	Cordierite ($2MgO\ 2Al_2O_3\ 5SiO_2$) (ρ=2.3g/cm^3)	0.0062
300	Boron Nitride (BN) parallel to a axis	0.0362
300	Aluminum Oxide (Al_2O_3) (single crystal)	0.047
300	Boron Nitride (BN) parallel to c axis	0.0687
315	Aluminum Oxide (Al_2O_3)	0.037

Source: data compiled by J.S. Park from *No. 1 Materials Index*, Peter T.B. Shaffer, Plenum Press, New York, (1964); *Smithells Metals Reference Book*, Eric A. Brandes, ed., in association with Fulmer Research Institute Ltd. 6th ed. London, Butterworths, Boston, (1983); and *Ceramic Source*, American Ceramic Society (1986-1991)

Table 358. SELECTING THERMAL CONDUCTIVITY OF CERAMICS AT TEMPERATURE

(SHEET 8 OF 19)

Temperature (°C)	Ceramic	Thermal Conductivity (cal • cm^{-1} • sec^{-1} • K^{-1})
400	Silicon Dioxide (SiO_2)	0.003
400	Sillimanite (Al_2O_3 SiO_2) (0% porosity)	0.004
400	Zirconium Oxide (ZrO_2) (5-10% CaO stabilized)	0.0045
400	Zirconium Oxide (ZrO_2) (stabilized, 0% porosity)	0.005
400	Titanium Oxide (TiO_2) (0% porosity)	0.009
400	Mullite ($3Al_2O_3$ $2SiO_2$) (0% porosity)	0.011
400	Uranium Dioxide (UO_2)	0.012
400	Zircon (SiO_2 ZrO_2) (0% porosity)	0.012
400	Thorium Dioxide (ThO_2) (0% porosity)	0.014
400	Uranium Dioxide (UO_2) (0% porosity)	0.015
400	Nickel monoxide (NiO) (0% porosity)	0.017

Source: data compiled by J.S. Park from *No. 1 Materials Index*, Peter T.B. Shaffer, Plenum Press, New York, (1964); *Smithells Metals Reference Book*, Eric A. Brandes, ed., in association with Fulmer Research Institute Ltd. 6th ed. London, Butterworths, Boston, (1983); and *Ceramic Source*, American Ceramic Society (1986-1991)

Table 358. Selecting Thermal Conductivity of Ceramics at Temperature

(Sheet 9 of 19)

Temperature (°C)	Ceramic	Thermal Conductivity ($cal \cdot cm^{-1} \cdot sec^{-1} \cdot K^{-1}$)
400	Aluminum Oxide (Al_2O_3)	0.02-0.031
400	Calcium Oxide (CaO)	0.022
400	Spinel (Al_2O_3 MgO) (0% porosity)	0.024
400	Magnesium Oxide (MgO)	0.038-0.045
400	Aluminum Nitride (AlN)	0.053
400	Beryllium Oxide (BeO)	0.14-0.16
425	Zirconium Mononitride (TiN)	0.025
425	Boron Carbide (B_4C)	0.198
425	Molybdenum Disilicide ($MoSi_2$)	0.074
500	Cordierite (2MgO $2Al_2O_3$ $5SiO_2$) (ρ=2.1g/cm^3)	0.0040
500	Zirconium Oxide (ZrO_2) (stabilized)	0.0044
500	Cordierite (2MgO $2Al_2O_3$ $5SiO_2$) (ρ=2.3g/cm^3)	0.0055

Source: data compiled by J.S. Park from *No. 1 Materials Index*, Peter T.B. Shaffer, Plenum Press, New York, (1964); *Smithells Metals Reference Book*, Eric A. Brandes, ed., in association with Fulmer Research Institute Ltd. 6th ed. London, Butterworths, Boston, (1983); and *Ceramic Source*, American Ceramic Society (1986-1991)

Table 358. Selecting Thermal Conductivity of Ceramics at Temperature

(Sheet 10 of 19)

Temperature (°C)	Ceramic	Thermal Conductivity ($cal \cdot cm^{-1} \cdot sec^{-1} \cdot K^{-1}$)
500	Aluminum Oxide (Al_2O_3)	0.035
500	Trisilicon tetranitride (Si_3N_4) (pressureless sintered)	0.036-0.042
540	Molybdenum Disilicide ($MoSi_2$)	0.053
600	Titanium Oxide (TiO_2) (0% porosity)	0.008
600	Uranium Dioxide (UO_2)	0.008
600	Thorium Dioxide (ThO_2) (0% porosity)	0.010
600	Uranium Dioxide (UO_2) (0% porosity)	0.010
600	Mullite ($3Al_2O_3$ $2SiO_2$) (0% porosity)	0.010
600	Spinel (Al_2O_3 MgO) (0% porosity)	0.019
600	Calcium Oxide (CaO)	0.020
600	Aluminum Oxide (Al_2O_3)	0.021-0.022

Source: data compiled by J.S. Park from *No. 1 Materials Index*, Peter T.B. Shaffer, Plenum Press, New York, (1964); *Smithells Metals Reference Book*, Eric A. Brandes, ed., in association with Fulmer Research Institute Ltd. 6th ed. London, Butterworths, Boston, (1983); and *Ceramic Source*, American Ceramic Society (1986-1991)

Table 358. SELECTING THERMAL CONDUCTIVITY OF CERAMICS AT TEMPERATURE

(SHEET 11 OF 19)

Temperature (°C)	Ceramic	Thermal Conductivity ($cal \cdot cm^{-1} \cdot sec^{-1} \cdot K^{-1}$)
600	Aluminum Nitride (AlN)	0.048
600	Silicon Carbide (SiC) (cubic, CVD)	0.049-0.080
600	Zirconium Monocarbide (ZrC)	0.080
600	Beryllium Oxide (BeO)	0.089-0.1137
650	Zirconium Mononitride (TiN)	0.018
650	Titanium Mononitride (TiN)	0.027
650	Molybdenum Disilicide ($MoSi_2$)	0.057
700	Uranium Dioxide (UO_2)	0.008
700	Boron Nitride (BN) parallel to a axis	0.0318
700	Boron Nitride (BN) parallel to c axis	0.0646
800	Zirconium Oxide (ZrO_2) (plasma sprayed)	0.0019-0.0022
800	Oxide (ZrO_2) (plasma sprayed and coated with Cr_2O_3)	0.0033
800	Sillimanite (Al_2O_3 SiO_2) (0% porosity)	0.0035

Source: data compiled by J.S. Park from *No. 1 Materials Index*, Peter T.B. Shaffer, Plenum Press, New York, (1964); *Smithells Metals Reference Book*, Eric A. Brandes, ed., in association with Fulmer Research Institute Ltd. 6th ed. London, Butterworths, Boston, (1983); and *Ceramic Source*, American Ceramic Society (1986-1991)

Table 358. SELECTING THERMAL CONDUCTIVITY OF CERAMICS AT TEMPERATURE

(SHEET 12 OF 19)

Temperature (°C)	Ceramic	Thermal Conductivity ($cal \cdot cm^{-1} \cdot sec^{-1} \cdot K^{-1}$)
800	Cordierite ($2MgO\ 2Al_2O_3\ 5SiO_2$) (ρ=2.1g/cm^3)	0.0038
800	Silicon Dioxide (SiO_2)	0.004
800	Zirconium Oxide (ZrO_2) (5-10% CaO stabilized)	0.0049
800	Zirconium Oxide (ZrO_2) (Y_2O_3 stabilized)	0.0053
800	Zirconium Oxide (ZrO_2) (stabilized, 0% porosity)	0.0055
800	Cordierite ($2MgO\ 2Al_2O_3\ 5SiO_2$) (ρ=2.3g/cm^3)	0.0055
800	Zirconium Oxide (ZrO_2) (MgO stabilized)	0.0057
800	Thorium Dioxide (ThO_2) (0% porosity)	0.008
800	Titanium Oxide (TiO_2) (0% porosity)	0.008
800	Uranium Dioxide (UO_2) (0% porosity)	0.009
800	Mullite ($3Al_2O_3\ 2SiO_2$) (0% porosity)	0.0095
800	Zircon ($SiO_2\ ZrO_2$) (0% porosity)	0.010

Source: data compiled by J.S. Park from *No. 1 Materials Index*, Peter T.B. Shaffer, Plenum Press, New York, (1964); *Smithells Metals Reference Book*, Eric A. Brandes, ed., in association with Fulmer Research Institute Ltd. 6th ed. London, Butterworths, Boston, (1983); and *Ceramic Source*, American Ceramic Society (1986-1991)

Table 358. SELECTING THERMAL CONDUCTIVITY OF CERAMICS AT TEMPERATURE

(SHEET 13 OF 19)

Temperature (°C)	Ceramic	Thermal Conductivity ($cal \cdot cm^{-1} \cdot sec^{-1} \cdot K^{-1}$)
800	Nickel monoxide (NiO) (0% porosity)	0.012
800	Spinel (Al_2O_3 MgO) (0% porosity)	0.015
800	Aluminum Oxide (Al_2O_3)	0.015-0.017
800	Calcium Oxide (CaO)	0.019
800	Magnesium Oxide (MgO)	0.0198-0.026
800	Aluminum Oxide (Al_2O_3) (single crystal)	0.029
800	Aluminum Nitride (AlN)	0.042
800	Beryllium Oxide (BeO)	0.060-0.093
800	Silicon Carbide (SiC) (cubic, CVD)	0.061
800	Zirconium Monocarbide (ZrC)	0.083
875	Zirconium Mononitride (TiN)	0.016
875	Molybdenum Disilicide ($MoSi_2$)	0.046

Source: data compiled by J.S. Park from *No. 1 Materials Index*, Peter T.B. Shaffer, Plenum Press, New York, (1964); *Smithells Metals Reference Book*, Eric A. Brandes, ed., in association with Fulmer Research Institute Ltd. 6th ed. London, Butterworths, Boston, (1983); and *Ceramic Source*, American Ceramic Society (1986-1991)

Table 358. SELECTING THERMAL CONDUCTIVITY OF CERAMICS AT TEMPERATURE

(SHEET 14 OF 19)

Temperature (°C)	Ceramic	Thermal Conductivity ($cal \cdot cm^{-1} \cdot sec^{-1} \cdot K^{-1}$)
1000	Zirconium Oxide (ZrO_2) (stabilized)	0.0048-0.0055
1000	Uranium Dioxide (UO_2)	0.006
1000	Thorium Dioxide (ThO_2) (0% porosity)	0.007-0.0074
1000	Titanium Oxide (TiO_2) (0% porosity)	0.008
1000	Uranium Dioxide (UO_2) (0% porosity)	0.008
1000	Mullite ($3Al_2O_3$ $2SiO_2$) (0% porosity)	0.009
1000	Nickel monoxide (NiO) (0% porosity)	0.011
1000	Spinel (Al_2O_3 MgO) (0% porosity)	0.013-0.0138
1000	Titanium Monocarbide (TiC)	0.0135
1000	Aluminum Oxide (Al_2O_3)	0.014-0.016
1000	Magnesium Oxide (MgO)	0.016-0.020
1000	Calcium Oxide (CaO)	0.0186-0.019

Source: data compiled by J.S. Park from *No. 1 Materials Index*, Peter T.B. Shaffer, Plenum Press, New York, (1964); *Smithells Metals Reference Book*, Eric A. Brandes, ed., in association with Fulmer Research Institute Ltd. 6th ed. London, Butterworths, Boston, (1983); and *Ceramic Source*, American Ceramic Society (1986-1991)

Table 358. SELECTING THERMAL CONDUCTIVITY OF CERAMICS AT TEMPERATURE

(SHEET 15 OF 19)

Temperature (°C)	Ceramic	Thermal Conductivity ($cal \cdot cm^{-1} \cdot sec^{-1} \cdot K^{-1}$)
1000	Titanium Mononitride (TiN)	0.020
1000	Boron Nitride (BN) parallel to a axis	0.0295
1000	Trisilicon tetranitride (Si_3N_4) (pressureless sintered)	0.038
1000	Silicon Carbide (SiC) (cubic, CVD)	0.051
1000	Boron Nitride (BN) parallel to c axis	0.0637
1000	Zirconium Monocarbide (ZrC)	0.086
1100	Zirconium Mononitride (TiN)	0.015
1100	Molybdenum Disilicide ($MoSi_2$)	0.041
1100	Beryllium Oxide (BeO)	0.043
1200	Sillimanite (Al_2O_3 SiO_2) (0% porosity)	0.0035
1200	Zirconium Oxide (ZrO_2) (stabilized)	0.0049-0.0050
1200	Silicon Dioxide (SiO_2)	0.005
1200	Zirconium Oxide (ZrO_2) (5-10% CaO stabilized)	0.0057

Source: data compiled by J.S. Park from *No. 1 Materials Index*, Peter T.B. Shaffer, Plenum Press, New York, (1964); *Smithells Metals Reference Book*, Eric A. Brandes, ed., in association with Fulmer Research Institute Ltd. 6th ed. London, Butterworths, Boston, (1983); and *Ceramic Source*, American Ceramic Society (1986-1991)

Table 358. Selecting Thermal Conductivity of Ceramics at Temperature

(Sheet 16 of 19)

Temperature (°C)	Ceramic	Thermal Conductivity ($cal \cdot cm^{-1} \cdot sec^{-1} \cdot K^{-1}$)
1200	Uranium Dioxide (UO_2)	0.006
1200	Zirconium Oxide (ZrO_2) (stabilized, 0% porosity)	0.006
1200	Thorium Dioxide (ThO_2) (0% porosity)	0.006-0.0076
1200	Titanium Oxide (TiO_2) (0% porosity)	0.008
1200	Mullite ($3Al_2O_3$ $2SiO_2$) (0% porosity)	0.009
1200	Zircon (SiO_2 ZrO_2) (0% porosity)	0.0095
1200	Spinel (Al_2O_3 MgO) (0% porosity)	0.013
1200	Aluminum Oxide (Al_2O_3)	0.013-0.015
1200	Magnesium Oxide (MgO)	0.0139-0.0148
1200	Trisilicon tetranitride (Si_3N_4) (pressureless sintered)	0.033-0.034
1200	Beryllium Oxide (BeO)	0.041-0.054
1200	Zirconium Monocarbide (ZrC)	0.089

Source: data compiled by J.S. Park from *No. 1 Materials Index*, Peter T.B. Shaffer, Plenum Press, New York, (1964); *Smithells Metals Reference Book*, Eric A. Brandes, ed., in association with Fulmer Research Institute Ltd. 6th ed. London, Butterworths, Boston, (1983); and *Ceramic Source*, American Ceramic Society (1986-1991)

Table 358. SELECTING THERMAL CONDUCTIVITY OF CERAMICS AT TEMPERATURE

(SHEET 17 OF 19)

Temperature (°C)	Ceramic	Thermal Conductivity ($cal \cdot cm^{-1} \cdot sec^{-1} \cdot K^{-1}$)
1250	Silicon Carbide (SiC) (cubic, CVD)	0.0059
1300	Beryllium Oxide (BeO)	0.038
1327	Silicon Carbide (SiC) (cubic, CVD)	0.0827
1400	Thorium Dioxide (ThO_2) (0% porosity)	0.006
1400	Zirconium Oxide (ZrO_2) (stabilized, 0% porosity)	0.0065
1400	Mullite ($3Al_2O_3$ $2SiO_2$) (0% porosity)	0.009
1400	Zircon (SiO_2 ZrO_2) (0% porosity)	0.0095
1400	Magnesium Oxide (MgO)	0.012-0.014
1400	Aluminum Oxide (Al_2O_3)	0.013
1400	Beryllium Oxide (BeO)	0.036
1400	Zirconium Monocarbide (ZrC)	0.092

Source: data compiled by J.S. Park from *No. 1 Materials Index*, Peter T.B. Shaffer, Plenum Press, New York, (1964); *Smithells Metals Reference Book*, Eric A. Brandes, ed., in association with Fulmer Research Institute Ltd. 6th ed. London, Butterworths, Boston, (1983); and *Ceramic Source*, American Ceramic Society (1986-1991)

Table 358. SELECTING THERMAL CONDUCTIVITY OF CERAMICS AT TEMPERATURE

(SHEET 18 OF 19)

Temperature (°C)	Ceramic	Thermal Conductivity ($cal \cdot cm^{-1} \cdot sec^{-1} \cdot K^{-1}$)
1500	Sillimanite (Al_2O_3 SiO_2) (0% porosity)	0.003
1500	Beryllium Oxide (BeO)	0.034
1500	Titanium Mononitride (TiN)	0.162
1530	Silicon Carbide (SiC) (cubic, CVD)	0.0032
1600	Silicon Dioxide (SiO_2)	0.006
1600	Magnesium Oxide (MgO)	0.0108-0.016
1600	Aluminum Oxide (Al_2O_3)	0.014
1600	Beryllium Oxide (BeO)	0.033-0.039
1600	Zirconium Monocarbide (ZrC)	0.096
1700	Beryllium Oxide (BeO)	0.033

Source: data compiled by J.S. Park from *No. 1 Materials Index*, Peter T.B. Shaffer, Plenum Press, New York, (1964); *Smithells Metals Reference Book*, Eric A. Brandes, ed., in association with Fulmer Research Institute Ltd. 6th ed. London, Butterworths, Boston, (1983); and *Ceramic Source*, American Ceramic Society (1986-1991)

Table 358. SELECTING THERMAL CONDUCTIVITY OF CERAMICS AT TEMPERATURE

(SHEET 19 OF 19)

Temperature (°C)	Ceramic	Thermal Conductivity ($cal \cdot cm^{-1} \cdot sec^{-1} \cdot K^{-1}$)
1800	Magnesium Oxide (MgO)	0.0096-0.0191
1800	Aluminum Oxide (Al_2O_3)	0.017
1800	Beryllium Oxide (BeO)	0.036
1800	Zirconium Monocarbide (ZrC)	0.099
1900	Beryllium Oxide (BeO)	0.036
2000	Beryllium Oxide (BeO)	0.036
2000	Zirconium Monocarbide (ZrC)	0.103
2200	Zirconium Monocarbide (ZrC)	0.105
2300	Titanium Mononitride (TiN)	0.136

Source: data compiled by J.S. Park from *No. 1 Materials Index*, Peter T.B. Shaffer, Plenum Press, New York, (1964); *Smithells Metals Reference Book*, Eric A. Brandes, ed., in association with Fulmer Research Institute Ltd. 6th ed. London, Butterworths, Boston, (1983); and *Ceramic Source*, American Ceramic Society (1986-1991)

Table 359. SELECTING THERMAL CONDUCTIVITY OF POLYMERS
(SHEET 1 OF 4)

Polymer	Thermal Conductivity (ASTM C177) Btu / (hr • ft • °F)
ABS Resins; Molded, Extruded: Very high impact	0.01—0.14
Polystyrene: Medium impact	0.024—0.090
Polystyrene: High impact	0.024—0.090
Rubber Phenolic: Asbestos Filled	0.04
Rubber Phenolic: Chopped Fabric Filled	0.05
Vinylidene Chloride	0.053
Polystyrene: General purpose	0.058—0.090
Polyvinyl Chloride & Copolymers: Nonrigid—General	0.07—0.10
Polyvinyl Chloride & Copolymers: Nonrigid—Electrical	0.07—0.10
Polyvinyl Chloride & Copolymers: Rigid—Normal Impact	0.07—0.10
Silicone: Woven Glass Fabric/ Silicone Laminate	0.075—0.125
ABS Resins; Molded, Extruded: Low temperature impact	0.08—0.14
ABS Resins; Molded, Extruded: Medium impact	0.08—0.18
Phenolics; High Shock: Chopped Fabric or Cord Filled	0.097—0.170
Phenolics; Molded: General: Woodflour and Flock Filled	0.097—0.3
Polyester, Thermoset: Cast Rigid	0.10—0.12
Cellulose Acetate, ASTM Grade: H6—1	0.10—0.19
Cellulose Acetate, ASTM Grade: H4—1	0.10—0.19
Cellulose Acetate, ASTM Grade: H2—1	0.10—0.19
Cellulose Acetate, ASTM Grade: MH—1, MH—2	0.10—0.19
Cellulose Acetate, ASTM Grade: MS—1, MS—2	0.10—0.19
Cellulose Acetate, ASTM Grade: S2—1	0.10—0.19
Cellulose Acetate Butyrate; ASTM Grade: H4	0.10—0.19
Cellulose Acetate Butyrate; ASTM Grade: MH	0.10—0.19
Cellulose Acetate Butyrate; ASTM Grade: S2	0.10—0.19
Cellulose Acetate Propionate, ASTM Grade: 1	0.10—0.19
Cellulose Acetate Propionate, ASTM Grade: 3	0.10—0.19
Cellulose Acetate Propionate, ASTM Grade: 6	0.10—0.19

Source: data compiled by J.S. Park from Charles T. Lynch, *CRC Handbook of Materials Science, Vol. 3,* CRC Press, Boca Raton, Florida, 1975 and *Engineered Materials Handbook, Vol.2,* Engineering Plastics, ASM International, Metals Park, Ohio, 1988.

Table 359. SELECTING THERMAL CONDUCTIVITY OF POLYMERS

(SHEET 2 OF 4)

Polymer	Thermal Conductivity (ASTM C177) Btu / (hr • ft • °F)
Phenolics; Molded: Shock: Paper, Flock, or Pulp Filled	0.1—0.16
Epoxy, Standard: Cast rigid	0.1—0.3
Epoxy, Standard: Molded	0.1—0.5
Polycarbonate	0.11
Polystyrene: Glass fiber -30% reinforced	0.117
Acrylic Cast Resin Sheets, Rods: General purpose, type I	0.12
Acrylic Cast Resin Sheets, Rods: General purpose, type II	0.12
Acrylic Moldings: Grades 5, 6, 8	0.12
Acrylic Moldings: High impact grade	0.12
Fluorinated ethylene propylene(FEP)	0.12
Rubber Phenolic: Woodflour or Flock Filled	0.12
ABS Resins; Molded, Extruded: High impact	0.12—0.16
ABS Resins; Molded, Extruded: Heat resistant	0.12—0.20
Polycarbonate (40% glass fiber reinforced)	0.13
Polyacetal Homopolymer: Standard	0.13
Polytetrafluoroethylene (PTFE)	0.14
Polyvinylidene— fluoride (PVDF)	0.14
Polytrifluoro chloroethylene (PTFCE)	0.145
Polyacetal Copolymer: Standard	0.16
Melamine; Molded: Cellulose Electrical	0.17—0.20
Urea; Molded: Alpha—Cellulose Filled (ASTM Type l)	0.17—0.244
Silicone: Fibrous (Glass) Reinforced	0.18
Polyethylene; Molded, Extruded; Type I: Melt Index 0.3—3.6	0.19
Polyethylene; Molded, Extruded; Type I: Melt Index 6—26	0.19
Polyethylene; Molded, Extruded; Type I: Melt Index 200	0.19
Polyethylene; Molded, Extruded; Type II: Melt Index 20	0.19
Polyethylene; Molded, Extruded; Type II: Melt Index L.0—1.9	0.19
Polyethylene; Molded, Extruded; Type III: Melt Index 0.2—0.9	0.19

Source: data compiled by J.S. Park from Charles T. Lynch, *CRC Handbook of Materials Science, Vol. 3,* CRC Press, Boca Raton, Florida, 1975 and *Engineered Materials Handbook, Vol.2,* Engineering Plastics, ASM International, Metals Park, Ohio, 1988.

Table 359. SELECTING THERMAL CONDUCTIVITY OF POLYMERS

(SHEET 3 OF 4)

Polymer	Thermal Conductivity (ASTM C177) Btu / (hr • ft • °F)
Polyethylene; Type III: Melt Melt Index 0.1—12.0	0.19
Polyethylene; Molded, Extruded; Type III: Melt Index 1.5—15	0.19
Polyethylene; Molded, Extruded; Type III: High Molecular Weight	0.19
Phenolics; Molded: Very High Shock: Glass Fiber Filled	0.2
Alkyds; Molded: Glass reinforced (heavy duty parts)	0.20—0.30
Phenolics; Molded: Arc Resistant—Mineral	0.24—0.34
Silicone: Granular (Silica) Reinforced	0.25—0.5
Melamine; Molded: Glass Fiber Filled	0.28
Alkyds; Molded: Putty (encapsulating)	0.35—0.60
Alkyds; Molded: Rope (general purpose)	0.35—0.60
Alkyds; Molded: Granular (high speed molding)	0.35—0.60
Polyester Injection Moldings: General purpose grade	0.36—0.55
Chlorinated polyether	0.91
Chlorinated polyvinyl chloride	0.95
PVC–Acrylic Injection Molded	0.98
PVC–Acrylic Sheet	1.01
Phenylene Oxide: SE—100	1.1
Polyarylsulfone	1.1
Phenylene Oxide: Glass fiber reinforced	1.1–1.15
Nylon; Molded, Extruded Type 6: General purpose	1.2—1.69
Nylon; Type 6: Cast	1.2—1.7
Polypropylene: General Purpose	1.21—1.36
Polyester, Thermoset: High strength (glass fiber filled)	1.32—1.68
Thermoset Allyl diglycol Carbonate	1.45
Nylon: Type 11	1.5
6/10 Nylon: General purpose	1.5
Phenylene Oxide: SE—1	1.5
6/6 Nylon: Glass fiber reinforced	1.5— 3.3

Source: data compiled by J.S. Park from Charles T. Lynch, *CRC Handbook of Materials Science, Vol. 3,* CRC Press, Boca Raton, Florida, 1975 and *Engineered Materials Handbook, Vol.2*, Engineering Plastics, ASM International, Metals Park, Ohio, 1988.

Table 359. SELECTING THERMAL CONDUCTIVITY OF POLYMERS
(SHEET 4 OF 4)

Polymer	Thermal Conductivity (ASTM C177) Btu / (hr • ft • °F)
Polyacetal Copolymer: High flow	1.6
6/6 Nylon: General purpose molding	1.69—1.7
Nylon; Molded, Extruded Type 6: Glass fiber (30%) reinforced	1.69—3.27
Nylon: Type 12	1.7
6/6 Nylon: General purpose extrusion	1.7
Polypropylene: High Impact	1.72
Phenylene oxides (Noryl): Standard	1.8
Polyphenylene Sulfide: Standard	2
Polyphenylene Sulfide: 40% Glass Reinforced	2
Epoxy, Standard: High strength laminate	2.35
ABS–Polycarbonate Alloy	2.46
6/10 Nylon: Glass fiber (30%) reinforced	3.5
Polymide: Glass Reinforced	3.59
Polymide: Unreinforced	3.8–6.78

Source: data compiled by J.S. Park from Charles T. Lynch, *CRC Handbook of Materials Science, Vol. 3,* CRC Press, Boca Raton, Florida, 1975 and *Engineered Materials Handbook, Vol.2*, Engineering Plastics, ASM International, Metals Park, Ohio, 1988.

Table 360. SELECTING THERMAL EXPANSION OF TOOL STEELS

(SHEET 1 OF 2)

Type	Temperature Change from 20 •C to	Thermal Expansion mm/(m•K)
M2	260°C	9.4
T1	200 °C	9.7
T15	200 °C	9.9
M2	100 °C	10.1
H13	100 °C	10.4
W1	100 °C	10.4
A2	260°C	10.6
A2	100 °C	10.7
W1	200 °C	11
T15	425°C	11
M2	425°C	11.2
T1	425°C	11.2
L6	100 °C	11.3
H13	200 °C	11.5
T15	540°C	11.5
T1	540°C	11.7
H11	100 °C	11.9
M2	540°C	11.9
T1	600°C	11.9
H13	425°C	12.2
M2	600°C	12.2
H21	100 °C	12.4
S1	100 °C	12.4
H11	200 °C	12.4
H13	540°C	12.4
H26	540°C	12.4
H21	200 °C	12.6
L6	200 °C	12.6
S1	200 °C	12.6
S7	200 °C	12.6

Source: data from *ASM Metals Reference Book, Second Edition*, American Society for Metals, Metals Park, Ohio 44073, p242, (1984).

Table 360. SELECTING THERMAL EXPANSION OF TOOL STEELS
(SHEET 2 OF 2)

Type	Temperature Change from 20 •C to	Thermal Expansion mm/(m•K)
L6	425°C	12.6
S5	425°C	12.6
H11	425°C	12.8
A2	425°C	12.9
H21	425°C	12.9
H11	540°C	12.9
W1	425°C	13.1
H13	600°C	13.1
S7	425°C	13.3
S5	540°C	13.3
H11	600°C	13.3
S7	600°C	13.3
S1	425°C	13.5
H21	540°C	13.5
L6	540°C	13.5
S7	500°C	13.7
L6	600°C	13.7
S5	600°C	13.7
W1	500°C	13.8
S1	540°C	13.9
H21	600°C	13.9
A2	540°C	14
A2	600°C	14.2
S1	600°C	14.2
W1	600°C	14.2
L2	425°C	14.4
L2	540°C	14.6
L2	600°C	14.8

Source: data from *ASM Metals Reference Book, Second Edition*, American Society for Metals, Metals Park, Ohio 44073, p242, (1984).

Table 361. SELECTING THERMAL EXPANSION OF TOOL STEELS AT TEMPERATURE (SHEET 1 OF 2)

Temperature Change from 20 •C to	Type	Thermal Expansion mm/(m•K)
100 °C	M2	10.1
	H13	10.4
	W1	10.4
	A2	10.7
	L6	11.3
	H11	11.9
	H21	12.4
	S1	12.4
200 °C	T1	9.7
	T15	9.9
	W1	11
	H13	11.5
	H11	12.4
	H21	12.6
	L6	12.6
	S1	12.6
	S7	12.6
260°C	M2	9.4
	A2	10.6
425°C	T15	11
	M2	11.2
	T1	11.2
	H13	12.2
	L6	12.6
	S5	12.6
	H11	12.8
	A2	12.9
	H21	12.9
	W1	13.1

Source: data from *ASM Metals Reference Book, Second Edition,* American Society for Metals, Metals Park, Ohio 44073, p242, (1984).

Table 361. SELECTING THERMAL EXPANSION OF TOOL STEELS AT TEMPERATURE (SHEET 2 OF 2)

Temperature Change from 20 •C to	Type	Thermal Expansion mm/(m•K)
	S7	13.3
	S1	13.5
	L2	14.4
500°C	S7	13.7
	W1	13.8
540°C	T15	11.5
	T1	11.7
	M2	11.9
	H13	12.4
	H26	12.4
	H11	12.9
	S5	13.3
	H21	13.5
	L6	13.5
	S1	13.9
	A2	14
	L2	14.6
600°C	T1	11.9
	M2	12.2
	H13	13.1
	H11	13.3
	S7	13.3
	L6	13.7
	S5	13.7
	H21	13.9
	A2	14.2
	S1	14.2
	W1	14.2
	L2	14.8

Source: data from *ASM Metals Reference Book, Second Edition*, American Society for Metals, Metals Park, Ohio 44073, p242, (1984).

Table 362. SELECTING THERMAL EXPANSION OF ALLOY CAST IRONS

Description	Thermal Expansion Coefficient mm/(m • °C)
Abrasion–Resistant White Martensitic Nickel–Chromium Iron	8 to 9
Corrosion–Resistant High–Nickel Gray Iron	8.1 to 19.3
Heat–Resistant Gray High–Nickel Iron	8.1 to 19.3
Heat–Resistant Gray High–Chromium Iron	9.3 to 9.9
Corrosion–Resistant High–Chromium Iron	9.4 to 9.9
Heat–Resistant Gray Medium–Silicon Iron	10.8
Heat–Resistant Medium–Silicon Ductile Iron	10.8 to 13.5
Abrasion–Resistant Low–C White Irons	12
Corrosion–Resistant High– Silicon Iron	12.4 to 13.1
Heat–Resistant Gray Nickel–Chromium–Silicon Iron	12.6 to 16.2
Corrosion–Resistant High–Nickel Ductile Iron	12.6 to 18.7
Heat–Resistant Gray High–Aluminum Iron	15.3
Heat–Resistant High–Nickel Ductile (23 Ni)	18.4
Heat–Resistant High–Nickel Ductile (20 Ni)	18.7

Source: Data from *ASM Metals Reference Book, Second Edition*, American Society for Metals, Metals Park, Ohio 44073, p172, (1984).

Table 363. SELECTING THERMAL EXPANSION OF CERAMICS (SHEET 1 OF 16)

Ceramic	Thermal Expansion ($°C^{-1}$)
Hafnium Dioxide (HfO_2) monoclinic, parallel to b axis	0 for 28–262°C
Silicon Dioxide (SiO_2) Vitreous	0.5 x 10^{-6} for 20–1250°C
Silicon Dioxide (SiO_2) Vitreous	0.527 x 10^{-6} for 25–500°C
Silicon Dioxide (SiO_2) Vitreous	0.564 x 10^{-6} for 25–1000°C
Boron Nitride (BN) parallel to a axis	0.59 x 10^{-6} for 25 to 350°C
Cordierite (2MgO $2Al_2O_3$ $5SiO_2$) (ρ=1.8g/cm^3)	0.6 x 10^{-6} for 25 to 400°C
Boron Nitride (BN) parallel to a axis	0.77 x 10^{-6} for 25 to 1000°C
Boron Nitride (BN) parallel to a axis	0.89 x 10^{-6} for 25 to 700°C
Hafnium Dioxide (HfO_2) monoclinic, parallel to a axis	0.9x10^{-6} for 28–494°C
Zirconium Oxide (ZrO_2) tetragonal, parallel to b axis	1.1 x 10^{-6} for 27 to 759°C
Hafnium Dioxide (HfO_2) monoclinic, parallel to a axis	1.3x10^{-6} for 28–697°C
Hafnium Dioxide (HfO_2) — tetragonal polycrystalline	1.31 x 10^{-6} for 25–1700°C
Hafnium Dioxide (HfO_2) monoclinic, parallel to a axis	1.4x10^{-6} for 28–903°C
Cordierite (2MgO $2Al_2O_3$ $5SiO_2$) (ρ=1.8g/cm^3)	1.5 x 10^{-6} for 25 to 700°C
Zirconium Oxide (ZrO_2) tetragonal, parallel to b axis	1.5 x 10^{-6} for 27 to 964°C
Aluminum Oxide (Al_2O_3) perpendicular to c axis	1.65 x 10^{-6} for 0 to –273°C
Cordierite (2MgO $2Al_2O_3$ $5SiO_2$) (ρ=1.8g/cm^3)	1.7 x 10^{-6} for 25 to 900°C
Aluminum Oxide (Al_2O_3) — polycrystalline	1.89 x 10^{-6} for 0 to –273°C
Zirconium Oxide (ZrO_2) tetragonal, parallel to b axis	1.9 x 10^{-6} for 27 to 1110°C
Aluminum Oxide (Al_2O_3) parallel to c axis	1.95 x 10^{-6} for 0 to –273°C
Zirconium Oxide (ZrO_2) tetragonal, parallel to b axis	2 x 10^{-6} for 27 to 504°C

Source: data compiled by J.S. Park from *No. 1 Materials Index*, Peter T.B. Shaffer, Plenum Press, New York, (1964); *Smithells Metals Reference Book*, Eric A. Brandes, ed., in association with Fulmer Research Institute Ltd. 6th ed. London, Butterworths, Boston, (1983); and *Ceramic Source*, American Ceramic Society (1986-1991)

Table 363. SELECTING THERMAL EXPANSION OF CERAMICS
(SHEET 2 OF 16)

Ceramic	Thermal Expansion ($°C^{-1}$)
Hafnium Dioxide (HfO_2) monoclinic, parallel to a axis	2.1x10^{-6} for 28–1098°C
Trisilicon Tetranitride (Si_3N_4)	2.11 x 10^{-6} for 25 to 500°C
Cordierite (2MgO $2Al_2O_3$ $5SiO_2$) (ρ=2.1g/cm^3)	2.2 x 10^{-6} for 25 to 400°C
Cordierite (2MgO $2Al_2O_3$ $5SiO_2$) (ρ=2.3g/cm^3)	2.3 x 10^{-6} for 25 to 400°C
Beryllium Oxide (BeO) — polycrystalline	2.4 x 10^{-6} for 25–200°C
Aluminum Oxide (Al_2O_3) perpendicular to c axis	2.55 x 10^{-6} for 0 to –173°C
Cordierite (2MgO $2Al_2O_3$ $5SiO_2$) (ρ=2.51g/cm^3)	2.7 x 10^{-6} for 25 to 1100°C
Cordierite (2MgO $2Al_2O_3$ $5SiO_2$) (ρ=2.1g/cm^3)	2.8 x 10^{-6} for 25 to 700°C
Cordierite (2MgO $2Al_2O_3$ $5SiO_2$) (ρ=2.1g/cm^3)	2.8 x 10^{-6} for 25 to 900°C
Trisilicon Tetranitride (Si_3N_4)	2.87 x 10^{-6} for 25 to 1000°C
Trisilicon Tetranitride (Si_3N_4) (reaction sintered)	2.9 x 10^{-6} for 20 to 1000°C
Aluminum Oxide (Al_2O_3) — polycrystalline	2.91 x 10^{-6} for 0 to –173°C
Zirconium Oxide (ZrO_2) tetragonal, parallel to b axis	3 x 10^{-6} for 27 to 264°C
Trisilicon Tetranitride (Si_3N_4) (hot pressed)	3–3.9 x 10^{-6} for 20 to 1000°C
Aluminum Oxide (Al_2O_3) parallel to c axis	3.01 x 10^{-6} for 0 to –173°C
Hafnium Dioxide (HfO_2) — tetragonal polycrystalline	3.03 x 10^{-6} for 25–2000°C
Cordierite (2MgO $2Al_2O_3$ $5SiO_2$) (ρ=2.3g/cm^3)	3.3 x 10^{-6} for 25 to 700°C
Trisilicon Tetranitride (Si_3N_4) (sintered)	3.5 x 10^{-6} for 20 to 1000°C
Trisilicon Tetranitride (Si_3N_4)	3.66 x 10^{-6} for 25 to 1500°C
Thorium Dioxide (ThO_2)	3.67 x 10^{-6} for 0 to –273°C

Source: data compiled by J.S. Park from *No. 1 Materials Index*, Peter T.B. Shaffer, Plenum Press, New York, (1964); *Smithells Metals Reference Book*, Eric A. Brandes, ed., in association with Fulmer Research Institute Ltd. 6th ed. London, Butterworths, Boston, (1983); and *Ceramic Source*, American Ceramic Society (1986-1991)

Table 363. SELECTING THERMAL EXPANSION OF CERAMICS

(SHEET 3 OF 16)

Ceramic	Thermal Expansion ($°C^{-1}$)
Cordierite ($2MgO\ 2Al_2O_3\ 5SiO_2$) (ρ=2.3g/cm^3)	3.7 x 10^{-6} for 25 to 900°C
Trisilicon Tetranitride (Si_3N_4) (pressureless sintered)	3.7 x 10^{-6} for 40 to 1000°C
Cordierite ($2MgO\ 2Al_2O_3\ 5SiO_2$) (glass)	3.7–3.8 x 10^{-6} for 25 to 900°C
Aluminum Oxide (Al_2O_3) perpendicular to c axis	3.75 x 10^{-6} for 0 to –73°C
Zircon ($SiO_2\ ZrO_2$)	3.79 x 10^{-6} for 25 to 500°C
Zirconium Oxide (ZrO_2) — tetragonal	4.0 x 10^{-6} for 0 to 500°C
Aluminum Nitride (AlN)	4.03 x 10^{-6} for 25 to 200°C
Aluminum Oxide (Al_2O_3) — polycrystalline	4.10 x 10^{-6} for 0 to –73°C
Aluminum Oxide (Al_2O_3) parallel to c axis	4.39 x 10^{-6} for 0 to –73°C
Tungsten Monocarbide (WC)	4.42 x 10^{-6} for 25–500°C
Mullite ($3Al_2O_3\ 2SiO_2$)	4.5 x 10^{-6} for 20 to 1325°C
Boron Carbide (B_4C)	4.5 x 10^{-6} for room temp.–800°C
Chromium Diboride (CrB_2)	4.6–11.1 x 10^{-6} for 20–1000°C
Titanium Diboride (TiB_2)	4.6–8.1 x 10^{-6}
Zircon ($SiO_2\ ZrO_2$)	4.62 x 10^{-6} for 25 to 1000°C
Mullite ($3Al_2O_3\ 2SiO_2$)	4.63 x 10^{-6} for 25 to 500°C
Silicon Carbide (SiC)	4.63 x 10^{-6} for 25–500°C
Silicon Carbide (SiC)	4.70 x 10^{-6} for 0–1700°C
Silicon Carbide (SiC)	4.70 x 10^{-6} for 20–1500°C
Aluminum Oxide (Al_2O_3) perpendicular to c axis	4.78 x 10^{-6} for 0 to 27°C

Source: data compiled by J.S. Park from *No. 1 Materials Index*, Peter T.B. Shaffer, Plenum Press, New York, (1964); *Smithells Metals Reference Book*, Eric A. Brandes, ed., in association with Fulmer Research Institute Ltd. 6th ed. London, Butterworths, Boston, (1983); and *Ceramic Source*, American Ceramic Society (1986-1991)

Table 363. SELECTING THERMAL EXPANSION OF CERAMICS
(SHEET 4 OF 16)

Ceramic	Thermal Expansion ($°C^{-1}$)
Boron Carbide (B_4C)	4.78 x 10^{-6} for 25–500°C
Aluminum Nitride (AlN)	4.83 x 10^{-6} for 25 to 600°C
Aluminum Nitride (AlN)	4.84 x 10^{-6} for 25 to 500°C
Tungsten Monocarbide (WC)	4.84–4.92 x 10^{-6} for 25–1000°C
Zirconium Oxide (ZrO_2) — tetragonal	5.0 x 10^{-6} for 0 to 1400°C
Mullite ($3Al_2O_3$ $2SiO_2$)	5.0 x 10^{-6} for 25 to 800°C
Tantalum Diboride (TaB_2)	5.1 x 10^{-6} at room temp.
Silicon Carbide (SiC)	5.12 x 10^{-6} for 25–1000°C
Mullite ($3Al_2O_3$ $2SiO_2$)	5.13 x 10^{-6} for 25 to 1000°C
Aluminum Oxide (Al_2O_3) parallel to c axis	5.31 x 10^{-6} for 0 to 27°C
Thorium Dioxide (ThO_2)	5.32 x 10^{-6} for 0 to –173°C
Tungsten Monocarbide (WC)	5.35–5.8 x 10^{-6} for 25–1500°C
Hafnium Dioxide (HfO_2) — monoclinic polycrystalline	5.47 x 10^{-6} for 25–500°C
Silicon Carbide (SiC)	5.48 x 10^{-6} for 25–1500°C
Zircon (SiO_2 ZrO_2)	5.5 x 10^{-6} for 20 to 1200°C
Hafnium Diboride (HfB_2)	5.5 –5.54 x 10^{-6} for 20 to1000°C
Zirconium Oxide (ZrO_2) — tetragonal	5.5–5.58 x 10^{-6} for 20 to 1200°C
Zirconium Diboride (ZrB_2)	5.5–6.57 x 10^{-6} °C for 25–1000°C
Aluminum Oxide (Al_2O_3) perpendicular to c axis	5.51 x 10^{-6} for 0 to 127°C
Boron Carbide (B_4C)	5.54 x 10^{-6} for 25–1000°C
Aluminum Nitride (AlN)	5.54–5.64 x 10^{-6} for 25 to 1000°C
Aluminum Oxide (Al_2O_3) — polycrystalline	5.60 x 10^{-6} for 0 to 27°C
Mullite ($3Al_2O_3$ $2SiO_2$)	5.62 x 10^{-6} for 20 to 1500°C
Zircon (SiO_2 ZrO_2)	5.63 x 10^{-6} for 20 to 1500°C

Source: data compiled by J.S. Park from *No. 1 Materials Index*, Peter T.B. Shaffer, Plenum Press, New York, (1964); *Smithells Metals Reference Book*, Eric A. Brandes, ed., in association with Fulmer Research Institute Ltd. 6th ed. London, Butterworths, Boston, (1983); and *Ceramic Source*, American Ceramic Society (1986-1991)

Table 363. SELECTING THERMAL EXPANSION OF CERAMICS (SHEET 5 OF 16)

Ceramic	Thermal Expansion (°C^{-1})
Zirconium Diboride (ZrB_2)	5.69 x 10^{-6} for 25–500°C
Silicon Carbide (SiC)	5.77 x 10^{-6} for 25–2000°C
Hafnium Dioxide (HfO_2) — monoclinic polycrystalline	5.8 x 10^{-6} for 25–1300°C
Tungsten Monocarbide (WC)	5.82–7.4 x 10^{-6} for 25–2000°C
Hafnium Dioxide (HfO_2) — monoclinic polycrystalline	5.85 x 10^{-6} for 25–1000°C
Silicon Carbide (SiC)	5.94 x 10^{-6} for 25–2500°C
Boron Carbide (B_4C)	6.02 x 10^{-6} for 25–1500°C
Aluminum Oxide (Al_2O_3) — polycrystalline	6.03 x 10^{-6} for 0 to 127°C
Aluminum Nitride (AlN)	6.09 x 10^{-6} for 25 to 1350°C
Aluminum Oxide (Al_2O_3) perpendicular to c axis	6.10 x 10^{-6} for 0 to 227°C
Zirconium Monocarbide (ZrC)	6.10x 10^{-6} for 25–500°C
Zirconium Monocarbide (ZrC)	6.10–6.73 x 10^{-6} for 25–650°C
Zirconium Mononitride (TiN)	6.13 x 10^{-6} for 20–450°C
Hafnium Dioxide (HfO_2) monoclinic, parallel to a axis	6.2x10^{-6} for 28–494°C
Hafnium Monocarbide (HfC)	6.25 x 10^{-6} for 25–1000°C
Aluminum Oxide (Al_2O_3) parallel to c axis	6.26 x 10^{-6} for 0 to 127°C
Hafnium Monocarbide (HfC)	6.27–6.59 x 10^{-6} for 25–650°C
Tantalum Monocarbide (TaC)	6.29–6.32 x 10^{-6} for 25–500°C
Beryllium Oxide (BeO) parallel to c axis	6.3 x 10^{-6} for 28 to 252°C
Hafnium Dioxide (HfO_2) — monoclinic polycrystalline	6.30 x 10^{-6} for 25–1500°C
Beryllium Oxide (BeO) — polycrystalline	6.3–6.4 x 10^{-6} for 25–300°C
Zirconium Monocarbide (ZrC)	6.32x 10^{-6} for 0–750°C
Hafnium Dioxide (HfO_2) — monoclinic polycrystalline	6.45 x 10^{-6} for 20–1700°C
Zirconium Monocarbide (ZrC)	6.46–6.66x 10^{-6} for 0–1000°C

Source: data compiled by J.S. Park from *No. 1 Materials Index*, Peter T.B. Shaffer, Plenum Press, New York, (1964); *Smithells Metals Reference Book*, Eric A. Brandes, ed., in association with Fulmer Research Institute Ltd. 6th ed. London, Butterworths, Boston, (1983); and *Ceramic Source*, American Ceramic Society (1986-1991)

Table 363. SELECTING THERMAL EXPANSION OF CERAMICS
(SHEET 6 OF 16)

Ceramic	Thermal Expansion ($°C^{-1}$)
Thorium Dioxide (ThO_2)	6.47 x 10^{-6} for 0 to –73°C
Tantalum Monocarbide (TaC)	6.50 x 10^{-6} for 0–1000°C
Aluminum Oxide (Al_2O_3) perpendicular to c axis	6.52 x 10^{-6} for 0 to 327°C
Titanium Monocarbide (TiC)	6.52–7.15 x 10^{-6} for 25–500°C
Zirconium Oxide (ZrO_2) — monoclinic	6.53 x 10^{-6} for 25 to 500°C
Boron Carbide (B_4C)	6.53 x 10^{-6} for 25–2000°C
Aluminum Oxide (Al_2O_3) — polycrystalline	6.55 x 10^{-6} for 0 to 227°C
Zirconium Monocarbide (ZrC)	6.56x 10^{-6} for 25–1000°C
Sillimanite (Al_2O_3 SiO_2)	6.58 x 10^{-6} at 20°C
Tantalum Monocarbide (TaC)	6.64 x 10^{-6} for 0–1200°C
Zirconium Monocarbide (ZrC)	6.65x 10^{-6} for 25–800°C
Tantalum Monocarbide (TaC)	6.67 x 10^{-6} for 25–1000°C
Zirconium Monocarbide (ZrC)	6.68x 10^{-6} for 0–1275°C
Beryllium Oxide (BeO) parallel to c axis	6.7 x 10^{-6} for 28 to 474°C
Hafnium Dioxide (HfO_2) monoclinic, parallel to a axis	6.7x10^{-6} for 28–697°C
Zirconium Oxide (ZrO_2) tetragonal, parallel to a axis	6.8 x 10^{-6} for 27 to 759°C
Hafnium Dioxide (HfO_2) monoclinic, parallel to a axis	6.8x10^{-6} for 28–262°C
Beryllium Oxide (BeO) average for (2a+c)/3	6.83 x 10^{-6} for 28 to 252°C
Zirconium Monocarbide (ZrC)	6.83x 10^{-6} for 0–1525°C
Aluminum Oxide (Al_2O_3) parallel to c axis	6.86 x 10^{-6} for 0 to 227°C
Aluminum Oxide (Al_2O_3) perpendicular to c axis	6.88 x 10^{-6} for 0 to 427°C
Aluminum Oxide (Al_2O_3) — polycrystalline	6.93 x 10^{-6} for 0 to 327°C
Zirconium Diboride (ZrB_2)	6.98 x 10^{-6} for 20–1500°C
Zirconium Monocarbide (ZrC)	6.98x 10^{-6} for 0–1775°C
Zirconium Mononitride (TiN)	7.03 x 10^{-6} for 20–680°C

Source: data compiled by J.S. Park from *No. 1 Materials Index*, Peter T.B. Shaffer, Plenum Press, New York, (1964); *Smithells Metals Reference Book*, Eric A. Brandes, ed., in association with Fulmer Research Institute Ltd. 6th ed. London, Butterworths, Boston, (1983); and *Ceramic Source*, American Ceramic Society (1986-1991)

Table 363. SELECTING THERMAL EXPANSION OF CERAMICS
(SHEET 7 OF 16)

Ceramic	Thermal Expansion (°C^{-1})
Zirconium Monocarbide (ZrC)	7.06x 10^{-6} for 25–1500°C
Titanium Monocarbide (TiC)	7.08 x 10^{-6} for 0–750°C
Boron Carbide (B_4C)	7.08 x 10^{-6} for 25–2500°C
Beryllium Oxide (BeO) perpendicular to c axis	7.1 x 10^{-6} for 28 to 252°C
Tantalum Monocarbide (TaC)	7.12 x 10^{-6} for 25–1500°C
Aluminum Oxide (Al_2O_3) perpendicular to c axis	7.15 x 10^{-6} for 0 to 527°C
Boron Nitride (BN) parallel to c axis	7.15 x 10^{-6} for 25 to 1000°C
Titanium Monocarbide (TiC)	7.18–7.45 x 10^{-6} for 25–750°C
Zirconium Oxide (ZrO_2) — tetragonal	7.2 x 10^{-6} for –10 to 1000°C
Aluminum Oxide (Al_2O_3) — polycrystalline	7.24 x 10^{-6} for 0 to 427°C
Aluminum Oxide (Al_2O_3) parallel to c axis	7.31 x 10^{-6} for 0 to 327°C
Aluminum Oxide (Al_2O_3) perpendicular to c axis	7.35 x 10^{-6} for 0 to 627°C
Titanium Monocarbide (TiC)	7.40–8.82 x 10^{-6} for 25–1000°C
Beryllium Oxide (BeO) average for (2a+c)/3	7.43 x 10^{-6} for 28 to 474°C
Zirconium Oxide (ZrO_2) tetragonal, parallel to a axis	7.5 x 10^{-6} for 27 to 504°C
Aluminum Oxide (Al_2O_3) — polycrystalline	7.50 x 10^{-6} for 0 to 527°C
Hafnium Dioxide (HfO_2) monoclinic, parallel to a axis	7.5x10^{-6} for 28–903°C
Aluminum Oxide (Al_2O_3) perpendicular to c axis	7.53 x 10^{-6} for 0 to 727°C
Zirconium Oxide (ZrO_2) — monoclinic	7.59 x 10^{-6} for 25 to 1000°C
Beryllium Oxide (BeO) — polycrystalline	7.59 x 10^{-6} for 25–500°C
Tantalum Monocarbide (TaC)	7.64 x 10^{-6} for 25–2000°C
Zirconium Monocarbide (ZrC)	7.65x 10^{-6} for 25–650°C
Aluminum Oxide (Al_2O_3) perpendicular to c axis	7.67 x 10^{-6} for 0 to 827°C
Aluminum Oxide (Al_2O_3) parallel to c axis	7.68 x 10^{-6} for 0 to 427°C

Source: data compiled by J.S. Park from *No. 1 Materials Index*, Peter T.B. Shaffer, Plenum Press, New York, (1964); *Smithells Metals Reference Book*, Eric A. Brandes, ed., in association with Fulmer Research Institute Ltd. 6th ed. London, Butterworths, Boston, (1983); and *Ceramic Source*, American Ceramic Society (1986-1991)

Table 363. SELECTING THERMAL EXPANSION OF CERAMICS
(SHEET 8 OF 16)

Ceramic	Thermal Expansion (°C^{-1})
Aluminum Oxide (Al_2O_3) — polycrystalline	7.69 x 10^{-6} for 0 to 627°C
Zirconium Oxide (ZrO_2) — monoclinic	7.72 x 10^{-6} for 25 to 1050°C
Spinel (Al_2O_3 MgO)	7.79 x 10^{-6} for 25 to 500°C
Molybdenum Disilicide ($MoSi_2$)	7.79 x 10^{-6} for 25–500°C
Tungsten Disilicide (WSi_2)	7.79 x 10^{-6} for 25–500°C
Titanium Oxide (TiO_2) — polycrystalline	7.8 x 10^{-6} for 20–600°C
Thorium Dioxide (ThO_2)	7.8 x 10^{-6} for 27 to 223°C
Zirconium Oxide (ZrO_2) tetragonal, parallel to a axis	7.8 x 10^{-6} for 27 to 964°C
Beryllium Oxide (BeO) perpendicular to c axis	7.8 x 10^{-6} for 28 to 474°C
Beryllium Oxide (BeO) parallel to c axis	7.8 x 10^{-6} for 28 to 749°C
Aluminum Oxide (Al_2O_3) perpendicular to c axis	7.80 x 10^{-6} for 0 to 927°C
Aluminum Oxide (Al_2O_3) — polycrystalline	7.83 x 10^{-6} for 0 to 727°C
Titanium Monocarbide (TiC)	7.85–7.86 x 10^{-6} for 0–1000°C
Aluminum Oxide (Al_2O_3) perpendicular to c axis	7.88 x 10^{-6} for 0 to 1027°C
Titanium Oxide (TiO_2) perpendicular to a axis	7.9 x 10^{-6} for 26 to 240°C
Titanium Monocarbide (TiC)	7.90 x 10^{-6} for 0–2500°C
Hafnium Dioxide (HfO_2) monoclinic, parallel to a axis	7.9x10^{-6} for 28–1098°C
Aluminum Oxide (Al_2O_3) parallel to c axis	7.96 x 10^{-6} for 0 to 527°C
Aluminum Oxide (Al_2O_3) perpendicular to c axis	7.96 x 10^{-6} for 0 to 1127°C
Aluminum Oxide (Al_2O_3) — polycrystalline	7.97 x 10^{-6} for 0 to 827°C
Zirconium Oxide (ZrO_2) — monoclinic	8.0 x 10^{-6} for 25 to 1080°C
Trichromium Dicarbide (Cr_3C_2)	8.00 x 10^{-6} for 25–500°C
Titanium Monocarbide (TiC)	8.02 x 10^{-6} for 0–1275°C
Aluminum Oxide (Al_2O_3) perpendicular to c axis	8.05 x 10^{-6} for 0 to 1227°C

Source: data compiled by J.S. Park from *No. 1 Materials Index*, Peter T.B. Shaffer, Plenum Press, New York, (1964); *Smithells Metals Reference Book*, Eric A. Brandes, ed., in association with Fulmer Research Institute Ltd. 6th ed. London, Butterworths, Boston, (1983); and *Ceramic Source*, American Ceramic Society (1986-1991)

Table 363. SELECTING THERMAL EXPANSION OF CERAMICS
(SHEET 9 OF 16)

Ceramic	Thermal Expansion (°C^{-1})
Thorium Dioxide (ThO_2)	8.06 x 10^{-6} for 0 to 127°C
Boron Nitride (BN) parallel to c axis	8.06 x 10^{-6} for 25 to 700°C
Aluminum Oxide (Al_2O_3) — polycrystalline	8.08 x 10^{-6} for 0 to 927°C
Titanium Oxide (TiO_2) perpendicular to a axis	8.1 x 10^{-6} for 26 to 670°C
Thorium Dioxide (ThO_2)	8.10 x 10^{-6} for 0 to 27°C
Aluminum Oxide (Al_2O_3) perpendicular to c axis	8.12 x 10^{-6} for 0 to 1327°C
Titanium Monocarbide (TiC)	8.15–9.45 x 10^{-6} for 25–1500°C
Aluminum Oxide (Al_2O_3) perpendicular to c axis	8.16 x 10^{-6} for 0 to 1427°C
Aluminum Oxide (Al_2O_3) — polycrystalline	8.18 x 10^{-6} for 0 to 1027°C
Aluminum Oxide (Al_2O_3) parallel to c axis	8.19 x 10^{-6} for 0 to 627°C
Titanium Oxide (TiO_2) perpendicular to a axis	8.2 x 10^{-6} for 26 to 455°C
Titanium Oxide (TiO_2) perpendicular to a axis	8.2 x 10^{-6} for 26 to 940°C
Beryllium Oxide (BeO) parallel to c axis	8.2 x 10^{-6} for 28 to 872°C
Aluminum Oxide (Al_2O_3) perpendicular to c axis	8.20 x 10^{-6} for 0 to 1527°C
Tungsten Disilicide (WSi_2)	8.21 x 10^{-6} for 0–1000°C
Cerium Dioxide (CeO_2)	8.22 x 10^{-6} for 25–500°C
Titanium Oxide (TiO_2) — polycrystalline	8.22 x 10^{-6} for 25–500°C
Aluminum Oxide (Al_2O_3) — polycrystalline	8.25 x 10^{-6} for 0 to 1127°C
Aluminum Oxide (Al_2O_3) perpendicular to c axis	8.26 x 10^{-6} for 0 to 1627°C
Titanium Monocarbide (TiC)	8.26 x 10^{-6} for 0–1525°C
Beryllium Oxide (BeO) average for (2a+c)/3	8.27 x 10^{-6} for 28 to 749°C
Titanium Monocarbide (TiC)	8.29 x 10^{-6} for 0–1400°C
Titanium Oxide (TiO_2) perpendicular to a axis	8.3 x 10^{-6} for 26 to 1110°C
Aluminum Oxide (Al_2O_3) perpendicular to c axis	8.30 x 10^{-6} for 0 to 1727°C

Source: data compiled by J.S. Park from *No. 1 Materials Index*, Peter T.B. Shaffer, Plenum Press, New York, (1964); *Smithells Metals Reference Book*, Eric A. Brandes, ed., in association with Fulmer Research Institute Ltd. 6th ed. London, Butterworths, Boston, (1983); and *Ceramic Source*, American Ceramic Society (1986-1991)

Table 363. SELECTING THERMAL EXPANSION OF CERAMICS
(SHEET 10 OF 16)

Ceramic	Thermal Expansion ($°C^{-1}$)
Thorium Dioxide (ThO_2)	8.31 x 10^{-6} for 0 to 227°C
Tungsten Disilicide (WSi_2)	8.31 x 10^{-6} for 25–1000°C
Aluminum Oxide (Al_2O_3) — polycrystalline	8.32 x 10^{-6} for 0 to 1227°C
Aluminum Oxide (Al_2O_3) parallel to c axis	8.38 x 10^{-6} for 0 to 727°C
Aluminum Oxide (Al_2O_3) — polycrystalline	8.39 x 10^{-6} for 0 to 1327°C
Zirconium Oxide (ZrO_2) tetragonal, parallel to a axis	8.4 x 10^{-6} for 27 to 264°C
Titanium Monocarbide (TiC)	8.40 x 10^{-6} for 0–1775°C
Tantalum Monocarbide (TaC)	8.40 x 10^{-6} for 25–2500°C
Beryllium Oxide (BeO) — polycrystalline	8.4–8.5 x 10^{-6} for 25–800°C
Molybdenum Disilicide ($MoSi_2$)	8.41 x 10^{-6} for 0–1000°C
Spinel (Al_2O_3 MgO)	8.41 x 10^{-6} for 25 to 1000°C
Dichromium Trioxide (Cr_2O_3)	8.43 x 10^{-6} for 25–500°C
Aluminum Oxide (Al_2O_3) — polycrystalline	8.45 x 10^{-6} for 0 to 1427°C
Aluminum Oxide (Al_2O_3) — polycrystalline	8.49 x 10^{-6} for 0 to 1527°C
Beryllium Oxide (BeO) perpendicular to c axis	8.5 x 10^{-6} for 28 to 749°C
Molybdenum Disilicide ($MoSi_2$)	8.51 x 10^{-6} for 25–1000°C
Aluminum Oxide (Al_2O_3) parallel to c axis	8.52 x 10^{-6} for 0 to 827°C
Thorium Dioxide (ThO_2)	8.53 x 10^{-6} for 0 to 327°C
Aluminum Oxide (Al_2O_3) — polycrystalline	8.53 x 10^{-6} for 0 to 1627°C
Titanium Oxide (TiO_2) average for (2a+c)/3	8.53 x 10^{-6} for 26 to 240°C
Molybdenum Disilicide ($MoSi_2$)	8.56 x 10^{-6} for 0–1400°C
Aluminum Oxide (Al_2O_3) — polycrystalline	8.58 x 10^{-6} for 0 to 1727°C
Dichromium Trioxide (Cr_2O_3)	8.62 x 10^{-6} for 25–1000°C
Thorium Dioxide (ThO_2)	8.63 x 10^{-6} for 25 to 500°C

Source: data compiled by J.S. Park from *No. 1 Materials Index*, Peter T.B. Shaffer, Plenum Press, New York, (1964); *Smithells Metals Reference Book*, Eric A. Brandes, ed., in association with Fulmer Research Institute Ltd. 6th ed. London, Butterworths, Boston, (1983); and *Ceramic Source*, American Ceramic Society (1986-1991)

Table 363. SELECTING THERMAL EXPANSION OF CERAMICS
(SHEET 11 OF 16)

Ceramic	Thermal Expansion (°C^{-1})
Zirconium Oxide (ZrO_2) — tetragonal	8.64 x 10^{-6} for –20 to 600°C
Aluminum Oxide (Al_2O_3) parallel to c axis	8.65 x 10^{-6} for 0 to 927°C
Thorium Dioxide (ThO_2)	8.7 x 10^{-6} for 27 to 498°C
Zirconium Oxide (ZrO_2) tetragonal, parallel to a axis	8.7 x 10^{-6} for 27 to 1110°C
Thorium Dioxide (ThO_2)	8.71 x 10^{-6} for 0 to 427°C
Aluminum Oxide (Al_2O_3) parallel to c axis	8.75 x 10^{-6} for 0 to 1027°C
Trichromium Dicarbide (Cr_3C_2)	8.8 x 10^{-6} for 25–120°C
Tungsten Disilicide (WSi_2)	8.81 x 10^{-6} for 0–1400°C
Titanium Monocarbide (TiC)	8.81 x 10^{-6} for 25–2000°C
Dichromium Trioxide (Cr_2O_3)	8.82 x 10^{-6} for 25–1500°C
Titanium Oxide (TiO_2) — polycrystalline	8.83 x 10^{-6} for 25–1000°C
Aluminum Oxide (Al_2O_3) parallel to c axis	8.84 x 10^{-6} for 0 to 1127°C
Thorium Dioxide (ThO_2)	8.87 x 10^{-6} for 0 to 527°C
Beryllium Oxide (BeO) average for (2a+c)/3	8.87 x 10^{-6} for 28 to 872°C
Thorium Dioxide (ThO_2)	8.9 x 10^{-6} for 27 to 755°C
Beryllium Oxide (BeO) parallel to c axis	8.9 x 10^{-6} for 28 to 1132°C
Aluminum Oxide (Al_2O_3) parallel to c axis	8.92 x 10^{-6} for 0 to 1227°C
Cerium Dioxide (CeO_2)	8.92 x 10^{-6} for 25–1000°C
Titanium Oxide (TiO_2) average for (2a+c)/3	8.93 x 10^{-6} for 26 to 670°C
Thorium Dioxide (ThO_2)	8.96 x 10^{-6} for 0 to 1000°C
Titanium Oxide (TiO_2) average for (2a+c)/3	8.97 x 10^{-6} for 26 to 455°C
Titanium Oxide (TiO_2) average for (2a+c)/3	8.97 x 10^{-6} for 26 to 940°C
Aluminum Oxide (Al_2O_3) parallel to c axis	8.98 x 10^{-6} for 0 to 1327°C
Titanium Oxide (TiO_2) — polycrystalline	8.98 x 10^{-6} for 0–1000°C

Source: data compiled by J.S. Park from *No. 1 Materials Index*, Peter T.B. Shaffer, Plenum Press, New York, (1964); *Smithells Metals Reference Book*, Eric A. Brandes, ed., in association with Fulmer Research Institute Ltd. 6th ed. London, Butterworths, Boston, (1983); and *Ceramic Source*, American Ceramic Society (1986-1991)

Table 363. SELECTING THERMAL EXPANSION OF CERAMICS
(SHEET 12 OF 16)

Ceramic	Thermal Expansion (°C^{-1})
Spinel (Al_2O_3 MgO)	9.0 x 10^{-6} for 20 to 1250°C
Thorium Dioxide (ThO_2)	9.00 x 10^{-6} for 0 to 627°C
Molybdenum Disilicide ($MoSi_2$)	9.00–9.18 x 10^{-6} for 25–1500°C
Zirconium Monocarbide (ZrC)	9.0x 10^{-6} for 1000–2000°C
Aluminum Oxide (Al_2O_3) parallel to c axis	9.02 x 10^{-6} for 0 to 1427°C
Beryllium Oxide (BeO) — polycrystalline	9.03 x 10^{-6} for 25–1000°C
Uranium Dioxide (UO_2) (heating)	9.07 x 10^{-6} for 27 to 400°C
Aluminum Oxide (Al_2O_3) parallel to c axis	9.08 x 10^{-6} for 0 to 1527°C
Thorium Dioxide (ThO_2)	9.1 x 10^{-6} for 27 to 1087°C
Aluminum Oxide (Al_2O_3) parallel to c axis	9.13 x 10^{-6} for 0 to 1627°C
Titanium Oxide (TiO_2) average for (2a+c)/3	9.13 x 10^{-6} for 26 to 1110°C
Thorium Dioxide (ThO_2)	9.14 x 10^{-6} for 0 to 727°C
Spinel (Al_2O_3 MgO)	9.17 x 10^{-6} for 25 to 1500°C
Aluminum Oxide (Al_2O_3) parallel to c axis	9.18 x 10^{-6} for 0 to 1727°C
Beryllium Oxide (BeO) — polycrystalline	9.18 x 10^{-6} for 25–1250°C
Uranium Dioxide (UO_2)	9.18 x 10^{-6} for 27 to 400°C
Thorium Dioxide (ThO_2)	9.2 x 10^{-6} for 27 to 994°C
Beryllium Oxide (BeO) perpendicular to c axis	9.2 x 10^{-6} for 28 to 872°C
Thorium Dioxide (ThO_2)	9.24 x 10^{-6} for 0 to 827°C
Uranium Dioxide (UO_2) (cooling)	9.28 x 10^{-6} for 27 to 400°C
Titanium Monocarbide (TiC)	9.32 x 10^{-6} for 25–1250°C
Thorium Dioxide (ThO_2)	9.34 x 10^{-6} for 0 to 927°C
Titanium Mononitride (TiN)	9.35 x 10^{-6}
Thorium Dioxide (ThO_2)	9.35 x 10^{-6} for 0 to 1200°C

Source: data compiled by J.S. Park from *No. 1 Materials Index*, Peter T.B. Shaffer, Plenum Press, New York, (1964); *Smithells Metals Reference Book*, Eric A. Brandes, ed., in association with Fulmer Research Institute Ltd. 6th ed. London, Butterworths, Boston, (1983); and *Ceramic Source*, American Ceramic Society (1986-1991)

Table 363. SELECTING THERMAL EXPANSION OF CERAMICS
(SHEET 13 OF 16)

Ceramic	Thermal Expansion ($°C^{-1}$)
Beryllium Oxide (BeO) — polycrystalline	9.40 x 10^{-6} for 500–1200°C
Thorium Dioxide (ThO_2)	9.42 x 10^{-6} for 0 to 1027°C
Thorium Dioxide (ThO_2)	9.44 x 10^{-6} for 25 to 1000°C
Uranium Dioxide (UO_2)	9.47 x 10^{-6} for 25 to 500°C
Titanium Oxide (TiO_2) — polycrystalline	9.50 x 10^{-6} for 25–1500°C
Thorium Dioxide (ThO_2)	9.53 x 10^{-6} for 0 to 1127°C
Thorium Dioxide (ThO_2)	9.55 x 10^{-6} for 20 to 800°C
Thorium Dioxide (ThO_2)	9.55 x 10^{-6} for 20 to 1400°C
Dichromium Trioxide (Cr_2O_3)	9.55 x 10^{-6} for 20–1400°C
Beryllium Oxide (BeO) average for (2a+c)/3	9.57 x 10^{-6} for 28 to 1132°C
Thorium Dioxide (ThO_2)	9.60 x 10^{-6} for 0 to 1227°C
Thorium Dioxide (ThO_2)	9.68 x 10^{-6} for 0 to 1327°C
Thorium Dioxide (ThO_2)	9.76 x 10^{-6} for 0 to 1427°C
Titanium Oxide (TiO_2) parallel to c axis	9.8 x 10^{-6} for 26 to 240°C
Thorium Dioxide (ThO_2)	9.83 x 10^{-6} for 0 to 1527°C
Thorium Dioxide (ThO_2)	9.84 x 10^{-6} for 0 to 1400°C
Beryllium Oxide (BeO) perpendicular to c axis	9.9 x 10^{-6} for 28 to 1132°C
Thorium Dioxide (ThO_2)	9.91 x 10^{-6} for 0 to 1627°C
Trichromium Dicarbide (Cr_3C_2)	9.95 x 10^{-6} for 25–500°C
Thorium Dioxide (ThO_2)	9.97 x 10^{-6} for 0 to 1727°C
Boron Nitride (BN) parallel to c axis	10.15 x 10^{-6} for 25 to 350°C
Thorium Dioxide (ThO_2)	10.17 x 10^{-6} for 25 to 1500°C
Beryllium Oxide (BeO) — polycrystalline	10.3 x 10^{-6} for 25–1500°C
Thorium Dioxide (ThO_2)	10.43 x 10^{-6} for 25 to 1700°C

Source: data compiled by J.S. Park from *No. 1 Materials Index*, Peter T.B. Shaffer, Plenum Press, New York, (1964); *Smithells Metals Reference Book*, Eric A. Brandes, ed., in association with Fulmer Research Institute Ltd. 6th ed. London, Butterworths, Boston, (1983); and *Ceramic Source*, American Ceramic Society (1986-1991)

Table 363. SELECTING THERMAL EXPANSION OF CERAMICS
(SHEET 14 OF 16)

Ceramic	Thermal Expansion ($°C^{-1}$)
Silicon Dioxide (SiO_2) β_2 tridymite	10.45 x 10^{-6} for 25–1000°C
Zirconium Oxide (ZrO_2) — tetragonal	10.5 x 10^{-6} for 0 to 1000°C
Titanium Oxide (TiO_2) parallel to c axis	10.5 x 10^{-6} for 26 to 455°C
Titanium Oxide (TiO_2) parallel to c axis	10.5 x 10^{-6} for 26 to 940°C
Zirconium Oxide (ZrO_2) — tetragonal	10.52 x 10^{-6} for 0 to 1000°C (MgO)
Zirconium Oxide (ZrO_2) — tetragonal	10.6 x 10^{-6} for 0 to 1200°C (CaO)
Titanium Oxide (TiO_2) parallel to c axis	10.6 x 10^{-6} for 26 to 670°C
Titanium Oxide (TiO_2) parallel to c axis	10.8 x 10^{-6} for 26 to 1110°C
Uranium Dioxide (UO_2) (cooling)	10.8 x 10^{-6} for 400 to 800°C
Uranium Dioxide (UO_2) (cooling)	10.8 x 10^{-6} for 400 to 800°C
Hafnium Dioxide (HfO_2) monoclinic, parallel to c axis	10.8x10^{-6} for 28–697°C
Trichromium Dicarbide (Cr_3C_2)	10.9 x 10^{-6} for 150–980°C
Zirconium Oxide (ZrO_2) — tetragonal	11.0 x 10^{-6} for 0 to 1500°C
Hafnium Dioxide (HfO_2) monoclinic, parallel to c axis	11x10^{-6} for 28–262°C
Beryllium Oxide (BeO) — polycrystalline	11.1 x 10^{-6} for 25–2000°C
Uranium Dioxide (UO_2) (heating)	11.1 x 10^{-6} for 400 to 800°C
Uranium Dioxide (UO_2)	11.15 x 10^{-6} for 25 to 1750°C
Uranium Dioxide (UO_2)	11.19 x 10^{-6} for 25 to 1000°C
Hafnium Dioxide (HfO_2) monoclinic, parallel to c axis	11.4x10^{-6} for 28–494°C
Zirconium Oxide (ZrO_2) tetragonal, parallel to c axis	11.9 x 10^{-6} for 27 to 759°C
Hafnium Dioxide (HfO_2) monoclinic, parallel to c axis	11.9x10^{-6} for 28–903°C
Hafnium Dioxide (HfO_2) monoclinic, parallel to c axis	12.1x10^{-6} for 28–1098°C
Uranium Dioxide (UO_2)	12.19 x 10^{-6} for 25 to 1200°C
Boron Nitride (BN)	12.2 x 10^{-6} for 25 to 500°C

Source: data compiled by J.S. Park from *No. 1 Materials Index*, Peter T.B. Shaffer, Plenum Press, New York, (1964); *Smithells Metals Reference Book*, Eric A. Brandes, ed., in association with Fulmer Research Institute Ltd. 6th ed. London, Butterworths, Boston, (1983); and *Ceramic Source*, American Ceramic Society (1986-1991)

Table 363. SELECTING THERMAL EXPANSION OF CERAMICS
(SHEET 15 OF 16)

Ceramic	Thermal Expansion ($°C^{-1}$)
Uranium Dioxide (UO_2) (cooling)	12.6 x 10^{-6} for 800 to 1250°C
Zirconium Oxide (ZrO_2) tetragonal, parallel to c axis	12.8 x 10^{-6} for 27 to 964°C
Magnesium Oxide (MgO)	12.83 x 10^{-6} for 25–500°C
Uranium Dioxide (UO_2) (cooling)	12.9 x 10^{-6} for 800 to 1200°C
Zirconium Oxide (ZrO_2) tetragonal, parallel to c axis	13 x 10^{-6} for 27 to 504°C
Uranium Dioxide (UO_2) (heating)	13.0 x 10^{-6} for 800 to 1200°C
Magnesium Oxide (MgO)	13.3 x 10^{-6} for 20–1700°C
Boron Nitride (BN)	13.3 x 10^{-6} for 25 to 1000°C
Zirconium Oxide (ZrO_2) tetragonal, parallel to c axis	13.6 x 10^{-6} for 27 to 1110°C
Magnesium Oxide (MgO)	13.63 x 10^{-6} for 25–1000°C
Magnesium Oxide (MgO)	13.90 x 10^{-6} for 0–1000°C
Zirconium Oxide (ZrO_2) tetragonal, parallel to c axis	14 x 10^{-6} for 27 to 264°C
Magnesium Oxide (MgO)	14.0 x 10^{-6} for 20–1400°C
Magnesium Oxide (MgO)	14.2–14.9 x 10^{-6} for 20–1700°C
Magnesium Oxide (MgO)	14.46 x 10^{-6} for 0–1200°C
Silicon Dioxide (SiO_2) β quartz	14.58 x 10^{-6} for 25–1000°C
Magnesium Oxide (MgO)	15.06 x 10^{-6} for 0–1400°C
Magnesium Oxide (MgO)	15.11 x 10^{-6} for 25–1500°C
Magnesium Oxide (MgO)	15.89 x 10^{-6} for 25–1800°C
Silicon Dioxide (SiO_2) α tridymite	18.5 x 10^{-6} for 25–117°C
Silicon Dioxide (SiO_2) α quartz	19.35 x 10^{-6} for 25–500°C
Silicon Dioxide (SiO_2) $β_2$ tridymite	19.35 x 10^{-6} for 25–500°C
Silicon Dioxide (SiO_2) α quartz	22.2 x 10^{-6} for 25–575°C
Silicon Dioxide (SiO_2) $β_1$ tridymite	25.0 x 10^{-6} for 25–117°C

Source: data compiled by J.S. Park from *No. 1 Materials Index*, Peter T.B. Shaffer, Plenum Press, New York, (1964); *Smithells Metals Reference Book*, Eric A. Brandes, ed., in association with Fulmer Research Institute Ltd. 6th ed. London, Butterworths, Boston, (1983); and *Ceramic Source*, American Ceramic Society (1986-1991)

Table 363. SELECTING THERMAL EXPANSION OF CERAMICS
(SHEET 16 OF 16)

Ceramic	Thermal Expansion (°C^{-1})
Silicon Dioxide (SiO_2) β_1 tridymite	27.5 x 10^{-6} for 25–163°C
Silicon Dioxide (SiO_2) β quartz	27.8 x 10^{-6} for 25–575°C
Silicon Dioxide (SiO_2) β_2 tridymite	31.9 x 10^{-6} for 25–163°C

Source: data compiled by J.S. Park from *No. 1 Materials Index*, Peter T.B. Shaffer, Plenum Press, New York, (1964); *Smithells Metals Reference Book*, Eric A. Brandes, ed., in association with Fulmer Research Institute Ltd. 6th ed. London, Butterworths, Boston, (1983); and *Ceramic Source*, American Ceramic Society (1986-1991)

Table 364. SELECTING THERMAL EXPANSION OF GLASSES
(SHEET 1 OF 11)

Glass	Temperature Range of Validity	Thermal Expansion (K^{-1})
SiO_2 glass	–60—20°C	3.50x10^{-7}
SiO_2 glass	–40—20°C	3.80x10^{-7}
SiO_2 glass	–20—20°C	4.00x10^{-7}
SiO_2 glass	0–20°C	4.30x10^{-7}
SiO_2 glass	20–100°C	5.35x10^{-7}
SiO_2 glass	20–150°C	5.75x10^{-7}
SiO_2 glass	20–200°C	5.85x10^{-7}
SiO_2 glass	20–350°C	5.90x10^{-7}
SiO_2 glass	20–250°C	5.92x10^{-7}
SiO_2 glass	20–300°C	5.94x10^{-7}
SiO_2–Al_2O_3 glass (3.1% mol Al_2O_3, 1000°C for 115 hr)	20–980°C	6.2x10^{-7}
SiO_2–Al_2O_3 glass (3.1% mol Al_2O_3, water quenching)	20–980°C	6.2x10^{-7}
SiO_2–Al_2O_3 glass (8.2% mol Al_2O_3, water quenching)	20–800°C	8.8x10^{-7}
SiO_2–Al_2O_3 glass (5.4% mol Al_2O_3, 1130°C for 20 hr)	20–350°C	12.2x10^{-7}
SiO_2–Al_2O_3 glass (8.2% mol Al_2O_3, 1000°C for 115 hr)	20–950°C	14.5x10^{-7}
SiO_2–Al_2O_3 glass (13.9% mol Al_2O_3, water quenching)	20–600°C	17.2x10^{-7}
SiO_2–Al_2O_3 glass (17.4% mol Al_2O_3, water quenching)	20–700°C	20.7x10^{-7}

Source: data compiled by Jun S. Park from O. V. Mazurin, M. V. Streltsina and T. P. Shvaiko–Shvaikovskaya, *Handbook of Glass Data, Part A and Part B*, Elsevier, New York, 1983

Table 364. SELECTING THERMAL EXPANSION OF GLASSES

(SHEET 2 OF 11)

Glass	Temperature Range of Validity	Thermal Expansion (K^{-1})
SiO_2–Al_2O_3 glass (13.9% mol Al_2O_3, 1000°C for 115 hr)	20–900°C	$22.7x10^{-7}$
SiO_2–Al_2O_3 glass (17.4% mol Al_2O_3, 1000°C for 115 hr)	20–800°C	$28.3x10^{-7}$
SiO_2–B_2O_3 glass (39.2% mol B_2O_3)	100–200°C	$44.9x10^{-7}$
SiO_2–B_2O_3 glass (39.2% mol B_2O_3)	0–100°C	$47.5x10^{-7}$
SiO_2–B_2O_3 glass (44.2% mol B_2O_3)	0–100°C	$49.8x10^{-7}$
SiO_2–B_2O_3 glass (44.2% mol B_2O_3)	100–200°C	$50.8x10^{-7}$
SiO_2–PbO glass (25.7% mol PbO)	20–170°C	$51.45–52.23x10^{-7}$
SiO_2–B_2O_3 glass (50.8% mol B_2O_3)	100–200°C	$54.8x10^{-7}$
B_2O_3–CaO glass (29.3% mol CaO)	room temp. to 100°C	$54.9–56.4x10^{-7}$
B_2O_3–CaO glass (31.4% mol CaO)	room temp. to 100°C	$57.3–58.2x10^{-7}$
SiO_2–B_2O_3 glass (50.8% mol B_2O_3)	0–100°C	$57.6x10^{-7}$
SiO_2–PbO glass (30.0% mol PbO)	20–170°C	$57.68–59.08x10^{-7}$
B_2O_3–CaO glass (34.9% mol CaO)	room temp. to 100°C	$60.1–66.2x10^{-7}$
B_2O_3–CaO glass (29.3% mol CaO)	100–200°C	$60.2–60.8x10^{-7}$
SiO_2–PbO glass (32.5% mol PbO)	20–170°C	$60.62–62.31x10^{-7}$
SiO_2–PbO glass (33.2% mol PbO)	20–170°C	$61.58–63.33x10^{-7}$
B_2O_3–CaO glass (37.1% mol CaO)	room temp. to 100°C	$63.1–64.0x10^{-7}$
B_2O_3–CaO glass (31.4% mol CaO)	100–200°C	$63.5–65.1x10^{-7}$
B_2O_3–CaO glass (29.3% mol CaO)	200–300°C	$63.9–65.4x10^{-7}$
SiO_2–PbO glass (35.0% mol PbO)	20–170°C	$63.99–66.17x10^{-7}$
B_2O_3–Na_2O glass (16.2% mol Na_2O)	–196—25°C	$65.9x10^{-7}$
B_2O_3–Na_2O glass (15.8% mol Na_2O)	–196—25°C	$67.4x10^{-7}$

Source: data compiled by Jun S. Park from O. V. Mazurin, M. V. Streltsina and T. P. Shvaiko–Shvaikovskaya, *Handbook of Glass Data, Part A and Part B*, Elsevier, New York, 1983

Table 364. SELECTING THERMAL EXPANSION OF GLASSES

(SHEET 3 OF 11)

Glass	Temperature Range of Validity	Thermal Expansion (K^{-1})
B_2O_3–CaO glass (31.4% mol CaO)	200–300°C	67.4–68.1x10^{-7}
B_2O_3–CaO glass (34.9% mol CaO)	100–200°C	67.5–67.6x10^{-7}
B_2O_3–CaO glass (37.1% mol CaO)	100–200°C	68.4–70.4x10^{-7}
SiO_2–PbO glass (37.5% mol PbO)	20–170°C	68.75–71.44x10^{-7}
B_2O_3–Na_2O glass (15% mol Na_2O, T_g = 407°C)	below T_g	69x10^{-7}
B_2O_3–Na_2O glass (18.4% mol Na_2O)	–196—25°C	69.1x10^{-7}
B_2O_3–Na_2O glass (13.7% mol Na_2O)	–196—25°C	69.3x10^{-7}
SiO_2–B_2O_3 glass (58.4% mol B_2O_3)	100–200°C	70.1x10^{-7}
B_2O_3–CaO glass (29.3% mol CaO)	300–400°C	71.3–71.6x10^{-7}
B_2O_3–Na_2O glass (11.5% mol Na_2O)	–196—25°C	71.5x10^{-7}
SiO_2–B_2O_3 glass (58.4% mol B_2O_3)	0–100°C	71.9x10^{-7}
B_2O_3–Na_2O glass (22.5% mol Na_2O)	–196—25°C	71.9x10^{-7}
B_2O_3–CaO glass (37.1% mol CaO)	200–300°C	74.6–75.8x10^{-7}
B_2O_3–CaO glass (34.9% mol CaO)	200–300°C	74.7–75.2x10^{-7}
SiO_2–PbO glass (42.6% mol PbO)	20–170°C	75.16–78.58x10^{-7}
B_2O_3–CaO glass (31.4% mol CaO)	300–400°C	76.5–76.7x10^{-7}
B_2O_3–CaO glass (29.3% mol CaO)	400–500°C	76.9–77.1x10^{-7}
B_2O_3–Na_2O glass (10% mol Na_2O, T_g = 354°C)	below T_g	77x10^{-7}
B_2O_3–CaO glass (34.9% mol CaO)	300–400°C	77.8–78.5x10^{-7}
SiO_2–PbO glass (45.8% mol PbO)	20–170°C	78.85–82.60x10^{-7}
B_2O_3–CaO glass (31.4% mol CaO)	400–500°C	79.2–81.0x10^{-7}
B_2O_3–Na_2O glass (15.8% mol Na_2O)	20–50°C	80.7x10^{-7}

Source: data compiled by Jun S. Park from O. V. Mazurin, M. V. Streltsina and T. P. Shvaiko–Shvaikovskaya, *Handbook of Glass Data, Part A and Part B*, Elsevier, New York, 1983

Table 364. SELECTING THERMAL EXPANSION OF GLASSES
(SHEET 4 OF 11)

Glass	Temperature Range of Validity	Thermal Expansion (K^{-1})
B_2O_3–CaO glass (29.3% mol CaO)	500–600°C	80.9–86.8x10^{-7}
B_2O_3–Na_2O glass (28.9% mol Na_2O)	−196—25°C	81.4x10^{-7}
B_2O_3–CaO glass (37.1% mol CaO)	300–400°C	81.6–82.2x10^{-7}
SiO_2–PbO glass (47.8% mol PbO)	20–170°C	83.03–87.03x10^{-7}
B_2O_3–CaO glass (31.4% mol CaO)	500–600°C	83.1–88.5x10^{-7}
B_2O_3–CaO glass (34.9% mol CaO)	400–500°C	83.8–95.0x10^{-7}
SiO_2–PbO glass (49.8% mol PbO)	20–170°C	85.57–89.82x10^{-7}
B_2O_3–Na_2O glass (17.4% mol Na_2O)	20–50°C	85.6x10^{-7}
B_2O_3–Na_2O glass (20% mol Na_2O, T_g = 456°C)	below T_g	86x10^{-7}
B_2O_3–Na_2O glass (16.2% mol Na_2O)	20–50°C	86.0x10^{-7}
B_2O_3–Na_2O glass (18.4% mol Na_2O)	20–50°C	86.2x10^{-7}
B_2O_3–Na_2O glass (19.6% mol Na_2O)	20–50°C	86.8x10^{-7}
B_2O_3–CaO glass (37.1% mol CaO)	400–500°C	86.9–87.6x10^{-7}
SiO_2–B_2O_3 glass (72.7% mol B_2O_3)	0–100°C	87.0x10^{-7}
B_2O_3–Na_2O glass (13.7% mol Na_2O)	20–50°C	87.5x10^{-7}
B_2O_3–Na_2O glass (20.0% mol Na_2O)	20–50°C	87.6x10^{-7}
B_2O_3–Na_2O glass (16.2% mol Na_2O)	20–150°C	87.7x10^{-7}
B_2O_3–Na_2O glass (15.8% mol Na_2O)	20–150°C	87.8x10^{-7}
B_2O_3–Na_2O glass (11.5% mol Na_2O)	20–50°C	88.7x10^{-7}
B_2O_3–Na_2O glass (17.4% mol Na_2O)	20–150°C	89.1x10^{-7}
B_2O_3–Na_2O glass (18.4% mol Na_2O)	20–150°C	89.2x10^{-7}
SiO_2–B_2O_3 glass (72.7% mol B_2O_3)	100–200°C	89.7x10^{-7}
B_2O_3–Na_2O glass (22.5% mol Na_2O)	20–50°C	90.4x10^{-7}

Source: data compiled by Jun S. Park from O. V. Mazurin, M. V. Streltsina and T. P. Shvaiko–Shvaikovskaya, *Handbook of Glass Data, Part A and Part B*, Elsevier, New York, 1983

Table 364. SELECTING THERMAL EXPANSION OF GLASSES
(SHEET 5 OF 11)

Glass	Temperature Range of Validity	Thermal Expansion (K^{-1})
B_2O_3–Na_2O glass (23.6% mol Na_2O)	20–50°C	$90.4x10^{-7}$
SiO_2–PbO glass (53.8% mol PbO)	20–170°C	$90.62–95.25x10^{-7}$
B_2O_3–Na_2O glass (13.7% mol Na_2O)	20–250°C	$90.9x10^{-7}$
B_2O_3–Na_2O glass (16.2% mol Na_2O)	20–250°C	$90.9x10^{-7}$
B_2O_3–Na_2O glass (19.6% mol Na_2O)	20–150°C	$91.2x10^{-7}$
B_2O_3–Na_2O glass (20.0% mol Na_2O)	20–150°C	$91.6x10^{-7}$
B_2O_3–CaO glass (34.9% mol CaO)	500–600°C	$91.8–92.1x10^{-7}$
B_2O_3–Na_2O glass (13.7% mol Na_2O)	20–150°C	$92.3x10^{-7}$
B_2O_3–Na_2O glass (17.4% mol Na_2O)	20–250°C	$92.4x10^{-7}$
B_2O_3–Na_2O glass (15.8% mol Na_2O)	20–250°C	$93.3x10^{-7}$
B_2O_3–CaO glass (37.1% mol CaO)	500–600°C	$93.5–95.5x10^{-7}$
B_2O_3–Na_2O glass (18.4% mol Na_2O)	20–250°C	$94.1x10^{-7}$
B_2O_3–Na_2O glass (4.4% mol Na_2O)	–196—25°C	$94.6x10^{-7}$
B_2O_3–Na_2O glass (22.5% mol Na_2O)	20–150°C	$94.7x10^{-7}$
B_2O_3–Na_2O glass (11.5% mol Na_2O)	20–150°C	$94.9x10^{-7}$
B_2O_3–Na_2O glass (25% mol Na_2O, T_g = 466°C)	below T_g	$95x10^{-7}$
B_2O_3–Na_2O glass (19.6% mol Na_2O)	20–250°C	$95.3x10^{-7}$
SiO_2–PbO glass (57.5% mol PbO)	20–170°C	$95.64–100.45x10^{-7}$
B_2O_3–Na_2O glass (18.4% mol Na_2O)	20–350°C	$96.2x10^{-7}$
B_2O_3–Na_2O glass (17.4% mol Na_2O)	20–350°C	$96.3x10^{-7}$
B_2O_3–Na_2O glass (23.6% mol Na_2O)	20–150°C	$96.7x10^{-7}$
B_2O_3–Na_2O glass (16.2% mol Na_2O)	20–350°C	$96.9x10^{-7}$

Source: data compiled by Jun S. Park from O. V. Mazurin, M. V. Streltsina and T. P. Shvaiko–Shvaikovskaya, *Handbook of Glass Data, Part A and Part B*, Elsevier, New York, 1983

Table 364. SELECTING THERMAL EXPANSION OF GLASSES

(SHEET 6 OF 11)

Glass	Temperature Range of Validity	Thermal Expansion (K^{-1})
SiO_2–PbO glass (59.0% mol PbO)	20–170°C	97.00–101.90x10^{-7}
SiO_2–Na_2O glass (20.3% mol Na_2O)	room temp–100°C	97.5x10^{-7}
B_2O_3–Na_2O glass (20.0% mol Na_2O)	20–250°C	97.6x10^{-7}
B_2O_3–Na_2O glass (11.5% mol Na_2O)	20–250°C	97.9x10^{-7}
B_2O_3–Na_2O glass (15.8% mol Na_2O)	20–350°C	97.9x10^{-7}
B_2O_3–Na_2O glass (22.5% mol Na_2O)	20–250°C	98.7x10^{-7}
B_2O_3–Na_2O glass (8.7% mol Na_2O)	20–50°C	98.8x10^{-7}
SiO_2–Na_2O glass (20.3% mol Na_2O)	100–200°C	99.3x10^{-7}
B_2O_3–Na_2O glass (19.6% mol Na_2O)	20–350°C	99.6x10^{-7}
B_2O_3–Na_2O glass (8.7% mol Na_2O)	20–150°C	100.5x10^{-7}
SiO_2–Na_2O glass (20.3% mol Na_2O)	200–300°C	100.6x10^{-7}
SiO_2–PbO glass (61.0% mol PbO)	20–170°C	100.66–105.58x10^{-7}
B_2O_3–Na_2O glass (23.6% mol Na_2O)	20–250°C	101.2x10^{-7}
B_2O_3–Na_2O glass (20.0% mol Na_2O)	20–350°C	101.3x10^{-7}
SiO_2–PbO glass (61.75% mol PbO)	20–170°C	101.36–106.30x10^{-7}
B_2O_3–Na_2O glass (28.9% mol Na_2O)	20–50°C	102.1x10^{-7}
B_2O_3–Na_2O glass (4.4% mol Na_2O)	20–50°C	103.0x10^{-7}
B_2O_3–Na_2O glass (22.5% mol Na_2O)	20–350°C	104.0x10^{-7}
B_2O_3–Na_2O glass (8.7% mol Na_2O)	20–250°C	105.3x10^{-7}
B_2O_3–Na_2O glass (23.6% mol Na_2O)	20–350°C	106.5x10^{-7}
SiO_2–Na_2O glass (20.3% mol Na_2O)	300–400°C	106.9x10^{-7}
B_2O_3–Na_2O glass (28.9% mol Na_2O)	20–150°C	107.4x10^{-7}
SiO_2–Na_2O glass (24.0% mol Na_2O)	room temp–100°C	109.7x10^{-7}
B_2O_3–Na_2O glass (4.4% mol Na_2O)	20–150°C	109.9x10^{-7}

Source: data compiled by Jun S. Park from O. V. Mazurin, M. V. Streltsina and T. P. Shvaiko–Shvaikovskaya, *Handbook of Glass Data, Part A and Part B*, Elsevier, New York, 1983

Table 364. SELECTING THERMAL EXPANSION OF GLASSES
(SHEET 7 OF 11)

Glass	Temperature Range of Validity	Thermal Expansion (K^{-1})
SiO_2–PbO glass (67.7% mol PbO)	20–170°C	110.38–115.48x10^{-7}
SiO_2–B_2O_3 glass (83.2% mol B_2O_3)	0–100°C	111.4x10^{-7}
B_2O_3–Na_2O glass (28.9% mol Na_2O)	20–250°C	112.8x10^{-7}
SiO_2–Na_2O glass (24.0% mol Na_2O)	100–200°C	114.3x10^{-7}
B_2O_3–Na_2O glass (5% mol Na_2O, T_g = 318°C)	below T_g	115x10^{-7}
B_2O_3–Na_2O glass (4.4% mol Na_2O)	20–250°C	116.0x10^{-7}
SiO_2–B_2O_3 glass (83.2% mol B_2O_3)	100–200°C	116.6x10^{-7}
SiO_2–Na_2O glass (24.0% mol Na_2O)	200–300°C	116.6x10^{-7}
B_2O_3–Na_2O glass (28.9% mol Na_2O)	20–350°C	117.1x10^{-7}
SiO_2–B_2O_3 glass (88.6% mol B_2O_3)	0–100°C	118.1x10^{-7}
SiO_2–Na_2O glass (20% mol Na_2O, T_g = 478°C)	below T_g	120x10^{-7}
SiO_2–Na_2O glass (24.0% mol Na_2O)	300–400°C	121.7x10^{-7}
SiO_2–B_2O_3 glass (88.6% mol B_2O_3)	100–200°C	126.0x10^{-7}
B_2O_3–Na_2O glass (30% mol Na_2O, T_g = 468°C)	below T_g	128x10^{-7}
SiO_2–B_2O_3 glass (94.0% mol B_2O_3)	0–100°C	131.7x10^{-7}
SiO_2–Na_2O glass (31.1% mol Na_2O)	room temp–100°C	136.0x10^{-7}
B_2O_3–Na_2O glass (0.01% mol Na_2O)	–196—25°C	140x10^{-7}
SiO_2–B_2O_3 glass (94.0% mol B_2O_3)	100–200°C	141.9x10^{-7}
SiO_2–Na_2O glass (31.1% mol Na_2O)	100–200°C	142.5x10^{-7}
SiO_2–Na_2O glass (33.8% mol Na_2O)	room temp–100°C	143.9x10^{-7}

Source: data compiled by Jun S. Park from O. V. Mazurin, M. V. Streltsina and T. P. Shvaiko–Shvaikovskaya, *Handbook of Glass Data, Part A and Part B*, Elsevier, New York, 1983

Table 364. SELECTING THERMAL EXPANSION OF GLASSES
(SHEET 8 OF 11)

Glass	Temperature Range of Validity	Thermal Expansion (K^{-1})
SiO_2–Na_2O glass (31.1% mol Na_2O)	200–300°C	148.3x10^{-7}
B_2O_3–Na_2O glass (0.01% mol Na_2O)	20–150°C	149.0x10^{-7}
B_2O_3–Na_2O glass (0.01% mol Na_2O)	20–50°C	149.3x10^{-7}
B_2O_3 glass	20–200°C	150±3–158±3x10^{-7}
SiO_2–Na_2O glass (30% mol Na_2O, T_g = 455°C)	below T_g	152x10^{-7}
SiO_2–Na_2O glass (37.2% mol Na_2O)	room temp–100°C	152.1x10^{-7}
SiO_2–Na_2O glass (33.8% mol Na_2O)	100–200°C	153.6x10^{-7}
B_2O_3 glass	100–200°C	154.5–169x10^{-7}
B_2O_3 glass	0–100°C	154.5–183x10^{-7}
SiO_2–Na_2O glass (33.8% mol Na_2O)	200–300°C	159.1x10^{-7}
SiO_2–Na_2O glass (31.1% mol Na_2O)	300–400°C	160.0x10^{-7}
SiO_2–Na_2O glass (37.2% mol Na_2O)	100–200°C	160.9x10^{-7}
SiO_2–Na_2O glass (33% mol Na_2O, T_g = 445°C)	below T_g	165x10^{-7}
SiO_2–Na_2O glass (37.2% mol Na_2O)	200–300°C	171.6x10^{-7}
SiO_2–Na_2O glass (33.8% mol Na_2O)	300–400°C	173.6x10^{-7}
SiO_2–Na_2O glass (40% mol Na_2O, T_g = 421°C)	below T_g	179x10^{-7}
SiO_2–Na_2O glass (37.2% mol Na_2O)	300–400°C	187.7x10^{-7}
SiO_2–Na_2O glass (45% mol Na_2O, T_g = 417°C)	below T_g	219x10^{-7}
SiO_2–B_2O_3 glass (39.2% mol B_2O_3)	390–410°C	301x10^{-7}

Source: data compiled by Jun S. Park from O. V. Mazurin, M. V. Streltsina and T. P. Shvaiko–Shvaikovskaya, *Handbook of Glass Data, Part A and Part B*, Elsevier, New York, 1983

Table 364. SELECTING THERMAL EXPANSION OF GLASSES
(SHEET 9 OF 11)

Glass	Temperature Range of Validity	Thermal Expansion (K^{-1})
SiO_2–Na_2O glass (20% mol Na_2O, T_g = 478°C)	above T_g	$315x10^{-7}$
SiO_2–Na_2O glass (30% mol Na_2O, T_g = 455°C)	above T_g	$402x10^{-7}$
SiO_2–B_2O_3 glass (44.2% mol B_2O_3)	380–400°C	$450x10^{-7}$
SiO_2–Na_2O glass (33% mol Na_2O, T_g = 445°C)	above T_g	$465x10^{-7}$
SiO_2–Na_2O glass (40% mol Na_2O, T_g = 421°C)	above T_g	$500x10^{-7}$
SiO_2–CaO glass (35% mol CaO)	1700°C	$53{\pm}5x10^{-6}$
SiO_2–Na_2O glass (45% mol Na_2O, T_g = 417°C)	above T_g	$574x10^{-7}$
SiO_2–B_2O_3 glass (50.8% mol B_2O_3)	350–370°C	$579x10–^7$
B_2O_3–Na_2O glass (20% mol Na_2O, T_g = 456°C)	above T_g	$586x10^{-7}$
SiO_2–CaO glass (40% mol CaO)	1700°C	$64{\pm}4x10^{-6}$
SiO_2–CaO glass (30% mol CaO)	1700°C	$66{\pm}5x10^{-6}$
SiO_2–Na_2O glass (20% mol Na_2O)	liquidus temp. to 1400°C	$6.7x10^{-5}$
SiO_2–B_2O_3 glass (58.4% mol B_2O_3)	320–340°C	$694x10^{-7}$
SiO_2–PbO glass (50% mol PbO)	1100°C	$723x10^{-7}$
SiO_2–CaO glass (42.5% mol CaO)	1700°C	$76{\pm}4x10^{-6}$

Source: data compiled by Jun S. Park from O. V. Mazurin, M. V. Streltsina and T. P. Shvaiko–Shvaikovskaya, *Handbook of Glass Data, Part A and Part B*, Elsevier, New York, 1983

Table 364. SELECTING THERMAL EXPANSION OF GLASSES
(SHEET 10 OF 11)

Glass	Temperature Range of Validity	Thermal Expansion (K^{-1})
SiO_2–CaO glass (47.5% mol CaO)	1700°C	$76\pm4x10^{-6}$
SiO_2–CaO glass (52.5% mol CaO)	1700°C	$76–107\pm4x10^{-6}$
B_2O_3–Na_2O glass (15% mol Na_2O, T_g = 407°C)	above T_g	$761x10^{-7}$
B_2O_3–Na_2O glass (25% mol Na_2O, T_g = 466°C)	above T_g	$834x10^{-7}$
SiO_2–CaO glass (50% mol CaO)	1700°C	$84–85\pm4x10^{-6}$
SiO_2–CaO glass (45% mol CaO)	1700°C	$85–100\pm4x10^{-6}$
SiO_2–PbO glass (66.7% mol PbO)	1100°C	$867x10^{-7}$
SiO_2–B_2O_3 glass (72.7% mol B_2O_3)	300–320°C	$899x10^{-7}$
SiO_2–CaO glass (55% mol CaO)	1700°C	$94–95\pm4x10^{-6}$
SiO_2–CaO glass (57.5% mol CaO)	1700°C	$95\pm4x10^{-6}$
SiO_2–B_2O_3 glass (83.2% mol B_2O_3)	280–300°C	$970x10^{-7}$
SiO_2–B_2O_3 glass (88.6% mol B_2O_3)	280–300°C	$1023x10^{-7}$
SiO_2–CaO glass (60% mol CaO)	1700°C	$103\pm4x10^{-6}$
B_2O_3–Na_2O glass (30% mol Na_2O, T_g = 468°C)	above T_g	$1150x10^{-7}$
SiO_2–B_2O_3 glass (94.0% mol B_2O_3)	270–290°C	$1200x10^{-7}$
B_2O_3–Na_2O glass (10% mol Na_2O, T_g = 354°C)	above T_g	$1230x10^{-7}$
B_2O_3–Na_2O glass (5% mol Na_2O, T_g = 318°C)	above T_g	$1400x10^{-7}$

Source: data compiled by Jun S. Park from O. V. Mazurin, M. V. Streltsina and T. P. Shvaiko–Shvaikovskaya, *Handbook of Glass Data, Part A and Part B*, Elsevier, New York, 1983

Table 364. SELECTING THERMAL EXPANSION OF GLASSES
(SHEET 11 OF 11)

Glass	Temperature Range of Validity	Thermal Expansion (K^{-1})
SiO_2–Na_2O glass (33.3% mol Na_2O)	liquidus temp.to 1400°C	$17.2x10^{-5}$
SiO_2–Na_2O glass (40% mol Na_2O)	liquidus temp. to 1400°C	$20.0x10^{-5}$
SiO_2–Na_2O glass (50% mol Na_2O)	liquidus temp. to 1400°C	$23.7x10^{-5}$

Source: data compiled by Jun S. Park from O. V. Mazurin, M. V. Streltsina and T. P. Shvaiko–Shvaikovskaya, *Handbook of Glass Data, Part A and Part B*, Elsevier, New York, 1983

Table 365. SELECTING THERMAL EXPANSION OF POLYMERS
(SHEET 1 OF 5)

Polymer	Thermal Expansion Coefficient ASTM D696 ($\cdot F^{-1}$)
Polymides: Glass Reinforced	0.8×10^{-6}
Polycarbonate (40% Glass Fiber Reinforced)	$1.0—1.1 \times 10^{-6}$
Epoxy Novolacs: Cast, Rigid	$1.6—3.0 \times 10^{-6}$
Epoxies: High Performance Resins: Molded	$1.7—2.2 \times 10^{-6}$
Polymides: Unreinforced	$2.5—4.5 \times 10^{-6}$
ABS Resin; Molded, Extruded: Heat Resistant	$3.0—4.0 \times 10^{-6}$
Acrylic Moldings: Grades 5, 6, 8	$3—4 \times 10^{-6}$
ABS Resin; Molded, Extruded: Medium Impact	$3.2—4.8 \times 10^{-6}$
Standard Epoxies: General Purpose Glass Cloth Laminate	$3.3—4.8 \times 10^{-6}$
Standard Epoxies: High Strength Laminate	$3.3—4.8 \times 10^{-6}$
Polycarbonate	3.75×10^{-6}
Acrylic Moldings: High Impact Grade	$4—6 \times 10^{-6}$
Chlorinated Polyvinyl Chloride	4.4×10^{-6}
Acrylics; Cast Resin Sheets, Rods: General Purpose, Type I	4.5×10^{-6}
Acrylics; Cast Resin Sheets, Rods: General Purpose, Type II	4.5×10^{-6}
ABS Resin; Molded, Extruded: Very High Impact	$5.0—6.0 \times 10^{-6}$
ABS Resin; Molded, Extruded: Low Temperature Impact	$5.0—6.0 \times 10^{-6}$
ABS Resin; Molded, Extruded: High Impact	$5.5—6.0 \times 10^{-6}$
Chlorinated Polyether	6.6×10^{-6}
Melamines; Molded: Glass Fiber Filled	0.82×10^{-5}
Rubber Phenolic—Woodflour or Flock	$0.83—2.20 \times 10^{-5}$
Phenolics, Molded; General: Very High Shock: Glass Fiber Filled	0.88×10^{-5}
Standard Epoxies: Molded	$1—2 \times 10^{-5}$
Melamines; Molded: Cellulose Filled Electrical	$1.11—2.78 \times 10^{-5}$

Source: data compiled by J.S. Park from Charles T. Lynch, *CRC Handbook of Materials Science, Vol. 3*, CRC Press, Boca Raton, Florida and *Engineered Materials Handbook, Vol.2*, Engineering Plastics, ASM International, Metals Park, Ohio, 1988.

Table 365. SELECTING THERMAL EXPANSION OF POLYMERS

(SHEET 2 OF 5)

Polymer	Thermal Expansion Coefficient ASTM D696 (•F^{-1})
Nylon; Molded, Extruded; Type 6: Glass Fiber (30%) Reinforced	1.2 x 10^{-5}
Phenylene Oxides (Noryl): Glass Fiber Reinforced	1.2–1.6 x 10^{-5}
Ureas; Molded: Alpha—Cellulose Filled (ASTM Type l)	1.22—1 .50 x 10^{-5}
Alkyds; Molded: Putty (encapsulating)	1.3 x 10^{-5}
Alkyds; Molded: Rope (general purpose)	1.3 x 10^{-5}
Alkyds; Molded: Granular (high speed molding)	1.3 x 10^{-5}
Alkyds; Molded: Glass reinforced (heavy duty parts)	1.3 x 10^{-5}
Reinforced Polyester Moldings: High Strength (Glass Fibers)	13—19 x 10^{-6}
Phenylene Oxides: Glass Fiber Reinforced	1.4–2.0 x 10^{-5}
6/10 Nylon: General purpose	1.5 x 10^{-5}
6/6 Nylon; General Purpose Molding: Glass Fiber Reinforced	1.5—3.3 x 10^{-5}
Glass Fiber (30%) Reinforced SAN	1.6 x 10^{-5}
Phenolics, General: High Shock: Chopped Fabric or Cord Filled	1.60—2.22 x 10^{-5}
Phenolics, Molded; General: Shock: Paper, Flock, or Pulp	1.6—2.3 x 10^{-5}
Polypropylene: Glass Reinforced	1.6—2.4 x 10^{-5}
Phenolics, Molded; General: Woodflour And Flock Filled	1.66—2.50 x 10^{-5}
6/6 Nylon; General Purpose Molding	1.69—1.7 x 10^{-5}
6/6 Nylon; General Purpose Extrusion	1.7 x 10^{-5}
Rubber Phenolic—Chopped Fabric	1.7 x 10^{-5}
Polytetrafluoroethylene (PTFE), Ceramic Reinforced	1.7—2.0 x 10^{-5}
Polystyrenes; Molded: Glass Fiber -30% Reinforced	1.8 x 10^{-5}
Polymide Homopolymer: 20% Glass Reinforced	2.0—4.5 x 10^{-5}
Polypropylene: Asbestos Filled	2—3 x 10^{-5}
Standard Epoxies: Filament Wound Composite	2—6 x 10^{-5}

Source: data compiled by J.S. Park from Charles T. Lynch, *CRC Handbook of Materials Science, Vol. 3*, CRC Press, Boca Raton, Florida and *Engineered Materials Handbook, Vol.2*, Engineering Plastics, ASM International, Metals Park, Ohio, 1988.

Table 365. SELECTING THERMAL EXPANSION OF POLYMERS (SHEET 3 OF 5)

Polymer	Thermal Expansion Coefficient ASTM D696 (•F^{-1})
Diallyl Phthalates; Molded: Glass Fiber Filled	2.2.—2.6 x 10^{-5}
Rubber Phenolic—Asbestos	2.2 x 10^{-5}
Polymide Copolymer: 25% Glass Reinforced	2.2—4.7 x 10^{-5}
Polystyrenes; Molded: High Impact	2.2—5.6 x 10^{-5}
Silicones; Molded, Laminated: Granular (Silica) Reinforced	2.5—5.0 x 10^{-5}
Polyarylsulfone	2.6 x 10^{-5}
Polyester; Injection Moldings: Glass Reinforced Grades	2.7—3.3 x 10^{-5}
Polyvinyl Chloride; Molded, Extruded: Rigid—normal impact	2.8—3 .3 x 10^{-5}
Polyphenylene Sulfide: Standard	3.0—4.9 x 10^{-5}
Standard Epoxies: Cast Flexible	3—5 x 10^{-5}
Phenylene Oxides (Noryl): Standard	3.1 x 10^{-5}
Silicones; Molded, Laminated: Fibrous (Glass) Reinforced	3.17—3.23 x 10^{-5}
Standard Epoxies: Cast rigid	3.3 x 10^{-5}
Phenylene Oxides: SE—1	3.3 x 10^{-5}
Polystyrenes; Molded: Medium Impact	3.3—4.7 x 10^{-5}
Polystyrenes; Molded: General Purpose	3.3—4.8 x 10^{-5}
6/10 Nylon: Glass fiber (30%) reinforced	3.5 x 10^{-5}
PVC–Acrylic Alloy Sheet	3.5 x 10^{-5}
Polyester; Injection Moldings: Glass Reinforced Self Extinguishing	3.5 x 10^{-5}
Styrene Acrylonitrile (SAN)	3.6—3.7 x 10^{-5}
Phenylene Oxides: SE—100	3.8 x 10^{-5}
Polypropylene: General Purpose	3.8—5.8 x 10^{-5}
Polytrifluoro chloroethylene (PTFCE)	3.88 x 10^{-5}
Thermoset Cast Polyyester: Rigid	3.9—5.6 x 10^{-5}

Source: data compiled by J.S. Park from Charles T. Lynch, *CRC Handbook of Materials Science, Vol. 3*, CRC Press, Boca Raton, Florida and *Engineered Materials Handbook, Vol.2*, Engineering Plastics, ASM International, Metals Park, Ohio, 1988.

Table 365. SELECTING THERMAL EXPANSION OF POLYMERS (SHEET 4 OF 5)

Polymer	Thermal Expansion Coefficient ASTM D696 (•F^{-1})
Polyphenylene Sulfide: 40% Glass Reinforced	4 x 10^{-5}
Diallyl Phthalates; Molded: Asbestos Filled	4.0 x 10^{-5}
Polypropylene: High Impact	4.0—5.9 x 10^{-5}
Nylon; Type 6: Cast	4.4 x 10^{-5}
Cellulose Acetate; Molded, Extruded; ASTM Grade: H6—1	4.4—9.0 x 10^{-5}
Cellulose Acetate; Molded, Extruded; ASTM Grade: H4—1	4.4—9.0 x 10^{-5}
Cellulose Acetate; Molded, Extruded; ASTM Grade: H2—1	4.4—9.0 x 10^{-5}
Cellulose Acetate; ASTM Grade: MH—1, MH—2	4.4—9.0 x 10^{-5}
Cellulose Acetate; ASTM Grade: MS—1, MS—2	4.4—9.0 x 10^{-5}
Cellulose Acetate; Molded, Extruded; ASTM Grade: S2—1	4.4—9.0 x 10^{-5}
Polymide Homopolymer: Standard	4.5 x 10^{-5}
Polymide Homopolymer: 22% TFE Reinforced	4.5 x 10^{-5}
Polymide Copolymer: Standard	4.7 x 10^{-5}
Polymide Copolymer: High Flow	4.7 x 10^{-5}
Nylon; Molded, Extruded; Type 6: General Purpose	4.8 x 10^{-5}
Polyester; Injection Moldings: General Purpose Grade	4.9—13.0 x 10^{-5}
Diallyl Phthalates; Molded: Orlon Filled	5.0 x 10^{-5}
Diallyl Phthalates; Molded: Dacron Filled	5.2 x 10^{-5}
Polyester; Thermoplastic Injection Moldings: General Purpose Grade	5.3 x 10^{-5}
Nylon; Type 11	5.5 x 10^{-5}
Thermoset Carbonate: Allyl diglycol carbonate	6 x 10^{-5}
Cellulose Acetate Butyrate; Molded, Extruded; ASTM Grade: H4	6—9 x 10^{-5}
Cellulose Acetate Butyrate; Molded, Extruded; ASTM Grade: MH	6—9 x 10^{-5}
Cellulose Acetate Butyrate; Molded, Extruded; ASTM Grade: S2	6—9 x 10^{-5}

Source: data compiled by J.S. Park from Charles T. Lynch, *CRC Handbook of Materials Science, Vol. 3*, CRC Press, Boca Raton, Florida and *Engineered Materials Handbook, Vol.2*, Engineering Plastics, ASM International, Metals Park, Ohio, 1988.

Table 365. SELECTING THERMAL EXPANSION OF POLYMERS
(SHEET 5 OF 5)

Polymer	Thermal Expansion Coefficient ASTM D696 ($\cdot F^{-1}$)
Cellusose Acetate Propionate; Molded, Extruded; ASTM Grade: 1	6—9 x 10^{-5}
Cellusose Acetate Propionate; Molded, Extruded; ASTM Grade: 3	6—9 x 10^{-5}
Cellusose Acetate Propionate; Molded, Extruded; ASTM Grade: 6	6—9 x 10^{-5}
ABS–Polycarbonate Alloy	6.12 x 10^{-5}
Nylon; Type 12	7.2 x 10^{-5}
Fluorinated Ethylene Propylene(FEP)	8.3—10.5 x 10^{-5}
Polyethylene; Molded, Extruded; Type II: Melt Index 20	8.3—16.7 x 10^{-5}
Polyethylene; Molded, Extruded; Type II: Melt index l.0—1.9	8.3—16.7 x 10^{-5}
Polyethylene; Molded, Extruded; Type III: Melt Index 0.2—0.9	8.3—16.7 x 10^{-5}
Polyethylene; Type III: Melt Melt Index 0.l—12.0	8.3—16.7 x 10^{-5}
Polyethylene; Molded, Extruded; Type III: Melt Index 1.5—15	8.3—16.7 x 10^{-5}
Polyvinylidene— Fluoride (PVDF)	8.5 x 10^{-5}
Vinylidene chloride	8.78 x 10^{-5}
Polyethylene; Molded, Extruded; Type I: Melt Index 0.3—3.6	8.9—11.0 x 10^{-5}
Polyethylene; Molded, Extruded; Type I: Melt Index 6—26	8.9—11.0 x 10^{-5}
Polyethylene; Molded, Extruded; Type I: Melt Index 200	11 x 10^{-5}
Polytetrafluoroethylene (PTFE)	55 x 10^{-5}

Source: data compiled by J.S. Park from Charles T. Lynch, *CRC Handbook of Materials Science, Vol. 3,* CRC Press, Boca Raton, Florida and *Engineered Materials Handbook, Vol.2*, Engineering Plastics, ASM International, Metals Park, Ohio, 1988.

Table 366. SELECTING THERMAL EXPANSION COEFFICIENTS FOR MATERIALS USED IN INTEGRATED CIRCUITS (SHEET 1 OF 6)

Material	Temperature Range (K)	Linear Thermal Expansion Coefficient (K^{-1})
Vitreous silica	300	0.42×10^{-6}
Vitreous silica	700	0.54×10^{-6}
Vitreous silica	800	0.54×10^{-6}
Vitreous silica	600	0.55×10^{-6}
Vitreous silica	400	0.56×10^{-6}
Vitreous silica	500	0.56×10^{-6}
Vitreous silica	500	0.56×10^{-6}
Silicon nitride (β)	25–1,000	2.25×10^{-6}
Pyroceram cement (Devitrified)	25–300	2.4×10^{-6}
Silicon	300	2.5×10^{-6}
Silicon nitride (α)	25–1,000	2.9×10^{-6}
Silicon	400	3.1×10^{-6}
Pyrex glass	25–300	3.2×10^{-6}
Silicon	500	3.5×10^{-6}
Silicon	500	3.5×10^{-6}
Silicon	600	3.8×10^{-6}
Pyroceram cement (Vitreous #45)	0–300	4×10^{-6}
Silicon	700	4.1×10^{-6}
Silicon	800	4.3×10^{-6}
Tungsten	300	4.5×10^{-6}

Source: from Beadles, R. L., Interconnections and Encapsulation, *Integrated Silicon Device Technology,* Vol. 14, Research Triangle Institute, Research Triangle Park, N. C., 1967. in *CRC Handbook of Materials Science*, Charles T. Lynch, Ed., CRC Press, Cleveland, (1974).

Table 366. SELECTING THERMAL EXPANSION COEFFICIENTS FOR MATERIALS USED IN INTEGRATED CIRCUITS (SHEET 2 OF 6)

Material	Temperature Range (K)	Linear Thermal Expansion Coefficient (K^{-1})
Tungsten	400	4.6×10^{-6}
Tungsten	500	4.6×10^{-6}
Tungsten	500	4.6×10^{-6}
Beryllium oxide	300	4.7×10^{-6}
Tungsten	600	4.7×10^{-6}
Tungsten	700	4.7×10^{-6}
Tungsten	800	4.8×10^{-6}
Silicon carbide	0–1,000	4.8×10^{-6}
Molybdenum	300	5×10^{-6}
Kovar	25–300	5.0×10^{-6}
Molybdenum	400	5.2×10^{-6}
Molybdenum	500	5.3×10^{-6}
Molybdenum	500	5.3×10^{-6}
Molybdenum	600	5.4×10^{-6}
Molybdenum	700	5.5×10^{-6}
Germanium	300	5.7×10^{-6}
Molybdenum	800	5.7×10^{-6}
Beryllium oxide	500	6×10^{-6}
Beryllium oxide	500	6×10^{-6}
Aluminum oxide ceramic	25–300	$6.0–7.0 \times 10^{-6}$

Source: from Beadles, R. L., Interconnections and Encapsulation, *Integrated Silicon Device Technology,* Vol. 14, Research Triangle Institute, Research Triangle Park, N. C., 1967. in *CRC Handbook of Materials Science*, Charles T. Lynch, Ed., CRC Press, Cleveland, (1974).

Table 366. SELECTING THERMAL EXPANSION COEFFICIENTS FOR MATERIALS USED IN INTEGRATED CIRCUITS (SHEET 3 OF 6)

Material	Temperature Range (K)	Linear Thermal Expansion Coefficient (K^{-1})
Germanium	400	6.2×10^{-6}
Tantalum	300	6.5×10^{-6}
Germanium	500	6.5×10^{-6}
Germanium	500	6.5×10^{-6}
Tantalum	400	6.6×10^{-6}
Germanium	600	6.7×10^{-6}
Tantalum	500	6.8×10^{-6}
Tantalum	500	6.8×10^{-6}
Tantalum	600	6.9×10^{-6}
Germanium	700	6.9×10^{-6}
Beryllium oxide	700	7×10^{-6}
Tantalum	700	7×10^{-6}
Tantalum	800	7.1×10^{-6}
Germanium	800	7.2×10^{-6}
Pyroceram cement (#89, #95)	—	$8\text{–}10 \times 10^{-6}$
Platinum	300	8.9×10^{-6}
Platinum	400	9.2×10^{-6}
Platinum	500	9.5×10^{-6}
Platinum	500	9.5×10^{-6}
Platinum	600	9.7×10^{-6}

Source: from Beadles, R. L., Interconnections and Encapsulation, *Integrated Silicon Device Technology*, Vol. 14, Research Triangle Institute, Research Triangle Park, N. C., 1967. in *CRC Handbook of Materials Science*, Charles T. Lynch, Ed., CRC Press, Cleveland, (1974).

Table 366. SELECTING THERMAL EXPANSION COEFFICIENTS FOR MATERIALS USED IN INTEGRATED CIRCUITS (SHEET 4 OF 6)

Material	Temperature Range (K)	Linear Thermal Expansion Coefficient (K^{-1})
Platinum	700	10×10^{-6}
Platinum	800	10.2×10^{-6}
Nickel	300	12.7×10^{-6}
Nickel	400	13.8×10^{-6}
Kanthal A	20–900	$13.9–15.1 \times 10^{-6}$
Gold	300	14.1×10^{-6}
Gold	400	14.5×10^{-6}
Gold	500	15×10^{-6}
Gold	500	15×10^{-6}
Nickel	500	15.2×10^{-6}
Nickel	500	15.2×10^{-6}
Gold	600	15.4×10^{-6}
Gold	700	15.9×10^{-6}
Nickel	700	16.4×10^{-6}
Gold	800	16.5×10^{-6}
Copper	300	16.8×10^{-6}
Nickel	800	16.8×10^{-6}
Nickel	600	17.2×10^{-6}
Copper	400	17.7×10^{-6}
Brass	25–300	$17.7–21.2 \times 10^{-6}$

Source: from Beadles, R. L., Interconnections and Encapsulation, *Integrated Silicon Device Technology,* Vol. 14, Research Triangle Institute, Research Triangle Park, N. C., 1967. in *CRC Handbook of Materials Science*, Charles T. Lynch, Ed., CRC Press, Cleveland, (1974).

Table 366. SELECTING THERMAL EXPANSION COEFFICIENTS FOR MATERIALS USED IN INTEGRATED CIRCUITS (SHEET 5 OF 6)

Material	Temperature Range (K)	Linear Thermal Expansion Coefficient (K^{-1})
Copper	500	18.3×10^{-6}
Copper	500	18.3×10^{-6}
Copper	600	18.9×10^{-6}
Silver	300	19.2×10^{-6}
Copper	700	19.4×10^{-6}
Silver	400	20×10^{-6}
Copper	800	20×10^{-6}
Silver	500	20.6×10^{-6}
Silver	500	20.6×10^{-6}
Tin	300	21.2×10^{-6}
Silver	600	21.4×10^{-6}
Silver	700	22.3×10^{-6}
Aluminum	300	23.2×10^{-6}
Silver	800	23.4×10^{-6}
Tin	400	24.2×10^{-6}
Aluminum	400	24.9×10^{-6}
Aluminum	500	26.4×10^{-6}
Aluminum	500	26.4×10^{-6}
Tin	500	27.5×10^{-6}
Tin	500	27.5×10^{-6}

Source: from Beadles, R. L., Interconnections and Encapsulation, *Integrated Silicon Device Technology*, Vol. 14, Research Triangle Institute, Research Triangle Park, N. C., 1967. in *CRC Handbook of Materials Science*, Charles T. Lynch, Ed., CRC Press, Cleveland, (1974).

Table 366. SELECTING THERMAL EXPANSION COEFFICIENTS FOR MATERIALS USED IN INTEGRATED CIRCUITS (SHEET 6 OF 6)

Material	Temperature Range (K)	Linear Thermal Expansion Coefficient (K^{-1})
Aluminum	600	28.3×10^{-6}
Lead	300	28.9×10^{-6}
Lead	400	29.8×10^{-6}
Aluminum	700	30.7×10^{-6}
Indium	300	31.9×10^{-6}
Lead	500	32.1×10^{-6}
Lead	500	32.1×10^{-6}
Aluminum	800	33.8×10^{-6}
Indium	400	38.5×10^{-6}
Pyroceram (#9608)	25–300	420×10^{-6}
Solder glass (Kimble CV-101)	0–300	809×10^{-6}

Source: from Beadles, R. L., Interconnections and Encapsulation, *Integrated Silicon Device Technology,* Vol. 14, Research Triangle Institute, Research Triangle Park, N. C., 1967. in *CRC Handbook of Materials Science*, Charles T. Lynch, Ed., CRC Press, Cleveland, (1974).

Table 367. SELECTING THERMAL EXPANSION COEFFICIENTS FOR MATERIALS USED IN INTEGRATED CIRCUITS AT TEMPERATURE (SHEET 1 OF 5)

Temperature Range (K)	Material	Linear Thermal Expansion Coefficient (K^{-1})
25–300	Pyroceram cement (Devitrified)	2.4×10^{-6}
25–300	Pyrex glass	3.2×10^{-6}
0–300	Pyroceram cement (Vitreous #45)	4×10^{-6}
25–300	Kovar	5.0×10^{-6}
25–300	Aluminum oxide ceramic	$6.0–7.0 \times 10^{-6}$
25–300	Brass	$17.7–21.2 \times 10^{-6}$
25–300	Pyroceram (#9608)	420×10^{-6}
0–300	Solder glass (Kimble CV-101)	809×10^{-6}
300	Vitreous silica	0.42×10^{-6}
300	Silicon	2.5×10^{-6}
300	Tungsten	4.5×10^{-6}
300	Beryllium oxide	4.7×10^{-6}
300	Molybdenum	5×10^{-6}
300	Germanium	5.7×10^{-6}
300	Tantalum	6.5×10^{-6}
300	Platinum	8.9×10^{-6}
300	Nickel	12.7×10^{-6}
300	Gold	14.1×10^{-6}
300	Copper	16.8×10^{-6}
300	Silver	19.2×10^{-6}

Source: from Beadles, R. L., Interconnections and Encapsulation, *Integrated Silicon Device Technology*, Vol. 14, Research Triangle Institute, Research Triangle Park, N. C., 1967. in *CRC Handbook of Materials Science*, Charles T. Lynch, Ed., CRC Press, Cleveland, (1974).

Table 367. SELECTING THERMAL EXPANSION COEFFICIENTS FOR MATERIALS USED IN INTEGRATED CIRCUITS AT TEMPERATURE (SHEET 2 OF 5)

Temperature Range (K)	Material	Linear Thermal Expansion Coefficient (K^{-1})
300	Tin	21.2×10^{-6}
300	Aluminum	23.2×10^{-6}
300	Lead	28.9×10^{-6}
300	Indium	31.9×10^{-6}
400	Vitreous silica	0.56×10^{-6}
400	Silicon	3.1×10^{-6}
400	Tungsten	4.6×10^{-6}
400	Molybdenum	5.2×10^{-6}
400	Germanium	6.2×10^{-6}
400	Tantalum	6.6×10^{-6}
400	Platinum	9.2×10^{-6}
400	Nickel	13.8×10^{-6}
400	Gold	14.5×10^{-6}
400	Copper	17.7×10^{-6}
400	Silver	20×10^{-6}
400	Tin	24.2×10^{-6}
400	Aluminum	24.9×10^{-6}
400	Lead	29.8×10^{-6}
400	Indium	38.5×10^{-6}

Source: from Beadles, R. L., Interconnections and Encapsulation, *Integrated Silicon Device Technology,* Vol. 14, Research Triangle Institute, Research Triangle Park, N. C., 1967. in *CRC Handbook of Materials Science*, Charles T. Lynch, Ed., CRC Press, Cleveland, (1974).

Table 367. SELECTING THERMAL EXPANSION COEFFICIENTS FOR MATERIALS USED IN INTEGRATED CIRCUITS AT TEMPERATURE (SHEET 3 OF 5)

Temperature Range (K)	Material	Linear Thermal Expansion Coefficient (K^{-1})
500	Vitreous silica	0.56×10^{-6}
500	Silicon	3.5×10^{-6}
500	Tungsten	4.6×10^{-6}
500	Molybdenum	5.3×10^{-6}
500	Beryllium oxide	6×10^{-6}
500	Germanium	6.5×10^{-6}
500	Tantalum	6.8×10^{-6}
500	Platinum	9.5×10^{-6}
500	Gold	15×10^{-6}
500	Nickel	15.2×10^{-6}
500	Copper	18.3×10^{-6}
500	Silver	20.6×10^{-6}
500	Aluminum	26.4×10^{-6}
500	Tin	27.5×10^{-6}
500	Lead	32.1×10^{-6}
600	Vitreous silica	0.55×10^{-6}
600	Silicon	3.8×10^{-6}
600	Tungsten	4.7×10^{-6}
600	Molybdenum	5.4×10^{-6}

Source: from Beadles, R. L., Interconnections and Encapsulation, *Integrated Silicon Device Technology,* Vol. 14, Research Triangle Institute, Research Triangle Park, N. C., 1967. in *CRC Handbook of Materials Science*, Charles T. Lynch, Ed., CRC Press, Cleveland, (1974).

Table 367. SELECTING THERMAL EXPANSION COEFFICIENTS FOR MATERIALS USED IN INTEGRATED CIRCUITS AT TEMPERATURE (SHEET 4 OF 5)

Temperature Range (K)	Material	Linear Thermal Expansion Coefficient (K^{-1})
600	Germanium	6.7×10^{-6}
600	Tantalum	6.9×10^{-6}
600	Platinum	9.7×10^{-6}
600	Gold	15.4×10^{-6}
600	Nickel	17.2×10^{-6}
600	Copper	18.9×10^{-6}
600	Silver	21.4×10^{-6}
600	Aluminum	28.3×10^{-6}
700	Vitreous silica	0.54×10^{-6}
700	Silicon	4.1×10^{-6}
700	Tungsten	4.7×10^{-6}
700	Molybdenum	5.5×10^{-6}
700	Germanium	6.9×10^{-6}
700	Beryllium oxide	7×10^{-6}
700	Tantalum	7×10^{-6}
700	Platinum	10×10^{-6}
700	Gold	15.9×10^{-6}
700	Nickel	16.4×10^{-6}
700	Copper	19.4×10^{-6}
700	Silver	22.3×10^{-6}
700	Aluminum	30.7×10^{-6}

Source: from Beadles, R. L., Interconnections and Encapsulation, *Integrated Silicon Device Technology*, Vol. 14, Research Triangle Institute, Research Triangle Park, N. C., 1967. in *CRC Handbook of Materials Science*, Charles T. Lynch, Ed., CRC Press, Cleveland, (1974).

Table 367. SELECTING THERMAL EXPANSION COEFFICIENTS FOR MATERIALS USED IN INTEGRATED CIRCUITS AT TEMPERATURE (SHEET 5 OF 5)

Temperature Range (K)	Material	Linear Thermal Expansion Coefficient (K^{-1})
800	Vitreous silica	0.54×10^{-6}
800	Silicon	4.3×10^{-6}
800	Tungsten	4.8×10^{-6}
800	Molybdenum	5.7×10^{-6}
800	Tantalum	7.1×10^{-6}
800	Germanium	7.2×10^{-6}
800	Platinum	10.2×10^{-6}
800	Gold	16.5×10^{-6}
800	Nickel	16.8×10^{-6}
800	Copper	20×10^{-6}
800	Silver	23.4×10^{-6}
800	Aluminum	33.8×10^{-6}
0–1,000	Silicon carbide	4.8×10^{-6}
25–1,000	Silicon nitride (α)	2.9×10^{-6}
25–1,000	Silicon nitride (β)	2.25×10^{-6}

Source: from Beadles, R. L., Interconnections and Encapsulation, *Integrated Silicon Device Technology*, Vol. 14, Research Triangle Institute, Research Triangle Park, N. C., 1967. in *CRC Handbook of Materials Science*, Charles T. Lynch, Ed., CRC Press, Cleveland, (1974).

CHAPTER 13

Selecting Mechanical Properties

List of Tables

Tensile Strength

Compressive Strength

Yield Strength

List of Tables (Continued)

Flexural Strength

Shear

Hardness and Microhardness

Friction

Abrasion Resistance

Fatigue

Impact Energy and Impact Strength

List of Tables (Continued)

Moduli

Poisson's Ratio

Elongation

Area Reduction

Table 368. SELECTING TENSILE STRENGTH OF TOOL STEELS

Type	Condition	Tensile Strength (MPa)
S7	Annealed	640
L6	Annealed	655
S1	Annealed	690
L2	Annealed	710
S5	Annealed	725
L2	Oil quenched from 855 •C and single tempered at: 650 •C	930
L6	Oil quenched from 845 •C and single tempered at: 650 •C	965
S5	Oil quenched from 870 •C and single tempered at: 650 •C	1035
S7	Fan cooled from 940 •C and single tempered at: 650 •C	1240
L2	Oil quenched from 855 •C and single tempered at: 540 •C	1275
L6	Oil quenched from 845 •C and single tempered at: 540 •C	1345
S1	Oil quenched from 845 •C and single tempered at: 650 •C	1345
S5	Oil quenched from 870 •C and single tempered at: 540 •C	1520
L2	Oil quenched from 855 •C and single tempered at: 425 •C	1550
L6	Oil quenched from 845 •C and single tempered at: 425 •C	1585
S1	Oil quenched from 845 •C and single tempered at: 540 •C	1680
L2	Oil quenched from 855 •C and single tempered at: 315 •C	1790
S1	Oil quenched from 845 •C and single tempered at: 425 •C	1790
S7	Fan cooled from 940 •C and single tempered at: 540 •C	1820
S5	Oil quenched from 870 •C and single tempered at: 425 •C	1895
S7	Fan cooled from 940 •C and single tempered at: 425 •C	1895
S7	Fan cooled from 940 •C and single tempered at: 315 •C	1965
L2	Oil quenched from 855 •C and single tempered at: 205 •C	2000
L6	Oil quenched from 845 •C and single tempered at: 315 •C	2000
S1	Oil quenched from 845 •C and single tempered at: 315 •C	2030
S1	Oil quenched from 845 •C and single tempered at: 205 •C	2070
S7	Fan cooled from 940 •C and single tempered at: 205 •C	2170
S5	Oil quenched from 870 •C and single tempered at: 315 •C	2240
S5	Oil quenched from 870 •C and single tempered at: 205 •C	2345

Source: data from ASM *Metals Reference Book, Second Edition*, American Society for Metals, Metals Park, Ohio 44073, p241, (1984).

Table 369. SELECTING TENSILE STRENGTH OF GRAY CAST IRONS

SAE grade	Maximum Tensile Strength (MPa)
G1800	118
G2500	173
G2500a	173
G3000	207
C3500	241
G4000	276
G3500b	1241
G3500c	1241
G4000d	1276
Grey Cast Iron Bars ASTM Class	**Tensile Strength (MPa)**
20	152
25	179
30	214
35	252
40	293
50	362
60	431

Source: data from *ASM Metals Reference Book*, Second Edition, American Society for Metals, Metals Park, Ohio 44073, p166-167, (1984).

Table 370. SELECTING TENSILE STRENGTH OF DUCTILE IRONS

Specification Number	Grade or Class	Tensile Strength (MPa)
MIL-I-24137(Ships)	Class C	345
MIL-I-24137(Ships)	Class B	379
ASTM A395-76; ASME SA395	60-40-18	414
ASTM A536-72; MIL-1-11466B(MR)	60-40-18	414
SAE J434c	D4018	414
MIL-I-24137(Ships)	Class A	414
ASTM A536-72; MIL-1-11466B(MR)	65-45-12	448
SAE J434c	D4512	448
ASTM A476-70(d); SAE AMS5316	80-60-03	552
ASTM A536-72; MIL-1-11466B(MR)	80-55-06	552
SAE J434c	D5506	552
ASTM A536-72; MIL-1-11466B(MR)	100-70-03	689
SAE J434c	D7003	689
ASTM A536-72; MIL-1-11466B(MR)	120-90-02	827

Source: data from *ASM Metals Reference Book, Second Edition*, American Society for Metals, Metals Park, Ohio 44073, p169, (1984).

Table 371. SELECTING TENSILE STRENGTHS OF MALLEABLE IRON CASTINGS

Specification Number	grade or class	Tensile Strength (MPa)
ASTM A197		276
ASTM A47, A338; ANSI G48.1; FED QQ–I–666c	32510	345
ASTM A602; SAE J158	M3210	345
ASTM A47, A338; ANSI G48.1; FED QQ–I–666c	35018	365
ASTM A220; ANSI C48.2; MIL–I–11444B	40010	414
ASTM A220; ANSI C48.2; MIL–I–11444B	45008	448
ASTM A220; ANSI C48.2; MIL–I–11444B	45006	448
ASTM A602; SAE J158	M4504(a)	448
ASTM A220; ANSI C48.2; MIL–I–11444B	50005	483
ASTM A602; SAE J158	M5003(a)	517
ASTM A602; SAE J158	M5503(b)	517
ASTM A220; ANSI C48.2; MIL–I–11444B	60004	552
ASTM A220; ANSI C48.2; MIL–I–11444B	70003	586
ASTM A602; SAE J158	M7002(b)	621
ASTM A220; ANSI C48.2; MIL–I–11444B	80002	655
ASTM A220; ANSI C48.2; MIL–I–11444B	90001	724
ASTM A602; SAE J158	M8501(b)	724

[a] Air quenched and tempered

[b] Liquid quenched and tempered

Source: data from ASM *Metals Reference Book, Second Edition*, American Society for Metals, Metals Park, Ohio 44073, p171, (1984).

Table 372. SELECTING TENSILE STRENGTHS OF ALUMINUM CASTING ALLOYS (SHEET 1 OF 3)

Alloy AA No.	Temper	Tensile Strength (MPa)
443.0	F	130
208.0	F	145
B443.0	F	159
850.0	T5	160
514.0	F	170
355.0	T71	175
356.0	T51	175
A390.0	F,T5	180
242.0	T21	185
319.0	F	185
308.0	F	195
355.0	T51	195
356.0	T71	195
A390.0	F,T5	200
242.0	T77	205
355.0	T51	210
713.0	T5	210
242.0	T571	220
295.0	T4	220
356.0	T7	220
713.0	T5	220
C443.0	F	228
356.0	T6	230
319.0	F	235
356.0	T7	235
355.0	T6	240
712.0	F	240
295.0	T6	250

Source: data from *ASM Metals Reference Book, Second Edition*, American Society for Metals, Metals Park, Ohio 44073, (1984).

Table 372. SELECTING TENSILE STRENGTHS OF ALUMINUM CASTING ALLOYS (SHEET 2 OF 3)

Alloy AA No.	Temper	Tensile Strength (MPa)
319.0	T6	250
336.0	T551	250
355.0	T71	250
A390.0	T7	250
296.0	T4	255
A390.0	T7	260
355.0	T7	265
356.0	T6	265
296.0	T7	270
355.0	T61	270
242.0	T571	275
296.0	T6	275
535.0	F	275
319.0	T6	280
355.0	T7	280
390.0	F	280
A390.0	T6	280
295.0	T62	285
355.0	T6	290
A413.0	F	290
390.0	T5	300
413.0	F	300
355.0	T62	310
383.0	F	310
A390.0	T6	310
518.0	F	310
A360.0	F	320
242.0	T61	325

Source: data from *ASM Metals Reference Book, Second Edition*, American Society for Metals, Metals Park, Ohio 44073, (1984).

Table 372. SELECTING TENSILE STRENGTHS OF ALUMINUM CASTING ALLOYS (SHEET 3 OF 3)

Alloy AA No.	Temper	Tensile Strength (MPa)
336.0	T65	325
360.0	F	325
359.0	T61	330
380.0	F	330
384.0, A384.0	F	330
520.0	T4	330
359.0	T62	345
771.0	T6	345
357.0, A357.0	T62	360
201.0	T4	365
354.0	T61	380
206.0, A206.0	T7	435
201.0	T7	460
201.0	T6	485

Source: data from *ASM Metals Reference Book, Second Edition*, American Society for Metals, Metals Park, Ohio 44073, (1984).

Table 373. SELECTING TENSILE STRENGTHS OF WROUGHT ALUMINUM ALLOYS (SHEET 1 OF 7)

Alloy	Temper	Tensile Strength (MPa)
1060	0	69
1050	0	76
1060	H12	83
1350	0	83
1100	0	90
6063	0	90
1060	H14	97
1350	H12	97
6101	H111	97
1050	H14	110
1060	H16	110
1100	H12	110
1350	H14	110
3003	0	110
3105	0	115
Alclad 6061	0	115
1100	H14	125
1350	H16	125
5005	0	125
6061	0	125
1050	H16	130
1060	H18	130
Alclad	H12	130
5457	0	130
5005	H12	140
5005	H32	140
1100	H16	145
4043	0	145

Source: data from *ASM Metals Reference Book, Second Edition*, American Society for Metals, Metals Park, Ohio 44073, p.299—302, (1984).

Table 373. SELECTING TENSILE STRENGTHS OF WROUGHT ALUMINUM ALLOYS (SHEET 2 OF 7)

Alloy	Temper	Tensile Strength (MPa)
5050	0	145
6070	0	145
3003	H14	150
3105	H12	150
6063	TI	150
6066	0	150
6463	Tl	150
1050	H18	160
5005	H14	160
5005	H34	160
5657	H25	160
1100	H18	165
Alclad 2014	0	170
2219	0	170
3105	H14	170
5050	H32	170
6005	T1	170
6063	T4	170
Alclad 2024	0	180
3003	H16	180
3004	0	180
3105	H25	180
5005	H16	180
5005	H36	180
5457	H25	180
1350	H19	185
2014	0	185
2024	0	185

Source: data from *ASM Metals Reference Book, Second Edition*, American Society for Metals, Metals Park, Ohio 44073, p.299—302, (1984).

Table 373. SELECTING TENSILE STRENGTHS OF WROUGHT ALUMINUM ALLOYS (SHEET 3 OF 7)

Alloy	Temper	Tensile Strength (MPa)
6063	T5	185
6463	T5	185
7005	0	193
3105	H16	195
5050	H34	195
5052	0	195
5652	0	195
5657	H28, H38	195
3003	H18	200
5005	H18	200
5005	H38	200
5050	H36	205
5457	H28, H38	205
6063	T831	205
Alclad	H32	215
3105	H18	215
5050	H38	220
6151	T6	220
Alclad 7075	0	220
5052	H32	230
5652	H32	230
Alclad 6061	T4, T451	230
7075	0	230
5252	H25	235
6009	T4	235
3004	H34	240
5154	0	240
5154	H112	240

Source: data from *ASM Metals Reference Book, Second Edition*, American Society for Metals, Metals Park, Ohio 44073, p.299—302, (1984).

Table 373. SELECTING TENSILE STRENGTHS OF WROUGHT ALUMINUM ALLOYS (SHEET 4 OF 7)

Alloy	Temper	Tensile Strength (MPa)
5254	0	240
5254	H112	240
6061	T4, T451	240
6063	T6	240
6463	T6	240
5454	0	250
5454	H112	250
6351	T4	250
6010	T4	255
6063	T83	255
3004	H36	260
5052	H34	260
5086	0	260
5454	H111	260
5454	H311	260
5652	H34	260
6005	T5	260
6205	Tl	260
5086	H112	270
5154	H32	270
5254	H32	270
5052	H36	275
5182	0	275
5454	H32	275
5652	H36	275
3004	H38	285
4043	H18	285
5252	H28, H38	285

Source: data from *ASM Metals Reference Book, Second Edition*, American Society for Metals, Metals Park, Ohio 44073, p.299—302, (1984).

Table 373. SELECTING TENSILE STRENGTHS OF WROUGHT ALUMINUM ALLOYS (SHEET 5 OF 7)

Alloy	Temper	Tensile Strength (MPa)
5052	H38	290
5056	0	290
5083	0	290
5086	H32, H116, H117	290
5154	H34	290
5254	H34	290
5652	H38	290
Alclad 6061	T6, T651	290
6063	T832	290
5083	H112	305
5454	H34	305
5154	H36	310
5254	H36	310
5456	0	310
5456	H112	310
6061	T6, T651	310
6205	T5	310
6351	T6	310
5083	H113	315
5083	H321	315
5182	H32	315
6070	T4	315
5083	H323, H32	325
5086	H34	325
5456	H111	325
2218	T72	330
5154	H38	330
5254	H38	330

Source: data from *ASM Metals Reference Book, Second Edition*, American Society for Metals, Metals Park, Ohio 44073, p.299—302, (1984).

Table 373. SELECTING TENSILE STRENGTHS OF WROUGHT ALUMINUM ALLOYS (SHEET 6 OF 7)

Alloy	Temper	Tensile Strength (MPa)
6201	T6	330
6201	T81	330
2036	T4	340
5182	H34	340
5454	H36	340
2218	T71	345
5083	H343, H34	345
6009	T6	345
5456	H321, H116	350
2219	T42	360
2219	T31, T351	360
6066	T4, T451	360
5454	H38	370
7005	T6,T63,T6351	372
2011	T3	380
4032	T6	380
6070	T6	380
7005	T53	393
2219	T37	395
6066	T6, T651	395
6262	T9	400
2011	T8	405
2218	T61	405
2219	T62	415
5056	H38	415
Alclad 2014	T4	420
5182	H19(n)	420
2014	T4	425

Source: data from *ASM Metals Reference Book, Second Edition*, American Society for Metals, Metals Park, Ohio 44073, p.299—302, (1984).

Table 373. SELECTING TENSILE STRENGTHS OF WROUGHT ALUMINUM ALLOYS (SHEET 7 OF 7)

Alloy	Temper	Tensile Strength (MPa)
Alclad 2014	T3	435
5056	H18	435
Alclad 2024	T4, T351	440
2618	All	440
Alclad 2024	T	450
Alclad 2024	T81, T851	450
2048		455
2219	T81, T851	455
Alclad 2024	T361	460
Alclad 2014	T6	470
2024	T4, T351	470
2219	T87	475
2014	T6	485
2024	T3	485
Alclad 2024	T861	485
2124	T851	490
2024	T361	495
7075	T73	505
7050	T736	515
Alclad 7075	T6,T651	525
7175	T736	525
7475	T61	525
7075	T6,T651	570
7175	T66	595

Source: data from *ASM Metals Reference Book, Second Edition*, American Society for Metals, Metals Park, Ohio 44073, p.299—302, (1984).

Table 374. SELECTING TENSILE STRENGTHS OF CERAMICS (SHEET 1 OF 4)

Ceramic	Temperature	Tensile Strength (psi)
Boron Nitride (BN)	1000°C	0.35×10^3
Boron Nitride (BN)	1500°C	0.35×10^3
Beryllium Oxide (BeO)	1300°C	0.6×10^3
Spinel (Al_2O_3 MgO)	1300°C	1.1×10^3
Boron Nitride (BN)	1800°C	1.15×10^3
Aluminum Oxide (Al_2O_3)	1460°C	1.5×10^3
Tantalum Monocarbide (TaC)		$2\text{-}42 \times 10^3$
Beryllium Oxide (BeO)	1140°C	2.0×10^3
Boron Nitride (BN)	2000°C	2.25×10^3
Cordierite ($2MgO\ 2Al_2O_3\ 5SiO_2$)(ρ=1.8g/cm^3)	1200°C	2.5×10^3
Cordierite ($2MgO\ 2Al_2O_3\ 5SiO_2$)(ρ=2.1g/cm^3)	800°C	3.5×10^3
Zircon (SiO_2 ZrO_2)	1200°C	3.6×10^3
Aluminum Oxide (Al_2O_3)	1400°C	4.3×10^3
Silicon Carbide (SiC)	25°C	$5\text{-}20 \times 10^3$
Beryllium Oxide (BeO)	1000°C	5.0×10^3
Silicon Carbide (SiC) (hot pressed)	1400°C	$5.75\text{-}21.75 \times 10^3$
Magnesium Oxide (MgO)	1300°C	6×10^3
Spinel (Al_2O_3 MgO)	1150°C	6.1×10^3
Aluminum Oxide (Al_2O_3)	1300°C	6.4×10^3
Zirconium Oxide (ZrO_2)	1000°C	$6.75\text{-}17.0 \times 10^3$

To convert **psi** to **MPa**, multiply by **145**.

Source: data compiled by J.S. Park from *No. 1 Materials Index*, Peter T.B. Shaffer, Plenum Press, New York, (1964); *Smithells Metals Reference Book*, Eric A. Brandes, ed., in association with Fulmer Research Institute Ltd. 6th ed. London, Butterworths, Boston, (1983); and *Ceramic Source*, American Ceramic Society (1986-1991).

Table 374. SELECTING TENSILE STRENGTHS OF CERAMICS
(SHEET 2 OF 4)

Ceramic	Temperature	Tensile Strength (psi)
Boron Nitride (BN)	2400°C	6.80 x10^3
Beryllium Oxide (BeO)	900°C	7.0 x10^3
Cordierite ($2MgO\ 2Al_2O_3\ 5SiO_2$)(ρ=2.51g/cm^3)	25°C	7.8 x10^3
Magnesium Oxide (MgO)	1200°C	8 x10^3
Zircon ($SiO_2\ ZrO_2$)	1050°C	8.7 x10^3
Magnesium Oxide (MgO)	1100°C	10 x10^3
Zirconium Oxide (ZrO_2)	1300°C	10.2 x10^3
Chromium Diboride (CrB_2)		10.6 x10^4
Beryllium Oxide (BeO)	500°C	11.1 x10^3
Silicon Carbide (SiC) (reaction bonded)	20°C	11.17 x10^3
Magnesium Oxide (MgO)	1000°C	11.5 x10^3
Zirconium Monocarbide (ZrC)	980°C	11.7-14.45 x10^3
Zirconium Oxide (ZrO_2)	1200°C	12.1 x10^3
Zircon ($SiO_2\ ZrO_2$)	room temp.	12.7 x10^3
Zirconium Monocarbide (ZrC)	1250°C	12.95-15.85 x10^3
Zirconium Oxide (ZrO_2)	1100°C	13.0-13.5 x10^3
Beryllium Oxide (BeO)	room temp.	13.5-20 x10^3
Spinel (Al_2O_3 MgO)	550°C	13.7 x10^3
Magnesium Oxide (MgO)	room temp.	14 x10^3
Magnesium Oxide (MgO)	200°C	14 x10^3

To convert **psi** to **MPa**, multiply by **145**.

Source: data compiled by J.S. Park from *No. 1 Materials Index*, Peter T.B. Shaffer, Plenum Press, New York, (1964); *Smithells Metals Reference Book*, Eric A. Brandes, ed., in association with Fulmer Research Institute Ltd. 6th ed. London, Butterworths, Boston, (1983); and *Ceramic Source*, American Ceramic Society (1986-1991).

Table 374. SELECTING TENSILE STRENGTHS OF CERAMICS
(SHEET 3 OF 4)

Ceramic	Temperature	Tensile Strength (psi)
Thorium Dioxide (ThO_2)	room temp.	14 x10^3
Magnesium Oxide (MgO)	400°C	15.2 x10^3
Magnesium Oxide (MgO)	800°C	16 x10^3
Zirconium Monocarbide (ZrC)	room temp.	16.0 x10^3
Zirconium Oxide (ZrO_2)	800°C	16.0 x10^3
Mullite ($3Al_2O_3$ $2SiO_2$)	25°C	16 x10^3
Zirconium Oxide (ZrO_2)	200°C	16.8 x10^3
Titanium Monocarbide (TiC)	1000°C	17.2 x10^3
Zirconium Oxide (ZrO_2)	400°C	17.5 x10^3
Zirconium Oxide (ZrO_2)	600°C	17.6 x10^3
Zirconium Oxide (ZrO_2)	room temp.	17.9-20 x10^3
Titanium Diboride (TiB_2)		18.4 x10^3
Aluminum Oxide (Al_2O_3)	1200°C	18.5-20 x10^3
Spinel (Al_2O_3 MgO)	room temp.	19.2 x10^3
Zirconium Oxide (ZrO_2)	500°C	20.0 x10^3
Trisilicon tetranitride (Si_3N_4) (reaction bonded)	1400°C	20.3 x10^3
Trisilicon tetranitride (Si_3N_4) (hot pressed)	1400°C	21.8 x10^3
Boron Carbide (B_4C)	980°C	22.5 x10^3
Trisilicon tetranitride (Si_3N_4) (reaction bonded)	20°C	24.7 x10^3

To convert **psi** to **MPa**, multiply by **145**.

Source: data compiled by J.S. Park from *No. 1 Materials Index*, Peter T.B. Shaffer, Plenum Press, New York, (1964); *Smithells Metals Reference Book*, Eric A. Brandes, ed., in association with Fulmer Research Institute Ltd. 6th ed. London, Butterworths, Boston, (1983); and *Ceramic Source*, American Ceramic Society (1986-1991).

Table 374. SELECTING TENSILE STRENGTHS OF CERAMICS
(SHEET 4 OF 4)

Ceramic	Temperature	Tensile Strength (psi)
Zirconium Diboride (ZrB_2)		28.7 x10^3
Silicon Carbide (SiC) (hot pressed)	20°C	29 x10^3
Aluminum Oxide (Al_2O_3)	1140°C	31.4 x10^3
Aluminum Oxide (Al_2O_3)	300°C	33.6 x10^3
Aluminum Oxide (Al_2O_3)	1050°C	33.9 x10^3
Aluminum Oxide (Al_2O_3)	800°C	34.6 x10^3
Aluminum Oxide (Al_2O_3)	1000°C	35 x10^3
Aluminum Oxide (Al_2O_3)	room temp.	37-37.8 x10^3
Aluminum Oxide (Al_2O_3)	500°C	40 x10^3
Molybdenum Disilicide ($MoSi_2$)	980°C	40 x10^3
Molybdenum Disilicide ($MoSi_2$)	1300°C	41.07 x10^3
Molybdenum Disilicide ($MoSi_2$)	1090°C	42.16 x10^3
Molybdenum Disilicide ($MoSi_2$)	1200°C	42.8 x10^3
Tungsten Monocarbide (WC)		50 x10^3
Trisilicon tetranitride (Si_3N_4) (hot pressed)	20°C	54.4 x10^3
Spinel (Al_2O_3 MgO)	900°C	110.8 x10^3

To convert psi to MPa, multiply by 145.

Source: data compiled by J.S. Park from *No. 1 Materials Index*, Peter T.B. Shaffer, Plenum Press, New York, (1964); *Smithells Metals Reference Book*, Eric A. Brandes, ed., in association with Fulmer Research Institute Ltd. 6th ed. London, Butterworths, Boston, (1983); and *Ceramic Source*, American Ceramic Society (1986-1991).

Table 375. SELECTING TENSILE STRENGTHS OF GLASS (SHEET 1 OF 2)

Glass	Tensile Strength ($Kg \cdot mm^{-2}$)
(Corning 7940 silica glass @ 100°C)	5.6
SiO_2 glass (1.5 mm diameter rod, 0.5 $g/mm^2 \cdot s$ stress rate)	5.84–7.08
(Corning 7940 silica glass @ 300°C)	6.2
(Corning 7940 silica glass @ 500°C)	6.6
(Corning 7940 silica glass @ 700°C)	7.1
(Corning 7940 silica glass @ 900°C)	7.6
SiO_2 glass (1.5 mm diameter rod, 54 $g/mm^2 \cdot s$ stress rate)	8.52±2.52
SiO_2 glass (1.5 mm diameter rod, 50 $g/mm^2 \cdot s$ stress rate)	9.73±2.13
SiO_2–Na_2O glass (5 mm diameter rod, 20% mol Na_2O)	15
SiO_2 glass (112 μm diameter fiber)	28.3
SiO_2 glass (108 μm diameter fiber)	28.8
SiO_2 glass (78 μm diameter fiber)	35.8
SiO_2 glass (74 μm diameter fiber)	36.5
SiO_2 glass (65 μm diameter fiber)	39.7
SiO_2 glass (60 μm diameter fiber)	42.3
SiO_2–PbO glass (17.2 μm diameter fiber, 50% mol PbO)	43–51.6
SiO_2 glass (56 μm diameter fiber)	44.3
SiO_2 glass (48 μm diameter fiber)	49.6
SiO_2–PbO glass (11.4 μm diameter fiber, 50% mol PbO)	51.9–56
B_2O_3 glass (10–30 μm diameter fiber)	60
SiO_2–PbO glass (7.1 μm diameter fiber, 50% mol PbO)	62–71.3
SiO_2–PbO glass (4.3 μm diameter fiber, 50% mol PbO)	64
SiO_2–PbO glass (8.0 μm diameter fiber, 50% mol PbO)	64.5
SiO_2–PbO glass (5.7 μm diameter fiber, 50% mol PbO)	66–67.2

Source: data compiled by J.S. Park from O. V. Mazurin, M. V. Streltsina and T. P. Shvaiko–Shvaikovskaya, *Handbook of Glass Data, Part A and Part B*, Elsevier, New York, 1983

Table 375. SELECTING TENSILE STRENGTHS OF GLASS
(SHEET 2 OF 2)

Glass	Tensile Strength ($Kg \cdot mm^{-2}$)
SiO_2-PbO glass (3.0 mm diameter fiber, 50% mol PbO)	70.8
SiO_2-Na_2O glass (11.4mm diameter fiber, 36.3% mol Na_2O)	91.2±1.480
SiO_2-Na_2O glass (25.7mm diameter fiber, 19.5% mol Na_2O)	92.5±10.08
SiO_2-Na_2O glass (8.6mm diameter fiber, 36.3% mol Na_2O)	98.0±0.344
B_2O_3-Na_2O glass (10-30 mm diameter fiber, 10% mol Na_2O)	102
SiO_2-Na_2O glass (12.8mm diameter fiber, 25.5% mol Na_2O)	103±1.020
SiO_2-Na_2O glass (5.4mm diameter fiber, 36.3% mol Na_2O)	107.6±0.308
SiO_2-Na_2O glass (6.3mm diameter fiber, 25.5% mol Na_2O)	127±0.259
SiO_2-Na_2O glass (8.6mm diameter fiber, 19.5% mol Na_2O)	134±1.34
B_2O_3-Na_2O glass (10-30 mm diameter fiber, 20% mol Na_2O)	137
SiO_2-Na_2O glass (3.6mm diameter fiber, 25.5% mol Na_2O)	142±0.189
B_2O_3-Na_2O glass (10-30 mm diameter fiber, 30% mol Na_2O)	152
SiO_2-Na_2O glass (6.0mm diameter fiber, 19.5% mol Na_2O)	173±1.36

Source: data compiled by J.S. Park from O. V. Mazurin, M. V. Streltsina and T. P. Shvaiko–Shvaikovskaya, *Handbook of Glass Data, Part A and Part B*, Elsevier, New York, 1983

Table 376. SELECTING TENSILE STRENGTHS OF POLYMERS
(SHEET 1 OF 5)

Polymer	Tensile Strength (ASTM D638) (10^3 psi)
Olefin Copolymer: EEA (ethylene ethyl acrylate)	0.2
Olefin Copolymer: Ethylene butene	0.35
Olefin Copolymer: EVA (ethylene vinyl acetate)	0.36
Propylene–ethylene	0.4
Ethylene Ionomer	0.4
Fluorocarbons: Ceramic reinforced (PTFE)	0.75—2.5
Polyethylene, Type I, low density: Melt index 200	0.9—1.1 (ASTM D412)
Polyvinyl Chloride & Copolymer: Nonrigid—general	1—3.5 (ASTM D412)
Polyesters, cast thermoset: Flexible	1—8
6/6 Nylon: General purpose extrusion	1.26, 8.6
Polyethylene, Type I, low density: Melt index 6—26	1.4—2.0 (ASTM D412)
Polyethylene, Type I, low density: Melt index 0.3—3.6	1.4—2.5 (ASTM D412)
Standard Epoxy: Cast flexible	1.4—7.6
Polyethylene, Type II, medium density: Melt index 20	2
Polyvinyl Chloride & Copolymer: Nonrigid—electrical	2—3.2 (ASTM D412)
Polyethylene, Type II, medium density: Melt index l.0—1.9	2.3—2.4
Fluorocarbons: Fluorinated ethylene propylene(FEP)	2.5—4.0
Fluorocarbons: Polytetrafluoroethylene (PTFE)	2.5—6.5
Polyethylene, Type III, higher density: Melt Melt index 0.l—12.0	2.9—4.0
Cellulose Acetate Butyrate, ASTM Grade: S2	3.0—4.0 at Fracture
Cellulose Acetate; ASTM Grade: S2—1	3.0—4.4 at Fracture
Alkyd; Molded: Granular (high speed molding)	3—4
Ethylene Polyallomer	3—4.3
Phenolics: Rubber phenolic—chopped fabric	3—5 (ASTM D651)

To convert **psi** to **MPa**, multiply by 145.

Source: data compiled by J.S. Park from Charles T. Lynch, *CRC Handbook of Materials Science, Vol. 3*, CRC Press, Boca Raton, Florida, 1975 and *Engineered Materials Handbook, Vol.2*, Engineering Plastics, ASM International, Metals Park, Ohio, 1988.

Table 376. SELECTING TENSILE STRENGTHS OF POLYMERS

(SHEET 2 OF 5)

Polymer	Tensile Strength (ASTM D638) (10^3 psi)
Polystyrene: High impact	3.3—5.1
Cellulose Acetate; ASTM Grade: MS—1, MS—2	3.9—5.3 at Fracture
Cellulose Acetate Propionate, ASTM Grade: 6	4
Phenolics: Rubber phenolic—asbestos	4 (ASTM D651)
Polystyrene: Medium impact	4.0—6.0
Alkyd; Molded: Putty (encapsulating)	4—5
ABS Resin; Molded, Extruded: Low temperature impact	4—6
Reinforced polyester moldings: Heat & chemical resistant (asbestos)	4—6
Silicone: Granular (silica) reinforced	4—6 (ASTM D651)
Diallyl Phthalates, Molded: Asbestos filled	4—6.5
Polyvinyl Chloride & Copolymer: Vinylidene chloride	4—8,15—40 (ASTM D412)
Polyethylene, Type III, higher density: Melt index 0.2—0.9	4.4
Polyethylene, Type III, higher density: Melt index 1.5—15	4.4
Diallyl Phthalates, Molded: Orlon filled	4.5—6
ABS Resin; Molded, Extruded: Very high impact	4.5—6.0
Polypropylene: general purpose	4.5—6.0
Phenolics: Rubber phenolic—woodflour or flock	4.5—9 (ASTM D651)
Fluorocarbons: Polytrifluoro chloroethylene (PTFCE)	4.6—5.7
Diallyl Phthalates, Molded: Dacron filled	4.6—6.2
Cellulose Acetate; ASTM Grade: MH—1, MH—2	4.8—6.3 at Fracture
Polystyrene: General purpose	5.0—10
ABS Resin; Molded, Extruded: High impact	5.0—6.0
Cellulose Acetate Butyrate, ASTM Grade: MH	5.0—6.0 at Fracture
Phenolics, General: woodflour and flock filler	5.0—8.5 (ASTM D651)

To convert psi to MPa, multiply by 145.

Source: data compiled by J.S. Park from Charles T. Lynch, *CRC Handbook of Materials Science, Vol. 3*, CRC Press, Boca Raton, Florida, 1975 and *Engineered Materials Handbook, Vol.2*, Engineering Plastics, ASM International, Metals Park, Ohio, 1988.

Table 376. SELECTING TENSILE STRENGTHS OF POLYMERS

(SHEET 3 OF 5)

Polymer	Tensile Strength (ASTM D638) (10^3 psi)
Phenolics, Shock: paper, flock, or pulp filler	5.0—8.5 (ASTM D651)
Reinforced polyester moldings: High strength (glass fibers)	5—10
Urea: Alpha, cellulose filled (ASTM Type l)	5—10
Phenolics, Very high shock: glass fiber filler	5—10 (ASTM D651)
Polyesters, cast thermoset: Rigid	5—15
Allyl diglycol carbonate (thermoset)	5—6
Melamine, molded: Alpha cellulose and mineral filler	5—8
Alkyd; Molded: Glass reinforced (heavy duty parts)	5—9
Melamine, molded: Cellulose electrical filler	5—9
Phenolics, High shock: chopped fabric or cord filler	5—9 (ASTM D651)
Cellulose Acetate Propionate, ASTM Grade: 3	5.1—5.9
Epoxiy, (cycloaliphatic diepoxides): Molded	5.2—5.3
Fluorocarbons: Polyvinylidene— fluoride (PVDF)	5.2—8.6
Polyethylene, Type III, higher density, high molecular weight	5.4
Diallyl Phthalates, Molded: Glass fiber filled	5.5—11
Polyvinyl Chloride & Copolymer: Rigid—normal impact	5.5—8 (ASTM D412)
Acrylic Moldings: High impact grade	5.5—8.0
Cellulose Acetate; ASTM Grade: H2—1	5.8—7.2 at Fracture
Cellulose Acetate Propionate, ASTM Grade: 1	5.9—6.5
Chlorinated polyether	6
Phenolics: Arc resistant—mineral	6 (ASTM D651)
Acrylic Cast Resin Sheets, Rods: General purpose, type I	6—9
Melamine, molded: Glass fiber filler	6—9
ABS Resin; Molded, Extruded: Medium impact	6.3—8.0
Silicone: Fibrous (glass) reinforced	6.5 (ASTM D651)
Polyacetal homopolymer: 22% TFE reinforced	6.9
Cellulose Acetate Butyrate, ASTM Grade: H4	6.9 at Fracture
ABS Resin; Molded, Extruded: Heat resistant	7.0—8.0

To convert **psi** to **MPa**, multiply by 145.

Source: data compiled by J.S. Park from Charles T. Lynch, *CRC Handbook of Materials Science, Vol. 3*, CRC Press, Boca Raton, Florida, 1975 and *Engineered Materials Handbook, Vol.2*, Engineering Plastics, ASM International, Metals Park, Ohio, 1988.

Table 376. SELECTING TENSILE STRENGTHS OF POLYMERS (SHEET 4 OF 5)

Polymer	Tensile Strength (ASTM D638) (10^3 psi)
Alkyd; Molded: Rope (general purpose)	7—8
Cellulose Acetate; ASTM Grade: H4—1	7—8 at Fracture
Nylon, Type 12	7.1—8.5
6/10 Nylon: General purpose	7.1—8.5
Chlorinated polyvinyl chloride	7.3
Nylon, Type 6: Flexible copolymers	7.5—10.0
Acrylic Cast Resin Sheets, Rods: General purpose, type II	8—10
Standard Epoxy: Molded	8—11
Epoxiy, (cycloaliphatic diepoxides): Cast, rigid	8—12
ABS–Polycarbonate Alloy	8.2
Polystyrene: Styrene acrylonitrile (SAN)	8.3—12.0
Polyacetal homopolymer: 20% glass reinforced	8.5
Polyacetal copolymer: Standard	8.8
Polyacetal copolymer: High flow	8.8
Acrylic Moldings: Grades 5, 6, 8	8.8—10.5
Polycarbonate	9.5
Standard Epoxy: Cast rigid	9.5-11.5
Nylon, Type 6: General purpose	9.5—12.5
Epoxy novolacs: Cast, rigid	9.6—12.0
Polyacetal homopolymer: Standard	10
6/6 Nylon: General purpose molding	11.2—11.8
Nylon, Type 6: Cast	12.8
Polyarylsulfone	13
Polystyrene: Glass fiber -30% reinforced	14
Reinforced polyester: Sheet molding, general purpose	15—17
Polycarbonate (40% glass fiber reinforced)	18
Polystyrene: Glass fiber (30%) reinforced SAN	18
Polyacetal copolymer: 25% glass reinforced	18.5

To convert **psi** to **MPa**, multiply by 145.

Source: data compiled by J.S. Park from Charles T. Lynch, *CRC Handbook of Materials Science, Vol. 3*, CRC Press, Boca Raton, Florida, 1975 and *Engineered Materials Handbook, Vol.2*, Engineering Plastics, ASM International, Metals Park, Ohio, 1988.

Table 376. SELECTING TENSILE STRENGTHS OF POLYMERS
(SHEET 5 OF 5)

Polymer	Tensile Strength (ASTM D638) (10^3 psi)
6/10 Nylon: Glass fiber (30%) reinforced	19
6/6 Nylon: Glass fiber Molybdenum disulfide filled	19—22
Nylon, Type 6: Glass fiber (30%) reinforced	21—24
6/6 Nylon: Glass fiber reinforced	25—30
Silicone: Woven glass fabric / silicone laminate	30—35 (ASTM D651)
Epoxy: Glass cloth laminate	50-58
Epoxiy, (cycloaliphatic diepoxides): Glass cloth laminate	50—52
Epoxy novolacs: Glass cloth laminate	59.2
Epoxy: Glass cloth: High strength laminate	160
Epoxy: Glass cloth laminate: Filament wound composite	230-240 (hoop)

To convert **psi** to **MPa**, multiply by **145**.

Source: data compiled by J.S. Park from Charles T. Lynch, *CRC Handbook of Materials Science, Vol. 3*, CRC Press, Boca Raton, Florida, 1975 and *Engineered Materials Handbook, Vol.2*, Engineering Plastics, ASM International, Metals Park, Ohio, 1988.

Table 377. SELECTING COMPRESSIVE STRENGTHS OF GRAY CAST IRON BARS

ASTM Class	Compressive Strength (MPa)
20	572
25	669
30	752
35	855
40	965
50	1130
60	1293

Source: data from ASM *Metals Reference Book, Second Edition*, American Society for Metals, Metals Park, Ohio 44073, p166-167, (1984).

Table 378. SELECTING COMPRESSIVE STRENGTHS OF CERAMICS
(SHEET 1 OF 3)

Ceramic	Temperature (°C)	Compressive Strength (psi)
Thorium Dioxide (ThO_2)	1500	1.5 x10^3
Zirconium Oxide (ZrO_2)	1500	2.8 x10^3
Thorium Dioxide (ThO_2)	1400	5.7 x10^3
Aluminum Oxide (Al_2O_3)	1600	7 x10^3
Beryllium Oxide (BeO)	1600	7 x10^3
Spinel (Al_2O_3 MgO)	1600	8.5 x10^3
Trisilicon tetranitride (Si_3N_4)	25	10-100 x10^3
Trisilicon tetranitride (Si_3N_4)	1000	10-30 x10^3
Aluminum Oxide (Al_2O_3)	1500	14 x10^3
Beryllium Oxide (BeO)	1500	17 x10^3
Zirconium Oxide (ZrO_2)	1400	18.5 x10^3
Cordierite (2MgO $2Al_2O_3$ $5SiO_2$) (ρ=1.8g/cm^3)	1200	18.5 x10^3
Spinel (Al_2O_3 MgO)	1400	21.4 x10^3
Beryllium Oxide (BeO)	1400	24 x10^3
Beryllium Oxide (BeO)	1145	28.5 x10^3
Thorium Dioxide (ThO_2)	1200	28.5 x10^3
Cordierite (2MgO $2Al_2O_3$ $5SiO_2$) (ρ=2.1g/cm^3)	800	30 x10^3
Boron Nitride (BN), parallel to c axis		34.0 x10^3
Beryllium Oxide (BeO)	1000	35.5-40 x10^3
Aluminum Oxide (Al_2O_3)	1400	35.6 x10^3

To convert **psi** to **MPa**, multiply by **145.**

Source: data compiled by J.S. Park from *No. 1 Materials Index*, Peter T.B. Shaffer, Plenum Press, New York, (1964); *Smithells Metals Reference Book*, Eric A. Brandes, ed., in association with Fulmer Research Institute Ltd. 6th ed. London, Butterworths, Boston, (1983); and *Ceramic Source*, American Ceramic Society (1986-1991).

Table 378. SELECTING COMPRESSIVE STRENGTHS OF CERAMICS
(SHEET 2 OF 3)

Ceramic	Temperature (°C)	Compressive Strength (psi)
Boron Nitride (BN), parallel to a axis		45×10^3
Titanium Diboride (TiB_2)		$47\text{-}97 \times 10^3$
Cordierite ($2MgO\ 2Al_2O_3\ 5SiO_2$) (ρ=2.51g/cm^3)	25	50×10^3
Cordierite ($2MgO\ 2Al_2O_3\ 5SiO_2$) (ρ=2.3g/cm^3)	400	50×10^3
Thorium Dioxide (ThO_2)	1000	51×10^3
Beryllium Oxide (BeO)	800	64×10^3
Aluminum Oxide (Al_2O_3)	1200	71×10^3
Beryllium Oxide (BeO)	500	71×10^3
Thorium Dioxide (ThO_2)	800	71×10^3
Spinel (Al_2O_3 MgO)	1200	71×10^3
Mullite ($3Al_2O_3\ 2SiO_2$)	25	$80\text{-}190 \times 10^3$
Silicon Carbide (SiC)	25	$82\text{-}200 \times 10^3$
Aluminum Oxide (Al_2O_3)	1100	85×10^3
Thorium Dioxide (ThO_2)	600	85×10^3
Spinel (Al_2O_3 MgO)	1100	85.5×10^3
Magnesium Oxide (MgO)	room temp.	112×10^3
Beryllium Oxide (BeO)	room temp.	$114\text{-}310 \times 10^3$
Zirconium Oxide (ZrO_2)	1200	114×10^3
Aluminum Oxide (Al_2O_3)	1000	128×10^3
Titanium mononitride (TiN)		141×10^3

To convert **psi** to **MPa**, multiply by **145**.

Source: data compiled by J.S. Park from *No. 1 Materials Index*, Peter T.B. Shaffer, Plenum Press, New York, (1964); *Smithells Metals Reference Book*, Eric A. Brandes, ed., in association with Fulmer Research Institute Ltd. 6th ed. London, Butterworths, Boston, (1983); and *Ceramic Source*, American Ceramic Society (1986-1991).

Table 378. SELECTING COMPRESSIVE STRENGTHS OF CERAMICS
(SHEET 3 OF 3)

Ceramic	Temperature (°C)	Compressive Strength (psi)
Thorium Dioxide (ThO_2)	room temp.	146-214 x10^3
Thorium Dioxide (ThO_2)	400	156 x10^3
Zirconium Oxide (ZrO_2)	1000	171 x10^3
Spinel (Al_2O_3 MgO)	800	171 x10^3
Aluminum Oxide (Al_2O_3)	800	183 x10^3
Aluminum Oxide (Al_2O_3)	600	199 x10^3
Spinel (Al_2O_3 MgO)	500	199 x10^3
Zirconium Oxide (ZrO_2)	room temp.	205-300 x10^3
Aluminum Oxide (Al_2O_3)	400	214 x10^3
Zirconium Oxide (ZrO_2)	500	228 x10^3
Zirconium Monocarbide (ZrC)	room temp.	238 x10^3
Spinel (Al_2O_3 MgO)	room temp.	270 x10^3
Aluminum Oxide (Al_2O_3)	room temp.	427 x10^3
Titanium Monocarbide (TiC)	room temp.	10.9-19 x10^4
Boron Carbide (B_4C)	room temp.	41.4 x10^4
Trichromium Dicarbide (Cr_3C_2)		60 x10^4

To convert **psi** to **MPa**, multiply by 145.

Source: data compiled by J.S. Park from *No. 1 Materials Index*, Peter T.B. Shaffer, Plenum Press, New York, (1964); *Smithells Metals Reference Book*, Eric A. Brandes, ed., in association with Fulmer Research Institute Ltd. 6th ed. London, Butterworths, Boston, (1983); and *Ceramic Source*, American Ceramic Society (1986-1991).

Table 379. SELECTING COMPRESSIVE STRENGTHS OF POLYMERS (SHEET 1 OF 3)

Polymer	Compressive Strength (1000 psi)
ABS Resins; Molded, Extruded: Medium impact	0.5—11.0
Polyester, Cast Thermoset: Flexible	1—17
Styrene acrylonitrile (SAN), Glass fiber (30%) reinforced	2.3
Polystyrene, Molded: Medium impact	4—9
Polystyrene, Molded: High impact	4—9
PVC–acrylic injection molded	6.2
ABS Resins; Molded, Extruded: High impact	7.0—9.0
PVC–acrylic sheet	8.4
Chlorinated polyether	9
ABS Resins; Molded, Extruded: Heat resistant	9.3—11.0
Silicone, Molded: Fibrous (glass) reinforced silicones	10—12.5
Rubber phenolic, Molded: , chopped fabric filled	10—15
Rubber phenolic, Molded: , asbestos filled	10—20
Silicone, Molded: Granular (silica) reinforced silicones	10.6—17
Polyvinyl Chloride: Rigid—normal impact	11—12
ABS–Polycarbonate Alloy	11.1—11.8
Polystyrene, Molded: General purpose	11.5—16.0
Phenylene Oxide: SE—100	12
Rubber phenolic, Molded: woodflour or flock filled	12—20
Polyester, Cast Thermoset: Rigid	12—37
Polycarbonate	12.5
Polyester; Thermoplastic Moldings: General purpose grade	13
Phenylene oxide (Noryl): Standard	13.9—14
Silicone, Laminated with woven glass fabric	15—24
Phenolic; Molded: High shock, chopped fabric or cord filled	15—30
Polyester; Thermoplastic Moldings: Glass reinforced grades	16—18
Alkyds; Molded: Granular (high speed molding)	16—20
Phenylene Oxide: SE—1	16.4

To convert **psi** to **MPa**, multiply by **145**.

Source: data compiled by J.S. Park from Charles T. Lynch, *CRC Handbook of Materials Science, Vol. 3,* CRC Press, Boca Raton, Florida, 1975 and *Engineered Materials Handbook, Vol.2,* Engineering Plastics, ASM International, Metals Park, Ohio, 1988.

Table 379. SELECTING COMPRESSIVE STRENGTHS OF POLYMERS (SHEET 2 OF 3)

Polymer	Compressive Strength (1000 psi)
Epoxy, Standard : Cast rigid	16.5—24
Epoxy, High performance resins: Cast, rigid	17—19
Phenolic; Molded: Very high shock, glass fiber filled	17—30
Phenylene Oxide: Glass fiber reinforced	17.6—17.9
Polyarylsulfone	17.8
Polyester; Thermoplastic: Glass reinforced, self extinguishing	18
Diallyl Phthalate; Molded: Asbestos filled	18—25
Polymide: Unreinforced	18.4, 27.4
Polycarbonate (40% glass fiber reinforced)	18.5
Polystyrene, Molded: Glass fiber -30% reinforced	19
Alkyds; Molded: Putty (encapsulating)	20—25
Diallyl Phthalate; Molded: Orlon filled	20—25
Polyester: Heat and chemical resistsnt (asbestos reinforced)	20—25
Polyester: High strength, (glass fibers reinforced)	20—26
Diallyl Phthalate; Molded: Dacron filled	20—30
Phenolic, Molded: Arc resistant, mineral filled	20—30
Melamine; Molded: Glass fiber filled	20—42
Epoxy, High performance resins: Molded	22—26
Phenolic; Molded: General, woodflour and flock filled	22—36
Polyester: Sheet molding compounds, general purpose	22—36
Thermoset Carbonate: Allyl diglycol carbonate	22.5
Alkyds; Molded: Glass reinforced (heavy duty parts)	24—30
Phenolic; Molded: Shock, paper, flock, or pulp filled	24—35
Diallyl Phthalate; Molded: Glass fiber filled	25
Melamine; Molded: Cellulose electrical filled	25—35
Urea, Molded: Woodflour filled	25—35
Urea, Molded: Alpha—cellulose filled (ASTM Type l)	25—38
Melamine; Molded: Mineral filled	26—30

To convert **psi** to **MPa**, multiply by 145.

Source: data compiled by J.S. Park from Charles T. Lynch, *CRC Handbook of Materials Science, Vol. 3,* CRC Press, Boca Raton, Florida, 1975 and *Engineered Materials Handbook, Vol.2*, Engineering Plastics, ASM International, Metals Park, Ohio, 1988.

Table 379. SELECTING COMPRESSIVE STRENGTHS OF POLYMERS (SHEET 3 OF 3)

Polymer	Compressive Strength (1000 psi)
Alkyds; Molded: Rope (general purpose)	28
Epoxy novolac: Cast, rigid	30—50
Epoxy, Standard : Molded	34-38
Melamine; Molded: Unfilled	40—45
Melamine; Molded: Alpha cellulose filled	40—45
Polymide: Glass reinforced	42
Epoxy novolac: Glass cloth laminate	48—57
Epoxy, Standard : General purpose glass cloth laminate	50-60
Epoxy, High performance resins: Glass cloth laminate	67—71
Epoxy, Standard : High strength laminate	80-90 (edgewise)

To convert psi to MPa, multiply by 145.

Source: data compiled by J.S. Park from Charles T. Lynch, *CRC Handbook of Materials Science, Vol. 3*, CRC Press, Boca Raton, Florida, 1975 and *Engineered Materials Handbook, Vol.2*, Engineering Plastics, ASM International, Metals Park, Ohio, 1988.

Table 380. SELECTING YIELD STRENGTHS OF TOOL STEELS

Type	Condition	0.2% Yield Strength (MPa)
L6	Annealed	380
S7	Annealed	380
S1	Annealed	415
S5	Annealed	440
L2	Annealed	510
L2	Oil quenched from 855 •C and single tempered at 650 •C	760
L6	Oil quenched from 845 •C and single tempered at 650 •C	830
S7	Fan cooled from 940 •C and single tempered a 650 •C	1035
L6	Oil quenched from 845 •C and single tempered at 540 •C	1100
L2	Oil quenched from 855 •C and single tempered at 540 •C	1170
S5	Oil quenched from 870 •C and single tempered a 650 •C	1170
S1	Oil quenched from 930 •C and single tempered at 650 •C	1240
L2	Oil quenched from 855 •C and single tempered at 425 •C	1380
L6	Oil quenched from 845 •C and single tempered at 425 •C	1380
S5	Oil quenched from 870 •C and single tempered a 540 •C	1380
S7	Fan cooled from 940 •C and single tempered a 540 •C	1380
S7	Fan cooled from 940 •C and single tempered a 425 •C	1410
S7	Fan cooled from 940 •C and single tempered a 205 •C	1450
S1	Oil quenched from 930 •C and single tempered at 540 •C	1525
S7	Fan cooled from 940 •C and single tempered a 315 •C	1585
L2	Oil quenched from 855 •C and single tempered at 315 •C	1655
S1	Oil quenched from 930 •C and single tempered at 425 •C	1690
S5	Oil quenched from 870 •C and single tempered a 425 •C	1690
L2	Oil quenched from 855 •C and single tempered at 205 •C	1790
L6	Oil quenched from 845 •C and single tempered at 315 •C	1790
S1	Oil quenched from 930 •C and single tempered at 315 •C	1860
S5	Oil quenched from 870 •C and single tempered a 315 •C	1860
S1	Oil quenched from 930 •C and single tempered at 205 •C	1895
S5	Oil quenched from 870 •C and single tempered a 205 •C	1930

Source: Data from *ASM Metals Reference Book, Second Edition*, American Society for Metals, Metals Park, Ohio 44073, p241, (1984).

Table 381. SELECTING YIELD STRENGTHS OF DUCTILE IRONS

Specification Number	Grade or Class	Yield Strength (MPa)
MIL-I-24137(Ships)	Class C	172
MIL-I-24137(Ships)	Class B	207
ASTM A395-76; ASME SA395	60-40-18	276
ASTM A536-72, MIL-1-11466B(MR)	60-40-18	276
SAE J434c	D4018	276
ASTM A536-72, MIL-1-11466B(MR)	65-45-12	310
SAE J434c	D4512	310
MIL-I-24137(Ships)	Class A	310
ASTM A536-72, MIL-1-11466B(MR)	80-55-06	379
SAE J434c	D5506	379
ASTM A476-70(d); SAE AMS5316	80-60-03	414
ASTM A536-72, MIL-1-11466B(MR)	100-70-03	483
SAE J434c	D7003	483
ASTM A536-72, MIL-1-11466B(MR)	120-90-02	621

Source: data from *ASM Metals Reference Book, Second Edition*, American Society for Metals, Metals Park, Ohio 44073, p169, (1984).

Table 382. SELECTING YIELD STRENGTHS OF MALLEABLE IRON CASTINGS

Specification Number	Grade or Class	Yield Strength (MPa)
ASTM A197		207
ASTM A47, A338; ANSI G48.1; FED QQ–I–666c	32510	224
ASTM A602; SAE J158	M3210	224
ASTM A47, A338; ANSI G48.1; FED QQ–I–666c	35018	241
ASTM A220; ANSI C48.2; MIL–I–11444B	40010	276
ASTM A220; ANSI C48.2; MIL–I–11444B	45008	310
ASTM A220; ANSI C48.2; MIL–I–11444B	45006	310
ASTM A602; SAE J158	M4504(a)	310
ASTM A220; ANSI C48.2; MIL–I–11444B	50005	345
ASTM A602; SAE J158	M5003(a)	345
ASTM A602; SAE J158	M5503(b)	379
ASTM A220; ANSI C48.2; MIL–I–11444B	60004	414
ASTM A220; ANSI C48.2; MIL–I–11444B	70003	483
ASTM A602; SAE J158	M7002(b)	483
ASTM A220; ANSI C48.2; MIL–I–11444B	80002	552
ASTM A602; SAE J158	M8501(b)	586
ASTM A220; ANSI C48.2; MIL–I–11444B	90001	621

(a) Air quenched and tempered
(b) Liquid quenched and tempered

Source: data from *ASM Metals Reference Book, Second Edition*, American Society for Metals, Metals Park, Ohio 44073, p171, (1984).

Table 383. SELECTING YIELD STRENGTHS OF CAST ALUMINUM ALLOYS (SHEET 1 OF 3)

Alloy AA No.	Temper	Yield Strength (MPa)
443.0	F	55
B443.0	F	62
850.0	T5	75
514.0	F	85
208.0	F	97
295.0	T4	110
308.0	F	110
C443.0	F	110
242.0	T21	125
319.0	F	125
296.0	T4	130
319.0	F	130
A413.0	F	130
296.0	T7	140
356.0	T51	140
413.0	F	140
535.0	F	140
356.0	T71	145
383.0	F	150
713.0	T5	150
713.0	T5	150
242.0	T77	160
355.0	T51	160
295.0	T6	165
319.0	T6	165
355.0	T51	165
356.0	T6	165
356.0	T7	165

Source: data from *ASM Metals Reference Book, Second Edition*, American Society for Metals, Metals Park, Ohio 44073, (1984).

Table 383. SELECTING YIELD STRENGTHS OF CAST ALUMINUM ALLOYS (SHEET 2 OF 3)

Alloy AA No.	Temper	Yield Strength (MPa)
A360.0	F	165
380.0	F	165
384.0, A384.0	F	165
360.0	F	170
712.0	F	170
355.0	T6	175
296.0	T6	180
A390.0	F,T5	180
520.0	T4	180
319.0	T6	185
356.0	T6	185
355.0	T6	190
518.0	F	190
336.0	T551	195
355.0	T71	200
A390.0	F,T5	200
242.0	T571	205
355.0	T7	210
356.0	T7	210
201.0	T4	215
355.0	T71	215
295.0	T62	220
242.0	T571	235
355.0	T61	240
390.0	F	240
355.0	T7	250
A390.0	T7	250
359.0	T61	255

Source: data from *ASM Metals Reference Book, Second Edition*, American Society for Metals, Metals Park, Ohio 44073, (1984).

Table 383. SELECTING YIELD STRENGTHS OF CAST ALUMINUM ALLOYS (SHEET 3 OF 3)

Alloy AA No.	Temper	Yield Strength (MPa)
390.0	T5	260
A390.0	T7	260
771.0	T6	275
355.0	T62	280
A390.0	T6	280
354.0	T61	285
242.0	T61	290
357.0, A357.0	T62	290
359.0	T62	290
336.0	T65	295
A390.0	T6	310
206.0, A206.0	T7	345
201.0	T7	415
201.0	T6	435

Source: data from *ASM Metals Reference Book, Second Edition*, American Society for Metals, Metals Park, Ohio 44073, (1984).

Table 384. SELECTING YIELD STRENGTHS OF WROUGHT ALUMINUM ALLOYS (SHEET 1 OF 7)

Alloy	Yield Strength Temper	(MPa)
1050	0	28
1060	0	28
1350	0	28
1100	0	34
5005	0	41
3003	0	42
5457	0	48
Alclad 6061	0	48
6063	0	48
3105	0	55
5050	0	55
6061	0	55
Alclad 2014	0	69
3004	0	69
4043	0	69
6070	0	69
1060	H12	76
2024	0	76
Alclad 2024	0	76
2219	0	76
6101	Hlll	76
1350	H12	83
6066	0	83
7005	0	83
1060	H14	90
5052	0	90
5652	0	90
6063	T1	90

Source: data from *ASM Metals Reference Book, Second Edition*, American Society for Metals, Metals Park, Ohio 44073, p.299–302, (1984).

Table 384. SELECTING YIELD STRENGTHS OF WROUGHT ALUMINUM ALLOYS (SHEET 2 OF 7)

Alloy	Yield Strength Temper	(MPa)
6063	T4	90
6463	Tl	90
Alclad 7075	0	95
1350	H14	97
2014	0	97
1050	H14	105
1060	H16	105
1100	H12	105
6005	T1	105
7075	0	105
1350	H16	110
1100	H14	115
5005	H32	115
5086	0	115
5154	0	115
5154	H112	115
5254	0	115
5254	H112	115
5454	0	115
1050	H16	125
1060	H18	125
Alclad	H12	125
5454	H112	125
3105	H12	130
5005	H12	130
5086	H112	130
6009	T4	130
Alclad 6061	T4, T451	130

Source: data from *ASM Metals Reference Book, Second Edition*, American Society for Metals, Metals Park, Ohio 44073, p.299–302, (1984).

Table 384. SELECTING YIELD STRENGTHS OF WROUGHT ALUMINUM ALLOYS (SHEET 3 OF 7)

Alloy	Yield Strength Temper	(MPa)
1100	H16	140
5005	H34	140
5182	0	140
5657	H25	140
6205	Tl	140
1050	H18	145
3003	H14	145
5050	H32	145
5083	0	145
6061	T4, T451	145
6063	T5	145
6463	T5	145
1100	H18	150
3105	H14	150
5005	H14	150
5056	0	150
6351	T4	150
3105	H25	160
5456	0	160
5457	H25	160
1350	H19	165
5005	H36	165
5050	H34	165
5456	H112	165
5657	H28, H38	165
3003	H16	170
Alclad	H32	170
3105	H16	170

Source: data from *ASM Metals Reference Book, Second Edition*, American Society for Metals, Metals Park, Ohio 44073, p.299–302, (1984).

Table 384. SELECTING YIELD STRENGTHS OF WROUGHT ALUMINUM ALLOYS (SHEET 4 OF 7)

Alloy	Yield Strength Temper	(MPa)
5005	H16	170
5252	H25	170
6010	T4	170
6070	T4	170
5050	H36	180
5454	H111	180
5454	H311	180
2219	T42	185
3003	H18	185
5005	H38	185
5457	H28, H38	185
6063	T831	185
2036	T4	195
3105	H18	195
5005	H18	195
5052	H32	195
5083	H112	195
5652	H32	195
6151	T6	195
3004	H34	200
5050	H38	200
5086	H32, H116, H117	205
5154	H32	205
5254	H32	205
5454	H32	205
6066	T4, T451	205
5052	H34	215
5652	H34	215

Source: data from *ASM Metals Reference Book, Second Edition*, American Society for Metals, Metals Park, Ohio 44073, p.299–302, (1984).

Table 384. SELECTING YIELD STRENGTHS OF WROUGHT ALUMINUM ALLOYS (SHEET 5 OF 7)

Alloy	Yield Strength Temper	(MPa)
6063	T6	215
6463	T6	215
3004	H36	230
5083	H113	230
5083	H321	230
5154	H34	230
5254	H34	230
5456	H111	230
5182	H32	235
5052	H36	240
5252	H28, H38	240
5454	H34	240
5652	H36	240
6005	T5	240
6063	T83	240
2219	T31, T351	250
3004	H38	250
5083	H323, H32	250
5154	H36	250
5254	H36	250
Alclad 2014	T4	255
2218	T72	255
5052	H38	255
5086	H34	255
5456	H321, H116	255
5652	H38	255
Alclad 6061	T6, T651	255
4043	H18	270

Source: data from *ASM Metals Reference Book, Second Edition*, American Society for Metals, Metals Park, Ohio 44073, p.299–302, (1984).

Table 384. SELECTING YIELD STRENGTHS OF WROUGHT ALUMINUM ALLOYS (SHEET 6 OF 7)

Alloy	Yield Strength Temper	(MPa)
5154	H38	270
5254	H38	270
6063	T832	270
Alclad 2014	T3	275
2218	T71	275
5454	H36	275
6061	T6, T651	275
5083	H343, H34	285
5182	H34	285
6351	T6	285
2014	T4	290
Alclad 2024	T4, T351	290
2219	T62	290
6205	T5	290
2011	T3	295
6201	T6	300
2218	T61	305
2011	T8	310
Alclad 2024	T	310
5454	H38	310
6201	T81	310
2219	T37	315
4032	T6	315
7005	T6,T63,T6351	315
2024	T4, T351	325
6009	T6	325
2024	T3	345
5056	H38	345

Source: data from *ASM Metals Reference Book, Second Edition*, American Society for Metals, Metals Park, Ohio 44073, p.299–302, (1984).

Table 384. SELECTING YIELD STRENGTHS OF WROUGHT ALUMINUM ALLOYS (SHEET 7 OF 7)

Alloy	Yield Strength Temper	(MPa)
7005	T53	345
2219	T81, T851	350
6070	T6	350
6066	T6, T651	360
Alclad 2024	T361	365
2618	All	370
6262	T9	380
2024	T361	395
2219	T87	395
5182	H19(n)	395
5056	H18	405
2014	T6	415
Alclad 2014	T6	415
Alclad 2024	T81, T851	415
2048		415
7075	T73	435
2124	T851	440
Alclad 2024	T861	455
7050	T736	455
7175	T736	455
Alclad 7075	T6,T651	460
7475	T61	460
7075	T6,T651	505
7175	T66	525

Source: data from *ASM Metals Reference Book, Second Edition*, American Society for Metals, Metals Park, Ohio 44073, p.299–302, (1984).

Table 385. SELECTING YIELD STRENGTHS OF POLYMERS
(SHEET 1 OF 2)

Polymer	Yield Strength, (ASTM D638) (10^3 psi)
Polypropylene: High impact	2.8—4.3
Polystyrene, Molded: High impact	2.8—5.3
Polypropylene: Asbestos filled	3.3—8.2
Polypropylene: Flame retardant	3.6—4.2
Polystyrene, Molded: Medium impact	3.7—6.0
Nylon; Molded or Extruded: Type 8	3.9
Polypropylene: General purpose	4.5—6.0
Polystyrene, Molded: General purpose	5.0—10
Polymide: Unreinforced	5—7.5
PVC–acrylic injection molded	5.5
Nylon; Molded or Extruded: Type 12	5.5—6.5
Chlorinated Polyether	5.9
PVC–acrylic sheet	6.5
Polypropylene: Glass reinforced	7—11
Nylon, Type 6/10; Molded or Extruded: General purpose	7.1—8.5
Nylon; Molded or Extruded: Flexible copolymers	7.5—10.0
Polyester Injection Moldings: General purpose grade	7.5—8
Phenylene Oxide: SE—100	7.8
Nylon, Type 6/6: General purpose molding	8.0—11.8
Polyarylsulfone	8—12
ABS–Polycarbonate Alloy	8.2
Polyester: General purpose grade	8.2
Polycarbonate	8.5
Nylon; Molded or Extruded: Type 11	8.5
Nylon; Molded or Extruded: General purpose	8.5—12.5
Nylon, Type 6/6: General purpose extrusion	8.6—12.6
Polyacetal Copolymer: Standard	8.8
Polyacetal Copolymer: High flow	8.8

To convert **psi** to **MPa**, multiply by **145**.

Source: data compiled by J.S. Park from Charles T. Lynch, *CRC Handbook of Materials Science, Vol. 3*, CRC Press, Boca Raton, Florida, 1975 and *Engineered Materials Handbook, Vol.2*, Engineering Plastics, ASM International, Metals Park, Ohio, 1988.

Table 385. SELECTING YIELD STRENGTHS OF POLYMERS
(SHEET 2 OF 2)

Polymer	Yield Strength, (ASTM D638) (10^3 psi)
Polyphenylene sulfide: Standard	9.511
Phenylene Oxide: SE—1	9.6
Polyacetal Homopolymer: Standard	10
Phenylene oxide (Noryl): Standard	10.2
Polyester: Asbestos filled grade	12
Nylon; Molded or Extruded: Cast	12.8
Polyester: Glass reinforced grade	14
Polystyrene, Molded: Glass fiber 30% reinforced	14
Phenylene Oxide: Glass fiber reinforced	14.5—17.0
Polyester Moldings: Glass reinforced self extinguishing	17
Phenylene oxide (Noryl): Glass fiber reinforced	17—19
Polyester Injection Moldings: Glass reinforced grades	17—25
Styrene acrylonitrile (SAN): Glass fiber (30%) reinforced	18
Polyacetal Copolymer: 25% glass reinforced	18.5
Polyphenylene sulfide: 40% glass reinforced	20—21
Nylon, Type 6/6; Molded or Extruded: Glass fiber reinforced	25
Polymide: Glass reinforced	28

To convert **psi** to **MPa**, multiply by 145.

Source: data compiled by J.S. Park from Charles T. Lynch, *CRC Handbook of Materials Science, Vol. 3*, CRC Press, Boca Raton, Florida, 1975 and *Engineered Materials Handbook, Vol.2*, Engineering Plastics, ASM International, Metals Park, Ohio, 1988.

Table 386. SELECTING COMPRESSIVE YIELD STRENGTHS OF POLYMERS (SHEET 1 OF 2)

Polymer	Compressive Yield Strength (ASTM D690 or D695) (0.1% offset, 1000 psi)
Polytetrafluoroethylene (PTFE)	0.7—1.8
Ceramic reinforced (PTFE)	1.4—1.8
Fluorinated ethylene propylene(FEP)	1.6
Polytrifluoro chloroethylene (PTFCE)	2
Cellulose Acetate Butyrate, ASTM Grade: S2	2.6—4.3
6/10 Nylon: General purpose	3.0
Cellulose Acetate, ASTM Grade: S2—1	3.15—6.1
Cellulose Acetate, ASTM Grade: MS—1, MS—2	3.2—7.2
Cellulose Acetate, ASTM Grade: H2—1	4.3—9.6
Polypropylene: High impact	4.4
Cellulose Acetate, ASTM Grade: MH—1, MH—2	4.4—8.4
Polyacetal Homopolymer: 22% TFE reinforced	4.5
Polyacetal Copolymer: Standard	4.5
Polyacetal Copolymer: High flow	4.5
6/6 Nylon: General purpose molding	4.9
6/6 Nylon: General purpose extrusion	4.9
Cellusose Acetate Propionate, ASTM Grade: 3	4.9—5.8
Polyacetal Homopolymer: Standard	5.2
Polyacetal Homopolymer: 20% glass reinforced	5.2
Cellulose Acetate Butyrate, ASTM Grade: MH	5.3—7.1
Polypropylene: General purpose	5.5—6.5
Cellusose Acetate Propionate, ASTM Grade: 1	6.2—7.3
Cellulose Acetate, ASTM Grade: H4—1	6.5—10.6
Polypropylene: Glass reinforced	6.5—7
Polypropylene: Asbestos filled	7
Acrylic Moldings: High impact grade	7.3—12.0
Cellulose Acetate Butyrate, ASTM Grade: H4	8.8
Nylon, Type 6: General purpose	9.7

To convert from **psi** to **MPa**, multiply by **145**.

Source: data compiled by J.S. Park from Charles T. Lynch, CRC *Handbook of Materials Science, Vol. 3,* CRC Press, Boca Raton, Florida, 1975 and *Engineered Materials Handbook, Vol.2,* Engineering Plastics, ASM International, Metals Park, Ohio, 1988.

Table 386. SELECTING COMPRESSIVE YIELD STRENGTHS OF POLYMERS (SHEET 2 OF 2)

Polymer	Compressive Yield Strength (ASTM D690 or D695) (0.1% offset, 1000 psi)
Polyvinyl Chloride: Rigid—normal impact	10—11
Acrylic Cast Resin Sheets, Rods: General purpose, type I	12—14
Polyvinylidene— fluoride (PVDF)	12.8—14.2
Nylon, Type 6: Cast	14
Acrylic Cast Resin Sheets, Rods: General purpose, type II	14—18
Acrylic Moldings: Grades 5, 6, 8	14.5—17
6/10 Nylon: Glass fiber (30%) reinforced	18
Nylon, Type 6: Glass fiber (30%) reinforced	19—20
6/6 Nylon: Glass fiber reinforced	20—24
Vinylidene chloride	75—85

To convert from psi to MPa, multiply by 145.

Source: data compiled by J.S. Park from Charles T. Lynch, CRC *Handbook of Materials Science, Vol. 3,* CRC Press, Boca Raton, Florida, 1975 and *Engineered Materials Handbook, Vol.2,* Engineering Plastics, ASM International, Metals Park, Ohio, 1988.

Table 387. SELECTING FLEXURAL STRENGTHS OF POLYMERS

(SHEET 1 OF 4)

Polymer	Flexural Strength (ASTM D790) (10^3 psi)
Epoxy, Standard: Cast flexible	1.2—12.7
Cellulose Acetate Butyrate, ASTM Grade: S2	2.5—3.95 (yield)
Fluorinated ethylene propylene(FEP)	3 (0.1% offset)
Nylon, Type 6: Flexible copolymers	3.4—16.4
Polytrifluoro chloroethylene (PTFCE)	3.5 (0.1% offset)
Cellulose Acetate, ASTM Grade: S2—1	3.5—5.7 (yield)
Cellulose Acetate, ASTM Grade: MS—1, MS—2	3.8—7.1 (yield)
Polyesters, Cast Thermoset: Flexible	4—16
Polypropylene: High impact	4.1 (yield)
Cellulose Acetate, ASTM Grade: MH—1, MH—2	4.4—8.65 (yield)
Chlorinated polyether	5 (0.1% offset)
ABS Resins; Molded or Extruded: Low temperature impact	5—8
Cellusose Acetate Propionate, ASTM Grade: 3	5.6—6.2 (yield)
Cellulose Acetate Butyrate, ASTM Grade: MH	5.6—6.7 (yield)
Cellulose Acetate, ASTM Grade: H2—1	6.0—10.0 (yield)
ABS Resins; Molded or Extruded: Very high impact	6.0—9.8
Silicone: Granular (silica) reinforced	6—10
Melamines, Molded: Cellulose filled, electrical	6—15
Reinforced polyester: High strength (glass fibers)	6—26
Polypropylene: General purpose	6—7 (yield)
Polymide: Unreinforced	6.6—11
Cellusose Acetate Propionate, ASTM Grade: 1	6.8—7.9 (yield)
Rubber phenolic—chopped fabric filled	7
Rubber phenolic—asbestos filled	7
Alkyd, Molded: Granular (high speed molding)	7—10
Rubber phenolic—woodflour or flock filled	7—12
Diallyl Phthalate, Molded: Orlon filled	7.5—10.5
Urea, Molded: Woodflour filled	7.5—12.0

To convert from **psi** to **MPa**, multiply by **145**.

Source: data compiled by J.S. Park from Charles T. Lynch, CRC *Handbook of Materials Science, Vol. 3,* CRC Press, Boca Raton, Florida, 1975 and *Engineered Materials Handbook, Vol.2,* Engineering Plastics, ASM International, Metals Park, Ohio, 1988.

Table 387. Selecting Flexural Strengths of Polymers

(Sheet 2 of 4)

Polymer	Flexural Strength (ASTM D790) (10^3 psi)
Urea, Molded: Cellulose filled (ASTM Type 2)	7.5—13
Polypropylene: Asbestos filled	7.5—9 (yield)
ABS Resins; Molded or Extruded: High impact	7.5—9.5
6/10 Nylon: General purpose	8
Phenolic: Shock: paper, flock, or pulp filled	8.0—11.5
Diallyl Phthalate, Molded: Asbestos filled	8—10
Alkyd, Molded: Putty (encapsulating)	8—11
Polypropylene: Glass reinforced	8—11 (yield)
Phenolic: High shock, chopped fabric or cord filled	8—15
Urea, Molded: Alpha—cellulose filled (ASTM Type l)	8—18
Polyesters, Cast Thermoset: Rigid	8—24
Cellulose Acetate, ASTM Grade: H4—1	8.1—11.15 (yield)
Phenolic: General, woodflour and flock filled	8.5—12
Polyvinylidene— fluoride (PVDF)	8.6—10.8 (0.1% offset)
PVC–acrylic injection molded	8.7
Acrylic Moldings: High impact grade	8.7—12.0
Cellulose Acetate Butyrate, ASTM Grade: H4	9 (yield)
Diallyl Phthalate, Molded: Dacron filled	9—11.5
Melamines, Molded: Unfilled	9.5—14
ABS Resins; Molded or Extruded: Medium impact	9.9—11.8
Epoxy, High performance resins: Molded	10—12
Phenolic: Arc resistant—mineral filled	10—13
Reinforced polyester: Heat and chemical resistant (asbestos)	10—13
Polystyrene: General purpose	10—15
Diallyl Phthalate, Molded: Glass fiber filled	10—18
Phenolic: Very high shock, glass fiber filled	10—45
PVC–acrylic sheet	10.7
ABS Resins; Molded or Extruded: Heat resistant	11.0—12.0

To convert from **psi** to **MPa**, multiply by 145.

Source: data compiled by J.S. Park from Charles T. Lynch, CRC *Handbook of Materials Science, Vol. 3,* CRC Press, Boca Raton, Florida, 1975 and *Engineered Materials Handbook, Vol.2,* Engineering Plastics, ASM International, Metals Park, Ohio, 1988.

Table 387. SELECTING FLEXURAL STRENGTHS OF POLYMERS

(SHEET 3 OF 4)

Polymer	Flexural Strength (ASTM D790) (10^3 psi)
Epoxy, High performance resins: Cast, rigid	11—16
Melamines, Molded: Alpha cellulose filled	11—16
Polyvinyl Chloride And Copolymers: Rigid—normal impact	11—16
Polyester Injection Moldings: General purpose grade	12
Epoxy novolacs: Cast, rigid	12—13
Acrylic, Cast Resin Sheets, Rods: General purpose, type I	12—14
Alkyd, Molded: Glass reinforced (heavy duty parts)	12—17
Polyester Injection Moldings: General purpose grade	12.8
Phenylene Oxide: SE—100	12.8
Polyacetal Copolymer: Standard	13
Polyacetal Copolymer: High flow	13
Polycarbonate	13.5
Phenylene Oxide: SE—1	13.5
Epoxy, Standard: Cast rigid	14—18
Melamines, Molded: Glass fiber filled	14—18
Polyacetal Homopolymer: Standard	14.1
ABS–Polycarbonate Alloy	14.3
Chlorinated polyvinyl chloride	14.5
Acrylic Moldings: Grades 5, 6, 8	15—16
Acrylic, Cast Resin Sheets, Rods: General purpose, type II	15—17
Vinylidene chloride	15—17
Phenylene oxides (Noryl): Standard	15.4
Silicone: Fibrous (glass) reinforced	16—19
Polyarylsulfone	16.1—17.2
Nylon, Type 6: Cast	16.5
Polystyrene: Glass fiber —30% reinforced	17
Melamines, Molded: Alpha mineral filled	18—10
Polyester Injection Moldings: Glass reinforced grade	19

To convert from **psi** to **MPa**, multiply by **145.**

Source: data compiled by J.S. Park from Charles T. Lynch, CRC *Handbook of Materials Science, Vol. 3,* CRC Press, Boca Raton, Florida, 1975 and *Engineered Materials Handbook, Vol.2*, Engineering Plastics, ASM International, Metals Park, Ohio, 1988.

Table 387. SELECTING FLEXURAL STRENGTHS OF POLYMERS
(SHEET 4 OF 4)

Polymer	Flexural Strength (ASTM D790) (10^3 psi)
Polyester Injection Moldings: Asbestos—filled grade	19
Alkyd, Molded: Rope (general purpose)	19—20
Epoxy, Standard: Molded	19—22
Polyphenylene sulfide: Standard	20
Phenylene Oxide: Glass fiber reinforced	20.5—22
Styrene acrylonitrile (SAN): Glass fiber (30%) reinforced	22
Polyester Injection Moldings: Glass reinforced grades	22—24
6/10 Nylon: Glass fiber (30%) reinforced	23
Polyester Injection Moldings: Glass reinforced self extinguishing	23
Phenylene oxides (Noryl): Glass fiber reinforced	25—28
6/6 Nylon: Glass fiber Molybdenum disulfide filled	26—28
Reinforced polyester sheet molding: general purpose	26—32
Nylon, Type 6: Glass fiber (30%) reinforced	26—34
6/6 Nylon: Glass fiber reinforced	26—35
Polycarbonate (40% glass fiber reinforced)	27
Polyacetal Copolymer: 25% glass reinforced	28
Silicone: Woven glass fabric/ silicone laminate	33—47
Polyphenylene sulfide: 40% glass reinforced	37
Polymide: Glass reinforced	56
Epoxy, High performance resins: Glass cloth laminate	70—72
Epoxy, Standard: General purpose glass cloth laminate	80—90
Epoxy novolacs: Glass cloth laminate	84—89
Epoxy, Standard: High strength laminate	165—177
Epoxy, Standard: Filament wound composite	170—180
Nylon, Type 6: General purpose	Unbreakable
6/6 Nylon: General purpose molding	Unbreakable

To convert from **psi** to **MPa**, multiply by 145.

Source: data compiled by J.S. Park from Charles T. Lynch, CRC *Handbook of Materials Science, Vol. 3,* CRC Press, Boca Raton, Florida, 1975 and *Engineered Materials Handbook, Vol.2,* Engineering Plastics, ASM International, Metals Park, Ohio, 1988.

Table 388. SELECTING SHEAR STRENGTHS OF WROUGHT ALUMINUM ALLOYS (SHEET 1 OF 6)

Alloy AA No.	Temper	Shear Strength (MPa)
1060	0	48
1060	H12	55
1350	0	55
7072	0	55
1050	0	62
1060	H14	62
1100	0	62
1350	H12	62
7072	H12	62
1050	H14	69
1060	H16	69
1100	H12	69
1350	H14	69
6063	0	69
7072	H14	69
1050	H16	76
1060	H18	76
1100	H14	76
1350	H16	76
3003	0	76
5005	0	76
Alclad 6061	0	76
1050	H18	83
1100	H16	83
Alclad	H12	83
3105	0	83
5457	0	83
6061	0	83

Source: Data from *ASM Metals Reference Book, Second Edition*, American Society for Metals, Metals Park, Ohio 44073, (1984).

Table 388. SELECTING SHEAR STRENGTHS OF WROUGHT ALUMINUM ALLOYS (SHEET 2 OF 6)

Alloy AA No.	Temper	Shear Strength (MPa)
1100	H18	90
3003	H14	97
3105	H12	97
5005	H12	97
5005	H14	97
5005	H32	97
5005	H34	97
5657	H25	97
6063	T1	97
6066	0	97
6070	0	97
6463	T1	97
1350	H19	105
3003	H16	105
3105	H14	105
3105	H25	105
5005	H16	105
5005	H36	105
5050	0	105
5657	H28, H38	105
3003	H18	110
3004	0	110
3105	H16	110
5005	H18	110
5005	H38	110
5457	H25	110
Alclad	H32	115
3105	H18	115

Source: Data from *ASM Metals Reference Book, Second Edition*, American Society for Metals, Metals Park, Ohio 44073, (1984).

Table 388. SELECTING SHEAR STRENGTHS OF WROUGHT ALUMINUM ALLOYS (SHEET 3 OF 6)

Alloy AA No.	Temper	Shear Strength (MPa)
5050	H32	115
6063	T5	115
6463	T5	115
7005	0	117
2014	0	125
Alclad 2014	0	125
2024	0	125
Alclad 2024	0	125
3004	H34	125
5050	H34	125
5052	0	125
5457	H28, H38	125
5652	0	125
6063	T831	125
5050	H36	130
3004	H36	140
5050	H38	140
5052	H32	140
5652	H32	140
6151	T6	140
3004	H38	145
5052	H34	145
5252	H25	145
5652	H34	145
5154	0	150
5154	H32	150
5182	0	150
5254	0	150

Source: Data from *ASM Metals Reference Book, Second Edition*, American Society for Metals, Metals Park, Ohio 44073, (1984).

Table 388. SELECTING SHEAR STRENGTHS OF WROUGHT ALUMINUM ALLOYS (SHEET 4 OF 6)

Alloy AA No.	Temper	Shear Strength (MPa)
5254	H32	150
6009	T4	150
Alclad 6061	T4, T451	150
6063	T6	150
6063	T83	150
6463	T6	150
7075	0	150
Alclad 7075	0	150
5052	H36	160
5086	0	160
5252	H28, H38	160
5454	0	160
5454	H111	160
5454	H112	160
5454	H311	160
5652	H36	160
5052	H38	165
5154	H34	165
5254	H34	165
5454	H32	165
5652	H38	165
6061	T4, T451	165
5083	0	170
5056	0	180
5154	H36	180
5254	H36	180
5454	H34	180
5086	H34	185

Source: Data from *ASM Metals Reference Book, Second Edition*, American Society for Metals, Metals Park, Ohio 44073, (1984).

Table 388. SELECTING SHEAR STRENGTHS OF WROUGHT ALUMINUM ALLOYS (SHEET 5 OF 6)

Alloy AA No.	Temper	Shear Strength (MPa)
Alclad 6061	T6, T651	185
6063	T832	185
5154	H38	195
5254	H38	195
6066	T4, T451	200
6351	T6	200
2218	T72	205
5456	H321, H116	205
6005	T5	205
6061	T6, T651	205
6070	T4	205
6205	T5	205
7005	T6,T63,T6351	214
2011	T3	220
5056	H38	220
7005	T53	221
5056	H18	235
6066	T6, T651	235
6070	T6	235
2011	T8	240
6262	T9	240
Alclad 2014	T3	255
Alclad 2014	T4	255
2014	T4	260
2618	All	260
4032	T6	260
7475	T7351	270
7475	T7651	270

Source: Data from *ASM Metals Reference Book, Second Edition*, American Society for Metals, Metals Park, Ohio 44073, (1984).

Table 388. SELECTING SHEAR STRENGTHS OF WROUGHT ALUMINUM ALLOYS (SHEET 6 OF 6)

Alloy AA No.	Temper	Shear Strength (MPa)
Alclad 2024	T	275
Alclad 2024	T4, T351	275
Alclad 2024	T81, T851	275
Alclad 2014	T6	285
2024	T3	285
2024	T4, T351	285
Alclad 2024	T361	285
2014	T6	290
2024	T361	290
Alclad 2024	T861	290
7175	T736	290
7475	T651	295
Alclad 7075	T6,T651	315
7175	T66	325
7075	T6,T651	330

Source: Data from *ASM Metals Reference Book, Second Edition,* American Society for Metals, Metals Park, Ohio 44073, (1984).

Table 389. SELECTING TORSIONAL SHEAR STRENGTHS OF GRAY CAST IRON BARS

ASTM Class	Torsional Shear Strength (MPa)
20	179
25	220
30	276
35	334
40	393
50	503
60	610

Source: data from *ASM Metals Reference Book, Second Edition*, American Society for Metals, Metals Park, Ohio 44073, p166-167, (1984).

Table 390. SELECTING HARDNESS OF TOOL STEELS

Type	Condition	Hardness (HRC)
S7	Annealed	95 HRB
L2	Annealed	96 HRB
S1	Annealed	96 HRB
S5	Annealed	96 HRB
L2	Oil quenched from 855 •C and single tempered at 650 •C	30
L6	Oil quenched from 845 •C and single tempered at 315 •C 650 •C	32
S5	Oil quenched from 870 •C and single tempered at 650 •C	37
S7	Fan cooled from 940 •C and single tempered at 650 •C	39
L2	Oil quenched from 855 •C and single tempered at 540 •C	41
L6	Oil quenched from 845 •C and single tempered at 315 •C 540 •C	42
S1	Oil quenched from 930 •C and single tempered at 650 •C	42
L6	Oil quenched from 845 •C and single tempered at 315 •C 425 •C	46
L2	Oil quenched from 855 •C and single tempered at 425 •C	47
S1	Oil quenched from 930 •C and single tempered at 540 •C	47.5
S5	Oil quenched from 870 •C and single tempered at 540 •C	48
S1	Oil quenched from 930 •C and single tempered at 425 •C	50.5
S7	Fan cooled from 940 •C and single tempered at 540 •C	51
L2	Oil quenched from 855 •C and single tempered at 315 •C	52
S5	Oil quenched from 870 •C and single tempered at 425 •C	52
S7	Fan cooled from 940 •C and single tempered at 425 •C	53
L2	Oil quenched from 855 •C and single tempered at 205 •C	54
L6	Oil quenched from 845 •C and single tempered at 315 •C	54
S1	Oil quenched from 930 •C and single tempered at 315 •C	54
S7	Fan cooled from 940 •C and single tempered at 315 •C	55
S1	Oil quenched from 930 •C and single tempered at 205 •C	57.5
S5	Oil quenched from 870 •C and single tempered at 315 •C	58
S7	Fan cooled from 940 •C and single tempered at 205 •C	58
S5	Oil quenched from 870 •C and single tempered at 205 •C	59

Source: Data from ASM *Metals Reference Book, Second Edition*, American Society for Metals, Metals Park, Ohio 44073, p241, (1984).

Table 391. SELECTING HARDNESS OF GRAY CAST IRONS

SAE grade	Hardness (HB)
G2500	170 to 229
G2500a	170 to 229
G1800	187 max
G3000	187 to 241
C3500	207 to 255
G3500b	207 to 255
G3500c	207 to 255
G4000	217 to 269
G4000d	241 to 321

Source: data from *ASM Metals Reference Book, Second Edition*, American Society for Metals, Metals Park, Ohio 44073, p166-167, (1984).

Table 392. SELECTING HARDNESS OF GRAY CAST IRON BARS

Grey Cast Iron Bars ASTM Class	Hardness (HB)
20	156
25	174
30	210
35	212
40	235
50	262
60	302

Source: data from *ASM Metals Reference Book, Second Edition*, American Society for Metals, Metals Park, Ohio 44073, p166-167, (1984).

Table 393. SELECTING HARDNESS OF DUCTILE IRONS

Specification Number	Grade or Class	Hardness (HB)
ASTM A395-76; ASME SA395	60-40-18	143-187
SAE J434c	D4512	156-217
SAE J434c	D4018	170 max
MIL-I-24137(Ships)	Class C	175 max
SAE J434c	D5506	187-255
MIL-I-24137(Ships)	Class A	190 max
MIL-I-24137(Ships)	Class B	190 max
ASTM A476-70(d); SAE AMS5316	80-60-03	201 min
SAE J434c	D7003	241-302

Source: data from *ASM Metals Reference Book, Second Edition*, American Society for Metals, Metals Park, Ohio 44073, p169, (1984).

Table 394. SELECTING HARDNESS OF MALLEABLE IRON CASTINGS

Specification Number	Grade or Class	Hardness (HB)
ASTM A220; ANSI C48.2; MIL–I–11444B	40010	149–197
ASTM A47, A338; ANSI G48.1; FED QQ–I–666c	32510	156 max
ASTM A47, A338; ANSI G48.1; FED QQ–I–666c	35018	156 max
ASTM A197		156 max
ASTM A602; SAE J158	M3210	156 max
ASTM A220; ANSI C48.2; MIL–I–11444B	45008	156–197
ASTM A220; ANSI C48.2; MIL–I–11444B	45006	156–207
ASTM A602; SAE J158	M4504(a)	163–217
ASTM A220; ANSI C48.2; MIL–I–11444B	50005	179–229
ASTM A602; SAE J158	M5003(a)	187–241
ASTM A602; SAE J158	M5503(b)	187–241
ASTM A220; ANSI C48.2; MIL–I–11444B	60004	197–241
ASTM A220; ANSI C48.2; MIL–I–11444B	70003	217–269
ASTM A602; SAE J158	M7002(b)	229–269
ASTM A220; ANSI C48.2; MIL–I–11444B	80002	241–285
ASTM A602; SAE J158	M8501(b)	269–302
ASTM A220; ANSI C48.2; MIL–I–11444B	90001	269–321

(a) Air quenched and tempered
(b) Liquid quenched and tempered

Source: data from *ASM Metals Reference Book, Second Edition*, American Society for Metals, Metals Park, Ohio 44073, p171, (1984).

Table 395. SELECTING HARDNESS OF WROUGHT ALUMINUM ALLOYS (SHEET 1 OF 5)

Alloy AA No.	Temper	Hardness (BHN)
1060	0	19
7072	0	20
1060	H12	23
1100	0	23
6063	0	25
1060	H14	26
1100	H12	28
3003	0	28
5005	0	28
7072	H12	28
1060	H16	30
6061	0	30
1100	H14	32
5457	0	32
7072	H14	32
1060	H18	35
Alclad	H12	35
6070	0	35
5005	H32	36
5050	0	36
1100	H16	38
3003	H14	40
5657	H25	40
5005	H34	41
6063	T1	42
6463	T1	42
6066	0	43
1100	H18	44

Source: data from *ASM Metals Reference Book, Second Edition*, American Society for Metals, Metals Park, Ohio 44073, (1984).

Table 395. SELECTING HARDNESS OF WROUGHT ALUMINUM ALLOYS (SHEET 2 OF 5)

Alloy AA No.	Temper	Hardness (BHN)
2014	0	45
3004	0	45
5005	H36	46
5050	H32	46
2024	0	47
3003	H16	47
5052	0	47
5652	0	47
5457	H25	48
5657	H28, H38	50
5005	H38	51
Alclad	H32	52
5050	H34	53
3003	H18	55
5457	H28, H38	55
5050	H36	58
5154	0	58
5182	0	58
5254	0	58
5052	H32	60
5652	H32	60
6063	T5	60
6463	T5	60
7075	0	60
5454	0	62
5454	H112	62
3004	H34	63
5050	H38	63

Source: data from *ASM Metals Reference Book, Second Edition,* American Society for Metals, Metals Park, Ohio 44073, (1984).

Table 395. SELECTING HARDNESS OF WROUGHT ALUMINUM ALLOYS (SHEET 3 OF 5)

Alloy AA No.	Temper	Hardness (BHN)
5154	H112	63
5254	H112	63
5056	0	65
6061	T4, T451	65
6205	T1	65
5154	H32	67
5254	H32	67
5052	H34	68
5252	H25	68
5652	H34	68
3004	H36	70
5454	H111	70
5454	H311	70
6009	T4	70
6063	T831	70
6151	T6	71
5052	H36	73
5154	H34	73
5254	H34	73
5454	H32	73
5652	H36	73
6063	T6	73
6463	T6	74
5252	H28, H38	75
6010	T4	76
3004	H38	77
5052	H38	77
5652	H38	77

Source: data from *ASM Metals Reference Book, Second Edition*, American Society for Metals, Metals Park, Ohio 44073, (1984).

Table 395. SELECTING HARDNESS OF WROUGHT ALUMINUM ALLOYS (SHEET 4 OF 5)

Alloy AA No.	Temper	Hardness (BHN)
5154	H36	78
5254	H36	78
5154	H38	80
5254	H38	80
5454	H34	81
6063	T83	82
5456	H321, H116	90
6066	T4, T451	90
6070	T4	90
6201	T6	90
2011	T3	95
2218	T72	95
6005	T5	95
6061	T6, T651	95
6063	T832	95
6205	T5	95
6351	T6	95
2011	T8	100
5056	H38	100
2014	T4	105
2218	T71	105
5056	H18	105
2218	T61	115
2024	T3	120
2024	T4, T351	120
4032	T6	120
6066	T6, T651	120
6070	T6	120

Source: data from *ASM Metals Reference Book, Second Edition*, American Society for Metals, Metals Park, Ohio 44073, (1984).

Table 395. SELECTING HARDNESS OF WROUGHT ALUMINUM ALLOYS (SHEET 5 OF 5)

Alloy AA No.	Temper	Hardness (BHN)
6262	T9	120
2024	T361	130
2014	T6	135
7049	T73	135
7175	T736	145
7075	T6,T651	150
7175	T66	150

Source: data from *ASM Metals Reference Book, Second Edition*, American Society for Metals, Metals Park, Ohio 44073, (1984).

Table 396. SELECTING HARDNESS OF CERAMICS
(SHEET 1 OF 6)

Ceramic	Hardness
Tantalum Monocarbide (TaC)	Brinell: 840
Titanium Oxide (TiO_2)	Knoop: 713-1121 kg/mm^2
Trisilicon tetranitride (Si_3N_4) (α)	Knoop: 815-1936kg/mm^2
Zirconium Oxide (ZrO_2) (partially stabilized)	Knoop: 1019-1121 kg/mm^2
Zirconium Oxide (ZrO_2)(fully stabilized)	Knoop: 1019-1529 kg/mm^2
Trichromium Dicarbide (Cr_3C_2)	Knoop: 1019-1834 kg/mm^2
Hafnium Monocarbide (HfC)	Knoop: 1790-1870 kg/mm^2
Zirconium Monocarbide (ZrC)	Knoop: 2138 kg/mm^2
Silicon Carbide (SiC) (cubic, CVD)	Knoop: 2853-4483 kg/mm^2
Dichromium Trioxide (Cr_2O_3)	Knoop: 2955 kg/mm^2
Zirconium Mononitride (ZrN)	Knoop 30g: 1983 kg/mm^2
Titanium mononitride (TiN)	Knoop 30g: 2160 kg/mm^2
Tantalum Diboride (TaB_2)	Knoop 30g: 2537 kg/mm^2
Titanium Diboride (TiB_2)	Knoop 30g: 3370 kg/mm^2
Tantalum Monocarbide (TaC)	Knoop 50g: 1800-1952 kg/mm^2
Calcium Oxide (CaO)	Knoop 100g: 560 kg/mm^2
Uranium Dioxide (UO_2)	Knoop 100g: 600 kg/mm^2
Silicon Dioxide (SiO_2) (parallel to optical axis)	Knoop 100g: 710 kg/mm^2
Silicon Dioxide (SiO_2) (normal to optical axis)	Knoop 100g: 790 kg/mm^2
Tantalum Monocarbide (TaC)	Knoop 100g: 825 kg/mm^2
Thorium Dioxide (ThO_2)	Knoop 100g: 945 kg/mm^2
Tungsten Disilicide (WSi_2)	Knoop 100g: 1090 kg/mm^2
Zirconium Oxide (ZrO_2)	Knoop 100g: 1200 kg/mm^2

Source: data compiled by J.S. Park from *No. 1 Materials Index*, Peter T.B. Shaffer, Plenum Press, New York, (1964); *Smithells Metals Reference Book*, Eric A. Brandes, ed., in association with Fulmer Research Institute Ltd. 6th ed. London, Butterworths, Boston, (1983); and *Ceramic Source*, American Ceramic Society (1986-1991).

Table 396. SELECTING HARDNESS OF CERAMICS
(SHEET 2 OF 6)

Ceramic	Hardness
Aluminum Nitride (AlN)	Knoop 100g: 1225-1230 kg/mm^2
Molybdenum Disilicide ($MoSi_2$)	Knoop 100g: 1257 kg/mm^2
Beryllium Oxide (BeO)	Knoop 100g: 1300 kg/mm^2
Zirconium Mononitride (ZrN)	Knoop 100g: 1510 kg/mm^2
Zirconium Diboride (ZrB_2)	Knoop 100g: 1560 kg/mm^2
Chromium Diboride (CrB2)	Knoop 100g: 1700 kg/mm^2
Titanium mononitride (TiN)	Knoop 100g: 1770 kg/mm^2
Tungsten Monocarbide (WC)	Knoop 100g: 1870-1880 kg/mm^2
Aluminum Oxide (Al_2O_3)	Knoop 100g: 2000-2050 kg/mm^2
Titanium Monocarbide (TiC)	Knoop 100g: 2470 kg/mm^2
Silicon Carbide (SiC)	Knoop 100g: 2500-2550 kg/mm^2
Tantalum Diboride (TaB_2)	Knoop 100g: 2615 ± 120 kg/mm^2
Titanium Diboride (TiB_2)	Knoop 100g: 2710-3000 kg/mm^2
Silicon Carbide (SiC)	Knoop 100g: 2745 kg/mm^2 (green)
Boron Carbide (B_4C)	Knoop 100g: 2800 kg/mm^2
Silicon Carbide (SiC)	Knoop 100g: 2960 kg/mm^2 (black)
Titanium Diboride (TiB_2) (single crystal)	Knoop 100g: 3250±100 kg/mm^2
Zirconium Diboride (ZrB_2) (single crystal)	Knoop 160g: 2000 kg/mm^2
Zirconium Diboride (ZrB_2)	Knoop 160g: 2100 kg/mm^2
Hafnium Diboride (HfB2) (polycrystalline)	Knoop 160g: 2400kg/mm at 24 °C
Titanium Diboride (TiB_2)	Knoop 160g: 3500 kg/mm^2
Hafnium Diboride (HfB2) (single crystal)	Knoop 160g: 3800kg/mm at 24 °C

Source: data compiled by J.S. Park from *No. 1 Materials Index*, Peter T.B. Shaffer, Plenum Press, New York, (1964); *Smithells Metals Reference Book*, Eric A. Brandes, ed., in association with Fulmer Research Institute Ltd. 6th ed. London, Butterworths, Boston, (1983); and *Ceramic Source*, American Ceramic Society (1986-1991).

Table 396. SELECTING HARDNESS OF CERAMICS
(SHEET 3 OF 6)

Ceramic	Hardness
Titanium Monocarbide (TiC)	Knoop 1000g: 1905 kg/mm^2
Boron Carbide (B_4C)	Knoop 1000g: 2230 kg/mm^2
Tantalum Diboride (TaB_2)	Micro: 1700 kg/mm^2
Zirconium Monocarbide (ZrC)	Micro: 2090 kg/mm^2
Titanium Monocarbide (TiC)	Micro 20g: 3200 kg/mm^2
Molybdenum Disilicide ($MoSi_2$)	Micro 50g: 1200 kg/mm^2
Tungsten Disilicide (WSi_2)	Micro 50g: 1260 kg/mm^2
Molybdenum Disilicide ($MoSi_2$)	Micro 100g: 1290 kg/mm^2
Chromium Diboride (CrB2)	Micro 100g: 1800 kg/mm^2
Boron Nitride (BN) (hexagonal)	Mohs: 2
Aluminum Nitride (AlN)	Mohs: 5-5.5
Magnesium Oxide (MgO)	Mohs: 5.5
Uranium Dioxide (UO_2)	Mohs: 6-7
Sillimanite (Al_2O_3 SiO_2)	Mohs: 6-7
Thorium Dioxide (ThO_2)	Mohs: 6.5
Zirconium Oxide (ZrO_2)	Mohs: 6.5
Mullite ($3Al_2O_3$ $2SiO_2$)	Mohs: 7.5
Zircon (SiO_2 ZrO_2)	Mohs: 7.5
Zirconium Mononitride (ZrN)	Mohs: 8+
Titanium mononitride (TiN)	Mohs: 8-10
Aluminum Oxide (Al_2O_3) (single crystal)	Mohs: 9
Trisilicon tetranitride (Si_3N_4)	Mohs: 9+
Silicon Carbide (SiC)	Mohs: 9.2

Source: data compiled by J.S. Park from *No. 1 Materials Index*, Peter T.B. Shaffer, Plenum Press, New York, (1964); *Smithells Metals Reference Book*, Eric A. Brandes, ed., in association with Fulmer Research Institute Ltd. 6th ed. London, Butterworths, Boston, (1983); and *Ceramic Source*, American Ceramic Society (1986-1991).

Table 396. SELECTING HARDNESS OF CERAMICS
(SHEET 4 OF 6)

Ceramic	Hardness
Beryllium Oxide (BeO)	R45N: 64-67
Mullite ($3Al_2O_3$ $2SiO_2$)	R45N: 71
Aluminum Oxide (Al_2O_3)	R45N: 78-90
Aluminum Nitride (AlN) (thin film)	Rockwell 15N: 94.0
Aluminum Nitride (AlN) (thick film)	Rockwell 15N: 94.5
Tungsten Monocarbide (WC) (6% Co, 1-3μm grain size)	Rockwell A: 81.4 ± 0.4
Tungsten Monocarbide (WC) (24% Co, 1-3μm grain size)	Rockwell A: 86.9 ± 0.6
Zirconium Diboride (ZrB_2)	Rockwell A: 87-89
Tungsten Monocarbide (WC) (6% Co, 3-6μm grain size)	Rockwell A: 87.3 ± 0.5
Titanium Monocarbide (TiC) (98.6% density)	Rockwell A: 88-89
Tungsten Monocarbide (WC) (6% Co, 2-4μm grain size)	Rockwell A: 88.6 ± 0.5
Tantalum Diboride (TaB_2)	Rockwell A: 89
Tantalum Monocarbide (TaC)	Rockwell A: 89
Tungsten Monocarbide (WC) (12% Co, 1-3μm grain size)	Rockwell A: 89.4 ± 0.5
Titanium Monocarbide (TiC) (99.5% density)	Rockwell A: 91-93.5
Titanium Monocarbide (TiC) (100% density)	Rockwell A: 91-93.5
Tungsten Monocarbide (WC)	Rockwell A: 92
Zirconium Monocarbide (ZrC)	Rockwell A: 92.5
Trisilicon tetranitride (Si_3N_4)	Rockwell A: 99

Source: data compiled by J.S. Park from *No. 1 Materials Index*, Peter T.B. Shaffer, Plenum Press, New York, (1964); *Smithells Metals Reference Book*, Eric A. Brandes, ed., in association with Fulmer Research Institute Ltd. 6th ed. London, Butterworths, Boston, (1983); and *Ceramic Source*, American Ceramic Society (1986-1991).

Table 396. SELECTING HARDNESS OF CERAMICS

(SHEET 5 OF 6)

Ceramic	Hardness
Cordierite ($2MgO\ 2Al_2O_3\ 5SiO_2$) (glass)	Vickers: 672.5 kg/mm^2
Titanium Oxide (TiO_2)	Vickers: 713-1121 kg/mm^2
Trisilicon tetranitride (Si_3N_4) (α)	Vickers: 815-1936kg/mm^2
Cordierite ($2MgO\ 2Al_2O_3\ 5SiO_2$)	Vickers: 835.6 kg/mm^2
Zirconium Oxide (ZrO_2) (partially stabilized)	Vickers: 1019-1121 kg/mm^2
Zirconium Oxide (ZrO_2)(fully stabilized)	Vickers: 1019-1529 kg/mm^2
Trichromium Dicarbide (Cr_3C_2)	Vickers: 1019-1834 kg/mm^2
Mullite ($3Al_2O_3\ 2SiO_2$)	Vickers: 1120 kg/mm^2
Boron Carbide (B_4C)	Vickers: 2400 kg/mm^2
Silicon Carbide (SiC) (cubic, CVD)	Vickers: 2853-4483 kg/mm^2
Dichromium Trioxide (Cr_2O_3)	Vickers: 2955 kg/mm^2
Tungsten Disilicide (WSi_2)	Vickers 10g: 1632 kg/mm^2
Aluminum Oxide (Al_2O_3)	Vickers 20g: 2600 kg/mm^2
Silicon Carbide (SiC)	Vickers 25g: 3000-3500 kg/mm^2
Chromium Diboride (CrB2)	Vickers 50g: 1800 kg/mm^2
Tantalum Monocarbide (TaC)	Vickers 50g: 1800 kg/mm^2
Zirconium Diboride (ZrB_2)	Vickers 50g: 2200 kg/mm^2
Tungsten Monocarbide (WC)	Vickers 50g: 2400 kg/mm^2
Hafnium Monocarbide (HfC)	Vickers 50g: 2533-3202 kg/mm^2
Zirconium Monocarbide (ZrC)	Vickers 50g: 2600 kg/mm^2
Aluminum Oxide (Al_2O_3)	Vickers 50g: 2720 kg/mm^2
Titanium Monocarbide (TiC)	Vickers 50g: 2900-3200 kg/mm^2
Titanium Diboride (TiB_2)	Vickers 50g: 3400 kg/mm^2

Source: data compiled by J.S. Park from *No. 1 Materials Index*, Peter T.B. Shaffer, Plenum Press, New York, (1964); *Smithells Metals Reference Book*, Eric A. Brandes, ed., in association with Fulmer Research Institute Ltd. 6th ed. London, Butterworths, Boston, (1983); and *Ceramic Source*, American Ceramic Society (1986-1991).

Table 396. SELECTING HARDNESS OF CERAMICS
(SHEET 6 OF 6)

Ceramic	Hardness
Tungsten Disilicide (WSi_2)	Vickers 100g: 1090 kg/mm^2
Molybdenum Disilicide ($MoSi_2$)	Vickers 100g: 1290-1550 kg/mm^2
Tungsten Monocarbide (WC)	Vickers 100g: 1730 kg/mm^2
Zirconium Monocarbide (ZrC)	Vickers 100g: 2836-3840 kg/mm^2
Titanium Monocarbide (TiC)	Vickers 100g: 2850-3390 kg/mm^2
Silicon Dioxide (SiO_2) (1011 face) 10 μm diagonal	Vickers 500g: 1040-1130 kg/mm^2
Silicon Dioxide (SiO_2) (normal to optical axis)	Vickers 500g: 1103 kg/mm^2
Silicon Dioxide (SiO_2)	Vickers 500g: 1120 kg/mm^2
Silicon Dioxide (SiO_2) (parallel to optical axis)	Vickers 500g: 1260 kg/mm^2
Silicon Dioxide (SiO_2) (polished 1010 face) 10 μm diagonal	Vickers 500g: 1300 kg/mm^2
Silicon Dioxide (SiO_2) (1010 face) 10 μm diagonal	Vickers 500g:1120-1230 kg/mm^2

Source: data compiled by J.S. Park from *No. 1 Materials Index*, Peter T.B. Shaffer, Plenum Press, New York, (1964); *Smithells Metals Reference Book*, Eric A. Brandes, ed., in association with Fulmer Research Institute Ltd. 6th ed. London, Butterworths, Boston, (1983); and *Ceramic Source*, American Ceramic Society (1986-1991).

Table 397. SELECTING MICROHARDNESS OF GLASS

Glass	Test	Microhardness
SiO_2 glass	Knoop	500–679
B_2O_3 glass	Vickers	194–205
SiO_2–B_2O_3 glass (95% mol B_2O_3)	Vickers	227–253
SiO_2–B_2O_3 glass (90% mol B_2O_3)	Vickers	231–257
SiO_2–B_2O_3 glass (75% mol B_2O_3)	Vickers	237–269–345
SiO_2–B_2O_3 glass (85% mol B_2O_3)	Vickers	239–267
SiO_2–B_2O_3 glass (80% mol B_2O_3)	Vickers	239–271
SiO_2–B_2O_3 glass (70% mol B_2O_3)	Vickers	251–279
B_2O_3–Na_2O glass (5% mol Na_2O)	Vickers	276
B_2O_3–Na_2O glass (10% mol Na_2O)	Vickers	292
SiO_2–B_2O_3 glass (65% mol B_2O_3)	Vickers	293–297
B_2O_3–Na_2O glass (15% mol Na_2O)	Vickers	297
SiO_2–B_2O_3 glass (60% mol B_2O_3)	Vickers	328–345
SiO_2–Na_2O glass (45% mol Na_2O)	Vickers	378±2
B_2O_3–Na_2O glass (20% mol Na_2O)	Vickers	380
SiO_2–Na_2O glass (40% mol Na_2O)	Vickers	394±2
SiO_2–Na_2O glass (30% mol Na_2O)	Vickers	413±3
SiO_2–Na_2O glass (35% mol Na_2O)	Vickers	414±4
SiO_2–Na_2O glass (25% mol Na_2O)	Vickers	423±4
B_2O_3–Na_2O glass (25% mol Na_2O)	Vickers	460
B_2O_3–Na_2O glass (30% mol Na_2O)	Vickers	503

Source: data compiled by J.S. Park from O. V. Mazurin, M. V. Streltsina and T. P. Shvaiko–Shvaikovskaya, *Handbook of Glass Data, Part A and Part B*, Elsevier, New York, 1983

Table 398. SELECTING HARDNESS OF POLYMERS
(SHEET 1 OF 5)

Polymer	Hardness, (ASTM D785) (Rockwell)
Polyester, Thermoset: Flexible	6—40 (Barcol)
Polyester, Thermoset: Rigid	35—50 (Barcol)
Polyester: Heat & chemical resistant (asbestos reinforced)	40—70 (Barcol)
Polyester: Sheet molding compounds, general purpose	45—60 (Barcol)
Cellusose Acetate Propionate, ASTM Grade: 6	57
Polyethylene, Type III: High molecular weight	60—65 (Shore)
Alkyd, Molded: Putty (encapsulating)	60—70 (Barcol)
Alkyd, Molded: Granular (high speed molding)	60—70 (Barcol)
Polyester moldings: High strength (glass fibers) Reinforced	60—80 (Barcol)
Epoxy, Standard: Cast High strength laminate	70—72 (Barcol)
Alkyd, Molded: Rope (general purpose)	70—75 (Barcol)
Alkyd, Molded: Glass reinforced (heavy duty parts)	70—80 (Barcol)
Silicone: Woven glass fabric/ silicone laminate	75 (Barcol)
Epoxy, Standard: Cast Molded	75-80 (Barcol)
Epoxy, High performance resins: Glass cloth laminate	75—80
Cellusose Acetate Propionate, ASTM Grade: 3	92—96
Cellusose Acetate Propionate, ASTM Grade: 1	100—109
Epoxy, High performance resins: Cast, rigid	107—112
Polyvinyl Chloride: Nonrigid—general	A50—100 (Shore, ASTM D676)
Polyvinyl Chloride: Nonrigid—electrical	A78—100 (Shore, ASTM D676)
Vinylidene chloride	>A95 (Shore, ASTM D676)
Polyethylene, Type I: Melt index 6—26	C73, D47—53 (Shore)
Polyethylene, Type I: Melt index 0.3—3.6	C73, D50—52 (Shore)
Olefin Copolymer, Molded: EEA (ethylene ethyl acrylate)	D35 (Shore)
Olefin Copolymer, Molded: EVA (ethylene vinyl acetate)	D36 (Shore)
Polyethylene, Type I: Melt index 200	D45 (Shore)
Polytetrafluoroethylene (PTFE)	D52

Source: data compiled by J.S. Park from Charles T. Lynch, *CRC Handbook of Materials Science, Vol. 3*, CRC Press, Boca Raton, Florida, 1975 and *Engineered Materials Handbook, Vol.2*, Engineering Plastics, ASM International, Metals Park, Ohio, 1988.

Table 398. SELECTING HARDNESS OF POLYMERS

(SHEET 2 OF 5)

Polymer	Hardness, (ASTM D785) (Rockwell)
Polyethylene, Type II: Melt index 20	D55 (Shore)
Polyethylene, Type II: Melt index l.0—1.9	D55—D56 (Shore)
Fluorinated ethylene propylene(FEP)	D57—58
Olefin Copolymer, Molded: Propylene—ethylene ionomer	D60 (Shore)
Polyethylene, Type III: Melt Melt index 0.l—12.0	D60—70 (Shore)
Olefin Copolymer, Molded: Ethylene butene	D65 (Shore)
Polyethylene, Type III: Melt index 0.2—0.9	D68—70 (Shore)
Polyethylene, Type III: Melt index 1.5—15	D68—70 (Shore)
Polyvinyl Chloride: Rigid—normal impact	D70—85 (Shore, ASTM D676)
Epoxy, High performance resins: Molded	D94—96
6/10 Nylon: Glass fiber (30%) reinforced	E40—50
Phenolic, Molded: Very high shock: glass fiber filled	E50—70
6/6 Nylon: Glass fiber reinforced	E60—E80
Phenolic, Molded: High shock: chopped fabric or cord filled	E80—90
Phenolic, Molded: General: woodflour and flock filled	E85—100
Phenolic, Molded: Shock: paper, flock, or pulp filled	E85—95
Urea, Molded: Alpha—cellulose filled (ASTM Type l)	E94—97
Melamine, Molded: Unfilled	E110
Polymide: Glass reinforced	E114
Phenylene Oxide: Glass fiber reinforced	L106, L108
Polystyrene, Molded: High impact	M3—43
Acrylic Moldings: High impact grade	M38—45
Rubber phenolic—woodflour or flock filled	M40—90
Polystyrene, Molded: Medium impact	M47—65

Source: data compiled by J.S. Park from Charles T. Lynch, *CRC Handbook of Materials Science, Vol. 3*, CRC Press, Boca Raton, Florida, 1975 and *Engineered Materials Handbook, Vol.2*, Engineering Plastics, ASM International, Metals Park, Ohio, 1988.

Table 398. SELECTING HARDNESS OF POLYMERS
(SHEET 3 OF 5)

Polymer	Hardness, (ASTM D785) (Rockwell)
Rubber phenolic—asbestos filled	M50
Epoxy, Standard: Cast Cast flexible	M50-100
Polyvinyl Chloride & Copolymers: Vinylidene chloride	M50—65
Rubber phenolic—chopped fabric filled	M57
Polycarbonate	M70
Silicone: Granular (silica) reinforced	M71—95
Polystyrene, Molded: General purpose	M72
Styrene acrylonitrile (SAN)	M75—85
Polyacetal Homopolymer: 22% TFE reinforced	M78
Polyacetal Copolymer: 25% glass reinforced	M79
Polyacetal Copolymer: Standard	M80
Polyacetal Copolymer: High flow	M80
Acrylic Moldings: Grades 5, 6, 8	M80—103
Acrylic Cast Resin Sheets, Rods: General purpose, type I	M80—90
Phenylene oxides (Noryl): Glass fiber reinforced	M84
Polyester, Thermoplastic Moldings: Asbestos—filled grade	M85
Polyarylsulfone	M85—110
Polystyrene, Molded: Glass fiber -30% reinforced	M85—95
Silicone: Fibrous (glass) reinforced	M87
Polyacetal Homopolymer: 20% glass reinforced	M90
Glass fiber (30%) reinforced Styrene acrylonitrile (SAN)	M90—123
Polyacetal Homopolymer: Standard	M94
6/6 Nylon: Glass fiber Molybdenum disulfide filled	M95—100
Thermoset Carbonate: Allyl diglycol carbonate	M95—M100 (Barcol)
Acrylic Cast Resin Sheets, Rods: General purpose, type II	M96—102
Polycarbonate (40% glass fiber reinforced)	M97
Epoxy, Standard: Cast Filament wound composite	M98-120
Phenolic, Molded: Arc resistant—mineral	M105—115

Source: data compiled by J.S. Park from Charles T. Lynch, *CRC Handbook of Materials Science, Vol. 3*, CRC Press, Boca Raton, Florida, 1975 and *Engineered Materials Handbook, Vol.2*, Engineering Plastics, ASM International, Metals Park, Ohio, 1988.

Table 398. SELECTING HARDNESS OF POLYMERS
(SHEET 4 OF 5)

Polymer	Hardness, (ASTM D785) (Rockwell)
Epoxy, Standard: Cast rigid	M106
Cellusose Acetate Propionate, ASTM Grade: Asbestos filled	M107
Cellusose Acetate Propionate, ASTM Grade: Orlon filled	M108
Cellusose Acetate Propionate, ASTM Grade: Glass fiber filled	M108
Epoxy, Standard: Cast General purpose glass cloth laminate	M115—117
Melamine, Molded: Cellulose filled electrical	M115—125
Urea, Molded: Alpha—cellulose filled (ASTM Type l)	M116—120
Urea, Molded: Woodflour filled	M116—120
Cellulose Acetate Butyrate, ASTM Grade: S2	R23—42
Polypropylene: High impact	R28—95
Polytetrafluoroethylene (PTFE): Ceramic reinforced	R35—55
Cellulose Acetate, ASTM Grade: S2—1	R49—88
Cellulose Acetate, ASTM Grade: MS—1, MS—2	R54—96
Polypropylene: Flame retardant	R60—R105
Nylon, Type 6: Flexible copolymers	R72—Rll9
Cellulose Acetate, ASTM Grade: MH—1, MH—2	R74—104
ABS Resin; Molded, Extruded: Low temperature impact	R75—95
Cellulose Acetate Butyrate, ASTM Grade: MH	R80—100
Polypropylene: General purpose	R80—R100
ABS Resin; Molded, Extruded: Very high impact	R85—105
Cellulose Acetate, ASTM Grade: H2—1	R89—112
Polypropylene: Asbestos filled	R90—R110
Polypropylene: Glass reinforced	R90—R115
Nylon, Type 6: Glass fiber (30%) reinforced	R93—121
ABS Resin; Molded, Extruded: High impact	R95—113
Chlorinated polyether	R100
Nylon, Type 11	R100—R108
Cellulose Acetate, ASTM Grade: H4—1	R103—120

Source: data compiled by J.S. Park from Charles T. Lynch, *CRC Handbook of Materials Science, Vol. 3*, CRC Press, Boca Raton, Florida, 1975 and *Engineered Materials Handbook, Vol.2*, Engineering Plastics, ASM International, Metals Park, Ohio, 1988.

Table 398. SELECTING HARDNESS OF POLYMERS
(SHEET 5 OF 5)

Polymer	Hardness, (ASTM D785) (Rockwell)
PVC–acrylic injection molded	R104
PVC–acrylic sheet	R105
Nylon, Type 12	R106
ABS Resin; Molded, Extruded: Heat resistant	R107—116
ABS Resin; Molded, Extruded: Medium impact	R108—115
Polyvinylidene— fluoride (PVDF)	R109—110
Polytrifluoro chloroethylene (PTFCE)	R110—115
Polyvinyl Chloride & Copolymers: Rigid—normal impact	R110—120
6/10 Nylon: General purpose	R111
Cellulose Acetate Butyrate, ASTM Grade: H4	R114
Phenylene Oxide: SE—100	R115
Nylon, Type 6: Cast	R116
Polyester, Thermoplastic Moldings: General purpose grade	R117
Polyester, Thermoplastic Moldings: General purpose grade	R117
Polyester, Thermoplastic Moldings: Glass reinforced grade	R117—M85
Chlorinated polyvinyl chloride	R118
ABS–Polycarbonate Alloy	R118
6/6 Nylon: General purpose extrusion	R118—108
Nylon, Type 6: General purpose	R118—R120
6/6 Nylon: General purpose molding	R118—120, R108
Polyester, Thermoplastic Moldings: Glass reinforced grades	R118—M90
Polyester, Thermoplastic: Glass reinforced self extinguishing	R119
Phenylene Oxide: SE—1	R119
Phenylene oxides (Noryl): Standard	R120
Polyphenylene sulfide: Standard	R120—124
Polyphenylene sulfide: 40% glass reinforced	R123

Source: data compiled by J.S. Park from Charles T. Lynch, *CRC Handbook of Materials Science, Vol. 3*, CRC Press, Boca Raton, Florida, 1975 and *Engineered Materials Handbook, Vol.2*, Engineering Plastics, ASM International, Metals Park, Ohio, 1988.

Table 399. SELECTING COEFFICIENTS OF STATIC FRICTION FOR POLYMERS

Polymer	Coefficient of Static Friction (Against Self) (Dimensionless)
6/6 Nylon: General purpose molding	0.04—0.13
Polyacetal Homopolymer: 22% TFE reinforced	0.05—0.15 (against steel)
Polyarylsulfone	0.1—0.3
Polyacetal Homopolymer: Standard	0.1—0.3 (against steel)
Polyacetal Homopolymer: 20% glass reinforced	0.1—0.3 (against steel)
Polyester; Thermoplastic Moldings: General purpose grade	0.13 (against steel)
Polyester; Thermoplastic Moldings: Glass reinforced grades	0.14 (against steel)
Polyester; Thermoplastic : Glass reinforced self extinguishing	0.14 (against steel)
Polyacetal Copolymer: Standard	0.15 (against steel)
Polyacetal Copolymer: 25% glass reinforced	0.15 (against steel)
Polyacetal Copolymer: High flow	0.15 (against steel)
Polyester; Thermoplastic Moldings: Glass reinforced grades	0.16 (ASTM D1894)
Polyester; Thermoplastic: Glass reinforced self extinguishing	0.16 (ASTM D1894)
Polyester; Thermoplastic Moldings: General purpose grade	0.17 (ASTM D1894)
ABS–Polycarbonate Alloy	0.2
Nylon, Type 6: Cast	0.32 (dynamic)
Polycarbonate	0.52
Phenylene oxides (Noryl): Standard	0.67

Source: data compiled by J.S. Park from Charles T. Lynch, *CRC Handbook of Materials Science, Vol. 3,* CRC Press, Boca Raton, Florida, 1975 and Engineered Materials Handbook, Vol.2, Engineering Plastics, ASM International, Metals Park, Ohio, 1988.

Table 400. SELECTING ABRASION RESISTANCE OF POLYMERS

Polymer	Abrasion Resistance (Taber, CS—17 wheel, ASTM D1044) (mg / 1000 cycles)
Polymide: Unreinforced	0.004—0.08
PVC–acrylic injection molded	0.0058 (CS—10 wheel)
PVC–acrylic sheet	0.073 (CS—10 wheel)
Nylon, Type 6: Cast	2.7
6/6 Nylon: General purpose extrusion	3—5
6/6 Nylon: General purpose molding	3—8
Nylon, Type 6: General purpose	5
Polyester Injection Moldings:General purpose grade	6.5
Polyacetal Homopolymer: 22% TFE reinforced	9
Polyester Injection Moldings:Glass reinforced grades	9—50
Polycarbonate	10
Polyester Injection Moldings:Glass reinforced self extinguishing	11
Polyacetal Copolymer:Standard	14
Polyacetal Copolymer:High flow	14
Polyacetal Homopolymer: Standard	14—20
Polymide: Glass reinforced	20
Phenylene Oxide: SE—1	20
Phenylene oxides (Noryl): Standard	20
Polyacetal Homopolymer: 20% glass reinforced	33
Phenylene Oxide: Glass fiber reinforced	35
Polycarbonate (40% glass fiber reinforced)	40
Polyacetal Copolymer:25% glass reinforced	40
Polyarylsulfone	40
Phenylene Oxide: SE—100	100
Polystyrene, Molded: Glass fiber -30% reinforced	164
Polyvinylidene— fluoride (PVDF)	600—1200
Polytrifluoro chloroethylene (PTFCE)	8000

Source: data compiled by J.S. Park from Charles T. Lynch, *CRC Handbook of Materials Science, Vol. 3*, CRC Press, Boca Raton, Florida, 1975 and *Engineered Materials Handbook, Vol.2*, Engineering Plastics, ASM International, Metals Park, Ohio, 1988.

Table 401. SELECTING FATIGUE STRENGTHS OF WROUGHT ALUMINUM ALLOYS (SHEET 1 OF 4)

Alloy AA No.	Temper	Fatigue Strength (MPa)
1060	0	21
1060	H12	28
1060	H14	34
1100	0	34
1100	H12	41
1060	H16	45
1060	H18	45
1100	H14	48
1350	H19	48
3003	0	48
Alclad	H12	55
6063	0	55
1100	H16	62
1100	H18	62
3003	H14	62
6061	0	62
6063	T1	62
6070	0	62
3003	H16	69
3003	H18	69
6063	T5	69
6063	T6	69
6463	T1	69
6463	T5	69
6463	T6	69
5050	0	83
2014	0	90
2024	0	90

Source: data from *ASM Metals Reference Book, Second Edition*, American Society for Metals, Metals Park, Ohio 44073, (1984).

Table 401. SELECTING FATIGUE STRENGTHS OF WROUGHT ALUMINUM ALLOYS (SHEET 2 OF 4)

Alloy AA No.	Temper	Fatigue Strength (MPa)
5050	H32	90
5050	H34	90
6070	T4	90
6262	T9	90
6351	T6	90
3004	0	97
5050	H36	97
5050	H38	97
6005	T1	97
6005	T5	97
6061	T4, T451	97
6061	T6, T651	97
6070	T6	97
2219	T62	105
2219	T81, T851	105
2219	T87	105
Alclad	H32	105
3004	H34	105
6205	T5	105
3004	H36	110
3004	H38	110
4032	T6	110
5052	0	110
5652	0	110
6066	T6, T651	110
5052	H32	115
5154	0	115
5154	H112	115

Source: data from *ASM Metals Reference Book, Second Edition*, American Society for Metals, Metals Park, Ohio 44073, (1984).

Table 401. SELECTING FATIGUE STRENGTHS OF WROUGHT ALUMINUM ALLOYS (SHEET 3 OF 4)

Alloy AA No.	Temper	Fatigue Strength (MPa)
5254	0	115
5254	H112	115
5652	H32	115
6009	T4	115
6010	T4	115
2011	T3	125
2011	T8	125
2014	T6	125
2024	T361	125
2036	T4	125
2618	All	125
5052	H34	125
5154	H32	125
5254	H32	125
5652	H34	125
7005	T6,T63,T6351	125
5052	H36	130
5154	H34	130
5254	H34	130
5652	H36	130
2014	T4	140
2024	T3	140
2024	T4, T351	140
5052	H38	140
5056	0	140
5154	H36	140
5182	0	140
5254	H36	140

Source: data from *ASM Metals Reference Book, Second Edition*, American Society for Metals, Metals Park, Ohio 44073, (1984).

Table 401. SELECTING FATIGUE STRENGTHS OF WROUGHT ALUMINUM ALLOYS (SHEET 4 OF 4)

Alloy AA No.	Temper	Fatigue Strength (MPa)
5652	H38	140
7005	T53	140
5154	H38	145
5254	H38	145
5056	H18	150
5056	H38	150
5083	H321	160
7075	T6,T651	160
7175	T66	160
7175	T736	160
2048		220
7475	T7351	220
7050	T736	240
7049	T73	295

Source: data from *ASM Metals Reference Book, Second Edition*, American Society for Metals, Metals Park, Ohio 44073, (1984).

Table 402. SELECTING REVERSED BENDING FATIGUE LIMITS OF GRAY CAST IRON BARS

ASTM Class	Reversed Bending Fatigue Limit (MPa)
20	69
25	79
30	97
35	110
40	128
50	148
60	169

Source: data from *ASM Metals Reference Book, Second Edition*, American Society for Metals, Metals Park, Ohio 44073, p166-167, (1984).

Table 403. SELECTING IMPACT ENERGY OF TOOL STEELS

Type	Condition	Impact Energy (J)
L6	Oil quenched from 845 °C and single tempered at 315 °C	12(a)
L6	Oil quenched from 845 °C and single tempered at 425 °C	18(a)
L2	Oil quenched from 855 °C and single tempered at 315 °C	19(a)
L6	Oil quenched from 845 °C and single tempered at 540 °C	23(a)
L2	Oil quenched from 855 °C and single tempered at 425 °C	26(a)
L2	Oil quenched from 855 °C and single tempered at 205 °C	28(a)
L2	Oil quenched from 855 °C and single tempered at 540 °C	39(a)
L6	Oil quenched from 845 °C and single tempered at 650 °C	81(a)
L2	Oil quenched from 855 °C and single tempered at 650 °C	125(a)
S5	Oil quenched from 870 °C and single tempered at 540 °C	188(b)
S1	Oil quenched from 930 °C and single tempered at 425 °C	203(b)
S5	Oil quenched from 870 °C and single tempered at 205 °C	206(b)
S1	Oil quenched from 930 °C and single tempered at 540 °C	230(b)
S5	Oil quenched from 870 °C and single tempered at 315 °C	232(b)
S1	Oil quenched from 930 °C and single tempered at 315 °C	233(b)
S5	Oil quenched from 870 °C and single tempered at 425 °C	243(b)
S7	Fan cooled from 940 °C and single tempered at 425 °C	243(b)
S7	Fan cooled from 940 °C and single tempered at 205 °C	244(b)
S1	Oil quenched from 930 °C and single tempered at 205 °C	249(b)
S7	Fan cooled from 940 °C and single tempered at 315 °C	309(b)
S7	Fan cooled from 940 °C and single tempered at 540 °C	324(b)
S7	Fan cooled from 940 °C and single tempered at 650 °C	358(b)

(a) Charpy V-notch.
(b) Charpy unnotched.

Source: Data from *ASM Metals Reference Book, Second Edition*, American Society for Metals, Metals Park, Ohio 44073, p241, (1984).

Table 404. SELECTING IMPACT STRENGTHS OF POLYMERS

(SHEET 1 OF 5)

Polymer	Impact Strength (Izod notched, ASTM D256) (ft—lb / in.)
Thermoset Cast Polyyester: Rigid	0.18—0.40
Melamine, Molded: mineral filled	0.2
Urea, Molded: Cellulose filled (ASTM Type 2)	0.20—0.275
Urea, Molded: Alpha—cellulose filled (ASTM Type l)	0.20—0.35
Acrylic Moldings: Grades 5, 6, 8	0.2—0.4
Thermoset Allyl diglycol carbonate	0.2—0.4
Polystyrene, Molded: General purpose	0.2—0.4 (ASTM D638)
Epoxy, Standard: Cast rigid	0.2—0.5
Phenolic, Molded: General, woodflour and flock filled	0.24—0.50
Alkyd, Molded: Putty (encapsulating)	0.25—0.35
Urea, Molded: Woodflour filled	0.25—0.35
Melamine, Molded: Cellulose filled electrical	0.27—0.36
Styrene acrylonitrile (SAN)	0.29—0.54
Polyphenylene sulfide: Standard	0.3
Alkyd, Molded: Granular (high speed molding)	0.30—0.35
Melamine, Molded: Alpha cellulose filled	0.30—0.35
Phenolic, Molded: Arc resistant—mineral filled	0.30—0.45
Diallyl Phthalate, Molded: Asbestos filled	0.30—0.50
Epoxy, Standard: Cast flexible	0.3—0.2
Rubber phenolic—asbestos filled	0.3—0.4
Epoxy, High performance: Molded	0.3—0.5
Silicone, Molded: Granular (silica) reinforced	0.34
Rubber phenolic—woodflour or flock filled	0.34—1.0
Acrylic Cast Resin Sheets, Rods: General purpose, type I	0.4

To convert **ft–lb / in.** to **N•m/m**, multiply by **53.38**

Source: data compiled by J.S. Park from Charles T. Lynch, *CRC Handbook of Materials Science, Vol. 3*, CRC Press, Boca Raton, Florida, 1975 and *Engineered Materials Handbook, Vol.2*, Engineering Plastics, ASM International, Metals Park, Ohio, 1988.

Table 404. SELECTING IMPACT STRENGTHS OF POLYMERS
(SHEET 2 OF 5)

Polymer	Impact Strength (Izod notched, ASTM D256) (ft—lb / in.)
Acrylic Cast Resin Sheets, Rods: General purpose, type II	0.4
Olefin Copolymers, Molded: Ethylene butene	0.4
Chlorinated polyether	0.4 (D758)
Epoxy, Standard: Molded	0.4—0.5
Phenolic, Molded: Shock: paper, flock, or pulp filled	0.4—1.0
Polypropylene: General purpose	0.4—2.2
Polyethylene, Type III: Melt Melt index 0.l—12.0	0.4—6.0
Reinforced polyester: Heat and chemical resistsnt (asbestos)	0.45—1.0
Epoxy, High performance: Cast, rigid	0.5
Polymide: Unreinforced	0.5
Polyester; Thermoplastic Moldings: Asbestos—filled grade	0.5
Diallyl Phthalate, Molded: Orlon filled	0.5—1.2
Polystyrene, Molded: Medium impact	0.5—1.2 (ASTM D638)
Polypropylene: Asbestos filled	0.5—1.5
Polyvinyl Chloride And Copolymers: Rigid—normal impact	0.5—10
Melamine, Molded: Glass fiber filled	0.5—12.0
Diallyl Phthalate, Molded: Glass fiber filled	0.5—15.0
Polypropylene: Glass reinforced	0.5—2
6/6 Nylon: General purpose molding	0.55—2.0 (ASTM D638)
Nylon Type 6: General purpose	0.6—1.2
6/10 Nylon: General purpose	0.6—1.6
Phenolic, Molded: High shock: chopped fabric or cord filled	0.6—8.0
Polyacetal Homopolymer: 22% TFE reinforced	0.7 (ASTM D638)
Polyacetal Homopolymer: 20% glass reinforced	0.8 (ASTM D638)

To convert **ft–lb / in.** to **N•m/m**, multiply by **53.38**

Source: data compiled by J.S. Park from Charles T. Lynch, *CRC Handbook of Materials Science, Vol. 3*, CRC Press, Boca Raton, Florida, 1975 and *Engineered Materials Handbook, Vol.2*, Engineering Plastics, ASM International, Metals Park, Ohio, 1988.

Table 404. SELECTING IMPACT STRENGTHS OF POLYMERS

(SHEET 3 OF 5)

Polymer	Impact Strength (Izod notched, ASTM D256) (ft—lb / in.)
Polystyrene, Molded: High impact	0.8—1.8 (ASTM D638)
Acrylic Moldings: High impact grade	0.8—2.3
Polyacetal Copolymer: High flow	1
Polyester; Thermoplastic Moldings: General purpose grade	1.0—1.2
Polyester; Thermoplastic Moldings: Glass reinforced grades	1.0—2.2
Reinforced polyester moldings: High strength (glass fibers)	1—10
Polyphenylene sulfide: 40% glass reinforced	1.09
Olefin Copolymers, Molded: Propylene—ethylene	1.1
Nylon Type 6: Cast	1.2
Phenylene oxides (Noryl): Standard	1.2—1.3
Polyethylene, Type III: Melt index 1.5—15	1.2—2.5
Nylon: Type 12	1.2—4.2
Polyacetal Copolymer: Standard	1.3
6/6 Nylon: General purpose extrusion	1.3 (ASTM D638)
Glass fiber (30%) reinforced Styrene acrylonitrile (SAN)	1.35—3.0
Polyacetal Homopolymer: Standard	1.4 (ASTM D638)
Olefin Copolymers, Molded: Polyallomer	1.5
Polypropylene: High impact	1.5—12
Nylon Type 6: Flexible copolymers	1.5—19
Polyarylsulfone	1.6—5.0
Cellusose Acetate Propionate, ASTM Grade: 1	1.7—2.7
Diallyl Phthalate, Molded: Dacron filled	1.7—5.0
Polyacetal Copolymer: 25% glass reinforced	1.8
Polyester; Moldings: Glass reinforced self extinguishing	1.8

To convert **ft–lb / in.** to **N•m/m**, multiply by **53.38**

Source: data compiled by J.S. Park from Charles T. Lynch, *CRC Handbook of Materials Science, Vol. 3*, CRC Press, Boca Raton, Florida, 1975 and *Engineered Materials Handbook, Vol.2*, Engineering Plastics, ASM International, Metals Park, Ohio, 1988.

Table 404. SELECTING IMPACT STRENGTHS OF POLYMERS
(SHEET 4 OF 5)

Polymer	Impact Strength (Izod notched, ASTM D256) (ft—lb / in.)
Phenylene oxides (Noryl): Glass fiber reinforced	1.8—2.0
Rubber phenolic—chopped fabric filled	2.0—2.3
ABS Resin: Medium impact	2.0—4.0
ABS Resin: Heat resistant	2.0—4.0
Polytetrafluoroethylene (PTFE)	2.0—4.0
Vinylidene chloride	2—8
Alkyd, Molded: Rope (general purpose)	2.2
Polypropylene: Flame retardant	2.2
Nylon Type 6: Glass fiber (30%) reinforced	2.2—3.4
Phenylene Oxide: Glass fiber reinforced	2.3 (ASTM D638)
Polystyrene, Molded: Glass fiber —30% reinforced	2.5
6/6 Nylon: Glass fiber reinforced	2.5—3.4 (ASTM D638)
Cellulose Acetate Butyrate, ASTM Grade: H4	3
Polyvinylidene— fluoride (PVDF)	3.0—10.3
ABS Resin: High impact	3.0—5.0
Nylon: Type 11	3.3—3.6
6/10 Nylon: Glass fiber (30%) reinforced	3.4
Polytrifluoro chloroethylene (PTFCE)	3.50—3.62
Cellusose Acetate Propionate, ASTM Grade: 3	3.5—5.6
Thermoset Cast Polyyester: Flexible	4
Polyethylene, Type III: Melt index 0.2—0.9	4.0—14
Cellulose Acetate Butyrate, ASTM Grade: MH	4.4—6.9
Phenylene Oxide: SE—100	5 (ASTM D638)
Phenylene Oxide: SE—1	5 (ASTM D638)

To convert ft–lb / in. to N•m/m, multiply by 53.38

Source: data compiled by J.S. Park from Charles T. Lynch, *CRC Handbook of Materials Science, Vol. 3*, CRC Press, Boca Raton, Florida, 1975 and *Engineered Materials Handbook, Vol.2*, Engineering Plastics, ASM International, Metals Park, Ohio, 1988.

Table 404. SELECTING IMPACT STRENGTHS OF POLYMERS
(SHEET 5 OF 5)

Polymer	Impact Strength (Izod notched, ASTM D256) (ft—lb / in.)
ABS Resin: Very high impact	5.0—7.5
Reinforced polyester Sheet molding, general purpose	5—15
ABS Resin: Low temperature impact	6—10
Chlorinated polyvinyl chloride	6.3
Cellulose Acetate Butyrate, ASTM Grade: S2	7.5—10.0
Alkyd, Molded: Glass reinforced (heavy duty parts)	8—12
Olefin Copolymers, Molded: Ionomer	9—14
Cellusose Acetate Propionate, ASTM Grade: 6	9.4
Silicone, Molded: Fibrous (glass) reinforced	10
ABS–Polycarbonate Alloy	10 (ASTM D638)
Silicone: Woven glass fabric/ silicone laminate	10—25
Phenolic, Molded: Very high shock: glass fiber filled	10—33
Epoxy, Standard: General purpose glass cloth laminate	12—15
Polycarbonate	12—16
Epoxy novolacs: Cast, rigid	13—17
PVC–acrylic sheet	15
PVC–acrylic injection molded	15
Polymide: Glass reinforced	17
Epoxy, Standard: High strength laminate	60—61
Nylon: Type 8	>16
Polyethylene, Type III: High molecular weight	>20
Fluorinated ethylene propylene(FEP)	No break
Polyvinyl Chloride And Copolymers: Nonrigid—general	Variable
Polyvinyl Chloride And Copolymers: Nonrigid—electrical	Variable

To convert **ft–lb / in.** to **N•m/m**, multiply by **53.38**

Source: data compiled by J.S. Park from Charles T. Lynch, *CRC Handbook of Materials Science, Vol. 3*, CRC Press, Boca Raton, Florida, 1975 and *Engineered Materials Handbook, Vol.2*, Engineering Plastics, ASM International, Metals Park, Ohio, 1988.

Table 405. SELECTING TENSILE MODULI OF GRAY CAST IRONS

ASTM Class	Tensile Modulus (GPa)
20	66 to 97
25	79 to 102
30	90 to 113
35	100 to 119
40	110 to 138
50	130 to 157
60	141 to 162

Source: data from *ASM Metals Reference Book, Second Edition*, American Society for Metals, Metals Park, Ohio 44073, p166-167, (1984).

Table 406. SELECTING TENSILE MODULI OF TREATED DUCTILE IRONS

Treatment	Tension Modulus (GPa)
120 90-02	164
65-45-12	168
80-55-06	168
60-40-18	169

Source: data from *ASM Metals Reference Book, Second Edition*, American Society for Metals, Metals Park, Ohio 44073, p169-170, (1984).

Table 407. SELECTING YOUNG'S MODULI OF CERAMICS
(SHEET 1 OF 6)

Ceramic	Temperature	Young's Modulus (psi)
Boron Nitride (BN), parallel to c axis	700°C	$0.51x10^6$
Boron Nitride (BN), parallel to a axis	700°C	$1.54x10^6$
Boron Nitride (BN), parallel to a axis	1000°C	$1.65x10^6$
Zirconium Oxide (ZrO_2) (plasma sprayed)	500°C	$2x10^6$
Zirconium Oxide (ZrO_2) (plasma sprayed)	1100°C	$3.05x10^6$
Zirconium Diboride (ZrB_2) (22.4% density, foam)		$3.305x10^6$
Boron Nitride (BN), parallel to c axis	300°C	$3.47x10^6$
Magnesium Oxide (MgO)	1300°C	$4 x10^6$
Mullite ($3Al_2O_3$ $2SiO_2$) (ρ=2.77 g/cm^3)	1200°C	$4.00x10^6$
Boron Nitride (BN), parallel to c axis	23°C	$4.91x10^6$
Titanium Diboride (TiB_2) (12.0 μm grain size, ρ=4.66g/cm^3, 9.6wt% Ni)		$6.29x10^6$
Zirconium Oxide (ZrO_2) (plasma sprayed)	room temp.	$6.96x10^6$
Hafnium Dioxide (HfO_2)		$8.2x10^6$
Boron Nitride (BN), parallel to a axis	300°C	$8.79x10^6$
Magnesium Oxide (MgO)	1200°C	$10 x10^6$
Titanium Mononitride (TiN)		$11.47\text{-}36.3x10^6$
Boron Nitride (BN), parallel to a axis	23°C	$12.46x10^6$
Thorium Dioxide (ThO_2)	1200°C	$12.8x10^6$
Zirconium Oxide (ZrO_2) (plasma sprayed)	1500°C	$12.8x10^6$

To convert from **psi** to **MPa**, multiply by **145.**

Source: data compiled by J.S. Park from *No. 1 Materials Index*, Peter T.B. Shaffer, Plenum Press, New York, (1964); *Smithells Metals Reference Book*, Eric A. Brandes, ed., in association with Fulmer Research Institute Ltd. 6th ed. London, Butterworths, Boston, (1983); and *Ceramic Source*, American Ceramic Society (1986-1991)

Table 407. SELECTING YOUNG'S MODULI OF CERAMICS
(SHEET 2 OF 6)

Ceramic	Temperature	Young's Modulus (psi)
Cordierite ($2MgO\ 2Al_2O_3\ 5SiO_2$) glass		$13.92x10^6$
Zirconium Oxide (ZrO_2) (fully stabilized)	room temp.	$14.1\text{-}30.0x10^6$
Zirconium Oxide (ZrO_2) (plasma sprayed)	1400˚C	$14.2x10^6$
Trisilicon tetranitride (Si_3N_4) (reaction sintered)	20˚C	$14.5\text{-}31.9x10^6$
Mullite ($3Al_2O_3\ 2SiO_2$) (ρ=2.77 g/cm^3)	800˚C	$14.79x10^6$
Dichromium Trioxide (Cr_2O_3)		$>14.9x10^6$
Zirconium Oxide (ZrO_2) (plasma sprayed)	1200˚C	$17.1\text{-}18.0x10^6$
Thorium Dioxide (ThO_2)	1000˚C	$17.1x10^6$
Trisilicon tetranitride (Si_3N_4) (reaction sintered)	1400˚C	$17.4\text{-}29.0x10^6$
Thorium Dioxide (ThO_2)	room temp.	$17.9\text{-}34.87x10^6$
Thorium Dioxide (ThO_2)	800˚C	$18\text{-}18.5x10^6$
Mullite ($3Al_2O_3\ 2SiO_2$) (ρ=2.77 g/cm^3)	25˚C	$18.42x10^6$
Zirconium Oxide (ZrO_2) (plasma sprayed)	1000˚C	$18.5\text{-}25x10^6$
Mullite ($3Al_2O_3\ 2SiO_2$) (ρ=2.77 g/cm^3)	400˚C	$18.89x10^6$
Zirconium Oxide (ZrO_2) (plasma sprayed)	800˚C	$18.9x10^6$
Zirconium Oxide (ZrO_2) (stabilized, ρ=5.634 g/cm^3)	room temp.	$19.96x10^6$
Beryllium Oxide (BeO)	1145˚C	$20\ x10^6$
Spinel (Al_2O_3 MgO)	1300˚C	$20.1x10^6$

To convert from **psi** to **MPa**, multiply by **145**.

Source: data compiled by J.S. Park from *No. 1 Materials Index*, Peter T.B. Shaffer, Plenum Press, New York, (1964); *Smithells Metals Reference Book*, Eric A. Brandes, ed., in association with Fulmer Research Institute Ltd. 6th ed. London, Butterworths, Boston, (1983); and *Ceramic Source*, American Ceramic Society (1986-1991)

Table 407. SELECTING YOUNG'S MODULI OF CERAMICS

(SHEET 3 OF 6)

Ceramic	Temperature	Young's Modulus (psi)
Cordierite (2MgO $2Al_2O_3$ $5SiO_2$)		$20.16x10^6$
Mullite ($3Al_2O_3$ $2SiO_2$) (ρ=2.779 g/cm^3)	room temp.	$20.75x10^6$
Magnesium Oxide (MgO)	1000°C	21 $x10^6$
Uranium Dioxide (UO_2)	0-1000°C	$21x10^6$
Zircon (SiO_2 ZrO_2)	room temp.	$24x10^6$
Zirconium Oxide (ZrO_2) (plasma sprayed)	room temp.	$24.8\text{-}27x10^6$
Cerium Dioxide (CeO_2)		$24.9x10^6$
Spinel (Al_2O_3 MgO)	1200°C	$25.0x10^6$
Uranium Dioxide (UO_2)	20°C	$25x10^6$
Trisilicon tetranitride (Si_3N_4) (hot pressed)	1400°C	$25.38\text{-}36.25x10^6$
Aluminum Oxide (Al_2O_3)	1500°C	25.6 $x10^6$
Uranium Dioxide (UO_2) (ρ=10.37 g/cm^3)	room temp.	$27.98x10^6$
Trisilicon tetranitride (Si_3N_4) (sintered)	20°C	$28.28\text{-}45.68x10^6$
Zirconium Monocarbide (ZrC)	room temp.	$28.3\text{-}69.6x10^6$
Silicon Carbide (SiC) (reaction sintered)	1400°C	$29\text{-}46.4x10^6$
Magnesium Oxide (MgO)	600°C	29.5 $x10^6$
Zirconium Oxide (ZrO_2) (partially stabilized)	room temp.	$29.7x10^6$
Spinel (Al_2O_3 MgO)	1000°C	$30.4x10^6$
Magnesium Oxide (MgO)	room temp.	$30.5\text{-}36.3x10^6$
Chromium Diboride (CrB_2)		$30.6x10^6$

To convert from **psi** to **MPa**, multiply by **145.**

Source: data compiled by J.S. Park from *No. 1 Materials Index*, Peter T.B. Shaffer, Plenum Press, New York, (1964); *Smithells Metals Reference Book*, Eric A. Brandes, ed., in association with Fulmer Research Institute Ltd. 6th ed. London, Butterworths, Boston, (1983); and *Ceramic Source*, American Ceramic Society (1986-1991)

Table 407. SELECTING YOUNG'S MODULI OF CERAMICS
(SHEET 4 OF 6)

Ceramic	Temperature	Young's Modulus (psi)
Aluminum Oxide (Al_2O_3)	1250°C	32 x10^6
Aluminum Oxide (Al_2O_3)	1400°C	32.7 x10^6
Spinel (Al_2O_3 MgO)	800°C	32.9x10^6
Beryllium Oxide (BeO)	1000°C	33 x10^6
Mullite ($3Al_2O_3$ $2SiO_2$) (full density)	room temp.	33.35x10^6
Spinel (Al_2O_3 MgO)	600°C	34x10^6
Spinel (Al_2O_3 MgO)	200°C	34.4x10^6
Spinel (Al_2O_3 MgO)	room temp.	34.5x10^6
Spinel (Al_2O_3 MgO)	400°C	34.5x10^6
Zirconium Oxide (ZrO_2) (plasma sprayed)	20°C	36x10^6
Trisilicon tetranitride (Si_3N_4) (hot pressed)	20°C	36.25-47.13x10^6
Tantalum Diboride (TaB_2)		37 x10^6
Spinel (Al_2O_3 MgO) (ρ=3.510 g/cm^3)	room temp.	38.23x10^6
Molybdenum Disilicide ($MoSi_2$)	room temp.	39.3-56.36x10^6
Aluminum Oxide (Al_2O_3)	1200°C	39.8-53.65 x10^6
Beryllium Oxide (BeO)	800°C	40 x10^6
Aluminum Nitride (AlN)	1400°C	40x10^6
Titanium Oxide (TiO_2)		41x10^6
Tantalum Monocarbide (TaC)	room temp.	41.3-91.3x10^6
Boron Carbide (B_4C)	room temp.	42-65.2x10^6

To convert from **psi** to **MPa**, multiply by **145**.

Source: data compiled by J.S. Park from *No. 1 Materials Index*, Peter T.B. Shaffer, Plenum Press, New York, (1964); *Smithells Metals Reference Book*, Eric A. Brandes, ed., in association with Fulmer Research Institute Ltd. 6th ed. London, Butterworths, Boston, (1983); and *Ceramic Source*, American Ceramic Society (1986-1991)

Table 407. SELECTING YOUNG'S MODULI OF CERAMICS

(SHEET 5 OF 6)

Ceramic	Temperature	Young's Modulus (psi)
Magnesium Oxide (MgO) ($\rho = 3.506\ g/cm^3$)	room temp.	$42.74x10^6$
Beryllium Oxide (BeO)	room temp.	$42.8-45.5x10^6$
Silicon Carbide (SiC) (sintered)	1400°C	$43.5-58.0x10^6$
Silicon Carbide (SiC) (pressureless sintered)	room temp.	$43.9x10^6$
Titanium Monocarbide (TiC)	1000°C	$45-55x10^6$
Aluminum Oxide (Al_2O_3)	1000°C	$45.5-50\ x10^6$
Aluminum Nitride (AlN)	1000°C	$46x10^6$
Zirconium Diboride (ZrB_2)		$49.8-63.8x10^6$
Aluminum Oxide (Al_2O_3)	500°C	$50-57.275\ x10^6$
Aluminum Oxide (Al_2O_3)	room temp.	$50-59.3x10^6$
Aluminum Nitride (AlN)	25°C	$50x10^6$
Silicon Carbide (SiC) (reaction sintered)	20°C	$50.75-54.38x10^6$
Silicon Carbide (SiC) (reaction sintered)	1200°C	$51x10^6$
Aluminum Oxide (Al_2O_3)	800°C	$51.2\ x10^6$
Silicon Carbide (SiC) (reaction sintered)	800°C	$53x10^6$
Titanium Diboride (TiB_2)		$53.2x10^6$
Trichromium Dicarbide (Cr_3C_2)		$54.1x10^6$
Silicon Carbide (SiC) (sintered)	20°C	$54.38-60.9x10^6$
Silicon Carbide (SiC) (reaction sintered)	400°C	$55x10^6$
Silicon Carbide (SiC) (hot presses)	1400°C	$55.1x10^6$

To convert from **psi** to **MPa**, multiply by **145.**

Source: data compiled by J.S. Park from *No. 1 Materials Index*, Peter T.B. Shaffer, Plenum Press, New York, (1964); *Smithells Metals Reference Book*, Eric A. Brandes, ed., in association with Fulmer Research Institute Ltd. 6th ed. London, Butterworths, Boston, (1983); and *Ceramic Source*, American Ceramic Society (1986-1991)

Table 407. SELECTING YOUNG'S MODULI OF CERAMICS

(SHEET 6 OF 6)

Ceramic	Temperature	Young's Modulus (psi)
Silicon Carbide (SiC) ($\rho = 3.128$ g/cm^3)	room temp.	58.2×10^6
Silicon Carbide (SiC) (self bonded)	room temp.	59.5×10^6
Silicon Carbide (SiC) ($\rho = 3.120$ g/cm^3)	room temp.	59.52×10^6
Silicon Carbide (SiC) (cubic, CVD)	room temp.	$60.2\text{-}63.9 \times 10^6$
Hafnium Monocarbide (HfC) ($\rho = 11.94$ g/cm^3)	room temp.	61.55×10^6
Silicon Carbide (SiC) (hot pressed)	20°C	$62.4\text{-}65.3 \times 10^6$
Titanium Monocarbide (TiC)	room temp.	63.715×10^6
Silicon Carbide (SiC) (hot pressed)	room temp.	63.8×10^6
Titanium Diboride (TiB_2) (3.5 μm grain size, ρ=4.37g/cm^3, 0.8wt% Ni)		75.0×10^6
Titanium Diboride (TiB_2) (6.0 μm grain size, ρ=4.56g/cm^3, 0.16wt% Ni)		77.9×10^6
Titanium Diboride (TiB_2) (6.0 μm grain size, ρ=4.46g/cm^3)		81.6×10^6
Tungsten Monocarbide (WC)	room temp.	$96.91\text{-}103.5 \times 10^6$

To convert from **psi to MPa**, multiply by **145.**

Source: data compiled by J.S. Park from *No. 1 Materials Index*, Peter T.B. Shaffer, Plenum Press, New York, (1964); *Smithells Metals Reference Book*, Eric A. Brandes, ed., in association with Fulmer Research Institute Ltd. 6th ed. London, Butterworths, Boston, (1983); and *Ceramic Source*, American Ceramic Society (1986-1991)

Table 408. SELECTING YOUNG'S MODULI OF GLASS
(SHEET 1 OF 2)

Glass	Temperature	Young's Modulus (GPa)
B_2O_3 glass	room temp.	17.2–17.7
SiO_2–B_2O_3 glass (90% mol B_2O_3)		20.9
SiO_2–B_2O_3 glass (85% mol B_2O_3)		21.2
SiO_2–B_2O_3 glass (95% mol B_2O_3)		21.2
SiO_2–B_2O_3 glass (65% mol B_2O_3)		22.5
SiO_2–B_2O_3 glass (80% mol B_2O_3)		22.8
SiO_2–B_2O_3 glass (60% mol B_2O_3)		23.3
SiO_2–B_2O_3 glass (70% mol B_2O_3)		23.5
SiO_2–B_2O_3 glass (75% mol B_2O_3)		24.1
B_2O_3–Na_2O glass (10% mol Na_2O)	15°C	31.4
SiO_2–PbO glass (65.0% mol PbO)		41.2
B_2O_3–Na_2O glass (20% mol Na_2O)	15°C	43.2
SiO_2–PbO glass (60.0% mol PbO)		43.6
SiO_2–PbO glass (50.0% mol PbO)		44.1
SiO_2–Na_2O glass (40% mol Na_2O)	200–250°C	46.1
SiO_2–PbO glass (35.7% mol PbO)		46.3
SiO_2–PbO glass (24.6% mol PbO)		47.1
SiO_2–PbO glass (55.0% mol PbO)		49.3
SiO_2–PbO glass (30.0% mol PbO)		50.1
SiO_2–Na_2O glass (33% mol Na_2O)	200–250°C	51.0
SiO_2–PbO glass (45.0% mol PbO)		51.7
SiO_2–Na_2O glass (40% mol Na_2O)	–196°C	51.9
SiO_2–PbO glass (38.4% mol PbO)		52.8
B_2O_3–Na_2O glass (25% mol Na_2O)	15°C	53.7

Source: data compiled by J.S. Park from O. V. Mazurin, M. V. Streltsina and T. P. Shvaiko–Shvaikovskaya, *Handbook of Glass Data, Part A and Part B*, Elsevier, New York, 1983.

Table 408. SELECTING YOUNG'S MODULI OF GLASS (SHEET 2 OF 2)

Glass	Temperature	Young's Modulus (GPa)
SiO_2–Na_2O glass (25% mol Na_2O)	200–250°C	53.9
SiO_2–Na_2O glass (33% mol Na_2O)	–196°C	54.9
SiO_2–Na_2O glass (25% mol Na_2O)	–196°C	56.9
B_2O_3–Na_2O glass (37% mol Na_2O)	15°C	57.1
B_2O_3–Na_2O glass (33.3% mol Na_2O)	15°C	59.4
SiO_2–Na_2O glass (35% mol Na_2O)	room temp.	60.2
SiO_2–Na_2O glass (33% mol Na_2O)	room temp.	60.3
SiO_2–Na_2O glass (30% mol Na_2O)	room temp.	60.5
SiO_2–Na_2O glass (25% mol Na_2O)	room temp.	61.4
SiO_2–Na_2O glass (20% mol Na_2O)	room temp.	62.0
SiO_2–Na_2O glass (15% mol Na_2O)	room temp.	64.4
SiO_2 glass	20°C	72.76–74.15
SiO_2 glass	998°C (annealing point)	79.87
SiO_2 glass	1096°C (straining point)	80.80

Source: data compiled by J.S. Park from O. V. Mazurin, M. V. Streltsina and T. P. Shvaiko–Shvaikovskaya, *Handbook of Glass Data, Part A and Part B*, Elsevier, New York, 1983.

Table 409. SELECTING MODULI OF ELASTICITY IN TENSION FOR POLYMERS (SHEET 1 OF 3)

Polymer	Modulus of Elasticity in Tension (ASTM D638) (10^5 psi)
Polyester, Cast Thermoset: Flexible	0.001—0.10
Polyvinyl Chloride & Copolymers: Nonrigid—general	0.004—0.03 (ASTM D412)
Polyvinyl Chloride & Copolymers: Nonrigid—electrical	0.01—0.03 (ASTM D412)
Polyethylene, Type I: Melt index 6—26	0.20—0.24
Polyethylene, Type I: Melt index 0.3—3.6	0.21—0.27
Polytetrafluoroethylene (PTFE)	0.38—0.65
Fluorinated ethylene propylene(FEP)	0.5—0.7
Epoxy, Standard: Cast flexible	0.5—2.5
Vinylidene chloride	0.7—2.0 (ASTM D412)
Chlorinated polyether	1.5
Polystyrene, Molded: High impact	1.50—3.80 (D638)
Ceramic reinforced (PTFE)	1.5—2.0
Polyester, Cast Thermoset: Rigid	1.5—6.5
Polyvinylidene— fluoride (PVDF)	1.7—2
Polytrifluoro chloroethylene (PTFCE)	1.9—3.0
ABS Resin: Very high impact	2.0—3.1
ABS Resin: Low temperature impact	2.0—3.1
Acrylic Cast Resin Moldings: High impact grade	2.3—3.3
ABS Resin: High impact	2.6—3.2
Polystyrene, Molded: Medium impact	2.6—4.7 (D638)
Polyvinyl Chloride & Copolymers: Rigid—normal impact	3 5—4.0 (ASTM D412)
ABS Resin: Medium impact	3.3—4.0
Polycarbonate	3.45
ABS Resin: Heat resistant	3.5—4.2

To convert from **psi** to **MPa,** multiply by **145.**

Source: data compiled by J.S. Park from Charles T. Lynch, *CRC Handbook of Materials Science*, Vol. 3, CRC Press, Boca Raton, Florida, 1975 and *Engineered Materials Handbook*, Vol.2, Engineering Plastics, ASM International, Metals Park, Ohio, 1988.

Table 409. SELECTING MODULI OF ELASTICITY IN TENSION FOR POLYMERS (SHEET 2 OF 3)

Polymer	Modulus of Elasticity in Tension (ASTM D638) (10^5 psi)
Acrylic Cast Resin Sheets, Rods: General purpose, type I	3.5—4.5
Acrylic Cast Resin Moldings: Grades 5, 6, 8	3.5—5.0
Rubber phenolic—chopped fabric filled	3.5—6
Chlorinated polyvinyl chloride	3.7
Acrylic Cast Resin Sheets, Rods: General purpose, type II	4.0—5.0
Styrene acrylonitrile (SAN)	4.0—5.2
Epoxy, High performance: Cast, rigid	4—5
Rubber phenolic—woodflour or flock filled	4—6
Epoxy, Standard: Cast rigid	4.5
Polystyrene, Molded: General purpose	4.6—5.0 (D638)
Epoxy novolacs: Cast, rigid	4.8—5.0
Rubber phenolic—asbestos filled	5—9
Diallyl Phthalate, Molded: Orlon filled	6
Phenolic, Shock: paper, flock, or pulp filled	8—12
Phenolic, General: woodflour and flock filled	8—13
Phenolic, High shock: chopped fabric or cord filled	9—14
Melamine; Molded: Cellulose filled electrical	10—11
Phenolic, Molded: Arc resistant—mineral filled	10—30
Urea, Molded: Woodflour filled	11—14
Diallyl Phthalate, Molded: Asbestos filled	12
Reinforced polyester moldings: Heat & chemical resistsnt (asbestos)	12—15
Polystyrene, Molded: Glass fiber -30% reinforced	12.1 (D638)
Urea, Molded: Alpha—cellulose filled (ASTM Type l)	13—16
Reinforced polyester Sheet molding: general purpose	15—20

To convert from **psi** to **MPa**, multiply by 145.

Source: data compiled by J.S. Park from Charles T. Lynch, *CRC Handbook of Materials Science*, Vol. 3, CRC Press, Boca Raton, Florida, 1975 and *Engineered Materials Handbook*, Vol.2, Engineering Plastics, ASM International, Metals Park, Ohio, 1988.

Table 409. SELECTING MODULI OF ELASTICITY IN TENSION FOR POLYMERS (SHEET 3 OF 3)

Polymer	Modulus of Elasticity in Tension (ASTM D638) (10^5 psi)
Reinforced polyester moldings: High strength (glass fibers)	16—20
Polycarbonate (40% glass fiber reinforced)	17
Glass fiber (30%) reinforced Styrene acrylonitrile (SAN)	17.5
Epoxy novolacs: Glass cloth laminate	27.5
Silicone: Woven glass fabric/ silicone laminate	28 (ASTM D651)
Phenolic, Very high shock: glass fiber filled	30—33
Epoxy, High performance Molded: Glass cloth laminate	32—33
Epoxy, Standard, Molded: General purpose glass cloth laminate	33—36
Epoxy, Standard, Molded: High strength laminate	57—58
Epoxy, Standard, Molded: Filament wound composite	72—64

To convert from **psi** to **MPa**, multiply by **145**.

Source: data compiled by J.S. Park from Charles T. Lynch, *CRC Handbook of Materials Science*, Vol. 3, CRC Press, Boca Raton, Florida, 1975 and *Engineered Materials Handbook*, Vol.2, Engineering Plastics, ASM International, Metals Park, Ohio, 1988.

Table 410. SELECTING COMPRESSION MODULI OF TREATED DUCTILE IRONS

Treatment	Compression Modulus (GPa)
65-45-12	163
60-40-18	164
120 90-02	164
80-55-06	165

Source: data from *ASM Metals Reference Book*, Second Edition, American Society for Metals, Metals Park, Ohio 44073, p169-170, (1984).

Table 411. SELECTING MODULUS OF ELASTICITY IN COMPRESSION FOR POLYMERS

Polymer	Modulus of Elasticity in Compression (ASTM D638) (10^5 psi)
Polytetrafluoroethylene (PTFE)	0 70—0.90
Fluorinated ethylene propylene(FEP)	0.6—0.8
Ceramic reinforced (PTFE)	1.5—2.0
Polyvinylidene— fluoride (PVDF)	1.7—2
Polytrifluoro chloroethylene (PTFCE)	1.8

To convert from **psi** to **MPa**, multiply by **145**.

Source: data compiled by J.S. Park from Charles T. Lynch, *CRC Handbook of Materials Science*, Vol. 3, CRC Press, Boca Raton, Florida, 1975 and *Engineered Materials Handbook*, Vol.2, Engineering Plastics, ASM International, Metals Park, Ohio, 1988.

Table 412. SELECTING BULK MODULI OF GLASS

Glass	Temperature	Bulk Modulus (GPa)
B_2O_3-Na_2O glass (10% mol Na_2O)	15°C	23.2
SiO_2-PbO glass (38.4% mol PbO)		25.1
SiO_2-PbO glass (30.0% mol PbO)		25.6
SiO_2-PbO glass (55.0% mol PbO)		29.5
SiO_2-PbO glass (50.0% mol PbO)		30.5
SiO_2-PbO glass (45.0% mol PbO)		30.6
SiO_2 glass		31.01-37.62
SiO_2-PbO glass (35.7% mol PbO)		31.1
SiO_2-PbO glass (65.0% mol PbO)		31.6
SiO_2-PbO glass (60.0% mol PbO)		33.1
B_2O_3-Na_2O glass (20% mol Na_2O)	15°C	33.6
SiO_2-Na_2O glass (15% mol Na_2O)	room temp.	33.8
SiO_2-PbO glass (24.6% mol PbO)		33.9
SiO_2-Na_2O glass (20% mol Na_2O)	room temp.	34.8
SiO_2-Na_2O glass (25% mol Na_2O)	room temp.	36.5
SiO_2-Na_2O glass (30% mol Na_2O)	room temp.	38.2
B_2O_3-Na_2O glass (25% mol Na_2O)	15°C	39.2
SiO_2-Na_2O glass (35% mol Na_2O)	room temp.	39.8
SiO_2-Na_2O glass (33% mol Na_2O)	room temp.	40.1
B_2O_3-Na_2O glass (37% mol Na_2O)	15°C	42.1
B_2O_3-Na_2O glass (33.3% mol Na_2O)	15°C	44.4

Source: data compiled by J.S. Park from O. V. Mazurin, M. V. Streltsina and T. P. Shvaiko-Shvaikovskaya, *Handbook of Glass Data*, Part A and Part B, Elsevier, New York, 1983

Table 413. SELECTING MODULI OF ELASTICITY IN FLEXURE OF POLYMERS (SHEET 1 OF 6)

Polymer	Modulus of Elasticity in Flexure (ASTM D790) (10^5 psi)
Polyester, Thermoset Cast: Flexible	0.001—0.39
Olefin Copolymer, Molded: Propylene—ethylene	0.00140
Olefin Copolymer, Molded: Ethylene butene	0.00165
Polyethylene, Type I: Melt index 200	0.1 (ASTM D747)
Polyethylene, Type I: Melt index 6—26	0.12—0.3 (ASTM D747)
Polyethylene, Type I: Melt index 0.3—3.6	0.13—0.27 (ASTM D747)
Polyethylene, Type II: Melt index 20	0.35—0.5 (ASTM D747)
Polyethylene, Type II: Melt index l.0—1.9	0.35—0.5 (ASTM D747)
Epoxy, Standard: Cast flexible	0.36—3.9
Nylon, Type 8	0.4
Polytetrafluoroethylene (PTFE)	0.6—1.1
Cellulose Acetate Butyrate, ASTM Grade: S2	0.70—0.90 (ASTM D747)
Olefin Copolymer, Molded: Polyallomer	0.7—1.3
Polyethylene, Type III: High molecular weight	0.75 (ASTM D747)
Fluorinated ethylene propylene(FEP)	0.8
Polyethylene, Type III: Melt Melt index 0.l—12.0	0.9—0.25 (ASTM D747)
Nylon, Type 6: Flexible copolymers	0.92—3.2
Nylon, Type 6: Glass fiber (30%) reinforced	1.0—1.4
Polypropylene: High impact	1.0—2.0
Polyester, Thermoset Cast: Rigid	1—9
Cellulose Acetate, ASTM Grade: S2—1	1.05—1.65 (ASTM D747)
Cellusose Acetate Propionate, ASTM Grade: 6	1.1
Cellulose Acetate Butyrate, ASTM Grade: MH	1.20—1.40 (ASTM D747)
Cellulose Acetate, ASTM Grade: MS—1, MS—2	1.25—1.90 (ASTM D747)

To convert from **psi to MPa, multiply by 145.**

Source: data compiled by J.S. Park from Charles T. Lynch, *CRC Handbook of Materials Science*, Vol. 3, CRC Press, Boca Raton, Florida, 1975 and *Engineered Materials Handbook*, Vol.2, Engineering Plastics, ASM International, Metals Park, Ohio, 1988.

Table 413. SELECTING MODULI OF ELASTICITY IN FLEXURE OF POLYMERS (SHEET 2 OF 6)

Polymer	Modulus of Elasticity in Flexure (ASTM D790) (10^5 psi)
Chlorinated polyether	1.3 (0.1% offset)
Polyethylene, Type III: Melt index 0.2—0.9	1.3—1.5 (ASTM D747)
Nylon, Type 6: General purpose	1.4—3.9
Cellusose Acetate Propionate, ASTM Grade: 3	1.45—1.55
Polyethylene, Type III: Melt index 1.5—15	1.5 (ASTM D747)
Cellulose Acetate, ASTM Grade: MH—1, MH—2	1.50—2.15 (ASTM D747)
Cellulose Acetate, ASTM Grade: H2—1	1.50—2.35 (ASTM D747)
Nylon, Type 11	1.51
6/10 Nylon: General purpose	1.6—2.8
Cellusose Acetate Propionate, ASTM Grade: 1	1.7—1.8
Polypropylene: General purpose	1.7—2.5
Polyvinylidene— fluoride (PVDF)	1.75—2.0
6/6 Nylon: General purpose extrusion	1.75—4.1
Cellulose Acetate Butyrate, ASTM Grade: H4	1.8 (ASTM D747)
Polypropylene: Flame retardant	1.9—6.1
Polytrifluoro chloroethylene (PTFCE)	2.0—2.5
Cellulose Acetate, ASTM Grade: H4—1	2.0—2.55 (ASTM D747)
ABS Resins; Molded, Extruded: Very high impact	2.0—3.2
ABS Resins; Molded, Extruded: Low temperature impact	2.0—3.2
Polystyrene; Molded: High impact	2.3—4.0
ABS Resins; Molded, Extruded: High impact	2.5—3.2
Thermoset Allyl diglycol carbonate	2.5—3.3
Acrylic Moldings: High impact grade	2.7—3.6
PVC–acrylic injection molded	3

To convert from **psi** to **MPa**, multiply by **145.**

Source: data compiled by J.S. Park from Charles T. Lynch, *CRC Handbook of Materials Science*, Vol. 3, CRC Press, Boca Raton, Florida, 1975 and *Engineered Materials Handbook*, Vol.2, Engineering Plastics, ASM International, Metals Park, Ohio, 1988.

Table 413. SELECTING MODULI OF ELASTICITY IN FLEXURE OF POLYMERS (SHEET 3 OF 6)

Polymer	Modulus of Elasticity in Flexure (ASTM D790) (10^5 psi)
Polycarbonate	3.4
Polyester, Injection Moldings: General purpose grade	3.4
Polypropylene: Asbestos filled	3.4—6.5
Rubber phenolic—chopped fabric filled	3.5
ABS Resins; Molded, Extruded: Medium impact	3.5—4.0
ABS Resins; Molded, Extruded: Heat resistant	3.5—4.2
Acrylic Cast Resin Sheets, Rods: General purpose, type I	3.5—4.5
Acrylic Moldings: Grades 5, 6, 8	3.5—5.0
Polystyrene; Molded: Medium impact	3.5—5.0
Phenylene Oxide: SE—100	3.6
Phenylene Oxide: SE—1	3.6
Polyacetal Copolymer: Standard	3.75
Polyacetal Copolymer: High flow	3.75
Polyvinyl Chloride And Copolymers: Rigid—normal impact	3.8—5.4
Chlorinated polyvinyl chloride	3.85
Phenylene oxides (Noryl): Standard	3.9
ABS–Polycarbonate Alloy	4
PVC–acrylic sheet	4
Polyacetal Homopolymer: 22% TFE reinforced	4
Polyarylsulfone	4
Acrylic Cast Resin Sheets, Rods: General purpose, type II	4.0—5.0
Epoxy, High performance: Cast, rigid	4—5
Polystyrene; Molded: General purpose	4—5
Rubber phenolic—woodflour or flock filled	4—6

To convert from **psi** to **MPa**, multiply by 145.

Source: data compiled by J.S. Park from Charles T. Lynch, *CRC Handbook of Materials Science*, Vol. 3, CRC Press, Boca Raton, Florida, 1975 and *Engineered Materials Handbook*, Vol.2, Engineering Plastics, ASM International, Metals Park, Ohio, 1988.

Table 413. SELECTING MODULI OF ELASTICITY IN FLEXURE OF POLYMERS (SHEET 4 OF 6)

Polymer	Modulus of Elasticity in Flexure (ASTM D790) (10^5 psi)
Polypropylene: Glass reinforced	4—8.2
Polyacetal Homopolymer: Standard	4.1
6/6 Nylon: General purpose molding	4.1—4.5, 1.75
Epoxy novolacs: Cast, rigid	4.4—4.8
Epoxy, Standard: Cast rigid	4.5—5.4
Ceramic reinforced (PTFE)	4.64
Rubber phenolic—asbestos filled	5
Polymide: Unreinforced	5—7
Nylon, Type 6: Cast	5.05
Polyphenylene sulfide: Standard	5.5—6.0
Phenylene Oxide: Glass fiber reinforced	7.4—10.4
Phenolic, Molded: General: woodflour and flock filled	8—12
Phenolic, Molded: Shock: paper, flock, or pulp filled	8—12
6/10 Nylon: Glass fiber (30%) reinforced	8.5
Polyacetal Homopolymer: 20% glass reinforced	8.8
Phenolic, Molded: High shock: chopped fabric or cord filled	9—13
Melamine, Molded: Unfilled	10—13
Melamine, Molded: Cellulose filled electrical	10—13
6/6 Nylon: Glass fiber reinforced	10—18
Phenolics: Molded: Arc resistant—mineral	10—30
Polyacetal Copolymer: 25% glass reinforced	11
6/6 Nylon: Glass fiber Molybdenum disulfide filled	11—13
Polycarbonate (40% glass fiber reinforced)	12
Polyester, Moldings: Glass reinforced self extinguishing	12

To convert from **psi** to **MPa**, multiply by **145**.

Source: data compiled by J.S. Park from Charles T. Lynch, *CRC Handbook of Materials Science*, Vol. 3, CRC Press, Boca Raton, Florida, 1975 and *Engineered Materials Handbook*, Vol.2, Engineering Plastics, ASM International, Metals Park, Ohio, 1988.

Table 413. SELECTING MODULI OF ELASTICITY IN FLEXURE OF POLYMERS (SHEET 5 OF 6)

Polymer	Modulus of Elasticity in Flexure (ASTM D790) (10^5 psi)
Polystyrene; Molded: Glass fiber -30% reinforced	12
Phenylene oxides (Noryl): Glass fiber reinforced	12, 15.5
Polyester, Thermoplastic Moldings: Glass reinforced grades	12—15
Silicone, Molded: Granular (silica) reinforced	14—17
Glass fiber (30%) reinforced Styrene acrylonitrile (SAN)	14.5
Reinforced polyester sheet molding: general purpose	15—18
Epoxy, Standard: Molded	15—25
Reinforced polyester moldings: High strength (glass fibers)	15—25
Polyphenylene sulfide: 40% glass reinforced	17—22
Alkyds, Molded Rope (general purpose)	22—27
Alkyds, Molded: Granular (high speed molding)	22—27
Alkyds, Molded: Glass reinforced (heavy duty parts)	22—28
Melamine, Molded: Glass fiber filled	24
Silicone, Molded: Fibrous (glass) reinforced	25
Silicone, Molded: Woven glass fabric/ silicone laminate	26—32
Epoxy, High performance: Glass cloth laminate	28—31
Phenolic, Molded: Very high shock: glass fiber filled	30—33
Epoxy novolacs: Glass cloth laminate	32—35
Polyester, Thermoplastic Moldings: General purpose grade	33
Epoxy, Standard: General purpose glass cloth laminate	36—39

To convert from **psi** to **MPa**, multiply by **145.**

Source: data compiled by J.S. Park from Charles T. Lynch, *CRC Handbook of Materials Science*, Vol. 3, CRC Press, Boca Raton, Florida, 1975 and *Engineered Materials Handbook*, Vol.2, Engineering Plastics, ASM International, Metals Park, Ohio, 1988.

Table 413. SELECTING MODULI OF ELASTICITY IN FLEXURE OF POLYMERS (SHEET 6 OF 6)

Polymer	Modulus of Elasticity in Flexure (ASTM D790) (10^5 psi)
Polyimide: Glass reinforced	38.4
Epoxy, Standard: High strength laminate	53—55
Epoxy, Standard: Filament wound composite	69—75
Polyester, Thermoplastic Moldings: Glass reinforced grade	87
Polyester, Thermoplastic Moldings: Asbestos—filled grade	90

To convert from psi to MPa, multiply by 145.

Source: data compiled by J.S. Park from Charles T. Lynch, *CRC Handbook of Materials Science*, Vol. 3, CRC Press, Boca Raton, Florida, 1975 and *Engineered Materials Handbook*, Vol.2, Engineering Plastics, ASM International, Metals Park, Ohio, 1988.

Table 414. SELECTING SHEAR MODULI OF GLASS
(SHEET 1 OF 2)

Glass	Temperature	Shear Modulus (GPa)
B_2O_3 glass	300°C	4.75
B_2O_3 glass	290°C	5.15
B_2O_3 glass	280°C	5.49
B_2O_3 glass	270°C	5.78
B_2O_3 glass	260°C	6.07
B_2O_3 glass	250°C	6.29
B_2O_3 glass	room temp.	6.55
B_2O_3–Na_2O glass (10% mol Na_2O)	15°C	12.3
SiO_2–PbO glass (65.0% mol PbO)		16.1
B_2O_3–Na_2O glass (20% mol Na_2O)	15°C	16.8
SiO_2–PbO glass (60.0% mol PbO)		17.0
SiO_2–PbO glass (50.0% mol PbO)		17.5
SiO_2–PbO glass (35.7% mol PbO)		18.5
SiO_2–PbO glass (55.0% mol PbO)		20.2
SiO_2–PbO glass (24.6% mol PbO)		20.4
B_2O_3–Na_2O glass (25% mol Na_2O)	15°C	21.1
SiO_2–PbO glass (45.0% mol PbO)		21.2
SiO_2–PbO glass (30.0% mol PbO)		21.4
B_2O_3–Na_2O glass (37% mol Na_2O)	15°C	22.4
SiO_2–PbO glass (38.4% mol PbO)		23.0
B_2O_3–Na_2O glass (33.3% mol Na_2O)	15°C	23.2
SiO_2–Na_2O glass (35% mol Na_2O)	room temp.	24.1
SiO_2–Na_2O glass (18% mol Na_2O)	160°C	24.2
SiO_2–Na_2O glass (33% mol Na_2O)	room temp.	24.2

Source: data compiled by J.S. Park from O. V. Mazurin, M. V. Streltsina and T. P. Shvaiko–Shvaikovskaya, *Handbook of Glass Data*, Part A and Part B, Elsevier, New York, 1983.

Table 414. SELECTING SHEAR MODULI OF GLASS

(SHEET 2 OF 2)

Glass	Temperature	Shear Modulus (GPa)
SiO_2–Na_2O glass (30% mol Na_2O)	room temp.	24.5
SiO_2–Na_2O glass (18% mol Na_2O)	80°C	24.8
SiO_2–Na_2O glass (18% mol Na_2O)	0°C	25.0
SiO_2–Na_2O glass (25% mol Na_2O)	room temp.	25.2
SiO_2–Na_2O glass (18% mol Na_2O)	–100°C	25.8
SiO_2–Na_2O glass (20% mol Na_2O)	room temp.	25.8
SiO_2–Na_2O glass (7.5% mol Na_2O)	–100—160°C	26.9
SiO_2–Na_2O glass (5% mol Na_2O)	–100°C	27.2
SiO_2–Na_2O glass (5% mol Na_2O)	160°C	27.2
SiO_2–Na_2O glass (15% mol Na_2O)	room temp.	27.2
SiO_2–Na_2O glass (5% mol Na_2O)	0°C	27.4
SiO_2–Na_2O glass (5% mol Na_2O)	80°C	27.6
SiO_2 glass	20°C	31.38
SiO_2 glass	998°C (annealing point)	33.57
SiO_2 glass	1096°C (straining point)	34.15

Source: data compiled by J.S. Park from O. V. Mazurin, M. V. Streltsina and T. P. Shvaiko–Shvaikovskaya, *Handbook of Glass Data*, Part A and Part B, Elsevier, New York, 1983.

Table 415. SELECTING TORSIONAL MODULI OF GRAY CAST IRONS

ASTM Class	Torsional Modulus (GPa)
20	27 to 39
25	32 to 41
30	36 to 45
35	40 to 48
40	44 to 54
50	50 to 55
60	54 to 59

Source: data from *ASM Metals Reference Book*, Second Edition, American Society for Metals, Metals Park, Ohio 44073, p166-167, (1984).

Table 416. SELECTING TORSIONAL MODULI OF TREATED DUCTILE IRONS

Treatment	Torsion Modulus (GPa)
80-55-06	62
60-40-18	63
120-90-02	63.4
65-45-12	64

Source: data from *ASM Metals Reference Book*, Second Edition, American Society for Metals, Metals Park, Ohio 44073, p169-170, (1984).

Table 417. SELECTING MODULI OF RUPTURE FOR CERAMICS
(SHEET 1 OF 5)

Ceramic	Temperature (°C)	Modulus of Rupture (psi)
Boron Nitride (BN) parallel to c axis	1000	$1.08x10^3$
Boron Nitride (BN) parallel to c axis	1500	$1.25x10^3$
Boron Nitride (BN) parallel to c axis	1800	$1.50x10^3$
Boron Nitride (BN) parallel to c axis	700	$1.90x10^3$
Boron Nitride (BN) parallel to a axis	1000	$2.18x10^3$
Boron Nitride (BN) parallel to c axis	2000	$2.45x10^3$
Zirconium Monocarbide (ZrC)	2000	$2.5x10^3$
Cordierite ($2MgO\ 2Al_2O_3\ 5SiO_2$) (ρ=1.8g/cm^3)	1200	$3.4x10^3$
Boron Nitride (BN) parallel to a axis	700	$3.84x10^3$
Hafnium Monocarbide (HfC) (ρ = 11.9 g/cm^3)	2200	$4.78x10^3$
Zirconium Monocarbide (ZrC)	1750	$5.14x10^3$
Titanium Diboride (TiB_2) (98% dense)		$5.37x10^3$
Titanium Diboride (TiB_2) (3.5 μm grain size, ρ=4.37g/cm^3, 0.8wt% Ni)		$5.7x10^3$
Mullite ($3Al_2O_3\ 2SiO_2$)	25	$6\text{-}27x10^3$
Titanium Diboride (TiB_2) (6.0 μm grain size, ρ=4.46g/cm^3)		$6.2x10^3$
Titanium Diboride (TiB_2) (12.0 μm grain size, ρ=4.66g/cm^3, 9.6wt% Ni)		$6.29x10^3$
Boron Nitride (BN) parallel to c axis	300	$7.03x10^3$
Trisilicon Tetranitride (Si_3N_4) (reaction sintered)	20	$7.25\text{-}43.5x10^3$

To convert from **psi** to **MPa**, multiply by 145.

Source: data compiled by J.S. Park from *No. 1 Materials Index*, Peter T.B. Shaffer, Plenum Press, New York, (1964); *Smithells Metals Reference Book*, Eric A. Brandes, ed., in association with Fulmer Research Institute Ltd. 6th ed. London, Butterworths, Boston, (1983); and *Ceramic Source*, American Ceramic Society (1986-1991).

Table 417. SELECTING MODULI OF RUPTURE FOR CERAMICS
(SHEET 2 OF 5)

Ceramic	Temperature (°C)	Modulus of Rupture (psi)
Boron Nitride (BN) parallel to c axis	25	$7.28\text{-}13.2x10^3$
Cordierite ($2MgO\ 2Al_2O_3\ 5SiO_2$) (ρ=2.1g/cm^3)	800	$8x10^3$
Zirconium Monocarbide (ZrC)	1250	$8.3x10^3$
Mullite ($3Al_2O_3\ 2SiO_2$) (ρ=2.77g/cm^3)	25	$8.5x10^3$
Titanium Oxide (TiO_2)	room temp.	$10\text{-}14.9x10^3$
Hafnium Dioxide (HfO_2)		$10x10^3$
Titanium Diboride (TiB_2) (6.0 μm grain size, ρ=4.56g/cm^3, 0.16wt% Ni)		$11.0x10^3$
Silicon Carbide (SiC)	1400	$11x10^3$
Mullite ($3Al_2O_3\ 2SiO_2$) (ρ=2.77g/cm^3)	1200	$11.5x10^3$
Hafnium Monocarbide (HfC) (ρ = 11.9 g/cm^3)	2000	$12.64x10^3$
Titanium mononitride (TiN) (10wt% AlO & 10wt% AlN)		$13.34x10^3$
Mullite ($3Al_2O_3\ 2SiO_2$) (ρ=2.77g/cm^3)	400	$13.5x10^3$
Titanium Monocarbide (TiC) (ρ = 4.85 g/cm^3)	2000	$13.6x10^3$
Silicon Carbide (SiC)	1800	$15x10^3$
Cordierite ($2MgO\ 2Al_2O_3\ 5SiO_2$) (ρ=2.3g/cm^3)	400	$15x10^3$
Boron Nitride (BN) parallel to a axis	300	$15.14x10^3$
Boron Nitride (BN) parallel to a axis	25	$15.88x10^3$
Cordierite ($2MgO\ 2Al_2O_3\ 5SiO_2$) (ρ=2.51g/cm^3)	25	$16x10^3$
Zirconium Monocarbide (ZrC)	room temp.	$16.6\text{-}22.5x10^3$

To convert from **psi** to **MPa**, multiply by **145**.

Source: data compiled by J.S. Park from *No. 1 Materials Index*, Peter T.B. Shaffer, Plenum Press, New York, (1964); *Smithells Metals Reference Book*, Eric A. Brandes, ed., in association with Fulmer Research Institute Ltd. 6th ed. London, Butterworths, Boston, (1983); and *Ceramic Source*, American Ceramic Society (1986-1991).

Table 417. SELECTING MODULI OF RUPTURE FOR CERAMICS
(SHEET 3 OF 5)

Ceramic	Temperature (°C)	Modulus of Rupture (psi)
Mullite ($3Al_2O_3$ $2SiO_2$) (ρ=2.77g/cm^3)	800	16.7x10^3
Aluminum Nitride (AlN)	1400	18.1x10^3
Molybdenum Disilicide ($MoSi_2$) (ρ = 5.57 g/cm^3)	room temp.	18.57x10^3
Titanium Diboride (TiB_2)		19x10^3
Zirconium Oxide (ZrO_2) (5-10 CaO stabilized)	room temp.	20-35x10^3
Titanium mononitride (TiN) (30wt% AlO & 10wt% AlN)		23.93x10^3
Beryllium Oxide (BeO)	room temp.	24-29 x10^3
Silicon Carbide (SiC)	1300	25x10^3
Silicon Carbide (SiC)	room temp.	27x10^3
Aluminum Nitride (AlN)	1000	27x10^3
Aluminum Oxide (Al_2O_3) (80% dense, 20μm grain size)	600	28x10^3
Aluminum Oxide (Al_2O_3) (80% dense, 20μm grain size)	20	30x10^3
Aluminum Oxide (Al_2O_3) (80% dense, 20μm grain size)	1100	30x10^3
Zirconium Oxide (ZrO_2) (MgO stabilized)	room temp.	30x10^3
Aluminum Oxide (Al_2O_3) (80% dense, 20μm grain size)	900	31x10^3
Titanium Monocarbide (TiC) (ρ = 4.85 g/cm^3)	room temp.	32.67x10^3
Titanium mononitride (TiN) (30wt% AlO & 30wt% AlN)		33.25x10^3
Titanium mononitride (TiN)		34x10^3
Hafnium Monocarbide (HfC) (ρ = 11.9 g/cm^3)	room temp.	34.67x10^3
Molybdenum Disilicide ($MoSi_2$) (hot pressed)	room temp.	36-57x10^3

To convert from **psi** to **MPa**, multiply by 145.

Source: data compiled by J.S. Park from *No. 1 Materials Index*, Peter T.B. Shaffer, Plenum Press, New York, (1964); *Smithells Metals Reference Book*, Eric A. Brandes, ed., in association with Fulmer Research Institute Ltd. 6th ed. London, Butterworths, Boston, (1983); and *Ceramic Source*, American Ceramic Society (1986-1991).

Table 417. SELECTING MODULI OF RUPTURE FOR CERAMICS
(SHEET 4 OF 5)

Ceramic	Temperature (°C)	Modulus of Rupture (psi)
Dichromium Trioxide (Cr_2O_3)		>38x10^3
Aluminum Nitride (AlN) (hot pressed)	25	38.5x10^3
Trisilicon Tetranitride (Si_3N_4) (sintered)	20	39.9-121.8x10^3
Silicon Carbide (SiC) (with 1wt% B addictive)		42x10^3
Aluminum Oxide (Al_2O_3) (80% dense, 3μm grain size)	1100	42x10^3
Molybdenum Disilicide ($MoSi_2$) (sintered)	room temp.	50.7x10^3
Molybdenum Disilicide ($MoSi_2$) (hot pressed)	1200	55.00x10^3
Tungsten Monocarbide (WC)	room temp.	55.65-84x10^3
Aluminum Oxide (Al_2O_3) (80% dense, 3μm grain size)	20	56x10^3
Silicon Carbide (SiC) (with 1 wt% Be addictive)		58x10^3
Aluminum Oxide (Al_2O_3) (80% dense, 3μm grain size)	900	58x10^3
Aluminum Oxide (Al_2O_3)	room temp.	60 x10^3
Aluminum Oxide (Al_2O_3) (80% dense, 3μm grain size)	600	62x10^3
Trisilicon Tetranitride (Si_3N_4) (hot pressed)	20	65.3-159.5x10^3
Molybdenum Disilicide ($MoSi_2$) (sintered)	980	67.25x10^3
Molybdenum Disilicide ($MoSi_2$) (hot pressed)	1090	72.00x10^3
Molybdenum Disilicide ($MoSi_2$) (sintered)	1090	86.00x10^3
Aluminum Oxide (Al_2O_3) (single crystal)		131 x10^3
Silicon Carbide (SiC) (with 1wt% Al addictive)		136x10^3
Aluminum Oxide (Al_2O_3) (zirconia toughened alumina, 15 vol% ZrO_2)		137x10^3

To convert from **psi** to **MPa**, multiply by **145**.

Source: data compiled by J.S. Park from *No. 1 Materials Index*, Peter T.B. Shaffer, Plenum Press, New York, (1964); *Smithells Metals Reference Book*, Eric A. Brandes, ed., in association with Fulmer Research Institute Ltd. 6th ed. London, Butterworths, Boston, (1983); and *Ceramic Source*, American Ceramic Society (1986-1991).

Table 417. SELECTING MODULI OF RUPTURE FOR CERAMICS
(SHEET 5 OF 5)

Ceramic	Temperature (°C)	Modulus of Rupture (psi)
Aluminum Oxide (Al_2O_3) (zirconia toughened alumina, 25 vol% ZrO_2)		$139x10^3$
Aluminum Oxide (Al_2O_3) (zirconia toughened alumina, 50 vol% ZrO_2)		$145x10^3$
Zirconium Oxide (ZrO_2) (sintered yittria doped zirconia)		$148x10^3$
Zirconium Oxide (ZrO_2) (hot pressed yittria doped zirconia)		$222x10^3$

To convert from **psi** to **MPa**, multiply by 145.

Source: data compiled by J.S. Park from *No. 1 Materials Index*, Peter T.B. Shaffer, Plenum Press, New York, (1964); *Smithells Metals Reference Book*, Eric A. Brandes, ed., in association with Fulmer Research Institute Ltd. 6th ed. London, Butterworths, Boston, (1983); and *Ceramic Source*, American Ceramic Society (1986-1991).

Table 418. SELECTING POISSON'S RATIOS FOR CERAMICS

(SHEET 1 OF 2)

Ceramic	Poisson's Ratio
Titanium Diboride (TiB_2)	0.09—0.28
Titanium Diboride (TiB_2) (6.0 μm grain size, ρ=4.46g/cm^3)	0.10
Titanium Diboride (TiB_2) (6.0 μm grain size, ρ=4.56g/cm^3, 0.16wt% Ni)	0.11
Titanium Diboride (TiB_2) (3.5 μm grain size, ρ=4.37g/cm^3, 0.8wt% Ni)	0.12
Zirconium Diboride (ZrB_2)	0.144
Titanium Diboride (TiB_2) (12.0 μm grain size, ρ=4.66g/cm^3, 9.6wt% Ni)	0.15
Molybdenum Disilicide ($MoSi_2$)	0.158—0.172
Magnesium Oxide (MgO) (ρ = 3.506 g/cm^3) (room temp)	0.163
Hafnium Monocarbide (HfC)	0.166
Cordierite (2MgO $2Al_2O_3$ $5SiO_2$) (ρ=2.1g/cm^3)	0.17
Tantalum Monocarbide (TaC)	0.1719—0.24
Silicon Carbide (SiC) (ρ = 3.128 g/cm^3) (room temp)	0.183—0.192
Titanium Monocarbide (TiC)	0.187—189
Boron Carbide (B_4C)	0.207
Cordierite (2MgO $2Al_2O_3$ $5SiO_2$) (ρ=2.3g/cm^3)	0.21
Aluminum Oxide (Al_2O_3)	0.21—0.27
Trisilicon tetranitride (Si_3N_4) (presureless sintered)	0.22—0.27
Zirconium Oxide (ZrO_2) (partially stabilized)	0.23
Zirconium Oxide (ZrO_2) (fully stabilized)	0.23—0.32

Source: data compiled by J.S. Park from *No. 1 Materials Index*, Peter T.B. Shaffer, Plenum Press, New York, (1964); *Smithells Metals Reference Book*, Eric A. Brandes, ed., in association with Fulmer Research Institute Ltd. 6th ed. London, Butterworths, Boston, (1983); and *Ceramic Source*, American Ceramic Society (1986-1991)

Table 418. SELECTING POISSON'S RATIOS FOR CERAMICS

(SHEET 2 OF 2)

Ceramic	Poisson's Ratio
Mullite ($3Al_2O_3$ $2SiO_2$) (ρ=2.779 g/cm^3)	0.238
Tungsten Monocarbide (WC)	0.24
Trisilicon tetranitride (Si_3N_4)	0.24
Zirconium Oxide (ZrO_2) (plasma sprayed)	0.25
Zirconium Monocarbide (ZrC) (ρ = 6.118 g/cm^3)	0.257
Cordierite ($2MgO$ $2Al_2O_3$ $5SiO_2$) (glass)	0.26
Beryllium Oxide (BeO)	0.26—0.34
Cerium Dioxide (CeO_2)	0.27—0.31
Thorium Dioxide (ThO_2) (ρ=9.722 g/cm^3)	0.275
Titanium Oxide (TiO_2)	0.28
Spinel (Al_2O_3 MgO) (ρ=3.510 g/cm^3)	0.294
Uranium Dioxide (UO_2) (ρ=10.37 g/cm^3)	0.302
Zirconium Oxide (ZrO_2) (room temp)	0.324—0.337

Source: data compiled by J.S. Park from *No. 1 Materials Index*, Peter T.B. Shaffer, Plenum Press, New York, (1964); *Smithells Metals Reference Book*, Eric A. Brandes, ed., in association with Fulmer Research Institute Ltd. 6th ed. London, Butterworths, Boston, (1983); and *Ceramic Source*, American Ceramic Society (1986-1991)

Table 419. SELECTING POISSON'S RATIOS OF GLASS
(SHEET 1 OF 2)

Glass	Temperature (°C)	Poisson's Ratio
SiO_2–PbO glass (38.4% mol PbO)		0.150
SiO_2 glass	room temp.	0.166–0.177
SiO_2–PbO glass (30.0% mol PbO)		0.174
SiO_2–Na_2O glass (15% mol Na_2O)	room temp.	0.183
SiO_2–Na_2O glass (20% mol Na_2O)	room temp.	0.203
SiO_2–Na_2O glass (25% mol Na_2O)	room temp.	0.219
SiO_2–PbO glass (45.0% mol PbO)		0.219
SiO_2–PbO glass (55.0% mol PbO)		0.222
SiO_2–Na_2O glass (30% mol Na_2O)	room temp.	0.236
SiO_2–Na_2O glass (35% mol Na_2O)	room temp.	0.248
SiO_2–Na_2O glass (33% mol Na_2O)	room temp.	0.249
SiO_2–PbO glass (24.6% mol PbO)		0.249
SiO_2–PbO glass (35.7% mol PbO)		0.252
SiO_2–PbO glass (50.0% mol PbO)		0.259
B_2O_3–Na_2O glass (15.4% mol Na_2O)		0.271
B_2O_3–Na_2O glass (25% mol Na_2O)	15	0.2713
B_2O_3–Na_2O glass (22.8% mol Na_2O)		0.272
B_2O_3–Na_2O glass (37% mol Na_2O)	15	0.2739
B_2O_3–Na_2O glass (29.8% mol Na_2O)		0.274
B_2O_3–Na_2O glass (10% mol Na_2O)	15	0.2740

Source: data compiled by J.S. Park from O. V. Mazurin, M. V. Streltsina and T. P. Shvaiko–Shvaikovskaya, *Handbook of Glass Data*, Part A and Part B, Elsevier, New York, 1983

Table 419. SELECTING POISSON'S RATIOS OF GLASS
(SHEET 2 OF 2)

Glass	Temperature (°C)	Poisson's Ratio
B_2O_3–Na_2O glass (33.3% mol Na_2O)	15	0.2771
B_2O_3–Na_2O glass (5.5% mol Na_2O)		0.279
SiO_2–PbO glass (60.0% mol PbO)		0.281
SiO_2–PbO glass (65.0% mol PbO)		0.283
B_2O_3–Na_2O glass (20% mol Na_2O)	15	0.2860
B_2O_3 glass	room temp.	0.288–0.309
B_2O_3–Na_2O glass (37.25% mol Na_2O)		0.292

Source: data compiled by J.S. Park from O. V. Mazurin, M. V. Streltsina and T. P. Shvaiko–Shvaikovskaya, *Handbook of Glass Data*, Part A and Part B, Elsevier, New York, 1983

Table 420. SELECTING COMPRESSION POISSON'S RATIOS OF TREATED DUCTILE IRONS

Treatment	Compression Poisson's Ratio
60-40-18	0.26
120 90-02	0.27
65-45-12	0.31
80-55-06	0.31

Source: data from *ASM Metals Reference Book*, Second Edition, American Society for Metals, Metals Park, Ohio 44073, p169-170, (1984).

Table 421. SELECTING TORSION POISSON'S RATIOS OF TREATED DUCTILE IRONS

Treatment	Torsion Poisson's Ratio
120 90-02	0.28
60-40-18	0.29
65-45-12	0.29
80-55-06	0.31

Source: data from *ASM Metals Reference Book*, Second Edition, American Society for Metals, Metals Park, Ohio 44073, p169-170, (1984).

Table 422. SELECTING ELONGATION OF TOOL STEELS

Type	Condition	Elongation (%)
L6	Oil quenched from 845 •C and single tempered at 315 •C	4
S1	Oil quenched from 930 •C and single tempered at 315 •C	4
L2	Oil quenched from 855 •C and single tempered at 205 •C	5
S1	Oil quenched from 930 •C and single tempered at 425 •C	5
S5	Oil quenched from 870 •C and single tempered at 205 •C	5
S5	Oil quenched from 870 •C and single tempered at 315 •C	7
S7	Fan cooled from 940 •C and single tempered at 205 •C	7
L6	Oil quenched from 845 •C and single tempered at 425 •C	8
S1	Oil quenched from 930 •C and single tempered at 540 •C	9
S5	Oil quenched from 870 •C and single tempered at 425 •C	9
S7	Fan cooled from 940 •C and single tempered at 315 •C	9
L2	Oil quenched from 855 •C and single tempered at 315 •C	10
S5	Oil quenched from 870 •C and single tempered at 540 •C	10
S7	Fan cooled from 940 •C and single tempered at 425 •C	10
S7	Fan cooled from 940 •C and single tempered at 540 •C	10
L2	Oil quenched from 855 •C and single tempered at 425 •C	12
L6	Oil quenched from 845 •C and single tempered at 540 •C	12
S1	Oil quenched from 930 •C and single tempered at 650 •C	12
S7	Fan cooled from 940 •C and single tempered at 650 •C	14
L2	Oil quenched from 855 •C and single tempered at 540 •C	15
S5	Oil quenched from 870 •C and single tempered at 650 •C	15
L6	Oil quenched from 845 •C and single tempered at 650 •C	20
S1	Annealed	24
L2	Annealed	25
L2	Oil quenched from 855 •C and single tempered at 650 •C	25
L6	Annealed	25
S5	Annealed	25
S7	Annealed	25

Source: Data from *ASM Metals Reference Book*, Second Edition, American Society for Metals, Metals Park, Ohio 44073, p241, (1984).

Table 423. SELECTING ELONGATION OF DUCTILE IRONS

Specification Number	Grade or Class	Elongation (%)
ASTM A536-72; MIL-1-11466B(MR)	120-90-02	2
ASTM A476-70(d); SAE AMS5316	80-60-03	3
ASTM A536-72; MIL-1-11466B(MR)	100-70-03	3
SAE J434c	D7003	3
ASTM A536-72; MIL-1-11466B(MR)	80-55-06	6
SAE J434c	D5506	6
MIL-I-24137(Ships)	Class B	7
ASTM A536-72; MIL-1-11466B(MR)	65-45-12	12
SAE J434c	D4512	12
MIL-I-24137(Ships)	Class A	15
ASTM A395-76; ASME SA395	60-40-18	18
ASTM A536-72; MIL-1-11466B(MR)	60-40-18	18
SAE J434c	D4018	18
MIL-I-24137(Ships)	Class C	20

Source: data from *ASM Metals Reference Book*, Second Edition, American Society for Metals, Metals Park, Ohio 44073, p169, (1984).

Table 424. SELECTING ELONGATION OF MALLEABLE IRON CASTINGS

Specification Number	Grade or Class	Elongation (%)
ASTM A220; ANSI C48.2; MIL-I-11444B	90001	1
ASTM A602; SAE J158	M8501(b)	1
ASTM A220; ANSI C48.2; MIL-I-11444B	80002	2
ASTM A602; SAE J158	M7002(b)	2
ASTM A220; ANSI C48.2; MIL-I-11444B	70003	3
ASTM A602; SAE J158	M5003(a)	3
ASTM A602; SAE J158	M5503(b)	3
ASTM A220; ANSI C48.2; MIL-I-11444B	60004	4
ASTM A602; SAE J158	M4504(a)	4
ASTM A197		5
ASTM A220; ANSI C48.2; MIL-I-11444B	50005	5
ASTM A220; ANSI C48.2; MIL-I-11444B	45006	6
ASTM A220; ANSI C48.2; MIL-I-11444B	45008	8
ASTM A47, A338; ANSI G48.1; FED QQ-I-666c	32510	10
ASTM A220; ANSI C48.2; MIL-I-11444B	40010	10
ASTM A602; SAE J158	M3210	10
ASTM A47, A338; ANSI G48.1; FED QQ-I-666c	35018	18

(a) Air quenched and tempered
(b) Liquid quenched and tempered

Source: data from *ASM Metals Reference Book*, Second Edition, American Society for Metals, Metals Park, Ohio 44073, p171, (1984).

Table 425. SELECTING TOTAL ELONGATION OF CAST ALUMINUM ALLOYS (SHEET 1 OF 3)

Alloy AA No.	Temper	Elongation (in 2 in.) (%)
242.0	T571	0.5
242.0	T61	0.5
336.0	T551	0.5
336.0	T65	0.5
355.0	T7	0.5
A390.0	F,T5	<1.0
A390.0	T6	<1.0
A390.0	T7	<1.0
A390.0	T6	<1.0
A390.0	T7	<1.0
242.0	T21	1.0
242.0	T571	1.0
355.0	T61	1.0
390.0	F	1.0
390.0	T5	1.0
A390.0	F,T5	1.0
355.0	T51	1.5
355.0	T71	1.5
355.0	T62	1.5
242.0	T77	2.0
295.0	T62	2.0
308.0	F	2.0
319.0	F	2.0
319.0	T6	2.0
355.0	T51	2.0
355.0	T7	2.0
356.0	T51	2.0
356.0	T7	2.0

Source: data from *ASM Metals Reference Book*, Second Edition, American Society for Metals, Metals Park, Ohio 44073, (1984).

Table 425. SELECTING TOTAL ELONGATION OF CAST ALUMINUM ALLOYS (SHEET 2 OF 3)

Alloy AA No.	Temper	Elongation (in 2 in.) (%)
208.0	F	2.5
319.0	F	2.5
384.0, A384.0	F	2.5
413.0	F	2.5
319.0	T6	3.0
355.0	T6	3.0
355.0	T71	3.0
360.0	F	3.0
380.0	F	3.0
713.0	T5	3.0
356.0	T6	3.5
356.0	T71	3.5
383.0	F	3.5
A413.0	F	3.5
355.0	T6	4.0
713.0	T5	4.0
201.0	T7	4.5
296.0	T7	4.5
295.0	T6	5.0
296.0	T6	5.0
356.0	T6	5.0
A360.0	F	5.0
712.0	F	5.0
518.0	F	5.0—8.0
359.0	T62	5.5
354.0	T61	6.0
356.0	T7	6.0
359.0	T61	6.0

Source: data from *ASM Metals Reference Book,* Second Edition, American Society for Metals, Metals Park, Ohio 44073, (1984).

Table 425. SELECTING TOTAL ELONGATION OF CAST ALUMINUM ALLOYS (SHEET 3 OF 3)

Alloy AA No.	Temper	Elongation (in 2 in.) (%)
201.0	T6	7
357.0, A357.0	T62	8.0
443.0	F	8.0
295.0	T4	8.5
296.0	T4	9.0
C443.0	F	9.0
514.0	F	9.0
771.0	T6	9.0
B443.0	F	10.0
850.0	T5	10.0
206.0, A206.0	T7	11.7
535.0	F	13
520.0	T4	16
201.0	T4	20

Source: data from *ASM Metals Reference Book,* Second Edition, American Society for Metals, Metals Park, Ohio 44073, (1984).

Table 426. SELECTING TOTAL ELONGATION OF POLYMERS
(SHEET 1 OF 4)

Polymer	Elongation (in 2 in.), (ASTM D638) (%)
Polycarbonate (40% glass fiber reinforced)	0—5
Phenolic, Molded, Very high shock: glass fiber filled	0.2
Reinforced polyester moldings: High strength (glass fibers)	0.3—0.5
Phenolic, Molded, High shock: chopped fabric or cord filled	0.37—0.57
Phenolic, Molded, General: woodflour and flock filled	0.4—0.8
Styrene acrylonitrile (SAN)	0.5—4.5
Melamine, Molded: Cellulose electrical	0.6
Rubber phenolic—woodflour or flock filled	0.75—2.25
Polymide: Glass reinforced	<1
Polymide: Unreinforced	<1—1.2
Ureas; Molded: Alpha—cellulose filled (ASTM Type l)	1
Polystyrenes, Molded: General purpose	1.0—2.3
Polyvinyl Chloride & Copolymers: Rigid—normal impact	1—10
Polyester, Thermoplastic Moldings: Glass reinforced grades	1—5
Polystyrenes, Molded: Glass fiber -30% reinforced	1.1
Glass fiber (30%) reinforced Styrene acrylonitrile (SAN)	1.4—1.6
Epoxy, Standard: Cast flexible	1.5-60
Polyester, Thermoset Cast: Rigid	1.7—2.6
6/6 Nylon, Molded, Extruded: Glass fiber reinforced	1.8—2.2
6/10 Nylon: Glass fiber (30%) reinforced	1.9
Polypropylene: Glass reinforced	2—4
Epoxy, High performance: Cast, rigid	2—5
Acrylic Cast Resin Sheets, Rods: General purpose, type I	2—7
Acrylic Cast Resin Sheets, Rods: General purpose, type II	2—7
Nylon, Type 6: Glass fiber (30%) reinforced	2.2—3.6
Epoxy novolacs: Glass cloth laminate	2.2—4.8
Silicones: Fibrous (glass) reinforced silicones	<3 (ASTM D651)
Silicone: Granular (silica) reinforced	<3 (ASTM D651)

Source: data compiled by J.S. Park from Charles T. Lynch, *CRC Handbook of Materials Science*, Vol. 3, CRC Press, Boca Raton, Florida, 1975 and *Engineered Materials Handbook*, Vol.2, Engineering Plastics, ASM International, Metals Park, Ohio, 1988.

Table 426. SELECTING TOTAL ELONGATION OF POLYMERS
(SHEET 2 OF 4)

Polymer	Elongation (in 2 in.), (ASTM D638) (%)
6/6 Nylon: Glass fiber Molybdenum disulfide filled	3
Polyacetal Copolymer: 25% glass reinforced	3
Polyphenylene sulfide: Standard	3
Polystyrenes, Molded: Medium impact	3.0—40
Polypropylene: Flame retardant	3—15
Polypropylene: Asbestos filled	3—20
Acrylic Moldings: Grades 5, 6, 8	3—5
Polyphenylene sulfide: 40% glass reinforced	3—9
Phenylene Oxides: Glass fiber reinforced	4—6
Epoxy, Standard: Cast rigid	4.4
Polyester, Thermoplastic Moldings: Glass reinforced grade	<5
Polyester, Thermoplastic Moldings: Asbestos—filled grade	<5
Polyester, Thermoplastic: Glass reinforced self extinguishing	5
ABS Resins: Medium impact	5—20
ABS Resins: High impact	5—50
Polyacetal Homopolymer: 20% glass reinforced	7
Ceramic reinforced (PTFE)	10—200
Polyacetal Homopolymer: 22% TFE reinforced	12
Polyvinyl Chloride & Copolymers: Vinylidene chloride	15—30
Polyarylsulfone	15—40
6/6 Nylon: General purpose molding	15—60, 300
ABS Resins: Heat resistant	20
Nylon, Molded, Extruded Type 6: Cast	20
Olefin Copolymers, Molded: Ethylene butene	20
ABS Resins: Very high impact	20—50
Polyacetal Homopolymer: Standard	25
Acrylic Moldings: High impact grade	>25
Polyester, Thermoset Cast: Flexible	25—300

Source: data compiled by J.S. Park from Charles T. Lynch, *CRC Handbook of Materials Science*, Vol. 3, CRC Press, Boca Raton, Florida, 1975 and *Engineered Materials Handbook*, Vol.2, Engineering Plastics, ASM International, Metals Park, Ohio, 1988.

Table 426. SELECTING TOTAL ELONGATION OF POLYMERS
(SHEET 3 OF 4)

Polymer	Elongation (in 2 in.), (ASTM D638) (%)
Nylon, Molded, Extruded Type 6: General purpose	30—100
ABS Resins: Low temperature impact	30—200
Polypropylene: High impact	30—>200
Polyacetal Copolymer: High flow	40
Phenylene Oxides: SE—100	50
Phenylene oxides (Noryl): Standard	50—100
Polyethylene, Type III: Melt Melt index 0.l—12.0	50—l,000
Phenylene Oxides: SE—1	60
Polyacetal Copolymer: Standard	60—75
Polyethylene, Type I: Melt index 200	80—100 (ASTM D412)
6/10 Nylon: General purpose	85—220
6/6 Nylon: General purpose extrusion	90—240
PVC–Acrylic Alloy: sheet	>100
Nylon, Type 11	100—120
Polypropylene: General purpose	100—600
Polyethylene, Type III: Melt index 1.5—15	100—700
Polycarbonate	110
ABS–Polycarbonate Alloy	110
Nylon, Type 12	120—350
Polytrifluoro chloroethylene (PTFCE)	125—175
Polyethylene, Type I: Melt index 6—26	125—675 (ASTM D412)
Chlorinated polyether	130
PVC–Acrylic Alloy: injection molded	150
Polyethylene, Type II: Melt index 20	200
Polyvinylidene— fluoride (PVDF)	200—300
Nylon, Molded, Extruded Type 6: Flexible copolymers	200—320
Polyethylene, Type II: Melt index l.0—1.9	200—425
Polyvinyl Chloride & Copolymers: Nonrigid—general	200—450

Source: data compiled by J.S. Park from Charles T. Lynch, *CRC Handbook of Materials Science*, Vol. 3, CRC Press, Boca Raton, Florida, 1975 and *Engineered Materials Handbook*, Vol.2, Engineering Plastics, ASM International, Metals Park, Ohio, 1988.

Table 426. SELECTING TOTAL ELONGATION OF POLYMERS
(SHEET 4 OF 4)

Polymer	Elongation (in 2 in.), (ASTM D638) (%)
Polyvinyl Chloride & Copolymers: Nonrigid—electrical	220—360
Polyester, Thermoplastic Moldings: General purpose grade	250
Fluorinated ethylene propylene (FEP)	250—330
Polytetrafluoroethylene (PTFE)	250—350
Polyester, Thermoplastic Moldings: General purpose grade	300
Olefin Copolymers, Molded: Polyallomer	300—400
Nylon, Type 8	400
Polyethylene, Type III: High molecular weight	400
Olefin Copolymers, Molded: Ionomer	450
Polyethylene, Type I: Melt index 0.3—3.6	500—725 (ASTM D412)
Olefin Copolymers, Molded: EEA (ethylene ethyl acrylate)	650
Olefin Copolymers, Molded: EVA (ethylene vinyl acetate)	650
Polyethylene, Type III: Melt index 0.2—0.9	700—1,000

Source: data compiled by J.S. Park from Charles T. Lynch, *CRC Handbook of Materials Science*, Vol. 3, CRC Press, Boca Raton, Florida, 1975 and *Engineered Materials Handbook*, Vol.2, Engineering Plastics, ASM International, Metals Park, Ohio, 1988.

Table 427. SELECTING ELONGATION AT YIELD OF POLYMERS

Polymer	Elongation at Yield, (ASTM D638) (%)
Polystyrene: General purpose	1.0—2.3
Polystyrene: Glass fiber -30% reinforced	1.1
Polystyrene: Medium impact	1.2—3.0
Polyphenylene sulfide: 40% glass reinforced	1.25
Polystyrene: Glass fiber (30%) reinforced SAN	1.4—1.6
Polystyrene: High impact	1.5—2.0
Polyphenylene sulfide: Standard	1.6
Phenylene oxides (Noryl): Glass fiber reinforced	2—1.6
Polyacetal Copolymer: 25% glass reinforced	3
Polycarbonate	5
Nylon, Type 6: Cast	5
Polypropylene: Asbestos filled	5
6/6 Nylon: General purpose molding	5—25
6/6 Nylon: General purpose extrusion	5—30
6/10 Nylon: General purpose	5—30
Phenylene oxides (Noryl): Standard	5.6
Nylon, Type 12	5.8
Polyarylsulfone	6.5—13
Polypropylene: High impact	7—13
Polypropylene: General purpose	9—15
Polyacetal Homopolymer: Standard	12
Polyacetal Copolymer: Standard	12
Polyacetal Copolymer: High flow	12
Chlorinated polyether	15

Source: data compiled by J.S. Park from Charles T. Lynch, *CRC Handbook of Materials Science*, Vol. 3, CRC Press, Boca Raton, Florida, 1975 and *Engineered Materials Handbook*, Vol.2, Engineering Plastics, ASM International, Metals Park, Ohio, 1988.

Table 428. SELECTING AREA REDUCTION OF TOOL STEELS

Type	Condition	Area Reduction (%)
L6	Oil quenched from 845 •C and single tempered at 315 •C	9
S1	Oil quenched from 930 •C and single tempered at 315 •C	12
L2	Oil quenched from 855 •C and single tempered at 205 •C	15
S1	Oil quenched from 930 •C and single tempered at 425 •C	17
L6	Oil quenched from 845 •C and single tempered at 425 •C	20
S5	Oil quenched from 870 •C and single tempered at 205 •C	20
S7	Fan cooled from 940 •C and single tempered at 205 •C	20
S1	Oil quenched from 930 •C and single tempered at 540 •C	23
S5	Oil quenched from 870 •C and single tempered at 315 •C	24
S7	Fan cooled from 940 •C and single tempered at 315 •C	25
S5	Oil quenched from 870 •C and single tempered at 425 •C	28
S7	Fan cooled from 940 •C and single tempered at 425 •C	29
L2	Oil quenched from 855 •C and single tempered at 315 •C	30
L6	Oil quenched from 845 •C and single tempered at 540 •C	30
S5	Oil quenched from 870 •C and single tempered at 540 •C	30
S7	Fan cooled from 940 •C and single tempered at 540 •C	33
L2	Oil quenched from 855 •C and single tempered at 425 •C	35
S1	Oil quenched from 930 •C and single tempered at 650 •C	37
S5	Oil quenched from 870 •C and single tempered at 650 •C	40
L2	Oil quenched from 855 •C and single tempered at 540 •C	45
S7	Fan cooled from 940 •C and single tempered at 650 •C	45
L6	Oil quenched from 845 •C and single tempered at 650 •C	48
L2	Annealed	50
S5	Annealed	50
S1	Annealed	52
L2	Oil quenched from 855 •C and single tempered at 650 •C	55
L6	Annealed	55
S7	Annealed	55

Area Reduction in 50 mm or 2 in.

Source: data from ASM *Metals Reference Book*, Second Edition, American Society for Metals, Metals Park, Ohio 44073, p241, (1984).

CHAPTER 14

Selecting Electrical Properties

List of Tables

(continued)

List of Tables (Continued)

Table 429. SELECTING ELECTRICAL RESISTIVITY OF ALLOY CAST IRONS

Description	Electrical Resistivity ($\mu\Omega \cdot m$)
Corrosion–Resistant High– Silicon iron	0.50
Abrasion–Resistant Low–C White Iron	0.53
Heat–Resistant Medium–silicon Ductile Iron	0.58 to 0.87
Abrasion–Resistant Martensitic nickel–chromium White Iron	0.80
Corrosion–Resistant High–nickel gray iron	1.0[a]
Corrosion–Resistant High–nickel ductile iron	1.0[a]
Heat–Resistant High–nickel Ductile Iron (23 Ni)	1.0[a]
Heat–Resistant High–nickel Ductile Iron (20 Ni)	1.02
Heat–Resistant Gray High–nickel Iron	1.4 to 1.7
Heat–Resistant Nickel–chromium–silicon Gray Iron	1.5 to 1.7
Heat–Resistant High–aluminum Gray Iron	2.4

[a] Estimated.

Source: data from ASM *Metals Reference Book*, Second Edition, American Society for Metals, Metals Park, Ohio 44073, (1984).

Table 430. SELECTING RESISTIVITY OF CERAMICS

(SHEET 1 OF 5)

Ceramic	Temperature Range of Validity	Resistivity (Ω–cm)
Boron Carbide (B_4C)		0.3–0.8
Titanium Monocarbide (TiC)		0.3–0.8
Zircoium Oxide (ZrO_2) (stabilized)	2200°C	0.37
Zircoium Oxide (ZrO_2) (stabilized)	2000°C	0.59
Silicon Carbide (SiC) (with 1 wt% Al additive)		0.8
Zircoium Oxide (ZrO_2) (stabilized)	1700°C	1.6
Zircoium Oxide (ZrO_2) (stabilized)	1300°C	9.4
Zircoium Oxide (ZrO_2) (stabilized)	1200°C	77
Silicon Carbide (SiC)	20°C	10^2 –10^{12}
Magnesium Oxide (MgO)	1727°C	4x10^2
Zircoium Oxide (ZrO_2) (stabilized)	700°C	2300
Cordierite (2MgO $2Al_2O_3$ $5SiO_2$) (ρ=2.3g/cm^3)	900°C	1.9x10^4
Silicon Carbide (SiC) (with 1 wt% B additive)		2x10^4
Boron Nitride (BN)	1000°C	3.1x10^4
Cordierite (2MgO $2Al_2O_3$ $5SiO_2$) (ρ=2.3g/cm^3)	700°C	8.0x10^4
Cordierite (2MgO $2Al_2O_3$ $5SiO_2$) (ρ=2.1g/cm^3)	900°C	3.5x10^5
Cordierite (2MgO $2Al_2O_3$ $5SiO_2$) (ρ=1.8g/cm^3)	900°C	7.0x10^5
Cordierite (2MgO $2Al_2O_3$ $5SiO_2$) (ρ=2.3g/cm^3)	500°C	7.7x10^5
Zirconium Diboride (ZrB_2)	liquid air temperature	1.8x10^6
Aluminum Oxide (Al_2O_3)	1000°C	2x10^6

Source: data compiled by J.S. Park from *No. 1 Materials Index*, Peter T.B. Shaffer, Plenum Press, New York, (1964); *Smithells Metals Reference Book*, Eric A. Brandes, ed., in association with Fulmer Research Institute Ltd. 6th ed. London, Butterworths, Boston, (1983); and *Ceramic Source*, American Ceramic Society (1986–1991).

Table 430. SELECTING RESISTIVITY OF CERAMICS
(SHEET 2 OF 5)

Ceramic	Temperature Range of Validity	Resistivity (Ω–cm)
Cordierite (2MgO $2Al_2O_3$ $5SiO_2$) (ρ=2.1g/cm^3)	700°C	$3.0x10^6$
Titanium Diboride (TiB_2) (polycrystalline) (100% dense, extrapolated)	liquid air temp.	$3.7x10^6$
Zirconium Mononitirde (TiN)	liquid air	$3.97x10^6$
Cordierite (2MgO $2Al_2O_3$ $5SiO_2$) (ρ=1.8g/cm^3)	700°C	$4.7x10^6$
Titanium Diboride (TiB_2) (monocrystalline) (crystal length 5 cm, 39 deg. and 59 deg. orientation with respect to growth axis)	room temp.	$6.6\pm0.2x10^6$
Titanium Diboride (TiB_2) (monocrystalline) (crystal length 1.5 cm, 16.5 deg. and 90 deg. orientation with respect to growth axis)	room temp.	$6.7\pm0.2x10^6$
Tantalum Monocarbide (TaC) (80% dense)	4.2K	$8x10^6$
Titanium Mononitirde (TiN)	liquid air	$8.13x10^6$
Titanium Diboride (TiB_2) (polycrystalline) (100% dense, extrapolated)	room temp.	$8.7–14.1x10^6$
Titanium Diboride (TiB_2) (polycrystalline) (85% dense)	room temp.	$9.0x10^6$
Zirconium Diboride (ZrB_2)	20 °C	$9.2x10^6$
Tantalum Monocarbide (TaC) (80% dense)	80K	$10x10^6$
Hafnium Diboride (HfB_2)	room temp.	10–12 x 10^6
Titanium Mononitirde (TiN)	room temp.	$11.07–130x10^6$
Zirconium Mononitirde (TiN)	room temp.	$11.52–160x10^6$
Tantalum Monocarbide (TaC) (80% dense)	160K	$15x10^6$

Source: data compiled by J.S. Park from *No. 1 Materials Index*, Peter T.B. Shaffer, Plenum Press, New York, (1964); *Smithells Metals Reference Book*, Eric A. Brandes, ed., in association with Fulmer Research Institute Ltd. 6th ed. London, Butterworths, Boston, (1983); and *Ceramic Source*, American Ceramic Society (1986–1991).

Table 430. SELECTING RESISTIVITY OF CERAMICS
(SHEET 3 OF 5)

Ceramic	Temperature Range of Validity	Resistivity (Ω–cm)
Molybdenum Disilicide ($MoSi_2$)	–80˚C	$18.9x10^6$
Magnesium Oxide (MgO)	1000˚C	$0.2–1x10^8$
Tantalum Monocarbide (TaC) (80% dense)	240K	$20x10^6$
Chromium Diboride (CrB_2)		$21x10^6$
Molybdenum Disilicide ($MoSi_2$)	22˚C	$21.5x10^6$
Tantalum Monocarbide (TaC) (80% dense)	300K	$25x10^6$
Titanium Diboride (TiB_2) (polycrystalline) (85% dense)	room temp.	$26.5–28.4x10^6$
Cordierite (2MgO $2Al_2O_3$ $5SiO_2$) (ρ=2.3g/cm^3)	300˚C	$3.3x10^7$
Tungsten Disilicide (WSi_2)		$33.4–54.9x10^6$
Hafnium Monocarbide (HfC)	4.2K	$41x10^6$
Hafnium Monocarbide (HfC)	80K	$41x10^6$
Zirconium Monocarbide (ZrC)	4.2K	$41x10^6$
Hafnium Monocarbide (HfC)	160K	$45x10^6$
Zirconium Monocarbide (ZrC)	80K	$45x10^6$
Zirconium Monocarbide (ZrC)	160K	$47x10^6$
Cordierite (2MgO $2Al_2O_3$ $5SiO_2$) (ρ=1.8g/cm^3)	500˚C	$4.9x10^7$
Hafnium Monocarbide (HfC)	240K	$49x10^6$
Zirconium Monocarbide (ZrC)	240K	$53x10^6$
Hafnium Monocarbide (HfC)	300K	$60x10^6$
Zirconium Monocarbide (ZrC)	300K	$61–64x10^6$

Source: data compiled by J.S. Park from *No. 1 Materials Index*, Peter T.B. Shaffer, Plenum Press, New York, (1964); *Smithells Metals Reference Book*, Eric A. Brandes, ed., in association with Fulmer Research Institute Ltd. 6th ed. London, Butterworths, Boston, (1983); and *Ceramic Source*, American Ceramic Society (1986–1991).

Table 430. SELECTING RESISTIVITY OF CERAMICS
(SHEET 4 OF 5)

Ceramic	Temperature Range of Validity	Resistivity (Ω–cm)
Tantalum Diboride (TaB_2)		$68 x10^6$
Molybdenum Disilicide ($MoSi_2$)	1600°C	$75–80x10^6$
Cordierite ($2MgO\ 2Al_2O_3\ 5SiO_2$) (ρ=2.1g/cm^3)	500°C	$9.0x10^7$
Zirconium Monocarbide (ZrC)	773K	$97x10^6$
Mullite ($3Al_2O_3\ 2SiO_2$)	500°C	10^8
Zirconium Monocarbide (ZrC)	1273K	$137x10^6$
Zirconium Mononitirde (TiN)	melting temp.	$320x10^6$
Titanium Mononitirde (TiN)	melting temp.	$340x10^6$
Aluminum Oxide (Al_2O_3)	700°C	$5.0x10^8$
Cordierite ($2MgO\ 2Al_2O_3\ 5SiO_2$) (ρ=1.8g/cm^3)	300°C	$3.0x10^9$
Boron Nitride (BN) (90% humidity)	25°C	$5.0x10^9$
Mullite ($3Al_2O_3\ 2SiO_2$)	300°C	10^{10}
Cordierite ($2MgO\ 2Al_2O_3\ 5SiO_2$) (ρ=2.1g/cm^3)	300°C	$2.0x10^{10}$
Boron Nitride (BN)	480°C	$2.3x10^{10}$
Aluminum Oxide (Al_2O_3)	500°C	$6.3x10^{10}$
Boron Nitride (BN) (50% humidity)	25°C	$7.0x10^{10}$
Silicon Carbide (SiC) (with 2.0 wt% BN additive)		$1x10^{11}$
Aluminum Nitride (AlN)	room temp.	$2x10^{11}–10^{13}$
Cordierite ($2MgO\ 2Al_2O_3\ 5SiO_2$) (ρ=2.3g/cm^3)	100°C	$2.5x10^{11}$
Boron Nitride (BN) (20% humidity)	25°C	$1.0x10^{12}$

Source: data compiled by J.S. Park from *No. 1 Materials Index*, Peter T.B. Shaffer, Plenum Press, New York, (1964); *Smithells Metals Reference Book*, Eric A. Brandes, ed., in association with Fulmer Research Institute Ltd. 6th ed. London, Butterworths, Boston, (1983); and *Ceramic Source*, American Ceramic Society (1986–1991).

Table 430. SELECTING RESISTIVITY OF CERAMICS
(SHEET 5 OF 5)

Ceramic	Temperature Range of Validity	Resistivity (Ω–cm)
Cordierite ($2MgO\ 2Al_2O_3\ 5SiO_2$) (ρ=1.8g/cm^3)	100°C	$1.0x10^{13}$
Aluminum Oxide (Al_2O_3)	300°C	$1x10^{13}$
Silicon Carbide (SiC) (with 1.6 wt% BeO additive)		$>10^{13}$
Trisilicon tetranitride (Si_3N_4)		$>10^{13}$
Boron Nitride (BN)	25°C	$1.7x10^{13}$
Aluminum Oxide (Al_2O_3)	100°C	$2x10^{13}$
Cordierite ($2MgO\ 2Al_2O_3\ 5SiO_2$) (ρ=2.1g/cm^3)	100°C	$3.0x10^{13}$
Silicon Carbide (SiC) (with 1 wt% Be additive)		$3x10^{13}$
Silicon Carbide (SiC) (with 3.2 wt% BeO additive)		$4x10^{13}$
Cordierite ($2MgO\ 2Al_2O_3\ 5SiO_2$) (ρ=1.8g/cm^3)	25°C	$1.0x10^{14}$
Cordierite ($2MgO\ 2Al_2O_3\ 5SiO_2$) (ρ=2.3g/cm^3)	25°C	$1x10^{14}$
Mullite ($3Al_2O_3\ 2SiO_2$)	25°C	$>10^{14}$
Cordierite ($2MgO\ 2Al_2O_3\ 5SiO_2$) (ρ=2.1g/cm^3)	25°C	$>1x10^{14}$
Beryllium Oxide (BeO)	500°C	$1–5x10^{15}$
Beryllium Oxide (BeO)	300°C	$>10^{15}$
Aluminum Oxide (Al_2O_3)	25°C	$>10x10^{14}$
Magnesium Oxide (MgO)	27°C	$1.3x10^{15}$
Beryllium Oxide (BeO)	700°C	$1.5–2x10^{15}$
Beryllium Oxide (BeO)	1000°C	$4–7x10^{15}$
Beryllium Oxide (BeO)	25°C	$>10^{17}$
Silicon Dioxide (SiO_2)	room temp.	10^{18}

Source: data compiled by J.S. Park from *No. 1 Materials Index*, Peter T.B. Shaffer, Plenum Press, New York, (1964); *Smithells Metals Reference Book*, Eric A. Brandes, ed., in association with Fulmer Research Institute Ltd. 6th ed. London, Butterworths, Boston, (1983); and *Ceramic Source*, American Ceramic Society (1986–1991).

Table 431. SELECTING VOLUME RESISTIVITY OF GLASS

(SHEET 1 OF 13)

Glass	Temperature (°C)	Resistivity (log Ω cm)
SiO_2–Na_2O glass (57.5% mol Na_2O)	1300	–0.67
SiO_2–Na_2O glass (49.3% mol Na_2O)	1300	–0.61
SiO_2–Na_2O glass (57.5% mol Na_2O)	1200	–0.61
SiO_2–Na_2O glass (49.3% mol Na_2O)	1200	–0.56
SiO_2–Na_2O glass (44.5% mol Na_2O)	1300	–0.52
SiO_2–Na_2O glass (57.5% mol Na_2O)	1100	–0.52
SiO_2–Na_2O glass (49.3% mol Na_2O)	1100	–0.47
SiO_2–Na_2O glass (44.5% mol Na_2O)	1200	–0.46
SiO_2–Na_2O glass (39.5% mol Na_2O)	1400	–0.45
SiO_2–Na_2O glass (39.5% mol Na_2O)	1300	–0.39
SiO_2–Na_2O glass (44.5% mol Na_2O)	1100	–0.38
SiO_2–Na_2O glass (34.7% mol Na_2O)	1400	–0.33
SiO_2–Na_2O glass (39.5% mol Na_2O)	1200	–0.32
SiO_2–Na_2O glass (34.7% mol Na_2O)	1300	–0.27
SiO_2–Na_2O glass (39.5% mol Na_2O)	1100	–0.24
SiO_2–Na_2O glass (34.7% mol Na_2O)	1200	–0.20
SiO_2–Na_2O glass (29.7% mol Na_2O)	1400	–0.16
SiO_2–Na_2O glass (39.5% mol Na_2O)	1000	–0.13
SiO_2–Na_2O glass (34.7% mol Na_2O)	1100	–0.11
SiO_2–Na_2O glass (29.7% mol Na_2O)	1300	–0.10
SiO_2–Na_2O glass (29.7% mol Na_2O)	1200	–0.02
SiO_2–Na_2O glass (34.7% mol Na_2O)	1000	0.00
SiO_2–Na_2O glass (39.5% mol Na_2O)	900	0.00
SiO_2–Na_2O glass (29.7% mol Na_2O)	1100	0.08

Source: data compiled by J. S. Park from O. V. Mazurin, M. V. Streltsina and T. P. Shvaiko–Shvaikovskaya, *Handbook of Glass Data*, Part A and Part B, Elsevier, New York, 1983

Table 431. SELECTING VOLUME RESISTIVITY OF GLASS
(SHEET 2 OF 13)

Glass	Temperature (°C)	Resistivity (log Ω cm)
SiO_2–Na_2O glass (34.7% mol Na_2O)	900	0.12
SiO_2–Na_2O glass (39.5% mol Na_2O)	800	0.13
SiO_2–Na_2O glass (24.8% mol Na_2O)	1200	0.17
SiO_2–Na_2O glass (29.7% mol Na_2O)	1000	0.20
SiO_2–Na_2O glass (24.8% mol Na_2O)	1100	0.26
SiO_2–PbO glass (66.7% mol PbO)	1000	0.26
SiO_2–Na_2O glass (19.9% mol Na_2O)	1300	0.30
SiO_2–Na_2O glass (39.5% mol Na_2O)	700	0.33
SiO_2–Na_2O glass (29.7% mol Na_2O)	900	0.34
SiO_2–CaO glass (55.2% mol CaO)	1600	0.34
SiO_2–Na_2O glass (19.9% mol Na_2O)	1200	0.38
SiO_2–Na_2O glass (24.8% mol Na_2O)	1000	0.38
SiO_2–CaO glass (51.4% mol CaO)	1618	0.38
SiO_2–PbO glass (60% mol PbO)	1000	0.40
B_2O_3–Na_2O glass (32.8% mol Na_2O)	900	0.40
SiO_2–CaO glass (55.2% mol CaO)	1550	0.42–0.43
SiO_2–CaO glass (51.4% mol CaO)	1560	0.47
SiO_2–Na_2O glass (19.9% mol Na_2O)	1100	0.48
SiO_2–CaO glass (51.4% mol CaO)	1500	0.48–0.49
SiO_2–PbO glass (66.7% mol PbO)	900	0.50
SiO_2–CaO glass (55.2% mol CaO)	1499	0.51–0.53
SiO_2–Na_2O glass (24.8% mol Na_2O)	900	0.52
SiO_2–Na_2O glass (29.7% mol Na_2O)	800	0.52
SiO_2–CaO glass (45.4% mol CaO)	1622	0.52

Source: data compiled by J. S. Park from O. V. Mazurin, M. V. Streltsina and T. P. Shvaiko–Shvaikovskaya, *Handbook of Glass Data*, Part A and Part B, Elsevier, New York, 1983

Table 431. SELECTING VOLUME RESISTIVITY OF GLASS

(SHEET 3 OF 13)

Glass	Temperature (°C)	Resistivity (log Ω cm)
SiO_2–PbO glass (51.6% mol PbO)	1200	0.54
SiO_2–Na_2O glass (15% mol Na_2O)	1500	0.56
SiO_2–CaO glass (45.4% mol CaO)	1585	0.58–0.59
SiO_2–PbO glass (50.0% mol PbO)	1200	0.60
B_2O_3–Na_2O glass (32.8% mol Na_2O)	800	0.60
SiO_2–Na_2O glass (15% mol Na_2O)	1400	0.61
SiO_2–Na_2O glass (19.9% mol Na_2O)	1000	0.61
SiO_2–CaO glass (45.4% mol CaO)	1550	0.65
B_2O_3–Na_2O glass (21.9% mol Na_2O)	1000	0.65
SiO_2–Na_2O glass (39.5% mol Na_2O)	600	0.67
SiO_2–CaO glass (41.3% mol CaO)	1600	0.67–0.68
SiO_2–PbO glass (51.6% mol PbO)	1100	0.70
B_2O_3–Na_2O glass (27.5% mol Na_2O)	900	0.70
B_2O_3–CaO glass (40.0% mol CaO)	1250	0.75
SiO_2–Na_2O glass (19.9% mol Na_2O)	900	0.76
SiO_2–PbO glass (60% mol PbO)	900	0.76
SiO_2–CaO glass (41.3% mol CaO)	1550	0.76
SiO_2–Na_2O glass (29.7% mol Na_2O)	700	0.78
SiO_2–CaO glass (33.6% mol CaO)	1600	0.79–0.80
SiO_2–PbO glass (50.0% mol PbO)	1100	0.80
SiO_2–PbO glass (44.7% mol PbO)	1300	0.82
SiO_2–PbO glass (66.7% mol PbO)	800	0.82
SiO_2–CaO glass (41.3% mol CaO)	1519	0.82
B_2O_3–CaO glass (33.3% mol CaO)	1250	0.85

Source: data compiled by J. S. Park from O. V. Mazurin, M. V. Streltsina and T. P. Shvaiko–Shvaikovskaya, *Handbook of Glass Data*, Part A and Part B, Elsevier, New York, 1983

Table 431. SELECTING VOLUME RESISTIVITY OF GLASS
(SHEET 4 OF 13)

Glass	Temperature (°C)	Resistivity (log Ω cm)
B_2O_3–Na_2O glass (17.3% mol Na_2O)	1000	0.89
SiO_2–Na_2O glass (39.5% mol Na_2O)	550	0.91
SiO_2–Na_2O glass (10% mol Na_2O)	1600	0.92
SiO_2–PbO glass (51.6% mol PbO)	1000	0.92
SiO_2–CaO glass (33.6% mol CaO)	1560	0.93–0.94
B_2O_3–Na_2O glass (21.9% mol Na_2O)	900	0.94
SiO_2–Na_2O glass (19.9% mol Na_2O)	800	0.96
SiO_2–CaO glass (33.6% mol CaO)	1500	0.97
SiO_2–PbO glass (44.7% mol PbO)	1200	0.98
B_2O_3–CaO glass (40.0% mol CaO)	1150	0.98
B_2O_3–Na_2O glass (27.5% mol Na_2O)	800	1.00
SiO_2–PbO glass (50.0% mol PbO)	1000	1.02
B_2O_3–Na_2O glass (32.8% mol Na_2O)	700	1.02
SiO_2–Na_2O glass (10% mol Na_2O)	1500	1.03
SiO_2–PbO glass (38.5% mol PbO)	1300	1.04
SiO_2–PbO glass (60% mol PbO)	800	1.07
B_2O_3–CaO glass (33.3% mol CaO)	1150	1.10
SiO_2–PbO glass (44.7% mol PbO)	1100	1.15
SiO_2–Na_2O glass (29.7% mol Na_2O)	600	1.16
B_2O_3–Na_2O glass (17.3% mol Na_2O)	900	1.18
SiO_2–PbO glass (51.6% mol PbO)	900	1.20
B_2O_3–CaO glass (55.4% mol CaO)	1150	1.22
SiO_2–PbO glass (38.5% mol PbO)	1200	1.26
B_2O_3–Na_2O glass (21.9% mol Na_2O)	800	1.29

Source: data compiled by J. S. Park from O. V. Mazurin, M. V. Streltsina and T. P. Shvaiko–Shvaikovskaya, *Handbook of Glass Data*, Part A and Part B, Elsevier, New York, 1983

Table 431. SELECTING VOLUME RESISTIVITY OF GLASS
(SHEET 5 OF 13)

Glass	Temperature (°C)	Resistivity (log Ω cm)
SiO_2–Na_2O glass (29.7% mol Na_2O)	550	1.31
SiO_2–PbO glass (66.7% mol PbO)	700	1.32
SiO_2–Na_2O glass (19.9% mol Na_2O)	700	1.34
SiO_2–PbO glass (50.0% mol PbO)	900	1.36
B_2O_3–Na_2O glass (17.3% mol Na_2O)	850	1.39
SiO_2–PbO glass (44.7% mol PbO)	1000	1.40
B_2O_3–CaO glass (40.0% mol CaO)	1050	1.40
B_2O_3–Na_2O glass (12.1% mol Na_2O)	900	1.48
B_2O_3–CaO glass (33.3% mol CaO)	1050	1.52
SiO_2–PbO glass (38.5% mol PbO)	1100	1.56
SiO_2–PbO glass (51.6% mol PbO)	800	1.62
SiO_2–Na_2O glass (19.9% mol Na_2O)	600	1.68
B_2O_3–CaO glass (55.4% mol CaO)	1050	1.70
SiO_2–PbO glass (60% mol PbO)	650	1.72
SiO_2–PbO glass (60% mol PbO)	700	1.74
SiO_2–PbO glass (44.7% mol PbO)	900	1.82
B_2O_3–Na_2O glass (12.1% mol Na_2O)	800	1.89
SiO_2–PbO glass (50.0% mol PbO)	800	1.90
SiO_2–PbO glass (38.5% mol PbO)	1000	1.94
B_2O_3–Na_2O glass (3.63% mol Na_2O)	1000	2.00
B_2O_3–CaO glass (40.0% mol CaO)	950	2.06
B_2O_3–CaO glass (33.3% mol CaO)	950	2.25
SiO_2 glass (0.5 atm Ar pressure)	2100	2.30
B_2O_3–Na_2O glass (3.63% mol Na_2O)	900	2.30

Source: data compiled by J. S. Park from O. V. Mazurin, M. V. Streltsina and T. P. Shvaiko–Shvaikovskaya, *Handbook of Glass Data*, Part A and Part B, Elsevier, New York, 1983

Table 431. SELECTING VOLUME RESISTIVITY OF GLASS
(SHEET 6 OF 13)

Glass	Temperature (°C)	Resistivity (log Ω cm)
SiO_2–Na_2O glass (45% mol Na_2O)	350	2.35
SiO_2–PbO glass (44.7% mol PbO)	800	2.38
B_2O_3–Na_2O glass (12.1% mol Na_2O)	700	2.43
B_2O_3–CaO glass (55.4% mol CaO)	950	2.46
SiO_2–PbO glass (38.5% mol PbO)	900	2.47
SiO_2–Na_2O glass (48% mol Na_2O)	300	2.58
SiO_2–Na_2O glass (40% mol Na_2O)	350	2.66
SiO_2–Na_2O glass (45% mol Na_2O)	300	2.69
SiO_2 glass (0.5 atm Ar pressure)	2000	2.70
B_2O_3–Na_2O glass (3.63% mol Na_2O)	800	2.70
SiO_2–Na_2O glass (35% mol Na_2O)	350	2.92
SiO_2–Na_2O glass (40% mol Na_2O)	300	2.97
B_2O_3–CaO glass (40.0% mol CaO)	850	2.97
SiO_2 glass (0.5 atm Ar pressure)	1900	3.00
B_2O_3–CaO glass (33.3% mol CaO)	850	3.10
SiO_2–PbO glass (38.5% mol PbO)	800	3.20
SiO_2–Al_2O_3 glass (5.51% wt Al_2O_3)	1900	3.20
SiO_2–Al_2O_3 glass (10.86% wt Al_2O_3)	1900	3.20
SiO_2–Na_2O glass (36% mol Na_2O)	300	3.22
SiO_2–Al_2O_3 glass (2.83% wt Al_2O_3)	1900	3.28
SiO_2–Na_2O glass (45% mol Na_2O)	250	3.30
SiO_2–Na_2O glass (33.3% mol Na_2O)	300	3.34
SiO_2–Al_2O_3 glass (10.86% wt Al_2O_3)	1700	3.34
SiO_2–Al_2O_3 glass (5.51% wt Al_2O_3)	1700	3.36

Source: data compiled by J. S. Park from O. V. Mazurin, M. V. Streltsina and T. P. Shvaiko–Shvaikovskaya, *Handbook of Glass Data*, Part A and Part B, Elsevier, New York, 1983

Table 431. SELECTING VOLUME RESISTIVITY OF GLASS
(SHEET 7 OF 13)

Glass	Temperature (°C)	Resistivity (log Ω cm)
SiO_2–Na_2O glass (30% mol Na_2O)	350	3.46
SiO_2–Al_2O_3 glass (2.83% wt Al_2O_3)	1700	3.46
SiO_2 glass (0.5 atm Ar pressure)	1800	3.48
SiO_2–Na_2O glass (25% mol Na_2O)	350	3.52
SiO_2–Al_2O_3 glass (10.86% wt Al_2O_3)	1500	3.52
SiO_2–B_2O_3 glass (2.74% wt B_2O_3)	1900	3.56
SiO_2–Al_2O_3 glass (5.51% wt Al_2O_3)	1500	3.56
SiO_2–Na_2O glass (40% mol Na_2O)	250	3.59
SiO_2–Na_2O glass (30% mol Na_2O)	300	3.64–3.78
SiO_2–Al_2O_3 glass (2.83% wt Al_2O_3)	1500	3.67
SiO_2–Al_2O_3 glass (10.86% wt Al_2O_3)	1300	3.74
SiO_2–B_2O_3 glass (2.74% wt B_2O_3)	1700	3.76
SiO_2–Al_2O_3 glass (5.51% wt Al_2O_3)	1300	3.76
SiO_2–Na_2O glass (20% mol Na_2O)	350	3.80
SiO_2–B_2O_3 glass (19.37% wt B_2O_3)	1900	3.84
SiO_2–Na_2O glass (35% mol Na_2O)	250	3.85
B_2O_3–CaO glass (55.4% mol CaO)	850	3.86
SiO_2–Na_2O glass (27% mol Na_2O)	300	3.94
SiO_2–B_2O_3 glass (5.48% wt B_2O_3)	1900	3.94
SiO_2–Al_2O_3 glass (2.83% wt Al_2O_3)	1300	3.94
SiO_2–B_2O_3 glass (10.75% wt B_2O_3)	1900	3.98
SiO_2 glass (0.5 atm Ar pressure)	1700	4.00
SiO_2–B_2O_3 glass (19.37% wt B_2O_3)	1700	4.00
SiO_2–B_2O_3 glass (2.74% wt B_2O_3)	1500	4.02

Source: data compiled by J. S. Park from O. V. Mazurin, M. V. Streltsina and T. P. Shvaiko–Shvaikovskaya, *Handbook of Glass Data*, Part A and Part B, Elsevier, New York, 1983

Table 431. SELECTING VOLUME RESISTIVITY OF GLASS (SHEET 8 OF 13)

Glass	Temperature (°C)	Resistivity (log Ω cm)
SiO_2–Al_2O_3 glass (10.86% wt Al_2O_3)	1100	4.02
SiO_2–Na_2O glass (25% mol Na_2O)	300	4.03
SiO_2–Na_2O glass (48% mol Na_2O)	150	4.09
SiO_2–B_2O_3 glass (5.48% wt B_2O_3)	1700	4.10
SiO_2–Al_2O_3 glass (5.51% wt Al_2O_3)	1100	4.15
SiO_2–B_2O_3 glass (10.75% wt B_2O_3)	1700	4.16
SiO_2–B_2O_3 glass (19.37% wt B_2O_3)	1500	4.22
SiO_2–Al_2O_3 glass (2.83% wt Al_2O_3)	1100	4.29
SiO_2–B_2O_3 glass (5.48% wt B_2O_3)	1500	4.30
SiO_2–Na_2O glass (15% mol Na_2O)	350	4.32
SiO_2–Na_2O glass (45% mol Na_2O)	150	4.33
SiO_2–Na_2O glass (20% mol Na_2O)	300	4.36–4.64
SiO_2 glass (0.5 atm Ar pressure)	1600	4.40
SiO_2–PbO glass (38.5% mol PbO)	700	4.40
SiO_2–B_2O_3 glass (2.74% wt B_2O_3)	1300	4.40
SiO_2–B_2O_3 glass (10.75% wt B_2O_3)	1500	4.40
SiO_2–Na_2O glass (30% mol Na_2O)	250	4.42
SiO_2–B_2O_3 glass (19.37% wt B_2O_3)	1300	4.48
SiO_2–Na_2O glass (25% mol Na_2O)	250	4.50
SiO_2–Al_2O_3 glass (10.86% wt Al_2O_3)	900	4.54
SiO_2–B_2O_3 glass (5.48% wt B_2O_3)	1300	4.56
SiO_2–Na_2O glass (40% mol Na_2O)	150	4.58
SiO_2–Al_2O_3 glass (5.51% wt Al_2O_3)	900	4.65
SiO_2 glass (0.5 atm Ar pressure)	1500	4.66

Source: data compiled by J. S. Park from O. V. Mazurin, M. V. Streltsina and T. P. Shvaiko–Shvaikovskaya, *Handbook of Glass Data*, Part A and Part B, Elsevier, New York, 1983

Table 431. SELECTING VOLUME RESISTIVITY OF GLASS
(SHEET 9 OF 13)

Glass	Temperature (°C)	Resistivity (log Ω cm)
SiO_2–B_2O_3 glass (10.75% wt B_2O_3)	1300	4.69
SiO_2–B_2O_3 glass (2.74% wt B_2O_3)	1100	4.72
SiO_2–Na_2O glass (13% mol Na_2O)	300	4.77–4.79
SiO_2–B_2O_3 glass (19.37% wt B_2O_3)	1100	4.82
SiO_2–Al_2O_3 glass (2.83% wt Al_2O_3)	900	4.82
SiO_2–Na_2O glass (20% mol Na_2O)	250	4.85
SiO_2–Na_2O glass (36% mol Na_2O)	150	4.89
SiO_2 glass	1500	4.90
SiO_2–Na_2O glass (10% mol Na_2O)	350	4.96
SiO_2 glass	1400	5.00
SiO_2–Na_2O glass (33.3% mol Na_2O)	150	5.06
SiO_2–B_2O_3 glass (10.75% wt B_2O_3)	1100	5.08
SiO_2 glass	1300	5.15
SiO_2–Na_2O glass (44.2% mol Na_2O)	100	5.15
SiO_2–B_2O_3 glass (5.48% wt B_2O_3)	1100	5.16
SiO_2–Na_2O glass (10% mol Na_2O)	300	5.18
SiO_2 glass	1200	5.30
SiO_2–Na_2O glass (7.5% mol Na_2O)	300	5.30
SiO_2–B_2O_3 glass (2.74% wt B_2O_3)	900	5.30
SiO_2–Al_2O_3 glass (5.51% wt Al_2O_3)	700	5.34
SiO_2–Al_2O_3 glass (10.86% wt Al_2O_3)	700	5.38
SiO_2–Na_2O glass (15% mol Na_2O)	250	5.44
SiO_2 glass	1100	5.46
SiO_2–Na_2O glass (30% mol Na_2O)	150	5.48–5.75

Source: data compiled by J. S. Park from O. V. Mazurin, M. V. Streltsina and T. P. Shvaiko–Shvaikovskaya, *Handbook of Glass Data*, Part A and Part B, Elsevier, New York, 1983

Table 431. SELECTING VOLUME RESISTIVITY OF GLASS (SHEET 10 OF 13)

Glass	Temperature (°C)	Resistivity (log Ω cm)
B_2O_3 glass	840	5.5
SiO_2–B_2O_3 glass (5.48% wt B_2O_3)	900	5.64
SiO_2–B_2O_3 glass (19.37% wt B_2O_3)	900	5.65
SiO_2 glass	1000	5.66
SiO_2–B_2O_3 glass (10.75% wt B_2O_3)	900	5.74
SiO_2–Al_2O_3 glass (2.83% wt Al_2O_3)	700	5.74
B_2O_3 glass	780	5.8
SiO_2–Na_2O glass (27% mol Na_2O)	150	5.87
SiO_2 glass	900	5.90
SiO_2–Na_2O glass (25% mol Na_2O)	150	6.05
B_2O_3–CaO glass (55.4% mol CaO)	750	6.13
SiO_2–Na_2O glass (10% mol Na_2O)	250	6.14
B_2O_3 glass	730	6.2
SiO_2 glass	800	6.20
SiO_2–Na_2O glass (5% mol Na_2O)	350	6.37
SiO_2–Na_2O glass (20% mol Na_2O)	150	6.45–6.80
SiO_2 glass	700	6.56
SiO_2–Na_2O glass (30.2% mol Na_2O)	100	6.58
B_2O_3 glass	680	6.6
B_2O_3 glass	640	6.9
SiO_2–Na_2O glass (13% mol Na_2O)	150	6.90–6.96
SiO_2 glass	600	7.00
B_2O_3 glass	600	7.3
SiO_2–Na_2O glass (5% mol Na_2O)	300	7.33–8.25

Source: data compiled by J. S. Park from O. V. Mazurin, M. V. Streltsina and T. P. Shvaiko–Shvaikovskaya, *Handbook of Glass Data*, Part A and Part B, Elsevier, New York, 1983

Table 431. SELECTING VOLUME RESISTIVITY OF GLASS
(SHEET 11 OF 13)

Glass	Temperature (°C)	Resistivity (log Ω cm)
SiO_2–Na_2O glass (10% mol Na_2O)	150	7.35
SiO_2–Na_2O glass (7.5% mol Na_2O)	150	7.59
B_2O_3 glass	560	7.6
SiO_2–Na_2O glass (5% mol Na_2O)	250	7.63
SiO_2 glass	500	7.80
SiO_2–PbO glass (65% mol PbO)	300	7.81
SiO_2–PbO glass (60% mol PbO)	300	8.11
SiO_2–Na_2O glass (15.1% mol Na_2O)	100	8.15
B_2O_3–Na_2O glass (40% mol Na_2O)	100	8.46
SiO_2 glass	400	8.5–10.80
SiO_2–CaO glass (50% mol CaO)	400	8.70
SiO_2–PbO glass (50% mol PbO)	300	8.80–9.2
B_2O_3–Na_2O glass (30% mol Na_2O)	100	8.82
B_2O_3–Na_2O glass (40% mol Na_2O)	80	9.08
B_2O_3–Na_2O glass (30% mol Na_2O)	80	9.43
SiO_2–PbO glass (40% mol PbO)	300	9.48
B_2O_3–Na_2O glass (40% mol Na_2O)	60	9.73
SiO_2–PbO glass (65% mol PbO)	200	9.76
SiO_2–Na_2O glass (7.8% mol Na_2O)	100	9.89
SiO_2–PbO glass (35% mol PbO)	300	9.89
SiO_2–PbO glass (60% mol PbO)	200	10.04
SiO_2–PbO glass (57.1% mol PbO)	172	10.14
B_2O_3–Na_2O glass (30% mol Na_2O)	60	10.14
SiO_2–PbO glass (63.2% mol PbO)	159	10.34

Source: data compiled by J. S. Park from O. V. Mazurin, M. V. Streltsina and T. P. Shvaiko–Shvaikovskaya, *Handbook of Glass Data*, Part A and Part B, Elsevier, New York, 1983

Table 431. SELECTING VOLUME RESISTIVITY OF GLASS

(SHEET 12 OF 13)

Glass	Temperature (°C)	Resistivity (log Ω cm)
SiO_2–PbO glass (30% mol PbO)	300	10.44
SiO_2–Na_2O glass (5% mol Na_2O)	150	10.45–11.71
B_2O_3–Na_2O glass (40% mol Na_2O)	40	10.48
SiO_2–PbO glass (50% mol PbO)	200	10.69
SiO_2 glass	250	11.0–13.6
B_2O_3–Na_2O glass (20% mol Na_2O)	100	11.28
SiO_2–PbO glass (40% mol PbO)	200	11.54
SiO_2–PbO glass (51.4% mol PbO)	139	11.59
B_2O_3–Na_2O glass (10% mol Na_2O)	100	11.61
SiO_2–PbO glass (40.2% mol PbO)	175	11.70
SiO_2–PbO glass (47.3% mol PbO)	149	11.74
B_2O_3–Na_2O glass (30% mol Na_2O)	40	11.90
B_2O_3–Na_2O glass (20% mol Na_2O)	80	12.05
SiO_2–PbO glass (35% mol PbO)	200	12.10
SiO_2–CaO glass (50% mol CaO)	300	12.2
B_2O_3–Na_2O glass (10% mol Na_2O)	80	12.40
B_2O_3–Na_2O glass (20% mol Na_2O)	60	12.91
SiO_2–PbO glass (30% mol PbO)	200	12.94
B_2O_3–CaO glass (33.3% mol CaO)	300	13.16
B_2O_3–Na_2O glass (10% mol Na_2O)	60	13.21
B_2O_3–CaO glass (33.3% mol CaO)	250	13.50
B_2O_3–Na_2O glass (16% mol Na_2O)	100	13.58
SiO_2–PbO glass (33.8% mol PbO)	135	13.68
SiO_2–PbO glass (57.1% mol PbO)	77	13.70

Source: data compiled by J. S. Park from O. V. Mazurin, M. V. Streltsina and T. P. Shvaiko–Shvaikovskaya, *Handbook of Glass Data*, Part A and Part B, Elsevier, New York, 1983

Table 431. SELECTING VOLUME RESISTIVITY OF GLASS (SHEET 13 OF 13)

Glass	Temperature (°C)	Resistivity (log Ω cm)
B_2O_3–Na_2O glass (20% mol Na_2O)	40	13.86
B_2O_3–CaO glass (33.3% mol CaO)	200	13.92
B_2O_3–Na_2O glass (10% mol Na_2O)	40	14.20
SiO_2–PbO glass (63.2% mol PbO)	57	14.29
B_2O_3–Na_2O glass (16% mol Na_2O)	80	14.32
B_2O_3–CaO glass (33.3% mol CaO)	150	14.40
SiO_2–PbO glass (47.3% mol PbO)	79	14.48
SiO_2–PbO glass (51.4% mol PbO)	65	14.52
SiO_2–PbO glass (40.2% mol PbO)	78	14.85
B_2O_3–Na_2O glass (16% mol Na_2O)	60	15.08
B_2O_3–Na_2O glass (16% mol Na_2O)	40	15.89
SiO_2–PbO glass (33.8% mol PbO)	66	16.14

Source: data compiled by J. S. Park from O. V. Mazurin, M. V. Streltsina and T. P. Shvaiko–Shvaikovskaya, *Handbook of Glass Data*, Part A and Part B, Elsevier, New York, 1983

Table 432. SELECTING VOLUME RESISTIVITY OF POLYMERS
(SHEET 1 OF 6)

Polymer	Volume Resistivity (ASTM D257) (Ω • cm)
Diallyl Phthalates; Molded: Dacron Filled	10^2—2.5×10^4
Diallyl Phthalates; Molded: Asbestos Filled	10^2—5×10^3
Diallyl Phthalates; Molded: Glass Fiber Filled	10^4—5×10^4
Diallyl Phthalates; Molded: Orlon Filled	6×10^4—6×10^6
Standard Epoxies: Cast Flexible	9.1×10^5—6.7×10^9
Standard Epoxies; Reinforced: High Strength Laminate	6.6×10^7—10^9
Molded Rubber Phenolic—Woodflour or Flock Filled	10^8—10^{11}
Phenolics; Molded: General: Woodflour and Flock Filled	10^9—10^{13}
Cellulose Acetate; Molded, Extruded; ASTM Grade: H6—1	10^{10}—10^{13}
Cellulose Acetate; Molded, Extruded; ASTM Grade: H4—1	10^{10}—10^{13}
Cellulose Acetate; Molded, Extruded; ASTM Grade: H2—1	10^{10}—10^{13}
Cellulose Acetate; ASTM Grade: MH—1, MH—2	10^{10}—10^{13}
Cellulose Acetate; ASTM Grade: MS—1, MS—2	10^{10}—10^{13}
Cellulose Acetate; Molded, Extruded; ASTM Grade: S2—1	10^{10}—10^{13}
Phenolics; Molded: High Shock: Chopped Fabric or Cord Filled	$>10^{10}$
Phenolics; Molded: Very High Shock: Glass Fiber Filled	10^{10}—10^{11}
Phenolics: Molded: Arc Resistant—Mineral Filled	10^{10}—10^{12}
Ureas; Molded: Cellulose Filled (ASTM Type 2)	5—8×10^{10}
Cellulose Acetate Butyrate; ASTM Grade: H4	10^{11}—10^{14}
Cellulose Acetate Butyrate; ASTM Grade: MH	10^{11}—10^{14}

Source: data compiled by J.S. Park from Charles T. Lynch, *CRC Handbook of Materials Science*, Vol. 3, CRC Press, Boca Raton, Florida, 1975 and *Engineered Materials Handbook*, Vol.2, Engineering Plastics, ASM International, Metals Park, Ohio, 1988.

Table 432. SELECTING VOLUME RESISTIVITY OF POLYMERS

(SHEET 2 OF 6)

Polymer	Volume Resistivity (ASTM D257) ($\Omega \cdot$ cm)
Cellulose Acetate Butyrate; ASTM Grade: S2	10^{11}—10^{14}
Cellusose Acetate Propionate; ASTM Grade: 1	10^{11}—10^{14}
Cellusose Acetate Propionate; ASTM Grade: 3	10^{11}—10^{14}
Cellusose Acetate Propionate; ASTM Grade: 6	10^{11}—10^{14}
Phenolics: Molded: Rubber Phenolic—Chopped Fabric Filled	10^{11}
Phenolics: Molded: Rubber Phenolic—Asbestos Filled	10^{11}
Ureas; Molded: Alpha—Cellulose filled (ASTM Type l)	0.5—5 x 10^{11}
Melamines; Molded: Glass Fiber Filled	1—7 x 10^{11}
Phenolics; Molded: Shock: Paper, Flock, or Pulp Filled	1—50 x 10^{11}
Nylons: Type 8	1.5 x 10^{11}
Polyvinyl Chloride & Copolymers: Nonrigid—Electrical	4—300 x 10^{11}
Melamines; Molded: Alpha Cellulose And Mineral Filled	10^{12}
Polyesters, Thermosets; Cast polyyester: Flexible	10^{12}
Melamines; Molded: Cellulose Electrical Filled	10^{12}—10^{13}
Reinforced Polyester: High Strength (Glass Fibers)	1 x 10^{12} —1 x 10^{13}
Reinforced Polyester: Heat & Chemical Resistant (Asbestos)	1 x 10^{12} —1 x 10^{13}
Polyvinyl Chloride & Copolymers: Nonrigid—General	1—700 x 10^{12}
Polyesters, Thermosets; Cast polyyester: Rigid	10^{13}
PVC–Acrylic Alloy: PVC–Acrylic Sheet	1—5 x 10^{13}
Nylons: Type 11	2 x 10^{13}

Source: data compiled by J.S. Park from Charles T. Lynch, *CRC Handbook of Materials Science*, Vol. 3, CRC Press, Boca Raton, Florida, 1975 and *Engineered Materials Handbook*, Vol.2, Engineering Plastics, ASM International, Metals Park, Ohio, 1988.

Table 432. SELECTING VOLUME RESISTIVITY OF POLYMERS
(SHEET 3 OF 6)

Polymer	Volume Resistivity (ASTM D257) ($\Omega \cdot$ cm)
Nylons; Molded, Extruded; Type 6: General purpose	4.5×10^{13}
Alkyds; Molded: Putty (Encapsulating)	10^{14}
Alkyds; Molded: Rope (General Purpose)	10^{14}
Alkyds; Molded: Glass reinforced (heavy duty parts)	10^{14}
Acrylics; Moldings: Grades 5, 6, 8	$>10^{14}$
Alkyds; Molded: Granular (high speed molding)	10^{14}—10^{15}
Nylons: Type 12	10^{14}—10^{15}
6/6 Nylon: General purpose molding	10^{14}—10^{15}
Polyacetal Copolymer: Standard	1×10^{14}
Polyacetal Copolymer: High Flow	1.0×10^{14}
Polyacetal Copolymer: 25% Glass Reinforced	1.2×10^{14}
High Performance Epoxies: Molded	1.4—5.5×10^{14}
Woven Glass Fabric/ Silicone Laminate	2—5×10^{14}
High Performance Epoxies: Cast, rigid	2.10×10^{14}
Nylons; Type 6: Cast	2.6×10^{14}
Nylons; Type 6: Glass fiber (30%) Reinforced	2.8×10^{14}—1.5×10^{15}
Polyester; Thermoplastic Moldings: Asbestos—Filled Grade	3×10^{14}
Thermoset Carbonate: Allyl Diglycol Carbonate	4×10^{14}
Polyphenylene sulfide: 40% Glass Reinforced	4.5×10^{14}
Polyvinylidene— fluoride (PVDF)	5×10^{14}
Polyacetal Homopolymer: 20% Glass Reinforced	5×10^{14}
Granular (Silica) Reinforced Silicones	5×10^{14}
Fibrous (Glass) Reinforced Silicones	9×10^{14}
Polyvinyl Chloride & Copolymers: Rigid—Normal Impact	10^{14}—10^{16}

Source: data compiled by J.S. Park from Charles T. Lynch, *CRC Handbook of Materials Science*, Vol. 3, CRC Press, Boca Raton, Florida, 1975 and *Engineered Materials Handbook*, Vol.2, Engineering Plastics, ASM International, Metals Park, Ohio, 1988.

Table 432. SELECTING VOLUME RESISTIVITY OF POLYMERS
(SHEET 4 OF 6)

Polymer	Volume Resistivity (ASTM D257) ($\Omega \cdot$ cm)
Vinylidene chloride	10^{14}—10^{16}
Ceramic Reinforced (PTFE)	10^{15}
6/6 Nylon: General Purpose Extrusion	10^{15}
6/10 Nylon: General purpose	10^{15}
Acrylics; Cast Resin Sheets, Rods: General purpose, type II	$>10^{15}$
Acrylics; Cast Resin Sheets, Rods: General purpose, type I	$>10^{15}$
Polyethylenes; Molded, Extruded; Type II: Melt Index 20	$>10^{15}$
Polyethylenes; Molded, Extruded; Type II: Melt Index 1.0—1.9	$>10^{15}$
Polyethylenes; Molded, Extruded; Type III: Melt Index 0.2—0.9	$>10^{15}$
Polyethylenes; Type III: Melt Melt Index 0.1—12.0	$>10^{15}$
Polyethylenes; Molded, Extruded; Type III: Melt Index 1.5—15	$>10^{15}$
Polyethylenes; Molded, Extruded; Type III: High Molecular Weight	$>10^{15}$
Olefin Copolymers; Molded: EVA (ethylene vinyl acetate)	0.15×10^{15}
Chlorinated Polyvinyl Chloride	1×10^{15}—2×10^{16}
Standard Epoxies: Molded	1—5×10^{15}
Polyacetal Homopolymer: Standard	1×10^{15}
ABS Resins; Molded, Extruded: High impact	1—4×10^{15}
ABS Resins; Molded, Extruded: Very high impact	1—4×10^{15}
ABS Resins; Molded, Extruded: Low temperature impact	1—4×10^{15}
ABS Resins; Molded, Extruded: Heat resistant	1—5×10^{15}
Polycarbonate (40% Glass Fiber Reinforced)	1.4×10^{15}
Polypropylene: Asbestos Filled	1.5×10^{15}
Polyester; Thermoplastic Moldings: General Purpose Grade	2×10^{15}
ABS Resins; Molded, Extruded: Medium impact	2—4×10^{15}

Source: data compiled by J.S. Park from Charles T. Lynch, *CRC Handbook of Materials Science*, Vol. 3, CRC Press, Boca Raton, Florida, 1975 and *Engineered Materials Handbook*, Vol.2, Engineering Plastics, ASM International, Metals Park, Ohio, 1988.

Table 432. SELECTING VOLUME RESISTIVITY OF POLYMERS
(SHEET 5 OF 6)

Polymer	Volume Resistivity (ASTM D257) ($\Omega \cdot cm$)
Olefin Copolymers; Molded: EEA (ethylene ethyl acrylate)	2.4×10^{15}
6/6 Nylon; Molded, Extruded: Glass Fiber Reinforced	$2.6—5.5 \times 10^{15}$
Polymides: Unreinforced	4×10^{15}
PVC–Acrylic Alloy: PVC–Acrylic Injection Molded	5×10^{15}
Standard Epoxies: Cast Rigid	6.1×10^{15}
Reinforced Polyester Sheet Molding, General Purpose	$6.4 \times 10^{15}—2.2 \times 10^{16}$
Polymides: Glass Reinforced	9.2×10^{15}
Olefin Copolymers; Molded: Ionomer	10×10^{15}
Styrene Acrylonitrile (SAN)	$>10^{16}$
Epoxy Novolacs: Cast, rigid	$>10^{16}$
Olefin Copolymers; Molded: Polyallomer	$>10^{16}$
Polystyrenes; Molded: General Purpose	$>10^{16}$
Polystyrenes; Molded: Medium Impact	$>10^{16}$
Polystyrenes; Molded: High Impact	$>10^{16}$
Polyester; Thermoplastic Moldings: General Purpose Grade	$1—4 \times 10^{16}$
Chlorinated Polyether	1.5×10^{16}
Polypropylene: Glass Reinforced	1.7×10^{16}
Acrylics; Moldings: High Impact Grade	2.0×10^{16}
Polycarbonate	2.1×10^{16}
ABS–Polycarbonate Alloy	2.2×10^{16}
Polyester; Thermoplastic Moldings: Glass Reinforced Grades	$3.2—3.3 \times 10^{16}$
Polyarylsulfone	$3.2—7.71 \times 10^{16}$
Polyester Moldings: Glass Reinforced Self Extinguishing	3.4×10^{16}
Polystyrenes; Molded: Glass Fiber -30% Reinforced	3.6×10^{16}

Source: data compiled by J.S. Park from Charles T. Lynch, *CRC Handbook of Materials Science*, Vol. 3, CRC Press, Boca Raton, Florida, 1975 and *Engineered Materials Handbook*, Vol.2, Engineering Plastics, ASM International, Metals Park, Ohio, 1988.

Table 432. SELECTING VOLUME RESISTIVITY OF POLYMERS
(SHEET 6 OF 6)

Polymer	Volume Resistivity (ASTM D257) ($\Omega \cdot$ cm)
Polypropylene: Flame Retardant	4×10^{16}—10^{17}
Glass Fiber (30%) Reinforced SAN	4.4×10^{16}
Phenylene Oxides (Noryl): Standard	5×10^{16}
Phenylene Oxides: SE—100	10^{17}
Phenylene Oxides: SE—1	10^{17}
Phenylene Oxides: Glass Fiber Reinforced	10^{17}
Phenylene Oxides (Noryl): Glass Fiber Reinforced	10^{17}
Polypropylene: High Impact	10^{17}
Polypropylene: General Purpose	$>10^{17}$
Polyethylenes; Molded, Extruded; Type I: Melt Index 0.3—3.6	10^{17}—10^{19}
Polyethylenes; Molded, Extruded; Type I: Melt Index 6—26	10^{17}—10^{19}
Polyethylenes; Molded, Extruded; Type I: Melt Index 200	10^{17}—10^{19}
Polytrifluoro Chloroethylene (PTFCE), Molded,Extruded	10^{18}
Polytetrafluoroethylene (PTFE), Molded,Extruded	$>10^{18}$
Fluorinated Ethylene Propylene (FEP)	$>2 \times 10^{18}$

Source: data compiled by J.S. Park from Charles T. Lynch, *CRC Handbook of Materials Science*, Vol. 3, CRC Press, Boca Raton, Florida, 1975 and *Engineered Materials Handbook*, Vol.2, Engineering Plastics, ASM International, Metals Park, Ohio, 1988.

Table 433. SELECTING CRITICAL TEMPERATURE OF SUPERCONDUCTIVE ELEMENTS (SHEET 1 OF 2)

Element	T_c(K)
W	0.0154
Be	0.026
Ir	0.11-0.14
Ti	0.39
Ru	0.493
Cd	0.518-0.52
Zr	0.53
Zr (ω)	0.65
Os	0.655
Zn	0.875
Mo	0.916
Ga	1.0833
Al	1.175
Th	1.39
Pa	1.4
Re	1.697
Ti	2.332-2.39
Sb	2.6-2.7[a]
In	3.405
Sn	3.721
Hg (β)	3.949
Hg (α)	4.154
Ta	4.47
La (α)	4.88

[a] Metastable.

Source: data from Roberts, B. W., *Properties of Selected Superconductive Materials* - 1974 Supplement, NBS Technical Note 825, National Bureau of Standards, U.S. Government Printing Office, Washington,D.C., 1974, 10.

Table 433. SELECTING CRITICAL TEMPERATURE OF SUPERCONDUCTIVE ELEMENTS (SHEET 2 OF 2)

Element	T_c(K)
V	5.43-5.31
Ga (β)	5.90-6.2
La (β)	6.00
Pb	7.23
Ga (γ)	7.62
Tc	7.73-7.78
Ga (δ)	7.85
Nb	9.25

[a] Metastable.

Source: data from Roberts, B. W., *Properties of Selected Superconductive Materials* - 1974 Supplement, NBS Technical Note 825, National Bureau of Standards, U.S. Government Printing Office, Washington,D.C., 1974, 10.

Table 434. SELECTING DISSIPATION FACTOR FOR POLYMERS AT 60 HZ (SHEET 1 OF 5)

Polymer	Dissipation Factor (ASTM D150) @ 60 Hz
Polystyrenes; Molded: General purpose	0.0001–0.0003
Fluorocarbons; Molded,Extruded: Polytetrafluoroethylene (PTFE)	0.0002
Fluorocarbons; Molded,Extruded: Fluorinated ethylene propylene (FEP)	0.0003
Polystyrenes; Molded: Medium impact	0.0004–0.002
Polystyrenes; Molded: High impact	0.0004–0.002
Polyethylenes; Molded, Extruded: Type I: Melt index 0.3—3.6	<0.0005
Polyethylenes; Molded, Extruded: Type I: Melt index 6—26	<0.0005
Polyethylenes; Molded, Extruded: Type I: Melt index 200	<0.0005
Polyethylenes; Molded, Extruded: Type II: Melt index 20	<0.0005
Polyethylenes; Molded, Extruded: Type II: Melt index l.0—1.9	<0.0005
Polyethylenes; Molded, Extruded: Type III: Melt index 0.2—0.9	<0.0005
Polyethylenes; Molded, Extruded: Type III: Melt Melt index 0.l—12.0	<0.0005
Polyethylenes; Molded, Extruded: Type III: Melt index 1.5—15	<0.0005
Polyethylenes; Molded, Extruded: Type III: High molecular weight	<0.0005
Olefin Copolymers; Molded: Polyallomer	>0.0005
Polypropylene: General purpose	0.0005–0.0007
Fluorocarbons; Molded,Extruded: Ceramic reinforced (PTFE)	0.0005–0.0015
Phenylene Oxides: SE—100	0.0007
Phenylene Oxides: SE—1	0.0007
Polypropylene: Flame retardant	0.0007–0.017
Phenylene oxides (Noryl): Standard	0.0008
Polycarbonate	0.0009
Phenylene Oxides: Glass fiber reinforced	0.0009
Olefin Copolymers; Molded: EEA (ethylene ethyl acrylate)	0.001
Epoxy novolacs: Cast, rigid	0.001—0.007
Polypropylene: High impact	<0.0016
Polyarylsulfone	0.0017—0.003
Phenylene oxides (Noryl): Glass fiber reinforced	0.0019

Source: data compiled by J.S. Park from Charles T. Lynch, *CRC Handbook of Materials Science*, Vol. 3, CRC Press, Boca Raton, Florida, 1975 and *Engineered Materials Handbook*, Vol.2, Engineering Plastics, ASM International, Metals Park, Ohio, 1988.

Table 434. SELECTING DISSIPATION FACTOR FOR POLYMERS AT 60 HZ (SHEET 2 OF 5)

Polymer	Dissipation Factor (ASTM D150) @ 60 Hz
Polypropylene: Glass reinforced	0.002
Silicones; Molded, Laminated: Granular (silica) reinforced silicones	0.002—0.004
ABS–Polycarbonate Alloy	0.0026
Polymides: Unreinforced	0.003
Olefin Copolymers; Molded: EVA (ethylene vinyl acetate)	0.003
Olefin Copolymers; Molded: Ionomer	0.003
ABS Resins; Molded, Extruded: Medium impact	0.003—0.006
Polyester;: Thermosets: Cast Rigid	0.003—0.04
Polymides: Glass Reinforced	0.0034
Standard Epoxies: General Purpose Glass Cloth Laminate	0.004-0.006
Diallyl Phthalates; Molded: Glass Fiber Filled	0.004—0.015 (Dry)
Diallyl Phthalates; Molded: Dacron Filled	0.004—0.016 (Dry)
Polyacetal Homopolymer: 20% glass reinforced	0.0047
Polyacetal Homopolymer: Standard	0.0048
Standard Epoxies: Cast Flexible	0.0048-0.0380
Polystyrenes; Molded: Glass fiber -30% reinforced	0.005
Polystyrenes; Molded: Glass fiber (30%) reinforced SAN	0.005
ABS Resins; Molded, Extruded: High impact	0.005—0.007
ABS Resins; Molded, Extruded: Low temperature impact	0.005—0.01
ABS Resins; Molded, Extruded: Very high impact	0.005—0.010
Epoxies; High Performance Resins: Cast, Rigid	0.0055—0.0074
Polycarbonate (40% glass fiber reinforced)	0.006
Polystyrenes; Molded: Styrene acrylonitrile (SAN)	>0.006
Polypropylene: Asbestos filled	0.007
Nylons; Molded, Extruded; Type 6: Flexible Copolymers	0.007—0.010
Epoxies; High Performance Resins: Molded	0.0071—0.025
Standard Epoxies: Cast Rigid	0.0074
Reinforced Polyester Sheet molding compounds, general purpose	0.0087—0.04

Source: data compiled by J.S. Park from Charles T. Lynch, *CRC Handbook of Materials Science*, Vol. 3, CRC Press, Boca Raton, Florida, 1975 and *Engineered Materials Handbook*, Vol.2, Engineering Plastics, ASM International, Metals Park, Ohio, 1988.

Table 434. SELECTING DISSIPATION FACTOR FOR POLYMERS AT 60 HZ (SHEET 3 OF 5)

Polymer	Dissipation Factor (ASTM D150) @ 60 Hz
Silicones; Molded, Laminated: Fibrous (glass) reinforced silicones	0.01
Cellulose Acetate Butyrate; Molded, Extruded; ASTM Grade: H4	0.01—0.04
Cellulose Acetate Butyrate; Molded, Extruded; ASTM Grade: MH	0.01—0.04
Cellulose Acetate Butyrate; Molded, Extruded; ASTM Grade: S2	0.01—0.04
Cellusose Acetate Propionate; Molded, Extruded; ASTM Grade: 1	0.01—0.04
Cellusose Acetate Propionate; Molded, Extruded; ASTM Grade: 3	0.01—0.04
Cellusose Acetate Propionate; Molded, Extruded; ASTM Grade: 6	0.01—0.04
Cellulose Acetate; Molded, Extruded; ASTM Grade: H4—1	0.01—0.06
Cellulose Acetate; Molded, Extruded; ASTM Grade: H2—1	0.01—0.06
Cellulose Acetate; Molded, Extruded; ASTM Grade: MH—1, MH—2	0.01—0.06
Cellulose Acetate; Molded, Extruded; ASTM Grade: MS—1, MS—2	0.01—0.06
Cellulose Acetate; Molded, Extruded; ASTM Grade: S2—1	0.01—0.06
Polyester;: Thermosets: Flexible	0.01—0.18
Chlorinated polyether	0.011
Standard Epoxies: Molded	0.011-0.018
Nylons; Molded, Extruded; 6/6 Nylon: General purpose molding	0.014—0.04
Nylons; Type 6: Cast	0.015
Nylons; Molded, Extruded; 6/6 Nylon: Glass fiber reinforced	0.018—0.009
Chlorinated polyvinyl chloride	0.0189—0.0208
Alkyds; Molded: Rope (general purpose)	0.019
Fluorocarbons; Molded,Extruded: Polytrifluoro chloroethylene (PTFCE)	0.02
Silicones; Molded, Laminated: Woven glass fabric/ silicone laminate	0.02
Polyvinyl Chloride & Copolymers: Rigid—normal impact	0.020—0.03
Alkyds; Molded: Glass reinforced (heavy duty parts)	0.02—0.03
Phenolics; Molded; Very High Shock: Glass Fiber Filled	0.02—0.03
Nylons; Molded, Extruded; Type 6: Glass fiber (30%) reinforced	0.022—0.008
Diallyl Phthalates; Molded: Orlon Filled	0.023—0.015 (Dry)
Melamines; Molded: Cellulose Electrical Filled	0.026—0.192

Source: data compiled by J.S. Park from Charles T. Lynch, *CRC Handbook of Materials Science*, Vol. 3, CRC Press, Boca Raton, Florida, 1975 and *Engineered Materials Handbook*, Vol.2, Engineering Plastics, ASM International, Metals Park, Ohio, 1988.

Table 434. SELECTING DISSIPATION FACTOR FOR POLYMERS AT 60 HZ (SHEET 4 OF 5)

Polymer	Dissipation Factor (ASTM D150) @ 60 Hz
Nylons; Molded, Extruded: Type 11	0.03
ABS Resins; Molded, Extruded: Heat resistant	0.030—0.040
Alkyds; Molded: Granular (high speed molding)	0.030—0.040
Alkyds; Molded: Putty (encapsulating)	0.030—0.045
Acrylics; Moldings: High impact grade	0.03—0.04
Thermoset Carbonate: Allyl diglycol carbonate	0.03—0.04
Polyvinyl Chloride & Copolymers: Vinylidene chloride	0.03—0.15
Ureas; Molded: Woodflour filled	0.035—0.040
Ureas; Molded: Alpha—cellulose filled (ASTM Type l)	0.035—0.043
PVC–Acrylic Injection Molded	0.037
Nylons; Molded, Extruded; 6/10 Nylon: General purpose	0.04
Acrylics; Moldings: Grades 5, 6, 8	0.04—0.06
Ureas; Molded: Cellulose filled (ASTM Type 2)	0.042—0.044
Melamines; Molded: Unfilled	0.048—0.162
Fluorocarbons; Molded,Extruded: Polyvinylidene—fluoride (PVDF)	0.05
Diallyl Phthalates; Molded: Asbestos Filled	0.05—0.03 (Dry)
Acrylics; Cast Resin Sheets, Rods: General Purpose, Type I	0.05—0.06
Acrylics; Cast Resin Sheets, Rods: General Purpose, Type II	0.05—0.06
Polyvinyl Chloride & Copolymers; Molded, Extruded: Nonrigid—general	0.05—0.15
Phenolics; Molded; General: Woodflour & Flock Filled	0.05—0.30
Nylons; Molded, Extruded; Type 6: General Purpose	0.06—0.014
PVC–Acrylic Sheet	0.076
Polyvinyl Chloride & Copolymers: Nonrigid—electrical	0.08—0.11
Phenolics; Molded; Shock: Paper, Flock, or Pulp Filled	0.08—0.35

Source: data compiled by J.S. Park from Charles T. Lynch, *CRC Handbook of Materials Science*, Vol. 3, CRC Press, Boca Raton, Florida, 1975 and *Engineered Materials Handbook*, Vol.2, Engineering Plastics, ASM International, Metals Park, Ohio, 1988.

Table 434. SELECTING DISSIPATION FACTOR FOR POLYMERS AT 60 HZ (SHEET 5 OF 5)

Polymer	Dissipation Factor (ASTM D150) @ 60 Hz
Phenolics; Molded; High Shock: Chopped Fabric or Cord Filled	0.08—0.45
Phenolics: Molded: Arc resistant—Mineral Filled	0.13—0.16
Melamines; Molded: Glass Fiber Filled	0.14—0.23
Rubber Phenolic—Asbestos Filled	0.15
Phenolics: Molded: Rubber Phenolic—Woodflour or Flock Filled	0.15—0.60
Nylons; Molded, Extruded: Type 8	0.19
Rubber Phenolic—Chopped Fabric Filled	0.5

Source: data compiled by J.S. Park from Charles T. Lynch, *CRC Handbook of Materials Science*, Vol. 3, CRC Press, Boca Raton, Florida, 1975 and *Engineered Materials Handbook*, Vol.2, Engineering Plastics, ASM International, Metals Park, Ohio, 1988.

Table 435. SELECTING DISSIPATION FACTOR FOR POLYMERS AT 1 MHZ (SHEET 1 OF 4)

Polymer	Dissipation Factor (ASTM D150) @ 10^6 Hz
Polystyrenes; Molded: General purpose	0.0001–0.0005
Fluorocarbons; Molded,Extruded: Polytetrafluoroethylene (PTFE)	0.0002
Polypropylene: General purpose	0.0002–0.0003
Polypropylene: High impact	0.0002—0.0003
Molded,Extruded Fluorinated ethylene propylene (FEP)	0.0003
Polystyrenes; Molded: Medium impact	0.0004–0.002
Polystyrenes; Molded: High impact	0.0004–0.002
Fluorocarbons; Molded,Extruded: Ceramic reinforced (PTFE)	0.0005–0.0015
Polypropylene: Flame retardant	0.0006–0.003
Polyphenylene sulfide: Standard	0.0007
Silicones; Molded, Laminated: Granular (silica) reinforced silicones	0.001—0.004
Polyphenylene sulfide: 40% glass reinforced	0.0014—0.0041
Phenylene Oxides: Glass fiber reinforced	0.0015
Polypropylene: Asbestos filled	0.002
Polystyrenes; Molded: Glass fiber -30% reinforced	0.002
Silicones; Molded, Laminated: Woven glass fabric/ silicone laminate	0.002
Phenylene Oxides: SE—100	0.0024
Phenylene Oxides: SE—1	0.0024
Polypropylene: Glass reinforced	0.003
Phenylene oxides (Noryl): Standard	0.0034
Polyacetal Homopolymer: 20% glass reinforced	0.0036
Silicones; Molded, Laminated: Fibrous (glass) reinforced silicones	0.004
Polyacetal Homopolymer: Standard	0.0048
Phenylene oxides (Noryl): Glass fiber reinforced	0.0049
ABS Resins; Molded, Extruded: Heat resistant	0.005—0.015
Polymides: Glass Reinforced	0.0055
Polyarylsulfone	0.0056—0.012
ABS–Polycarbonate Alloy	0.0059

Source: data compiled by J.S. Park from Charles T. Lynch, CRC *Handbook of Materials Science*, Vol. 3, CRC Press, Boca Raton, Florida, 1975 and *Engineered Materials Handbook*, Vol.2, Engineering Plastics, ASM International, Metals Park, Ohio, 1988.

Table 435. SELECTING DISSIPATION FACTOR FOR POLYMERS AT 1 MHZ (SHEET 2 OF 4)

Polymer	Dissipation Factor (ASTM D150) @ 10^6 Hz
Polyester;: Thermosets: Cast Rigid	0.006—0.04
Polycarbonate (40% glass fiber reinforced)	0.007
Polystyrenes; Molded: Styrene acrylonitrile (SAN)	0.007–0.010
Fluorocarbons; Molded,Extruded: Polytrifluoro chloroethylene (PTFCE)	0.007—0.010
ABS Resins; Molded, Extruded: High impact	0.007—0.015
ABS Resins; Molded, Extruded: Medium impact	0.008—0.009
ABS Resins; Molded, Extruded: Very high impact	0.008—0.016
ABS Resins; Molded, Extruded: Low temperature impact	0.008—0.016
Reinforced Polyester Sheet molding compounds, general purpose	0.0086—0.022
Polystyrenes; Molded: Glass fiber (30%) reinforced SAN	0.009
Diallyl Phthalates; Molded: Dacron Filled	0.009—0.017 (Wet)
Polycarbonate	0.01
Standard Epoxies: High Strength Laminate	0.010-0.017
Nylons; Molded, Extruded; Type 6: Flexible Copolymers	0.010—0.015
Acrylics; Moldings: High impact grade	0.01—0.02
Cellulose Acetate; Molded, Extruded; ASTM Grade: H4—1	0.01—0.10
Cellulose Acetate; Molded, Extruded; ASTM Grade: H2—1	0.01—0.10
Cellulose Acetate; Molded, Extruded; ASTM Grade: MH—1, MH—2	0.01—0.10
Cellulose Acetate; Molded, Extruded; ASTM Grade: MS—1, MS—2	0.01—0.10
Cellulose Acetate; Molded, Extruded; ASTM Grade: S2—1	0.01—0.10
Chlorinated polyether	0.011
Polymides: Unreinforced	0.011
Diallyl Phthalates; Molded: Glass Fiber Filled	0.012—0.020 (Wet)
Standard Epoxies: Molded	0.013—0.020
Alkyds; Molded: Glass reinforced (heavy duty parts)	0.015—0.022
Epoxies; High Performance Resins: Glass Cloth Laminate	0.0158
Alkyds; Molded: Putty (encapsulating)	0.016—0.020
Nylons; Molded, Extruded; 6/6 Nylon: Glass fiber reinforced	0.017—0.018

Source: data compiled by J.S. Park from Charles T. Lynch, CRC *Handbook of Materials Science*, Vol. 3, CRC Press, Boca Raton, Florida, 1975 and *Engineered Materials Handbook*, Vol.2, Engineering Plastics, ASM International, Metals Park, Ohio, 1988.

Table 435. SELECTING DISSIPATION FACTOR FOR POLYMERS AT 1 MHZ (SHEET 3 OF 4)

Polymer	Dissipation Factor (ASTM D150) @ 10^6 Hz
Alkyds; Molded: Granular (high speed molding)	0.017—0.020
Nylons; Molded, Extruded; Type 6: Glass fiber (30%) reinforced	0.019—0.015
Chlorinated polyvinyl chloride	0.02
Nylons; Molded, Extruded: Type 11	0.02
Phenolics; Molded; Very High Shock: Glass Fiber Filled	0.02
Melamines; Molded: Glass Fiber Filled	0.020—0.03
Acrylics; Cast Resin Sheets, Rods: General Purpose, Type I	0.02—0.03
Acrylics; Cast Resin Sheets, Rods: General Purpose, Type II	0.02—0.03
Acrylics; Moldings: Grades 5, 6, 8	0.02—0.03
Cellulose Acetate Butyrate; Molded, Extruded; ASTM Grade: H4	0.02—0.05
Cellulose Acetate Butyrate; Molded, Extruded; ASTM Grade: MH	0.02—0.05
Cellulose Acetate Butyrate; Molded, Extruded; ASTM Grade: S2	0.02—0.05
Cellusose Acetate Propionate; Molded, Extruded; ASTM Grade: 1	0.02—0.05
Cellusose Acetate Propionate; Molded, Extruded; ASTM Grade: 3	0.02—0.05
Cellusose Acetate Propionate; Molded, Extruded; ASTM Grade: 6	0.02—0.05
Polyester;: Thermosets: Flexible	0.02—0.06
Alkyds; Molded: Rope (general purpose)	0.023
Standard Epoxies: General Purpose Glass Cloth Laminate	0.024—0.026
Ureas; Molded: Cellulose filled (ASTM Type 2)	0.027—0.029
Melamines; Molded: Alpha Cellulose Filled	0.028
Ureas; Molded: Alpha—cellulose filled (ASTM Type l)	0.028—0.032
Ureas; Molded: Woodflour filled	0.028—0.032
Epoxies; High Performance Resins: Cast, Rigid	0.029—0.028
Melamines; Molded: Alpha Cellulose Mineral Filled	0.030
Nylons; Molded, Extruded; Type 6: General Purpose	0.03—0.04
Phenolics; Molded; General: Woodflour & Flock Filled	0.03—0.07
Phenolics; Molded; Shock: Paper, Flock, or Pulp Filled	0.03—0.07
Phenolics; Molded; High Shock: Chopped Fabric or Cord Filled	0.03—0.09

Source: data compiled by J.S. Park from Charles T. Lynch, CRC *Handbook of Materials Science*, Vol. 3, CRC Press, Boca Raton, Florida, 1975 and *Engineered Materials Handbook*, Vol.2, Engineering Plastics, ASM International, Metals Park, Ohio, 1988.

Table 435. SELECTING DISSIPATION FACTOR FOR POLYMERS AT 1 MHZ (SHEET 4 OF 4)

Polymer	Dissipation Factor (ASTM D150) @ 10^6 Hz
PVC–Acrylic Injection Molded	0.031
Melamines; Molded: Unfilled	0.031—0.040
Standard Epoxies: Cast Rigid	0.032
Melamines; Molded: Cellulose Electrical Filled	0.032—0.12
Standard Epoxies: Cast Flexible	0.0369-0.0622
Nylons; Molded, Extruded; 6/6 Nylon: General purpose molding	0.04
Diallyl Phthalates; Molded: Orlon Filled	0.045—0.040 (Wet)
Nylons; Type 6: Cast	0.05
Nylons; Molded, Extruded: Type 8	0.08
Rubber Phenolic—Chopped Fabric Filled	0.09
PVC–Acrylic Sheet	0.094
Phenolics: Molded: Arc resistant—Mineral Filled	0.1
Thermoset Carbonate: Allyl diglycol carbonate	0.1—0.2
Phenolics: Molded: Rubber Phenolic—Woodflour or Flock Filled	0.1—0.2
Rubber Phenolic—Asbestos Filled	0.13
Diallyl Phthalates; Molded: Asbestos Filled	0.154—0.050 (Wet)
Fluorocarbons; Molded,Extruded: Polyvinylidene—fluoride (PVDF)	0.184

Source: data compiled by J.S. Park from Charles T. Lynch, CRC *Handbook of Materials Science*, Vol. 3, CRC Press, Boca Raton, Florida, 1975 and *Engineered Materials Handbook*, Vol.2, Engineering Plastics, ASM International, Metals Park, Ohio, 1988.

Table 436. SELECTING DIELECTRIC STRENGTH OF POLYMERS
(SHEET 1 OF 5)

Polymer	Dielectric Strength (Short Time, ASTM D149) (V / mil)
Polyvinyl Chloride & Copolymers: Nonrigid–electrical	24—500
Phenolics; Molded: High shock: chopped fabric or cord filled	200—350
Reinforced polyester moldings: High strength (glass fibers)	200—400
Phenolics; Molded: General: woodflour and flock filled	200—425
Phenolics; Molded: Rubber phenolic—chopped fabric filled	250
Melamines; Molded: Glass fiber filled	250 —300
Phenolics; Molded: Shock: paper, flock, or pulp filled	250—350
Phenolics; Molded: Rubber phenolic—woodflour or flock filled	250—375
Cellulose Acetate Butyrate; Molded, Extruded; ASTM Grade: H4	250—400
Cellulose Acetate Butyrate; Molded, Extruded; ASTM Grade: MH	250—400
Cellulose Acetate Butyrate; Molded, Extruded; ASTM Grade: S2	250—400
Cellulose Acetate; Molded, Extruded; ASTM Grade: H6—1	250—600
Cellulose Acetate; Molded, Extruded; ASTM Grade: H4—1	250—600
Cellulose Acetate; Molded, Extruded; ASTM Grade: H2—1	250—600
Cellulose Acetate; ASTM Grade: MH—1, MH—2	250—600
Cellulose Acetate; ASTM Grade: MS—1, MS—2	250—600
Cellulose Acetate; Molded, Extruded; ASTM Grade: S2—1	250—600
Polyvinylidene— fluoride (PVDF): Molded,Extruded	260
Silicones: Fibrous (glass) reinforced silicones	280 (in oil)
Epoxies; High performance resins: Molded	280—400 (step)
Polymides: Glass reinforced	300—310
Resins; Molded, Extruded: Very high impact	300—375
Ceramic reinforced (PTFE): Molded,Extruded	300—400
6/6 Nylon: Glass fiber Molybdenum disulfide filled	300—400
Polyesters: Cast Thermosets: Rigid	300—400
Polyesters: Cast Thermosets: Flexible	300—400
Ureas; Molded: Alpha–cellulose filled (ASTM Type l)	300—400
Ureas; Molded: Woodflour filled	300—400

Source: data compiled by J.S. Park from Charles T. Lynch, *CRC Handbook of Materials Science*, Vol. 3, CRC Press, Boca Raton, Florida, 1975 and *Engineered Materials Handbook*, Vol.2, Engineering Plastics, ASM International, Metals Park, Ohio, 1988.

Table 436. SELECTING DIELECTRIC STRENGTH OF POLYMERS

(SHEET 2 OF 5)

Polymer	Dielectric Strength (Short Time, ASTM D149) (V / mil)
Diallyl Phthalates; Molded: Asbestos filled	300—400 (wet)
Resins; Molded, Extruded: Low temperature impact	300—415
Diallyl Phthalates; Molded: Glass fiber filled	300—420 (wet)
Cellusose Acetate Propionate; Molded, Extruded' ASTM Grade: 1	300—450
Cellusose Acetate Propionate; Molded, Extruded' ASTM Grade: 3	300—450
Cellusose Acetate Propionate; Molded, Extruded' ASTM Grade: 6	300—450
Polystyrenes; Molded: High impact	300—650
Polypropylene: Glass reinforced	317—475
Nylons; Molded, Extruded: Type 8	340
Ureas; Molded: Cellulose filled (ASTM Type 2)	340—370
Phenolics; Molded: Rubber phenolic—asbestos filled	350
Reinforced polyester moldings: Heat & chemical resistant (asbestos)	350
Polyarylsulfone	350—383
Melamines; Molded: Cellulose electrical	350—400
Phenolics; Molded: Arc resistant—mineral filled	350—425
Diallyl Phthalates; Molded: Glass fiber filled	350—430 (dry)
Resins; Molded, Extruded: High impact	350—440
Diallyl Phthalates; Molded: Asbestos filled	350—450 (dry)
Diallyl Phthalates; Molded: Dacron filled	360—391 (wet)
Resins; Molded, Extruded: Heat resistant	360—400
Melamines; Molded: Alpha cellulose and mineral filled	375
Diallyl Phthalates; Molded: Orlon filled	375 (wet)
Phenolics; Molded: Very high shock: glass fiber filled	375—425
Diallyl Phthalates; Molded: Dacron filled	376—400 (dry)
Nylons; Molded, Extruded; Type 6: Cast	380
Silicones: Granular (silica) reinforced silicones	380 (in oil)
ABS Resins; Molded, Extruded: Medium impact	385
6/6 Nylon: General purpose molding	385

Source: data compiled by J.S. Park from Charles T. Lynch, *CRC Handbook of Materials Science*, Vol. 3, CRC Press, Boca Raton, Florida, 1975 and *Engineered Materials Handbook*, Vol.2, Engineering Plastics, ASM International, Metals Park, Ohio, 1988.

Table 436. SELECTING DIELECTRIC STRENGTH OF POLYMERS
(SHEET 3 OF 5)

Polymer	Dielectric Strength (Short Time, ASTM D149) (V / mil)
Nylons; Molded, Extruded; Type 6: General purpose	385—400
Polystyrenes; Molded: Glass fiber -30% reinforced	396
Acrylics; Moldings: Grades 5, 6, 8	400
Chlorinated polyether	400
Polycarbonate	400
PVC–Acrylic Alloy: PVC–acrylic injection molded	400
Phenylene Oxides: SE—100	400 (1/8 in.)
Diallyl Phthalates; Molded: Orlon filled	400 (dry)
Reinforced polyester: Sheet molding compounds, general purpose	400—440
Nylons; Molded, Extruded; Type 6: Glass fiber (30%) reinforced	400—450
6/6 Nylon; Molded, Extruded: Glass fiber reinforced	400—480
Acrylics; Moldings: High impact grade	400—500
Styrene acrylonitrile (SAN)	400—500
Polyester; Thermoplastic Moldings: General purpose grade	420—540
Nylons; Molded, Extruded: Type 11	425
Phenylene oxides (Noryl): Standard	425
Polystyrenes; Molded: Medium impact	>425
PVC–Acrylic Alloy: PVC–acrylic sheet	>429
Nylons; Molded, Extruded; Type 6: Flexible copolymers	440
Epoxy novolacs: Cast, rigid	444
Polypropylene: Asbestos filled	450
Acrylics; Cast Resin Sheets, Rods: General purpose, type II	450—500
Acrylics; Cast Resin Sheets, Rods: General purpose, type I	450—530
Polyphenylene sulfide: Standard	450—595
Polypropylene: High impact	450—650
6/10 Nylon: General purpose extrusion	470
Polycarbonate (40% glass fiber reinforced)	475
Phenylene oxides (Noryl): Glass fiber reinforced	480

Source: data compiled by J.S. Park from Charles T. Lynch, *CRC Handbook of Materials Science*, Vol. 3, CRC Press, Boca Raton, Florida, 1975 and *Engineered Materials Handbook*, Vol.2, Engineering Plastics, ASM International, Metals Park, Ohio, 1988.

Table 436. SELECTING DIELECTRIC STRENGTH OF POLYMERS
(SHEET 4 OF 5)

Polymer	Dielectric Strength (Short Time, ASTM D149) (V / mil)
Polyethylenes; Molded, Extruded; Type I: Melt index 0.3—3.6	480
Polyethylenes; Molded, Extruded; Type I: Melt index 6—26	480
Polyethylenes; Molded, Extruded; Type I: Melt index 200	480
Polyethylenes; Molded, Extruded; Type II: Melt index 20	480
Polyethylenes; Molded, Extruded; Type II: Melt index l.0—1.9	480
Polyethylenes; Molded, Extruded; Type III: Melt index 0.2—0.9	480
Polyethylenes; Molded, Extruded; Type III: Melt Melt index 0.l—12.0	480
Polyethylenes; Molded, Extruded; Type III: Melt index 1.5—15	480
Polyethylenes; Molded, Extruded; Type III: High molecular weight	480
Polypropylene: Flame retardant	485—700
Polyphenylene sulfide: 40% glass reinforced	490
ABS–Polycarbonate Alloy	500
Polyacetal Homopolymer: Standard	500
Polyacetal Homopolymer: 20% glass reinforced	500
Polyacetal Copolymer: Standard	500
Polyacetal Copolymer: High flow	500
Phenylene Oxides: SE—1	500 (1/8 in.)
Polystyrenes; Molded: General purpose	>500
Olefin Copolymers; Molded: Polyallomer	500—650
Glass fiber (30%) reinforced SAN	515
Olefin Copolymers; Molded: EVA (ethylene vinyl acetate)	525
Polytrifluoro chloroethylene (PTFCE): Molded,Extruded	530—600
Olefin Copolymers; Molded: EEA (ethylene ethyl acrylate)	550
Polyester; Thermoplastic Moldings: Glass reinforced grades	560—750
Polyacetal Copolymer: 25% glass reinforced	580
Polyester; Thermoplastic Moldings: Asbestos—filled grade	580
Polyester; Thermoplastic Moldings: General purpose grade	590
Polypropylene: General purpose	650 (125 mil)

Source: data compiled by J.S. Park from Charles T. Lynch, *CRC Handbook of Materials Science*, Vol. 3, CRC Press, Boca Raton, Florida, 1975 and *Engineered Materials Handbook*, Vol.2, Engineering Plastics, ASM International, Metals Park, Ohio, 1988.

Table 436. SELECTING DIELECTRIC STRENGTH OF POLYMERS
(SHEET 5 OF 5)

Polymer	Dielectric Strength (Short Time, ASTM D149) (V / mil)
Silicones: Woven glass fabric/ silicone laminate	725
Polyvinyl Chloride & Copolymers: Rigid–normal impact	725—1,400
Polyester; Thermoplastic Moldings: Glass reinforced self extinguishing	750
Nylons; Molded, Extruded: Type 12	840
Olefin Copolymers; Molded: Ionomer	1,000
Polytetrafluoroethylene (PTFE): Molded,Extruded	1,000—2,000
Phenylene Oxides: Glass fiber reinforced	1,020 (1/32 in.)
Chlorinated polyvinyl chloride	1,250—1,550
Fluorinated ethylene propylene(FEP): Molded,Extruded	2,100

Source: data compiled by J.S. Park from Charles T. Lynch, *CRC Handbook of Materials Science*, Vol. 3, CRC Press, Boca Raton, Florida, 1975 and *Engineered Materials Handbook*, Vol.2, Engineering Plastics, ASM International, Metals Park, Ohio, 1988.

Table 437. SELECTING DIELECTRIC CONSTANTS OF POLYMERS AT 60 HZ (SHEET 1 OF 5)

Polymer	Dielectric Constant (ASTM D150) 60 Hz
Polytetrafluoroethylene (PTFE) (0.01 in thickness)	2.1
Fluorinated ethylene propylene(FEP) (0.01 in thickness)	2.1
Polypropylene: General purpose	2.20—2.28
Polypropylene: High impact	2.20—2.28
Polyethylenes; Molded, Extruded; Type I: Melt index 0.3—3.6	2.3
Polyethylenes; Molded, Extruded; Type I: Melt index 6—26	2.3
Polyethylenes; Molded, Extruded; Type I: Melt index 200	2.3
Polyethylenes; Molded, Extruded; Type II: Melt index 20	2.3
Polyethylenes; Molded, Extruded; Type II: Melt index l.0—1.9	2.3
Polyethylenes; Molded, Extruded; Type III: Melt index 0.2—0.9	2.3
Polyethylenes; Molded, Extruded; Type III: Melt Melt index 0.l—12.0	2.3
Polyethylenes; Molded, Extruded; Type III: Melt index 1.5—15	2.3
Polyethylenes; Molded, Extruded; Type III: High molecular weight	2.3
Polyallomer	2.3
Polypropylene: Glass reinforced	2.3—2.5
Polyvinyl Chloride & Copolymers: Rigid—normal impact	2.3—3.7
Olefin Copolymers; Molded: Ionomer	2.4
Polystyrenes; Molded: General purpose	2.45—2.65
Polystyrenes; Molded: Medium impact	2.45—4.75
Polystyrenes; Molded: High impact	2.45—4.75
Polypropylene: Flame retardant	2.46—2.79
ABS Resins; Molded, Extruded: Low temperature impact	2.5—3.5
Polytrifluoro chloroethylene (PTFCE)	2.6—2.7
Styrene acrylonitrile (SAN)	2.6—3.4
Phenylene Oxides: SE—100	2.65
Phenylene Oxides: SE—1	2.69
ABS Resins; Molded, Extruded: Heat resistant	2.7—3.5
ABS–Polycarbonate Alloy	2.74

Source: data compiled by J.S. Park from Charles T. Lynch, CRC *Handbook of Materials Science*, Vol. 3, CRC Press, Boca Raton, Florida, 1975 and *Engineered Materials Handbook*, Vol.2, Engineering Plastics, ASM International, Metals Park, Ohio, 1988.

Table 437. SELECTING DIELECTRIC CONSTANTS OF POLYMERS AT 60 HZ (SHEET 2 OF 5)

Polymer	Dielectric Constant (ASTM D150) 60 Hz
Polypropylene: Asbestos filled	2.75
Olefin Copolymers; Molded: EEA (ethylene ethyl acrylate)	2.8
ABS Resins; Molded, Extruded: Medium impact	2.8—3.2
ABS Resins; Molded, Extruded: High impact	2.8—3.2
ABS Resins; Molded, Extruded: Very high impact	2.8—3.5
Polyesters Cast Thermosets: Rigid	2.8—4.4
Ceramic reinforced (PTFE)	2.9—3.6
Phenylene Oxides: Glass fiber reinforced	2.93
Polyvinyl Chloride & Copolymers: Vinylidene chloride	3—5
Phenylene oxides (Noryl): Standard	3.06—3.15
Chlorinated polyvinyl chloride	3.08
Chlorinated polyether	3.1
Polystyrenes; Molded: Glass fiber -30% reinforced	3.1
Polyester; Thermoplastic Moldings: General purpose grade	3.1—3.3
Polyester; Thermoplastic Moldings: General purpose grade	3.16
Olefin Copolymers; Molded: EVA (ethylene vinyl acetate)	3.16
Polycarbonate	3.17
Polyesters Cast Thermosets: Flexible	3.18—7.0
Nylons; Molded, Extruded Type 6: Flexible copolymers	3.2—4.0
Nylons: Type 11	3.3 (10^3 Hz)
Diallyl Phthalates; Molded: Orlon filled	3.3—3.9 (Dry)
Epoxy novolacs: Cast, rigid	3.34—3.39
Glass fiber (30%) reinforced Styrene acrylonitrile (SAN)	3.5
Diallyl Phthalates; Molded: Dacron filled	3.5—3.8 (Dry)
Acrylics; Moldings: Grades 5, 6, 8	3.5—3.9
Acrylics; Moldings: High impact grade	3.5—3.9
Polyester; Thermoplastic Moldings: Asbestos—filled grade	3.5—4.2
Acrylics; Cast Resin Sheets, Rods: General purpose, type I	3.5—4.5

Source: data compiled by J.S. Park from Charles T. Lynch, CRC *Handbook of Materials Science*, Vol. 3, CRC Press, Boca Raton, Florida, 1975 and *Engineered Materials Handbook*, Vol.2, Engineering Plastics, ASM International, Metals Park, Ohio, 1988.

Table 437. SELECTING DIELECTRIC CONSTANTS OF POLYMERS AT 60 HZ (SHEET 3 OF 5)

Polymer	Dielectric Constant (ASTM D150) 60 Hz
Acrylics; Cast Resin Sheets, Rods: General purpose, type II	3.5—4.5
Diallyl Phthalates; Molded: Glass fiber filled	3.5—4.5 (Dry)
Cellulose Acetate Butyrate; Molded, Extruded; ASTM Grade: H4	3.5—6.4
Cellulose Acetate Butyrate; Molded, Extruded; ASTM Grade: MH	3.5—6.4
Cellulose Acetate Butyrate; Molded, Extruded; ASTM Grade: S2	3.5—64
Cellulose Acetate; Molded, Extruded; ASTM Grade: H6—1	3.5—7.5
Cellulose Acetate; Molded, Extruded; ASTM Grade: H4—1	3.5—7.5
Cellulose Acetate; Molded, Extruded; ASTM Grade: H2—1	3.5—7.5
Cellulose Acetate; Molded, Extruded; ASTM Grade: MH—1, MH—2	3.5—7.5
Cellulose Acetate; Molded, Extruded; ASTM Grade: MS—1, MS—2	3.5—7.5
Cellulose Acetate; Molded, Extruded; ASTM Grade: S2—1	3.5—7.5
Polyarylsulfone	3.51—3.94
Phenylene oxides (Noryl): Glass fiber reinforced	3.55
Nylons: Type 12	3.6 (10^3 Hz)
Polyacetal Homopolymer: Standard	3.7
Polyacetal Copolymer: Standard	3.7 (100 Hz)
Polyacetal Copolymer: High flow	3.7 (100 Hz)
Polyester Moldings: Glass reinforced self extinguishing	3.7—3.8
Cellusose Acetate Propionate; Molded, Extruded; ASTM Grade: 1	3.7—4.0
Cellusose Acetate Propionate; Molded, Extruded; ASTM Grade: 3	3.7—4.0
Cellusose Acetate Propionate; Molded, Extruded; ASTM Grade: 6	3.7—4.0
Polyester; Thermoplastic Moldings: Glass reinforced grades	3.7—4.2
Polycarbonate (40% glass fiber reinforced)	3.8
PVC–Acrylic Alloy: PVC–acrylic sheet	3.86
6/10 Nylon: General purpose	3.9
Polyacetal Copolymer: 25% glass reinforced	3.9 (100 Hz)
Silicones; Molded, Laminated: Woven glass fabric/ silicone laminate	3.9—4.2
High performance Epoxies: Cast, rigid	3.96—4.02

Source: data compiled by J.S. Park from Charles T. Lynch, CRC *Handbook of Materials Science*, Vol. 3, CRC Press, Boca Raton, Florida, 1975 and *Engineered Materials Handbook*, Vol.2, Engineering Plastics, ASM International, Metals Park, Ohio, 1988.

Table 437. SELECTING DIELECTRIC CONSTANTS OF POLYMERS AT 60 HZ (SHEET 4 OF 5)

Polymer	Dielectric Constant (ASTM D150) 60 Hz
Nylons; Type 6: Cast	4
6/6 Nylon: General purpose molding	4
PVC–Acrylic Alloy: PVC–acrylic injection molded	4
Polyacetal Homopolymer: 20% glass reinforced	4
Nylons; Molded, Extruded Type 6: General purpose	4.0—5.3
Standard Epoxies: Cast rigid	4.02
Silicones; Molded, Laminated: Granular (silica) reinforced silicones	4.1—4.5
Polymides: Unreinforced	4.12
Silicones; Molded, Laminated: Fibrous (glass) reinforced silicones	4.34
Thermoset Carbonate: Allyl diglycol carbonate	4.4
Standard Epoxies: Molded	4.4-5.4
Epoxy novolacs: Glass cloth laminate	4.41—4.43
Standard Epoxies: Cast flexible	4.43-4.79
Diallyl Phthalates; Molded: Asbestos filled	4.5—5.2 (Dry)
Nylons; Molded, Extruded Type 6: Glass fiber (30%) reinforced	4.6—5.6
Polyester Thermosets: Sheet molding compounds, general purpose	4.62—5.0
High performance Epoxies: Molded	4.7—5.7
Polymides: Glass reinforced	4.84
Phenolics; Molded; General: woodflour and flock filled	5.0—9.0
Alkyds; Molded: Glass reinforced (heavy duty parts)	5.2—6.0
Standard Epoxies: General purpose glass cloth laminate	5.3-5.4
Alkyds; Molded: Putty (encapsulating)	5.4—5.9
Polyvinyl Chloride & Copolymers: Nonrigid—general	5.5—9.1
Phenolics; Molded; Shock: paper, flock, or pulp filled	5.6—11.0
Alkyds; Molded: Granular (high speed molding)	5.7—6.3
Polyvinyl Chloride & Copolymers: Nonrigid—electrical	6.0—8.0
Melamines; Molded: Cellulose electrical	6.2—7.7
Phenolics; Molded; High shock: chopped fabric or cord filled	6.5—15.0

Source: data compiled by J.S. Park from Charles T. Lynch, CRC *Handbook of Materials Science*, Vol. 3, CRC Press, Boca Raton, Florida, 1975 and *Engineered Materials Handbook*, Vol.2, Engineering Plastics, ASM International, Metals Park, Ohio, 1988.

Table 437. SELECTING DIELECTRIC CONSTANTS OF POLYMERS AT 60 HZ (SHEET 5 OF 5)

Polymer	Dielectric Constant (ASTM D150) 60 Hz
Melamines; Molded: Glass fiber filled	7.0—11.1
Ureas; Molded: Alpha—cellulose filled (ASTM Type l)	7.0—9.5
Ureas; Molded: Woodflour filled	7.0—9.5
Phenolics; Molded; Very high shock: glass fiber filled	7.1—7.2
Ureas; Molded: Cellulose filled (ASTM Type 2)	7.2—7.3
Alkyds; Molded: Rope (general purpose)	7.4
Phenolics; Molded: Arc resistant—mineral	7.4
Melamines; Molded: Unfilled	7.9—11.0
Phenolics; Molded: Rubber phenolic—woodflour or flock	9—16
Nylons: Type 8	9.3
Polyvinylidene— fluoride (PVDF) (0.125 in thickness)	10
Rubber phenolic—chopped fabric	15
Rubber phenolic—asbestos	15
6/6 Nylon; Molded, Extruded:Glass fiber reinforced	40—44

Source: data compiled by J.S. Park from Charles T. Lynch, CRC *Handbook of Materials Science*, Vol. 3, CRC Press, Boca Raton, Florida, 1975 and *Engineered Materials Handbook*, Vol.2, Engineering Plastics, ASM International, Metals Park, Ohio, 1988.

Table 438. SELECTING DIELECTRIC CONSTANTS OF POLYMERS AT 1 MHZ (SHEET 1 OF 4)

Polymer	Dielectric Constant (ASTM D150) 10^6 Hz
Polypropylene: Glass reinforced	2—2.25
Polypropylene: General purpose	2.23—2.24
Polypropylene: High impact	2.23—2.27
ABS Resins; Molded, Extruded: Very high impact	2.4—3.0
ABS Resins; Molded, Extruded: Low temperature impact	2.4—3.0
Polystyrenes; Molded: Medium impact	2.4—3.8
Polystyrenes; Molded: General purpose	2.45—2.65
Polypropylene: Flame retardant	2.45—2.70
Acrylics; Moldings: High impact grade	2.5—3.0
Polystyrenes; Molded: High impact	2.5—4.0
Styrene acrylonitrile (SAN)	2.6—3.02
Polypropylene: Asbestos filled	2.6—3.17
Phenylene Oxides: SE—100	2.64
Phenylene Oxides: SE—1	2.68
ABS–Polycarbonate Alloy	2.69
Acrylics; Moldings: Grades 5, 6, 8	2.7—2.9
ABS Resins; Molded, Extruded: High impact	2.7—3.0
Acrylics; Cast Resin Sheets, Rods: General purpose, type I	2.7—3.2
Acrylics; Cast Resin Sheets, Rods: General purpose, type II	2.7—3.2
ABS Resins; Molded, Extruded: Medium impact	2.75—3.0
Standard Epoxies: Cast flexible	2.78-3.52
ABS Resins; Molded, Extruded: Heat resistant	2.8—3.2
Polyesters Cast Thermosets: Rigid	2.8—4.4
Chlorinated polyether	2.92
Phenylene Oxides: Glass fiber reinforced	2.92
Polycarbonate	2.96
Polystyrenes; Molded: Glass fiber -30% reinforced	3
Nylons; Molded, Extruded Type 6: Flexible copolymers	3.0—3.6

Source: data compiled by J.S. Park from Charles T. Lynch, CRC *Handbook of Materials Science*, Vol. 3, CRC Press, Boca Raton, Florida, 1975 and *Engineered Materials Handbook*, Vol.2, Engineering Plastics, ASM International, Metals Park, Ohio, 1988.

Table 438. SELECTING DIELECTRIC CONSTANTS OF POLYMERS AT 1 MHZ (SHEET 2 OF 4)

Polymer	Dielectric Constant (ASTM D150) 10^6 Hz
Polyacetal Copolymer: Standard	3—7
Polyacetal Copolymer: High flow	3—7
Polyacetal Copolymer: 25% glass reinforced	3—9
Phenylene oxides (Noryl): Standard	3.03—3.10
Chlorinated polyvinyl chloride	3.2—3.6
Cellulose Acetate Butyrate; Molded, Extruded; ASTM Grade: H4	3.2—6.2
Cellulose Acetate Butyrate; Molded, Extruded; ASTM Grade: MH	3.2—6.2
Cellulose Acetate Butyrate; Molded, Extruded; ASTM Grade: S2	3.2—6.2
Cellulose Acetate; Molded, Extruded; ASTM Grade: H6—1	3.2—7.0
Cellulose Acetate; Molded, Extruded; ASTM Grade: H4—1	3.2—7.0
Cellulose Acetate; Molded, Extruded; ASTM Grade: H2—1	3.2—7.0
Cellulose Acetate; Molded, Extruded; ASTM Grade: MH—1, MH—2	3.2—7.0
Cellulose Acetate; Molded, Extruded; ASTM Grade: MS—1, MS—2	3.2—7.0
Cellulose Acetate; Molded, Extruded; ASTM Grade: S2—1	3.2—7.0
Polyphenylene sulfide: Standard	3.22—3.8
Nylons; Type 6: Cast	3.3
PVC–Acrylic Alloy: PVC–acrylic injection molded	3.4
Silicones; Molded, Laminated: Granular (silica) reinforced silicones	3.4 —4.3
Glass fiber (30%) reinforced Styrene acrylonitrile (SAN)	3.4—3.6
Cellusose Acetate Propionate; Molded, Extruded; ASTM Grade: 1	3.4—3.7
Cellusose Acetate Propionate; Molded, Extruded; ASTM Grade: 3	3.4—3.7
Phenylene oxides (Noryl): Glass fiber reinforced	3.41
Standard Epoxies: Cast rigid	3.42
PVC–Acrylic Alloy: PVC–acrylic sheet	3.44
6/10 Nylon: General purpose	3.5
Thermoset Carbonate: Allyl diglycol carbonate	3.5—3.8
6/6 Nylon; Molded, Extruded:Glass fiber reinforced	3.5—4.1
High performance Epoxies: Cast, rigid	3.53—3.58

Source: data compiled by J.S. Park from Charles T. Lynch, CRC *Handbook of Materials Science*, Vol. 3, CRC Press, Boca Raton, Florida, 1975 and *Engineered Materials Handbook*, Vol.2, Engineering Plastics, ASM International, Metals Park, Ohio, 1988.

Table 438. SELECTING DIELECTRIC CONSTANTS OF POLYMERS AT 1 MHZ (SHEET 3 OF 4)

Polymer	Dielectric Constant (ASTM D150) 10^6 Hz
Polyarylsulfone	3.54—3.7
Polycarbonate (40% glass fiber reinforced)	3.58
6/6 Nylon: General purpose molding	3.6
Nylons; Molded, Extruded Type 6: General purpose	3.6—3.8
Polyacetal Homopolymer: Standard	3.7
Cellusose Acetate Propionate; Molded, Extruded; ASTM Grade: 6	3.7—3.4
Diallyl Phthalates; Molded: Dacron filled	3.7—3.9 (Wet)
Polyesters Cast Thermosets: Flexible	3.7—6.1
Silicones; Molded, Laminated: Woven glass fabric/ silicone laminate	3.8—397
Polyphenylene sulfide: 40% glass reinforced	3.88
Nylons; Molded, Extruded Type 6: Glass fiber (30%) reinforced	3.9—5.4
Polymides: Unreinforced	3.96
Nylons: Type 8	4
Phenolics; Molded; General: woodflour and flock filled	4.0—7.0
Polyacetal Homopolymer: 20% glass reinforced	4—0
Standard Epoxies: Molded	4.1-4.6
Diallyl Phthalates; Molded: Orlon filled	4.1—3.4 (Wet)
Silicones; Molded, Laminated: Fibrous (glass) reinforced silicones	4.28
High performance Epoxies: Molded	4.3—4.8
Diallyl Phthalates; Molded: Glass fiber filled	4.4—4.6 (Wet)
Alkyds; Molded: Putty (encapsulating)	4.5—4.7
Alkyds; Molded: Glass reinforced (heavy duty parts)	4.5—5.0
Phenolics; Molded; Shock: paper, flock, or pulp filled	4.5—7.0
Phenolics; Molded; High shock: chopped fabric or cord filled	4.5—7.0
Polyester Thermosets: Sheet molding compounds, general purpose	4.55—4.75
Phenolics; Molded; Very high shock: glass fiber filled	4.6—6.6
Standard Epoxies: General purpose glass cloth laminate	4.7-4.8
Polymides: Glass reinforced	4.74

Source: data compiled by J.S. Park from Charles T. Lynch, CRC *Handbook of Materials Science*, Vol. 3, CRC Press, Boca Raton, Florida, 1975 and *Engineered Materials Handbook*, Vol.2, Engineering Plastics, ASM International, Metals Park, Ohio, 1988.

Table 438. SELECTING DIELECTRIC CONSTANTS OF POLYMERS AT 1 MHZ (SHEET 4 OF 4)

Polymer	Dielectric Constant (ASTM D150) 10^6 Hz
Standard Epoxies: High strength laminate	4.8-5.2
Alkyds; Molded: Granular (high speed molding)	4.8—5.1
Diallyl Phthalates; Molded: Asbestos filled	4.8—6.5 (Wet)
Phenolics; Molded: Arc resistant—mineral	5
Phenolics; Molded: Rubber phenolic—woodflour or flock	5
Rubber phenolic—chopped fabric	5
Rubber phenolic—asbestos	5
High performance Epoxies: Glass cloth laminate	5.1
Melamines; Molded: Cellulose electrical	5.2—6.0
Melamines; Molded: Alpha cellulose mineral filled	5.6
Melamines; Molded: Glass fiber filled	6.0—7.9
Melamines; Molded: Unfilled	6.3—7.3
Ureas; Molded: Cellulose filled (ASTM Type 2)	6.4—6.5
Ureas; Molded: Alpha—cellulose filled (ASTM Type l)	6.4—6.9
Ureas; Molded: Woodflour filled	6.4—6.9
Melamines; Molded: Alpha cellulose filled	6.4—8.1
Alkyds; Molded: Rope (general purpose)	6.8

Source: data compiled by J.S. Park from Charles T. Lynch, CRC *Handbook of Materials Science*, Vol. 3, CRC Press, Boca Raton, Florida, 1975 and *Engineered Materials Handbook*, Vol.2, Engineering Plastics, ASM International, Metals Park, Ohio, 1988.

Table 439. SELECTING TANGENT LOSS IN GLASS

(SHEET 1 OF 5)

Glass	Frequency (Hz)	Temperature	Tangent Loss (tan δ)
SiO_2 glass	100 Hz	25°C	0.00002
SiO_2 glass	1 kHz	25°C	0.00002
SiO_2 glass	10 kHz	25°C	0.00002
SiO_2 glass	10 kHz	200°C	0.00004
B_2O_3 glass	32 kHz	50K	0.00005
B_2O_3 glass	32 kHz	100K	0.00011
SiO_2 glass	1 kHz	200°C	0.00012
B_2O_3 glass	32 kHz	300K	0.0003
B_2O_3-Na_2O glass (10% mol Na_2O)	1 kHz	134.5°C	0.0003
B_2O_3 glass	1 MHz	100°C	0.0004
B_2O_3 glass	1 MHz	200°C	0.0005
B_2O_3-Na_2O glass (12.5% mol Na_2O)	1 kHz	134.5°C	0.0005
SiO_2 glass	100 Hz	200°C	0.00052
B_2O_3 glass	32 kHz	150K	0.0007
SiO_2 glass	10 kHz	300°C	0.00072
B_2O_3 glass	32 kHz	250K	0.0008
B_2O_3 glass	1 MHz	300°C	0.0009
B_2O_3-Na_2O glass (10% mol Na_2O)	1 kHz	214°C	0.0009
B_2O_3-Na_2O glass (20% mol Na_2O)	1 kHz	16°C	0.0009
B_2O_3-CaO glass (33.3% mol CaO)	2 MHz	25°C	0.001
B_2O_3 glass	32 kHz	200K	0.0010
SiO_2-B_2O_3 glass (46.3% mol B_2O_3)	10 GHz		0.0014
SiO_2 glass	9.4 GHz	20°C	0.0015
B_2O_3-Na_2O glass (15% mol Na_2O)	1 kHz	134.5°C	0.0015

Source: data compiled by J.S. Park from O. V. Mazurin, M. V. Streltsina and T. P. Shvaiko-Shvaikovskaya, *Handbook of Glass Data*, Part A and Part B, Elsevier, New York, 1983.

Table 439. SELECTING TANGENT LOSS IN GLASS

(SHEET 2 OF 5)

Glass	Frequency (Hz)	Temperature	Tangent Loss (tan δ)
SiO_2 glass	9.4 GHz	200°C	0.0018
SiO_2 glass	9.4 GHz	400°C	0.002
B_2O_3-CaO glass (33.3% mol CaO)	2 MHz	100°C	0.002
SiO_2-Al_2O_3 glass (0.5% mol Al_2O_3)	100 K	100 K	0.0021
B_2O_3-Na_2O glass (10% mol Na_2O)	1MHz	room temp.	0.0022
B_2O_3-Na_2O glass (12.5% mol Na_2O)	1 kHz	214°C	0.0022
B_2O_3-Na_2O glass (25% mol Na_2O)	1 kHz	16°C	0.0022
SiO_2-Al_2O_3 glass (0.5% mol Al_2O_3)	50 K	50 K	0.0025
B_2O_3-Na_2O glass (8% mol Na_2O)	1MHz	room temp.	0.0025
B_2O_3-CaO glass (33.3% mol CaO)	2 MHz	200°C	0.0025
SiO_2-Al_2O_3 glass (0.5% mol Al_2O_3)	150 K	150 K	0.0026
B_2O_3-Na_2O glass (20% mol Na_2O)	1 kHz	90.5°C	0.0026
SiO_2 glass	9.4 GHz	600°C	0.0029
B_2O_3-Na_2O glass (16% mol Na_2O)	1MHz	room temp.	0.0031
B_2O_3-CaO glass (33.3% mol CaO)	2 MHz	300°C	0.0035
B_2O_3-Na_2O glass (10% mol Na_2O)	1 kHz	277°C	0.0038
B_2O_3-CaO glass (33.3% mol CaO)	2 MHz	400°C	0.0045
SiO_2 glass	9.4 GHz	800°C	0.0048
SiO_2-PbO glass (40% mol PbO)	100 GHz	room temp.	0.005
B_2O_3-CaO glass (33.3% mol CaO)	2 MHz	500°C	0.0055
SiO_2-Na_2O glass (16% mol Na_2O)	$4.5x10^8$ Hz	20°C	0.0058
B_2O_3-Na_2O glass (25% mol Na_2O)	1MHz	room temp.	0.0063
B_2O_3-Na_2O glass (15% mol Na_2O)	1 kHz	214°C	0.0064
B_2O_3-Na_2O glass (10% mol Na_2O)	1 kHz	298°C	0.0066

Source: data compiled by J.S. Park from O. V. Mazurin, M. V. Streltsina and T. P. Shvaiko-Shvaikovskaya, *Handbook of Glass Data*, Part A and Part B, Elsevier, New York, 1983.

Table 439. SELECTING TANGENT LOSS IN GLASS

(SHEET 3 OF 5)

Glass	Frequency (Hz)	Temperature	Tangent Loss (tan δ)
B_2O_3-CaO glass (33.3% mol CaO)	2 MHz	550°C	0.007
SiO_2 glass	1 kHz	300°C	0.0072
SiO_2-Na_2O glass (20% mol Na_2O)	4.5x10^8 Hz	20°C	0.0073
SiO_2-Na_2O glass (22.2% mol Na_2O)	4.5x10^8 Hz	20°C	0.0081
B_2O_3-Na_2O glass (28% mol Na_2O)	1MHz	room temp.	0.0081
B_2O_3-Na_2O glass (12.5% mol Na_2O)	1 kHz	277°C	0.0100
SiO_2-Na_2O glass (28.6% mol Na_2O)	4.5x10^8 Hz	20°C	0.0102
SiO_2 glass	9.4 GHz	1000°C	0.011
B_2O_3-Na_2O glass (20% mol Na_2O)	1 kHz	157°C	0.0149
SiO_2-PbO glass (40% mol PbO)	32 GHz	-150°C	0.015
B_2O_3-Na_2O glass (25% mol Na_2O)	1 kHz	90.5°C	0.0150
SiO_2-Na_2O glass (36% mol Na_2O)	4.5x10^8 Hz	20°C	0.0162
B_2O_3-Na_2O glass (12.5% mol Na_2O)	1 kHz	298°C	0.0170
SiO_2-PbO glass (40% mol PbO)	32 GHz	-100°C	0.018
SiO_2-PbO glass (40% mol PbO)	32 GHz	-50°C	0.020
SiO_2 glass	10 kHz	400°C	0.022
SiO_2-PbO glass (40% mol PbO)	32 GHz	0°C	0.022
SiO_2-PbO glass (40% mol PbO)	32 GHz	50°C	0.024
SiO_2 glass	9.4 GHz	1200°C	0.025
SiO_2-Na_2O glass (19.5% mol Na_2O)	300 kHz	room temp.	0.0295
B_2O_3-Na_2O glass (15% mol Na_2O)	1 kHz	277°C	0.0296
SiO_2-Na_2O glass (19.5% mol Na_2O)	100 kHz	room temp.	0.0364
SiO_2-Na_2O glass (24.4% mol Na_2O)	300 kHz	room temp.	0.0369
SiO_2-Na_2O glass (19.5% mol Na_2O)	50 kHz	room temp.	0.0428

Source: data compiled by J.S. Park from O. V. Mazurin, M. V. Streltsina and T. P. Shvaiko-Shvaikovskaya, *Handbook of Glass Data*, Part A and Part B, Elsevier, New York, 1983.

Table 439. SELECTING TANGENT LOSS IN GLASS

(SHEET 4 OF 5)

Glass	Frequency (Hz)	Temperature	Tangent Loss (tan δ)
SiO_2-Na_2O glass (24.4% mol Na_2O)	100 kHz	room temp.	0.0456
SiO_2 glass	9.4 GHz	1400°C	0.046
B_2O_3-Na_2O glass (15% mol Na_2O)	1 kHz	298°C	0.0477
SiO_2-Na_2O glass (19.5% mol Na_2O)	30 kHz	room temp.	0.0492
SiO_2-PbO glass (40% mol PbO)	1000 GHz	room temp.	0.050
SiO_2-Na_2O glass (24.4% mol Na_2O)	50 kHz	room temp.	0.0563
SiO_2-Na_2O glass (29.4% mol Na_2O)	300 kHz	room temp.	0.0568
SiO_2-Na_2O glass (24.4% mol Na_2O)	30 kHz	room temp.	0.0652
SiO_2-Na_2O glass (19.5% mol Na_2O)	10 kHz	room temp.	0.0656
SiO_2-Na_2O glass (29.4% mol Na_2O)	100 kHz	room temp.	0.0758
SiO_2 glass	100 Hz	300°C	0.080
SiO_2-Na_2O glass (19.5% mol Na_2O)	5 kHz	room temp.	0.0832
B_2O_3-Na_2O glass (20% mol Na_2O)	1 kHz	219°C	0.0890
SiO_2-Na_2O glass (24.4% mol Na_2O)	10 kHz	room temp.	0.0916
SiO_2-Na_2O glass (34.3% mol Na_2O)	300 kHz	room temp.	0.0936
SiO_2-Na_2O glass (29.4% mol Na_2O)	50 kHz	room temp.	0.0972
SiO_2-Na_2O glass (19.5% mol Na_2O)	3 kHz	room temp.	0.0984
SiO_2-Na_2O glass (34.3% mol Na_2O)	1kHz	room temp.	0.10324
B_2O_3-Na_2O glass (25% mol Na_2O)	1 kHz	157°C	0.1080
SiO_2-Na_2O glass (29.4% mol Na_2O)	30 kHz	room temp.	0.1172
SiO_2-Na_2O glass (24.4% mol Na_2O)	5 kHz	room temp.	0.1194
SiO_2-Na_2O glass (34.3% mol Na_2O)	100 kHz	room temp.	0.1388
SiO_2-Na_2O glass (39.3% mol Na_2O)	300 kHz	room temp.	0.1402
SiO_2-Na_2O glass (19.5% mol Na_2O)	1kHz	room temp.	0.144

Source: data compiled by J.S. Park from O. V. Mazurin, M. V. Streltsina and T. P. Shvaiko-Shvaikovskaya, *Handbook of Glass Data*, Part A and Part B, Elsevier, New York, 1983.

Table 439. SELECTING TANGENT LOSS IN GLASS

(SHEET 5 OF 5)

Glass	Frequency (Hz)	Temperature	Tangent Loss (tan δ)
SiO_2-Na_2O glass (24.4% mol Na_2O)	3 kHz	room temp.	0.1455
SiO_2-Na_2O glass (29.4% mol Na_2O)	10 kHz	room temp.	0.1764
SiO_2-Na_2O glass (34.3% mol Na_2O)	50 kHz	room temp.	0.1864
SiO_2 glass	1 kHz	400°C	0.2
SiO_2-Na_2O glass (39.3% mol Na_2O)	100 kHz	room temp.	0.2144
SiO_2-Na_2O glass (24.4% mol Na_2O)	1kHz	room temp.	0.2207
SiO_2-Na_2O glass (34.3% mol Na_2O)	30 kHz	room temp.	0.2314
SiO_2-Na_2O glass (29.4% mol Na_2O)	5 kHz	room temp.	0.2426
B_2O_3-Na_2O glass (20% mol Na_2O)	1 kHz	274°C	0.2480
SiO_2-Na_2O glass (29.4% mol Na_2O)	3 kHz	room temp.	0.3027
SiO_2-Na_2O glass (39.3% mol Na_2O)	50 kHz	room temp.	0.3032
SiO_2-Na_2O glass (34.3% mol Na_2O)	10 kHz	room temp.	0.3752
SiO_2-Na_2O glass (39.3% mol Na_2O)	30 kHz	room temp.	0.3835
SiO_2-Na_2O glass (29.4% mol Na_2O)	1kHz	room temp.	0.4923
SiO_2-Na_2O glass (34.3% mol Na_2O)	5 kHz	room temp.	0.5280
SiO_2-Na_2O glass (39.3% mol Na_2O)	10 kHz	room temp.	0.6338
SiO_2-Na_2O glass (34.3% mol Na_2O)	3 kHz	room temp.	0.6520
SiO_2 glass	100 Hz	400°C	1.0

Source: data compiled by J.S. Park from O. V. Mazurin, M. V. Streltsina and T. P. Shvaiko-Shvaikovskaya, *Handbook of Glass Data*, Part A and Part B, Elsevier, New York, 1983.

Table 440. SELECTING TANGENT LOSS IN GLASS BY TEMPERATURE (SHEET 1 OF 5)

Temperature	Glass	Frequency (Hz)	Tangent Loss (tan δ)
-100°C	SiO_2-PbO glass (40% mol PbO)	32 GHz	0.018
-150°C	SiO_2-PbO glass (40% mol PbO)	32 GHz	0.015
-50°C	SiO_2-PbO glass (40% mol PbO)	32 GHz	0.020
0°C	SiO_2-PbO glass (40% mol PbO)	32 GHz	0.022
16°C	B_2O_3-Na_2O glass (20% mol Na_2O)	1 kHz	0.0009
16°C	B_2O_3-Na_2O glass (25% mol Na_2O)	1 kHz	0.0022
20°C	SiO_2 glass	9.4 GHz	0.0015
20°C	B_2O_3-Na_2O glass (10% mol Na_2O)	1MHz	0.0022
20°C	B_2O_3-Na_2O glass (8% mol Na_2O)	1MHz	0.0025
20°C	B_2O_3-Na_2O glass (16% mol Na_2O)	1MHz	0.0031
20°C	SiO_2-PbO glass (40% mol PbO)	100 GHz	0.005
20°C	SiO_2-Na_2O glass (16% mol Na_2O)	4.5×10^8 Hz	0.0058
20°C	B_2O_3-Na_2O glass (25% mol Na_2O)	1MHz	0.0063
20°C	SiO_2-Na_2O glass (20% mol Na_2O)	4.5×10^8 Hz	0.0073
20°C	B_2O_3-Na_2O glass (28% mol Na_2O)	1MHz	0.0081
20°C	SiO_2-Na_2O glass (22.2% mol Na_2O)	4.5×10^8 Hz	0.0081
20°C	SiO_2-Na_2O glass (28.6% mol Na_2O)	4.5×10^8 Hz	0.0102
20°C	SiO_2-Na_2O glass (36% mol Na_2O)	4.5×10^8 Hz	0.0162
20°C	SiO_2-Na_2O glass (19.5% mol Na_2O)	300 kHz	0.0295
20°C	SiO_2-Na_2O glass (19.5% mol Na_2O)	100 kHz	0.0364
20°C	SiO_2-Na_2O glass (24.4% mol Na_2O)	300 kHz	0.0369
20°C	SiO_2-Na_2O glass (19.5% mol Na_2O)	50 kHz	0.0428

Source: data compiled by J.S. Park from O. V. Mazurin, M. V. Streltsina and T. P. Shvaiko-Shvaikovskaya, *Handbook of Glass Data*, Part A and Part B, Elsevier, New York, 1983.

Table 440. SELECTING TANGENT LOSS IN GLASS BY TEMPERATURE (SHEET 2 OF 5)

Temperature	Glass	Frequency (Hz)	Tangent Loss (tan δ)
20°C	SiO_2-Na_2O glass (24.4% mol Na_2O)	100 kHz	0.0456
20°C	SiO_2-Na_2O glass (19.5% mol Na_2O)	30 kHz	0.0492
20°C	SiO_2-PbO glass (40% mol PbO)	1000 GHz	0.050
20°C	SiO_2-Na_2O glass (24.4% mol Na_2O)	50 kHz	0.0563
20°C	SiO_2-Na_2O glass (29.4% mol Na_2O)	300 kHz	0.0568
20°C	SiO_2-Na_2O glass (24.4% mol Na_2O)	30 kHz	0.0652
20°C	SiO_2-Na_2O glass (19.5% mol Na_2O)	10 kHz	0.0656
20°C	SiO_2-Na_2O glass (29.4% mol Na_2O)	100 kHz	0.0758
20°C	SiO_2-Na_2O glass (19.5% mol Na_2O)	5 kHz	0.0832
20°C	SiO_2-Na_2O glass (24.4% mol Na_2O)	10 kHz	0.0916
20°C	SiO_2-Na_2O glass (34.3% mol Na_2O)	300 kHz	0.0936
20°C	SiO_2-Na_2O glass (29.4% mol Na_2O)	50 kHz	0.0972
20°C	SiO_2-Na_2O glass (19.5% mol Na_2O)	3 kHz	0.0984
20°C	SiO_2-Na_2O glass (34.3% mol Na_2O)	1kHz	0.10324
20°C	SiO_2-Na_2O glass (29.4% mol Na_2O)	30 kHz	0.1172
20°C	SiO_2-Na_2O glass (24.4% mol Na_2O)	5 kHz	0.1194
20°C	SiO_2-Na_2O glass (34.3% mol Na_2O)	100 kHz	0.1388
20°C	SiO_2-Na_2O glass (39.3% mol Na_2O)	300 kHz	0.1402
20°C	SiO_2-Na_2O glass (19.5% mol Na_2O)	1kHz	0.144
20°C	SiO_2-Na_2O glass (24.4% mol Na_2O)	3 kHz	0.1455
20°C	SiO_2-Na_2O glass (29.4% mol Na_2O)	10 kHz	0.1764
20°C	SiO_2-Na_2O glass (34.3% mol Na_2O)	50 kHz	0.1864
20°C	SiO_2-Na_2O glass (39.3% mol Na_2O)	100 kHz	0.2144
20°C	SiO_2-Na_2O glass (24.4% mol Na_2O)	1kHz	0.2207

Source: data compiled by J.S. Park from O. V. Mazurin, M. V. Streltsina and T. P. Shvaiko-Shvaikovskaya, *Handbook of Glass Data*, Part A and Part B, Elsevier, New York, 1983.

Table 440. SELECTING TANGENT LOSS IN GLASS BY TEMPERATURE (SHEET 3 OF 5)

Temperature	Glass	Frequency (Hz)	Tangent Loss (tan δ)
20°C	SiO_2-Na_2O glass (34.3% mol Na_2O)	30 kHz	0.2314
20°C	SiO_2-Na_2O glass (29.4% mol Na_2O)	5 kHz	0.2426
20°C	SiO_2-Na_2O glass (29.4% mol Na_2O)	3 kHz	0.3027
20°C	SiO_2-Na_2O glass (39.3% mol Na_2O)	50 kHz	0.3032
20°C	SiO_2-Na_2O glass (34.3% mol Na_2O)	10 kHz	0.3752
20°C	SiO_2-Na_2O glass (39.3% mol Na_2O)	30 kHz	0.3835
20°C	SiO_2-Na_2O glass (29.4% mol Na_2O)	1kHz	0.4923
20°C	SiO_2-Na_2O glass (34.3% mol Na_2O)	5 kHz	0.5280
20°C	SiO_2-Na_2O glass (39.3% mol Na_2O)	10 kHz	0.6338
20°C	SiO_2-Na_2O glass (34.3% mol Na_2O)	3 kHz	0.6520
25°C	SiO_2 glass	100 Hz	0.00002
25°C	SiO_2 glass	1 kHz	0.00002
25°C	SiO_2 glass	10 kHz	0.00002
25°C	B_2O_3-CaO glass (33.3% mol CaO)	2 MHz	0.001
50°C	SiO_2-PbO glass (40% mol PbO)	32 GHz	0.024
90.5°C	B_2O_3-Na_2O glass (20% mol Na_2O)	1 kHz	0.0026
90.5°C	B_2O_3-Na_2O glass (25% mol Na_2O)	1 kHz	0.0150
100°C	B_2O_3 glass	1 MHz	0.0004
100°C	B_2O_3-CaO glass (33.3% mol CaO)	2 MHz	0.002
134.5°C	B_2O_3-Na_2O glass (10% mol Na_2O)	1 kHz	0.0003
134.5°C	B_2O_3-Na_2O glass (12.5% mol Na_2O)	1 kHz	0.0005
134.5°C	B_2O_3-Na_2O glass (15% mol Na_2O)	1 kHz	0.0015
157°C	B_2O_3-Na_2O glass (20% mol Na_2O)	1 kHz	0.0149
157°C	B_2O_3-Na_2O glass (25% mol Na_2O)	1 kHz	0.1080

Source: data compiled by J.S. Park from O. V. Mazurin, M. V. Streltsina and T. P. Shvaiko-Shvaikovskaya, *Handbook of Glass Data*, Part A and Part B, Elsevier, New York, 1983.

Table 440. SELECTING TANGENT LOSS IN GLASS BY TEMPERATURE (SHEET 4 OF 5)

Temperature	Glass	Frequency (Hz)	Tangent Loss (tan δ)
200°C	SiO_2 glass	10 kHz	0.00004
200°C	SiO_2 glass	1 kHz	0.00012
200°C	B_2O_3 glass	1 MHz	0.0005
200°C	SiO_2 glass	100 Hz	0.00052
200°C	SiO_2 glass	9.4 GHz	0.0018
200°C	B_2O_3-CaO glass (33.3% mol CaO)	2 MHz	0.0025
214°C	B_2O_3-Na_2O glass (10% mol Na_2O)	1 kHz	0.0009
214°C	B_2O_3-Na_2O glass (12.5% mol Na_2O)	1 kHz	0.0022
214°C	B_2O_3-Na_2O glass (15% mol Na_2O)	1 kHz	0.0064
219°C	B_2O_3-Na_2O glass (20% mol Na_2O)	1 kHz	0.0890
274°C	B_2O_3-Na_2O glass (20% mol Na_2O)	1 kHz	0.2480
277°C	B_2O_3-Na_2O glass (10% mol Na_2O)	1 kHz	0.0038
277°C	B_2O_3-Na_2O glass (12.5% mol Na_2O)	1 kHz	0.0100
277°C	B_2O_3-Na_2O glass (15% mol Na_2O)	1 kHz	0.0296
298°C	B_2O_3-Na_2O glass (10% mol Na_2O)	1 kHz	0.0066
298°C	B_2O_3-Na_2O glass (12.5% mol Na_2O)	1 kHz	0.0170
298°C	B_2O_3-Na_2O glass (15% mol Na_2O)	1 kHz	0.0477
300°C	SiO_2 glass	10 kHz	0.00072
300°C	B_2O_3 glass	1 MHz	0.0009
300°C	B_2O_3-CaO glass (33.3% mol CaO)	2 MHz	0.0035
300°C	SiO_2 glass	1 kHz	0.0072
300°C	SiO_2 glass	100 Hz	0.080

Source: data compiled by J.S. Park from O. V. Mazurin, M. V. Streltsina and T. P. Shvaiko-Shvaikovskaya, *Handbook of Glass Data*, Part A and Part B, Elsevier, New York, 1983.

Table 440. SELECTING TANGENT LOSS IN GLASS BY TEMPERATURE (SHEET 5 OF 5)

Temperature	Glass	Frequency (Hz)	Tangent Loss (tan δ)
323°C	B_2O_3 glass	32 kHz	0.00005
373°C	B_2O_3 glass	32 kHz	0.00011
373°C	SiO_2-Al_2O_3 glass (0.5% mol Al_2O_3)	100 K	0.0021
323°C	SiO_2-Al_2O_3 glass (0.5% mol Al_2O_3)	50 K	0.0025
400°C	SiO_2 glass	9.4 GHz	0.002
400°C	B_2O_3-CaO glass (33.3% mol CaO)	2 MHz	0.0045
400°C	SiO_2 glass	10 kHz	0.022
400°C	SiO_2 glass	1 kHz	0.2
400°C	SiO_2 glass	100 Hz	1.0
423°C	SiO_2-Al_2O_3 glass (0.5% mol Al_2O_3)	150 K	0.0026
423°C	B_2O_3 glass	32 kHz	0.0007
473°C	B_2O_3 glass	32 kHz	0.0010
500°C	B_2O_3-CaO glass (33.3% mol CaO)	2 MHz	0.0055
523°C	B_2O_3 glass	32 kHz	0.0008
550°C	B_2O_3-CaO glass (33.3% mol CaO)	2 MHz	0.007
573°C	B_2O_3 glass	32 kHz	0.0003
600°C	SiO_2 glass	9.4 GHz	0.0029
800°C	SiO_2 glass	9.4 GHz	0.0048
1000°C	SiO_2 glass	9.4 GHz	0.011
1200°C	SiO_2 glass	9.4 GHz	0.025
1400°C	SiO_2 glass	9.4 GHz	0.046

Source: data compiled by J.S. Park from O. V. Mazurin, M. V. Streltsina and T. P. Shvaiko-Shvaikovskaya, *Handbook of Glass Data*, Part A and Part B, Elsevier, New York, 1983.

Table 441. SELECTING TANGENT LOSS IN GLASS BY FREQUENCY
(SHEET 1 OF 5)

Frequency (Hz)	Glass	Temperature	Tangent Loss (tan δ)
100 Hz	SiO_2 glass	25°C	0.00002
100 Hz	SiO_2 glass	200°C	0.00052
100 Hz	SiO_2 glass	300°C	0.080
100 Hz	SiO_2 glass	400°C	1.0
1 kHz	SiO_2 glass	25°C	0.00002
1 kHz	SiO_2 glass	200°C	0.00012
1 kHz	B_2O_3-Na_2O glass (10% mol Na_2O)	134.5°C	0.0003
1 kHz	B_2O_3-Na_2O glass (12.5% mol Na_2O)	134.5°C	0.0005
1 kHz	B_2O_3-Na_2O glass (10% mol Na_2O)	214°C	0.0009
1 kHz	B_2O_3-Na_2O glass (20% mol Na_2O)	16°C	0.0009
1 kHz	B_2O_3-Na_2O glass (15% mol Na_2O)	134.5°C	0.0015
1 kHz	B_2O_3-Na_2O glass (12.5% mol Na_2O)	214°C	0.0022
1 kHz	B_2O_3-Na_2O glass (25% mol Na_2O)	16°C	0.0022
1 kHz	B_2O_3-Na_2O glass (20% mol Na_2O)	90.5°C	0.0026
1 kHz	B_2O_3-Na_2O glass (10% mol Na_2O)	277°C	0.0038
1 kHz	B_2O_3-Na_2O glass (15% mol Na_2O)	214°C	0.0064
1 kHz	B_2O_3-Na_2O glass (10% mol Na_2O)	298°C	0.0066
1 kHz	SiO_2 glass	300°C	0.0072
1 kHz	B_2O_3-Na_2O glass (12.5% mol Na_2O)	277°C	0.0100
1 kHz	B_2O_3-Na_2O glass (20% mol Na_2O)	157°C	0.0149
1 kHz	B_2O_3-Na_2O glass (25% mol Na_2O)	90.5°C	0.0150
1 kHz	B_2O_3-Na_2O glass (12.5% mol Na_2O)	298°C	0.0170
1 kHz	B_2O_3-Na_2O glass (15% mol Na_2O)	277°C	0.0296
1 kHz	B_2O_3-Na_2O glass (15% mol Na_2O)	298°C	0.0477

Source: data compiled by J.S. Park from O. V. Mazurin, M. V. Streltsina and T. P. Shvaiko-Shvaikovskaya, *Handbook of Glass Data*, Part A and Part B, Elsevier, New York, 1983.

Table 441. SELECTING TANGENT LOSS IN GLASS BY FREQUENCY
(SHEET 2 OF 5)

Frequency (Hz)	Glass	Temperature	Tangent Loss (tan δ)
1 kHz	B_2O_3-Na_2O glass (20% mol Na_2O)	219°C	0.0890
1 kHz	SiO_2-Na_2O glass (34.3% mol Na_2O)	room temp.	0.10324
1 kHz	B_2O_3-Na_2O glass (25% mol Na_2O)	157°C	0.1080
1 kHz	SiO_2-Na_2O glass (19.5% mol Na_2O)	room temp.	0.144
1 kHz	SiO_2 glass	400°C	0.2
1 kHz	SiO_2-Na_2O glass (24.4% mol Na_2O)	room temp.	0.2207
1 kHz	B_2O_3-Na_2O glass (20% mol Na_2O)	274°C	0.2480
1 kHz	SiO_2-Na_2O glass (29.4% mol Na_2O)	room temp.	0.4923
3 kHz	SiO_2-Na_2O glass (19.5% mol Na_2O)	room temp.	0.0984
3 kHz	SiO_2-Na_2O glass (24.4% mol Na_2O)	room temp.	0.1455
3 kHz	SiO_2-Na_2O glass (29.4% mol Na_2O)	room temp.	0.3027
3 kHz	SiO_2-Na_2O glass (34.3% mol Na_2O)	room temp.	0.6520
5 kHz	SiO_2-Na_2O glass (19.5% mol Na_2O)	room temp.	0.0832
5 kHz	SiO_2-Na_2O glass (24.4% mol Na_2O)	room temp.	0.1194
5 kHz	SiO_2-Na_2O glass (29.4% mol Na_2O)	room temp.	0.2426
5 kHz	SiO_2-Na_2O glass (34.3% mol Na_2O)	room temp.	0.5280
10 kHz	SiO_2 glass	25°C	0.00002
10 kHz	SiO_2 glass	200°C	0.00004
10 kHz	SiO_2 glass	300°C	0.00072
10 kHz	SiO_2 glass	400°C	0.022
10 kHz	SiO_2-Na_2O glass (19.5% mol Na_2O)	room temp.	0.0656
10 kHz	SiO_2-Na_2O glass (24.4% mol Na_2O)	room temp.	0.0916
10 kHz	SiO_2-Na_2O glass (29.4% mol Na_2O)	room temp.	0.1764
10 kHz	SiO_2-Na_2O glass (34.3% mol Na_2O)	room temp.	0.3752
10 kHz	SiO_2-Na_2O glass (39.3% mol Na_2O)	room temp.	0.6338

Source: data compiled by J.S. Park from O. V. Mazurin, M. V. Streltsina and T. P. Shvaiko-Shvaikovskaya, *Handbook of Glass Data*, Part A and Part B, Elsevier, New York, 1983.

Table 441. SELECTING TANGENT LOSS IN GLASS BY FREQUENCY
(SHEET 3 OF 5)

Frequency (Hz)	Glass	Temperature	Tangent Loss (tan δ)
30 kHz	SiO_2-Na_2O glass (19.5% mol Na_2O)	room temp.	0.0492
30 kHz	SiO_2-Na_2O glass (24.4% mol Na_2O)	room temp.	0.0652
30 kHz	SiO_2-Na_2O glass (29.4% mol Na_2O)	room temp.	0.1172
30 kHz	SiO_2-Na_2O glass (34.3% mol Na_2O)	room temp.	0.2314
30 kHz	SiO_2-Na_2O glass (39.3% mol Na_2O)	room temp.	0.3835
32 kHz	B_2O_3 glass	50K	0.00005
32 kHz	B_2O_3 glass	100K	0.00011
32 kHz	B_2O_3 glass	300K	0.0003
32 kHz	B_2O_3 glass	150K	0.0007
32 kHz	B_2O_3 glass	250K	0.0008
32 kHz	B_2O_3 glass	200K	0.0010
50 kHz	SiO_2-Na_2O glass (19.5% mol Na_2O)	room temp.	0.0428
50 kHz	SiO_2-Na_2O glass (24.4% mol Na_2O)	room temp.	0.0563
50 kHz	SiO_2-Na_2O glass (29.4% mol Na_2O)	room temp.	0.0972
50 kHz	SiO_2-Na_2O glass (34.3% mol Na_2O)	room temp.	0.1864
50 kHz	SiO_2-Na_2O glass (39.3% mol Na_2O)	room temp.	0.3032
100 kHz	SiO_2-Na_2O glass (19.5% mol Na_2O)	room temp.	0.0364
100 kHz	SiO_2-Na_2O glass (24.4% mol Na_2O)	room temp.	0.0456
100 kHz	SiO_2-Na_2O glass (29.4% mol Na_2O)	room temp.	0.0758
100 kHz	SiO_2-Na_2O glass (34.3% mol Na_2O)	room temp.	0.1388
100 kHz	SiO_2-Na_2O glass (39.3% mol Na_2O)	room temp.	0.2144
300 kHz	SiO_2-Na_2O glass (19.5% mol Na_2O)	room temp.	0.0295
300 kHz	SiO_2-Na_2O glass (24.4% mol Na_2O)	room temp.	0.0369
300 kHz	SiO_2-Na_2O glass (29.4% mol Na_2O)	room temp.	0.0568

Source: data compiled by J.S. Park from O. V. Mazurin, M. V. Streltsina and T. P. Shvaiko-Shvaikovskaya, *Handbook of Glass Data*, Part A and Part B, Elsevier, New York, 1983.

Table 441. SELECTING TANGENT LOSS IN GLASS BY FREQUENCY
(SHEET 4 OF 5)

Frequency (Hz)	Glass	Temperature	Tangent Loss (tan δ)
300 kHz	SiO_2-Na_2O glass (34.3% mol Na_2O)	room temp.	0.0936
300 kHz	SiO_2-Na_2O glass (39.3% mol Na_2O)	room temp.	0.1402
1 MHz	B_2O_3 glass	100°C	0.0004
1 MHz	B_2O_3 glass	200°C	0.0005
1 MHz	B_2O_3 glass	300°C	0.0009
1 MHz	B_2O_3-Na_2O glass (10% mol Na_2O)	room temp.	0.0022
1 MHz	B_2O_3-Na_2O glass (8% mol Na_2O)	room temp.	0.0025
1 MHz	B_2O_3-Na_2O glass (16% mol Na_2O)	room temp.	0.0031
1 MHz	B_2O_3-Na_2O glass (25% mol Na_2O)	room temp.	0.0063
1 MHz	B_2O_3-Na_2O glass (28% mol Na_2O)	room temp.	0.0081
2 MHz	B_2O_3-CaO glass (33.3% mol CaO)	25°C	0.001
2 MHz	B_2O_3-CaO glass (33.3% mol CaO)	100°C	0.002
2 MHz	B_2O_3-CaO glass (33.3% mol CaO)	200°C	0.0025
2 MHz	B_2O_3-CaO glass (33.3% mol CaO)	300°C	0.0035
2 MHz	B_2O_3-CaO glass (33.3% mol CaO)	400°C	0.0045
2 MHz	B_2O_3-CaO glass (33.3% mol CaO)	500°C	0.0055
2 MHz	B_2O_3-CaO glass (33.3% mol CaO)	550°C	0.007
4.5×10^8 Hz	SiO_2-Na_2O glass (16% mol Na_2O)	20°C	0.0058
4.5×10^8 Hz	SiO_2-Na_2O glass (20% mol Na_2O)	20°C	0.0073
4.5×10^8 Hz	SiO_2-Na_2O glass (22.2% mol Na_2O)	20°C	0.0081
4.5×10^8 Hz	SiO_2-Na_2O glass (28.6% mol Na_2O)	20°C	0.0102
4.5×10^8 Hz	SiO_2-Na_2O glass (36% mol Na_2O)	20°C	0.0162

Source: data compiled by J.S. Park from O. V. Mazurin, M. V. Streltsina and T. P. Shvaiko-Shvaikovskaya, *Handbook of Glass Data,* Part A and Part B, Elsevier, New York, 1983.

Table 441. SELECTING TANGENT LOSS IN GLASS BY FREQUENCY

(SHEET 5 OF 5)

Frequency (Hz)	Glass	Temperature	Tangent Loss (tan δ)
9.4 GHz	SiO_2 glass	20°C	0.0015
9.4 GHz	SiO_2 glass	200°C	0.0018
9.4 GHz	SiO_2 glass	400°C	0.002
9.4 GHz	SiO_2 glass	600°C	0.0029
9.4 GHz	SiO_2 glass	800°C	0.0048
9.4 GHz	SiO_2 glass	1000°C	0.011
9.4 GHz	SiO_2 glass	1200°C	0.025
9.4 GHz	SiO_2 glass	1400°C	0.046
10 GHz	SiO_2-B_2O_3 glass (46.3% mol B_2O_3)		0.0014
32 GHz	SiO_2-PbO glass (40% mol PbO)	-150°C	0.015
32 GHz	SiO_2-PbO glass (40% mol PbO)	-100°C	0.018
32 GHz	SiO_2-PbO glass (40% mol PbO)	-50°C	0.020
32 GHz	SiO_2-PbO glass (40% mol PbO)	0°C	0.022
32 GHz	SiO_2-PbO glass (40% mol PbO)	50°C	0.024
100 GHz	SiO_2-PbO glass (40% mol PbO)	room temp.	0.005
1000 GHz	SiO_2-PbO glass (40% mol PbO)	room temp.	0.050

Source: data compiled by J.S. Park from O. V. Mazurin, M. V. Streltsina and T. P. Shvaiko-Shvaikovskaya, *Handbook of Glass Data*, Part A and Part B, Elsevier, New York, 1983.

Table 442. SELECTING ELECTRICAL PERMITTIVITY OF GLASS

(SHEET 1 OF 6)

Glass	Frequency (Hz)	Temperature (°C)	Electrical Permittivity
B_2O_3 glass	50 kHz	800	3.04
B_2O_3 glass	50 kHz	620	3.05
B_2O_3 glass	50 kHz	750	3.06
B_2O_3 glass	50 kHz	700	3.09
B_2O_3 glass	50 kHz	500	3.10
B_2O_3 glass	50 kHz	650	3.10
B_2O_3 glass	50 kHz	580	3.115
B_2O_3 glass	50 kHz	550	3.12
B_2O_3 glass	10 kHz	500	3.13
B_2O_3 glass	10 kHz	550	3.14
B_2O_3 glass	10 kHz	580	3.145
B_2O_3 glass	3 kHz	500	3.15
B_2O_3 glass	10 kHz	620	3.15
B_2O_3 glass	10 kHz	650	3.15
B_2O_3 glass	10 kHz	700	3.16
B_2O_3 glass	1 kHz	500	3.17
B_2O_3 glass	3 kHz	550	3.17
B_2O_3 glass	3 kHz	580	3.18
B_2O_3 glass	1 kHz	550	3.21
B_2O_3 glass	3 kHz	620	3.21
B_2O_3 glass	3 kHz	650	3.25
B_2O_3 glass	1 kHz	580	3.27
SiO_2–Al_2O_3 glass (46.3% mol B_2O_3)	10 GHz		3.55
B_2O_3–Na_2O glass (4.08% mol Na_2O)	56.8 MHz	room temp.	3.72

Source: data compiled by J.S. Park from O. V. Mazurin, M. V. Streltsina and T. P. Shvaiko-Shvaikovskaya, *Handbook of Glass Data*, Part A and Part B, Elsevier, New York, 1983.

Table 442. SELECTING ELECTRICAL PERMITTIVITY OF GLASS
(SHEET 2 OF 6)

Glass	Frequency (Hz)	Temperature (°C)	Electrical Permittivity
SiO_2 glass	9.4 GHz	20	3.81
SiO_2 glass	10 GHz	20	3.82
SiO_2 glass	10 GHz	220	3.82
SiO_2 glass	9.4 GHz	200	3.83
SiO_2 glass	9.4 GHz	400	3.84
SiO_2 glass	9.4 GHz	600	3.86
SiO_2 glass	9.4 GHz	800	3.88
SiO_2 glass	9.4 GHz	1000	3.91
SiO_2 glass	10 GHz	888	3.91
SiO_2 glass	9.4 GHz	1200	3.93
SiO_2 glass	9.4 GHz	1400	3.96
SiO_2 glass	10 GHz	1170	3.98
SiO_2 glass	100 Hz	25	4.0
SiO_2 glass	100 Hz	200	4.0
SiO_2 glass	100 Hz	300	4.0
SiO_2 glass	1 kHz	25	4.0
SiO_2 glass	1 kHz	200	4.0
SiO_2 glass	1 kHz	300	4.0
SiO_2 glass	10 kHz	25	4.0
SiO_2 glass	10 kHz	200	4.0
SiO_2 glass	10 kHz	300	4.0
SiO_2 glass	10 kHz	400	4.0
SiO_2 glass	10 GHz	1764	4.04
SiO_2 glass	10 GHz	1335	4.05

Source: data compiled by J.S. Park from O. V. Mazurin, M. V. Streltsina and T. P. Shvaiko-Shvaikovskaya, *Handbook of Glass Data*, Part A and Part B, Elsevier, New York, 1983.

Table 442. SELECTING ELECTRICAL PERMITTIVITY OF GLASS

(SHEET 3 OF 6)

Glass	Frequency (Hz)	Temperature (°C)	Electrical Permittivity
SiO_2 glass	10 GHz	1764	4.05
SiO_2 glass	10 GHz	1420	4.07
SiO_2 glass	10 GHz	1480	4.09
SiO_2 glass	1 kHz	400	4.1
SiO_2 glass	10 GHz	1526	4.11
SiO_2 glass	10 GHz	1584	4.12
SiO_2 glass	10 GHz	1647	4.12
SiO_2 glass	10 GHz	1602	4.15
B_2O_3–Na_2O glass (7.35% mol Na_2O)	56.8 MHz	room temp.	4.20
SiO_2–PbO glass (40% mol PbO)	32 GHz	–150	4.25
SiO_2–PbO glass (40% mol PbO)	32 GHz	–100	4.30
SiO_2–PbO glass (40% mol PbO)	32 GHz	–50	4.40
SiO_2–PbO glass (40% mol PbO)	32 GHz	0	4.45
B_2O_3–Na_2O glass (14.15% mol Na_2O)	56.8 MHz	room temp.	4.94
SiO_2–PbO glass (40% mol PbO)	32 GHz	50	5.00
B_2O_3–Na_2O glass (10% mol Na_2O)	1 kHz	73	5.00
B_2O_3–Na_2O glass (10% mol Na_2O)	1 kHz	134.5	5.05
B_2O_3–Na_2O glass (10% mol Na_2O)	1 kHz	214	5.15
B_2O_3–Na_2O glass (17.31% mol Na_2O)	56.8 MHz	room temp.	5.27
B_2O_3–Na_2O glass (10% mol Na_2O)	1 kHz	277	5.45
B_2O_3–Na_2O glass (12.5% mol Na_2O)	1 kHz	73	5.45
SiO_2 glass	100 Hz	400	5.5
B_2O_3–Na_2O glass (10% mol Na_2O)	1 kHz	298	5.60
B_2O_3–Na_2O glass (12.5% mol Na_2O)	1 kHz	134.5	5.60

Source: data compiled by J.S. Park from O. V. Mazurin, M. V. Streltsina and T. P. Shvaiko-Shvaikovskaya, *Handbook of Glass Data*, Part A and Part B, Elsevier, New York, 1983.

Table 442. SELECTING ELECTRICAL PERMITTIVITY OF GLASS
(SHEET 4 OF 6)

Glass	Frequency (Hz)	Temperature (°C)	Electrical Permittivity
B_2O_3–Na_2O glass (12.5% mol Na_2O)	1 kHz	214	5.75
B_2O_3–Na_2O glass (15% mol Na_2O)	1 kHz	73	5.80
B_2O_3–Na_2O glass (15% mol Na_2O)	1 kHz	134.5	6.00
SiO_2–Na_2O glass (16% mol Na_2O)	$4.5x10^8$ Hz	20	6.01
B_2O_3–Na_2O glass (20% mol Na_2O)	1 kHz	16	6.15
B_2O_3–Na_2O glass (24.77% mol Na_2O)	56.8 MHz	room temp.	6.24
B_2O_3–Na_2O glass (12.5% mol Na_2O)	1 kHz	277	6.30
B_2O_3–Na_2O glass (20% mol Na_2O)	1 kHz	90.5	6.43
SiO_2–Na_2O glass (20% mol Na_2O)	$4.5x10^8$ Hz	20	6.48
B_2O_3–Na_2O glass (15% mol Na_2O)	1 kHz	214	6.50
B_2O_3–Na_2O glass (12.5% mol Na_2O)	1 kHz	298	6.65
SiO_2–Na_2O glass (22.2% mol Na_2O)	$4.5x10^8$ Hz	20	6.85
B_2O_3–Na_2O glass (31.98% mol Na_2O)	56.8 MHz	room temp.	7.03
B_2O_3–Na_2O glass (20% mol Na_2O)	1 kHz	157	7.45
B_2O_3–Na_2O glass (25% mol Na_2O)	1 kHz	16	7.50
SiO_2–Na_2O glass (19.5% mol Na_2O)	300 kHz	room temp.	7.62
SiO_2–Na_2O glass (28.6% mol Na_2O)	$4.5x10^8$ Hz	20	7.62
SiO_2–Na_2O glass (19.5% mol Na_2O)	100 kHz	room temp.	7.74
B_2O_3–Na_2O glass (15% mol Na_2O)	1 kHz	277	7.80
SiO_2–Na_2O glass (19.5% mol Na_2O)	50 kHz	room temp.	7.88
SiO_2–Na_2O glass (19.5% mol Na_2O)	30 kHz	room temp.	8.00
SiO_2–Na_2O glass (19.5% mol Na_2O)	10 kHz	room temp.	8.26
SiO_2–Na_2O glass (19.5% mol Na_2O)	5 kHz	room temp.	8.56
B_2O_3–Na_2O glass (15% mol Na_2O)	1 kHz	298	8.60

Source: data compiled by J.S. Park from O. V. Mazurin, M. V. Streltsina and T. P. Shvaiko-Shvaikovskaya, *Handbook of Glass Data*, Part A and Part B, Elsevier, New York, 1983.

Table 442. SELECTING ELECTRICAL PERMITTIVITY OF GLASS
(SHEET 5 OF 6)

Glass	Frequency (Hz)	Temperature (°C)	Electrical Permittivity
SiO_2–Na_2O glass (24.4% mol Na_2O)	300 kHz	room temp.	8.75
B_2O_3–Na_2O glass (25% mol Na_2O)	1 kHz	90.5	8.90
SiO_2–Na_2O glass (24.4% mol Na_2O)	100 kHz	room temp.	8.91
SiO_2–Na_2O glass (19.5% mol Na_2O)	3 kHz	room temp.	8.97
SiO_2–Na_2O glass (24.4% mol Na_2O)	50 kHz	room temp.	9.14
SiO_2–Na_2O glass (24.4% mol Na_2O)	30 kHz	room temp.	9.30
SiO_2–Na_2O glass (19.5% mol Na_2O)	1kHz	room temp.	9.40
SiO_2–Na_2O glass (36% mol Na_2O)	4.5×10^8 Hz	20	9.40
SiO_2–Na_2O glass (24.4% mol Na_2O)	10 kHz	room temp.	9.74
SiO_2–Na_2O glass (29.4% mol Na_2O)	300 kHz	room temp.	10.15
SiO_2–Na_2O glass (24.4% mol Na_2O)	5 kHz	room temp.	10.21
SiO_2–Na_2O glass (29.4% mol Na_2O)	100 kHz	room temp.	10.47
SiO_2–Na_2O glass (24.4% mol Na_2O)	3 kHz	room temp.	10.61
SiO_2–Na_2O glass (29.4% mol Na_2O)	50 kHz	room temp.	10.86
SiO_2–Na_2O glass (34.3% mol Na_2O)	300 kHz	room temp.	11.14
SiO_2–Na_2O glass (29.4% mol Na_2O)	30 kHz	room temp.	11.21
SiO_2–Na_2O glass (24.4% mol Na_2O)	1kHz	room temp.	11.62
SiO_2–Na_2O glass (34.3% mol Na_2O)	100 kHz	room temp.	11.78
B_2O_3–Na_2O glass (20% mol Na_2O)	1 kHz	219	11.85
SiO_2–Na_2O glass (29.4% mol Na_2O)	10 kHz	room temp.	12.08
SiO_2–Na_2O glass (39.3% mol Na_2O)	300 kHz	room temp.	12.43
SiO_2–Na_2O glass (34.3% mol Na_2O)	50 kHz	room temp.	12.57
SiO_2–Na_2O glass (29.4% mol Na_2O)	5 kHz	room temp.	13.19
SiO_2–Na_2O glass (34.3% mol Na_2O)	30 kHz	room temp.	13.28

Source: data compiled by J.S. Park from O. V. Mazurin, M. V. Streltsina and T. P. Shvaiko-Shvaikovskaya, *Handbook of Glass Data*, Part A and Part B, Elsevier, New York, 1983.

Table 442. SELECTING ELECTRICAL PERMITTIVITY OF GLASS

(SHEET 6 OF 6)

Glass	Frequency (Hz)	Temperature (°C)	Electrical Permittivity
SiO_2–Na_2O glass (39.3% mol Na_2O)	100 kHz	room temp.	13.55
SiO_2–Na_2O glass (29.4% mol Na_2O)	3 kHz	room temp.	14.23
SiO_2–Na_2O glass (39.3% mol Na_2O)	50 kHz	room temp.	15.06
SiO_2–Na_2O glass (34.3% mol Na_2O)	10 kHz	room temp.	15.22
SiO_2–Na_2O glass (39.3% mol Na_2O)	30 kHz	room temp.	16.56
B_2O_3–Na_2O glass (25% mol Na_2O)	1 kHz	157	17.30
SiO_2–Na_2O glass (29.4% mol Na_2O)	1kHz	room temp.	17.52
SiO_2–Na_2O glass (34.3% mol Na_2O)	5 kHz	room temp.	18.13
SiO_2–Na_2O glass (34.3% mol Na_2O)	3 kHz	room temp.	21.30
SiO_2–Na_2O glass (39.3% mol Na_2O)	10 kHz	room temp.	22.08
B_2O_3–Na_2O glass (20% mol Na_2O)	1 kHz	274	31.00
SiO_2–Na_2O glass (34.3% mol Na_2O)	1kHz	room temp.	38.61

Source: data compiled by J.S. Park from O. V. Mazurin, M. V. Streltsina and T. P. Shvaiko-Shvaikovskaya, *Handbook of Glass Data*, Part A and Part B, Elsevier, New York, 1983.

Table 443. SELECTING ELECTRICAL PERMITTIVITY OF GLASS BY FREQUENCY (SHEET 1 OF 6)

Frequency (Hz)	Glass	Temperature (°C)	Electrical Permittivity
100 Hz	SiO_2 glass	25	4.0
100 Hz	SiO_2 glass	200	4.0
100 Hz	SiO_2 glass	300	4.0
100 Hz	SiO_2 glass	400	5.5
1 kHz	B_2O_3 glass	500	3.17
1 kHz	B_2O_3 glass	550	3.21
1 kHz	B_2O_3 glass	580	3.27
1 kHz	SiO_2 glass	25	4.0
1 kHz	SiO_2 glass	200	4.0
1 kHz	SiO_2 glass	300	4.0
1 kHz	SiO_2 glass	400	4.1
1 kHz	B_2O_3–Na_2O glass (10% mol Na_2O)	73	5.00
1 kHz	B_2O_3–Na_2O glass (10% mol Na_2O)	134.5	5.05
1 kHz	B_2O_3–Na_2O glass (10% mol Na_2O)	214	5.15
1 kHz	B_2O_3–Na_2O glass (10% mol Na_2O)	277	5.45
1 kHz	B_2O_3–Na_2O glass (12.5% mol Na_2O)	73	5.45
1 kHz	B_2O_3–Na_2O glass (10% mol Na_2O)	298	5.60
1 kHz	B_2O_3–Na_2O glass (12.5% mol Na_2O)	134.5	5.60
1 kHz	B_2O_3–Na_2O glass (12.5% mol Na_2O)	214	5.75
1 kHz	B_2O_3–Na_2O glass (15% mol Na_2O)	73	5.80
1 kHz	B_2O_3–Na_2O glass (15% mol Na_2O)	134.5	6.00
1 kHz	B_2O_3–Na_2O glass (20% mol Na_2O)	16	6.15
1 kHz	B_2O_3–Na_2O glass (12.5% mol Na_2O)	277	6.30
1 kHz	B_2O_3–Na_2O glass (20% mol Na_2O)	90.5	6.43

Source: data compiled by J.S. Park from O. V. Mazurin, M. V. Streltsina and T. P. Shvaiko-Shvaikovskaya, *Handbook of Glass Data*, Part A and Part B, Elsevier, New York, 1983.

Table 443. SELECTING ELECTRICAL PERMITTIVITY OF GLASS BY FREQUENCY (SHEET 2 OF 6)

Frequency (Hz)	Glass	Temperature (°C)	Electrical Permittivity
1 kHz	B_2O_3–Na_2O glass (15% mol Na_2O)	214	6.50
1 kHz	B_2O_3–Na_2O glass (12.5% mol Na_2O)	298	6.65
1 kHz	B_2O_3–Na_2O glass (20% mol Na_2O)	157	7.45
1 kHz	B_2O_3–Na_2O glass (25% mol Na_2O)	16	7.50
1 kHz	B_2O_3–Na_2O glass (15% mol Na_2O)	277	7.80
1 kHz	B_2O_3–Na_2O glass (15% mol Na_2O)	298	8.60
1 kHz	B_2O_3–Na_2O glass (25% mol Na_2O)	90.5	8.90
1 kHz	SiO_2–Na_2O glass (19.5% mol Na_2O)	room temp.	9.40
1 kHz	SiO_2–Na_2O glass (24.4% mol Na_2O)	room temp.	11.62
1 kHz	B_2O_3–Na_2O glass (20% mol Na_2O)	219	11.85
1 kHz	B_2O_3–Na_2O glass (25% mol Na_2O)	157	17.30
1 kHz	SiO_2–Na_2O glass (29.4% mol Na_2O)	room temp.	17.52
1 kHz	B_2O_3–Na_2O glass (20% mol Na_2O)	274	31.00
1 kHz	SiO_2–Na_2O glass (34.3% mol Na_2O)	room temp.	38.61
3 kHz	B_2O_3 glass	500	3.15
3 kHz	B_2O_3 glass	550	3.17
3 kHz	B_2O_3 glass	580	3.18
3 kHz	B_2O_3 glass	620	3.21
3 kHz	B_2O_3 glass	650	3.25
3 kHz	SiO_2–Na_2O glass (19.5% mol Na_2O)	room temp.	8.97
3 kHz	SiO_2–Na_2O glass (24.4% mol Na_2O)	room temp.	10.61
3 kHz	SiO_2–Na_2O glass (29.4% mol Na_2O)	room temp.	14.23
3 kHz	SiO_2–Na_2O glass (34.3% mol Na_2O)	room temp.	21.30

Source: data compiled by J.S. Park from O. V. Mazurin, M. V. Streltsina and T. P. Shvaiko-Shvaikovskaya, *Handbook of Glass Data*, Part A and Part B, Elsevier, New York, 1983.

Table 443. SELECTING ELECTRICAL PERMITTIVITY OF GLASS BY FREQUENCY (SHEET 3 OF 6)

Frequency (Hz)	Glass	Temperature (°C)	Electrical Permittivity
5 kHz	SiO_2–Na_2O glass (19.5% mol Na_2O)	room temp.	8.56
5 kHz	SiO_2–Na_2O glass (24.4% mol Na_2O)	room temp.	10.21
5 kHz	SiO_2–Na_2O glass (29.4% mol Na_2O)	room temp.	13.19
5 kHz	SiO_2–Na_2O glass (34.3% mol Na_2O)	room temp.	18.13
10 kHz	B_2O_3 glass	500	3.13
10 kHz	B_2O_3 glass	550	3.14
10 kHz	B_2O_3 glass	580	3.145
10 kHz	B_2O_3 glass	620	3.15
10 kHz	B_2O_3 glass	650	3.15
10 kHz	B_2O_3 glass	700	3.16
10 kHz	SiO_2 glass	25	4.0
10 kHz	SiO_2 glass	200	4.0
10 kHz	SiO_2 glass	300	4.0
10 kHz	SiO_2 glass	400	4.0
10 kHz	SiO_2–Na_2O glass (19.5% mol Na_2O)	room temp.	8.26
10 kHz	SiO_2–Na_2O glass (24.4% mol Na_2O)	room temp.	9.74
10 kHz	SiO_2–Na_2O glass (29.4% mol Na_2O)	room temp.	12.08
10 kHz	SiO_2–Na_2O glass (34.3% mol Na_2O)	room temp.	15.22
10 kHz	SiO_2–Na_2O glass (39.3% mol Na_2O)	room temp.	22.08
30 kHz	SiO_2–Na_2O glass (19.5% mol Na_2O)	room temp.	8.00
30 kHz	SiO_2–Na_2O glass (24.4% mol Na_2O)	room temp.	9.30
30 kHz	SiO_2–Na_2O glass (29.4% mol Na_2O)	room temp.	11.21
30 kHz	SiO_2–Na_2O glass (34.3% mol Na_2O)	room temp.	13.28
30 kHz	SiO_2–Na_2O glass (39.3% mol Na_2O)	room temp.	16.56

Source: data compiled by J.S. Park from O. V. Mazurin, M. V. Streltsina and T. P. Shvaiko-Shvaikovskaya, *Handbook of Glass Data*, Part A and Part B, Elsevier, New York, 1983.

Table 443. **SELECTING ELECTRICAL PERMITTIVITY OF GLASS BY FREQUENCY** (SHEET 4 OF 6)

Frequency (Hz)	Glass	Temperature (°C)	Electrical Permittivity
50 kHz	B_2O_3 glass	800	3.04
50 kHz	B_2O_3 glass	620	3.05
50 kHz	B_2O_3 glass	750	3.06
50 kHz	B_2O_3 glass	700	3.09
50 kHz	B_2O_3 glass	500	3.10
50 kHz	B_2O_3 glass	650	3.10
50 kHz	B_2O_3 glass	580	3.115
50 kHz	B_2O_3 glass	550	3.12
50 kHz	SiO_2–Na_2O glass (19.5% mol Na_2O)	room temp.	7.88
50 kHz	SiO_2–Na_2O glass (24.4% mol Na_2O)	room temp.	9.14
50 kHz	SiO_2–Na_2O glass (29.4% mol Na_2O)	room temp.	10.86
50 kHz	SiO_2–Na_2O glass (34.3% mol Na_2O)	room temp.	12.57
50 kHz	SiO_2–Na_2O glass (39.3% mol Na_2O)	room temp.	15.06
100 kHz	SiO_2–Na_2O glass (19.5% mol Na_2O)	room temp.	7.74
100 kHz	SiO_2–Na_2O glass (24.4% mol Na_2O)	room temp.	8.91
100 kHz	SiO_2–Na_2O glass (29.4% mol Na_2O)	room temp.	10.47
100 kHz	SiO_2–Na_2O glass (34.3% mol Na_2O)	room temp.	11.78
100 kHz	SiO_2–Na_2O glass (39.3% mol Na_2O)	room temp.	13.55
300 kHz	SiO_2–Na_2O glass (19.5% mol Na_2O)	room temp.	7.62
300 kHz	SiO_2–Na_2O glass (24.4% mol Na_2O)	room temp.	8.75
300 kHz	SiO_2–Na_2O glass (29.4% mol Na_2O)	room temp.	10.15
300 kHz	SiO_2–Na_2O glass (34.3% mol Na_2O)	room temp.	11.14
300 kHz	SiO_2–Na_2O glass (39.3% mol Na_2O)	room temp.	12.43

Source: data compiled by J.S. Park from O. V. Mazurin, M. V. Streltsina and T. P. Shvaiko-Shvaikovskaya, *Handbook of Glass Data*, Part A and Part B, Elsevier, New York, 1983.

Table 443. SELECTING ELECTRICAL PERMITTIVITY OF GLASS BY FREQUENCY (SHEET 5 OF 6)

Frequency (Hz)	Glass	Temperature (°C)	Electrical Permittivity
56.8 MHz	B_2O_3–Na_2O glass (4.08% mol Na_2O)	room temp.	3.72
56.8 MHz	B_2O_3–Na_2O glass (7.35% mol Na_2O)	room temp.	4.20
56.8 MHz	B_2O_3–Na_2O glass (14.15% mol Na_2O)	room temp.	4.94
56.8 MHz	B_2O_3–Na_2O glass (17.31% mol Na_2O)	room temp.	5.27
56.8 MHz	B_2O_3–Na_2O glass (24.77% mol Na_2O)	room temp.	6.24
56.8 MHz	B_2O_3–Na_2O glass (31.98% mol Na_2O)	room temp.	7.03
$4.5x10^8$ Hz	SiO_2–Na_2O glass (16% mol Na_2O)	20	6.01
$4.5x10^8$ Hz	SiO_2–Na_2O glass (20% mol Na_2O)	20	6.48
$4.5x10^8$ Hz	SiO_2–Na_2O glass (22.2% mol Na_2O)	20	6.85
$4.5x10^8$ Hz	SiO_2–Na_2O glass (28.6% mol Na_2O)	20	7.62
$4.5x10^8$ Hz	SiO_2–Na_2O glass (36% mol Na_2O)	20	9.40
9.4 GHz	SiO_2 glass	20	3.81
9.4 GHz	SiO_2 glass	200	3.83
9.4 GHz	SiO_2 glass	400	3.84
9.4 GHz	SiO_2 glass	600	3.86
9.4 GHz	SiO_2 glass	800	3.88
9.4 GHz	SiO_2 glass	1000	3.91
9.4 GHz	SiO_2 glass	1200	3.93
9.4 GHz	SiO_2 glass	1400	3.96
10 GHz	SiO_2–Al_2O_3 glass (46.3% mol B_2O_3)		3.55
10 GHz	SiO_2 glass	20	3.82
10 GHz	SiO_2 glass	220	3.82
10 GHz	SiO_2 glass	888	3.91

Source: data compiled by J.S. Park from O. V. Mazurin, M. V. Streltsina and T. P. Shvaiko-Shvaikovskaya, *Handbook of Glass Data*, Part A and Part B, Elsevier, New York, 1983.

Table 443. SELECTING ELECTRICAL PERMITTIVITY OF GLASS BY FREQUENCY (SHEET 6 OF 6)

Frequency (Hz)	Glass	Temperature (°C)	Electrical Permittivity
10 GHz	SiO_2 glass	1170	3.98
10 GHz	SiO_2 glass	1764	4.04
10 GHz	SiO_2 glass	1335	4.05
10 GHz	SiO_2 glass	1764	4.05
10 GHz	SiO_2 glass	1420	4.07
10 GHz	SiO_2 glass	1480	4.09
10 GHz	SiO_2 glass	1526	4.11
10 GHz	SiO_2 glass	1584	4.12
10 GHz	SiO_2 glass	1647	4.12
10 GHz	SiO_2 glass	1602	4.15
32 GHz	SiO_2–PbO glass (40% mol PbO)	–150	4.25
32 GHz	SiO_2–PbO glass (40% mol PbO)	–100	4.30
32 GHz	SiO_2–PbO glass (40% mol PbO)	–50	4.40
32 GHz	SiO_2–PbO glass (40% mol PbO)	0	4.45
32 GHz	SiO_2–PbO glass (40% mol PbO)	50	5.00

Source: data compiled by J.S. Park from O. V. Mazurin, M. V. Streltsina and T. P. Shvaiko-Shvaikovskaya, *Handbook of Glass Data*, Part A and Part B, Elsevier, New York, 1983.

Table 444. SELECTING ARC RESISTANCE OF POLYMERS
(SHEET 1 OF 3)

Polymer	Arc Resistance (ASTM D495) (seconds)
Rubber phenolic—asbestos filled	5—20
Phenolics; Molded; General: woodflour and flock filled	5—60
Phenolics; Molded; Shock: paper, flock, or pulp filled	5—60
Phenolics; Molded; High shock: chopped fabric or cord filled	5—60
Rubber phenolic—woodflour or flock filled	7—20
Rubber phenolic—chopped fabric filled	10—20
Polypropylene: Flame retardant	15—40
Polystyrenes; Molded: High impact	20—100
Polystyrenes; Molded: Medium impact	20—135
PVC–Acrylic Alloy: PVC–acrylic injection molded	25
Polystyrenes; Molded: Glass fiber -30% reinforced	28
Polyphenylene sulfide: 40% glass reinforced	34
Polymides: Glass reinforced	50—180
Phenolics; Molded; Very high shock: glass fiber filled	60
Polystyrenes; Molded: General purpose	60—135
Glass fiber (30%) reinforced SAN	65
Polyarylsulfone	67—81
Melamines; Molded: Cellulose electrical filled	70—135
Polypropylene: Glass reinforced	73—77
Phenylene Oxides: SE—100	75
Phenylene Oxides: SE—1	75
Standard Epoxies: Cast flexible	75—98
PVC–Acrylic Alloy: PVC–acrylic sheet	80
Polyester; Thermoplastic Moldings: Glass reinforced self extinguishing	80
Ureas; Molded: Woodflour filled	80—110
Ureas; Molded: Cellulose filled (ASTM Type 2)	85—110
Diallyl Phthalates; Molded: Orlon filled	85—115
Nylons; Molded, Extruded Type 6: Glass fiber (30%) reinforced	92—81

Source: data compiled by J.S. Park from Charles T. Lynch, *CRC Handbook of Materials Science*, Vol. 3, CRC Press, Boca Raton, Florida, 1975 and *Engineered Materials Handbook*, Vol.2, Engineering Plastics, ASM International, Metals Park, Ohio, 1988.

Table 444. SELECTING ARC RESISTANCE OF POLYMERS
(SHEET 2 OF 3)

Polymer	Arc Resistance (ASTM D495) (seconds)
ABS–Polycarbonate Alloy	96
Standard Epoxies: Cast rigid	100
Ureas; Molded: Alpha—cellulose filled (ASTM Type l)	100—135
Melamines; Molded: Unfilled	100—145
Styrene acrylonitrile (SAN)	100—150
Diallyl Phthalates; Molded: Dacron filled	105—125
Polyester; Thermoplastic Moldings: Asbestos—filled grade	108
Phenylene oxides (Noryl): Glass fiber reinforced	114
Polyesters Cast Thermosets: Rigid	115—135
Epoxy novolacs: Cast, rigid	120
6/6 Nylon; Molded, Extruded: General purpose molding	120
6/6 Nylon; Molded, Extruded: General purpose extrusion	120
6/10 Nylon: General purpose	120
Phenylene Oxides: Glass fiber reinforced	120
Polycarbonate	120 (tungsten electrode)
Polycarbonate (40% glass fiber reinforced)	120 (tungsten electrode)
Polypropylene: Asbestos filled	121—125
Phenylene oxides (Noryl): Standard	122
Polypropylene: High impact	123—140
Melamines; Molded: Alpha cellulose and mineral filled	125
Polyester; Thermoplastic Moldings: General purpose grade	125
Polypropylene: General purpose	125—136
Diallyl Phthalates; Molded: Asbestos filled	125—140
Diallyl Phthalates; Molded: Glass fiber filled	125—140
Polyesters Cast Thermosets: Flexible	125—145
Polyacetal Homopolymer: Standard	129
Polyester; Thermoplastic Moldings: Glass reinforced grades	130
Reinforced polyester moldings: High strength (glass fibers)	130—170

Source: data compiled by J.S. Park from Charles T. Lynch, *CRC Handbook of Materials Science*, Vol. 3, CRC Press, Boca Raton, Florida, 1975 and *Engineered Materials Handbook*, Vol.2, Engineering Plastics, ASM International, Metals Park, Ohio, 1988.

Table 444. SELECTING ARC RESISTANCE OF POLYMERS

(SHEET 3 OF 3)

Polymer	Arc Resistance (ASTM D495) (seconds)
Standard Epoxies: General purpose glass cloth laminate	130—180
Reinforced polyester: Sheet molding compounds, general purpose	130—180
6/6 Nylon; Molded, Extruded: Glass fiber Molybdenum disulfide filled	135
Standard Epoxies: Molded	135—190
Polyacetal Copolymer: 25% glass reinforced	136
6/6 Nylon; Molded, Extruded: Glass fiber reinforced	148—100
Polymides: Unreinforced	152
Alkyds; Molded: Putty (encapsulating)	180
Alkyds; Molded: Rope (general purpose)	180
Alkyds; Molded: Granular (high speed molding)	180
Alkyds; Molded: Glass reinforced (heavy duty parts)	180
Phenolics; Molded: Arc resistant—mineral	180
High performance Epoxies: Molded	180—185
Melamines; Molded: Glass fiber filled	180—186
Thermoset Carbonate: Allyl diglycol carbonate	185
Polyacetal Homopolymer: 20% glass reinforced	188
Polyester; Thermoplastic Moldings: General purpose grade	190
Molded,Extruded Polytetrafluoroethylene (PTFE)	>200
Silicones; Molded, Laminated: Woven glass fabric/ silicone laminate	225—250
Polyacetal Copolymer: Standard	240
Polyacetal Copolymer: High flow	240
Silicones; Molded, Laminated: Fibrous (glass) reinforced silicones	240
Silicones; Molded, Laminated: Granular (silica) reinforced silicones	250—310
Molded,Extruded Polytrifluoro chloroethylene (PTFCE)	>360
Acrylics; Cast Resin Sheets, Rods: General purpose, type I	No track
Acrylics; Cast Resin Sheets, Rods: General purpose, type II	No track
Acrylic Moldings: Grades 5, 6, 8	No track
Acrylic Moldings: High impact grade	No track

Source: data compiled by J.S. Park from Charles T. Lynch, *CRC Handbook of Materials Science*, Vol. 3, CRC Press, Boca Raton, Florida, 1975 and *Engineered Materials Handbook*, Vol.2, Engineering Plastics, ASM International, Metals Park, Ohio, 1988.

CHAPTER 15 Selecting Optical Properties

List of Tables

Table 445. SELECTING TRANSMISSION RANGE OF OPTICAL MATERIALS (SHEET 1 OF 2)

Material & Crystal Structure	Transmission Region (μm, at 298 K)
Magnesium Fluoride (Single Crystal)	0.1 – 9.7
Silica (High Purity Crystalline)	0.12 – 4.5
Silica (High Purity Fused)	0.12 – 4.5
Lithium Fluoride (Single Crystal)	0.12 – 9.0
Ammonium Dihydrogen Phosphate (ADP, Single Crystal)	0.13 – 1.7
Calcium Fluoride (Single Crystal)	0.13 – 12
Alumina (Sapphire, Single Crystal)	0.15 – 6.5
Sodium Fluoride (Single Crystal)	0.19 – 15
Magnesium Fluoride (Film)	0.2 – 5.0
Calcium Carbonate (Calcite, Single Crystal)	0.2 – 5.5
Thallium Chloribromide (KRS–6, Mixed Crystal)	0.21 – 35
Magnesium Oxide (Single Crystal)	0.25 – 8.5
Barium Fluoride (Single Crystal)	0.25 – 15
Potassium Bromide (Single Crystal)	0.25 – 35
Potassium Iodide (Single Crystal)	0.25 – 45
Cesium Iodide (Single Crystal)	0.25 – 80
Cesium Bromide (Single Crystal)	0.3 – 55
Lithium Niobate (Single Crystal)	0.33 – 5.2
Strontium Titanate (Single Crystal)	0.39 – 6.8
Silver Chloride (Single Crystal)	0.4 – 2.8
Cuprous Chloride (Single Crystal)	0.4 – 19
Titanium Dioxide (Rutile, Single Crystal)	0.43 – 6.2
Silver Bromide (Single Crystal)	0.45 – 35
Cadmium Sulfide (Bulk and Hexagonal Single Crystal)	0.5 – 16
Zinc Selenide (Single Crystal, Cubic)	~0.5 – 22
Arsenic Trisulfade (Glass)	0.6 – 13
Zinc Sulfide (Single Crystal, Cubic)	~0.6 – 15.6
Thallium Bromoiodide (KRS–5, Mixed Crystal)	0.6 – 40

External transmittance ≥ 10% with 2.0 mm thickness.

Source: Data compiled by J.S. Park from various sources.

Table 445. SELECTING TRANSMISSION RANGE OF OPTICAL MATERIALS (SHEET 2 OF 2)

Material & Crystal Structure	Transmission Region (μm, at 298 K)
Cadmium Telluride (Hot Pressed Polycrystalline)	0.9 – 16
Gallium Arsenide (Intrinsic Single Crystal)	1.0 – 15
Selenium (Amorphous)	1.0 – 20
Silicon (Single Crystal)	1.2 – 15
Germanium (Intrinsic Single Crystal)	1.8 – 23
Lead Sulfide (Single Crystal)	3.0 – 7.0
Tellurium (Polycrystalline Film)	3.5 – 8.0
Tellurium (Single Crystal)	3.5 – 8.0
Indium Arsenide (Single Crystal)	3.8 – 7.0

External transmittance ≥ 10% with 2.0 mm thickness.

Source: Data compiled by J.S. Park from various sources.

Table 446. SELECTING TRANSPARENCY OF POLYMERS

(SHEET 1 OF 3)

Polymer	Transparency (visible light) (ASTM D791) (%)
Alkyds; Molded: Putty (encapsulating)	Opaque
Alkyds; Molded: Rope (general purpose)	Opaque
Alkyds; Molded: Granular (high speed molding)	Opaque
Alkyds; Molded: Glass reinforced (heavy duty parts)	Opaque
Chlorinated polyether	Opaque
Chlorinated polyvinyl chloride	Opaque
Standard Epoxies: General purpose glass cloth laminate	Opaque
Standard Epoxies: High strength laminate	Opaque
Standard Epoxies: Filament wound composite	Opaque
High performance Epoxies: Molded	Opaque
High performance Epoxies: Glass cloth laminate	Opaque
Epoxy novolacs: Glass cloth laminate	Opaque
Melamines; Molded: Cellulose electrical	Opaque
6/6 Nylon; Molded, Extruded: Glass fiber reinforced	Opaque
6/6 Nylon; Molded, Extruded: Glass fiber Molybdenum disulfide filled	Opaque
6/6 Nylon; Molded, Extruded: General purpose extrusion	Opaque
6/10 Nylon: General purpose	Opaque
6/10 Nylon: Glass fiber (30%) reinforced	Opaque
ABS–Polycarbonate Alloy	Opaque
PVC–Acrylic Alloy: PVC–acrylic injection molded	Opaque
Polymides: Unreinforced	Opaque
Polymides: Glass reinforced	Opaque
Reinforced polyester moldings: High strength (glass fibers)	Opaque
Reinforced polyester moldings: Heat & chemical resistsnt (asbestos)	Opaque
Reinforced polyester: Sheet molding compounds, general purpose	Opaque
Phenylene Oxides: SE—100	Opaque
Phenylene Oxides: SE—1	Opaque
Phenylene Oxides: Glass fiber reinforced	Opaque

Source: data compiled by J.S. Park from Charles T. Lynch, *CRC Handbook of Materials Science*, Vol. 3, CRC Press, Boca Raton, Florida, 1975 and *Engineered Materials Handbook*, Vol.2, Engineering Plastics, ASM International, Metals Park, Ohio, 1988.

Table 446. SELECTING TRANSPARENCY OF POLYMERS

(SHEET 2 OF 3)

Polymer	Transparency (visible light) (ASTM D791) (%)
Phenylene oxides (Noryl): Glass fiber reinforced	Opaque
Polypropylene: Asbestos filled	Opaque
Polypropylene: Glass reinforced	Opaque
Polypropylene: Flame retardant	Opaque
Polyphenylene sulfide: Standard	Opaque
Polyphenylene sulfide: 40% glass reinforced	Opaque
Polystyrenes; Molded: Medium impact	Opaque
Polystyrenes; Molded: High impact	Opaque
Polystyrenes; Molded: Glass fiber -30% reinforced	Opaque
Glass fiber (30%) reinforced Styrene acrylonitrile (SAN)	Opaque
Silicones; Molded, Laminated: Fibrous (glass) reinforced silicones	Opaque
Silicones; Molded, Laminated: Granular (silica) reinforced silicones	Opaque
Silicones; Molded, Laminated: Woven glass fabric/ silicone laminate	Opaque
Ureas; Molded: Cellulose filled (ASTM Type 2)	Opaque
Ureas; Molded: Woodflour filled	Opaque
PVC–Acrylic Alloy: PVC–acrylic sheet	Opaque
Polypropylene: General purpose	Translucent—opaque
Polypropylene: High impact	Translucent—opaque
Polycarbonate (40% glass fiber reinforced)	Translucent
6/6 Nylon; Molded, Extruded: General purpose molding	Translucent
Polystyrenes; Molded: General purpose	Transparent
Styrene acrylonitrile (SAN)	Transparent
Ureas; Molded: Alpha—cellulose filled (ASTM Type 1)	21.8
Polycarbonate	75—85
Cellulose Acetate; Molded, Extruded; ASTM Grade: H6—1	75—90
Cellulose Acetate; Molded, Extruded; ASTM Grade: H4—1	75—90
Cellulose Acetate Butyrate; Molded, Extruded; ASTM Grade: H4	75—92
Cellulose Acetate; Molded, Extruded; ASTM Grade: H2—1	80—90

Source: data compiled by J.S. Park from Charles T. Lynch, *CRC Handbook of Materials Science*, Vol. 3, CRC Press, Boca Raton, Florida, 1975 and *Engineered Materials Handbook*, Vol.2, Engineering Plastics, ASM International, Metals Park, Ohio, 1988.

Table 446. SELECTING TRANSPARENCY OF POLYMERS

(SHEET 3 OF 3)

Polymer	Transparency (visible light) (ASTM D791) (%)
Cellulose Acetate; Molded, Extruded; ASTM Grade: MH—1, MH—2	80—90
Cellulose Acetate; Molded, Extruded; ASTM Grade: MS—1, MS—2	80—90
Cellulose Acetate Butyrate; Molded, Extruded; ASTM Grade: MH	80—92
Cellusose Acetate Propionate; Molded, Extruded; ASTM Grade: 1	80—92
Cellusose Acetate Propionate; Molded, Extruded; ASTM Grade: 3	80—92
Cellusose Acetate Propionate; Molded, Extruded; ASTM Grade: 6	80—92
Polytrifluoro chloroethylene (PTFCE) Molded, Extruded	80—92
Cellulose Acetate; Molded, Extruded; ASTM Grade: S2—1	80—95
Standard Epoxies: Molded	85
Cellulose Acetate Butyrate; Molded, Extruded; ASTM Grade: S2	85—95
Thermoset Carbonate: Allyl diglycol carbonate	89—92
Acrylic Moldings: High impact grade	90
Standard Epoxies: Cast flexible	90
Acrylics; Cast Resin Sheets, Rods: General purpose, type I	91—92 (0.125 in.)
Acrylics; Cast Resin Sheets, Rods: General purpose, type II	91—92 (0.125 in.)
Acrylic Moldings: Grades 5, 6, 8	>92

Source: data compiled by J.S. Park from Charles T. Lynch, *CRC Handbook of Materials Science*, Vol. 3, CRC Press, Boca Raton, Florida, 1975 and *Engineered Materials Handbook*, Vol.2, Engineering Plastics, ASM International, Metals Park, Ohio, 1988.

Table 447. SELECTING REFRACTIVE INDICES OF GLASSES

(SHEET 1 OF 6)

Glass	Wavelength (λ)	Temperature (°C)	Refractive Index (n_D)
B_2O_3 glass	5461 Å	700	1.4130
SiO_2 glass	3.245 μm	26	1.41353
B_2O_3 glass	5461 Å	650	1.4155
B_2O_3 glass	5461 Å	600	1.4180
B_2O_3 glass	5461 Å	550	1.4210
SiO_2 glass	3.245 μm	828	1.42243
B_2O_3 glass	5461 Å	500	1.4240
B_2O_3 glass	5461 Å	450	1.4270
SiO_2 glass	2.553 μm	26	1.42949
B_2O_3 glass	5461 Å	400	1.4315
SiO_2 glass	2.553 μm	471	1.43450
B_2O_3 glass	5461 Å	350	1.4365
SiO_2 glass	2.553 μm	828	1.43854
SiO_2 glass	1.981 μm	26	1.43863
B_2O_3 glass	5461 Å	300	1.4420
SiO_2 glass	1.660 μm	26	1.44307
SiO_2 glass	1.981 μm	471	1.44361
SiO_2 glass	1.470 μm	26	1.44524
SiO_2 glass	1.981 μm	828	1.44734
SiO_2 glass	1.254 μm	26	1.44772
SiO_2 glass	1.660 μm	471	1.44799
SiO_2–B_2O_3 glass (quenched, 13.5% mol B_2O_3)	1.002439 μm	23	1.4485
SiO_2–B_2O_3 glass (annealed, 13.5% mol B_2O_3)	1.002439 μm	23	1.4493
SiO_2 glass	1.470 μm	471	1.45031

Source: data compiled by J.S. Park from O. V. Mazurin, M. V. Streltsina and T. P. Shvaiko–Shvaikovskaya, *Handbook of Glass Data*, Part A and Part B, Elsevier, New York, 1983.

Table 447. SELECTING REFRACTIVE INDICES OF GLASSES

(SHEET 2 OF 6)

Glass	Wavelength (λ)	Temperature (°C)	Refractive Index (n_D)
SiO_2 glass	1.01398 μm	26	1.45039
B_2O_3 glass	5461 Å	250	1.4505
SiO_2–B_2O_3 glass (quenched, 13.5% mol B_2O_3)	0.852111 μm	23	1.4507
SiO_2–B_2O_3 glass (annealed, 13.5% mol B_2O_3)	0.852111 μm	23	1.4515
SiO_2 glass	1.660 μm	828	1.45174
SiO_2–B_2O_3 glass (quenched, 13.5% mol B_2O_3)	0.734620 μm	23	1.4528
SiO_2 glass	1.254 μm	471	1.45283
SiO_2–B_2O_3 glass (annealed, 13.5% mol B_2O_3)	0.734620 μm	23	1.4537
SiO_2 glass	1.470 μm	828	1.45440
SiO_2 glass	1.01398 μm	471	1.45562
SiO_2–B_2O_3 glass (quenched, 13.5% mol B_2O_3)	0.589263 μm	23	1.4570
SiO_2 glass	1.254 μm	828	1.45700
SiO_2–B_2O_3 glass (annealed, 13.5% mol B_2O_3)	0.589263 μm	23	1.4579
SiO_2–B_2O_3 glass (20% mol B_2O_3)	5145 Å		1.4582
SiO_2–B_2O_3 glass (15% mol B_2O_3)	5145 Å		1.4584
SiO_2–B_2O_3 glass (30% mol B_2O_3)	5145 Å		1.4588
SiO_2–B_2O_3 glass (10% mol B_2O_3)	5145 Å		1.4592
SiO_2–Al_2O_3 glass (1.4% mol Al_2O_3)	589.262 nm		1.4595
SiO_2 glass	1.01398 μm	828	1.45960
SiO_2 glass	0.54607 μm	26	1.46028
SiO_2–B_2O_3 glass (50% mol B_2O_3)	5145 Å		1.4604
B_2O_3 glass	5461 Å	200	1.4605
SiO_2–B_2O_3 glass (quenched, 13.5% mol B_2O_3)	0.508582 μm	23	1.4606
SiO_2–B_2O_3 glass (75% mol B_2O_3)	5145 Å		1.4612

Source: data compiled by J.S. Park from O. V. Mazurin, M. V. Streltsina and T. P. Shvaiko–Shvaikovskaya, *Handbook of Glass Data*, Part A and Part B, Elsevier, New York, 1983.

Table 447. SELECTING REFRACTIVE INDICES OF GLASSES
(SHEET 3 OF 6)

Glass	Wavelength (λ)	Temperature (°C)	Refractive Index (n_D)
SiO_2–B_2O_3 glass (annealed, 13.5% mol B_2O_3)	0.508582 μm	23	1.4615
SiO_2–B_2O_3 glass (90% mol B_2O_3)	5145 Å		1.4617
B_2O_3 glass	5461 Å	150	1.4625
SiO_2–Al_2O_3 glass (3.1% mol Al_2O_3)	589.262 nm		1.4630
B_2O_3 glass	5461 Å	100	1.4635
B_2O_3 glass	5461 Å	20	1.4650
SiO_2–Al_2O_3 glass (3.7% mol Al_2O_3)	589.262 nm		1.4652–1.4667
B_2O_3–Na_2O glass (0.01% mol Na_2O)		25	1.46536
SiO_2–B_2O_3 glass (quenched, 13.5% mol B_2O_3)	0.435833 μm	23	1.4657
SiO_2 glass	0.54607 μm	471	1.46575
SiO_2–B_2O_3 glass (annealed, 13.5% mol B_2O_3)	0.435833 μm	23	1.4665
B_2O_3 glass	5461 Å	0	1.467
B_2O_3 glass	5461 Å	–100	1.469
SiO_2 glass	0.40466 μm	26	1.46978
SiO_2 glass	0.54607 μm	828	1.47004
SiO_2 glass	0.40466 μm	471	1.47575
SiO_2 glass	0.33415 μm	26	1.48000
SiO_2 glass	0.40466 μm	828	1.48033
SiO_2–Na_2O glass (15% mol Na_2O)			1.4822
B_2O_3–Na_2O glass (4.4% mol Na_2O)		25	1.48387
SiO_2 glass	0.33415 μm	471	1.48633
SiO_2 glass	0.30215 μm	26	1.48738
SiO_2 glass	3.245 μm	471	1.4893
SiO_2–Na_2O glass (20% mol Na_2O)			1.4906

Source: data compiled by J.S. Park from O. V. Mazurin, M. V. Streltsina and T. P. Shvaiko–Shvaikovskaya, *Handbook of Glass Data*, Part A and Part B, Elsevier, New York, 1983.

Table 447. SELECTING REFRACTIVE INDICES OF GLASSES
(SHEET 4 OF 6)

Glass	Wavelength (λ)	Temperature (°C)	Refractive Index (n_D)
SiO_2 glass	0.28936 μm	26	1.49121
SiO_2 glass	0.33415 μm	828	1.49135
SiO_2 glass	0.30215 μm	471	1.49407
B_2O_3–Na_2O glass (8.7% mol Na_2O)		25	1.49442
SiO_2 glass	0.27528 μm	26	1.49615
B_2O_3–Na_2O glass (11.5% mol Na_2O)		25	1.49662
SiO_2 glass	0.28936 μm	471	1.49818
SiO_2–Na_2O glass (25% mol Na_2O)			1.4983
B_2O_3–Na_2O glass (13.7% mol Na_2O)		25	1.49841
SiO_2 glass	0.30215 μm	828	1.49942
B_2O_3–Na_2O glass (16.2% mol Na_2O)		25	1.49984
B_2O_3–Na_2O glass (15.8% mol Na_2O)		25	1.50024
B_2O_3–Na_2O glass (17.4% mol Na_2O)		25	1.50155
B_2O_3–Na_2O glass (18.4% mol Na_2O)		25	1.50210
SiO_2 glass	0.27528 μm	471	1.50327
SiO_2 glass	0.28936 μm	828	1.50358
SiO_2–Na_2O glass (30% mol Na_2O)			1.5041
B_2O_3–Na_2O glass (19.6% mol Na_2O)		25	1.50468
B_2O_3–Na_2O glass (20.0% mol Na_2O)		25	1.50500
SiO_2–Na_2O glass (33.3% mol Na_2O)			1.5061
B_2O_3–Na_2O glass (22.5% mol Na_2O)		25	1.50806
SiO_2 glass	0.24827 μm	26	1.50865
SiO_2 glass	0.27528 μm	828	1.50889
B_2O_3–Na_2O glass (23.6% mol Na_2O)		25	1.50979

Source: data compiled by J.S. Park from O. V. Mazurin, M. V. Streltsina and T. P. Shvaiko–Shvaikovskaya, *Handbook of Glass Data*, Part A and Part B, Elsevier, New York, 1983.

Table 447. SELECTING REFRACTIVE INDICES OF GLASSES
(SHEET 5 OF 6)

Glass	Wavelength (λ)	Temperature (°C)	Refractive Index (n_D)
SiO_2–Na_2O glass (39.3% mol Na_2O)			1.5099
SiO_2 glass	0.2407 μm	26	1.51361
SiO_2–Na_2O glass (45.1% mol Na_2O)			1.5137
B_2O_3–Na_2O glass (28.9% mol Na_2O)		25	1.51611
SiO_2 glass	0.24827 μm	471	1.51665
SiO_2–Na_2O glass (50% mol Na_2O)			1.517
SiO_2 glass	0.23021 μm	26	1.52034
SiO_2 glass	0.2407 μm	471	1.52201
SiO_2 glass	0.24827 μm	828	1.52289
SiO_2 glass	0.2407 μm	828	1.52832
SiO_2 glass	0.23021 μm	471	1.52908
SiO_2 glass	0.23021 μm	828	1.53584
SiO_2–CaO glass (39.0% mol CaO)			1.5905
B_2O_3–CaO glass (35% mol CaO)			1.6021
SiO_2–CaO glass (44.6% mol CaO)			1.6120
SiO_2–PbO glass (20.78% mol PbO)			1.6174
SiO_2–Al_2O_3 glass (70.2% mol Al_2O_3)			1.629
SiO_2–CaO glass (50.0% mol CaO)			1.6295
SiO_2–Al_2O_3 glass (77.0% mol Al_2O_3)			1.634
SiO_2–CaO glass (52.9% mol CaO)			1.6350
SiO_2–CaO glass (57.5% mol CaO)			1.6455
SiO_2–PbO glass (24.90% mol PbO)			1.6509
B_2O_3–CaO glass (64.1% mol CaO)			1.6525
SiO_2–PbO glass (29.71% mol PbO)			1.6948

Source: data compiled by J.S. Park from O. V. Mazurin, M. V. Streltsina and T. P. Shvaiko–Shvaikovskaya, *Handbook of Glass Data*, Part A and Part B, Elsevier, New York, 1983.

Table 447. SELECTING REFRACTIVE INDICES OF GLASSES
(SHEET 6 OF 6)

Glass	Wavelength (λ)	Temperature (°C)	Refractive Index (n_D)
SiO_2–Al_2O_3 glass (84.1% mol Al_2O_3)			1.720
SiO_2–PbO glass (33.01% mol PbO)			1.7270
SiO_2–Al_2O_3 glass (91.8% mol Al_2O_3)			1.728
SiO_2–PbO glass (36.64% mol PbO)			1.7632
SiO_2–PbO glass (40.80% mol PbO)			1.8092
SiO_2–PbO glass (44.07% mol PbO)			1.8457
SiO_2–PbO glass (47.83% mol PbO)			1.8865
SiO_2–PbO glass (50.50% mol PbO)			1.9189
SiO_2–PbO glass (53.46% mol PbO)			1.9545
SiO_2–PbO glass (56.43% mol PbO)			1.9894
SiO_2–PbO glass (61.38% mol PbO)			2.0460–2.0512
SiO_2–PbO glass (65.97% mol PbO)			2.1030

Source: data compiled by J.S. Park from O. V. Mazurin, M. V. Streltsina and T. P. Shvaiko–Shvaikovskaya, *Handbook of Glass Data*, Part A and Part B, Elsevier, New York, 1983.

Table 448. SELECTING REFRACTIVE INDICES OF POLYMERS

(SHEET 1 OF 2)

Polymer	Refractive Index (ASTM D542) (n_D)
Fluorinated ethylene propylene(FEP) Molded, Extruded	1.34
Polytetrafluoroethylene (PTFE) Molded, Extruded	1.35
Polyvinylidene— fluoride (PVDF) Molded, Extruded	1.42
Polytrifluoro chloroethylene (PTFCE) Molded, Extruded	1.43
Cellusose Acetate Propionate; Molded, Extruded; ASTM Grade: 1	1.46—1.49
Cellusose Acetate Propionate; Molded, Extruded; ASTM Grade: 3	1.46—1.49
Cellusose Acetate Propionate; Molded, Extruded; ASTM Grade: 6	1.46—1.49
Cellulose Acetate Butyrate; Molded, Extruded; ASTM Grade: H4	1.46—1.49 (D543)
Cellulose Acetate Butyrate; Molded, Extruded; ASTM Grade: MH	1.46—1.49 (D543)
Cellulose Acetate Butyrate; Molded, Extruded; ASTM Grade: S2	1.46—1.49 (D543)
Cellulose Acetate; Molded, Extruded; ASTM Grade: H6—1	1.46—1.50
Cellulose Acetate; Molded, Extruded; ASTM Grade: H4—1	1.46—1.50
Cellulose Acetate; Molded, Extruded; ASTM Grade: H2—1	1.46—1.50
Cellulose Acetate; Molded, Extruded; ASTM Grade: MH—1, MH—2	1.46—1.50
Cellulose Acetate; Molded, Extruded; ASTM Grade: MS—1, MS—2	1.46—1.50
Cellulose Acetate; Molded, Extruded; ASTM Grade: S2—1	1.46—1.50
Acrylics; Cast Resin Sheets, Rods: General purpose, type II	1.485—1.495
Acrylics; Cast Resin Sheets, Rods: General purpose, type I	1.485—1.500
Acrylic Moldings: Grades 5, 6, 8	1.489—1.493
Acrylic Moldings: High impact grade	1.49
Thermoset Carbonate: Allyl diglycol carbonate	1.5
Polyesters Cast Thermosets: Flexible	1.50—1.57
Polyethylenes; Molded, Extruded; Type I: Melt index 0.3—3.6	1.51
Polyethylenes; Molded, Extruded; Type I: Melt index 6—26	1.51
Polyethylenes; Molded, Extruded; Type I: Melt index 200	1.51
Polyethylenes; Molded, Extruded; Type II: Melt index 20	1.51
Polyethylenes; Molded, Extruded; Type II: Melt index l.0—1.9	1.51
Polyesters Cast Thermosets: Rigid	1.53—1.58

Source: data compiled by J.S. Park from Charles T. Lynch, *CRC Handbook of Materials Science*, Vol. 3, CRC Press, Boca Raton, Florida, 1975 and *Engineered Materials Handbook*, Vol.2, Engineering Plastics, ASM International, Metals Park, Ohio, 1988.

Table 448. SELECTING REFRACTIVE INDICES OF POLYMERS

(SHEET 2 OF 2)

Polymer	Refractive Index (ASTM D542) (n_D)
Polyethylenes; Molded, Extruded; Type III: Melt index 0.2—0.9	1.54
Polyethylenes; Molded, Extruded; Type III: Melt Melt index 0.l—12.0	1.54
Polyethylenes; Molded, Extruded; Type III: Melt index 1.5—15	1.54
Styrene acrylonitrile (SAN)	1.565—1.569
Polycarbonate	1.586
Polystyrenes; Molded: General purpose	1.6
Polyvinyl Chloride & Copolymers: Vinylidene chloride	1.60—1.63
Standard Epoxies: Cast flexible	1.61
Standard Epoxies: Molded	1.61
Phenylene oxides (Noryl): Standard	1.63
Polyarylsulfone	1.651
Polyacetal Homopolymer: Standard	Opaque
Polyacetal Homopolymer: 20% glass reinforced	Opaque
Polyacetal Homopolymer: 22% TFE reinforced	Opaque
Polyacetal Copolymer: Standard	Opaque
Polyacetal Copolymer: 25% glass reinforced	Opaque
Polyacetal Copolymer: High flow	Opaque
Polystyrenes; Molded: Medium impact	Opaque
Polystyrenes; Molded: High impact	Opaque
Polystyrenes; Molded: Glass fiber -30% reinforced	Opaque
Glass fiber (30%) reinforced Styrene acrylonitrile (SAN)	Opaque

Source: data compiled by J.S. Park from Charles T. Lynch, *CRC Handbook of Materials Science*, Vol. 3, CRC Press, Boca Raton, Florida, 1975 and *Engineered Materials Handbook*, Vol.2, Engineering Plastics, ASM International, Metals Park, Ohio, 1988.

CHAPTER 16 Selecting Chemical Properties

List of Tables

Table 449. SELECTING WATER ABSORPTION OF POLYMERS
(SHEET 1 OF 5)

Polymer	Water Absorption in 24 hr (ASTM D570) (%)
Polytrifluoro chloroethylene (PTFCE); Molded, Extruded	0
Alkyds; Molded: Glass reinforced (heavy duty parts)	0.007—0.10
Fluorinated ethylene propylene(FEP)	<0.01
Polyethylenes; Molded, Extruded; Type I: Melt index 0.3—3.6	<0.01
Polyethylenes; Molded, Extruded; Type I: Melt index 6—26	<0.01
Polyethylenes; Molded, Extruded; Type I: Melt index 200	<0.01
Polyethylenes; Molded, Extruded; Type II: Melt index 20	<0.01
Polyethylenes; Molded, Extruded; Type II: Melt index l.0—1.9	<0.01
Polyethylenes; Molded, Extruded; Type III: Melt index 0.2—0.9	<0.01
Polyethylenes; Molded, Extruded; Type III: Melt Melt index 0.l—12.0	<0.01
Polyethylenes; Molded, Extruded; Type III: Melt index 1.5—15	<0.01
Polyethylenes; Molded, Extruded; Type III: High molecular weight	<0.01
Polypropylene: High impact	<0.01—0.02
Polypropylene: General purpose	<0.01—0.03
Chlorinated polyether	0.01
Polytetrafluoroethylene (PTFE); Molded, Extruded	0.01
Polyvinyl Chloride & Copolymers: Vinylidene chloride	>0.1 (ASTM D635)
Polypropylene: Flame retardant	0.02—0.03
Polypropylene: Asbestos filled	0.02—0.04
Polypropylene: Glass reinforced	0.02—0.05
Silicones: Woven glass fabric/ silicone laminate	0.03—0.05
Polyvinylidene— fluoride (PVDF)	0.03—0.06
Polystyrenes; Molded: Medium impact	0.03—0.09
Polyvinyl Chloride & Copolymers: Rigid—normal impact	0.03—0.40 (ASTM D635)
High performance Epoxies; Glass cloth laminate	0.04—0.06
Standard Epoxies; High strength laminate	0.05
Standard Epoxies; General purpose glass cloth laminate	0.05—0.07
Standard Epoxies; Filament wound composite	0.05—0.07

Source: data compiled by J.S. Park from Charles T. Lynch, *CRC Handbook of Materials Science,* Vol. 3, CRC Press, Boca Raton, Florida, 1975 and *Engineered Materials Handbook*, Vol.2, Engineering Plastics, ASM International, Metals Park, Ohio, 1988.

Table 449. SELECTING WATER ABSORPTION OF POLYMERS
(SHEET 2 OF 5)

Polymer	Water Absorption in 24 hr (ASTM D570) (%)
Alkyds; Molded: Rope (general purpose)	0.05—0.08
Polystyrenes; Molded: High impact	0.05—0.22
PVC–Acrylic Alloy: PVC–acrylic sheet	0.06
Phenylene Oxides: Glass fiber reinforced	0.06
Polyester; Thermoplastic Moldings: Glass reinforced grades	0.06—0.07
Polyester; Moldings: Glass reinforced self extinguishing	0.07
Polyester; Thermoplastic Moldings: Glass reinforced grade	0.07
Phenylene Oxides: SE—100	0.07
Phenylene Oxides: SE—1	0.07
Polystyrenes; Molded: Glass fiber –30% reinforced	0.07
Polycarbonate (40% glass fiber reinforced)	0.08
Polyester; Thermoplastic Moldings: General purpose grade	0.08
Silicones; Molded, Laminated: Granular (silica) reinforced	0.08—0.1
Alkyds; Molded: Granular (high speed molding)	0.08—0.12
Polyester; Thermoplastic Moldings: General purpose grade	0.09
Melamines; Molded: Glass fiber filled	0.09—0.60
Polyester; Thermoplastic Moldings: Asbestos—filled grade	0.1
Alkyds; Molded: Putty (encapsulating)	0.10—0.15
Rubber phenolic—asbestos filled	0.10—0.50
Silicones; Molded, Laminated: Fibrous (glass) reinforced	0.1—0.15
Standard Epoxies; Cast rigid	0.1—0.2
Epoxy novolacs: Cast, rigid	0.1—0.7
Phenolics; Molded; Very high shock: glass fiber filled	0.1—1.0
Chlorinated polyvinyl chloride	0.11
High performance Epoxies; Molded	0.11—0.2
Polyesters: Cast Thermosets: Flexible	0.12—2.5
PVC–Acrylic Alloy: PVC–acrylic injection molded	0.13
Polycarbonate	0.15

Source: data compiled by J.S. Park from Charles T. Lynch, *CRC Handbook of Materials Science,* Vol. 3, CRC Press, Boca Raton, Florida, 1975 and *Engineered Materials Handbook*, Vol.2, Engineering Plastics, ASM International, Metals Park, Ohio, 1988.

Table 449. SELECTING WATER ABSORPTION OF POLYMERS

(SHEET 3 OF 5)

Polymer	Water Absorption in 24 hr (ASTM D570) (%)
Styrene acrylonitrile (SAN): Glass fiber (30%) reinforced	0.15
Polyester: Sheet molding compounds, general purpose	0.15—0.25
Phenylene oxides (Noryl): Glass fiber reinforced	0.18—0.22
Thermoset Carbonate: Allyl diglycol carbonate	0.2
6/10 Nylon: Glass fiber (30%) reinforced	0.2
Polymides: Glass reinforced	0.2
Polyacetal Homopolymer: 22% TFE reinforced	0.2
Ceramic reinforced (PTFE)	>0.2
Styrene acrylonitrile (SAN)	0.20—0.35
Polyesters: Cast Thermosets: Rigid	0.20—0.60
ABS Resins; Molded, Extruded: Medium impact	0.2—0.4
ABS Resins; Molded, Extruded: Heat resistant	0.2—0.4
Acrylics; Cast Resin Sheets, Rods: General purpose, type II	0.2—0.4
Acrylics; Moldings: High impact grade	0.2—0.4
ABS Resins; Molded, Extruded: High impact	0.2—0.45
ABS Resins; Molded, Extruded: Very high impact	0.2—0.45
ABS Resins; Molded, Extruded: Low temperature impact	0.2—0.45
Melamines; Molded: Unfilled	0.2—0.5
Polyvinyl Chloride & Copolymers: Nonrigid—general	0.2—1.0 (ASTM D635)
ABS–Polycarbonate Alloy	0.21
Polyacetal Copolymer: Standard	0.22
Polyacetal Copolymer: High flow	0.22
Phenylene oxides (Noryl): Standard	0.22
Polymides: Unreinforced	0.24—0.47
Nylons; Type 12	0.25
Polyacetal Homopolymer: Standard	0.25
Polyacetal Homopolymer: 20% glass reinforced	0.25
Ppolyester moldings: Heat & chemical resistant (asbestos)	0.25—0.50

Source: data compiled by J.S. Park from Charles T. Lynch, *CRC Handbook of Materials Science*, Vol. 3, CRC Press, Boca Raton, Florida, 1975 and *Engineered Materials Handbook*, Vol.2, Engineering Plastics, ASM International, Metals Park, Ohio, 1988.

Table 449. SELECTING WATER ABSORPTION OF POLYMERS
(SHEET 4 OF 5)

Polymer	Water Absorption in 24 hr (ASTM D570) (%)
Melamines; Molded: Cellulose electrical filled	0.27—0.80
Polyacetal Copolymer: 25% glass reinforced	0.29
Polystyrenes; Molded: General purpose	0.30—0.2
Acrylics; Cast Resin Sheets, Rods: General purpose, type I	0.3—0.4
Acrylics; Moldings: Grades 5, 6, 8	0.3—0.4
Melamines; Molded: Alpha cellulose and mineral filled	0.3—0.5
Standard Epoxies; Molded	0.3—0.8
Phenolics; Molded; General: woodflour and flock filled	0.3—0.8
Nylons; Type 11	0.4
6/10 Nylon: General purpose	0.4
Polyarylsulfone	0.4
Polyvinyl Chloride & Copolymers: Nonrigid—electrical	0.40—0.75 (ASTM D635)
Standard Epoxies; Cast flexible	0.4—0.1
Ureas; Molded: Alpha—cellulose filled (ASTM Type l)	0.4—0.8
Phenolics; Molded; Shock: paper, flock, or pulp filled	0.4—1.5
Phenolics; Molded; High shock: chopped fabric or cord filled	0.4—1.75
Nylons; 6/6 Nylon: Glass fiber Molybdenum disulfide filled	0.5—0.7
Phenolics; Molded; Arc resistant—mineral filled	0.5—0.7
Reinforced polyester moldings: High strength (glass fibers)	0.5—0.75
Rubber phenolic—woodflour or flock filled	0.5—2.0
Rubber phenolic—chopped fabric filled	0.5—2.0
Nylons; Type 6: Cast	0.6
Nylons; Molded, Extruded; 6/6 Nylon: Glass fiber reinforced	0.8—0.9
Nylons; Molded, Extruded; Type 6: Flexible copolymers	0.8—1.4
Nylons; Molded, Extruded; Type 6: Glass fiber (30%) reinforced	0.9—1.2
Cellulose Acetate Butyrate; ASTM Grade: S2	0.9—1.3
Cellulose Acetate Butyrate; ASTM Grade: MH	1.3—1.6
Cellusose Acetate Propionate; ASTM Grade: 3	1.3—1.8

Source: data compiled by J.S. Park from Charles T. Lynch, *CRC Handbook of Materials Science*, Vol. 3, CRC Press, Boca Raton, Florida, 1975 and *Engineered Materials Handbook*, Vol.2, Engineering Plastics, ASM International, Metals Park, Ohio, 1988.

Table 449. SELECTING WATER ABSORPTION OF POLYMERS
(SHEET 5 OF 5)

Polymer	Water Absorption in 24 hr (ASTM D570) (%)
Nylons; Molded, Extruded; Type 6: General purpose	1.3—1.9
Nylons; Molded, Extruded; 6/6 Nylon: General purpose molding	1.5
Nylons; Molded, Extruded; 6/6 Nylon: General purpose extrusion	1.5
Cellusose Acetate Propionate; ASTM Grade: 6	1.6
Cellusose Acetate Propionate; ASTM Grade: 1	1.6—2.0
Cellulose Acetate; Molded, Extruded; ASTM Grade: H4—1	1.7—2.7
Cellulose Acetate; Molded, Extruded; ASTM Grade: H2—1	1.7—2.7
Cellulose Acetate; ASTM Grade: MH—1, MH—2	1.8—4.0
Cellulose Acetate Butyrate; ASTM Grade: H4	2
Cellulose Acetate; ASTM Grade: MS—1, MS—2	2.1—4.0
Cellulose Acetate; Molded, Extruded; ASTM Grade: S2—1	2.3—4.0
Nylons; Type 8	9.5

Source: data compiled by J.S. Park from Charles T. Lynch, *CRC Handbook of Materials Science*, Vol. 3, CRC Press, Boca Raton, Florida, 1975 and *Engineered Materials Handbook*, Vol.2, Engineering Plastics, ASM International, Metals Park, Ohio, 1988.

Table 450. SELECTING IRON ALLOYS IN 10% CORROSIVE MEDIUM

(SHEET 1 OF 16)

Corrosive Medium	Corrosion Rate* at 70°F in a 10% Corrosive Medium†							
	1020 Steel	Grey Cast Iron	Ni–Resist Cast Iron	12% Cr Steel	17% Cr Steel	Stainless Steel 301	Stainless Steel 316	14% Si Iron
Acetaldehyde	<0.05	<0.05	—	—	—	—	—	<0.002
Acetic Acid (Aerated)	>0.05	>0.05	<0.02	<0.02	<0.002	<0.002	<0.002	<0.002
Acetic Acid (Air Free)	>0.05	>0.05	<0.02	<0.02	<0.02	<0.02	<0.002	<0.002
Acetic Anhydride	—	—	—	—	—	—	—	<0.002
Acetoacetic Acid	>0.05	>0.05	—	—	<0.02	<0.02	<0.02	<0.02
Acetone	<0.05	—	—	<0.02	<0.02	<0.02	<0.02	<0.002
Acrolein	<0.02	—	—	<0.02	<0.02	<0.02	<0.02	<0.02
Alcohol (Ethyl)	<0.02	<0.02	<0.02	<0.02	<0.02	<0.02	<0.002	<0.002
Alcohol (Methyl)	<0.02	<0.02	<0.02	<0.02	<0.02	<0.02	<0.002	<0.002
Alcohol (Allyl)	—	—	—	—	—	—	—	<0.02
Allylamine	<0.02 (30%)	—	—	—	—	<0.002 (30%)	<0.002 (30%)	<0.002 (30%)
Aluminum Acetate	>0.05	>0.05	—	<0.02	—	<0.02	<0.02	<0.02

Source: data compiled by J.S. Park from Earl R. Parker, *Materials Data Book for Engineers and Scientists*, McGraw-Hill Book Company, New York, 1967.

Table 450. SELECTING IRON ALLOYS IN 10% CORROSIVE MEDIUM

(SHEET 2 OF 16)

Corrosive Medium	Corrosion Rate* at 70°F in a 10% Corrosive Medium†							
	1020 Steel	Grey Cast Iron	Ni–Resist Cast Iron	12% Cr Steel	17% Cr Steel	Stainless Steel 301	Stainless Steel 316	14% Si Iron
Aluminum Chlorate	—	—	—	—	<0.002	<0.002	—	<0.02
Aluminum Chloride	>0.05	>0.05	>0.05	>0.05	>0.05	>0.05	<0.05	<0.002
Aluminum Fluoride	<0.02	<0.02	—	>0.05	>0.05	>0.05	—	>0.05
Aluminum Formate	<0.05	—	—	<0.02	<0.02	<0.02	<0.02	<0.02
Aluminum Hydroxide	<0.02	<0.02	<0.02	<0.02	<0.02	<0.02	<0.02	<0.02
Aluminum Nitrate	>0.05	>0.05	—	<0.02	<0.02	<0.02	<0.02	—
Aluminum Potassium Sulfate	>0.05	>0.05	>0.05	>0.05	<0.05	<0.02	<0.02	—
Aluminum Sulfate	>0.05	>0.05	<0.02	>0.05	—	<0.02	<0.02	<0.002
Ammonia	<0.002	<0.002	<0.002	<0.002	<0.002	<0.002	<0.002	<0.02
Ammonium Acetate	—	—	<0.002	<0.002	<0.002	<0.002	<0.002	<0.002
Ammonium Bicarbonate	<0.02	<0.02	<0.02	<0.02	<0.02	<0.02	<0.02	<0.002
Ammonium Bromide	>0.05	>0.05	—	<0.05	<0.05	<0.05	<0.02	<0.002

Source: data compiled by J.S. Park from Earl R. Parker, *Materials Data Book for Engineers and Scientists*, McGraw-Hill Book Company, New York, 1967.

Table 450. SELECTING IRON ALLOYS IN 10% CORROSIVE MEDIUM

(SHEET 3 OF 16)

Corrosive Medium	Corrosion Rate* at 70°F in a 10% Corrosive Medium†							
	1020 Steel	Grey Cast Iron	Ni–Resist Cast Iron	12% Cr Steel	17% Cr Steel	Stainless Steel 301	Stainless Steel 316	14% Si Iron
Ammonium Carbonate	<0.02	<0.02	<0.02	<0.02	<0.02	<0.02	<0.02	<0.002
Ammonium Chloride	<0.05	>0.05	<0.02	<0.05	<0.05	<0.02	<0.02	<0.002
Ammonium Citrate	>0.05	>0.05	>0.05	—	<0.02	<0.02	<0.02	—
Ammonium Formate	—	—	—	—	—	<0.02	<0.02	<0.02
Ammonium Nitrate	<0.002	<0.02	<0.02	<0.02	<0.002	<0.002	<0.002	<0.002
Ammonium Sulfate	<0.02	<0.05	>0.05	>0.05	<0.05	<0.05	<0.02	<0.002
Ammonium Sulfite	>0.05	>0.05	>0.05	>0.05	>0.05	<0.05	<0.02	<0.02
Ammonium Thiocyanate	<0.02	<0.02	<0.02	—	<0.02	<0.02	<0.02	<0.02
Amyl Acetate	<0.002	—	—	—	—	<0.002	<0.002	<0.002
Amyl Chloride	>0.05	—	—	—	—	>0.05	—	<0.02
Aniline	—	—	<0.02	<0.02	<0.02	<0.02	<0.02	<0.002
Aniline Hydro-chloride	>0.05	>0.05	>0.05	>0.05	>0.05	>0.05	>0.05	<0.02

Source: data compiled by J.S. Park from Earl R. Parker, *Materials Data Book for Engineers and Scientists*, McGraw-Hill Book Company, New York, 1967.

Table 450. SELECTING IRON ALLOYS IN 10% CORROSIVE MEDIUM

(SHEET 4 OF 16)

Corrosive Medium	Corrosion Rate* at 70°F in a 10% Corrosive Medium†							
	1020 Steel	Grey Cast Iron	Ni–Resist Cast Iron	12% Cr Steel	17% Cr Steel	Stainless Steel 301	Stainless Steel 316	14% Si Iron
Antimony Trichloride	>0.05	>0.05	>0.05	>0.05	>0.05	>0.05	>0.05	<0.002
Barium Carbonate	<0.02	<0.02	<0.02	<0.02	<0.02	<0.02	<0.02	<0.02
Barium Chloride	<0.02	>0.05	<0.02	<0.05	<0.02	<0.02	<0.02	<0.02
Barium Nitrate	<0.02	—	—	—	<0.02	<0.02	<0.02	<0.02
Barium Peroxide	<0.05	—	—	>0.05	—	<0.02	<0.02	<0.02
Benzal-dehyde	>0.05	>0.05	<0.02	—	—	<0.02	—	<0.02
Benzene	—	—	—	<0.02	<0.02	<0.02	<0.02	<0.002
Benzoic Acid	>0.05	>0.05	—	<0.02	<0.02	<0.02	<0.02	<0.02
Boric Acid	<0.05	>0.05	<0.002	<0.02	<0.02	<0.002	<0.002	<0.02
Bromic Acid	>0.05	>0.05	—	>0.05	>0.05	>0.05	>0.05	—
Butyric Acid	<0.05	>0.05	>0.05	<0.05	<0.05	<0.02	<0.02	<0.002
Cadmium Chloride	>0.05	>0.05	>0.05	>0.05	>0.05	<0.02	<0.02	<0.02

Source: data compiled by J.S. Park from Earl R. Parker, *Materials Data Book for Engineers and Scientists*, McGraw-Hill Book Company, New York, 1967.

Table 450. SELECTING IRON ALLOYS IN 10% CORROSIVE MEDIUM

(SHEET 5 OF 16)

Corrosive Medium	Corrosion Rate* at 70°F in a 10% Corrosive Medium†							
	1020 Steel	Grey Cast Iron	Ni–Resist Cast Iron	12% Cr Steel	17% Cr Steel	Stainless Steel 301	Stainless Steel 316	14% Si Iron
Cadmium Sulfate	<0.02	<0.02	—	—	<0.002	<0.002	<0.002	<0.002
Calcium Acetate	<0.02	<0.05	—	<0.02	<0.02	<0.02	<0.02	<0.02
Calcium Bicarbonate	<0.02	—	—	—	—	—	—	—
Calcium Bromide	—	—	—	<0.02	<0.02	<0.02	<0.02	—
Calcium Chlorate	<0.002	<0.02	<0.05	<0.02	<0.02	<0.02	<0.02	<0.02
Calcium Chloride	<0.002	<0.02	<0.02	<0.02	<0.05	<0.02	<0.02	<0.002
Calcium Hydroxide	<0.02	<0.02	<0.02	<0.02	<0.02	<0.02	<0.02	<0.02
Calcium Hypochlorite	<0.05	<0.05	<0.02	>0.05	>0.05	<0.05	<0.05	<0.02
Carbon Tetrachloride	—	—	—	>0.05	<0.002	>0.05	<0.02	<0.002
Carbon Acid (Air Free)	<0.02	—	—	—	—	<0.02	<0.02	<0.02
Chloroacetic Acid	>0.05	>0.05	>0.05	>0.05	>0.05	>0.05	>0.05	>0.05
Chlorine Gas	>0.05	>0.05	>0.05	>0.05	>0.05	—	—	—

Source: data compiled by J.S. Park from Earl R. Parker, *Materials Data Book for Engineers and Scientists*, McGraw-Hill Book Company, New York, 1967.

Table 450. SELECTING IRON ALLOYS IN 10% CORROSIVE MEDIUM

(SHEET 6 OF 16)

Corrosive Medium	Corrosion Rate* at 70°F in a 10% Corrosive Medium†							
	1020 Steel	Grey Cast Iron	Ni–Resist Cast Iron	12% Cr Steel	17% Cr Steel	Stainless Steel 301	Stainless Steel 316	14% Si Iron
Chromic Acid	>0.05	<0.05	<0.05	>0.05	<0.02	<0.02	<0.02	<0.002
Chromic Sulfates	>0.05	—	—	>0.05	>0.05	<0.02	<0.02	<0.002
Citric Acid	>0.05	>0.05	>0.05	<0.05	<0.02	<0.02	<0.02	<0.002
Copper Nitrate	>0.05	>0.05	>0.05	<0.02	<0.02	<0.02	<0.002	<0.002
Copper Sulfate	>0.05	>0.05	>0.05	<0.02	<0.02	<0.02	<0.02	<0.002
Diethylene Glycol	<0.002 (60%)	—	—	—	—	—	—	—
Ethyl Chloride	>0.05 (90%)	—	—	>0.05 (90%)	>0.05 (90%)	>0.05 (90%)	—	—
Ethylene Glycol	<0.02	—	—	—	—	—	—	<0.02
Ferric Chloride	>0.05	>0.05	>0.05	>0.05	>0.05	>0.05	>0.05	>0.05
Ferric Nitrate	>0.05	>0.05	—	<0.02	<0.02	<0.02	<0.02	<0.02
Ferrous Chloride	>0.05	>0.05	>0.05	>0.05	>0.05	>0.05	>0.05	>0.05
Ferrous Sulfate	>0.05	>0.05	—	<0.02	<0.02	<0.02	<0.02	<0.02

Source: data compiled by J.S. Park from Earl R. Parker, *Materials Data Book for Engineers and Scientists*, McGraw-Hill Book Company, New York, 1967.

Table 450. SELECTING IRON ALLOYS IN 10% CORROSIVE MEDIUM

(SHEET 7 OF 16)

Corrosive Medium	Corrosion Rate* at 70°F in a 10% Corrosive Medium†							
	1020 Steel	Grey Cast Iron	Ni–Resist Cast Iron	12% Cr Steel	17% Cr Steel	Stainless Steel 301	Stainless Steel 316	14% Si Iron
Formaldehyde	<0.05 (40%)	<0.05 (40%)	<0.05 (40%)	<0.02	<0.002	<0.002 (20%)	<0.02	<0.002
Formic Acid	>0.05	>0.05	>0.05	<0.05	<0.05	<0.02	<0.002	<0.002
Furfural	<0.02 (30%)	—	<0.02 (30%)	<0.02 (80%)	<0.002 (30%)	<0.002 (30%)	<0.002	<0.02 (20%)
Hydrazine	>0.05	>0.05	—	—	—	<0.002	<0.002	—
Hydrobromic Acid	>0.05	>0.05	—	>0.05	>0.05	>0.05	>0.05	>0.05
Hydrochloric Acid (Areated)	>0.05	>0.05	>0.05	>0.05	>0.05	>0.05	>0.05	<0.02
Hydrochloric Acid (Air Free)	>0.05	>0.05	<0.05	>0.05	>0.05	>0.05	>0.05	<0.02
Hydrofluoric Acid (Areated)	>0.05	>0.05	<0.002	—	—	<0.002	<0.002	>0.05

Source: data compiled by J.S. Park from Earl R. Parker, *Materials Data Book for Engineers and Scientists*, McGraw-Hill Book Company, New York, 1967.

Table 450. SELECTING IRON ALLOYS IN 10% CORROSIVE MEDIUM

(SHEET 8 OF 16)

Corrosive Medium	Corrosion Rate* at 70°F in a 10% Corrosive Medium†							
	1020 Steel	Grey Cast Iron	Ni–Resist Cast Iron	12% Cr Steel	17% Cr Steel	Stainless Steel 301	Stainless Steel 316	14% Si Iron
Hydrofluoric Acid (Air Free)	>0.05	>0.05	<0.002	>0.05	>0.05	>0.05	>0.05	>0.05
Hydrogen Chloride	>0.05 90	>0.05 90	—	>0.05 90	>0.05 90	>0.05 90	—	<0.02 90
Hydrogen Iodide	<0.05 (1%)	>0.05	—	<0.05	—	<0.02 1%)	—	>0.05
Hydrogen Peroxide	>0.05 (20%)	>0.05 (20%)	—	<0.02 (20%)	<0.02 (20%)	<0.02 (20%)	<0.02 (20%)	<0.02 (20%)
Hydrogen Sulfide	<0.02	<0.02	<0.02	<0.02	<0.02	>0.05	<0.002	—
Lactic Acid	>0.05	>0.05	>0.05	>0.05	>0.05	<0.02	<0.02	<0.002
Lead Acetate	>0.05 (20%)	>0.05	—	<0.02	<0.02	<0.02	<0.02	<0.02
Lead Nitrate	>0.05	>0.05	—	<0.02	<0.02	<0.02	<0.02	<0.002

Source: data compiled by J.S. Park from Earl R. Parker, *Materials Data Book for Engineers and Scientists*, McGraw-Hill Book Company, New York, 1967.

Table 450. SELECTING IRON ALLOYS IN 10% CORROSIVE MEDIUM

(SHEET 9 OF 16)

Corrosive Medium	Corrosion Rate* at 70°F in a 10% Corrosive Medium†							
	1020 Steel	Grey Cast Iron	Ni–Resist Cast Iron	12% Cr Steel	17% Cr Steel	Stainless Steel 301	Stainless Steel 316	14% Si Iron
Lithium Chloride	<0.02 (30%)	<0.02 (30%)	<0.002 (30%)	—	—	<0.002 (30%)	<0.002 (30%)	<0.02 (30%)
Lithium Hydroxide	<0.02	<0.02	<0.02	<0.02	<0.02	<0.02	<0.02	>0.05
Magnesium Chloride	<0.02	<0.02	<0.02	<0.05	<0.05	<0.05	<0.02	<0.002
Magnesium Hydroxide	<0.02	<0.02	<0.02	<0.02	<0.02	<0.02	<0.02	<0.02
Magnesium Sulfate	<0.02	>0.05	<0.02	>0.05	<0.002	<0.002	<0.002	<0.002
Maleic Acid	>0.05	>0.05	>0.05	—	<0.02	<0.02	<0.02	<0.02
Malic Acid	>0.05	>0.05	—	<0.02	<0.02	<0.002	<0.002	—
Maganous Chloride	>0.05 (40%)	>0.05 (40%)	<0.05 (40%)	—	—	<0.02 (40%)	<0.02 (40%)	—
Mercuric Chloride	>0.05	>0.05	>0.05	>0.05	>0.05	>0.05	>0.05	<0.02
Mercurous Nitrate	—	—	—	<0.02	<0.02	<0.02	<0.02	<0.02
Methallylamine	<0.02	—	<0.02	<0.02	<0.02	<0.02	<0.02	<0.02
Methanol	<0.02	<0.02	<0.02	<0.02	<0.02	<0.02	<0.02	<0.002

Source: data compiled by J.S. Park from Earl R. Parker, *Materials Data Book for Engineers and Scientists*, McGraw-Hill Book Company, New York, 1967.

Table 450. SELECTING IRON ALLOYS IN 10% CORROSIVE MEDIUM

(SHEET 10 OF 16)

Corrosive Medium	Corrosion Rate* at 70°F in a 10% Corrosive Medium†							
	1020 Steel	Grey Cast Iron	Ni–Resist Cast Iron	12% Cr Steel	17% Cr Steel	Stainless Steel 301	Stainless Steel 316	14% Si Iron
Methyl Ethyl Ketone	<0.02	<0.02	<0.02	<0.02	<0.02	<0.02	<0.02	<0.02
Methyl Isobutyl Ketone	<0.02	<0.02	<0.02	<0.02	<0.02	<0.02	<0.02	<0.02
Methylamine	<0.02	<0.02	<0.02	<0.02	<0.02	<0.02	<0.02	<0.02
Methylene Chloride	—	—	—	—	—	<0.02	<0.02	—
Monochloroacetic Acid	>0.05	>0.05	—	>0.05	>0.05	<0.05	<0.05	<0.02
Monorthanolamine	<0.02	—	—	<0.02	<0.002	<0.002	<0.02	—
Monoethalamine	<0.02	—	<0.02	<0.02	<0.02	<0.02	<0.02	<0.02
Monoethylamine	<0.02	<0.02	<0.02	<0.02	<0.02	<0.02	<0.02	<0.02
Monosodium Phosphate	>0.05	>0.05	>0.05	>0.05	>0.05	<0.02	<0.02	<0.02
Nickel Chloride	>0.05	>0.05	>0.05	>0.05	>0.05	>0.05	>0.05	<0.02
Nickel Nitrate	<0.02	<0.02	<0.02	<0.02	<0.02	<0.02	<0.02	<0.002
Nickel Sulfate	>0.05	>0.05	—	—	—	<0.002	<0.02	<0.002

Source: data compiled by J.S. Park from Earl R. Parker, *Materials Data Book for Engineers and Scientists*, McGraw-Hill Book Company, New York, 1967.

Table 450. SELECTING IRON ALLOYS IN 10% CORROSIVE MEDIUM

(SHEET 11 OF 16)

Corrosive Medium	Corrosion Rate* at 70°F in a 10% Corrosive Medium†							
	1020 Steel	Grey Cast Iron	Ni–Resist Cast Iron	12% Cr Steel	17% Cr Steel	Stainless Steel 301	Stainless Steel 316	14% Si Iron
Nitric Acid	>0.05	>0.05	>0.05	<0.02	<0.02	<0.002	<0.002	<0.002
Nitric + Sulfuric Acid	—	—	—	—	—	—	—	<0.02
Nitrous Acid	—	—	—	<0.05	<0.02	<0.02	<0.02	<0.002
Oleic Acid	—	—	—	<0.02	<0.02	<0.02	<0.02	<0.002
Oxalic Acid	>0.05	>0.05	>0.05	>0.05	>0.05	<0.02	<0.02	<0.02
Phosphoric Acid (Areated)	>0.05	>0.05	>0.05	<0.02	<0.02	<0.02	<0.002	<0.002
Phosphoric Acid (Air Free)	>0.05	>0.05	>0.05	>0.05	>0.05	<0.02	<0.02	<0.02
Picric Acid	>0.05	>0.05	—	<0.02	<0.02	<0.02	<0.02	<0.02
Potassium Bicarbonate	<0.02	<0.02	<0.02	<0.02	<0.02	<0.02	<0.02	<0.02
Potassium Bromide	<0.05	<0.05	<0.02	<0.02	<0.02	<0.02	<0.02	<0.02
Potassium Carbonate	<0.02	<0.02	<0.02	<0.02	<0.02	<0.02	<0.02	<0.02
Potassium Chlorate	<0.02	—	<0.02	<0.02	<0.02	<0.02	<0.02	<0.02

Source: data compiled by J.S. Park from Earl R. Parker, *Materials Data Book for Engineers and Scientists*, McGraw-Hill Book Company, New York, 1967.

Table 450. Selecting Iron Alloys in 10% Corrosive Medium

(Sheet 12 of 16)

Corrosive Medium	Corrosion Rate* at 70°F in a 10% Corrosive Medium†							
	1020 Steel	Grey Cast Iron	Ni–Resist Cast Iron	12% Cr Steel	17% Cr Steel	Stainless Steel 301	Stainless Steel 316	14% Si Iron
Potassium Chromate	<0.02	<0.02	<0.02	<0.02	<0.02	<0.02	<0.02	<0.02
Potassium Cyanide	<0.02	>0.05	<0.02	<0.02	<0.02	<0.02	<0.02	<0.02
Potassium Dichromate	<0.02	<0.02	<0.02	<0.02	<0.02	<0.002	<0.002	<0.002
Potassium Ferricyanide	<0.02	<0.02	<0.02	<0.02	<0.02	<0.02	<0.02	<0.02
Potassium Ferrocyanide	>0.05	>0.05	<0.02	>0.05	<0.02	<0.02	<0.02	<0.02
Potassium Hydroxide	<0.02	<0.02	<0.02	<0.02	<0.02	<0.02	<0.02	>0.05
Potassium Hypochlorite	>0.05	>0.05	>0.05	>0.05	>0.05	>0.05	<0.05	<0.002
Potassium Iodide	<0.02	—	<0.02	>0.05	>0.05	<0.02	<0.02	<0.02
Potassium Nitrate	<0.02	<0.02	<0.02	<0.02	<0.02	<0.02	<0.02	<0.002
Potassium Nitrite	<0.02	<0.02	<0.02	<0.02	<0.02	<0.02	<0.02	<0.02
Potassium Permanganate	<0.02	<0.02	<0.02	<0.002	<0.02	<0.02	<0.02	<0.02
Potassium Silicate	<0.02	<0.02	<0.02	<0.02	<0.02	<0.02	<0.02	<0.02

Source: data compiled by J.S. Park from Earl R. Parker, *Materials Data Book for Engineers and Scientists*, McGraw-Hill Book Company, New York, 1967.

Table 450. SELECTING IRON ALLOYS IN 10% CORROSIVE MEDIUM

(SHEET 13 OF 16)

Corrosive Medium	Corrosion Rate* at 70°F in a 10% Corrosive Medium†							
	1020 Steel	Grey Cast Iron	Ni–Resist Cast Iron	12% Cr Steel	17% Cr Steel	Stainless Steel 301	Stainless Steel 316	14% Si Iron
Propionic Acid	>0.05	>0.05	—	—	—	—	—	<0.02
Pyridine	<0.02	<0.02	<0.02	<0.02	<0.02	<0.02	<0.02	<0.02
Quinine Sulfate	>0.05	>0.05	<0.02	—	<0.02	<0.02	<0.02	<0.02
Silver Bromide	>0.05	>0.05	>0.05	>0.05	>0.05	>0.05	>0.05	—
Silver Chloride	>0.05	>0.05	—	>0.05	>0.05	>0.05	>0.05	—
Silver Nitrate	>0.05	>0.05	—	<0.02	<0.02	<0.02	<0.002	<0.002
Sodium Acetate	<0.02	—	<0.02	<0.02	<0.02	<0.02	<0.02	<0.002
Sodium Bicarbonate	<0.02	<0.02	<0.02	<0.02	<0.02	<0.02	<0.02	<0.002
Sodium Bisulfate	>0.05	>0.05	<0.002	<0.002	<0.002	<0.002	<0.002	<0.002
Sodium Bromide	<0.02	—	<0.02	<0.05	<0.05	<0.05	<0.05	<0.05
Sodium Carbonate	<0.002	<0.002	<0.002	<0.02	<0.02	<0.02	<0.02	<0.02
Sodium Chloride	<0.02	<0.02	<0.02	<0.02	<0.02	<0.02	<0.02	<0.02

Source: data compiled by J.S. Park from Earl R. Parker, *Materials Data Book for Engineers and Scientists*, McGraw-Hill Book Company, New York, 1967.

Table 450. SELECTING IRON ALLOYS IN 10% CORROSIVE MEDIUM

(SHEET 14 OF 16)

Corrosive Medium	Corrosion Rate* at 70°F in a 10% Corrosive Medium†							
	1020 Steel	Grey Cast Iron	Ni–Resist Cast Iron	12% Cr Steel	17% Cr Steel	Stainless Steel 301	Stainless Steel 316	14% Si Iron
Sodium Chromate	<0.02	<0.02	<0.02	<0.02	<0.02	<0.02	<0.02	<0.02
Sodium Hydroxide	<0.002	<0.02	<0.002	<0.002	<0.002	<0.002	<0.002	>0.05
Sodium Hypochlorite	>0.05	>0.05	>0.05	>0.05	>0.05	>0.05	>0.05	—
Sodium Metasilicate	<0.02	<0.02	<0.002	<0.002	<0.002	<0.002	<0.002	<0.02
Sodium Nitrate	<0.02	<0.02	<0.02	<0.02	<0.02	<0.002	<0.002	<0.002
Sodium Nitrite	<0.02	<0.02	<0.02	<0.02	<0.02	<0.02	<0.02	<0.02
Sodium Phosphate	<0.02	<0.02	<0.02	<0.02	<0.02	<0.02	<0.02	<0.02
Sodium Silicate	<0.02	<0.02	<0.02	<0.02	<0.02	<0.02	<0.02	<0.02
Sodium Sulfate	<0.02	<0.02	<0.02	<0.05	<0.05	<0.02	<0.002	<0.002
Sodium Sulfide	<0.05	<0.05	—	>0.05	>0.05	<0.02	>0.05	<0.02
Sodium Sulfite	<0.02	>0.05	<0.02	<0.02	<0.02	<0.002	<0.002	<0.002
Stannic Chloride	>0.05	>0.05	>0.05	>0.05	>0.05	>0.05	>0.05	>0.05

Source: data compiled by J.S. Park from Earl R. Parker, *Materials Data Book for Engineers and Scientists*, McGraw-Hill Book Company, New York, 1967.

Table 450. SELECTING IRON ALLOYS IN 10% CORROSIVE MEDIUM

(SHEET 15 OF 16)

Corrosive Medium	Corrosion Rate* at 70°F in a 10% Corrosive Medium†							
	1020 Steel	Grey Cast Iron	Ni–Resist Cast Iron	12% Cr Steel	17% Cr Steel	Stainless Steel 301	Stainless Steel 316	14% Si Iron
Stannous Chloride	>0.05	>0.05	>0.05	>0.05	>0.05	>0.05	<0.02	<0.002
Strontium Nitrate	>0.05	>0.05	<0.02	<0.02	<0.02	<0.02	<0.02	<0.02
Succinic Acid	<0.02	<0.02	<0.02	<0.02	<0.02	<0.02	<0.02	<0.02
Sulfur Dioxide	>0.05	—	—	>0.05	>0.05	>0.05	<0.002	—
Sulfuric Acid (Areated)	>0.05	>0.05	<0.02	<0.05	<0.05	>0.05	<0.002	<0.002
Sulfuric Acid (Air Free)	>0.05	>0.05	<0.02	>0.05	>0.05	>0.05	<0.05	
Sulfurous Acid	<0.05	—	<0.05	>0.05	>0.05	<0.02	<0.02	<0.02
Tannic Acid	>0.05	—	—	<0.02	<0.02	<0.02	<0.02	<0.002
Tartaric Acid	>0.05	>0.05	<0.02	<0.02	<0.02	<0.002	<0.02	<0.02
Tetraphosphoric Acid	>0.05	>0.05	>0.05	>0.05	>0.05	—	—	—
Trichloroacetic Acid	>0.05	>0.05	>0.05	>0.05	>0.05	>0.05	>0.05	<0.002
Urea	<0.05	—	—	<0.02	<0.02	<0.02	<0.02	<0.02

Source: data compiled by J.S. Park from Earl R. Parker, *Materials Data Book for Engineers and Scientists*, McGraw-Hill Book Company, New York, 1967.

Table 450. SELECTING IRON ALLOYS IN 10% CORROSIVE MEDIUM

(SHEET 16 OF 16)

Corrosive Medium	Corrosion Rate[*] at 70°F in a 10% Corrosive Medium[†]							
	1020 Steel	Grey Cast Iron	Ni–Resist Cast Iron	12% Cr Steel	17% Cr Steel	Stainless Steel 301	Stainless Steel 316	14% Si Iron
Zinc Chloride	>0.05	>0.05	<0.02	—	—	—	—	—
Zinc Sulfate	>0.05	>0.05	<0.02	<0.05	<0.05	<0.002	<0.02	<0.002

Source: data compiled by J.S. Park from Earl R. Parker, *Materials Data Book for Engineers and Scientists*, McGraw-Hill Book Company, New York, 1967.

* <0.002 means that corrosion rate is likely to be less than 0.002 inch per year (Excellent).
<0.02 means that corrosion rate is likely to be less than about 0.02 inch per year (Good).
<0.05 means that corrosion rate is likely to be less than about 0.05 inch per year (Fair).
>0.05 means that corrosion rate is likely to be more than 0.05 inch per year (Poor).

† 10% corrosive medium in 90% water at 70°F

Table 451. SELECTING IRON ALLOYS IN 100% CORROSIVE MEDIUM

(SHEET 1 OF 17)

Corrosive Medium	Corrosion Rate* at 70°F in a 100% Corrosive Medium†							
	1020 Steel	Grey Cast Iron	Ni–Resist Cast Iron	12% Cr Steel	17% Cr Steel	Stainless Steel 301	Stainless Steel 316	14% Si Iron
Acetaldehyde	<0.002	<0.002	<0.002	<0.002	<0.002	<0.002	<0.002	<0.002
Acetic Acid (Aerated)	>0.05	>0.05	>0.05	>0.05	<0.002	<0.002	<0.002	<0.002
Acetic Acid (Air Free)	>0.05	>0.05	>0.05	>0.05	<0.05	<0.002	<0.02	<0.002
Acetic Anhydride	>0.05	>0.05	<0.02	<0.05	<0.05	<0.02	<0.02	<0.002
Acetoacetic Acid	>0.05	>0.05	—	—	<0.02	<0.02	<0.02	<0.02
Acetone	<0.002	<0.002	<0.002	<0.002	<0.002	<0.002	<0.002	<0.002
Acetylene	<0.002	<0.002	<0.002	<0.002	<0.002	<0.002	<0.002	<0.002
Acrolein	<0.02	<0.02	<0.02	<0.02	<0.02	<0.002	<0.02	<0.02
Acrylonitril	<0.002	<0.002	<0.002	<0.002	<0.002	<0.002	<0.002	<0.002
Alcohol (Ethyl)	<0.002	<0.02	<0.02	<0.02	<0.02	<0.02	<0.002	<0.002
Alcohol (Methyl)	<0.002	<0.002	<0.002	<0.02	<0.02	<0.02	<0.002	<0.002
Alcohol (Allyl)	<0.002	<0.02	<0.02	<0.02	<0.02	<0.02	<0.02	<0.02

Source: data compiled by J.S. Park from Earl R. Parker, *Materials Data Book for Engineers and Scientists*, McGraw-Hill Book Company, New York, 1967.

Table 451. SELECTING IRON ALLOYS IN 100% CORROSIVE MEDIUM

(SHEET 2 OF 17)

Corrosive Medium	Corrosion Rate* at 70°F in a 100% Corrosive Medium†							
	1020 Steel	Grey Cast Iron	Ni–Resist Cast Iron	12% Cr Steel	17% Cr Steel	Stainless Steel 301	Stainless Steel 316	14% Si Iron
Alcohol (Amyl)	<0.02	<0.02	<0.02	<0.02	<0.02	<0.02	<0.02	<0.02
Alcohol (Benzyl)	<0.002	—	—	<0.02	<0.02	<0.02	<0.02	<0.02
Alcohol (Butyl)	<0.002	<0.002	—	<0.002	<0.002	<0.002	<0.002	<0.002
Alcohol (Cetyl)	<0.02	—	—	<0.02	<0.02	<0.02	<0.02	<0.02
Alcohol (Isopropyl)	<0.002	<0.02	<0.02	<0.02	<0.02	<0.02	<0.02	<0.02
Allylamine	<0.02	<0.02	<0.02	<0.02	<0.02	<0.02	<0.02	<0.02
Allyl Chloride	<0.002	<0.02	—	<0.02	<0.02	<0.02	<0.002	<0.002
Allyl Sulfide	<0.02	<0.02	<0.02	<0.02	<0.02	<0.02	<0.02	<0.02
Aluminum Acetate	—	—	<0.02	<0.02	—	<0.02	<0.02	<0.002
Aluminum Chlorate	—	—	—	—	—	—	—	<0.002
Aluminum Chloride	<0.002	>0.05	>0.05	<0.002	<0.002	<0.002	—	<0.02
Aluminum Fluoride	—	—	—	>0.05	>0.05	>0.05	<0.05	>0.05

Source: data compiled by J.S. Park from Earl R. Parker, *Materials Data Book for Engineers and Scientists*, McGraw-Hill Book Company, New York, 1967.

Table 451. SELECTING IRON ALLOYS IN 100% CORROSIVE MEDIUM

(SHEET 3 OF 17)

Corrosive Medium	Corrosion Rate* at 70°F in a 100% Corrosive Medium†							
	1020 Steel	Grey Cast Iron	Ni–Resist Cast Iron	12% Cr Steel	17% Cr Steel	Stainless Steel 301	Stainless Steel 316	14% Si Iron
Aluminum Fluosilicate	>0.05	>0.05	—	<0.02	<0.02	<0.02	<0.02	<0.02
Aluminum Formate	>0.05	—	—	<0.02	<0.02	<0.02	<0.02	<0.02
Aluminum Hydroxide	—	—	—	—	<0.02	<0.02	<0.02	—
Aluminum Nitrate	—	—	—	<0.02	<0.02	<0.02	<0.02	—
Aluminum Potassium Sulfate	—	—	—	<0.05	>0.05	<0.02	—	<0.002
Aluminum Sulfate	—	—	—	>0.05	>0.05	<0.02	<0.02	<0.02
Ammonia	<0.002	<0.002	<0.002	<0.002	<0.002	<0.002	<0.002	<0.02
Ammonium Acetate	<0.002	<0.02	<0.002	<0.002	<0.002	<0.002	<0.002	<0.02
Ammonium Bicarbonate	<0.002	<0.02	<0.02	—	—	<0.05	<0.02	<0.002
Ammonium Bromide	>0.05	>0.05	—	>0.05	—	<0.05	—	—
Ammonium Carbonate	<0.002	<0.02	<0.02	<0.02	<0.02	<0.02	<0.02	<0.02
Ammonium Chloride	<0.02	—	—	>0.05	>0.05	>0.05	—	<0.02

Source: data compiled by J.S. Park from Earl R. Parker, *Materials Data Book for Engineers and Scientists*, McGraw-Hill Book Company, New York, 1967.

Table 451. Selecting Iron Alloys in 100% Corrosive Medium
(Sheet 4 of 17)

Corrosive Medium	Corrosion Rate* at 70°F in a 100% Corrosive Medium†							
	1020 Steel	Grey Cast Iron	Ni–Resist Cast Iron	12% Cr Steel	17% Cr Steel	Stainless Steel 301	Stainless Steel 316	14% Si Iron
Ammonium Citrate	<0.002	—	—	—	—	—	—	—
Ammonium Formate	—	—	—	—	—	<0.02	<0.02	<0.02
Ammonium Nitrate	<0.02	<0.05	—	<0.02	<0.02	<0.002	<0.002	—
Ammonium Sulfate	—	<0.02	<0.02	—	—	—	—	<0.002
Ammonium Sulfite	—	—	—	—	—	<0.05	<0.02	—
Amyl Acetate	<0.02	<0.02	<0.002	<0.002	<0.02	<0.002	<0.002	<0.002
Amyl Chloride	<0.02	<0.02	—	<0.05	<0.05	<0.002	<0.002	<0.02
Aniline	<0.002	<0.002	<0.02	<0.02	<0.02	<0.02	<0.02	<0.002
Aniline Hydrochloride	>0.05	>0.05	>0.05	>0.05	>0.05	>0.05	>0.05	<0.02
Anthracine	<0.02	<0.02	<0.02	<0.02	<0.02	<0.02	<0.02	<0.02
Antimony Trichloride	<0.05	—	—	>0.05	>0.05	>0.05	—	—
Barium Carbonate	<0.02	<0.02	<0.02	<0.02	<0.02	<0.02	<0.02	<0.02

Source: data compiled by J.S. Park from Earl R. Parker, *Materials Data Book for Engineers and Scientists*, McGraw-Hill Book Company, New York, 1967.

Table 451. SELECTING IRON ALLOYS IN 100% CORROSIVE MEDIUM

(SHEET 5 OF 17)

Corrosive Medium	Corrosion Rate* at 70°F in a 100% Corrosive Medium†							
	1020 Steel	Grey Cast Iron	Ni–Resist Cast Iron	12% Cr Steel	17% Cr Steel	Stainless Steel 301	Stainless Steel 316	14% Si Iron
Barium Chloride	<0.002	<0.02	—	—	<0.02	<0.05	<0.02	—
Barium Hydroxide	<0.02	<0.02	—	<0.02	<0.02	<0.02	<0.02	<0.02
Barium Nitrate	<0.02	—	—	—	—	<0.02	<0.02	<0.02
Barium Oxide	<0.002	—	—	<0.02	<0.02	<0.02	<0.02	<0.02
Barium Peroxide	<0.002	—	—	—	—	—	—	—
Benzaldehyde	<0.002	>0.05	<0.002	<0.02	<0.02	<0.02	<0.02	<0.02
Benzene	<0.02	<0.02	<0.02	<0.02	<0.02	<0.02	<0.02	<0.002
Benzoic Acid	>0.05	>0.05	<0.02	<0.02	<0.02	<0.02	<0.02	<0.02
Boric Acid	—	—	<0.02	<0.02	<0.02	<0.02	<0.02	<0.02
Bromic Acid	>0.05	>0.05	—	>0.05	>0.05	—	—	—
Bromine (Dry)	<0.05	>0.05	<0.02	>0.05	>0.05	>0.05	>0.05	>0.05
Bromine (Wet)	>0.05	>0.05	>0.05	>0.05	>0.05	>0.05	>0.05	>0.05

Source: data compiled by J.S. Park from Earl R. Parker, *Materials Data Book for Engineers and Scientists*, McGraw-Hill Book Company, New York, 1967.

Table 451. Selecting Iron Alloys in 100% Corrosive Medium

(Sheet 6 of 17)

Corrosive Medium	Corrosion Rate* at 70°F in a 100% Corrosive Medium†							
	1020 Steel	Grey Cast Iron	Ni–Resist Cast Iron	12% Cr Steel	17% Cr Steel	Stainless Steel 301	Stainless Steel 316	14% Si Iron
Butyric Acid	>0.05	—	>0.05	—	<0.05	<0.02	<0.02	<0.002
Cadmium Chloride	<0.002	—	—	—	—	—	—	—
Cadmium Sulfate	<0.02	<0.02	—	—	—	—	—	—
Calcium Acetate	<0.05	<0.05	—	<0.02	<0.02	<0.02	<0.02	<0.02
Calcium Bicarbonate	<0.02	<0.02	—	<0.02	<0.02	<0.02	<0.02	<0.02
Calcium Bromide	<0.05	<0.05	—	<0.02	<0.02	<0.02	<0.02	<0.02
Calcium Chlorate	<0.02	<0.02	<0.02	—	—	—	—	—
Calcium Chloride	<0.002	<0.002	—	—	<0.02	<0.02	<0.002	<0.02
Calcium Hydroxide	<0.02	<0.02	—	<0.02	<0.02	—	—	—
Calcium Hypochlorite	<0.02	<0.02	—	>0.05	>0.05	—	—	<0.05
Carbon Dioxide	<0.002	<0.002	<0.002	<0.002	<0.002	<0.002	<0.002	<0.002
Carbon Monoxide	<0.002	<0.002	<0.002	<0.002	<0.002	<0.002	<0.002	<0.002

Source: data compiled by J.S. Park from Earl R. Parker, *Materials Data Book for Engineers and Scientists*, McGraw-Hill Book Company, New York, 1967.

Table 451. SELECTING IRON ALLOYS IN 100% CORROSIVE MEDIUM

(SHEET 7 OF 17)

Corrosive Medium	Corrosion Rate* at 70°F in a 100% Corrosive Medium†							
	1020 Steel	Grey Cast Iron	Ni–Resist Cast Iron	12% Cr Steel	17% Cr Steel	Stainless Steel 301	Stainless Steel 316	14% Si Iron
Carbon Tetrachloride	<0.002	<0.05	<0.02	<0.02	<0.002	<0.02	<0.02	<0.002
Carbon Acid (Air Free)	<0.02	<0.05	<0.002	<0.002	<0.002	<0.02	<0.02	<0.002
Chloroacetic Acid	>0.05	>0.05	>0.05	>0.05	>0.05	—	—	>0.05
Chlorine Gas	<0.02	<0.02	<0.02	<0.05	<0.05	<0.002	<0.02	<0.02
Chlorine Liquid	<0.02	—	—	—	—	—	—	—
Chloroform (Dry)	<0.002	<0.002	—	<0.002	<0.02	<0.002	<0.002	—
Chromic Acid	<0.002	<0.02	<0.02	<0.02	—	—	—	<0.02
Chromic Hydroxide	<0.02	—	<0.02	<0.02	<0.02	<0.02	<0.02	<0.02
Chromic Sulfates	>0.05	—	—	>0.05	>0.05	<0.05	—	<0.02
Citric Acid	<0.002	—	>0.05	—	—	<0.02	<0.02	<0.002
Diethylene Glycol	<0.002	—	—	—	<0.002	<0.002	<0.002	<0.002
Ethyl Chloride	<0.002	—	—	<0.002	<0.002	<0.002	<0.002	<0.002

Source: data compiled by J.S. Park from Earl R. Parker, *Materials Data Book for Engineers and Scientists*, McGraw-Hill Book Company, New York, 1967.

Table 451. SELECTING IRON ALLOYS IN 100% CORROSIVE MEDIUM

(SHEET 8 OF 17)

Corrosive Medium	Corrosion Rate* at 70°F in a 100% Corrosive Medium†							
	1020 Steel	Grey Cast Iron	Ni–Resist Cast Iron	12% Cr Steel	17% Cr Steel	Stainless Steel 301	Stainless Steel 316	14% Si Iron
Ethylene Glycol	<0.002	<0.02	<0.02	<0.02	<0.02	<0.02	<0.02	<0.02
Ethylene Oxide	<0.002	<0.02	—	<0.02	<0.02	<0.02	<0.02	<0.02
Fatty Acids	>0.05	>0.05	<0.02	<0.02	<0.02	<0.02	<0.002	<0.002
Ferric Chloride	<0.02	—	—	—	—	—	—	—
Fluorine	<0.002	>0.05	—	>0.05	<0.002	<0.002	<0.002	>0.05
Formaldehyde	<0.002	<0.02	—	<0.02	<0.002	<0.002	<0.002	<0.002
Formic Acid	>0.05	>0.05	>0.05	<0.02	<0.05	<0.02	<0.002	<0.002
Furfural	<0.02	<0.02	<0.02	—	—	<0.02	<0.02	<0.02
Hydrazine	>0.05	—	—	—	—	—	—	—
Hydrobromic Acid	<0.02	<0.02	>0.05	—	—	>0.05	—	>0.05
Hydrocyanic Acid	<0.002	<0.02	<0.02	>0.05	<0.05	<0.02	<0.02	<0.02
Hydrofluoric Acid (Areated)	<0.02	>0.05	<0.02	—	—	<0.02	<0.02	>0.05

Source: data compiled by J.S. Park from Earl R. Parker, *Materials Data Book for Engineers and Scientists*, McGraw-Hill Book Company, New York, 1967.

Table 451. SELECTING IRON ALLOYS IN 100% CORROSIVE MEDIUM

(SHEET 9 OF 17)

Corrosive Medium	Corrosion Rate* at 70°F in a 100% Corrosive Medium†							
	1020 Steel	Grey Cast Iron	Ni–Resist Cast Iron	12% Cr Steel	17% Cr Steel	Stainless Steel 301	Stainless Steel 316	14% Si Iron
Hydrofluoric Acid (Air Free)	<0.05	>0.05	<0.02	>0.05	>0.05	>0.05	<0.02	>0.05
Hydrogen Chloride	<0.002	<0.02	<0.002	>0.05	>0.05	<0.002	<0.002	<0.02
Hydrogen Fluoride	<0.002	—	<0.02	<0.02	<0.02	<0.002	<0.002	—
Hydrogen Iodide	<0.02	<0.02	<0.02	>0.05	>0.05	<0.02	<0.02	<0.02
Hydrogen Peroxide	—	—	—	<0.02	<0.02	<0.02	<0.02	<0.02
Hydrogen Sulfide	<0.02	<0.02	<0.02	<0.02	<0.05	<0.05	<0.02	<0.02
Lactic Acid	>0.05	>0.05	>0.05	—	—	<0.02	<0.02	<0.02
Lead Acetate	<0.002	—	—	<0.02	<0.02	<0.02	<0.02	<0.05
Lead Chromate	<0.02	<0.02	<0.02	<0.02	<0.02	<0.02	<0.02	<0.02
Lead Nitrate	<0.02	<0.02	—	—	—	<0.02	<0.02	<0.002
Lead Sulfate	<0.02	<0.02	<0.02	<0.02	<0.02	<0.02	<0.02	<0.02
Lithium Chloride	<0.002	<0.002	—	—	—	<0.002	<0.002	<0.02

Source: data compiled by J.S. Park from Earl R. Parker, *Materials Data Book for Engineers and Scientists*, McGraw-Hill Book Company, New York, 1967.

Table 451. SELECTING IRON ALLOYS IN 100% CORROSIVE MEDIUM

(SHEET 10 OF 17)

Corrosive Medium	Corrosion Rate* at 70°F in a 100% Corrosive Medium†							
	1020 Steel	Grey Cast Iron	Ni–Resist Cast Iron	12% Cr Steel	17% Cr Steel	Stainless Steel 301	Stainless Steel 316	14% Si Iron
Lithium Hydroxide	<0.002	—	—	—	—	—	—	—
Magnesium Chloride	<0.002	<0.02	<0.02	—	—	—	—	>0.05
Magnesium Hydroxide	<0.002	—	<0.02	<0.02	<0.02	<0.02	<0.02	—
Magnesium Sulfate	<0.02	<0.02	<0.02	<0.05	<0.02	<0.02	<0.02	<0.002
Maleic Acid	<0.002	—	—	<0.05	<0.02	<0.02	<0.02	<0.02
Malic Acid	—	—	—	—	—	<0.002	<0.002	—
Mercuric Chloride	—	—	—	>0.05	>0.05	>0.05	—	<0.02
Mercurous Nitrate	<0.02	—	—	<0.02	—	<0.02	<0.02	<0.002
Methallylamine	<0.02	<0.02	<0.02	<0.02	<0.02	<0.02	<0.02	<0.002
Methanol	<0.002	<0.002	<0.002	<0.002	<0.002	<0.002	<0.002	<0.002
Methyl Ethyl Ketone	<0.002	<0.002	<0.002	<0.002	<0.002	<0.002	<0.002	<0.002
Methyl Isobutyl Ketone	<0.02	<0.02	<0.02	<0.02	<0.02	<0.02	<0.02	<0.02

Source: data compiled by J.S. Park from Earl R. Parker, *Materials Data Book for Engineers and Scientists*, McGraw-Hill Book Company, New York, 1967.

Table 451. SELECTING IRON ALLOYS IN 100% CORROSIVE MEDIUM

(SHEET 11 OF 17)

Corrosive Medium	Corrosion Rate* at 70°F in a 100% Corrosive Medium†							
	1020 Steel	Grey Cast Iron	Ni–Resist Cast Iron	12% Cr Steel	17% Cr Steel	Stainless Steel 301	Stainless Steel 316	14% Si Iron
Methylamine	<0.02	<0.02	<0.02	<0.02	<0.02	<0.02	<0.02	<0.02
Methylene Chloride	<0.02	<0.02	<0.02	<0.02	<0.02	<0.02	<0.02	<0.02
Monochloroacetic Acid	<0.002	>0.05	<0.05	>0.05	>0.05	<0.02	<0.02	<0.02
Monorthanolamine	<0.02	<0.02	<0.02	—	—	<0.02	<0.02	—
Monoethalamine	<0.02	—	<0.02	<0.02	<0.02	<0.02	<0.02	<0.02
Monoethylamine	<0.02	<0.02	<0.02	<0.02	<0.02	<0.02	<0.02	<0.02
Nitric Acid	>0.05	>0.05	>0.05	>0.05	<0.05	<0.002	<0.002	<0.002
Nitric Acid (Red Fuming)	<0.05	>0.05	>0.05	<0.002	<0.002	<0.002	<0.002	<0.002
Nitric + Hydrochloric Acid	>0.05	>0.05	>0.05	>0.05	>0.05	>0.05	>0.05	<0.05
Nitric + Hydrofluoric Acid	>0.05	>0.05	>0.05	>0.05	>0.05	>0.05	>0.05	>0.05
Nitric + Sulfuric Acid	>0.05	>0.05	>0.05	>0.05	>0.05	>0.05	>0.05	<0.02
Nitrobenzene	<0.002	<0.02	<0.02	<0.02	<0.02	<0.02	<0.02	<0.002

Source: data compiled by J.S. Park from Earl R. Parker, *Materials Data Book for Engineers and Scientists*, McGraw-Hill Book Company, New York, 1967.

Table 451. SELECTING IRON ALLOYS IN 100% CORROSIVE MEDIUM

(SHEET 12 OF 17)

Corrosive Medium	Corrosion Rate* at 70°F in a 100% Corrosive Medium†							
	1020 Steel	Grey Cast Iron	Ni–Resist Cast Iron	12% Cr Steel	17% Cr Steel	Stainless Steel 301	Stainless Steel 316	14% Si Iron
Nitrocelluolose	<0.02	<0.02	<0.02	<0.02	<0.02	<0.02	<0.02	<0.02
Nitroglycerine	<0.05	<0.05	<0.02	<0.02	<0.02	<0.02	<0.02	<0.05
Nitrotolune	<0.02	<0.02	<0.02	<0.02	<0.02	<0.02	<0.02	<0.02
Nitrous Acid	>0.05	—	—	—	—	<0.02	<0.02	<0.002
Oleic Acid	<0.02	<0.02	<0.002	<0.02	<0.02	<0.02	<0.02	<0.002
Oxalic Acid	>0.05	>0.05	<0.02	>0.05	>0.05	>0.05	>0.05	<0.02
Phenol	<0.002	<0.02	<0.02	<0.02	<0.02	<0.02	<0.02	<0.002
Phosphoric Acid (Areated)	>0.05	>0.05	>0.05	—	—	>0.05	<0.02	<0.002
Phosphoric Acid (Air Free)	>0.05	>0.05	>0.05	>0.05	>0.05	—	—	<0.02
Picric Acid	>0.05	>0.05	>0.05	<0.02	<0.02	<0.02	<0.02	<0.02
Potassium Bicarbonate	<0.002	—	—	—	<0.02	<0.02	<0.02	—
Potassium Bromide	>0.05	>0.05	<0.02	<0.002	<0.02	<0.05	—	<0.02

Source: data compiled by J.S. Park from Earl R. Parker, *Materials Data Book for Engineers and Scientists*, McGraw-Hill Book Company, New York, 1967.

Table 451. SELECTING IRON ALLOYS IN 100% CORROSIVE MEDIUM

(SHEET 13 OF 17)

Corrosive Medium	Corrosion Rate* at 70°F in a 100% Corrosive Medium†							
	1020 Steel	Grey Cast Iron	Ni–Resist Cast Iron	12% Cr Steel	17% Cr Steel	Stainless Steel 301	Stainless Steel 316	14% Si Iron
Potassium Carbonate	<0.02	<0.02	<0.02	<0.02	<0.02	<0.02	<0.02	<0.02
Potassium Chlorate	<0.002	—	—	<0.02	<0.02	<0.02	<0.02	<0.02
Potassium Chromate	—	—	—	<0.02	<0.02	<0.02	<0.02	—
Potassium Cyanide	<0.002	<0.02	—	<0.02	<0.02	<0.02	<0.02	<0.02
Potassium Dichromate	—	—	<0.02	<0.02	<0.02	<0.02	<0.02	—
Potassium Ferricyanide	<0.02	<0.02	<0.02	—	<0.02	<0.02	<0.02	—
Potassium Hydroxide	<0.002	<0.02	—	<0.002	<0.002	<0.002	—	>0.05
Potassium Hypochlorite	<0.002	—	—	—	—	—	<0.02	<0.002
Potassium Iodide	<0.02	—	—	—	—	<0.02	<0.02	<0.02
Potassium Nitrate	<0.002	<0.02	—	<0.02	<0.02	<0.02	—	<0.002
Potassium Nitrite	<0.02	<0.02	<0.02	<0.02	<0.02	<0.02	<0.02	<0.02
Potassium Permanganate	<0.002	<0.02	—	<0.02	<0.02	<0.02	—	—

Source: data compiled by J.S. Park from Earl R. Parker, *Materials Data Book for Engineers and Scientists*, McGraw-Hill Book Company, New York, 1967.

Table 451. SELECTING IRON ALLOYS IN 100% CORROSIVE MEDIUM

(SHEET 14 OF 17)

Corrosive Medium	Corrosion Rate* at 70°F in a 100% Corrosive Medium†							
	1020 Steel	Grey Cast Iron	Ni–Resist Cast Iron	12% Cr Steel	17% Cr Steel	Stainless Steel 301	Stainless Steel 316	14% Si Iron
Potassium Silicate	<0.02	<0.02	<0.02	<0.02	<0.02	<0.02	<0.02	<0.02
Propionic Acid	<0.02	—	—	—	—	—	<0.02	<0.02
Pyridine	<0.02	<0.02	<0.02	<0.02	<0.02	<0.02	<0.02	<0.02
Quinine Sulfate	>0.05	>0.05	<0.02	—	<0.02	<0.02	<0.02	<0.02
Salicylic Acid	>0.05	>0.05	<0.02	<0.02	<0.02	<0.02	<0.02	<0.02
Silicon Tetrachloride (Dry)	<0.002	<0.002	<0.002	<0.002	<0.002	<0.002	<0.002	<0.002
Silicon Tetrachloride (Wet)	>0.05	>0.05	>0.05	>0.05	>0.05	>0.05	—	<0.002
Silver Bromide	>0.05	>0.05	>0.05	>0.05	>0.05	<0.05	—	<0.02
Silver Chloride	>0.05	>0.05	—	>0.05	>0.05	>0.05	—	<0.02
Silver Nitrate	—	—	—	—	—	—	<0.02	—
Sodium Acetate	<0.002	<0.002	—	<0.02	<0.02	<0.02	<0.02	<0.02
Sodium Bicarbonate	<0.05	<0.05	<0.02	—	<0.02	—	—	—

Source: data compiled by J.S. Park from Earl R. Parker, *Materials Data Book for Engineers and Scientists*, McGraw-Hill Book Company, New York, 1967.

Table 451. SELECTING IRON ALLOYS IN 100% CORROSIVE MEDIUM

(SHEET 15 OF 17)

Corrosive Medium	Corrosion Rate* at 70°F in a 100% Corrosive Medium†							
	1020 Steel	Grey Cast Iron	Ni–Resist Cast Iron	12% Cr Steel	17% Cr Steel	Stainless Steel 301	Stainless Steel 316	14% Si Iron
Sodium Bisulfate	<0.002	—	<0.002	>0.05	—	>0.05	—	<0.002
Sodium Bromide	<0.02	<0.05	<0.02	—	—	—	—	—
Sodium Carbonate	<0.02	<0.02	<0.02	<0.02	<0.02	<0.02	<0.02	<0.02
Sodium Chloride	<0.002	<0.02	<0.02	—	—	—	—	—
Sodium Chromate	<0.02	<0.02	<0.02	<0.02	<0.02	<0.02	<0.02	<0.02
Sodium Hydroxide	<0.02	—	<0.02	—	—	—	—	—
Sodium Hypochlorite	>0.05	—	—	>0.05	>0.05	>0.05	>0.05	—
Sodium Metasilicate	<0.002	<0.02	<0.02	<0.002	<0.002	<0.002	<0.002	<0.02
Sodium Nitrate	<0.02	<0.02	<0.02	<0.02	<0.002	<0.02	<0.02	<0.002
Sodium Nitrite	<0.002	—	—	<0.002	—	<0.02	—	—
Sodium Phosphate	<0.02	<0.02	<0.02	<0.02	<0.02	<0.02	<0.02	<0.02
Sodium Silicate	<0.02	<0.02	<0.02	<0.02	<0.02	<0.02	<0.02	<0.02

Source: data compiled by J.S. Park from Earl R. Parker, *Materials Data Book for Engineers and Scientists*, McGraw-Hill Book Company, New York, 1967.

Table 451. SELECTING IRON ALLOYS IN 100% CORROSIVE MEDIUM

(SHEET 16 OF 17)

Corrosive Medium	Corrosion Rate* at 70°F in a 100% Corrosive Medium†							
	1020 Steel	Grey Cast Iron	Ni–Resist Cast Iron	12% Cr Steel	17% Cr Steel	Stainless Steel 301	Stainless Steel 316	14% Si Iron
Sodium Sulfate	<0.02	<0.02	<0.02	>0.05	>0.05	<0.002	<0.002	<0.002
Sodium Sulfide	<0.02	<0.02	—	<0.02	>0.05	>0.05	—	<0.02
Stannic Chloride	<0.002	—	—	—	—	—	—	—
Stannous Chloride	<0.02	<0.02	<0.02	—	<0.05	<0.05	—	—
Strontium Nitrate	>0.05	>0.05	—	—	<0.02	<0.02	<0.02	<0.02
Succinic Acid	<0.02	<0.02	<0.02	<0.02	<0.02	<0.02	<0.02	—
Sulfur Dioxide	<0.002	<0.02	<0.02	<0.02	<0.02	<0.02	<0.02	>0.05
Sulfur Trioxide	<0.02	<0.02	<0.02	<0.02	<0.02	<0.02	<0.02	>0.05
Sulfuric Acid (Areated)	<0.02	<0.02	<0.02	>0.05	>0.05	<0.02	<0.02	<0.02
Sulfuric Acid (Air Free)	<0.02	<0.02	<0.02	<0.05	<0.05	<0.05	<0.02	
Sulfuric Acid (Fuming)	<0.02	<0.02	<0.05	<0.002	<0.002	<0.02	<0.02	<0.02
Sulfurous Acid	>0.05	>0.05	>0.05	>0.05	>0.05	>0.05	<0.002	<0.02

Source: data compiled by J.S. Park from Earl R. Parker, *Materials Data Book for Engineers and Scientists,* McGraw-Hill Book Company, New York, 1967.

Table 451. Selecting Iron Alloys in 100% Corrosive Medium

(Sheet 17 of 17)

Corrosive Medium	Corrosion Rate* at 70°F in a 100% Corrosive Medium†							
	1020 Steel	Grey Cast Iron	Ni–Resist Cast Iron	12% Cr Steel	17% Cr Steel	Stainless Steel 301	Stainless Steel 316	14% Si Iron
Tannic Acid	<0.002	<0.02	—	<0.02	<0.02	<0.02	<0.02	<0.002
Tartaric Acid	<0.05	>0.05	—	—	—	—	—	<0.02
Tetraphosphoric Acid	>0.05	>0.05	<0.05	>0.05	>0.05	<0.02	<0.02	<0.05
Trichloroacetic Acid	>0.05	>0.05	>0.05	>0.05	>0.05	>0.05	>0.05	<0.002
Trichloroethylene	<0.002	<0.02	<0.02	<0.02	<0.02	<0.02	<0.02	<0.002
Zinc Chloride	<0.002	<0.02	<0.02	>0.05	>0.05	—	—	—

Source: data compiled by J.S. Park from Earl R. Parker, *Materials Data Book for Engineers and Scientists*, McGraw-Hill Book Company, New York, 1967.

* <0.002 means that corrosion rate is likely to be less than 0.002 inch per year (Excellent).
<0.02 means that corrosion rate is likely to be less than about 0.02 inch per year (Good).
<0.05 means that corrosion rate is likely to be less than about 0.05 inch per year (Fair).
>0.05 means that corrosion rate is likely to be more than 0.05 inch per year (Poor).

† Water-free, Dry or Maximum concentration of corrosive medium. Quantitatively

Table 452. SELECTING NONFERROUS METALS FOR USE IN A 10% CORROSIVE MEDIUM

(SHEET 1 OF 18)

Corrosive Medium	Corrosion Rate* at 70°F in a 10% Corrosive Medium†									
	Copper, Sn-Braze, Al-Braze	70-30 Brass	Silicon Bronze	Monel	Nickel	Inconel	Hastelloy	Aluminum	Lead	Titanium
Acetaldehyde	<0.002	<0.02	<0.02	<0.002	<0.002	—	—	<0.02	<0.02	—
Acetic Acid (Aerated)	>0.05	>0.05	>0.05	<0.02	<0.05	<0.02	<0.002	<0.02	>0.05	<0.002
Acetic Acid (Air Free)	<0.002	>0.05	>0.05	<0.02	<0.02	<0.02	<0.002	<0.002	>0.05	<0.002
Acetic Anhydride	—	—	—	—	—	—	<0.002	—	—	—
Acetoacetic Acid	—	—	—	<0.02	<0.02	—	<0.02	<0.02	—	—
Acetone	<0.002	<0.002	<0.002	<0.002	<0.002	<0.002	<0.002	<0.02	<0.002	<0.002
Acrolein	<0.02	<0.02	<0.02	—	—	—	—	<0.02	<0.02	—
Alcohol (Ethyl)	<0.002	<0.002	<0.002	<0.002	<0.002	<0.002	<0.002	<0.02	<0.002	<0.002
Alcohol (Methyl)	<0.02	<0.02	<0.02	<0.002	<0.002	<0.002	<0.002	—	<0.02	—
Alcohol (Benzyl)	—	—	—	—	—	—	<0.02	—	—	—
Alcohol (Butyl)	—	—	—	—	—	—	—	<0.002	—	—
Aluminum Acetate	<0.02	—	<0.02	<0.02	<0.02	<0.02	<0.02	<0.002	<0.002	—

Source: data compiled by J.S. Park from Earl R. Parker, *Materials Data Book for Engineers and Scientists*, McGraw-Hill Book Company, New York, 1967.

Table 452. SELECTING NONFERROUS METALS FOR USE IN A 10% CORROSIVE MEDIUM

(SHEET 2 OF 18)

Corrosive Medium	Corrosion Rate* at 70°F in a 10% Corrosive Medium†									
	Copper, Sn-Braze, Al-Braze	70-30 Brass	Silicon Bronze	Monel	Nickel	Inconel	Hastelloy	Aluminum	Lead	Titanium
Aluminum Chlorate	—	—	—	<0.02	<0.02	<0.02	<0.02	—	<0.02	<0.002
Aluminum Chloride	<0.02	>0.05	<0.02	<0.02	<0.05	>0.05	<0.002	>0.05	>0.05	>0.05
Aluminum Fluoride	<0.02	>0.05	<0.02	<0.002	<0.02	—	<0.02	<0.002	<0.02	—
Aluminum Formate	—	—	<0.02	<0.02	<0.02	<0.02	<0.02	<0.02	—	—
Aluminum Hydroxide	<0.02	<0.02	<0.02	<0.02	<0.02	—	<0.02	<0.02	<0.02	<0.002
Aluminum Nitrate	—	—	—	<0.02	<0.02	<0.02	<0.02	<0.02	<0.02	<0.002
Aluminum Potassium Sulfate	<0.02	>0.05	<0.02	<0.02	<0.02	—	<0.02	<0.02	<0.002	—
Aluminum Sulfate	<0.02	<0.02	<0.02	<0.02	<0.02	<0.02	<0.002	<0.002	<0.02	<0.002

Source: data compiled by J.S. Park from Earl R. Parker, *Materials Data Book for Engineers and Scientists*, McGraw-Hill Book Company, New York, 1967.

Table 452. SELECTING NONFERROUS METALS FOR USE IN A 10% CORROSIVE MEDIUM

(SHEET 3 OF 18)

Corrosive Medium	Corrosion Rate* at 70°F in a 10% Corrosive Medium†									
	Copper, Sn-Braze, Al-Braze	70-30 Brass	Silicon Bronze	Monel	Nickel	Inconel	Hastelloy	Aluminum	Lead	Titanium
Ammonia	>0.05	>0.05	>0.05	>0.05	>0.05	<0.002	<0.002	<0.002	<0.02	<0.002
Ammonium Acetate	—	—	—	<0.002	<0.002	<0.002	<0.002	<0.002	—	—
Ammonium Bicarbonate	>0.05	>0.05	>0.05	—	—	—	—	<0.02	<0.02	—
Ammonium Bromide	>0.05	>0.05	>0.05	<0.02	<0.02	—	<0.02	>0.05	>0.05	—
Ammonium Carbonate	>0.05	>0.05	>0.05	<0.02	>0.05	>0.05	>0.05	<0.02	<0.02	—
Ammonium Chloride	>0.05	>0.05	>0.05	<0.02	<0.02	<0.02	<0.002	>0.05	>0.05	<0.002
Ammonium Citrate	>0.05	>0.05	>0.05	<0.02	<0.02	<0.02	<0.02	<0.02	—	<0.002
Ammonium Formate	—	—	—	<0.02	<0.02	<0.02	<0.002	<0.02	—	<0.002

Source: data compiled by J.S. Park from Earl R. Parker, *Materials Data Book for Engineers and Scientists*, McGraw-Hill Book Company, New York, 1967.

Table 452. Selecting Nonferrous Metals for Use in a 10% Corrosive Medium

(Sheet 4 of 18)

Corrosive Medium	Corrosion Rate* at 70°F in a 10% Corrosive Medium†									
	Copper, Sn-Braze, Al-Braze	70-30 Brass	Silicon Bronze	Monel	Nickel	Inconel	Hastelloy	Aluminum	Lead	Titanium
Ammonium Nitrate	>0.05	>0.05	>0.05	>0.05	<0.02	—	<0.02	<0.02	>0.05	<0.05
Ammonium Sulfate	<0.05	>0.05	<0.02	<0.02	<0.02	<0.02	<0.02	>0.05	<0.02	<0.002
Ammonium Sulfite	>0.05	>0.05	>0.05	>0.05	>0.05	>0.05	—	—	—	—
Ammonium Thiocyanate	>0.05	>0.05	>0.05	<0.02	<0.02		—	—	—	—
Amyl Acetate	<0.02	<0.02	<0.02	<0.02	—	—	<0.002	—	—	—
Amyl Chloride	<0.02	—	—	<0.02	<0.02	—	—	—	—	—
Aniline	—	—	—	<0.02	<0.02	—	—	—	—	—
Aniline Hydrochloride	>0.05	>0.05	>0.05	>0.05	<0.05	>0.05	<0.02	>0.05	>0.05	<0.002

Source: data compiled by J.S. Park from Earl R. Parker, *Materials Data Book for Engineers and Scientists*, McGraw-Hill Book Company, New York, 1967.

Table 452. SELECTING NONFERROUS METALS FOR USE IN A 10% CORROSIVE MEDIUM

(SHEET 5 OF 18)

Corrosive Medium	Corrosion Rate* at 70°F in a 10% Corrosive Medium†									
	Copper, Sn-Braze, Al-Braze	70-30 Brass	Silicon Bronze	Monel	Nickel	Inconel	Hastelloy	Aluminum	Lead	Titanium
Antimony Trichloride	>0.05	>0.05	>0.05	>0.05	>0.05	—	>0.05	>0.05	<0.02	—
Barium Carbonate	<0.02	<0.02	<0.02	<0.02	<0.02	—	—	—	—	—
Barium Chloride	<0.02	>0.05	<0.02	<0.02	<0.02	<0.02	<0.02	<0.02	<0.02	<0.002
Barium Hydroxide	>0.05	>0.05	>0.05	<0.02	<0.002	<0.02	<0.02	>0.05	>0.05	—
Barium Nitrate	>0.05	>0.05	>0.05	—	<0.02	<0.02	<0.02	<0.02	<0.02	—
Barium Peroxide	>0.05	>0.05	>0.05	<0.02	<0.02	—	—	>0.05	>0.05	—
Benzaldehyde	>0.05	>0.05	>0.05	—	—	—	—	<0.02	>0.05	—
Benzene	<0.002	<0.02	<0.02	<0.002	<0.002	<0.002	<0.02	<0.02	<0.02	<0.002
Benzoic Acid	<0.02	<0.02	<0.02	<0.02	<0.02	<0.02	<0.002	<0.02	>0.05	<0.002
Boric Acid	<0.02	<0.02	<0.02	<0.02	<0.02	<0.02	<0.002	<0.05	<0.02	<0.002
Bromic Acid	>0.05	>0.05	>0.05	>0.05	>0.05	>0.05	—	>0.05	<0.02	—
Butyric Acid	<0.05	<0.05	<0.02	<0.05	<0.05	<0.05	<0.002	<0.02	>0.05	<0.002

Source: data compiled by J.S. Park from Earl R. Parker, *Materials Data Book for Engineers and Scientists*, McGraw-Hill Book Company, New York, 1967.

Table 452. SELECTING NONFERROUS METALS FOR USE IN A 10% CORROSIVE MEDIUM

(SHEET 6 OF 18)

	Corrosion Rate* at 70°F in a 10% Corrosive Medium†									
Corrosive Medium	**Copper, Sn-Braze, Al-Braze**	**70-30 Brass**	**Silicon Bronze**	**Monel**	**Nickel**	**Inconel**	**Hastelloy**	**Aluminum**	**Lead**	**Titanium**
Cadmium Chloride	<0.02	>0.05	<0.02	<0.02	<0.02	—	<0.02	>0.05	—	—
Cadmium Sulfate	<0.02	<0.02	<0.02	<0.002	<0.002	<0.002	<0.002	<0.02	<0.002	—
Calcium Acetate	<0.02	<0.02	<0.02	<0.02	<0.02	<0.02	<0.02	—	<0.02	<0.002
Calcium Bromide	<0.02	<0.02	<0.02	<0.02	<0.02	<0.02	<0.02	<0.05	<0.02	—
Calcium Chlorate	<0.02	>0.05	<0.02	<0.02	<0.02	<0.02	<0.02	<0.02	<0.02	—
Calcium Chloride	<0.002	<0.02	<0.02	<0.002	<0.002	<0.002	<0.002	<0.002	>0.05	<0.002
Calcium Hydroxide	<0.02	<0.02	<0.02	<0.02	<0.02	<0.02	<0.002	>0.05	>0.05	—
Calcium Hypochlorite	<0.02	<0.02	<0.02	>0.05	>0.05	>0.05	<0.02	>0.05	<0.05	<0.002

Source: data compiled by J.S. Park from Earl R. Parker, *Materials Data Book for Engineers and Scientists*, McGraw-Hill Book Company, New York, 1967.

Table 452. SELECTING NONFERROUS METALS FOR USE IN A 10% CORROSIVE MEDIUM

(SHEET 7 OF 18)

Corrosive Medium	Corrosion Rate* at 70°F in a 10% Corrosive Medium†									
	Copper, Sn-Braze, Al-Braze	70-30 Brass	Silicon Bronze	Monel	Nickel	Inconel	Hastelloy	Aluminum	Lead	Titanium
Carbon Tetrachloride	—	—	—	<0.02	<0.02	<0.002	<0.002	—	—	—
Carbon Acid (Air Free)	<0.02	—	<0.02	<0.02	<0.02	<0.02	<0.002	<0.02	—	—
Chloroacetic Acid	>0.05	>0.05	—	<0.02	—	—	<0.02	>0.05	>0.05	—
Chromic Acid	>0.05	>0.05	>0.05	>0.05	>0.05	<0.02	<0.02	>0.05	<0.02	<0.002
Chromic Sulfates	<0.02	<0.02	<0.02	—	—	—	<0.02	—	<0.02	—
Citric Acid	<0.05	>0.05	<0.05	<0.02	<0.02	<0.02	<0.002	<0.02	<0.02	<0.002
Copper Nitrate	>0.05	>0.05	>0.05	>0.05	>0.05	>0.05	<0.02	>0.05	—	—
Copper Sulfate	>0.05	>0.05	<0.02	<0.02	<0.02	<0.02	<0.002	>0.05	<0.02	—
Ethyl Chloride	<0.02	—	—	<0.02	—	—	—	—	—	—
Ethylene Glycol	<0.02	—	—	—	—	—	—	<0.002	—	—
Ferric Chloride	>0.05	>0.05	>0.05	>0.05	>0.05	<0.05	<0.002	>0.05	>0.05	<0.002
Ferric Nitrate	>0.05	>0.05	>0.05	>0.05	>0.05	>0.05	<0.002	>0.05	<0.002	<0.002

Source: data compiled by J.S. Park from Earl R. Parker, *Materials Data Book for Engineers and Scientists*, McGraw-Hill Book Company, New York, 1967.

Table 452. Selecting Nonferrous Metals for Use in a 10% Corrosive Medium

(Sheet 8 of 18)

Corrosive Medium	Corrosion Rate* at 70°F in a 10% Corrosive Medium†									
	Copper, Sn-Braze, Al-Braze	70-30 Brass	Silicon Bronze	Monel	Nickel	Inconel	Hastelloy	Aluminum	Lead	Titanium
Ferrous Chloride	<0.02	>0.05	<0.05	>0.05	<0.05	>0.05	<0.02	>0.05	>0.05	<0.002
Ferrous Sulfate	<0.02	>0.05	<0.02	—	>0.05	<0.02	<0.02	<0.002	<0.02	<0.002
Formaldehyde	<0.002	<0.002	<0.002	<0.002	<0.002	<0.002	<0.02	<0.02	<0.02	<0.002
Formic Acid	<0.02	<0.05	<0.02	<0.02	<0.02	<0.02	<0.002	<0.02	>0.05	<0.02
Furfural	<0.02	<0.02	<0.02	<0.02	<0.02	<0.02	<0.02	—	—	—
Hydrazine	>0.05	>0.05	>0.05	—	—	—	—	—	>0.05	—
Hydrobromic Acid	>0.05	>0.05	<0.02	>0.05	>0.05	—	<0.02	>0.05	>0.05	—
Hydrochloric Acid (Areated)	>0.05	>0.05	>0.05	>0.05	>0.05	>0.05	<0.02	>0.05	<0.02	<0.02
Hydrochloric Acid (Air Free)	>0.05	>0.05	<0.02	>0.05	>0.05	>0.05	<0.02	>0.05	<0.02	<0.02
Hydrocyanic Acid	>0.05	>0.05	>0.05	>0.05	—	—	—	<0.02	>0.05	—
Hydrofluoric Acid (Areated)	<0.02	>0.05	>0.05	<0.02	<0.02	<0.02	<0.02	>0.05	>0.05	>0.05
Hydrofluoric Acid (Air Free)	<0.02	>0.05	<0.02	<0.02	<0.02	<0.02	<0.02	>0.05	<0.002	>0.05

Source: data compiled by J.S. Park from Earl R. Parker, *Materials Data Book for Engineers and Scientists*, McGraw-Hill Book Company, New York, 1967.

Table 452. SELECTING NONFERROUS METALS FOR USE IN A 10% CORROSIVE MEDIUM

(SHEET 9 OF 18)

Corrosive Medium	Corrosion Rate* at 70°F in a 10% Corrosive Medium†									
	Copper, Sn-Braze, Al-Braze	70-30 Brass	Silicon Bronze	Monel	Nickel	Inconel	Hastelloy	Aluminum	Lead	Titanium
Hydrogen Iodide	—	—	—	<0.02	—	—	—	—	—	—
Hydrogen Peroxide	>0.05	>0.05	>0.05	<0.02	<0.02	<0.02	<0.002	<0.002	>0.05	<0.002
Hydrogen Sulfide	<0.02	<0.02	<0.02	—	—	<0.02	—	—	—	—
Lactic Acid	<0.002	<0.05	<0.05	>0.05	<0.02	<0.02	<0.02	<0.02	>0.05	<0.002
Lead Acetate	<0.05		—	<0.02	<0.02	<0.02	<0.02	—	—	<0.002
Lead Chromate	—	—	—	—	—	—	—	>0.05	—	—
Lead Nitrate	—	—	—	—	<0.02	—	—	>0.05	—	—
Lead Sulfate	—	—	—	—	<0.02	—	—	>0.05	—	—

Source: data compiled by J.S. Park from Earl R. Parker, *Materials Data Book for Engineers and Scientists*, McGraw-Hill Book Company, New York, 1967.

Table 452. SELECTING NONFERROUS METALS FOR USE IN A 10% CORROSIVE MEDIUM

(SHEET 10 OF 18)

Corrosive Medium	Corrosion Rate* at 70°F in a 10% Corrosive Medium†									
	Copper, Sn-Braze, Al-Braze	70-30 Brass	Silicon Bronze	Monel	Nickel	Inconel	Hastelloy	Aluminum	Lead	Titanium
Lithium Chloride	<0.02 30	<0.02 30	<0.02 30	<0.002 30	<0.002 30	<0.002 30	<0.002 30	<0.05	<0.02	—
Lithium Hydroxide	>0.05	>0.05	>0.05	<0.02	<0.02	<0.02	<0.02	>0.05	>0.05	—
Magnesium Chloride	<0.02	<0.02	<0.02	<0.002	<0.002	<0.002	<0.002	>0.05	>0.05	<0.002
Magnesium Hydroxide	<0.02	<0.02	<0.02	<0.02	—	—	<0.02	>0.05	>0.05	—
Magnesium Sulfate	<0.002	<0.02	<0.002	<0.02	<0.02	<0.02	<0.002	<0.02	<0.02	—
Maleic Acid	<0.02	<0.02	<0.02	<0.05	<0.02	<0.02	<0.002	<0.02	—	—
Malic Acid	—	—	—	<0.02	<0.02	<0.002	—	<0.02	—	—
Maganous Chloride	—	—	—	—	—	—	<0.02	—	—	<0.002
Mercuric Chloride	>0.05	>0.05	>0.05	>0.05	<0.05	>0.05	<0.02	>0.05	<0.05	<0.002
Mercurous Nitrate	>0.05	>0.05	>0.05	<0.02	—	—	<0.02	>0.05	—	—
Methanol	<0.02	<0.02	<0.02	<0.002	<0.002	<0.002	<0.002	—	<0.02	—
Methyl Ethyl Ketone	<0.02	<0.02	<0.02	<0.02	<0.02	<0.02	<0.02	<0.02	<0.02	<0.002

Source: data compiled by J.S. Park from Earl R. Parker, *Materials Data Book for Engineers and Scientists*, McGraw-Hill Book Company, New York, 1967.

Table 452. SELECTING NONFERROUS METALS FOR USE IN A 10% CORROSIVE MEDIUM
(SHEET 11 OF 18)

Corrosive Medium	Corrosion Rate* at 70°F in a 10% Corrosive Medium†									
	Copper, Sn-Braze, Al-Braze	70-30 Brass	Silicon Bronze	Monel	Nickel	Inconel	Hastelloy	Aluminum	Lead	Titanium
Methyl Isobutyl Ketone	<0.02	<0.02	<0.02	<0.02	<0.02	<0.02	<0.02	<0.02	<0.02	<0.002
Methylamine	—	—	—	—	—	—	—	<0.02	—	—
Methylene Chloride	<0.02	—	<0.02	—	—	—	<0.02	>0.05	—	—
Monochloroacetic Acid	>0.05	>0.05	>0.05	—	<0.02	<0.02	—	>0.05	>0.05	—
Monoethalamine	—	—	—	—	—	—	—	<0.02	—	—
Monoethylamine	—	—	—	—	—	—	—	<0.02	—	—
Monosodium Phosphate	<0.02	<0.02	<0.02	<0.02	<0.02	<0.02	<0.02	>0.05	<0.02	—
Nickel Chloride	>0.05	>0.05	>0.05	<0.02	—	—	<0.002	>0.05	—	<0.02

Source: data compiled by J.S. Park from Earl R. Parker, *Materials Data Book for Engineers and Scientists*, McGraw-Hill Book Company, New York, 1967.

Table 452. SELECTING NONFERROUS METALS FOR USE IN A 10% CORROSIVE MEDIUM

(SHEET 12 OF 18)

Corrosive Medium	Corrosion Rate* at 70°F in a 10% Corrosive Medium†									
	Copper, Sn-Braze, Al-Braze	70-30 Brass	Silicon Bronze	Monel	Nickel	Inconel	Hastelloy	Aluminum	Lead	Titanium
Nickel Nitrate	<0.05	<0.05	<0.05	>0.05	>0.05	>0.05	<0.02	>0.05	—	—
Nickel Sulfate	<0.02	<0.05	<0.02	—	<0.02	<0.02	<0.02	>0.05	<0.02	—
Nitric Acid	>0.05	>0.05	>0.05	>0.05	>0.05	<0.02	<0.002	>0.05	>0.05	<0.002
Nitric + Sulfuric Acid	>0.05	>0.05	>0.05	>0.05	>0.05	>0.05	—	>0.05	>0.05	—
Nitrous Acid	—	—	—	—	>0.05	—	—	<0.05	—	—
Oleic Acid	—	>0.05	—	—	—	—	—	—	—	—
Oxalic Acid	<0.02	<0.02	<0.02	<0.02	<0.02	<0.02	<0.02	<0.02	>0.05	<0.02
Phenol	—	—	—	<0.002	—	—	—	—	—	—

Source: data compiled by J.S. Park from Earl R. Parker, *Materials Data Book for Engineers and Scientists*, McGraw-Hill Book Company, New York, 1967.

Table 452. SELECTING NONFERROUS METALS FOR USE IN A 10% CORROSIVE MEDIUM

(SHEET 13 OF 18)

Corrosive Medium	Corrosion Rate* at 70°F in a 10% Corrosive Medium†									
	Copper, Sn-Braze, Al-Braze	70-30 Brass	Silicon Bronze	Monel	Nickel	Inconel	Hastelloy	Aluminum	Lead	Titanium
Phosphoric Acid (Areated)	>0.05	>0.05	>0.05	<0.05	<0.05	<0.02	<0.002	>0.05	<0.02	<0.02
Phosphoric Acid (Air Free)	<0.02	<0.02	<0.02	<0.02	<0.02	<0.02	<0.002	>0.05	<0.002	—
Picric Acid	>0.05	>0.05	>0.05	<0.05	>0.05	—	<0.02	>0.05	>0.05	—
Potassium Bicarbonate	<0.02	<0.02	<0.02	<0.02	<0.02	<0.02	<0.02	>0.05	>0.05	—
Potassium Bromide	<0.02	<0.02	<0.02	<0.02	<0.02	<0.02	<0.002	<0.02	<0.02	<0.002
Potassium Carbonate	<0.02	<0.02	<0.02	<0.02	<0.02	<0.02	<0.02	>0.05	>0.05	<0.002
Potassium Chlorate	<0.02	<0.02	<0.02	<0.05	<0.02	<0.05	<0.02	<0.02	<0.02	<0.002
Potassium Chromate	<0.02	<0.02	<0.02	<0.02	<0.002	<0.002	<0.002	<0.02	<0.02	—

Source: data compiled by J.S. Park from Earl R. Parker, *Materials Data Book for Engineers and Scientists*, McGraw-Hill Book Company, New York, 1967.

Table 452. SELECTING NONFERROUS METALS FOR USE IN A 10% CORROSIVE MEDIUM

(SHEET 14 OF 18)

	Corrosion Rate* at 70°F in a 10% Corrosive Medium†									
Corrosive Medium	Copper, Sn-Braze, Al-Braze	70-30 Brass	Silicon Bronze	Monel	Nickel	Inconel	Hastelloy	Aluminum	Lead	Titanium
Potassium Cyanide	>0.05	>0.05	>0.05	<0.02	<0.02	<0.02	<0.02	>0.05	>0.05	—
Potassium Dichromate	<0.02	<0.02	<0.02	<0.02	<0.02	<0.02	<0.02	<0.002	<0.02	<0.002
Potassium Ferricyanide	<0.02	<0.02	<0.02	<0.02	<0.02	—	<0.02	<0.02	<0.02	—
Potassium Ferrocyanide	<0.02	<0.02	<0.02	<0.02	<0.02	<0.02	<0.02	<0.002	<0.02	—
Potassium Hydroxide	<0.02	<0.02	<0.02	<0.002	<0.002	<0.02	<0.02	>0.05	>0.05	<0.002
Potassium Hypochlorite	<0.02	>0.05	>0.05	<0.05	<0.05	<0.05	<0.02	>0.05	<0.02	<0.002
Potassium Iodide	<0.02	—	<0.02	<0.02	<0.02	<0.02	<0.02	<0.02	>0.05	<0.002
Potassium Nitrate	<0.02	<0.02	<0.02	<0.02	<0.02	<0.02	<0.02	<0.002	<0.02	<0.002

Source: data compiled by J.S. Park from Earl R. Parker, *Materials Data Book for Engineers and Scientists*, McGraw-Hill Book Company, New York, 1967.

Table 452. SELECTING NONFERROUS METALS FOR USE IN A 10% CORROSIVE MEDIUM

(SHEET 15 OF 18)

Corrosive Medium	Corrosion Rate* at 70°F in a 10% Corrosive Medium†									
	Copper, Sn-Braze, Al-Braze	70-30 Brass	Silicon Bronze	Monel	Nickel	Inconel	Hastelloy	Aluminum	Lead	Titanium
Potassium Nitrite	<0.02	<0.02	<0.02	<0.02	<0.02	<0.02	<0.02	<0.02	<0.02	<0.002
Potassium Permanganate	<0.02	<0.02	<0.02	<0.05	<0.02	<0.02	<0.002	<0.02	<0.05	—
Potassium Silicate	<0.02	<0.02	<0.02	<0.02	<0.02	<0.02	<0.02	>0.05	—	—
Propionic Acid	<0.02	<0.02	<0.02	<0.02	<0.02	—	—	<0.02	>0.05	—
Pyridine	<0.02	<0.02	<0.02	<0.02	<0.02	<0.02	<0.02	<0.02	<0.02	—
Quinine Sulfate	<0.02	<0.02	<0.02	<0.02	<0.02	<0.02	<0.02	—	—	—
Salicylic Acid	—	—	—	<0.02	<0.02	—	—	>0.05	—	—
Silver Bromide	>0.05	>0.05	>0.05	—	—	—	<0.002	>0.05	—	—
Silver Chloride	>0.05	>0.05	>0.05	—	—	—	<0.02	>0.05	—	<0.002
Silver Nitrate	>0.05	>0.05	>0.05	>0.05	>0.05	<0.02	<0.002	>0.05	>0.05	—
Sodium Acetate	<0.02	<0.02	<0.02	<0.05	<0.02	<0.02	<0.02	<0.02	—	—
Sodium Bicarbonate	<0.02	<0.02	<0.02	<0.02	<0.02	<0.02	<0.02	>0.05	<0.02	—

Source: data compiled by J.S. Park from Earl R. Parker, *Materials Data Book for Engineers and Scientists*, McGraw-Hill Book Company, New York, 1967.

Table 452. SELECTING NONFERROUS METALS FOR USE IN A 10% CORROSIVE MEDIUM

(SHEET 16 OF 18)

Corrosive Medium	Corrosion Rate* at 70°F in a 10% Corrosive Medium†									
	Copper, Sn-Braze, Al-Braze	70-30 Brass	Silicon Bronze	Monel	Nickel	Inconel	Hastelloy	Aluminum	Lead	Titanium
Sodium Bisulfate	—	>0.05	<0.02	<0.02	<0.02	<0.02	<0.02	>0.05	<0.02	—
Sodium Bromide	<0.02	<0.05	<0.02	<0.02	<0.02	<0.02	<0.02	<0.05	—	—
Sodium Carbonate	<0.02	>0.05	<0.02	<0.02	<0.02	<0.02	<0.02	>0.05	<0.02	—
Sodium Chloride	<0.02	<0.05	<0.02	<0.002	<0.002	<0.002	<0.02	<0.05	<0.02	<0.002
Sodium Chromate	<0.02	<0.02	<0.02	<0.02	<0.02	<0.02	<0.02	<0.02	<0.02	—
Sodium Hydroxide	<0.002	>0.05	<0.02	<0.002	<0.002	<0.002	<0.002	>0.05	<0.02	<0.002
Sodium Hypochlorite	>0.05	>0.05	<0.02	>0.05	>0.05	>0.05	<0.002	>0.05	>0.05	<0.002
Sodium Metasilicate	<0.02	<0.02	<0.02	<0.002	<0.002	<0.002	<0.002	>0.05	—	—
Sodium Nitrate	<0.02	<0.05	<0.02	<0.02	<0.02	<0.002	<0.02	<0.002	>0.05	—
Sodium Nitrite	<0.02	<0.02	<0.02	<0.02	<0.02	<0.02	<0.02	<0.02	<0.02	<0.002
Sodium Phosphate	<0.02	<0.02	<0.02	<0.02	<0.02	<0.02	<0.02	>0.05	<0.02	—
Sodium Silicate	<0.02	<0.02	<0.02	<0.02	<0.02	<0.02	<0.02	>0.05	>0.05	—

Source: data compiled by J.S. Park from Earl R. Parker, *Materials Data Book for Engineers and Scientists*, McGraw-Hill Book Company, New York, 1967.

Table 452. SELECTING NONFERROUS METALS FOR USE IN A 10% CORROSIVE MEDIUM

(SHEET 17 OF 18)

Corrosive Medium	Corrosion Rate* at 70°F in a 10% Corrosive Medium†									
	Copper, Sn-Braze, Al-Braze	70-30 Brass	Silicon Bronze	Monel	Nickel	Inconel	Hastelloy	Aluminum	Lead	Titanium
Sodium Sulfate	<0.02	<0.02	<0.02	<0.02	<0.02	<0.02	<0.02	<0.002	<0.02	—
Sodium Sulfide	>0.05	<0.05	>0.05	<0.02	<0.02	<0.02	<0.02	>0.05	<0.002	<0.002
Sodium Sulfite	<0.02	>0.05	<0.02	<0.02	<0.02	<0.02	<0.02	<0.02	<0.02	—
Stannic Chloride	>0.05	>0.05	>0.05	>0.05	>0.05	>0.05	<0.02	>0.05	>0.05	<0.002
Stannous Chloride	>0.05	>0.05	<0.02	>0.05	<0.05	>0.05	<0.02	>0.05	>0.05	—
Strontium Nitrate	<0.02	<0.02	<0.02	<0.02	<0.02	<0.02	<0.02	<0.02	—	—
Succinic Acid	<0.02	<0.02	<0.02	<0.02	<0.02	<0.02	<0.02	<0.02	<0.02	<0.002
Sulfur Dioxide	<0.02	>0.05	—	>0.05	>0.05	<0.02	<0.002	>0.05	—	—
Sulfuric Acid (Areated)	>0.05	>0.05	>0.05	<0.05	<0.05	>0.05	<0.002	>0.05	<0.002	<0.02
Sulfuric Acid (Air Free)	<0.02	<0.05	<0.02	<0.002	<0.02	<0.05	<0.002	>0.05	<0.002	—
Sulfurous Acid	<0.02	<0.02	<0.02	>0.05	<0.05	<0.05	<0.02	<0.02	<0.02	<0.002
Tannic Acid	<0.02	—	<0.02	<0.02	—	—	<0.02	<0.02	>0.05	<0.002

Source: data compiled by J.S. Park from Earl R. Parker, *Materials Data Book for Engineers and Scientists*, McGraw-Hill Book Company, New York, 1967.

Table 452. SELECTING NONFERROUS METALS FOR USE IN A 10% CORROSIVE MEDIUM

(SHEET 18 OF 18)

Corrosive Medium	Corrosion Rate* at 70°F in a 10% Corrosive Medium†									
	Copper, Sn-Braze, Al-Braze	70-30 Brass	Silicon Bronze	Monel	Nickel	Inconel	Hastelloy	Aluminum	Lead	Titanium
Tartaric Acid	<0.02	<0.05	<0.05	<0.02	<0.02	<0.02	<0.02	<0.02	<0.02	<0.002
Tetraphosphoric Acid	—	>0.05	>0.05	—	—	—	—	>0.05	>0.05	—
Trichloroacetic Acid	>0.05	>0.05	—	—	—	—	<0.02	>0.05	>0.05	<0.002
Urea	<0.02	<0.02	<0.02	<0.02	<0.02	<0.02	<0.02	<0.02	—	—
Zinc Chloride	<0.02	>0.05	<0.02	<0.02	<0.02	—	<0.02	>0.05	<0.02	<0.002
Zinc Sulfate	<0.02	<0.05	<0.02	<0.02	<0.02	<0.002	<0.02	<0.05	<0.02	—

Source: data compiled by J.S. Park from Earl R. Parker, *Materials Data Book for Engineers and Scientists*, McGraw-Hill Book Company, New York, 1967.

* <0.002 means that corrosion rate is likely to be less than 0.002 inch per year (Excellent).
<0.02 means that corrosion rate is likely to be less than about 0.02 inch per year (Good).
<0.05 means that corrosion rate is likely to be less than about 0.05 inch per year (Fair).
>0.05 means that corrosion rate is likely to be more than 0.05 inch per year (Poor).

† 10% corrosive medium in 90% water

Table 453. SELECTING NONFERROUS METALS FOR USE IN A 100% CORROSIVE MEDIUM

(SHEET 1 OF 18)

Corrosive Medium	Corrosion Rate* at 70°F in a 100% Corrosive Medium†									
	Copper, Sn-Braze, Al-Braze	70-30 Brass	Silicon Bronze	Monel	Nickel	Inconel	Hastelloy	Aluminum	Lead	Titanium
Acetaldehyde	<0.002	<0.002	<0.002	<0.002	<0.002	<0.002	<0.002	<0.002	<0.002	<0.002
Acetic Acid (Aerated)	<0.02	>0.05	>0.05	<0.02	>0.05	<0.02	<0.002	<0.002	<0.05	<0.002
Acetic Acid (Air Free)	<0.002	>0.05	<0.02	<0.02	<0.02	<0.02	<0.002	<0.002	<0.02	<0.002
Acetic Anhydride	<0.02	>0.05	<0.02	<0.02	<0.02	<0.02	<0.002	<0.002	<0.002	<0.002
Acetoacetic Acid	—	—	—	<0.02	<0.02	—	<0.02	<0.02	<0.02	—
Acetone	<0.002	<0.002	<0.002	<0.002	<0.002	<0.002	<0.002	<0.002	<0.02	<0.002
Acetylene	<0.002	<0.002	<0.002	<0.002	<0.002	<0.002	<0.002	<0.002	<0.002	<0.002
Acrolein	<0.02	<0.02	<0.02	<0.02	<0.02	<0.02	<0.02	<0.02	—	<0.02
Acrylonitril	<0.002	<0.002	<0.002	<0.002	<0.002	<0.002	<0.002	<0.002	<0.002	<0.002
Alcohol (Ethyl)	<0.002	<0.002	<0.002	<0.002	<0.002	<0.002	<0.002	<0.02	<0.002	<0.002
Alcohol (Methyl)	<0.02	<0.02	<0.02	<0.002	<0.002	<0.002	<0.002	<0.02	<0.02	—
Alcohol (Allyl)	<0.02	<0.02	<0.02	<0.02	<0.02	<0.02	<0.02	<0.02	<0.02	<0.002

Source: data compiled by J.S. Park from Earl R. Parker, *Materials Data Book for Engineers and Scientists*, McGraw-Hill Book Company, New York, 1967.

Table 453. SELECTING NONFERROUS METALS FOR USE IN A 100% CORROSIVE MEDIUM
(SHEET 2 OF 18)

Corrosive Medium	Corrosion Rate* at 70°F in a 100% Corrosive Medium†									
	Copper, Sn-Braze, Al-Braze	70-30 Brass	Silicon Bronze	Monel	Nickel	Inconel	Hastelloy	Aluminum	Lead	Titanium
Alcohol (Amyl)	<0.002	—	<0.02	—	—	—	—	<0.002	—	<0.002
Alcohol (Benzyl)	<0.02	<0.02	<0.02	<0.02	<0.02	<0.02	<0.02	<0.02	<0.02	<0.002
Alcohol (Butyl)	<0.002	<0.002	<0.002	<0.002	<0.002	<0.002	—	<0.002	—	<0.002
Alcohol (Cetyl)	<0.02	—	—	<0.02	<0.02	<0.02	—	<0.02	<0.02	<0.002
Alcohol (Isopropyl)	<0.02	<0.02	<0.02	<0.02	<0.02	<0.02	<0.02	<0.02	<0.002	—
Allylamine	>0.05	>0.05	>0.05	—	—	—	—	—	—	—
Allyl Chloride	<0.02	<0.02	<0.02	<0.02	<0.02	<0.02	<0.02	>0.05	<0.05	—
Allyl Sulfide	>0.05	>0.05	>0.05	—	—	—	—	<0.02	>0.05	—
Aluminum Acetate	<0.02	<0.02	<0.02	—	—	—	<0.02	<0.002	<0.002	<0.002
Aluminum Chlorate	—	—	—	<0.02	<0.02	<0.02	<0.02	—	<0.02	—
Aluminum Chloride	<0.02	>0.05	<0.02	—	<0.02	—	<0.002	<0.02	—	—
Aluminum Fluosilicate	<0.02	<0.02	<0.02	<0.02	<0.02	<0.02	<0.02	—	<0.02	—

Source: data compiled by J.S. Park from Earl R. Parker, *Materials Data Book for Engineers and Scientists*, McGraw-Hill Book Company, New York, 1967.

Table 453. SELECTING NONFERROUS METALS FOR USE IN A 100% CORROSIVE MEDIUM

(SHEET 3 OF 18)

Corrosive Medium	Corrosion Rate* at 70°F in a 100% Corrosive Medium†									
	Copper, Sn-Braze, Al-Braze	70-30 Brass	Silicon Bronze	Monel	Nickel	Inconel	Hastelloy	Aluminum	Lead	Titanium
Aluminum Formate	<0.02	—	<0.02	<0.02	<0.02	<0.02	<0.02	<0.02	<0.02	<0.002
Aluminum Hydroxide	—	—	<0.02	—	—	—	—	—	—	<0.002
Aluminum Nitrate	—	—	—	—	—	—	—	<0.02	—	<0.002
Aluminum Potassium Sulfate	<0.02	>0.05	<0.02	—	—	—	—	<0.02	<0.02	<0.002
Aluminum Sulfate	<0.002	<0.05	<0.02	<0.02	<0.02	—	<0.02	>0.05	—	—
Ammonia	<0.002	<0.002	<0.002	<0.002	<0.002	<0.002	<0.002	<0.002	<0.02	<0.002
Ammonium Acetate	>0.05	>0.05	>0.05	<0.002	<0.002	<0.002	<0.002	<0.002	—	—
Ammonium Bicarbonate	—	—	—	—	—	—	—	<0.02	—	—
Ammonium Carbonate	—	—	<0.02	<0.02	<0.02	<0.02	—	<0.02	—	—
Ammonium Chloride	>0.05	>0.05	>0.05	<0.02	<0.02	<0.02	<0.02	<0.02	<0.02	—
Ammonium Citrate	—	—	—	—	—	<0.02	—	<0.02	—	<0.002
Ammonium Formate	—	—	—	—	—	<0.02	—	—	—	<0.002

Source: data compiled by J.S. Park from Earl R. Parker, *Materials Data Book for Engineers and Scientists*, McGraw-Hill Book Company, New York, 1967.

Table 453. SELECTING NONFERROUS METALS FOR USE IN A 100% CORROSIVE MEDIUM

(SHEET 4 OF 18)

Corrosive Medium	Corrosion Rate* at 70°F in a 100% Corrosive Medium†									
	Copper, Sn-Braze, Al-Braze	70-30 Brass	Silicon Bronze	Monel	Nickel	Inconel	Hastelloy	Aluminum	Lead	Titanium
Ammonium Nitrate	>0.05	>0.05	>0.05	<0.02	<0.02	—	—	<0.02	—	—
Ammonium Sulfate	<0.02	<0.02	<0.02	<0.02	<0.02	—	<0.02	<0.02	<0.02	—
Ammonium Sulfite	>0.05	>0.05	>0.05	—	—	—	—	—	—	—
Ammonium Thiocyanate	—	—	—	<0.02	<0.02	—	—	—	—	—
Amyl Acetate	<0.02	<0.02	<0.02	<0.02	<0.02	<0.02	<0.002	<0.002	<0.02	<0.002
Amyl Chloride	<0.002	<0.02	<0.002	<0.02	<0.02	—	<0.02	<0.02	>0.05	—
Aniline	>0.05	>0.05	—	<0.02	<0.02	—	<0.02	<0.02	>0.05	—
Aniline Hydrochloride	—	—	—	—	—	—	<0.05	>0.05	—	—
Anthracine	<0.02	<0.02	<0.02	<0.02	<0.02	<0.02	<0.02	<0.02	<0.02	<0.002
Antimony Trichloride	<0.05	—	—	—	<0.02	—	<0.002	<0.02	<0.002	—
Barium Carbonate	<0.02	<0.02	<0.02	<0.02	<0.02	—	<0.02	>0.05	>0.05	—
Barium Chloride	<0.02	<0.02	<0.02	<0.02	<0.02	<0.02	<0.02	>0.05	—	—

Source: data compiled by J.S. Park from Earl R. Parker, *Materials Data Book for Engineers and Scientists*, McGraw-Hill Book Company, New York, 1967.

Table 453. SELECTING NONFERROUS METALS FOR USE IN A 100% CORROSIVE MEDIUM

(SHEET 5 OF 18)

Corrosive Medium	Corrosion Rate* at 70°F in a 100% Corrosive Medium†									
	Copper, Sn-Braze, Al-Braze	70-30 Brass	Silicon Bronze	Monel	Nickel	Inconel	Hastelloy	Aluminum	Lead	Titanium
Barium Hydroxide	—	—	—	<0.02	<0.02	<0.02	<0.02	>0.05	>0.05	—
Barium Nitrate	—	—	—	—	—	<0.02	<0.02	—	—	—
Barium Oxide	—	—	—	<0.02	—	<0.02	<0.02	—	—	—
Benzaldehyde	<0.02	<0.02	<0.02	<0.02	<0.02	<0.02	<0.02	<0.002	>0.05	—
Benzene	<0.02	<0.02	<0.02	<0.02	<0.02	<0.02	<0.02	<0.02	<0.02	<0.002
Benzoic Acid	<0.02	<0.02	<0.02	<0.02	<0.02	—	—	<0.02	>0.05	<0.002
Boric Acid	<0.02	<0.02	<0.02	<0.02	<0.02	<0.02	<0.002	<0.02	<0.02	—
Bromic Acid	>0.05	>0.05	>0.05	>0.05	>0.05	>0.05	—	—	<0.02	—
Bromine (Dry)	<0.02	<0.02	<0.02	<0.002	<0.002	<0.002	<0.002	<0.02	<0.002	>0.05
Bromine (Wet)	>0.05	>0.05	>0.05	>0.05	>0.05	>0.05	<0.002	>0.05	>0.05	>0.05
Butyric Acid	<0.02	—	<0.02	<0.02	<0.05	<0.05	<0.002	<0.002	>0.05	<0.002
Calcium Acetate	<0.02	<0.02	<0.02	<0.02	<0.02	<0.02	<0.02	<0.05	<0.02	<0.002

Source: data compiled by J.S. Park from Earl R. Parker, *Materials Data Book for Engineers and Scientists*, McGraw-Hill Book Company, New York, 1967.

Table 453. SELECTING NONFERROUS METALS FOR USE IN A 100% CORROSIVE MEDIUM

(SHEET 6 OF 18)

Corrosive Medium	Corrosion Rate* at 70°F in a 100% Corrosive Medium†									
	Copper, Sn-Braze, Al-Braze	70-30 Brass	Silicon Bronze	Monel	Nickel	Inconel	Hastelloy	Aluminum	Lead	Titanium
Calcium Bicarbonate	<0.02	<0.02	<0.02	<0.02	<0.02	<0.02	<0.02	<0.02	<0.05	<0.002
Calcium Bromide	<0.02	<0.02	<0.02	<0.02	<0.02	<0.02	<0.02	<0.05	<0.02	<0.05
Calcium Chlorate	—	<0.02	—	—	—	—	<0.02	—	—	<0.002
Calcium Chloride	<0.02	<0.02	<0.02	<0.02	<0.02	<0.02	<0.002	>0.05	—	—
Calcium Hydroxide	—	—	—	<0.02	<0.02	<0.02	—	>0.05	—	—
Calcium Hypochlorite	—	—	—	—	—	—	<0.02	—	<0.002	—
Carbon Dioxide	<0.002	<0.002	<0.002	<0.002	<0.002	<0.002	<0.002	<0.002	<0.002	<0.002
Carbon Monoxide	<0.002	<0.002	<0.002	<0.002	<0.002	<0.002	<0.002	<0.002	<0.002	<0.002
Carbon Tetrachloride	<0.002	<0.05	<0.002	<0.002	<0.002	<0.002	<0.002	<0.02	<0.002	<0.002
Carbon Acid (Air Free)	<0.02	>0.05	<0.02	<0.05	<0.02	<0.002	<0.002	<0.002	>0.05	—
Chloroacetic Acid	>0.05	>0.05	<0.05	<0.05	<0.02	<0.05	<0.002	>0.05	>0.05	<0.002
Chlorine Gas	<0.02	>0.05	<0.02	<0.02	<0.002	<0.02	<0.02	<0.02	<0.02	>0.05

Source: data compiled by J.S. Park from Earl R. Parker, *Materials Data Book for Engineers and Scientists*, McGraw-Hill Book Company, New York, 1967.

Table 453. SELECTING NONFERROUS METALS FOR USE IN A 100% CORROSIVE MEDIUM

(SHEET 7 OF 18)

Corrosive Medium	Corrosion Rate* at 70°F in a 100% Corrosive Medium†									
	Copper, Sn-Braze, Al-Braze	70-30 Brass	Silicon Bronze	Monel	Nickel	Inconel	Hastelloy	Aluminum	Lead	Titanium
Chlorine Liquid	—	—	—	<0.02	—	—	—	—	<0.02	—
Chloroform (Dry)	<0.002	<0.02	<0.02	<0.002	<0.002	<0.002	<0.02	<0.02	<0.02	—
Chromic Acid	—	>0.05	—	—	—	—	<0.02	>0.05	—	—
Chromic Hydroxide	<0.02	<0.02	<0.02	<0.02	<0.02	<0.02	<0.02	<0.02	<0.02	—
Chromic Sulfates	<0.05	—	—	<0.05	—	—	<0.02	<0.05	<0.02	—
Citric Acid	<0.02	<0.02	<0.02	<0.02	<0.02	<0.02	<0.002	<0.02	>0.05	—
Copper Nitrate	>0.05	>0.05	<0.05	—	—	—	<0.02	—	—	—
Copper Sulfate	>0.05	>0.05	>0.05	—	—	—	<0.002	>0.05	<0.02	—
Diethylene Glycol	<0.002	<0.002	<0.002	<0.02	<0.02	<0.02	<0.02	<0.02	<0.02	<0.002
Ethyl Chloride	<0.002	<0.002	<0.002	<0.02	<0.002	<0.002	<0.02	<0.002	<0.02	<0.002
Ethylene Glycol	<0.02	<0.02	<0.02	<0.02	<0.02	<0.02	—	<0.002	<0.05	—
Ethylene Oxide	>0.05	>0.05	>0.05	<0.02	<0.02	<0.02	<0.002	<0.002	<0.02	<0.002

Source: data compiled by J.S. Park from Earl R. Parker, *Materials Data Book for Engineers and Scientists*, McGraw-Hill Book Company, New York, 1967.

Table 453. SELECTING NONFERROUS METALS FOR USE IN A 100% CORROSIVE MEDIUM
(SHEET 8 OF 18)

Corrosive Medium	Corrosion Rate* at 70°F in a 100% Corrosive Medium†									
	Copper, Sn-Braze, Al-Braze	70-30 Brass	Silicon Bronze	Monel	Nickel	Inconel	Hastelloy	Aluminum	Lead	Titanium
Fatty Acids	<0.05	<0.05	<0.05	<0.02	<0.02	<0.02	<0.002	<0.002	>0.05	<0.002
Ferric Chloride	<0.02	<0.02	<0.02	>0.05	—	>0.05	<0.02	>0.05	—	—
Ferric Nitrate	—	—	—	—	—	—	—	—	<0.002	—
Ferrous Chloride	<0.02	—	<0.02	—	—	—	<0.02	—	—	—
Ferrous Sulfate	<0.02	<0.05	<0.02	<0.02	<0.02	—	<0.02	—	—	—
Fluorine	<0.002	<0.02	>0.05	<0.002	<0.002	<0.002	<0.02	>0.05	<0.02	—
Formaldehyde	<0.002	<0.02	<0.02	<0.002	<0.002	<0.02	<0.02	<0.002	<0.02	<0.002
Formic Acid	<0.02	<0.02	<0.02	—	<0.02	<0.02	<0.002	<0.02	>0.05	<0.02
Furfural	<0.02	<0.02	<0.02	<0.02	<0.02	<0.02	<0.02	<0.02	<0.02	<0.002
Hydrazine	—	—	—	>0.05	<0.002	<0.002	<0.002	<0.002	>0.05	—
Hydrobromic Acid	<0.02	>0.05	<0.02	—	<0.02	—	—	>0.05	—	—
Hydrocyanic Acid	<0.02	<0.02	<0.02	<0.02	<0.02	<0.02	<0.02	<0.002	<0.02	—

Source: data compiled by J.S. Park from Earl R. Parker, *Materials Data Book for Engineers and Scientists*, McGraw-Hill Book Company, New York, 1967.

Table 453. SELECTING NONFERROUS METALS FOR USE IN A 100% CORROSIVE MEDIUM

(SHEET 9 OF 18)

Corrosive Medium	Corrosion Rate* at 70°F in a 100% Corrosive Medium†									
	Copper, Sn-Braze, Al-Braze	70-30 Brass	Silicon Bronze	Monel	Nickel	Inconel	Hastelloy	Aluminum	Lead	Titanium
Hydrofluoric Acid (Areated)	<0.02	—	—	<0.02	<0.02	<0.02	<0.02	—	—	—
Hydrofluoric Acid (Air Free)	<0.02	<0.02	<0.02	<0.02	<0.02	<0.02	<0.05	—	>0.05	>0.05
Hydrogen Chloride	<0.02	<0.02	<0.02	<0.002	<0.002	<0.002	<0.002	>0.05	<0.02	—
Hydrogen Fluoride	<0.02	<0.02	<0.02	<0.02	<0.002	<0.02	<0.02	<0.02	>0.05	<0.002
Hydrogen Iodide	<0.02	>0.05	<0.02	—	<0.02	—	<0.02	>0.05	—	—
Hydrogen Peroxide	>0.05	>0.05	>0.05	<0.002	<0.02	<0.02	<0.002	<0.002	<0.002	>0.05
Hydrogen Sulfide	<0.02	<0.02	<0.02	<0.02	<0.02	<0.02	<0.002	<0.002	<0.02	<0.002
Lactic Acid	<0.02	<0.05	<0.02	—	—	—	<0.02	<0.02	>0.05	<0.002
Lead Acetate	—	<0.05	<0.02	<0.02	—	—	>0.05	>0.05	—	
Lead Chromate	<0.02	<0.02	<0.02	<0.02	<0.02	<0.02	<0.02	—	<0.02	—
Lead Nitrate	—	—	—	<0.02	<0.02	<0.02	<0.02	—	<0.02	—
Lead Sulfate	<0.02	<0.02	<0.02	<0.02	<0.02	<0.02	<0.02	—	<0.02	—

Source: data compiled by J.S. Park from Earl R. Parker, *Materials Data Book for Engineers and Scientists*, McGraw-Hill Book Company, New York, 1967.

Table 453. SELECTING NONFERROUS METALS FOR USE IN A 100% CORROSIVE MEDIUM
(SHEET 10 OF 18)

Corrosive Medium	Corrosion Rate* at 70°F in a 100% Corrosive Medium†									
	Copper, Sn-Braze, Al-Braze	70-30 Brass	Silicon Bronze	Monel	Nickel	Inconel	Hastelloy	Aluminum	Lead	Titanium
Lithium Chloride	—	—	—	<0.002	—	—	—	—	<0.02	—
Lithium Hydroxide	—	—	—	<0.02	<0.02	<0.02	<0.02	>0.05	—	—
Magnesium Chloride	<0.02	—	<0.02	<0.02	<0.02	<0.02	<0.002	—	>0.05	<0.002
Magnesium Hydroxide	<0.02	<0.02	<0.02	<0.02	<0.02	—	—	>0.05	—	—
Magnesium Sulfate	<0.02	<0.02	<0.02	<0.02	<0.02	<0.02	<0.002	<0.02	—	—
Maleic Acid	<0.02	—	—	—	—	—	<0.02	—	—	—
Malic Acid	—	—	—	—	<0.02	<0.02	—	<0.002	—	<0.002
Mercuric Chloride	>0.05	>0.05	>0.05	—	—	—	—	—	—	—
Mercurous Nitrate	>0.05	>0.05	—	—	—	—	<0.02	>0.05	>0.05	—
Mercury	>0.05	>0.05	>0.05	<0.02	<0.02	<0.02	<0.02	>0.05	>0.05	—
Methallylamine	>0.05	>0.05	>0.05	<0.05	<0.02	<0.02	<0.02	<0.02	—	—
Methanol	<0.02	<0.02	<0.02	<0.002	<0.002	<0.002	<0.02	<0.02	<0.02	—

Source: data compiled by J.S. Park from Earl R. Parker, *Materials Data Book for Engineers and Scientists*, McGraw-Hill Book Company, New York, 1967.

Table 453. SELECTING NONFERROUS METALS FOR USE IN A 100% CORROSIVE MEDIUM

(SHEET 11 OF 18)

Corrosive Medium	Corrosion Rate* at 70°F in a 100% Corrosive Medium†									
	Copper, Sn-Braze, Al-Braze	70-30 Brass	Silicon Bronze	Monel	Nickel	Inconel	Hastelloy	Aluminum	Lead	Titanium
Methyl Ethyl Ketone	<0.002	<0.002	<0.002	<0.002	<0.002	<0.002	<0.002	<0.002	<0.002	<0.002
Methyl Isobutyl Ketone	<0.02	<0.02	<0.02	<0.02	<0.02	<0.02	<0.002	<0.002	<0.002	<0.002
Methylamine	>0.05	>0.05	>0.05	—	—	—	—	<0.02	—	—
Methylene Chloride	<0.002	<0.002	<0.02	<0.002	<0.02	<0.02	—	<0.002	<0.02	—
Monochloroacetic Acid	>0.05	>0.05	>0.05	<0.05	<0.02	<0.02	<0.002	>0.05	>0.05	<0.002
Monorthanolamine	>0.05	>0.05	>0.05	<0.02	<0.02	<0.02	—	<0.02	—	—
Monoethalamine	>0.05	>0.05	>0.05	—	—	—	—	<0.02	—	—
Monoethylamine	>0.05	>0.05	>0.05	—	—	—	—	<0.02	—	—
Nickel Chloride	—	—	<0.02	<0.02	—	<0.02	<0.002	>0.05	<0.02	—
Nickel Nitrate	—	—	—	<0.02	<0.02	<0.02	<0.02	—	<0.02	—
Nickel Sulfate	<0.02	<0.02	<0.02	<0.02	—	<0.02	<0.02	>0.05	<0.02	—
Nitric Acid	>0.05	>0.05	>0.05	>0.05	>0.05	—	—	<0.02	>0.05	—
Nitric Acid (Red Fuming)	>0.05	>0.05	>0.05	>0.05	>0.05	<0.02	<0.02	<0.002	—	<0.002

Source: data compiled by J.S. Park from Earl R. Parker, *Materials Data Book for Engineers and Scientists*, McGraw-Hill Book Company, New York, 1967.

Table 453. SELECTING NONFERROUS METALS FOR USE IN A 100% CORROSIVE MEDIUM

(SHEET 12 OF 18)

Corrosive Medium	Corrosion Rate* at 70°F in a 100% Corrosive Medium†									
	Copper, Sn-Braze, Al-Braze	70-30 Brass	Silicon Bronze	Monel	Nickel	Inconel	Hastelloy	Aluminum	Lead	Titanium
Nitric + Hydrochloric Acid	>0.05	>0.05	>0.05	>0.05	>0.05	>0.05	>0.05	>0.05	>0.05	<0.02
Nitric + Hydrofluoric Acid	>0.05	—	—	—	—	—	<0.05	—	—	>0.05
Nitric + Sulfuric Acid	>0.05	>0.05	>0.05	>0.05	>0.05	>0.05	—	>0.05	>0.05	—
Nitrobenzene	<0.02	<0.02	<0.02	<0.02	<0.02	<0.02	<0.02	<0.02	<0.02	—
Nitrocelluolose	—	<0.02	<0.02	<0.002	<0.02	<0.02	—	<0.002	<0.002	—
Nitroglycerine	<0.02	<0.02	<0.02	<0.02	—	<0.02	—	<0.002	<0.05	—
Nitrotolune	<0.02	<0.02	<0.02	<0.02	<0.02	<0.02	—	<0.02	<0.02	—
Nitrous Acid	>0.05	>0.05	>0.05	>0.05	>0.05	—	—	—	>0.05	—
Oleic Acid	<0.002	<0.02	<0.02	<0.002	<0.002	<0.002	<0.02	<0.002	>0.05	<0.002
Oxalic Acid	<0.05	<0.05	<0.02	<0.02	<0.05	<0.02	<0.02	<0.02	>0.05	—
Phenol	<0.002	<0.002	<0.002	<0.002	<0.002	<0.002	<0.002	<0.002	<0.02	—
Phosphoric Acid (Areated)	>0.05	>0.05	>0.05	—	>0.05	>0.05	<0.002	<0.02	<0.02	>0.05
Phosphoric Acid (Air Free)	—	>0.05	—	—	—	—	<0.002	>0.05	<0.02	>0.05

Source: data compiled by J.S. Park from Earl R. Parker, *Materials Data Book for Engineers and Scientists*, McGraw-Hill Book Company, New York, 1967.

Table 453. SELECTING NONFERROUS METALS FOR USE IN A 100% CORROSIVE MEDIUM

(SHEET 13 OF 18)

Corrosive Medium	Corrosion Rate* at 70°F in a 100% Corrosive Medium†									
	Copper, Sn-Braze, Al-Braze	70-30 Brass	Silicon Bronze	Monel	Nickel	Inconel	Hastelloy	Aluminum	Lead	Titanium
Picric Acid	>0.05	>0.05	>0.05	>0.05	<0.02	<0.02	<0.02	<0.02	<0.02	—
Potassium Bicarbonate	<0.02	<0.02	<0.02	—	—	—	—	<0.02	—	—
Potassium Bromide	<0.02	<0.02	<0.02	<0.02	<0.02	<0.02	<0.02	—	<0.02	—
Potassium Carbonate	<0.02	<0.02	<0.02	<0.02	<0.02	<0.02	<0.02	>0.05	>0.05	—
Potassium Chlorate	<0.05	<0.05	<0.05	—	—	—	—	<0.02	—	—
Potassium Chromate	—	<0.02	<0.02	—	—	—	—	<0.02	—	—
Potassium Cyanide	>0.05	>0.05	>0.05	<0.02	<0.02	<0.02	—	—	—	>0.05
Potassium Dichromate	—	—	—	—	—	—	—	<0.02	—	—
Potassium Ferricyanide	<0.02	—	—	—	—	—	—	—	—	—
Potassium Ferrocyanide	—	—	—	—	—	—	—	<0.02	—	—
Potassium Hydroxide	—	—	>0.05	—	—	—	—	—	>0.05	—
Potassium Hypochlorite	—	—	—	—	—	—	<0.02	—	—	—
Potassium Iodide	<0.02	—	<0.02	<0.02	<0.02	<0.02	<0.02	—	—	<0.002

Source: data compiled by J.S. Park from Earl R. Parker, *Materials Data Book for Engineers and Scientists*, McGraw-Hill Book Company, New York, 1967.

Table 453. Selecting Nonferrous Metals for Use in a 100% Corrosive Medium

(Sheet 14 of 18)

Corrosive Medium	Corrosion Rate* at 70°F in a 100% Corrosive Medium†									
	Copper, Sn-Braze, Al-Braze	70-30 Brass	Silicon Bronze	Monel	Nickel	Inconel	Hastelloy	Aluminum	Lead	Titanium
Potassium Nitrate	<0.002	<0.02	<0.02	<0.02	<0.02	—	—	<0.02	—	—
Potassium Nitrite	<0.02	<0.02	<0.02	<0.02	<0.02	<0.02	<0.02	<0.02	<0.02	<0.002
Potassium Permanganate	<0.02	—	<0.02	—	—	—	<0.002	<0.02	>0.05	—
Potassium Silicate	<0.02	<0.02	<0.02	<0.02	<0.02	<0.02	<0.02	<0.02	—	—
Propionic Acid	<0.02	—	—	<0.02	—	—	—	<0.02	—	>0.05
Pyridine	<0.02	<0.02	<0.02	<0.02	<0.02	<0.02	<0.02	<0.02	<0.02	—
Quinine Sulfate	<0.02	<0.02	<0.02	<0.02	<0.02	<0.02	<0.02	—	—	<0.002
Salicylic Acid	<0.02	—	<0.02	<0.02	<0.02	<0.02	<0.02	<0.02	<0.02	—
Silicon Tetrachloride (Dry)	<0.002	<0.002	<0.002	<0.002	<0.002	<0.002	<0.02	<0.02	<0.02	—
Silicon Tetrachloride (Wet)	>0.05	>0.05	>0.05	>0.05	>0.05	—	<0.02	>0.05	—	—
Silver Bromide	—	—	—	<0.02	<0.02	—	—	—	—	<0.002
Silver Chloride	<0.02	—	—	—	—	—	—	—	—	—

Source: data compiled by J.S. Park from Earl R. Parker, *Materials Data Book for Engineers and Scientists*, McGraw-Hill Book Company, New York, 1967.

Table 453. SELECTING NONFERROUS METALS FOR USE IN A 100% CORROSIVE MEDIUM
(SHEET 15 OF 18)

Corrosive Medium	Corrosion Rate* at 70°F in a 100% Corrosive Medium†									
	Copper, Sn-Braze, Al-Braze	70-30 Brass	Silicon Bronze	Monel	Nickel	Inconel	Hastelloy	Aluminum	Lead	Titanium
Sodium Acetate	<0.02	—	—	<0.02	<0.02	<0.02	—	<0.002	<0.02	—
Sodium Bicarbonate	<0.02	—	—	—	—	—	—	<0.02	—	—
Sodium Bisulfate	<0.02	<0.05	<0.02	<0.02	<0.02	<0.02	<0.02	—	—	—
Sodium Bromide	<0.05	—	—	—	—	—	—	—	—	—
Sodium Carbonate	—	—	<0.02	<0.02	<0.02	<0.02	<0.02	—	—	—
Sodium Chromate	<0.02	<0.02	<0.02	<0.02	<0.02	<0.02	<0.02	<0.02	<0.02	—
Sodium Hydroxide	—	—	—	<0.002	<0.002	<0.002	<0.002	—	—	—
Sodium Hypochlorite	—	>0.05	>0.05	<0.02	—	—	<0.05	>0.05	>0.05	<0.002
Sodium Metasilicate	<0.02	<0.02	<0.02	<0.002	<0.002	<0.002	<0.002	<0.02	—	—
Sodium Nitrate	<0.05	<0.05	<0.02	<0.02	<0.02	—	—	<0.02	—	—
Sodium Nitrite	—	—	—	<0.002	<0.02	<0.02	—	—	—	—
Sodium Phosphate	<0.02	<0.02	<0.02	<0.02	<0.02	<0.02	<0.02	—	<0.02	—

Source: data compiled by J.S. Park from Earl R. Parker, *Materials Data Book for Engineers and Scientists*, McGraw-Hill Book Company, New York, 1967.

Table 453. SELECTING NONFERROUS METALS FOR USE IN A 100% CORROSIVE MEDIUM

(SHEET 16 OF 18)

Corrosive Medium	Corrosion Rate* at 70°F in a 100% Corrosive Medium†									
	Copper, Sn-Braze, Al-Braze	70-30 Brass	Silicon Bronze	Monel	Nickel	Inconel	Hastelloy	Aluminum	Lead	Titanium
Sodium Silicate	<0.02	<0.02	<0.02	<0.02	<0.02	<0.02	<0.02	<0.002	—	—
Sodium Sulfate	<0.02	>0.05	<0.02	<0.02	<0.02	<0.02	<0.002	—	<0.02	—
Sodium Sulfide	>0.05	>0.05	>0.05	—	—	—	—	>0.05	<0.002	—
Sodium Sulfite	<0.05	>0.05	<0.02	<0.02	—	<0.02	—	—	<0.02	—
Stannic Chloride	—	—	>0.05	—	—	—	<0.02	—	—	—
Stannous Chloride	—	—	<0.02	<0.02	<0.02	<0.02	<0.02	—	—	—
Strontium Nitrate	<0.02	<0.02	<0.02	<0.02	<0.02	<0.02	<0.02	<0.02	—	—
Succinic Acid	<0.02	<0.02	<0.02	<0.02	<0.02	<0.02	—	<0.02	<0.02	<0.002
Sulfur Dioxide	<0.02	<0.05	<0.02	<0.02	<0.02	<0.02	<0.02	<0.02	<0.02	—
Sulfur Trioxide	<0.02	<0.02	<0.02	<0.02	<0.02	<0.02	<0.02	<0.02	<0.02	—
Sulfuric Acid (Areated)	>0.05	>0.05	>0.05	>0.05	>0.05	>0.05	<0.02	>0.05	>0.05	>0.05
Sulfuric Acid (Air Free)	—	—	—	>0.05	>0.05	—	<0.02	>0.05	>0.05	>0.05

Source: data compiled by J.S. Park from Earl R. Parker, *Materials Data Book for Engineers and Scientists*, McGraw-Hill Book Company, New York, 1967.

Table 453. Selecting Nonferrous Metals for Use in a 100% Corrosive Medium

(Sheet 17 of 18)

Corrosive Medium	Corrosion Rate* at 70°F in a 100% Corrosive Medium†									
	Copper, Sn-Braze, Al-Braze	70-30 Brass	Silicon Bronze	Monel	Nickel	Inconel	Hastelloy	Aluminum	Lead	Titanium
Sulfuric Acid (Fuming)	>0.05	>0.05	>0.05	>0.05	>0.05	<0.02	<0.002	<0.02	>0.05	—
Sulfurous Acid	<0.05	>0.05	<0.02	>0.05	>0.05	<0.02	<0.02	<0.02	<0.02	<0.002
Tannic Acid	<0.02	<0.05	<0.02	<0.02	<0.02	<0.02	—	>0.05	>0.05	<0.002
Tartaric Acid	<0.02	—	<0.02	—	—	—	<0.02	—	>0.05	<0.002
Tetraphosphoric Acid	<0.05	<0.05	<0.05	<0.05	>0.05	<0.02	<0.02	>0.05	>0.05	—
Trichloroacetic Acid	>0.05	>0.05	<0.05	>0.05	<0.02	—	<0.02	>0.05	>0.05	>0.05
Trichloroethylene	<0.002	<0.02	<0.02	<0.002	<0.002	<0.02	<0.002	<0.002	>0.05	<0.002
Urea	—	—	—	—	—	—	—	<0.02	—	—

Source: data compiled by J.S. Park from Earl R. Parker, *Materials Data Book for Engineers and Scientists*, McGraw-Hill Book Company, New York, 1967.

Table 453. SELECTING NONFERROUS METALS FOR USE IN A 100% CORROSIVE MEDIUM

(SHEET 18 OF 18)

Corrosive Medium	Corrosion Rate* at 70°F in a 100% Corrosive Medium†									
	Copper, Sn-Braze, Al-Braze	70-30 Brass	Silicon Bronze	Monel	Nickel	Inconel	Hastelloy	Aluminum	Lead	Titanium
Zinc Chloride	—	—	>0.05	<0.02	<0.02	<0.02	<0.02	—	<0.02	—
Zinc Sulfate	<0.02	<0.02	<0.02	—	—	—	—	—	—	—

Source: data compiled by J.S. Park from Earl R. Parker, *Materials Data Book for Engineers and Scientists*, McGraw-Hill Book Company, New York, 1967.

* <0.002 means that corrosion rate is likely to be less than 0.002 inch per year (Excellent).
<0.02 means that corrosion rate is likely to be less than about 0.02 inch per year (Good).
<0.05 means that corrosion rate is likely to be less than about 0.05 inch per year (Fair).
>0.05 means that corrosion rate is likely to be more than 0.05 inch per year (Poor).

† Water-free, Dry or Maximum concentration of corrosive medium. Quantitatively

Table 454. SELECTING CORROSION RATES OF METALS

(SHEET 1 OF 5)

Metal	Corrosive Environment	Corrosion Rate* (Mils Penetration per Year)
Silicon iron	Acetic, 5% (Non–oxidizing)	0–0.2
Iron	Sodium Hydroxide, 5%	0–0.2
Nickel alloys	Sodium Hydroxide, 5%	0–0.2
Stainless steel	Sodium Hydroxide, 5%	0–0.2
Nickel alloys	Fresh Water	0–0.2
Silicon iron	Fresh Water	0–0.2
Stainless steel	Fresh Water	0–0.2
Copper alloys	Normal Outdoor Air (Urban Exposure)	0–0.2
Lead	Normal Outdoor Air (Urban Exposure)	0–0.2
Nickel alloys	Normal Outdoor Air (Urban Exposure)	0–0.2
Silicon iron	Normal Outdoor Air (Urban Exposure)	0–0.2
Stainless steel	Normal Outdoor Air (Urban Exposure)	0–0.2
Tin	Normal Outdoor Air (Urban Exposure)	0–0.2
Stainless steel	Acetic, 5% (Non–oxidizing)	0–0.5
Tin	Fresh Water	0–0.5
Aluminum	Normal Outdoor Air (Urban Exposure)	0–0.5
Zinc	Normal Outdoor Air (Urban Exposure)	0–0.5
Copper alloys	Fresh Water	0–1
Nickel alloys	Sea Water	0–1
Silicon iron	Sodium Hydroxide, 5%	0–10
Stainless steel	Sulfuric, 5% (Non–oxidizing)	0–2
Stainless steel	Nitric, 5% (Oxidizing)	0–2
Silicon iron	Sulfuric, 5% (Non–oxidizing)	0–20
Silicon iron	Nitric, 5% (Oxidizing)	0–20
Stainless steel	Sea Water	0–200***
Silicon iron	Sea Water	0–3
Gold	Sulfuric, 5% (Non–oxidizing)	<0.1
Platinum	Sulfuric, 5% (Non–oxidizing)	<0.1

Source: data compiled by J.S. Park from R. E. Bolz and G. L. Tuve, *CRC Handbook of Tables for Applied Engineering Science, 2nd edition*, CRC Press, Inc., Boca Raton, Florida, (1987).

Table 454. SELECTING CORROSION RATES OF METALS
(SHEET 2 OF 5)

Metal	Corrosive Environment	Corrosion Rate* (Mils Penetration per Year)
Tantalum	Sulfuric, 5% (Non–oxidizing)	<0.1
Zirconium	Sulfuric, 5% (Non–oxidizing)	<0.1
Gold	Acetic, 5% (Non–oxidizing)	<0.1
Molybdenum	Acetic, 5% (Non–oxidizing)	<0.1
Platinum	Acetic, 5% (Non–oxidizing)	<0.1
Silver	Acetic, 5% (Non–oxidizing)	<0.1
Tantalum	Acetic, 5% (Non–oxidizing)	<0.1
Titanium	Acetic, 5% (Non–oxidizing)	<0.1
Zirconium	Acetic, 5% (Non–oxidizing)	<0.1
Gold	Nitric, 5% (Oxidizing)	<0.1
Platinum	Nitric, 5% (Oxidizing)	<0.1
Tantalum	Nitric, 5% (Oxidizing)	<0.1
Zirconium	Nitric, 5% (Oxidizing)	<0.1
Gold	Sodium Hydroxide, 5%	<0.1
Molybdenum	Sodium Hydroxide, 5%	<0.1
Platinum	Sodium Hydroxide, 5%	<0.1
Silver	Sodium Hydroxide, 5%	<0.1
Zirconium	Sodium Hydroxide, 5%	<0.1
Gold	Fresh Water	<0.1
Molybdenum	Fresh Water	<0.1
Platinum	Fresh Water	<0.1
Silver	Fresh Water	<0.1
Tantalum	Fresh Water	<0.1
Titanium	Fresh Water	<0.1
Zirconium	Fresh Water	<0.1
Aluminum	Fresh Water	0.1
Gold	Sea Water	<0.1
Molybdenum	Sea Water	<0.1

Source: data compiled by J.S. Park from R. E. Bolz and G. L. Tuve, *CRC Handbook of Tables for Applied Engineering Science, 2nd edition*, CRC Press, Inc., Boca Raton, Florida, (1987).

Table 454. SELECTING CORROSION RATES OF METALS

(SHEET 3 OF 5)

Metal	Corrosive Environment	Corrosion Rate* (Mils Penetration per Year)
Platinum	Sea Water	<0.1
Silver	Sea Water	<0.1
Tantalum	Sea Water	<0.1
Titanium	Sea Water	<0.1
Zirconium	Sea Water	<0.1
Tin	Sea Water	0.1
Gold	Normal Outdoor Air (Urban Exposure)	<0.1
Molybdenum	Normal Outdoor Air (Urban Exposure)	<0.1
Platinum	Normal Outdoor Air (Urban Exposure)	<0.1
Silver	Normal Outdoor Air (Urban Exposure)	<0.1
Tantalum	Normal Outdoor Air (Urban Exposure)	<0.1
Titanium	Normal Outdoor Air (Urban Exposure)	<0.1
Zirconium	Normal Outdoor Air (Urban Exposure)	<0.1
Titanium	Sulfuric, 5% (Non–oxidizing)	0.1–1
Titanium	Nitric, 5% (Oxidizing)	0.1–1
Iron	Fresh Water	0.1–10**
Iron	Sea Water	0.1–10**
Nickel alloys	Sulfuric, 5% (Non–oxidizing)	0.1–1500
Nickel alloys	Nitric, 5% (Oxidizing)	0.1–1500
Lead	Fresh Water	0.1–2
Titanium	Sodium Hydroxide, 5%	<0.2
Lead	Sea Water	0.2–15
Copper alloys	Sea Water	0.2–15**
Zinc	Fresh Water	0.5–10
Zinc	Sea Water	0.5–10**
Aluminum	Acetic, 5% (Non–oxidizing)	0.5–5
Tantalum	Sodium Hydroxide, 5%	<1
Aluminum	Sea Water	1–50

Source: data compiled by J.S. Park from R. E. Bolz and G. L. Tuve, *CRC Handbook of Tables for Applied Engineering Science, 2nd edition*, CRC Press, Inc., Boca Raton, Florida, (1987).

Table 454. SELECTING CORROSION RATES OF METALS
(SHEET 4 OF 5)

Metal	Corrosive Environment	Corrosion Rate* (Mils Penetration per Year)
Iron	Normal Outdoor Air (Urban Exposure)	1–8
Nickel alloys	Acetic, 5% (Non–oxidizing)	2–10**
Copper alloys	Acetic, 5% (Non–oxidizing)	2–15**
Copper alloys	Sodium Hydroxide, 5%	2–5
Tin	Acetic, 5% (Non–oxidizing)	2–500**
Tin	Sodium Hydroxide, 5%	5–20
Lead	Sodium Hydroxide, 5%	5–500**
Lead	Acetic, 5% (Non–oxidizing)	10–150**
Iron	Acetic, 5% (Non–oxidizing)	10–400
Zinc	Sodium Hydroxide, 5%	15–200
Aluminum	Sulfuric, 5% (Non–oxidizing)	15–80
Aluminum	Nitric, 5% (Oxidizing)	15–80
Tin	Sulfuric, 5% (Non–oxidizing)	100–400
Tin	Nitric, 5% (Oxidizing)	100–400
Lead	Sulfuric, 5% (Non–oxidizing)	100–6000
Lead	Nitric, 5% (Oxidizing)	100–6000
Copper alloys	Sulfuric, 5% (Non–oxidizing)	150–1500
Copper alloys	Nitric, 5% (Oxidizing)	150–1500
Zinc	Acetic, 5% (Non–oxidizing)	600–800
Iron	Sulfuric, 5% (Non–oxidizing)	1000–10000
Iron	Nitric, 5% (Oxidizing)	1000–10000
Aluminum	Sodium Hydroxide, 5%	13000
Molybdenum	Sulfuric, 5% (Non–oxidizing)	high
Silver	Sulfuric, 5% (Non–oxidizing)	high

Source: data compiled by J.S. Park from R. E. Bolz and G. L. Tuve, *CRC Handbook of Tables for Applied Engineering Science, 2nd edition*, CRC Press, Inc., Boca Raton, Florida, (1987).

Table 454. SELECTING CORROSION RATES OF METALS
(SHEET 5 OF 5)

Metal	Corrosive Environment	Corrosion Rate* (Mils Penetration per Year)
Zinc	Sulfuric, 5% (Non–oxidizing)	high
Molybdenum	Nitric, 5% (Oxidizing)	high
Silver	Nitric, 5% (Oxidizing)	high
Zinc	Nitric, 5% (Oxidizing)	high

Source: data compiled by J.S. Park from R. E. Bolz and G. L. Tuve, *CRC Handbook of Tables for Applied Engineering Science, 2nd edition*, CRC Press, Inc., Boca Raton, Florida, (1987).

* Corrosion Rate Ranges Expressed in Mils Penetration per Year (1 Mil = 0.001 in)
Note: The corrosion–rate ranges for the solutions are based on temperature up to 212 °F.

** Aeration leads to the higher rates in the range.

*** Aeration leads to passivity, scarcity of dissolved air to activity.

Table 455. SELECTING CORROSION RATES OF METALS IN CORROSIVE ENVIRONMENTS (SHEET 1 OF 5)

Corrosive Environment	Metal	Corrosion Rate* (Mils Penetration per Year)
Sulfuric, 5% (Non–oxidizing)	Stainless steel	0–2
	Silicon iron	0–20
	Gold	<0.1
	Platinum	<0.1
	Tantalum	<0.1
	Zirconium	<0.1
	Titanium	0.1–1
	Nickel alloys	0.1–1500
	Aluminum	15–80
	Tin	100–400
	Lead	100–6000
	Copper alloys	150–1500
	Iron	1000–10000
	Molybdenum	high
	Silver	high
	Zinc	high
Acetic, 5% (Non–oxidizing)	Gold	<0.1
	Molybdenum	<0.1
	Platinum	<0.1
	Silver	<0.1
	Tantalum	<0.1
	Titanium	<0.1
	Zirconium	<0.1
	Silicon iron	0–0.2
	Stainless steel	0–0.5
	Aluminum	0.5–5
	Nickel alloys	2–10**
	Copper alloys	2–15**

Source: data compiled by J.S. Park from R. E. Bolz and G. L. Tuve, *CRC Handbook of Tables for Applied Engineering Science, 2nd edition*, CRC Press, Inc., Boca Raton, Florida, (1987).

Table 455. SELECTING CORROSION RATES OF METALS IN CORROSIVE ENVIRONMENTS (SHEET 2 OF 5)

Corrosive Environment	Metal	Corrosion Rate* (Mils Penetration per Year)
	Tin	2–500**
	Lead	10–150**
	Iron	10–400
	Zinc	600–800
Nitric, 5% (Oxidizing)	Stainless steel	0–2
	Silicon iron	0–20
	Gold	<0.1
	Platinum	<0.1
	Tantalum	<0.1
	Zirconium	<0.1
	Titanium	0.1–1
	Nickel alloys	0.1–1500
	Aluminum	15–80
	Tin	100–400
	Lead	100–6000
	Copper alloys	150–1500
	Iron	1000–10000
	Molybdenum	high
	Silver	high
	Zinc	high
Sodium Hydroxide, 5%	Iron	0–0.2
	Nickel alloys	0–0.2
	Stainless steel	0–0.2
	Silicon iron	0–10
	Gold	<0.1
	Molybdenum	<0.1
	Platinum	<0.1
	Silver	<0.1

Source: data compiled by J.S. Park from R. E. Bolz and G. L. Tuve, *CRC Handbook of Tables for Applied Engineering Science, 2nd edition*, CRC Press, Inc., Boca Raton, Florida, (1987).

Table 455. SELECTING CORROSION RATES OF METALS IN CORROSIVE ENVIRONMENTS (SHEET 3 OF 5)

Corrosive Environment	Metal	Corrosion Rate* (Mils Penetration per Year)
	Zirconium	<0.1
	Titanium	<0.2
	Tantalum	<1
	Copper alloys	2–5
	Tin	5–20
	Lead	5–500**
	Zinc	15–200
	Aluminum	13000
Fresh Water	Nickel alloys	0–0.2
	Silicon iron	0–0.2
	Stainless steel	0–0.2
	Tin	0–0.5
	Gold	<0.1
	Molybdenum	<0.1
	Platinum	<0.1
	Silver	<0.1
	Tantalum	<0.1
	Titanium	<0.1
	Zirconium	<0.1
	Copper alloys	0–1
	Aluminum	0.1
	Iron	0.1–10**
	Lead	0.1–2
	Zinc	0.5–10
Sea Water	Nickel alloys	0–1
	Stainless steel	0–200***
	Silicon iron	0–3
	Gold	<0.1

Source: data compiled by J.S. Park from R. E. Bolz and G. L. Tuve, *CRC Handbook of Tables for Applied Engineering Science, 2nd edition*, CRC Press, Inc., Boca Raton, Florida, (1987).

Table 455. SELECTING CORROSION RATES OF METALS IN CORROSIVE ENVIRONMENTS (SHEET 4 OF 5)

Corrosive Environment	Metal	**Corrosion Rate*** **(Mils Penetration per Year)**
Normal Outdoor Air (Urban Exposure)	Molybdenum	<0.1
	Platinum	<0.1
	Silver	<0.1
	Tantalum	<0.1
	Titanium	<0.1
	Zirconium	<0.1
	Tin	0.1
	Iron	0.1–10**
	Lead	0.2–15
	Copper alloys	0.2–15**
	Zinc	0.5–10**
	Aluminum	1–50
	Copper alloys	0–0.2
	Lead	0–0.2
	Nickel alloys	0–0.2
	Silicon iron	0–0.2
	Stainless steel	0–0.2
	Tin	0–0.2
	Aluminum	0–0.5
	Zinc	0–0.5
	Gold	<0.1
	Molybdenum	<0.1
	Platinum	<0.1
	Silver	<0.1

Source: data compiled by J.S. Park from R. E. Bolz and G. L. Tuve, *CRC Handbook of Tables for Applied Engineering Science, 2nd edition*, CRC Press, Inc., Boca Raton, Florida, (1987).

Table 455. SELECTING CORROSION RATES OF METALS IN CORROSIVE ENVIRONMENTS (SHEET 5 OF 5)

Corrosive Environment	Metal	Corrosion Rate* (Mils Penetration per Year)
	Tantalum	<0.1
	Titanium	<0.1
	Zirconium	<0.1
	Iron	1–8

Source: data compiled by J.S. Park from R. E. Bolz and G. L. Tuve, *CRC Handbook of Tables for Applied Engineering Science, 2nd edition*, CRC Press, Inc., Boca Raton, Florida, (1987).

* Corrosion Rate Ranges Expressed in Mils Penetration per Year (1 Mil = 0.001 in)
Note: The corrosion–rate ranges for the solutions are based on temperature up to 212 °F.

** Aeration leads to the higher rates in the range.

*** Aeration leads to passivity, scarcity of dissolved air to activity.

Table 456. SELECTING FLAMMABILITY OF POLYMERS

(SHEET 1 OF 5)

Polymer	Flammability (ASTM D635) (ipm)
Alkyds; Molded: Glass reinforced (heavy duty parts)	Nonburning
Alkyds; Molded: Putty (encapsulating)	Nonburning
Ceramic reinforced (PTFE)	Noninflammable
Chlorinated polyvinyl chloride	Nonburning
Fibrous (glass) reinforced silicones	Nonburning
Fluorinated ethylene propylene (FEP)	Noninflammable
Granular (silica) reinforced silicones	Nonburning
Polyphenylene sulfide: 40% glass reinforced	Non—burning
Polyphenylene sulfide: Standard	Non—burning
Polytetrafluoroethylene (PTFE); Molded,Extruded	Nonintlammable
Polytrifluoro chloroethylene (PTFCE); Molded,Extruded	Noninflammable
PVC–Acrylic Alloy: PVC–acrylic injection molded	Nonburning
PVC–Acrylic Alloy: PVC–acrylic sheet	Nonburning
Alkyds; Molded: Granular (high speed molding)	Self extinguishing
Alkyds; Molded: Rope (general purpose)	Self extinguishing
Chlorinated polyether	Self extinguishing
Epoxies; High performance resins: Cast, rigid	Self extinguishing
Epoxies; High performance resins: Glass cloth laminate	Self extinguishing
Epoxies; High performance resins: Molded	Self extinguishing
Melamines; Molded: Alpha cellulose and mineral filled	Self extinguishing
Melamines; Molded: Cellulose electrical filled	Self extinguishing
Melamines; Molded: Glass fiber filled	Self extinguishing
Melamines; Molded: Unfilled	Self extinguishing
Nylons; Molded, Extruded; Type 6: Cast	Self extinguishing
Nylons; Molded, Extruded; Type 6: General purpose	Self extinguishing
Nylons; Molded, Extruded; Type 8	Self extinguishing
Nylons; Molded, Extruded; Type 11	Self extinguishing
6/6 Nylon: General purpose extrusion	Self extinguishing
6/6 Nylon: General purpose molding	Self extinguishing

Source: data compiled by J.S. Park from Charles T. Lynch, *CRC Handbook of Materials Science, Vol. 3*, CRC Press, Boca Raton, Florida, 1975 and *Engineered Materials Handbook, Vol.2*, Engineering Plastics, ASM International, Metals Park, Ohio, 1988.

Table 456. SELECTING FLAMMABILITY OF POLYMERS
(SHEET 2 OF 5)

Polymer	Flammability (ASTM D635) (ipm)
6/10 Nylon: General purpose	Self extinguishing
Phenolics: Molded: Arc resistant—mineral filled	Self extinguishing
Phenolics; Molded; General: woodflour and flock filled	Self extinguishing
Phenolics; Molded; High shock: chopped fabric or cord filled	Self extinguishing
Phenolics; Molded; Shock: paper, flock, or pulp filled	Self extinguishing
Phenolics; Molded; Very high shock: glass fiber filled	Self extinguishing
Phenylene oxides (Noryl): Glass fiber reinforced	Self extinguishing
Phenylene oxides (Noryl): Standard	Self extinguishing
Phenylene Oxides: Glass fiber reinforced	Self extinguishing
Phenylene Oxides: SE—1	Self extinguishing
Phenylene Oxides: SE—100	Self extinguishing
Polyarylsulfone	Self extinguishing
Polycarbonate	Self extinguishing
Polycarbonate (40% glass fiber reinforced)	Self extinguishing
Polyester; Moldings: Glass reinforced self extinguishing	Self extinguishing
Polypropylene: Flame retardant	Self extinguishing
Polyvinyl Chloride & Copolymers: Nonrigid—electrical	Self extinguishing
Polyvinyl Chloride & Copolymers: Nonrigid—general	Self extinguishing
Polyvinyl Chloride & Copolymers: Rigid—normal impact	Self extinguishing
Polyvinyl Chloride & Copolymers: Vinylidene chloride	Self extinguishing
Polyvinylidene— fluoride (PVDF)	Self extinguishing
Reinforced polyester moldings: High strength (glass fibers)	Self extinguishing
Reinforced polyester: Heat and chemical resistant (asbestos)	Self extinguishing
Reinforced polyester: Sheet molding compounds, general purpose	Self extinguishing
Rubber phenolic—asbestos filled	Self extinguishing
Rubber phenolic—chopped fabric filled	Self extinguishing
Rubber phenolic—woodflour or flock filled	Self extinguishing
Standard Epoxies: Filament wound composite	Self extinguishing

Source: data compiled by J.S. Park from Charles T. Lynch, *CRC Handbook of Materials Science, Vol. 3*, CRC Press, Boca Raton, Florida, 1975 and *Engineered Materials Handbook, Vol.2*, Engineering Plastics, ASM International, Metals Park, Ohio, 1988.

Table 456. SELECTING FLAMMABILITY OF POLYMERS
(SHEET 3 OF 5)

Polymer	Flammability (ASTM D635) (ipm)
Standard Epoxies: High strength laminate	Self extinguishing
Standard Epoxies: Molded	Self extunguishing
Ureas; Molded: Alpha—cellulose filled (ASTM Type l)	Self extinguishing
Ureas; Molded: Cellulose filled (ASTM Type 2)	Self extinguishing
Ureas; Molded: Woodflour filled	Self extinguishing
Standard Epoxies: General purpose glass cloth laminate	Slow burn to Self extinguishing
Polyester; Thermoset: Cast polyyester: Flexible	Slow burn to self extinguishing
Nylons; Molded, Extruded; Type 6: Glass fiber (30%) reinforced	Slow burn
6/6 Nylon: Glass fiber reinforced	Slow burn
6/6 Nylon: Glass fiber Molybdenum disulfide filled	Slow burn
6/10 Nylon: Glass fiber (30%) reinforced	Slow burn
Polyester; Thermoplastic Injection Moldings: General purpose grade	Slow burn
Polyester; Thermoplastic Injection Moldings: Glass reinforced grades	Slow burn
Polyester; Thermoplastic Moldings: General purpose grade	Slow burn
Polyester; Thermoplastic Moldings: Glass reinforced grade	Slow burn
Silicones; Woven glass fabric/ silicone laminate	0.12
Standard Epoxies: Cast rigid	0.3-0.34
Thermoset Carbonate: Allyl diglycol carbonate	0.35
Cellulose Acetate Butyrate; Molded, Extruded; ASTM Grade: H4	0.5—1.5
Cellulose Acetate Butyrate; Molded, Extruded; ASTM Grade: MH	0.5—1.5
Cellulose Acetate Butyrate; Molded, Extruded; ASTM Grade: S2	0.5—1.5
Cellusose Acetate Propionate; Molded, Extruded; ASTM Grade: 1	0.5—1.5
Cellusose Acetate Propionate; Molded, Extruded; ASTM Grade: 3	0.5—1.5
Cellusose Acetate Propionate; Molded, Extruded; ASTM Grade: 6	0.5—1.5
Polystyrenes; Molded: High impact	0.5—1.5
Cellulose Acetate; Molded, Extruded; ASTM Grade: H6—1	0.5—2.0
Cellulose Acetate; Molded, Extruded; ASTM Grade: H4—1	0.5—2.0

Source: data compiled by J.S. Park from Charles T. Lynch, *CRC Handbook of Materials Science, Vol. 3*, CRC Press, Boca Raton, Florida, 1975 and *Engineered Materials Handbook, Vol.2*, Engineering Plastics, ASM International, Metals Park, Ohio, 1988.

Table 456. SELECTING FLAMMABILITY OF POLYMERS
(SHEET 4 OF 5)

Polymer	Flammability (ASTM D635) (ipm)
Cellulose Acetate; Molded, Extruded; ASTM Grade: H2—1	0.5—2.0
Cellulose Acetate; Molded, Extruded; ASTM Grade: MH—1, MH—2	0.5—2.0
Cellulose Acetate; Molded, Extruded; ASTM Grade: MS—1, MS—2	0.5—2.0
Cellulose Acetate; Molded, Extruded; ASTM Grade: S2—1	0.5—2.0
Polystyrenes; Molded: Medium impact	0.5—2.0
Nylons; Molded, Extruded; Type 6: Flexible copolymers	Slow burn, 0.6
Polypropylene: General purpose	0.7—1
Polyacetal Homopolymer: 20% glass reinforced	0.8
Polyacetal Homopolymer: 22% TFE reinforced	0.8
Polystyrenes; Molded: Styrene acrylonitrile (SAN)	0.8
Polyester; Thermoset: Cast polyyester: Rigid	0.87 to self extinguishing
ABS–Polycarbonate Alloy	0.9
Polyacetal Copolymer: 25% glass reinforced	1
Polypropylene: High impact	1
Polypropylene: Asbestos filled	1
Polypropylene: Glass reinforced	1
Polyethylenes; Molded, Extruded; Type I: Melt index 0.3—3.6	1
Polyethylenes; Molded, Extruded; Type I: Melt index 6—26	1
Polyethylenes; Molded, Extruded; Type I: Melt index 200	1
Polyethylenes; Molded, Extruded; Type II: Melt index 20	1
Polyethylenes; Molded, Extruded; Type II: Melt index l.0—1.9	1
Polyethylenes; Molded, Extruded; Type III: Melt index 0.2—0.9	1
Polyethylenes; Molded, Extruded; Type III: Melt Melt index 0.l—12.0	1
Polyethylenes; Molded, Extruded; Type III: Melt index 1.5—15	1
Polyethylenes; Molded, Extruded; Type III: High molecular weight	1
ABS Resins; Molded, Extruded: Low temperature impact	1.0—1.5
Polystyrenes; Molded: General purpose	1.0—1.5
ABS Resins; Molded, Extruded: Medium impact	1.0—1.6

Source: data compiled by J.S. Park from Charles T. Lynch, *CRC Handbook of Materials Science, Vol. 3*, CRC Press, Boca Raton, Florida, 1975 and *Engineered Materials Handbook, Vol.2*, Engineering Plastics, ASM International, Metals Park, Ohio, 1988.

Table 456. SELECTING FLAMMABILITY OF POLYMERS
(SHEET 5 OF 5)

Polymer	Flammability (ASTM D635) (ipm)
Polyacetal Homopolymer: Standard	1.1
Polyacetal Copolymer: Standard	1.1
Polyacetal Copolymer: High flow	1.1
ABS Resins; Molded, Extruded: High impact	1.3—1.5
ABS Resins; Molded, Extruded: Very high impact	1.3—1.5
ABS Resins; Molded, Extruded: Heat resistant	1.3—2.0

Source: data compiled by J.S. Park from Charles T. Lynch, *CRC Handbook of Materials Science, Vol. 3*, CRC Press, Boca Raton, Florida, 1975 and *Engineered Materials Handbook, Vol.2*, Engineering Plastics, ASM International, Metals Park, Ohio, 1988.

Index

KEY WORD IN CONTEXT (SHEET 1 OF 54)

KEY WORD IN CONTEXT (SHEET 2 OF 54)

KEY WORD IN CONTEXT (SHEET 3 OF 54)

KEY WORD IN CONTEXT (SHEET 4 OF 54)

KEY WORD IN CONTEXT (SHEET 5 OF 54)

KEY WORD IN CONTEXT (SHEET 6 OF 54)

KEY WORD IN CONTEXT (SHEET 7 OF 54)

KEY WORD IN CONTEXT (SHEET 8 OF 54)

KEY WORD IN CONTEXT (SHEET 9 OF 54)

KEY WORD IN CONTEXT (SHEET 10 OF 54)

KEY WORD IN CONTEXT (SHEET 11 OF 54)

KEY WORD IN CONTEXT (SHEET 12 OF 54)

KEY WORD IN CONTEXT (SHEET 13 OF 54)

KEY WORD IN CONTEXT (SHEET 14 OF 54)

KEY WORD IN CONTEXT (SHEET 15 OF 54)

KEY WORD IN CONTEXT (SHEET 16 OF 54)

Key Word in Context (Sheet 17 of 54)

Key Word in Context (Sheet 18 of 54)

KEY WORD IN CONTEXT (SHEET 19 OF 54)

KEY WORD IN CONTEXT (SHEET 20 OF 54)

KEY WORD IN CONTEXT (SHEET 21 OF 54)

Key Word in Context (Sheet 22 of 54)

KEY WORD IN CONTEXT (SHEET 23 OF 54)

KEY WORD IN CONTEXT (SHEET 24 OF 54)

KEY WORD IN CONTEXT (SHEET 25 OF 54)

Key Word in Context (Sheet 26 of 54)

KEY WORD IN CONTEXT (SHEET 27 OF 54)

Key Word in Context (Sheet 28 of 54)

KEY WORD IN CONTEXT (SHEET 29 OF 54)

KEY WORD IN CONTEXT (SHEET 30 OF 54)

Key Word in Context (Sheet 31 of 54)

KEY WORD IN CONTEXT (SHEET 32 OF 54)

KEY WORD IN CONTEXT (SHEET 33 OF 54)

KEY WORD IN CONTEXT (SHEET 34 OF 54)

KEY WORD IN CONTEXT (SHEET 35 OF 54)

KEY WORD IN CONTEXT (SHEET 36 OF 54)

KEY WORD IN CONTEXT (SHEET 37 OF 54)

KEY WORD IN CONTEXT (SHEET 38 OF 54)

KEY WORD IN CONTEXT (SHEET 39 OF 54)

KEY WORD IN CONTEXT (SHEET 40 OF 54)

KEY WORD IN CONTEXT (SHEET 41 OF 54)

KEY WORD IN CONTEXT (SHEET 42 OF 54)

KEY WORD IN CONTEXT (SHEET 43 OF 54)

KEY WORD IN CONTEXT (SHEET 44 OF 54)

KEY WORD IN CONTEXT (SHEET 45 OF 54)

Key Word in Context (Sheet 46 of 54)

Key Word in Context (Sheet 47 of 54)

KEY WORD IN CONTEXT (SHEET 48 OF 54)

KEY WORD IN CONTEXT (SHEET 49 OF 54)

KEY WORD IN CONTEXT (SHEET 50 OF 54)

KEY WORD IN CONTEXT (SHEET 51 OF 54)

KEY WORD IN CONTEXT (SHEET 52 OF 54)

KEY WORD IN CONTEXT (SHEET 53 OF 54)

KEY WORD IN CONTEXT (SHEET 54 OF 54)